Important Concepts Guide

CHAPTER 1   A REVIEW OF BASIC ALGEBRA

Natural numbers: 1, 2, 3, 4, 5, 6, . . .

Prime numbers: 2, 3, 5, 7, 11, 13, 17, . . .

Composite numbers: 4, 6, 8, 9, 10, 12, . . .

Whole numbers: 0, 1, 2, 3, 4, 5, 6, . . .

Integers: . . . , −3, −2, −1, 0, 1, 2, 3, . . .

Rational numbers: Any number that can be written in the form $\frac{a}{b}$ $(b \neq 0)$, where a and b are integers

Irrational numbers: Any decimal that cannot be written in the form $\frac{a}{b}$ $(b \neq 0)$, where a and b are integers

Real numbers: The union of the rational and irrational numbers

Properties of real numbers:
If a, b, and c are real numbers, then

Closure properties:
$\begin{cases} a + b \text{ is a real number.} \\ a - b \text{ is a real number.} \\ ab \text{ is a real number.} \\ \dfrac{a}{b} \text{ is a real number} \quad (b \neq 0). \end{cases}$

Associative properties: $\begin{cases} (a + b) + c = a + (b + c) \\ (ab)c = a(bc) \end{cases}$

Commutative properties: $\begin{cases} a + b = b + a \\ ab = ba \end{cases}$

Distributive property: $a(b + c) = ab + ac$

0 is the **additive identity.**

1 is the **multiplicative identity.**

a and $-a$ are **additive inverses.**

a and $\dfrac{1}{a}$ $(a \neq 0)$ are **multiplicative inverses.**

Double-negative rule: $-(-a) = a$

Properties of exponents:
If there are no divisions by 0, then

$x^m x^n = x^{m+n}$ $\qquad (x^m)^n = x^{mn}$

$(xy)^n = x^n y^n$ $\qquad \left(\dfrac{x}{y}\right)^n = \dfrac{x^n}{y^n}$

$x^0 = 1 \quad (x \neq 0)$ $\qquad x^{-n} = \dfrac{1}{x^n}$

$\dfrac{x^m}{x^n} = x^{m-n}$ $\qquad \left(\dfrac{x}{y}\right)^{-n} = \left(\dfrac{y}{x}\right)^n$

Formulas for area and volume:

Figure	Area	Figure	Volume
Square	$A = s^2$	Cube	$V = s^3$
Rectangle	$A = lw$	Rectangular solid	$V = lwh$
Circle	$A = \pi r^2$	Sphere	$V = \frac{4}{3}\pi r^3$
Triangle	$A = \frac{1}{2}bh$	Cylinder	$V = Bh*$
Trapezoid	$A = \frac{1}{2}h(b_1 + b_2)$	Cone	$V = \frac{1}{3}Bh*$
		Pyramid	$V = \frac{1}{3}Bh*$

*B is the area of the base.

CHAPTER 2   GRAPHS, EQUATIONS OF LINES, AND FUNCTIONS

Midpoint formula: If $P(x_1, y_1)$ and $Q(x_2, y_2)$, the midpoint of segment PQ is

$$M\left(\frac{x_1 + x_2}{2}, \frac{y_1 + y_2}{2}\right)$$

Slope of a nonvertical line: If $x_1 \neq x_2$, then

$$m = \frac{\Delta y}{\Delta x} = \frac{y_2 - y_1}{x_2 - x_1}$$

Equations of a line:

Point-slope form: $y - y_1 = m(x - x_1)$

Slope-intercept form: $y = mx + b$

General form: $Ax + By = C$ $\quad$ (A and B not both 0)

Horizontal line: $y = b$

Vertical line: $x = a$

Function notation:
$f(a)$ is the value of $y = f(x)$ when $x = a$.

CHAPTER 3   SYSTEMS OF LINEAR EQUATIONS

Determinants:

$$\begin{vmatrix} a & b \\ c & d \end{vmatrix} = ad - bc$$

Cramer's rule:

The solution of $\begin{cases} ax + by = e \\ cx + dy = f \end{cases}$ is

$$x = \frac{D_x}{D} \quad \text{and} \quad y = \frac{D_y}{D}, \text{ where}$$

$$D = \begin{vmatrix} a & b \\ c & d \end{vmatrix}, D_x = \begin{vmatrix} e & b \\ f & d \end{vmatrix}, D_y = \begin{vmatrix} a & e \\ c & f \end{vmatrix}$$

(continues on the back endsheet)

10th Edition

Intermediate Algebra

A Guided Approach

Rosemary M. Karr
Collin College

Marilyn B. Massey
Collin College

R. David Gustafson
Rock Valley College

Australia • Brazil • Canada • Mexico • Singapore • United Kingdom • United States

Cengage

Intermediate Algebra: A Guided Approach,
Tenth Edition
Rosemary M. Karr, Marilyn B. Massey,
R. David Gustafson

Product Manager: Marc Bove

Content Developer: Stefanie Beeck, Cynthia Ashton

Content Coordinator: Lauren Crosby

Product Assistant: Kathryn Clark

Media Developer: Bryon Spencer

Brand Manager: Gordon Lee

Market Development Manager: Mark Linton

Content Project Manager: Jennifer Risden

Art Director: Vernon Boes

Manufacturing Planner: Becky Cross

Rights Acquisitions Specialist: Tom McDonough

Production Service: Lachina Publishing Services

Photo Researcher: Q2A/Bill Smith Group

Text Researcher: Pablo D'Stair

Copy Editor: Lachina Publishing Services

Illustrator: Lori Heckelman; Lachina Publishing Services

Text Designer: Terri Wright

Cover Designer: Terri Wright

Cover Image: Kevin Twomey

Compositor: Lachina Publishing Services

Design image credits: Fig leaf (chapter opener): © Michael Breuer/Masterfile; Leaf (front matter): ©JoLin/Shutterstock.com; Green background: © John Fox/Getty Images

Library of Congress Control Number: 2013932722

ISBN-13: 978-1-4354-6250-2
ISBN-10: 1-4354-6250-5

Cengage
200 Pier 4 Boulevard
Boston, MA 02210
USA

Cengage is a leading provider of customized learning solutions with employees residing in nearly 40 different countries and sales in more than 125 countries around the world. Find your local representative at: **www.cengage.com.**

To learn more about Cengage platforms and services, register or access your online learning solution, or purchase materials for your course, visit **www.cengage.com.**

Printed in the United States of America
Print Number: 06   Print Year: 2022

About the Authors

ROSEMARY M. KARR graduated from Eastern Kentucky University (EKU) with a Bachelor's degree in Mathematics, attained her Master of Arts degree at EKU in Mathematics Education, and earned her Ph.D. from the University of North Texas. After two years of teaching high school mathematics, she joined the faculty at Eastern Kentucky University, where she earned tenure as Assistant Professor of Mathematics. A professor at Collin College in Plano, Texas, since 1990, Professor Karr has written more than 10 solutions manuals, presented numerous papers, and been an active member in several educational associations including President of the National Association for Developmental Education. She has been honored several times by Collin College, and has received such national recognitions as U.S. Professor of the Year (2007) and induction as Fellow by the Council of Learning Assistance and Developmental Education Associations (2012).

MARILYN B. MASSEY teaches mathematics at Collin College in McKinney, Texas, where she joined the faculty in 1991. She has been President of the Texas Association for Developmental Education, served as academic chair of the Developmental Mathematics Department, and received an Excellence in Teaching Award from the National Conference for College Teaching and Learning. Professor Massey has presented at numerous state and national conferences. She earned her Bachelor's degree from the University of North Texas and Master of Arts degree in Mathematics Education from the University of Texas at Dallas.

R. DAVID GUSTAFSON is Professor Emeritus of Mathematics at Rock Valley College in Illinois and also has taught extensively at Rockford College and Beloit College. He is coauthor of several best-selling mathematics textbooks, including Gustafson/Frisk/Hughes, *College Algebra*; Gustafson/Karr/Massey, *Beginning Algebra, Intermediate Algebra, Beginning and Intermediate Algebra: A Combined Approach*; and the Tussy/Gustafson and Tussy/Gustafson/Koenig developmental mathematics series. His numerous professional honors include Rock Valley Teacher of the Year and Rockford's Outstanding Educator of the Year. He has been very active in AMATYC as a Midwest Vice-president and has been President of IMACC, AMATYC's Illinois affiliate. He earned a Master of Arts degree in Mathematics from Rockford College in Illinois, as well as a Master of Science degree from Northern Illinois University.

To my husband and family, for their
unwavering support of my work
—R.M.K.

To YoYo, Danielle, and Ethan
—M.B.M.

To Craig, Jeremy, Paula, Gary, Bob, Jennifer,
John-Paul, Gary, and Charlie
—R.D.G.

Contents

1   A Review of Basic Algebra   1

REACH FOR SUCCESS   2
1.1   The Real Number System   3
1.2   Arithmetic and Properties of Real Numbers   15
1.3   Exponents   29
1.4   Scientific Notation   40
1.5   Solving Linear Equations in One Variable   48
1.6   Using Linear Equations to Solve Applications   59
1.7   More Applications of Linear Equations in One Variable   70
■   *Projects*   77
REACH FOR SUCCESS EXTENSION   78
CHAPTER REVIEW   79
CHAPTER TEST   87

2   Graphs, Equations of Lines, and Functions   89

REACH FOR SUCCESS   90
2.1   Graphing Linear Equations   91
2.2   Slope of a Line   105
2.3   Writing Equations of Lines   117
2.4   Introduction to Functions   129
2.5   Graphing Other Functions   142
■   *Projects*   152
REACH FOR SUCCESS EXTENSION   154
CHAPTER REVIEW   155
CHAPTER TEST   160
CUMULATIVE REVIEW FOR CHAPTERS 1–2   161

3 Systems of Linear Equations   163

REACH FOR SUCCESS   164

3.1 Solving Systems of Two Linear Equations in Two Variables by Graphing   165

3.2 Solving Systems of Two Linear Equations in Two Variables by Substitution and Elimination (Addition)   175

3.3 Solving Systems of Three Linear Equations in Three Variables   189

3.4 Solving Systems of Linear Equations Using Matrices   199

3.5 Solving Systems of Linear Equations Using Determinants   207

■ *Projects*   218

REACH FOR SUCCESS EXTENSION   220

CHAPTER REVIEW   221

CHAPTER TEST   226

4 Inequalities   229

REACH FOR SUCCESS   230

4.1 Solving Linear Inequalities in One Variable   231

4.2 Solving Absolute Value Equations and Inequalities in One Variable   241

4.3 Solving Linear Inequalities in Two Variables   254

4.4 Solving Systems of Linear and Quadratic Inequalities in Two Variables by Graphing   262

4.5 Solving Systems Using Linear Programming   269

■ *Projects*   279

REACH FOR SUCCESS EXTENSION   280

CHAPTER REVIEW   281

CHAPTER TEST   286

CUMULATIVE REVIEW FOR CHAPTERS 1–4   287

5 Polynomials, Polynomial Functions, and Equations   289

REACH FOR SUCCESS   290

5.1 Polynomials and Polynomial Functions   291

5.2 Adding and Subtracting Polynomials   300

5.3 Multiplying Polynomials   306

5.4 The Greatest Common Factor and Factoring by Grouping   316

5.5 The Difference of Two Squares; the Sum and Difference of Two Cubes   324

5.6 Factoring Trinomials   332

5.7 Summary of Factoring Techniques   344
5.8 Solving Equations by Factoring   347
◼ *Projects*   355
REACH FOR SUCCESS EXTENSION   357
CHAPTER REVIEW   358
CHAPTER TEST   363

6 ∽ **Rational Expressions**   365

REACH FOR SUCCESS   366
6.1 Finding the Domain of Rational Functions and Simplifying Rational Expressions   367
6.2 Multiplying and Dividing Rational Expressions   377
6.3 Adding and Subtracting Rational Expressions   385
6.4 Simplifying Complex Fractions   395
6.5 Solving Equations Containing Rational Expressions   404
6.6 Dividing Polynomials   414
6.7 Synthetic Division   421
6.8 Proportion and Variation   428
◼ *Projects*   441
REACH FOR SUCCESS EXTENSION   443
CHAPTER REVIEW   444
CHAPTER TEST   450
CUMULATIVE REVIEW FOR CHAPTERS 1–6   451

7 ∽ **Radical Expressions and Equations**   453

REACH FOR SUCCESS   454
7.1 Radical Expressions   455
7.2 Applications of the Pythagorean Theorem and the Distance Formula   467
7.3 Rational Exponents   474
7.4 Simplifying and Combining Radical Expressions   482
7.5 Multiplying Radical Expressions and Rationalizing   492
7.6 Radical Equations   501
7.7 Complex Numbers   510
◼ *Projects*   520
REACH FOR SUCCESS EXTENSION   522
CHAPTER REVIEW   523
CHAPTER TEST   530

8   Quadratic Equations and Functions; Inequalities; and Algebra, Composition, and Inverses of Functions   533

REACH FOR SUCCESS   534

8.1 Solving Quadratic Equations Using the Square-Root Property and by Completing the Square   535

8.2 Solving Quadratic Equations Using the Quadratic Formula   546

8.3 The Discriminant and Equations That Can Be Written in Quadratic Form   553

8.4 Graphs of Quadratic Functions   562

8.5 Quadratic and Other Nonlinear Inequalities   578

8.6 Algebra and Composition of Functions   588

8.7 Inverses of Functions   595

■ *Projects*   605

REACH FOR SUCCESS EXTENSION   607

CHAPTER REVIEW   608

CHAPTER TEST   615

CUMULATIVE REVIEW FOR CHAPTERS 1–8   616

9   Exponential and Logarithmic Functions   619

REACH FOR SUCCESS   620

9.1 Exponential Functions   621

9.2 Base-*e* Exponential Functions   632

9.3 Logarithmic Functions   640

9.4 Natural Logarithms   650

9.5 Properties of Logarithms   656

9.6 Exponential and Logarithmic Equations   666

■ *Projects*   677

REACH FOR SUCCESS EXTENSION   679

CHAPTER REVIEW   680

CHAPTER TEST   686

10   Conic Sections and More Graphing   687

REACH FOR SUCCESS   688
10.1   The Circle and the Parabola   689
10.2   The Ellipse   702
10.3   The Hyperbola   712
10.4   Systems Containing Second-Degree Equations   723
10.5   Piecewise-Defined Functions and the Greatest Integer Function   729
    ■ *Projects*   736
REACH FOR SUCCESS EXTENSION   737
CHAPTER REVIEW   738
CHAPTER TEST   745
CUMULATIVE REVIEW FOR CHAPTERS 1–10   746

11   Miscellaneous Topics   749

REACH FOR SUCCESS   750
11.1   The Binomial Theorem   751
11.2   Arithmetic Sequences and Series   759
11.3   Geometric Sequences and Series   768
11.4   Infinite Geometric Sequences and Series   776
11.5   Permutations and Combinations   782
11.6   Probability   791
    ■ *Projects*   797
REACH FOR SUCCESS EXTENSION   799
CHAPTER REVIEW   800
CHAPTER TEST   805
CUMULATIVE REVIEW FOR CHAPTERS 1–11   805

GLOSSARY   G-1

APPENDIX 1   *Symmetries of Graphs*   A-1

APPENDIX 2   *Sample Final Examination*   A-7

APPENDIX 3   *Tables*   A-11

APPENDIX 4   *Answers to Selected Exercises*   A-15

INDEX   I-1

Preface

This tenth edition of *Intermediate Algebra: A Guided Approach* is an exciting and innovative revision. The new edition reflects a thorough update, has new pedagogical features that make the text easier to read, and has an entirely new and fresh interior design. This series is known for its integrated approach, for the clarity of its writing, for making algebra relevant and engaging, and for developing student skills. The revisions to this already successful text will further promote student achievement. Coauthors Rosemary Karr and Marilyn Massey joined David Gustafson and have now assumed primary responsibility, bringing more experience in developmental education.

This new edition has expanded on the learning plan that helps students transition to the next level, teaching them the problem-solving strategies that will serve them well in their everyday lives. Most textbooks share the goals of clear writing, well-developed examples, and ample exercises, whereas the Karr/Massey/Gustafson series develops student success beyond the demands of traditional required coursework. The tenth edition's learning tools have been developed with your students in mind.

Through their collective teaching experience, the authors have developed an acute awareness of students' approaches to homework and have determined that exercise sets should serve as more than just a group of problems to work. The authors' philosophy is to guide the student through new material in a gentle progression of thought development that slowly reduces the student's dependence on external factors and relates new concepts to previously learned material. They have written the textbook to guide students through the material while providing a decreasing level of support throughout each section. Initially, the authors provide a map to the content through learning objectives that serve as advanced organizers for what students can expect to learn that day. The vocabulary encourages students to speak the language of mathematics. *Getting Ready* exercises at the beginning of each section prepares students for the upcoming concepts by reviewing relevant previous skills. The instructor may guide the students through the examples, but the students will independently attempt the Self Checks. Students will begin to use their "mathematical voice" to explain problems to one another or work collectively to find a solution to a problem not previously encountered, a primary goal of the *Now Try This* feature. The guidance shifts from instructor to fellow students to individual through carefully designed exercise sets.

■ New to This Edition

The major changes to this edition are the inclusion of *Self Checks* for ALL examples, study skills strategies for each chapter, modification of the *Warm-Up* exercises, and the revision of many exercises.

- The *Warm-Ups* have been significantly revised to reflect the original intention of this group of exercises. We wanted the *Warm-Ups* to be just that, warm-ups. Thus, most have been revised to include skills needed for the development of section concepts.

- A set of interactive study skills strategies, called *Reach for Success* have been incorporated throughout the textbook for a more holistic learning experience.

The authors feel that most students already know effective study strategies and could list many of the successful techniques, but it is the authors' intent to encourage the student's thoughtful consideration and implementation of these skills. The interactive approach engages the student, encouraging an active participation to develop and reinforce the skills. The incorporation of student advising and the development of study skills is critical for academic success. This is why this feature is titled *REACH for Success*. The authors hope to help students do just that. These worksheets (and others) are available online for flexibility in the order of assignment.

- *Everyday Connections* have been revised to more accurately reflect a topic of the selected chapter.

- *Self Checks* have been included for EVERY example in the textbook, especially significant for applications where students need to develop problem solving strategies.

- Additional *Teaching Tips* have been included at the request of our reviewers.

Calculators

The use of calculators is assumed throughout the text although the calculator icon makes the exercises easily identifiable for possible omission if the instructor chooses to do so. We believe that students should learn calculator skills in the mathematics classroom. They will then be prepared to use calculators in science and business classes and for nonacademic purposes.

Since most intermediate algebra students now have graphing calculators, keystrokes are given for both scientific and graphing calculators. Removable cards for the *Basic Calculator Keystroke Guide for the TI-83/84 Family of Calculator* and *Basic Calculator Keystroke Guide for the fx-9750GII CASIO* are bound into the book as resources for those students learning how to use a graphing calculator.

A Guided Tour

Chapter Openers showcase the variety of career paths available in the world of mathematics. We include a brief overview of each career, as well as job outlook statistics from the U.S. Department of Labor, including potential job growth and annual earning potential.

Graphs, Equations of Lines, and Functions

2

Careers and Mathematics

ATMOSPHERIC SCIENTISTS, INCLUDING METEOROLOGISTS

Atmospheric science is the study of the atmosphere—the blanket of air covering the Earth. Atmospheric scientists, including meteorologists, study the atmosphere's physical characteristics, motions, and processes. The best-known application of this knowledge is in forecasting the weather. A bachelor's degree in meteorology, which generally includes advanced courses in mathematics, is usually the minimum educational requirement for an entry-level position in the field.

Job Outlook:
Employment of atmospheric scientists is projected to increase about as fast as the average for all occupations through 2020. Opportunities will be greater in private industry than in the federal government.

Annual Earnings:
$48,090–$132,130

For More Information:
http://www.bls.gov/oco/ocos051.htm

For a Sample Application:
See Example 1 in Section 2.1

REACH FOR SUCCESS
2.1 Graphing Linear Equations
2.2 Slope of a Line
2.3 Writing Equations of Lines
2.4 Introduction to Functions
2.5 Graphing Other Functions
■ *Projects*
REACH FOR SUCCESS EXTENSION
CHAPTER REVIEW
CHAPTER TEST
CUMULATIVE REVIEW

In this chapter
it is often said that "A picture is worth a thousand words." In this chapter, we will show how numerical relationships can be described by using mathematical pictures called graphs.

89

Reach for Success   Analyzing Your Time

We can all agree that we only have 24 hours in a day, right? The expression "My, how time flies!" is certainly relevant. It might surprise you where the time goes when you begin to monitor the hours.

How can you schedule your time to allow for a successful semester? Begin by completing the pie chart to analyze a typical day.

HOW MANY HOURS A DAY DO YOU SPEND:
For every hour spent on each activity shade that many pie segments. (It might be helpful to use a different color to shade for each question.) There are no right or wrong answers. You just need an honest assessment of your time.

_____ working at outside employment (include commute time)?

_____ sleeping?

_____ preparing meals, eating, and exercising?

_____ getting ready for work/class?

_____ in class (include lab hours, commuting, or tutoring time)?

_____ with family and friends?

_____ on the Internet, phone, playing video games, texting, watching TV, going to movies, or other entertainment?

24-Hour Pie Chart

Now we have a mathematics problem! How many hours are left for studying? Remember, the rule of thumb is for every **1** hour in class, you will need at least **2** hours of studying. For mathematics, you could need even more. The key is to spend as much time as you need to understand the material. There is no magic formula!

A Successful Study Strategy . . .

Adjust your schedule as needed to support your educational goals.

At the end of the chapter you will find an additional exercise to guide you in planning for a successful semester.

Reach for Success is a new feature where study skills have been written for inclusion in every chapter. This feature is designed as an opening activity to help the student prepare for a successful semester.

Appearing at the beginning of each section, *Learning Objectives* are mapped to the appropriate content, as well as to relevant exercises in the *Guided Practice* section. Measurable objectives allow students to identify specific mathematical processes that may need additional reinforcement. For the instructor, homework assignments can be developed more easily with problems keyed to objectives, thus facilitating identification of appropriate exercises.

In order to work mathematics, one must be able to speak the language. Not only are *Vocabulary* words identified at the beginning of each section, these words are also bolded within the section. Exercises include questions on the vocabulary words, and a glossary has been included to facilitate the students' reference to these words. An optional Spanish glossary is available upon request.

Getting Ready questions appear at the beginning of each section, linking past concepts to the upcoming material.

Solving Systems of Two Linear Equations in Two Variables by Graphing

Objectives
1. Solve a system of two linear equations by graphing.
2. Recognize that an inconsistent system has no solution.
3. Express the infinitely many solutions of a dependent system as a general ordered pair.
4. Use a system of linear equations in two variables to solve a linear equation in one variable graphically.

Vocabulary
system of equations   inconsistent system   independent equations
consistent system   distinct lines   dependent equations
equivalent systems

Getting Ready
Let $y = -3x + 2$. Find the value of y when

1. $x = 0$.   **2.** $x = 3$.   **3.** $x = -3$.   **4.** $x = -\frac{1}{3}$.

Find five pairs of numbers

5. with a sum of 12.   **6.** with a difference of 3.

Examples are worked out in each chapter, highlighting the concept being discussed. We include Author notes in many of the text's examples, giving students insight into the thought process one goes through when approaching a problem and working toward a solution.

All examples end with a **Self Check**, so that students can measure their reading comprehension. Answers to each section's Self Checks are found at the end of that section for ease of reference.

Each section ends with **Now Try This** exercises intended to increase conceptual understanding through active classroom participation and involvement. These problems transition to the Exercise Sets, as well as to material in future sections and can be worked independently or in small groups. The exercises reinforce topics, digging a little deeper than the examples. To discourage a student from simply looking up the answer and trying to find the process that will produce the answer, answers to these problems will be provided only in the *Annotated Instructor's Edition* of the text.

> **NOW TRY THIS**
>
> A vertical asymptote appears at any value for which a simplified rational function is undefined. A hole in the graph of a function occurs at any value resulting from a factor common to the numerator and denominator of the original function.
>
> *For each of the functions below, determine where the vertical asymptote(s) occurs (if any) and where the hole(s) occurs (if any). Graph the original function and any simplified function and compare their graphs. What do you notice?*
>
> **1.** $f(x) = \dfrac{x^2 + 4}{x^2 - 9}$ **2.** $g(x) = \dfrac{x^3 - 3x^2}{15x - 5x^2}$

The *Exercise Sets* transition students through progressively more difficult homework problems. Students are initially asked to work quick, basic problems on their own, then proceed to work exercises keyed to examples and/or objectives, and finally to complete applications and critical-thinking questions on their own.

> ### 7.2 Exercises
>
> **WARM-UPS** *Evaluate each expression.*
>
> **1.** $\sqrt{625}$ **2.** $\sqrt{289}$
> **3.** $\sqrt{7^2 + 24^2}$ **4.** $\sqrt{15^2 + 8^2}$
> **5.** $\sqrt{5^2 - 3^2}$ **6.** $\sqrt{5^2 - 4^2}$
>
> **REVIEW** *Find each product.*
>
> **7.** $(2x + 5)(3x - 4)$ **8.** $(6y - 7)(4y + 5)$
>
> **9.** $(4a - 3b)(5a - 2b)$ **10.** $(4r - 3)(2r^2 + 3r - 4)$
>
> **VOCABULARY AND CONCEPTS** *Fill in the blanks.*
>
> **11.** In a right triangle, the side opposite the 90° angle is called the _____.
>
> **12.** In a right triangle, the two shorter sides are called ____.
>
> **19.** $b = 18$ m and $c = 82$ m
> **20.** $b = 7$ ft and $c = 25$ ft
> **21.** $a = 14$ in. and $c = 50$ in.
> **22.** $a = 8$ cm and $b = 15$ cm
> **23.** $a = \sqrt{8}$ mi and $b = \sqrt{8}$ mi
> **24.** $a = \sqrt{11}$ ft and $b = \sqrt{38}$ ft
>
> *Find the distance between the given points. If an answer is not exact, use a calculator and give an approximation to the nearest tenth.*
> **SEE EXAMPLE 2. (OBJECTIVE 2)**
>
> **25.** $(0, 0), (3, -4)$ **26.** $(0, 0), (-6, 8)$
> **27.** $(10, 8), (-2, 3)$ **28.** $(5, 9), (8, 13)$
> **29.** $(-7, 12), (-1, 4)$ **30.** $(-5, -2), (7, 3)$
> **31.** $(12, -6), (-3, 2)$ **32.** $(-14, -9), (10, -2)$
> **33.** $(-3, 5), (-5, -5)$ **34.** $(2, -3), (4, -8)$
> **35.** $(-9, 3), (4, 7)$ **36.** $(-1, -3), (-5, 8)$

- **Warm-Ups** get students into the homework mind-set, asking quick memory-testing questions.

- **Review** exercises are included to remind students of previous skills.

- **Vocabulary and Concepts** exercises emphasize the main concepts taught in this section.

- **Guided Practice** exercises are keyed to the objectives to increase student success by directing students to the concept covered in that group of exercises. Should a student encounter difficulties working a problem, a specific example within the objective is also cross-referenced.

- **Additional Practice** problems are mixed and not linked to objectives or examples, providing the student with the opportunity to distinguish among problem types and to select an appropriate problem-solving strategy. This will help students prepare for the format generally seen on exams.

- *Applications* ask students to apply their new skills in real-life situations, *Writing About Math* problems build students' mathematical communication skills, and *Something to Think About* encourages students to consider what they have learned in a section, and apply those concepts to a new situation.

- Many exercises are available online through Enhanced WebAssign®. These homework problems are algorithmic, ensuring that your students will learn mathematical processes, not just how to work with specific numbers.

Chapter-ending *Projects* encourage in-depth exploration of key concepts.

A *Chapter Review* grid presents the material cleanly and simply, giving students an efficient means of reviewing the chapter.

Chapter Tests allow students to pinpoint their strengths and challenges with the material. Answers are included at the back of the book.

Cumulative Review Exercises follow the end of chapter material for every even-numbered chapter, and keep students' skills current before moving on to the next topic.

A *Glossary* has been included to function as a student reference to the vocabulary words and definitions.

■ Additional Features

Everyday Connections boxes reveal the real-world power of mathematics. Each Everyday Connections box invites students to see how the material covered in the chapter is relevant to their lives.

Perspective boxes highlight interesting facts from mathematics history or important mathematicians, past and present. These brief but interesting biographies connect students to discoveries of the past and their importance to the present.

Perspective

THE FIBONACCI SEQUENCE AND THE GOLDEN RATIO

Perhaps one of the most intriguing examples of how a mathematical idea can represent natural phenomena is the *Fibonacci Sequence*, a list of whole numbers that is generated by a very simple rule. This sequence was first developed by the Italian mathematician Leonardo da Pisa, more commonly known as Fibonacci. The Fibonacci Sequence is the following list of numbers

$$1, 1, 2, 3, 5, 8, 13, 21, \ldots$$

where each successive number in the list is obtained by adding the two preceding numbers. Although Fibonacci originally developed this sequence to solve a mathematical puzzle, subsequent study of the numbers in this sequence has uncovered many examples in the natural world in which this sequence emerges. For example, the arrangement of the seeds on the face of a sunflower, the hibernation periods of certain insects, and the branching patterns of many plants all give rise to Fibonacci numbers.

Among the many special properties of these numbers is the fact that, as we generate more and more numbers in the list, the ratio of successive numbers approaches a constant value. This value is designated by the symbol ϕ and often is referred to as the "Golden Ratio." One way to calculate the value of ϕ is to solve the quadratic equation $\phi^2 - \phi - 1 = 0$.

Leonardo Fibonacci

1. Using the quadratic formula, find the exact value of ϕ.

2. Using a calculator, find a decimal approximation of ϕ, correct to three decimal places.

Note: Negative solutions are not applicable in this context.

Accent
on technology

▶ Generating Tables
of Solutions

If an equation in x and y is solved for y, we can use a graphing calculator to generate a table of solutions. The instructions in this discussion are for a TI-84 graphing calculator.

To construct a table of solutions for $2x + 5y = 10$, we first solve for y.

$$2x + 5y = 10$$
$$5y = -2x + 10 \quad \text{Subtract } 2x \text{ from both sides.}$$
$$y = -\frac{2}{5}x + 2 \quad \text{Divide both sides by 5 and simplify.}$$

To enter $y = -\frac{2}{5}x + 2$, we press **Y=** and enter (–) (2 ÷ 5) x, T, θ, n (x) + 2, as shown in Figure 2-9(a).

To enter the x-values that are to appear in the table, we press **2ND WINDOW** (TBLSET) and enter the first value for x on the line labeled TblStart=. In Figure 2-9(b), -5 has been entered on this line. Other values for x that are to appear in the table are determined by setting an increment value on the line labeled ΔTbl=. Figure 2-9(b) shows that an increment of 1 was entered. This means that each x-value in the table will be 1 unit larger than the previous x-value.

The final step is to press the keys **2ND GRAPH** (TABLE). This displays a table of solutions, as shown in Figure 2-9(c).

Accent on Technology boxes teach students the calculator skills to prepare them for using these tools in science and business classes, as well as for nonacademic purposes. Calculator examples are given in these boxes, and keystrokes are given for both scientific and graphing calculators. For instructors who do not use calculators in the classroom, the material on calculators is easily omitted without interrupting the flow of ideas.

Comment notations alert students to common errors as well as provide helpful and pertinent information about the concepts they are learning.

COMMENT If a and b are both negative, then $\sqrt{ab} \neq \sqrt{a}\sqrt{b}$. Recall, the simplified form of a radical expression does not contain a negative radicand. Therefore the $\sqrt{-16}$ and the $\sqrt{-4}$ must be simplified prior to multiplying. For example, if $a = -16$ and $b = -4$, we have

$$\sqrt{(-16)}\sqrt{(-4)} = (4i)(2i) = 8i^2 = 8(-1) = -8$$

The imaginary numbers are a subset of a set of numbers called the *complex numbers*.

For the instructor, *Teaching Tips* are provided in the margins of the *Annotated Instructor's Edition* as interesting historical information, alternative approaches for teaching the material, and class activities.

A *Spanish Glossary* is available for inclusion upon request.

Teaching Tip

If your class is full and the chairs are arranged in rows and columns, introduce graphing this way:
1. Number the rows and columns.
2. Ask row 3, column 1 to stand. This is the point (3, 1).
3. Ask (1, 3) to stand. That this is a different point emphasizes the ordered pair idea.
4. Repeat with more examples.

Supplements

FOR THE STUDENT	FOR THE INSTRUCTOR
	### Annotated Instructor Edition (ISBN: 978-1-4354-6251-9) The **Annotated Instructor Edition** features answers to all problems in the book.
### Student Solutions Manual (ISBN: 978-1-285-84624-8) *Author: Michael Welden, Mt. San Jacinto College* The **Student Solutions Manual** provides worked-out solutions to the odd-numbered problems in the book.	### Complete Solutions Manual (ISBN: 978-1-285-84623-1) *Author: Michael Welden, Mt. San Jacinto College* The **Complete Solutions Manual** provides worked-out solutions to all of the problems in the text.
### Student Workbook (ISBN: 978-1-285-84625-5) *Author: Maria H. Andersen, former math faculty at Muskegon Community College and now working in the learning software industry* The **Student Workbook** contains all of the assessments, activities, and worksheets from the **Instructor's Resource Binder** for classroom discussions, in-class activities, and group work.	### Instructor's Resource Binder (ISBN: 978-0-538-73675-6) *Author: Maria H. Andersen, former math faculty at Muskegon Community College and now working in the learning software industry* The **Instructor's Resource Binder** contains uniquely designed Teaching Guides, which include instruction tips, examples, activities worksheets, overheads, and assessments, with answers provided.
	### Instructor Companion Website Everything you need for your course in one place! This collection of book-specific lecture and class tools is available online via **www.cengage.com/login**. Formerly found on the PowerLecture, access and download PowerPoint® presentations, images, and more.
	### Solution Builder This online instructor database offers complete worked solutions to all exercises in the text, allowing you to create customized, secure solutions printouts (in PDF format) matched exactly to the problems you assign in class. For more information, visit **www.cengage.com/solutionbuilder**.
ENHANCED Web**Assign** ### Enhanced WebAssign® (Printed Access Card ISBN: 978-1-285-85770-1 Online Access Code ISBN: 978-1-285-85773-2) **Enhanced WebAssign** (assigned by the instructor) provides instant feedback on homework assignments to students. This online homework system is easy to use and includes a multimedia eBook, video examples, and problem-specific tutorials.	**ENHANCED** Web**Assign** ### Enhanced WebAssign® (Printed Access Card ISBN: 978-1-285-85770-1 Online Access Code ISBN: 978-1-285-85773-2) Instant feedback and ease of use are just two reasons why **WebAssign** is the most widely used homework system in higher education. **WebAssign's** homework delivery system allows you to assign, collect, grade, and record homework assignments via the web. And now this proven system has been enhanced to include a multimedia eBook, video examples, and problem-specific tutorials. **Enhanced WebAssign** is more than a homework system—it is a complete learning system for math students.
	### Text-Specific Videos *Author: Rena Petrello* These videos are available at no charge to qualified adopters of the text and feature 10–20 minute problem-solving lessons that cover each section of every chapter.

To the Student

Congratulations! You now own a state-of-the-art textbook that has been written especially for you. We have tried to write a book that you can read and understand. The text includes carefully written narrative and an extensive number of worked examples with corresponding Self Checks. **Now Try This** exercises can be worked with your classmates, and **Guided Practice** exercises tell you exactly which example to use as a resource for each question. These are just a few of the many features included in this text with your success in mind. Perhaps the biggest change is the inclusion of Study Skill strategies at the start and end of each chapter. We urge you to take the time to complete these worksheets to increase your understanding of what it takes to be a successful student.

To get the most out of this course, you must read and study the textbook carefully. We recommend that you work the examples on paper first, and then work the Self Checks. Only after you thoroughly understand the concepts taught in the examples should you attempt to work the exercises. A **Student Solutions Manual** is available, which contains the worked-out solutions to the odd-numbered exercises.

Since the material presented in **Intermediate Algebra, Tenth Edition**, may be of value to you in later years, we suggest that you keep this text. It will be a good reference in the future and will keep at your fingertips the material that you have learned here.

■ Hints on Studying Algebra

The phrase "practice makes perfect" is not quite true. It is "*perfect* practice that makes perfect." For this reason, it is important that you learn how to study algebra to get the most out of this course.

Although we all learn differently, here are some hints on studying algebra that most students find useful.

Planning a Strategy for Success To get where you want to be, you need a goal and a plan. Your goal should be to pass this course with a grade of A or B. To earn one of these grades, you must have a plan to achieve it. A good plan involves several points:

- Getting ready for class,
- Attending class,
- Doing homework,
- Making use of the extensive extra help available, including WebAssign if your instructor has set up a course, and
- Having a strategy for taking tests.

Getting Ready for Class   To get the most out of every class period, you will need to prepare for class. One of the best things you can do is to preview the material in the text that your instructor will be discussing in class. Perhaps you will not understand all of what you read, but you will be better able to understand your instructor when he or she discusses the material in class.

Do your work every day. If you fall behind, you will become frustrated and discouraged. Make a promise that you will always prepare for class, and then keep that promise.

Attending Class   The classroom experience is your opportunity to learn from your instructor and interact with your classmates. Make the most of it by attending every class. Sit near the front of the room where you can see and hear easily. Remember that it is your responsibility to follow the discussion, even though it takes concentration and hard work.

Pay attention to your instructor, and jot down the important things that he or she says. However, do not spend so much time taking notes that you fail to concentrate on what your instructor is explaining. Listening and understanding the big picture is much more important than just copying solutions to problems.

Do not be afraid to ask questions when your instructor asks for them. Asking questions will make you an active participant in the class. This will help you pay attention and keep you alert and involved.

Doing Homework   It requires practice to excel at tennis, master a musical instrument, or learn a foreign language. In the same way, it requires practice to learn mathematics. Since practice in mathematics is homework, homework is your opportunity to practice your skills and experiment with ideas. Consider creating note cards for important concepts.

It is important for you to pick a definite time to study and do homework. Set a formal schedule and stick to it. Try to study in a place that is comfortable and quiet. If you can, do some homework shortly after class, or at least before you forget what was discussed in class. This quick follow-up will help you remember the skills and concepts your instructor taught that day.

Each formal study session should include three parts:

1. Begin every study session with a review period. Look over previous chapters and see if you can do a few problems from previous sections. Keeping old skills alive will greatly reduce the amount of time you will need to prepare for tests.

2. After reviewing, read the assigned material. Resist the temptation to dive into the problems without reading and understanding the examples. Instead, work the examples and Self Checks with pencil and paper. Only after you completely understand the underlying principles behind them should you try to work the exercises.

   Once you begin to work the exercises, check your answers with the printed answers in the back of your text. If one of your answers differs from the printed answer, see if the two can be reconciled. Sometimes answers have more than one form. If you decide that your answer is incorrect, compare your work to the example in the text that most closely resembles the exercise, and try to find your mistake. If you cannot find an error, consult the *Student Solutions Manual*. If nothing works, mark the problem and ask about it in your next class meeting.

3. After completing the written assignment, preview the next section. This preview will be helpful when you hear the material discussed during the next class period.

You probably already know the general rule of thumb for college homework: two to three hours of practice for every hour you spend in class. If mathematics is difficult for you, plan on spending even more time on homework.

To make doing homework more enjoyable, study with one or more friends. The interaction will clarify ideas and help you remember them. If you choose to study alone, a good study technique is to explain the material to yourself out loud or use a white board where you can stand back and look at the big picture.

Accessing Additional Help   Access any help that is available from your instructor. Often, your instructor can clear up difficulties in a short period of time. Find out whether your college has a free tutoring program. Peer tutors often can be a great help or consider setting up your own study group.

Taking Tests   Students often become nervous before taking a test because they are afraid that they will do poorly.

To build confidence in your ability to take tests, rework many of the problems in the exercise sets, work the exercises in the Chapter Reviews, and take the Chapter Tests. Check all answers with the answers printed at the back of your text.

Guess what the instructor will ask, build your own tests, and work them. Once you know your instructor, you will be surprised at how good you can get at selecting test questions. With this preparation, you will have some idea of what will be on the test, and you will have more confidence in your ability to do well.

When you take a test, work slowly and deliberately. Write down any formulas at the top of the page. Scan the test and work the problems you find easiest first. Tackle the hardest problems last.

We wish you well.

Acknowledgements

We are grateful to the following people who reviewed the new editions of this series of texts. They all had valuable suggestions that have been incorporated into the texts.

Rob Farinelli, *Community College of Allegheny County*
Kathleen Kane, *Community College of Allegheny County- Allegheny Campus*
Marsha Lake, *Brevard (FL) Community College*
Carla A. Monticelli, *Camden County College*
Donna Szott, *Community College of Allegheny County- South Campus*
Yoshiyuki Yamamoto, *Brevard Community College*

Additional Acknowledgments

We also thank the following people who reviewed previous editions.
Kent Aeschliman, *Oakland Community College*
Carol Anderson, *Rock Valley College*
Cynthia Broughtou, *Arizona Western College*
David Byrd, *Enterprise State Junior College*
Pablo Chalmeta, *New River Community College*
Jerry Chen, *Suffolk County Community College*
Michael F. Cullinan, *Glendale Community College*
Amy Cuneo, *Santiago Canyon College*
Lou D'Alotto, *York College-CUNY*
Thomas DeAgostino, *Jackson Community College*
Kristin Dillard, *San Bernardino Valley College*
Kirsten Dooley, *Midlands Technical College*
Karen Driskell, *Calhoun Community College*
Joan Evans, *Texas Southern University*
Hamidullah Farhat, *Hampton University*
Harold Farmer, *Wallace Community College-Hanceville*
Mark Fitch, *University of Alaska, Anchorage*
Mark Foster, *Santa Monica College*
Tom Fox, *Cleveland State Community College*
Jeremiah Gilbert, *San Bernardino Valley College*
Harvey Hanna, *Ferris State University*
Jonathan P. Hexter, *Piedmont Virginia Community College*
Kathy Holster, *South Plains College*
Dorothy K. Holtgrefe, *Seminole Community College*
Mark Hopkins, *Oakland Community College*
Mike Judy, *Fullerton College*
Lynette King, *Gadsden State Community College*
Janet Mazzarella, *Southwestern College*
Donald J. McCarthy, *Glendale Community College*
Robert McCoy, *University of Alaska Anchorage*
Andrew P. McKintosh, *Glendale Community College*
Christian R. Miller, *Glendale Community College*
Feridoon Moinian, *Cameron University*
Brent Monte, *Irvine Valley College*

Daniel F. Mussa, *Southern Illinois University*
Joanne Peeples, *El Paso Community College*
Mary Ann Petruska, *Pensacola Junior College*
Linda Pulsinelli, *Western Kentucky University*
Kimberly Ricketts, *Northwest-Shoals Community College*
Janet Ritchie, *SUNY-Old Westbury*
Joanne Roth, *Oakland Community College*
Richard Rupp, *Del Mar College*
Rebecca Sellers, *Jefferson State Community College*
Kathy Spradlin, *Liberty University*
John Squires, *Cleveland State Community College*
April D. Strom, *Glendale Community College*
Robert Vilardi, *Troy University*
Victoria Wacek, *Missouri Western State College*
Judy Wells, *University of Southern Indiana*
Hattie White, *St. Phillip's College*
George J. Witt, *Glendale Community College*
Margaret Yoder, *Eastern Kentucky University*

We are grateful to the staff at Cengage Learning, especially our publisher Charlie Van Wagner and our editor Marc Bove. We also thank Vernon Boes, Jennifer Risden, Stefanie Beeck, and Bryon Spencer.

We are indebted to Lachina Publishing Services, our production service; to Rhoda Oden, who read the entire manuscript and worked every problem; and to Michael Welden, who prepared the *Student Solutions Manual.*

—ROSEMARY M. KARR
  MARILYN B. MASSEY
  R. DAVID GUSTAFSON

Index of Applications

Examples that are applications are shown with boldface numbers.
Exercises that are applications are shown with regular face numbers.

Architecture

Architecture, 354
Calculating clearance, 712
Designing a patio, 767–768
Designing a swimming pool, 355
Designing an underpass, 712
Drafting, 438
Finding the dimensions of a truss, **351–352**
Gateway Arch, **696–697**
House construction, 355
Landscape design, **708–709**

Business

Analyzing ads, 28
Annual rate of depreciation, 128
Appreciation equations, 128
Art, 128
Bank accounts, 40
Bookstores, 76
Building shelves, 76
Car repair, 128, 174
Carpentry, 86, 305, 355, 472
Cell phone growth, **626**
Cell phone usage, 631
Choosing salary plans, 189
Computing salaries, 69
Computing selling price, 75
Cost of electricity, 59
Cost of water, 59
Costs of a trucking company, 440
Crafts, 278
Customer service, 300
Cutting beams, 67
Cutting boards, 67
Cutting cables, 67
Cutting hair, **137–138**
Demand equations, 104
Depreciating a lawn mower, 128
Depreciation, **99**, 649
Depreciation equations, 128
Diet supplements, **274–275**
Directory costs, 376–377
Discount buying, 413
Earning money, 27, **257–258**
Fertilizer, 286
Figuring taxes, 261
Finding the percent of markdown, **62**
Fine arts, 354
Food service, 174
Furnace equipment, 269
Furniture, 269, 278

Housing, 141
Income, **273–274**
Inventories, 217, 278
Long-distance telephone call, **237**
Making bicycles, 187
Making clothing, 198
Making furniture, 76
Making gutters, 300
Making sporting goods, 261
Making statues, 197
Making water pumps, 188
Managing a beauty shop, 188
Managing a computer store, 188
Manufacturing footballs, 197
Manufacturing hammers, **194–195**
Merchandising, 187
Milling brass plates, 188
Operating costs, 577
Price increases, 413
Printer charges, 128
Printing books, 188
Printing stationery, **732**
Rate of decrease, 116
Rate of growth, 116
Reading proof, 412
Real estate, 128
Real estate listings, 128
Renting a rototiller, 240
Renting a truck, 240
Retail sales, **181**
Retailing, 174
Running a record company, 188
Salvage value, 128, 631
Scheduling equipment, 240
Selling apples, 480
Selling calculators, 67
Selling clothes, 28
Selling grass seed, 67
Selling hot dogs, 141
Selling shirts, 438
Selling tires, 141
Selling toys, 67
Selling trees, 261
Selling TVs, 315
Selling vacuum cleaners, **312**
Small businesses, 238
Supply equations, 104
Time for money to grow, 649
Transportation, 403
Trucking, 75
Value of a boat, 439

Value of a lathe, **122–123**
Work rates, 394

Education
Averaging grades, 28, 240
Calculating grades, 28
Computing grades, 76
History, 14
Planning a work schedule, 240
Raising grades, 75
Saving for college, **627**
School plays, 76
Selecting committees, **787**
Staffing, 438
Taking a test, 791

Electronics
Buying a computer, 240
Buying CDs, 240
CDs, 268
Computers, 790
DVDs, **543**
Electronics, 187, **321**, 403, 440, 466, 520
Finding the gain of an amplifier, 649
Production, 278–279
Selling DVD players, 141
TV translators, **693–694**

Entertainment
Buying tickets, 104
Concert receipts, 68
Concert tickets, 198, 262, 552
Juggling, 299
Movie stunts, 331
Production schedules, **275–276**
Theater seating, 206
TV coverage, 104
TV programming, **785**

Farming
Farming, 105, 187, 439
Fencing a field, 577
Fencing pastures, 68
Fencing pens, 68
Filling a pond, 412
Gardening, 305, 354
Grooming horses, 28
Making whole milk, 76
Milk production, **73–74**
Planting corn, 802
Production, 279
Storing corn, 412

Finance
Amount of an annuity, **773–774**
Annuities, 776
Boat depreciation, 806
Break-point analysis, **183–184**, 188
Car depreciation, 802
Comparing interest rates, 631
Comparing savings plans, 631

Compound interest, 631, 677
Compounding methods, 639
Computing taxes, **732**
Continuous compound interest, **633–634**, 638, 677
Cost and revenue, 174
Declining savings, 776
Depreciation, 158, 639
Depreciation rates, **506–507**
Determining a previous balance, 638
Determining the initial deposit, 638
Doubling money, 58, 655
Earning interest, 59, 103
Financial planning, **70–71**, 279
Frequency of compounding, 631
House appreciation, 776
Inheriting money, 75
Installment loans, 767
Interest rates, 552
Investing, 75, 88, 217, 238, 240, 262, 282, 313, 451, 575, 686
Investing money, 728
Investment, 87
Investment income, 187
Managing a checkbook, 28
Maximizing revenue, 577
Motorboat depreciation, 776
Pension funds, 67
Portfolio analysis, **62–63**
Profit, 552
Rule of seventy, 677
Saving money, **543**, 545, 767
Savings growth, 776
Stock appreciation, 802
Stock averages, 28
Supplemental income, 75
Tripling money, 655
Value of an IRA, 67

Geometry
Angles in a triangle, 64
Angles of a polygon, 59
Area of a circle, 439
Area of a triangle, 384
Area of an ellipse, 712
Area of many cubes, 473
Base of a triangle, 552
Circumference of a circle, 28
Complementary angles, 68
Complementary angles in a right triangle, 68
Curve fitting, **195–196**, 197, 198, 206
Diagonal of a baseball diamond, 466
Dimensions of a rectangle, **550**, 552, 609
Equilateral triangles, 67
Finding dimensions, 68
Geometry, 86, 88, 187, 188, 197, 206, 207, 305, 315, 316, 354, 439, 440, 472, 473, **487–489**, 728
Height of a triangle, 552
Inscribed squares, 776
Integers, 197, 354, 364, 552, 728
Interior angles, 768
Parallelograms, **185**
Perimeter of a rectangle, 394, 552

Perimeter of a square, 28, 355
Perimeter of a trapezoid, 27
Perimeter of a triangle, 27
Quadrilaterals, 67
Radius of a circle, 466
Side of a square, 552
Supplementary angles, 68
Supplementary angles and parallel lines, 69
Surface area of a cube, 473
Triangles, 69
Vertical angles, 68
Volume of a cone, **435–436**
Volume of a pyramid, 363
Volume of a rectangular solid, 363

Home Management and Shopping
Buying a TV and a Blu-ray player, 69
Buying a washer and dryer, 67
Buying books, 67
Buying furniture, 67
Buying roses, 68
Car depreciation, 104
Choosing housekeepers, 261
Cooking, 438
Cost of carpet, **111**
Dog runs, **64–65**
Draining a pool, 27, 412
Drywalling a house, **408**
Enclosing a swimming pool, 69
Filling a pool, 27, 412
Finding the percent of markdown, **62**
Framing pictures, 69
Grocery shopping, **430**
House appreciation, 104
House painting, 412
Housecleaning, 412
Landscaping, 174, **265–266**
Renting trucks, 68
Roofing a house, 412
Siding a house, 448
Slope of a ladder, 115
Slope of a roof, 115
Telephone costs, 105
Value of a car, 305
Value of a house, 305
Value of two cars, 305
Value of two houses, 305

Medicine
Body mass, 438, 473
Choosing a medical plan, 240–241
Diet, 278
Epidemics, 639
Forensic medicine, 656
Losing weight, 27
Medicine, 466, 545, 639, 676, 677, 796
Nutritional planning, 197
Physical therapy, 188, 206
Pulse rates, 501
Recommended dosage, 438
Sizes of viruses, 28

Miscellaneous
Accidents, 545
Arranging an evening, 790
Arranging appointments, 790
Arranging books, 790
Artillery, 728
Balancing a seesaw, 86
Balancing seesaws, 68, 69
Bouncing balls, 781
Broadcast ranges, 701
Building construction, 440
Buying gasoline, 26
Call letters, 791
Candy, 331
Cards, 806
Cats and dogs, 206
Chainsaw sculpting, 198
Checkers, 343
Choosing books, 791
Choosing clothes, 791
Choosing people, 804
Combination locks, 790
Communications, 525
Controlling moths, 781
Cutting ropes, 67
Darts, 331
Dimensions of a window, 552
Discharging a battery, 631
Dolphins, 141
Doubling time, **653–654**
Draining a tank, 448
Earthquakes, 649, 683
Electric service, 472
Elevators, 240
Falling object, 545
Falling objects, 768
Finding the endpoint of a line segment, 104
Finding the variance, 577
Fitness equipment, 711
Forming a committee, 791
Forming committees, 791
Framing a picture, 323, 552
Free-falling objects, 639
Genealogy, 776
Generating power, 509
Hardware, 491
Height of a bridge, **503**
Height of a flagpole, 439
Height of a tree, **431–432**, 439
Highway design, 701
Hobbies, 438
Horizon distance, 508
Hourly wages, 103
Ice cubes, 343
Information access, 735
Ironing boards, 492
Labor force, 553
Lifting cars, 68
Lining up, 790, 804, 806
Making Jell-O, 656
Maximizing area, **572–573**, 577

Meshing gears, 701
Metal fabrication, 552
Mixing candies, 75
Mixing candy, 187
Mixing fuel, 438
Mixing nuts, **72–73**, 198
Mixing solutions, 87, 88, 480
Model railroading, 438
Monuments, 394
Moving stones, 68
Numbers, 728
Office furniture, 343
Organ pipes, 440
Packing a tennis racket, 473
Palindromes, 790
Paying taxes, 26
Pendulum, 530, 545
Period of a pendulum, **462**, 466
Photography, **498–499**, 501
Piggy banks, 206
Planning a picnic, 791
Pool tables, 712
Preformed concrete, 364
Projectiles, 701
Quality control, 796
Reach of a ladder, 472
Renting a jet ski, 735
Riding in a taxi, 735
Roller coasters, 300
Royalties, 735
Satellite antennas, 701
Selecting committees, 805
Selecting people, 805
Shipping crates, 530
Shipping packages, 473
Signaling, 217
Sonic boom, 722
Springs, 253–254
Statistics, 466
Storage, 343
Storage capacity, 385
Storing oil, 440
Stretching a vacation, 413
Supporting a weight, 472
Targets, 501
Telephone service, 472
Value of a car, 103
Variance, **573**
View from a submarine, 525
Walking for charity, 412
Water usage, 84
Wheelchair ramps, 116
Width of a ring, 394
Width of a river, 439
Width of a walkway, 701
Winning a lottery, 791
Winter recreation, 355

Politics, Government, and the Military
Aircraft carriers, 323
Building a freeway, **470**

Building highways, **436**
City planning, **635–636**
Congress, **787**
Crime prevention, 105
Displaying the flag, 27
Environmental cleanup, 376
Fighting fires, **468**
Flags, 545
Forming committees, 804, 806
Growth of a town, **772–773**
Highway design, 508
Law enforcement, 466, 545
Louisiana Purchase, 631
Making a ballot, 790
Making license plates, 790
Marine Corps, 188
Military, 28
Phone numbers, 790
Police investigation, 141
Police investigations, 577
Population decline, 639, 775
Population growth, **636**, 638–639, 655, **673**, 676, 775
Predicting burglaries, 128
Rodent control, 676
School enrollment, 577
Space program, 553
Taxes, 450
Town population, 631
U.S. population, 682
Water usage, 577
World population growth, 638

Science
Alcohol absorption, 639
Alloys, 595
Alpha particles, 722
Angstroms per inch, 47
Aquariums, 665
Artillery fire, 141
Astronomy, **44**, 314
Atomic structure, **719–720**
Bacteria cultures, 631
Bacterial cultures, 676
Bacterial growth, 676
Balancing levers, 69
Ballistics, 141, **352**, 355, **571–572**, 576, 577, 609
Carbon-14 dating, **672**, 676, 686
Change in intensity, 666
Change in loudness, 666, 685
Changing temperatures, 27
Chemistry, 553
Choosing sample size, 241
Climate change, 115, 116
Conversion from degrees Celsius to degrees Fahrenheit, 141
Conversion from degrees Fahrenheit to degrees Celsius, 141
Converting temperatures, 58
Data analysis, 403
dB gain, 683
dB gain of an amplifier, 649
Diluting solutions, 76
Distance to Alpha Centauri, 47

Distance to the Moon, 47
Distance to the Sun, 47
Earth's atmosphere, 198
Electrostatic repulsion, 722
Engineering, **65–66**, 403, 413
Engineering designs, 69
Establishing equilibrium, 68
Falling objects, 439, 466, 802
Finding the constant of variation, 440
Finding the pH of a solution, **663**, 665
Flight to Pluto, 47
Focal length, 413
Force of gravity, 58
Gas pressure, 439, 440
Generation time, **673–674**
Half-life, 676
Half-life of radon-22, **671**
Heaviside unit step function, 734
Height of a rocket, **294–295**
Hurricanes, **93–94**
Hydrogen ion concentration, 665
Increasing concentration, 76
Lab experiments, 385
Lead decay, 677
Lens maker's formula, 413
Life of a comet, 47
Light intensity, **434–435**
Light year, 47
Mass of protons, 47
Measuring earthquakes, **646**
Mixing solutions, 76, **181–183**, 187
Oceanography, 677
Ohm's Law, 59
Operating temperatures, 254
Path of a comet, 701
pH of grapefruit, 685
pH of pickles, 666
Physics, 331
Power loss, 59
Radio frequencies, 189
Radioactive decay, 631, **637**, 639, 676, 682, 686
Range of a comet, 47
Range of camber angles, 254
Reading temperatures, 14
Research, 288
Signum function, 734
Solar flares, 47
Solar heating, 69
Speed of light, 47
Speed of sound, 47
Storm research, 75
Temperature change, **593**
Temperature ranges, 254
Temperature scales, 69
Tension, 440
Thermodynamics, 58
Thorium decay, 677
Tritium decay, 676
Volume of a tank, 47
Volume of sphere, **36**
Wavelengths, 47

Weather forecasting, 595
Weber–Fechner law, **663**

Sports
Area of a track, 712
Averaging weights, 28
Baseball, 472
Boating, 240, 413
Boats, 268
Bowling, 473
Buying golf clubs, 69
Cutting a rope, **61**
Cycling, 75
Football, **21**, 268
Mountain climbing, 27
NFL records, 198
Rate of descent, **112**
Rowing a boat, 413
Running a race, 75
Sailing, 472, 525
Ski runs, 439
Skydiving, 639
Track and field, 268
U.S. sports participation, 105

Travel
Aviation, 254, 413
Braking distance, 300
Bus fares, 552
Car rentals, 68
Computing distance, 75
Computing time, 75
Computing travel time, 76
Distance, rate, and time, 385
Distance traveled, 103
Driving rates, 728
Driving to a convention, **408–409**
Finding distance, 439
Finding rates, 413
Flight paths, 439
Flying speed, 448
Gas consumption, 438
Gas mileage, **98–99**
Grade of a road, 115
Motion, 87
Navigation, 174
Planning a trip, 804
Rates, 552
Riding a jet ski, 76
Riding a motorboat, 76
River cruise, **409–410**
Sailing time, 451
Stowing baggage, 187
Taking a walk, 76
Time of flight, 355
Time on the road, 413
Train travel, 413
Travel, 187, 575
Travel choices, 790
Travel times, **71–72**
Trip length, 448

A Review of Basic Algebra

Careers and Mathematics

FARMERS, RANCHERS, AND AGRICULTURAL MANAGERS

American farmers and ranchers direct the activities of one of the world's largest and most productive agricultural sectors. They produce enough food to meet the needs of the United States and produce a surplus for export.

Modern farming requires increasingly complex scientific, business, and financial decisions. Therefore, even people who were raised on farms must acquire the appropriate education.

Job Outlook:
The long-term trend toward consolidation of farms into fewer and larger farms is expected to continue through 2020. However, many small-scale farmers have developed successful market niches that involve direct contact with their customers, such as organic food production.

Since incomes of farmers vary greatly from year to year, many farmers have incomes from off-farm business activities.

Annual Earnings:
$29,280–$106,980

For More Information:
http://www.bls.gov/oco/ocos176.htm

For a Sample Application:
See Problem 45 in Section 1.6.

REACH FOR SUCCESS

1.1 The Real Number System

1.2 Arithmetic and Properties of Real Numbers

1.3 Exponents

1.4 Scientific Notation

1.5 Solving Linear Equations in One Variable

1.6 Using Linear Equations to Solve Applications

1.7 More Applications of Linear Equations in One Variable

■ *Projects*

REACH FOR SUCCESS EXTENSION

CHAPTER REVIEW

CHAPTER TEST

In this chapter

The language of mathematics is algebra. The word algebra comes from a book written by the Arabian mathematician Al-Khowarazmi around A.D. 800. Its title, Ihm al-jabr wa'l muqabalah, means restoration and reduction, a process then used to solve equations.

Reach for Success   Analyzing Your Time

We can all agree that we only have 24 hours in a day, right? The expression "My, how time flies!" is certainly relevant. It might surprise you where the time goes when you begin to monitor the hours.

How can you schedule your time to allow for a successful semester? Begin by completing the pie chart to analyze a typical day.

©Brad Collett/Shutterstock.com

HOW MANY HOURS A DAY DO YOU SPEND:

For every hour spent on each activity shade that many pie segments. (It might be helpful to use a different color to shade for each question.) There are no right or wrong answers. You just need an honest assessment of your time.

_____ working at outside employment (include commute time)?

_____ sleeping?

_____ preparing meals, eating, and exercising?

_____ getting ready for work/class?

_____ in class (include lab hours, commuting, or tutoring time)?

_____ with family and friends?

_____ on the Internet, phone, playing video games, texting, watching TV, going to movies, or other entertainment?

24-Hour Pie Chart

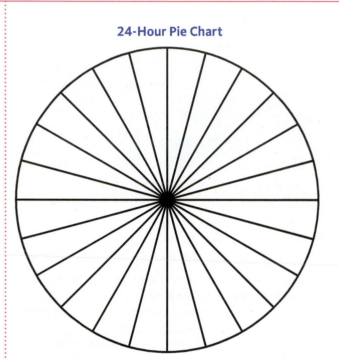

Now we have a mathematics problem! How many hours are left for studying? Remember, the rule of thumb is for every **1** hour in class, you will need at least **2** hours of studying. For mathematics, you could need even more. The key is to spend as much time as you need to understand the material. There is no magic formula!

A Successful Study Strategy . . .

Adjust your schedule as needed to support your educational goals.

At the end of the chapter you will find an additional exercise to guide you in planning for a successful semester.

Section 1.1

The Real Number System

Objectives

1. Identify any number in a set of real numbers as composite, whole, irrational, natural, prime, rational, and/or an integer.
2. Insert a symbol $<$, $>$, or $=$ to define the relationship between two rational numbers.
3. Graph a real number, a finite set of real numbers, or an inequality on the number line, and write the set in interval notation where appropriate.
4. Graph a compound inequality.
5. Find the absolute value of a real number.

Vocabulary

variables	prime numbers	real numbers
set	composite numbers	intervals
elements of a set	even integers	interval notation
finite set	odd integers	compound inequalities
natural numbers	roster method	union
whole numbers	set-builder notation	intersection
integers	rational numbers	absolute value
subset	irrational numbers	trichotomy property
coordinate		

Getting Ready

1. Name some different kinds of numbers.
2. Is there a largest number?

In algebra, we will work with expressions that contain numbers and letters. The letters that represent numbers are called **variables**. Some examples of algebraic expressions are

$$x + 45, \qquad \frac{8x}{3} - \frac{y}{2}, \qquad \frac{1}{2}bh, \qquad \frac{a}{b} + \frac{c}{d}, \qquad \text{and} \qquad 3x + 2y - 12$$

It is the use of variables that distinguishes algebra from arithmetic.

We begin the study of algebra by reviewing various subsets of the real numbers. We also will show how to graph these sets on a number line.

 1

Identify any number in a set of real numbers as composite, whole, irrational, natural, prime, rational, and/or an integer.

A **set** is a collection of objects. To denote a set, we can enclose a list of its **elements** with braces. If we can count the number of elements, we have a **finite set**. For example,

$\{1, 2, 3, 4\}$ denotes the finite set with elements 1, 2, 3, and 4.

$\left\{\frac{1}{2}, \frac{2}{3}, \frac{3}{4}\right\}$ denotes the finite set with elements $\frac{1}{2}$, $\frac{2}{3}$, and $\frac{3}{4}$.

Three infinite sets of numbers commonly used in algebra are as follows:

NATURAL NUMBERS	The set of **natural numbers** includes the numbers we use for counting:
	$\{1, 2, 3, 4, 5, 6, 7, 8, 9, \ldots\}$
WHOLE NUMBERS	The set of **whole numbers** includes the natural numbers together with 0:
	$\{0, 1, 2, 3, 4, 5, 6, 7, 8, 9, \ldots\}$
INTEGERS	The set of **integers** includes the natural numbers, 0, and the negatives of the natural numbers:
	$\{\ldots, -5, -4, -3, -2, -1, 0, 1, 2, 3, 4, 5, \ldots\}$

Each group of three dots, called *ellipses,* indicates that the numbers continue forever.

In Figure 1-1, we graph each of these sets from -6 to 6 on a number line.

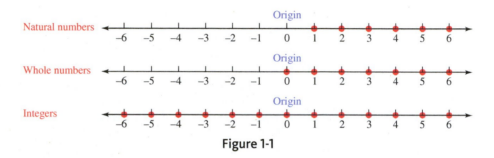

Figure 1-1

Since every natural number is also a whole number, we say that the set of natural numbers is a **subset** of the set of whole numbers. Note that the set of natural numbers and the set of whole numbers are also subsets of the integers.

To each number x in Figure 1-1, there corresponds a point on the number line called its *graph*, and to each point there corresponds a number x called its **coordinate**. Numbers to the left of 0 are *negative numbers*, and numbers to the right of 0 are *positive numbers*.

COMMENT   0 is neither positive nor negative.

There are two important subsets of the natural numbers.

PRIME NUMBERS	The **prime numbers** are the natural numbers greater than 1 that are divisible only by 1 and themselves.
	$\{2, 3, 5, 7, 11, 13, 17, 19, 23, \ldots\}$
COMPOSITE NUMBERS	The **composite numbers** are the natural numbers greater than 1 that are not prime.
	$\{4, 6, 8, 9, 10, 12, 14, 15, 16, 18, 20, 21, \ldots\}$

Figure 1-2 shows the graphs of the primes and composites that are less than or equal to 14.

Figure 1-2

COMMENT   1 is the only natural number that is neither prime nor composite.

There are two important subsets of the integers.

EVEN INTEGERS	The **even integers** are the integers that are exactly divisible by 2.
	$\{\ldots, -8, -6, -4, -2, 0, 2, 4, 6, 8, \ldots\}$
ODD INTEGERS	The **odd integers** are the integers that are not even.
	$\{\ldots, -9, -7, -5, -3, -1, 1, 3, 5, 7, 9, \ldots\}$

Figure 1-3 shows the graphs of the even and odd integers from -6 to 6.

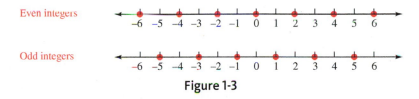

Figure 1-3

So far, we have written sets by listing their elements within braces. This method is called the **roster method**. A set also can be written in **set-builder notation**. In this method, we write a rule that describes which elements are in a set. For example, the set of even integers can be written in set-builder notation as follows:

$$\{x \mid x \text{ is an integer that can be divided exactly by 2}\}$$

The set of all x such that   a rule that describes membership in the set

To find coordinates of more points on the number line, we need the *rational numbers*.

RATIONAL NUMBERS	The set of **rational numbers** is
	$\left\{\dfrac{a}{b} \mid a \text{ is an integer and } b \text{ is a nonzero integer}\right\}.$

Each of the following numbers is an example of a rational number.

$$\frac{8}{5}, \quad \frac{2}{3}, \quad -\frac{44}{23}, \quad \frac{-315}{476}, \quad 0 = \frac{0}{7}, \quad \text{and} \quad 17 = \frac{17}{1}$$

Because each number has an integer numerator and a nonzero integer denominator

COMMENT   Note that $\frac{0}{5} = 0$, because $5 \cdot 0 = 0$. However, $\frac{5}{0}$ is undefined, because there is no number that, when multiplied by 0, gives 5. The fraction $\frac{0}{0}$ is indeterminate, because all numbers, when multiplied by 0, give 0. Remember that the denominator of a fraction cannot be 0.

EXAMPLE 1   Explain why each number is a rational number: **a.** -7 **b.** $-0.66\ldots$ **c.** 0.125

Solution   **a.** The integer -7 is a rational number because it can be written as $\frac{-7}{1}$, where -7 and 1 are integers and the denominator is not 0. Note that all integers are rational numbers.

        b. The decimal $-0.666\ldots$ is a rational number because it can be written as $\frac{-2}{3}$, where -2 and 3 are integers and the denominator is not 0.

        c. The decimal 0.125 is a rational number because it can be written as $\frac{125}{1,000}$, where 125 and $1,000$ are integers and the denominator is not 0. The fraction simplifies to $\frac{1}{8}$.

SELF CHECK 1   Explain why each number is a rational number:

     a. 4 **b.** $0.33\ldots$ **c.** 0.5

The next example illustrates that a rational number can be written as a decimal that either terminates or repeats a block of digits.

EXAMPLE 2   Change each fraction to decimal form and determine whether the decimal terminates or repeats: **a.** $\frac{3}{4}$ **b.** $\frac{421}{990}$

Solution   **a.** To change $\frac{3}{4}$ to a decimal, we divide 3 by 4 to obtain 0.75. This is a terminating decimal.

        b. To change $\frac{421}{990}$ to a decimal, we divide 421 by 990 to obtain $0.4252525\ldots$. This is a repeating decimal, because the block of digits 25 repeats forever. This decimal can be written as $0.4\overline{25}$, where the overbar indicates the repeating block of digits.

SELF CHECK 2   Change each fraction to decimal form and determine whether the decimal terminates or repeats:

     a. $\dfrac{5}{11}$ **b.** $\dfrac{2}{5}$

The rational numbers provide coordinates for many points on the number line that lie between the integers (see Figure 1-4). Note that the integers are a subset of the rational numbers.

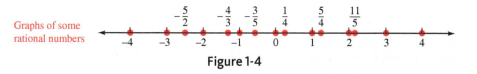

Graphs of some rational numbers

Figure 1-4

Points on the number line whose coordinates are nonterminating, nonrepeating decimals have coordinates that are called *irrational numbers*.

IRRATIONAL NUMBERS   The set of **irrational numbers** is

$$\{x \mid x \text{ is a nonterminating, nonrepeating decimal}\}.$$

Some examples of irrational numbers are

$$0.313313331\ldots \qquad \sqrt{2} \approx 1.414213562\ldots \qquad \pi \approx 3.141592653\ldots$$

If we combine the set of rational numbers (the terminating and repeating decimals) and the set of irrational numbers (the nonterminating, nonrepeating decimals), we obtain the set of *real numbers*.

REAL NUMBERS

The set of **real numbers** is

$\{x \mid x$ is a terminating decimal; a repeating decimal; or a nonterminating, nonrepeating decimal$\}$.

The number line in Figure 1-5 shows several points on the number line and their real-number coordinates. The points whose coordinates are real numbers fill up the number line.

Figure 1-5

Figure 1-6 illustrates how various sets of numbers are interrelated.

François Vieta (Viête)
1540–1603
By using letters in place of unknown numbers, Vieta simplified the subject of algebra and brought its notation closer to the notation that we use today. One symbol he didn't use was the equal sign.

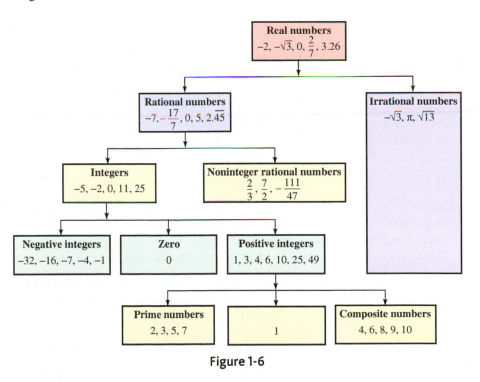

Figure 1-6

EXAMPLE 3   Which numbers in the set $\left\{-2, 0, \frac{5}{9}, 1.25, \sqrt{8}, 17\right\}$ are

a. natural numbers   **b.** whole numbers   **c.** integers
d. rational numbers   **e.** irrational numbers   **f.** real numbers?

Solution   **a.** The only natural number is 17.

b. The whole numbers are 0 and 17.

c. The integers are -2, 0, and 17.

d. The rational numbers are -2, 0, $\frac{5}{9}$, 1.25, and 17. $\left(1.25 \text{ is rational, because } 1.25 \text{ can be written in the form } \frac{5}{4}.\right)$

e. The only irrational number is $\sqrt{8}$. It is irrational because $\sqrt{8} = 2.828427125\ldots$ and this decimal is a nonterminating, nonrepeating decimal.

f. All of the numbers are real numbers.

SELF CHECK 3   Which numbers in the set $\left\{-2, 0, 1.5, \sqrt{5}, 7\right\}$ are

a. positive integers   **b.** rational numbers?

---

2   Insert a symbol $<$, $>$, or $=$ to define the relationship between two rational numbers.

To indicate that two quantities are not equal, we can use an inequality symbol, as summarized in Table 1-1.

TABLE 1-1

Symbol	Read as	Examples
$\neq$	"is not equal to"	$6 \neq 9$ and $0.33 \neq \frac{3}{5}$
$<$	"is less than"	$22 < 40$ and $0.27 < 3.1$
$>$	"is greater than"	$19 > 5$ and $\frac{1}{2} > 0.3$
$\leq$	"is less than or equal to"	$1.8 \leq 3.5$ and $35 \leq 35$
$\geq$	"is greater than or equal to"	$25.2 \geq 23.7$ and $29 \geq 29$
$\approx$	"is approximately equal to"	$\sqrt{2} \approx 1.414$ and $\sqrt{3} \approx 1.732$

We can always write an inequality with the inequality symbol pointing in the opposite direction. For example,

$17 < 25$ is equivalent to $25 > 17$

$5.3 \geq -2.9$ is equivalent to $-2.9 \leq 5.3$

COMMENT   Note that the inequality symbol always points to the smaller number.

On the number line, the coordinates of points get larger as we move from left to right, as shown in Figure 1-7. Thus, if a and b are the coordinates of two points, the one to the right is the greater.

This suggests the following general principle:

If $a > b$, point a lies to the right of point b on a number line.

If $a < c$, point a lies to the left of point c on a number line.

Figure 1-7

---

3   Graph a real number, a finite set of real numbers, or an inequality on the number line, and write the set in interval notation where appropriate.

Graphs of some sets of real numbers are portions of a number line called **intervals**. For example, the graph shown in Figure 1-8(a) is the interval that includes all real numbers x that are greater than -5. Since these numbers make the inequality $x > -5$ true, we say that they satisfy the inequality. The parenthesis on the graph indicates that -5 is not included in the interval.

To express this interval in **interval notation**, we write $(-5, \infty)$. Once again, the parentheses indicate that the endpoints are not included.

The interval shown in Figure 1-8(b) is the graph of the inequality $x \le 7$. It contains all real numbers that are less than or equal to 7. The bracket at 7 indicates that 7 is included in the interval. To express this interval in interval notation, we write $(-\infty, 7]$. The bracket indicates that 7 is in the interval.

COMMENT   The symbol ∞ (infinity) is not a real number. It is used in Figure 1-8(a) to indicate that the graph extends infinitely far to the right. The $-\infty$ in Figure 1-8(b) indicates that the graph extends infinitely far to the left.

When using interval notation, $-\infty$ and ∞ will always be preceded or followed by a parenthesis.

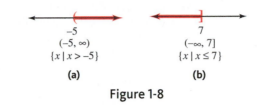

Figure 1-8

The graphs in Figure 1-8 also can be drawn using open and filled circles. An open circle indicates that an endpoint is not included, and a closed circle indicates that an endpoint is included.

EXAMPLE 4   Graph each set on the number line and then write the set in interval notation:

    a. $\{x \mid x < 9\}$ **b.** $\{x \mid x \ge 6\}$

Solution   **a.** $\{x \mid x < 9\}$ includes all real numbers that are less than 9. The graph is shown in Figure 1-9(a). This is the interval $(-\infty, 9)$.

    b. $\{x \mid x \ge 6\}$ includes all real numbers that are greater than or equal to 6. The graph is shown in Figure 1-9(b). This is the interval $[6, \infty)$.

Figure 1-9

SELF CHECK 4   Graph $\{x \mid x \ge 5\}$ and then write the set in interval notation.

4   Graph a compound inequality.

COMMENT   In a compound inequality involving *or*, the entire statement is true if either part is true.

Expressions that involve more than one inequality are called **compound inequalities**. One type of compound inequality involves the word *or*. For example, the graph of the set

    $\{x \mid x < -2 \text{ or } x \ge 3\}$ Read as "the set of all real numbers x such that x is less than -2 or x is greater than or equal to 3."

is shown in Figure 1-10. This graph is called the **union** of two intervals and can be written in interval notation as

    $(-\infty, -2) \cup [3, \infty)$ Read the symbol $\cup$ as "union."

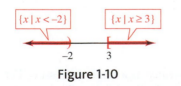

Figure 1-10

EXAMPLE 5   Graph each set on the number line and write it in interval notation:

    a. $\{x \mid x \le -4 \text{ or } x > 5\}$ **b.** $\{x \mid x \ge -4 \text{ or } x > 5\}$

Solution   **a.** The set $\{x \mid x \le -4 \text{ or } x > 5\}$ includes all real numbers less than or equal to -4 together with all real numbers greater than 5, as shown in Figure 1-11(a). This is the interval $(-\infty, -4] \cup (5, \infty)$.

    b. The set $\{x \mid x \ge -4 \text{ or } x > 5\}$ includes all real numbers that are greater than or equal to -4 together with all real numbers that are greater than 5. This includes all numbers x that are greater than or equal to -4 or greater than 5. (See Figure 1-11(b).) This is the interval $[-4, \infty)$.

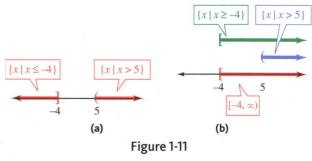

(a)   (b)

Figure 1-11

SELF CHECK 5   Graph $\{x \mid x < -1 \text{ or } x > 5\}$ and write it in interval notation.

Another type of compound inequality involves the word *and*. For example, the graph of the set

    $\{x \mid x \ge -5 \text{ and } x \le 6\}$ Read as "the set of all real numbers x such that x is greater than or equal to -5 and x is less than or equal to 6."

is shown in red in Figure 1-12(a). It is obtained by finding the overlap (or **intersection**) of the graphs of $\{x \mid x \ge -5\}$ and $\{x \mid x \le 6\}$. Using interval notation, this could be written as

    $(-\infty, 6] \cap [-5, \infty)$ Read the symbol $\cap$ as "intersection."

The overlap shown in red is the intersection of two intervals.

    From the graph, we see that the interval can be written more compactly as $[-5, 6]$.

    To graph the set $\{x \mid x \ge -5 \text{ and } x > 6\}$, we find the overlap of the graphs of $\{x \mid x \ge -5\}$ and $\{x \mid x > 6\}$. This overlap is the interval $(6, \infty)$ as shown in Figure 1-12(b). In set-builder notation, the solution is $\{x \mid x > 6\}$.

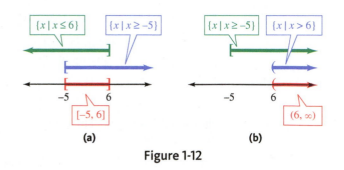

(a)   (b)

Figure 1-12

COMMENT   In an intersection, the statement is true only when both statements are true.

Two inequalities involving the word *and* often can be written as a single expression. For example, the compound inequality $x \ge -5$ and $x \le 6$ previously discussed can be written as

$-5 \leq x$ and $x \leq 6$ From Figure 1-12(a), note that this includes all numbers from -5 to 6, including -5 and 6.

or as the single expression $-5 \leq x \leq 6$.

EXAMPLE 6   Graph each set on the number line and then write it in interval notation:

a. $\{x \mid -5 \leq x \leq 6\}$ **b.** $\{x \mid 2 < x \leq 8\}$ **c.** $\{x \mid -4 \leq x < 3\}$
d. $\{x \mid 0 < x < 7\}$

Solution   **a.** The set $\{x \mid -5 \leq x \leq 6\}$ includes all real numbers from -5 to 6, as shown in Figure 1-13(a). This is the interval $[-5, 6]$.

b. The set $\{x \mid 2 < x \leq 8\}$ includes all real numbers between 2 and 8, including 8, as shown in Figure 1-13(b). This is the interval $(2, 8]$.

c. The set $\{x \mid -4 \leq x < 3\}$ includes all real numbers between -4 and 3, including -4, as shown in Figure 1-13(c). This is the interval $[-4, 3)$.

COMMENT   In part **d**, $(0, 7)$ is an interval and not a point.

d. The set $\{x \mid 0 < x < 7\}$ includes all real numbers between 0 and 7 as shown in Figure 1-13(d). This is the interval $(0, 7)$.

(a)   (b)   (c)   (d)

Figure 1-13

SELF CHECK 6   Graph $\{x \mid -6 < x \leq 10\}$ and then write it in interval notation.

Table 1-2 illustrates three ways to describe an interval.

TABLE 1-2		
Set notation	**Graph**	**Interval notation**
$\{x \mid x > a\}$		(a, ∞)
$\{x \mid x < a\}$		$(-\infty, a)$
$\{x \mid x \geq a\}$		$[a, \infty)$
$\{x \mid x \leq a\}$		$(-\infty, a]$
$\{x \mid a < x < b\}$		(a, b)
$\{x \mid a \leq x < b\}$		$[a, b)$
$\{x \mid a < x \leq b\}$		$(a, b]$
$\{x \mid a \leq x \leq b\}$		$[a, b]$
$\{x \mid x < a \text{ or } x > b\}$		$(-\infty, a) \cup (b, \infty)$

5 Find the absolute value of a real number.

The **absolute value** of a real number a, denoted as $|a|$, is the distance on a number line between 0 and the point with coordinate a. For example, the points shown in Figure 1-14 with coordinates of 3 and -3 both lie 3 units from 0. Thus, $|3| = |-3| = 3$.

3 units   3 units

-3 0   3

Figure 1-14

In general, for any real number a, $|a| = |-a|$.

The absolute value of a number can be defined more formally.

ABSOLUTE VALUE

For any real number x, $\begin{cases} \text{if } x \geq 0, \text{ then } |x| = x. \\ \text{if } x < 0, \text{ then } |x| = -x. \end{cases}$

If x is positive or 0, then x is its own absolute value. However, if x is negative, then $-x$, read as "the opposite of x" (which is a positive number) is the absolute value of x. Thus, $|x| \geq 0$ for all real numbers x.

EXAMPLE 7   Find each absolute value.

a. $|3| = 3$ **b.** $|-4| = 4$

c. $|0| = 0$ **d.** $-|\mathbf{-8}| = -(8) = -8$ Note that $|-8| = 8$.

SELF CHECK 7   Find each absolute value: **a.** $|-9|$ **b.** $-|-12|$

SELF CHECK ANSWERS

1. a. $4 = \frac{4}{1}$ **b.** $0.33\ldots = \frac{1}{3}$ **c.** $0.5 = \frac{1}{2}$ **2. a.** $0.\overline{45}$, repeating decimal   **b.** 0.4, terminating decimal   **3. a.** 7
b. $-2, 0, 1.5, 7$ **4.** $[5, \infty)$ **5.** $(-\infty, -1) \cup (5, \infty)$

6. $(-6, 10]$ **7. a.** 9   **b.** -12

NOW TRY THIS

1. Which numbers in the set $\left\{-\frac{5}{8}, \sqrt{3}, 6.2222\ldots, 0, \sqrt{25}, 4, -9.01020304\ldots\right\}$ are
    a. integers   **b.** irrational numbers
    c. rational numbers?

2. Describe each graph in set-builder and interval notation.

    (a)   (b)   (c)

1.1 Exercises

WARM-UPS *Identify two numbers on a number line whose distance is*

1. 5 units from zero.
2. 8 units from zero.

Determine whether the given number is divisible only by 1 and itself.

3. 7
4. 6

REVIEW *To simplify a fraction, factor the numerator and the denominator and divide out common factors. For example,* $\frac{12}{18} = \frac{6 \cdot 2}{6 \cdot 3} = \frac{\cancel{6} \cdot 2}{\cancel{6} \cdot 3} = \frac{2}{3}$. *Simplify each fraction.*

5. $\frac{6}{8}$
6. $\frac{20}{24}$

7. $\frac{32}{40}$
8. $\frac{54}{60}$

To multiply fractions, multiply the numerators and multiply the denominators. To divide fractions, invert the divisor and multiply. Always simplify the result if possible.

9. $\frac{1}{4} \cdot \frac{3}{5}$
10. $\frac{2}{7} \cdot \frac{21}{26}$

11. $\frac{2}{7} \div \frac{7}{4}$
12. $\frac{3}{5} \div \frac{9}{15}$

To add (or subtract) fractions, write each fraction with a common denominator and add (or subtract) the numerators and keep the same denominator. Always simplify the result if possible.

13. $\frac{5}{9} + \frac{4}{9}$
14. $\frac{16}{7} - \frac{2}{7}$

15. $\frac{4}{5} + \frac{3}{4}$
16. $\frac{6}{7} - \frac{2}{5}$

VOCABULARY AND CONCEPTS *Fill in the blanks.*

17. A letter that represents a number is called a _____.
18. A __ is a collection of objects called _____.
19. The numbers 1, 2, 3, 4, 5, 6, . . . form the set of _____ numbers.
20. The natural numbers together with 0 is called the set of _____ numbers.
21. The natural numbers, 0, and the negatives of the natural numbers form the set of _____.
22. The union of the rational and irrational numbers is called the set of ___ numbers.
23. An ____ integer can be divided exactly by 2.
24. An ___ integer cannot be divided exactly by 2.
25. A prime number is a _____ number that is larger than _ and can only be divided exactly by ____ and 1.
26. A _____ number is a natural number greater than _ that is not _____.

27. _ is neither positive nor negative.
28. The denominator of a fraction can never be _.
29. A repeating decimal represents a _____ number.
30. A nonrepeating, nonterminating decimal represents an _____ number.
31. The symbol __ means "is less than."
32. The symbol __ means "is greater than or equal to."
33. The symbol _ means "is approximately equal to."
34. If x is negative, $|x| = $ __.
35. $x \geq 5$ is called an _____ while "$x < -2$ or $x \geq 9$" is a _____.
36. Integers are a _____ of the rational numbers.
37. The symbol $\cup$ means _____ and the symbol $\cap$ means _____.
38. An answer such as $\{0, 1, 2, 3, \ldots\}$ is in _____ notation, $(-2, \infty)$ is in _____ notation and $\{x \mid x > -2\}$ is in _____ notation.

GUIDED PRACTICE *Change each fraction into a decimal and classify the result as a terminating or a repeating decimal.* SEE EXAMPLE 2. (OBJECTIVE 1)

39. $\frac{7}{8}$
40. $\frac{5}{3}$

41. $-\frac{7}{15}$
42. $-\frac{19}{16}$

List the elements in the set $\left\{-3, 0, \frac{2}{3}, 1, \sqrt{3}, 2, 9\right\}$ *that satisfy the given condition.* SEE EXAMPLES 1 AND 3. (OBJECTIVE 1)

43. natural number
44. whole number

45. integer
46. rational number

47. irrational number
48. real number

49. even natural number
50. odd integer

51. prime number
52. composite number

53. odd composite number
54. even prime number

Insert a $<$ or a $>$ symbol to make a true statement. (OBJECTIVE 2)

55. 4 ▯ 7
56. −2 ▯ −3

57. −5 ▯ −10
58. −6 ▯ 2

59. −7 ▯ 7
60. 0 ▯ −5

61. 6 ▯ −6
62. −8 ▯ −1

Write each statement with the inequality symbol pointing in the opposite direction. (OBJECTIVE 2)

63. 19 > 12
64. −7 ≥ −10

65. −6 ≤ −5
66. −10 < 13

67. $5 \geq -3$ **68.** $0 \geq -1$

69. $-3 < 0$ **70.** $-2 > -5$

Graph each set on the number line. SEE EXAMPLE 4. (OBJECTIVE 3)

71. the set of prime numbers less than 8

0 1 2 3 4 5 6 7 8

72. the set of integers between -7 and 0

−7 −6 −5 −4 −3 −2 −1 0

73. the set of odd integers between 10 and 18

10 11 12 13 14 15 16 17 18

74. the set of composite numbers less than 10

0 1 2 3 4 5 6 7 8 9 10

75. $\{x \mid x > 3\}$ **76.** $\{x \mid x < 0\}$

77. $[-5, \infty)$ **78.** $(-\infty, 9]$

Graph each set on the number line. SEE EXAMPLE 5. (OBJECTIVE 4)

79. $\{x \mid x < -3 \text{ or } x > 3\}$ **80.** $(\infty, -4] \cup (2, \infty)$

81. $(-\infty, -6] \cup [5, \infty)$ **82.** $\{x \mid x < -2 \text{ or } x \geq 3\}$

Graph each set on the number line. SEE EXAMPLE 6. (OBJECTIVE 4)

83. $\{x \mid 6 < x \leq 10\}$ **84.** $\{x \mid -3 \leq x \leq 8\}$

85. $\{x \mid -1 < x \leq 3\}$ **86.** $\{x \mid 2 < x < 5\}$

Write each expression without using absolute value symbols. Simplify the result when possible. SEE EXAMPLE 7. (OBJECTIVE 5)

87. $|25|$ **88.** $|-14|$

89. $-|-5|$ **90.** $-|8|$

ADDITIONAL PRACTICE *Simplify.*

91. $|-5| + |-2|$ **92.** $|-8| + |5|$

93. $|-5| \cdot |4|$ **94.** $|-2| \cdot |3|$

95. Find the value of x if $|x| = 10$.

96. Find the value of x if $|x| = 7$.

97. What numbers x are equal to their own absolute values?

98. What numbers x when added to their own absolute values give a sum of 0?

Graph each set on the number line.

99. $\{x \mid x \leq 7\}$ **100.** $\{x \mid x \geq -2\}$

101. $[0, 5)$ **102.** $[-6, 9]$

103. $(6, \infty) \cup [9, \infty)$ **104.** $\{x \mid x > -8 \text{ and } x > 0\}$

105. $\{x \mid x > 7 \text{ and } x \geq 11\}$ **106.** $\{x \mid x < -2 \text{ and } x \leq -3\}$

APPLICATIONS

107. Reading temperatures On the thermometer in the illustration, graph each of the following temperature readings: $12°, 8°, 0°, -6°$

108. History On the number line in the illustration, the origin is the point B.C./A.D.
   a. What happened in the year 1441?
   b. What happened in the year 500 B.C.? (This would correspond to -500 on a number line.)

MAYA CIVILIZATION

Based on data from People in Time and Place, Western Hemisphere *(Silver Burdett & Ginn Inc., 1991), p. 129*

WRITING ABOUT MATH

109. Explain why the integers are a subset of the rational numbers.

110. Explain why every integer is a rational number, but not every rational number is an integer.

111. Explain why the set of primes together with the set of composites is not the set of natural numbers.

112. Is the absolute value of a number always positive? Explain.

SOMETHING TO THINK ABOUT

113. How many integers have an absolute value that is less than 50?

114. How many odd integers have an absolute value between 20 and 40?

115. The **trichotomy property** of real numbers states that

If a and b are two real numbers, then

$$a < b \quad \text{or} \quad a = b \quad \text{or} \quad a > b$$

Explain why this is true.

116. Which of the following statements are always true?
  a. $|a + b| = |a| + |b|$
  b. $|a \cdot b| = |a| \cdot |b|$
  c. $|a + b| \leq |a| + |b|$

Section 1.2

Arithmetic and Properties of Real Numbers

Objectives

1. Add real numbers.
2. Subtract real numbers.
3. Multiply real numbers.
4. Divide real numbers.
5. Evaluate a numeric expression following the rules for order of operations.
6. Find the mean, median, and mode when given a set of values.
7. Evaluate an algebraic expression when given values for its variables.
8. Identify the property of real numbers that justifies a given statement.

Vocabulary

addend
sum
difference
factor
product
quotient
grouping symbols

mean
median
mode
expression
perimeter
circumference
commutative properties

associative properties
distributive property
additive identity
multiplicative identity
additive inverse
multiplicative inverse

Getting Ready

Perform each operation.

1. $5 + 4$ **2.** $4 + 5$ **3.** $3 \cdot 4$ **4.** $4 \cdot 3$

5. $12 - 7$ **6.** $15 \div 3$ **7.** $21 \div 7$ **8.** $25 - 19$

We will begin this section by discussing how to add, subtract, multiply, and divide real numbers.

1 Add real numbers.

When two numbers—each number called an **addend**—are added, we call the result their **sum**. To find the sum of $+2$ and $+3$, we can use a number line and represent the numbers with arrows, as shown in Figure 1-15(a). Since the endpoint of the second arrow is at $+5$, we have $+2 + (+3) = +5$.

To add -2 and -3, we can draw arrows as shown in Figure 1-15(b). Since the endpoint of the second arrow is at -5, we have $(-2) + (-3) = -5$.

COMMENT A negative number moves to the left and a positive number moves to the right on a number line.

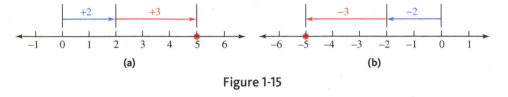

(a)   (b)

Figure 1-15

To add -6 and $+2$, we can draw arrows as shown in Figure 1-16(a). Since the endpoint of the second arrow is at -4, we have $(-6) + (+2) = -4$.

To add $+7$ and -4, we can draw arrows as shown in Figure 1-16(b). Since the endpoint of the final arrow is at $+3$, we have $(+7) + (-4) = +3$.

(a)   (b)

Figure 1-16

These examples suggest the following rules.

ADDING TWO REAL NUMBERS	With like signs: Add the absolute values of the numbers and use the common sign.
	With unlike signs: Subtract the absolute values of the numbers (the smaller from the larger) and use the sign of the number with the larger absolute value.

EXAMPLE 1 Add:

  a. $+4 + (+6) = +10$ Add the absolute values and use the common sign: $4 + 6 = +10$

  b. $-5 + (-3) = -8$ Add the absolute values and use the common sign: $-(5 + 3) = -8$

  c. $+9 + (-5) = +4$ Subtract the absolute values and use a $+$ sign: $+(9 - 5) = +4$

  d. $-12 + (+5) = -7$ Subtract the absolute values and use a $-$ sign: $-(12 - 5) = -7$

SELF CHECK 1 Add: **a.** $-7 + (-2)$ **b.** $-7 + 2$ **c.** $7 + 2$ **d.** $7 + (-2)$

2 Subtract real numbers.

When one number is subtracted from another number, we call the result their **difference**. To find the difference between two real numbers, we can write the subtraction as an equivalent addition. For example, the subtraction $7 - 4$ is equivalent to the addition $7 + (-4)$, because they have the same result:

$$7 - 4 = 3 \qquad \text{and} \qquad 7 + (-4) = 3$$

This suggests that to subtract two numbers, we can change the sign of the number being subtracted and add.

SUBTRACTING TWO REAL NUMBERS	If a and b are real numbers, then $a - b = a + (-b)$.

EXAMPLE 2   Subtract:

a. $12 - \mathbf{4} = 12 + (\mathbf{-4})$ Add the opposite of 4.
$= 8$

b. $-13 - \mathbf{5} = -13 + (\mathbf{-5})$ Add the opposite of 5.
$= -18$

c. $-14 - (\mathbf{-6}) = -14 + (\mathbf{+6})$ Add the opposite of -6.
$= -8$

SELF CHECK 2   Subtract: **a.** $-15 - 4$ **b.** $8 - 5$ **c.** $-12 - (-7)$

3   Multiply real numbers.

When two numbers—each number called a **factor**—are multiplied, we call the result their **product**. We can find the product of 5 and 4 by using 4 in an addition five times:

five addends of 4

$$5(4) = 4 + 4 + 4 + 4 + 4 = 20$$

We can find the product of 5 and -4 by using -4 in an addition five times:

five addends of -4

$$5(-4) = (-4) + (-4) + (-4) + (-4) + (-4) = -20$$

Since multiplication by a negative number can be defined as repeated subtraction, we can find the product of -5 and 4 by using 4 in a subtraction five times:

$$-5(4) = -4 - 4 - 4 - 4 - 4$$
$$= -4 + (-4) + (-4) + (-4) + (-4) \quad \text{Change the sign of each 4 and add.}$$
$$= -20$$

We can find the product of -5 and -4 by using -4 in a subtraction five times:

$$-5(-4) = -(-4) - (-4) - (-4) - (-4) - (-4)$$
$$= 4 + 4 + 4 + 4 + 4 \quad \text{Change the sign of each } -4 \text{ and add.}$$
$$= 20$$

The products $5(4)$ and $-5(-4)$ both equal $+20$, and the products $5(-4)$ and $-5(4)$ both equal -20. These results suggest the first two of the following rules.

MULTIPLYING TWO REAL NUMBERS	With like signs: Multiply their absolute values. The product is positive.
	With unlike signs: Multiply their absolute values. The product is negative.
	Multiplication by 0: If x is any real number, then $x \cdot 0 = 0 \cdot x = 0$.

COMMENT   The product of an even number of negative values produces a positive result and the product of an odd number of negative values produces a negative result.

EXAMPLE 3   Multiply:

a. $4(-7) = -28$ Multiply the absolute values: $4 \cdot 7 = 28$. Since the signs are unlike, the product is negative.

b. $-5(-6) = +30$ Multiply the absolute values: $5 \cdot 6 = 30$. Since the signs are alike, the product is positive.

c. $-7(6) = -42$ Multiply the absolute values: $7 \cdot 6 = 42$. Since the signs are unlike, the product is negative.

d. $8(6) = +48$ Multiply the absolute values: $8 \cdot 6 = 48$. Since the signs are alike, the product is positive.

SELF CHECK 3   Multiply:
a. $(-6)(5)$ **b.** $(-4)(-8)$ **c.** $(17)(-2)$ **d.** $(12)(6)$

4   **Divide real numbers.**

When two numbers are divided, we call the result their **quotient**. In the division $x \div y = q$ or $\frac{x}{y} = q$ $(y \neq 0)$, the quotient q is a number such that $y \cdot q = x$. We can use this relationship to find rules for dividing real numbers. We consider four divisions:

$$\frac{+10}{+2} = +5, \text{ because } +2(+5) = +10 \qquad \frac{-10}{-2} = +5, \text{ because } -2(+5) = -10$$

$$\frac{-10}{+2} = -5, \text{ because } +2(-5) = -10 \qquad \frac{+10}{-2} = -5, \text{ because } -2(-5) = +10$$

These results suggest the first two rules for dividing real numbers.

DIVIDING TWO REAL NUMBERS

With like signs: Divide their absolute values. The quotient is positive.

With unlike signs: Divide their absolute values. The quotient is negative.

Division by 0: Division by 0 is undefined.

COMMENT   If $x \neq 0$, then $\frac{0}{x} = 0$. However, $\frac{x}{0}$ is undefined for any value of x.

EXAMPLE 4   Divide:

a. $\frac{36}{18} = +2$ Divide the absolute values: $\frac{36}{18} = 2$. Since the signs are alike, the quotient is positive.

b. $\frac{-44}{11} = -4$ Divide the absolute values: $\frac{44}{11} = 4$. Since the signs are unlike, the quotient is negative.

c. $\frac{27}{-9} = -3$ Divide the absolute values: $\frac{27}{9} = 3$. Since the signs are unlike, the quotient is negative.

d. $\dfrac{-64}{-8} = +8$ Divide the absolute values: $\dfrac{64}{8} = 8$. Since the signs are alike, the quotient is positive.

 SELF CHECK 4   Divide: **a.** $\dfrac{55}{-5}$ **b.** $\dfrac{-72}{-6}$ **c.** $\dfrac{-100}{10}$ **d.** $\dfrac{50}{25}$

5 Evaluate a numeric expression following the rules for order of operations.

Suppose you are asked to contact a friend if you see a rug for sale while traveling in Turkey. After locating a nice one, you send the following text to your friend.

The next day, you receive this response.

The first statement in your friend's text says to buy the rug at any price. The second says not to buy it, because it is too expensive. The placement of the exclamation point makes these statements read differently, resulting in different interpretations.

When reading mathematical statements, the same kind of confusion is possible. To illustrate, we consider the expression $5 + 3 \cdot 7$, which contains the operations of addition and multiplication. To avoid the possibility of obtaining two different results when performing this type of calculation, we will agree to do the multiplications first.

Correct choice: Multiply first	*Incorrect choice: Add first*
$5 + \mathbf{3 \cdot 7} = 5 + \mathbf{21}$ Multiply 3 and 7.	$\mathbf{5 + 3} \cdot 7 = \mathbf{8} \cdot 7$ Add 5 and 3.
$= 26$ Add 5 and 21.	$= 56$ Multiply 8 and 7.

Different results

If we want to do additions before multiplications, we must use **grouping symbols** such as parentheses ( ), brackets [ ], or braces { }. You will also see radical symbols, absolute value bars, and the fraction bar used as grouping symbols. In the expression $(5 + 3)7$, the parentheses indicate that the addition is to be done first:

$$(\mathbf{5 + 3})7 = \mathbf{8} \cdot 7$$
$$= 56$$

To guarantee that calculations will have one correct result, we always will do calculations in the following order.

RULES FOR THE ORDER OF OPERATIONS FOR EXPRESSIONS WITHOUT EXPONENTS

Use the following steps to perform all calculations within each pair of grouping symbols, working from the innermost pair to the outermost pair.

1. Perform all multiplications and divisions, working from left to right.
2. Perform all additions and subtractions, working from left to right.
3. Because a fraction bar is a grouping symbol, simplify the numerator and the denominator separately and simplify the fraction, whenever possible.

When all grouping symbols have been removed, repeat the rules above to complete the calculation.

EXAMPLE 5   Evaluate each expression:

a. $4 + 2 \cdot 3 = 4 + 6$ Do the multiplication first.
$= 10$ Then do the addition.

b. $2(3 + 4) = 2 \cdot 7$ Do the addition within the parentheses first.

$= 14$ Then do the multiplication.

c. $5(3 - 6) \div 3 + 1 = 5(-3) \div 3 + 1$ Do the subtraction within parentheses.
$= -15 \div 3 + 1$ Then do the multiplication: $5(-3) = -15$
$= -5 + 1$ Then do the division: $-15 \div 3 = -5$
$= -4$ Finally, do the addition.

d. $5[3 - 2(6 \div 3 + 1)] = 5[3 - 2(2 + 1)]$ Do the division within parentheses.
$= 5[3 - 2(3)]$ Do the addition: $2 + 1 = 3$
$= 5(3 - 6)$ Do the multiplication: $2(3) = 6$
$= 5(-3)$ Do the subtraction: $3 - 6 = -3$
$= -15$ Do the multiplication.

e. $\dfrac{4 + 8(3 - 4)}{6 - 2(2)} = \dfrac{-4}{2}$ Simplify the numerator and denominator separately.

$= -2$ Do the division: $-4 \div 2 = -2$

SELF CHECK 5   Evaluate: **a.** $5 + 3 \cdot 4$ **b.** $(5 + 3) \cdot 4$ **c.** $3(5 - 7) \div 6 + 3$
d. $\dfrac{5 - 2(4 - 6)}{9 - 2 \cdot 3}$

6 Find the mean, median, and mode when given a set of values.

Three measures of central tendency are commonly used in newspapers and magazines: the *mean*, the *median*, and the *mode*.

MEAN

The **mean** of several values is the sum of those values divided by the number of values.

$$\text{Mean} = \frac{\text{Sum of the values}}{\text{Number of values}}$$

EXAMPLE 6   **FOOTBALL**   Figure 1-17 shows the gains and losses made by a running back on seven plays. Find the mean number of yards per carry.

Solution   To find the mean number of yards per carry, we add the numbers and divide by 7.

$$\frac{-8 + (+2) + (-6) + (+6) + (+4) + (-7) + (-5)}{7} = \frac{-14}{7} = -2$$

The running back averaged -2 yards (or lost 2 yards) per carry.

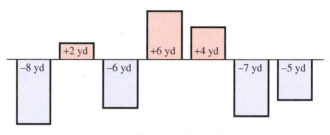

Figure 1-17

🌿 **SELF CHECK 6**   Find the mean if the running back's seven plays were an 8-yard gain followed by a 5-yard gain, 2-yard loss, 7-yard gain, 12-yard loss, 3-yard loss, and an 11-yard gain.

MEDIAN	The **median** of several values is the middle value. To find the median,
	1. arrange the values in increasing order;
	2. if there is an odd number of values, choose the middle value; and
	3. if there is an even number of values, find the mean of the middle two values.
MODE	The **mode** of several values is the value that occurs most often.

EXAMPLE 7   Ten workers in a small business have monthly salaries of

$2,500, $1,750, $2,415, $3,240, $2,790,
$3,240, $2,650, $2,415, $2,415, $2,650

Find: **a.** the median   **b.** the mode of the distribution

Solution   **a.** To find the median, we first arrange the salaries in increasing order:

$1,750, $2,415, $2,415, $2,415, $2,500,
$2,650, $2,650, $2,790, $3,240, $3,240

Because there is an even number of salaries, the median will be the mean of the middle two scores, $2,500 and $2,650.

$$\text{Median} = \frac{\$2,500 + \$2,650}{2} = \$2,575$$

b. Since the salary $2,415 occurs most often, it is the mode.

🌿 **SELF CHECK 7**   Given salaries of $3,275, $2,295, $4,025, $3,500, $3,015, $2,985, $3,275, and $3,000, find   **a.** the median and   **b.** the mode.

If two different numbers in a distribution tie for occurring most often, the distribution is called *bimodal*.

COMMENT   Although the mean is probably the most common measure of central tendency, the median and the mode are used frequently. For example, workers' salaries often are compared to the median salary. To say that the modal shoe size is 9 means that a shoe size of 9 occurs more often than any other size.

7   Evaluate an algebraic expression when given values for its variables.

Variables and numbers can be combined with the operations of arithmetic to produce an *algebraic* **expression**. To evaluate algebraic expressions, we substitute numbers for the variables and simplify.

EXAMPLE 8   If $x = 2$, $y = -3$, and $z = -5$, evaluate: **a.** $x + yz$ **b.** $\dfrac{xy + 3z}{y(z - x)}$

Solution   We substitute 2 for x, -3 for y, and -5 for z and simplify.

a. $x + yz = 2 + (-3)(-5)$
$$= 2 + (15)$$
$$= 17$$

b. $\dfrac{xy + 3z}{y(z - x)} = \dfrac{2(-3) + 3(-5)}{-3(-5 - 2)}$
$$= \dfrac{-6 + (-15)}{-3(-7)}$$
$$= \dfrac{-21}{21}$$
$$= -1$$

SELF CHECK 8   If $x = 2$, $y = -5$, and $z = 3$, evaluate: **a.** $y - xz$ **b.** $\dfrac{xy - 2z}{xz + 2y}$

Table 1-3 shows the formulas for the **perimeters** (the distance around a figure) of several geometric figures. The distance around a circle is called a **circumference**.

TABLE 1-3

Figure	Name	Perimeter	Figure	Name	Perimeter/circumference
	Square	$P = 4s$		Trapezoid	$P = a + b + c + d$
	Rectangle	$P = 2l + 2w$		Circle	$C = \pi D$ $C = 2\pi r$ (π is approximately 3.1416)
	Triangle	$P = a + b + c$			

EXAMPLE 9   Find the perimeter of the rectangle shown in Figure 1-18.

Solution   We substitute 2.75 for l and 1.25 for w into the formula $P = 2l + 2w$ and simplify.

$$P = 2l + 2w$$
$$P = 2(2.75) + 2(1.25)$$
$$= 5.50 + 2.50$$
$$= 8.00$$

The perimeter is 8 meters.

Figure 1-18

2.75 m

1.25 m

SELF CHECK 9   Find the perimeter of a rectangle with a length of 8 meters and a width of 5 meters.

---

8   Identify the property of real numbers that justifies a given statement.

When we work with real numbers, we will use the following properties.

PROPERTIES OF REAL NUMBERS

If a, b, and c are real numbers, the following properties apply.

The commutative properties of addition and multiplication

$$a + b = b + a \qquad ab = ba$$

The associative properties of addition and multiplication

$$(a + b) + c = a + (b + c) \qquad (ab)c = a(bc)$$

The distributive property of multiplication over addition

$$a(b + c) = ab + ac$$

The *commutative properties* enable us to add or multiply two numbers in either order and obtain the same result. For example,

$$2 + 3 = 5 \qquad \text{and} \qquad 3 + 2 = 5$$
$$7 \cdot 9 = 63 \qquad \text{and} \qquad 9 \cdot 7 = 63$$

COMMENT   Subtraction and division are not commutative, because doing these operations in different orders will give different results. For example,

$$8 - 4 = 4 \qquad \text{but} \qquad 4 - 8 = -4$$
$$8 \div 4 = 2 \qquad \text{but} \qquad 4 \div 8 = \frac{1}{2}$$

The *associative properties* enable us to group the numbers in a sum or a product in any way that we wish and still get the same result. For example,

$$(2 + 3) + 4 = 5 + 4 \qquad\qquad 2 + (3 + 4) = 2 + 7$$
$$= 9 \qquad\qquad\qquad\qquad = 9$$

$$(2 \cdot 3) \cdot 4 = 6 \cdot 4 \qquad\qquad 2 \cdot (3 \cdot 4) = 2 \cdot 12$$
$$= 24 \qquad\qquad\qquad\qquad = 24$$

COMMENT   Subtraction and division are not associative, because different groupings give different results. For example,

$$(\mathbf{8 - 4}) - 2 = \mathbf{4} - 2 = 2 \qquad \text{but} \qquad 8 - (\mathbf{4 - 2}) = 8 - \mathbf{2} = 6$$
$$(\mathbf{8 \div 4}) \div 2 = \mathbf{2} \div 2 = 1 \qquad \text{but} \qquad 8 \div (\mathbf{4 \div 2}) = 8 \div \mathbf{2} = 4$$

The *distributive property* enables us to evaluate many expressions involving a multiplication over an addition. We can add first inside the parentheses and then multiply, or multiply over the addition first and then add.

$$2(\mathbf{3 + 7}) = 2 \cdot \mathbf{10} \qquad\qquad 2(3 + 7) = \mathbf{2} \cdot \mathbf{3} + \mathbf{2} \cdot \mathbf{7}$$
$$= 20 \qquad\qquad\qquad\qquad = 6 + 14$$
$$\qquad\qquad\qquad\qquad\qquad = 20$$

We can interpret the distributive property geometrically. Since the area of the largest rectangle in Figure 1-19 is the product of its width a and its length $b + c$, its area is $a(b + c)$. The areas of the two smaller rectangles are ab and ac. Since the area of the largest rectangle is equal to the sum of the areas of the smaller rectangles, we have $a(b + c) = ab + ac$.

Figure 1-19

The distributive property can be extended to any number of addends.

$$a(\mathbf{b + c + d + e + \cdots}) = ab + ac + ad + ae + \cdots$$

EXAMPLE 10   Use the distributive property to write each expression without parentheses.
a. $2(x + 3)$ **b.** $2(x + y - 7)$

Solution   **a.** $2(x + 3) = \mathbf{2}x + \mathbf{2} \cdot 3$
$$= 2x + 6$$

b. $2(x + y - 7) = \mathbf{2}x + \mathbf{2}y - \mathbf{2} \cdot 7$
$$= 2x + 2y - 14$$

SELF CHECK 10   Use the distributive property to write $-5(a - 2b + 3c)$ without parentheses.

The real numbers 0 and 1 have important special properties.

PROPERTIES OF 0 AND 1

Additive identity (0): The sum of 0 and any number is the number itself.

$$0 + a = a + 0 = a$$

Multiplication property of 0: The product of any number and 0 is 0.

$$a \cdot 0 = 0 \cdot a = 0$$

Multiplicative identity (1): The product of 1 and any number is the number itself.

$$1 \cdot a = a \cdot 1 = a$$

For example,

$$7 + 0 = 7, \qquad 7(0) = 0, \qquad 1(5) = 5, \qquad \text{and} \qquad (-7)1 = -7$$

If the sum of two numbers is 0, the numbers are called **additive inverses**, **negatives**, or **opposites** of each other. For example, 6 and -6 are negatives, because $6 + (-6) = 0$.

THE ADDITIVE INVERSE PROPERTY

For every real number a, there is a real number $-a$ such that

$$a + (-a) = -a + a = 0$$

The symbol $-(-6)$ means "the negative of negative 6" or "the opposite of negative 6." Because the sum of two numbers that are negatives is 0, we have

$$-6 + [-(-6)] = 0 \quad \text{and} \quad -6 + 6 = 0$$

Because -6 has only one additive inverse, it follows that $-(-6) = 6$. In general, the following rule applies.

THE DOUBLE NEGATIVE RULE	If a represents any real number, then $-(-a) = a$.

If the product of two numbers is 1, the numbers are called **multiplicative inverses** or **reciprocals** of each other.

THE MULTIPLICATIVE INVERSE PROPERTY	For every nonzero real number a, there exists a real number $\frac{1}{a}$ such that $$a \cdot \frac{1}{a} = \frac{1}{a} \cdot a = 1 \quad (a \neq 0)$$

Some examples of reciprocals are

- 5 and $\frac{1}{5}$ are reciprocals, because $5\left(\frac{1}{5}\right) = 1$.
- $\frac{-3}{2}$ and $\frac{-2}{3}$ are reciprocals, because $\frac{-3}{2}\left(\frac{-2}{3}\right) = 1$.

The reciprocal of 0 does not exist, because $\frac{1}{0}$ is undefined.

SELF CHECK ANSWERS

1. a. -9 b. -5 c. 9 d. 5 2. a. -19 b. 3 c. -5 3. a. -30 b. 32 c. -34 d. 72 4. a. -11 b. 12 c. -10 d. 2 5. a. 17 b. 32 c. 2 d. 3 6. 2-yard gain 7. a. \$3,145 b. \$3,275 8. a. -11 b. 4 9. 26 m 10. $-5a + 10b - 15c$

NOW TRY THIS

1. Insert the correct operations, including any grouping symbols needed to make each expression equal to the given value.

   a. $3 \quad 5 = -2$
   b. $-8 \quad 4 = -4$
   c. $3 \quad 5 \quad 2 = 4$
   d. $3 \quad 5 \quad 2 = 9$

2. Evaluate each of the following for $x = -2$, $y = 5$, and $z = -4$.

   a. $x - y - z$
   b. $x - y(-z)$

1.2   Exercises

WARM-UPS *Identify the operation to be performed in simplifying each expression.*

1. $-5 + (-3)$

2. $-4 - (-5)$

3. $-7 - 4$

4. $-5(-6)$

5. $\dfrac{-10}{2}$

6. $-\dfrac{7}{8} + \dfrac{3}{4}$

REVIEW *Graph each set on the number line.*

7. $\{x \mid x > 4\}$

8. $(-\infty, -5]$

9. $(2, 10]$

10. $\{x \mid -4 \le x \le 4\}$

11. Buying gasoline   A man bought 32 gallons of gasoline at $2.29 per gallon and 3 quarts of oil at $1.35 per quart. The sales tax was included in the price of the gasoline, but 5% sales tax was added to the cost of the oil. Find the total cost.

12. Paying taxes   On an adjusted income of $57,760, a woman must pay taxes according to the schedule shown in the table. Compute the tax bill.

2012 Tax Rate Schedules			
Schedule X—If your filing status is Single			
If your taxable income is:		The tax is:	
Over—	**But not over—**		**Of the amount over—**
$0	$8,700	.....10%	$0
8,700	35,350	$870.00 + 15%	8,700
35,350	85,650	$4,867.50 + 25%	35,350
85,650	178,650	$17,442.50 + 28%	85,650
178,650	388,350	$43,482.50 + 33%	178,650
388,350	.....	$112,683.50 + 35%	388,350

VOCABULARY AND CONCEPTS   *Fill in the blanks.*

13. To add two numbers with like signs, we add their _____ values and keep the _____ sign.

14. To add two numbers with unlike signs, we _____ their absolute values and keep the sign of the number with the larger absolute value.

15. To subtract one number from another, we _____ the sign of the number that is being subtracted and ___.

16. The product of two real numbers with like signs is _____.

17. The quotient of two real numbers with unlike signs is _____.

18. The result of addition is a ___ and the result of subtraction is a _____.

19. The result of multiplication is a _____ and the result of division is a _____.

20. To simplify a numerical expression, we follow the _____.

21. The denominator of a fraction can never be _.

22. The three measures of central tendency are the _____, _____, and _____.

23. Write the formula for the circumference of a circle. _____ or _____

24. Write the associative property of multiplication. _____

25. Write the commutative property of addition. _____

26. Write the distributive property of multiplication over addition. _____

27. Multiplicative inverses are also called _____.

28. What is the additive identity? _ What is the multiplicative identity? _

29. The additive inverse of -5 is _ and the multiplicative inverse of 7 is __.

30. $-(-a) =$ _

GUIDED PRACTICE   *Add.* **SEE EXAMPLE 1. (OBJECTIVE 1)**

31. $-6 + (-3)$

32. $2 + (+8)$

33. $-7 + 2$

34. $6 + (-3)$

35. $\dfrac{1}{2} + \left(-\dfrac{1}{3}\right)$

36. $-\dfrac{3}{4} + \left(-\dfrac{1}{5}\right)$

37. $-\dfrac{5}{9} + \left(\dfrac{3}{4}\right)$

38. $\dfrac{7}{8} + \left(-\dfrac{6}{11}\right)$

Subtract. **SEE EXAMPLE 2. (OBJECTIVE 2)**

39. $-5 - 7$

40. $-11 - (-17)$

41. $-33 - (-33)$

42. $18 - (-2)$

43. $\dfrac{1}{2} - \left(-\dfrac{3}{5}\right)$

44. $\dfrac{1}{26} - \dfrac{11}{13}$

45. $\dfrac{1}{3} - \dfrac{1}{2}$

46. $\dfrac{7}{8} - \left(-\dfrac{3}{4}\right)$

Multiply. **SEE EXAMPLE 3. (OBJECTIVE 3)**

47. $-2(6)$

48. $4(-8)$

49. $-9(-3)$

50. $-2(-5)$

51. $\left(-\dfrac{2}{9}\right)\left(\dfrac{18}{7}\right)$

52. $\left(-\dfrac{6}{7}\right)\left(-\dfrac{5}{12}\right)$

53. $\left(\dfrac{9}{8}\right)\left(-\dfrac{6}{7}\right)$

54. $\left(\dfrac{8}{3}\right)\left(-\dfrac{15}{16}\right)$

Divide. **SEE EXAMPLE 4. (OBJECTIVE 4)**

55. $\dfrac{-8}{4}$

56. $\dfrac{36}{-4}$

57. $\dfrac{-49}{-7}$

58. $\dfrac{-5}{-25}$

59. $\dfrac{3}{4} \div \left(-\dfrac{3}{8}\right)$

60. $-\dfrac{3}{5} \div \dfrac{7}{10}$

61. $-\dfrac{16}{5} \div \left(-\dfrac{10}{3}\right)$

62. $\dfrac{7}{22} \div \left(-\dfrac{14}{11}\right)$

Evaluate each expression following the order of operations. **SEE EXAMPLE 5. (OBJECTIVE 5)**

63. $3 + 4 \cdot 5$

64. $5 + 5(3)$

65. $2 - 3 \cdot 5$

66. $5 \cdot 3 - 6 \cdot 4$

67. $8 - 4 - 3$

68. $5 - 3 - 1$

69. $8 - (4 - 3)$

70. $5 - (3 - 1)$

71. $18 \div 6 \div 3$

72. $100 \div 10 \div 5$

73. $18 \div (6 \div 3)$

74. $100 \div (10 \div 5)$

Use the distribution: 7, 5, 9, 10, 8, 6, 6, 7, 9, 12, 9. *You may use a calculator.* SEE EXAMPLES 6–7. (OBJECTIVE 6)

75. Find the mean.   **76.** Find the median.
77. Find the mode.

Use the distribution: 8, 12, 23, 12, 10, 16, 26, 12, 14, 8, 16, 23.

78. Find the median.   **79.** Find the mode.
80. Find the mean.

Let $a = 3$, $b = -2$, $c = -1$, and $d = 2$ and evaluate each expression. SEE EXAMPLE 8. (OBJECTIVE 7)

81. $ab + cd$ **82.** $ac - bd$

83. $\dfrac{ad + c}{cd + b}$ **84.** $\dfrac{ab + d}{bd + a}$

Sorting records is a common task in data processing. A selection sort requires C comparisons to sort N records, where C and N are related by the formula $C = \dfrac{N(N - 1)}{2}$. SEE EXAMPLES 8–9. (OBJECTIVE 7)

85. How many comparisons are needed to sort 200 records?

86. How many comparisons are needed to sort 10,000 records?

87. Perimeter of a triangle Find the perimeter of a triangle with sides that are 23.5, 37.2, and 39.7 feet long.

88. Perimeter of a trapezoid Find the perimeter of a trapezoid with sides that are 43.27, 47.37, 50.21, and 52.93 centimeters long.

Determine which property of real numbers justifies each statement. SEE EXAMPLE 10. (OBJECTIVE 8)

89. $5 + 8 = 8 + 5$

90. $2 \cdot (9 \cdot 13) = (2 \cdot 9) \cdot 13$

91. $3(2 + 5) = 3 \cdot 2 + 3 \cdot 5$

92. $4 \cdot 3 = 3 \cdot 4$

93. $81 + 0 = 81$

94. $6(7 + 1) = 6 \cdot 7 + 6 \cdot 1$

95. $5 \cdot \dfrac{1}{5} = 1$

96. $3 + (9 + 0) = (9 + 0) + 3$

ADDITIONAL PRACTICE *Determine which property of real numbers justifies each statement.*

97. $a + (7 + 8) = (a + 7) + 8$

98. $1 \cdot 3 = 3$

99. $(2 \cdot 3) \cdot 4 = 4 \cdot (2 \cdot 3)$

100. $5 + (-5) = 0$

Use a calculator to verify each statement and then identify the property of real numbers that is being illustrated.

101. $(37.9 + 25.2) + 14.3 = 37.9 + (25.2 + 14.3)$

102. $7.1(3.9 + 8.8) = 7.1 \cdot 3.9 + 7.1 \cdot 8.8$

103. $2.73(4.534 + 57.12) = 2.73 \cdot 4.534 + 2.73 \cdot 57.12$

104. $(6.789 + 345.1) + 27.347 = (345.1 + 6.789) + 27.347$

Evaluate each expression.

105. $98 - 14 - 14$ **106.** $-43 - 19 + 59$
107. $24 - (-3) - 5$ **108.** $46 + (-21) - 31$
109. $-2(21)(-4)$ **110.** $-5(-12)(-3)$
111. $56 \div 7 \div 2$ **112.** $96 \div (-4) \div (-3)$
113. $2 + 6 \div 3 - 5$ **114.** $6 - 8 \div 4 - 2$
115. $(3 + 9) \div (4 - 8)$ **116.** $(6 - 8) \div (4 - 2)$

117. $\dfrac{3(8 + 4)}{2 \cdot 3 - 9}$ **118.** $\dfrac{5(4 - 1)}{3 \cdot 2 + 5 \cdot 3}$

119. $\dfrac{100(2 - 4)}{1{,}000 \div 10 \div 10}$ **120.** $\dfrac{8(3) - 4(6)}{5(3) + 3(-7)}$

Let $a = 3$, $b = -2$, $c = -1$, and $d = 2$ and evaluate each expression.

121. $2b - 5ac$ **122.** $4a(6c - b) + d$
123. $c - d(2 - a)$ **124.** $(a - c) - (b - d)$
125. $a(b + c)$ **126.** $d(b + a)$

127. $\dfrac{ad + bc}{cd - ad}$ **128.** $\dfrac{bc - ad}{bd + ac}$

APPLICATIONS

129. Earning money One day Scott earned $22.25 tutoring mathematics and $39.75 tutoring physics. How much did he earn that day?

130. Losing weight During an illness, Wendy lost 13.5 pounds. By the time she recovered, she had lost another 11.5 pounds. What integer represents her change in weight?

131. Changing temperatures The temperature rose 17° in the afternoon and then dropped 13° overnight. Find the overall change in temperature.

132. Displaying the flag Before the American flag is displayed at half-mast, it should first be raised to the top of the flagpole. How far has the flag in the illustration traveled?

38 ft

133. Changing temperatures If the temperature has been dropping 4° each hour, how much warmer was it 3 hours ago?

134. Mountain climbing A team of mountaineers climbed 2,347 feet one day and then came down 597 feet to a base camp. What signed number represents their net change in altitude?

135. Filling a pool The flow of water from a pipe is filling a pool at the rate of 23 gallons per minute. How much less water was in the pool 5 hours ago?

136. Draining a pool If a drain is emptying a pool at the rate of 12 gallons per minute, how much more water was in the pool 2 hours ago?

Use a calculator to help solve the following.

137. Military   An army retreated 2,300 meters. After regrouping, it moved forward 1,750 meters. The next day, it gained another 1,875 meters. What integer represents the army's net gain (or loss)?

138. Grooming horses   John earned $10 an hour for grooming horses. After working for 12 hours, he had $154. How much did he have before he started work?

139. Managing a checkbook   Sally started with $437.37 in a checking account. One month, she had deposits of $125.18, $137.26, and $145.56. That same month, she had withdrawals of $117.11, $183.49, and $122.89. Find her ending balance.

140. Stock averages   The illustration shows the daily advances and declines of the Dow Jones average for one week. What integer represents the total gain or loss for the week?

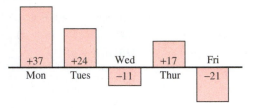

141. Selling clothes   If a clerk had the sales shown in the table for one week, find the mean of daily sales.

Monday	$1,525
Tuesday	$ 785
Wednesday	$1,628
Thursday	$1,214
Friday	$ 917
Saturday	$1,197

142. Sizes of viruses   The table gives the approximate lengths (in centimicrons) of the viruses that cause five common diseases. Find the mean length of the viruses.

Polio	2.5
Influenza	105.1
Pharyngitis	74.9
Chicken pox	137.4
Yellow fever	52.6

143. Calculating grades   A student has scores of 77, 79, 89, 84, and 86 on five exams. Find his average (mean) score.

144. Averaging weights   The offensive line of a football team has two guards, two tackles, and a center. If the guards weigh 298 and 287 pounds, the tackles 310 and 302 pounds, and the center 303 pounds, find the average (mean) weight of the offensive line.

145. Analyzing ads   The businessman who ran the following ad earns $100,000 and employs four students who earn $10,000 each. Is the ad misleading?

146. Averaging grades   A student has grades of 78%, 85%, 88%, and 96%. There is one test left, and the student needs to average 90% to earn an A. Does he have a chance?

147. Perimeter of a square   Find the perimeter of the square shown in the illustration.

148. Circumference of a circle   To the nearest hundredth, find the circumference of the circle shown in the illustration.

WRITING ABOUT MATH

149. The symmetric property of equality states that if $a = b$, then $b = a$. Explain why this property is often confused with the commutative properties. Why do you think this is so?

150. Explain why the mean of two numbers is halfway between the two numbers.

SOMETHING TO THINK ABOUT

151. Pick five numbers and find their mean. Add 7 to each of the numbers to get five new numbers and find their mean. What do you discover? Is this property always true?

152. Take the original five numbers in Exercise 151 and multiply each one by 7 to get five new numbers and find their mean. What do you discover? Is this property always true?

153. Give three applications in which the median would be the most appropriate average to use.

154. Give three applications in which the mode would be the most appropriate average to use.

Section 1.3

Exponents

In this section, we will review exponents, a shortcut way of indicating repeated multiplication. The rules for the order of operations will be modified to include exponents.

1 Identify the base and the exponent to simplify an exponential expression.

Exponents indicate repeated multiplication. For example,

$y^2 = y \cdot y$ Read y^2 as "y to the second power" or "y squared."

$z^3 = z \cdot z \cdot z$ Read z^3 as "z to the third power" or "z cubed."

$x^4 = x \cdot x \cdot x \cdot x$ Read x^4 as "x to the fourth power."

These examples suggest the following definition.

NATURAL-NUMBER EXPONENTS

If n is a natural number, then

$$x^n = \overbrace{x \cdot x \cdot x \cdots \cdot x}^{n \text{ factors of } x}$$

COMMENT A natural-number exponent tells how many times the base of an exponential expression is to be used as a **factor** in a product.

The exponential expression x^n is called a **power of x**, and we read it as "x to the nth power." In this expression, x is called the **base**, and n is called the **exponent**.

Base $\rightarrow x^n \leftarrow$ Exponent

EXAMPLE 1   Write each expression without exponents:

a. $2^5 = 2 \cdot 2 \cdot 2 \cdot 2 \cdot 2$
$= 32$

b. $(-2)^5 = (-2)(-2)(-2)(-2)(-2)$
$= -32$

c. $-4^4 = -(4^4)$ The base is 4.
$= -(4 \cdot 4 \cdot 4 \cdot 4)$
$= -256$

d. $(-4)^4 = (-4)(-4)(-4)(-4)$ The base is -4.
$= 256$

e. $\left(\dfrac{1}{2}a\right)^3 = \left(\dfrac{1}{2}a\right)\left(\dfrac{1}{2}a\right)\left(\dfrac{1}{2}a\right)$
$= \dfrac{1}{8}a^3$

f. $\left(-\dfrac{1}{5}b\right)^2 = \left(-\dfrac{1}{5}b\right)\left(-\dfrac{1}{5}b\right)$
$= \dfrac{1}{25}b^2$

 SELF CHECK 1   Write each expression without exponents:

a. 3^4 **b.** $(-5)^3$ **c.** $\left(-\dfrac{3}{4}a\right)^2$

COMMENT   Note the difference between $-x^n$ and $(-x)^n$.

$$\overbrace{}^{n \text{ factors of } x} \qquad \overbrace{}^{n \text{ factors of } (-x)}$$
$$-x^n = -(x \cdot x \cdot x \cdot \cdots \cdot x) \qquad (-x)^n = (-x)(-x)(-x) \cdots \cdot (-x)$$

Also, note the difference between ax^n and $(ax)^n$.

$$\overbrace{}^{n \text{ factors of } x} \qquad \overbrace{}^{n \text{ factors of } (ax)}$$
$$ax^n = a \cdot x \cdot x \cdot x \cdot \cdots \cdot x \qquad (ax)^n = (ax)(ax)(ax) \cdots \cdot (ax)$$

2   Simplify an expression by applying properties of exponents.

Since x^5 means that x is to be used as a factor five times, and since x^3 means that x is to be used as a factor three times, $x^5 \cdot x^3$ means that x will be used as a factor eight times.

$$\overbrace{}^{5 \text{ factors of } x} \quad \overbrace{}^{3 \text{ factors of } x} \qquad \overbrace{}^{8 \text{ factors of } x}$$
$$x^5 x^3 = x \cdot x \cdot x \cdot x \cdot x \quad \cdot \quad x \cdot x \cdot x \quad = \quad x \cdot x \cdot x \cdot x \cdot x \cdot x \cdot x \cdot x$$

In general,

$$\overbrace{}^{m \text{ factors of } x} \quad \overbrace{}^{n \text{ factors of } x} \qquad \overbrace{}^{m + n \text{ factors of } x}$$
$$x^m x^n = x \cdot x \cdot x \cdot x \cdot \cdots \cdot x \quad \cdot \quad x \cdot x \cdot \cdots \cdot x \quad = \quad x \cdot x \cdot x \cdot x \cdot x \cdot x \cdot x \cdot x \cdot \cdots \cdot x$$

Thus, *to multiply exponential expressions with the same base, we keep the same base and add the exponents.*

THE PRODUCT RULE OF EXPONENTS   If m and n are natural numbers, then
$$x^m x^n = x^{m+n}$$

COMMENT   The product rule of exponents applies only to exponential expressions with the same base. The expression $x^5 y^3$, for example, cannot be simplified, because the bases of the exponential expressions are different.

EXAMPLE 2 Simplify each expression:

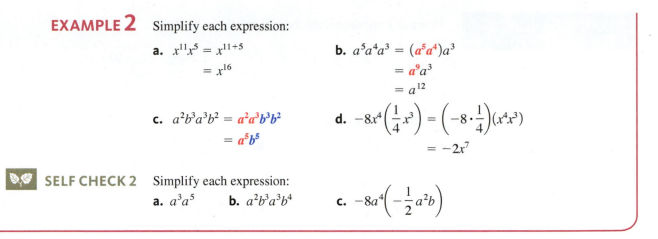

a. $x^{11}x^5 = x^{11+5}$
   $= x^{16}$

b. $a^5a^4a^3 = (\mathbf{a^5a^4})a^3$
   $= \mathbf{a^9}a^3$
   $= a^{12}$

c. $a^2b^3a^3b^2 = \mathbf{a^2a^3}\mathbf{b^3b^2}$
   $= \mathbf{a^5b^5}$

d. $-8x^4\left(\dfrac{1}{4}x^3\right) = \left(-8\cdot\dfrac{1}{4}\right)(x^4x^3)$
   $= -2x^7$

SELF CHECK 2 Simplify each expression:

a. a^3a^5 **b.** $a^2b^3a^3b^4$ **c.** $-8a^4\left(-\dfrac{1}{2}a^2b\right)$

To find another property of exponents, we simplify $(x^4)^3$, which means x^4 cubed or $x^4 \cdot x^4 \cdot x^4$.

$$(x^4)^3 = \overbrace{x^4}\cdot\overbrace{x^4}\cdot\overbrace{x^4} = x\cdot x\cdot x\cdot x\cdot x\cdot x\cdot x\cdot x\cdot x\cdot x\cdot x\cdot x = x^{12}$$

In general, we have

$$(x^m)^n = \overbrace{x^m\cdot x^m\cdot x^m\cdot\cdots\cdot x^m}^{n\text{ factors of }x^m} = \overbrace{x\cdot x\cdot x\cdot x\cdot x\cdot\cdots\cdot x}^{mn\text{ factors of }x} = x^{mn}$$

Thus, *to raise an exponential expression to a power, we keep the same base and multiply the exponents.*

To find a third property of exponents, we square $3x$ and get

$$(3x)^2 = (3x)(3x) = 3\cdot 3\cdot x\cdot x = 3^2x^2 = 9x^2$$

In general, we have

$$(xy)^n = \overbrace{(xy)(xy)(xy)\cdots\cdots(xy)}^{n\text{ factors of }xy} = \overbrace{xxx\cdots\cdots x}^{n\text{ factors of }x}\cdot\overbrace{yyy\cdots\cdots y}^{n\text{ factors of }y} = x^ny^n$$

To find a fourth property of exponents, we cube $\dfrac{x}{3}$ to get

$$\left(\frac{x}{3}\right)^3 = \frac{x}{3}\cdot\frac{x}{3}\cdot\frac{x}{3} = \frac{x\cdot x\cdot x}{3\cdot 3\cdot 3} = \frac{x^3}{3^3} = \frac{x^3}{27}$$

In general, we have

$$\left(\frac{x}{y}\right)^n = \overbrace{\left(\frac{x}{y}\right)\left(\frac{x}{y}\right)\left(\frac{x}{y}\right)\cdots\cdots\left(\frac{x}{y}\right)}^{n\text{ factors of }x/y} \quad (y \neq 0)$$

$$= \frac{\overbrace{xxx\cdots\cdots x}^{n\text{ factors of }x}}{\underbrace{yyy\cdots\cdots y}_{n\text{ factors of }y}}$$

Multiply the numerators and multiply the denominators.

$$= \frac{x^n}{y^n}$$

The previous three results are called the *power rules of exponents.*

THE POWER RULES OF EXPONENTS

If m and n are natural numbers, then

$$(x^m)^n = x^{mn} \qquad (xy)^n = x^n y^n \qquad \left(\frac{x}{y}\right)^n = \frac{x^n}{y^n} \quad (y \neq 0)$$

EXAMPLE 3   Simplify each expression:

a. $(3^2)^3 = 3^{2 \cdot 3}$

$\qquad = 3^6$

$\qquad = 729$

b. $(x^{11})^5 = x^{11 \cdot 5}$

$\qquad = x^{55}$

c. $(x^2 x^3)^6 = (x^5)^6$

$\qquad = x^{30}$

d. $(x^2)^4 (x^3)^2 = x^8 x^6$

$\qquad = x^{14}$

SELF CHECK 3   Simplify each expression: **a.** $(a^5)^8$ **b.** $(a^4 a^3)^3$ **c.** $(a^3)^3 (a^2)^3$

EXAMPLE 4   Simplify each expression. Assume that no denominators are zero.

a. $(x^2 y)^3 = (x^2)^3 y^3$

$\qquad = x^6 y^3$

b. $(x^3 y^4)^4 = (x^3)^4 (y^4)^4$

$\qquad = x^{12} y^{16}$

c. $\left(\dfrac{x}{y^2}\right)^4 = \dfrac{x^4}{(y^2)^4}$

$\qquad\quad = \dfrac{x^4}{y^8}$

d. $\left(\dfrac{x^3}{y^4}\right)^2 = \dfrac{(x^3)^2}{(y^4)^2}$

$\qquad\quad = \dfrac{x^6}{y^8}$

SELF CHECK 4   Simplify each expression. Assume that no denominators are zero.

a. $(a^4 b^5)^2$ **b.** $\left(\dfrac{a^5}{b^7}\right)^3$

We can expand the rules for exponents to hold for exponents of 0.

$$x^0 x^n = x^{0+n} = x^n = 1x^n$$

Therefore, since $x^0 x^n = 1 x^n$, it follows that $x^0 = 1$ $(x \neq 0)$.

ZERO EXPONENTS

If $x \neq 0$, then $x^0 = 1$.

COMMENT
0^0 is indeterminant.

Because of the previous definition, any nonzero base raised to the 0 power is 1. For example, if no variables are zero, then

$$5^0 = 1, \qquad (-7)^0 = 1, \qquad (3ax^3)^0 = 1, \qquad \left(\tfrac{1}{2} x^5 y^7 z^9\right)^0 = 1$$

We can expand the rules for exponents to include negative integer exponents.

$$x^{-n} x^n = x^{-n+n} = x^0 = 1 \quad (x \neq 0)$$

Therefore, since $x^{-n} \cdot x^n = 1$ and $\dfrac{1}{x^n} \cdot x^n = 1$, we define x^{-n} to be the reciprocal of x^n.

NEGATIVE EXPONENTS

If n is an integer and $x \neq 0$, then

$$x^{-n} = \frac{1}{x^n} \qquad \text{and} \qquad \frac{1}{x^{-n}} = x^n$$

COMMENT By the definition of negative exponents, the base cannot be 0. Thus, an expression such as 0^{-5} is undefined.

Because of this definition, we can write expressions containing negative exponents as expressions without negative exponents. For example,

$$5^{-2} = \frac{1}{5^2} = \frac{1}{25} \qquad\qquad 10^{-3} = \frac{1}{10^3} = \frac{1}{1,000}$$

and if $x \neq 0$, we have

$$(2x)^{-3} = \frac{1}{(2x)^3} = \frac{1}{8x^3} \qquad 3x^{-1} = 3 \cdot \frac{1}{x} = \frac{3}{x}$$

EXAMPLE 5 Write each expression without negative exponents. Assume that $x \neq 0$.

a. $x^{-5}x^3 = x^{-5+3}$
$= x^{-2}$
$= \dfrac{1}{x^2}$

b. $(x^{-3})^{-2} = x^{(-3)(-2)}$
$= x^6$

c. $\dfrac{4}{x^{-3}} = 4x^3$

SELF CHECK 5 Write each expression without negative exponents. Assume that $a \neq 0$.

a. $a^{-7}a^3$ **b.** $(a^{-5})^{-3}$ **c.** $\dfrac{2}{a^{-1}}$

To develop a rule for dividing exponential expressions, we proceed as follows:

$$\frac{x^m}{x^n} = x^m\left(\frac{1}{x^n}\right) = x^m x^{-n} = x^{m+(-n)} = x^{m-n}$$

Thus, *to divide exponential expressions with the same nonzero base, we keep the same base and subtract the exponent in the denominator from the exponent in the numerator.*

THE QUOTIENT RULE

If m and n are integers, then

$$\frac{x^m}{x^n} = x^{m-n} \quad (x \neq 0)$$

EXAMPLE 6 Simplify each expression. Assume no variable is zero.

a. $\dfrac{a^5}{a^3} = a^{5-3}$
$= a^2$

b. $\dfrac{x^{-5}}{x^{11}} = x^{-5-11}$
$= x^{-16}$
$= \dfrac{1}{x^{16}}$

c. $\dfrac{x^4 x^3}{x^{-5}} = \dfrac{x^7}{x^{-5}}$
$= x^{7-(-5)}$
$= x^{12}$

d. $\dfrac{(x^2)^3}{(x^3)^2} = \dfrac{x^6}{x^6}$
$= x^{6-6}$
$= x^0$
$= 1$

e. $\dfrac{x^2 y^3}{x y^4} = x^{2-1} y^{3-4}$

$\qquad = x y^{-1}$

$\qquad = \dfrac{x}{y}$

f. $\left(\dfrac{a^{-2} b^3}{a^2 a^3 b^4}\right)^3 = \left(\dfrac{a^{-2} b^3}{a^5 b^4}\right)^3$

$\qquad = (a^{-2-5} b^{3-4})^3$

$\qquad = (a^{-7} b^{-1})^3$

$\qquad = \left(\dfrac{1}{a^7 b}\right)^3$

$\qquad = \dfrac{1}{a^{21} b^3}$

SELF CHECK 6 Simplify each expression. Assume that no variable is zero.

a. $\dfrac{(a^{-2})^3}{(a^2)^{-3}}$ **b.** $\left(\dfrac{a^{-2} b^5}{b^8}\right)^{-3}$

To illustrate one more property of exponents, we consider the simplification of $\left(\frac{2}{3}\right)^{-4}$.

$$\left(\frac{2}{3}\right)^{-4} = \frac{1}{\left(\dfrac{2}{3}\right)^4} = \frac{1}{\dfrac{2^4}{3^4}} = 1 \div \frac{2^4}{3^4} = 1 \cdot \frac{3^4}{2^4} = \frac{3^4}{2^4} = \left(\frac{3}{2}\right)^4$$

This example suggests that to raise a fraction to a negative power, we can invert the fractional base (take the reciprocal) and then raise it to a positive power.

FRACTIONS TO NEGATIVE POWERS

If n is an integer, then

$$\left(\frac{x}{y}\right)^{-n} = \left(\frac{y}{x}\right)^n = \frac{y^n}{x^n} \quad (x \neq 0, y \neq 0)$$

EXAMPLE 7 Write each expression without using parentheses. Write answers without negative exponents.

a. $\left(\dfrac{3}{5}\right)^{-4} = \left(\dfrac{5}{3}\right)^4$

$\qquad = \dfrac{625}{81}$

b. $\left(\dfrac{y^2}{x^3}\right)^{-3} = \left(\dfrac{x^3}{y^2}\right)^3$

$\qquad = \dfrac{x^9}{y^6}$

c. $\left(\dfrac{2x^2}{3y^{-3}}\right)^{-4} = \left(\dfrac{3y^{-3}}{2x^2}\right)^4$

$\qquad = \dfrac{81 y^{-12}}{16 x^8}$

$\qquad = \dfrac{81}{16 x^8} \cdot y^{-12}$

$\qquad = \dfrac{81}{16 x^8} \cdot \dfrac{1}{y^{12}}$

$\qquad = \dfrac{81}{16 x^8 y^{12}}$

d. $\left(\dfrac{a^{-2} b^3}{a^2 a^3 b^4}\right)^{-3} = \left(\dfrac{a^2 a^3 b^4}{a^{-2} b^3}\right)^3$

$\qquad = \left(\dfrac{a^5 b^4}{a^{-2} b^3}\right)^3$

$\qquad = (a^{5-(-2)} b^{4-3})^3$

$\qquad = (a^7 b)^3$

$\qquad = a^{21} b^3$

SELF CHECK 7 Write each expression without using parentheses. Write answers without negative exponents.

a. $\left(\dfrac{2}{5}\right)^{-3}$ **b.** $\left(\dfrac{3a^3}{2b^{-2}}\right)^{-5}$

We summarize the rules of exponents as follows.

PROPERTIES OF EXPONENTS

If there are no divisions by 0, then for all integers m and n,

$$x^m x^n = x^{m+n} \qquad (x^m)^n = x^{mn} \qquad (xy)^n = x^n y^n \qquad \left(\frac{x}{y}\right)^n = \frac{x^n}{y^n}$$

$$x^0 = 1 \quad (x \neq 0) \qquad x^{-n} = \frac{1}{x^n} \qquad \frac{x^m}{x^n} = x^{m-n} \qquad \left(\frac{x}{y}\right)^{-n} = \left(\frac{y}{x}\right)^n = \frac{y^n}{x^n}$$

The same properties apply to exponents that are variables.

EXAMPLE 8 Simplify each expression. Assume that $a \neq 0$ and $x \neq 0$.

a. $\dfrac{a^n a}{a^2} = a^{n+1-2}$

$= a^{n-1}$

b. $\dfrac{x^3 x^2}{x^n} = x^{3+2-n}$

$= x^{5-n}$

c. $\left(\dfrac{x^n}{x^2}\right)^2 = \dfrac{x^{2n}}{x^4}$

$= x^{2n-4}$

d. $\dfrac{a^n a^{-3}}{a^{-1}} = a^{n+(-3)-(-1)}$

$= a^{n-3+1}$

$= a^{n-2}$

 SELF CHECK 8 Simplify each expression. Assume no division by 0.

a. $\dfrac{t^n t^2}{t^3}$ **b.** $\left(\dfrac{r^5 r^7}{r^n}\right)^3$ **c.** $\left(\dfrac{a^n}{a^5}\right)^3$ **d.** $\dfrac{x^n x^{-4}}{x^{-7}}$

Accent on technology

▶ Finding Powers

To find powers of numbers with many calculators, we use the $\boxed{y^x}$ key. For example, to find 5.37^4, we enter these numbers and press these keys:

5.37 $\boxed{y^x}$ 4 $\boxed{=}$ Some calculators have an $\boxed{x^y}$ key.

The display will read **831.5668016** .

To use a TI84 graphing calculator, we enter these numbers and press these keys:

5.37 $\boxed{\wedge}$ 4 **ENTER**

The display will read **5.37^4.**

831.5668016

If these methods do not work, consult your owner's manual.

For instructions regarding the use of a Casio graphing calculator, please refer to the Casio Keystroke Guide in the back of the book.

---

3 **Evaluate an expression using the rules for the order of operations.**

When simplifying expressions containing exponents, we find powers before performing additions and multiplications. Below we repeat the rules for order of operations, first discussed in the previous section, with a step included for exponents.

RULES FOR THE ORDER OF OPERATIONS

Use the following steps to perform all calculations within each pair of grouping symbols, working from the innermost pair to the outermost pair.

1. Find the values of any exponential expressions.
2. Perform all multiplications and divisions, working from left to right.

3. Perform all additions and subtractions, working from left to right.

4. Because a fraction bar is a grouping symbol, simplify the numerator and the denominator separately and simplify the fraction, whenever possible.

EXAMPLE 9   If $x = 2$ and $y = -3$, find the value of $3x + 2y^3$.

Solution
$$
\begin{aligned}
3x + 2y^3 &= 3(2) + 2(-3)^3 &&\text{Substitute 2 for } x \text{ and } -3 \text{ for } y. \\
&= 3(2) + 2(-27) &&\text{First find the power: } (-3)^3 = -27 \\
&= 6 - 54 &&\text{Then do the multiplications.} \\
&= -48 &&\text{Then do the subtraction.}
\end{aligned}
$$

SELF CHECK 9   Evaluate $-2a^2 - 3b$ if $a = -4$ and $b = 2$.

4   Apply the correct geometric formula to solve an application.

Table 1-4 shows the formulas used to compute the areas and volumes of many geometric figures. Remember that **area** is defined in square units and **volume** is defined in cubic units.

EXAMPLE 10   **VOLUME OF SPHERE**   Find the volume of the sphere shown in Figure 1-20 as an approximation correct to 4 decimal places.

Solution   The formula for the volume of a sphere is $V = \frac{4}{3}\pi r^3$. Since a radius is half as long as a diameter, the radius of the sphere is half of 20 centimeters, or 10 centimeters.

$$
V = \frac{4}{3}\pi r^3
$$

$$
V = \frac{4}{3}\pi(10)^3 \qquad \text{Substitute 10 for } r.
$$

$$
\approx 4188.790205 \qquad \text{Use a calculator.}
$$

20 cm

Figure 1-20   To four decimal places, the volume is $4{,}188.7902$ cm³.

SELF CHECK 10   Find the volume of a pyramid with a square base, 20 meters on each side, and a height of 21 meters.

TABLE 1-4

Figure	Name	Area	Figure	Name	Volume
s s	Square	$A = s^2$	s s s	Cube	$V = s^3$
l w	Rectangle	$A = lw$	h l w	Rectangular solid	$V = lwh$

Figure	Name	Area	Figure	Name	Volume
	Circle	$A = \pi r^2$		Sphere	$V = \frac{4}{3}\pi r^3$
	Triangle	$A = \frac{1}{2}bh$		Cylinder	$V = Bh^*$
	Trapezoid	$A = \frac{1}{2}h(b_1 + b_2)$		Cone	$V = \frac{1}{3}Bh^*$
				Pyramid	$V = \frac{1}{3}Bh^*$

*B represents the area of the base.

NOW TRY THIS

Let $a = -3$, $b = -2$, $c = 5$, and $d = -4$. Evaluate each expression.

1. $\dfrac{a - b}{c - d}$ 2. $b^2 - 4ac$ 3. $(c - a)^2 + (d - b)^2$

1.3 Exercises

WARM-UPS *Write each product using exponents.*

1. $(5)(5)(5)$
2. $(-2)(-2)(-2)(-2)$
3. $(-y)(-y)$
4. $(x)(x)(x)(x)(x)$
5. $(a)(a)(b)(b)(b)$
6. $(x)(x)(x)(y)(y)(y)(y)$
7. $(x^2)(x^2)(x^2)$
8. $(y^3)(y^3)$
9. $\left(\dfrac{2}{3}\right)\left(\dfrac{2}{3}\right)\left(\dfrac{2}{3}\right)\left(\dfrac{2}{3}\right)$
10. $\left(\dfrac{a}{b^2}\right)\left(\dfrac{a}{b^2}\right)\left(\dfrac{a}{b^2}\right)$
11. $-(ab)(ab)(ab)$
12. $-3(x^2y)(x^2y)$

REVIEW *If $a = 4$, $b = -2$, and $c = 5$, find each value.*

13. $a + b + c$
14. $a - 2b - c$
15. $\dfrac{ab + 2c}{a + b}$
16. $\dfrac{ac - bc}{6ab + b}$

VOCABULARY AND CONCEPTS *Fill in the blanks.*

17. In the exponential expression x^n, x is called the _____, and n is called the _____ or _____.
18. A natural-number exponent tells how many times the base is used as a _____.
19. $x^m x^n =$ _____
20. $(x^m)^n =$ _____

21. $(xy)^n =$ ___

22. $\left(\dfrac{x}{y}\right)^n =$ ___ $(y \neq 0)$

23. If $a \neq 0$, then $a^0 =$ _.

24. If $a \neq 0$, then $a^{-1} =$ _.

25. If $x \neq 0$, then $\dfrac{x^m}{x^n} =$ ___.

26. $\left(\dfrac{4}{5}\right)^{-3} = \left(\ \ \right)^3$

Write the formula to find each quantity.

27. Area of a square: _____

28. Area of a rectangle: _____

29. Area of a triangle: _____

30. Area of a trapezoid: _____

31. Area of a circle: _____

32. Volume of a cube: _____

33. Volume of a rectangular solid: _____

34. Volume of a sphere: _____

35. Volume of a cylinder: _____

36. Volume of a cone: _____

37. Volume of a pyramid: _____

38. In Exercises 35–37, B represents the area of the ___ of a solid.

Identify the base and the exponent of each expression.

39. 5^3

40. -7^2

41. $-x^5$

42. $(-t)^4$

43. $2b^6$

44. $(3xy)^5$

45. $(-mn^2)^3$

46. $(-p^2q)^2$

GUIDED PRACTICE *Simplify each expression. SEE EXAMPLE 1.*
(OBJECTIVE 1)

47. 3^2

48. 3^4

49. -3^2

50. -3^4

51. $(-3)^2$

52. $(-3)^3$

53. $\left(-\dfrac{2}{5}x\right)^4$

54. $(-3a)^3$

Simplify each expression. Assume that no variables are zero. SEE EXAMPLE 2. (OBJECTIVE 2)

55. x^3x^5

56. y^6y^4

57. $x^4y^2x^5y^6$

58. $a^5b^3a^9b^4$

59. $-12x^6\left(\dfrac{1}{3}x^3y^5\right)$

60. $\dfrac{1}{6}a^3b^2(-18a^4b^6)$

61. p^9pp^0

62. z^7z^0z

Simplify each expression. Assume that no variables are zero. SEE EXAMPLE 3. (OBJECTIVE 2)

63. $(x^6)^3$

64. $(y^8)^5$

65. $(-x)^2y^4x^4$

66. $-x^2y^3(y^4)^5$

67. $(a^2a^3)^4$

68. $(bb^2b^3)^4$

69. $(x^3)^4(x^5)^2$

70. $(a^3)^2(a^8)^3$

Simplify each expression. Assume that no denominators are zero. SEE EXAMPLE 4. (OBJECTIVE 2)

71. $(-2x)^6$

72. $(-3y)^5$

73. $(x^3y^2)^4$

74. $(x^2y^5)^2$

75. $\left(\dfrac{a^4}{b^5}\right)^7$

76. $\left(\dfrac{a^2}{b^3}\right)^4$

77. $\left(\dfrac{1}{2}a^2b^5\right)^4$

78. $\left(-\dfrac{1}{3}mn^2\right)^6$

Simplify each expression. Assume that no variables are zero. (OBJECTIVE 2)

79. 8^0

80. -8^0

81. $-(5x^3y^6)^0$

82. $(-3a^2b)^0$

83. $7x^0$

84. $-3x^0$

85. $7^0 + 7^2$

86. $5^2 - 5^0$

Simplify each expression. Assume that no variables are zero. SEE EXAMPLE 5. (OBJECTIVE 2)

87. 5^{-2}

88. 4^{-3}

89. $(3x)^{-4}$

90. $3x^{-4}$

91. $x^{-7}x^3$

92. $(a^{-5})^{-2}$

93. $\dfrac{1}{x^{-5}}$

94. $\dfrac{5}{x^{-6}}$

Simplify each expression. Assume that no denominators are zero. SEE EXAMPLE 6. (OBJECTIVE 2)

95. $\dfrac{a^8}{a^3}$

96. $\dfrac{c^2}{c^7}$

97. $\dfrac{x^{-6}x^5}{x^{10}}$

98. $\dfrac{(a^2)^{-3}}{a^{10}a^{-1}}$

99. $\dfrac{5a^7b^{-4}}{a^{-2}b^{-6}}$

100. $\dfrac{(2a^{-2})^3}{a^3a^{-4}}$

101. $\dfrac{(x^{-2})^9}{(x^4)^{-2}}$

102. $\left(\dfrac{a^4b^{-4}}{a^{-5}b^6}\right)^2$

Simplify each expression. Assume that no variables are zero. SEE EXAMPLE 7. (OBJECTIVE 2)

103. $\left(\dfrac{a^{-3}}{b^{-2}}\right)^{-2}$

104. $\left(\dfrac{k^{-3}}{k^{-4}}\right)^{-1}$

105. $\left(\dfrac{3x^5y^2}{6x^5y^{-2}}\right)^{-4}$

106. $\left(\dfrac{2a^3b^2}{3a^{-3}b^2}\right)^{-3}$

Simplify each expression. Assume that no variables are zero. SEE EXAMPLE 8. (OBJECTIVE 2)

107. $\dfrac{a^na^3}{a^4}$

108. $\dfrac{b^9b^7}{b^n}$

109. $\left(\dfrac{b^n}{b^3}\right)^3$

110. $\left(\dfrac{a^2}{a^n}\right)^4$

Evaluate each expression when $x = -2$ and $y = 3$. SEE EXAMPLE 9. (OBJECTIVE 3)

111. x^2y^3

112. x^3y^2

113. $\dfrac{x^{-3}}{y^3}$

114. $\dfrac{x^2}{y^{-3}}$

115. $(xy^2)^{-2}$ **116.** $-y^3x^{-2}$

117. $(-yx^{-1})^3$ **118.** $(-y)^3x^{-2}$

Find the area of each figure. Round all answers to the nearest unit.
SEE EXAMPLE 10. (OBJECTIVE 4)

119.   **120.**

121.   **122.**

Find the volume of each figure. Round all answers to the nearest unit.
SEE EXAMPLE 10. (OBJECTIVE 4)

123.   **124.**

125.   **126.**

ADDITIONAL PRACTICE *Simplify each expression. Assume that no variables are zero.*

127. aba^3b^4 **128.** $x^2y^3x^3y^2$

129. $(-3p^2q^3)^5$ **130.** $(3x^3y^4)^3$

131. $(x^2)^3(x^3)^2$ **132.** $(z^{12})^2$

133. $(r^{-3}s)^3$ **134.** $(m^5n^2)^{-3}$

135. $(b^{-8})^9$ **136.** $(a^{-2})^4(a^3)^2$

137. $(-d^2)^3(d^{-3})^3$ **138.** $(c^3)^2(c^4)^{-2}$

139. $\left(\dfrac{3a^{-2}b^2}{17a^2b^3}\right)^0$ **140.** $\dfrac{a^0 + b^0}{2(a+b)^0}$

141. $\dfrac{(3x^2)^{-2}}{x^3x^{-4}x^0}$ **142.** $\dfrac{y^{-3}y^{-4}y^0}{(2y^{-2})^3}$

143. $\dfrac{a^{-n}a^2}{a^3}$ **144.** $\dfrac{a^na^{-2}}{a^4}$

145. $\dfrac{a^{-n}a^{-2}}{a^{-4}}$ **146.** $\dfrac{a^n}{a^{-3}a^5}$

147. $\left(\dfrac{4a^{-2}b}{3ab^{-3}}\right)^3$ **148.** $\left(\dfrac{2ab^{-3}}{3a^{-2}b^2}\right)^2$

149. $\left(\dfrac{-2a^4b}{a^{-3}b^2}\right)^{-3}$ **150.** $\left(\dfrac{-3x^4y^2}{-9x^5y^{-2}}\right)^{-2}$

Find the area of each figure. Round all answers to the nearest unit.

151.   **152.**

153.   **154.**

Find the volume of each figure. Round all answers to the nearest unit.

155.   **156.**

157.   **158.**

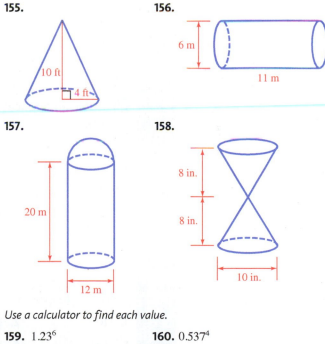

📟 *Use a calculator to find each value.*

159. 1.23^6 **160.** 0.537^4

161. -6.25^3 **162.** $(-25.1)^5$

📟 *Use a calculator to verify that each statement is true.*

163. $(3.68)^0 = 1$ **164.** $(2.1)^4(2.1)^3 = (2.1)^7$

165. $(7.2)^2(2.7)^2 = [(7.2)(2.7)]^2$

166. $(3.7)^2 + (4.8)^2 \neq (3.7 + 4.8)^2$

167. $(3.2)^2(3.2)^{-2} = 1$ **168.** $[(5.9)^3]^2 = (5.9)^6$

169. $(7.23)^{-3} = \dfrac{1}{(7.23)^3}$ **170.** $\left(\dfrac{5.4}{2.7}\right)^{-4} = \left(\dfrac{2.7}{5.4}\right)^4$

APPLICATIONS *Use a calculator to find each value.*

171. Bank accounts The formula $A = P(1 + i)^n$ gives the amount A in an account when P dollars is the amount originally deposited (the principal), i is the annual interest rate, and n is the number of years. If $5,000 is deposited in an account paying 11% compounded annually, how much will be in the account in 50 years?

172. Bank accounts The formula $P = A(1 + i)^{-n}$ gives the principal P that must be deposited at an annual rate i to grow to A dollars in n years. How much must be invested at 9% annual interest to have $1 million in 50 years?

WRITING ABOUT MATH

173. Explain why a positive number raised to a negative power is positive.

174. Explain the rules that determine the order in which operations are performed.

175. In the definition of x^{-1}, x cannot be 0. Why not?

176. Explain why $(xyz)^2 = x^2y^2z^2$.

SOMETHING TO THINK ABOUT

177. Find the value: $2^{-1} + 3^{-1} - 4^{-1}$

178. Simplify: $(3^{-1} + 4^{-1})^{-2}$

179. Construct an example using numbers to show that $x^m + x^n \neq x^{m+n}$.

180. Construct an example using numbers to show that $x^m + y^m \neq (x + y)^m$.

Section 1.4

Scientific Notation

Objectives

1. Write a number in scientific notation given a number in standard notation.
2. Write a number in standard notation given a number in scientific notation.
3. Use scientific notation to simplify a computation.
4. Solve an application requiring answers rounded to the proper significant digit.

Vocabulary

standard notation   scientific notation

Getting Ready

Evaluate each expression.

1. 10^1 **2.** 10^2 **3.** 10^3 **4.** 10^4

5. 10^{-2} **6.** 10^{-4} **7.** $4(10^3)$ **8.** $7(10^{-4})$

Very large and very small numbers occur often in science. For example, the speed of light written in **standard notation** is approximately 29,980,000,000 centimeters per second, and the mass of a hydrogen atom is approximately 0.00000000000000000000001673 gram.

Because these numbers contain a large number of zeros, they are difficult to read and remember. In this section, we will discuss a notation that will enable us to express these numbers in a more compact form.

1 Write a number in scientific notation given a number in standard notation.

With exponents, we can write very large and very small numbers in a form called *scientific notation*.

SCIENTIFIC NOTATION

A number is written in **scientific notation** when it is written in the form $N \times 10^n$, where $1 \le |N| < 10$ and n is an integer.

EXAMPLE 1 Write 29,980,000,000 in scientific notation.

Solution The number 2.998 is between 1 and 10. To get 29,980,000,000, the decimal point in 2.998 must be moved ten places to the right. We can do this by multiplying 2.998 by 10^{10}.

$$29,980,000,000 = 2.998 \times 10^{10}$$

 SELF CHECK 1 Write 150,000,000 in scientific notation.

EXAMPLE 2 Write 0.000000000000000000000001673 in scientific notation.

Solution The number 1.673 is between 1 and 10. To get 0.000000000000000000000001673, the decimal point in 1.673 must be moved 24 places to the left. We can do this by multiplying 1.673 by 10^{-24}.

$$0.000000000000000000000001673 = 1.673 \times 10^{-24}$$

 SELF CHECK 2 Write 0.000025 in scientific notation.

EXAMPLE 3 Write -0.0013 in scientific notation.

Solution The absolute value of -1.3 is between 1 and 10. To get -0.0013, we move the decimal point in -1.3 three places to the left by multiplying by 10^{-3}.

$$-0.0013 = -1.3 \times 10^{-3}$$

 SELF CHECK 3 Write $-45,700$ in scientific notation.

2 Write a number in standard notation given a number in scientific notation.

We can write a number written in scientific notation in standard notation. For example, to write 9.3×10^7 in standard notation, we multiply 9.3×10^7.

$$9.3 \times 10^7 = 9.3 \times 10,000,000 = 93,000,000$$

EXAMPLE 4 Write each number in standard notation: **a.** 3.7×10^5 **b.** -1.1×10^{-3}

Solution **a.** Since multiplication by 10^5 moves the decimal point five places to the right,

$$3.7 \times 10^5 = 370,000$$

b. Since multiplication by 10^{-3} moves the decimal point three places to the left,

$$-1.1 \times 10^{-3} = -0.0011$$

SELF CHECK 4   Write each number in standard notation.
   a. -9.6×10^{4} **b.** 5.62×10^{-3}

Each of the following numbers is written in both scientific and standard notation. In each case, the exponent gives the number of places that the decimal point moves, and the sign of the exponent indicates the direction that it moves:

$5.32 \times 10^{4} = 5\,3\,2\,0\,0.$ $6.45 \times 10^{7} = 6\,4\,5\,0\,0\,0\,0\,0.$
      4 places to the right   7 places to the right

$2.37 \times 10^{-4} = 0.0\,0\,0\,2\,3\,7$ $9.234 \times 10^{-2} = 0.0\,9\,2\,3\,4$
      4 places to the left   2 places to the left

$4.89 \times 10^{0} = 4.8\,9$
      No movement of the decimal point

Perspective

The ancient Egyptians developed two systems of writing. In hieroglyphics, each symbol was a picture of an object. Because hieroglyphic writing usually was inscribed in stone, many examples still survive today. For daily life, Egyptians used hieratic writing. Similar to hieroglyphics, hieratic writing was done with ink on papyrus sheets.

One papyrus that survives, the Rhind Papyrus, was discovered in 1858 by a British archaeologist, Henry Rhind. Also known as the Ahmes Papyrus after its ancient author, it begins with a description of its contents: *Directions for Obtaining the Knowledge of All Dark Things.*

The Ahmes Papyrus and another, the Moscow Papyrus, together contain 110 mathematical problems and their solutions. Many of these were probably for education,

The Ahmes Papyrus

because they represented situations that scribes, priests, and other government and temple administration workers were expected to be able to solve.

3   ## Use scientific notation to simplify a computation.

COMMENT   Numbers such as 47.2×10^{3} and 0.063×10^{-2} appear to be written in scientific notation, because they are the product of a number and a power of 10. However, they are not in scientific notation, because 47.2 and 0.063 are not between 1 and 10.

EXAMPLE 5   Change each number to scientific notation: **a.** 47.2×10^{3} **b.** 0.063×10^{-2}

Solution   Since the first factors are not between 1 and 10, neither number is in scientific notation. However, we can change them to scientific notation as follows:

a. $47.2 \times 10^{3} = (4.72 \times 10^{1}) \times 10^{3}$ Write 47.2 in scientific notation.
$= 4.72 \times (10^{1} \times 10^{3})$
$= 4.72 \times 10^{4}$

b. $0.063 \times 10^{-2} = (6.3 \times 10^{-2}) \times 10^{-2}$ Write 0.063 in scientific notation.
$= 6.3 \times (10^{-2} \times 10^{-2})$
$= 6.3 \times 10^{-4}$

SELF CHECK 5 Write each number in scientific notation.

a. 27.3×10^2 b. 0.0025×10^{-3}

Scientific notation is useful when simplifying expressions containing very large or very small numbers.

EXAMPLE 6 Use scientific notation to simplify $\dfrac{(0.00000064)(24,000,000,000)}{(400,000,000)(0.0000000012)}$.

Solution After writing each number in scientific notation, we can perform the arithmetic on the numbers and the exponential expressions separately.

$$\frac{(0.00000064)(24,000,000,000)}{(400,000,000)(0.0000000012)} = \frac{(6.4 \times 10^{-7})(2.4 \times 10^{10})}{(4 \times 10^8)(1.2 \times 10^{-9})}$$

$$= \frac{(6.4)(2.4)}{(4)(1.2)} \times \frac{10^{-7}10^{10}}{10^8 10^{-9}}$$

$$= 3.2 \times 10^4 \quad \text{The result in scientific notation.}$$

In standard notation, the result is 32,000.

 SELF CHECK 6 Use scientific notation to simplify: $\dfrac{(320)(25,000)}{0.00004}$

Accent
on technology

▶ **Using Scientific Notation**

Gottfried Wilhelm Leibniz
1646–1716
Leibniz, a German philosopher and logician, is principally known as one of the inventors of calculus, along with Newton. He also developed the binary numeration system, which is basic to modern computers.

Scientific and graphing calculators often give answers in scientific notation. For example, if we use a calculator to find 301.2^8, the display will read

     6.77391496 ^{19} Using a scientific calculator

     301.2 ∧8
              6.773914961E19   Using a TI84 graphing calculator

In either case, the answer is given in scientific notation and is to be interpreted as

     $6.77391496 \times 10^{19}$

Numbers also can be entered into a calculator in scientific notation. For example, to enter 24,000,000,000 (which is 2.4×10^{10} in scientific notation), we enter these numbers and press these keys:

     2.4 **EXP** 10
     2.4 **2ND** , (EE) 10 **ENTER**   }   Whichever of these keys is on your calculator

To use a calculator to simplify

     $\dfrac{(24,000,000,000)(0.00000006495)}{0.00000004824}$

we can enter each number in scientific notation. In scientific notation, the three numbers are

     2.4×10^{10} 6.495×10^{-8} 4.824×10^{-8}

The display will read **3.231343284E10**. In standard notation, the answer is 32,313,432,840.

For instructions regarding the use of a Casio graphing calculator, please refer to the Casio Keystroke Guide in the back of the book.

4 Solve an application requiring answers rounded to the proper significant digit.

If we measure the length of a rectangle and report the length to be 45 centimeters, we have rounded to the nearest centimeter. If we measure more carefully and find the length to be 45.2 centimeters, we have rounded to the nearest tenth of a centimeter. We say that the second measurement is more accurate than the first, because 45.2 has three *significant digits* but 45 has only two.

It is not always easy to know how many significant digits a number has. For example, 270 might be accurate to two or three significant digits. If 270 is rounded to the nearest ten, the number has two significant digits. If 270 is rounded to the nearest unit, it has three significant digits. This ambiguity does not occur when a number is written in scientific notation.

FINDING SIGNIFICANT DIGITS	If a number M is written in scientific notation as $N \times 10^n$, where $1 \le |N| < 10$ and n is an integer, the number of significant digits in M is the same as the number of digits in N.

In an application in which measurements are multiplied or divided, the final result should be rounded so that the answer has the same number of significant digits as the least accurate measurement.

EXAMPLE 7   **ASTRONOMY**   Earth is approximately 93,000,000 miles from the Sun, and Jupiter is approximately 484,000,000 miles from the Sun. Assuming the alignment shown in Figure 1-21, how long would it take a spaceship traveling at 7,500 mph to fly from Earth to Jupiter?

Solution   When the planets are aligned as shown in Figure 1-21, the distance between Earth and Jupiter is $(484,000,000 - 93,000,000)$ miles or 391,000,000 miles. To find the length of time in hours for the trip, we divide the distance by the rate.

$$\frac{391,000,000 \text{ mi}}{7,500 \frac{\text{mi}}{\text{hr}}} = \frac{3.91 \times 10^8 \text{ mi}}{7.5 \times 10^3 \frac{\text{mi}}{\text{hr}}}$$

There are three significant digits in the numerator and two in the denominator.

$$\approx 0.5213333 \times 10^5 \text{ mi} \cdot \frac{\text{hr}}{\text{mi}}$$

$$\approx 52,133.33 \text{ hr}$$

Since there are 24×365 hours in a year, we can change this result from hours to years by dividing 52,133.33 by (24×365).

$$\frac{52,133.33 \text{ hr}}{(24 \times 365) \frac{\text{hr}}{\text{yr}}} \approx 5.951293379 \text{ hr} \cdot \frac{\text{yr}}{\text{hr}} \approx 5.951293379 \text{ yr}$$

Rounding to two significant digits, the trip will take about 6.0 years.

Figure 1-21

SELF CHECK 7   How long would it take if the spaceship could travel at 12,000 mph?

Everyday connections
United States Federal Deficit

United States Federal Deficit
Fiscal[a] Years 1970–2016

Year	Federal deficit ($ billion[b])	Year	Federal deficit ($ billion)	Year	Federal deficit ($ billion)	Year	Federal deficit ($ billion)	Year	Federal deficit ($ billion)
1970	2.84	1980	73.83	1990	221.17	2000	−235.97	2010	1293.49
1971	23.03	1981	78.97	1991	269.34	2001	−127.89	2011	1645.12*
1972	23.37	1982	127.98	1992	290.44	2002	158.01	2012	1101.24*
1973	14.91	1983	207.8	1993	255.18	2003	377.81	2013	767.54*
1974	6.14	1984	185.42	1994	203.34	2004	412.9	2014	644.55*
1975	53.24	1985	212.36	1995	164.09	2005	318.59	2015	606.73*
1976	73.73	1986	221.29	1996	107.56	2006	248.57	2016	648.71*
1977	53.66	1987	149.79	1997	22.08	2007	160.96		
1978	59.19	1988	155.24	1998	−69.05	2008	458.55		
1979	40.73	1989	152.73	1999	−125.41	2009	1412.69		

Source: http://www.usgovernmentspending.com (customized question)

[a]The U.S. fiscal year runs from Oct. 1 through Sept. 30.   [b]1 billion = 1×10^9 *Values are estimates.

The table shows data for the federal deficit in billions of dollars for fiscal years 1970 through 2016. Use the table to answer each of the following questions.

1. Write the deficit for 1970 in standard notation.

2. Write the deficit for 2001 in standard notation and interpret the result.

3. Write the deficit for 2009 in standard notation.

4. What happened between October 1, 2008 and October 1, 2009 to result in such a dramatic increase in the deficit?

5. Given the 2010 federal deficit and that the total population that year was 308,745,538, what was each person's debt portion?

SELF CHECK ANSWERS   1. 1.5×10^8 2. 2.5×10^{-5} 3. -4.57×10^4 4. a. −96,000   b. 0.00562   5. a. 2.73×10^3 b. 2.5×10^{-6}
6. 200,000,000,000   7. about 3.7 years

NOW TRY THIS

1. An **angstrom** is a unit of length used to specify radiation wavelength. One angstrom is equal to one hundred-millionth of a centimeter.
   a. Write this number in standard notation.
   b. Write this number in scientific notation.
2. One micron is equal to 10,000 angstroms. Write this number in scientific notation.

1.4 Exercises

WARM-UPS *Multiply.*

1. $2.3(10,000)$ **2.** $-1.5(100)$

3. $-5.7(1,000)$ **4.** $6.6(100,000)$

Divide.

5. $\dfrac{3.2}{100}$ **6.** $\dfrac{-8.5}{10,000}$

7. $\dfrac{-5.1}{1,000}$ **8.** $\dfrac{8.9}{10}$

REVIEW *Write each fraction as a terminating or a repeating decimal.*

9. $\dfrac{3}{4}$ **10.** $\dfrac{4}{5}$

11. $\dfrac{13}{9}$ **12.** $\dfrac{14}{11}$

13. A man raises 3 to the second power, 4 to the third power, and 2 to the fourth power and then finds their sum. What number does he obtain?

14. If $a = -2$, $b = -3$, and $c = 4$, evaluate

$$\frac{5ab - 4ac - 2}{3bc + abc}$$

VOCABULARY AND CONCEPTS *Fill in the blanks.*

15. A number is written in _____ notation when it is written in the form $N \times$ ___, where $1 \le |N| < 10$ and n is an integer.

16. To change 6.31×10^4 to _____ notation, we move the decimal point in 6.31 ___ places to the right.

17. To change 6.31×10^{-4} to standard notation, we move the decimal point four places to the ___.

18. The number 6.7×10^3 ___ (insert > or <) the number $6,700,000 \times 10^{-4}$.

GUIDED PRACTICE *Write each number in scientific notation. SEE EXAMPLE 1. (OBJECTIVE 1)*

19. 5,200   **20.** 34,000

21. 17,600,000   **22.** 89,800,000

Write each number in scientific notation. SEE EXAMPLE 2. (OBJECTIVE 1)

23. 0.0059   **24.** 0.071

25. 0.0000096   **26.** 0.000046

Write each number in scientific notation. SEE EXAMPLE 3. (OBJECTIVE 1)

27. $-45,000$ **28.** $-547,000$

29. -0.000037 **30.** -0.00078

Write each number in standard notation. SEE EXAMPLE 4. (OBJECTIVE 2)

31. 4.6×10^2 **32.** 7.2×10^3

33. 7.96×10^5 **34.** 9.67×10^6

Write each number in scientific notation. SEE EXAMPLE 5. (OBJECTIVE 3)

35. $6,000 \times 10^{-7}$ **36.** 765×10^{-5}

37. 0.0527×10^5 **38.** 0.0298×10^3

Write each number in scientific notation and perform the operations. Give all answers in scientific notation. SEE EXAMPLE 6. (OBJECTIVE 3)

39. $\dfrac{(4,000)(30,000)}{(0.0006)}$ **40.** $\dfrac{(0.0006)(0.00007)}{21,000}$

41. $\dfrac{(640,000)(2,700,000)}{120,000}$ **42.** $\dfrac{(0.0000013)(0.000090)}{0.00039}$

ADDITIONAL PRACTICE *Write each number in standard notation.*

43. 5.26×10^{-3} **44.** 7.65×10^{-2}

45. 3.7×10^{-4} **46.** 4.12×10^{-5}

47. 23.65×10^6 **48.** 75.6×10^{-5}

49. 323×10^5 **50.** 689×10^6

Write each number in scientific notation.

51. 0.0317×10^{-2} **52.** 0.0012×10^{-3}

53. 52.3×10^0 **54.** 867×10^0

55. $\dfrac{(0.006)(0.008)}{0.0012}$ **56.** $\dfrac{(600)(80,000)}{120,000}$

57. $\dfrac{(220,000)(0.000009)}{0.00033}$ **58.** $\dfrac{(0.00024)(96,000,000)}{640,000,000}$

Use a calculator to evaluate each expression. Round all answers to the proper number of significant digits.

59. $23,437^3$ **60.** 0.00034^4

61. $(63,480)(893,322)$ **62.** $(0.0000413)(0.0000049)^2$

63. $\dfrac{(320,000)^2(0.0009)}{12,000^2}$ **64.** $\dfrac{(0.000012)^2(49,000)^2}{0.021}$

65. $\dfrac{(69.4)^8(73.1)^2}{(0.0043)^3}$ **66.** $\dfrac{(0.0031)^4(0.0012)^5}{(0.0456)^{-7}}$

APPLICATIONS *Use scientific notation to find each answer. Round all answers to the proper number of significant digits.* SEE EXAMPLE 7. *(OBJECTIVE 4)*

67. **Wavelengths** Transmitters, vacuum tubes, and lights emit energy that can be modeled as a wave. List the wavelengths shown in the table in order, from shortest to longest.

Type	Use	Wavelength (m)
visible light	lighting	9.3×10^{-6}
infrared	photography	3.7×10^{-5}
x-ray	medical	2.3×10^{-11}
radio wave	communication	3.0×10^{2}
gamma ray	treating cancer	8.9×10^{-14}

68. **Distance to the Sun** The Sun is about 93 million miles from Earth. Find this distance in feet. (*Hint*: 1 mi = 5,280 ft.)

69. **Speed of sound** The speed of sound in air is 3.31×10^{4} centimeters per second. Find the speed of sound in centimeters per hour.

70. **Volume of a tank** Find the volume of the tank shown in the illustration.

4,000 mm
7,000 mm
3,000 mm

71. **Mass of protons** If the mass of 1 proton is 0.000000000000000000000000167248 gram, find the mass of 1 million protons.

72. **Speed of light** The speed of light in a vacuum is about 30,000,000,000 centimeters per second. Find the speed of light in mph. (*Hint*: 160,000 cm ≈ 1 mile.)

73. **Distance to the Moon** The Moon is about 239,000 miles from Earth. Find this distance in inches.

74. **Distance to the Sun** The Sun is about 149,700,000 kilometers from Earth. Find this distance in miles. (*Hint*: 1 km ≈ 0.6214 mile.)

75. **Solar flares** Solar flares often produce immense loops of glowing gas ejected from the Sun's surface. The flare in the illustration extends about 95,000 kilometers into space. Express this distance in miles. (*Hint*: 1 km ≈ 0.6214 mile.)

76. **Distance to the Moon** The Moon is about 378,196 kilometers from Earth. Express this distance in inches. (*Hint*: 1 km ≈ 0.6214 mile.)

77. **Angstroms per inch** One **angstrom** is exactly 0.0000001 millimeter, and one inch is 25.4 millimeters. Find the number of angstroms in one inch.

78. **Range of a comet** One **astronomical unit** (AU) is the distance from Earth to the Sun—about 9.3×10^{7} miles. Halley's comet ranges from 0.6 to 18 AU from the Sun. Express this range in miles.

79. **Flight to Pluto** The planet Pluto is approximately 3,574,000,000 miles from Earth. If a spaceship can travel 18,000 mph, how long will it take to reach Pluto?

80. **Light year** Light travels about 300,000,000 meters per second. A **light year** is the distance that light can travel in one year. How many meters are in one light year?

81. **Distance to Alpha Centauri** Light travels about 186,000 miles per second. A **parsec** is 3.26 light years. The star Alpha Centauri is 1.3 parsecs from Earth. Express this distance in miles.

82. **Life of a comet** The mass of the comet shown in the illustration is about 10^{16} grams. When the comet is close to the Sun, matter evaporates at the rate of 10^{7} grams per second. Calculate the life of the comet if it appears every 50 years and spends ten days close to the Sun.

WRITING ABOUT MATH

83. Explain how to change a number from standard notation to scientific notation.

84. Explain how to change a number from scientific notation to standard notation.

SOMETHING TO THINK ABOUT

85. Find the highest power of 2 that can be evaluated with a scientific calculator.

86. Find the highest power of 7 that can be evaluated with a scientific calculator.

Section 1.5

Solving Linear Equations in One Variable

Objectives

1. Determine whether a number is a solution of an equation.
2. Solve a linear equation in one variable by applying the properties of equality.
3. Simplify an expression using the rules for order of operations and combining like terms.
4. Solve a linear equation in one variable requiring simplifying one or both sides.
5. Solve a linear equation in one variable that is an identity or a contradiction.
6. Solve a formula for an indicated variable.

Vocabulary

equation
solution set
solutions
roots
equivalent equations
linear equations

algebraic terms
constants
numerical coefficients
like terms
similar terms
conditional equations

identity
contradiction
empty set
literal equation

Getting Ready

Fill in the blanks.

1. ▢ + 3 = 5

2. 8 − ▢ = 4

3. $\dfrac{12}{▢} = 4$

4. ▢ · 5 = 30

In this section, we show how to solve equations, one of the most important concepts in algebra. Then, we will apply these equation-solving techniques to solve formulas for various variables.

1 Determine whether a number is a solution of an equation.

An **equation** is a statement indicating that two quantities are equal. The equation $2 + 4 = 6$ is true, and the equation $2 + 4 = 7$ is false. If an equation has a variable (say, x) it can be either true or false, depending on the value of x. For example, if $x = 1$, the equation $7x - 3 = 4$ is true.

$$7(\mathbf{1}) - 3 = 4 \quad \text{Substitute 1 for } x.$$
$$7 - 3 = 4$$
$$4 = 4$$

However, the equation is false for all other values of x. Since 1 makes the equation true, we say that 1 *satisfies* the equation.

The set of numbers that satisfies an equation is called its **solution set**. The elements of the solution set are called **solutions** or **roots** of the equation. Finding the solution set of an equation is called *solving the equation*.

EXAMPLE 1   Determine whether 3 is a solution of $2x + 4 = 10$.

Solution   We substitute 3 for x and see whether it satisfies the equation.

$$2\mathbf{x} + 4 = 10$$
$$2(\mathbf{3}) + 4 \stackrel{?}{=} 10 \quad \text{Substitute 3 for } x.$$
$$6 + 4 \stackrel{?}{=} 10 \quad \text{First do the multiplication on the left side.}$$
$$10 = 10 \quad \text{Then do the addition.}$$

Since $10 = 10$, the number 3 satisfies the equation. It is a solution.

 SELF CHECK 1   Is -5 a solution of $2x - 3 = -13$?

2   Solve a linear equation in one variable by applying the properties of equality.

To solve an equation we replace the equation with simpler ones, all having the same solution set. Such equations are called *equivalent equations*.

EQUIVALENT EQUATIONS   Equations with the same solution set are called **equivalent equations**.

We continue to replace each resulting equation with an equivalent one until we have isolated the variable on one side of an equation. To isolate the variable, we can use the following properties.

ADDITION, SUBTRACTION, MULTIPLICATION, AND DIVISION PROPERTIES OF EQUALITY

If a, b, and c are real numbers and $a = b$, then

$$a + c = b + c \quad \text{and} \quad a - c = b - c$$

and if $c \neq 0$, then

$$ac = bc \quad \text{and} \quad \frac{a}{c} = \frac{b}{c}$$

In words, we can say: *If any quantity is added to (or subtracted from) both sides of an equation, a new equation is formed that is equivalent to the original equation.*

If both sides of an equation are multiplied (or divided) by the same nonzero quantity, a new equation is formed that is equivalent to the original equation.

The most basic equations that we will solve are *linear equations in one variable.*

LINEAR EQUATIONS

A **linear equation** in one variable (say, x) is any equation that can be written in the form

$$ax + c = 0 \quad (a \text{ and } c \text{ are real numbers and } a \neq 0)$$

EXAMPLE 2   Solve: $2x + 8 = 0$

Solution   To solve the equation, we will isolate x on the left side.

$$2x + 8 = 0$$

$2x + 8 - 8 = 0 - 8$ To eliminate 8 from the left side, subtract 8 from both sides.

$2x = -8$ Simplify.

$\dfrac{2x}{2} = \dfrac{-8}{2}$ To eliminate 2 from the left side, divide both sides by 2.

$x = -4$ Simplify.

Check: We substitute -4 for x to verify that it satisfies the original equation.

$$2x + 8 = 0$$

$2(-4) + 8 \stackrel{?}{=} 0$ Substitute -4 for x.

$-8 + 8 \stackrel{?}{=} 0$

$0 = 0$

Since -4 satisfies the original equation, it is the solution.

🌿 **SELF CHECK 2**   Solve: $-3a + 15 = 0$

EXAMPLE 3   Solve: $3(x - 2) = 20$

Solution   We isolate x on the left side.

$$3(x - 2) = 20$$

$3x - 6 = 20$ Use the distributive property to remove parentheses.

$3x - 6 + 6 = 20 + 6$ To eliminate -6 from the left side, add 6 to both sides.

$3x = 26$ Simplify.

$\dfrac{3x}{3} = \dfrac{26}{3}$ To eliminate 3 from the left side, divide both sides by 3.

$x = \dfrac{26}{3}$ Simplify.

Check: $3(x - 2) = 20$

$3\left(\dfrac{26}{3} - 2\right) \stackrel{?}{=} 20$ Substitute $\frac{26}{3}$ for x.

$3\left(\dfrac{26}{3} - \dfrac{6}{3}\right) \stackrel{?}{=} 20$ Get a common denominator: $2 = \frac{6}{3}$.

$$3\left(\frac{20}{3}\right) \stackrel{?}{=} 20 \qquad \text{Combine the fractions: } \frac{26}{3} - \frac{6}{3} = \frac{20}{3}$$

$$20 = 20 \qquad \text{Simplify.}$$

Since $\frac{26}{3}$ satisfies the equation, it is the solution.

SELF CHECK 3   Solve: $-2(a + 3) = 18$

3 Simplify an expression using the rules for order of operations and combining like terms.

To solve many equations, we will need to combine like terms. An **algebraic term** is either a number (called a **constant**) or the product of numbers and variables. Some examples of terms are $3x$, $-7y$, y^2, and 8. The **numerical coefficients** of these terms are 3, -7, 1, and 8 (8 can be written as $8x^0$), respectively.

In algebraic expressions, terms are separated by $+$ and $-$ signs. For example, the expression $3x^2 + 2x - 4$ has three terms, and the expression $3x + 7y$ has two terms.

Terms with the same variables with the same exponents are called **like terms** or **similar terms**:

$5x$ and $6x$ are like terms.

$27x^2y^3$ and $-326x^2y^3$ are like terms.

$4x$ and $-17y$ are unlike terms.   The terms have different variables.

$15x^2y$ and $6xy^2$ are unlike terms.   The variables have different exponents.

By using the distributive law, we can combine like terms. For example,

$$\mathbf{5}x + \mathbf{6}x = (\mathbf{5 + 6})x = 11x \qquad \text{and} \qquad \mathbf{32}y - \mathbf{16}y = (\mathbf{32 - 16})y = 16y$$

This suggests that to combine like terms, *we add or subtract their numerical coefficients and keep the same variables with the same exponents.*

EXAMPLE 4   Simplify: $2(x + 3) + 8(x - 2)$

Solution   To simplify the expression, we will use the distributive property to remove parentheses and then combine like terms.

$$\mathbf{2}(x + 3) + \mathbf{8}(x - 2) = \mathbf{2}x + \mathbf{2} \cdot 3 + \mathbf{8}x - \mathbf{8} \cdot 2 \qquad \text{Use the distributive property to}$$
$$\text{remove parentheses.}$$
$$= 2x + 6 + 8x - 16 \qquad 2 \cdot 3 = 6 \text{ and } 8 \cdot 2 = 16$$
$$= 2x + 8x + 6 - 16 \qquad \text{Use the commutative property of}$$
$$\text{addition: } 6 + 8x = 8x + 6$$
$$= 10x - 10 \qquad \text{Combine like terms}$$

SELF CHECK 4   Simplify: $-3(a + 5) + 5(a - 2)$

COMMENT   In this book, we will simplify expressions and solve equations. Recognizing when to do which is a skill that we will apply throughout this book. Solving an equation means finding the values of the variable that make the equation true. Remember that

Expressions are to be simplified.   *Equations are to be solved.*

4 Solve a linear equation in one variable requiring simplifying one or both sides.

EXAMPLE 5 Solve: $3(2x - 1) = 2x + 9$

Solution

$$3(2x - 1) = 2x + 9$$

$6x - 3 = 2x + 9$ Use the distributive property to remove parentheses.

$6x - 3 \; \mathbf{+ \; 3} = 3x + 9 \; \mathbf{+ \; 3}$ To eliminate -3 from the left side, add 3 to both sides.

$6x = 2x + 12$ Combine like terms.

$6x \; \mathbf{- \; 2x} = 2x \; \mathbf{- \; 2x} + 12$ To eliminate $2x$ from the right side, subtract $2x$ from both sides.

$4x = 12$ Combine like terms.

$x = 3$ To eliminate 4 from the left side, divide both sides by 4.

Check:

$$3(2x - 1) = 2x + 9$$

$3(2 \cdot \mathbf{3} - 1) \stackrel{?}{=} 2 \cdot \mathbf{3} + 9$ Substitute 3 for x.

$3(5) \stackrel{?}{=} 6 + 9$

$15 = 15$

Since 3 satisfies the equation, it is the solution.

SELF CHECK 5 Solve: $-4(3a + 4) = 2a - 4$

To solve more complicated linear equations, we will follow these steps.

SOLVING LINEAR EQUATIONS

1. If the equation contains fractions, multiply both sides of the equation by a number that will eliminate the denominators. (This is referred to as clearing the fractions.)

2. Use the distributive property to remove all sets of parentheses and combine like terms on each side of the equation.

3. Use the addition and subtraction properties of equality to group all variables on one side of the equation and all numbers on the other side. Combine like terms, if necessary.

4. Use the multiplication and division properties of equality to make the coefficient of the variable equal to 1.

5. Check the result by replacing the variable with the possible solution and verifying that the number satisfies the equation.

EXAMPLE 6 Solve: $\dfrac{5}{3}(x - 3) = \dfrac{3}{2}(x - 2) + 2$

Solution **Step 1:** Since 6 is the smallest number that can be divided by both 2 and 3, we multiply both sides of the equation by 6 to clear the fractions.

$$\frac{5}{3}(x - 3) = \frac{3}{2}(x - 2) + 2$$

$\mathbf{6}\left[\dfrac{5}{3}(x - 3)\right] = \mathbf{6}\left[\dfrac{3}{2}(x - 2) + 2\right]$ To clear the fractions, multiply both sides by 6.

$\mathbf{6} \cdot \dfrac{5}{3}(x - 3) = \mathbf{6} \cdot \dfrac{3}{2}(x - 2) + \mathbf{6} \cdot 2$ Use the distributive property on the right side.

$10(x - 3) = 9(x - 2) + 12$ Simplify.

Step 2:   We use the distributive property to remove parentheses and then combine like terms.

$$10x - 30 = 9x - 18 + 12$$
$$10x - 30 = 9x - 6$$

Step 3:   We use the addition and subtraction properties of equality by adding 30 to both sides and subtracting $9x$ from both sides.

$$10x - 30 \ \mathbf{-\ 9x + 30} = 9x - 6 \ \mathbf{-\ 9x + 30}$$
$$x = 24 \qquad\qquad \text{Combine like terms.}$$

Since the coefficient of x in the above equation is 1, Step 4 is unnecessary.

Step 5:   We check by substituting 24 for x in the original equation and simplifying:

$$\frac{5}{3}(x - 3) = \frac{3}{2}(x - 2) + 2$$

$$\frac{5}{3}(\mathbf{24} - 3) \stackrel{?}{=} \frac{3}{2}(\mathbf{24} - 2) + 2$$

$$\frac{5}{3}(21) \stackrel{?}{=} \frac{3}{2}(22) + 2$$

$$5(7) \stackrel{?}{=} 3(11) + 2$$

$$35 = 35$$

Since 24 satisfies the equation, it is the solution.

SELF CHECK 6   Solve: $\dfrac{3}{2}(x - 2) = \dfrac{5}{3}(x - 3) + 2$

EXAMPLE 7   Solve: $\dfrac{x + 2}{5} - 4x = \dfrac{8}{5} - \dfrac{x + 9}{2}$

Solution

$$\frac{x + 2}{5} - 4x = \frac{8}{5} - \frac{x + 9}{2}$$

$$\mathbf{10}\left(\frac{x + 2}{5} - 4x\right) = \mathbf{10}\left(\frac{8}{5} - \frac{x + 9}{2}\right) \qquad \text{To clear the fractions, multiply both sides by 10.}$$

$$10\left(\frac{x + 2}{5}\right) - 10(4x) = 10\left(\frac{8}{5}\right) - 10\left(\frac{x + 9}{2}\right) \qquad \text{Use the distributive property to remove parentheses.}$$

$$2(x + 2) - 40x = 2(8) - 5(x + 9) \qquad \text{Simplify.}$$

$$2x + 4 - 40x = 16 - 5x - 45 \qquad \text{Use the distributive property to remove parentheses.}$$

$$-38x + 4 = -5x - 29 \qquad \text{Combine like terms.}$$

$$-38x + 4 + 5x - 4 = -5x - 29 + 5x - 4 \qquad \text{Add } 5x \text{ and } -4 \text{ to both sides.}$$

$$-33x = -33 \qquad \text{Simplify.}$$

$$\frac{-33x}{-33} = \frac{-33}{-33} \qquad \text{Divide both sides by } -33.$$

$$x = 1 \qquad \text{Simplify.}$$

Check:

$$\frac{x + 2}{5} - 4x = \frac{8}{5} - \frac{x + 9}{2}$$

$$\frac{\mathbf{1} + 2}{5} - 4(\mathbf{1}) \stackrel{?}{=} \frac{8}{5} - \frac{\mathbf{1} + 9}{2} \qquad \text{Substitute 1 for } x.$$

$$\frac{3}{5} - 4 \stackrel{?}{=} \frac{8}{5} - 5$$

$$\frac{3}{5} - \frac{20}{5} \overset{?}{=} \frac{8}{5} - \frac{25}{5}$$

$$-\frac{17}{5} = -\frac{17}{5}$$

Since 1 satisfies the equation, it is the solution.

SELF CHECK 7   Solve: $\dfrac{a+5}{5} + 2a = \dfrac{a}{2} - \dfrac{a+14}{5}$

EXAMPLE 8   Solve: $0.06x + 0.07(15{,}000 - x) = 990$

Solution

$$0.06x + 0.07(15{,}000 - x) = 990$$

$$\mathbf{100}\left[0.06x + 0.07(15{,}000 - x)\right] = \mathbf{100}\left[990\right]$$ To remove the decimals, multiply both sides by 100.

$$6x + 7(15{,}000 - x) = 99{,}000$$ Use the distributive property to remove the brackets.

$$6x + 105{,}000 - 7x = 99{,}000$$ Use the distributive property to remove the parentheses.

$$-x = -6{,}000$$ Combine like terms and subtract 105,000 from both sides.

$$x = 6{,}000$$ Multiply both sides by -1.

Check:
$$0.06x + 0.07(15{,}000 - x) = 990$$
$$0.06(\mathbf{6{,}000}) + 0.07(15{,}000 - \mathbf{6{,}000}) \overset{?}{=} 990$$
$$360 + 0.07(9{,}000) \overset{?}{=} 990$$
$$360 + 630 \overset{?}{=} 990$$
$$990 = 990$$

Since 6,000 satisfies the equation, the solution is 6,000.

SELF CHECK 8   Solve: $0.05(8{,}000 - x) + 0.06x = 450$

5   **Solve a linear equation in one variable that is an identity or a contradiction.**

The linear equations discussed so far are called **conditional equations**. These equations have exactly one solution. An **identity** is an equation that is satisfied by every number x for which both sides of the equation are defined. The solution of an identity is the set of *all real numbers* and is denoted by the symbol $\mathbb{R}$.

EXAMPLE 9   Solve: $2(x - 1) + 4 = 4(1 + x) - (2x + 2)$

Solution
$$2(x - 1) + 4 = 4(1 + x) - (2x + 2)$$
$$2x - 2 + 4 = 4 + 4x - 2x - 2$$ Use the distributive property to remove parentheses.
$$2x + 2 = 2x + 2$$ Combine like terms.

The result $2x + 2 = 2x + 2$ is true for every value of x. Since every number x satisfies the equation, it is an identity. The solution set is $\mathbb{R}$.

SELF CHECK 9   Solve: $3(a + 4) + 5 = 2(a - 1) + a + 19$

A **contradiction** is an equation that has no solution, as in the next example.

EXAMPLE 10 Solve: $\dfrac{x-1}{3} + 4x = \dfrac{3}{2} + \dfrac{13x-2}{3}$

Solution

$$\dfrac{x-1}{3} + 4x = \dfrac{3}{2} + \dfrac{13x-2}{3}$$

$$6\left(\dfrac{x-1}{3} + 4x\right) = 6\left(\dfrac{3}{2} + \dfrac{13x-2}{3}\right) \qquad \text{To clear the fractions, multiply both sides by 6.}$$

$$2(x-1) + 6(4x) = 3(3) + 2(13x-2) \qquad \text{Use the distributive property to remove parentheses.}$$

$$2x - 2 + 24x = 9 + 26x - 4 \qquad \text{Remove parentheses.}$$

$$26x - 2 = 26x + 5 \qquad \text{Combine like terms.}$$

$$-2 = 5 \qquad \text{Subtract } 26x \text{ from both sides.}$$

Since $-2 = 5$ is false, no number x can satisfy the equation. The solution set is the **empty set**, denoted as $\varnothing$.

SELF CHECK 10 Solve: $\dfrac{x+5}{5} = \dfrac{1}{5} + \dfrac{x}{5}$

Every linear equation in one variable will be one of these types, as summarized in Table 1-5.

TABLE 1-5				
Type of equation	**Examples**		**Solution set**	
Conditional	$2x + 4 = 8$	$\dfrac{x}{2} - 4 = 12$	$\{2\}$	$\{32\}$
Identity	$x + x = 2x$	$2(x+3) = 2x+6$	$\mathbb{R}$	$\mathbb{R}$
Contradiction	$x = x - 1$	$2(x+3) = 2x+5$	$\varnothing$	$\varnothing$

6 Solve a formula for an indicated variable.

Suppose we want to find the heights of several triangles whose areas and bases are known. It would be tedious to substitute values of A and b into the formula $A = \frac{1}{2}bh$ and then repeatedly solve the formula for h. It is more efficient to solve for h first and then substitute values for A and b and compute h directly.

To solve a formula, sometimes called a **literal equation**, for an indicated variable means to isolate that variable on one side of the equation.

EXAMPLE 11 Solve $A = \dfrac{1}{2}bh$ for h.

Solution

$$A = \dfrac{1}{2}bh$$

$$2 \cdot A = 2\left(\dfrac{1}{2}bh\right) \qquad \text{To clear the fraction, multiply both sides by 2.}$$

$$2A = bh \qquad \text{Multiply.}$$

$$\dfrac{2A}{b} = h \qquad \text{To isolate } h, \text{ divide both sides by } b.$$

$$h = \dfrac{2A}{b} \qquad \text{Write } h \text{ on the left side.}$$

🌿 **SELF CHECK 11** Solve $A = \frac{1}{2}bh$ for b.

EXAMPLE 12 For simple interest, the formula $A = p + prt$ represents the amount of money in an account at the end of a specific time. A represents the amount, p the principal, r the rate of interest, and t the time. We can solve the formula for t as follows:

Solution

$$A = p + prt$$

$$A - p = prt \qquad \text{To isolate the term involving } t, \text{ subtract } p \text{ from both sides.}$$

$$\frac{A - p}{pr} = t \qquad \text{To isolate } t, \text{ divide both sides by } pr.$$

$$t = \frac{A - p}{pr} \qquad \text{Write } t \text{ on the left side.}$$

🌿 **SELF CHECK 12** Solve $A = p + prt$ for r.

EXAMPLE 13 The formula $F = \frac{9}{5}C + 32$ converts degrees Celsius to degrees Fahrenheit. Solve the formula for C.

Solution

$$F = \frac{9}{5}C + 32$$

$$F - 32 = \frac{9}{5}C \qquad \text{To isolate the term involving } C, \text{ subtract 32 from both sides.}$$

$$\frac{5}{9}(F - 32) = \frac{5}{9}\left(\frac{9}{5}C\right) \qquad \text{To isolate } C, \text{ multiply both sides by } \frac{5}{9}.$$

$$\frac{5}{9}(F - 32) = C \qquad \frac{5}{9} \cdot \frac{9}{5} = 1$$

$$C = \frac{5}{9}(F - 32)$$

To convert degrees Fahrenheit to degrees Celsius, we can use the formula $C = \frac{5}{9}(F - 32)$.

🌿 **SELF CHECK 13** Solve $S = \dfrac{180(n - 2)}{7}$ for n.

🌿 **SELF CHECK ANSWERS**
1. yes **2.** 5 **3.** -12 **4.** $2a - 25$ **5.** $-\frac{6}{7}$ **6.** 0 **7.** -2 **8.** 5,000 **9.** $\mathbb{R}$ **10.** $\varnothing$
11. $b = \frac{2A}{h}$ **12.** $r = \frac{A - p}{pt}$ **13.** $n = \frac{7S}{180} + 2$ or $n = \frac{7S + 360}{180}$

NOW TRY THIS

Solve each equation for the indicated variable.

1. $y - y_1 = m(x - x_1)$ for m

2. $A = p + prt$ for p

1.5 Exercises

WARM-UPS *Evaluate each expression when $x = -2$.*

1. $3x + 1$

2. $7 + 4x$

3. $-x + 5$

4. $6 - x$

5. $3 - x$

6. $-3(x - 1)$

Multiply each expression by 4.

7. $\dfrac{1}{4}x$

8. $\dfrac{-3}{4}x$

9. $\dfrac{x + 1}{4}$

10. $\dfrac{-x + 2}{4}$

11. $\dfrac{x}{4} - \dfrac{3}{2}$

12. $\dfrac{x}{2} + \dfrac{5}{4}$

REVIEW *Simplify each expression.*

13. $\left(\dfrac{x + y}{x - y}\right)^0$

14. $(-2x^5)^4$

15. $\left(\dfrac{x^2x^5}{x^3}\right)^2$

16. $\left(\dfrac{x^4y^3}{x^5y}\right)^3$

17. $(-4x)^{-3}$

18. $\left(\dfrac{x^3}{y^4}\right)^{-5}$

VOCABULARY AND CONCEPTS *Fill in the blanks.*

19. An _____ is a statement that two quantities are equal.

20. If a number is substituted for a variable in an equation and the equation is true, we say that the number _____ the equation. The numbers that satisfy an equation form its _____ set.

21. If two equations have the same solution set, they are called _____ equations.

22. The solutions of an equation are also called _____.

23. If a, b, and c are real numbers and $a = b$, then $a + c = b + \_$ and $a - c = b - \_$, using the addition and subtraction properties of equality.

24. If a, b, and c are real numbers and $a = b$, then $a \cdot c = b \cdot \_$ and $\dfrac{a}{c} = \dfrac{}{\_}$ $(c \neq 0)$, using the multiplication and division properties of equality.

25. A number or the product of numbers and variables is called an algebraic _____.

26. _____ terms (similar terms) are terms with the same variables with the same exponents.

27. To combine like terms, add their _____ and keep the same _____ and exponents.

28. An _____ is an equation that is true for all values of its variable and its solution set is denoted by $\mathbb{R}$. A contradiction is true for ___ values of its variable and its solution set is denoted by $\varnothing$.

GUIDED PRACTICE *Determine whether 5 is a solution of each equation.* **SEE EXAMPLE 1. (OBJECTIVE 1)**

29. $3x + 2 = 17$

30. $7 + 5x = 32$

31. $\dfrac{3}{5}x - 5 = -2$

32. $\dfrac{2}{5}x + 12 = 8$

Solve each equation. **SEE EXAMPLE 2. (OBJECTIVE 2)**

33. $x + 6 = 8$

34. $y - 7 = 3$

35. $-5y = 15$

36. $3v = 12$

37. $3a - 27 = 0$

38. $4b + 28 = 0$

39. $\dfrac{x}{4} = 7$

40. $\dfrac{x}{-8} = 3$

Solve each equation. **SEE EXAMPLE 3. (OBJECTIVE 2)**

41. $3x + 1 = 3$

42. $7x + 5 = 20$

43. $6y - 10 = 14$

44. $2x - 4 = 16$

45. $3(x - 4) = -36$

46. $4(x + 6) = 84$

47. $-3(r + 6) = -2$

48. $-4(s - 4) = -1$

Simplify each expression. **SEE EXAMPLE 4. (OBJECTIVE 3)**

49. $6(2x - 5) + 3(x + 1)$

50. $7(x + 3) + 4(3x - 8)$

51. $3(x + 4) + 2(x - 1)$

52. $2(x - 3) + 4(x + 5)$

53. $8 + 2(x - 4) - (6x - 9)$

54. $9 - 3(6x + 2) - (4x + 3)$

55. $6x - 3(9 - 5x) + 8$

56. $-8x - 7(5 - 6x) - 9$

Solve each equation. **SEE EXAMPLE 5. (OBJECTIVE 4)**

57. $2(a - 5) - (3a + 1) = 0$

58. $8(3a - 5) - 4(2a + 3) = 12$

59. $3(y - 5) + 10 = 2(y + 4)$

60. $2(5x + 2) = 3(3x - 2)$

61. $9(x + 2) = -6(4 - x) + 18$

62. $3(x + 2) - 2 = -(5 + x) + x$

63. $-4p - 2(3p + 5) = -6p + 2(p + 2)$

64. $2q - 3(q - 5) = 5(q + 2) - 7$

Solve each equation. **SEE EXAMPLE 6. (OBJECTIVE 4)**

65. $\dfrac{x}{6} + 1 = \dfrac{x}{3}$

66. $\dfrac{3}{2}(y + 4) = \dfrac{20 - y}{2}$

67. $5 - \dfrac{x + 2}{3} = 7 - x$

68. $3x - \dfrac{2(x + 3)}{3} = 16 - \dfrac{x + 2}{2}$

69. $\dfrac{4x - 2}{2} = \dfrac{3x + 6}{3}$

70. $\dfrac{t + 4}{2} = \dfrac{2t - 3}{3}$

71. $\dfrac{a + 1}{3} + \dfrac{a - 1}{5} = \dfrac{2}{15}$

72. $\dfrac{2z + 3}{3} + \dfrac{3z - 4}{6} = \dfrac{z - 2}{2}$

Solve each equation. SEE EXAMPLE 7. (OBJECTIVE 4)

73. $\frac{1}{2}x - 4 = -1 + 2x$

74. $2x + 3 = \frac{2}{3}x - 1$

75. $\frac{x}{2} - \frac{x}{3} = 4$

76. $\frac{x}{2} + \frac{x}{3} = 10$

77. $\frac{5a}{2} - 12 = \frac{a}{3} + 1$

78. $\frac{5a}{6} - \frac{5}{2} = -\frac{1}{2} - \frac{a}{6}$

79. $\frac{a+1}{4} + \frac{2a-3}{4} = \frac{a}{2} - 2$

80. $\frac{y-8}{5} + 2 = \frac{2}{5} - \frac{y}{3}$

Solve each equation. SEE EXAMPLE 8. (OBJECTIVE 4)

81. $0.45 = 16.95 - 0.25(75 - 3a)$

82. $3.2 + x = 0.25(x + 32)$

83. $0.09x + 0.14(10,000 - x) = 1,275$

84. $0.04(20) + 0.01l = 0.02(20 + l)$

Solve each equation and state whether the equation is an identity or a contradiction. Then give the solution set. SEE EXAMPLES 9–10. (OBJECTIVE 5)

85. $4(2 - 3t) + 6t = -6t + 8$

86. $2(x - 3) = \frac{3}{2}(x - 4) + \frac{x}{2}$

87. $2x - 6 = -2x + 4(x - 2)$

88. $3(x - 4) + 6 = -2(x + 4) + 5x$

Solve each formula for the indicated variable. SEE EXAMPLE 11. (OBJECTIVE 6)

89. $A = lw$ for w

90. $p = 4s$ for s

91. $V = \frac{1}{3}Bh$ for B

92. $b = \frac{2A}{h}$ for A

93. $I = prt$ for t

94. $I = prt$ for r

95. $p = 2l + 2w$ for w

96. $p = 2l + 2w$ for l

Solve each formula for the indicated variable. SEE EXAMPLE 12. (OBJECTIVE 6)

97. $A = \frac{1}{2}h(B + b)$ for B

98. $A = \frac{1}{2}h(B + b)$ for b

99. $y = mx + b$ for x

100. $y = mx + b$ for m

Solve each formula for the indicated variable. SEE EXAMPLE 13. (OBJECTIVE 6)

101. $S = \frac{a - lr}{1 - r}$ for l

102. $C = \frac{5}{9}(F - 32)$ for F

103. $S = \frac{n(a + l)}{2}$ for l

104. $S = \frac{n(a + l)}{2}$ for n

ADDITIONAL PRACTICE *Simplify each expression.*

105. $4(x + 1) + (x + 6)$

106. $2(x + 2) + 4(x + 3)$

107. $-2(x + 1) - (x + 3)$

108. $-(x - 2) - (-3x + 1)$

Solve each equation. If the equation is an identity or a contradiction, so indicate.

109. $5(2x - 3) = 3x + 6$

110. $4(3x + 2) = 2x - 12$

111. $-2(4x - 5) = 4x - 3$

112. $-3(2x + 3) = 8x - 6$

113. $3a - 22 = -2a - 7$

114. $a + 18 = 6a - 3$

115. $3(y - 4) - 6 = y$

116. $2x + (2x - 3) = 5$

117. $5(5 - a) = 37 - 2a$

118. $4a + 17 = 7(a + 2)$

119. $4(y + 1) = -2(4 - y)$

120. $5(r + 4) = -2(r - 3)$

121. $2(2x + 1) = 15 + 3x$

122. $-2(x + 5) = 30 - x$

123. $9.8 - 15z = -15.7$

124. $0.05a + 0.25 = 0.77$

125. $-3x = -2x + 1 - (5 + x)$

126. $4 + 4(n + 2) = 3n - 2(n - 5)$

127. $4x - 2(3x + 2) = 2(x + 3)$

128. $2(y + 2) + 1 = 2y + 5$

129. $l = a + (n - 1)d$ for n

130. $l = a + (n - 1)d$ for d

APPLICATIONS

131. Force of gravity The masses of the two objects in the illustration are m and M. The force of gravitation F between the masses is

$$F = \frac{GmM}{d^2}$$

where G is a constant and d is the distance between them. Solve for m.

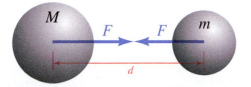

132. Thermodynamics The Gibbs free-energy formula is $G = U - TS + pV$. Solve for S.

133. Converting temperatures Solve the formula $F = \frac{9}{5}C + 32$ for C and find the Celsius temperatures that correspond to Fahrenheit temperatures of 32°, 70°, and 212°.

134. Doubling money A man intends to invest \$1,000 at simple interest. Solve the formula $A = p + prt$ for t and find how long it will take to double his money at the rates of 5%, 7%, and 10%.

135. Cost of electricity   The cost of electricity in a certain city is given by the formula $C = 0.07n + 6.50$, where C is the cost and n is the number of kilowatt hours used. Solve for n and find the number of kWh used for costs of $49.97, $76.50, and $125.

136. Cost of water   A monthly water bill in a certain city is calculated by using the formula $n = \frac{5{,}000C - 17{,}500}{6}$, where n is the number of gallons used and C is the monthly cost. Solve for C and compute the bill for quantities used of 500, 1,200, and 2,500 gallons.

137. Ohm's Law   The formula $E = IR$, called **Ohm's Law**, is used in electronics. Solve for R and then calculate the resistance R if the voltage E is 56 volts and the current I is 7 amperes. (Resistance has units of **ohms**.)

138. Earning interest   An amount P, invested at a simple interest rate r, will grow to an amount A in t years, according to the formula $A = P(1 + rt)$. Solve for P. Suppose a man invested some money at 5.5%. If after 5 years, he had $6,693.75 on deposit, what amount did he originally invest?

139. Angles of a polygon   A regular polygon has n equal sides and n equal angles. The measure a of an interior angle is given by $a = 180°\left(1 - \frac{2}{n}\right)$. Solve for n. Find the number of sides of the regular polygon in the illustration if an interior angle is 135°.

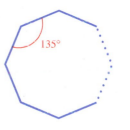

140. Power loss   The illustration is the schematic diagram of a resistor connected to a voltage source of 60 volts. As a result, the resistor dissipates power in the form of heat. The power P lost when a voltage V is placed across a resistance R is given by the formula

$$P = \frac{E^2}{R}$$

Solve for R. If P is 4.8 watts and E is 60 volts, find R.

WRITING ABOUT MATH

141. Explain the difference between a conditional equation, an identity, and a contradiction.

142. Explain how you would solve an equation.

SOMETHING TO THINK ABOUT   *Find the mistake.*

143. $3(x - 2) + 4 = 14$
$$3x - 2 + 4 = 14$$
$$3x + 2 = 14$$
$$3x = 12$$
$$x = 4$$

144.
$$A = p + prt$$
$$A - p = prt$$
$$A - p - pr = t$$
$$t = A - p - pr$$

Section 1.6

Using Linear Equations to Solve Applications

Objectives

1 Solve a recreation application by setting up and solving a linear equation in one variable.

2 Solve a business application by setting up and solving a linear equation in one variable.

3 Solve a geometry application by setting up and solving a linear equation in one variable.

4 Solve a lever application by setting up and solving a linear equation in one variable.

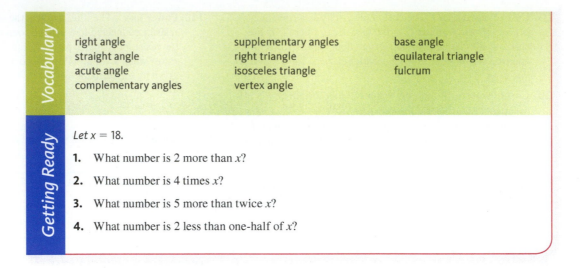

right angle	supplementary angles	base angle
straight angle	right triangle	equilateral triangle
acute angle	isosceles triangle	fulcrum
complementary angles	vertex angle	

Getting Ready

Let x = 18.

1. What number is 2 more than *x*?

2. What number is 4 times *x*?

3. What number is 5 more than twice *x*?

4. What number is 2 less than one-half of *x*?

In this section, we will apply equation-solving skills to solve applications. When we translate the words of a problem into mathematics, we are creating a *mathematical model* of the situation. To create these models, we can use Table 1-6 to translate certain words into mathematical operations.

TABLE 1-6			
Addition (+)	**Subtraction (−)**	**Multiplication (·)**	**Division (÷)**
added to	subtracted from	multiplied by	divided by
plus	difference	product	quotient
the sum of	less than	times	ratio
more than	less	of	half
increased by	decreased by	twice	

We can then change English phrases into algebraic expressions, as in Table 1-7.

TABLE 1-7	
English phrase	**Algebraic expression**
2 added to some number	$x + 2$
the difference between some number and 2	$x - 2$
5 times some number	$5x$
the product of 925 and some number	$925x$
5% of some number	$0.05x$
the sum of twice a number and 10	$2x + 10$
the quotient (or ratio) of 5 and some number	$\frac{5}{x}$
half of a number	$\frac{x}{2}$

Once we know how to change phrases into algebraic expressions, we can solve many problems. The following list of steps provides a strategy for solving applications.

PROBLEM SOLVING

1. **Analyze the problem** and identify a variable by asking yourself "What am I asked to find?" Choose a variable to represent the quantity to be found and then express all other unknown quantities in the problem as expressions involving that variable.

2. **Form an equation** by expressing a quantity in two different ways. This may require reading the problem several times to understand the given facts. What information is given? Is there a formula that applies to this situation? Often a sketch, chart, or diagram will help you visualize the facts of the problem.

3. **Solve the equation** found in Step 2.

4. **State the conclusion**, including proper units.

5. **Check the result** to be certain it satisfies the given conditions.

1 Solve a recreation application by setting up and solving a linear equation in one variable.

EXAMPLE 1 **CUTTING A ROPE**   A mountain climber wants to cut a rope 213 feet long into three pieces. If each piece is to be 2 feet longer than the previous one, where should he make the cuts?

Analyze the problem   We are asked to find the lengths of three pieces of rope. The information is given in terms of the length of the shortest piece. Therefore, we let x represent the length of the shortest piece in feet and express the other lengths in terms of x. Then $x + 2$ represents the length of the second piece in feet and $x + 2 + 2$, or $x + 4$, represents the length of the longest piece in feet.

Form an equation   The climber knows that the sum of these three lengths can be expressed in two ways: as $x + (x + 2) + (x + 4)$ and as 213. (See Figure 1-22.)

Figure 1-22

From Figure 1-22, we can set up the equation

The length of the first piece	plus	the length of the second piece	plus	the length of the third piece	equals	the total length of the rope.
x	$+$	$x + 2$	$+$	$x + 4$	$=$	213

Solve the equation   We can now solve the equation.

$$x + x + 2 + x + 4 = 213 \quad \text{This is the equation to solve.}$$
$$3x + 6 = 213 \quad \text{Combine like terms.}$$
$$3x = 207 \quad \text{Subtract 6 from both sides.}$$
$$x = 69 \quad \text{Divide both sides by 3. This is the length of the first piece in feet.}$$
$$x + 2 = 71 \quad \text{This is the length of the second piece in feet.}$$
$$x + 4 = 73 \quad \text{This is the length of the third piece in feet.}$$

COMMENT   Remember to include any units (feet, inches, pounds, etc.) when stating the conclusion to an application.

State the conclusion   He should make one cut 69 feet from one end. Then he should make a second cut 71 feet from one end of the remaining rope. This will leave a length of 73 feet.

Check the result   Each length is 2 feet longer than the previous length, and the sum of the lengths is 213 feet.

🌿 **SELF CHECK 1**   If the rope in the example is 291 feet long, where should he make the cuts?

---

2 Solve a business application by setting up and solving a linear equation in one variable.

When the regular price of merchandise is reduced, the amount of reduction is called the *markdown* (or *discount*).

> Sale price   =   regular price   −   markdown.

Usually, the markdown is expressed as a percent of the regular price.

> Markdown   =   percent of markdown   ·   regular price.

EXAMPLE 2   **FINDING THE PERCENT OF MARKDOWN**   A home theater system is on sale for $777. If the list price was $925, find the percent of markdown.

Analyze the problem   We are asked to find the percent of markdown, so we let p represent the percent of markdown, expressed as a decimal. In this case, $777 is the sale price, and $925 is the regular price.

Form an equation   The amount of the markdown is found by multiplying the regular price by the *percent of markdown*. If we subtract this amount from the regular price, we will have the sale price. This can be expressed as the equation

Sale price	equals	regular price	minus	markdown.
777	=	925	−	$p \cdot 925$

Solve the equation   We can now solve the equation.

$$777 = 925 - p \cdot 925 \qquad \text{This is the equation to solve.}$$
$$777 = 925 - 925p \qquad \text{Use the commutative property of multiplication.}$$
$$-148 = -925p \qquad \text{Subtract 925 from both sides.}$$
$$0.16 = p \qquad \text{Divide both sides by } -925.$$

State the conclusion   Because 0.16 equals 16%, the theater system is on sale at a 16% markdown.

Check the result   Since the markdown is 16% of $925, or $148, the sale price is $925 − $148, or $777.

🌿 **SELF CHECK 2**   If the theater system in the example is on sale for $740, find the price of the markdown.

EXAMPLE 3   **PORTFOLIO ANALYSIS**   A college foundation owns stock in IBC (selling at $54 per share), GS (selling at $65 per share), and ATB (selling at $105 per share). The foundation owns equal shares of GS and IBC, but five times as many shares of ATB.

  If this portfolio is worth $450,800, how many shares of each type does the foundation own?

Analyze the problem   We must find how many shares of each type of stock is owned. The given information states that there are the same number of shares for both IBC and GS. If we will let x represent the number of shares of IBC, then x also represents the number of shares of GS. There are five times as many shares of ATB, so the number of shares of ATB will be represented by $5x$.

Form an equation The value of the IBC stock plus the value of the GS stock plus the value of the ATB stock must equal $450,800.

- Since each of the x shares of IBC stock is worth $54, the value of this stock is $54x$.
- Since each share of x shares of GS stock is worth $65, the value of this stock is $65x$.
- Since each share of the $(5x)$ shares of ATB stock is worth $105, the value of this stock is $105(5x)$.

We set the sum of these values equal to $450,800.

The value of IBC stock	plus	the value of GS stock	plus	the value of ATB stock	equals	the total value of the portfolio.
$54x$	$+$	$65x$	$+$	$105(5x)$	$=$	$450{,}800$

Solve the equation We can now solve the equation.

$$54x + 65x + 105(5x) = 450{,}800 \quad \text{This is the equation to solve.}$$
$$54x + 65x + 525x = 450{,}800 \quad \text{Multiply.}$$
$$644x = 450{,}800 \quad \text{Combine like terms.}$$
$$x = 700 \quad \text{Divide both sides by 644.}$$

State the conclusion The foundation owns 700 shares of IBC, 700 shares of GS, and $5(700)$ or 3,500 shares of ABT.

Check the result The value of 700 shares of IBC at $54 per share is $37,800. The value of 700 shares of GS at $65 per share is $45,500. The value of 3,500 shares of ATB at $105 per share is $367,500. The sum of these values is $450,800.

SELF CHECK 3 If the portfolio is worth $209,300, how many shares of each type does the foundation own?

---

3 Solve a geometry application by setting up and solving a linear equation in one variable.

Figure 1-23 illustrates several geometric figures. A **right angle** is an angle whose measure is 90°. A **straight angle** is an angle whose measure is 180°. An **acute angle** is an angle whose measure is greater than 0° but less than 90°.

Figure 1-23

If the sum of two angles equals 90°, the angles are called **complementary angles**, and each angle is called the *complement* of the other. If the sum of two angles equals 180°, the angles are called **supplementary angles**, and each angle is called the *supplement* of the other.

A **right triangle** is a triangle with one right angle. In Figure 1-24(a) on the next page, $\angle C$ (read as "angle C") is a right angle. An **isosceles triangle** is a triangle with two sides of equal measure that meet to form the **vertex angle**. The angles opposite the sides of equal measure, called the **base angles**, are also equal. An **equilateral triangle** is a triangle with three sides of equal measure and three equal angles.

Figure 1-24

EXAMPLE 4   **ANGLES IN A TRIANGLE**   If the vertex angle of the isosceles triangle shown in Figure 1-24(b) measures 64°, find the measure of each base angle.

Analyze the problem   We are asked to find the measure of each base angle in degrees. If we let x represent the measure of one base angle in degrees, the measure of the other base angle is also x.

Form an equation   From geometry, we know that the sum of the measures of the three angles in any triangle is 180°. Thus, the sum of the angles in this triangle, $x + x + 64°$, is equal to 180°. We can form the equation

The measure of one base angle in degrees	plus	the measure of the other base angle in degrees	plus	the measure of the vertex angle in degrees	equals	180°.
x	$+$	x	$+$	64	$=$	180

Solve the equation   We now can solve the equation.

$x + x + 64 = 180$ This is the equation to solve.

$2x + 64 = 180$ Combine like terms.

$2x = 116$ Subtract 64 from both sides.

$x = 58$ Divide both sides by 2.

State the conclusion   The measure of each base angle is 58°.

Check the result   The sum of the measures of each base angle and the vertex angle is 180°:

$$58° + 58° + 64° = 180°$$

SELF CHECK 4   If the vertex angle of an isosceles triangle measures 58°, find the measure of each base angle.

EXAMPLE 5   **DOG RUNS**   A man has 28 meters of fencing to make a rectangular kennel. If the kennel is to be 6 meters longer than it is wide, find its dimensions.

Analyze the problem   We are asked to find the dimensions of a rectangle. Therefore, we will need to find both the length and the width. If we let w represent the width of the rectangle in meters, the length will be represented by $w + 6$ meters.

Form an equation   The perimeter P of a rectangle is the distance around it. Because this is a rectangle, opposite sides have the same measure. (See Figure 1-25.) The perimeter can be expressed as either $w + (w + 6) + w + (w + 6) = P$ or as $2w + 2(w + 6) = P$.

Figure 1-25

We can form the equation as follows.

Two widths	plus	two lengths	equals	the perimeter.
$2 \cdot w$	$+$	$2 \cdot (w + 6)$	$=$	28

Solve the equation   We now can solve the equation.

$$2 \cdot w + 2 \cdot (w + 6) = 28 \qquad \text{This is the equation to solve.}$$
$$2w + 2w + 12 = 28 \qquad \text{Use the distributive property to remove parentheses.}$$
$$4w + 12 = 28 \qquad \text{Combine like terms.}$$
$$4w = 16 \qquad \text{Subtract 12 from both sides.}$$
$$w = 4 \qquad \text{Divide both sides by 4.}$$
$$w + 6 = 10$$

State the conclusion   The dimensions of the kennel are 4 meters by 10 meters.

Check the result   If a kennel has a width of 4 meters and a length of 10 meters, its length is 6 meters longer than its width, and the perimeter is $2(4)$ meters $+ 2(10)$ meters $= 28$ meters.

 SELF CHECK 5   Find the dimensions of the kennel if there are 60 meters of fencing available.

---

4   Solve a lever application by setting up and solving a linear equation in one variable.

EXAMPLE 6   **ENGINEERING**   Design engineers must position two hydraulic cylinders as in Figure 1-26 to balance a 9,500-pound force at point A. The first cylinder at the end of the lever exerts a 3,500-pound force. Where should the engineers position the second cylinder, which is capable of exerting a 5,500-pound force?

Figure 1-26

Analyze the problem   We must find the distance in feet from the fulcrum that the second cylinder must be placed in order to balance the two sides. We will let x represent this distance in feet.

Form an equation   From a principle in physics, the lever will be in balance when the force of the first cylinder multiplied by its distance from the pivot (also called the **fulcrum**) added to the force of the second cylinder multiplied by its distance from the fulcrum is equal to the 9,500-pound force multiplied by its distance from the fulcrum.

Using this information, we can form the equation

Force of cylinder 1 times the distance from the fulcrum	plus	force of cylinder 2 times the distance from the fulcrum	equals	force to be balanced times the distance from the fulcrum.
$3{,}500 \cdot 5$	$+$	$5{,}500x$	$=$	$9{,}500 \cdot 3$

Solve the equation   We can now solve the equation.

$$3{,}500 \cdot 5 + 5{,}500x = 9{,}500 \cdot 3 \qquad \text{This is the equation to solve.}$$
$$17{,}500 + 5{,}500x = 28{,}500 \qquad \text{Multiply.}$$
$$5{,}500x = 11{,}000 \qquad \text{Subtract 17,500 from both sides.}$$
$$x = 2 \qquad \text{Divide both sides by 5,500.}$$

State the conclusion   The design must specify that the second cylinder be positioned 2 feet from the fulcrum.

Check the result $3{,}500 \cdot 5 + 5{,}500 \cdot 2 = 17{,}500 + 11{,}000 = 28{,}500$
$9{,}500 \cdot 3 = 28{,}500$

SELF CHECK 6   Find the distance from the fulcrum the 5,500-pound force in Example 6 should be placed to balance a 12,250-pound force.

SELF CHECK ANSWERS

1. He should make one cut 95 feet from one end and then a second cut 97 feet from one end of the remaining rope. This will leave a length of 99 feet.   **2.** The theater system is on sale at a 20% markdown.   **3.** The portfolio contains 325 shares each of IBC and GS and 1,625 shares of ATB.   **4.** The measure of each base angle is 61°.   **5.** The dimensions are 12 meters by 18 meters.   **6.** The 5,500-pound force should be 3.5 feet from the fulcrum.

NOW TRY THIS

The sum of two numbers is 92.

1. If one of the numbers is 27, what is the other?

2. If one of the numbers is -56, what is the other?

3. If one of the numbers is x, what expression represents the other number?

1.6   Exercises

WARM-UPS *Find each value.*

1. 30% of 650

2. $33\frac{1}{3}\%$ of 600

If a stock costs $54, find the cost of

3. 14 shares.

4. x shares.

Find the area of the rectangle with the given dimensions.

5. 8 meters long, 5 meters wide

6. l meters long, $l - 8$ meters wide

REVIEW *Simplify each expression. Assume no variables are 0.*

7. $\left(\dfrac{3x^{-3}}{4x^2}\right)^{-4}$

8. $\left(\dfrac{r^{-3}s^2}{r^2r^3s^{-4}}\right)^{-5}$

9. $\dfrac{a^n a^5}{a^6}$

10. $\left(\dfrac{b^n}{b^3}\right)^3$

VOCABULARY AND CONCEPTS *Fill in the blanks.*

11. The expression _____ represents the phrase "4 more than 5 times x."

12. The expression _____ represents the phrase "9 percent of x."

13. The expression ___ represents the phrase "the value of x shares priced at $89 per share."

14. A right angle is an angle whose measure is ___. A straight angle measures ____.

15. An _____ angle measures greater than 0° but less than 90°.

16. If the sum of the measures of two angles equals 90°, the angles are called _____ angles.

17. If the sum of the measures of two angles equals ____, the angles are called supplementary angles.

18. If a triangle has a 90° angle, it is called a ____ triangle.

19. If a triangle has two sides with equal measure, it is called an _____ triangle.

20. The sides of equal measure of an isosceles triangle form the _____ angle and the remaining angles are called _____.

21. An _____ triangle has three sides of equal measure and three equal angles.

22. The pivot point of a lever is called the _____.

APPLICATIONS *Set up an equation to solve. SEE EXAMPLE 1. (OBJECTIVE 1)*

23. Cutting ropes A 90-foot rope is cut into four pieces, with each successive piece twice as long as the previous one. Find the length of the longest piece.

24. Cutting cables A 186-foot cable is to be cut into four pieces. Find the length of each piece if each successive piece is 3 feet longer than the previous one.

25. Cutting boards The carpenter in the illustration saws a board into two pieces. He wants one piece to be 1 foot longer than twice the length of the shorter piece. Find the length of each piece.

26. Cutting beams A 30-foot steel beam is to be cut into two pieces. The longer piece is to be 2 feet more than three times as long as the shorter piece. Find the length of each piece.

Set up an equation to solve. SEE EXAMPLE 2. (OBJECTIVE 2)

27. Buying a washer and dryer Find the percent of markdown of the sale in the following ad.

28. Buying furniture A bedroom set regularly sells for $983. If it is on sale for $737.25, what is the percent of markdown?

29. Buying books When the price of merchandise is increased, the amount of increase is called the *markup*. If a bookstore buys a used calculus book for $18 and sells it for $78, find the percent of markup.

30. Selling toys When the price of merchandise is increased, the amount of increase is called the *markup*. If the owner of a gift shop buys stuffed animals for $18 and sells them for $30, find the percent of markup.

Set up an equation to solve. SEE EXAMPLE 3. (OBJECTIVE 2)

31. Value of an IRA In an Individual Retirement Account (IRA) valued at $53,900, a student has 500 shares of stock, some in Big Bank Corporation and some in Safe Savings and Loan. If Big Bank sells for $115 per share and Safe Savings sells for $97 per share, how many shares of each does the student own?

32. Pension funds A pension fund owns 12,000 shares in a stock mutual fund and a bond mutual fund. Currently, the stock fund sells for $12 per share, and the bond fund sells for $15 per share. How many shares of each does the pension fund own if the value of the securities is $165,000?

33. Selling calculators Last month, a bookstore ran the following ad and sold 85 calculators, generating $3,875 in sales. How many of each type of calculator did the bookstore sell?

34. Selling grass seed A seed company sells two grades of grass seed. A 100-pound bag of a mixture of rye and bluegrass sells for $245, and a 100-pound bag of bluegrass sells for $347. How many bags of each are sold in a week when the receipts for 19 bags are $5,369?

Set up an equation to solve. SEE EXAMPLE 4. (OBJECTIVE 3)

35. Equilateral triangles Find the measure of each angle in an equilateral triangle.

36. Quadrilaterals The sum of the angles of any four-sided figure (called a *quadrilateral*) is 360°. The following quadrilateral has two equal base angles. Find the measure of x.

37. Vertical angles   When two lines intersect as in the illustration, four angles are formed. Angles that are side-by-side, such as $\angle 1$ and $\angle 2$, are called **adjacent angles**. Angles that are nonadjacent, such as $\angle 1$ and $\angle 3$ or $\angle 2$ and $\angle 4$, are called **vertical angles**. From geometry, we know that if two lines intersect, vertical angles have the same measure. If $m(\angle 1) = (3x + 10)°$ and $m(\angle 3) = (5x - 10)°$, find x. Read $m(\angle 1)$ as "the measure of $\angle 1$."

38. If $m(\angle 2) = (6x + 20)°$ in the illustration of Exercise 37 and $m(\angle 4) = (8x - 20)°$, find $m(\angle 1)$.

39. Supplementary angles   If one of two supplementary angles is 35° larger than the other, find the measure of the smaller angle.

40. Supplementary angles   Refer to the illustration and find x.

$(2x + 30)°$ $(2x - 10)°$

41. Complementary angles   If one of two complementary angles is 22° greater than the other, find the measure of the larger angle.

42. Complementary angles in a right triangle   Explain why the acute angles in the following right triangle are complementary. Then find the measure of angle A.

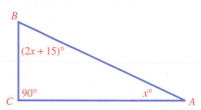

B
$(2x + 15)°$
$90°$ $x°$
C A

Set up an equation to solve. *SEE EXAMPLE 5. (OBJECTIVE 3)*

43. Finding dimensions   The rectangular garden shown in the illustration is twice as long as it is wide. Find its dimensions.

72 m

44. Finding dimensions   The width of a rectangular swimming pool is one-half its length. If its perimeter is 90 feet, find the dimensions of the pool.

45. Fencing pastures   A farmer has 624 feet of fencing to enclose the rectangular pasture shown in the illustration. Because a river runs along one side, fencing will be needed on only three sides. Find the dimensions of the pasture if its length is double its width.

624 ft

46. Fencing pens   A man has 150 feet of fencing to build the pen shown in the illustration. If one end is a square, find the outside dimensions.

x ft

x ft $(x + 5)$ ft

Set up an equation to solve. Ignore the weight of the lever. *SEE EXAMPLE 6. (OBJECTIVE 4)*

47. Balancing a seesaw   A seesaw is 20 feet long, and the fulcrum is in the center. If an 80-pound boy sits at one end, how far will the boy's 160-pound father have to sit from the fulcrum to balance the seesaw?

48. Establishing equilibrium   Two forces—110 pounds and 88 pounds—are applied to opposite ends of an 18-foot lever. How far from the greater force must the fulcrum be placed so that the lever is balanced?

49. Moving stones   A woman uses a 10-foot bar to lift a 210-pound stone. If she places another rock 3 feet from the stone to act as the fulcrum, how much force must she exert to move the stone?

50. Lifting cars   A 350-pound football player brags that he can lift a 2,500-pound car. If he uses a 12-foot bar with the fulcrum placed 3 feet from the car, will he be able to lift the car?

ADDITIONAL APPLICATIONS   *Set up an equation to solve.*

51. Buying roses   A man with $27.50 stops after work to order some roses for his wife's birthday. If each rose costs $1.50 and there is a delivery charge of $9.50, how many roses can he buy?

52. Renting trucks   To move to Wisconsin, a man can rent a truck for $69.95 per day plus 45¢ per mile. If he keeps the truck for one day, how many miles can he drive for a cost of $170.75?

53. Car rentals   While waiting for his car to be repaired, a man rents a car for $12 per day plus 30¢ per mile. If he keeps the car for 2 days, how many miles can he drive for a total cost of $42? How many miles can he drive for a total cost of $60?

54. Computing salaries A student earns $17 per day for delivering overnight packages. She is paid $5 per day plus 60¢ for each package delivered. How many more deliveries must she make each day to increase her daily earnings to $23?

55. Buying a TV and a Blu-ray player See the following ad. If the TV costs $501 more than the Blu-ray player, how much does the TV cost?

BUY *BOTH*
FOR
$655

56. Buying golf clubs The cost of a set of golf clubs is $750. If the irons cost $90 more than the woods, find the cost of the irons.

57. Triangles If the height of a triangle with a base of 12 inches is tripled, its area is increased by 84 square inches. Find the height of the triangle.

58. Engineering designs The width, *w*, of the flange in the following engineering drawing has not yet been determined. Find *w* so that the area of the 7-in.-by-12-in. rectangular portion is exactly one-half of the total area.

7 in.

w

12 in.   3 in.

59. Supplementary angles and parallel lines In the illustration, lines *r* and *s* are cut by a third line *l* to form ∠1 and ∠2. When lines *r* and *s* are parallel, ∠1 and ∠2 are supplementary. If m(∠1) = (*x* + 50)° and m(∠2) = (2*x* − 20)° and lines *r* and *s* are parallel, find *x*.

60. In the illustration of Exercise 59, find m(∠3).

61. Enclosing a swimming pool A woman wants to enclose the swimming pool shown in the illustration and have a walkway of uniform width all the way around. How wide will the walkway be if the woman uses 180 feet of fencing?

62. Framing pictures An artist wants to frame the following picture with a frame 2 inches wide. How wide will the framed picture be if the artist uses 70 inches of framing material?

x in.

x + 5 in.

63. Balancing levers Forces are applied to a lever as indicated in the illustration. Find *x*, the distance of the smallest force from the fulcrum. Ignore the weight of the lever.

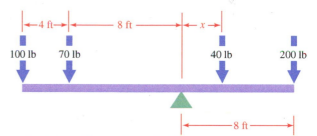

4 ft   8 ft   *x*

100 lb   70 lb   40 lb   200 lb

8 ft

64. Balancing seesaws Jim and Bob sit at opposite ends of an 18-foot seesaw, with the fulcrum at its center. Jim weighs 160 pounds, and Bob weighs 200 pounds. When Kim sits 4 feet in front of Jim, the seesaw balances. How much does Kim weigh?

65. Temperature scales The Celsius and Fahrenheit temperature scales are related by the equation $C = \frac{5}{9}(F - 32)$. At what temperature will a Fahrenheit and a Celsius thermometer give the same reading?

66. Solar heating One solar panel in the illustration is to be 3 feet wider than the other, but to be equally efficient, they must have the same area. Find the width of each.

11 ft

8 ft

w

WRITING ABOUT MATH

67. Explain the steps for solving an application.

68. Explain how to check the solution of an application.

SOMETHING TO THINK ABOUT

69. Find the distance x required to balance the lever in the illustration.

70. Interpret the answer to Exercise 69.

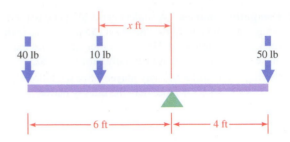

Section 1.7

More Applications of Linear Equations in One Variable

Objectives

1 Solve an investment application by setting up and solving a linear equation in one variable.

2 Solve a uniform motion application by setting up and solving a linear equation in one variable.

3 Solve a mixture application by setting up and solving a linear equation in one variable.

Getting Ready

1. Find 8% of $500.

2. Write an expression for 8% of x.

3. If $9,000 of $15,000 is invested at 7%, how much is left to invest at another rate?

4. If x of $15,000 is invested at 7%, write an expression for how much is left to be invested at another rate.

5. If a car travels 50 mph for 4 hours, how far will it go?

6. If coffee sells for $4 per pound, how much will 5 pounds cost?

In this section, we will continue our investigation of problem solving by considering investment, motion, and mixture applications.

Solve an investment application by setting up and solving a linear equation in one variable.

EXAMPLE 1 **FINANCIAL PLANNING**   A professor has $15,000 to invest for one year, some at 8% and the rest at 7% annual interest. If she will earn $1,110 from these investments, how much did she invest at each rate?

Analyze the problem We must find how much money the professor has invested in each of the two accounts. If we let x represent the amount she has invested in the account paying 8% interest, then the rest, or $15,000 - x$, is invested in the account paying 7% interest.

Form an equation Simple interest is computed by the formula $i = prt$, where i is the interest earned, p is the principal, r is the annual interest rate, and t is the length of time the principal is invested.

In this problem, $t = 1$ year. Thus, if $\$x$ is invested at 8% for one year, the interest earned is $i = \$x(8\%)(1) = \$0.08x$. If the remaining $\$(15{,}000 - x)$ is invested at 7%, the amount earned on that investment is $\$0.07(15{,}000 - x)$. See Figure 1-27. The sum of these amounts should equal the total earned interest of $\$1{,}110$.

	i	$=$	p	$\cdot$	r	$\cdot$	t
8% investment	$0.08x$		x		0.08		1
7% investment	$0.07(15{,}000 - x)$		$15{,}000 - x$		0.07		1

Figure 1-27

Using this information, we can form the equation

The interest earned at 8%	plus	the interest earned at 7%	equals	the total interest.
$0.08x$	$+$	$0.07(15{,}000 - x)$	$=$	$1{,}110$

Solve the equation   We can now solve the equation.

$0.08x + 0.07(15{,}000 - x) = 1{,}110$	This is the equation to solve.
$8x + 7(15{,}000 - x) = 111{,}000$	To eliminate the decimals, multiply both sides by 100.
$8x + 105{,}000 - 7x = 111{,}000$	Use the distributive property to remove parentheses.
$x + 105{,}000 = 111{,}000$	Combine like terms.
$x = 6{,}000$	Subtract 105,000 from both sides.
$15{,}000 - x = 9{,}000$	

State the conclusion   She invested $\$6{,}000$ at 8% and $\$9{,}000$ at 7%.

Check the result   The interest on $\$6{,}000$ is $0.08(\$6{,}000) = \$480$. The interest earned on $\$9{,}000$ is $0.07(\$9{,}000) = \$630$. The total interest is $\$1{,}110$.

 SELF CHECK 1   If the professor earned $\$1{,}170$ in interest, how much did she invest at each rate?

---

2   Solve a uniform motion application by setting up and solving a linear equation in one variable.

EXAMPLE 2   **TRAVEL TIMES**   A car leaves Rockford traveling toward Wausau at the rate of 55 mph. At the same time, another car leaves Wausau traveling toward Rockford at the rate of 50 mph. How long will it take them to meet if the cities are 157.5 miles apart?

Analyze the problem   Since we must find the time it takes for the two cars to meet, we will let t represent the time in hours.

Form an equation   Uniform motion applications are based on the formula $d = r \cdot t$, where d is distance, r is rate, and t is time. In this case, the cars are traveling toward each other as shown in Figure 1-28(a) on the next page and we can organize the given information in a chart as shown in Figure 1-28(b).

   We know that one car is traveling at 55 mph and that the other is going 50 mph. We also know that they travel for the same amount of time, t hours. Thus, the distance that the faster car travels is $55t$ miles, and the distance that the slower car travels is $50t$ miles. The sum of these distances equals 157.5 miles, the distance between the cities.

	Rate	·	Time	=	Distance
Faster car	55		t		$55t$
Slower car	50		t		$50t$

(a)   (b)

Figure 1-28

Using this information, we can form the equation

The distance the faster car goes	plus	the distance the slower car goes	equals	the distance between the cities.
$55t$	$+$	$50t$	$=$	157.5

Solve the equation   We can now solve the equation.

$55t + 50t = 157.5$	This is the equation to solve.
$105t = 157.5$	Combine like terms.
$t = 1.5$	Divide both sides by 105.

State the conclusion   The two cars will meet in $1\frac{1}{2}$ hours.

Check the result   The faster car travels $1.5(55) = 82.5$ miles. The slower car travels $1.5(50) = 75$ miles. The total distance traveled is 157.5 miles.

SELF CHECK 2   How long will it take the cars to meet if the rate of the first car is 60 mph and the rate of the second is 66 mph?

3   Solve a mixture application by setting up and solving a linear equation in one variable.

EXAMPLE 3   **MIXING NUTS**   The owner of a candy store notices that 20 pounds of gourmet cashews are not selling at $12 per pound. The store owner decides to mix peanuts with the cashews to lower the price per pound. If peanuts sell for $3 per pound, how many pounds of peanuts must be mixed with the cashews to make a mixture that could be sold for $6 per pound?

Analyze the problem   We must find the number of pounds of peanuts needed to mix with the cashews, so we will let x represent the number of pounds of peanuts. We will be adding the x pounds of peanuts to the 20 pounds of cashews, so the total number of pounds of the mixture will be $(20 + x)$.

Form an equation   This application is based on the formula $V = pn$, where V represents the value, p represents the price per pound, and n represents the number of pounds. We can enter the known information in a table as shown in Figure 1-29. The *value* of the cashews plus the *value* of the peanuts will be equal to the *value* of the mixture.

	Price	·	Number of pounds	=	Value
Cashews	12		20		240
Peanuts	3		x		$3x$
Mixture	6		$20 + x$		$6(20 + x)$

Figure 1-29

Using this information, we can form the equation

The value of the cashews	plus	the value of the peanuts	equals	the value of the mixture.
240	+	$3x$	=	$6(20 + x)$

Solve the equation   We can now solve the equation.

$240 + 3x = 6(20 + x)$ This is the equation to solve.

$240 + 3x = 120 + 6x$ Use the distributive property to remove parentheses.

$120 = 3x$ Subtract $3x$ and 120 from both sides.

$40 = x$ Divide both sides by 3.

State the conclusion   The store owner should mix 40 pounds of peanuts with the 20 pounds of cashews.

Check the result   The cashews are valued at $\$12(20) = \$240$ and the peanuts are valued at $\$3(40) = \$120$, so the mixture is valued at $\$6(60) = \$360$.

The value of the cashews plus the value of the peanuts equals the value of the mixture.

SELF CHECK 3   If the mixture is to be sold for $8 per pound, how many pounds of peanuts must be mixed with the cashews?

EXAMPLE 4   **MILK PRODUCTION**   A container is partially filled with 12 liters of whole milk containing 4% butterfat. How much 1% butterfat must be added to get a mixture that is 2% butterfat?

Analyze the problem   We will let x represent the number of liters of the 1% butterfat milk to be added. If there are already 12 liters of milk in the container, the total amount of milk will be $(12 + x)$ liters.

Form an equation   Since the first container shown in Figure 1-30(a) contains 12 liters of 4% butterfat, it contains $0.04(12)$ liters of butterfat. To this container we will add the contents of the second container, which holds $0.01x$ liters of butterfat. The sum of these two amounts of butterfat $[0.04(12) + 0.01x]$ will be equal to the amount of butterfat in the third container, which is $0.02(12 + x)$ liters of butterfat. This information is presented in table form in Figure 1-30(b).

	Amount of butterfat	=	Amount of milk	·	Percent of butterfat
Whole milk	$0.04(12)$		12		0.04
1% milk	$0.01(x)$		x		0.01
2% milk	$0.02(12 + x)$		$12 + x$		0.02

(a)   (b)

Figure 1-30

Using this information, we can form the equation

The amount of butterfat in 12 liters of 4% milk	plus	the amount of butterfat in x liters of 1% milk	equals	the amount of butterfat in $(12 + x)$ liters of 2% mixture.
$0.04(12)$	+	$0.01x$	=	$0.02(12 + x)$

Solve the equation   We now solve the equation.

$$0.04(12) + 0.01x = 0.02(12 + x) \quad \text{This is the equation to solve.}$$
$$4(12) + 1x = 2(12 + x) \quad \text{Multiply both sides by 100.}$$
$$48 + x = 24 + 2x \quad \text{Use the distributive property to remove parentheses.}$$
$$24 = x \quad \text{Subtract 24 and } x \text{ from both sides.}$$

State the conclusion   Thus, 24 liters of 1% butterfat milk should be added to get a mixture that is 2% butterfat.

Check the result   12 liters of 4% milk contains 0.48 liters of butterfat and 24 liters of 1% milk contains 0.24 liters of butterfat. This gives a total of 36 liters of a mixture that contains 0.72 liters of butterfat. This is 2% butterfat.

SELF CHECK 4   A container is partially filled with 15 gallons of whole milk containing 4% butter-fat. How much 1% butterfat milk must be added to get a mixture that is 2% butterfat?

SELF CHECK ANSWERS

1. She invested $12,000 at 8% interest and $3,000 at 7% interest. 2. The cars will meet in 1.25 hours (1 hour 15 minutes). 3. The owner should mix 16 pounds of peanuts with the 20 pounds of cashews. 4. There should be 30 gallons of 1% butterfat milk added.

NOW TRY THIS

Let s represent the speed of a plane in still air in mph. If it flies *into* the wind (headwind), its speed is decreased by the speed of the wind, and if it flies *with* the wind (tailwind), its speed is increased by the speed of the wind.

a. Represent the speed of a plane flying into a 40 mph headwind.

b. Represent the speed of a plane flying with a 40 mph tailwind.

c. If a plane flies 480 miles in 4 hours into a 40 mph headwind, find its speed in still air.

d. If a plane flies 480 miles in 2 hours with a 40 mph tailwind, find its speed in still air.

1.7   Exercises

WARM-UPS *For Exercises 1–4, use the formula i = prt to find i when*

1. $p = \$1,000, r = 3\%, t = 1$

2. $p = \$5,000, r = 2\%, t = 1$

3. $p = \$(12,000 - x), r = 4\%, t = 1$

4. $p = \$(20,000 - x), r = 3\%, t = 1$

5. How much interest will $2,500 earn if invested at 3% for one year?

6. Express the amount of interest $x will earn if invested at 4% for one year.

REVIEW *Solve each equation.*

7. $10x - 6 = 4x$

8. $4a + 1 = 15 - 6(a - 1)$

9. $\dfrac{8(y - 5)}{3} = 2(y - 4)$

10. $\dfrac{t - 1}{3} = \dfrac{t + 2}{6} + 2$

VOCABULARY AND CONCEPTS *Fill in the blanks.*

11. The formula for simple interest i is $i = prt$, where p is the _____, r is the ___, and t is the ____.

12. Uniform motion problems are based on the formula _____, where d is the _____, r is the ___ of speed, and t is the ____.

13. Dry mixture problems are based on the formula $V = pn$, where V is the _____, p is the _____ per unit, and n is the _____ of units.

14. The liquid mixture application in Example 4 is based on the idea that the amount of _____ in the first container plus the amount of _____ in the second container is equal to the amount of _____ in the mixture.

APPLICATIONS *Set up an equation to solve. SEE EXAMPLE 1.* (*OBJECTIVE 1*)

15. Investing Lured by the following ad, a woman invested $12,000, some in a money market account and the rest in a 5-year CD. How much was invested in each account if the income from both investments is $680 per year?

First Republic Savings and Loan	
Account	**Rate**
NOW	2.5%
Savings	3.5%
Money market	4.0%
Checking	2.0%
5-year CD	6.0%

16. Investing A man invested $18,000, some at 4% and some at 6% annual interest. The annual income from these investments was $960. How much did he invest at each rate?

17. Supplemental income A teacher wants to earn $1,500 per year in supplemental income from a cash gift of $16,000. She puts $6,000 in a credit union that pays 7% annual interest. What rate must she earn on the remainder to achieve her goal?

18. Inheriting money Paul split an inheritance between two investments, one paying 7% annual interest and the other 10%. He invested twice as much in the 10% investment as he did in the 7% investment. If his combined annual income from the two investments was $4,050, how much did he inherit?

19. Investing Kyoko has some money to invest. If she invests $3,000 more, she can qualify for an 11% investment. Otherwise, she can invest the money at 7.5% annual interest. If the 11% investment yields twice as much annual income as the 7.5% investment, how much does she have on hand to invest?

20. Supplemental income A bus driver wants to earn $3,500 per year in supplemental income from an inheritance of $40,000. If the driver invests $10,000 in a mutual fund paying 8%, what rate must he earn on the remainder to achieve his goal?

Set up an equation to solve. SEE EXAMPLE 2. (OBJECTIVE 2)

21. Computing time One car leaves Chicago headed for Cleveland, a distance of 343 miles. At the same time, a second car leaves Cleveland headed toward Chicago. If the first car averages 50 mph and the second car averages 48 mph, how long will it take the cars to meet?

22. Storm research During a storm, two teams of scientists leave a university at the same time to search for tornados. The first team travels east at 20 mph, and the second travels west at 25 mph, as shown in the illustration in the next column. If their radios have a range of 90 miles, how long will it be before they lose radio contact?

23. Cycling A cyclist leaves Las Vegas riding at the rate of 18 mph. One hour later, a car leaves Las Vegas going 45 mph in the same direction. How long will it take the car to overtake the cyclist?

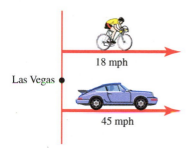

24. Computing distance At 2 P.M., two cars leave Eagle River, WI, one headed north and one headed south. If the car headed north averages 50 mph and the car headed south averages 60 mph, when will the cars be 165 miles apart?

25. Running a race Two runners leave the starting gate, one running 12 mph and the other 10 mph. If they maintain the pace, how long will it take for them to be one-quarter of a mile apart?

26. Trucking Two truck drivers leave a warehouse traveling in opposite directions, one driving at 50 mph and one driving at 56 mph. If the slower driver has a 2-hour head start, how long has she been on the road when the two drivers are 683 miles apart?

Set up an equation to solve. SEE EXAMPLE 3. (OBJECTIVE 3)

27. Mixing candies The owner of a store wants to make a 30-pound mixture of two candies to sell for $3 per pound. If one candy sells for $2.95 per pound and the other for $3.10 per pound, how many pounds of each should be used?

28. Computing selling price A mixture of candy is made to sell for $3.89 per pound. If 32 pounds of candy, selling for $3.80 per pound, are used along with 12 pounds of a more expensive candy, find the price per pound of the better candy.

29. Raising grades A student scores 70% on a test that contains 30 questions. To improve his score, the instructor agrees to let him work 15 additional questions. How many must he get right to raise his grade to 80%?

30. Raising grades On a second exam, the student in Exercise 29 earns a score of 60% on a 20-question test. This time, the instructor allows him to work 20 extra problems to improve his score. How many must he get right to raise his grade to 70%?

Set up an equation to solve each problem. SEE EXAMPLE 4.
(OBJECTIVE 3)

31. **Making whole milk**   Cream is approximately 22% butterfat. How many gallons of cream must be mixed with milk testing at 1% butterfat to get 42 gallons of milk containing 2% butterfat?

32. **Mixing solutions**   How much acid must be added to 60 grams of a solution that is 65% acid to obtain a new solution that is 75% acid?

33. **Computing grades**   Before the final, Estela had earned a total of 375 points on four tests. To receive an A in the course, she must have 90% of a possible total of 450 points. Find the lowest number of points that she can earn on the final exam and still receive an A.

34. **Computing grades**   A student has earned a total of 435 points on five tests. To receive a B in the course, he must have 80% of a possible total of 600 points. Find the lowest number of points that the student can make on the final exam and still receive a B.

ADDITIONAL APPLICATIONS   *Set up an equation to solve.*

35. **Bookstores**   A bookstore sells a mathematics book for $108. If the bookstore makes a profit of 35% on each book, what does the bookstore pay the publisher for each book?   (*Hint:* The retail price = the wholesale price + the markup.)

36. **Bookstores**   A bookstore sells a textbook for $39.20. If the bookstore makes a profit of 40% on each sale, what does the bookstore pay the publisher for each book?   (*Hint:* The retail price = the wholesale price + the markup.)

37. **Riding a jet ski**   A jet ski can go 12 mph in still water. If a rider goes upstream for 3 hours against a current of 4 mph, how long will it take the rider to return?

38. **Riding a motorboat**   The motorboat in the illustration can go 18 mph in still water. If a trip downstream takes 4 hours and the return trip takes 5 hours, find the speed of the current.

39. **Concert receipts**   For a jazz concert, student tickets are $4 each, and adult tickets are $8 each. If 225 tickets were sold and the total receipts are $1,300 how many student tickets were sold?

40. **School plays**   At a school play, 140 tickets are sold, with total receipts of $290. If adult tickets cost $2.50 each and student tickets cost $1.50 each, how many adult tickets were sold?

41. **Taking a walk**   Sarah walked north at the rate of 3 mph and returned at the rate of 4 mph. How many miles did she walk if the round trip took 3.5 hours?

42. **Computing travel time**   Jamal traveled a distance of 520 miles in 9 hours. Part of the time, his rate of speed was 55 mph. The rest of the time, his rate of speed was 60 mph. How long did Jamal travel at each rate?

43. **Diluting solutions**   In the illustration, how much water should be added to 20 ounces of a 15% solution of alcohol to dilute it to a 10% solution?

44. **Increasing concentration**   How much water must be boiled away to increase the concentration of 300 gallons of a salt solution from 2% to 3%?

45. **Making furniture**   A woodworker wants to put two partitions crosswise in the drawer shown in the illustration. He wants to place the partitions so that the spaces created increase by 3 inches from front to back. If the thickness of each partition is $\frac{1}{2}$ inch, how far from the front end should he place the first partition?

46. **Building shelves**   A carpenter wants to put four shelves on an 8-foot wall so that the five spaces created decrease by 6 inches as we move up the wall. (See the illustration.) If the thickness of each shelf is $\frac{3}{4}$ inch, how far will the bottom shelf be from the floor?

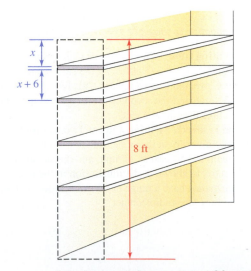

WRITING ABOUT MATH

47. What do you find most difficult in solving applications, and why?

48. Which type of application do you find easiest, and why?

SOMETHING TO THINK ABOUT

49. Discuss the difficulties in solving this application:

A man drives 100 miles at 30 mph. How fast should he drive on the return trip to average 60 mph for the entire trip?

50. What difficulties do you encounter when solving this application?

Adult tickets cost $4, and student tickets cost $2. Sales of 71 tickets bring in $245. How many of each were sold?

Projects

PROJECT 1

Theresa manages the local outlet of Hook n' Slice, a sporting goods retailer. She has hired you to help solve some problems. Be sure to provide explanations of how you arrive at your solutions.

a. The company recently directed that all shipments of golf clubs be sold at a retail price of three times what they cost the company. A shipment of golf clubs arrived yesterday, and they have not yet been labeled. This morning, the main office of Hook n' Slice called to say that the clubs should be sold at 35% off their retail prices. Rather than label each club with its retail price, calculate the sale price, and then relabel the club, Theresa wonders how to go directly from the cost of the club to the sale price. That is, if C is the cost to the company of a certain club, what is the sale price for that club?

1. Develop a formula and answer this question for her.

2. Now try out your formula on a club that cost the company $26. Compare your answer with what you find by following the company procedure of first computing the retail price and then reducing that by 35%.

b. All golf bags have been sale-priced at 20% off for the past few weeks. Theresa has all of them in a large rack, marked with a sign that reads "20% off retail price." The company has now directed that this sale price be reduced by 35%. Rather than relabel every bag with its new sale price, Theresa would like to simply change the sign to say " _____% off retail price."

1. Determine what number Theresa should put in the blank.

2. Check your answer by computing the final sale price for a golf bag with a $100 retail price in two ways:

- by using the percentage you told Theresa should go on the sign, and
- by doing the two separate price reductions.

3. Will the sign read the same if the retail price is first reduced by 35%, and then the sale price reduced by 20%? Explain.

PROJECT 2

The Parchdale City Planning Council is deciding whether or not the town should build an emergency reservoir for use in especially dry periods. The proposed site has room enough for a conical reservoir of diameter 141 feet and depth 85.3 feet.

As long as Parchdale's main water supply is working, the reservoir will be kept full. When a water emergency occurs, however, the reservoir will lose water to evaporation as well as supplying the town with water. The company that has designed the reservoir has given the town the following information.

If D equals the number of consecutive days the reservoir of volume V is used as Parchdale's water supply, then the total amount of water lost to evaporation after D days will be

$$0.1V\left(\frac{D - 0.7}{D}\right)^2$$

So the volume of water left in the reservoir after it has been used for D days is

$$\text{Volume} = V - (\text{usage per day}) \cdot D - (\text{evaporation})$$

Parchdale uses about 57,500 cubic feet of water per day under emergency conditions. The majority of the council members think that building the reservoir is a good idea if it will supply the town with water for a week. Otherwise, they will vote against building the reservoir.

a. Calculate the volume of water the reservoir will hold. Remember to use significant digits correctly. Express the answer in scientific notation.

b. Show that the reservoir will supply Parchdale with water for a full week.

c. How much water per day could Parchdale use if the proposed reservoir had to supply the city with water for ten days?

Reach for Success
EXTENSION OF ANALYZING YOUR TIME

Now that you've analyzed a day, let's move on to take a look at a typical week. Fill in the chart below to account for every hour of every day. To simplify the process, you can use the following abbreviations:

W (work time including commute), **S** (sleeping), **P** (preparing for work; preparing meals, eating, exercising), **C** (class time including commute), **F** (time with family and friends), **ST** (study time), and **E** (entertainment)

Remember, there are no right or wrong answers. This information is to give you a complete picture of how you are spending your time in a typical week.

	SUNDAY	MONDAY	TUESDAY	WEDNESDAY	THURSDAY	FRIDAY	SATURDAY
6:00–7:00							
7:00–8:00							
8:00–9:00							
9:00–10:00							
10:00–11:00							
11:00–Noon							
Noon–1:00							
1:00–2:00							
2:00–3:00							
3:00–4:00							
4:00–5:00							
5:00–6:00							
6:00–7:00							
7:00–8:00							
8:00–9:00							
9:00–10:00							
10:00–11:00							
11:00–12:00							
12:00–1:00							
1:00–2:00							
2:00–3:00							
3:00–4:00							
4:00–5:00							
5:00–6:00							

After reviewing your weekly schedule, do you have enough time to meet the "rule of thumb" of studying at least 2 hours per week for every 1 hour you are in class? If yes, congratulations!

If not, is it possible for you to find the additional hours to help you be successful in this and all your other classes?

Are you able/willing to adjust your schedule to find this time? _____ Explain your answer.

1   Review

SECTION 1.1   The Real Number System

DEFINITIONS AND CONCEPTS	EXAMPLES
A **set** is a collection of objects.	Which numbers in the set
Natural numbers:	$$\left\{-5, 0, \frac{2}{3}, 1.75, \sqrt{49}, \pi, 6\right\} \text{ are}$$
$\{1, 2, 3, 4, 5, \ldots\}$	
Whole numbers:	**a.** natural numbers, **b.** whole numbers, **c.** integers, **d.** rational numbers, **e.** irrational numbers, **f.** real numbers, **g.** prime numbers, **h.** composite numbers, **i.** even integers, **j.** odd integers?
$\{0, 1, 2, 3, 4, 5, \ldots\}$	
Integers:	**a.** $\sqrt{49}, 6$ ($\sqrt{49}$ is a natural number since $\sqrt{49} = 7$)
$\{\ldots, -3, -2, -1, 0, 1, 2, 3, \ldots\}$	**b.** $0, \sqrt{49}, 6$
Even integers:	**c.** $-5, 0, \sqrt{49}, 6$
$\{\ldots, -6, -4, -2, 0, 2, 4, 6, \ldots\}$	**d.** $-5, 0, \frac{2}{3}, 1.75, \sqrt{49}, 6$
Odd integers:	**e.** π
$\{\ldots, -5, -3, -1, 1, 3, 5, \ldots\}$	**f.** $-5, 0, \frac{2}{3}, 1.75, \sqrt{49}, \pi, 6$
Prime numbers:	**g.** $\sqrt{49}$
$\{2, 3, 5, 7, 11, 13, \ldots\}$	**h.** 6
Composite numbers:	**i.** $0, 6$
$\{4, 6, 8, 9, 10, 12, \ldots\}$	**j.** $-5, \sqrt{49}$
Rational numbers: numbers that can be written as	
$$\left\{\frac{a}{b} \,\middle|\, a \text{ and } b \text{ are integers and } b \neq 0\right\}$$	
Irrational numbers: numbers that can be written as nonterminating, nonrepeating decimals	
Real numbers: numbers that can be written as decimals	

Graphing:

The following table shows three ways to describe an interval.

Set Notation	Graph	Interval Notation
$\{x \mid x > a\}$		(a, ∞)
$\{x \mid x < a\}$		$(-\infty, a)$
$\{x \mid x \geq a\}$		$[a, \infty)$
$\{x \mid x \leq a\}$		$(-\infty, a]$
$\{x \mid a < x < b\}$		(a, b)
$\{x \mid a \leq x < b\}$		$[a, b)$
$\{x \mid a < x \leq b\}$		$(a, b]$
$\{x \mid a \leq x \leq b\}$		$[a, b]$
$\{x \mid x < a \text{ or } x > b\}$		$(-\infty, a) \cup (b, \infty)$

The graph of $\{x \mid x < 7\}$ includes all real numbers that are less than 7 as shown in the figure below. This is the interval $(-\infty, 7)$.

The graph of $\{x \mid x \geq -5\}$ includes all real numbers that are greater than or equal to -5 as shown in the figure below. This is the interval $[-5, \infty)$.

The set $\{x \mid 0 < x \leq 9\}$ includes all real numbers from 0 to 9, including 9, as shown in the figure below. This is the interval $(0, 9]$.

The set $\{x \mid -2 < x < 5\}$ includes all real numbers between -2 and 5, as shown in the figure below. This is the interval $(-2, 5)$.

Absolute value:

$$\begin{cases} \text{If } x \geq 0, \text{then } |x| = x. \\ \text{If } x < 0, \text{then } |x| = -x. \end{cases}$$

Evaluate: $-|-26|$

$$-|-26| = -(26) = -26$$

REVIEW EXERCISES

List the numbers in $\left\{-4, -\frac{2}{3}, 0, 1, 2, \pi, 4\right\}$ *that satisfy the given condition.*

1. whole number
2. natural number
3. rational number
4. integer
5. irrational number
6. real number
7. negative number
8. positive number
9. prime number
10. composite number
11. even integer
12. odd integer
13. Graph the set of prime numbers between 20 and 30.

       20 21 22 23 24 25 26 27 28 29 30

14. Graph the set of composite numbers between 5 and 13.

        5   6   7   8   9   10 11 12 13

Graph each set on the number line.

15. $\{x \mid x \geq -2\}$

16. $\{x \mid -2 < x \leq 6\}$

17. $[-4, 1)$

18. $(-3, 5]$

19. $\{x \mid x > 2\}$

20. $(-\infty, -1)$

21. $(-\infty, 0] \cup (2, \infty)$

Write each expression without absolute value symbols.

22. $|0|$ 23. $|-1|$

24. $-|-5|$ 25. $-|6|$

SECTION 1.2   Arithmetic and Properties of Real Numbers

DEFINITIONS AND CONCEPTS	EXAMPLES
Adding two real numbers:	
With like signs: Add their absolute values and keep the same sign.	$(+1) + (+6) = +7 \qquad (-1) + (-6) = -7$
With unlike signs: Subtract their absolute values and keep the sign of the number with the greater absolute value.	$(-1) + (+6) = +5 \qquad (+1) + (-6) = -5$
Subtracting two real numbers: $x - y$ is equivalent to $x + (-y)$	$-8 - 2 = -8 + (-2) \quad$ To subtract 2, add -2. $= -10$
Multiplying and dividing two real numbers:	
With like signs: Multiply (or divide) their absolute values. The sign is positive.	$3(7) = 21 \qquad -3(-7) = 21$ $\dfrac{+6}{+2} = +3 \qquad \dfrac{-6}{-2} = +3$
With unlike signs: Multiply (or divide) their absolute values. The sign is negative.	$-3(7) = -21 \qquad 3(-7) = -21$ $\dfrac{+6}{-2} = -3 \qquad \dfrac{-6}{+2} = -3$
Zero divided by any nonzero number is 0.	$\dfrac{0}{2} = 0$ because $(2)(0) = 0$.
Division by zero is undefined.	$\dfrac{2}{0}$ is undefined because no number multiplied by 0 gives 2.
Order of operations without exponents:	Simplify:
Use the following steps to perform all calculations within each pair of grouping symbols, working from the inner-most pair to the outermost pair.	$5(4 - 7) \div 5 + 1 = 5(-3) \div 5 + 1 \quad$ Do the subtraction within parentheses.
1. Perform all multiplications and divisions, working from left to right.	$= -15 \div 5 + 1 \quad$ Multiply.
2. Perform all additions and subtractions, working from left to right.	$= -3 + 1 \quad$ Divide.
3. Because a fraction bar is a grouping symbol, simplify the numerator and the denominator separately and then simplify the fraction, whenever possible.	$= -2 \quad$ Add.
When all grouping symbols have been removed, repeat the rules above to finish the calculation.	
Evaluating algebraic expressions:	If $x = 2$, $y = -3$, and $z = -5$, evaluate:
To evaluate algebraic expressions, we substitute numbers for the variables and simplify.	$\dfrac{xy - 3z}{y(z + x)}$
	To evaluate the expression, we substitute 2 for x, -3 for y, and -5 for z and simplify.
	$\dfrac{xy - 3z}{y(z + x)} = \dfrac{2(-3) - 3(-5)}{-3(-5 + 2)}$
	$= \dfrac{-6 + 15}{-3(-3)}$
	$= \dfrac{9}{9}$
	$= 1$

Commutative properties of addition and multiplication:	
$a + b = b + a$	The commutative property of addition justifies the statement $x + 3 = 3 + x$.
$ab = ba$	The commutative property of multiplication justifies the statement $x \cdot 3 = 3 \cdot x$.
Associative properties of addition and multiplication:	
$(a + b) + c = a + (b + c)$	The associative property of addition justifies the statement $(x + 3) + 4 = x + (3 + 4)$.
$(ab)c = a(bc)$	The associative property of multiplication justifies the statement $(x \cdot 3) \cdot 4 = x \cdot (3 \cdot 4)$.
Distributive property of multiplication over addition:	
$a(b + c) = ab + ac$	The distributive property justifies the statement $7(2x - 8) = 7(2x) - 7(8) = 14x - 56$

Additive and multiplicative identities:	
$a + 0 = 0 + a = a$ (Additive identity is 0.)	$5 + 0 = 0 + 5 = 5$
$a \cdot 1 = 1 \cdot a = a$ (Multiplicative identity is 1.)	$5 \cdot 1 = 1 \cdot 5 = 5$

Additive and multiplicative inverses:	
$a + (-a) = 0$ $(a, -a$ are additive inverses.)	The additive inverse of -2 is 2 because $-2 + 2 = 0$.
If $a \neq 0$, then	
$a\left(\dfrac{1}{a}\right) = \dfrac{1}{a} \cdot a = 1$ $\left(a, \dfrac{1}{a}\right.$ are multiplicative inverses or reciprocals.$\left.\right)$	The multiplicative inverse (reciprocal) of -2 is $-\frac{1}{2}$ because $-2\left(-\frac{1}{2}\right) = 1$.

Double negative rule:	
$-(-a) = a$	$-(-17) = 17$

Measures of central tendency:	Given the values 725, 650, 800, 500, and 725, find the
The **mean** of several values is the sum of those values divided by the number of values.	**a.** mean
$\text{Mean} = \dfrac{\text{sum of the values}}{\text{number of values}}$	$\dfrac{725 + 650 + 800 + 500 + 725}{5} = 680$
The **median** of several values is the middle value. To find the median,	**b.** median
1. Arrange the values in increasing order.	725 because it is the middle value
2. If there is an odd number of values, choose the middle value.	
3. If there is an even number of values, find the mean of the middle two values.	
The **mode** of several values is the value that occurs most often.	**c.** mode
	725 because it occurs more often than any other value

REVIEW EXERCISES

Perform the operations and simplify when possible.

26. $3 + (+5)$ **27.** $17 + (+5)$

28. $-15 + (-13)$ **29.** $-9 + (-18)$

30. $-2 + 5$ **31.** $3 + (-12)$

32. $8 + (-3)$ **33.** $11 + (-15)$

34. $-25 + 12$ **35.** $-30 + 35$

36. $-3 - 10$ **37.** $-7 - 8$

38. $27 - (-12)$ **39.** $38 - (-15)$

40. $(-3)(-7)$ **41.** $(-6)(-7)$

42. $\dfrac{-16}{-4}$ **43.** $\dfrac{-25}{-5}$

44. $4(-3)$ **45.** $-3(8)$

46. $\dfrac{-8}{2}$ **47.** $\dfrac{12}{-3}$

Simplify each expression.

48. $-4(3 - 6)$ **49.** $5[9 - (-2)]$

50. $-[4 - 2(6 - 4)]$ **51.** $3[-5 + 3(2 - 7)]$

52. $\dfrac{3 - 8}{10 - 5}$ **53.** $\dfrac{-32 - 8}{6 - 16}$

Consider the numbers 14, 12, 13, 14, 15, 20, 15, 17, 19, 15.

54. Find the mean.

55. Find the median.

56. Find the mode.

57. Can the mean, median, and mode of a group of numbers be the same?

Evaluate when a = 5, b = −2, c = −3, and d = 2.

58. $\dfrac{3a - 2b}{cd}$

59. $\dfrac{3b + 2d}{ac}$

60. $\dfrac{ab + cd}{c(b - d)}$

61. $\dfrac{ac - bd}{a(d + c)}$

Determine which property justifies each statement.

62. $3(4 + 2) = 3 \cdot 4 + 3 \cdot 2$

63. $3 + (x + 7) = (x + 7) + 3$

64. $3 + (x + 7) = (3 + x) + 7$

65. $-4 + 0 = -4$

66. $3 + (-3) = 0$

67. $4(-3) = (-3)4$

68. $3(xy) = (3x)y$

69. $5x \cdot 1 = 5x$

70. $a\left(\dfrac{1}{a}\right) = 1 \quad (a \ne 0)$

71. $-(-x) = x$

SECTION 1.3   Exponents

DEFINITIONS AND CONCEPTS	EXAMPLES

Properties of exponents:

If no variable base is 0,

n factors of x

$$x^n = \overbrace{x \cdot x \cdot x \cdots \cdots x}$$

$$x^m x^n = x^{m+n}$$

$$(x^m)^n = x^{mn}$$

$$(xy)^n = x^n y^n$$

$$\left(\dfrac{x}{y}\right)^n = \dfrac{x^n}{y^n}$$

$$x^0 = 1$$

$$x^{-n} = \dfrac{1}{x^n}$$

$$\dfrac{x^m}{x^n} = x^{m-n}$$

$$\left(\dfrac{x}{y}\right)^{-n} = \left(\dfrac{y}{x}\right)^n = \dfrac{y^n}{x^n}$$

5 factors of x

$$x^5 = \overbrace{x \cdot x \cdot x \cdot x \cdot x}$$

$$x^2 \cdot x^7 = x^{2+7} = x^9$$

$$(x^2)^7 = x^{2 \cdot 7} = x^{14}$$

$$(xy)^3 = x^3 y^3$$

$$\left(\dfrac{x}{y}\right)^3 = \dfrac{x^3}{y^3} \quad (y \ne 0)$$

$$(2x)^0 = 1 \quad (x \ne 0)$$

$$x^{-3} = \dfrac{1}{x^3} \quad (x \ne 0)$$

$$\dfrac{x^7}{x^2} = x^{7-2} = x^5 \quad (x \ne 0)$$

$$\left(\dfrac{5}{4}\right)^{-2} = \left(\dfrac{4}{5}\right)^2 = \dfrac{4^2}{5^2} = \dfrac{16}{25}$$

Order of operations:

Use the following steps to perform all calculations within each pair of grouping symbols, working from the innermost pair to the outermost pair.

1. Find the values of any exponential expressions.

2. Perform all multiplications and divisions, working from left to right.

3. Perform all additions and subtractions, working from left to right.

4. Because a fraction bar is a grouping symbol, simplify the numerator and the denominator separately and simplify the fraction, whenever possible.

When all grouping symbols have been removed, repeat the rules above to finish the calculation.

To simplify $2^3 - 5(6 - 13) - 8^2$, we first perform the subtraction within the parentheses.

$$2^3 - 5(-7) - 8^2$$

and then follow the steps in the rules for order of operations.

$8 - 5(-7) - 64$ Evaluate exponential expressions.

$= 8 + 35 - 64$ Perform the multiplication.

$= 43 - 64$ Perform addition/subtraction left to right.

$= -21$

REVIEW EXERCISES

Simplify each expression and write all answers without negative exponents.

72. 5^4

73. -3^5

74. $(-4)^3$

75. -5^{-4}

76. $(3x^4)(-2x^2)$

77. $(-x^5)(3x^3)$

78. x^7x^{-10}

79. $y^{-12}y^7$

80. $(3x^2)^3$

81. $(4x^4)^4$

82. $(-2x^2)^5$

83. $-(-3x^3)^5$

84. $(x^2)^{-5}$

85. $(x^{-4})^{-5}$

86. $(-4x^{-5})^2$

87. $(-3x^{-6})^{-4}$

88. $\dfrac{x^6}{x^4}$

89. $\dfrac{x^{12}}{x^7}$

90. $\dfrac{a^7}{a^{12}}$

91. $\dfrac{a^4}{a^7}$

92. $\dfrac{y^{-3}}{y^4}$

93. $\dfrac{y^5}{y^{-4}}$

94. $\dfrac{x^{-5}}{x^{-4}}$

95. $\dfrac{x^{-6}}{x^{-9}}$

Simplify each expression and write all answers without negative exponents.

96. $(-3a^3b^2)^{-4}$

97. $\left(\dfrac{4y^{-2}}{5y^{-3}}\right)^3$

98. $4^2 - 3(2) + 5^3$

99. $-2^2 - 4(5) - 3(7 - 12)$

SECTION 1.4   Scientific Notation

DEFINITIONS AND CONCEPTS	**EXAMPLES**
Scientific notation: $N \times 10^n$, where $1 \le |N| < 10$ and n is an integer.	To write 352,000 in scientific notation, we note that 3.52 is between 1 and 10. To obtain 352,000, the decimal point in 3.52 must be moved five places to the right. We can do this by multiplying 3.52 by 10^5. $$352{,}000 = 3.52 \times 10^5$$ To write 0.0001593 in scientific notation, we note that 1.593 is between 1 and 10. To obtain 0.0001593, the decimal point in 1.593 must be moved four places to the left. We can do this by multiplying 1.593 by 10^{-4}. $$0.0001593 = 1.593 \times 10^{-4}$$
Significant digits: If a number M is written in scientific notation as $N \times 10^n$, where $1 \le |N| < 10$ and n is an integer, the number of significant digits in M is the same as the number of digits in N.	There are two significant digits in 23,000 and four significant digits in 126,200. The product $(23{,}000)(126{,}200) = 2.9026 \times 10^9$ should be rounded to two significant digits: 2.9×10^9

REVIEW EXERCISES

Write each numeral in scientific notation.

100. 19,300,000,000

101. 0.0000000273

Write each numeral in standard notation.

102. 8.4×10^6

103. 8.3×10^{-9}

104. Water usage   Each person in the United States uses approximately 1,640 gallons of water per day. If the population of the United States is 306 million, how many gallons of water are used each day?

SECTION 1.5   Solving Linear Equations in One Variable

DEFINITIONS AND CONCEPTS	**EXAMPLES**
If a and b are real numbers and $a = b$, then $a + c = b + c \qquad a - c = b - c$ $ac = bc \quad (c \ne 0) \qquad \dfrac{a}{c} = \dfrac{b}{c} \quad (c \ne 0)$	If $a = b$, then $a + 3 = b + 3 \qquad a - 5 = b - 5$ $4a = 4b \qquad \dfrac{a}{7} = \dfrac{b}{7}$

Solving linear equations:

1. If the equation contains fractions, multiply both sides of the equation by a number that will clear the denominators. (This is referred to as "clearing the fractions.")

2. Use the distributive property to remove all sets of parentheses and combine like terms.

3. Use the addition and subtraction properties to group all variables on one side of the equation and all numbers on the other side. Combine like terms, if necessary.

4. Use the multiplication and division properties to make the coefficient of the variable equal to 1.

5. Check the result.

To solve $\frac{x-2}{5} - x = \frac{8}{5} - x + 2$, we first clear the fractions by multiplying both sides by 5 and then proceed as follows:

$$5\left(\frac{x-2}{5} - x\right) = 5\left(\frac{8}{5} - x + 2\right)$$

$$x - 2 - 5x = 8 - 5x + 10 \qquad \text{Use the distributive property to remove parentheses.}$$

$$-4x - 2 = 18 - 5x \qquad \text{Combine like terms.}$$

$$x = 20 \qquad \text{Add } 5x \text{ and 2 to both sides.}$$

Show that 20 satisfies the equation.

An **identity** is an equation that is satisfied by every number x for which both sides of the equation are defined. The solution of an identity is the set of *all real numbers* and is denoted by $\mathbb{R}$.

Show that the equation is an identity:

$$-5(2x + 3) + 8x - 9 = 6x - 8(x + 3)$$

$$-10x - 15 + 8x - 9 = 6x - 8x - 24 \qquad \text{Use the distributive property to remove parentheses.}$$

$$-2x - 24 = -2x - 24 \qquad \text{Combine like terms.}$$

Since the left side of the equation is the same as the right side, every number x will satisfy the equation. The equation is an identity and its solution set is all real numbers, $\mathbb{R}$.

A **contradiction** is an equation that has no solution. Its solution set is the **empty set**, $\varnothing$.

Show that the equation is a contradiction:

$$5(2x + 3) - 8x - 9 = 8x - 6(x + 3)$$

$$10x + 15 - 8x - 9 = 8x - 6x - 18 \qquad \text{Use the distributive property to remove parentheses.}$$

$$2x + 6 = 2x - 18 \qquad \text{Combine like terms.}$$

$$6 = -18 \qquad \text{Subtract } 2x \text{ from both sides.}$$

Because $6 = -18$ is a false statement, there is no number that will satisfy the equation. Therefore, it is a contradiction and the solution set is the empty set, $\varnothing$.

To solve a formula for an indicated variable means to isolate that variable on one side of the equation.

To solve $ax + by = c$ for y, we proceed as follows:

$$ax + by = c$$

$$by = c - ax \qquad \text{Subtract } ax \text{ from both sides.}$$

$$\frac{by}{b} = \frac{c - ax}{b} \qquad \text{Divide both sides by } b.$$

$$y = \frac{c - ax}{b} \qquad \frac{b}{b} = 1$$

REVIEW EXERCISES

Solve each equation.

105. $7x + 30 = 2$

106. $-5x - 8 = 17$

107. $4(y - 1) = 28$

108. $6(x + 2) = -36$

109. $13(x - 9) - 2 = 7x - 5$

110. $\dfrac{8(x - 5)}{3} = 2(x - 4)$

111. $\dfrac{3y}{4} - 13 = -\dfrac{y}{3}$

112. $\dfrac{2y}{5} + 5 = \dfrac{4y}{10}$

113. $0.07a = 0.10 - 0.04(a - 3)$

114. $0.12x + 0.06(50{,}000 - 2x) = 3{,}000$

Solve for the indicated quantity.

115. $V = \dfrac{4}{3}\pi r^3$ for r^3

116. $V = \dfrac{1}{3}\pi r^2 h$ for h

117. $v = \dfrac{1}{6}ab(x + y)$ for x

118. $V = \pi h^2\left(r - \dfrac{h}{3}\right)$ for r

SECTION 1.6   Using Linear Equations to Solve Applications

DEFINITIONS AND CONCEPTS	EXAMPLES
To solve applications, use the following strategy:	A home theater system is on sale for $966. If the list price was $1,288, find the percent of markdown.
1. *Analyze the problem* and identify a variable by asking yourself "What am I asked to find"? Choose a variable to represent the quantity to be found and then express all other unknown quantities in the problem as expressions involving that variable.	*Analyze the problem.* We must find the percent of markdown, so we let p represent the percent of markdown, expressed as a decimal. $966 is the sale price and $1,288 is the regular price.
2. *Form an equation* by expressing a quantity in two different ways. This may require reading the problem several times to understand the given facts. What information is given? Is there a formula that applies to this situation? Often a sketch, chart, or diagram will help you visualize the facts of the problem.	*Form an equation.* The amount of the markdown is found by multiplying the regular price by the *percent* of markdown. If we subtract this amount from the regular price, we will have the sale price. This can be expressed as the equation

Sale price	equals	regular price	minus	markdown.
966	=	1,288	−	$p \cdot 1{,}288$

3. *Solve the equation* found in Step 2.	*Solve the equation.*

$$966 = 1{,}288 - p \cdot 1{,}288 \qquad \text{This is the equation to solve.}$$
$$-322 = -1{,}288p \qquad \text{Subtract 1,288 from both sides.}$$
$$0.25 = r \qquad \text{Divide both sides by } -1{,}288.$$

4. *State the conclusion* including proper units.	*State the conclusion.* The theater system has been marked down 25%.
5. *Check the result* to be certain it satisfies the given conditions.	*Check the result.* A markdown of 25% from $1,288 is $322. The sale price is $1,288 − $322 or $966.

REVIEW EXERCISES

119. Carpentry   A carpenter wants to cut a 20-foot rafter so that one piece is three times as long as the other. Where should he cut the board?

120. Geometry   A rectangle is 8 meters longer than it is wide. If the perimeter of the rectangle is 60 meters, find its area.

121. Balancing a seesaw   Sue weighs 48 pounds, and her father weighs 180 pounds. If Sue sits on one end of a 20-foot-long seesaw with the fulcrum in the middle, how far from the fulcrum should her father sit to balance the seesaw?

SECTION 1.7   More Applications of Linear Equations in One Variable

DEFINITIONS AND CONCEPTS	EXAMPLES
Use the strategy given in the Section 1.6 review to solve applications.	A car leaves Dallas traveling toward Austin at the rate of 65 mph. At the same time, another car leaves Austin traveling toward Dallas at the rate of 55 mph. How long will it take them to meet if the cities are 192 miles apart?
	Analyze the problem. We must find the time it takes for the two cars to meet, so we will let t represent the time in hours.
	Form an equation. We will use the formula $d = rt$. We know that one car is traveling at 65 mph and that the other is going 55 mph. We also know that they travel for the same amount of time, t hours. Thus, the distance that the faster car travels is $65t$ miles, and the distance that the slower car travels is $55t$ miles. The sum of these distances must be 192 miles.
	We can form the equation
	$$65t + 55t = 192$$

Solve the equation.

$$65t + 55t = 192 \qquad \text{\color{red}{This is the equation to solve.}}$$
$$120t = 192 \qquad \text{\color{red}{Combine like terms.}}$$
$$t = 1.6 \qquad \text{\color{red}{Divide both sides by 120.}}$$

State the conclusion. The two cars will meet in 1.6 hours or 1 hour 36 minutes.

Check the result. The faster car travels $1.6(65) = 104$ miles. The slower car travels $1.6(55) = 88$ miles. The total distance traveled is 192 miles.

REVIEW EXERCISES

122. Investment Sally has $25,000 to invest. She invests some money at 10% interest and the rest at 9%. If her total annual income from these two investments is $2,430, how much does she invest at each rate?

123. Mixing solutions How much water must be added to 25 liters of a 14% alcohol solution to dilute it to a 10% solution?

124. Motion A car and a motorcycle both leave from the same point and travel in the same direction. (See the illustration.) The car travels at an average rate of 55 mph and the motorcycle at an average rate of 40 mph. How long will it take before the vehicles are 5 miles apart?

<table>
<tr><td>**1**</td><td>**Test**</td></tr>
</table>

Let $A = \left\{ -2, 0, 1, \frac{6}{5}, 2, \sqrt{7}, 5 \right\}$.

1. What numbers in A are natural numbers?

2. What numbers in A are irrational numbers?

Graph each set on the number line.

3. The set of odd integers between -4 and 6

$$\text{—} | \; | \; | \; | \; | \; | \; | \; | \; | \; | \; | \text{—}$$
$$-4 \;\; -3 \;\; -2 \;\; -1 \;\; 0 \;\; 1 \;\; 2 \;\; 3 \;\; 4 \;\; 5 \;\; 6$$

4. The set of prime numbers less than 12

$$\text{—} | \; | \; | \; | \; | \; | \; | \; | \; | \; | \; | \; | \; | \text{—}$$
$$0 \;\; 1 \;\; 2 \;\; 3 \;\; 4 \;\; 5 \;\; 6 \;\; 7 \;\; 8 \;\; 9 \;\; 10 \;\; 11 \;\; 12$$

Graph each set on the number line.

5. $\{x \mid x > 4\}$

6. $[-3, \infty)$

7. $\{x \mid -2 \le x < 4\}$

8. $(-\infty, -1] \cup [2, \infty)$

Write each expression without using absolute value symbols.

9. $-|8|$

10. $|-5|$

Perform the operations.

11. $7 + (-5)$

12. $-5(-4)$

13. $\dfrac{12}{-3}$

14. $-4 - \dfrac{-15}{3}$

Consider the numbers $-2, 0, -2, 2, 3, -1, -1, 1, 1, 2$.

15. Find the mean.

16. Find the median.

Let $a = 2$, $b = -3$, and $c = 4$ and evaluate each expression.

17. ab

18. $a + bc$

19. $ab - bc$

20. $\dfrac{-3b + a}{ac - b}$

Determine which property of real numbers justifies each statement.

21. $3 + 5 = 5 + 3$

22. $a(b + c) = ab + ac$

Simplify. Write all answers without using negative exponents. Assume that no denominators are zero.

23. $x^3 x^5$

24. $(2x^2 y^3)^3$

25. $(m^{-4})^2$

26. $\left(\dfrac{m^2 n^3}{m^4 n^{-2}} \right)^{-2}$

Write each numeral in scientific notation.

27. 4,700,000

28. 0.00000023

Write in standard notation.

29. 6.53×10^5

Simplify.

30. $-3 - 12 \div 6 - (-4)^2$

Solve each equation.

31. $9(x + 4) + 4 = 4(x - 5)$

32. $\dfrac{2y + 3}{3} + \dfrac{3y - 4}{6} = \dfrac{y - 2}{2}$

33. $\dfrac{y - 1}{5} + 2 = \dfrac{2y - 3}{3}$

34. $400 + 1.5x = 0.5(200 + 3x)$

35. Solve $P = L + \dfrac{s}{f}i$ for i.

36. Geometry   A rectangle has a perimeter of 26 centimeters and is 5 centimeters longer than it is wide. Find its area.

37. Investing   Jamie invests part of $10,000 at 9% annual interest and the rest at 8%. If his annual income from these investments is $860, how much does he invest at 8%?

38. Mixing solutions   How many liters of water are needed to dilute 20 liters of a 5% salt solution to a 1% salt solution?

Graphs, Equations of Lines, and Functions

2

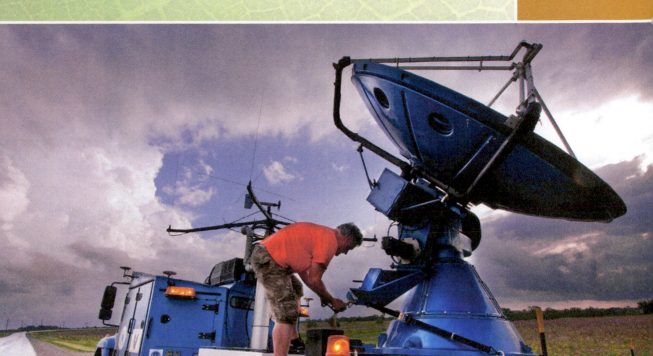

© Ryan McGinnis/Alamy

Careers and Mathematics

ATMOSPHERIC SCIENTISTS, INCLUDING METEOROLOGISTS

Atmospheric science is the study of the atmosphere—the blanket of air covering the Earth. Atmospheric scientists, including meteorologists, study the atmosphere's physical characteristics, motions, and processes. The best-known application of this knowledge is in forecasting the weather. A bachelor's degree in meteorology, which generally includes advanced courses in mathematics, is usually the minimum educational requirement for an entry-level position in the field.

Job Outlook:
Employment of atmospheric scientists is projected to increase about as fast as the average for all occupations through 2020. Opportunities will be greater in private industry than in the federal government.

Annual Earnings:
$45,050–$132,130

For More Information:
http://www.bls.gov/oco/ocos051.htm

For a Sample Application:
See Example 1 in Section 2.1.

REACH FOR SUCCESS
2.1 Graphing Linear Equations
2.2 Slope of a Line
2.3 Writing Equations of Lines
2.4 Introduction to Functions
2.5 Graphing Other Functions
■ *Projects*
REACH FOR SUCCESS EXTENSION
CHAPTER REVIEW
CHAPTER TEST
CUMULATIVE REVIEW

In this chapter

It is often said that "A picture is worth a thousand words." In this chapter, we will show how numerical relationships can be described by using mathematical pictures called graphs.

Reach for Success Reading This Mathematics Textbook

You might be asking yourself, "Who reads their math book anyway?" The better question might be, "Why should I read it?" or "How do I read it?" Why is easy to answer . . . to improve understanding and, thus, your grade! How is not as easy to answer. It takes practice.

Get acquainted with some of the features of this textbook.

Note that the objectives are provided at the beginning of each section. Why do you think these are listed?

_____

Turn to the first page of Section 2.2 in this chapter. On which page did you find this? _____

Note the vocabulary list at the beginning of each section. Where might you find the definitions?

1. _____ 2. _____

What is the purpose of the self-check exercises that directly follow each section's worked-out examples?

_____

A mathematics textbook is structured differently than texts in other subjects. Usually, each chapter is divided into 5 to 7 sections, consisting of 6 to 10 pages each. Thus, you only need to study a few pages at a time. Can you see an advantage to this structure? _____

_____

_____

One advantage to fewer pages is that you can study a smaller "chunk" of mathematics at a time. Use your syllabus (or ask your instructor) to identify which objectives are covered in each section for your course.

Which objectives are you responsible for learning in the second section of this chapter? _____

_____

Did you know that the answers to the odd-numbered exercises are in the back of the book? _____
Why could this be important to you? _____

_____

GLOSSARY

What is the purpose of the glossary? _____

When would you use it?

_____

INDEX

What is the purpose of the index? _____

When would you use it?

_____

A Successful Study Strategy . . .

Read each textbook section prior to the classroom discussion. Even having an idea of the topic for the day will better prepare you for learning.

At the end of the chapter you will find an additional exercise to help guide you to planning for a successful semester.

90

Section 2.1

Graphing Linear Equations

Objectives

1. Plot an ordered pair on a coordinate plane and identify the coordinates of a point plotted on a coordinate plane.
2. Graph a linear equation in two variables by plotting points.
3. Identify the *x*-intercept and *y*-intercept given a linear equation and use them to graph the line.
4. Graph a horizontal line and a vertical line.
5. Apply the skills of graphing to model applications.
6. Find the midpoint of a line segment.

Vocabulary

rectangular coordinate system	quadrants	*y*-intercept
perpendicular	*x*-coordinate	*x*-intercept
x-axis	*y*-coordinate	intercept method
y-axis	graph	vertical line
origin	ordered pair	horizontal line
coordinate plane	linear equation	midpoint

Getting Ready

Graph each set of numbers on the number line.

1. $\{-2, 1, 3\}$

(number line marked $-3\ -2\ -1\ 0\ 1\ 2\ 3\ 4$)

2. $\{x \,|\, x > -2\}$

3. $\{x \,|\, x \leq 3\}$

4. $\{x \,|\, -3 < x < 2\}$

Many cities are laid out on a rectangular grid. For example, on the east side of Rockford, all streets run north and south and all avenues run east and west. (See Figure 2-1 on the next page.) If we agree to list the street numbers first, every location can be identified by using an ordered pair of numbers. Jose lives on the corner of Third Street and Sixth Avenue, so his location is given by the ordered pair $(3, 6)$.

Lisa has a location of $(6, 3)$. We know that she lives on the corner of Sixth Street and Third Avenue. From the figure, we can see that

- Bob's location is $(4, 1)$.
- Rosa's location is $(7, 5)$.
- The location of the store is $(8, 2)$.

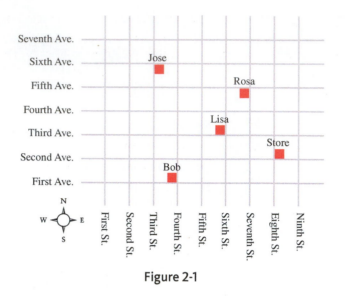

Figure 2-1

The idea of associating an ordered pair of numbers with points on a grid is attributed to the 17th-century French mathematician René Descartes. The grid often is called a **rectangular coordinate system**, or Cartesian coordinate system, after its inventor.

1 **Plot an ordered pair on a coordinate plane and identify the coordinates of a point plotted on a coordinate plane.**

A rectangular coordinate system (see Figure 2-2) is formed by two intersecting **perpendicular** (meeting at right angles) number lines.

- The horizontal number line usually is called the **x-axis**.
- The vertical number line usually is called the **y-axis**.

The positive direction on the *x*-axis is to the right, and the positive direction on the *y*-axis is upward. If no scale is indicated on the axes, we assume that the axes are scaled in units of 1.

The point where the axes cross is called the **origin**. The coordinates of the origin are $(0, 0)$. The two axes form a **coordinate plane** and divide it into four regions called **quadrants**, which are numbered as in Figure 2-2.

René Descartes
1596–1650
Descartes is famous for his work in philosophy as well as in mathematics. His philosophy is expressed in the words "I think, therefore I am." He is best known in mathematics for his invention of a coordinate system and his work with conic sections.

Figure 2-2

Every point in a coordinate plane can be identified by a pair of real numbers *x* and *y*, written as (x, y). The first number in the pair is the **x-coordinate**, and the second number is the **y-coordinate**. The numbers are called the *coordinates* of the point.

$(-4, 6)$

The *x*-coordinate is listed first.   The *y*-coordinate is listed second.

COMMENT   Note that $(-4, 6)$ and $(2, 3)$ could define intervals or points depending on the context.

The process of locating a point in the coordinate plane is called *graphing* or *plotting* the point. In Figure 2-3(a), we show how to graph the point Q with coordinates of $(-4, 6)$. Since the x-coordinate is negative, we start at the origin and move 4 units to the left along the x-axis. Since the y-coordinate is positive, we then move up 6 units to locate point Q. Point Q is the **graph** of the point with coordinates $(-4, 6)$ and lies in quadrant II.

To plot the point $P(2, 3)$, we start at the origin, move 2 units to the right along the x-axis, and then move up 3 units to locate point P. Point P lies in quadrant I. To plot point $R(6, -4)$, we start at the origin and move 6 units to the right and then 4 units down. Point R lies in quadrant IV.

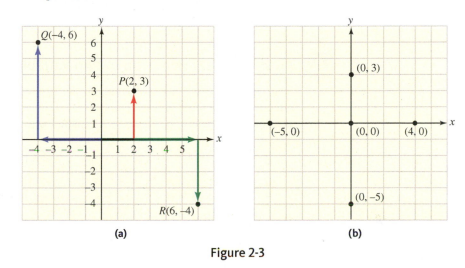

Figure 2-3

COMMENT   Note that point Q with coordinates of $(-4, 6)$ is not the same as point R with coordinates $(6, -4)$. Since the order of the coordinates of a point is important, we call the pairs **ordered pairs**.

In Figure 2-3(b), we see that the points $(-5, 0)$, $(0, 0)$, and $(4, 0)$ all lie on the x-axis. In fact, every point with a y-coordinate of 0 will lie on the x-axis. We also see that the points $(0, -5)$, $(0, 0)$, and $(0, 3)$ all lie on the y-axis. In fact, every point with an x-coordinate of 0 will lie on the y-axis.

EXAMPLE 1

HURRICANES   The map shown in Figure 2-4 shows the path of a hurricane that is currently located at 85° longitude and 25° latitude. If we agree to list longitude first, the location of the hurricane is given by the ordered pair $(85, 25)$. Use the map to answer the following questions.

a. What are the coordinates of New Orleans?

b. What are the estimated coordinates of Houston?

c. What are the estimated coordinates of the landfall of the hurricane?

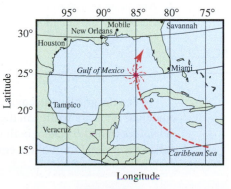

Longitude

Figure 2-4

Solution   a. Since New Orleans is at 90° longitude and 30° latitude, its coordinates are $(90, 30)$.

b. Since Houston has a longitude that is a little more than 95° and a latitude that is a little less than 30°, its estimated coordinates are $(96, 29)$.

c. The hurricane appears to be headed for the point with coordinates of $(83, 29)$.

SELF CHECK 1   What are the approximate coordinates of Veracruz?

Many times we want to show a graphical representation of an algebraic equation. We will do this first by plotting points.

2   Graph a linear equation in two variables by plotting points.

The equation $y = -\frac{1}{2}x + 4$ contains the two variables x and y. The solutions of this equation form a set of ordered pairs of real numbers. For example, the ordered pair $(-4, 6)$ is a solution, because the equation is satisfied when $x = -4$ and $y = 6$.

$y = -\frac{1}{2}x + 4$

x	y	(x, y)
-4	6	$(-4, 6)$
-2	5	$(-2, 5)$
0	4	$(0, 4)$
2	3	$(2, 3)$
4	2	$(4, 2)$

Figure 2-5

$$y = -\frac{1}{2}x + 4$$

$$6 = -\frac{1}{2}(-4) + 4 \quad \text{Substitute } -4 \text{ for } x \text{ and } 6 \text{ for } y.$$

$$6 = 2 + 4$$

$$6 = 6$$

This pair and some others that satisfy the equation are listed in the table shown in Figure 2-5.

The graph of the equation $y = -\frac{1}{2}x + 4$ is the graph of all points (x, y) on the rectangular coordinate system whose coordinates satisfy the equation.

EXAMPLE 2   Graph: $y = -\frac{1}{2}x + 4$

Solution   To graph the equation, we plot the ordered pairs listed in the table shown in Figure 2-5. These points appear to lie on a line, as shown in Figure 2-6. This line is the graph of the equation. Note that we write the equation of the line next to the graphed line.

When we say that the graph of an equation is a line, we imply two things:

1. Every point with coordinates that satisfy the equation will lie on the line.

2. Any point on the line will have coordinates that satisfy the equation.

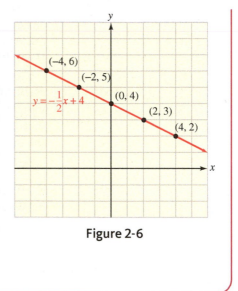

Figure 2-6

SELF CHECK 2   Graph: $y = 2x - 3$

When the graph of an equation is a line, we call the equation a **linear equation**. Linear equations often are written in the form $Ax + By = C$, where A, B, and C represent numerical values (A and B cannot both be zero) and x and y are variables. This form is sometimes called the general form of the equation of a line.

EXAMPLE 3   Graph: $3x + 2y = 12$

Solution   We can select values for either x or y, substitute them into the equation, and solve for the other variable. For example, if $x = 2$,

$$3x + 2y = 12$$
$$3(2) + 2y = 12 \qquad \text{Substitute 2 for } x.$$
$$6 + 2y = 12 \qquad \text{Simplify.}$$
$$2y = 6 \qquad \text{Subtract 6 from both sides.}$$
$$y = 3 \qquad \text{Divide both sides by 2.}$$

The ordered pair $(2, 3)$ satisfies the equation. If $y = 6$,

$$3x + 2y = 12$$
$$3x + 2(6) = 12 \qquad \text{Substitute 6 for } y.$$
$$3x + 12 = 12 \qquad \text{Simplify.}$$
$$3x = 0 \qquad \text{Subtract 12 from both sides.}$$
$$x = 0 \qquad \text{Divide both sides by 3.}$$

A second ordered pair that satisfies the equation is $(0, 6)$.

These pairs and others that satisfy the equation lie on a line, as shown in Figure 2-7. The graph of the equation is the line shown in the figure.

$3x + 2y = 12$

x	y	(x, y)
-2	9	$(-2, 9)$
0	6	$(0, 6)$
2	3	$(2, 3)$
4	0	$(4, 0)$
6	-3	$(6, -3)$

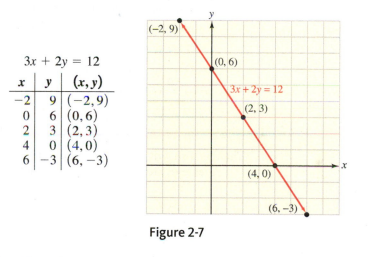

Figure 2-7

🌿 **SELF CHECK 3**   Graph: $3x - 2y = 6$

3   Identify the *x*-intercept and *y*-intercept given a linear equation and use them to graph the line.

In Example 3, the graph intersects the *y*-axis at the point with coordinates $(0, 6)$, called the *y*-intercept, and intersects the *x*-axis at the point with coordinates $(4, 0)$, called the *x*-intercept. In general, we have the following definitions.

INTERCEPTS OF A LINE   The **y-intercept** of a line is the point $(0, b)$ where the line intersects the *y*-axis. To find the value of b, substitute 0 for x in the equation of the line and solve for y.

The **x-intercept** of a line is the point $(a, 0)$ where the line intersects the *x*-axis. To find the value of a, substitute 0 for y in the equation of the line and solve for x.

Any two points will determine the graph of a line. However, many times we are asked to graph using the x- and y-intercepts as the two points. The method of graphing a line using the intercepts is called the **intercept method**.

EXAMPLE 4   Use the x- and y-intercepts to graph $2x + 5y = 10$.

Solution   To find the y-intercept, we substitute 0 for x and solve for y:

$$2x + 5y = 10$$
$$2(0) + 5y = 10 \qquad \text{Substitute 0 for } x.$$
$$5y = 10 \qquad \text{Simplify.}$$
$$y = 2 \qquad \text{Divide both sides by 5.}$$

The y-intercept is the point $(0, 2)$.

To find the x-intercept, we substitute 0 for y and solve for x:

$$2x + 5y = 10$$
$$2x + 5(0) = 10 \qquad \text{Substitute 0 for } y.$$
$$2x = 10 \qquad \text{Simplify.}$$
$$x = 5 \qquad \text{Divide both sides by 2.}$$

The x-intercept is the point $(5, 0)$.

Although the intercepts are sufficient to graph this line, we plot a third point as a check. To find the coordinates of a third point, we can substitute any convenient number for x or y and solve for the other variable. If we let $x = -5$, we will obtain

$$2x + 5y = 10$$
$$2(-5) + 5y = 10 \qquad \text{Substitute } -5 \text{ for } x.$$
$$-10 + 5y = 10 \qquad \text{Simplify.}$$
$$5y = 20 \qquad \text{Add 10 to both sides.}$$
$$y = 4 \qquad \text{Divide both sides by 5.}$$

The line also will pass through the point $(-5, 4)$.

A table of ordered pairs and the graph are shown in Figure 2-8.

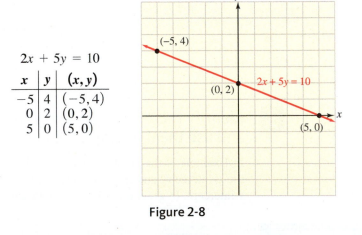

$2x + 5y = 10$

x	y	(x, y)
-5	4	$(-5, 4)$
0	2	$(0, 2)$
5	0	$(5, 0)$

Figure 2-8

SELF CHECK 4   Use the x- and y-intercepts to graph $3x - y = -4$.

Accent
on technology

▸ Generating Tables
  of Solutions

If an equation in x and y is solved for y, we can use a graphing calculator to generate a table of solutions. The instructions in this discussion are for a TI-84 graphing calculator.

To construct a table of solutions for $2x + 5y = 10$, we first solve for y.

$$2x + 5y = 10$$

$$5y = -2x + 10 \qquad \text{Subtract } 2x \text{ from both sides.}$$

$$y = -\frac{2}{5}x + 2 \qquad \text{Divide both sides by 5 and simplify.}$$

To enter $y = -\frac{2}{5}x + 2$, we press **Y=** and enter **(–) (2 ÷ 5) x, T, θ, n** (x) **+** 2, as shown in Figure 2-9(a).

To enter the x-values that are to appear in the table, we press **2ND WINDOW** (TBLSET) and enter the first value for x on the line labeled TblStart=. In Figure 2-9(b), -5 has been entered on this line. Other values for x that are to appear in the table are determined by setting an increment value on the line labeled ΔTbl=. Figure 2-9(b) shows that an increment of 1 was entered. This means that each x-value in the table will be 1 unit larger than the previous x-value.

The final step is to press the keys **2ND GRAPH** (TABLE). This displays a table of solutions, as shown in Figure 2-9(c).

(a)   (b)   (c)

Figure 2-9

For instructions regarding the use of a Casio graphing calculator, please see the Casio Keystroke Guide in the back of the book.

4 Graph a horizontal line and a vertical line.

Equations such as $y = 3$ and $x = -2$ are linear equations in two variables, because they can be written in the form $Ax + By = C$.

$$y = 3 \qquad \text{is equivalent to} \qquad 0x + 1y = 3$$

$$x = -2 \qquad \text{is equivalent to} \qquad 1x + 0y = -2$$

In the next example, we will discuss how to graph these equations.

EXAMPLE 5 Graph: **a.** $y = 3$ **b.** $x = -2$

Solution   **a.** Since the equation $y = 3$ does not contain x, the numbers chosen for x have no effect on y. The value of y is always 3. After plotting the pairs (x, y) shown in Figure 2-10 on the next page, we see that the graph is a horizontal line with a y-intercept of $(0, 3)$. The line has no x-intercept.

b. Since the equation $x = -2$ does not contain y, y can be any number. After plotting the pairs (x, y) shown in Figure 2-10, we see that the graph is a vertical line with an x-intercept of $(-2, 0)$. The line has no y-intercept.

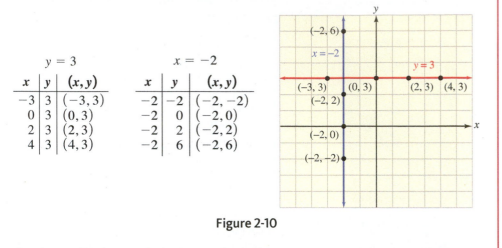

$y = 3$		
x	y	(x, y)
-3	3	$(-3, 3)$
0	3	$(0, 3)$
2	3	$(2, 3)$
4	3	$(4, 3)$

$x = -2$		
x	y	(x, y)
-2	-2	$(-2, -2)$
-2	0	$(-2, 0)$
-2	2	$(-2, 2)$
-2	6	$(-2, 6)$

Figure 2-10

SELF CHECK 5   Graph $x = 4$ and $y = -3$ on one set of coordinate axes.

The results of Example 5 suggest the following facts.

HORIZONTAL AND VERTICAL LINES

If a and b are real numbers, then

The graph of $x = a$ is a **vertical line** with x-intercept at $(a, 0)$. If $a = 0$, the line is the y-axis.

The graph of $y = b$ is a **horizontal line** with y-intercept at $(0, b)$. If $b = 0$, the line is the x-axis.

5   Apply the skills of graphing to model applications.

EXAMPLE 6   **GAS MILEAGE**   The following table gives the number of miles (y) that a bus can travel on x gallons of gas. Plot the ordered pairs and estimate how far the bus can travel on 9 gallons.

x	2	3	4	5	6
y	12	18	24	30	36

Solution   Since the distances driven are rather large numbers, we plot the points on a coordinate system where the unit distance on the y-axis is larger than the unit distance on the x-axis. After plotting each ordered pair as in Figure 2-11, we see that the points lie on a line.

To estimate how far the bus can travel on 9 gallons, we find 9 on the x-axis, move up to the graph, and then move to the left to locate a y-value of 54. Thus, the bus can travel approximately 54 miles on 9 gallons of gas.

Figure 2-11

🌿 **SELF CHECK 6**   How far can the bus travel on 10 gallons?

EXAMPLE 7   **DEPRECIATION**   A copy machine purchased for $6,750 is expected to depreciate according to the formula $y = -950x + 6,750$, where y is the value of the copier after x years. When will the copier have no monetary value?

Solution   The copier will have no monetary value when its value (y) is 0. To find x when $y = 0$, we substitute 0 for y and solve for x.

$$y = -950x + 6,750$$
$$0 = -950x + 6,750$$
$$-6,750 = -950x \qquad \text{Subtract 6,750 from both sides.}$$
$$7.105263158 \approx x \qquad \text{Divide both sides by } -950.$$

The copier will have no monetary value in about 7 years.

🌿 **SELF CHECK 7**   When will the copier be worth $2,000?

6   ## Find the midpoint of a line segment.

COMMENT   Segment PQ can be written as $\overline{PQ}$.

To distinguish between the coordinates of two points on a line, we often use subscript notation. Point $P(x_1, y_1)$ is read as "point P with coordinates of x sub 1 and y sub 1." Point $Q(x_2, y_2)$ is read as "point Q with coordinates of x sub 2 and y sub 2."

If point M in Figure 2-12 lies midway between points $P(x_1, y_1)$ and $Q(x_2, y_2)$, point M is called the **midpoint** of segment PQ. To find the coordinates of M, we find the mean of the x-coordinates and the mean of the y-coordinates of P and Q.

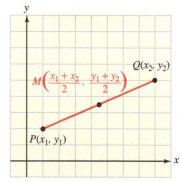

Figure 2-12

THE MIDPOINT FORMULA   The midpoint of the line segment with endpoints at $P(x_1, y_1)$ and $Q(x_2, y_2)$ is the point M with coordinates of

$$\left(\frac{x_1 + x_2}{2}, \frac{y_1 + y_2}{2}\right).$$

EXAMPLE 8   Find the midpoint of the line segment joining $P(-2, 3)$ and $Q(7, -5)$.

Solution   To find the midpoint, we find the mean of the x-coordinates and the mean of the y-coordinates to obtain

$$\frac{x_1 + x_2}{2} = \frac{-2 + 7}{2} \qquad\qquad \frac{y_1 + y_2}{2} = \frac{3 + (-5)}{2}$$

$$= \frac{5}{2} \qquad\qquad\qquad\qquad\qquad\quad = -1$$

The midpoint of segment PQ is the point $M\left(\frac{5}{2}, -1\right)$.

SELF CHECK 8   Find the midpoint of the line segment joining $P(5, -4)$ and $Q(-3, 5)$.

Accent on technology

▶ Graphing Lines

We have graphed linear equations by finding ordered pairs, plotting points, and drawing a line through those points. Graphing is much easier if we use a graphing calculator.

Graphing calculators have a window to display graphs. To see the complete graph, we must decide on the minimum and maximum values for the x- and y-coordinates. A window with standard settings of

Xmin = -10 Xmax = 10 Ymin = -10 Ymax = 10

will produce a graph in which the value of x is in the interval $[-10, 10]$ and y is in the interval $[-10, 10]$.

To graph $3x + 2y = 12$, we must first solve the equation for y.

$$3x + 2y = 12$$

$$2y = -3x + 12 \qquad \text{Subtract } 3x \text{ from both sides.}$$

$$y = -\frac{3}{2}x + 6 \qquad \text{Divide both sides by 2.}$$

To graph the equation, press **Y=** and enter the right side of the equation. The display should look like

$$\backslash Y_1 = -(3/2)X + 6$$

We then press the **GRAPH** key to obtain the graph shown in Figure 2-13(a). To show more detail, we can draw the graph in a different window. A window with settings of $[-1, 5]$ for x and $[-2, 7]$ for y will give the graph shown in Figure 2-13(b).

$$y = -\frac{3}{2}x + 6$$

$$y = -\frac{3}{2}x + 6$$

(a)   (b)

Figure 2-13

We can use the VALUE command found in the CALC menu on the TI-84 calculator to find the coordinates of any point on a graph. For example, to find the y-intercept of the graph of $y = -\frac{3}{2}x + 6$, we press **2ND TRACE** (CALC) **ENTER** to obtain Figure 2-14(a). Press 0 and then **ENTER** to obtain Figure 2-14(b).

Figure 2-14

We can obtain the x-coordinate by using the zero command found in the CALC menu. We press **2ND TRACE** (CALC) 2 (ZERO) to obtain Figure 2-15(a). We then enter a left-bound guess (a number that lies to the left of the intercept) such as 2, press **ENTER**, enter a right-bound guess (a number that lies to the right of the intercept) such as 5, press **ENTER**, and press **ENTER** again to obtain Figure 2-15(b). From the figure, we see that the x-intercept is the point $(4, 0)$.

Figure 2-15

For instructions regarding the use of a Casio graphing calculator, please see the Casio Keystroke Guide in the back of the book.

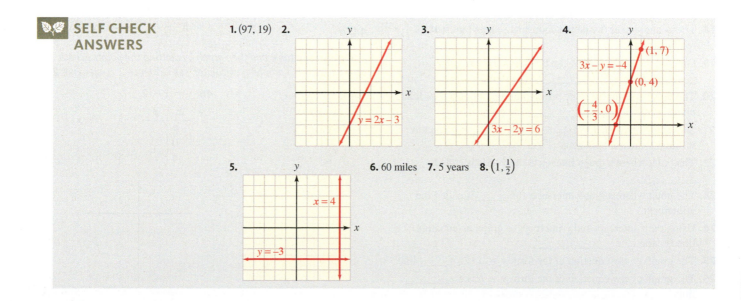

SELF CHECK ANSWERS

1. $(97, 19)$ 2. 3. 4.

5. 6. 60 miles 7. 5 years 8. $\left(1, \frac{1}{2}\right)$

NOW TRY THIS

1. Find the center of the circle with endpoints of its diameter at $(-5, 2)$ and $(-9, -7)$.

2. Find the midpoint of the line segment joining the points with coordinates
   a. $(p - 2, p)$ and $(4 - p, 5p - 2)$.
   b. $(p + 5, 3p - 1)$ and $(6p, p + 9)$.

2.1 Exercises

WARM-UPS *Given x = 2, find the value of y for each equation.*

1. $-2x - 4y = 12$

2. $3x - 5y = -4$

3. $y = \dfrac{5}{2}x - 3$

4. $y = -\dfrac{3}{2}x + 6$

5. $y = 3$

6. $y = 7$

7. $8y = 12$

8. $-10y = 5$

REVIEW

9. Evaluate: $-5 - 2(-7)$

10. Evaluate: $(-7)^2 + (-7)$

11. Simplify: $\dfrac{-3 + 5(2)}{9 + 5}$

12. Simplify: $|-7 - 5|$

13. Solve: $-3x + 5 = 26$

14. Solve $P = 2s + l$ for s.

VOCABULARY AND CONCEPTS *Fill in the blanks.*

15. The pair of numbers $(6, -2)$ is called an _____.

16. In the ordered pair $(-5, 7)$, -5 is called the _-coordinate and 7 is called the _-coordinate.

17. The point with coordinates $(0, 0)$ is called the _____.

18. The x- and y-axes divide the coordinate plane into four regions called _____.

19. Ordered pairs of numbers can be graphed on a _____ system.

20. The process of locating the position of a point on a coordinate plane is called _____ the point.

21. Any equation in the form $Ax + By = C$ is a _____ equation if A and B are not both zero.

22. The point where a graph intersects the x-axis is called _____.

23. The point where a graph intersects the _____ is called the y-intercept.

24. Using the x-intercept and y-intercept to graph an equation of a line is called the _____.

25. The graph of any equation of the form $x = a$ is a _____ line.

26. The graph of any equation of the form $y = b$ is a _____ line.

27. The symbol x_1 is read as "x _____."

28. The midpoint of a line segment joining (a, b) and (c, d) is given by the formula _____.

GUIDED PRACTICE *Plot each point on the rectangular coordinate system.* (OBJECTIVE 1)

29. $A(4, 3)$

30. $B(-2, 1)$

31. $C(3, -2)$

32. $D(-2, -3)$

33. $E(0, 5)$

34. $F(-4, 0)$

35. $G(2, 0)$

36. $H(0, 3)$

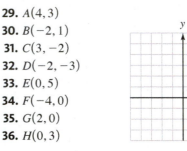

Give the coordinates of each point. SEE EXAMPLE 1. (OBJECTIVE 1)

37. A

38. B

39. C

40. D

41. E

42. F

43. G

44. H

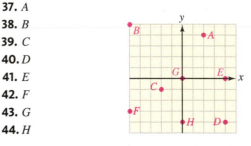

Complete each table and then graph by plotting the points. Check your work with a graphing calculator. SEE EXAMPLE 2. (OBJECTIVE 2)

45. $y = -x + 4$

x	y
-1	
0	
2	

46. $y = x - 2$

x	y
-2	
0	
4	

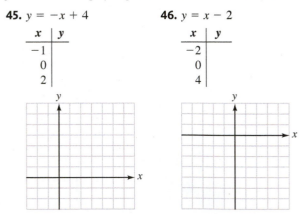

47. $y = 2x - 3$

x	y
−1	
0	
3	

48. $y = -\dfrac{1}{2}x + \dfrac{5}{2}$

x	y
−3	
−1	
3	

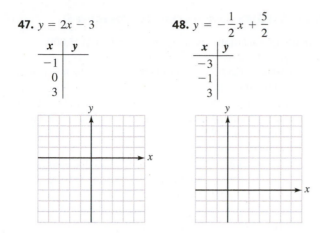

Graph each equation. Check your work with a graphing calculator. SEE EXAMPLE 3. *(OBJECTIVE 2)*

49. $3x + 4y = 12$

50. $4x - 3y = 12$

51. $6x - 3y = 12$

52. $3x + 6y = 12$

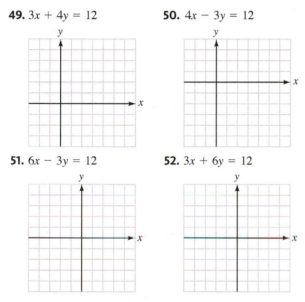

Graph each equation using the intercept method. Check your work with a graphing calculator. SEE EXAMPLE 4. *(OBJECTIVE 3)*

53. $3y = 6x - 9$

54. $2x = 4y - 10$

55. $3x + 4y - 8 = 0$

56. $-2y - 3x + 9 = 0$

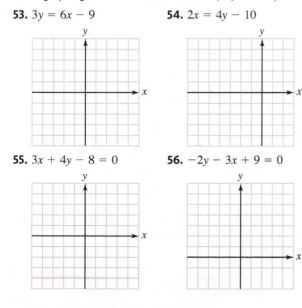

Graph each equation. SEE EXAMPLE 5. *(OBJECTIVE 4)*

57. $x = 3$

58. $y = -4$

59. $-3y + 2 = 5$

60. $-2x + 3 = 11$

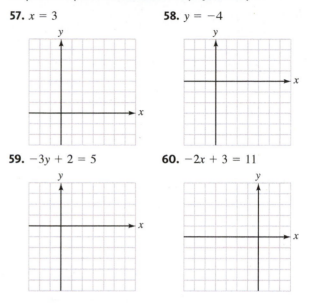

Plot the ordered pairs on a coordinate plane to find the answer. SEE EXAMPLE 6. *(OBJECTIVE 5)*

61. Hourly wages   The table gives the amount y (in dollars) that a student can earn for working x hours. Plot the ordered pairs and estimate how much the student will earn for working 8 hours.

x	2	4	5	6
y	12	24	30	36

62. Distance traveled   The table shows how far y (in miles) a biker can travel in x hours. Plot the ordered pairs and estimate how far the biker can travel in 8 hours.

x	2	4	5	6
y	30	60	75	90

63. Value of a car   The table shows the value y (in dollars) of a car that is x years old. Plot the ordered pairs and estimate the value of the car when it is 4 years old.

x	0	1	3
y	15,000	12,000	6,000

64. Earning interest   The table shows the amount y (in dollars) in a bank account drawing simple interest left on deposit for x years. Plot the ordered pairs and estimate the value of the account in 6 years.

x	0	1	4
y	1,000	1,050	1,200

Find the midpoint of $\overline{PQ}$. SEE EXAMPLE 8. (OBJECTIVE 6)

65. $P(0,0), Q(2,4)$

66. $P(10,12), Q(0,0)$

67. $P(4,6), Q(14,20)$

68. $P(10,4), Q(2,-2)$

69. $P(6,7), Q(5,1)$

70. $P(8,11), Q(3,9)$

71. $P(-10,-2), Q(3,12)$

72. $P(-5,-2), Q(7,3)$

ADDITIONAL PRACTICE *Graph each equation. Check your work with a graphing calculator.*

73. $y = -3x + 2$

74. $y = 2x + 3$

75. $y = \dfrac{3}{2}x$

76. $y = -\dfrac{2}{3}x$

Find the midpoint of $\overline{PQ}$.

77. $Q(-3,5), P(-5,-5)$

78. $Q(2,-3), P(4,-8)$

79. Finding the endpoint of a line segment   If $M(7,-4)$ is the midpoint of $\overline{PQ}$ and the coordinates of P are $(5,-6)$, find the coordinates of Q.

80. Finding the endpoint of a line segment   If $M(6,-5)$ is the midpoint of $\overline{PQ}$ and the coordinates of Q are $(-5,-8)$, find the coordinates of P.

Use a graphing calculator to create a table of solutions for each equation. Set the first value of x to be −2.5 and the increment value to be 0.5. Find the value of y that corresponds to x = −1. If an answer is not exact, round to the nearest thousandth.

81. $y = 2.5x + 1.5$

82. $y = 0.6x - 3.2$

83. $3.2x - 1.5y = 2.7$

84. $-1.7x + 3.7y = -2.8$

Use a graphing calculator to graph each equation, and then identify the x-coordinate of the x-intercept to the nearest hundredth.

85. $y = 3.7x - 4.5$

86. $y = 0.6x + 1.25$

87. $1.5x - 3y = 7$

88. $0.3x + y = 7.5$

APPLICATIONS *Solve. SEE EXAMPLE 7. (OBJECTIVE 5)*

89. House appreciation   A house purchased for $125,000 is expected to appreciate according to the formula $y = 7,500x + 125,000$, where y is the value of the house after x years. Find the value of the house 5 years later.

90. Car depreciation   A car purchased for $17,000 is expected to depreciate according to the formula $y = -1,360x + 17,000$, where y is the value of the car after x years. When will the car have no monetary value?

91. Demand equations   The number of HDTVs that consumers buy depends on price. The higher the price, the fewer people will buy. The equation that relates price to the number of HDTVs sold at that price is called a **demand equation**. If the demand equation for a 27-inch HDTV is $p = -\dfrac{1}{10}q + 170$, where p is the price and q is the number sold at that price, how many HDTVs will be sold at a price of $150?

92. Supply equations   The number of HDTVs that manufacturers produce depends on price. The higher the price, the more HDTVs manufacturers will produce. The equation that relates price to the number produced at that price is called a **supply equation**. If the supply equation for a 32-inch HDTV is $p = \dfrac{1}{10}q + 130$, where p is the price and q is the number produced for sale at that price, how many HDTVs will be produced if the price is $150?

93. TV coverage   In the illustration, a TV camera is located at point $(-3,0)$. The camera is to follow the launch of the first private industry rocket. As the rocket rises vertically, the camera can tilt back to a line of sight given by $y = 2x + 6$. Estimate how many miles the rocket will remain in the camera's view.

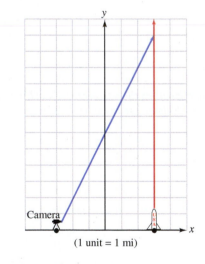

(1 unit = 1 mi)

94. Buying tickets   Children's tickets to the circus cost $5 each plus a $2 service fee for each block of tickets.

   a. Write a linear equation that gives the cost y for a student buying x tickets.

   b. Complete the table in the illustration and graph the equation.

   c. Use the graph to find the cost of buying 4 tickets.

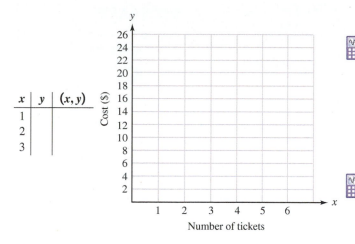

x	y	(x, y)
1		
2		
3		

Number of tickets

95. Telephone costs   In one community, the monthly cost of telephone service is $50 per month, plus 25¢ per text.

   a. Write a linear equation that gives the cost y for a person sending x texts.

   b. Complete the table in the illustration and graph the equation.

   c. Use the graph to estimate the cost of service in a month when 20 texts were sent.

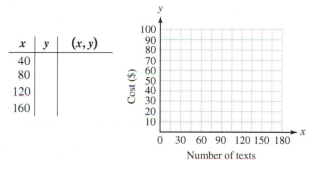

x	y	(x, y)
40		
80		
120		
160		

Number of texts

96. Crime prevention   The number n of incidents of daytime robberies appears to be related to d, the money spent for patrol officers, by the equation $n = 430 - 0.0005d$.

What expenditure would reduce the number of incidents to 350?

97. U.S. sports participation   The equation $s = -0.9t + 65.5$ estimates the number of people 7 years of age and older who went swimming during a given year, where s is the annual number of swimmers (in millions) and t is the number of years since 1990. (*Source*: National Sporting Goods Association)

   a. What information can be obtained from the s-intercept of the graph?

   b. Estimate the number of swimmers in 2015.

98. Farming   The equation

$$a = -3,700,000t + 983,000,000$$

estimates the number of acres a of farmland in the United States, t years after 1990. (*Source*: U.S. Department of Agriculture)

   a. What information can be obtained from the a-intercept of the graph?

   b. To the nearest million acres, estimate the number of acres of farmland in 2020.

WRITING ABOUT MATH

99. Explain how to graph a line using the intercept method.

100. Explain how to determine the quadrant in which the point $P(a, b)$ lies.

SOMETHING TO THINK ABOUT

101. If the line $y = ax + b$ passes only through quadrants I and II, what can be known about the constants a and b?

102. What are the coordinates of the three points that divide the line segment joining $P(a, b)$ and $Q(c, d)$ into four equal parts?

Section 2.2

Slope of a Line

Objectives

1. Find the slope of a line given a graph.
2. Find the slope of a line passing through two given points.
3. Find the slope of a line given its equation.
4. Identify the slope of a horizontal line and a vertical line.
5. Determine whether two lines are parallel, perpendicular, or neither.
6. Interpret slope in an application.

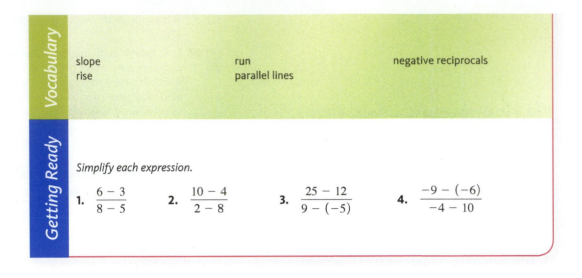

Vocabulary

slope   run   negative reciprocals
rise   parallel lines

Getting Ready

Simplify each expression.

1. $\dfrac{6-3}{8-5}$ **2.** $\dfrac{10-4}{2-8}$ **3.** $\dfrac{25-12}{9-(-5)}$ **4.** $\dfrac{-9-(-6)}{-4-10}$

In Section 2.1, we graphed equations of lines by plotting points. Later, we will show that we can graph a line if we know the coordinates of one point on the line and the slant (steepness) of the line. We will now discuss a measure of this slant called the *slope* of the line.

1 Find the slope of a line given a graph.

A service offered by an online research company costs \$2 per month plus \$3 for each hour of connect time. The table in Figure 2-16(a) gives the cost y for certain numbers of hours x of connect time. If we construct a graph from these data, we obtain the line shown in Figure 2-16(b).

Grace Murray Hopper
1906–1992

Grace Hopper graduated from Vassar College in 1928 and obtained a master's degree from Yale in 1930. In 1943, she entered the U.S. Naval Reserve. While in the Navy, she became a programmer of the Mark I, the world's first large computer. She is credited for first using the word "bug" to refer to a computer problem. The first bug was actually a moth that flew into one of the relays of the Mark II. From then on, locating computer problems was called "debugging" the system.

x	0	1	2	3	4	5
y	2	5	8	11	14	17

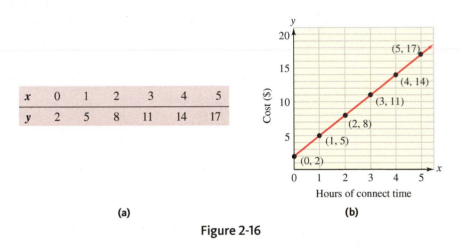

(a)

(b)

Figure 2-16

From the graph, we can see that if x changes from 0 to 1, y changes from 2 to 5. As x changes from 1 to 2, y changes from 5 to 8, and so on. The ratio of the change in y divided by the change in x is the constant 3.

$$\frac{\text{Change in } y}{\text{Change in } x} = \frac{5-2}{1-0} = \frac{8-5}{2-1} = \frac{11-8}{3-2} = \frac{14-11}{4-3} = \frac{17-14}{5-4} = \frac{3}{1} = 3$$

The ratio of the change in y divided by the change in x between any two points on any line is always a constant. This constant rate of change is called the **slope of the line**.

SLOPE OF A NONVERTICAL LINE	The slope m of the nonvertical line passing through points (x_1, y_1) and (x_2, y_2) is

$$m = \frac{\text{change in } y}{\text{change in } x} = \frac{y_2 - y_1}{x_2 - x_1}$$

2   Find the slope of a line passing through two given points.

EXAMPLE 1   Use the two points shown in Figure 2-17 to find the slope of the line passing through them.

Solution   We can let $(x_1, y_1) = (-2, 4)$ and $(x_2, y_2) = (3, -4)$. Then

$$m = \frac{\text{change in } y}{\text{change in } x}$$

$$= \frac{y_2 - y_1}{x_2 - x_1}$$

$$= \frac{-4 - 4}{3 - (-2)}$$

Substitute -4 for y_2, 4 for y_1, 3 for x_2, and -2 for x_1.

$$= -\frac{8}{5}$$

COMMENT   The slope $-\frac{8}{5}$ can be viewed as a change in y of -8 and change in x of 5 or a change in y of 8 and change in x of -5.

Figure 2-17

The slope of the line is $-\frac{8}{5}$. We would obtain the same result if we let $(x_1, y_1) = (3, -4)$ and $(x_2, y_2) = (-2, 4)$.

 SELF CHECK 1   Find the slope of the line passing through the points $(-3, 6)$ and $(4, -8)$.

COMMENT   If the ordered pairs are integer values, an alternative is to "count" the change in x and y. If you do, be careful of the direction. For example, to move from $(-2, 4)$ to $(3, -4)$, go down (negative) 8 units and to the right (positive) 5 units.

COMMENT   When calculating slope, always subtract the y-values and the x-values from the ordered pairs in the same order.

$$m = \frac{y_2 - y_1}{x_2 - x_1} \quad \text{or} \quad m = \frac{y_1 - y_2}{x_1 - x_2}$$

The change in y (often denoted as Δy) is the **rise** of the line between two points. The change in x (often denoted as Δx) is the **run**. Using this terminology, we can define slope to be the ratio of the rise to the run:

$$m = \frac{\Delta y}{\Delta x} = \frac{\text{rise}}{\text{run}} \quad (\Delta x \neq 0)$$

3   Find the slope of a line given its equation.

Previously, to find the slope of a line from a given equation, we graphed the equation and counted squares on the resulting line graph to determine the rise and the run. Another way is to find the x- and y-intercepts of the graph and use the slope formula.

EXAMPLE 2   Find the slope of the line determined by $3x - 4y = 12$.

Solution   We first find the coordinates of two points on the line.

- If $x = 0$, then $y = -3$, and the point $(0, -3)$ is on the line.
- If $y = 0$, then $x = 4$, and the point $(4, 0)$ is on the line.

We then refer to Figure 2-18 and find the slope of the line between $(0, -3)$ and $(4, 0)$ by substituting 0 for y_2, -3 for y_1, 4 for x_2, and 0 for x_1 in the formula for slope.

$$m = \frac{\Delta y}{\Delta x}$$

$$= \frac{y_2 - y_1}{x_2 - x_1}$$

$$= \frac{0 - (-3)}{4 - 0}$$

$$= \frac{3}{4}$$

The slope of the line is $\frac{3}{4}$.

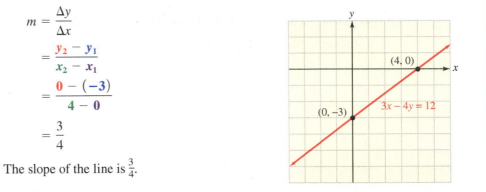

Figure 2-18

SELF CHECK 2   Find the slope of the line determined by $2x + 5y = 12$.

---

4   Identify the slope of a horizontal line and a vertical line.

If $P(x_1, y_1)$ and $Q(x_2, y_2)$ are points on the horizontal line shown in Figure 2-19(a), then $y_1 = y_2$, and the numerator of the fraction

$$\frac{y_2 - y_1}{x_2 - x_1}$$ On a horizontal line, $x_2 \neq x_1$.

is 0. Thus, the value of the fraction is 0, and the slope of the horizontal line is 0.

If $P(x_1, y_1)$ and $Q(x_2, y_2)$ are two points on the vertical line shown in Figure 2-19(b), then $x_1 = x_2$, and the denominator of the fraction

$$\frac{y_2 - y_1}{x_2 - x_1}$$ On a vertical line, $y_2 \neq y_1$.

is 0. Since the denominator cannot be 0, a vertical line has an undefined slope.

(a)   (b)

Figure 2-19

SLOPES OF HORIZONTAL AND VERTICAL LINES	All horizontal lines (lines with equations of the form $y = b$) have a slope of 0.
	All vertical lines (lines with equations of the form $x = a$) have an undefined slope.

If a line rises as we follow it from left to right, as in Figure 2-20(a), its slope is positive. If a line drops as we follow it from left to right, as in Figure 2-20(b), its slope is negative. If a line is horizontal, as in Figure 2-20(c), its slope is 0. If a line is vertical, as in Figure 2-20(d), it has an undefined slope.

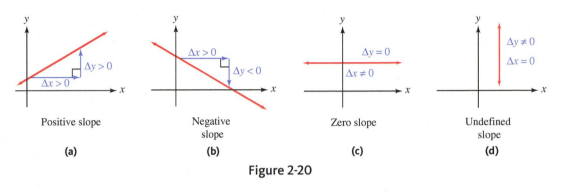

Positive slope	Negative slope	Zero slope	Undefined slope
(a)	(b)	(c)	(d)

Figure 2-20

5   Determine whether two lines are parallel, perpendicular, or neither.

To see a relationship between **parallel lines** (two lines in a plane that never intersect) and their slopes, we refer to the parallel lines l_1 and l_2 shown in Figure 2-21, with slopes of m_1 and m_2, respectively. Because right triangles ABC and DEF are similar, it follows that

$$m_1 = \frac{\Delta y \text{ of } l_1}{\Delta x \text{ of } l_1}$$
$$= \frac{\Delta y \text{ of } l_2}{\Delta x \text{ of } l_2}$$
$$= m_2$$

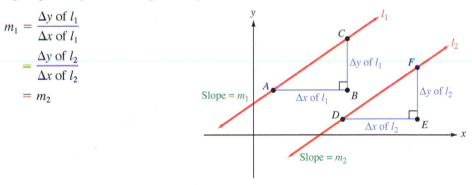

Figure 2-21

Thus, if two nonvertical lines are parallel, they have the same slope. It is also true that when two lines have the same slope, they are parallel.

SLOPES OF PARALLEL LINES	Nonvertical parallel lines have the same slope, and lines having the same slope are parallel.
	Since vertical lines are parallel, lines with an undefined slope are parallel.

EXAMPLE 3 The lines in Figure 2-22 are parallel. Find the slope of l_2.

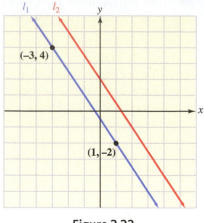

Figure 2-22

Solution From the figure, we can find the slope of line l_1. Since the lines are parallel, they will have equal slopes. Therefore, the slope of line l_2 will be equal to the slope of line l_1. We can use the points with coordinates $(-3, 4)$ and $(1, -2)$ on line l_1 to find the slope of l_1 as follows:

$$m = \frac{y_2 - y_1}{x_2 - x_1}$$

$$= \frac{-2 - 4}{1 - (-3)} \quad \text{Substitute } -2 \text{ for } y_2, 4 \text{ for } y_1, 1 \text{ for } x_2, \text{ and } -3 \text{ for } x_1.$$

$$= \frac{-6}{4}$$

The slope of l_1 is $-\frac{3}{2}$ and because the lines are parallel, the slope of l_2 is also $-\frac{3}{2}$.

SELF CHECK 3 Find the slope of any line parallel to a line with a slope of -3.

Two real numbers a and b are called **negative reciprocals** if $ab = -1$. For example,

$$-\frac{4}{3} \quad \text{and} \quad \frac{3}{4}$$

are negative reciprocals, because $-\frac{4}{3}\left(\frac{3}{4}\right) = -1$.

The following relates perpendicular lines and their slopes.

SLOPES OF PERPENDICULAR LINES

If two nonvertical lines are perpendicular, their slopes are negative reciprocals.

If the slopes of two lines are negative reciprocals, the lines are perpendicular.

Because a horizontal line is perpendicular to a vertical line, a line with a slope of 0 is perpendicular to a line with an undefined slope.

EXAMPLE 4   Are the lines shown in Figure 2-23 perpendicular?

Solution   We find the slopes of the lines and see whether they are negative reciprocals.

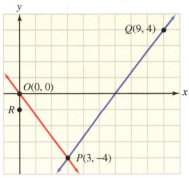

Figure 2-23

$$\text{Slope of line } OP = \frac{\Delta y}{\Delta x}$$

$$= \frac{y_2 - y_1}{x_2 - x_1}$$

$$= \frac{-4 - 0}{3 - 0}$$

$$= -\frac{4}{3}$$

$$\text{Slope of line } PQ = \frac{\Delta y}{\Delta x}$$

$$= \frac{y_2 - y_1}{x_2 - x_1}$$

$$= \frac{4 - (-4)}{9 - 3}$$

$$= \frac{8}{6}$$

$$= \frac{4}{3}$$

Since their slopes $-\frac{4}{3}$ and $\frac{4}{3}$ are not negative reciprocals, the lines are not perpendicular.

SELF CHECK 4   In Figure 2-23, is line PR perpendicular to line PQ?

---

6   Interpret slope in an application.

Many applications involve equations of lines and their slopes.

EXAMPLE 5   **COST OF CARPET**   A store sells a carpet for \$25 per square yard, plus a \$20 delivery charge. The total cost c of n square yards is given by the following formula.

c	equals	cost per square yard	times	the number of square yards	plus	the delivery charge.
c	$=$	25	$\cdot$	n	$+$	20

Graph the equation $c = 25n + 20$ and interpret the slope of the line.

Solution   We can graph the equation on a coordinate system with a vertical c-axis and a horizontal n-axis. Figure 2-24 shows a table of ordered pairs and the graph.

$$c = 25n + 20$$

n	c	(n, c)
10	270	$(10, 270)$
20	520	$(20, 520)$
30	770	$(30, 770)$
40	1,020	$(40, 1{,}020)$
50	1,270	$(50, 1{,}270)$

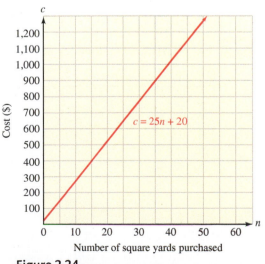

Figure 2-24

COMMENT Recall that any two points on a line can be used to find the slope. In this example, we chose $(30, 770)$ and $(50, 1{,}270)$, but we could have chosen any two points and obtained the same result.

If we choose the points $(30, 770)$ and $(50, 1{,}270)$ to find the slope, we have

$$
\begin{aligned}
m &= \frac{\Delta c}{\Delta n} \\[4pt]
&= \frac{c_2 - c_1}{n_2 - n_1} \\[4pt]
&= \frac{1{,}270 - 770}{50 - 30} \qquad \text{Substitute } 1{,}270 \text{ for } c_2,\ 770 \text{ for } c_1,\ 50 \text{ for } n_2,\ \text{and } 30 \text{ for } n_1. \\[4pt]
&= \frac{500}{20} \\[4pt]
&= 25
\end{aligned}
$$

The slope of 25 is the cost of the carpet in dollars per square yard.

SELF CHECK 5 Interpret the y-intercept of the graph in Figure 2-24.

EXAMPLE 6

RATE OF DESCENT It takes a skier 25 minutes to complete the course shown in Figure 2-25. Find his average rate of descent in feet per minute.

Solution To find the average rate of descent, we must find the ratio of the change in altitude to the change in time. To find this ratio, we calculate the slope of the line passing through the points with coordinates $(0, 12{,}000)$ and $(25, 8{,}500)$.

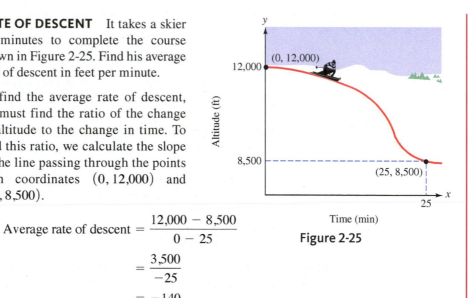

Figure 2-25

$$
\begin{aligned}
\text{Average rate of descent} &= \frac{12{,}000 - 8{,}500}{0 - 25} \\[4pt]
&= \frac{3{,}500}{-25} \\[4pt]
&= -140
\end{aligned}
$$

The slope is -140. Thus, the rate of change of descent is 140 ft/min.

SELF CHECK 6 Find the average rate of descent if the skier completes the course in 20 minutes.

Everyday connections
Labor Force

We can approximate the **rate of growth (or decrease)** of a quantity during a given time interval by calculating the slope of the line segment that connects the endpoints of the graph on the given interval.

Use the data from the graph to compute the rate of growth (or decrease) of the

1. labor force from 2010 to 2011.

2. labor force from 2005 to 2006.

3. During which 1-year interval was a positive rate of growth of the labor force the smallest?

Source: http://www.bls.gov/eag/eag.nc_durham _msa.htm

SELF CHECK ANSWERS

1. -2 **2.** $-\frac{2}{5}$ **3.** -3 **4.** no **5.** The y-coordinate of the y-intercept is the delivery charge. **6.** 175 ft/min

---

NOW TRY THIS

Find the slope of a line

1. parallel to $3x - 2 = 0$.

2. perpendicular to $6x = 4$.

3. perpendicular to $5y + 4 = 0$.

2.2 Exercises

WARM-UPS *Evaluate* $\dfrac{a - b}{c - d}$ *when*

1. $a = 2, b = 3, c = -1, d = 4.$

2. $a = -3, b = 5, c = 7, d = -2.$

3. $a = -2, b = -2, c = -3, d = 4.$

4. $a = 1, b = 1, c = -5, d = -2.$

5. $a = 4, b = -3, c = 6, d = 6.$
6. $a = 2, b = 3, c = -5, d = -5.$
7. Find the negative reciprocal of -0.2.

REVIEW *Simplify each expression. Write all answers without negative exponents. Assume no variable is zero.*

8. $(x^{-4}y^5)^{-3}$

9. $(x^{-3}y^2)^{-4}$

10. $\left(\dfrac{x^{-6}}{y^3}\right)^{-4}$

11. $\left(\dfrac{3x^2y^3}{8}\right)^0$

VOCABULARY AND CONCEPTS *Fill in the blanks.*

12. Slope is defined as the change in _ divided by the change in _.
13. A slope is a rate of _____.
14. The formula to compute slope is $m =$ _____.
15. The change in y (denoted as Δy) is the ___ of the line between two points.
16. The change in x (denoted as Δx) is the ___ of the line between two points.
17. The slope of a _____ line is 0.
18. The slope of a _____ line is undefined.
19. If a line rises as x increases, its slope is _____.
20. _____ lines have the same slope.
21. The slopes of nonvertical _____ lines are negative _____.
22. A line with an undefined slope and a line with a slope of _ are perpendicular.

GUIDED PRACTICE *Find the slope of the line passing through the given points, if possible. SEE EXAMPLE 1. (OBJECTIVES 1–2)*

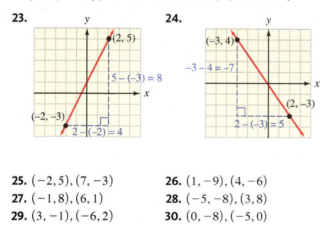

25. $(-2, 5), (7, -3)$ 26. $(1, -9), (4, -6)$
27. $(-1, 8), (6, 1)$ 28. $(-5, -8), (3, 8)$
29. $(3, -1), (-6, 2)$ 30. $(0, -8), (-5, 0)$

Find the slope of the line determined by each equation.
SEE EXAMPLE 2. (OBJECTIVE 3)

31. $5x + 3y = 15$ 32. $2x - y = 6$
33. $3x = 4y - 2$ 34. $x = y$

Determine whether the slope of the line in each graph is positive, negative, 0, or undefined. (OBJECTIVE 4)

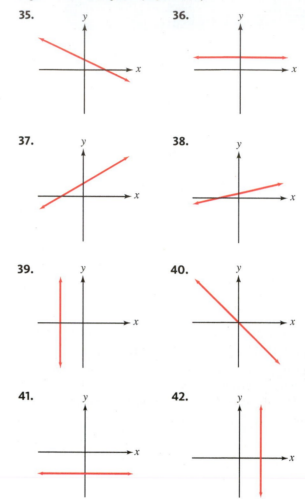

Determine whether the lines with the given slopes are parallel, perpendicular, or neither. SEE EXAMPLES 3–4. (OBJECTIVE 5)

43. $m_1 = 5, m_2 = -\dfrac{1}{5}$

44. $m_1 = \dfrac{1}{4}, m_2 = 4$

45. $m_1 = -0.5, m_2 = -2$

46. $m_1 = -5, m_2 = \dfrac{1}{-0.2}$

Determine whether the line PQ is parallel or perpendicular or neither parallel nor perpendicular to a line with a slope of -2.
SEE EXAMPLES 3–4. (OBJECTIVE 5)

47. $P(-2, 5), Q(6, -11)$
48. $P(6, 4), Q(8, 5)$
49. $P(-2, 1), Q(6, 5)$
50. $P(-6, 1), Q(-2, 9)$

ADDITIONAL PRACTICE *Find the slope of the line passing through the given points, if possible.*

51. $(4, 9), (-8, 9)$

52. $(2, -8), (3, -8)$

53. $(-7, -5), (-7, -2)$

54. $(7, 5), (7, -3)$

Find the slope of the line determined by each equation.

55. $y = \dfrac{3x - 6}{2}$

56. $x = \dfrac{3 - y}{4}$

57. $4y = 3(y + 2)$

58. $x + y = \dfrac{2 - 3y}{3}$

59. Are the lines passing through the points $(2.5, 3.7), (3.7, 2.5)$ and $(1.7, -2.3), (2.3, -1.7)$ parallel, perpendicular, or neither parallel nor perpendicular?

60. Are the lines with slopes $m_1 = \frac{2.3}{1.7}$ and $m_2 = \left(\frac{1.7}{2.3}\right)^{-1}$ parallel, perpendicular, or neither parallel nor perpendicular ?

61. Are the lines with slopes $m_1 = \frac{3.2}{-9.1}$ and $m_2 = \frac{-9.1}{3.2}$ parallel, perpendicular, or neither parallel nor perpendicular?

62. Is the line passing through the points $P(5, 4)$ and $Q(6, 6)$ parallel, perpendicular, or neither parallel nor perpendicular to a line with a slope of -2?

63. Is the line passing through the points $P(-2, 3)$ and $Q(4, -9)$ parallel, perpendicular, or neither parallel nor perpendicular to a line with a slope of -2?

Find the slopes of lines PQ and PR and tell whether the points P, Q, and R lie on the same line. (Hint: Two lines with the same slope and a point in common must be the same line.)

64. $P(-2, 4), Q(4, 8), R(8, 12)$

65. $P(0, 8), Q(1, 6), R(3, 2)$

66. $P(-4, 10), Q(-6, 0), R(-1, 5)$

67. $P(-10, -13), Q(-8, -10), R(-12, -16)$

68. $P(-2, 4), Q(0, 8), R(2, 12)$

69. $P(8, -4), Q(0, -12), R(8, -20)$

70. Find the equation of the x-axis and its slope.

71. Find the equation of the y-axis and its slope.

72. Show that points with coordinates of $(-3, 4), (4, 1)$, and $(-1, -1)$ are the vertices of a right triangle.

73. Show that a triangle with vertices at $(0, 0), (12, 0)$, and $(13, 12)$ is not a right triangle.

74. A square has vertices at points $(a, 0), (0, a), (-a, 0)$, and $(0, -a)$, where $a \neq 0$. Show that its adjacent sides are perpendicular.

75. If a and b are not both 0, show that the points $(2b, a), (b, b)$, and $(a, 0)$ are the vertices of a right triangle.

76. Show that the points $(0, 0), (0, a), (b, c)$, and $(b, a + c)$ are the vertices of a parallelogram. (*Hint*: Opposite sides of a parallelogram are parallel.)

77. If $b \neq 0$, show that the points $(0, 0), (0, b), (8, b + 2)$, and $(12, 3)$ are the vertices of a trapezoid. (*Hint*: A **trapezoid** is a four-sided figure with exactly two sides parallel.)

APPLICATIONS *SEE EXAMPLES 5–6. (OBJECTIVE 6)*

78. Grade of a road Find the slope of the road. (*Hint*: 1 mi = 5,280 ft.)

32 ft

1 mi

79. Slope of a roof Find the slope of the roof.

3 ft

24 ft

80. Slope of a ladder A ladder reaches 18 feet up the side of a building with its base 5 feet from the building. Find the slope of the ladder.

81. Physical fitness Find the slope of the treadmill for each setting listed in the table.

Height setting
2 in.
5 in.
8 in.

Height setting

50 in.

82. Climate change The following line graphs estimate the global temperature rise between the years of 1990 and 2040. Find the average rate of temperature change (the slope) of Model A: Status quo.

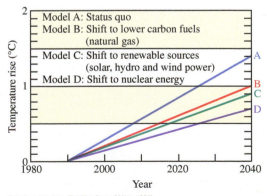

Based on data from *The Blue Planet* (Wiley, 1995)

83. Climate change   Find the average rate of temperature change of Model D: Shift to nuclear energy.

84. Rate of growth   When a college started an aviation program, the administration agreed to predict enrollments using a straight-line method. If the enrollment during the first year was 8, and the enrollment during the fifth year was 20, find the rate of growth per year. (See the illustration.)

85. Wheelchair ramps   The illustration shows two designs for a ramp to make a platform wheelchair accessible.

a. Find the slope of the ramp shown in design 1.

b. Find the slope of each part of the ramp shown in design 2.

c. Give one advantage and one disadvantage of each design.

86. Rate of growth   A small business predicts sales according to a straight-line method. If sales were $110,000 in the first year and $200,000 in the fourth year, find the rate of growth in sales per year.

87. Rate of decrease   The price of computer technology has been dropping for the past ten years. If a desktop PC cost $5,700 ten years ago, and the same computing power cost $1,499 two years ago, find the rate of decrease per year. (Assume a straight-line model.)

WRITING ABOUT MATH

88. Explain why a vertical line has no defined slope.

89. Explain how to determine from their slopes whether two lines are parallel, perpendicular, or neither.

SOMETHING TO THINK ABOUT

90. Find the slope of the line $Ax + By = C$. Follow the procedure of Example 2.

91. Follow Example 2 to find the slope of the line $y = mx + b$.

92. The points $(3, a)$, $(5, 7)$, and $(7, 10)$ lie on a line. Find a.

93. The line passing through points $(1, 3)$ and $(-2, 7)$ is perpendicular to the line passing through points $(4, b)$ and $(8, -1)$. Find b.

Section 2.3

Writing Equations of Lines

Objectives

1. Find the point-slope equation of a line with a given slope and passing through a given point.
2. Write the equation in slope-intercept form of a line with a given slope and passing through a given point.
3. Graph a linear equation using the slope and y-intercept.
4. Determine whether two linear equations define lines that are parallel, perpendicular, or neither.
5. Write an equation of the line passing through a given point and parallel or perpendicular to a given line.
6. Write an equation of a line representing real-world data.

Vocabulary

point-slope form   slope-intercept form

Getting Ready

Solve each equation.

1. $3 = \dfrac{x - 2}{4}$
2. $-2 = 3(x + 1)$

3. Solve $y - 2 = 3(x - 2)$ for y.

4. Solve $Ax + By + 3 = 0$ for x.

We now apply the concept of slope to write the equation of a line given sufficient information. We also will use slope as an aid in graphing lines and writing equations of lines representing real-world data.

 Find the point-slope equation of a line with a given slope and passing through a given point.

Suppose that the line shown in Figure 2-26 on the next page has a slope of m and passes through the point (x_1, y_1). If (x, y) is a second point on the line, we have

$$m = \frac{y - y_1}{x - x_1}$$

and if we multiply both sides by $(x - x_1)$, the denominator, we have

$$y - y_1 = m(x - x_1)$$

Figure 2-26

Because this new equation displays the coordinates of the point (x_1, y_1) on the line and the slope m of the line, it is called the **point-slope form** of the equation of a line.

POINT-SLOPE FORM	The point-slope form of the line passing through $P(x_1, y_1)$ and with slope m is
	$$y - y_1 = m(x - x_1)$$

EXAMPLE 1   Write the point-slope equation of the line with a slope of $-\frac{2}{3}$ and passing through $(-4, 5)$.

Solution   We substitute $-\frac{2}{3}$ for m, -4 for x_1, and 5 for y_1 into the point-slope form and simplify.

$$y - y_1 = m(x - x_1)$$

$$y - 5 = -\frac{2}{3}[x - (-4)] \quad \text{Substitute.}$$

$$y - 5 = -\frac{2}{3}(x + 4) \qquad -(-4) = 4$$

The point-slope equation of the line is $y - 5 = -\frac{2}{3}(x + 4)$.

SELF CHECK 1   Write the point-slope equation of the line with slope of $\frac{5}{4}$ and passing through $(0, 5)$.

EXAMPLE 2   Write the point-slope equation of the line passing through $(-5, 4)$ and $(8, -6)$. Then solve the equation for y.

Solution   First we find the slope of the line.

$$m = \frac{y_2 - y_1}{x_2 - x_1}$$

$$= \frac{-6 - 4}{8 - (-5)} \quad \text{Substitute } -6 \text{ for } y_2, \text{ 4 for } y_1, \text{ 8 for } x_2, \text{ and } -5 \text{ for } x_1.$$

$$= -\frac{10}{13} \qquad \text{Simplify.}$$

Because the line passes through both points, we can choose either one and substitute its coordinates into the point-slope form. If we choose $(-5, 4)$, we substitute -5 for x_1, 4 for y_1, and $-\frac{10}{13}$ for m and proceed as follows.

$$y - y_1 = m(x - x_1)$$

$$y - 4 = -\frac{10}{13}[x - (-5)] \quad \text{Substitute.}$$

$$y - 4 = -\frac{10}{13}(x + 5) \qquad -(-5) = 5$$

To solve the equation for y, we proceed as follows:

$$y - 4 = -\frac{10}{13}x - \frac{50}{13} \quad \text{Use the distributive property to remove parentheses.}$$

$$y = -\frac{10}{13}x + \frac{2}{13} \quad \text{Add 4 to both sides and simplify. } 4 = \frac{52}{13}, \frac{52}{13} - \frac{50}{13} = \frac{2}{13}$$

The equation of the line is $y = -\frac{10}{13}x + \frac{2}{13}$.

SELF CHECK 2   Write the equation of the line passing through the points $(-2, 5)$ and $(4, -3)$ in point-slope form, and then solve for y.

2   Write the equation in slope-intercept form of a line with a given slope and passing through a given point.

Since the y-intercept of the line shown in Figure 2-27 is the point $(0, b)$, we can write the equation of the line by substituting 0 for x_1 and b for y_1 in the point-slope form and then simplifying.

$$y - y_1 = m(x - x_1) \quad \text{This is the point-slope form of the equation of a line.}$$

$$y - b = m(x - 0) \quad \text{Substitute } b \text{ for } y_1 \text{ and 0 for } x_1.$$

$$y - b = mx \quad \text{Multiply.}$$

$$y = mx + b \quad \text{Add } b \text{ to both sides.}$$

Because $y = mx + b$ displays the slope m and the y-coordinate b of the y-intercept, it is called the **slope-intercept form** of the equation of a line.

Figure 2-27

SLOPE-INTERCEPT FORM   The slope-intercept form of a line with slope m and y-intercept $(0, b)$ is

$$y = mx + b.$$

EXAMPLE 3   Use the slope-intercept form to write an equation of the line with slope 4 that passes through the point $(5, 9)$.

Solution   Since we are given that $m = 4$ and that the ordered pair $(5, 9)$ satisfies the equation, we can substitute 5 for x, 9 for y, and 4 for m in the equation $y = mx + b$ and solve for b.

$$y = mx + b$$

$$9 = 4(5) + b \quad \text{Substitute.}$$

$$9 = 20 + b \quad \text{Multiply.}$$

$$-11 = b \quad \text{Subtract 20 from both sides.}$$

Because $m = 4$ and $b = -11$, the equation is $y = 4x - 11$.

SELF CHECK 3   Write the slope-intercept equation of the line with slope -2 and passing through the point $(-2, 8)$.

3 Graph a linear equation using the slope and *y*-intercept.

It is sometimes easier to graph a linear equation when it is written in slope-intercept form. For example, to graph $y = \frac{4}{3}x - 2$, we note that $b = -2$ so the *y*-intercept is $(0, b) = (0, -2)$. (See Figure 2-28.)

Because the slope of the line is $\frac{\Delta y}{\Delta x} = \frac{4}{3}$, we can locate another point on the line by starting at the point $(0, -2)$ and counting 3 units to the right and 4 units up. The line joining the two points is the graph of the equation.

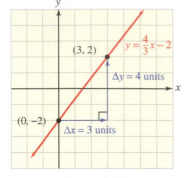

Figure 2-28

EXAMPLE 4 Find the slope and the *y*-intercept of the line with the equation $2(x - 3) = -3(y + 5)$. Then graph the line.

Solution We write the equation in the form $y = mx + b$ to find the slope m and the *y*-intercept $(0, b)$.

$$2(x - 3) = -3(y + 5)$$
$$2x - 6 = -3y - 15 \qquad \text{Use the distributive property to remove parentheses.}$$
$$2x + 3y - 6 = -15 \qquad \text{Add } 3y \text{ to both sides.}$$
$$3y - 6 = -2x - 15 \qquad \text{Subtract } 2x \text{ from both sides.}$$
$$3y = -2x - 9 \qquad \text{Add 6 to both sides.}$$
$$y = -\frac{2}{3}x - 3 \qquad \text{Divide both sides by 3 and simplify.}$$

Figure 2-29

The slope is $-\frac{2}{3}$, and the *y*-intercept is $(0, -3)$. To draw the graph, we plot the *y*-intercept $(0, -3)$ and then locate a second point on the line by moving 3 units to the right and 2 units down. We draw a line through the two points to obtain the graph shown in Figure 2-29.

SELF CHECK 4 Find the slope and the *y*-intercept of the line with the equation $2(y - 1) = 3x + 2$ and graph the line.

4 Determine whether two linear equations define lines that are parallel, perpendicular, or neither.

EXAMPLE 5 Show that the lines represented by $4x + 8y = 10$ and $2x = 12 - 4y$ are parallel.

Solution In the previous section, we saw that distinct lines are parallel when their slopes are equal. To see whether this is true in this case, we can solve each equation for *y*.

$$4x + 8y = 10 \qquad\qquad 2x = 12 - 4y$$
$$8y = -4x + 10 \qquad\qquad 4y = -2x + 12$$
$$y = -\frac{1}{2}x + \frac{5}{4} \qquad\qquad y = -\frac{1}{2}x + 3$$

Because the y-intercepts $\left(0, \frac{5}{4}\right)$ and $(0, 3)$ are different, the lines are distinct. Since the lines are distinct and have the same slope $\left(-\frac{1}{2}\right)$, they are parallel.

SELF CHECK 5   Are lines represented by $3x - 2y = 4$ and $6x = 4(y + 1)$ parallel?

EXAMPLE 6   Show that the lines represented by $4x + 8y = 10$ and $4x - 2y = 21$ are perpendicular.

Solution   Since two lines are perpendicular when their slopes are negative reciprocals, we solve each equation for y to show that the slopes of their graphs are negative reciprocals.

$$4x + 8y = 10 \qquad\qquad\qquad 4x - 2y = 21$$
$$8y = -4x + 10 \qquad\qquad -2y = -4x + 21$$
$$y = -\frac{1}{2}x + \frac{5}{4} \qquad\qquad y = 2x - \frac{21}{2}$$

Since the slopes are $-\frac{1}{2}$ and 2 (which are negative reciprocals), the lines are perpendicular.

SELF CHECK 6   Are lines represented by $3x + 2y = 6$ and $2x - 3y = 6$ perpendicular?

---

5   Write an equation of the line passing through a given point and parallel or perpendicular to a given line.

We will now use the slope properties of parallel and perpendicular lines to write equations of lines.

EXAMPLE 7   Write the equation of the line passing through $(-2, 5)$ and parallel to the line $y = 8x - 3$.

Solution   Since the equation is solved for y, the slope of the line given by $y = 8x - 3$ is the coefficient of x, which is 8. Since the desired equation is to have a graph that is parallel to the graph of $y = 8x - 3$, its slope also must be 8.

We substitute -2 for x_1, 5 for y_1, and 8 for m in the point-slope form and simplify.

$$y - y_1 = m(x - x_1)$$
$$y - 5 = 8[x - (-2)] \qquad \text{Substitute 5 for } y_1, \text{ 8 for } m, \text{ and } -2 \text{ for } x_1.$$
$$y - 5 = 8(x + 2) \qquad -(-2) = 2$$
$$y - 5 = 8x + 16 \qquad \text{Use the distributive property to remove parentheses.}$$
$$y = 8x + 21 \qquad \text{Add 5 to both sides.}$$

The equation is $y = 8x + 21$.

SELF CHECK 7   Write the equation of the line passing through the origin and parallel to the line $y = 8x - 3$.

EXAMPLE 8   Write the equation of the line passing through $(-2, 5)$ and perpendicular to the line $y = 8x - 3$.

Solution   The slope of the given line is 8. Thus, the slope of the desired line must be $-\frac{1}{8}$, which is the negative reciprocal of 8.

We substitute -2 for x_1, 5 for y_1, and $-\frac{1}{8}$ for m into the point-slope form and simplify:

$$y - \mathbf{y_1} = \mathbf{m}(x - \mathbf{x_1})$$

$$y - \mathbf{5} = -\frac{1}{8}[x - (\mathbf{-2})] \quad \text{Substitute.}$$

$$y - 5 = -\frac{1}{8}(x + 2) \quad \quad -(-2) = 2$$

$$y - 5 = -\frac{1}{8}x - \frac{1}{4} \quad \quad \text{Use the distributive property to remove parentheses.}$$

$$y = -\frac{1}{8}x - \frac{1}{4} + 5 \quad \text{Add 5 to both sides.}$$

$$y = -\frac{1}{8}x + \frac{19}{4} \quad \quad \text{Combine like terms.}$$

The equation is $y = -\frac{1}{8}x + \frac{19}{4}$.

SELF CHECK 8   Write the equation of the line passing through $(2, 4)$ and perpendicular to the line $y = 8x - 3$.

We summarize the various forms for the equation of a line in Table 2-1.

TABLE 2-1	
Point-slope form of a linear equation	$y - y_1 = m(x - x_1)$ The slope is m, and the line passes through (x_1, y_1).
Slope-intercept form of a linear equation	$y = mx + b$ The slope is m, and the y-intercept is $(0, b)$.
General form of a linear equation	$Ax + By = C$ A and B cannot both be 0.
A horizontal line	$y = b$ The slope is 0, and the y-intercept is $(0, b)$.
A vertical line	$x = a$ The slope is undefined, and the x-intercept is $(a, 0)$.

6   Write an equation of a line representing real-world data.

For tax purposes, many businesses use straight-line depreciation to find the declining value of aging equipment.

EXAMPLE 9   **VALUE OF A LATHE**   The owner of a machine shop buys a lathe for $1,970 and expects it to last for ten years. It can then be sold as scrap for an estimated **salvage value** of $270. If y represents the value of the lathe after x years of use, and y and x are related by the equation of a line,

a. Write the equation of the line.

b. Find the value of the lathe after $2\frac{1}{2}$ years.

c. State the economic meaning of the y-intercept of the line.

d. State the economic meaning of the slope of the line.

Solution   **a.** To find the equation of the line, we need to find
its slope and a point on the line.

COMMENT   When the size
of data is large, we can insert a
⫽ (break) symbol on the *x*- or
y-axis to indicate that the scale
does not begin until the first
value is listed.

When the lathe is new, its age *x* is 0, and its
value *y* is \$1,970. When the lathe is 10 years old,
$x = 10$ and its value is $y = \$270$. Since the line
passes through the points $(0, 1{,}970)$ and $(10, 270)$,
as shown in Figure 2-30, the slope of the line is

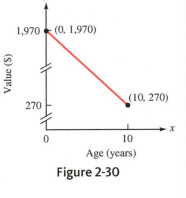

Figure 2-30

$$m = \frac{y_2 - y_1}{x_2 - x_1}$$
$$= \frac{270 - 1{,}970}{10 - 0}$$
$$= \frac{-1{,}700}{10}$$
$$= -170$$

To find the equation of the line, we substitute -170 for m, 0 for x_1, and 1,970 for
y_1 into the point-slope form and solve for *y*.

$$y - y_1 = m(x - x_1)$$
$$y - 1{,}970 = -170(x - 0)$$
$$y = -170x + 1{,}970$$

The current value *y* of the lathe is related to its age *x* by the equation
$y = -170x + 1{,}970$.

b. To find the value of the lathe after $2\frac{1}{2}$ years, we substitute 2.5 for *x* in the equation
and solve for *y*.

$$y = -170x + 1{,}970$$
$$= -170(2.5) + 1{,}970$$
$$= -425 + 1{,}970$$
$$= 1{,}545$$

After $2\frac{1}{2}$ years, the lathe will be worth \$1,545.

c. The *y*-intercept of the graph is $(0, b)$, where *b* is the value of *y* when $x = 0$.

$$y = -170x + 1{,}970$$
$$y = -170(0) + 1{,}970$$
$$y = 1{,}970$$

Thus, the *y* coordinate of the *y* intercept represents the lathe's original cost, \$1,970.

d. Each year, the value of the lathe decreases by \$170, because the slope of the line is
-170. The slope of the line is the annual depreciation rate.

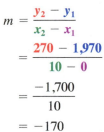 **SELF CHECK 9**   Find the value of the lathe after 6 years.

In statistics, the process of using one variable to predict another is called *regression*.
For example, if we know a man's height, we can make a good prediction about his weight,
because taller men usually weigh more than shorter men.

Figure 2-31 on the next page shows the result of sampling ten men at random and recording their heights and weights. The graph of the ordered pairs (h, w) is called a *scattergram*.

Man	Height in inches	Weight in pounds
1	66	140
2	68	150
3	68	165
4	70	180
5	70	165
6	71	175
7	72	200
8	74	190
9	75	210
10	75	215

(a)

(b)

Figure 2-31

To write a prediction equation (sometimes called a *regression equation*), we must find the equation of the line that comes closer to all of the points in the scattergram than any other possible line. There are exact methods to find this equation, but we can only approximate it here.

To write an approximation of the regression equation, we place a straightedge on the scattergram shown in Figure 2-31 and draw the line joining two points that seems to best fit all the points (the line that appears to minimize the distances between the data points and the line). In the figure, line PQ is drawn, where point P has coordinates of $(66, 140)$ and point Q has coordinates of $(75, 210)$.

Our approximation of the regression equation will be the equation of the line passing through points P and Q. To find the equation of this line, we first find its slope.

$$m = \frac{y_2 - y_1}{x_2 - x_1}$$

$$= \frac{210 - 140}{75 - 66}$$

$$= \frac{70}{9}$$

We can then use point-slope form to find its equation.

$$y - y_1 = m(x - x_1)$$

$$y - 140 = \frac{70}{9}(x - 66) \qquad \text{Choose } (66, 140) \text{ for } (x_1, y_1).$$

$$y = \frac{70}{9}x - \frac{4{,}620}{9} + 140 \qquad \text{Use the distributive property to remove parentheses and add 140 to both sides.}$$

$$y = \frac{70}{9}x - \frac{1{,}120}{3} \qquad \text{Combine like terms.}$$

Our approximation of the regression equation is $y = \frac{70}{9}x - \frac{1{,}120}{3}$.

To predict the weight of a man who is 73 inches tall, for example, we substitute 73 for x in the regression equation and simplify.

$$y = \frac{70}{9}\,x - \frac{1{,}120}{3}$$

$$y = \frac{70}{9}\,(73) - \frac{1{,}120}{3}$$

$$y \approx 194.4$$

We would predict that a 73-inch-tall man chosen at random will weigh about 194 pounds.

SELF CHECK ANSWERS

1. $y - 5 = \frac{5}{4}x$ **2.** $y = -\frac{4}{3}x + \frac{7}{3}$ **3.** $y = -2x + 4$
4. $m = \frac{3}{2}, (0, 2)$

5. yes   **6.** yes   **7.** $y = 8x$ **8.** $y = -\frac{1}{8}x + \frac{17}{4}$ **9.** After 6 years the lathe will be worth $950.

NOW TRY THIS

1. Which of the following graphs could be the graph of $y = -px - q$ if p and q are positive values?

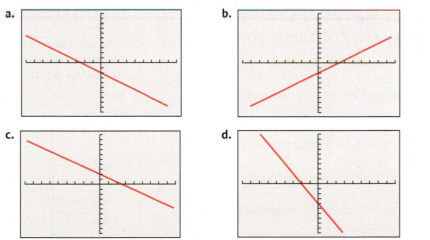

2. Write an equation of a line in slope-intercept form (if possible) with the given information.

   a. parallel to $y = -3$ through $(4, 6)$

   b. perpendicular to $x = 4$ through $(0, 0)$

   c. parallel to $3x = 2y + 8$ through $(-4, -6)$

2.3   Exercises

WARM-UPS   *Solve for y.*

1. $2x - 3y = 6$ **2.** $5x + 2y = 10$

Find the slope of the line passing through the given two points.

3. $(-1, 5)$ and $(6, 8)$

4. $(4, -3)$ and $(-2, 7)$

State the slope of each line whose equation is

5. $x = 3$ **6.** $y = -2$

REVIEW EXERCISES   *Solve each equation.*

7. $4x + 3 = 2(x - 5) + 7$ **8.** $12b + 6(3 - b) = b + 3$

9. $\dfrac{5(2 - x)}{3} - 1 = x + 5$ **10.** $\dfrac{r - 1}{3} = \dfrac{r + 2}{6} + 2$

11. Mixing alloys   In 60 ounces of alloy for watch cases, there are 20 ounces of gold. How much copper must be added to the alloy so that a watch case weighing 4 ounces, made from the new alloy, will contain exactly 1 ounce of gold?

12. Mixing coffee   To make a mixture of 80 pounds of coffee worth $272, a grocer mixes coffee worth $3.25 a pound with coffee worth $3.85 a pound. How many pounds of the cheaper coffee should the grocer use?

VOCABULARY AND CONCEPTS   *Fill in the blanks.*

13. The point-slope form of the equation of a line is
_____.

14. The slope-intercept form of the equation of a line is
_____.

15. The general form of the equation of a line is _____.

16. Two nonvertical lines are parallel when they have the _____ slope.

17. If the slopes of two lines are negative reciprocals, the lines are _____.

18. The process that recognizes that equipment loses value with age is called _____.

GUIDED PRACTICE   *Use point-slope form to write the equation of the line with the given properties. SEE EXAMPLE 1. (OBJECTIVE 1)*

19. $m = \dfrac{1}{2}$, passing through $(1, 5)$

20. $m = -\dfrac{2}{5}$, passing through $(-7, 3)$

21. $m = -3$, passing through $(2, 0)$

22. $m = -\dfrac{1}{3}$, passing through $(-4, -2)$

Use point-slope form to write the equation of the line passing through the two given points and solve for y. SEE EXAMPLE 2. (OBJECTIVE 1)

23. $P(2, 5), Q(4, 8)$
24. $P(-3, -7), Q(-6, -2)$
25. $P(3, 4), Q(0, -3)$
26. $P(4, 0), Q(6, -8)$

Use slope-intercept form to write the equation of the line with the given properties. SEE EXAMPLE 3. (OBJECTIVE 2)

27. $m = 5, b = -12$
28. $m = -2, b = 11$
29. $m = -7$, passing through $(7, 5)$
30. $m = 3$, passing through $(-2, -5)$
31. $m = 0$, passing through $(2, -4)$
32. $m = -7$, passing through the origin
33. passing through $(3, -7)$ and $(-5, 2)$
34. passing through $(-4, 5)$ and $(2, -6)$

Find the slope and the y-intercept of the line determined by the given equation. SEE EXAMPLE 4. (OBJECTIVE 3)

35. $5x + 2y = 10$ **36.** $3x - 4y = 8$

37. $-2(x + 3y) = 5$ **38.** $5(2x - 3y) = 4$

Write each equation in slope-intercept form to find the slope and the y-intercept. Then use the slope and y-intercept to graph the line. SEE EXAMPLE 4. (OBJECTIVE 3)

39. $y + 1 = x$ **40.** $x + y = 2$

41. $x = \dfrac{3}{2}y - 3$ **42.** $x = -\dfrac{4}{5}y + 2$

Determine whether the graphs of each pair of equations are parallel, perpendicular, or neither parallel nor perpendicular. SEE EXAMPLES 5–6. (OBJECTIVE 4)

43. $y = 3x + 4$, $y = 3x - 7$

44. $y = \frac{2}{3}x - 5$, $y = \frac{3}{2}x + 7$

45. $x + y = 2$, $y = x + 5$

46. $x = y + 2$, $y = x + 3$

47. $y = 3x + 7$, $2y = 6x - 9$

48. $2x + 3y = 9$, $3x - 2y = 5$

49. $x = 3y + 4$, $y = -3x + 7$

50. $2x + 5y = 3$, $y = \frac{2}{5}x$

Write the equation of the line that passes through the given point and is parallel or perpendicular to the given line. Write the answer in slope-intercept form. SEE EXAMPLES 7–8. (OBJECTIVE 5)

51. $(0, 0)$, parallel to $y = 4x - 7$

52. $(0, 0)$, perpendicular to $x = -6y + 5$

53. $(2, 5)$, parallel to $4x - y = 7$

54. $(-6, 3)$, parallel to $y + 3x = -12$

55. $(1, -3)$, perpendicular to $y = 7x - 2$

56. $(-4, 6)$, parallel to $x = 4y - 5$

57. $(2, 5)$, perpendicular to $4x - y = 7$

58. $(-6, 3)$, perpendicular to $y + 3x = -12$

ADDITIONAL PRACTICE *Determine whether the graphs of each pair of equations are parallel, perpendicular, or neither parallel nor perpendicular.*

59. $4x + 5y = 20$, $5x - 4y = 20$

60. $9x - 12y = 17$, $3x - 4y = 17$

61. $2x + 3y = 12$, $6x + 9y = 32$

62. $5x + 6y = 30$, $6x + 5y = 24$

63. $y = -5$, $x = 2$

64. $y = -3$, $y = -7$

65. $x = \frac{y - 2}{3}$, $3(y - 3) + x = 0$

66. $2y = 8$, $3(2 + x) = 2(x + 2)$

Write the equation of the line that passes through the given point and is parallel or perpendicular to the given line. Write the answer in slope-intercept form.

67. $(4, -2)$, parallel to $x = \frac{5}{4}y - 2$

68. $(1, -5)$, parallel to $x = -\frac{3}{4}y + 5$

69. $(4, -2)$, perpendicular to $x = \frac{5}{4}y - 2$

70. $(1, -5)$, perpendicular to $x = -\frac{3}{4}y + 5$

Find the slope and y-intercept and then graph the line.

71. $x = \frac{2y - 4}{7}$

72. $3x + 4 = -\frac{2(y - 3)}{5}$

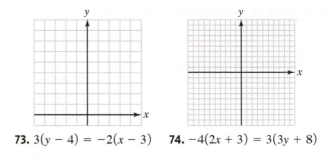

73. $3(y - 4) = -2(x - 3)$ **74.** $-4(2x + 3) = 3(3y + 8)$

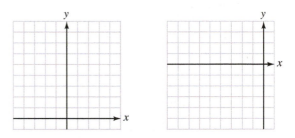

Write the equation of each line in slope-intercept form.

75.
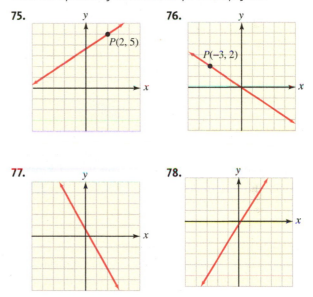
$P(2, 5)$

76.
$P(-3, 2)$

77.

78.

79. Write the equation of the line perpendicular to the line $y = 3$ and passing through the midpoint of the segment joining $(2, 4)$ and $(-6, 10)$.

80. Write the equation of the line parallel to the line $y = -8$ and passing through the midpoint of the segment joining $(-4, 2)$ and $(-2, 8)$.

81. Write the equation of the line parallel to the line $x = 3$ and passing through the midpoint of the segment joining $(2, -4)$ and $(8, 12)$.

82. Write the equation of the line perpendicular to the line $x = 3$ and passing through the midpoint of the segment joining $(-2, 2)$ and $(4, -8)$.

83. Solve $Ax + By = C$ for y and thereby show that the slope of its graph is $-\frac{A}{B}$ and its y-intercept is $\left(0, \frac{C}{B}\right)$.

84. Show that the *x*-intercept of the graph of $Ax + By = C$ is $\left(\frac{C}{A}, 0\right)$.

APPLICATIONS *For problems involving depreciation or appreciation, assume straight-line depreciation or straight-line appreciation.* SEE EXAMPLE 9. (OBJECTIVE 6)

85. Depreciation equations   A truck was purchased for $19,984. Its salvage value at the end of 8 years is expected to be $1,600. Find the depreciation equation.

86. Depreciation equations   A business purchased the computer shown. It will be depreciated over a 5-year period, when it will probably be worth $200. Find the depreciation equation.

$2,350

87. Art   In 1987, the painting *Rising Sunflowers* by Vincent van Gogh sold for $36,225,000. Suppose that an appraiser expected the painting to double in value in 20 years. Let *x* represent the time in years after 1987. Find the straight-line appreciation equation.

88. Real estate listings   Use the information given in the following description of the property to write a straight-line appreciation equation for the house.

Vacation Home
$122,000
Only 2 years old

• Great investment property!
• Expected to appreciate $4,000/yr

Sq ft: 1,635	Fam rm: yes	Den: no
Bdrm: 3	Ba: 1.5	Gar: enclosed
A/C: yes	Firepl: yes	Kit: built-ins

89. Appreciation equations   A famous oil painting was purchased for $250,000 and is expected to double in value in 5 years. Find the appreciation equation.

90. Appreciation equations   A house purchased for $142,000 is expected to double in value in 8 years. Find its appreciation equation.

91. Depreciation equations   Find the depreciation equation for the TV in the want ad in the illustration.

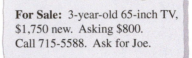

For Sale: 3-year-old 65-inch TV, $1,750 new. Asking $800. Call 715-5588. Ask for Joe.

92. Depreciating a lawn mower   A lawn mower cost $450 when new and is expected to last 10 years. What will it be worth in $6\frac{1}{2}$ years?

93. Salvage value   A copy machine that cost $1,750 when new will be depreciated at the rate of $180 per year. If the useful life of the copier is 7 years, find its salvage value.

94. Annual rate of depreciation   A machine that cost $47,600 when new will have a salvage value of $500 after its useful life of 15 years. Find its annual rate of depreciation.

95. Real estate   A vacation home is expected to appreciate about $4,000 a year. If the home will be worth $122,000 in 2 years, what will it be worth in 10 years?

96. Car repair   A garage charges a fixed amount, plus an hourly rate, to service a car. Use the information in the table to find the hourly rate.

A-1 Car Repair Typical charges	
2 hours	$143
5 hours	$320

97. Printer charges   A printer charges a fixed setup cost, plus $15 for every 100 copies. If 300 copies cost $75, how much will 1,000 copies cost?

98. Predicting burglaries   A police department knows that city growth and the number of burglaries are related by a linear equation. City records show that 575 burglaries were reported in a year when the local population was 77,000, and 675 were reported in a year when the population was 87,000. How many burglaries can be expected when the population reaches 110,000?

WRITING ABOUT MATH

99. Explain how to find the equation of a line passing through two given points.

100. In straight-line depreciation, explain why the slope of the line is called the *rate of depreciation*.

SOMETHING TO THINK ABOUT *Investigate the properties of the slope and the y-intercept by experimenting with the following problems.*

101. Graph $y = mx + 2$ for several positive values of *m*. What do you notice?

102. Graph $y = mx + 2$ for several negative values of *m*. What do you notice?

103. Graph $y = 2x + b$ for several increasing positive values of b. What do you notice?

104. Graph $y = 2x + b$ for several decreasing negative values of b. What do you notice?

105. How will the graph of $y = \frac{1}{2}x + 5$ compare to the graph of $y = \frac{1}{2}x - 5$?

106. How will the graph of $y = \frac{1}{2}x - 5$ compare to the graph of $y = \frac{1}{2}x$?

107. If the graph of $y = ax + b$ passes through quadrants I, II, and IV, what can be known about the constants a and b?

108. The graph of $Ax + By = C$ passes only through quadrants I and IV. What is known about the constants A, B, and C?

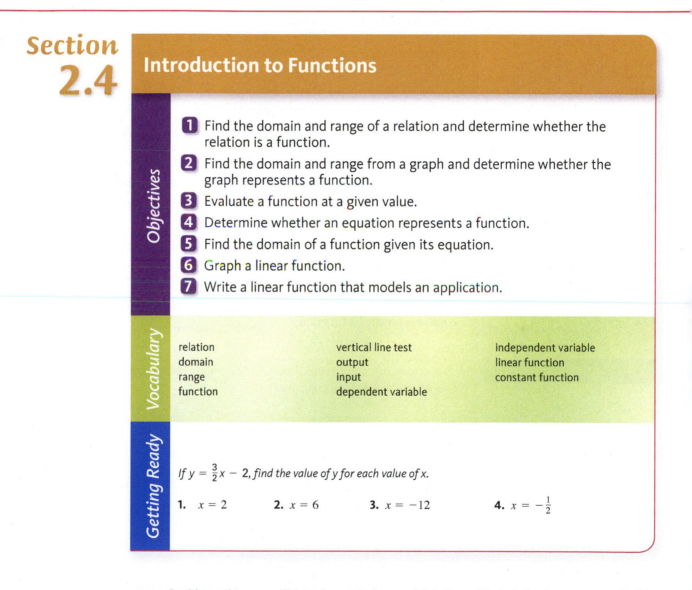

Section 2.4

Introduction to Functions

Objectives

1. Find the domain and range of a relation and determine whether the relation is a function.
2. Find the domain and range from a graph and determine whether the graph represents a function.
3. Evaluate a function at a given value.
4. Determine whether an equation represents a function.
5. Find the domain of a function given its equation.
6. Graph a linear function.
7. Write a linear function that models an application.

Vocabulary

relation	vertical line test	independent variable
domain	output	linear function
range	input	constant function
function	dependent variable	

Getting Ready

If $y = \frac{3}{2}x - 2$, find the value of y for each value of x.

1. $x = 2$ **2.** $x = 6$ **3.** $x = -12$ **4.** $x = -\frac{1}{2}$

In this section, we will introduce *relations* and *functions*. We include these concepts in this chapter because they involve ordered pairs. The concept of *function* is one of the most important ideas in mathematics. The farther you go in mathematics, the more you will study functions.

 1 ### Find the domain and range of a relation and determine whether the relation is a function.

Table 2-2 on the next page shows the number of women serving in the U.S. House of Representatives for several recent sessions of Congress.

TABLE 2-2						
Women in the U.S. House of Representatives						
Session of Congress	107th	108th	109th	110th	111th	112th
Number of Female Representatives	59	59	68	71	76	75

We can display the data in the table as sets of ordered pairs, where the *first component* represents the session of Congress and the *second component* represents the number of women serving in that session.

$$(107, 59) \quad (108, 59) \quad (109, 68) \quad (110, 71) \quad (111, 76) \quad (112, 75)$$

Sets of ordered pairs like these are called **relations**. The set of all first components $\{107, 108, 109, 110, 111, 112\}$ is called the **domain of the relation**, and the set of all second components $\{59, 68, 71, 75, 76\}$ is called the **range of the relation**. Although 59 occurs twice as a second component, we list it only once in the range.

EXAMPLE 1 Find the domain and range of the relation $\{(3, 2), (5, -7), (-8, 2), (-9, -12)\}$.

Solution Since the set of first components is the domain, the domain can be written as

D: $\{3, 5, -8, -9\}$

Since the set of second components is the range, the range can be written as

R: $\{2, -7, -12\}$ The second component 2 is listed only once in the range.

SELF CHECK 1 Find the domain and range of the relation $\{(5, 6), (-12, 4), (8, 6), (5, 4)\}$.

When each first component in a relation determines exactly one second component, the relation is called a **function**.

FUNCTION

A function is any set of ordered pairs (a relation) in which each first component determines exactly one second component.

EXAMPLE 2 Determine whether each relation defines a function.

a.
4 → 3
6 → 5
12 → 9

b.
x	y
8	2
1	4
8	6

c. $\{(-2, 3), (-1, 3), (0, 3), (1, 3)\}$

Solution **a.** The arrow diagram defines a function because each value in the first oval determines exactly one value in the second oval.

- $4 \rightarrow 3$ To the 4 in the first oval, there corresponds exactly one value, 3, in the second oval.

- $6 \rightarrow 5$ To the 6 in the first oval, there corresponds exactly one value, 5, in the second oval.

- $12 \rightarrow 9$ To the 12 in the first oval, there corresponds exactly one value, 9, in the second oval.

b. The table does not define a function, because the value 8 in the first column determines more than one value in the second column.

• In the first row, to the number 8, there corresponds the value 2.

• In the third row, to the number 8, there corresponds the value 6.

c. Since the first component of each ordered pair determines exactly one value for the second component, the set of ordered pairs is a function.

• $(-2, 3)$ To the number -2, there corresponds exactly one value, 3.

• $(-1, 3)$ To the number -1, there corresponds exactly one value, 3.

• $(0, 3)$ To the number 0, there corresponds exactly one value, 3.

• $(1, 3)$ To the number 1, there corresponds exactly one value, 3.

The results from parts (b) and (c) illustrate an important fact:

> *Two different ordered pairs of a function can have the same second component, but they cannot have the same first component.*

SELF CHECK 2   Determine whether each relation is a function.

a.
$$\begin{array}{c} 0 \rightarrow 2 \\ 9 \rightarrow 3 \\ \searrow 4 \end{array}$$

b.
x	y
-1	-60
0	55
3	0

c. $\{(4, -1), (9, 2), (16, 15), (4, -4)\}$

---

2   Find the domain and range from a graph and determine whether the graph represents a function.

A **vertical line test** can be used to determine whether the graph of an equation represents a function. If every vertical line that intersects a graph does so exactly once, the graph represents a function, because every number x determines a single value of y. If any vertical line that intersects a graph does so more than once, the graph cannot represent a function, because some number x determines more than one value of y.

The graph in Figure 2-32(a) represents a function, because every vertical line that intersects the graph does so exactly once. The graph in Figure 2-32(b) does not represent a function, because some vertical lines intersect the graph more than once.

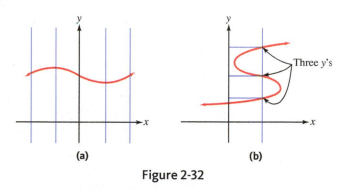

(a)   (b)

Figure 2-32

EXAMPLE 3   Find the domain and range of the relation determined by each graph and then tell whether the graph defines a function.

a. To find the domain, we look at the leftmost point on the graph in Figure 2-33 and identify -3 as its x-coordinate. Since the graph continues forever to the right, there is no rightmost point. Therefore, the domain $[-3, \infty)$ is expressed in interval notation.

   To find the range, we look for the lowest point on the graph and identify -4 as its y-coordinate. Since the

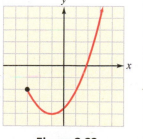

Figure 2-33

graph continues upward forever, there is no highest point. Therefore, the range is $[-4, \infty)$.

Since every vertical line that intersects the graph will do so exactly once, the vertical line test indicates that the graph is a function.

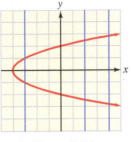

b. To find the domain, we note that the x-coordinate of the leftmost point in Figure 2-34 is -4 and that there is no rightmost point. Therefore, the domain is $[-4, \infty)$.

To find the range, we note that there is no lowest point or highest point. Therefore, the range is $(-\infty, \infty)$.

Since many vertical lines that intersect the graph will do so more than once, the vertical line test indicates that the graph is not a function.

Figure 2-34

 SELF CHECK 3   Find the domain and range of the relation determined by the graph and then tell whether the graph defines a function.

---

3 **Evaluate a function at a given value.**

We can use a special notation to denote functions.

The notation $y = f(x)$ denotes that the variable y is a function of x.

COMMENT The notation $f(x)$ does not mean "f times x."

The notation $y = f(x)$ is read as "y equals f of x." Note that y and $f(x)$ are two notations for the same quantity. Thus, the equations $y = 4x + 3$ and $f(x) = 4x + 3$ are equivalent.

To see why function notation is helpful, consider the following equivalent statements:

1. In the equation $y = 4x + 3$, find the value of y when x is 3.
2. If $f(x) = 4x + 3$, find $f(3)$.

Statement 2, which uses $f(x)$ notation, is much more concise.

The notation $y = f(x)$ provides a way of denoting the value of y that corresponds to some number x. Since the value of y usually depends on the number x, we call y the **dependent variable** and x the **independent variable**. For example, if $y = f(x)$, the value of y that is determined by $x = 3$ is denoted by $f(3)$.

EXAMPLE 4   Let $f(x) = 4x + 3$. Find: **a.** $f(3)$ **b.** $f(-1)$ **c.** $f(0)$
d. the value of x for which $f(x) = 7$

Solution   **a.** We replace x with 3.

$$f(\boldsymbol{x}) = 4\boldsymbol{x} + 3$$
$$f(\boldsymbol{3}) = 4(\boldsymbol{3}) + 3$$
$$= 12 + 3$$
$$= 15$$

b. We replace x with -1.

$$f(\boldsymbol{x}) = 4\boldsymbol{x} + 3$$
$$f(\boldsymbol{-1}) = 4(\boldsymbol{-1}) + 3$$
$$= -4 + 3$$
$$= -1$$

c. We replace x with 0.

$$f(x) = 4x + 3$$
$$f(0) = 4(0) + 3$$
$$= 3$$

d. We replace $f(x)$ with 7.

$$f(x) = 4x + 3$$
$$7 = 4x + 3$$
$$4 = 4x \qquad \text{Subtract 3 from both sides.}$$
$$x = 1 \qquad \text{Divide each side by 4.}$$

SELF CHECK 4   Let $f(x) = -2x - 1$. Find: **a.** $f(2)$ **b.** $f(-3)$ **c.** the value of x for which $f(x) = 5$

We can think of a function as a machine that takes some input x and turns it into some output $f(x)$, as shown in Figure 2-35(a). The machine shown in Figure 2-35(b) turns the input number 2 into the output value -3 and turns the input number 6 into the output value -11. The set of numbers that we can put into the machine is the domain of the function, and the set of numbers that comes out is the range.

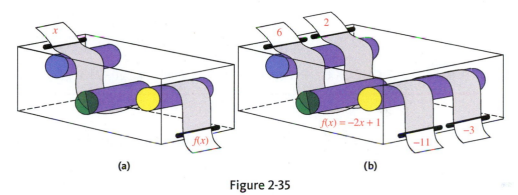

(a)   **(b)**

Figure 2-35

The letter f used in the notation $y = f(x)$ represents the word *function*. However, other letters can be used to represent functions. The notations $y = g(x)$ and $y = h(x)$ also denote functions involving the independent variable x.

In Example 5, the equation $y = g(x) = x^2 - 2x$ determines a function, because every possible value of x gives a single value of $g(x)$.

EXAMPLE 5   Let $g(x) = x^2 - 2x$. Find: **a.** $g\left(\frac{2}{5}\right)$ **b.** $g(s)$ **c.** $g(s^2)$ **d.** $g(-t)$

Solution   **a.** We replace x with $\frac{2}{5}$.

$$g(x) = x^2 - 2x$$
$$g\left(\frac{2}{5}\right) = \left(\frac{2}{5}\right)^2 - 2\left(\frac{2}{5}\right)$$
$$= \frac{4}{25} - \frac{4}{5}$$
$$= -\frac{16}{25}$$

b. We replace x with s.

$$g(x) = x^2 - 2x$$
$$g(s) = s^2 - 2s$$
$$= s^2 - 2s$$

c. We replace x with s^2.

$$g(x) = x^2 - 2x$$
$$g(s^2) = (s^2)^2 - 2s^2$$
$$= s^4 - 2s^2$$

d. We replace x with $-t$.

$$g(x) = x^2 - 2x$$
$$g(-t) = (-t)^2 - 2(-t)$$
$$= t^2 + 2t$$

SELF CHECK 5   Let $h(x) = -x^2 + 3$. Find: **a.** $h\left(\frac{1}{2}\right)$ **b.** $h(-a)$

EXAMPLE 6   Let $f(x) = 4x - 1$. Find: **a.** $f(3) + f(2)$ **b.** $f(a) - f(b)$

Solution   **a.** We find $f(3)$ and $f(2)$ separately.

$$f(x) = 4x - 1 \qquad\qquad f(x) = 4x - 1$$
$$f(3) = 4(3) - 1 \qquad\quad f(2) = 4(2) - 1$$
$$= 12 - 1 \qquad\qquad\quad = 8 - 1$$
$$= 11 \qquad\qquad\qquad\quad = 7$$

We then add the results to obtain $f(3) + f(2) = 11 + 7 = 18$.

b. We find $f(a)$ and $f(b)$ separately.

$$f(x) = 4x - 1 \qquad\qquad f(x) = 4x - 1$$
$$f(a) = 4a - 1 \qquad\qquad f(b) = 4b - 1$$

We then subtract the results to obtain

$$f(a) - f(b) = (4a - 1) - (4b - 1)$$
$$= 4a - 1 - 4b + 1$$
$$= 4a - 4b$$

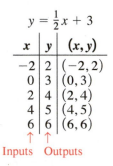 **SELF CHECK 6**   Let $g(x) = -2x + 3$. Find: **a.** $g(-2) + g(3)$ **b.** $g(a) - g(b)$

4   ## Determine whether an equation represents a function.

A function can be defined by an equation. For example, the equation $y = \frac{1}{2}x + 3$ sets up a rule in which each number x determines exactly one value of y. To find the value of y (called an **output value**) that corresponds to $x = 4$ (called an **input value**), we substitute 4 for x and simplify.

$$y = \frac{1}{2}x + 3$$

$$y = \frac{1}{2}(4) + 3 \qquad \text{Substitute the input value of 4 for } x.$$

$$= 2 + 3$$
$$= 5 \qquad\qquad \text{This is the output value.}$$

The ordered pair $(4, 5)$ satisfies the equation and shows that a y-value of 5 corresponds to an x-value of 4. This ordered pair and others that satisfy the equation appear in the table shown in Figure 2-36. The graph of the equation also appears in the figure.

$$y = \frac{1}{2}x + 3$$

x	y	(x, y)	
-2	2	$(-2, 2)$	A y-value of 2 corresponds to an x-value of -2.
0	3	$(0, 3)$	A y-value of 3 corresponds to an x-value of 0.
2	4	$(2, 4)$	A y-value of 4 corresponds to an x-value of 2.
4	5	$(4, 5)$	A y-value of 5 corresponds to an x-value of 4.
6	6	$(6, 6)$	A y-value of 6 corresponds to an x-value of 6.

Inputs   Outputs

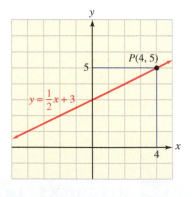

Figure 2-36

COMMENT Use the vertical line test to see that all linear equations represent functions except those of the form $x = c$.

To see how the table in Figure 2-36 determines the correspondence, we find an input in the x-column and then read across to find the corresponding output in the y-column. For example, if we select 2 as an input value, we get 4 as an output value. Thus, a y-value of 4 corresponds to an x-value of 2.

To see how the graph in Figure 2-36 determines the correspondence, we draw a vertical and a horizontal line through any point (say, point P) on the graph, as shown in the figure. Because these lines intersect the x-axis at 4 and the y-axis at 5, the point $P(4, 5)$ associates 5 on the y-axis with 4 on the x-axis. This shows that a y-value of 5 corresponds to an x-value of 4.

In this example, the set of all inputs x is the set of real numbers, $(-\infty, \infty)$. The set of all outputs y is also the set of real numbers, $(-\infty, \infty)$.

y IS A FUNCTION OF x	An equation, table, or graph in x and y in which each value of x (the *input*) determines exactly one value of y (the *output*) is a *function*. In this case, we say that y *is a function of x.*
	The set of all input values x is the domain of the function, and the set of all output values y is the range.

EXAMPLE 7 Does $y^2 = x$ define y to be a function of x? Find its domain and range.

Solution For a function to exist, each value of x must determine exactly one value of y. If we let $x = 16$, for example, y could be either 4 or -4, because $4^2 = 16$ and $(-4)^2 = 16$. Since more than one value of y is determined when $x = 16$, the equation does not represent a function.

A table of values and the graph appear in Figure 2-37.

COMMENT Notice that the graph does not pass the vertical line test. Thus, the equation $y^2 = x$ does not define a function.

$$y^2 = x$$

x	y	(x, y)
1	1	$(1, 1)$
0	0	$(0, 0)$
1	-1	$(1, -1)$
4	2	$(4, 2)$
4	-2	$(4, -2)$
16	4	$(16, 4)$
16	-4	$(16, -4)$

The inputs can be any nonnegative number. The outputs can be any real number.

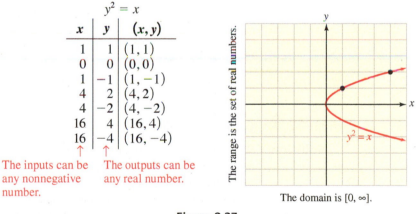

The range is the set of real numbers.

The domain is $[0, \infty]$.

Figure 2-37

Since the input x must be nonnegative, the domain is $[0, \infty)$. Since the output y can be any real number, the range is $(-\infty, \infty)$.

 SELF CHECK 7 Does $|y| = x$ define y to be a function of x? Find its domain and range.

5 **Find the domain of a function given its equation.**

The domain of a function that is defined by an equation is the set of all numbers that are permissible replacements for its independent variable.

EXAMPLE 8   Find the domain of the functions defined by

   a. $f(x) = x^2 + 8x - 3$ **b.** $f(x) = \dfrac{1}{x - 2}$.

Solution   **a.** Since any real number can be substituted for x in the function $f(x) = x^2 + 8x - 3$ to obtain a single value y, the domain is $(-\infty, \infty)$.

   b. The number 2 cannot be substituted for x in the function $f(x) = \dfrac{1}{x - 2}$, because that would make the denominator 0. However, any real number, except 2, can be substituted for x to obtain a single value y. Therefore, the domain is the set of all real numbers except 2. This is the interval $(-\infty, 2) \cup (2, \infty)$.

SELF CHECK 8   Find the domain of the function defined by the equation $y = \dfrac{2}{x + 3}$.

6   Graph a linear function.

The **graph of a function** is the graph of the ordered pairs $(x, f(x))$ that define the function. For the graph of the function shown in Figure 2-38, the domain is shown on the x-axis, and the range is shown on the y-axis. For any x in the domain, there corresponds one value $y = f(x)$ in the range.

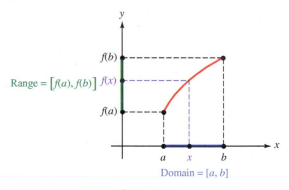

Figure 2-38

EXAMPLE 9   Graph the function $f(x) = -2x + 1$ and find its domain and range.

Solution   We graph the equation as in Figure 2-39. Since every real number x on the x-axis determines a corresponding value of y, the domain is the interval $(-\infty, \infty)$ shown on the x-axis. Since the values of y can be any real number, the range is the interval $(-\infty, \infty)$ shown on the y-axis.

Figure 2-39

SELF CHECK 9   Graph the function $f(x) = 3x - 1$ and find its domain and range.

In Section 2.1, we graphed equations whose graphs were lines. These equations define basic functions, called *linear functions*.

LINEAR FUNCTION	A **linear function** is a function defined by an equation that can be written in the form

$$f(x) = mx + b \qquad \text{or} \qquad y = mx + b$$

where m is the slope of the line graph and $(0, b)$ is the y-intercept.

EXAMPLE 10 Solve the equation $3x + 2y = 10$ for y to show that it defines a linear function. Then graph the function and find its domain and range.

Solution We solve the equation for y as follows:

$$3x + 2y = 10$$
$$2y = -3x + 10 \qquad \text{Subtract } 3x \text{ from both sides.}$$
$$y = -\frac{3}{2}x + 5 \qquad \text{Divide both sides by 2.}$$

Because the given equation can be written in the form $f(x) = mx + b$, it defines a linear function. The slope of its line graph is $-\frac{3}{2}$, and the y-intercept is $(0, 5)$. The graph appears in Figure 2-40. From the graph, we can see that both the domain and the range are the interval $(-\infty, \infty)$.

$3x + 2y = 10$

Figure 2-40

 SELF CHECK 10 Solve the equation $5x - 2y = 20$ for y to show that it defines a linear function.

A special case of a linear function is the **constant function**, defined by the equation $f(x) = b$, where b is a constant. Its graph, domain, and range are shown in Figure 2-41.

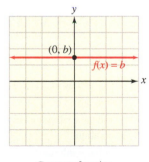

$(0, b)$

$f(x) = b$

Constant function
Domain: $(-\infty, \infty)$
Range: $\{b\}$

Figure 2-41

7 Write a linear function that models an application.

EXAMPLE 11 **CUTTING HAIR** A barber earns $26 per day plus $6.50 for each haircut she gives that day. Write a linear function that describes her daily income if she gives x haircuts each day.

Solution   The barber earns \$6.50 per haircut, so if she serves x customers a day, her earnings for haircuts will be \$6.50x. To find her total daily income, we must add \$26 to \$6.50x. Thus, her total daily income $I(x)$ is described by the function

$$I(x) = 6.50x + 26$$

🌱 **SELF CHECK 11**   Write a linear function that describes her daily income if she gets a raise of \$0.50 per haircut.

🌱 **SELF CHECK ANSWERS**

1. D: $\{5, -12, 8\}$; R: $\{6, 4\}$ **2. a.** no **b.** yes **c.** no   **3.** D: $(-\infty, \infty)$; R: $(-\infty, 0]$; yes   **4. a.** -5
b. 5 **c.** -3 **5. a.** $\frac{11}{4}$ **b.** $-a^2 + 3$ **6. a.** 4 **b.** $-2a + 2b$ **7.** no; domain $[0, \infty)$, range $(-\infty, \infty)$
8. $(-\infty, -3) \cup (-3, \infty)$ **9.**

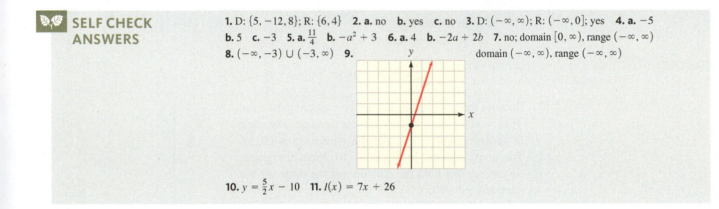

domain $(-\infty, \infty)$, range $(-\infty, \infty)$

10. $y = \frac{5}{2}x - 10$ **11.** $I(x) = 7x + 26$

NOW TRY THIS

Given $f(x) = 3x - 4$, *find:*

a. $f(x - 2)$ **b.** $f(x + h)$ **c.** $f(x + h) - f(x)$

2.4 Exercises 🌿

WARM-UPS *Given the table,*

x	y
-2	0
0	3
3	9

1. List the x-values in set notation.

2. List the y-values in set notation.

3. Write the table of values as ordered pairs in set notation.

4. Identify the y-intercept.

Solve each equation for y.

5. $3x - 2y = 8$

6. $4x - 3y = 9$

REVIEW *Solve each equation.*

7. $\dfrac{y + 2}{2} = 4(y + 2)$

8. $\dfrac{3z - 1}{6} - \dfrac{3z + 4}{3} = \dfrac{z + 3}{2}$

9. $\dfrac{2a}{3} + \dfrac{1}{2} = \dfrac{6a - 1}{6}$

10. $\dfrac{2x + 3}{5} - \dfrac{3x - 1}{3} = \dfrac{x - 1}{15}$

VOCABULARY AND CONCEPTS *Fill in the blanks.*

11. Any set of ordered pairs is called a _____.

12. A _____ is a correspondence between a set of input values and a set of output values, where each _____ value determines one _____ value.

13. In a function, the set of all input values is called the _____ of the function.

14. In a function, the set of all output values is called the _____ of the function.

15. The denominator of a fraction can never be _.

16. To decide whether a graph determines a function, use the _____.

17. If any vertical line intersects a graph more than once, the graph _____ represent a function.

18. A linear function is any function that can be written in the form _____.

19. In the function $f(x) = mx + b$, m is the _____ of its graph, and b is the y-coordinate of the _____.

Consider the function $f(x) = mx + b$. Fill in the blanks.

20. Any substitution for x is called an _____ value.

21. The value _ is called an output value.

22. The independent variable is _.

23. The dependent variable is _.

24. The notation $f(5)$ is the value of _ when $x = 5$.

GUIDED PRACTICE *State the domain and range of each relation and determine if it is a function.* **SEE EXAMPLES 1–2. (OBJECTIVE 1)**

25. $\{(6, -4), (-1, 3), (-2, 1), (5, 3)\}$

26. $\{(-2, 8), (-8, 4), (3, 8), (5, 4)\}$

27. $\{(-2, 3), (6, 8), (-2, 5), (5, 4)\}$

28. $\{(3, -2), (5, 2), (4, 5), (3, 0)\}$

State the domain and range of the relation determined by each graph in interval notation and determine whether it represents a function. **SEE EXAMPLE 3. (OBJECTIVE 2)**

29.

30.

31.

32.

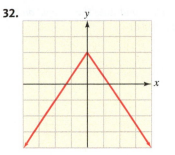

Find $f(3)$, $f(-1)$, and all values of x for which $f(x) = 0$. **SEE EXAMPLE 4. (OBJECTIVE 3)**

33. $f(x) = 3x$ **34.** $f(x) = -5x$

35. $f(x) = 2x - 3$ **36.** $f(x) = 3x - 5$

Find $f(2)$ and $f(3)$. **SEE EXAMPLE 4. (OBJECTIVE 3)**

37. $f(x) = x^2$ **38.** $f(x) = x^2 - 2$

39. $f(x) = x^3 - 3$ **40.** $f(x) = x^3$

Find $f(2)$ and $f(-2)$. **SEE EXAMPLE 4. (OBJECTIVE 3)**

41. $f(x) = |x| - 3$ **42.** $f(x) = |x| + 6$

43. $f(x) = x^2 - 2$ **44.** $f(x) = x^2 + 3$

Find $g(w)$ and $g(w + 1)$. **SEE EXAMPLE 5. (OBJECTIVE 3)**

45. $g(x) = 5x$ **46.** $g(x) = -4x$

47. $g(x) = 3x - 5$ **48.** $g(x) = 2x - 7$

Find each value given that $f(x) = 2x + 1$. **SEE EXAMPLE 6. (OBJECTIVE 3)**

49. $f(3) + f(2)$ **50.** $f(-5) - f(4)$

51. $f(b) - f(a)$ **52.** $f(b) + f(a)$

Determine whether the equation determines y to be a function of x. **SEE EXAMPLE 7. (OBJECTIVE 4)**

53. $y = 2x + 3$ **54.** $y = 3$

55. $y = 2x^2$ **56.** $y = 5x^2 - 2$

57. $y^2 = x + 1$ **58.** $y^2 = 3 - 2x$

59. $x = |y|$ **60.** $y = |x|$

State the domain of each function. **SEE EXAMPLE 8. (OBJECTIVE 5)**

61. $f(x) = x^2 - 4$ **62.** $f(x) = x^2 + 3x - 4$

63. $f(x) = \dfrac{1}{x + 5}$ **64.** $f(x) = \dfrac{5}{x + 1}$

65. $f(x) = \dfrac{1}{x + 3}$ **66.** $f(x) = \dfrac{3}{x - 4}$

67. $f(x) = \dfrac{x}{x^2 + 2}$ **68.** $f(x) = \dfrac{x}{x - 3}$

Draw the graph of each linear function and state the domain and range. **SEE EXAMPLES 9–10.** *(OBJECTIVE 6)*

69. $f(x) = 2x - 1$

70. $f(x) = -x + 2$

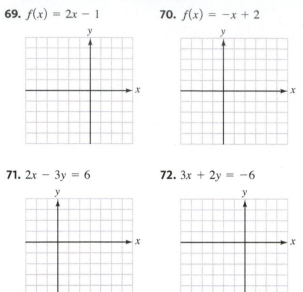

71. $2x - 3y = 6$

72. $3x + 2y = -6$

Determine whether each equation defines a linear function. **SEE EXAMPLES 9–10.** *(OBJECTIVE 6)*

73. $y = 4x^2 + 3$

74. $y = \dfrac{x - 3}{2}$

75. $x = 4y - 3$

76. $x = \dfrac{8}{y}$

ADDITIONAL PRACTICE *For each function, find $f(2)$ and $f(3)$.*

77. $f(x) = (x + 1)^2$

78. $f(x) = (x - 3)^2$

79. $f(x) = 2x^2 - x$

80. $f(x) = 5x^2 + 2x$

For each function, find $f(3)$ and $f(-1)$.

81. $f(x) = 7 + 5x$

82. $f(x) = 3 + 3x$

83. $f(x) = 9 - 2x$

84. $f(x) = 12 + 3x$

Find each value given that $f(x) = 2x + 1$.

85. $f(b) - 1$

86. $f(b) - f(1)$

87. $f(0) + f\left(-\tfrac{1}{2}\right)$

88. $f(a) + f(2a)$

Determine whether each relation defines a function.

89.

90.

91.

x	y
-1	4
0	6
1	3

92.

x	y
4	5
2	-1
4	3

State the domain and range of the relation determined by each graph in interval notation (if possible) and determine whether it represents a function.

93.

94.

95.

96.

97.

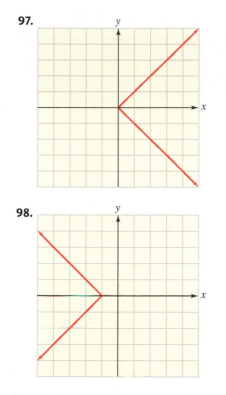

98.

APPLICATIONS *For Exercises 101 and 102, set up a linear equation to solve. SEE EXAMPLE 11. (OBJECTIVE 7)*

99. Selling DVD players An electronics firm manufactures portable DVD players, receiving $120 for each unit it makes. If x represents the number of units produced, the income received is determined by the *revenue function* $R(x) = 120x$. The manufacturer has fixed costs of $12,000 per month and variable costs of $57.50 for each unit manufactured. Thus, the *cost function* is $C(x) = 57.50x + 12,000$. How many DVD players must the company sell for revenue to equal cost?

100. Selling tires A tire company manufactures premium tires, receiving $130 for each tire it makes. If the manufacturer has fixed costs of $15,512.50 per month and variable costs of $93.50 for each tire manufactured, how many tires must the company sell for revenue to equal cost? (*Hint:* See Exercise 99.)

101. Selling hot dogs At a football game, a vendor sells hot dogs. He earns $50 per game plus $0.10 for each hot dog sold.

    a. Write a linear function that describes the vendor's income if he sells h hot dogs.

    b. Find his income if he sells 115 hot dogs.

102. Housing A housing contractor lists the following costs.

Fees and permits	$14,000
Cost per square foot	$102

    a. Write a linear function that describes the cost of building a house with s square feet.

    b. Find the cost to build a house having 1,800 square feet.

103. Ballistics A bullet shot straight up is s feet above the ground after t seconds, where $s = f(t) = -16t^2 + 256t$. Find the height of the bullet 3 seconds after it is shot.

104. Artillery fire A mortar shell is s feet above the ground after t seconds, where $s = f(t) = -16t^2 + 512t + 64$. Find the height of the shell 20 seconds after it is fired.

105. Dolphins See the illustration. The height h in feet reached by a dolphin t seconds after breaking the surface of the water is given by

$$h = -16t^2 + 32t$$

How far above the water will the dolphin be 1.5 seconds after a jump?

106. Police investigation The kinetic energy E of a moving object is given by $E = \frac{1}{2}mv^2$, where m is the mass of the object (in kilograms) and v is the object's velocity (in meters per second). Kinetic energy is measured in joules. Examining the damage done to a vehicle, a police investigator estimates that the velocity of a club with a 3-kilogram mass was 6 meters per second. Find the kinetic energy of the club.

107. Conversion from degrees Celsius to degrees Fahrenheit The temperature in degrees Fahrenheit that is equivalent to a temperature in degrees Celsius is given by the function $F(C) = \frac{9}{5}C + 32$. Find the Fahrenheit temperature that is equivalent to 25°C.

108. Conversion from degrees Fahrenheit to degrees Celsius The temperature in degrees Celsius that is equivalent to a temperature in degrees Fahrenheit is given by the function $C(F) = \frac{5}{9}F - \frac{160}{9}$. Find the Celsius temperature that is equivalent to 14°F.

WRITING ABOUT MATH

109. Explain the concepts of function, domain, and range.

110. Explain why the constant function is a special case of a linear function.

SOMETHING TO THINK ABOUT *Let $f(x) = 2x + 1$ and $g(x) = x^2$. Assume that $f(x) \neq 0$ and $g(x) \neq 0$.*

111. Is $f(x) + g(x)$ equal to $g(x) + f(x)$?

112. Is $f(x) - g(x)$ equal to $g(x) - f(x)$?

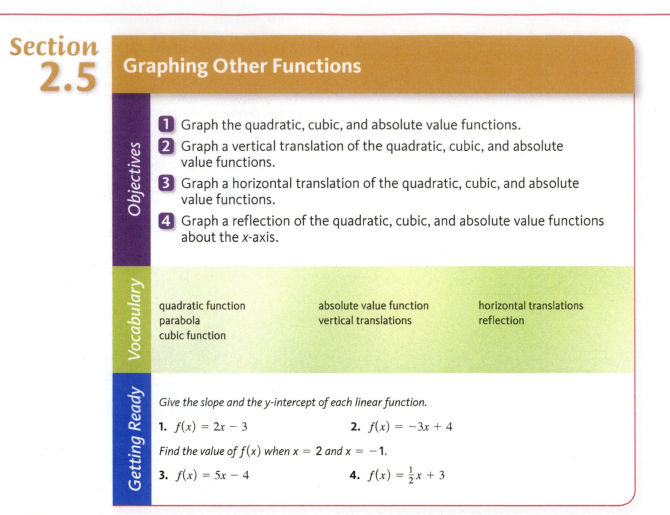

Graphing Other Functions

Objectives

1. Graph the quadratic, cubic, and absolute value functions.
2. Graph a vertical translation of the quadratic, cubic, and absolute value functions.
3. Graph a horizontal translation of the quadratic, cubic, and absolute value functions.
4. Graph a reflection of the quadratic, cubic, and absolute value functions about the *x*-axis.

Vocabulary

quadratic function
parabola
cubic function

absolute value function
vertical translations

horizontal translations
reflection

Getting Ready

Give the slope and the y-intercept of each linear function.

1. $f(x) = 2x - 3$ **2.** $f(x) = -3x + 4$

Find the value of $f(x)$ when $x = 2$ and $x = -1$.

3. $f(x) = 5x - 4$ **4.** $f(x) = \frac{1}{2}x + 3$

In the previous sections, we discussed *linear functions*, functions whose graphs are straight lines. We now extend the discussion to include nonlinear functions, functions whose graphs are not straight lines.

1   Graph the quadratic, cubic, and absolute value functions.

If f is a function whose domain and range are sets of real numbers, its graph is the set of all points $(x, f(x))$ in the xy-plane. In other words, the graph of f is the graph of the equation $y = f(x)$. In this section, we will graph many basic functions. The first is $f(x) = x^2$ (or $y = x^2$), sometimes called the squaring function. This function is a special case of a broader set of functions called **quadratic functions**. We will discuss these functions in detail in Chapter 8.

EXAMPLE 1   Graph the function: $f(x) = x^2$

Solution   We substitute values for x in the equation and compute the corresponding values of $f(x)$. For example, if $x = -3$, we have

$$f(x) = x^2$$
$$f(-3) = (-3)^2 \quad \text{Substitute } -3 \text{ for } x.$$
$$= 9$$

The ordered pair $(-3, 9)$ satisfies the equation and will lie on the graph. We list this pair and others that satisfy the equation in the table shown in Figure 2-42. We plot the points and draw a smooth curve through them to get the graph, called a **parabola**.

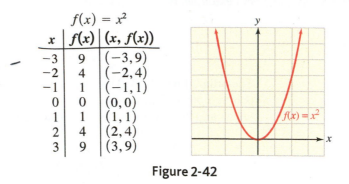

$f(x) = x^2$

x	$f(x)$	$(x, f(x))$
-3	9	$(-3, 9)$
-2	4	$(-2, 4)$
-1	1	$(-1, 1)$
0	0	$(0, 0)$
1	1	$(1, 1)$
2	4	$(2, 4)$
3	9	$(3, 9)$

Figure 2-42

SELF CHECK 1   Find the domain and range of $f(x) = x^2$.

The second basic function is $f(x) = x^3$ (or $y = x^3$), called the **cubic function**.

EXAMPLE 2   Graph the function: $f(x) = x^3$

Solution   We substitute values for x in the equation and compute the corresponding values of $f(x)$. For example, if $x = -2$, we have

$$f(x) = x^3$$
$$f(-2) = (-2)^3 \quad \text{Substitute } -2 \text{ for } x.$$
$$= -8$$

The ordered pair $(-2, -8)$ satisfies the equation and will lie on the graph. We list this pair and others that satisfy the equation in the table shown in Figure 2-43. We plot the points and draw a smooth curve through them to obtain the graph.

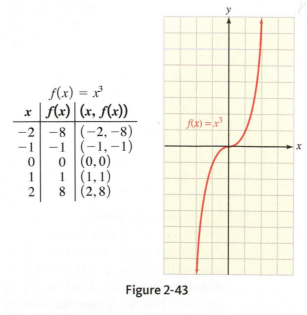

$f(x) = x^3$

x	$f(x)$	$(x, f(x))$
-2	-8	$(-2, -8)$
-1	-1	$(-1, -1)$
0	0	$(0, 0)$
1	1	$(1, 1)$
2	8	$(2, 8)$

Figure 2-43

SELF CHECK 2   Find the domain and range of $f(x) = x^3$.

The third basic function is $f(x) = |x|$ (or $y = |x|$), called the **absolute value function**.

EXAMPLE 3   Graph the function: $f(x) = |x|$

Solution   We substitute values for x in the equation and compute the corresponding values of $f(x)$. For example, if $x = -3$, we have

$$f(x) = |x|$$
$$f(-3) = |-3| \quad \text{Substitute } -3 \text{ for } x.$$
$$= 3$$

The ordered pair $(-3, 3)$ satisfies the equation and will lie on the graph. We list this pair and others that satisfy the equation in the table shown in Figure 2-44. We plot the points and draw a V-shaped graph through them.

x	$f(x)$	$(x, f(x))$
-3	3	$(-3, 3)$
-2	2	$(-2, 2)$
-1	1	$(-1, 1)$
0	0	$(0, 0)$
1	1	$(1, 1)$
2	2	$(2, 2)$
3	3	$(3, 3)$

$f(x) = |x|$

Figure 2-44

 SELF CHECK 3   Find the domain and range of $f(x) = |x|$.

Accent on technology

▸ Graphing Functions

We can graph nonlinear functions with a TI84 graphing calculator. For example, to graph $f(x) = x^2$ in a standard window of $[-10, 10]$ for x and $[-10, 10]$ for y, we enter the function by entering **X, T, θ, n** (x) **x²** (^{2}) and press the **GRAPH** key. We will obtain the graph shown in Figure 2-45(a).

To graph $f(x) = x^3$, we enter the function by entering **X, T, θ, n** (x) **^** 3 and press the **GRAPH** key to obtain the graph in Figure 2-45(b). To graph $f(x) = |x|$, we enter the function by selecting "abs" from the MATH NUM menu, typing x, **)** and pressing the **GRAPH** key to obtain the graph in Figure 2-45(c).

| (a) | (b) | (c) |

$f(x) = x^2$ $f(x) = x^3$ $f(x) = |x|$

Figure 2-45

When using a graphing calculator, we must be sure that the viewing window does not show a misleading graph. For example, if we graph $f(x) = |x|$ in the window $[0, 10]$ for x and $[0, 10]$ for y, we will obtain a misleading graph that looks like a line. (See Figure 2-46.) This is not true. The complete graph is the V-shaped graph shown in Figure 2-45(c).

Figure 2-46

For instructions regarding the use of a Casio graphing calculator, please refer to the Casio Keystroke Guide in the back of the book.

Graph a vertical translation of the quadratic, cubic, and absolute value functions.

The graphs of different functions may be identical except for their positions in the xy-plane. For example, Figure 2-47 shows the graph of $f(x) = x^2 + k$ for three different values of k. If $k = 0$, we have the graph of $f(x) = x^2$. If $k = 3$, we obtain the graph $f(x) = x^2 + 3$, which is identical to the graph of $f(x) = x^2$, except that it is shifted 3 units upward. If $k = -4$, we obtain the graph $f(x) = x^2 - 4$, which is identical to the graph of $f(x) = x^2$, except that it is shifted 4 units downward. These shifts are called **vertical translations**.

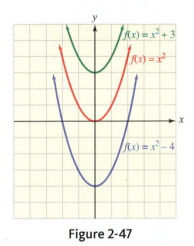

Figure 2-47

In general, we can make these observations.

VERTICAL TRANSLATIONS	If f is a function and k is a positive number, then

- The graph of $f(x) + k$ is identical to the graph of $f(x)$, except that it is translated k units upward.
- The graph of $f(x) - k$ is identical to the graph of $f(x)$, except that it is translated k units downward.

EXAMPLE 4   Graph: $f(x) = |x| + 2$

Solution   The graph of $f(x) = |x| + 2$ will be the same V-shaped graph as $f(x) = |x|$, except that it is shifted 2 units upward. The graph appears in Figure 2-48.

Figure 2-48

 SELF CHECK 4   Graph: $f(x) = |x| - 3$

3   Graph a horizontal translation of the quadratic, cubic, and absolute value functions.

Figure 2-49 shows the graph of $f(x) = (x + h)^2$ for three different values of h. If $h = 0$, we have the graph of $f(x) = x^2$. The graph of $f(x) = (x - 3)^2$ is identical to the graph of $f(x) = x^2$, except that it is shifted 3 units to the right. The graph of $f(x) = (x + 2)^2$ is identical to the graph of $f(x) = x^2$, except that it is shifted 2 units to the left. These shifts are called **horizontal translations**.

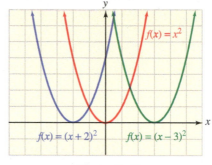

Figure 2-49

In general, we can make these observations.

<table>
<tr><td>

HORIZONTAL TRANSLATIONS

</td><td>

If f is a function and k is a positive number, then

- The graph of $f(x - k)$ is identical to the graph of $f(x)$, except that it is translated k units to the right.
- The graph of $f(x + k)$ is identical to the graph of $f(x)$, except that it is translated k units to the left.

</td></tr>
</table>

EXAMPLE 5   Graph: $f(x) = (x - 2)^3$

Solution   The graph of $f(x) = (x - 2)^3$ will be the same shape as the graph of $f(x) = x^3$, except that it is shifted 2 units to the right. The graph appears in Figure 2-50.

$f(x) = (x - 2)^3$

Figure 2-50

SELF CHECK 5   Graph: $f(x) = (x + 3)^3$

The next example will contain both a horizontal and a vertical translation.

EXAMPLE 6   Graph: $f(x) = (x - 3)^2 + 2$

Solution   We can graph this function by translating the graph of $f(x) = x^2$ to the right 3 units and then upward 2 units, as shown in Figure 2-51.

COMMENT   In Example 6, we could have translated the graph 2 units upward and then 3 units to the right to obtain the same graph.

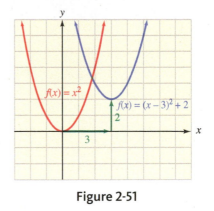

Figure 2-51

🌿 **SELF CHECK 6**   Graph: $f(x) = |x + 2| - 3$

4   Graph a reflection of the quadratic, cubic, and absolute value functions about the *x*-axis.

Figure 2-52 shows tables of solutions for $f(x) = x^2$ and for $f(x) = -x^2$. We note that for a given value of *x*, the corresponding *y*-values in the tables are *opposites*. When the values are graphed, we see that the $(-)$ in $f(x) = -x^2$ has the effect of "flipping" the graph of $f(x) = x^2$ over the *x*-axis, so that the parabola opens downward. We say that the graph of $f(x) = -x^2$ is a **reflection** of the graph of $f(x) = x^2$ about the *x*-axis.

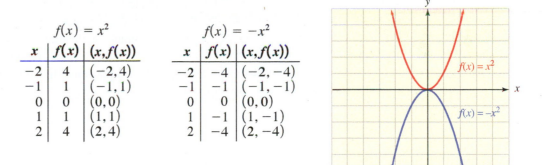

$f(x) = x^2$

x	$f(x)$	$(x, f(x))$
-2	4	$(-2, 4)$
-1	1	$(-1, 1)$
0	0	$(0, 0)$
1	1	$(1, 1)$
2	4	$(2, 4)$

$f(x) = -x^2$

x	$f(x)$	$(x, f(x))$
-2	-4	$(-2, -4)$
-1	-1	$(-1, -1)$
0	0	$(0, 0)$
1	-1	$(1, -1)$
2	-4	$(2, -4)$

Figure 2-52

EXAMPLE 7   Graph: $f(x) = -x^3$

Solution   To graph $f(x) = -x^3$, we use the graph of $f(x) = x^3$ from Example 2. First, we reflect the portion of the graph of $f(x) = x^3$ in quadrant I to quadrant IV, as shown in Figure 2-53 on the next page. Then we reflect the portion of the graph of $f(x) = x^3$ that is in quadrant III to quadrant II.

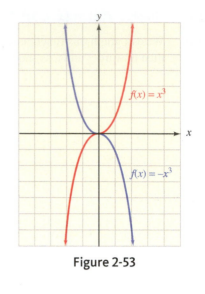

Figure 2-53

🌿 **SELF CHECK 7**   Graph: $f(x) = -|x|$

REFLECTION OF A GRAPH   The graph of $y = -f(x)$ is the graph of $f(x)$ reflected about the x-axis.

Perspective

GRAPHS IN SPACE

In an xy-coordinate system, graphs of equations containing the two variables x and y are lines or curves. Other equations have more than two variables, and graphing them often requires some ingenuity and perhaps the aid of a computer. Graphs of equations with the three variables x, y, and z are viewed in a three-dimensional coordinate system with three axes. The coordinates of points in a three-dimensional coordinate system are ordered triples (x, y, z). For example, the points $P(2, 3, 4)$ and $Q(-1, 2, 3)$ are plotted in Illustration 1.

Graphs of equations in three variables are not lines or curves, but flat planes or curved surfaces. Only the simplest of these equations can be conveniently graphed by hand; a computer provides the best images of others. The graph in Illustration 2 is called a **paraboloid**; it is the three-dimensional version of a parabola. Illustration 3 models a portion of the vibrating surface of a drum head.

Illustration 1   **Illustration 2**   **Illustration 3**

1. D: $(-\infty, \infty)$; R: $[0, \infty)$ **2.** D: $(-\infty, \infty)$; R: $(-\infty, \infty)$ **3.** D: $(-\infty, \infty)$; R: $[0, \infty)$

4.

$f(x) = |x| - 3$

5.

$f(x) = (x + 3)^3$

6.

$f(x) = |x + 2| - 3$

7.

$f(x) = -|x|$

NOW TRY THIS

Given the graph of $f(x)$ below, sketch a graph of each translation or reflection.

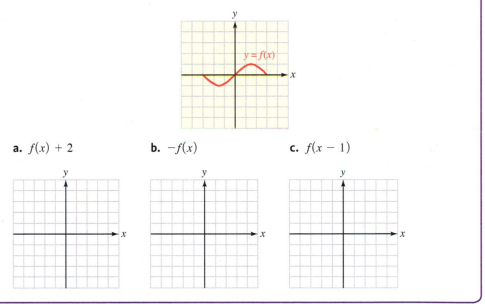

$y = f(x)$

a. $f(x) + 2$ **b.** $-f(x)$ **c.** $f(x - 1)$

2.5 Exercises

WARM-UPS *Find the value of f(x) when x = 3 and x = −2.*

1. $f(x) = x^2 - 4$ **2.** $f(x) = x^2 + 1$

3. $f(x) = x^3 - 1$ **4.** $f(x) = x^3 + 5$

5. $f(x) = |x| - 2$ **6.** $f(x) = |x| + 4$

Find the value of g(a) when a = −3.

7. $g(a) = -a^2 - 4$

8. $g(a) = -|a| + 2$

REVIEW

9. List the prime numbers between 20 and 30.
10. State the associative property of addition.

11. State the commutative property of multiplication.
12. What is the additive identity element?
13. What is the multiplicative identity element?
14. Find the multiplicative inverse of $-\frac{4}{7}$.

VOCABULARY AND CONCEPTS *Fill in the blanks.*

15. The function $f(x) = x^2$ is called a _____ function. Its graph is a _____.
16. The function $f(x) = x^3$ is called a _____ function.
17. The function $f(x) = |x|$ is called an _____ function.
18. Shifting the graph of an equation upward or downward is called a _____ translation.
19. Shifting the graph of an equation to the left or to the right is called a _____ translation.
20. The graph of $f(x) = x^2 + 3$ is the same as the graph of $f(x) = x^2$, except that it is shifted _ units _____.
21. The graph of $f(x) = x^3 - 2$ is the same as the graph of $f(x) = x^3$, except that it is shifted _ units _____.
22. The graph of $f(x) = (x - 5)^3$ is the same as the graph of $f(x) = x^3$, except that it is shifted _ units _____.
23. The graph of $f(x) = (x + 7)^3$ is the same as the graph of $f(x) = x^3$, except that it is shifted _ units _____.
24. The graph of $f(x) = (x - 2)^2 - 4$ is the same as the graph of $f(x) = x^2$, except that it is shifted 2 units _____ and 4 units _____.
25. The graph of $f(x) = -|x + 5|$ is the same as the graph of $f(x) = |x + 5|$, except that it is _____ about the x-axis.
26. The graph of $f(x) = -(x + 1)^2 + 2$ is the same as the graph of $f(x) = x^2$ except that it includes a _____ and _____ translation and a _____.

GUIDED PRACTICE *Graph each function by plotting points. Check your work with a graphing calculator. SEE EXAMPLES 1–3. (OBJECTIVE 1)*

27. $f(x) = x^2 - 3$

28. $f(x) = x^2 + 2$

29. $f(x) = (x - 1)^3$

30. $f(x) = (x + 1)^3$

31. $f(x) = |x| - 2$

32. $f(x) = |x| + 1$

Graph each function using a vertical translation of the graph of $f(x) = x^2, f(x) = x^3$, or $f(x) = |x|$. SEE EXAMPLE 4. (OBJECTIVE 2)

33. $f(x) = x^2 - 5$

34. $f(x) = x^3 + 4$

35. $f(x) = |x| + 6$

36. $f(x) = |x| - 4$

Graph each function using a horizontal translation of the graph of $f(x) = x^2, f(x) = x^3$, or $f(x) = |x|$. SEE EXAMPLE 5. (OBJECTIVE 3)

37. $f(x) = (x - 1)^3$

38. $f(x) = (x + 4)^2$

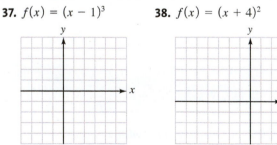

39. $f(x) = (x - 3)^2$

40. $f(x) = |x + 2|$

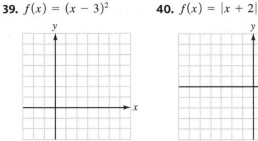

Graph each function using translations of the graph of
$f(x) = x^2, f(x) = x^3, or f(x) = |x|.$ *SEE EXAMPLE 6.*
(OBJECTIVES 2–3)

41. $f(x) = |x - 2| - 1$

42. $f(x) = (x + 2)^2 - 1$

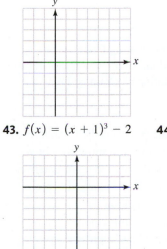

43. $f(x) = (x + 1)^3 - 2$

44. $f(x) = |x + 4| + 3$

Graph each function using a reflection and a translation of the
graph of $f(x) = x^2, f(x) = x^3, or f(x) = |x|.$ *SEE EXAMPLE 7.*
(OBJECTIVE 4)

45. $f(x) = -|x| + 1$

46. $f(x) = -x^3 - 2$

47. $f(x) = -(x - 1)^2$

48. $f(x) = -(x + 2)^2$

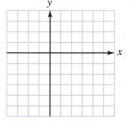

ADDITIONAL PRACTICE *Graph each function and state*
the domain and range.

49. $f(x) = |x - 1|$

50. $f(x) = |x + 2|$

51. $f(x) = x^2 + 8$

52. $f(x) = x^3 - 8$

53. $f(x) = |x + 5|$

54. $f(x) = |x - 5|$

55. $f(x) = (x - 6)^2$

56. $f(x) = (x + 9)^2$

Use a graphing calculator to graph each function, using values
of $[-4, 4]$ *for x and* $[-4, 4]$ *for y. The initial graph is not what*
it appears to be. Select a better viewing window to find the
complete graph.

57. $f(x) = x^3 + 8$

58. $f(x) = x^3 - 12$

WRITING ABOUT MATH

59. Explain how to graph an equation by plotting points.

60. Explain why the correct choice of window settings is important when using a graphing calculator.

⟰ SOMETHING TO THINK ABOUT *Use a graphing calculator.*

61. Use a graphing calculator with settings of $[-10, 10]$ for x and $[-10, 10]$ for y to graph   **a.** $y = -x^2$,   **b.** $y = -x^2 + 1$, and   **c.** $y = -x^2 + 2$. What do you notice?

62. Use a graphing calculator with settings of $[-10, 10]$ for x and $[-10, 10]$ for y to graph   **a.** $y = -|x|$,   **b.** $y = -|x| - 1$, and   **c.** $y = -|x| - 2$. What do you notice?

63. Graph $y = (x + k)^2 + 1$ for several positive values of k. What do you notice?

64. Graph $y = (x + k)^2 + 1$ for several negative values of k. What do you notice?

65. Graph $y = (kx)^2 + 1$ for several values of k, where $k > 1$. What do you notice?

66. Graph $y = (kx)^2 + 1$ for several values of k, where $0 < k < 1$. What do you notice?

67. Graph $y = (kx)^2 + 1$ for several values of k, where $k < -1$. What do you notice?

68. Graph $y = (kx)^2 + 1$ for several values of k, where $-1 < k < 0$. What do you notice?

Projects

PROJECT 1

The Board of Administrators of a county has hired your consulting firm to plan a highway. The new highway is to be built in the outback section of the county.

The two main roads in the outback section are Highway N, running in a straight line north and south, and Highway E, running in a straight line east and west. These two highways meet at an intersection that the locals call Four Corners. The only other county road in the area is Slant Road, which runs in a straight line from northwest to southeast, cutting across Highway N north of Four Corners and Highway E east of Four Corners.

The county clerk is unable to find an official map of the area, but there is an old sketch made by the original road designer. It shows that if a rectangular coordinate system is set up using Highways N and E as the axes and Four Corners as the origin, then the equation of the line representing Slant Road is

$$2x + 3y = 12 \quad \text{(where the unit length is 1 mile)}$$

Given this information, the county wants you to do the following:

a. Update the current information by giving the coordinates of the intersections of Slant Road with Highway N and Highway E.

b. Plan a new highway, Country Drive, that will begin 1 mile north of Four Corners and run in a straight line in a generally northeasterly direction, intersecting Slant Road at right angles. The county wants to know

the equation of the line representing Country Drive. You also should state the domain on which this equation is valid as a representation of Country Drive.

PROJECT 2

You represent your branch of the Buy-from-Us Corporation. At the company's regional meeting, you must present your revenue and cost reports to the other branch representatives. But disaster strikes! The graphs you had planned to present, containing cost and revenue information for this year and last year, are unlabeled! You cannot immediately recognize which graphs represent costs, which represent revenues, and which represent which year. Without these graphs, your presentation will not be effective.

The only other information you have with you is in the notes you made for your talk. From these you glean the following financial data about your branch.

1. All cost and revenue figures on the graphs are rounded to the nearest $50,000.

2. Costs for the fourth quarter of last year were $400,000.

3. Revenue was not above $400,000 for any quarter last year.

4. Last year, your branch lost money during the first quarter.

5. This year, your branch made money during three of the four quarters.

6. Profit during the second quarter of this year was $150,000.

And, of course, you know that profit = revenue − cost.

With this information, you must match each of the graphs (Illustrations 1–4) with one of the following titles:

Costs, This Year   Costs, Last Year

Revenues, This Year   Revenues, Last Year

You should be sure to have sound reasons for your choices—reasons ensuring that no other arrangement of the titles will fit the data. The *last* thing you want to do is present incorrect information.

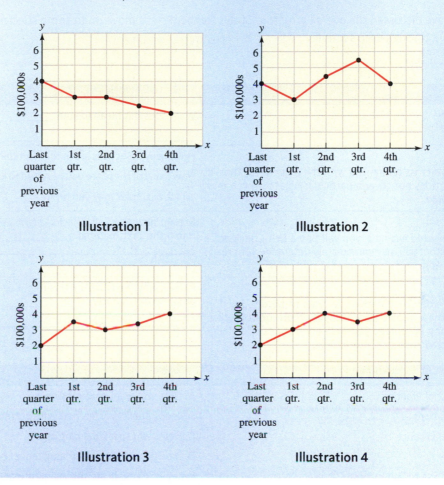

Illustration 1

Illustration 2

Illustration 3

Illustration 4

Reach for Success

EXTENSION OF READING THIS MATHEMATICS TEXTBOOK

READING THIS MATHEMATICS TEXTBOOK

Now that we have discussed the overall structure of this textbook, let's move on to some special features designed to help you work through each section.

Prior to the homework exercise set, you will find the following features:	
• Getting Ready	What does this mean to you? _____ _____
• Examples All examples are cross-referenced with the objectives. Thus, if your instructor does not cover a particular objective, you may omit that example.	How many objectives are in this section? _____ How many of these are you required to know? _____ Which examples may you omit? _____
• Self Checks After each numbered example, you will find a Self Check practice exercise so that you may immediately determine your reading comprehension.	Explain how the Self Checks might help you understand and retain the text material. _____ _____
• Now Try This Created to provide collaborative exercises to be discussed in small groups, these problems are slightly more difficult than the examples and encourage you to deepen your understanding or provide transition to a future topic.	Explain why the Now Try This might be considered transitional group-work exercises. _____ _____ _____
Match each exercise set feature with its description.	
_____ 1. Additional Practice	A. Gets you ready for homework
_____ 2. Applications	B. Keeps previously learned skills current
_____ 3. Guided Practice	C. Helps you speak the language
_____ 4. Review	D. Provides references for assistance
_____ 5. Something to Think About	E. Begins to duplicate the exam format
_____ 6. Vocabulary	F. Shows relevance of the topics to the world
_____ 7. Warm-up	G. Increases communication skills
_____ 8. Writing About Math	H. Transitions students' concepts
When beginining the exercises, be sure to *read the instructions* carefully.	1. Find the perimeter of the figure. 2. Find the area of the figure. 7 in. ☐ 21 in. What is the importance of understanding the instructions in the two exercises? _____ _____

Don't forget the computer software, Enhanced WebAssign. The exercises there can be done on the computer with additional help options, an excellent resource particularly during late evening when your other resources are not available!

2   Review

SECTION 2.1   Graphing Linear Equations

DEFINITIONS AND CONCEPTS	EXAMPLES
Any ordered pair of real numbers represents a point on the rectangular coordinate system.	On the coordinate axis, plot $(0, -2)$, $(2, -4)$, and $(-3, 5)$.

On the coordinate axis, plot $(0, -2)$, $(2, -4)$, and $(-3, 5)$.

The point where the axes cross is called the **origin**.	The origin is represented by the ordered pair $(0, 0)$.
The four regions of a coordinate plane are called **quadrants**.	The point with coordinates $(5, 3)$ is in quadrant I.
	The point with coordinates $(-5, 3)$ is in quadrant II.
	The point with coordinates $(-5, -3)$ is in quadrant III.
	The point with coordinates $(5, -3)$ is in quadrant IV.

An ordered pair of real numbers is a **solution** of an equation in two variables if it satisfies the equation.	The ordered pair $(-1, 5)$ is a solution of the equation $x - 2y = -11$ because it satisfies the equation.

$$x - 2y = -11$$
$$(-1) - 2(5) \stackrel{?}{=} -11 \quad \text{Substitute } -1 \text{ for } x \text{ and } 5 \text{ for } y.$$
$$-1 - 10 \stackrel{?}{=} -11$$
$$-11 = -11 \quad \text{True.}$$

Since the results are equal, $(-1, 5)$ is a solution.

General form of an equation of a line in two variables:

$$Ax + By = C \quad (A \text{ and } B \text{ are not both } 0.)$$

To graph a linear equation,

1. Find two ordered pairs (x, y) that satisfy the equation.
2. Plot each pair on the rectangular coordinate system.
3. Draw a line passing through the two points.
4. Use a third point as a check.

The equation $x + y = -2$ is written in general form. To graph it, we find two ordered pairs that satisfy the equation. A third point is included as a check.

$x + y = -2$

x	y	(x, y)
1	-3	$(1, -3)$
2	-4	$(2, -4)$
3	-5	$(3, -5)$

We then plot the points and draw a line passing through them.

The **y-intercept** of a line is the point where the line intersects the y-axis.

The y-intercept of the graph on the previous page is the point with coordinates $(0, -2)$.

The **x-intercept** of a line is the point where the line intersects the x-axis.

The x-intercept of the graph on the previous page is the point with coordinates $(-2, 0)$.

Vertical line:

The equation is

$$x = a$$

x-intercept at $(a, 0)$

Horizontal line:

The equation is

$$y = b$$

y-intercept at $(0, b)$

The graph of $x = 2$ is a vertical line passing through $(2, 0)$.

The graph of $y = -3$ is a horizontal line passing through $(0, -3)$.

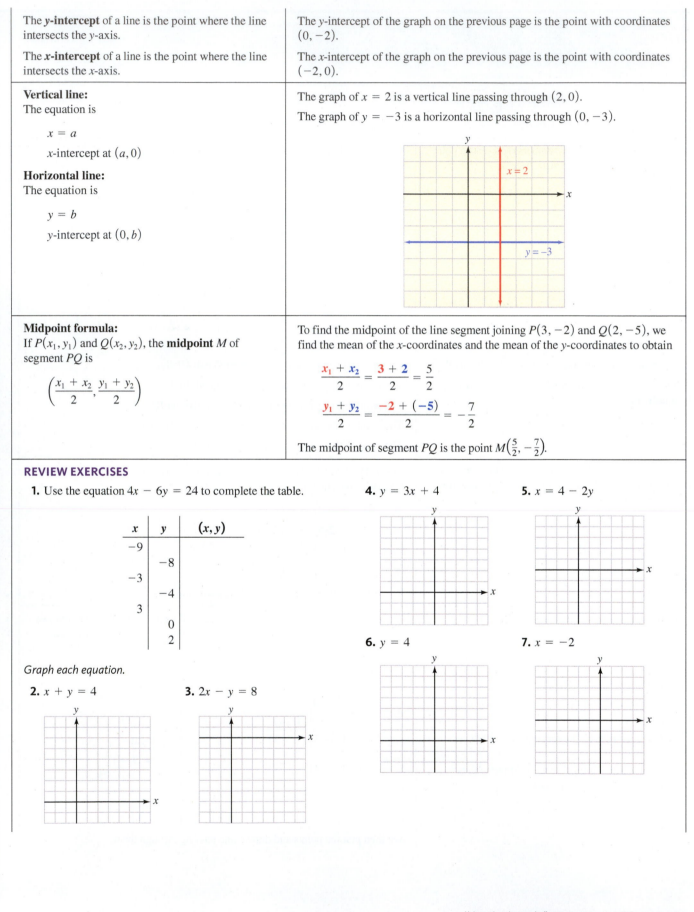

Midpoint formula:

If $P(x_1, y_1)$ and $Q(x_2, y_2)$, the **midpoint M** of segment PQ is

$$\left(\frac{x_1 + x_2}{2}, \frac{y_1 + y_2}{2} \right)$$

To find the midpoint of the line segment joining $P(3, -2)$ and $Q(2, -5)$, we find the mean of the x-coordinates and the mean of the y-coordinates to obtain

$$\frac{x_1 + x_2}{2} = \frac{3 + 2}{2} = \frac{5}{2}$$

$$\frac{y_1 + y_2}{2} = \frac{-2 + (-5)}{2} = -\frac{7}{2}$$

The midpoint of segment PQ is the point $M\left(\frac{5}{2}, -\frac{7}{2} \right)$.

REVIEW EXERCISES

1. Use the equation $4x - 6y = 24$ to complete the table.

x	y	(x, y)
−9		
	−8	
−3		
	−4	
3		
	0	
	2	

Graph each equation.

2. $x + y = 4$

3. $2x - y = 8$

4. $y = 3x + 4$

5. $x = 4 - 2y$

6. $y = 4$

7. $x = -2$

8. $2(x + 3) = x + 2$ **9.** $3y = 2(y - 1)$

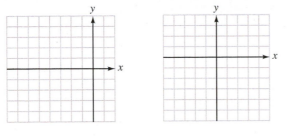

10. Find the midpoint of the line segment joining $P(-4, 3)$ and $Q(9, -7)$.

SECTION 2.2   Slope of a Line

DEFINITIONS AND CONCEPTS	EXAMPLES
Slope of a nonvertical line: If $x_2 \neq x_1$, $$m = \frac{\Delta y}{\Delta x} = \frac{y_2 - y_1}{x_2 - x_1}$$	The slope of a line passing through $(-1, 4)$ and $(5, -3)$ is given by $$m = \frac{y_2 - y_1}{x_2 - x_1} = \frac{-3 - 4}{5 - (-1)} = \frac{-7}{6}$$
Horizontal lines have a slope of 0.	$y = 5$ is a horizontal line with slope 0.
Vertical lines have an undefined slope.	$x = 5$ is a vertical line with an undefined slope.
Parallel lines have the same slope.	Two distinct lines with slopes $\frac{3}{2}$ and $\frac{3}{2}$ are parallel.
The slopes of two nonvertical perpendicular lines are negative reciprocals.	Two lines with slopes 5 and $-\frac{1}{5}$ are perpendicular.

REVIEW EXERCISES

Find the slope of the line passing through points P and Q, if possible.

11. $P(2, 5)$ and $Q(5, 8)$ **12.** $P(-3, -2)$ and $Q(6, 12)$

13. $P(-3, 4)$ and $Q(-5, -6)$ **14.** $P(5, -4)$ and $Q(-6, -9)$

15. $P(-2, 4)$ and $Q(8, 4)$ **16.** $P(-5, -4)$ and $Q(-5, 8)$

Find the slope of the graph of each equation, if one exists.

17. $2x - 3y = 18$ **18.** $4x + 5y = 10$

19. $-2(x - 3) = 10$ **20.** $2y - 7 = 1$

Determine whether the lines with the given slopes are parallel, perpendicular, or neither.

21. $m_1 = 4, m_2 = -\frac{1}{4}$ **22.** $m_1 = 0.5, m_2 = \frac{1}{2}$

23. $m_1 = 0.5, m_2 = -\frac{1}{2}$ **24.** $m_1 = 5, m_2 = -0.2$

25. Sales growth If the sales of a new business were $65,000 in its first year and $130,000 in its fourth year, find the rate of growth in sales per year.

SECTION 2.3   Writing Equations of Lines

DEFINITIONS AND CONCEPTS	EXAMPLES
Equations of a line: *Point-slope form:* $$y - y_1 = m(x - x_1)$$	Write the point-slope form of a line passing through $(-3, 5)$ with slope $\frac{1}{2}$. $y - y_1 = m(x - x_1)$ This is point-slope form. $y - 5 = \frac{1}{2}[x - (-3)]$ Substitute. $y - 5 = \frac{1}{2}(x + 3)$

Slope-intercept form: $y = mx + b$	Write the slope-intercept form of a line with slope $\frac{1}{2}$ and y-intercept $(0, -4)$. $y = \mathbf{m}x + \mathbf{b}$ This is slope-intercept form. $y = \dfrac{1}{2}x - 4$ Substitute. $y = \dfrac{1}{2}x - 4$
To find the slope of a linear equation written in general form, solve for y.	Find the slope of the line $3x + 5y = 7$. $5y = -3x + 7$ $y = -\dfrac{3}{5}x + \dfrac{7}{5}$ The slope is $-\dfrac{3}{5}$.

REVIEW EXERCISES

Write the equation of the line with the given properties in slope-intercept form.

26. slope of 3; passing through $P(-8, 5)$

27. passing through $(-2, 4)$ and $(6, -9)$

28. passing through $(-3, -5)$; parallel to the graph of

$3x - 2y = 7$

29. passing through $(-3, -5)$; perpendicular to the graph of

$3x - 2y = 7$

30. Depreciation   A business purchased a copy machine for $8,700 and will depreciate it on a straight-line basis over the next 5 years. At the end of its useful life, it will be sold as scrap for $100. Find its depreciation equation.

SECTION 2.4   Introduction to Functions

DEFINITIONS AND CONCEPTS	**EXAMPLES**
A **relation** is any set of ordered pairs.	Find the domain and range of the relation $\{(2, -1), (6, 3), (2, 5)\}$ and determine whether it defines a function.
The **domain** is the set of first components of the ordered pairs.	**Domain:** $\{2, 6\}$ because 2 and 6 are the first components in the ordered pairs.
The **range** is the set of second components of the ordered pairs.	**Range:** $\{-1, 3, 5\}$ because $-1, 3,$ and 5 are the second components in the ordered pairs.
A **function** is any set of ordered pairs (a relation) in which each first component determines exactly one second component.	The relation does not define a function because the first component 2 is paired with two different second components.
	Find the domain and range of the graph and determine whether the graph represents a function.
The **domain** of a function is the set of input values.	**Domain:** Since the graph extends forever to the left, and stops at $x = 3$ on the right, the domain is $(-\infty, 3]$ in interval notation.
The **range** is the set of output values.	**Range:** Since the graph extends forever downward and ends at $y = 4$, the range is $(-\infty, 4]$ in interval notation.
The **vertical line test** can be used to determine whether a graph represents a function.	Since every vertical line that intersects the graph will do so exactly once, the vertical line test indicates that the graph is a function.
$f(k)$ represents the value of $f(x)$ when $x = k$.	If $f(x) = -7x + 4$, find $f(2)$. $f(\mathbf{2}) = -7(\mathbf{2}) + 4$ Substitute 2 for x. $= -14 + 4$ $= -10$ In ordered pair form, we can write $(2, -10)$.

The domain of a function that is defined by an equation is the set of all numbers that are permissible replacements for its variable.	Find the domain of $f(x) = \frac{-5}{x+7}$. The number -7 cannot be substituted for x in the function $f(x) = \frac{-5}{x+7}$ because that would make the denominator 0. However, any real number, except -7, can be substituted for x to obtain a single value y. Therefore, the domain is the set of all real numbers except -7. This is the interval $(-\infty, -7) \cup (-7, \infty)$.
A **linear function** is a function defined by an equation that can be written in the form $f(x) = mx + b$ or $y = mx + b$ where m is the slope of the line graph and $(0, b)$ is the y-intercept.	The graph of $y = 4x - 1$ is a line with slope 4 and y-intercept $(0, -1)$. The domain is the set of real numbers $\mathbb{R}$, and the range is the set of real numbers $\mathbb{R}$.

REVIEW EXERCISES

Determine whether each equation determines y to be a function of x.

31. $y = 6x - 4$ **32.** $y = 4 - x$

33. $y^2 = x$ **34.** $|y| = x$

Assume that $f(x) = 3x + 2$ and $g(x) = x^2 - 4$ and find each value.

35. $f(6)$ **36.** $g(-5)$

37. $g(-2)$ **38.** $f(5)$

Find the domain of each function.

39. $f(x) = 6x - 5$ **40.** $f(x) = -2x + 3$

41. $f(x) = x^2 + 1$ **42.** $f(x) = \dfrac{4}{2 - x}$

43. $f(x) = \dfrac{7}{x - 3}$ **44.** $y = -4$

Use the vertical line test to determine whether each graph represents a function. State the domain and range.

45.   **46.**

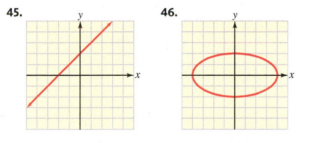

47.   **48.**

Determine whether each equation defines a linear function.

49. $y = 4x + 2$ **50.** $y = \dfrac{x + 5}{4}$

51. $4x - 3y = 12$ **52.** $y = x^2 - 10$

SECTION 2.5   Graphing Other Functions

DEFINITIONS AND CONCEPTS	EXAMPLES
Vertical translations: If f is a function and k is a positive number, then • The graph of $f(x) + k$ is identical to the graph of $f(x)$, except that it is translated k units upward.	 The graph of the absolute value function $f(x) = |x| + 4$ will be the same shape as $f(x) = |x|$, but shifted upward 4 units.
• The graph of $f(x) - k$ is identical to the graph of $f(x)$, except that it is translated k units downward.	The graph of the absolute value function $f(x) = |x| - 4$ will be the same shape as $f(x) = |x|$, but shifted downward 4 units.

Horizontal translations:

If f is a function and k is a positive number, then

- The graph of $f(x - k)$ is identical to the graph of $f(x)$, except that it is translated k units to the right.

- The graph of $f(x + k)$ is identical to the graph of $f(x)$, except that it is translated k units to the left.

The graph of the quadratic function $f(x) = (x - 3)^2$ will be the same shape as $f(x) = x^2$, but shifted 3 units to the right.

The graph of the quadratic function $f(x) = (x + 4)^2$ will be the same shape as $f(x) = x^2$, but shifted 4 units to the left.

Reflections:

The graph of $y = -f(x)$ is identical to the graph of $f(x)$ except that it is reflected about the x-axis.

The graph of the cubic function $f(x) = -x^3$ is identical to the graph of $f(x) = x^3$ except that it is reflected about the x-axis.

REVIEW EXERCISES

Graph each function.

53. $f(x) = x^2 - 3$

54. $f(x) = |x| - 4$

Use a graphing calculator to graph each function. Compare the results to the answers to Exercises 53–56.

57. $f(x) = x^2 - 3$

58. $f(x) = |x| - 4$

59. $f(x) = (x - 2)^3$

60. $f(x) = (x + 4)^2 - 3$

55. $f(x) = (x - 2)^3$

56. $f(x) = (x + 4)^2 - 3$

61. Graph $f(x) = -|x - 3|$ and state the domain and range.

---

2   Test

1. Graph the equation $2x - 5y = 10$.

3. Find the midpoint of the line segment.

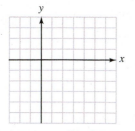

2. Find the x- and y-intercepts of the graph of $y = \dfrac{x + 2}{7}$.

Find the slope of each line, if possible.

4. the line through $(-2, 4)$ and $(6, 8)$

5. the graph of $2x - 3y = 8$

6. the graph of $x = 5$

7. the graph of $y = 5$

8. Write the equation of the line with slope of $\frac{2}{3}$ and passing through $(4, -5)$. Give the answer in slope-intercept form.

9. Write the equation of the line passing through $(-2, 6)$ and $(-4, -10)$. Give the answer in slope-intercept form.

10. Find the slope and the y-intercept of the graph of $-2(x - 3) = 3(2y + 5)$.

11. Are the graphs of $x - 3y = 9$ and $y = \frac{1}{3}x + 2$ parallel, perpendicular, or neither?

12. Determine whether the graphs of $y = -\frac{2}{3}x + 4$ and $2y = 3x - 3$ are parallel, perpendicular, or neither.

13. Write the equation of the line that passes through the origin and is parallel to the graph of $y = \frac{3}{2}x - 7$.

14. Write the equation of the line that passes through $P(-3, 6)$ and is perpendicular to the graph of $y = -\frac{2}{3}x - 7$.

15. Does $|y| = x$ define y to be a function of x?

16. State the domain and range of the function $f(x) = |x|$.

17. State the domain and range of the function $f(x) = x^3$.

Let $f(x) = 3x + 1$ and $g(x) = x^2 - 2$. Find each value.

18. $f(3)$

19. $g(-3)$

20. $f(a)$

21. $g(-x)$

Determine whether each graph represents a function.

22.
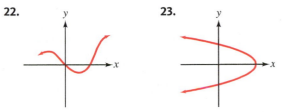

23.

Graph the function and state the domain and range.

24. $f(x) = x^2 - 1$

25. $f(x) = -|x + 2|$

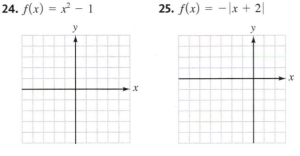

Cumulative Review

Determine which numbers in the set $\left\{-2, 0, 1, 2, \frac{13}{12}, 6, 7, \sqrt{5}, \pi\right\}$ are in each category.

1. natural numbers

2. whole numbers

3. rational numbers

4. irrational numbers

5. negative numbers

6. real numbers

7. prime numbers

8. composite numbers

9. even numbers

10. odd numbers

Graph each set on the number line.

11. $\{x \mid -2 < x \leq 5\}$

12. $[-5, 0) \cup [3, 6]$

Simplify each expression.

13. $-|-4| + |6|$

14. $\dfrac{|-5| + |-3|}{-|4|}$

Perform the operations.

15. $3 + 5 \cdot 4$

16. $\dfrac{9 - (6 - 3)}{6 - 9}$

17. $20 \div (-10 \div 2)$

18. $\dfrac{6 + 3(6 + 4)}{2(3 - 9)}$

Evaluate each expression when $x = 2$ and $y = -3$.

19. $-3x - y$

20. $\dfrac{x^2 - y^2}{2x + y}$

Determine the property of real numbers that justifies each statement.

21. $(a + b) + c = a + (b + c)$

22. $5(a + b) = 5a + 5b$

23. $(a + b) + c = c + (a + b)$

24. $(ab)c = a(bc)$

Simplify each expression. Assume that all variables are positive numbers and write all answers without negative exponents.

25. $(-2x^4y)^3$

26. $\dfrac{c^4c^8}{(c^5)^2}$

27. $\left(-\dfrac{a^3b^{-2}}{ab}\right)^{-1}$

28. $\left(\dfrac{-2a^{-3}b^2}{8a^4b^{-5}}\right)^{-2}$

29. Write 0.0000382 in scientific notation.

30. Write 9.32×10^8 in standard notation.

Solve each equation.

31. $4x - 8 = 20$

32. $\dfrac{2x - 6}{3} = x + 7$

33. $4(y - 3) + 4 = -3(y + 5)$

34. $2x - \dfrac{3(x - 2)}{2} = 7 - \dfrac{x - 3}{3}$

Solve each formula for the indicated variable.

35. $S = \dfrac{n(a + l)}{2}$ for a

36. $A = \frac{1}{2}h(b_1 + b_2)$ for h

37. The sum of three consecutive even integers is 90. Find the integers.

38. A rectangle is three times as long as it is wide. If its perimeter is 112 centimeters, find its dimensions.

39. Determine whether the graph of $2x - 3y = 6$ defines a function.

40. Find the slope of the line passing through $P(-2, 5)$ and $Q(8, -9)$.

41. Write the equation of the line passing through $P(-2, 5)$ and $Q(8, -9)$.

42. Write the equation of the line that passes through $P(-2, 3)$ and is parallel to the graph of $3x + y = 8$.

Evaluate each expression, given that $f(x) = 3x^2 + 2$ and $g(x) = 2x - 1$.

43. $f(-1)$

44. $g(8)$

45. $g(t)$

46. $f(-r)$

Graph each equation and determine if it is a function. State the domain and range.

47. $y = -x^2 + 1$

48. $y = \left|\dfrac{1}{2}x - 3\right|$

Systems of Linear Equations

Careers and Mathematics

COSMETOLOGISTS

Cosmetologists offer services such as hair care, make-up, and manicures and pedicures. Make-up artists' skills can range from providing a personal service for special occasions to those required for the unique needs of the television, movie, and photography industries.

Almost half of all cosmetologists are self-employed and all are required to be licensed, although the qualifications vary from state to state.

Job Outlook:
Overall employment of barbers and cosmetologists is projected to grow about as fast as the average for all occupations through 2020. Opportunities will be best for those licensed to provide a broad range of services.

Annual Earnings:
$20,490–$120,050

For More Information:
http://www.bls.gov/ooh/personal-care
-and-service/barbers-hairdressers-and
-cosmetologists.htm

For a Sample Application:
See Problem 80 in Section 3.2.

REACH FOR SUCCESS

3.1 Solving Systems of Two Linear Equations in Two Variables by Graphing

3.2 Solving Systems of Two Linear Equations in Two Variables by Substitution and Elimination (Addition)

3.3 Solving Systems of Three Linear Equations in Three Variables

3.4 Solving Systems of Linear Equations Using Matrices

3.5 Solving Systems of Linear Equations Using Determinants

■ *Projects*

REACH FOR SUCCESS EXTENSION

CHAPTER REVIEW

CHAPTER TEST

In this chapter

We have considered linear equations with two variables, say x and y. We found that each equation had infinitely many solutions (x, y), and that we can graph each equation on the rectangular coordinate system. In this chapter, we will discuss different methods for solving systems of linear equations involving two or three equations.

Reach for Success Understanding Your Syllabus

When you go on a trip, you have a plan, an itinerary. You might even have a travel representative along on the trip as a guide to handle the details and any unexpected situations that might occur. Your course syllabus is similar to your itinerary and your instructor is similar to the travel guide.

A syllabus is mandatory for all college courses and serves as a contract between the student and the instructor. Examine the syllabus you received for this course to identify important information.

Course Syllabus

Course Title: Beginning Algebra
Instructor's Office: B232 University Building
Office Hours: MTWR: 8:30am - 10:00am plus other times by appointment
Course Resources: The college provides a Math Lab at no charge to support student success.
Supplies: Textbook: Beginning and Intermediate Algebra: A Guided Approach, 7th edition by Karr/Massey/Gustafson
A graphing calculator is recommended. The TI 84 and Casio fx-9760GII are each supported by a tear-out card in the back of the textbook which provide keystrokes for calculator concepts covered in the text.
Access to homework will be through WebAssign. The course code required to register for this site will be provided the first day of the semester.

Method of Evaluation:

4 Tests (10% each)	40%
20 Homework Assignments	10%
10 Online Labs	10%
Midterm	20%
Comprehensive Final Exam	20%

90 - 100	80 - 89	70 - 79	60 - 69	Below 60
A	B	C	D	F

Tests
Tests count 40% of your final grade.
Tests are given in class on the date indicated in your schedule. There are no make-ups. However, if you miss a test, your Final Exam grade will be used to replace it. If you take all tests, the Final Exam grade, if higher, can be used to replace a lower test grade.

Homework
Homework counts 10% of your final grade.
You have a 3 calendar day "extension" after the deadline in which to submit homework. There is, however, a 10% penalty each day but ONLY for the problems that were not completed by the deadline.

LABS
Labs count 10% of your final grade.
Late labs are not accepted, i.e., there is no 3-day "extension."

MIDTERM
The Midterm counts 20% of your final grade. If you miss the midterm, your Final Exam grade will be used to replace it.

FINAL EXAM
The Final Exam counts 20% of your final grade.
The final exam is cumulative, meaning that it includes all topics covered during the semester.

Your syllabus will include your instructor's contact information, office location, and e-mail address. It typically contains the instructor's office hours, which are used to meet with and help students.

Write the information regarding your instructor:

Office location: _____

E-mail address: _____

Office hours: _____

Write down at least one listed office hour during which you could meet with your instructor if necessary. _____

Does your instructor include a statement such as "and other times by appointment" with the office hours?

If your schedule absolutely prohibits you from meeting with your instructor during office hours, write down at least two times during the week that you are available and could request an appointment.

_____

Most instructors prefer that students be proactive. For example, it is important to contact your instructor prior to any known life event that may interfere with meeting due dates.

Write down any issues that might interfere with your work this semester and that you might wish to discuss with your instructor.

_____

Your syllabus will include requirements of the course and how each will be used to determine your grade. Knowing exactly how your grade will be computed, you can estimate your average at any time.

Write down how the following will be used to determine your grade.

Tests (how many?) _____ _____ points or _____ percentage
Homework/daily work _____ points or _____ percentage
Attendance _____ points or _____ percentage
Quizzes _____ points or _____ percentage
Other _____ points or _____ percentage

How would you use this information to calculate your grade?

_____

A Successful Study Strategy . . .

Know your instructor's office hours so you can have a plan of action if it becomes necessary to meet. In addition, knowing how your grade will be determined early in the semester can eliminate surprises at the end of the semester.

At the end of the chapter you will find additional exercises to guide you to planning for a successful semester.

Solving Systems of Two Linear Equations in Two Variables by Graphing

Objectives

1. Solve a system of two linear equations by graphing.
2. Recognize that an inconsistent system has no solution.
3. Express the infinitely many solutions of a dependent system as a general ordered pair.
4. Use a system of linear equations in two variables to solve a linear equation in one variable graphically.

Vocabulary

system of equations   inconsistent system   independent equations
consistent system   distinct lines   dependent equations
equivalent systems

Getting Ready

Let $y = -3x + 2$. Find the value of y when

1. $x = 0$.   2. $x = 3$.   3. $x = -3$.   4. $x = -\frac{1}{3}$.

Find five pairs of numbers with

5. a sum of 12.   6. a difference of 3.

In the pair of linear equations

$$\begin{cases} x + 2y = 4 \\ 2x - y = 3 \end{cases}$$

(called a **system of equations**) there are infinitely many ordered pairs (x, y) that satisfy the first equation and infinitely many ordered pairs (x, y) that satisfy the second equation. However, there is only one ordered pair (x, y) that satisfies both equations at the same time. The process of finding this ordered pair is called *solving the system.*

 1 Solve a system of two linear equations by graphing.

In general, we follow these steps.

THE GRAPHING METHOD

1. On a single set of coordinate axes, carefully graph each equation.
2. Find the coordinates of the point where the graphs intersect, if applicable.
3. Check the solution in both of the original equations, if applicable.
4. If the graphs have no point in common, the system has no solution.
5. If the graphs of the equations coincide (are the same), the system has infinitely many solutions that can be expressed as a general ordered pair.

When a system of equations has a solution (as in Example 1), the system is called a **consistent system**.

EXAMPLE 1 Solve the system by graphing: $\begin{cases} x + 2y = 4 \\ 2x - y = 3 \end{cases}$

Solution We graph both equations on one set of coordinates axes, as shown in Figure 3-1.

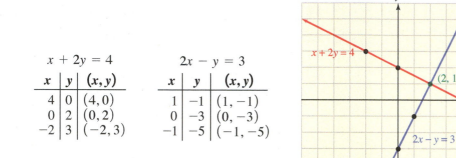

$x + 2y = 4$

x	y	(x, y)
4	0	$(4, 0)$
0	2	$(0, 2)$
-2	3	$(-2, 3)$

$2x - y = 3$

x	y	(x, y)
1	-1	$(1, -1)$
0	-3	$(0, -3)$
-1	-5	$(-1, -5)$

Figure 3-1

Although infinitely many ordered pairs (x, y) satisfy $x + 2y = 4$, and infinitely many ordered pairs (x, y) satisfy $2x - y = 3$, only the coordinates of the point where the graphs intersect satisfy both equations. Since the intersection point has coordinates of $(2, 1)$, the solution is the ordered pair $(2, 1)$, or $x = 2$ and $y = 1$.

When we check the solution, we substitute 2 for x and 1 for y in both equations and verify that $(2, 1)$ satisfies each one as shown below.

$$x + 2y = 4 \qquad\qquad 2x - y = 3$$
$$2 + 2(1) = 4 \qquad\qquad 2(2) - 1 = 3$$
$$2 + 2 \stackrel{?}{=} 4 \qquad\qquad 4 - 1 \stackrel{?}{=} 3$$
$$4 = 4 \qquad\qquad\qquad 3 = 3$$

SELF CHECK 1 Solve the system by graphing: $\begin{cases} 2x + y = 4 \\ x - 3y = -5 \end{cases}$

If the equations in two systems are equivalent, the systems are called **equivalent systems**. In Example 2, we solve a more difficult system by writing it as a simpler equivalent system.

EXAMPLE 2 Solve the system by graphing: $\begin{cases} \frac{3}{2}x - y = \frac{5}{2} \\ x + \frac{1}{2}y = 4 \end{cases}$

Solution We multiply both sides of $\frac{3}{2}x - y = \frac{5}{2}$ by 2 to clear the fractions and obtain the equation $3x - 2y = 5$. We multiply both sides of $x + \frac{1}{2}y = 4$ by 2 to clear the fraction and obtain the equation $2x + y = 8$. This will result in a new system

$$\begin{cases} 3x - 2y = 5 \\ 2x + y = 8 \end{cases}$$

which is equivalent to the original system but has no fractions. If we graph each equation in the new system, as in Figure 3-2, we see that the coordinates of the point where the two lines intersect are $(3, 2)$.

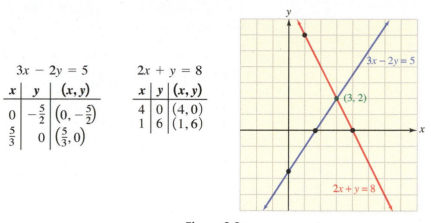

$$3x - 2y = 5$$

x	y	(x, y)
0	$-\frac{5}{2}$	$(0, -\frac{5}{2})$
$\frac{5}{3}$	0	$(\frac{5}{3}, 0)$

$$2x + y = 8$$

x	y	(x, y)
4	0	(4, 0)
1	6	(1, 6)

Figure 3-2

To check the solution, $(3, 2)$, we substitute 3 for x and 2 for y in the two original equations.

$$\frac{3}{2}x - y = \frac{5}{2}$$

$$\frac{3}{2}(3) - 2 \stackrel{?}{=} \frac{5}{2}$$

$$\frac{9}{2} - 2 \stackrel{?}{=} \frac{5}{2}$$

$$\frac{5}{2} = \frac{5}{2}$$

$$x + \frac{1}{2}y = 4$$

$$3 + \frac{1}{2}(2) \stackrel{?}{=} 4$$

$$3 + 1 \stackrel{?}{=} 4$$

$$4 = 4$$

SELF CHECK 2   Solve the system by graphing: $\begin{cases} \frac{5}{2}x - y = 2 \\ x + \frac{1}{3}y = 3 \end{cases}$

2   Recognize that an inconsistent system has no solution.

When a system has no solution (as in Example 3), it is called an **inconsistent system**. Since there is no solution, the solution set is $\varnothing$.

EXAMPLE 3   Solve the system by graphing: $\begin{cases} 2x + 3y = 6 \\ 4x + 6y = 24 \end{cases}$

Solution   We graph both equations on one set of coordinate axes, as shown in Figure 3-3 on the next page. In this example, the graphs appear to be parallel. We can show that this is true by writing each equation in slope-intercept form.

$$2x + 3y = 6$$
$$3y = -2x + 6$$
$$y = -\frac{2}{3}x + 2$$

$$4x + 6y = 24$$
$$6y = -4x + 24$$
$$y = -\frac{2}{3}x + 4$$

The slope of both lines is $-\frac{2}{3}$, but their y-intercepts, $(2, 0)$ and $(4, 0)$, are different, thus they are called **distinct lines**. The slopes are equal and the y-intercepts are different, therefore the lines are parallel.

Since the graphs are parallel lines, the lines do not intersect, and the system does not have a solution. The system is an *inconsistent system* and its solution set is $\varnothing$.

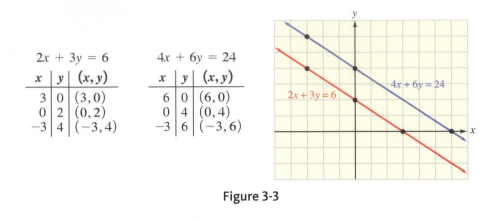

$$2x + 3y = 6$$

x	y	(x, y)
3	0	$(3, 0)$
0	2	$(0, 2)$
-3	4	$(-3, 4)$

$$4x + 6y = 24$$

x	y	(x, y)
6	0	$(6, 0)$
0	4	$(0, 4)$
-3	6	$(-3, 6)$

Figure 3-3

SELF CHECK 3   Solve the system by graphing: $\begin{cases} 2x + 3y = 6 \\ y = -\dfrac{2}{3}x - 3 \end{cases}$

3 Express the infinitely many solutions of a dependent system as a general ordered pair.

When the equations of a system have different graphs (as in Examples 1, 2, and 3), the equations are called **independent equations**. Two equations with the same graph are called **dependent equations**, as in Example 4.

EXAMPLE 4   Solve the system by graphing: $\begin{cases} 2y - x = 4 \\ 2x + 8 = 4y \end{cases}$

Solution   We graph each equation on one set of coordinate axes, as shown in Figure 3-4. Since the graphs are the same line (coincide), the system has infinitely many solutions. In fact, any ordered pair (x, y) that satisfies one equation satisfies the other.

$$2y - x = 4$$

x	y	(x, y)
-4	0	$(-4, 0)$
0	2	$(0, 2)$

$$2x + 8 = 4y$$

x	y	(x, y)
-4	0	$(-4, 0)$
0	2	$(0, 2)$

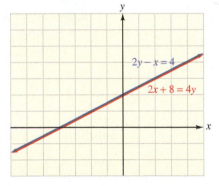

Figure 3-4

To find a general ordered pair solution, we can solve either equation of the system for either variable. If we solve one equation for y, we will obtain an expression for y in terms of x,

$$2y - x = 4$$
$$2y = x + 4$$
$$y = \frac{1}{2}x + 2$$

Substituting $\frac{1}{2}x + 2$ for y in the ordered pair (x, y), we obtain the general solution $\left(x, \frac{1}{2}x + 2\right)$. Since x is the independent variable, we can substitute any value for x to obtain a value for y to generate infinitely many ordered pair solutions.

SELF CHECK 4   Solve the system by graphing: $\begin{cases} 2x - y = 4 \\ x = \frac{1}{2}y + 2 \end{cases}$

We summarize the possibilities that can occur when we graph two linear equations, each with two variables.

SOLVING SYSTEMS OF TWO LINEAR EQUATIONS

If the lines are distinct and intersect, the equations are independent and the system is consistent. **One solution exists**, expressed as (x, y).

If the lines are distinct and parallel, the equations are independent and the system is inconsistent. **No solution exists**, expressed as $\varnothing$.

If the lines coincide (are the same), the equations are dependent and the system is consistent. **Infinitely many solutions exist**, expressed as a general ordered pair such as $(x, ax + b)$.

Accent on technology

▸ Solving Systems by Graphing

To solve the system

$$\begin{cases} 3x + 2y = 12 \\ 2x - 3y = 12 \end{cases}$$

using a graphing calculator, we first solve each equation for y so we can enter them into a graphing calculator. After solving for y, we obtain the following equivalent system. The graph is shown in Figure 3-5.

$$\begin{cases} y = -\frac{3}{2}x + 6 \\ y = \frac{2}{3}x - 4 \end{cases}$$

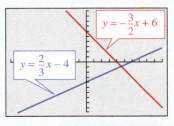

Figure 3-5

COMMENT   The intersect method will work only if the x-coordinate of the intersection point is visible on the screen. If it is not, you may need to adjust the viewing window.

We can solve this system using the INTERSECT feature found in the CALC menu on a TI-84 graphing calculator. From the graph, press **2nd TRACE** (CALC) and select "5: intersect" by using the down arrow to highlight and then press **ENTER** or by pressing 5 on the keyboard to obtain Figure 3-6(a) on the next page. We direct the calculator to look for the intersection by selecting a point on the first line and pressing **ENTER** and a point on the second line and pressing **ENTER**. We can move the cursor closer to the intersection point using the **TRACE** key and then press **ENTER** again. We obtain Figure 3-6(b). From the figure, we see that the solution is approximately $(4.6153846, -0.9230769)$.

(Continued)

(a)   (b)

Figure 3-6

Refer to the tear-out card for the method for converting decimal answers to fractions and verify that the exact solution is $x = \frac{60}{13}$ and $y = -\frac{12}{13}$ or $\left(\frac{60}{13}, \frac{-12}{13}\right)$.

For instructions regarding the use of a Casio graphing calculator, please refer to the Casio Keystroke Guide in the back of the book.

4   Use a system of linear equations in two variables to solve a linear equation in one variable graphically.

The graphing method discussed in this section can be used to solve equations in one variable.

EXAMPLE 5   Solve $2x + 4 = -2$ graphically.

Solution   To solve $2x + 4 = -2$ graphically, we can set the left and right sides of the equation equal to y. The graphs of $y = 2x + 4$ and $y = -2$ are shown in Figure 3-7. To solve $2x + 4 = -2$, we need to find the value of x that makes $2x + 4$ equal to -2. The point of intersection of the graphs is $(-3, -2)$. This indicates that if x is -3, the expression $2x + 4$ equals -2. So the solution of $2x + 4 = -2$ is -3. Check this result by substituting 3 for x in the original equation.

Figure 3-7

 SELF CHECK 5   Solve $2x + 4 = 2$ graphically.

Accent on technology

▶ Solving Equations Graphically

To solve $2(x - 3) + 3 = 7$ with a TI-84 graphing calculator, we graph the left and right sides of the equation in the same window by entering

$$Y_1 = 2(x - 3) + 3$$
$$Y_2 = 7$$

Figure 3-8(a) shows the graphs, generated using settings of $[-10, 10]$ for x and for y.

The coordinates of the point of intersection of the graphs can be determined using the INTERSECT feature outlined in the previous Accent on Technology.

COMMENT Notice that this is an equation in one variable. Although the intersection will display both an x-value and a y-value, only the value of x is the solution.

In Figure 3-8(b), we see that the point of intersection is $(5, 7)$, which indicates that 5 is the solution of $2(x - 3) + 3 = 7$.

(a)   (b)

Figure 3-8

For instructions regarding the use of a Casio graphing calculator, please refer to the Casio Keystroke Guide in the back of the book.

SELF CHECK ANSWERS

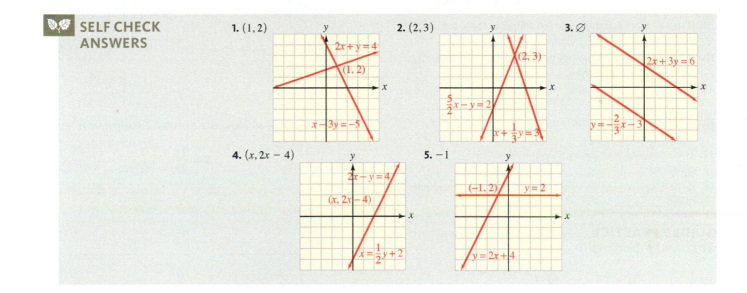

1. $(1, 2)$

2. $(2, 3)$

3. $\varnothing$

4. $(x, 2x - 4)$

5. -1

NOW TRY THIS

Solve each system by graphing.

a. $\begin{cases} y = x^2 - 4 \\ y = 2x - 1 \end{cases}$

b. $\begin{cases} y = -|x + 1| + 4 \\ y = -\frac{1}{3}x + 1 \end{cases}$

c. $\begin{cases} y = |x - 2| \\ y = x^2 \end{cases}$

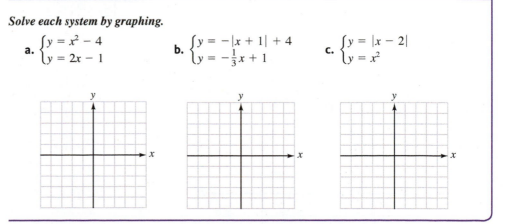

3.1   Exercises

WARM-UPS *Determine whether the ordered pair is a solution of the system of equations.*

1. $(1, 3)$; $\begin{cases} y = 3x \\ y = \frac{1}{3}x + \frac{8}{3} \end{cases}$ **2.** $(-1, 2)$; $\begin{cases} y = 2x + 4 \\ y = x + 2 \end{cases}$

3. $(2, -1)$; $\begin{cases} 2x - 4y = 8 \\ 4x + y = 9 \end{cases}$ **4.** $(-4, 3)$; $\begin{cases} 4x - y = -19 \\ 3x + 2y = -6 \end{cases}$

REVIEW *Write each number in scientific notation.*

5. 850,000,000   **6.** 0.000000479

7. 239×10^3 **8.** 465×10^{-4}

VOCABULARY AND CONCEPTS *Fill in the blanks.*

9. If two or more equations are considered at the same time, they are called a _____ of equations.

10. When a system of equations has one or more solutions, it is called a _____ system.

11. If a system has no solution, it is called an _____ system. Its solution set is $\varnothing$.

12. If two equations have different graphs, they are called _____ equations.

13. Two equations with the same graph are called _____ equations.

14. If the equations in two systems are equivalent, the systems are called _____ systems.

GUIDED PRACTICE *Solve each system by graphing.*
SEE EXAMPLE 1. (OBJECTIVE 1)

15. $\begin{cases} x + y = 6 \\ x - y = 2 \end{cases}$ **16.** $\begin{cases} x - y = 4 \\ 2x + y = 5 \end{cases}$

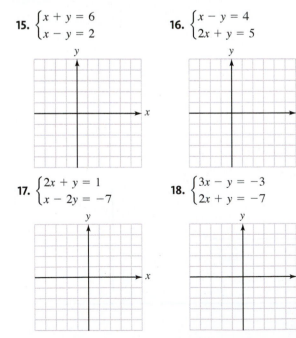

17. $\begin{cases} 2x + y = 1 \\ x - 2y = -7 \end{cases}$ **18.** $\begin{cases} 3x - y = -3 \\ 2x + y = -7 \end{cases}$

19. $\begin{cases} 2x + 3y = 0 \\ 2x + y = 4 \end{cases}$ **20.** $\begin{cases} 3x - 2y = 0 \\ 2x + 3y = 0 \end{cases}$

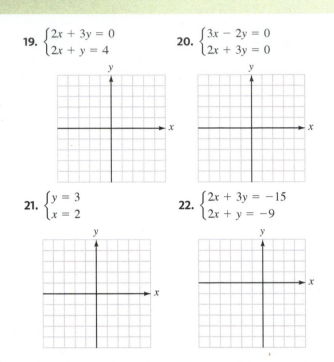

21. $\begin{cases} y = 3 \\ x = 2 \end{cases}$ **22.** $\begin{cases} 2x + 3y = -15 \\ 2x + y = -9 \end{cases}$

Solve each system by graphing. **SEE EXAMPLE 2. (OBJECTIVE 1)**

23. $\begin{cases} x = 2 \\ y = \frac{4 - x}{2} \end{cases}$ **24.** $\begin{cases} y = -2 \\ x = \frac{4 + 3y}{2} \end{cases}$

25. $\begin{cases} \frac{5}{2}x + y = \frac{1}{2} \\ 2x - \frac{3}{2}y = 5 \end{cases}$ **26.** $\begin{cases} x = \frac{5y - 4}{2} \\ x - \frac{5}{3}y + \frac{1}{3} = 0 \end{cases}$

Solve each system by graphing. State whether the system is inconsistent or if the equations are dependent. Write each solution set. SEE EXAMPLE 3. (OBJECTIVE 2)

27. $\begin{cases} 3x = 5 - 2y \\ 3x + 2y = 7 \end{cases}$

28. $\begin{cases} y = x \\ x - y = 7 \end{cases}$

29. $\begin{array}{l} x = 2y - 8 \\ y = \frac{1}{2}x - 5 \end{array}$

30. $\begin{cases} 8x - 2y = 7 \\ y = 4x + 3 \end{cases}$

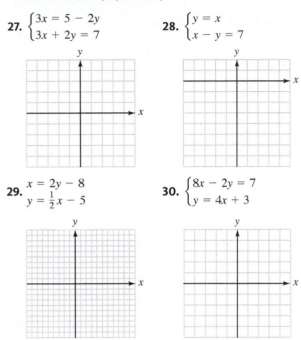

Solve each system by graphing. State whether the system is inconsistent or if the equations are dependent. Write each solution set. SEE EXAMPLE 4. (OBJECTIVE 3)

31. $\begin{cases} x = 3 - 2y \\ 2x + 4y = 6 \end{cases}$

32. $\begin{cases} x = y \\ y - x = 0 \end{cases}$

33. $\begin{cases} 6x + 3y = 9 \\ y + 2x = 3 \end{cases}$

34. $\begin{cases} x - 3y = 9 \\ 3x - 27 = 9y \end{cases}$

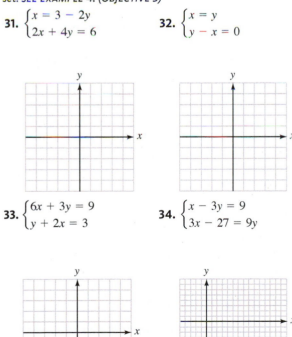

Solve each equation graphically. SEE EXAMPLE 5. (OBJECTIVE 4).

35. $\frac{3}{2}a + 4 = -5$

36. $-8(a - 1) = -8$

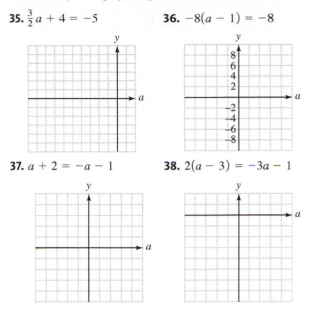

37. $a + 2 = -a - 1$

38. $2(a - 3) = -3a - 1$

ADDITIONAL PRACTICE *Determine whether the following systems will have one solution, no solution, or infinitely many solutions.*

39. $\begin{cases} y = 5x \\ y = 5x + 3 \end{cases}$

40. $\begin{cases} y = 4x \\ y = 3x + x \end{cases}$

41. $\begin{cases} y = 6x + 1 \\ y = -6x + 1 \end{cases}$

42. $\begin{cases} y = -3x + 2 \\ -3x = y \end{cases}$

Solve each system by graphing.

43. $\begin{cases} x = \frac{11 - 2y}{3} \\ y = \frac{11 - 6x}{4} \end{cases}$

44. $\begin{cases} x = \frac{1 - 3y}{4} \\ y = \frac{12 + 3x}{2} \end{cases}$

45. $\begin{cases} \frac{5}{2}x + 3y = 6 \\ y = \frac{24 - 10x}{12} \end{cases}$

46. $\begin{cases} 2x = 5y - 11 \\ 3x = 2y \end{cases}$

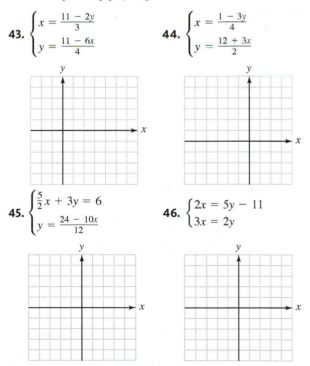

Use a graphing calculator to solve each system. Give the exact answer.

47. $\begin{cases} x = 13 - 4y \\ 3x = 4 + 2y \end{cases}$ **48.** $\begin{cases} 3x = 7 - 2y \\ 2x = 2 + 4y \end{cases}$

49. $\begin{cases} x = -\frac{3}{2}y \\ x = \frac{3}{2}y - 2 \end{cases}$ **50.** $\begin{cases} x = \frac{3y - 1}{4} \\ y = \frac{4 - 8x}{3} \end{cases}$

Use a graphing calculator to solve each system. Give all answers to the nearest hundredth.

51. $\begin{cases} y = 3.2x - 1.5 \\ y = -2.7x - 3.7 \end{cases}$ **52.** $\begin{cases} y = -0.45x + 5 \\ y = 5.55x - 13.7 \end{cases}$

53. $\begin{cases} 1.7x + 2.3y = 3.2 \\ y = 0.25x + 8.95 \end{cases}$ **54.** $\begin{cases} 2.75x = 12.9y - 3.79 \\ 7.1x - y = 35.76 \end{cases}$

APPLICATIONS

55. Retailing   The cost of manufacturing one type of camera and the revenue from the sale of those cameras are shown in the illustration.
  a. From the illustration, find the cost of manufacturing 15,000 cameras.
  b. From the illustration, find the revenue obtained by selling 20,000 cameras.
  c. For what number of cameras sold will the revenue equal the cost?

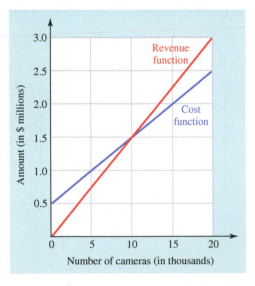

56. Food service
  a. Estimate the point of intersection of the two graphs shown in the illustration. Express your answer in the form (year, number of meals).
  b. What information about dining out does the point of intersection provide?

Number of Take-Out and On-Premise Meals Purchased at Commercial Restaurants per Person Annually

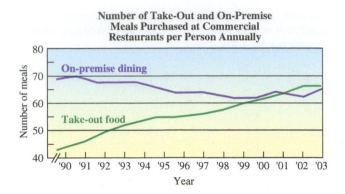

57. Navigation   Two ships are sailing on the same coordinate system. One ship is following a course described by $2x + 3y = 6$, and the other is following a course described by $2x - 3y = 9$.
  a. Is there a possibility of a collision?
  b. Find the coordinates of the danger point.
  c. Is a collision a certainty?

58. Navigation   Two airplanes are flying at the same altitude and in the same coordinate system. One plane is following a course described by $y = \frac{2}{5}x - 2$, and the other is following a course described by $x = \frac{5y + 7}{2}$. Is there a possibility of a collision?

59. Car repairs   Smith Chevrolet charges $50 per hour for labor on car repairs. Lopez Ford charges a diagnosis fee of $30 plus $40 per hour for labor. If the labor on an engine repair costs the same at either shop, how long does the repair take?

60. Landscaping   A landscaper installs some trees and shrubs at a bank. He installs 25 plants for a total cost of $1,500. How many trees ($t$) and how many shrubs ($s$) did he install if each tree costs $100 and each shrub costs $50?

61. Cost and revenue   The function $C(x) = 200x + 400$ gives the cost for a college to offer x sections of an introductory class in CPR (cardiopulmonary resuscitation). The function $R(x) = 280x$ gives the amount of revenue the college brings in when offering x sections of CPR. Find the point where the cost equals the revenue by graphing each function on the same coordinate system.

62. In Exercise 61, how many sections does the college need to offer to make a profit on the CPR training course?

WRITING ABOUT MATH

63. Explain how to solve a system of two equations in two variables.
64. Can a system of two equations in two variables have exactly two solutions? Why or why not?

SOMETHING TO THINK ABOUT

65. Form an independent system of equations with a solution of $(-5, 2)$.

66. Form a dependent system of equations with a solution of $(-5, 2)$.

Section 3.2

Solving Systems of Two Linear Equations in Two Variables by Substitution and Elimination (Addition)

Objectives

 1 Solve a system of two linear equations by substitution.

2 Solve a system of two linear equations by elimination (addition).

3 Identify an inconsistent linear system.

4 Identify a dependent linear system and express the answer as a general ordered pair.

5 Write a repeating decimal in fractional form.

6 Solve an application by setting up and solving a system of two linear equations.

Getting Ready

Remove parentheses by using the distributive property.

1. $3(2x - 7)$ **2.** $-4(3x + 5)$

Substitute x − 3 for y and remove parentheses by using the distributive property.

3. $3y$ **4.** $-2(y + 2)$

The graphing method discussed in the previous section provides a way to visualize the process of solving systems of equations. However, it can be difficult to use to solve systems of higher order, such as three equations, each with three variables. In this section, we will discuss algebraic methods that enable us to solve such systems.

1 Solve a system of two linear equations by substitution.

To solve a system of two linear equations (each with two variables) by substitution, we use the following steps.

THE SUBSTITUTION METHOD

1. If necessary, solve one equation for one of its variables, preferably a variable with a coefficient of 1.

2. Substitute the resulting expression for the variable obtained in Step 1 into the other equation and solve that equation.

3. Find the value of the other variable by substituting the value of the variable found in Step 2 into any equation containing both variables.

4. State the solution.

5. Check the solution in both of the original equations.

EXAMPLE 1   Solve the system by substitution: $\begin{cases} 4x + y = 13 \\ -2x + 3y = -17 \end{cases}$

Solution   **Step 1:**   We solve the first equation for y, because y has a coefficient of 1 and no fractions are introduced.

$$4x + y = 13$$

(1) $$y = -4x + 13 \quad \text{Subtract } 4x \text{ from both sides.}$$

Step 2:   We then substitute $-4x + 13$ for y in the second equation of the system and solve for x.

$$-2x + 3y = -17$$
$$-2x + 3(\mathbf{-4x + 13}) = -17 \quad \text{Substitute.}$$
$$-2x - 12x + 39 = -17 \quad \text{Use the distributive property to remove parentheses.}$$
$$-14x = -56 \quad \text{Combine like terms and subtract 39 from both sides.}$$
$$x = 4 \quad \text{Divide both sides by } -14.$$

Step 3:   To find the value of y, we substitute 4 for x in Equation 1 and simplify:

$$y = -4\mathbf{x} + 13$$
$$y = -4(\mathbf{4}) + 13 \quad \text{Substitute.}$$
$$= -3$$

Step 4:   The solution is $x = 4$ and $y = -3$, or $(4, -3)$. The graphs of these two equations would intersect at the point with coordinates $(4, -3)$.

Step 5:   To verify that this solution satisfies both equations, we substitute $x = 4$ and $y = -3$ into each equation in the original system and simplify.

$$4\mathbf{x} + \mathbf{y} = 13 \qquad\qquad -2\mathbf{x} + 3\mathbf{y} = -17$$
$$4(\mathbf{4}) + (\mathbf{-3}) \stackrel{?}{=} 13 \qquad\qquad -2(\mathbf{4}) + 3(\mathbf{-3}) \stackrel{?}{=} -17$$
$$16 - 3 \stackrel{?}{=} 13 \qquad\qquad -8 - 9 \stackrel{?}{=} -17$$
$$13 = 13 \qquad\qquad -17 = -17$$

Since the ordered pair $(4, -3)$ satisfies both equations of the system, the solution checks.

SELF CHECK 1   Solve the system by substitution: $\begin{cases} x + 3y = 9 \\ 2x - y = -10 \end{cases}$

EXAMPLE 2   Solve the system by substitution: $\begin{cases} \frac{4}{3}x + \frac{1}{2}y = -\frac{2}{3} \\ \frac{1}{2}x + \frac{2}{3}y = \frac{5}{3} \end{cases}$

Solution   First we find an equivalent system without fractions by multiplying both sides of each equation by 6, the LCD.

(1) $\begin{cases} 8x + 3y = -4 \\ 3x + 4y = 10 \end{cases}$
(2)

Because no variable in either equation has a coefficient of 1, it is impossible to avoid fractions when solving for a variable. We solve Equation 2 for x.

$$3x + 4y = 10$$
$$3x = -4y + 10 \quad \text{Subtract } 4y \text{ from both sides.}$$

(3) $$x = -\frac{4}{3}y + \frac{10}{3} \quad \text{Divide both sides by 3.}$$

We then substitute $-\frac{4}{3}y + \frac{10}{3}$ for x in Equation 1 and solve for y.

$$8x + 3y = -4$$

$$8\left(-\frac{4}{3}y + \frac{10}{3}\right) + 3y = -4 \qquad \text{Substitute.}$$

$$-\frac{32}{3}y + \frac{80}{3} + 3y = -4 \qquad \text{Use the distributive property to remove parentheses.}$$

$$-32y + 80 + 9y = -12 \qquad \text{Multiply both sides by 3 to clear fractions.}$$

$$-23y = -92 \qquad \text{Combine like terms and subtract 80 from both sides.}$$

$$y = 4 \qquad \text{Divide both sides by } -23.$$

We can find the value of x by substituting 4 for y in Equation 3 and simplifying:

$$x = -\frac{4}{3}y + \frac{10}{3}$$

$$= -\frac{4}{3}(4) + \frac{10}{3} \qquad \text{Substitute.}$$

$$= -\frac{6}{3} \qquad\qquad -\frac{16}{3} + \frac{10}{3} = -\frac{6}{3}$$

$$= -2$$

The solution is the ordered pair $(-2, 4)$. Verify that this solution checks in both of the equations in the original system.

SELF CHECK 2   Solve the system by substitution: $\begin{cases} \frac{2}{3}x + \frac{1}{2}y = 1 \\ \frac{1}{3}x - \frac{3}{2}y = 4 \end{cases}$

---

2 | Solve a system of two linear equations by elimination (addition).

Sometimes using the substitution method can be cumbersome, especially if none of the coefficients are 1. We have another algebraic method of solving systems that may be more efficient. In the elimination (addition) method, we combine the equations of the system in a way that will eliminate the terms involving one of the variables.

THE ELIMINATION (ADDITION) METHOD

1. If necessary, write both equations of the system in $Ax + By = C$ form.
2. If necessary, multiply the terms of one or both of the equations by constants chosen to make the coefficients of one of the variables opposites.
3. Add the equations and solve the resulting equation, if possible.
4. Substitute the value obtained in Step 3 into either of the original equations and solve for the remaining variable, if applicable.
5. State the solution obtained in Steps 3 and 4.
6. Check the solution in both equations of the original system.

EXAMPLE 3   Solve the system by elimination: $\begin{cases} 4x + y = 13 \\ -2x + 3y = -17 \end{cases}$

Solution   This is the same system we solved in Example 1. Since both equations are already written in $Ax + By = C$ form, Step 1 is unnecessary.

Step 2:   To solve the system by elimination, we multiply the second equation by 2 to make the coefficients of x opposites.

$$\begin{cases} 4x + y = 13 \\ -4x + 6y = -34 \end{cases}$$

Step 3:   When we add the equations, the terms involving x are eliminated because $4x$ and $-4x$ are additive inverses and their sum is 0. We will obtain

$$7y = -21$$
$$y = -3 \qquad \text{Divide both sides by 7.}$$

Step 4:   To find the value of x, we substitute -3 for y in either of the original equations and solve. If we use the first equation, we have

$$4x + y = 13$$
$$4x + (-3) = 13 \qquad \text{Substitute.}$$
$$4x = 16 \qquad \text{Add 3 to both sides.}$$
$$x = 4 \qquad \text{Divide both sides by 4.}$$

Step 5:   The solution is $x = 4$ and $y = -3$, or $(4, -3)$.

Step 6:   The check was completed in Example 1.

SELF CHECK 3   Solve the system by elimination: $\begin{cases} 3x + 2y = 0 \\ 2x - y = -7 \end{cases}$

EXAMPLE 4   Solve the system by elimination: $\begin{cases} \frac{4}{3}x + \frac{1}{2}y = -\frac{2}{3} \\ \frac{1}{2}x + \frac{2}{3}y = \frac{5}{3} \end{cases}$

Solution   This is the same system we solved in Example 2. To solve it by elimination, we may find an equivalent system with no fractions by multiplying both sides of each equation by 6 to obtain

(1) $\begin{cases} 8x + 3y = -4 \\ 3x + 4y = 10 \end{cases}$
(2)

We can solve for x by eliminating the terms involving y. To do so, we multiply both sides of Equation 1 by 4 and both sides of Equation 2 by -3 to obtain

$$\begin{cases} 32x + 12y = -16 \\ -9x - 12y = -30 \end{cases}$$

When these equations are added, the y-terms are eliminated and we obtain the result

$$23x = -46$$
$$x = -2 \qquad \text{Divide both sides by 23.}$$

To find the value of y, we substitute -2 for x in either Equation 1 or Equation 2. If we substitute -2 for x in Equation 2, we have

$$3x + 4y = 10$$
$$3(-2) + 4y = 10 \qquad \text{Substitute.}$$
$$-6 + 4y = 10 \qquad \text{Simplify.}$$
$$4y = 16 \qquad \text{Add 6 to both sides.}$$
$$y = 4 \qquad \text{Divide both sides by 4.}$$

The solution is $(-2, 4)$. The check was completed in Example 2.

SELF CHECK 4   Solve the system by elimination: $\begin{cases} \frac{4}{3}x + \frac{1}{2}y = -3 \\ \frac{1}{2}x + \frac{2}{3}y = -\frac{1}{6} \end{cases}$

3 Identify an inconsistent linear system.

In the next two examples, we encounter the situation in which both variables are simultaneously eliminated but with different interpretations.

EXAMPLE 5   Solve the system using any method: $\begin{cases} y = 2x + 4 \\ 8x - 4y = 7 \end{cases}$

Solution   Because the first equation is already solved for y, we will use the substitution method.

$$8x - 4\mathbf{y} = 7$$

$$8x - 4(\mathbf{2x + 4}) = 7 \quad \text{Substitute } 2x + 4 \text{ for } y \text{ in the second equation.}$$

We then solve this equation for x:

$$8x - 8x - 16 = 7 \quad \text{Use the distributive property to remove parentheses.}$$

$$-16 = 7 \quad \text{Combine like terms.}$$

This false statement indicates that the equations in the system are independent (they represent distinct lines), but that the system is inconsistent. If the equations of this system were graphed, the graphs would be parallel. Since the system has no point of intersection, its solution set is $\varnothing$.

SELF CHECK 5   Solve the system using any method: $\begin{cases} x = -\frac{5}{2}y + 5 \\ y = -\frac{2}{5}x + 5 \end{cases}$

4 Identify a dependent linear system and express the answer as a general ordered pair.

In the next example, both variables are eliminated but the result is different.

EXAMPLE 6   Solve the system using any method: $\begin{cases} 4x + 6y = 12 \\ -2x - 3y = -6 \end{cases}$

Solution   Since the equations are written in $Ax + By = C$ form, we will use the elimination (addition) method. To eliminate the x-terms when adding the equations, we multiply both sides of the second equation by 2 to obtain

$$\begin{cases} 4x + 6y = 12 \\ -4x - 6y = -12 \end{cases}$$

COMMENT When we solved a linear equation in one variable, a result of $0 = 0$ indicated a solution of all real numbers. When we solve a system of linear equations in two variables a result of $0 = 0$ indicates a solution of an infinite number of ordered pairs that could be written in the form $(x, ax + b)$.

After adding the left and right sides, we have

$$0x + 0y = 0$$

$$0 = 0$$

Here, both the x- and y-terms are eliminated. The true statement $0 = 0$ indicates that the equations in this system are dependent and that the system is consistent.

Note that the equations of the system are equivalent, because when the second equation is multiplied by -2, it becomes the first equation. The line graphs of these equations would coincide. Since any ordered pair that satisfies one of the equations also satisfies the other, there are infinitely many solutions.

To find a general solution, we can solve either equation of the system to obtain $y = -\frac{2}{3}x + 2$. Substituting $-\frac{2}{3}x + 2$ for y in the ordered pair (x, y), we obtain the general solution $\left(x, -\frac{2}{3}x + 2\right)$.

SELF CHECK 6   Solve the system using any method: $\begin{cases} x = -\frac{5}{2}y + 5 \\ y = -\frac{2}{5}x + 2 \end{cases}$

5   Write a repeating decimal in fractional form.

We have seen that to write a fraction in decimal form, we divide its numerator by its denominator. The result is often a repeating decimal. By using systems of equations, we can write a repeating decimal in fractional form.

EXAMPLE 7   Write $0.2\overline{54}$ as a fraction.

Solution   To write $0.2\overline{54}$ as a fraction, we note that the decimal has a repeating block of two digits and then form an equation by setting x equal to the decimal.

(1) $x = 0.2545454\ldots$

We then form another equation by multiplying both sides of Equation 1 by 10^2.

(2) $100x = 25.4545454\ldots$ $10^2 = 100$

We can *subtract* each side of Equation 1 from the corresponding side of Equation 2 to obtain

$$100x = 25.4\,54\,54\,54\ldots$$
$$\underline{\quad x = \ 0.2\,54\,54\,54\ldots}$$
$$99x = 25.2$$

Finally, we solve $99x = 25.2$ for x and simplify the fraction.

$$x = \frac{25.2}{99} = \frac{25.2 \cdot \mathbf{10}}{99 \cdot \mathbf{10}} = \frac{252}{990} = \frac{18 \cdot 14}{18 \cdot 55} = \frac{14}{55}$$

We can use a calculator to verify that the decimal representation of $\frac{14}{55}$ is $0.2\overline{54}$.

SELF CHECK 7   Write $0.3\overline{72}$ as a fraction.

The key step in solving Example 7 was multiplying both sides of Equation 1 by 10^2. If there had been n digits in the repeating block of the decimal, we would have multiplied both sides of Equation 1 by 10^n.

6   Solve an application by setting up and solving a system of two linear equations.

To solve applications using two variables, we follow the same problem-solving strategy discussed in Chapter 1, except that we use two variables and form two equations instead of one.

EXAMPLE 8   **RETAIL SALES**   A store advertises two types of cell phones, one selling for $67 and the other for $100. If the receipts from the sale of 36 phones totaled $2,940, how many of each type were sold?

Analyze the problem   We can let x represent the number of phones sold for $67 and let y represent the number of phones sold for $100.

Form two equations   Because a total of 36 phones were sold, we can form the equation

The number of lower-priced phones	plus	the number of higher-priced phones	equals	the total number of phones.
x	$+$	y	$=$	36

We know that the receipts for the sale of x of the $67 phones will be $67x$ and that the receipts for the sale of y of the $100 phones will be $100y$. Since the sum of these receipts is $2,940, the second equation is

The value of the lower-priced phones	plus	the value of the higher-priced phones	equals	the total receipts.
$67x$	$+$	$100y$	$=$	$2,940$

Solve the system   To find out how many of each type of phone were sold, we must solve the system

$$(1) \quad \begin{cases} x + y = 36 \\ 67x + 100y = 2{,}940 \end{cases}$$
$$(2)$$

We multiply both sides of Equation 1 by -100, add the resulting equation to Equation 2, and solve for x:

$$\begin{array}{r} -100x - 100y = -3{,}600 \\ \underline{67x + 100y = 2{,}940} \\ -33x = -660 \end{array}$$

$$x = 20 \qquad \text{Divide both sides by } -33.$$

To find the value of y, we substitute 20 for x in Equation 1 and solve.

$$x + y = 36$$
$$20 + y = 36 \qquad \text{Substitute.}$$
$$y = 16 \qquad \text{Subtract 20 from both sides.}$$

State the conclusion   The store sold 20 of the $67 phones and 16 of the $100 phones.

Check the result   If 20 of one type were sold and 16 of the other type were sold, a total of 36 phones were sold.

Since the value of the lower-priced phones is $20(\$67) = \$1{,}340$ and the value of the higher-priced phones is $16(\$100) = \$1{,}600$, the total receipts are $2,940.

SELF CHECK 8   If the sale of the 36 phones totaled $2,775, how many of each type were sold?

EXAMPLE 9   **MIXING SOLUTIONS**   How many ounces of a 5% saline solution and how many ounces of a 20% saline solution must be mixed together to obtain 50 ounces of a 15% saline solution?

Analyze the problem To find how many ounces of each type of saline solution should be mixed, we can let x represent the number of ounces of the 5% saline solution and y represent the number of ounces of the 20% saline solution. (See Figure 3-9.)

Figure 3-9

Form two equations Because a total of 50 ounces are needed, one of the equations will be

The number of ounces of 5% saline solution	plus	the number of ounces of 20% saline solution	equals	the total number of ounces in the mixture.
x	$+$	y	$=$	50

The amount of salt in the x ounces of 5% saline solution is $0.05x$, and the amount of salt in the y ounces of 20% saline solution is $0.20y$. The amount of salt in the 50 ounces of 15% saline solution will be $0.15(50)$. This gives the equation

The salt in the 5% solution	plus	the salt in the 20% solution	equals	the salt in the mixture.
$0.05x$	$+$	$0.20y$	$=$	$0.15(50)$

Solve the system To find out how many ounces of each are needed, we solve the following system:

(1) $\begin{cases} x + y = 50 \\ 0.05x + 0.20y = 7.5 \quad \text{\textcolor{red}{$0.15(50) = 7.5$}} \end{cases}$
(2)

To solve this system by substitution, we can solve Equation 1 for y

$x + y = 50$

(3) $\qquad y = 50 - x$ Subtract x from both sides.

and then substitute $50 - x$ for y in Equation 2.

$$0.05x + 0.20\textcolor{red}{y} = 7.5$$
$$0.05x + 0.20(\textcolor{red}{50 - x}) = 7.5 \qquad \text{Substitute.}$$
$$5x + 20(50 - x) = 750 \qquad \text{Multiply both sides by 100.}$$
$$5x + 1{,}000 - 20x = 750 \qquad \text{Use the distributive property to remove parentheses.}$$
$$-15x = -250 \qquad \text{Combine like terms and subtract 1,000 from both sides.}$$
$$x = \frac{-250}{-15} \qquad \text{Divide both sides by } -15.$$
$$x = \frac{50}{3} \qquad \text{Simplify.}$$

To find the value of y, we substitute $\frac{50}{3}$ for x in Equation 3:

$$y = 50 - x$$

$$= 50 - \frac{50}{3} \quad \text{Substitute.}$$

$$= \frac{100}{3}$$

State the conclusion   To obtain 50 ounces of a 15% saline solution, we must mix $\frac{50}{3}\left(16\frac{2}{3}\right)$ ounces of the 5% saline solution with $\frac{100}{3}\left(33\frac{1}{3}\right)$ ounces of the 20% saline solution.

Check the result   We note that $16\frac{2}{3}$ ounces of solution plus $33\frac{1}{3}$ ounces of solution equals the required 50 ounces of solution. We also note that 5% of $16\frac{2}{3} \approx 0.83$, and 20% of $33\frac{1}{3} \approx 6.67$, giving a total of 7.5, which is 15% of 50.

SELF CHECK 9   How many ounces of a 10% saline solution and how many ounces of a 15% saline solution must be mixed together to obtain 50 ounces of a 12% saline solution?

Systems of equations can be used to solve many applications. Sometimes the equations may involve more than two variables.

In manufacturing, running a machine involves both *setup costs* and *unit costs*. Setup costs include the cost of preparing a machine to do a certain job. Unit costs depend on the number of items to be manufactured, including costs of raw materials and labor.

Suppose that a certain machine has a setup cost of $600 and a unit cost of $3 per item. If x items will be manufactured using this machine, the cost will be

$$\text{Cost} = 600 + 3x \quad \text{Cost = setup cost + unit cost · the number of items}$$

Furthermore, suppose that a larger and more efficient machine has a setup cost of $800 and a unit cost of $2 per item. The cost of manufacturing x items using this machine is

$$\text{Cost on larger machine} = 800 + 2x$$

The *break point* is the number of units x that need to be manufactured to make the cost of running either machine the same. The break point can be found by setting the two costs equal to each other and solving for x.

$$600 + 3x = 800 + 2x$$

$$x = 200 \qquad \text{Subtract 600 and } 2x \text{ from both sides.}$$

The break point is 200 units, because the cost using either machine is $1,200 when $x = 200$.

Cost on small machine $= 600 + 3x$	Cost on large machine $= 800 + 2x$
$= 600 + 3(200)$	$= 800 + 2(200)$
$= 600 + 600$	$= 800 + 400$
$= \$1,200$	$= \$1,200$

EXAMPLE 10   **BREAK-POINT ANALYSIS**   One machine has a setup cost of $400 and a unit cost of $1.50, and another machine has a setup cost of $500 and a unit cost of $1.25. Find the break point.

Analyze the problem   The cost of manufacturing x units using machine 1 is x times $1.50, plus the setup cost of $400. The cost of manufacturing x units using machine 2 is x times $1.25, plus the setup cost of $500. The break point occurs when the costs are equal.

Form two equations The cost C_1 using machine 1 is

The cost of using machine 1	equals	the cost of manufacturing x units at \$1.50 per unit	plus	the setup cost.
C_1	=	$1.5x$	+	400

The cost C_2 using machine 2 is

The cost of using machine 2	equals	the cost of manufacturing x units at \$1.25 per unit	plus	the setup cost.
C_2	=	$1.25x$	+	500

Solve the system To find the break point, we must solve the system $\begin{cases} C_1 = 1.5x + 400 \\ C_2 = 1.25x + 500 \end{cases}$. Since the break point occurs when $C_1 = C_2$, we can substitute $1.5x + 400$ for C_2 to obtain

$$1.5x + 400 = 1.25x + 500$$

$1.5x = 1.25x + 100$ Subtract 400 from both sides.

$0.25x = 100$ Subtract $1.25x$ from both sides.

$x = 400$ Divide both sides by 0.25.

State the conclusion The break point is 400 units.

Check the result For 400 units, the cost using machine 1 is $400 + 1.5(400) = 400 + 600 = \$1,000$. The cost using machine 2 is $500 + 1.25(400) = 500 + 500 = \$1,000$. Since the costs are equal, the break point is 400 units.

SELF CHECK 10 One machine has a setup cost of \$300 and a unit cost of \$1.50 and another machine has a setup cost of \$700 and a unit cost of \$1.25. Find the break point.

Some applications include geometric figures. A *parallelogram* is a four-sided figure with its opposite sides parallel. (See Figure 3-10(a).) Here are some important facts about parallelograms.

1. Opposite sides of a parallelogram have the same measure.

2. Opposite angles of a parallelogram have the same measure.

3. Consecutive angles of a parallelogram are supplementary.

4. A diagonal of a parallelogram (see Figure 3-10(b)) divides the parallelogram into two *congruent triangles*—triangles with the same shape and same area.

5. In Figure 3-10(b), $\angle 1$ and $\angle 2$, and $\angle 3$ and $\angle 4$, are called pairs of *alternate interior angles*. When a diagonal intersects two parallel sides of a parallelogram, all pairs of alternate interior angles have the same measure.

(a)　　　　　　　　　　(b)

Figure 3-10

EXAMPLE 11 **PARALLELOGRAMS** Refer to the parallelogram shown in Figure 3-11 and find the values of x and y.

Figure 3-11

Solution   Since diagonal AC intersects two parallel sides, the alternate interior angles that are formed have the same measure. Thus, $(x - y)° = 30°$. Since opposite angles of a parallelogram have the same measure, we know that $(x + y)° = 110°$. We can form the following system of equations and solve it by addition.

$$(1) \quad \begin{cases} x - y = 30 \\ x + y = 110 \end{cases}$$
$$(2)$$

$$2x = 140 \quad \text{Add Equations 1 and 2 to eliminate } y.$$
$$x = 70 \quad \text{Divide both sides by 2.}$$

We can substitute 70 for x in Equation 2 and solve for y.

$$x + y = 110$$
$$70 + y = 110 \quad \text{Substitute 70 for } x.$$
$$y = 40 \quad \text{Subtract 70 from both sides.}$$

Thus, $x = 70$ and $y = 40$.

SELF CHECK 11   If the angle at B is 112° and the other angle symbol (DCA) is 42°, find the value of x and y.

SELF CHECK ANSWERS   **1.** $(-3, 4)$ **2.** $(3, -2)$ **3.** $(-2, 3)$ **4.** $(-3, 2)$ **5.** $\varnothing$ **6.** $\left(x, -\frac{2}{5}x + 2\right)$ **7.** $\frac{41}{110}$ **8.** The store sold 25 of the $67 phones and 11 of the $100 phones.   **9.** There should be 20 ounces of 15% saline solution and 30 ounces of the 10% saline solution.   **10.** The break point is 1,600 units.   **11.** $x = 77$ and $y = 35$

NOW TRY THIS

Use substitution to solve the system: $\begin{cases} t = r + 1 \\ r + s + t = 0 \\ r + s = -2 \end{cases}$

3.2 Exercises

WARM-UPS *Substitute $x - 2$ for y and solve for x.*

1. $3x - 4y = 7$ **2.** $5x + 3y = 18$

Add the left and right sides of the equations in each system and write the result.

3. $\begin{cases} 4x - 2y = 6 \\ 3x + 2y = 8 \end{cases}$ **4.** $\begin{cases} 5x + 3y = 10 \\ -5x - 7y = -8 \end{cases}$

REVIEW *Simplify each expression. Write all answers without using negative exponents.*

5. $(a^3b^4)^2(ab^2)^3$

6. $\left(\dfrac{a^2b^3c^4d}{ab^2c^3d^4}\right)^{-3}$

7. $\left(\dfrac{-3x^3y^4}{x^{-5}y^3}\right)^{-4}$

8. $\dfrac{5t^0 - 2t^0 + 3}{4t^0 + t^0}$

VOCABULARY AND CONCEPTS *Fill in the blanks.*

9. Running a machine involves both _____ costs and _____ costs.

10. The _____ point is the number of units that need to be manufactured to make the cost the same on either of two machines.

11. A _____ is a four-sided figure with both pairs of opposite sides parallel.

12. _____ sides of a parallelogram have the same length.

13. _____ angles of a parallelogram have the same measure.

14. _____ angles of a parallelogram are supplementary.

GUIDED PRACTICE *Solve each system by substitution.*
SEE EXAMPLE 1. (OBJECTIVE 1)

15. $\begin{cases} y = 3x \\ x + y = 8 \end{cases}$

16. $\begin{cases} y = x + 2 \\ x + 2y = 16 \end{cases}$

17. $\begin{cases} x - y = 2 \\ 2x + y = 13 \end{cases}$

18. $\begin{cases} x - y = -4 \\ 3x - 2y = -5 \end{cases}$

Solve each system by substitution. **SEE EXAMPLE 2. (OBJECTIVE 1)**

19. $\begin{cases} \dfrac{x}{2} + \dfrac{y}{2} = 6 \\ \dfrac{x}{2} - \dfrac{y}{2} = -2 \end{cases}$

20. $\begin{cases} \dfrac{x}{2} - \dfrac{y}{3} = -4 \\ \dfrac{x}{2} + \dfrac{y}{9} = 0 \end{cases}$

21. $\begin{cases} x = \dfrac{2}{3}y \\ y = 4x + 5 \end{cases}$

22. $\begin{cases} \dfrac{3x}{5} + \dfrac{5y}{3} = 2 \\ \dfrac{6x}{5} - \dfrac{5y}{3} = 1 \end{cases}$

Solve each system by elimination (addition). **SEE EXAMPLE 3. (OBJECTIVE 2)**

23. $\begin{cases} 2x - y = -5 \\ 2x + y = -3 \end{cases}$

24. $\begin{cases} x + 3y = 7 \\ x - 3y = -11 \end{cases}$

25. $\begin{cases} 3x + 5y = 2 \\ 4x - 3y = 22 \end{cases}$

26. $\begin{cases} 4x - 3y = 9 \\ 3x + 2y = 11 \end{cases}$

27. $\begin{cases} 5x - 2y = 19 \\ 3x + 4y = 1 \end{cases}$

28. $\begin{cases} 2y - 3x = -13 \\ 3x - 17 = 4y \end{cases}$

29. $\begin{cases} 4x + 6y = 5 \\ 8x - 9y = 3 \end{cases}$

30. $\begin{cases} 4x + 9y = 8 \\ 2x - 6y = -3 \end{cases}$

Solve each system by elimination (addition). **SEE EXAMPLE 4. (OBJECTIVE 2)**

31. $\begin{cases} \dfrac{5}{6}x + \dfrac{2}{3}y = \dfrac{7}{6} \\ \dfrac{10}{7}x - \dfrac{4}{9}y = \dfrac{17}{21} \end{cases}$

32. $\begin{cases} \dfrac{3x}{2} - \dfrac{2y}{3} = 0 \\ \dfrac{3x}{4} + \dfrac{4y}{3} = \dfrac{5}{2} \end{cases}$

33. $\begin{cases} \dfrac{3}{4}x + \dfrac{2}{3}y = 7 \\ \dfrac{3}{5}x - \dfrac{1}{2}y = 18 \end{cases}$

34. $\begin{cases} \dfrac{2}{3}x - \dfrac{1}{4}y = -8 \\ \dfrac{1}{2}x - \dfrac{3}{8}y = -9 \end{cases}$

Solve each system using any method. **SEE EXAMPLE 5. (OBJECTIVE 3)**

35. $\begin{cases} 3x = 2y - 4 \\ 6x - 4y = -4 \end{cases}$

36. $\begin{cases} x = \dfrac{3}{2}y + 5 \\ 2x - 3y = 8 \end{cases}$

37. $\begin{cases} x = 5y + 2 \\ 3x = 15y + 10 \end{cases}$

38. $\begin{cases} 3x - 6y = 8 \\ -6x + 12y = -10 \end{cases}$

Solve each system using any method. **SEE EXAMPLE 6. (OBJECTIVE 4)**

39. $\begin{cases} 12x = 4y + 6 \\ 6x - 2y = 3 \end{cases}$

40. $\begin{cases} 8x - 4y = 16 \\ 2x - 4 = y \end{cases}$

41. $\begin{cases} y - 2x = 6 \\ 4x + 12 = 2y \end{cases}$

42. $\begin{cases} 6x - 2y = 7 \\ 12x - 4y = 14 \end{cases}$

Write each repeating decimal as a fraction. Simplify the answer when possible. **SEE EXAMPLE 7. (OBJECTIVE 5)**

43. $0.\overline{6}$

44. $0.\overline{29}$

45. $-0.3\overline{489}$

46. $-2.3\overline{47}$

ADDITIONAL PRACTICE *Solve each system using any method.*

47. $\begin{cases} y = 2x \\ x + y = 6 \end{cases}$

48. $\begin{cases} y = -x \\ 2x + y = 4 \end{cases}$

49. $\begin{cases} 3x - 4y = 9 \\ x + 2y = 8 \end{cases}$

50. $\begin{cases} 3x - 2y = -10 \\ 6x + 5y = 25 \end{cases}$

51. $\begin{cases} x - y = 6 \\ x + y = 2 \end{cases}$

52. $\begin{cases} 4x + 2y = 4 \\ 2x + y = 5 \end{cases}$

53. $\begin{cases} 2x + 2y = -1 \\ 3x + 4y = 0 \end{cases}$

54. $\begin{cases} 5x + 3y = -7 \\ 3x - 3y = 7 \end{cases}$

55. $\begin{cases} \dfrac{2}{5}x - \dfrac{1}{6}y = \dfrac{7}{10} \\ \dfrac{3}{4}x - \dfrac{2}{3}y = \dfrac{19}{8} \end{cases}$

56. $\begin{cases} \dfrac{5}{6}x - y = \dfrac{-13}{6} \\ \dfrac{3}{2}x + y = \dfrac{-19}{2} \end{cases}$

57. $\begin{cases} 3x + 5y = -14 \\ 2x - 3y = 16 \end{cases}$

58. $\begin{cases} 2x + 3y = 8 \\ 6y + 4x = 16 \end{cases}$

Solve each system.

59. $\begin{cases} \dfrac{1}{x} + \dfrac{1}{y} = \dfrac{5}{6} \\ \dfrac{1}{x} - \dfrac{1}{y} = \dfrac{1}{6} \end{cases}$

60. $\begin{cases} \dfrac{1}{x} + \dfrac{1}{y} = \dfrac{9}{20} \\ \dfrac{1}{x} - \dfrac{1}{y} = \dfrac{1}{20} \end{cases}$

61. $\begin{cases} \dfrac{1}{x} + \dfrac{2}{y} = -1 \\ \dfrac{2}{x} - \dfrac{1}{y} = -7 \end{cases}$

62. $\begin{cases} \dfrac{3}{x} - \dfrac{2}{y} = -30 \\ \dfrac{2}{x} - \dfrac{3}{y} = -30 \end{cases}$

APPLICATIONS *Use a system of two linear equations to solve.*
SEE EXAMPLES 8–11. (OBJECTIVE 6)

63. Merchandising   A pair of shoes and a sweater cost $110. If the sweater cost $20 more than the shoes, how much did the sweater cost?

64. Merchandising   A sporting goods salesperson sells 2 fishing reels and 5 rods for $270. The next day, the salesperson sells 4 reels and 2 rods for $220. How much does each cost?

65. Electronics   Two resistors in the voltage divider circuit in the illustration have a total resistance of 1,375 ohms. To provide the required voltage, R_1 must be 125 ohms greater than R_2. Find both resistances.

66. Stowing baggage   A small aircraft can carry 950 pounds of baggage, distributed between two storage compartments. On one flight, the plane is fully loaded, with 150 pounds more baggage in one compartment than the other. How much is stowed in each compartment?

67. Geometry   The rectangular field in the illustration is surrounded by 72 meters of fencing. If the field is partitioned as shown, a total of 88 meters of fencing is required. Find the dimensions of the field.

68. Geometry   In a right triangle, one acute angle is 15° greater than two times the other acute angle. Find the difference between the angles.

69. Investment income   Part of $8,000 was invested at 10% interest and the rest at 12%. If the annual income from these investments was $900, how much was invested at each rate?

70. Investment income   Part of $12,000 was invested at 6% interest and the rest at 7.5%. If the annual income from these investments was $810, how much was invested at each rate?

71. Mixing solutions   How many ounces of the two alcohol solutions in the illustration must be mixed to obtain 100 ounces of a 12.2% solution?

72. Mixing candy   How many pounds each of candy shown in the illustration must be mixed to obtain 60 pounds of candy that is worth $3 per pound?

73. Travel   A car travels 50 miles in the same time that a plane travels 180 miles. The speed of the plane is 143 mph faster than the speed of the car. Find the speed of the car.

74. Travel   A car and a truck leave Rockford at the same time, heading in opposite directions. When they are 350 miles apart, the car has gone 70 miles farther than the truck. How far has the car traveled?

75. Making bicycles   A bicycle manufacturer builds racing bikes and mountain bikes, with the per-unit manufacturing costs shown in the table. The company has budgeted $15,900 for labor and $13,075 for materials. How many bicycles of each type can be built?

Model	Cost of materials	Cost of labor
Racing	$55	$60
Mountain	$70	$90

76. Farming   A farmer keeps some animals on a strict diet. Each animal is to receive 15 grams of protein and 7.5 grams of carbohydrates. The farmer uses two food mixes with nutrients shown in the illustration. How many grams of each mix should be used to provide the correct nutrients for each animal?

Mix	Protein	Carbohydrates
Mix A	12%	9%
Mix B	15%	5%

77. Milling brass plates Two machines can mill a brass plate. One machine has a setup cost of $300 and a cost per plate of $2. The other machine has a setup cost of $500 and a cost per plate of $1. Find the break point.

78. Printing books A printer has two presses. One has a setup cost of $210 and can print the pages of a certain book for $5.98. The other press has a setup cost of $350 and can print the pages of the same book for $5.95. Find the break point.

79. Managing a computer store The manager of a computer store knows that his fixed costs are $8,925 per month and that his unit cost is $850 for every computer sold. If he can sell all the computers he can get for $1,275 each, how many computers must he sell each month to break even?

80. Managing a makeup studio A makeup studio specializing in photo shoots has fixed costs of $23,600 per month. The owner estimates that the cost for each indoor shoot is $910. This cost covers labor and makeup. If her studio can book as many indoor shoots as she wants at a price of $1500 each, how many shoots will it take each month to break even?

81. Running a small business A person invests $18,375 to set up a small business that produces a piece of computer software that will sell for $29.95. If each piece can be produced for $5.45, how many pieces must be sold to break even?

82. Running a record company Three people invest $35,000 each to start a record company that will produce reissues of classic jazz. Each release will be a set of 3 CDs that will retail for $15 per disc. If each set can be produced for $18.95, how many sets must be sold for the investors to make a profit?

Break-point analysis *A paint manufacturer can choose between two processes for manufacturing house paint, with monthly costs shown in the table. Assume that the paint sells for $18 per gallon.*

Process	Fixed costs	Unit cost (per gallon)
A	$32,500	$13
B	$80,600	$ 5

83. For process A, how many gallons must be sold for the manufacturer to break even?

84. For process B, how many gallons must be sold for the manufacturer to break even?

85. If expected sales are 6,000 gallons per month, which process should the company use?

86. If expected sales are 7,000 gallons per month, which process should the company use?

Making water pumps *A manufacturer of automobile water pumps is considering retooling for one of two manufacturing processes, with monthly fixed costs and unit costs as indicated in the table. Each water pump can be sold for $50.*

Process	Fixed costs	Unit cost (per gallon)
A	$12,390	$29
B	$20,460	$17

87. For process A, how many water pumps must be sold for the manufacturer to break even?

88. For process B, how many water pumps must be sold for the manufacturer to break even?

89. If expected sales are 550 per month, which process should be used?

90. If expected sales are 600 per month, which process should be used?

91. If expected sales are 650 per month, which process should be used?

92. At what monthly sales level is process B better?

93. Geometry If two angles are supplementary, their sum is 180°. If the difference between two supplementary angles is 110°, find the measure of each angle.

94. Geometry If two angles are complementary, their sum is 90°. If one of two complementary angles is 16° greater than the other, find the measure of each angle.

95. Find the values of x and y in the parallelogram.

96. Find the values of x and y in the parallelogram.

97. Physical therapy To rehabilitate her knee, an athlete does leg extensions. Her goal is to regain a full 90° range of motion in this exercise. Use the information in the illustration to determine her current range of motion in degrees.

She needs to extend this much more.

Current range of motion. This angle is four times as large as the other.

98. The Marine Corps The Marine Corps War Memorial in Arlington, Virginia, portrays the raising of the U.S. flag on Iwo Jima during World War II. Find the measure of the two angles shown in the illustration if the measure of one of the angles is 15° less than twice the other.

99. Radio frequencies In a radio, an inductor and a capacitor are used in a resonant circuit to select a wanted radio station at a frequency f and reject all others. The inductance L and the capacitance C determine the inductance reactance X_L and the capacitive reactance X_C of that circuit, where

$$X_L = 2\pi fL \qquad \text{and} \qquad X_C = \frac{1}{2\pi fC}$$

The radio station selected will be at the frequency f where $X_L = X_C$. Write a formula for f^2 in terms of L and C.

100. Choosing salary plans A sales clerk can choose from two salary options:

1. a straight 7% commission
2. $150 + 2% commission

How much would the clerk have to sell for each plan to produce the same monthly paycheck?

WRITING ABOUT MATH

101. Which method would you use to solve the following system? Why?

$$\begin{cases} y = 3x + 1 \\ 3x + 2y = 12 \end{cases}$$

102. Which method would you use to solve the following system? Why?

$$\begin{cases} 2x + 4y = 9 \\ 3x - 5y = 20 \end{cases}$$

SOMETHING TO THINK ABOUT

103. Under what conditions will a system of two equations in two variables be inconsistent?

104. Under what conditions will the equations of a system of two equations in two variables be dependent?

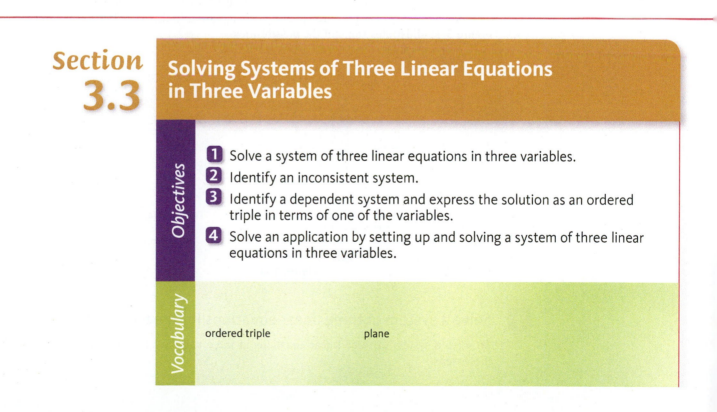

Section 3.3

Solving Systems of Three Linear Equations in Three Variables

Objectives

1. Solve a system of three linear equations in three variables.
2. Identify an inconsistent system.
3. Identify a dependent system and express the solution as an ordered triple in terms of one of the variables.
4. Solve an application by setting up and solving a system of three linear equations in three variables.

Vocabulary

ordered triple   plane

Determine whether the equation $x + 2y + 3z = 6$ is satisfied by the following values.

1. $x = 1, y = 1, z = 1$ **2.** $x = -2, y = 1, z = 2$

3. $x = 2, y = -2, z = -1$ **4.** $x = 2, y = 2, z = 0$

We now extend the definition of a linear equation to include equations of the form $ax + by + cz = d$ where $a, b, c,$ and d are numerical values (d is a constant). The solution of a system of three linear equations with three variables is an **ordered triple** of numbers if the equations are independent and the system consistent. For example, the solution of the system

$$\begin{cases} 2x + 3y + 4z = 20 \\ 3x + 4y + 2z = 17 \\ 3x + 2y + 3z = 16 \end{cases}$$ (x, y, z), which equals $(1, 2, 3)$

is the ordered triple $(1, 2, 3)$. Each equation is satisfied if $x = 1, y = 2,$ and $z = 3$.

$$2x + 3y + 4z = 20 \qquad\qquad 3x + 4y + 2z = 17 \qquad\qquad 3x + 2y + 3z = 16$$
$$2(1) + 3(2) + 4(3) \overset{?}{=} 20 \quad 3(1) + 4(2) + 2(3) \overset{?}{=} 17 \quad 3(1) + 2(2) + 3(3) \overset{?}{=} 16$$
$$2 + 6 + 12 \overset{?}{=} 20 \qquad\qquad 3 + 8 + 6 \overset{?}{=} 17 \qquad\qquad 3 + 4 + 9 \overset{?}{=} 16$$
$$20 = 20 \qquad\qquad\qquad\qquad 17 = 17 \qquad\qquad\qquad\qquad 16 = 16$$

The graph of an equation of the form $ax + by + cz = d$ is a flat surface called a **plane**. A system of three linear equations in three variables is consistent or inconsistent, depending on how the three planes corresponding to the three equations intersect. Figure 3-12 illustrates some of the possibilities.

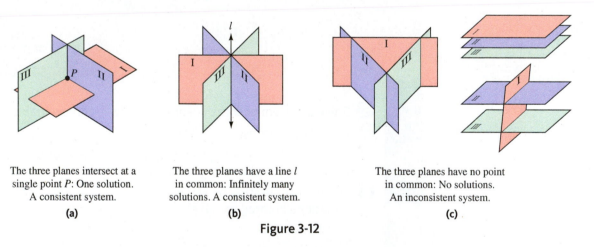

The three planes intersect at a single point P: One solution. A consistent system.

(a)

The three planes have a line l in common: Infinitely many solutions. A consistent system.

(b)

The three planes have no point in common: No solutions. An inconsistent system.

(c)

Figure 3-12

1   Solve a system of three linear equations in three variables.

To solve a system of three linear equations in three variables by elimination, we follow these steps.

SOLVING THREE LINEAR EQUATIONS IN THREE VARIABLES

1. Write all equations in the system in $ax + by + cz = d$ form.
2. Select any two equations and eliminate a variable.
3. Select a different pair of equations and eliminate the same variable.
4. Solve the resulting pair of two equations in two variables.
5. To find the value of the third variable, substitute the values of the two variables found in Step 4 into any equation containing all three variables and solve the equation.
6. Check the solution in all three of the original equations.

EXAMPLE 1 Solve the system: $\begin{cases} 2x + y + 4z = 12 \\ x + 2y + 2z = 9 \\ 3x - 3y - 2z = 1 \end{cases}$

Solution We are given the system

COMMENT To keep track of the equations it is helpful to number them.

(1) $\quad$ $\begin{cases} 2x + y + 4z = 12 \\ x + 2y + 2z = 9 \\ 3x - 3y - 2z = 1 \end{cases}$
(2)
(3)

If we select Equations 2 and 3 and add them, the variable z is eliminated:

(2) $\quad x + 2y + 2z = 9$
(3) $\quad \underline{3x - 3y - 2z = 1}$
(4) $\quad 4x - y = 10$

We now select a different pair of equations, 1 and 3 (we could have used Equations 1 and 2) and eliminate z again. If each side of Equation 3 is multiplied by 2 and the resulting equation is added to Equation 1, z is again eliminated:

(1) $\quad 2x + y + 4z = 12$
$\quad\quad \underline{6x - 6y - 4z = 2}$
(5) $\quad 8x - 5y = 14$

Equations 4 and 5 form a system of two equations in two variables:

(4) $\quad \begin{cases} 4x - y = 10 \\ 8x - 5y = 14 \end{cases}$
(5)

To solve this system, we multiply Equation 4 by -5 and add the resulting equation to Equation 5 to eliminate y:

$\quad -20x + 5y = -50$
(5) $\quad \underline{8x - 5y = 14}$
$\quad -12x = -36$
$\quad\quad\quad x = 3$ $\qquad$ Divide both sides by -12.

To find the value of y, we substitute 3 for x in any equation containing only x and y (such as Equation 5) and solve.

(5) $\quad 8x - 5y = 14$
$\quad 8(3) - 5y = 14$ $\qquad$ Substitute.
$\quad 24 - 5y = 14$ $\qquad$ Simplify.
$\quad\quad\quad -5y = -10$ $\qquad$ Subtract 24 from both sides.
$\quad\quad\quad\quad y = 2$ $\qquad$ Divide both sides by -5.

To find the value of z, we substitute 3 for x and 2 for y in an equation containing x, y, and z (such as Equation 1) and solve.

$$(1) \qquad 2\boldsymbol{x} + \boldsymbol{y} + 4z = 12$$

$$2(\boldsymbol{3}) + \boldsymbol{2} + 4z = 12 \qquad \color{red}{\text{Substitute.}}$$

$$8 + 4z = 12 \qquad \color{red}{\text{Simplify.}}$$

$$4z = 4 \qquad \color{red}{\text{Subtract 8 from both sides.}}$$

$$z = 1 \qquad \color{red}{\text{Divide both sides by 4.}}$$

The solution of the system is $(3, 2, 1)$. Verify that these values satisfy each equation in the original system.

 SELF CHECK 1   Solve the system: $\begin{cases} 2x + y + 4z = 16 \\ x + 2y + 2z = 11 \\ 3x - 3y - 2z = -9 \end{cases}$

2 Identify an inconsistent system.

The next example is an inconsistent system and has no solution.

EXAMPLE 2   Solve the system: $\begin{cases} 2x + y - 3z = -3 \\ 3x - 2y + 4z = 2 \\ 4x + 2y - 6z = -7 \end{cases}$

Solution   We are given the system of equations

$$\begin{array}{ll}(1) & \\ (2) & \begin{cases} 2x + y - 3z = -3 \\ 3x - 2y + 4z = 2 \\ 4x + 2y - 6z = -7 \end{cases} \\ (3) & \end{array}$$

We can multiply Equation 1 by 2 and add the resulting equation to Equation 2 to eliminate y.

$$\begin{array}{ll} & 4x + 2y - 6z = -6 \\ (2) & \underline{3x - 2y + 4z = 2} \\ (4) & 7x - 2z = -4 \end{array}$$

We now add Equations 2 and 3 to again eliminate y.

$$\begin{array}{ll} (2) & 3x - 2y + 4z = 2 \\ (3) & \underline{4x + 2y - 6z = -7} \\ (5) & 7x - 2z = -5 \end{array}$$

Equations 4 and 5 form the system

$$\begin{array}{ll}(4) & \\ (5) & \end{array} \begin{cases} 7x - 2z = -4 \\ 7x - 2z = -5 \end{cases}$$

Since $7x - 2z$ cannot equal both -4 and -5, this system is inconsistent; it has no solution. Its solution set is $\varnothing$.

 SELF CHECK 2   Solve the system: $\begin{cases} 2x + y - 3z = 8 \\ 3x - 2y + 4z = 10 \\ 4x + 2y - 6z = -5 \end{cases}$

3 Identify a dependent system and express the solution as an ordered triple in terms of one of the variables.

When the equations in a system of two equations in two variables were dependent, the system had infinitely many solutions. This is not always true for systems of three equations in three variables. In fact, a system can have dependent equations and still be inconsistent. Figure 3-13 illustrates the different possibilities.

When three planes coincide, the equations are dependent, and there are infinitely many solutions.

(a)

When three planes intersect in a common line, the equations are dependent, and there are infinitely many solutions.

(b)

When two planes coincide and are parallel to a third plane, the system is inconsistent, and there are no solutions.

(c)

Figure 3-13

EXAMPLE 3 Solve the system: $\begin{cases} 3x - 2y + z = -1 \\ 2x + y - z = 5 \\ 5x - y = 4 \end{cases}$

Solution We can add the first two equations to obtain

(1) $\quad 3x - 2y + z = -1$
(2) $\quad \underline{2x + y - z = 5}$
(4) $\quad 5x - y = 4$

Since Equation 4 is the same as the third equation of the system, the equations of the system are dependent, and there will be infinitely many solutions.

To write the general solution to this system, we can solve Equation 4 for y to obtain

$5x - y = 4$

$-y = -5x + 4$ Subtract $5x$ from both sides.

$y = 5x - 4$ Multiply both sides by -1.

We then can substitute $5x - 4$ for y in the first equation of the system and solve for z to obtain

$3x - 2\mathbf{y} + z = -1$

$3x - 2(\mathbf{5x - 4}) + z = -1$ Substitute.

$3x - 10x + 8 + z = -1$ Use the distributive property to remove parentheses.

$-7x + 8 + z = -1$ Combine like terms.

$z = 7x - 9$ Add $7x$ and -8 to both sides.

COMMENT The solution in Example 3 is called a *general* solution. If we had eliminated different variables, we could have expressed the general solution in terms of y or in terms of z.

Since we have found the values of y and z in terms of x, every solution to the system has the form $(x, 5x - 4, 7x - 9)$, where x can be any real number. For example,

If $x = 1$, a solution is $(1, 1, -2)$. $5(1) - 4 = 1$, and $7(1) - 9 = -2$

If $x = 2$, a solution is $(2, 6, 5)$. $5(2) - 4 = 6$, and $7(2) - 9 = 5$

If $x = 3$, a solution is $(3, 11, 12)$. $5(3) - 4 = 11$, and $7(3) - 9 = 12$

This system has infinitely many solutions of the form $(x, 5x - 4, 7x - 9)$.

🍃 **SELF CHECK 3** Solve the system: $\begin{cases} 3x + 2y + z = -1 \\ 2x - y - z = 5 \\ 5x + y = 4 \end{cases}$

Solve an application by setting up and solving a system of three linear equations in three variables.

EXAMPLE 4   **MANUFACTURING HAMMERS**   A company makes three types of hammers—good, better, and best. The cost of making each type of hammer is $4, $6, and $7, respectively, and the hammers sell for $6, $9, and $12. Each day, the cost of making 100 hammers is $520, and the daily revenue from their sale is $810. How many of each type are manufactured?

Analyze the problem   We need to know the number of each type of hammer manufactured, so we will let x represent the number of good hammers, y represent the number of better hammers, and z represent the number of best hammers.

Form three equations   Since x represents the number of good hammers, y represents the number of better hammers, and z represents the number of best hammers, we know that

> The total number of hammers is $x + y + z$.
> The cost of making good hammers is $4x$ ($4 times x hammers).
> The cost of making better hammers is $6y$ ($6 times y hammers).
> The cost of making best hammers is $7z$ ($7 times z hammers).
> The revenue received by selling good hammers is $6x$ ($6 times x hammers).
> The revenue received by selling better hammers is $9y$ ($9 times y hammers).
> The revenue received by selling best hammers is $12z$ ($12 times z hammers).

The information leads to three equations:

The number of good hammers	plus	the number of better hammers	plus	the number of best hammers	equals	the total number of hammers.
x	$+$	y	$+$	z	$=$	100

The cost of making good hammers	plus	the cost of making better hammers	plus	the cost of making best hammers	equals	the total cost.
$4x$	$+$	$6y$	$+$	$7z$	$=$	520

The revenue from the good hammers	plus	the revenue from the better hammers	plus	the revenue from the best hammers	equals	the total revenue.
$6x$	$+$	$9y$	$+$	$12z$	$=$	810

Solve the system   These three equations give the following system:

$$
\begin{cases}
(1) & x + y + z = 100 \\
(2) & 4x + 6y + 7z = 520 \\
(3) & 6x + 9y + 12z = 810
\end{cases}
$$

that we can solve as follows:

If we multiply Equation 1 by -7 and add the result to Equation 2, we obtain

$$
\begin{array}{r}
-7x - 7y - 7z = -700 \\
\underline{4x + 6y + 7z = 520} \\
(4) \quad -3x - y = -180
\end{array}
$$

If we multiply Equation 1 by -12 and add the result to Equation 3, we obtain

$$-12x - 12y - 12z = -1{,}200$$
$$\underline{\quad 6x + \ 9y + 12z = \quad\ 810}$$
(5) $$-6x - \ 3y \quad\quad = \ -390$$

If we multiply Equation 4 by -3 and add it to Equation 5, we obtain

$$9x + 3y = \quad 540$$
$$\underline{-6x - 3y = -390}$$
$$3x \quad\quad = \quad 150$$
$$\quad\quad x = 50 \qquad \text{Divide both sides by 3.}$$

To find y, we substitute 50 for x in Equation 4:

$$-3x - y = -180$$
$$-3(50) - y = -180 \qquad \text{Substitute.}$$
$$-y = -30 \qquad \text{Add 150 to both sides.}$$
$$y = 30 \qquad \text{Divide both sides by } -1.$$

To find z, we substitute 50 for x and 30 for y in Equation 1:

$$x + y + z = 100$$
$$50 + 30 + z = 100$$
$$z = 20 \qquad \text{Subtract 80 from both sides.}$$

State the conclusion   Each day, the company makes 50 good hammers, 30 better hammers, and 20 best hammers. This totals 100. The cost of producing the hammers is \$4(50) + \$6(30) + \$7(20) or \$520 and the revenue from the sale of the hammers is \$6(50) + \$9(30) + \$12(20) or \$810.

Check the result   Check the solution in each equation in the original system.

SELF CHECK 4   If the cost of the 100 hammers is \$560 and the daily revenue is \$870, how many of each type are manufactured?

EXAMPLE 5   **CURVE FITTING**   The equation of the parabola shown in Figure 3-14 is of the form $y = ax^2 + bx + c$. Find the equation of the parabola.

Solution   Since the parabola passes through the points shown in the figure, each pair of coordinates satisfies the equation $y = ax^2 + bx + c$. For example, $(-1, 5)$ substituted in $y = ax^2 + bx + c$ becomes $5 = a(-1)^2 + b(-1) + c$ or simplified as $a - b + c = 5$. If we substitute the x- and y-values of each point into the equation and simplify, we obtain the following system.

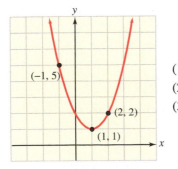

Figure 3-14

(1) $\begin{cases} a - b + c = 5 & \text{for point } (-1, 5) \\ a + b + c = 1 & \text{for point } (1, 1) \\ 4a + 2b + c = 2 & \text{for point } (2, 2) \end{cases}$
(2)
(3)

If we add Equations 1 and 2, we obtain $2a + 2c = 6$. If we multiply Equation 1 by 2 and add the result to Equation 3, we get $6a + 3c = 12$. We can then divide both sides of $2a + 2c = 6$ by 2 and divide both sides of $6a + 3c = 12$ by 3 to obtain the system

(4) $\begin{cases} a + c = 3 \\ 2a + c = 4 \end{cases}$
(5)

If we multiply Equation 4 by -1 and add the result to Equation 5, we get $a = 1$. To find the value of c, we can substitute 1 for a in Equation 4 and find that $c = 2$. To find the value of b, we can substitute 1 for a and 2 for c in Equation 2 and find that $b = -2$.

After we substitute these values of a, b, and c into the equation $y = ax^2 + bx + c$ we have the equation of the parabola.

$$y = ax^2 + bx + c$$
$$y = 1x^2 + (-2)x + 2$$
$$y = x^2 - 2x + 2$$

SELF CHECK 5   If a parabola passes through the points $(1, -10)$, $(-2, 8)$, and $(5, -6)$, find the equation of the parabola.

SELF CHECK ANSWERS   **1.** $(1, 2, 3)$ **2.** $\varnothing$ **3.** infinitely many solutions of the form $(x, 4 - 5x, -9 + 7x)$ **4.** The company makes 30 good hammers, 50 better hammers, and 20 best hammers.   **5.** $y = x^2 - 5x - 6$

NOW TRY THIS

The manager of a coffee bar wants to mix some Peruvian Organic coffee worth $15 per pound with some Colombian coffee worth $10 per pound and Indian Malabar coffee worth $18 per pound to obtain 50 pounds of a blend that he can sell for $17.50 per pound. He wants to use 10 fewer pounds of the Indian Malabar than Peruvian Organic. How many pounds of each should he use?   (*Hint*: This problem is based on the formula $V = np$, where V represents value, n represents the number of pounds, and p represents the price per pound.)

3.3 Exercises

WARM-UPS *Is the ordered triple a solution of the system?*

1. $(1, 1, 1)$; $\begin{cases} 2x + y - 3z = 0 \\ 3x - 2y + 4z = 5 \\ 4x + 2y - 6z = 0 \end{cases}$

2. $(2, 1, 1)$; $\begin{cases} 3x + 2y - z = 5 \\ 2x - 3y + 2z = 4 \\ 4x - 2y + 3z = 10 \end{cases}$

REVIEW *Consider the line passing through $(-1, -5)$ and $(2, 3)$.*

3. Find the slope of the line.

4. Write the equation of the line in slope-intercept form.

Find each value if $f(x) = 2x^2 + 1$.

5. $f(0)$ **6.** $f(-2)$

7. $f(3s)$ **8.** $f(4t)$

VOCABULARY AND CONCEPTS *Fill in the blanks.*

9. The graph of the equation $2x + 3y + 4z = 5$ is a flat surface called a ____.

10. When three planes coincide, the equations of the system are _____, and there are _____ many solutions.

11. When three planes intersect in a line, the system will have _____ many solutions.

12. When three planes are parallel, the system will have __ solutions.

Determine whether the ordered triple is a solution of the given system.

13. $(1, 2, 1)$, $\begin{cases} x + y - z = 2 \\ 2x + y + 3z = 7 \\ x - \frac{1}{3}y + \frac{2}{3}z = 1 \end{cases}$

14. $(-3, 2, -1)$, $\begin{cases} 2x + 2y + 3z = 5 \\ 3x + y = -6 \\ x + y + 2z = 1 \end{cases}$

GUIDED PRACTICE *Solve each system.* **SEE EXAMPLE 1.** **(OBJECTIVE 1)**

15. $\begin{cases} x + y + z = 4 \\ 2x + y - z = 1 \\ 2x - 3y + z = 1 \end{cases}$

16. $\begin{cases} x + y + z = 4 \\ x - y + z = 2 \\ x - y - z = 0 \end{cases}$

17. $\begin{cases} 4x + 3z = 4 \\ 2y - 6z = -1 \\ 8x + 4y + 3z = 9 \end{cases}$

18. $\begin{cases} 2x + 3y + 2z = 1 \\ 2x - 3y + 2z = -1 \\ 4x + 3y - 2z = 4 \end{cases}$

Solve each system. SEE EXAMPLE 2. (OBJECTIVE 2)

19. $\begin{cases} 3a - 2b - c = 4 \\ 6a - 4b - 2c = 10 \\ a + 3b + c = 2 \end{cases}$

20. $\begin{cases} 2x + y - z = 1 \\ x + 2y + 2z = 2 \\ 4x + 5y + 3z = 3 \end{cases}$

Solve each system. SEE EXAMPLE 3. (OBJECTIVE 3)

21. $\begin{cases} 7x - 2y - z = 1 \\ 9x - 6y + z = 7 \\ x - 2y + z = 3 \end{cases}$

22. $\begin{cases} x - 2y + 3z = 9 \\ -x + 3y = -4 \\ 2x - 5y + 3z = 13 \end{cases}$

23. $\begin{cases} x + 2y + z = 1 \\ 3x + y + 3z = 3 \\ -2x + y - 2z = -2 \end{cases}$

24. $\begin{cases} -x + y - 2z = -5 \\ 3x + 2y = 4 \\ 4x - 4y + 8z = 20 \end{cases}$

ADDITIONAL PRACTICE *Solve each system.*

25. $\begin{cases} a + 3b + c = 10 \\ 3a + b + c = 12 \\ a + b + 3c = 8 \end{cases}$

26. $\begin{cases} 3x - y - 2z = 12 \\ x + y + 6z = 8 \\ 2x - 2y - z = 11 \end{cases}$

27. $\begin{cases} x + \frac{1}{3}y + z = 13 \\ \frac{1}{2}x - y + \frac{1}{3}z = -2 \\ x + \frac{1}{2}y - \frac{1}{3}z = 2 \end{cases}$

28. $\begin{cases} x - \frac{1}{5}y - z = 9 \\ \frac{1}{4}x + \frac{1}{5}y - \frac{1}{2}z = 5 \\ 2x + y + \frac{1}{6}z = 12 \end{cases}$

29. $\begin{cases} x - 3y + z = 1 \\ 2x - y - 2z = 2 \\ x + 2y - 3z = -1 \end{cases}$

30. $\begin{cases} 2x + y - 3z = 5 \\ x - 2y + 4z = 9 \\ 4x + 2y - 6z = 1 \end{cases}$

31. $\begin{cases} 2x + 3y + 4z = 6 \\ 2x - 3y - 4z = -4 \\ 4x + 6y + 8z = 12 \end{cases}$

32. $\begin{cases} x - 3y + 4z = 2 \\ 2x + y + 2z = 3 \\ 4x - 5y + 10z = 7 \end{cases}$

33. $\begin{cases} 2x + 2y + 3z = 10 \\ 3x + y - z = 0 \\ x + y + 2z = 6 \end{cases}$

34. $\begin{cases} x - y + z = 4 \\ x + 2y - z = -1 \\ x + y - 3z = -2 \end{cases}$

APPLICATIONS *Use a system of three linear equations to solve.* SEE EXAMPLES 4–5. (OBJECTIVE 4)

35. Making statues An artist makes three types of ceramic statues at a monthly cost of $650 for 180 statues. The manufacturing costs for the three types are $5, $4, and $3. If the statues sell for $20, $12, and $9, respectively, how many of each type should be made to produce $2,100 in monthly revenue?

36. Manufacturing footballs A factory manufactures three types of footballs at a monthly cost of $2,425 for 1,125 footballs. The manufacturing costs for the three types of footballs are $4, $3, and $2. These footballs sell for $16, $12, and $10, respectively. How many of each type are

manufactured if the monthly profit is $9,275? (*Hint*: Profit = Income − Cost.)

37. Curve fitting Find the equation of the parabola shown in the illustration.

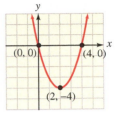

38. Curve fitting Find the equation of the parabola shown in the illustration.

39. Integers The sum of three integers is 19. The third integer is twice the second, and the second integer is 5 more than the first. Find the integers.

40. Integers The sum of three integers is 48. If the first integer is doubled, the sum is 60. If the second integer is doubled, the sum is 63. Find the integers.

41. Geometry The sum of the angles in any triangle is 180°. In triangle ABC, $\angle A$ is 80° less than the sum of $\angle B$ and $\angle C$, and $\angle C$ is 50° less than twice $\angle B$. Find the measure of each angle.

42. Geometry The sum of the angles of any four-sided figure is 360°. In the quadrilateral, $\angle A = \angle B$, $\angle C$ is 20° greater than $\angle A$, and $\angle D = 40°$. Find the measure of each angle.

43. Nutritional planning One unit of each of three foods contains the nutrients shown in the table. How many units of each must be used to provide exactly 11 grams of fat, 6 grams of carbohydrates, and 10 grams of protein?

Food	Fat	Carbohydrates	Protein
A	1	1	2
B	2	1	1
C	2	1	2

44. Nutritional planning One unit of each of three foods contains the nutrients shown in the table. How many units of each must be used to provide exactly 14 grams of fat, 9 grams of carbohydrates, and 9 grams of protein?

Food	Fat	Carbohydrates	Protein
A	2	1	2
B	3	2	1
C	1	1	2

45. Concert tickets Tickets for a concert cost $5, $3, and $2. Twice as many $5 tickets were sold as $2 tickets. The receipts for 750 tickets were $2,625. How many of each price ticket were sold?

46. Mixing nuts The owner of a candy store mixed some peanuts worth $3 per pound, some cashews worth $9 per pound, and some Brazil nuts worth $9 per pound to get 50 pounds of a mixture that would sell for $6 per pound. She used 15 fewer pounds of cashews than peanuts. How many pounds of each did she use?

47. Chainsaw sculpting A north woods sculptor carves three types of statues with a chainsaw. The times required for carving, sanding, and painting a totem pole, a bear, and a deer are shown in the table. How many of each should be produced to use all available labor hours?

	Totem pole	Bear	Deer	Time available
Carving	2 hours	2 hours	1 hour	14 hours
Sanding	1 hour	2 hours	2 hours	15 hours
Painting	3 hours	2 hours	2 hours	21 hours

48. Making clothing A clothing manufacturer makes coats, shirts, and slacks. The times required for cutting, sewing, and packaging each item are shown in the table. How many of each should be made to use all available labor hours?

	Coats	Shirts	Slacks	Time available
Cutting	20 min	15 min	10 min	115 hr
Sewing	60 min	30 min	24 min	280 hr
Packaging	5 min	12 min	6 min	65 hr

49. Earth's atmosphere Use the information in the graph to determine what percent of Earth's atmosphere is nitrogen, is oxygen, and is other gases.

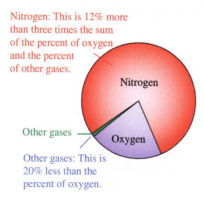

Nitrogen: This is 12% more than three times the sum of the percent of oxygen and the percent of other gases.

Other gases: This is 20% less than the percent of oxygen.

50. NFL records Jerry Rice, who played with the San Francisco 49ers and the Oakland Raiders, holds the all-time record for touchdown passes caught (197). Here are interesting facts about three quarterbacks who helped him achieve this record.

- He caught 17 more TD passes from Steve Young (SF) than he did from Joe Montana.
- He caught 50 more TD passes from Joe Montana (SF) than he did from Rich Gannon (Oak).
- He caught a total of 171 TD passes from Young, Montana, and Gannon.

Determine the number of touchdown passes Rice caught from Young, from Montana, and from Gannon.

The equation of a circle is of the form $x^2 + y^2 + cx + dy + e = 0$.

51. Curve fitting Find the equation of the circle shown in the illustration.

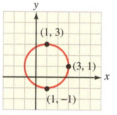

52. Curve fitting Find the equation of the circle shown in the illustration.

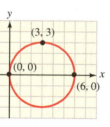

WRITING ABOUT MATH

53. What makes a system of three equations in three variables inconsistent?

54. What makes the equations of a system of three equations in three variables dependent?

SOMETHING TO THINK ABOUT

55. Solve the system:

$$\begin{cases} x + y + z + w = 3 \\ x - y - z - w = -1 \\ x + y - z - w = 1 \\ x + y - z + w = 3 \end{cases}$$

56. Solve the system:

$$\begin{cases} 2x + y + z + w = 3 \\ x - 2y - z + w = -3 \\ x - y - 2z - w = -3 \\ x + y - z + 2w = 4 \end{cases}$$

Section 3.4

Solving Systems of Linear Equations Using Matrices

Objectives

1 Solve a system of linear equations with the same number of equations as variables using Gaussian elimination.

2 Solve a system with more linear equations than variables using row operations on a matrix.

3 Solve a system with fewer linear equations than variables using row operations on a matrix.

Vocabulary

matrix
element of a matrix
square matrix

augmented matrix
coefficient matrix
Gaussian elimination

triangular form of a matrix
back substitution

Getting Ready

Multiply the first row by 2 and add the result to the second row.

1.
$\begin{matrix} 2 & 3 & 5 \\ 1 & 2 & 3 \end{matrix}$

2.
$\begin{matrix} -1 & 0 & 4 \\ 2 & 3 & -1 \end{matrix}$

Multiply the first row by -1 and add the result to the second row.

3.
$\begin{matrix} 2 & 3 & 5 \\ 1 & 2 & 3 \end{matrix}$

4.
$\begin{matrix} -1 & 0 & 4 \\ 2 & 3 & -1 \end{matrix}$

In this section, we will discuss an alternative method used for solving systems of linear equations. This method involves the use of *matrices*.

1 Solve a system of linear equations with the same number of equations as variables using Gaussian elimination.

MATRIX   A **matrix** is any rectangular array of numbers.

Some examples of matrices are

$$A = \begin{bmatrix} 1 & 2 & 3 \\ 4 & 5 & 6 \end{bmatrix} \qquad B = \begin{bmatrix} 1 & 2 \\ 3 & 4 \\ 5 & 6 \end{bmatrix} \qquad C = \begin{bmatrix} 2 & 4 & 6 \\ 8 & 10 & 12 \\ 14 & 16 & 18 \end{bmatrix}$$

The numbers in each matrix are called **elements**. Because matrix A has two rows and three columns, it is called a 2×3 matrix (read "2 by 3" matrix). Matrix B is a 3×2 matrix, because the matrix has three rows and two columns. Matrix C is a 3×3 matrix (three rows and three columns).

Arthur Cayley
1821–1895
Cayley taught mathematics at Cambridge University. When he refused to take religious vows, he was fired and became a lawyer. After 14 years, he returned to mathematics and to Cambridge. Cayley was a major force in developing the theory of matrices.

Any matrix with the same number of rows and columns, like matrix C, is called a **square matrix**.

To use matrices to solve systems of linear equations, we consider the system

$$\begin{cases} x - 2y - z = 6 \\ 2x + 2y - z = 1 \\ -x - y + 2z = 1 \end{cases}$$

which can be represented by the following matrix, called an **augmented matrix**:

$$\begin{bmatrix} 1 & -2 & -1 & 6 \\ 2 & 2 & -1 & 1 \\ -1 & -1 & 2 & 1 \end{bmatrix}$$

The first three columns of the augmented matrix form a 3×3 matrix called a **coefficient matrix**. It is determined by the coefficients of x, y, and z in the equations of the system. The 3×1 matrix to the right of the dashed line is determined by the constants in the equations.

Coefficient matrix   *Column of constants*

$$\begin{bmatrix} 1 & -2 & 1 \\ 2 & 2 & -1 \\ -1 & -1 & 2 \end{bmatrix} \qquad\qquad \begin{bmatrix} 6 \\ 1 \\ 1 \end{bmatrix}$$

Each row of the augmented matrix represents one equation of the system:

$$\begin{bmatrix} 1 & -2 & -1 & 6 \\ 2 & 2 & -1 & 1 \\ -1 & -1 & 2 & 1 \end{bmatrix} \quad \begin{matrix} \leftrightarrow \\ \leftrightarrow \\ \leftrightarrow \end{matrix} \quad \begin{cases} x - 2y - z = 6 \\ 2x + 2y - z = 1 \\ -x - y + 2z = 1 \end{cases}$$

To solve a 3×3 system of equations by **Gaussian elimination**, we transform an augmented matrix into the following matrix that has all 0's below its main diagonal, which is formed by the elements a, e, and h.

$$\begin{bmatrix} a & b & c & d \\ 0 & e & f & g \\ 0 & 0 & h & i \end{bmatrix} \qquad (a, b, c, \ldots, i \text{ are real numbers})$$

We often can write a matrix in this form, called **triangular form**, by using the following operations.

ELEMENTARY ROW OPERATIONS

1. Any two rows of a matrix can be interchanged.

2. Any row of a matrix can be multiplied by a nonzero constant.

3. Any row of a matrix can be changed by adding a nonzero constant multiple of another row to it.

- A type 1 row operation corresponds to interchanging two equations of a system.

- A type 2 row operation corresponds to multiplying both sides of an equation by a nonzero constant.

- A type 3 row operation corresponds to adding a nonzero multiple of one equation to another.

None of these operations will change the solution of the given system of equations.

After we have written the matrix in triangular form, we can solve the corresponding system of equations by a process called **back substitution**, as shown in Example 1.

EXAMPLE 1   Solve the system using matrices: $\begin{cases} x - 2y - z = 6 \\ 2x + 2y - z = 1 \\ -x - y + 2z = 1 \end{cases}$

Solution   We can represent the system with the following augmented matrix:

$$\left[\begin{array}{ccc|c} \mathbf{1} & -2 & -1 & 6 \\ 2 & 2 & -1 & 1 \\ -1 & -1 & 2 & 1 \end{array} \right]$$

To get 0's under the 1 in the first column, we multiply row 1 of the augmented matrix by -2 and add it to row 2 to get a new row 2. We then add row 1 to row 3 to obtain a new row 3.

$$\left[\begin{array}{ccc|c} 1 & -2 & -1 & 6 \\ 0 & \mathbf{6} & 1 & -11 \\ 0 & -3 & 1 & 7 \end{array} \right] \quad \begin{array}{l} -2R1 + R2 \rightarrow R2 \\ R1 + R3 \rightarrow R3 \end{array}$$

To get a 0 under the 6 in the second column of the previous matrix, we multiply row 2 by $\frac{1}{2}$ and add it to row 3.

$$\left[\begin{array}{ccc|c} 1 & -2 & 1 & 6 \\ 0 & 6 & 1 & -11 \\ 0 & 0 & \frac{3}{2} & \frac{3}{2} \end{array} \right] \quad \frac{1}{2}R2 + R3 \rightarrow R3$$

Finally, to clear the fraction in the third row, third column, we multiply row 3 by $\frac{2}{3}$, which is the reciprocal of $\frac{3}{2}$.

$$\left[\begin{array}{ccc|c} 1 & -2 & 1 & 6 \\ 0 & 6 & 1 & -11 \\ 0 & 0 & 1 & 1 \end{array} \right] \quad \frac{2}{3}R3 \rightarrow R3$$

The final matrix represents the system of equations

(1)
(2) $\begin{cases} x - 2y - z = 6 \\ 0x + 6y + z = -11 \\ 0x + 0y + z = 1 \end{cases}$
(3)

From Equation 3, we can see that $z = 1$. Now we begin the back substitution process. To find the value of y, we substitute 1 for z in Equation 2 and solve.

(2) $6y + z = -11$

$6y + \mathbf{1} = -11$ Substitute 1 for z.

$6y = -12$ Subtract 1 from both sides.

$y = -2$ Divide both sides by 6.

Thus, $y = -2$. To find the value of x, we substitute 1 for z and -2 for y in Equation 1 and solve.

(1) $x - 2\mathbf{y} - z = 6$

$x - 2(\mathbf{-2}) - \mathbf{1} = 6$ Substitute 1 for z and -2 for y.

$x + 3 = 6$ Simplify.

$x = 3$ Subtract 3 from both sides.

Thus, $x = 3$. The solution of the given system is $(3, -2, 1)$. Verify that this ordered triple satisfies each equation of the original system.

SELF CHECK 1   Solve the system using matrices: $\begin{cases} x - 2y - z = 2 \\ 2x + 2y - z = -5 \\ -x - y + 2z = 7 \end{cases}$

2 Solve a system with more linear equations than variables using row operations on a matrix.

We can use matrices to solve systems that have more equations than variables.

EXAMPLE 2 Solve the system using matrices: $\begin{cases} x + y = -1 \\ 2x - y = 7 \\ -x + 2y = -8 \end{cases}$

Solution This system can be represented by the following augmented matrix:

$$\left[\begin{array}{rr|r} 1 & 1 & -1 \\ 2 & -1 & 7 \\ -1 & 2 & -8 \end{array} \right]$$

To obtain 0's under the 1 in the first column, we multiply row 1 by -2 and add it to row 2. Then we can add row 1 to row 3.

$$\left[\begin{array}{rr|r} 1 & 1 & -1 \\ 0 & -3 & 9 \\ 0 & 3 & -9 \end{array} \right] \qquad \begin{array}{l} -2R1 + R2 \rightarrow R2 \\ R1 + R3 \rightarrow R3 \end{array}$$

To get a 0 under the -3 in the second column, we can add row 2 to row 3.

$$\left[\begin{array}{rr|r} 1 & 1 & -1 \\ 0 & -3 & 9 \\ 0 & 0 & 0 \end{array} \right] \qquad R2 + R3 \rightarrow R3$$

Finally, to get a 1 in the second row, second column, we multiply row 2 by $-\frac{1}{3}$.

$$\left[\begin{array}{rr|r} 1 & 1 & -1 \\ 0 & 1 & -3 \\ 0 & 0 & 0 \end{array} \right] \qquad -\frac{1}{3}R2 \rightarrow R2$$

The final matrix represents the system

$$\begin{cases} x + y = -1 \\ 0x + y = -3 \\ 0x + 0y = 0 \end{cases}$$

The third equation can be discarded, because $0x + 0y = 0$ for all x and y. From the second equation, we can read that $y = -3$. To find the value of x, we substitute -3 for y in the first equation and solve.

$$\begin{array}{ll} x + y = -1 & \\ x + (-3) = -1 & \text{Substitute } -3 \text{ for } y. \\ x = 2 & \text{Add 3 to both sides.} \end{array}$$

The solution is $(2, -3)$. Verify that this solution satisfies all three equations of the original system.

SELF CHECK 2 Solve the system using matrices: $\begin{cases} x + y = 1 \\ 2x - y = 8 \\ -x + 2y = -7 \end{cases}$

COMMENT If the last row of the final matrix in Example 2 had been representative of the form $0x + 0y = k$, where $k \neq 0$, the system would not have a solution. No values of x and y could make the expression $0x + 0y$ equal to a nonzero constant k.

Solve a system with fewer linear equations than variables using row operations on a matrix.

We also can solve many systems that have more variables than equations.

EXAMPLE 3   Solve the system using matrices: $\begin{cases} x + y - 2z = -1 \\ 2x - y + z = -3 \end{cases}$

Solution   This system can be represented by the following augmented matrix.

$$\left[\begin{array}{ccc|c} \mathbf{1} & 1 & -2 & -1 \\ 2 & -1 & 1 & -3 \end{array}\right]$$

To obtain a 0 under the 1 in the first column, we multiply row 1 by -2 and add it to row 2.

$$\left[\begin{array}{ccc|c} 1 & 1 & -2 & -1 \\ 0 & -3 & 5 & -1 \end{array}\right] \quad -2R1 + R2 \rightarrow R2$$

Then to get a 1 in the second row, second column, we multiply row 2 by $-\frac{1}{3}$.

$$\left[\begin{array}{ccc|c} 1 & 1 & -2 & -1 \\ 0 & 1 & -\frac{5}{3} & \frac{1}{3} \end{array}\right] \quad -\frac{1}{3}R2 \rightarrow R2$$

The final matrix represents the system

$$\begin{cases} x + y - 2z = -1 \\ y - \frac{5}{3}z = \frac{1}{3} \end{cases}$$

We add $\frac{5}{3}z$ to both sides of the second equation to obtain

$$y = \frac{1}{3} + \frac{5}{3}z$$

We have not found a specific value for y. However, we have found y in terms of z.

To find a value of x in terms of z, we substitute $\frac{1}{3} + \frac{5}{3}z$ for y in the first equation and simplify to obtain

$$x + y - 2z = -1$$
$$x + \mathbf{\frac{1}{3}} + \mathbf{\frac{5}{3}}z - 2z = -1 \qquad \text{\color{red}Substitute.}$$
$$x + \frac{1}{3} - \frac{1}{3}z = -1 \qquad \text{\color{red}Combine like terms.}$$
$$x - \frac{1}{3}z = -\frac{4}{3} \qquad \text{\color{red}Subtract } \tfrac{1}{3} \text{ from both sides.}$$
$$x = -\frac{4}{3} + \frac{1}{3}z \qquad \text{\color{red}Add } \tfrac{1}{3}z \text{ to both sides.}$$

A solution of this system must have the form

$$\left(-\frac{4}{3} + \frac{1}{3}z,\ \frac{1}{3} + \frac{5}{3}z,\ z\right) \quad \text{\color{red}This solution is a general solution of the system.}$$

for all values of z.

 SELF CHECK 3   Solve the system using matrices: $\begin{cases} x + y - 2z = 11 \\ 2x - y + z = -2 \end{cases}$

Everyday connections
Staffing

Suburban Office

	Male	Female
Day shift	14	12
Night shift	5	2

Matrices with the same number of rows and columns can be added. We simply add their corresponding elements. For example,

$$\begin{bmatrix} 2 & 3 & -4 \\ -1 & 2 & 5 \end{bmatrix} + \begin{bmatrix} 3 & -1 & 0 \\ 4 & 3 & 2 \end{bmatrix}$$

$$= \begin{bmatrix} 2+3 & 3+(-1) & -4+0 \\ -1+4 & 2+3 & 5+2 \end{bmatrix}$$

$$= \begin{bmatrix} 5 & 2 & -4 \\ 3 & 5 & 7 \end{bmatrix}$$

To multiply a matrix by a constant, we multiply each element of the matrix by the constant. For example,

$$5 \cdot \begin{bmatrix} 2 & 3 & -4 \\ -1 & 2 & 5 \end{bmatrix}$$

$$= \begin{bmatrix} 5 \cdot 2 & 5 \cdot 3 & 5 \cdot (-4) \\ 5 \cdot (-1) & 5 \cdot 2 & 5 \cdot 5 \end{bmatrix}$$

$$= \begin{bmatrix} 10 & 15 & -20 \\ -5 & 10 & 25 \end{bmatrix}$$

Since matrices provide a good way to store information in computers, they often are used in applied problems. For example, suppose there are 66 security officers employed at either the downtown office or the suburban office:

Downtown Office

	Male	Female
Day shift	12	18
Night shift	3	0

The information about the employees is contained in the following matrices.

$$D = \begin{bmatrix} 12 & 18 \\ 3 & 0 \end{bmatrix} \quad \text{and} \quad S = \begin{bmatrix} 14 & 12 \\ 5 & 2 \end{bmatrix}$$

The entry in the first row-first column in matrix D gives the information that 12 males work the day shift at the downtown office. Company management can add the matrices D and S to find corporate-wide totals:

$$D + S = \begin{bmatrix} 12 & 18 \\ 3 & 0 \end{bmatrix} + \begin{bmatrix} 14 & 12 \\ 5 & 2 \end{bmatrix}$$

$$= \begin{bmatrix} 26 & 30 \\ 8 & 2 \end{bmatrix}$$

We interpret the total to mean:

	Male	Female
Day shift	26	30
Night shift	8	2

If one-third of the force in each category at the downtown location retires, the downtown staff would be reduced to $\frac{2}{3} D$ people. We can compute $\frac{2}{3} D$ by multiplying each entry by $\frac{2}{3}$.

$$\frac{2}{3} D = \frac{2}{3} \begin{bmatrix} 12 & 18 \\ 3 & 0 \end{bmatrix}$$

$$= \begin{bmatrix} 8 & 12 \\ 2 & 0 \end{bmatrix}$$

After retirements, downtown staff would be

	Male	Female
Day shift	8	12
Night shift	2	0

SELF CHECK ANSWERS   **1.** $(1, -2, 3)$ **2.** $(3, -2)$ **3.** $\left(3 + \frac{1}{3}z, 8 + \frac{5}{3}z, z\right)$

NOW TRY THIS

A toy company builds authentic models of a compact car, a sedan, and a truck. The times required for preparation, assembly, and post-production are given below. Use matrices to help determine how many of each should be made in order to use all available labor hours.

	Compact	Sedan	Truck	Total labor hours
Preparation	1 hr	1 hr	2 hrs	60 hrs
Assembly	2 hrs	3 hrs	4 hrs	130 hrs
Post-production	2 hrs	2 hrs	3 hrs	100 hrs

3.4 Exercises

WARM-UPS *Write the system of equations that correspond to the following.*

1. $\begin{bmatrix} 2 & -1 & | & 3 \\ 1 & 5 & | & 7 \end{bmatrix}$

2. $\begin{bmatrix} 1 & -3 & | & 5 \\ 0 & 4 & | & -1 \end{bmatrix}$

3. $\begin{bmatrix} 3 & -2 & 4 & | & 1 \\ 5 & 2 & -3 & | & -7 \\ -1 & 9 & 8 & | & 0 \end{bmatrix}$

4. $\begin{bmatrix} 1 & 5 & -1 & | & 6 \\ 0 & 4 & -2 & | & 3 \\ -1 & -1 & 0 & | & 8 \end{bmatrix}$

REVIEW *Write each number in scientific notation.*

5. 470,000,000

6. 0.0000089

7. 75×10^4

8. 0.63×10^4

VOCABULARY AND CONCEPTS *Fill in the blanks.*

9. A _____ is a rectangular array of numbers.

10. The numbers in a matrix are called its _____.

11. A 3×4 matrix has _ rows and 4 _____.

12. A _____ matrix has the same number of rows as columns.

13. An _____ matrix of a system of equations includes the _____ matrix and the column of constants.

14. If a matrix has all 0's below its main diagonal, it is written in _____ form.

Consider the system $\begin{cases} 2x - 3y = 9 \\ 4x + 2y = 2 \end{cases}$

15. Find the coefficient matrix.

16. Find the augmented matrix.

Determine whether each matrix is in triangular form.

17. $\begin{bmatrix} 5 & 3 & -1 \\ 0 & 7 & 6 \\ 0 & 0 & 2 \end{bmatrix}$

18. $\begin{bmatrix} 7 & -2 & 3 \\ 0 & 1 & 4 \\ 0 & 6 & 0 \end{bmatrix}$

Use a row operation on the first matrix to find the missing number in the second matrix.

19. $\begin{bmatrix} 3 & 1 & 2 \\ 4 & 5 & 7 \end{bmatrix}$
$\begin{bmatrix} 3 & 1 & 2 \\ 1 & 4 & \blacksquare \end{bmatrix}$

20. $\begin{bmatrix} -1 & 3 & 2 \\ 1 & -2 & 3 \end{bmatrix}$
$\begin{bmatrix} -1 & 3 & 2 \\ \blacksquare & 1 & 5 \end{bmatrix}$

21. $\begin{bmatrix} -5 & -1 & 3 \\ -2 & 3 & 1 \end{bmatrix}$
$\begin{bmatrix} -5 & -1 & 3 \\ -4 & 6 & \blacksquare \end{bmatrix}$

22. $\begin{bmatrix} 2 & 1 & -3 \\ 2 & 6 & 1 \end{bmatrix}$
$\begin{bmatrix} 6 & 3 & \blacksquare \\ 2 & 6 & 1 \end{bmatrix}$

GUIDED PRACTICE *Use matrices to solve each system of equations. Give a general solution if necessary.* **SEE EXAMPLE 1.** (OBJECTIVE 1)

23. $\begin{cases} 2x - 3y = 3 \\ 2x - y = 5 \end{cases}$

24. $\begin{cases} x + y = 3 \\ x - y = -1 \end{cases}$

25. $\begin{cases} x + 2y = -2 \\ 3x - y = 8 \end{cases}$

26. $\begin{cases} 2x - 3y = 16 \\ -4x + y = -22 \end{cases}$

27. $\begin{cases} x + y + z = 6 \\ x + 2y + z = 8 \\ x + y + 2z = 9 \end{cases}$

28. $\begin{cases} x - y + z = 2 \\ x + 2y - z = 6 \\ 2x - y - z = 3 \end{cases}$

29. $\begin{cases} x - y = 1 \\ y + z = 1 \\ x + z = 2 \end{cases}$

30. $\begin{cases} x + z = 1 \\ x + y = 2 \\ 2x + y + z = 3 \end{cases}$

Use matrices to solve each system of equations. **SEE EXAMPLE 2.** (OBJECTIVE 2)

31. $\begin{cases} x + y = 3 \\ 3x - y = 1 \\ 2x + y = 4 \end{cases}$

32. $\begin{cases} x - y = -5 \\ 2x + 3y = 5 \\ x + y = 1 \end{cases}$

33. $\begin{cases} 2x + y = 7 \\ x - y = 2 \\ -x + 3y = -2 \end{cases}$

34. $\begin{cases} 3x - y = 2 \\ -6x + 3y = 0 \\ -x + 2y = -4 \end{cases}$

Use matrices to solve each system of equations. Give a general solution. **SEE EXAMPLE 3.** (OBJECTIVE 3)

35. $\begin{cases} x + 3y + 2z = 4 \\ -x - 2y - z = -2 \end{cases}$

36. $\begin{cases} 2x - 4y + 3z = 6 \\ -4x + 6y + 4z = -6 \end{cases}$

37. $\begin{cases} -3x - 2y + 2z = 0 \\ -x + y - z = -5 \end{cases}$

38. $\begin{cases} 4x - 3y + 5z = 0 \\ 4x + 3y - z = 0 \end{cases}$

ADDITIONAL PRACTICE *Use matrices to solve each system of equations.*

39. $\begin{cases} 5x - 4y = 8 \\ 10x + 3y = -6 \end{cases}$

40. $\begin{cases} 5x - 4y = 10 \\ x - 7y = 2 \end{cases}$

41. $\begin{cases} 5a = 24 + 2b \\ 5b = 3a + 16 \end{cases}$

42. $\begin{cases} 3m = 2n + 16 \\ 2m = -5n - 2 \end{cases}$

43. $\begin{cases} 3a + b - 3c = 5 \\ a - 2b + 4c = 10 \\ a + b + c = 13 \end{cases}$

44. $\begin{cases} 2a + b - 3c = -1 \\ 3a - 2b - c = -5 \\ a - 3b - 2c = -12 \end{cases}$

45. $\begin{cases} x + 2y + 2z = 2 \\ 2x + y - z = 1 \\ 4x + 5y + 3z = 3 \end{cases}$ **46.** $\begin{cases} x + 2y - z = 3 \\ 2x - y + 2z = 6 \\ x - 3y + 3z = 4 \end{cases}$

47. $\begin{cases} 2x - y = 4 \\ x + 3y = 2 \\ -x - 4y = -2 \end{cases}$ **48.** $\begin{cases} 3x - 2y = 5 \\ x + 2y = 7 \\ -3x - y = -11 \end{cases}$

49. $\begin{cases} x + 3y = 7 \\ x + y = 3 \\ 3x + y = 5 \end{cases}$ **50.** $\begin{cases} x + y = 3 \\ x - 2y = -3 \\ x - y = 1 \end{cases}$

51. $\begin{cases} 5x - 2y = 4 \\ 2x - 4y = -8 \end{cases}$ **52.** $\begin{cases} 2x - y = -1 \\ x - 2y = 1 \end{cases}$

53. $\begin{cases} 2x + y = -4 \\ 6x + 3y = 1 \end{cases}$ **54.** $\begin{cases} x - 4y = 6 \\ -3x + 12y = 10 \end{cases}$

55. $\begin{cases} 3x - 2y + 4z = 4 \\ x + y + z = 3 \\ 6x - 2y - 3z = 10 \end{cases}$ **56.** $\begin{cases} 2x + y - z = -6 \\ x - y - z = -2 \\ -4x + 3y + z = 6 \end{cases}$

57. $\begin{cases} 3x - y = 9 \\ -6x + 2y = -18 \end{cases}$ **58.** $\begin{cases} x - y = 1 \\ -3x + 3y = -3 \end{cases}$

59. $\begin{cases} x + y + z = 6 \\ x - y + z = 2 \end{cases}$ **60.** $\begin{cases} x - y = 0 \\ y + z = 3 \\ x + z = 3 \end{cases}$

61. $\begin{cases} x + 2y + z = 1 \\ 2x - y + 2z = 2 \\ 3x + y + 3z = 3 \end{cases}$ **62.** $\begin{cases} x - 2y + 3z = 9 \\ -x + 3y = -4 \\ 2x - 5y + 3z = 13 \end{cases}$

63. $\begin{cases} 2x + y + 3z = 3 \\ -2x - y + z = 5 \\ 4x - 2y + 2z = 2 \end{cases}$ **64.** $\begin{cases} 3x + 2y + z = 8 \\ 6x - y + 2z = 16 \\ -9x + y - z = -20 \end{cases}$

APPLICATIONS

65. Piggy banks   When a child breaks open her piggy bank, she finds a total of 64 coins, consisting of nickels, dimes, and quarters. The total value of the coins is $6. If the nickels were dimes, and the dimes were nickels, the value of the coins would be $5. How many nickels, dimes, and quarters were in the piggy bank?

66. Theater seating   The illustration shows the cash receipts and the ticket prices from two sold-out performances of a play. Find the number of seats in each of the three sections of the 800-seat theater.

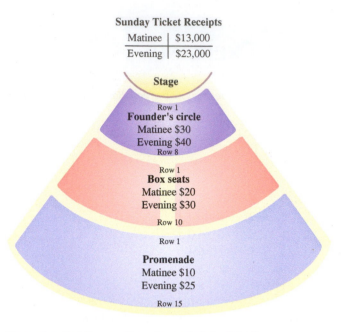

Sunday Ticket Receipts	
Matinee	$13,000
Evening	$23,000

Stage

Row 1
Founder's circle
Matinee $30
Evening $40
Row 8

Row 1
Box seats
Matinee $20
Evening $30
Row 10

Row 1
Promenade
Matinee $10
Evening $25
Row 15

Remember that the equation of a parabola is of the form
$y = ax^2 + bx + c.$

67. Curve fitting   Find the equation of the parabola passing through the points $(0, 1), (1, 2)$, and $(-1, 4)$.

68. Curve fitting   Find the equation of the parabola passing through the points $(0, 1), (1, 1)$, and $(-1, -1)$.

69. Physical therapy   After an elbow injury, a volleyball player has restricted movement of her arm. Her range of motion (the measure of $\angle 1$) is 28° less than the measure of $\angle 2$. Find the measure of each angle.

70. Cats and dogs   In 2011, there were approximately 164 million dogs and cats in the United States. If there were 8 million more cats than dogs, how many dogs and cats were there?

Remember these facts from geometry. Then solve each problem using two variables.

Two angles whose measures add up to 90° are complementary.
Two angles whose measures add up to 180° are supplementary.
The sum of the measures of the interior angles in a triangle is 180°.

71. Geometry   One angle is 28° larger than its complement. Find the measure of each angle.

72. Geometry   One angle is 46° larger than its supplement. Find the measure of each angle.

73. Geometry In the triangle below, $\angle B$ is 25° more than $\angle A$, and $\angle C$ is 5° less than twice $\angle A$. Find the measure of each angle in the triangle.

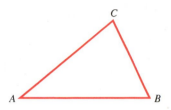

74. Geometry In the triangle below, $\angle A$ is 10° less than $\angle B$, and $\angle B$ is 10° less than $\angle C$. Find the measure of each angle in the triangle.

WRITING ABOUT MATH

75. Explain how to check the solution of a system of equations.

76. Explain how to perform a type 3 row operation.

SOMETHING TO THINK ABOUT

77. If the system represented by

$$\begin{bmatrix} 1 & 1 & 0 & | & 1 \\ 0 & 0 & 1 & | & 2 \\ 0 & 0 & 0 & | & k \end{bmatrix}$$

has no solution, what do you know about k?

78. Is it possible for a system with fewer equations than variables to have no solution? Illustrate.

Section 3.5
Solving Systems of Linear Equations Using Determinants

Objectives

1 Find the determinant of a 2 × 2 and a 3 × 3 matrix without a calculator.
2 Solve a system of linear equations using Cramer's rule.

Vocabulary

determinant   minors   Cramer's rule

Getting Ready

Find each value.

1. $3(-4) - 2(5)$ **2.** $5(2) - 3(-4)$

3. $2(2 - 5) - 3(5 - 2) + 2(4 - 3)$

4. $-3(5 - 2) + 2(3 + 1) - 2(5 + 1)$

We now discuss a final method for solving systems of linear equations. This method involves *determinants*, an idea related to the concept of matrices.

1 Find the determinant of a 2 × 2 and a 3 × 3 matrix without a calculator.

If a matrix A has the same number of rows as columns, it is called a *square matrix.* To each square matrix A, there is associated a number called its *determinant,* represented by the symbol $|A|$.

VALUE OF A 2 × 2 DETERMINANT

If a, b, c, and d are real numbers, the **determinant** of the matrix $\begin{bmatrix} a & b \\ c & d \end{bmatrix}$ is

$$\begin{vmatrix} a & b \\ c & d \end{vmatrix} = ad - bc$$

COMMENT Note that if A is a matrix, $|A|$ represents the determinant of A. If A is a number, $|A|$ represents the absolute value of A.

The determinant of a 2 × 2 matrix is the number that is equal to the product of the numbers on the major diagonal

$$\begin{vmatrix} a & b \\ c & d \end{vmatrix}$$

minus the product of the numbers on the other diagonal

$$\begin{vmatrix} a & b \\ c & d \end{vmatrix}$$

EXAMPLE 1 Evaluate the determinants: **a.** $\begin{vmatrix} 3 & 2 \\ 6 & 9 \end{vmatrix}$ **b.** $\begin{vmatrix} -5 & \frac{1}{2} \\ -1 & 0 \end{vmatrix}$

Solution **a.** $\begin{vmatrix} 3 & 2 \\ 6 & 9 \end{vmatrix} = 3(9) - 2(6)$ **b.** $\begin{vmatrix} -5 & \frac{1}{2} \\ -1 & 0 \end{vmatrix} = -5(0) - \frac{1}{2}(-1)$

$$= 27 - 12 \qquad\qquad\qquad = 0 + \frac{1}{2}$$

$$= 15 \qquad\qquad\qquad\qquad = \frac{1}{2}$$

🌿 **SELF CHECK 1** Evaluate the determinant: $\begin{vmatrix} 4 & -3 \\ 2 & 1 \end{vmatrix}$

A 3 × 3 determinant can be evaluated by expanding by **minors**.

VALUE OF A 3 × 3 DETERMINANT

$$\begin{vmatrix} a_1 & b_1 & c_1 \\ a_2 & b_2 & c_2 \\ a_3 & b_3 & c_3 \end{vmatrix} = a_1 \underset{\text{Minor of } a_1}{\begin{vmatrix} b_2 & c_2 \\ b_3 & c_3 \end{vmatrix}} - b_1 \underset{\text{Minor of } b_1}{\begin{vmatrix} a_2 & c_2 \\ a_3 & c_3 \end{vmatrix}} + c_1 \underset{\text{Minor of } c_1}{\begin{vmatrix} a_2 & b_2 \\ a_3 & b_3 \end{vmatrix}}$$

To find the minor of a_1, we find the determinant formed by crossing out the elements of the matrix that are in the same row and column as a_1:

$$\begin{vmatrix} a_1 & b_1 & c_1 \\ a_2 & b_2 & c_2 \\ a_3 & b_3 & c_3 \end{vmatrix} \qquad \text{The minor of } a_1 \text{ is } \begin{vmatrix} b_2 & c_2 \\ b_3 & c_3 \end{vmatrix}.$$

To find the minor of b_1, we cross out the elements of the matrix that are in the same row and column as b_1:

$$\begin{vmatrix} a_1 & b_1 & c_1 \\ a_2 & b_2 & c_2 \\ a_3 & b_3 & c_3 \end{vmatrix} \qquad \text{The minor of } b_1 \text{ is } \begin{vmatrix} a_2 & c_2 \\ a_3 & c_3 \end{vmatrix}.$$

To find the minor of c_1, we cross out the elements of the matrix that are in the same row and column as c_1:

$$\begin{vmatrix} a_1 & b_1 & c_1 \\ a_2 & b_2 & c_2 \\ a_3 & b_3 & c_3 \end{vmatrix} \qquad \text{The minor of } c_1 \text{ is } \begin{vmatrix} a_2 & b_2 \\ a_3 & b_3 \end{vmatrix}.$$

We can evaluate a 3×3 determinant by expanding it along any row or column. To determine the signs between the terms of the expansion of a 3×3 determinant, we use the following array of signs.

ARRAY OF SIGNS FOR A 3×3 DETERMINANT	$\begin{array}{ccc} + & - & + \\ - & + & - \\ + & - & + \end{array}$

The pattern above (alternating signs within the rows and columns beginning with a positive in row 1, column 1) holds true for any square matrix.

EXAMPLE 2 Evaluate the determinant: $\begin{vmatrix} 1 & 3 & -2 \\ 2 & 1 & 3 \\ 1 & 2 & 3 \end{vmatrix}$

Solution

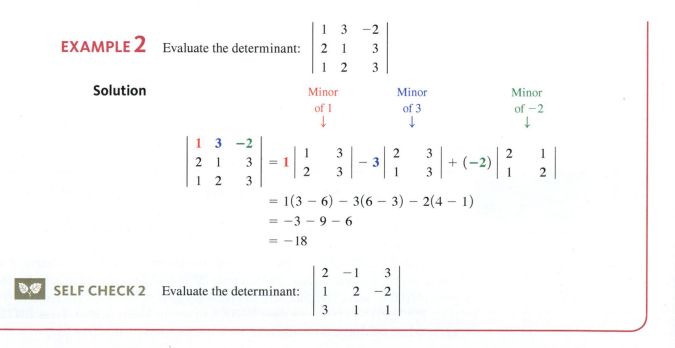

$$\begin{vmatrix} \mathbf{1} & \mathbf{3} & \mathbf{-2} \\ 2 & 1 & 3 \\ 1 & 2 & 3 \end{vmatrix} = \mathbf{1} \begin{vmatrix} 1 & 3 \\ 2 & 3 \end{vmatrix} - \mathbf{3} \begin{vmatrix} 2 & 3 \\ 1 & 3 \end{vmatrix} + (\mathbf{-2}) \begin{vmatrix} 2 & 1 \\ 1 & 2 \end{vmatrix}$$

$$= 1(3 - 6) - 3(6 - 3) - 2(4 - 1)$$
$$= -3 - 9 - 6$$
$$= -18$$

SELF CHECK 2 Evaluate the determinant: $\begin{vmatrix} 2 & -1 & 3 \\ 1 & 2 & -2 \\ 3 & 1 & 1 \end{vmatrix}$

Perspective

LEWIS CARROLL

One of the more amusing historical anecdotes concerning matrices and determinants involves the English mathematician Charles Dodgson, also known as Lewis Carroll. The anecdote describes how England's Queen Victoria so enjoyed reading Carroll's book *Alice in Wonderland* that she requested a copy of his next publication. To her great surprise, she received an autographed copy of a mathematics text titled *An Elementary Treatise on Determinants*. The story was repeated as fact so often that Carroll finally included an explicit disclaimer in his book *Symbolic Logic*, insisting that the incident never actually occurred.

© Morphart Creations Inc./Shutterstock.com

Evaluate each determinant.

1. $\begin{vmatrix} 3 & 4 \\ 2 & 1 \end{vmatrix}$

2. $\begin{vmatrix} 3 & 4 & 2 \\ 1 & -1 & 5 \\ 1 & 2 & -2 \end{vmatrix}$

Source: http://mathworld.wolfram.com/Determinant.html

EXAMPLE 3   Evaluate $\begin{vmatrix} 1 & 3 & -2 \\ 2 & 1 & 3 \\ 1 & 2 & 3 \end{vmatrix}$ by expanding on the middle column.

Solution   This is the determinant of Example 2. To expand it along the middle column, we use the signs of the middle column of the array of signs:

$$\begin{vmatrix} 1 & \textbf{3} & -2 \\ 2 & \textbf{1} & 3 \\ 1 & \textbf{2} & 3 \end{vmatrix} = -3 \begin{vmatrix} 2 & 3 \\ 1 & 3 \end{vmatrix} + 1 \begin{vmatrix} 1 & -2 \\ 1 & 3 \end{vmatrix} - 2 \begin{vmatrix} 1 & -2 \\ 2 & 3 \end{vmatrix}$$

Minor of 3   Minor of 1   Minor of 2

$$= -3(6 - 3) + 1[3 - (-2)] - 2[3 - (-4)]$$
$$= -3(3) + 1(5) - 2(7)$$
$$= -9 + 5 - 14$$
$$= -18$$

As expected, we get the same value as in Example 2.

SELF CHECK 3   Evaluate: $\begin{vmatrix} 2 & -1 & 3 \\ 1 & 2 & 2 \\ 3 & 1 & 1 \end{vmatrix}$

Accent on technology

▸ **Evaluating Determinants**

To use a TI-84 graphing calculator to evaluate the determinant in Example 3, $\begin{vmatrix} 1 & 3 & -2 \\ 2 & 1 & 3 \\ 1 & 2 & 3 \end{vmatrix}$, we first enter the matrix by pressing the **MATRX** key, selecting EDIT, and pressing the **ENTER** key. We then enter the dimensions and the elements of the matrix to obtain Figure 3-15(a). We then press **2ND QUIT** to clear the screen. We then press **MATRX**, select MATH(1 det), and press **ENTER** to obtain Figure 3-15(b). Next, press **MATRX** (NAMES) and press **ENTER** to obtain Figure 3-15(c). Press **ENTER** again to find the determinant. The value of the determinant is −18.

Figure 3-15

For instructions regarding the use of a Casio graphing calculator, please refer to the Casio Keystroke Guide in the back of the book.

2 Solve a system of linear equations using Cramer's rule.

The method of using determinants to solve systems of equations is called **Cramer's rule**, named after the 18th-century mathematician Gabriel Cramer. To develop Cramer's rule, we consider the system

$$\begin{cases} ax + by = e \\ cx + dy = f \end{cases}$$

where x and y are variables and $a, b, c, d, e,$ and f are numerical values.

If we multiply both sides of the first equation by d and multiply both sides of the second equation by $-b$, we can add the equations and eliminate y:

$$\begin{aligned} adx + bdy &= ed \\ \underline{-bcx - bdy} &= \underline{-bf} \\ adx - bcx &= ed - bf \end{aligned}$$

To solve for x, we use the distributive property to write $adx - bcx$ as $(ad - bc)x$ on the left side and divide each side by $ad - bc$:

$$(ad - bc)x = ed - bf$$

$$x = \frac{ed - bf}{ad - bc} \quad (ad - bc) \neq 0$$

We can find y in a similar manner. After eliminating the variable x, we have

$$y = \frac{af - ec}{ad - bc} \quad (ad - bc) \neq 0$$

Determinants provide a way of remembering these formulas. Note that the denominator for both x and y is

$$\begin{vmatrix} a & b \\ c & d \end{vmatrix} = ad - bc$$

Gabriel Cramer
1704–1752
Although other mathematicians had worked with determinants, it was the work of Cramer that popularized them.

The numerators can be expressed as determinants also:

$$x = \frac{ed - bf}{ad - bc} = \frac{\begin{vmatrix} e & b \\ f & d \end{vmatrix}}{\begin{vmatrix} a & b \\ c & d \end{vmatrix}} \quad \text{and} \quad x = \frac{af - ec}{ad - bc} = \frac{\begin{vmatrix} a & e \\ c & f \end{vmatrix}}{\begin{vmatrix} a & b \\ c & d \end{vmatrix}}$$

If we compare these formulas with the original system

$$\begin{cases} ax + by = e \\ cx + dy = f \end{cases}$$

we note that in the expressions for x and y above, the denominator determinant is formed by using the coefficients $a, b, c,$ and d of the variables in the equations. The numerator determinants are the same as the denominator determinant, except that the column of coefficients of the variable for which we are solving is replaced with the column of constants e and f.

CRAMER'S RULE FOR TWO LINEAR EQUATIONS IN TWO VARIABLES

The solution of the system $\begin{cases} ax + by = e \\ cx + dy = f \end{cases}$ is given by

$$x = \frac{D_x}{D} \quad \text{and} \quad y = \frac{D_y}{D}$$

where $D = \begin{vmatrix} a & b \\ c & d \end{vmatrix}, D_x = \begin{vmatrix} e & b \\ f & d \end{vmatrix},$ and $D_y = \begin{vmatrix} a & e \\ c & f \end{vmatrix}.$

If $D \neq 0$, the system is consistent and the equations are independent.

If $D = 0$ and D_x or D_y is nonzero, the system is inconsistent.

EXAMPLE 4   Use Cramer's rule to solve $\begin{cases} 4x - 3y = 6 \\ -2x + 5y = 4 \end{cases}.$

Solution   The value of x is the quotient of two determinants. The denominator determinant is made up of the coefficients of x and y:

$$D = \begin{vmatrix} 4 & -3 \\ -2 & 5 \end{vmatrix}$$

To solve for x, we form the numerator determinant from the denominator determinant by replacing its first column (the coefficients of x) with the column of constants (6 and 4).

To solve for y, we form the numerator determinant from the denominator determinant by replacing the second column (the coefficients of y) with the column of constants (6 and 4).

To find the values of x and y, we evaluate each determinant:

$$x = \frac{D_x}{D} = \frac{\begin{vmatrix} 6 & -3 \\ 4 & 5 \end{vmatrix}}{\begin{vmatrix} 4 & -3 \\ -2 & 5 \end{vmatrix}} = \frac{6(5) - (-3)(4)}{4(5) - (-3)(-2)} = \frac{30 + 12}{20 - 6} = \frac{42}{14} = 3$$

$$y = \frac{D_y}{D} = \frac{\begin{vmatrix} 4 & 6 \\ -2 & 4 \end{vmatrix}}{\begin{vmatrix} 4 & -3 \\ -2 & 5 \end{vmatrix}} = \frac{4(4) - 6(-2)}{4(5) - (-3)(-2)} = \frac{16 + 12}{20 - 6} = \frac{28}{14} = 2$$

The solution of this system is $(3, 2)$. Verify that $x = 3$ and $y = 2$ satisfy each equation in the given system.

SELF CHECK 4   Solve the system: $\begin{cases} 2x - 3y = -16 \\ 3x + 5y = 14 \end{cases}$

EXAMPLE 5   Use Cramer's rule to solve $\begin{cases} 7x = 8 - 4y \\ 2y = 3 - \frac{7}{2}x \end{cases}$.

Solution   We multiply both sides of the second equation by 2 to eliminate the fraction and write the system in the form

$$\begin{cases} 7x + 4y = 8 \\ 7x + 4y = 6 \end{cases}$$

When we attempt to use Cramer's rule to solve this system for x, we obtain

$$x = \frac{D_x}{D} = \frac{\begin{vmatrix} 8 & 4 \\ 6 & 4 \end{vmatrix}}{\begin{vmatrix} 7 & 4 \\ 7 & 4 \end{vmatrix}} = \frac{8}{0}, \text{ which is undefined.}$$

This system is inconsistent because the denominator determinant is 0 and the numerator determinant is not 0. Since this system has no solution, its solution set is $\varnothing$.

We can see directly from the system that it is inconsistent. For any values of x and y, it is impossible that 7 times x plus 4 times y could be both 8 and 6.

SELF CHECK 5   Solve the system: $\begin{cases} 3x = 8 - 4y \\ y = \frac{5}{2} - \frac{3}{4}x \end{cases}$

EXAMPLE 6   Use Cramer's rule to solve $\begin{cases} y = -3x + 1 \\ 6x + 2y = 2 \end{cases}$.

Solution   We first write the system in the form

$$\begin{cases} 3x + y = 1 \\ 6x + 2y = 2 \end{cases}$$

When we attempt to use Cramer's rule to solve this system for x, we obtain

$$x = \frac{D_x}{D} = \frac{\begin{vmatrix} 1 & 1 \\ 2 & 2 \end{vmatrix}}{\begin{vmatrix} 3 & 1 \\ 6 & 2 \end{vmatrix}} = \frac{0}{0}, \text{ which is indeterminate.}$$

When we attempt to use Cramer's rule to solve this system for y, we obtain

$$y = \frac{D_y}{D} = \frac{\begin{vmatrix} 3 & 1 \\ 6 & 2 \end{vmatrix}}{\begin{vmatrix} 3 & 1 \\ 6 & 2 \end{vmatrix}} = \frac{0}{0}, \text{ which is indeterminate.}$$

This system is consistent and its equations are dependent because every determinant is 0. Every solution of one equation is also a solution of the other. The solution is the general ordered pair $(x, -3x + 1)$.

SELF CHECK 6   Solve the system: $\begin{cases} y = 2x - 5 \\ 10x - 5y = 25 \end{cases}$

CRAMER'S RULE FOR THREE LINEAR EQUATIONS IN THREE VARIABLES

The solution of the system $\begin{cases} ax + by + cz = j \\ dx + ey + fz = k \\ gx + hy + iz = l \end{cases}$ is given by

$$x = \frac{D_x}{D}, \quad y = \frac{D_y}{D}, \quad \text{and} \quad z = \frac{D_z}{D}$$

where

$$D = \begin{vmatrix} a & b & c \\ d & e & f \\ g & h & i \end{vmatrix} \qquad D_x = \begin{vmatrix} j & b & c \\ k & e & f \\ l & h & i \end{vmatrix}$$

$$D_y = \begin{vmatrix} a & j & c \\ d & k & f \\ g & l & i \end{vmatrix} \qquad D_z = \begin{vmatrix} a & b & j \\ d & e & k \\ g & h & l \end{vmatrix}$$

If $D \neq 0$, the system is consistent and the equations are independent.

If $D = 0$ and D_x or D_y or D_z is nonzero, the system is inconsistent.

If the determinant of the coefficient, D, is 0, the system is inconsistent or has infinitely many solutions. Use another method to solve.

EXAMPLE 7   Use Cramer's rule to solve $\begin{cases} 2x + y + 4z = 12 \\ x + 2y + 2z = 9 \\ 3x - 3y - 2z = 1 \end{cases}$.

Solution   The denominator determinant is the determinant formed by the coefficients of the variables. To form the numerator determinants, we substitute the column of constants for the coefficients of the variable to be found. We form the quotients for $x, y,$ and z and evaluate the determinants:

$$x = \frac{D_x}{D} = \frac{\begin{vmatrix} 12 & 1 & 4 \\ 9 & 2 & 2 \\ 1 & -3 & -2 \end{vmatrix}}{\begin{vmatrix} 2 & 1 & 4 \\ 1 & 2 & 2 \\ 3 & -3 & -2 \end{vmatrix}} = \frac{12\begin{vmatrix} 2 & 2 \\ -3 & -2 \end{vmatrix} - 1\begin{vmatrix} 9 & 2 \\ 1 & -2 \end{vmatrix} + 4\begin{vmatrix} 9 & 2 \\ 1 & -3 \end{vmatrix}}{2\begin{vmatrix} 2 & 2 \\ -3 & -2 \end{vmatrix} - 1\begin{vmatrix} 1 & 2 \\ 3 & -2 \end{vmatrix} + 4\begin{vmatrix} 1 & 2 \\ 3 & -3 \end{vmatrix}} = \frac{12(2) - (-20) + 4(-29)}{2(2) - (-8) + 4(-9)} = \frac{-72}{-24} = 3$$

$$y = \frac{D_y}{D} = \frac{\begin{vmatrix} 2 & 12 & 4 \\ 1 & 9 & 2 \\ 3 & 1 & -2 \end{vmatrix}}{\begin{vmatrix} 2 & 1 & 4 \\ 1 & 2 & 2 \\ 3 & -3 & -2 \end{vmatrix}} = \frac{2\begin{vmatrix} 9 & 2 \\ 1 & -2 \end{vmatrix} - 12\begin{vmatrix} 1 & 2 \\ 3 & -2 \end{vmatrix} + 4\begin{vmatrix} 1 & 9 \\ 3 & 1 \end{vmatrix}}{-24} = \frac{2(-20) - 12(-8) + 4(-26)}{-24} = \frac{-48}{-24} = 2$$

$$z = \frac{D_z}{D} = \frac{\begin{vmatrix} 2 & 1 & 12 \\ 1 & 2 & 9 \\ 3 & -3 & 1 \end{vmatrix}}{\begin{vmatrix} 2 & 1 & 4 \\ 1 & 2 & 2 \\ 3 & -3 & -2 \end{vmatrix}} = \frac{2\begin{vmatrix} 2 & 9 \\ -3 & 1 \end{vmatrix} - 1\begin{vmatrix} 1 & 9 \\ 3 & 1 \end{vmatrix} + 12\begin{vmatrix} 1 & 2 \\ 3 & -3 \end{vmatrix}}{-24} = \frac{2(29) - 1(-26) + 12(-9)}{-24} = \frac{-24}{-24} = 1$$

The solution of this system is $(3, 2, 1)$.

SELF CHECK 7   Solve the system: $\begin{cases} x + y + 2z = 6 \\ 2x - y + z = 9 \\ x + y - 2z = -6 \end{cases}$

SELF CHECK ANSWERS

1. 10   **2.** 0   **3.** -20 **4.** $(-2, 4)$ **5.** $\varnothing$ **6.** $(x, 2x - 5)$ **7.** $(2, -2, 3)$

NOW TRY THIS

Solve for x.

1. $\begin{vmatrix} x & 2 \\ x & 3 \end{vmatrix} = 4$

2. $\begin{vmatrix} 2x & 3 \\ -5x & -3 \end{vmatrix} = 10(x - 1)$

3. $\begin{vmatrix} x + 4 & 3 \\ 2x - 5 & 2 \end{vmatrix} = 2x + 5$

4. $\begin{vmatrix} 2 & x & -1 \\ 1 & 2x & 4 \\ -4 & x & 1 \end{vmatrix} = 30$

3.5   Exercises

WARM-UPS *Evaluate $ad - bc$ for the following.*

1. $a = 4, b = 3, c = 2, d = -1$

2. $a = 2, b = 0, c = -1, d = 4$

3. $a = 0, b = 8, c = 2, d = 5$

When using Cramer's rule to solve the system $\begin{cases} 2x - y = -21 \\ 4x + 5y = 7 \end{cases}$

4. Set up the denominator determinant for x.

5. Set up the numerator determinant for x.

6. Set up the numerator determinant for y.

REVIEW *Solve each equation.*

7. $5(x - 2) - (3 - x) = x + 2$

8. $\frac{3}{7}x = 2(x + 11)$

9. $\frac{5}{3}(5x + 6) - 10 = 0$

10. $5 - 3(2x - 1) = 2(4 + 3x) - 24$

VOCABULARY AND CONCEPTS *Fill in the blanks.*

11. A determinant is a _____ that is associated with a _____ matrix.

12. The value of $\begin{vmatrix} a & b \\ c & d \end{vmatrix}$ is _____.

13. The minor of b_1 in $\begin{vmatrix} a_1 & b_1 & c_1 \\ a_2 & b_2 & c_2 \\ a_3 & b_3 & c_3 \end{vmatrix}$ is _____.

14. We can evaluate a determinant by expanding it along any ___ or ___.

15. The method of solving a system of linear equations using determinants is called _____.

16. The setup for the denominator determinant for the value of x in the system $\begin{cases} 5x + 3y = 6 \\ 4x - 2y = 7 \end{cases}$ is _____.

17. If $D \neq 0$, then the system is _____ and the equations are _____.

18. If the denominator determinant for y in a system of equations is zero, the equations of the system are _____ or the system is _____.

GUIDED PRACTICE
Evaluate each determinant. SEE EXAMPLE 1. (OBJECTIVE 1)

19. $\begin{vmatrix} 4 & 2 \\ -3 & 2 \end{vmatrix}$

20. $\begin{vmatrix} 3 & -2 \\ -2 & 4 \end{vmatrix}$

21. $\begin{vmatrix} -2 & 5 \\ 1 & -3 \end{vmatrix}$

22. $\begin{vmatrix} -1 & -2 \\ -3 & -4 \end{vmatrix}$

Evaluate each determinant. SEE EXAMPLES 2–3. (OBJECTIVE 1)

23. $\begin{vmatrix} 1 & 1 & 1 \\ 1 & 0 & 2 \\ 0 & 2 & 0 \end{vmatrix}$

24. $\begin{vmatrix} 1 & 2 & 0 \\ 0 & 1 & 2 \\ 0 & 0 & 1 \end{vmatrix}$

25. $\begin{vmatrix} -1 & 2 & 1 \\ 2 & 1 & -3 \\ 1 & 1 & 1 \end{vmatrix}$

26. $\begin{vmatrix} 1 & 2 & 3 \\ 1 & 2 & 3 \\ 1 & 2 & 3 \end{vmatrix}$

27. $\begin{vmatrix} 1 & -2 & 3 \\ -2 & 1 & 1 \\ -3 & -2 & 1 \end{vmatrix}$

28. $\begin{vmatrix} 1 & 1 & 2 \\ 2 & 1 & -2 \\ 3 & 1 & 3 \end{vmatrix}$

29. $\begin{vmatrix} 2 & 4 & 6 \\ 1 & 3 & 5 \\ 9 & 8 & 7 \end{vmatrix}$

30. $\begin{vmatrix} 1 & 4 & 7 \\ 2 & 5 & 8 \\ 3 & 6 & 9 \end{vmatrix}$

Use Cramer's rule to solve each system, if possible. If the equations of the system are dependent, give a general solution. SEE EXAMPLE 4. (OBJECTIVE 2)

31. $\begin{cases} 4x + 3y = 5 \\ 3x - 2y = -9 \end{cases}$

32. $\begin{cases} 3x - y = -3 \\ 2x + y = -7 \end{cases}$

33. $\begin{cases} x + y = 6 \\ x - y = 2 \end{cases}$

34. $\begin{cases} x - y = 4 \\ 2x + y = 5 \end{cases}$

Use Cramer's rule to solve each system, if possible. If the equations of the system are dependent, give a general solution. SEE EXAMPLE 5. (OBJECTIVE 2)

35. $\begin{cases} 4x = 2y - 5 \\ y = \dfrac{4x - 5}{2} \end{cases}$

36. $\begin{cases} y = \dfrac{11 - 3x}{2} \\ x = \dfrac{11 - 4y}{6} \end{cases}$

37. $\begin{cases} 2x = \dfrac{y + 5}{2} \\ y = 4x - 6 \end{cases}$

38. $\begin{cases} y = \dfrac{12 + 2x}{3} \\ x = \dfrac{3}{2}y + 3 \end{cases}$

Use Cramer's rule to solve each system, if possible. If the equations of the system are dependent, give a general solution. SEE EXAMPLE 6. (OBJECTIVE 2)

39. $\begin{cases} 2x + 3y = 9 \\ y = -\dfrac{2}{3}x + 3 \end{cases}$

40. $\begin{cases} x = \dfrac{12 - 6y}{5} \\ y = \dfrac{24 - 10x}{12} \end{cases}$

41. $\begin{cases} 4x - 3y = 6 \\ y = \dfrac{4x - 6}{3} \end{cases}$

42. $\begin{cases} 2x + 3y = 12 \\ x = \dfrac{12 - 3y}{2} \end{cases}$

Use Cramer's rule to solve each system, if possible. If the equations of the system are dependent, give a general solution. SEE EXAMPLE 7. (OBJECTIVE 2)

43. $\begin{cases} x + y + z = 4 \\ x + y - z = 0 \\ x - y + z = 2 \end{cases}$

44. $\begin{cases} x + y + z = 4 \\ x - y + z = 2 \\ x - y - z = 0 \end{cases}$

45. $\begin{cases} x + y + 2z = 7 \\ x + 2y + z = 8 \\ 2x + y + z = 9 \end{cases}$

46. $\begin{cases} x + 2y + 2z = 10 \\ 2x + y + 2z = 9 \\ 2x + 2y + z = 1 \end{cases}$

47. $\begin{cases} 2x + y - z = 1 \\ x + 2y + 2z = 2 \\ 4x + 5y + 3z = 3 \end{cases}$

48. $\begin{cases} 2x - y + 4z + 2 = 0 \\ 5x + 8y + 7z = -8 \\ x + 3y + z + 3 = 0 \end{cases}$

49. $\begin{cases} 2x + 3y + 4z = 6 \\ 2x - 3y - 4z = -4 \\ 4x + 6y + 8z = 12 \end{cases}$

50. $\begin{cases} x - 3y + 4z - 2 = 0 \\ 2x + y + 2z - 3 = 0 \\ 4x - 5y + 10z - 7 = 0 \end{cases}$

ADDITIONAL PRACTICE
Evaluate each determinant.

51. $\begin{vmatrix} 2a & b \\ b & 2a \end{vmatrix}$

52. $\begin{vmatrix} y & x \\ x + y & x - y \end{vmatrix}$

53. $\begin{vmatrix} a & 2a & -a \\ 2 & -1 & 3 \\ 1 & 2 & -3 \end{vmatrix}$

54. $\begin{vmatrix} 1 & 2b & -3 \\ 2 & -b & 2 \\ 1 & 3b & 1 \end{vmatrix}$

55. $\begin{vmatrix} 1 & a & b \\ 1 & 2a & 2b \\ 1 & 3a & 3b \end{vmatrix}$

56. $\begin{vmatrix} a & b & c \\ 0 & b & c \\ 0 & 0 & c \end{vmatrix}$

Use a graphing calculator to evaluate each determinant.

57. $\begin{bmatrix} 2 & -3 & 4 \\ -1 & 2 & 4 \\ 3 & -3 & 1 \end{bmatrix}$ **58.** $\begin{bmatrix} -3 & 2 & -5 \\ 3 & -2 & 6 \\ 1 & -3 & 4 \end{bmatrix}$

59. $\begin{bmatrix} 2 & 1 & -3 \\ -2 & 2 & 4 \\ 1 & -2 & 2 \end{bmatrix}$ **60.** $\begin{bmatrix} 4 & 2 & -3 \\ 2 & -5 & 6 \\ 2 & 5 & -2 \end{bmatrix}$

Use Cramer's rule to solve each system.

61. $\begin{cases} 2x + 3y = 0 \\ 4x - 6y = -4 \end{cases}$ **62.** $\begin{cases} 4x - 3y = -1 \\ 8x + 3y = 4 \end{cases}$

63. $\begin{cases} y = \dfrac{-2x + 1}{3} \\ 3x - 2y = 8 \end{cases}$ **64.** $\begin{cases} 2x + 3y = -1 \\ x = \dfrac{y - 9}{4} \end{cases}$

65. $\begin{cases} x = \dfrac{5y - 4}{2} \\ y = \dfrac{3x - 1}{5} \end{cases}$ **66.** $\begin{cases} y = \dfrac{1 - 5x}{2} \\ x = \dfrac{3y + 10}{4} \end{cases}$

67. $\begin{cases} 2x + y + z = 5 \\ x - 2y + 3z = 10 \\ x + y - 4z = -3 \end{cases}$ **68.** $\begin{cases} 3x + 2y - z = -8 \\ 2x - y + 7z = 10 \\ 2x + 2y - 3z = -10 \end{cases}$

69. $\begin{cases} 4x + 3z = 4 \\ 2y - 6z = -1 \\ 8x + 4y + 3z = 9 \end{cases}$ **70.** $\begin{cases} \frac{1}{2}x + y + z + \frac{3}{2} = 0 \\ x + \frac{1}{2}y + z - \frac{1}{2} = 0 \\ x + y + \frac{1}{2}z + \frac{1}{2} = 0 \end{cases}$

71. $\begin{cases} x + y = 1 \\ \frac{1}{2}y + z = \frac{5}{2} \\ x - z = -3 \end{cases}$ **72.** $\begin{cases} 3x + 4y + 14z = 7 \\ -\frac{1}{2}x - y + 2z = \frac{3}{2} \\ x + \frac{3}{2}y + \frac{5}{2}z = 1 \end{cases}$

Solve each equation.

73. $\begin{vmatrix} x & 2 \\ -3 & 1 \end{vmatrix} = 2$ **74.** $\begin{vmatrix} x & -x \\ 2 & -3 \end{vmatrix} = -5$

75. $\begin{vmatrix} x & -2 \\ 3 & 1 \end{vmatrix} = \begin{vmatrix} 4 & 2 \\ x & 3 \end{vmatrix}$ **76.** $\begin{vmatrix} x & 3 \\ x & 2 \end{vmatrix} = \begin{vmatrix} 3 & 2 \\ 1 & 1 \end{vmatrix}$

APPLICATIONS

77. Signaling A system of sending signals uses two flags held in various positions to represent letters of the alphabet. The illustration shows how the letter U is signaled. Find the measures x and y, if y is to be 30° more than x.

78. Inventories The table shows an end-of-the-year inventory report for a warehouse that supplies electronics stores. If the warehouse stocks two models of cordless telephones, one valued at $67 and the other at $100, how many of each model of phone did the warehouse have at the time of the inventory?

Item	Number	Merchandise value
Television	800	$1,005,450
Radios	200	$15,785
Cordless phones	360	$29,400

79. Investing A student wants to average a 6.6% return by investing $20,000 in the three stocks listed in the table. Because HiTech is considered to be a high-risk investment, he wants to invest three times as much in SaveTel and HiGas combined as he invests in HiTech. How much should he invest in each stock?

Stock	Rate of return
HiTech	10%
SaveTel	5%
HiGas	6%

80. Investing A woman wants to average a $7\frac{1}{3}\%$ return by investing $30,000 in three certificates of deposit. She wants to invest five times as much in the 8% CD as in the 6% CD. How much should she invest in each CD?

Type of CD	Rate of return
12 month	6%
24 month	7%
36 month	8%

WRITING ABOUT MATH

81. Explain how to find the minor of an element of a determinant.

82. Explain how to find the value of x when solving a system of linear equations by Cramer's rule.

SOMETHING TO THINK ABOUT

83. Show that

$$\begin{vmatrix} x & y & 1 \\ -2 & 3 & 1 \\ 3 & 5 & 1 \end{vmatrix} = 0$$

is the equation of the line passing through $(-2, 3)$ and $(3, 5)$.

84. Show that

$$\frac{1}{2}\begin{vmatrix} 0 & 0 & 1 \\ 3 & 0 & 1 \\ 0 & 4 & 1 \end{vmatrix}$$

is the area of the triangle with vertices at $(0, 0), (3, 0)$, and $(0, 4)$.

Determinants with more than 3 rows and 3 columns can be evaluated by expanding them by minors. The sign array for a 4×4 determinant is

$$\begin{matrix} + & - & + & - \\ - & + & - & + \\ + & - & + & - \\ - & + & - & + \end{matrix}$$

Evaluate each determinant.

85. $\begin{vmatrix} 1 & 0 & 2 & 1 \\ 2 & 1 & 1 & 3 \\ 1 & 1 & 1 & 1 \\ 2 & 1 & 1 & 1 \end{vmatrix}$

86. $\begin{vmatrix} 1 & 2 & -1 & 1 \\ -2 & 1 & 3 & -1 \\ 0 & 1 & 1 & 2 \\ 2 & 0 & 3 & 1 \end{vmatrix}$

Projects

PROJECT 1

The number of units of a product that will be produced depends on the unit price of the product. As the unit price gets higher, the product will be produced in greater quantity, because the producer will make more money on each item. The *supply* of the product will grow, and we say that supply *is a function of* (or *depends on*) the unit price. Furthermore, as the price rises, fewer consumers will buy the product, and the *demand* will decrease. The demand for the product is also a function of the unit price.

In this project, we will assume that both supply and demand are *linear* functions of the unit price. Thus, the graph of supply (the *y*-coordinate) versus price (the *x*-coordinate) is a line with positive slope. The graph of the demand function is a line with negative slope. Because these two lines cannot be parallel, they must intersect. The price at which supply equals demand is called the *market price:* At this price, the same number of units of the product will be sold as are manufactured.

You work for Soda Pop Inc. and have the task of analyzing the sales figures for the past year. You have been provided with the following supply and demand functions. (Supply and demand are measured in cases per week; p, the price per case, is measured in dollars.)

The demand for soda is $D(p) = 19{,}000 - 2{,}200p$.
The supply of soda is $S(p) = 3{,}000 + 1{,}080p$.

Both functions are true for values of p from \$3.50 to \$5.75.

Graph both functions on the same set of coordinate axes, being sure to label each graph, and include any other important information. Then write a report for your supervisor that answers the following questions.

a. Explain why producers will be able to sell all of the soda they make when the price is \$3.50 per case. How much money will the producers take in from these sales?

b. How much money will producers take in from sales when the price is \$5.75 per case? How much soda will not be sold?

c. Find the market price for soda (to the nearest cent). How many cases per week will be sold at this price? How much money will the producers take in from sales at the market price?

d. Explain why prices always tend toward the market price. That is, explain why the unit price will rise if the demand is greater than the supply, and why the unit price will fall if supply is greater than demand.

PROJECT 2

Goodstuff Produce Company has two large water canals that feed the irrigation ditches on its fruit farm. One of these canals runs directly north and south, and the other runs directly east and west. The canals cross at the center of the farm property (the origin) and divide the farm into four quadrants. The company is interested in digging some new irrigation ditches in a portion of the northeast quadrant. You have been hired to plan the layout of the new system.

Your design is to make use of ditch Z, which is already present. This ditch runs from a point 300 meters north of the origin to a point 400 meters east of the origin. The owners of Goodstuff want two new ditches.

- Ditch A is to begin at a point 100 meters north of the origin and follow a line that travels 3 meters north for every 7 meters it travels east until it intersects ditch Z.

- Ditch B is to run from the origin to ditch Z in such a way that it exactly bisects the area in the northeast quadrant that is south of both ditch Z and ditch A.

You are to provide the equations of the lines that the three ditches follow, as well as the exact location of the gates that will be installed where the ditches intersect one another. Be sure to provide explanations and organized work that will clearly display the desired information and assure the owners of Goodstuff that they will get exactly what they want.

Reach for Success
EXTENSION OF UNDERSTANDING YOUR SYLLABUS

Now that you know where your instructor's office is, how to get in touch with him or her, and how your grade will be determined, take a look at other information in the syllabus that may be important to know before you get too far into the semester.

On your syllabus, locate your instructor's make-up policy. If you know of any conflicts in meeting established deadlines, contact your instructor *prior* to the due date.	Does your instructor accept late work? _____
	
	If so, under what conditions? State the work that is accepted late and any penalty assessed.
	
	   Test make-up policy _____
	
	   Penalty assessed _____
	
	   Other requirements make-up policy _____
	
	   Penalty assessed _____
	
	If not, try to reschedule any personal obligations.

On your syllabus, locate your instructor's attendance policy.	What is your instructor's attendance policy? _____ What does the syllabus indicate about arriving late or leaving class early? _____

Understanding any penalties for late work can make a difference in the successful completion of the course. Do not hesitate to contact your instructor if you are unclear regarding any of the policies stated on the syllabus.

3 Review

SECTION 3.1 Solving Systems of Two Linear Equations in Two Variables by Graphing

DEFINITIONS AND CONCEPTS

1. On a single set of coordinate axes, carefully graph each equation.

2. Find the coordinates of the point where the graphs intersect, if applicable.

3. Check the solution in both of the original equations, if applicable.

4. If the graphs have no point in common, the system has no solution.

5. If the graphs of the equations coincide (are the same), the system has infinitely many solutions that can be expressed as a general ordered pair.

In a graph of two equations, each with two variables:

If the lines are distinct (are different) and intersect, the equations are *independent* and the system is *consistent*. **One solution exists**, expressed as (x, y).

If the lines are distinct and parallel, the equations are *independent* and the system is *inconsistent*. **No solution exists**, expressed as $\varnothing$.

If the lines coincide (are the same), the equations are *dependent* and the system is *consistent*. **Infinitely many solutions exist**, expressed as $(x, ax + b)$.

EXAMPLES

Solve each system by graphing.

a. $\begin{cases} x + y = 6 \\ x - y = 2 \end{cases}$ b. $\begin{cases} 2x + y = 6 \\ y = -2x - 3 \end{cases}$

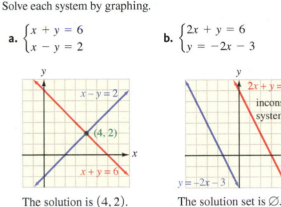

The solution is $(4, 2)$.   The solution set is $\varnothing$.

c. $\begin{cases} x - 3y = 6 \\ y = \dfrac{1}{3}x - 2 \end{cases}$

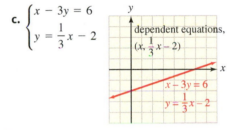

The equations of the system are dependent, and infinitely many solutions exist. A general solution is $\left(x, \dfrac{1}{3}x - 2\right)$.

REVIEW EXERCISES

Solve each system by the graphing method.

1. $\begin{cases} 2x + y = 11 \\ -x + 2y = 7 \end{cases}$

2. $\begin{cases} 3x + 2y = 0 \\ 2x - 3y = -13 \end{cases}$

3. $\begin{cases} \dfrac{1}{2}x + \dfrac{1}{3}y = 2 \\ y = 6 - \dfrac{3}{2}x \end{cases}$

4. $\begin{cases} \dfrac{1}{3}x - \dfrac{1}{2}y = 1 \\ 6x - 9y = 2 \end{cases}$

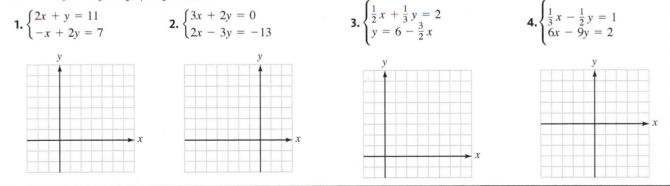

SECTION 3.2   Solving Systems of Two Linear Equations in Two Variables by Substitution and Elimination (Addition)

DEFINITIONS AND CONCEPTS	EXAMPLES
1. If necessary, solve one equation for one of its variables, preferably a variable with a coefficient of 1. **2.** Substitute the resulting expression for the variable obtained in Step 1 into the other equation and solve that equation. **3.** Find the value of the other variable by substituting the value of the variable found in Step 2 into any equation containing both variables. **4.** State the solution. **5.** Check the solution in both of the original equations.	Solve by substitution: $\begin{cases} x = 3y - 9 \\ 2x - y = 2 \end{cases}$ Since the first equation is already solved for x, we will substitute its right side for x in the second equation. $2(\mathbf{3y - 9}) - y = 2$ $6y - 18 - y = 2$ Use the distributive property to remove parentheses. $5y - 18 = 2$ Combine like terms. $5y = 20$ Add 18 to both sides. $y = 4$ Divide both sides by 5. To find x, we can substitute 4 for y in the first equation. $x = 3(\mathbf{4}) - 9$ $x = 3$ The solution is $(3, 4)$.
1. If necessary, write both equations of the system in $Ax + By = C$ form. **2.** If necessary, multiply the terms of one or both of the equations by constants chosen to make the coefficients of one of the variables opposites. **3.** Add the equations and solve the resulting equation, if possible. **4.** Substitute the value obtained in Step 3 into either of the original equations and solve for the remaining variable, if applicable. **5.** State the solution obtained in Steps 3 and 4. **6.** Check the solution in both equations of the original system.	Solve by elimination: $\begin{cases} 2x - 3y = 8 \\ x + 2y = 4 \end{cases}$ To eliminate x, we multiply the second equation by -2 and add the result to the first equation. $\begin{aligned} 2x - 3y &= 8 \\ -2x - 4y &= -8 \\ \hline -7y &= 0 \\ y &= 0 \end{aligned}$ Divide both sides by 7. To find x, we can substitute 0 for y in the first equation. $2x - 3(\mathbf{0}) = 8$ Substitute. $2x = 8$ Simplify. $x = 4$ Divide both sides by 2. The solution is $(4, 0)$.

REVIEW EXERCISES

Solve each system by substitution.

5. $\begin{cases} y = 2x + 5 \\ 3x + 4y = 9 \end{cases}$ **6.** $\begin{cases} y = 3x + 5 \\ 3x - y = -5 \end{cases}$

7. $\begin{cases} x + 2y = 11 \\ 2x - y = 2 \end{cases}$ **8.** $\begin{cases} 2x + 3y = -2 \\ 3x + 5y = -2 \end{cases}$

Solve each system by elimination.

9. $\begin{cases} x - y = 1 \\ 5x + 2y = -16 \end{cases}$ **10.** $\begin{cases} 3x + 2y = 1 \\ 2x - 3y = 5 \end{cases}$

11. $\begin{cases} 2x + 6 = -3y \\ -2x = 3y + 6 \end{cases}$ **12.** $\begin{cases} y = \dfrac{2x - 3}{2} \\ x = \dfrac{2y + 7}{2} \end{cases}$

SECTION 3.3   Solving Systems of Three Linear Equations in Three Variables

DEFINITIONS AND CONCEPTS	EXAMPLES

Strategy for solving three linear equations in three variables:

1. Write all equations in the system in $ax + by + cz = d$ form.

2. Select any two equations and eliminate a variable.

3. Select a different pair of equations and eliminate the same variable.

4. Solve the resulting pair of two equations in two variables.

5. To find the value of the third variable, substitute the values of the two variables found in Step 4 into any equation containing all three variables and solve the equation.

6. Check the solution in all three of the original equations.

To solve the system $\begin{cases} x + y + z = 4 \\ x - 2y - z = -9 \\ 2x - y + 2z = -1 \end{cases}$, we can add the first and second equations to obtain Equation 1:

$$\begin{aligned} x + y + z &= 4 \\ x - 2y - z &= -9 \\ \hline (1) \quad 2x - y \phantom{{}- z} &= -5 \end{aligned}$$

We now multiply the second equation by 2 and add it to the third equation to obtain Equation 2:

$$\begin{aligned} 2x - 4y - 2z &= -18 \\ 2x - y + 2z &= -1 \\ \hline (2) \quad 4x - 5y \phantom{{}+ 2z} &= -19 \end{aligned}$$

To solve the system $\begin{cases} 2x - y = -5 \\ 4x - 5y = -19 \end{cases}$, formed by Equations 1 and 2, we can multiply the first equation by -2 and add the result to the second equation to eliminate x.

$$\begin{aligned} -4x + 2y &= 10 \\ 4x - 5y &= -19 \\ \hline -3y &= -9 \\ y &= 3 \qquad \text{Divide both sides by } -3. \end{aligned}$$

We can substitute 3 into either equation of the system to find x.

$$\begin{aligned} 2x - y &= -5 && \text{This is the first equation of the system.} \\ 2x - 3 &= -5 && \text{Substitute 3 for } y. \\ 2x &= -2 && \text{Add 3 to both sides.} \\ x &= -1 && \text{Divide both sides by 2.} \end{aligned}$$

We now can substitute -1 for x and 3 for y into any of the equations in the original system and solve for z:

$$\begin{aligned} x + y + z &= 4 && \text{This is the first equation of the original system.} \\ -1 + 3 + z &= 4 && \text{Substitute.} \\ 2 + z &= 4 && \text{Simplify.} \\ z &= 2 && \text{Subtract 2 from both sides.} \end{aligned}$$

The solution is $(-1, 3, 2)$.

REVIEW EXERCISES

Solve each system.

13. $\begin{cases} x + y + z = 6 \\ x - y - z = -4 \\ -x + y - z = -2 \end{cases}$

14. $\begin{cases} 2x + 3y + z = -5 \\ -x + 2y - z = -6 \\ 3x + y + 2z = 4 \end{cases}$

15. $\begin{cases} -x + 5y + 2z = -6 \\ x - 5y - z = 6 \\ 3x - 15y - 6z = 18 \end{cases}$

SECTION 3.4 Solving Systems of Linear Equations Using Matrices

DEFINITIONS AND CONCEPTS	EXAMPLES

DEFINITIONS AND CONCEPTS

A **matrix** is any rectangular array of numbers.

Systems of linear equations can be solved using matrices and the method of **Gaussian elimination** and back substitution.

A matrix with m rows and n columns is called an $m \times n$ matrix.

$$\begin{bmatrix} a & b & c & | & d \\ 0 & e & f & | & g \\ 0 & 0 & h & | & i \end{bmatrix} \quad \begin{array}{l} (a, b, c, \ldots i \text{ are real} \\ \text{numbers)} \end{array}$$

This matrix is in triangular form.

EXAMPLES

To use matrices to solve the system

$$\begin{cases} x + y + z = 4 \\ 2x - y + 2z = -1 \\ x - 2y - z = -9 \end{cases}$$

we can represent it with the following augmented matrix:

$$\begin{bmatrix} \mathbf{1} & 1 & 1 & | & 4 \\ 2 & -1 & 2 & | & -1 \\ 1 & -2 & -1 & | & -9 \end{bmatrix}$$

To get 0's under the 1 in the first column, we multiply row 1 of the augmented matrix by -2 and add it to row 2 to get a new row 2. We then multiply row 1 by -1 and add it to row 3 to get a new row 3.

$$\begin{bmatrix} 1 & 1 & 1 & | & 4 \\ 0 & \mathbf{-3} & 0 & | & -9 \\ 0 & -3 & -2 & | & -13 \end{bmatrix} \quad \begin{array}{l} -2R1 + R2 \rightarrow R2 \\ -1R1 + R3 \rightarrow R3 \end{array}$$

To get a 0 under the -3 in the second column of the previous matrix, we multiply row 2 by -1 and add it to row 3.

$$\begin{bmatrix} 1 & 1 & 1 & | & 4 \\ 0 & -3 & 0 & | & -9 \\ 0 & 0 & \mathbf{-2} & | & -4 \end{bmatrix} \quad -1R2 + R3 \rightarrow R3$$

Finally, to obtain a 1 in the third row, third column, we multiply row 3 by $-\frac{1}{2}$.

$$\begin{bmatrix} 1 & 1 & 1 & | & 4 \\ 0 & -3 & 0 & | & -9 \\ 0 & 0 & 1 & | & 2 \end{bmatrix} \quad -\frac{1}{2}R3 \rightarrow R3$$

The final matrix represents the system

$$\begin{array}{l} (1) \\ (2) \\ (3) \end{array} \begin{cases} x + y + z = 4 \\ 0x - 3y + 0z = -9 \\ 0x + 0y + z = 2 \end{cases}$$

From Equation 3, we see that $z = 2$. From Equation 2, we see that $y = 3$.
To find the value of x, we use back substitution to substitute 2 for z and 3 for y in Equation 1 and solve for x:

$$\begin{array}{llll} (1) & x + \mathbf{y} + \mathbf{z} = 4 & & \\ & x + \mathbf{3} + \mathbf{2} = 4 & & \text{Substitute.} \\ & x + 5 = 4 & & \text{Simplify.} \\ & x = -1 & & \text{Subtract 5 from both sides.} \end{array}$$

Thus, $x = -1$. The solution of the given system is $(-1, 3, 2)$. Verify that this ordered triple satisfies each equation of the original system.

REVIEW EXERCISES

Solve each system by using matrices.

16. $\begin{cases} x + y + z = 6 \\ 2x - y + z = 1 \\ 4x + y - z = 5 \end{cases}$

17. $\begin{cases} x + y = 3 \\ x - 2y = -3 \\ 2x + y = 4 \end{cases}$

18. $\begin{cases} x - 3y + z = 4 \\ 2x - 5y + 3z = 6 \end{cases}$

SECTION 3.5   Solving Systems of Linear Equations Using Determinants

DEFINITIONS AND CONCEPTS

A **determinant of a square matrix** is a number.

$$\begin{vmatrix} a & b \\ c & d \end{vmatrix} = ad - bc$$

$$\begin{vmatrix} a_1 & b_1 & c_1 \\ a_2 & b_2 & c_2 \\ a_3 & b_3 & c_3 \end{vmatrix}$$

$$= a_1 \begin{vmatrix} b_2 & c_2 \\ b_3 & c_3 \end{vmatrix} - b_1 \begin{vmatrix} a_2 & c_2 \\ a_3 & c_3 \end{vmatrix} + c_1 \begin{vmatrix} a_2 & b_2 \\ a_3 & b_3 \end{vmatrix}$$

EXAMPLES

Find the determinant: $\begin{vmatrix} 8 & -3 \\ 2 & -1 \end{vmatrix}$

$$\begin{vmatrix} 8 & -3 \\ 2 & -1 \end{vmatrix} = 8(-1) - (-3)(2)$$

$$= -8 + 6$$

$$= -2$$

To evaluate the determinant $\begin{vmatrix} 1 & 3 & -2 \\ 1 & -2 & -1 \\ 2 & -1 & 3 \end{vmatrix}$, we can expand by minors:

Minor of 1 ↓   Minor of 3 ↓   Minor of −2 ↓

$$\begin{vmatrix} 1 & 3 & -2 \\ 1 & -2 & -1 \\ 2 & -1 & 3 \end{vmatrix} = 1 \begin{vmatrix} -2 & -1 \\ -1 & 3 \end{vmatrix} - 3 \begin{vmatrix} 1 & -1 \\ 2 & 3 \end{vmatrix} + (-2) \begin{vmatrix} 1 & -2 \\ 2 & -1 \end{vmatrix}$$

$$= 1(-6 - 1) - 3(3 + 2) - 2(-1 + 4)$$

$$= -7 - 15 - 6$$

$$= -28$$

Cramer's rule for two linear equations in two variables:

The solution of the system $\begin{cases} ax + by = e \\ cx + dy = f \end{cases}$ is given by

$$x = \frac{D_x}{D} = \frac{\begin{vmatrix} e & b \\ f & d \end{vmatrix}}{\begin{vmatrix} a & b \\ c & d \end{vmatrix}} \quad \text{and} \quad y = \frac{D_y}{D} = \frac{\begin{vmatrix} a & e \\ c & f \end{vmatrix}}{\begin{vmatrix} a & b \\ c & d \end{vmatrix}}$$

If $D \neq 0$, the system is consistent and the equations are independent.

If $D = 0$ and D_x or D_y is nonzero, the system is inconsistent.

If every determinant is 0, the system is consistent but the equations are dependent.

Solve using Cramer's rule: $\begin{cases} 2x - 4y = -14 \\ 3x + y = -7 \end{cases}$

$$x = \frac{D_x}{D} = \frac{\begin{vmatrix} -14 & -4 \\ -7 & 1 \end{vmatrix}}{\begin{vmatrix} 2 & -4 \\ 3 & 1 \end{vmatrix}} = \frac{-14 - 28}{2 - (-12)} = \frac{-42}{14} = -3$$

$$y = \frac{D_y}{D} = \frac{\begin{vmatrix} 2 & -14 \\ 3 & -7 \end{vmatrix}}{\begin{vmatrix} 2 & -4 \\ 3 & 1 \end{vmatrix}} = \frac{-14 - (-42)}{2 - (-12)} = \frac{28}{14} = 2$$

The solution is $(-3, 2)$.

Cramer's rule for three linear equations in three variables:

The solution of the system $\begin{cases} ax + by + cz = j \\ dx + ey + fz = k \\ gx + hy + iz = l \end{cases}$

is given by $x = \frac{D_x}{D}, y = \frac{D_y}{D}$, and $z = \frac{D_z}{D}$ where

$$D = \begin{vmatrix} a & b & c \\ d & e & f \\ g & h & i \end{vmatrix} \quad D_x = \begin{vmatrix} j & b & c \\ k & e & f \\ l & h & i \end{vmatrix}$$

$$D_y = \begin{vmatrix} a & j & c \\ d & k & f \\ g & l & i \end{vmatrix} \quad D_z = \begin{vmatrix} a & b & j \\ d & e & k \\ g & h & l \end{vmatrix}$$

To use Cramer's rule to solve $\begin{cases} x + y + z = 4 \\ 2x - y + 2z = -1 \\ x - 2y - z = -9 \end{cases}$, we can find D, D_x, D_y, and D_z and substitute these values into the formulas

$$x = \frac{D_x}{D}, \quad y = \frac{D_y}{D}, \quad \text{and} \quad z = \frac{D_z}{D}$$

After forming and evaluating the determinants, we will obtain

$$D = 6, \quad D_x = -6, \quad D_y = 18, \quad \text{and} \quad D_z = 12$$

and we have

$$x = \frac{D_x}{D} = \frac{-6}{6} = -1, y = \frac{D_y}{D} = \frac{18}{6} = 3, z = \frac{D_z}{D} = \frac{12}{6} = 2$$

The solution of this system is $(-1, 3, 2)$.

If $D \neq 0$, the system is consistent and the equations are independent.

If $D = 0$ and D_x or D_y or D_z is nonzero, the system is inconsistent.

If every determinant is 0, the system is consistent but the equations are dependent.

REVIEW EXERCISES

Evaluate each determinant.

19. $\begin{vmatrix} 3 & 1 \\ -2 & 4 \end{vmatrix}$

20. $\begin{vmatrix} -4 & 5 \\ 6 & -2 \end{vmatrix}$

21. $\begin{vmatrix} -1 & 2 & -1 \\ 2 & -1 & 3 \\ 1 & -2 & 2 \end{vmatrix}$

22. $\begin{vmatrix} 3 & -2 & 2 \\ 1 & -2 & -2 \\ 2 & 1 & -1 \end{vmatrix}$

Use Cramer's rule to solve each system.

23. $\begin{cases} 3x + 4y = 10 \\ 2x - 3y = 1 \end{cases}$

24. $\begin{cases} 2x - 5y = -17 \\ 3x + 2y = 3 \end{cases}$

25. $\begin{cases} x + 2y + z = 0 \\ 2x + y + z = 3 \\ x + y + 2z = 5 \end{cases}$

26. $\begin{cases} x + y + z = 1 \\ -x + y - z = 3 \\ -x - y + z = 5 \end{cases}$

3 Test

1. Solve $\begin{cases} 2x + y = 5 \\ y = 2x - 3 \end{cases}$ by graphing.

2. Use substitution to solve: $\begin{cases} 2x - 4y = 14 \\ x = -2y + 7 \end{cases}$

3. Use elimination to solve: $\begin{cases} 2x + 3y = -5 \\ 3x - 2y = 12 \end{cases}$

4. Use any method to solve: $\begin{cases} \frac{x}{2} - \frac{y}{4} = -4 \\ x + y = -2 \end{cases}$

Consider the system $\begin{cases} 3(x + y) = x - 3 \\ -y = \frac{2x + 3}{3} \end{cases}$.

5. Are the equations of the system dependent or independent?

6. Is the system consistent or inconsistent?

Use an elementary row operation to find the missing number in the second matrix.

7. $\begin{bmatrix} 1 & 3 & -2 \\ 4 & -2 & -1 \end{bmatrix}, \begin{bmatrix} 1 & 3 & -2 \\ -1 & -17 & \blacksquare \end{bmatrix}$

8. $\begin{bmatrix} -1 & 3 & 6 \\ 3 & -2 & 4 \end{bmatrix}, \begin{bmatrix} -1 & 3 & 6 \\ 5 & -8 & \blacksquare \end{bmatrix}$

Consider the system $\begin{cases} x + y + z = 4 \\ x + y - z = 6 \\ 2x - 3y + z = -1 \end{cases}$.

9. Write the augmented matrix that represents the system.

10. Solve for x. **11.** Solve for y.

12. Solve for z.

Use matrices to solve each system.

13. $\begin{cases} x + y = 4 \\ 2x - y = 2 \end{cases}$

14. $\begin{cases} x + y = 2 \\ x - y = -4 \\ 2x + y = 1 \end{cases}$

Evaluate each determinant.

15. $\begin{vmatrix} 2 & -3 \\ 4 & 5 \end{vmatrix}$

16. $\begin{vmatrix} -3 & -4 \\ -2 & 3 \end{vmatrix}$

17. $\begin{vmatrix} 1 & 2 & 0 \\ 2 & 0 & 3 \\ 1 & -2 & 2 \end{vmatrix}$

18. $\begin{vmatrix} 1 & -2 & 3 \\ 3 & 1 & -2 \\ 2 & -4 & 6 \end{vmatrix}$

Consider the system $\begin{cases} x - y = -6 \\ 3x + y = -6 \end{cases}$, *which is to be solved with Cramer's rule.*

19. When solving for x, what is the numerator determinant? (Do not evaluate it.)

20. When solving for y, what is the denominator determinant? (Do not evaluate it.)

21. Solve the system for x.

22. Solve the system for y.

23. Use a matrix to solve.
$$\begin{cases} 2x - 2y - 2z = 2 \\ 2x + 3y + z = 2 \\ 3x + 2y = 0 \end{cases}$$

24. Interpret the results of the matrix.
$$\left[\begin{array}{ccc|c} 1 & 0 & 4 & 4 \\ 0 & 1 & 3 & 2 \\ 0 & 0 & 0 & 0 \end{array}\right]$$

25. $\begin{cases} -2x - y + z = -4 \\ x - y - 3z = -1 \end{cases}$

Inequalities

Careers and Mathematics

DIESEL SERVICE TECHNICIANS AND MECHANICS

Not all diesel engine service technicians work on cars and trucks. Some service heavy equipment machinery, such as bulldozers and road graders, while more highly trained technicians might be found servicing engines on cruise ships, on offshore oil drilling platforms, or in underground mining operations.

Although on-the-job training is an option for meeting the requirements for some jobs, many employers seek applicants who have completed training programs in diesel engine repair. Some of these programs are offered through community and technical colleges.

Job Outlook:
Employment of diesel engine service technicians is expected to increase 15% by 2020. Employment growth will continue to be concentrated in automobile dealerships and independent automotive repair shops.

Annual Earnings:
$26,550–$60,830

For More Information:
http://www.bls.gov/ooh/installation-maintenance-and-repair/diesel-service-technicians-and-mechanics.htm

For a Sample Application:
See Problem 135 in Section 4.2.

REACH FOR SUCCESS

4.1 Solving Linear Inequalities in One Variable

4.2 Solving Absolute Value Equations and Inequalities in One Variable

4.3 Solving Linear Inequalities in Two Variables

4.4 Solving Systems of Linear and Quadratic Inequalities in Two Variables by Graphing

4.5 Solving Systems Using Linear Programming

■ *Projects*

REACH FOR SUCCESS EXTENSION

CHAPTER REVIEW

CHAPTER TEST

CUMULATIVE REVIEW

In this chapter

We previously have considered linear equations. In this chapter we will solve inequalities in one variable. We will also solve systems of linear inequalities in two variables and apply that technique to solving applications.

Reach for Success Setting Course Goals

Two professional football teams do not show up on a Sunday afternoon as if they were playing a game of sandlot football. Instead, before the game they've studied films, they've practiced the plays—they have a "game plan."

In this exercise you will set some realistic goals for this course and then establish a "game plan" that will help you achieve your goals.

Melinda Nagy/Shutterstock.com

We're going to assume one of your goals is to be successful in this course.

The grade I am willing to work to achieve is a/an _____.

Most college classes recommended 2–3 hours of outside time per week for every 1 hour in class if students want to be successful. With mathematics, you may find it takes more time.

Considering other course commitments as well as any work and family commitments, state the number of hours *outside of class* you realistically believe you can devote each week to this one course. _____

To help yourself be successful in this course,

- **complete all homework and other assignments on time.**
- take good notes in class.
- ask questions in class.
- study on a daily basis.
- participate in a study group.
- utilize the mathematics or tutor center.
- attend instructor reviews (if available) or office hours.
- access online tutoring services (if available).

List at least three things you are willing to add to your "game plan" and commit to, that will support your success in this course.

1. _____

2. _____

3. _____

Can you think of anything else you can do to improve your performance in this class?

_____

Consider your answers to the questions above. Is your plan sufficient to achieve your goal?

YES _____   NO _____

Please explain. _____

A Successful Study Strategy . . .

 Set a goal, outline a plan of action, and stick to the plan to achieve that goal.

At the end of the chapter you will find an additional exercise to guide you through a successful semester.

Section 4.1

Solving Linear Inequalities in One Variable

Objectives

1. Solve a linear inequality in one variable and express the result as a graph and in set-builder and interval notation.

2. Solve a compound inequality in one variable involving "and," and express the result as a graph and in set-builder and interval notation.

3. Solve a compound inequality in one variable involving "or," and express the result as a graph and in set-builder and interval notation.

4. Solve an application by setting up and solving a linear inequality in one variable.

Vocabulary

inequality	interval notation	closed interval
trichotomy property	unbounded interval	compound inequality
transitive property	open interval	intersection
linear inequality	half-open interval	union

Getting Ready

Graph each set on the number line.

1. $\{x \mid x > -3\}$ 2. $\{x \mid x < 4\}$

3. $\{x \mid x \le 5\}$ 4. $\{x \mid x \ge -1\}$

In this section, we will review the basics of linear inequalities, first discussed in Chapter 1. We will solve linear inequalities and compound inequalities involving "and" and "or" and conclude the section with applications.

 Solve a linear inequality in one variable and express the result as a graph and in set-builder and interval notation.

Inequalities are statements indicating that two quantities are unequal. Inequalities can be recognized because they contain one or more of the following symbols.

INEQUALITY SYMBOLS	Inequality	Read as	Example
	$a \ne b$	"a is not equal to b."	$5 \ne 9$
	$a < b$	"a is less than b."	$2 < 3$
	$a > b$	"a is greater than b."	$7 > -5$
	$a \le b$	"a is less than or equal to b."	$-5 \le 0$ or $-4 \le -4$
	$a \ge b$	"a is greater than or equal to b."	$9 \ge 6$ or $8 \ge 8$
	$a \approx b$	"a is approximately equal to b."	$1.14 \approx 1.1$

If a is to the left of b on the number line, then $a < b$. By definition, $a < b$ means that "a is less than b," but it also means that $b > a$.

There are several basic properties of inequalities.

TRICHOTOMY PROPERTY

For any real numbers a and b, exactly one of the following statements is true:

$$a < b, \qquad a = b, \qquad \text{or} \qquad a > b$$

The **trichotomy property** indicates that exactly one of the statements is true about any two real numbers. Either the *first is less than the second, the first is equal to the second, or the first is greater than the second.*

TRANSITIVE PROPERTY

If a, b, and c are real numbers with $a < b$ and $b < c$, then $a < c$.
If a, b, and c are real numbers with $a > b$ and $b > c$, then $a > c$.

The first part of the **transitive property** indicates that:

If a first number is less than a second number and the second is less than a third, then the first number is less than the third.

The second part is similar, with the words "is greater than" substituted for "is less than."

ADDITION PROPERTY OF INEQUALITIES

Any real number can be added to (or subtracted from) both sides of an inequality to produce another inequality with the same direction as the original.

To illustrate the addition property, we add 4 to both sides of the inequality $3 < 12$ to get

$$3 + 4 < 12 + 4$$
$$7 < 16$$

We note that the $<$ symbol is unchanged (has the same direction).

Subtracting 4 from both sides of $3 < 12$ does not change the direction of the inequality either.

$$3 - 4 < 12 - 4$$
$$-1 < 8$$

MULTIPLICATION PROPERTY OF INEQUALITIES

If both sides of an inequality are multiplied (or divided) by a positive number, another inequality results with the same relationship as the original inequality and the direction of the symbol remains the same. If both sides of an inequality are multiplied (or divided) by a negative number, another inequality results with the opposite relationship from the original inequality and the symbol must be reversed.

To illustrate the first part of the multiplication property, we multiply both sides of the inequality $-4 < 6$ by 2 to get

$$2(-4) < 2(6)$$
$$-8 < 12$$

The $<$ symbol is unchanged.

Dividing both sides by 2 does not change the direction of the inequality either.

$$\frac{-4}{2} < \frac{6}{2}$$

$$-2 < 3$$

To illustrate the second part of the multiplication property, we multiply both sides of the inequality $-4 < 6$ by -2 to obtain

$$-4 < 6$$

$$-2(-4) > -2(6) \quad \text{Change} < \text{to} >.$$

$$8 > -12$$

COMMENT Remember to change the direction of an inequality symbol every time you multiply or divide both sides by a negative number.

Here, the $<$ symbol changes to a $>$ symbol.

Dividing both sides by -2 also changes the direction of the inequality.

$$-4 < 6$$

$$\frac{-4}{-2} > \frac{6}{-2} \quad \text{Change} < \text{to} >.$$

$$2 > -3$$

LINEAR INEQUALITIES IN ONE VARIABLE	A **linear inequality** in one variable is any inequality that can be expressed in one of the following forms (with $a \neq 0$).

$$ax + c < 0 \qquad ax + c > 0 \qquad ax + c \leq 0 \qquad ax + c \geq 0$$

We solve linear inequalities just as we solve linear equations, but with one exception. If we multiply or divide both sides by a negative number, we must change the direction of the inequality.

EXAMPLE 1   Solve each linear inequality and graph its solution set.

a. $3(2x - 9) < 9$ **b.** $-4(3x + 2) \leq 16$

Solution   In each part, we use the same steps as for solving equations.

a. $3(2x - 9) < 9$

$6x - 27 < 9$ Use the distributive property to remove parentheses.

$6x < 36$ Add 27 to both sides.

$x < 6$ Divide both sides by 6.

The solution set is $\{x \mid x < 6\}$. The graph is shown in Figure 4-1(a) on the next page. Since we did not have equality in the original statement, the parenthesis at 6 indicates that 6 is not included in the solution set.

b. $-4(3x + 2) \leq 16$

$-12x - 8 \leq 16$ Use the distributive property to remove parentheses.

$-12x \leq 24$ Add 8 to both sides.

$x \geq -2$ Divide both sides by -12 and reverse the $\leq$ symbol.

The solution set is $\{x \mid x \geq -2\}$. The graph is shown in Figure 4-1(b). Since we had equality in the original statement, the bracket at -2 indicates that -2 is included in the solution set.

(a)   (b)

Figure 4-1

🌿 **SELF CHECK 1**   Solve the inequality and graph its solution set: $-3(2x - 4) \leq 24$

Amalie Noether
1882–1935
Albert Einstein described Noether as the most creative female mathematical genius since the beginning of higher education for women. Her work was in the area of abstract algebra. Although she received a doctoral degree in mathematics, she was denied a mathematics position in Germany because she was a woman.

In Chapter 1, we saw that **interval notation** is another way to express the solution set of an inequality in one variable. Recall that this notation uses parentheses and brackets to indicate whether endpoints of an interval are included or excluded from a solution set.

An interval is called a (an)

- **unbounded interval** if it extends forever in one or more directions. See Figure 4-2(a).
- **open interval** if it is bounded and has no endpoints. See Figure 4-2(b).
- **half-open interval** if it is bounded and has one endpoint. See Figure 4-2(c).
- **closed interval** if it is bounded and has two endpoints. See Figure 4-2(d).

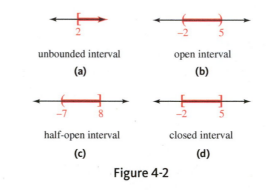

unbounded interval   open interval
(a)   (b)

half-open interval   closed interval
(c)   (d)

Figure 4-2

Table 4-1 shows the possible intervals that exist when a and b are real numbers.

TABLE 4-1			
Kind of interval	**Set notation**	**Graph**	**Interval**
Unbounded intervals	$\{x \mid x > a\}$		(a, ∞)
	$\{x \mid x < a\}$		$(-\infty, a)$
	$\{x \mid x \geq a\}$		$[a, \infty)$
	$\{x \mid x \leq a\}$		$(-\infty, a]$
Open interval	$\{x \mid a < x < b\}$		(a, b)
Half-open intervals	$\{x \mid a \leq x < b\}$		$[a, b)$
	$\{x \mid a < x \leq b\}$		$(a, b]$
Closed interval	$\{x \mid a \leq x \leq b\}$		$[a, b]$

EXAMPLE 2   Solve each linear inequality. Express the solution in set-builder and interval notation and graph it.

a. $5(x + 2) \geq 3x - 1$ **b.** $\dfrac{2}{3}(x + 2) > \dfrac{4}{5}(x - 3)$

Solution   In both parts, we use the same steps as for solving linear equations.

a. $5(x + 2) \geq 3x - 1$

$5x + 10 \geq 3x - 1$ Use the distributive property to remove parentheses.

$2x + 10 \geq -1$ Subtract $3x$ from both sides.

$2x \geq -11$ Subtract 10 from both sides.

$x \geq -\dfrac{11}{2}$ Divide both sides by 2.

The solution set is $\left\{x \mid x \geq -\frac{11}{2}\right\}$, which is the interval $\left[-\frac{11}{2}, \infty\right)$. Its graph is shown in Figure 4-3(a).

b. $\dfrac{2}{3}(x + 2) > \dfrac{4}{5}(x - 3)$

$\mathbf{15} \cdot \dfrac{2}{3}(x + 2) > \mathbf{15} \cdot \dfrac{4}{5}(x - 3)$ To clear fractions, multiply both sides by 15, the LCD of 3 and 5.

$10(x + 2) > 12(x - 3)$ Simplify.

$10x + 20 > 12x - 36$ Use the distributive property to remove parentheses.

$-2x + 20 > -36$ Subtract $12x$ from both sides.

$-2x > -56$ Subtract 20 from both sides.

$x < 28$ Divide both sides by -2 and reverse the $>$ symbol.

The solution set is $\{x \mid x < 28\}$, which is the interval $(-\infty, 28)$. Its graph is shown in Figure 4-3(b).

(a)   (b)

Figure 4-3

 SELF CHECK 2   Solve the linear inequality $\frac{3}{2}(x + 2) < \frac{3}{5}(x - 3)$. Express the solution in set-builder and interval notation and graph it.

2   Solve a compound inequality in one variable involving "and," and express the result as a graph and in set-builder and interval notation.

To say that x is between -3 and 8, we write the inequality

$-3 < x < 8$ Read as "-3 is less than x and x is less than 8."

This inequality is called a **compound inequality**, because it is a combination of two inequalities:

$-3 < x$ and $x < 8$

The word *and* indicates that both inequalities are true at the same time. The result is called the **intersection** of the two statements.

COMPOUND INEQUALITIES   The inequality $c < x < d$ is equivalent to $c < x$ and $x < d$.

EXAMPLE 3   Solve the inequality $-3 \le 2x + 5 < 7$. Express the solution in set-builder and interval notation and graph it.

Solution   This inequality means that $2x + 5$ is between -3 and 7. We can solve it by isolating x between the inequality symbols.

$$-3 \le 2x + 5 < 7$$
$$-8 \le 2x < 2 \qquad \text{Subtract 5 from all three parts.}$$
$$-4 \le x < 1 \qquad \text{Divide all three parts by 2.}$$

Figure 4-4

The solution set is $\{x \mid -4 \le x < 1\}$, which is the interval $[-4, 1)$. The graph is shown in Figure 4-4.

SELF CHECK 3   Solve $-5 \le 3x - 8 \le 7$. Express the solution in set-builder and interval notation and graph it.

EXAMPLE 4   Solve $x + 3 < 2x - 1 < 4x - 3$. Express the solution in set-builder and interval notation and graph it.

Solution   Since it is impossible to isolate x between the inequality symbols, we solve each of the linear inequalities separately.

$$
\begin{array}{ccc}
x + 3 < 2x - 1 & \text{and} & 2x - 1 < 4x - 3 \\
4 < x & & 2 < 2x \\
x > 4 & & 1 < x \\
& & x > 1
\end{array}
$$

Figure 4-5

Only those x values where $x > 4$ and $x > 1$ are in the solution set. Since all numbers greater than 4 are also greater than 1, the solutions are the numbers x where $x > 4$, $\{x \mid x > 4\}$. Thus, the solution set is the interval $(4, \infty)$, whose graph is shown in Figure 4-5.

SELF CHECK 4   Solve $x + 2 < 3x + 1 < 5x + 3$. Express the solution in set-builder and interval notation and graph it.

3   Solve a compound inequality in one variable involving "or," and express the result as a graph and in set-builder and interval notation.

To say that x is less than -5 or greater than 10, we write the inequality

$$x < -5 \qquad \text{or} \qquad x > 10 \qquad \text{Read as "}x\text{ is less than } -5 \text{ or } x \text{ is greater than 10."}$$

The word *or* indicates that only one of the inequalities needs to be true to make the entire statement true. The result is called the **union**, $\cup$, of the two statements.

EXAMPLE 5   Solve the compound inequality $x \le -3$ or $x \ge 8$. Express the solution in set-builder and interval notation and graph it.

Solution   The word *or* in the statement $x \le -3$ or $x \ge 8$ indicates that only one of the inequalities needs to be true to make the statement true. The graph of the inequality is shown in Figure 4-6. The solution set is $\{x \mid x \le -3 \text{ or } x \ge 8\}$, which is the union of two intervals $(-\infty, -3] \cup [8, \infty)$.

Figure 4-6

SELF CHECK 5   Solve $x < -5$ or $x \ge 3$. Express the solution in set-builder and interval notation and graph it.

COMMENT   In the statement $x \leq -3$ or $x \geq 8$, it is incorrect to string the inequalities together as $8 \leq x \leq -3$, because that would imply that $8 \leq -3$, which is false.

 Solve an application by setting up and solving a linear inequality in one variable.

EXAMPLE 6   **LONG-DISTANCE**   Suppose that a long-distance telephone call costs 36¢ for the first three minutes and 11¢ for each additional minute. For how many minutes can a person talk for less than $2?

Analyze the problem   We can let x represent the total number of minutes that the call can last.

Form an inequality   The cost of the call will be 36¢ for the first three minutes plus 11¢ times the number of additional minutes, where the number of additional minutes is $x - 3$ (the total number of minutes minus 3 minutes). The cost of the call is to be less than $2. With this information, we can form the inequality

The cost of the first three minutes	plus	the cost of the additional minutes	is less than	$2.
0.36	+	$0.11(x - 3)$	<	2

Solve the inequality   We can solve the inequality as follows:

$$0.36 + 0.11(x - 3) < 2$$
$$36 + 11(x - 3) < 200 \qquad \text{To eliminate the decimals, multiply both sides by 100.}$$
$$36 + 11x - 33 < 200 \qquad \text{Use the distributive property to remove parentheses.}$$
$$11x + 3 < 200 \qquad \text{Combine like terms.}$$
$$11x < 197 \qquad \text{Subtract 3 from both sides.}$$
$$x < 17.\overline{90} \qquad \text{Divide both sides by 11.}$$

State the conclusion   Since the phone company doesn't bill for part of a minute, the longest that the person can talk is 17 minutes.

Check the result   If the call lasts 17 minutes, the customer will be billed $0.36 + $0.11(14) = $1.90. If the call lasts 18 minutes, the customer will be billed $0.36 + $0.11(15) = $2.01.

SELF CHECK 6   If a long-distance telephone call costs 25¢ for the first 3 minutes and 8¢ for each additional minute, for how many minutes can a person talk for less than $2?

Accent on technology

▶ Solving Linear Inequalities

We can solve linear inequalities with a TI84 graphing calculator. For example, to solve the inequality $3(2x - 9) < 9$, we can graph $y = 3(2x - 9)$ and $y = 9$ using window settings of $[-10, 10]$ for x and $[-10, 10]$ for y to obtain Figure 4-7(a) on the next page. We use the INTERSECT feature to see that the graph of $y = 3(2x - 9)$ is below the graph of $y = 9$ for x-values in the interval $(-\infty, 6)$. See Figure 4-7(b). This interval is the solution, because in this interval, $3(2x - 9) < 9$.

(continued)

Figure 4-7

For instructions regarding the use of a Casio graphing calculator, please refer to the Casio Keystroke Guide in the back of the book.

SELF CHECK ANSWERS

1. $\{x \mid x \geq -2\}$
2. $\{x \mid x < -\frac{16}{3}\}; \left(-\infty, -\frac{16}{3}\right)$
3. $\{x \mid 1 \leq x \leq 5\}; [1, 5]$
4. $\{x \mid x > \frac{1}{2}\}; \left(\frac{1}{2}, \infty\right)$
5. $\{x \mid x < -5 \text{ or } x \geq 3\}; (-\infty, -5) \cup [3, \infty)$
6. The call can last for 24 minutes.

NOW TRY THIS

Solve the compound inequality and express the solution as a graph.

1. $6x - 2(x + 1) < 10$ or $-5x < -10$
2. $4x \geq 2(x - 6)$ and $-5x \geq 30$

Solve the compound inequality and express the solution in interval notation.

3. $x > 3$ or $x \leq -2$
4. $x > 3$ and $x \leq -2$

4.1 Exercises

WARM-UPS *Write each inequality in set-builder notation.*

1. $x < 2$
2. $x > -3$
3. $x \geq -5$
4. $x \geq 1$

Write each inequality in interval notation.

5. $x > -6$
6. $x \geq -4$
7. $x \leq 2$
8. $x < 7$

REVIEW *Simplify each expression.*

9. $\left(\dfrac{t^4 t^2 t^{-3}}{t^5 t^{-7}}\right)^{-2}$

10. $\left(\dfrac{a^{-2} b^3 a^5 b^{-2}}{a^6 b^{-5}}\right)^{-4}$

11. **Small businesses**   A man invests $1,200 in baking equipment to make pies. Each pie requires $4.20 in ingredients. If the man can sell all the pies he can make for $7.95 each, how many pies will he have to make to earn a profit?

12. **Investing**   A woman invested $15,000, part at 7% annual interest and the rest at 8%. If she earned $2,200 in income over a two-year period, how much did she invest at 7%?

VOCABULARY AND CONCEPTS *Fill in the blanks.*

13. The symbol for "is not equal to" is __.
14. The symbol for "is greater than" is __.
15. The symbol for "is less than" is __.
16. The symbol for "is less than or equal to" is __.
17. The symbol for "is greater than or equal to" is __.
18. The trichotomy property states if a and b are two numbers, then $a < b$, _____, or _____.
19. The transitive property states if $a < b$ and $b < c$, then _____.
20. If both sides of an inequality are multiplied by a _____ number, the relationship and direction of the inequality remains the same.
21. If both sides of an inequality are divided by a negative number, the relationship changes and the direction of the inequality symbol must be _____.

22. $3x + 2 > 7$ is an example of a _____ inequality and $3 < 2x < 6$ is a _____ inequality.

23. The inequality $c < x < d$ is equivalent to _____ and _____.

24. The word "or" between two inequality statements indicates that only ___ of the inequalities needs to be true for the entire statement to be true.

25. The interval $(2, 5)$ is called an _____ interval.

26. The interval $[-5, 3]$ is called a _____ interval.

GUIDED PRACTICE *Solve each inequality. Express each result as a graph and in set-builder notation.* **SEE EXAMPLE 1.** *(OBJECTIVE 1)*

27. $x + 4 < 5$

28. $x - 5 > 2$

29. $2(x + 4) \geq -2$

30. $3(x - 2) \leq 18$

31. $2x + 3 < 9$

32. $5x - 1 \geq 19$

33. $-2(4x + 1) \leq 14$

34. $-4(3x - 2) \geq 68$

Solve each inequality. Express each result as a graph and in interval notation. **SEE EXAMPLE 2.** *(OBJECTIVE 1)*

35. $3(z - 2) \leq 2(z + 7)$

36. $5(3 + z) > -3(z + 3)$

37. $-11(2 - b) < 4(2b + 2)$

38. $-9(h - 3) + 2h \leq 8(4 - h)$

39. $\frac{1}{2}y + 2 > \frac{1}{3}y - 4$

40. $\frac{1}{4}x - \frac{1}{3} \leq x + 2$

41. $\frac{2}{3}x + \frac{3}{2}(x - 5) \leq x$

42. $\frac{5}{9}(x + 3) - \frac{4}{3}(x - 3) \geq x - 1$

Solve each inequality. Express each result as a graph and in interval notation. **SEE EXAMPLE 3.** *(OBJECTIVE 2)*

43. $-4 < 2x < 8$

44. $-3 \leq 3x < 12$

45. $-2 < -b + 3 < 5$

46. $2 < -t - 2 < 9$

47. $15 > 2x - 7 > 9$

48. $14 > 2x - 4 > 2$

49. $-6 < -3(x - 4) \leq 24$

50. $-4 \leq -2(x + 8) < 8$

Solve each inequality. Express each result as a graph and in interval notation. **SEE EXAMPLE 4.** *(OBJECTIVE 2)*

51. $x + 3 < 3x - 1 < 2x + 2$

52. $x - 1 \leq 2x + 4 \leq 3x - 1$

53. $5(x + 1) \leq 4(x + 3) < 3(x - 1)$

54. $-5(2 + x) < 4x + 1 < 3x$

Solve each inequality. Express each result as a graph and in interval notation. **SEE EXAMPLE 5.** *(OBJECTIVE 3)*

55. $3x + 2 < 8$ or $2x - 3 > 11$

56. $4x + 5 < -3$ or $4x - 1 > 7$

57. $-4(x + 2) \geq 12$ or $3x + 8 < 11$

58. $x < 6$ or $x > -6$

ADDITIONAL PRACTICE *Solve each inequality. Express each result as a graph and in interval notation.*

59. $x < -3$ and $x > 3$

60. $x + 2 < -\frac{1}{3}x < \frac{1}{2}x$

61. $5(x - 2) \geq 0$ and $-3x < 9$

62. $4x \geq -x + 5 \geq 3x - 4$

63. $6x + 7 > 19$

64. $8x - 10 < 6$

65. $6x + 12 > 2x$

66. $-4x - 36 \leq -10x$

67. $-\dfrac{1}{3}x + 2 \geq 1$

68. $-\dfrac{1}{2}x + 4 < 32$

69. $4(x + 5) < 12$

70. $-5(x - 2) \geq 15$

71. $0.4x + 0.4 \leq 0.1x + 0.85$

72. $0.05 - 0.5x \geq -0.7 - 0.8x$

73. $0 \geq \dfrac{1}{2}x - 4 > 6$

74. $-6 \leq \dfrac{1}{3}a + 1 < 0$

75. $0 \leq \dfrac{4 - x}{3} \leq 2$

76. $-2 \leq \dfrac{5 - 3x}{2} \leq 2$

77. $x - 2 \leq 5x + 2$ or $5x + 2 \leq 2x - 4$

78. $2x - 4 < -x + 5$ or $-x + 5 \leq 4x - 10$

79. $3x - 2 < 4$ or $-6x - 1 \leq 5$

80. $-5x - 2 \geq 8$ or $7x - 3 > 4$

Use a graphing calculator to solve each inequality.

81. $2x + 3 < 5$

82. $3x - 2 > 4$

83. $5x + 2 \geq -18$

84. $3x - 4 \leq 20$

APPLICATIONS *Use an inequality to solve. Use a calculator if necessary.* SEE EXAMPLE 6. (OBJECTIVE 4)

85. Renting a rototiller The cost of renting a rototiller is $15.50 for the first hour and $7.95 for each additional hour. How long can a person have the rototiller if the cost is to be less than $50?

86. Renting a truck How long can a person rent the truck described in the ad if the cost is to be less than $110?

87. Boating A speedboat is rated to carry 750 lb. If the driver weighs 205 lb and a passenger weighs 175 lb, how many children can safely ride along if they average 90 lb each?

88. Elevators An elevator is rated to carry 900 lb. How many boxes of books can the elevator safely carry if each box weighs 80 lb and the operator weighs 165 lb?

89. Investing If a woman invests $10,000 at 8% annual interest, how much more must she invest at 9% so that her annual income will exceed $1,250?

90. Investing If a man invests $8,900 at 5.5% annual interest, how much more must he invest at 8.75% so that his annual income will be more than $1,500?

91. Buying CDs A student can afford to spend up to $330 on a stereo system and some CDs. If the stereo costs $175 and the CDs are $8.50 each, find the greatest number of CDs he can buy.

92. Buying a computer A student who can afford to spend up to $2,000 sees the following ad. If she buys a computer, find the greatest number of DVDs that she can buy.

Big Sale!!!!
$1,695.95
All DVDs
$19.95

93. Averaging grades A student has scores of 70, 77, and 85 on three exams. What score is needed on a fourth exam to make an average of 80 or better?

94. Averaging grades A student has scores of 70, 79, 85, and 88 on four exams. What score does she need on the fifth exam to keep her average above 80?

95. Planning a work schedule Nguyen can earn $5 an hour for working at the college library and $9 an hour for construction work. To save time for study, he wants to limit his work to 20 hours a week but still earn more than $125. How many hours can he work at the library?

96. Scheduling equipment An excavating company charges $300 an hour for the use of a backhoe and $500 an hour for the use of a bulldozer. (Part of an hour counts as a full hour.) The company employs one operator for 40 hours per week. If the company wants to take in at least $18,500 each week, how many hours per week can it schedule the operator to use a backhoe?

97. Choosing a medical plan A college provides its employees with a choice of the two medical plans shown in the table. For what size hospital bills is Plan 2 better than Plan 1? (*Hint*: The cost to the employee includes both the deductible payment and the employee's insurance copayment.)

Plan 1	Plan 2
Employee pays $100	Employee pays $200
Plan pays 70% of the rest	Plan pays 80% of the rest

98. Choosing a medical plan To save costs, the college in Exercise 97 raised the employee deductible, as shown in the following table. For what size hospital bills is Plan 2 better than Plan 1? (*Hint*: The cost to the employee includes both the deductible payment and the employee's insurance copayment.)

Plan 1	Plan 2
Employee pays $200 Plan pays 70% of the rest	Employee pays $400 Plan pays 80% of the rest

Statistics *A researcher wants to estimate the mean (average) real estate tax paid by homeowners in Rockford. To do so, he selects a random sample of homeowners and computes the mean tax paid by the homeowners in that sample. The formula to compute the sample size to be 95% certain that the sample mean will be within* E *of the true population mean is* $\dfrac{3.84\sigma^2}{N} < E^2$ *where* σ^2 *is the square of the standard deviation,* E *is the maximum acceptable error, and* N *is the sample size.*

99. Choosing sample size How large would a sample have to be for the researcher to be 95% certain that the true population mean would be within $20 of the sample mean if the standard deviation, σ, of all tax bills is $120?

100. Choosing sample size How large would the sample have to be for the researcher to be 95% certain that the true population mean would be within $10 of the sample mean if the standard deviation, σ, of all tax bills is $120?

WRITING ABOUT MATH

101. The techniques for solving linear equations and linear inequalities are similar, yet different. Explain.

102. Explain the concepts of *absolute* inequality and *conditional* inequality.

SOMETHING TO THINK ABOUT

103. If $x > -3$, must it be true that $x^2 > 9$?

104. If $x > 2$, must it be true that $x^2 > 4$?

105. Which of these relations is transitive?

a. $=$ **b.** $\leq$ **c.** $\not\geq$ **d.** $\neq$

106. The following solution is not correct. Why?

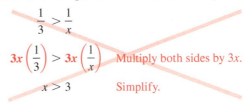

$$\frac{1}{3} > \frac{1}{x}$$

$$3x\left(\frac{1}{3}\right) > 3x\left(\frac{1}{x}\right) \quad \text{Multiply both sides by } 3x.$$

$$x > 3 \quad \text{Simplify.}$$

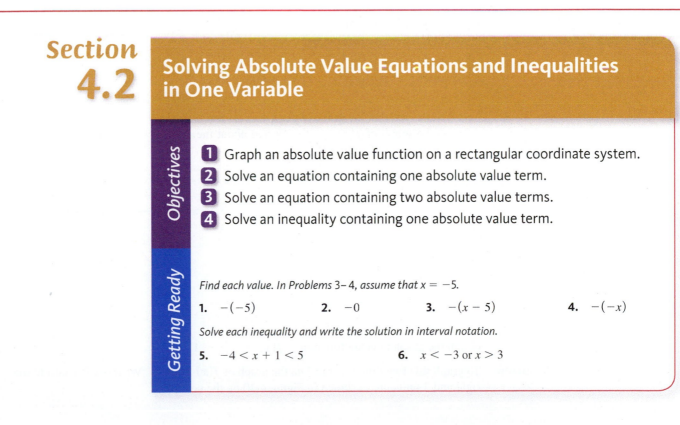

Section 4.2

Solving Absolute Value Equations and Inequalities in One Variable

Objectives

1. Graph an absolute value function on a rectangular coordinate system.
2. Solve an equation containing one absolute value term.
3. Solve an equation containing two absolute value terms.
4. Solve an inequality containing one absolute value term.

Getting Ready

Find each value. In Problems 3–4, assume that $x = -5$.

1. $-(-5)$ **2.** -0 **3.** $-(x - 5)$ **4.** $-(-x)$

Solve each inequality and write the solution in interval notation.

5. $-4 < x + 1 < 5$ **6.** $x < -3$ or $x > 3$

In this section, we review the definition of absolute value and the graphs of absolute value functions. Then we will show how to solve equations and inequalities that contain one or more absolute value expressions.

1   Graph an absolute value function on a rectangular coordinate system.

Earlier, we graphed the absolute value function $f(x) = |x|$ and considered many translations of its graph. The work in Example 1 reviews these concepts.

EXAMPLE 1   Graph the function $f(x) = |x - 1| + 3$.

Solution   To graph this function, we translate the graph of $f(x) = |x|$. We move it 1 unit to the right and 3 units up, as shown in Figure 4-8.

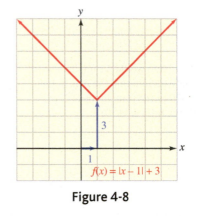

Figure 4-8

🌿 **SELF CHECK 1**   Graph: $f(x) = |x + 2| - 3$

COMMENT   Using a table of values would produce the same graph.

We also saw that the graph of $f(x) = -|x|$ is the same as the graph of $f(x) = |x|$ reflected about the x-axis, as shown in Figure 4-9.

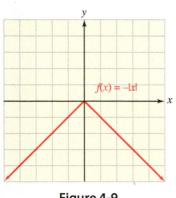

Figure 4-9

EXAMPLE 2   Graph the absolute value function: $f(x) = -|x - 1| + 3$

Solution   To graph this function, we translate the graph of $f(x) = -|x|$. We move it 1 unit to the right and 3 units up, as shown in Figure 4-10 on the next page.

Figure 4-10

 SELF CHECK 2   Graph: $f(x) = -|x + 2| - 3$

---

2 Solve an equation containing one absolute value term.

We discussed that the absolute value of a number is its distance from zero on the number line. The definition of $|x|$ is as follows:

ABSOLUTE VALUE

Assume x is any real number.

> If $x \geq 0$, then $|x| = x$.
> If $x < 0$, then $|x| = -x$.

This definition associates a nonnegative real number with any real number.

- If $x \geq 0$, then x (which is positive or 0) is its own absolute value.
- If $x < 0$, then $-x$ (which is positive) is the absolute value.

Either way, $|x|$ is positive or 0:

$|x| \geq 0$ for all real numbers x

Since $9 \geq 0$, 9 is its own absolute value: $|9| = 9$.

Since $-5 < 0$, the opposite of -5 is the absolute value:

$$|-5| = -(-5) = 5$$

Since $0 \geq 0$, 0 is its own absolute value: $|0| = 0$.

Since $\pi \approx 3.14$, it follows that $2 - \pi < 0$. Thus,

$$|2 - \pi| = -(2 - \pi) = \pi - 2$$

COMMENT   The placement of a $-$ sign in an expression containing an absolute value symbol is important. For example, $|-19| = 19$, but $-|19| = -19$.

In the equation $|x| = 5$, x can be either 5 or -5, because $|5| = 5$ and $|-5| = 5$. In the equation $|x| = 8$, x can be either 8 or -8. In general, the following is true.

ABSOLUTE VALUE EQUATIONS

If $k > 0$, then

$$|x| = k \quad \text{is equivalent to} \quad x = k \quad \text{or} \quad x = -k$$

Figure 4-11

Figure 4-12

Since the absolute value of a number represents its distance on the number line from zero, the solutions of $|x| = k$ are the coordinates of the two points that lie exactly k units from zero. See Figure 4-11.

The equation $|x - 3| = 7$ indicates that a point on the number line with a coordinate of $x - 3$ is 7 units from the origin. Thus, $x - 3$ can be either 7 or -7.

$$x - 3 = 7 \quad \text{or} \quad x - 3 = -7$$
$$x = 10 \qquad \qquad x = -4$$

The solutions of $|x - 3| = 7$ are 10 and -4, as shown in Figure 4-12.

COMMENT   The equation $|x - 3| = 7$ also can be solved using the definition of absolute value.

If $x - 3 \geq 0$, then

$|x - 3| = x - 3$, and

we have

$$x - 3 = 7$$
$$x = 10$$

If $x - 3 < 0$, then

$|x - 3| = -(x - 3)$, and

we have

$$-(x - 3) = 7$$
$$-x + 3 = 7$$
$$-x = 4$$
$$x = -4$$

EXAMPLE 3   Solve: $|3x - 2| = 5$

Solution   We can write $|3x - 2| = 5$ as

$$3x - 2 = 5 \quad \text{or} \quad -(3x - 2) = 5$$

and solve each equation for x:

$$3x - 2 = 5 \quad \text{or} \quad -(3x - 2) = 5$$
$$3x = 7 \qquad \qquad -3x + 2 = 5$$
$$x = \frac{7}{3} \qquad \qquad -3x = 3$$
$$\qquad \qquad x = -1$$

Verify that both solutions check.

SELF CHECK 3   Solve: $|2x - 3| = 7$

To solve more complicated equations containing a term involving an absolute value, we will isolate the absolute value on one side of the equation.

EXAMPLE 4   Solve: $|5x + 3| - 7 = 4$

Solution   We first add 7 to both sides of the equation to isolate the absolute value on the left side.

$$|5x + 3| - 7 = 4$$
$$|5x + 3| = 11 \qquad \text{Add 7 to both sides.}$$

We can now write this equation as

$$5x + 3 = 11 \quad \text{or} \quad -(5x + 3) = 11$$

and solve each equation for x.

$$
\begin{array}{c|c}
5x = 8 & -5x - 3 = 11 \\
x = \dfrac{8}{5} & -5x = 14 \\
& x = -\dfrac{14}{5}
\end{array}
$$

Verify that both solutions check.

SELF CHECK 4 Solve: $|7x - 2| + 4 = 5$

EXAMPLE 5 Solve: $\left| \dfrac{2}{3}x + 3 \right| + 4 = 10$

Solution We subtract 4 from both sides of the equation to isolate the absolute value on the left side.

$$\left| \dfrac{2}{3}x + 3 \right| + 4 = 10$$

$$\left| \dfrac{2}{3}x + 3 \right| = 6 \qquad \text{Subtract 4 from both sides.}$$

We now can write this equation as

$$\dfrac{2}{3}x + 3 = 6 \quad \text{or} \quad -\left(\dfrac{2}{3}x + 3 \right) = 6$$

and solve each equation for x:

$$
\begin{array}{c|c}
\dfrac{2}{3}x = 3 & -\dfrac{2}{3}x - 3 = 6 \\
2x = 9 & -\dfrac{2}{3}x = 9 \\
x = \dfrac{9}{2} & x = -\dfrac{27}{2}
\end{array}
$$

Verify that both solutions check.

SELF CHECK 5 Solve: $\left| \dfrac{3}{5}x - 2 \right| - 3 = 4$

EXAMPLE 6 Solve: $\left| \dfrac{1}{2}x - 5 \right| - 4 = -4$

Solution We first isolate the absolute value on the left side.

$$\left| \dfrac{1}{2}x - 5 \right| - 4 = -4$$

$$\left| \dfrac{1}{2}x - 5 \right| = 0 \qquad \text{Add 4 to both sides.}$$

Since 0 is the only number whose absolute value is 0, the expression $\frac{1}{2}x - 5$ must be equal to 0, and we have

$$\frac{1}{2}x - 5 = 0$$

$$\frac{1}{2}x = 5 \qquad \text{Add 5 to both sides.}$$

$$x = 10 \qquad \text{Multiply both sides by 2.}$$

Verify that 10 satisfies the original equation.

 SELF CHECK 6   Solve: $\left| \frac{2}{3}x + 4 \right| + 5 = 5$

COMMENT   Since the absolute value of a quantity cannot be negative, equations such as $\left| 7x + \frac{1}{2} \right| = -4$ have no solution. Since there are no solutions, the solution set is empty. Recall that an empty set is denoted by the symbol $\varnothing$.

3   Solve an equation containing two absolute value terms.

The equation $|a| = |b|$ is true when $a = b$ or when $a = -b$. For example,

$$|3| = |3| \quad \text{or} \quad |3| = |-3|$$
$$3 = 3 \qquad\quad\ \ 3 = 3$$

In general, the following statement is true.

EQUATIONS WITH TWO ABSOLUTE VALUES	If a and b represent algebraic expressions, the equation $|a| = |b|$ is equivalent to the pair of equations
	$a = b \qquad \text{or} \qquad a = -b$

COMMENT   Notice that $|a| = |b|$ is also equivalent to the pair of equations $b = a$ or $b = -a$.

EXAMPLE 7   Solve: $|5x + 3| = |3x + 25|$

Solution   This equation is true when $5x + 3 = 3x + 25$, or when $5x + 3 = -(3x + 25)$. We solve each equation for x.

$$5x + 3 = 3x + 25 \quad \text{or} \quad 5x + 3 = -(3x + 25)$$
$$2x = 22 \qquad\qquad\qquad 5x + 3 = -3x - 25$$
$$x = 11 \qquad\qquad\qquad\qquad 8x = -28$$
$$x = -\frac{28}{8}$$
$$x = -\frac{7}{2}$$

Verify that both solutions check.

 SELF CHECK 7   Solve: $|2x - 3| = |4x + 9|$

4 Solve an inequality containing one absolute value term.

We will first consider when the absolute value term is less than a number. The inequality $|x| < 5$ indicates that a point with a coordinate of x is less than 5 units from zero. (See Figure 4-13.) Thus, x is between -5 and 5, and

$$|x| < 5 \qquad \text{is equivalent to} \qquad -5 < x < 5$$

The solution set of the inequality $|x| < k$ $(k > 0)$ includes the coordinates of the points on the number line that are less than k units from zero. (See Figure 4-14.)

Figure 4-13   **Figure 4-14**

SOLVING $|x| < k$ AND $|x| \leq k$	If $k \geq 0$, then
	$|x| < k$ is equivalent to $-k < x < k$
	$|x| \leq k$ is equivalent to $-k \leq x \leq k$

COMMENT Since $|x|$ is always positive or 0, the inequality $|x| < k$ has no solutions when $k < 0$. For example, the solution set of $|x| < -5$ is $\varnothing$.

EXAMPLE 8 Solve: $|2x - 3| < 9$

Solution Since $|2x - 3| < 9$ is equivalent to $-9 < 2x - 3 < 9$, we proceed as follows to isolate x between the inequality symbols.

$$-9 < 2x - 3 < 9$$
$$-6 < 2x < 12 \qquad \text{Add 3 to all three parts.}$$
$$-3 < x < 6 \qquad \text{Divide all parts by 2.}$$

Any number between -3 and 6, not including either -3 or 6, is in the solution set. This is the open interval $(-3, 6)$, whose graph is shown in Figure 4-15.

Figure 4-15

SELF CHECK 8 Solve: $|3x + 2| < 4$

COMMENT The inequality $|2x - 3| < 9$ also can be solved using the definition of absolute value.

If $2x - 3 \geq 0$, then	If $2x - 3 < 0$, then
$|2x - 3| = 2x - 3$,	$|2x - 3| = -(2x - 3)$,
and we have	and we have
$2x - 3 < 9$	$-(2x - 3) < 9$
$2x < 12$	$-2x + 3 < 9$
$x < 6$	$-2x < 6$
	$x > -3$

These results indicate that $x < 6$ and $x > -3$, which can be written as $-3 < x < 6$.

EXAMPLE 9   Solve: $|2 - 3x| - 2 \le 3$

Solution   We first add 2 to both sides of the inequality to isolate the absolute value term

$$|2 - 3x| \le 5$$

Since $|2 - 3x| \le 5$ is equivalent to $-5 \le 2 - 3x \le 5$, we proceed as follows to isolate x between the inequality symbols.

$$-5 \le 2 - 3x \le 5$$

$$-7 \le -3x \le 3 \qquad \text{Subtract 2 from all parts.}$$

$$\frac{7}{3} \ge x \ge -1 \qquad \text{Divide all three parts by } -3 \text{ and reverse the directions of the inequality symbols.}$$

$$-1 \le x \le \frac{7}{3} \qquad \text{Write an equivalent inequality with the inequality symbols pointing in the opposite direction.}$$

The solution set is the closed interval $\left[-1, \frac{7}{3}\right]$, whose graph is shown in Figure 4-16.

Figure 4-16

🍃 **SELF CHECK 9**   Solve: $|3 - 2x| + 3 \le 8$

We will now consider when the absolute value term is greater than a number. The inequality $|x| > 5$ indicates that a point with a coordinate of x is more than 5 units from the origin. (See Figure 4-17.) Thus, $x < -5$ or $x > 5$.

Figure 4-17

Figure 4-18

In general, the inequality $|x| > k$ can be interpreted to mean that a point with coordinate x is more than k units from the origin. (See Figure 4-18.)

Thus,

$$|x| > k \qquad \text{is equivalent to} \qquad x < -k \quad \text{or} \quad x > k$$

The *or* indicates an either/or situation. To be in the solution set, x needs to satisfy only one of the two conditions.

SOLVING $|x| > k$ AND $|x| \ge k$

If $k \ge 0$, then

$$|x| > k \qquad \text{is equivalent to} \qquad x < -k \quad \text{or} \quad x > k$$
$$|x| \ge k \qquad \text{is equivalent to} \qquad x \le -k \quad \text{or} \quad x \ge k$$

EXAMPLE 10   Solve: $|5x - 10| > 20$

Solution   We write the inequality as two separate inequalities and solve each one for x. Since $|5x - 10| > 20$ is equivalent to $5x - 10 < -20$ or $5x - 10 > 20$, we have

$$5x - 10 < -20 \quad \text{or} \quad 5x - 10 > 20$$
$$5x < -10 \qquad\qquad 5x > 30 \qquad \text{Add 10 to both sides.}$$
$$x < -2 \qquad\qquad x > 6 \qquad \text{Divide both sides by 5.}$$

Thus, x is either less than -2 or greater than 6:

$$x < -2 \qquad \text{or} \qquad x > 6$$

These are the unbounded intervals $(-\infty, -2) \cup (6, \infty)$, whose graph appears in Figure 4-19.

Figure 4-19

SELF CHECK 10 Solve: $|3x + 4| > 13$

COMMENT The inequality $|5x - 10| > 20$ can also be solved using the definition of absolute value.

If $5x - 10 \geq 0$, then   or   If $5x - 10 < 0$, then

$|5x - 10| = 5x - 10,$ $|5x - 10| = -(5x - 10),$

and we have   and we have

$$5x - 10 > 20 \qquad\qquad\qquad -(5x - 10) > 20$$
$$5x > 30 \qquad\qquad\qquad\qquad -5x + 10 > 20$$
$$x > 6 \qquad\qquad\qquad\qquad\qquad -5x > 10$$
$$x < -2$$

Thus, $x < -2$ or $x > 6$.

EXAMPLE 11 Solve: $\left| \dfrac{3 - x}{5} \right| \geq 6$

Solution We write the inequality as two separate inequalities:

$$\left| \frac{3 - x}{5} \right| \geq 6 \quad \text{is equivalent to} \quad \frac{3 - x}{5} \leq -6 \quad \text{or} \quad \frac{3 - x}{5} \geq 6$$

Then we solve each one for x:

$$\frac{3 - x}{5} \leq -6 \quad \text{or} \quad \frac{3 - x}{5} \geq 6$$

$3 - x \leq -30$	$3 - x \geq 30$	Multiply both sides by 5.
$-x \leq -33$	$-x \geq 27$	Subtract 3 from both sides.
$x \geq 33$	$x \leq -27$	Divide both sides by -1 and reverse the direction of the inequality symbol.

Figure 4-20

The solution set is $(-\infty, -27] \cup [33, \infty)$, whose graph appears in Figure 4-20.

SELF CHECK 11 Solve: $\left| \dfrac{3 - x}{6} \right| \geq 5$

EXAMPLE 12 Solve: $\left| \dfrac{2}{3}x - 2 \right| - 3 > 6$

Solution We begin by adding 3 to both sides to isolate the absolute value on the left side. We then proceed as follows:

$$\left| \frac{2}{3}x - 2 \right| > 9 \quad \text{Add 3 to both sides to isolate the absolute value.}$$

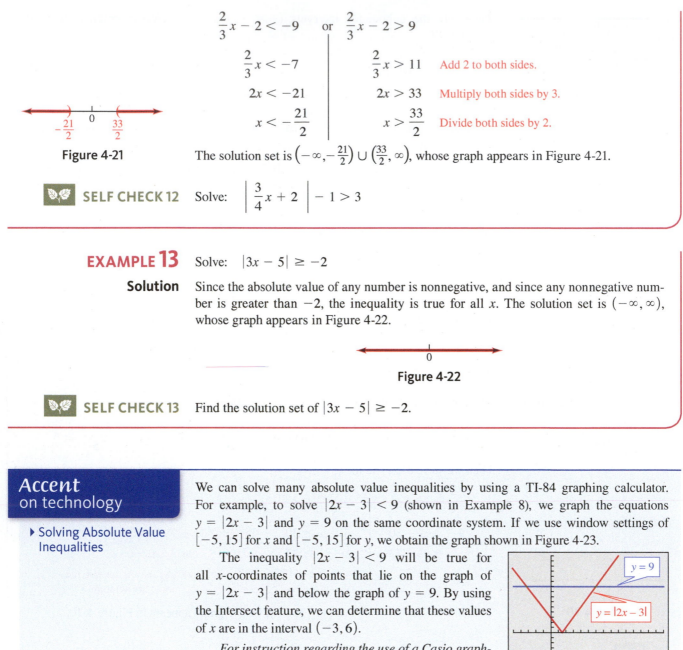

$$\frac{2}{3}x - 2 < -9 \quad \text{or} \quad \frac{2}{3}x - 2 > 9$$

$\frac{2}{3}x < -7$	$\frac{2}{3}x > 11$ Add 2 to both sides.
$2x < -21$	$2x > 33$ Multiply both sides by 3.
$x < -\frac{21}{2}$	$x > \frac{33}{2}$ Divide both sides by 2.

Figure 4-21

The solution set is $\left(-\infty, -\frac{21}{2}\right) \cup \left(\frac{33}{2}, \infty\right)$, whose graph appears in Figure 4-21.

SELF CHECK 12 Solve: $\left|\frac{3}{4}x + 2\right| - 1 > 3$

EXAMPLE 13 Solve: $|3x - 5| \geq -2$

Solution Since the absolute value of any number is nonnegative, and since any nonnegative number is greater than -2, the inequality is true for all x. The solution set is $(-\infty, \infty)$, whose graph appears in Figure 4-22.

Figure 4-22

SELF CHECK 13 Find the solution set of $|3x - 5| \geq -2$.

Accent
on technology

▶ Solving Absolute Value
Inequalities

We can solve many absolute value inequalities by using a TI-84 graphing calculator. For example, to solve $|2x - 3| < 9$ (shown in Example 8), we graph the equations $y = |2x - 3|$ and $y = 9$ on the same coordinate system. If we use window settings of $[-5, 15]$ for x and $[-5, 15]$ for y, we obtain the graph shown in Figure 4-23.

The inequality $|2x - 3| < 9$ will be true for all x-coordinates of points that lie on the graph of $y = |2x - 3|$ and below the graph of $y = 9$. By using the Intersect feature, we can determine that these values of x are in the interval $(-3, 6)$.

For instruction regarding the use of a Casio graphing calculator, please refer to the Casio Keystroke Guide in the back of the book.

Figure 4-23

EXAMPLE 14 Solve: $|3x - 5| < -2$

Solution Since the absolute value of any number is nonnegative, there are no values of x that would make the inequality true. The solution is $\varnothing$.

SELF CHECK 14 Solve: $|2x - 7| < -5$

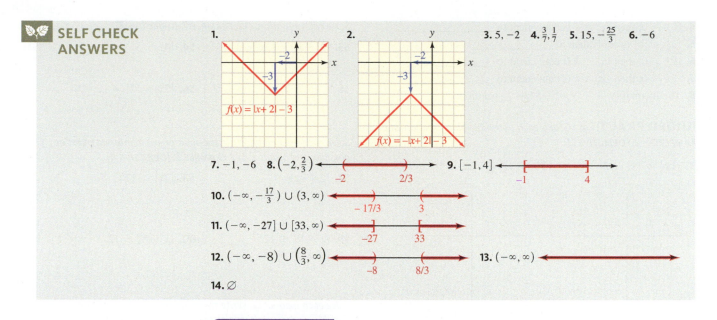

3. $5, -2$ 4. $\frac{3}{7}, \frac{1}{7}$ 5. $15, -\frac{25}{3}$ 6. -6

7. $-1, -6$ 8. $\left(-2, \frac{2}{3}\right)$ 9. $[-1, 4]$

10. $(-\infty, -\frac{17}{3}) \cup (3, \infty)$

11. $(-\infty, -27] \cup [33, \infty)$

12. $(-\infty, -8) \cup \left(\frac{8}{3}, \infty\right)$ 13. $(-\infty, \infty)$

14. $\varnothing$

NOW TRY THIS

1. Express the following as an absolute value equation. $x = 2$ or $x = -2$

2. Express each of the following as an absolute value inequality.
   a. $-5 \le x \le 5$
   b. $3x - 5 < -7$ or $3x - 5 > 7$

4.2 Exercises

WARM-UPS *Given the graphs below, state the solutions.*

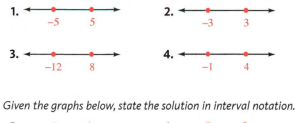

1. -5, 5
2. -3, 3
3. -12, 8
4. -1, 4

Given the graphs below, state the solution in interval notation.

5. -8, 8
6. -4, 4
7. -2, 2
8. -3, 3
9.
10. -5, 2
11. 7, 9
12.

REVIEW *Solve each equation or formula.*

13. $3(2a - 1) = 2a$
14. $\frac{t}{6} - \frac{t}{3} = -1$
15. $\frac{5x}{2} - 1 = \frac{x}{3} + 12$
16. $4b - \frac{b+9}{2} = \frac{b+2}{5} - \frac{8}{5}$
17. $A = p + prt$ for t
18. $P = s_1 + s_2 + 2s_3$ for s_3

VOCABULARY AND CONCEPTS *Fill in the blanks.*

19. If $x \ge 0$, $|x| = $ __.
20. If $x < 0$, $|x| = $ __.
21. $|x| \ge $ __ for all real numbers x.
22. The graph of the function $f(x) = |x - 2| - 4$ is the same as the graph of $f(x) = |x|$, except that it has been translated 2 units to the _____ and 4 units _____.
23. The graph of $f(x) = -|x|$ is the same as the graph of $f(x) = |x|$, except that it has been _____ about the x-axis.

24. If $k > 0$, then $|x| = k$ is equivalent to _____.
25. $|a| = |b|$ is equivalent to _____.
26. If $k > 0$, then $|x| < k$ is equivalent to _____.
27. If $k \geq 0$, then $|x| \geq k$ is equivalent to _____.
28. The inequality $|x - 3| < -5$ has __ solutions.

GUIDED PRACTICE *Graph each absolute value function.* SEE EXAMPLES 1–2. (OBJECTIVE 1)

29. $f(x) = |x| - 2$

30. $f(x) = -|x| + 1$

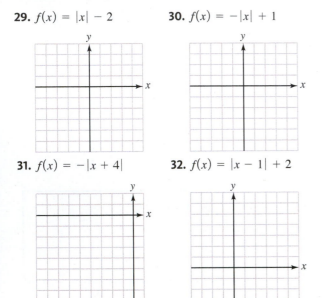

31. $f(x) = -|x + 4|$

32. $f(x) = |x - 1| + 2$

Solve each equation. SEE EXAMPLE 3. (OBJECTIVE 2)

33. $|x| = 4$
34. $|x| = 9$
35. $|x - 3| = 6$
36. $|x + 4| = 8$
37. $|2x - 3| = 5$
38. $|4x - 4| = 20$
39. $|3 - 4x| = 5$
40. $|8 - 5x| = 18$

Solve each equation. SEE EXAMPLE 4. (OBJECTIVE 2)

41. $|2x + 1| - 3 = 12$
42. $|3x - 2| + 1 = 11$

43. $|x + 3| + 7 = 10$
44. $|2 - x| + 3 = 5$

Solve each equation. SEE EXAMPLE 5. (OBJECTIVE 2)

45. $\left| \dfrac{5}{2}x - 3 \right| + 2 = 4$
46. $\left| \dfrac{1}{4}x + 2 \right| - 3 = 5$

47. $\left| \dfrac{2}{3}x + 4 \right| + 3 = 7$
48. $\left| \dfrac{1}{3}x - 1 \right| - 6 = 1$

Solve each equation. SEE EXAMPLE 6. (OBJECTIVE 2)

49. $|3x + 24| = 0$
50. $|2x + 18| + 6 = 6$

51. $|5 - x| + 7 = 7$
52. $|8 - 2x| + 3 = 3$

Solve each equation. SEE EXAMPLE 7. (OBJECTIVE 3)

53. $|2 - x| = |3x + 2|$
54. $|4x + 3| = |9 - 2x|$

55. $\left| \dfrac{x}{2} + 2 \right| = \left| \dfrac{x}{2} - 2 \right|$
56. $|7x + 12| = |x - 6|$

Solve each inequality. Write the solution set in interval notation, if possible, and graph it. SEE EXAMPLE 8. (OBJECTIVE 4)

57. $|2x| < 8$
58. $|3x| < 27$

59. $|x + 9| \leq 12$
60. $|x - 8| \leq 12$

Solve each inequality. Write the solution set in interval notation, if possible, and graph it. SEE EXAMPLE 9. (OBJECTIVE 4)

61. $\left| \dfrac{1}{2}x - 3 \right| - 4 < 2$

62. $|3x + 1| + 2 < 6$

63. $-|5x - 1| + 2 > 0$

64. $\left| \dfrac{1}{6}x + 6 \right| + 2 < 3$

Solve each inequality. Write the solution set in interval notation and graph it. SEE EXAMPLE 10. (OBJECTIVE 4)

65. $|5x| + 2 > 7$
66. $|7x| - 3 > 4$

67. $|2 - 3x| \geq 8$
68. $|-1 - 2x| > 5$

Solve each inequality. Write the solution set in interval notation and graph it. SEE EXAMPLE 11. (OBJECTIVE 4)

69. $\left| \dfrac{x - 2}{3} \right| > 4$
70. $\left| \dfrac{2 - x}{3} \right| \geq 5$

71. $\left| \dfrac{x - 7}{2} \right| > 4$
72. $\left| \dfrac{x + 1}{3} \right| > 1$

Solve each inequality. Write the solution set in interval notation and graph it. SEE EXAMPLE 12. (OBJECTIVE 4)

73. $\left| \dfrac{5 - x}{2} \right| - \dfrac{3}{2} > 3$ **74.** $\left| 1 - \dfrac{1}{2}x \right| - 5 > 2$

75. $\left| \dfrac{1}{3}x + 7 \right| + 5 > 6$ **76.** $\left| \dfrac{1}{5}x - 5 \right| + 4 > 4$

Solve each inequality. Write the solution set in interval notation and graph it. SEE EXAMPLE 13. (OBJECTIVE 4)

77. $|x - 7| \geq -3$ **78.** $|2x - 5| \geq -1$

79. $|4x + 3| > -5$ **80.** $|7x + 2| > -8$

Solve each inequality. Write the solution set in interval notation. SEE EXAMPLE 14. (OBJECTIVE 4)

81. $|3x - 30| + 10 \leq 4$ **82.** $|2x + 5| + 8 < 3$

83. $\left| \dfrac{2}{7}x - 3 \right| + 5 < 3$ **84.** $\left| \dfrac{2}{3}x + 1 \right| + 4 < 2$

ADDITIONAL PRACTICE *Solve each equation or inequality. Graph the inequalities and state the solution in interval notation.*

85. $\left| \dfrac{x}{2} - 1 \right| = 3$ **86.** $\left| \dfrac{4x - 64}{4} \right| = 32$

87. $\left| \dfrac{3x + 48}{3} \right| = 12$ **88.** $\left| \dfrac{x}{2} + 2 \right| = 4$

89. $|3x + 2| = |5x - 6|$ **90.** $|4x - 3| = |2x + 7|$

91. $|3x + 7| = -|8x - 2|$ **92.** $-|17x + 13| = |3x - 14|$

93. $|3x - 2| < 10$ **94.** $\left| \dfrac{1}{7}x + 1 \right| \leq 0$

95. $|5x - 12| < -5$ **96.** $|3 - 2x| < 7$

97. $|3x + 2| > 14$ **98.** $|2x - 5| > 25$

99. $|4x + 3| > 0$ **100.** $|8x - 3| > 0$

101. $|2x + 10| = 0$ **102.** $\left| \dfrac{7}{2}x + 3 \right| = -5$

103. $|3x + 2| = 16$ **104.** $|5x - 3| = 22$

105. $|3x - 1| = |x + 5|$ **106.** $|3x + 1| = |x - 5|$

107. $\left| x + \dfrac{1}{3} \right| = |x - 3|$ **108.** $\left| x - \dfrac{1}{4} \right| = |x + 4|$

109. $|4x - 1| \leq 7$ **110.** $-2|3x - 4| < 16$

111. $3|2x + 5| \geq 9$ **112.** $|4 - 3x| \leq 13$

113. $-|2x - 3| < -7$ **114.** $-|3x + 1| < -8$

115. $|3x - 2| + 2 \geq 0$ **116.** $|5x - 1| + 4 \leq 0$

117. $|x - 12| > 24$ **118.** $|x + 5| \geq 7$

119. $|3x + 2| \leq -3$ **120.** $\left| \dfrac{x - 2}{3} \right| \leq 4$

121. $\left| \dfrac{3}{5}x - 4 \right| - 2 = -2$ **122.** $\left| \dfrac{3}{4}x + 2 \right| + 4 = 4$

123. $\left| \dfrac{x - 5}{10} \right| + 5 \leq 5$ **124.** $\left| \dfrac{3}{5}x - 2 \right| + 3 \leq 3$

125. $|x - 21| = -8$ **126.** $|2x + 1| + 2 \leq 2$

Write each inequality as an inequality using absolute values.

127. $-4 < x < 4$
128. $x < -4 \text{ or } x > 4$
129. $x + 3 < -6 \text{ or } x + 3 > 6$
130. $-5 \leq x - 3 \leq 5$

APPLICATIONS

131. Springs The weight on the spring shown in the illustration oscillates up and down according to the formula $|d - 5| \leq 1$, where d is the distance of the weight above the ground. Solve the formula for d and give the range of heights that the weight is above the ground.

5 ft

132. Aviation An airplane is cruising at 30,000 ft and has been instructed to maintain an altitude that is described by the inequality $|a - 30,000| \leq 1,500$. Solve the inequality for a and give the range of altitudes at which the plane can fly.

133. Temperature ranges The temperatures on a summer day satisfied the inequality $|t - 78°| \leq 8°$, where t is a temperature in degrees Fahrenheit. Express the range of temperatures as a compound inequality.

134. Operating temperatures A car CD player has an operating temperature of $|t - 40°| < 80°$, where t is a temperature in degrees Fahrenheit. Express this range of temperatures as a compound inequality.

135. Range of camber angles The specifications for a car state that the camber angle c of its wheels should be $0.6° \pm 0.5°$. Express this range with an inequality containing absolute value symbols.

136. Tolerance of sheet steel A sheet of steel is to be 0.25 inch thick with a tolerance of 0.015 inch. Express this specification with an inequality containing absolute value symbols.

WRITING ABOUT MATH

137. Explain how to find the absolute value of a given number.

138. Explain why the equation $|x| + 5 = 0$ has no solution.

139. Explain the use of parentheses and brackets when graphing inequalities.

140. If $k > 0$, explain the differences between the solution sets of $|x| < k$ and $|x| > k$.

SOMETHING TO THINK ABOUT

141. For what values of k does $|x| + k = 0$ have exactly two solutions?

142. For what value of k does $|x| + k = 0$ have exactly one solution?

143. Under what conditions is $|x| + |y| > |x + y|$?

144. Under what conditions is $|x| + |y| = |x + y|$?

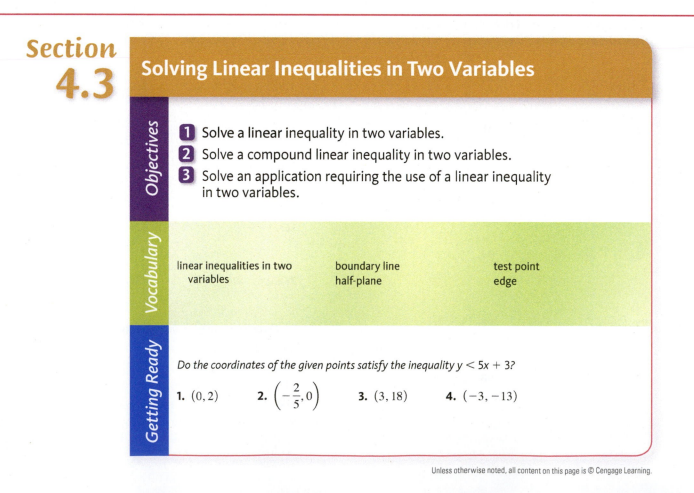

Section 4.3
Solving Linear Inequalities in Two Variables

Objectives

1. Solve a linear inequality in two variables.
2. Solve a compound linear inequality in two variables.
3. Solve an application requiring the use of a linear inequality in two variables.

Vocabulary

linear inequalities in two variables

boundary line
half-plane

test point
edge

Getting Ready

Do the coordinates of the given points satisfy the inequality $y < 5x + 3$?

1. $(0, 2)$ **2.** $\left(-\frac{2}{5}, 0\right)$ **3.** $(3, 18)$ **4.** $(-3, -13)$

In this section, we will show the solutions of linear inequalities in two variables. We begin by discussing their graphs.

 Solve a linear inequality in two variables.

Inequalities such as

$$2x + 3y < 6 \quad \text{or} \quad 3x - 4y \geq 9$$

are called **linear inequalities in two variables**. In general, we have the following definition:

<table>
<tr>
<td>**LINEAR INEQUALITIES IN TWO VARIABLES**</td>
<td>A linear inequality in x and y is any inequality that can be written in the form

$$Ax + By < C, \quad Ax + By > C, \quad Ax + By \leq C, \quad \text{or} \quad Ax + By \geq C$$

where A, B, and C are real numbers and A and B are not both 0.</td>
</tr>
</table>

The *graph of a linear inequality* in x and y is the graph of all ordered pairs (x, y) that satisfy the inequality.

The inequality $y > 3x + 2$ is an example of a linear inequality in two variables because it can be written in the form $-3x + y > 2$. To graph it, we first graph the related equation $y = 3x + 2$ as shown in Figure 4-24(a). This **boundary line**, sometimes called an **edge**, divides the coordinate plane into two **half-planes**, one on either side of the line.

To find which half-plane is the graph of $y > 3x + 2$, we select a **test point** not on the line and substitute the coordinates into the inequality to determine if it results in a true statement. In this case, a convenient test point is the origin, $(0, 0)$.

$$y > 3x + 2$$
$$0 \overset{?}{>} 3(0) + 2 \quad \text{Substitute 0 for } x \text{ and 0 for } y.$$
$$0 \not> 2$$

Since the coordinates do not satisfy $y > 3x + 2$, the origin is not part of the graph. Thus, the half-plane on the other side of the dashed line is the graph, which is shown in Figure 4-24(b). We use a *dashed* line to indicate that the boundary line is not included in the graph.

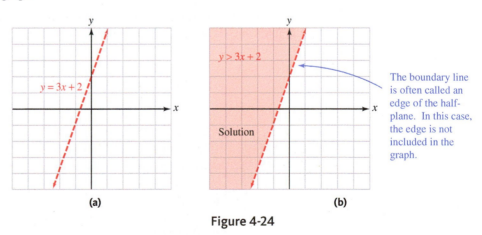

Figure 4-24

EXAMPLE 1   Solve: $2x - 3y \leq 6$

Solution   We start by graphing the related equation $2x - 3y = 6$ to find the boundary line. This time, we draw the *solid* line shown in Figure 4-25(a) on the next page, because equality is included in the original inequality. To determine which half-plane represents $2x - 3y < 6$, we select a test point, $(0, 0)$, to determine whether the coordinates satisfy the inequality.

$$2x - 3y < 6$$
$$2(\mathbf{0}) - 3(\mathbf{0}) < 6 \qquad \text{Substitute 0 for } x \text{ and 0 for } y.$$
$$0 < 6$$

Since the coordinates satisfy the inequality, the origin is in the half-plane that is the graph of $2x - 3y < 6$. The graph is shown in Figure 4-25(b).

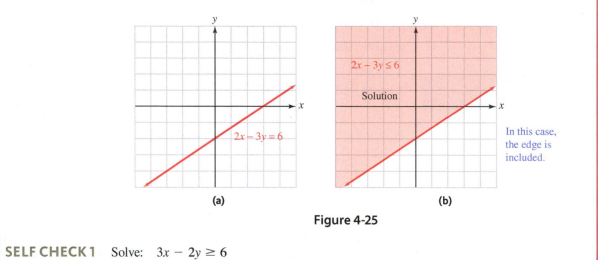

Figure 4-25

🌿 **SELF CHECK 1**   Solve: $3x - 2y \geq 6$

EXAMPLE 2   Solve: $y < 2x$

Solution   We start by graphing the related equation $y = 2x$, as shown in Figure 4-26(a). Because it is not part of the inequality, we draw the edge as a dashed line.

   To determine which half-plane is the graph of $y < 2x$, we select a test point. We cannot use the origin because the edge passes through it. We select a different test point—say, $(3, 1)$.

$$y < 2x$$
$$1 \overset{?}{<} 2(\mathbf{3}) \qquad \text{Substitute 1 for } y \text{ and 3 for } x.$$
$$1 < 6$$

Since $1 < 6$ is a true inequality, the point $(3, 1)$ satisfies the inequality and is in the graph, which is shown in Figure 4-26(b).

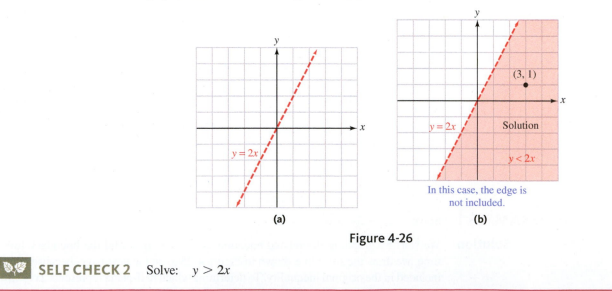

Figure 4-26

🌿 **SELF CHECK 2**   Solve: $y > 2x$

COMMENT   As the first two examples suggest, we draw a boundary line as a solid line when the inequality symbol is $\leq$ or $\geq$. We draw a dashed line when the inequality symbol is $<$ or $>$.

2   Solve a compound linear inequality in two variables.

EXAMPLE 3   Solve: $2 < x \leq 5$

Solution   The inequality $2 < x \leq 5$ is equivalent to the following two inequalities:

$$2 < x + 0y \qquad \text{and} \qquad x \leq 5 + 0y$$

Its graph will contain all points in the plane that satisfy the inequalities $2 < x + 0y$ and $x \leq 5 + 0y$ simultaneously. These points are in the shaded region of Figure 4-27.

Figure 4-27

SELF CHECK 3   Solve: $-2 \leq x < 3$

3   Solve an application requiring the use of a linear inequality in two variables.

EXAMPLE 4   **EARNING MONEY**   Rick has two part-time jobs, one paying \$7 per hour and the other paying \$5 per hour. He must earn at least \$140 per week to pay his expenses while attending college. Write and graph an inequality that shows the various ways he can schedule his time to achieve his goal.

Analyze the problem   If we let x represent the number of hours per week he works on the first job, he will earn \$7x per week on the first job. If we let y represent the number of hours per week he works on the second job, he will earn \$5y per week on the second job.

Form an inequality   To achieve his goal, the sum of these two incomes must be at least \$140.

The hourly rate on the first job	times	the hours worked on the first job	plus	the hourly rate on the second job	times	the hours worked on the second job	is greater than or equal to	$140.
\$7	·	x	+	\$5	·	y	$\geq$	\$140

Solve the inequality   The graph of $7x + 5y \geq 140$ is shown in Figure 4-28 on the next page. Any point in the shaded region indicates a way that he can schedule his time and earn \$140 or more per week. For example, if he works 10 hours on the first job and 15 hours on the second job, he will earn

$$\$7(10) + \$5(15) = \$70 + \$75$$
$$= \$145$$

COMMENT We could let x represent the number of hours he works on the second job and y represent the number of hours he works on the first job. The resulting graph would be different, but the combination of hours would remain the same.

If he works 5 hours on the first job and 25 hours on the second job, he will earn

$$\$7(5) + \$5(25) = \$35 + \$125$$
$$= \$160$$

Since Rick cannot work a negative number of hours, the graph has no meaning when x or y is negative, so only the first quadrant of the graph is shown.

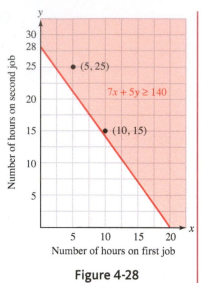

Figure 4-28

🌿 **SELF CHECK 4**   If Rick needs at least \$175 per week, write and graph an inequality that shows the various ways he can schedule his time.

Accent
on technology

▶ Graphing Inequalities

The TI-84 graphing calculator has a graphing-style icon in the Y = editor. See Figure 4-29(a). Some of the different graphing styles are as follows.

\	line	A straight line or curved graph is shown.
◥	above	Shading covers the area above a graph.
◣	below	Shading covers the area below a graph.

We can change the icon by placing the cursor on it and pressing **ENTER** until the desired style is visible.

Figure 4-29

To graph the inequality of Example 1, $2x - 3y \le 6$, using window settings of $x = [-10, 10]$ and $y = [-10, 10]$, we solve for y in the inequality $2x - 3y \le 6$, $-3y \le -2x + 6$, $y \ge \frac{2}{3}x - 2$. We enter the equation $2x - 3y = 6$ as $y = \frac{2}{3}x - 2$, change the graphing-style icon to "above" (◥), and press **GRAPH** to obtain Figure 4-29(b).

To graph the inequality of Example 2, $y < 2x$, we change the graphing-style icon to "below" (◣), enter the equation $y = 2x$, and press **GRAPH** to get Figure 4-29(c).

Note that graphing calculators do not distinguish between solid and dashed lines to show whether or not the edge of a region is included within the graph.

For instructions regarding the use of a Casio graphing calculator, please refer to the Casio Keystroke Guide in the back of the book.

SELF CHECK ANSWERS

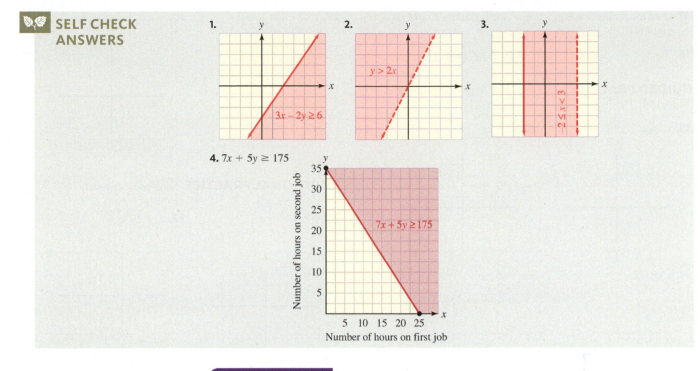

1.

2. $y > 2x$

3. $-2 \leq x < 3$

4. $7x + 5y \geq 175$

$7x + 5y \geq 175$

Number of hours on second job

Number of hours on first job

NOW TRY THIS

Solve each of the following by graphing.

1. $x \geq 2$ and $y < -1$

2. $x = 2$ and $y = -1$

4.3 Exercises

WARM-UPS *Graph* $2x + 5y = 10$*. State whether the points with the given coordinates lie* above *or* below *the graph of the line.*

1. $(0, 0)$ **2.** $(2, 3)$

3. $(6, 1)$ **4.** $(-1, 1)$

Graph $x = 3$*. State whether the points with the given coordinates lie to the* right *or* left *of the graph of the line.*

5. $(0, 0)$ **6.** $(4, 1)$

Graph $y = -2$*. State whether the points with the given coordinates lie* above *or* below *the graph of the line.*

7. $(0, 0)$ **8.** $(1, -3)$

REVIEW *Solve each system.*

9. $\begin{cases} x + y = 3 \\ x - y = 5 \end{cases}$ **10.** $\begin{cases} 5x + 3y = -2 \\ x = -3y + 2 \end{cases}$

11. $\begin{cases} 4x + 3y = 12 \\ 5x - 2y = 15 \end{cases}$ **12.** $\begin{cases} 3x + 5y = 21 \\ 4x + 2y = 14 \end{cases}$

VOCABULARY AND CONCEPTS *Fill in the blanks.*

13. $3x + 2y < 12$ is an example of a _____ inequality in two variables.

14. Graphs of linear inequalities in two variables are _____.

15. The boundary line of a half-plane is called an _____.

16. If an inequality involves $\leq$ or $\geq$, the boundary line of its graph will be _____.

17. If an inequality involves $<$ or $>$, the boundary line of its graph will be _____.

18. If $y < \frac{1}{2}x - 2$ and $y > \frac{1}{2}x - 2$ are false, then _____.

GUIDED PRACTICE *Solve each inequality. SEE EXAMPLE 1.*
(OBJECTIVE 1)

19. $y \geq x$

20. $y \leq 2x$

21. $y \geq -2$

22. $x \geq -3$

Solve each inequality. SEE EXAMPLE 2. (OBJECTIVE 1)

23. $y > x + 1$

24. $y < 2x - 1$

25. $x < 4$

26. $y < 3$

Solve each inequality. SEE EXAMPLE 3. (OBJECTIVE 2)

27. $-2 \leq x < 0$

28. $-3 < y \leq -1$

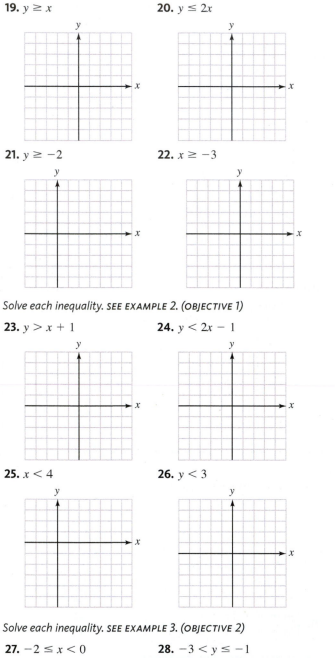

29. $y < -2$ or $y > 3$

30. $-x \leq 1$ or $x \geq 2$

ADDITIONAL PRACTICE *Solve each inequality.*

31. $y \geq 1 - \frac{3}{2}x$

32. $x - 2y \geq 4$

33. $0.5x + y > 1.5 + x$

34. $0.5x + 0.5y \leq 2$

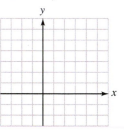

35. $y < \frac{1}{3}x - 1$

36. $2x + y \leq 6$

37. $3x \geq -y + 3$

38. $2x \leq -3y - 12$

Use a graphing calculator to solve each inequality.

39. $y < 0.27x - 1$ **40.** $y > -3.5x + 2.7$

41. $y \geq -2.37x + 1.5$ **42.** $y \leq 3.37x - 1.7$

Find the equation of the boundary line or lines. Then write the inequality whose graph is shown.

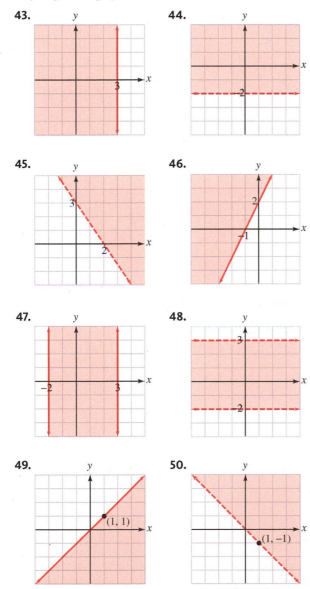

43.

44.

45.

46.

47.

48.

49.

50.

51.

52.

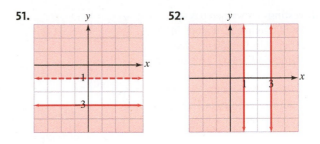

APPLICATIONS *Graph each inequality for nonnegative values of x and y. Then state three ordered pairs that satisfy the inequality.* SEE EXAMPLE 4. (OBJECTIVE 3)

53. Figuring taxes   On average, it takes an accountant 1 hour to complete a simple tax return and 3 hours to complete a complicated return. If the accountant wants to work no more than 9 hours per day, graph an inequality that shows the possible ways that the number of simple returns (x) and complicated returns (y) can be completed each day.

54. Selling trees   During a sale, a garden store sold more than $2,000 worth of trees. If a 6-foot maple costs $100 and a 5-foot pine costs $125, graph an inequality that shows the possible ways that the number of maple trees (x) and pine trees (y) were sold.

55. Choosing housekeepers One housekeeper charges $6 per hour, and another charges $7 per hour. If Sarah can afford no more than $42 per week to clean her house, graph an inequality that shows the possible ways that she can hire the first housekeeper for x hours and the second housekeeper for y hours.

56. Making sporting goods   A sporting goods manufacturer allocates at least 1,200 units of time per day to make fishing rods and reels. If it takes 10 units of time to make a rod and 15 units of time to make a reel, graph an inequality that shows the possible ways to schedule the time to make x rods and y reels.

57. Investing A woman has up to $6,000 to invest. If stock in Traffico sells for $50 per share and stock in Cleanco sells for $60 per share, graph an inequality that shows the possible ways that she can buy x shares of Traffico and y shares of Cleanco.

58. Concert tickets Tickets to a children's matinee concert cost $6 for reserved seats and $4 for general admission. If receipts must be at least $10,200 to meet expenses, graph an inequality that shows the possible ways that the box office can sell x reserved seats and y general admission tickets to meet expenses.

WRITING ABOUT MATH

59. Explain how to decide where to draw the boundary of the graph of a linear inequality, and whether to draw it as a solid or a broken line.

60. Explain how to decide which side of the boundary of the graph of a linear inequality should be shaded.

SOMETHING TO THINK ABOUT

61. Can an inequality be an identity, one that is satisfied by all (x, y) pairs? Illustrate.

62. Can an inequality have no solutions? Illustrate.

Section 4.4

Solving Systems of Linear and Quadratic Inequalities in Two Variables by Graphing

Objectives

1. Solve a system of linear inequalities in two variables.
2. Solve a system of quadratic inequalities in two variables.
3. Solve an application requiring the use of a system of inequalities in two variables.

Getting Ready

Do the given coordinates satisfy the inequalities $y > 2x - 3$ and $x - y > -1$?

1. $(0, 0)$ **2.** $(1, 1)$ **3.** $(-2, 0)$ **4.** $(0, 4)$

In the previous section, we learned how to graph the solution of linear inequalities in two variables, and in the previous chapter we learned how to solve systems of equations in two variables. In this section, we will combine these skills to solve systems of linear inequalities.

1 Solve a system of linear inequalities in two variables.

We now consider the graphs of systems of inequalities in the variables x and y. These graphs will usually be the intersection of half-planes.

1. Graph each inequality in the system on the same coordinate axes using solid or dashed lines as appropriate.

2. Identify the region where the graphs overlap.

3. Select a test point from the region to verify the solution.

4. Graph only the solution set, if there is one, on a new set of coordinate axes.

EXAMPLE 1   Graph the solution set: $\begin{cases} x + y \leq 1 \\ 2x - y > 2 \end{cases}$

Solution   On one set of coordinate axes, we graph each inequality as shown in Figure 4-30(a).

The graph of $x + y \leq 1$ includes the line graph of the equation $x + y = 1$ and all points below it. Since the edge is included, we draw it as a solid line.

The graph of $2x - y > 2$ contains the points below the graph of the equation $2x - y = 2$. Since the edge is not included, we draw it as a dashed line.

The area where the half-planes intersect represents the solution of the system of inequalities, because any point in that region has coordinates that will satisfy both inequalities.

The solution to the system is shown in Figure 4-30(b).

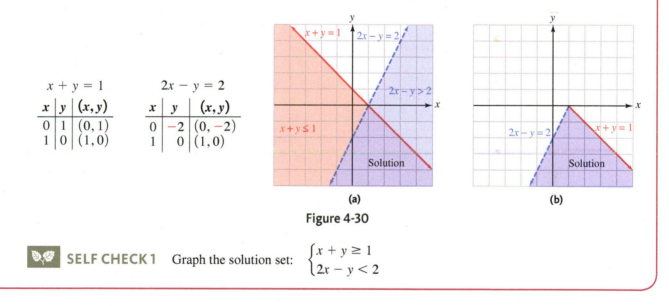

$x + y = 1$		
x	y	(x, y)
0	1	$(0, 1)$
1	0	$(1, 0)$

$2x - y = 2$		
x	y	(x, y)
0	-2	$(0, -2)$
1	0	$(1, 0)$

(a)

(b)

Figure 4-30

SELF CHECK 1   Graph the solution set: $\begin{cases} x + y \geq 1 \\ 2x - y < 2 \end{cases}$

EXAMPLE 2   Graph the solution set: $\begin{cases} x \geq 1 \\ y \geq x \\ 4x + 5y < 20 \end{cases}$

Solution   The graph of $x \geq 1$ is shown in Figure 4-31(a) on the next page.

Figure 4-31(b) includes the graph of $y \geq x$ and $x \geq 1$.

Figure 4-31(c) includes the graphs of $4x + 5y < 20$, $y \geq x$, and $x \geq 1$.

We see that the graph of the system includes the points that lie within a shaded triangle, together with the points on two of the three sides of the triangle, as shown in Figure 4-31(d).

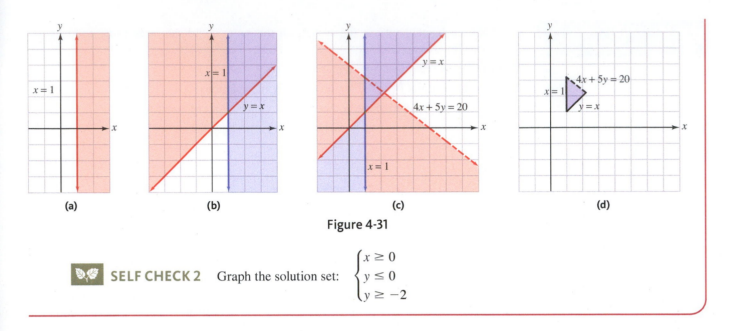

Figure 4-31

SELF CHECK 2   Graph the solution set: $\begin{cases} x \geq 0 \\ y \leq 0 \\ y \geq -2 \end{cases}$

2   Solve a system of quadratic inequalities in two variables.

EXAMPLE 3   Graph the solution set: $\begin{cases} y < x^2 \\ y > \dfrac{x^2}{4} - 2 \end{cases}$

Solution   The graph of $y = x^2$ is the red, dashed parabola shown in Figure 4-32(a). The points with coordinates that satisfy $y < x^2$ are the points that lie below the parabola.

The graph of $y = \dfrac{x^2}{4} - 2$ is the blue, dashed parabola. This time, the points that lie above the parabola satisfy the inequality, $y > \dfrac{x^2}{4} - 2$. Thus, the solution of the system is the area between the parabolas as shown in Figure 4-32(b).

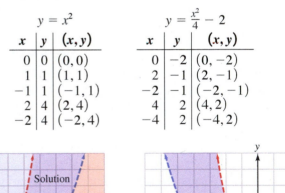

Figure 4-32

SELF CHECK 3 Graph the solution set: $\begin{cases} y \geq x^2 \\ y \leq x + 2 \end{cases}$

Accent on technology

▶ Solving Systems of Inequalities

To solve the system of Example 1 using a T184 graphing calculator, we use window settings of $x = [-10, 10]$ and $y = [-10, 10]$. To graph $x + y \leq 1$, we enter the equation $x + y = 1$ $(y = -x + 1)$ and change the graphing-style icon to below (◣). To graph $2x - y > 2$, we enter the equation $2x - y = 2$ $(y = 2x - 2)$ and change the graphing-style icon to below (◣). Finally, we press **GRAPH** to obtain Figure 4-33(a).

To solve the system of Example 3, we enter the equation $y = x^2$ and change the graphing-style icon to below (◣). We then enter the equation $y = \frac{x^2}{4} - 2$ and change the graphing-style icon to above (◥). Finally, we press **GRAPH** to obtain Figure 4-33(b).

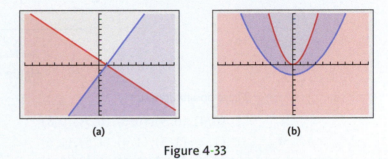

(a)   (b)

Figure 4-33

COMMENT When transferring your answer to paper, be sure to shade only the solution.

For instructions regarding the use of a Casio graphing calculator, please refer to the Casio Keystroke Guide in the back of the book.

3 Solve an application requiring the use of a system of inequalities in two variables.

EXAMPLE 4 **LANDSCAPING** A homeowner budgets from $300 to $600 for trees and shrubs to landscape his yard. After shopping around, he finds that good trees cost $150 and mature shrubs cost $75. What combinations of trees and shrubs can he afford?

Analyze the problem If x represents the number of trees purchased, $150x$ will be the cost of the trees. If y represents the number of shrubs purchased, $75y$ will be the cost of the shrubs.

Form two inequalities We know that the sum of these costs must be from $300 to $600. We can then form the following system of inequalities:

The cost of a tree	times	the number of trees purchased	plus	the cost of a shrub	times	the number of shrubs purchased	is greater than or equal to	$300.
$150	·	x	+	$75	·	y	≥	$300

The cost of a tree	times	the number of trees purchased	plus	the cost of a shrub	times	the number of shrubs purchased	is less than or equal to	$600.
$150	·	x	+	$75	·	y	≤	$600

Solve the system We graph the system

$$\begin{cases} 150x + 75y \geq 300 \\ 150x + 75y \leq 600 \end{cases}$$

as in Figure 4-34. The coordinates of each point shown in the graph give a possible combination of trees (x) and number of shrubs (y) that can be purchased.

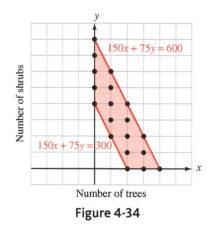

Figure 4-34

State the conclusion These possibilities are

$$(0,4), (0,5), (0,6), (0,7), (0,8)$$
$$(1,2), (1,3), (1,4), (1,5), (1,6)$$
$$(2,0), (2,1), (2,2), (2,3), (2,4)$$
$$(3,0), (3,1), (3,2), (4,0)$$

Only these points can be used, because the homeowner cannot buy a portion of a tree.

Check the result Check some of the ordered pairs to verify that they satisfy both inequalities.

SELF CHECK 4 If the homeowner can budget from $450 to $600, what combination of trees and shrubs can he afford?

SELF CHECK ANSWERS

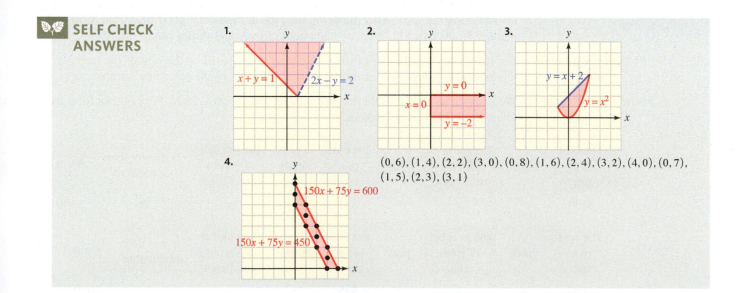

$(0,6), (1,4), (2,2), (3,0), (0,8), (1,6), (2,4), (3,2), (4,0), (0,7),$
$(1,5), (2,3), (3,1)$

NOW TRY THIS

Solve each system.

1. $\begin{cases} x - y > 0 \\ y \leq x + 2 \end{cases}$ 2. $\begin{cases} 4x - 3y \leq -4 \\ y < \frac{4}{3}x - \frac{5}{2} \end{cases}$ 3. $\begin{cases} y \leq x^2 - 6x + 7 \\ x - y < 3 \end{cases}$

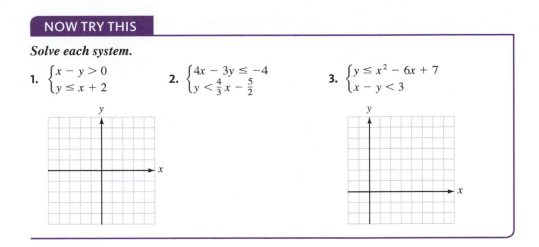

4.4 Exercises

WARM-UPS *Fill in the blanks with "<" or ">" so that the coordinates (1, 0) satisfy both inequalities.*

1. $\begin{cases} y \, \_ \, x + 2 \\ y \, \_ \, x - 2 \end{cases}$ 2. $\begin{cases} y \, \_ \, x + 3 \\ y \, \_ \, x + 1 \end{cases}$

REVIEW *Solve each formula for the given variable.*

3. $A = p + prt$ for t

4. $F = \frac{9}{5}C + 32$ for C

5. $z = \dfrac{x - \mu}{\sigma}$ for x 6. $3x - 4y = 8$ for y

7. $l = a + (n - 1)d$ for d 8. $z = \dfrac{x - \mu}{\sigma}$ for μ

VOCABULARY AND CONCEPTS *Fill in the blanks.*

9. To solve a system of inequalities by graphing, we graph each inequality. The solution is the region where the graphs _____.

10. If an edge is included in the graph of an inequality, we draw it as a ____ line and if not, it will be _____.

GUIDED PRACTICE *Graph the solution set of each system.*
SEE EXAMPLE 1. (OBJECTIVE 1)

11. $\begin{cases} y < 3x + 2 \\ y < -2x + 3 \end{cases}$ 12. $\begin{cases} y \leq x - 2 \\ y \geq 2x + 1 \end{cases}$

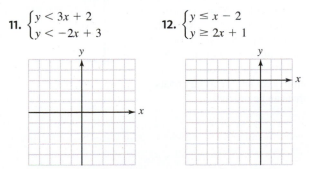

13. $\begin{cases} 3x + 2y > 6 \\ x + 3y \leq 2 \end{cases}$ 14. $\begin{cases} x + y < 2 \\ x + y \leq 1 \end{cases}$

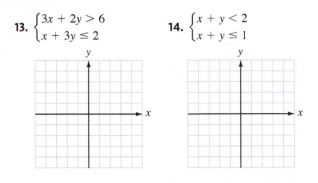

Graph the solution set of each system. SEE EXAMPLE 2. (OBJECTIVE 1)

15. $\begin{cases} 2x + 3y \leq 6 \\ 3x + y \leq 1 \\ x \leq 0 \end{cases}$ 16. $\begin{cases} 2x + y \leq 2 \\ y \geq x \\ x \geq 0 \end{cases}$

17. $\begin{cases} x - y < 4 \\ y \leq 0 \\ x \geq 0 \end{cases}$ 18. $\begin{cases} x + y \leq 4 \\ x \geq 0 \\ y \geq 0 \end{cases}$

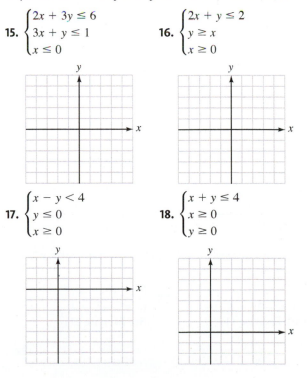

Graph the solution set of each system. SEE EXAMPLE 3. (OBJECTIVE 2)

19. $\begin{cases} 2x - y > 4 \\ y < -x^2 + 2 \end{cases}$ **20.** $\begin{cases} x \le y^2 \\ y \ge x \end{cases}$

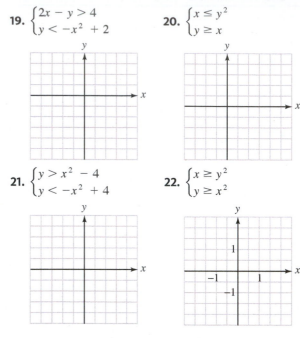

21. $\begin{cases} y > x^2 - 4 \\ y < -x^2 + 4 \end{cases}$ **22.** $\begin{cases} x \ge y^2 \\ y \ge x^2 \end{cases}$

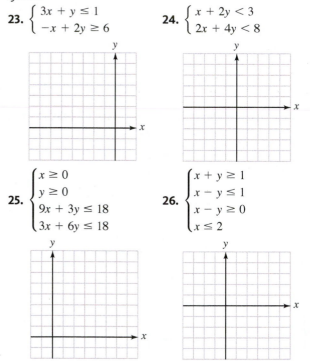

ADDITIONAL PRACTICE *Graph the solution set of each system.*

23. $\begin{cases} 3x + y \le 1 \\ -x + 2y \ge 6 \end{cases}$ **24.** $\begin{cases} x + 2y < 3 \\ 2x + 4y < 8 \end{cases}$

25. $\begin{cases} x \ge 0 \\ y \ge 0 \\ 9x + 3y \le 18 \\ 3x + 6y \le 18 \end{cases}$ **26.** $\begin{cases} x + y \ge 1 \\ x - y \le 1 \\ x - y \ge 0 \\ x \le 2 \end{cases}$

Use a graphing calculator to solve each system. SEE EXAMPLE 3. (OBJECTIVE 2)

27. $\begin{cases} y < 3x + 2 \\ y < -2x + 3 \end{cases}$ **28.** $\begin{cases} y > x^2 - 4 \\ y < -x^2 + 4 \end{cases}$
(See Exercise 11)   (See Exercise 21)

APPLICATIONS *Graph the solution.* SEE EXAMPLE 4. (OBJECTIVE 3)

29. Football In 2003, the Green Bay Packers scored either a touchdown or a field goal 65.4% of the time when their offense was in the red zone. This was the best record in the NFL! If x represents the yard line the football is on, a team's *red zone* is an area on their opponent's half of the field that can be described by the system

$$\begin{cases} x > 0 \\ x \le 20 \end{cases}$$

Shade the red zone on the field shown below.

← Packers moving this direction

30. Track and field In the shot put, the solid metal ball must land in a marked sector for it to be a fair throw. In the illustration, graph the system of inequalities that describes the region in which a shot must land.

$$\begin{cases} y \le \frac{3}{8}x \\ y \ge -\frac{3}{8}x \\ x \ge 1 \end{cases}$$

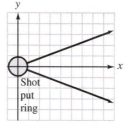

Graph a system of inequalities and give two possible solutions to each problem.

31. CDs Melodic Music has CDs on sale for either $10 or $15. If a customer wants to spend at least $30 but no more than $60 on CDs, graph a system of inequalities that will show the possible ways a customer can buy $10 CDs and $15 CDs.

32. Boats Dry Boat Works wholesales aluminum boats for $800 and fiberglass boats for $600. Northland Marina wants to order at least $2,400 worth, but no more than $4,800 worth of boats. Graph a system of inequalities that will show the possible combinations of aluminum boats and fiberglass boats that can be ordered.

33. Furniture A distributor wholesales desk chairs for $150 and side chairs for $100. Best Furniture wants to order no more than $900 worth of chairs, including more side chairs than desk chairs. Graph a system of inequalities that will show the possible combinations of desk chairs and side chairs that can be ordered.

34. Furnace equipment Bolden Heating Company wants to order no more than $2,000 worth of electronic air cleaners and humidifiers from a wholesaler that charges $500 for air cleaners and $200 for humidifiers. If Bolden wants more humidifiers than air cleaners, graph a system of inequalities that will show the possible combinations of air cleaners and humidifiers that can be ordered.

WRITING ABOUT MATH

35. When graphing a system of linear inequalities, explain how to decide which region to shade.

36. Explain how a system of two linear inequalities might have no solution.

SOMETHING TO THINK ABOUT

37. The solution of a system of inequalities in two variables is *bounded* if it is possible to draw a circle around it. Can a non-empty solution of two linear inequalities be bounded?

38. The solution of $\begin{cases} y \geq |x| \\ y \leq k \end{cases}$ has an area of 25. Find k.

Section 4.5

Solving Systems Using Linear Programming

Objectives

1 Find the maximum and minimum value of an equation in the form $P = ax + by$, subject to specific constraints.

2 Solve an application using linear programming.

Vocabulary

linear programming objective function | constraints | feasibility region

Getting Ready

Evaluate $2x + 3y$ for each pair of coordinates.

1. $(0, 0)$ **2.** $(3, 0)$ **3.** $(2, 2)$ **4.** $(0, 4)$

We now use our knowledge of solving systems of inequalities in two variables to solve linear programming applications.

1 Find the maximum and minimum value of an equation in the form $P = ax + by$, subject to specific constraints.

Linear programming is a mathematical technique used to find the optimal allocation of resources in the military, business, telecommunications, and other fields. It got its start during World War II when it became necessary to move huge quantities of people, materials, and supplies as efficiently and economically as possible.

To solve a linear program, we maximize (or minimize) a function (called the **objective function**) subject to given conditions on its variables. These conditions (called **constraints**) are usually given as a system of linear inequalities. For example, suppose that the annual profit (in millions of dollars) earned by a business is given by the equation $P = y + 2x$, where the profit is determined by the sale of two different items, x and y, and are subject to the following constraints:

$$\begin{cases} 3x + y \leq 120 \\ x + y \leq 60 \\ x \geq 0 \\ y \geq 0 \end{cases}$$

To find the maximum profit P that can be earned by the business, we solve the system of inequalities as shown in Figure 4-35(a) and find the coordinates of each *corner point* of the region R, called a **feasibility region**. We can then write the profit equation

$$P = y + 2x \qquad \text{in the form} \qquad y = -2x + P$$

The equation $y = -2x + P$ is the equation of a set of parallel lines, each with a slope of -2 and a y-intercept of P. To find the line that passes through region R and provides the maximum value of P, we refer to Figure 4-35(b) and locate the line with the greatest y-intercept. Since line l has the greatest y-intercept and intersects region R at the corner point $(30, 30)$, the maximum value of P (subject to the given constraints) is

$$P = y + 2x$$
$$= 30 + 2(30)$$
$$= 90$$

Thus, the maximum profit P that can be earned is \$90 million. This profit occurs when $x = 30$ and $y = 30$.

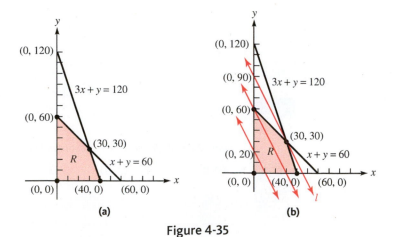

Figure 4-35

Perspective

HOW TO SOLVE IT

As a young student, George Polya (1888–1985) enjoyed mathematics and understood the solutions presented by his teachers. However, Polya had questions still asked by mathematics students today: "Yes, the solution works, but how is it possible to come up with such a solution? How could I discover such things by myself?" These questions still concerned him years later when, as Professor of Mathematics at Stanford University, he developed an

George Polya
1888–1985

approach to teaching mathematics that was very popular with faculty and students. His book, *How to Solve It*, became a bestseller.

Polya's problem-solving approach involves four steps.

- *Understand the problem.* What is the unknown? What information is known? What are the conditions?
- *Devise a plan.* Have you seen anything like it before? Do you know any related problems you have solved before? If you can't solve the proposed problem, can you solve a similar but easier problem?
- *Carry out the plan.* Check each step. Can you explain why each step is correct?
- *Look back.* Examine the solution. Can you check the result? Can you use the result, or the method, to solve any other problem?

The preceding discussion illustrates the following important fact.

MAXIMUM OR MINIMUM OF AN OBJECTIVE FUNCTION	If a linear function, subject to the constraints of a system of linear inequalities in two variables, attains a maximum or a minimum value, that value will occur at a corner point or along an entire edge of the region R that represents the solution of the system.

EXAMPLE 1   If $P = 2x + 3y$, find the maximum value of P subject to the following constraints:

$$\begin{cases} x + y \le 4 \\ 2x + y \le 6 \\ x \ge 0 \\ y \ge 0 \end{cases}$$

Solution   We solve the system of inequalities to find the feasibility region R shown in Figure 4-36. The coordinates of its corner points are $(0, 0)$, $(3, 0)$, $(0, 4)$, and $(2, 2)$.

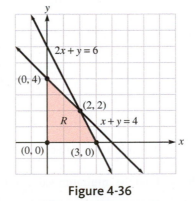

Figure 4-36

Since the maximum value of P will occur at a corner of R, we substitute the coordinates of each corner point into the objective function $P = 2x + 3y$ and find the one that gives the maximum value of P.

Point	$P = 2x + 3y$
$(0, 0)$	$P = 2(\mathbf{0}) + 3(\mathbf{0}) = 0$
$(3, 0)$	$P = 2(\mathbf{3}) + 3(\mathbf{0}) = 6$
$(2, 2)$	$P = 2(\mathbf{2}) + 3(\mathbf{2}) = 10$
$(0, 4)$	$P = 2(\mathbf{0}) + 3(\mathbf{4}) = 12$

The maximum value $P = 12$ occurs when $x = 0$ and $y = 4$.

SELF CHECK 1   Find the maximum value of $P = 4x + 3y$, subject to the constraints of Example 1.

EXAMPLE 2   If $P = 3x + 2y$, find the minimum value of P subject to the following constraints:

$$\begin{cases} x + y \geq 1 \\ x - y \leq 1 \\ x - y \geq 0 \\ x \leq 2 \end{cases}$$

Solution   We refer to the feasibility region shown in Figure 4-37 with corner points at $\left(\frac{1}{2}, \frac{1}{2}\right)$, $(2, 2)$, $(2, 1)$, and $(1, 0)$.

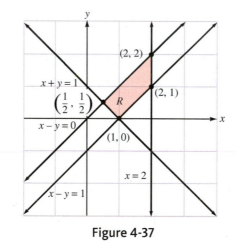

Figure 4-37

Since the minimum value of P occurs at a corner point of region R, we substitute the coordinates of each corner point into the objective function $P = 3x + 2y$ and find the one that gives the minimum value of P.

Point	$P = 3x + 2y$
$\left(\frac{1}{2}, \frac{1}{2}\right)$	$P = 3\left(\frac{1}{2}\right) + 2\left(\frac{1}{2}\right) = \frac{5}{2}$
$(2, 2)$	$P = 3(\mathbf{2}) + 2(\mathbf{2}) = 10$
$(2, 1)$	$P = 3(\mathbf{2}) + 2(\mathbf{1}) = 8$
$(1, 0)$	$P = 3(\mathbf{1}) + 2(\mathbf{0}) = 3$

The minimum value $P = \frac{5}{2}$ occurs when $x = \frac{1}{2}$ and $y = \frac{1}{2}$.

SELF CHECK 2   Find the minimum value of $P = 2x + y$, subject to the constraints of Example 2.

Solve an application using linear programming.

Linear programming applications can be complex and involve hundreds of variables. In this section, we will consider a few simpler situations. Since they involve only two variables, we can solve them using graphical methods.

EXAMPLE 3   **INCOME**   An accountant prepares tax returns for individuals and for small businesses. On average, each individual return requires 3 hours of her time and 1 hour of computer time. Each business return requires 4 hours of her time and 2 hours of computer time. Because of other business considerations, her time is limited to 240 hours, and the computer time is limited to 100 hours. If she earns a profit of $80 on each individual return and a profit of $150 on each business return, how many returns of each type should she prepare to maximize her profit?

Solution   First, we organize the given information into a table.

	Individual tax return	Business tax return	Time available
Accountant's time	3	4	240 hours
Computer time	1	2	100 hours
Profit	$80	$150	

Then we solve using the following steps.

Find the objective function   Suppose that x represents the number of individual returns to be completed and y represents the number of business returns to be completed. Because each of the x individual returns will earn an $80 profit, and each of the y business returns will earn a $150 profit, the total profit is given by the equation

$$P = 80x + 150y$$

Find the feasibility region   The number of individual returns and business returns cannot be negative, thus we know that $x \geq 0$ and $y \geq 0$.

Each of the x individual returns will take 3 hours of her time, each of the y business returns will take 4 hours of her time, and the total number of hours she will work will be $(3x + 4y)$ hours. This amount must be less than or equal to her available time, which is 240 hours. Thus, the inequality $3x + 4y \leq 240$ is a constraint on the accountant's time.

Each of the x individual returns will take 1 hour of computer time, each of the y business returns will take 2 hours of computer time, and the total number of hours of computer time will be $(x + 2y)$ hours. This amount must be less than or equal to the available computer time, which is 100 hours. Thus, the inequality $x + 2y \leq 100$ is a constraint on the computer time.

We have the following constraints on the values of x and y.

$$\begin{cases} x \geq 0 \\ y \geq 0 \\ 3x + 4y \leq 240 \\ x + 2y \leq 100 \end{cases}$$

The number of individual returns is nonnegative.
The number of business returns is nonnegative.
The accountant's time must be less than or equal to 240 hours.
The computer time must be less than or equal to 100 hours.

To find the feasibility region, we graph each of the constraints to find region R, as in Figure 4-38 on the next page. The four corner points of this region have coordinates of $(0, 0), (80, 0), (40, 30),$ and $(0, 50)$.

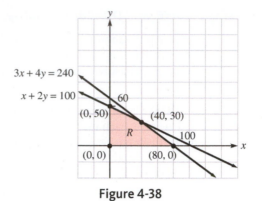

Figure 4-38

Find the maximum profit   To find the maximum profit, we substitute the coordinates of each corner point into the objective function $P = 80x + 150y$.

Point	$P = 80x + 150y$
$(0, 0)$	$P = 80(0) + 150(0) = 0$
$(80, 0)$	$P = 80(80) + 150(0) = 6{,}400$
$(40, 30)$	$P = 80(40) + 150(30) = 7{,}700$
$(0, 50)$	$P = 80(0) + 150(50) = 7{,}500$

From the table, we can see that the accountant will earn a maximum profit of \$7,700 if she prepares 40 individual returns and 30 business returns.

SELF CHECK 3   The accountant can make \$7,500 when $x = 0$ and $y = 50$. Interpret the meaning of these values.

EXAMPLE 4   **SUPPLEMENTS**   Healthtab and Robust are two diet supplements. Each Healthtab tablet costs 50¢ and contains 3 units of calcium, 20 units of Vitamin C, and 40 units of iron. Each Robust tablet costs 60¢ and contains 4 units of calcium, 40 units of Vitamin C, and 30 units of iron. At least 24 units of calcium, 200 units of Vitamin C, and 120 units of iron are required for the daily needs of one patient. How many tablets of each supplement should be taken daily for a minimum cost? Find the daily minimum cost.

Solution   First, we organize the given information into a table.

	Healthtab	Robust	Amount required
Calcium	3	4	24
Vitamin C	20	40	200
Iron	40	30	120
Cost	50¢	60¢	

Find the objective function   We let x represent the number of Healthtab tablets to be taken daily and y the corresponding number of Robust tablets. Because each of the x Healthtab tablets will cost 50¢, and each of the y Robust tablets will cost 60¢, the total cost will be given by the equation

$$C = 0.50x + 0.60y \quad \text{50¢ = \$0.50 and 60¢ = \$0.60}$$

Find the feasibility region   Since there are requirements for calcium, Vitamin C, and iron, there is a constraint for each. Note that neither x nor y can be negative.

$$\begin{cases} 3x + 4y \geq 24 & \text{The amount of calcium must be greater than or equal to 24 units.} \\ 20x + 40y \geq 200 & \text{The amount of Vitamin C must be greater than or equal to 200 units.} \\ 40x + 30y \geq 120 & \text{The amount of iron must be greater than or equal to 120 units.} \\ x \geq 0, y \geq 0 & \text{The number of tablets taken must be greater than or equal to 0.} \end{cases}$$

We graph the inequalities to find the feasibility region and the coordinates of its corner points, as in Figure 4-39.

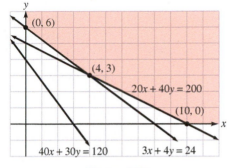

Figure 4-39

Find the minimum cost   In this case, the feasibility region is not bounded on all sides. The coordinates of the corner points are $(0, 6)$, $(4, 3)$, and $(10, 0)$. To find the minimum cost, we substitute each pair of coordinates into the objective function.

Point	$C = 0.50x + 0.60y$
$(0, 6)$	$C = 0.50(0) + 0.60(6) = 3.60$
$(4, 3)$	$C = 0.50(4) + 0.60(3) = 3.80$
$(10, 0)$	$C = 0.50(10) + 0.60(0) = 5.00$

A minimum cost will occur if no Healthtab and 6 Robust tablets are taken daily. The minimum daily cost is $3.60.

SELF CHECK 4   If the minimum daily cost is chosen, determine the number of units of calcium, Vitamin C, and iron that the patient will receive.

EXAMPLE 5   **PRODUCTION SCHEDULES**   A television program director must schedule comedy skits and musical numbers for prime-time variety shows. Each comedy skit requires 2 hours of rehearsal time, costs $3,000, and brings in $20,000 from the show's sponsors. Each musical number requires 1 hour of rehearsal time, costs $6,000, and generates $12,000. If 250 hours are available for rehearsal, and $600,000 is budgeted for comedy and music, how many segments of each type should be produced to maximize income? Find the maximum income.

Solution   First, we organize the given information into a table.

	Comedy	Musical	Available
Rehearsal time (hours)	2	1	250
Cost (in $1,000s)	3	6	600
Generated income (in $1,000s)	20	12	

Find the objective function   We let x represent the number of comedy skits and y the number of musical numbers to be scheduled. Because each of the x comedy skits generates $20 thousand, the income generated by the comedy skits is $20x$ thousand. The musical numbers produce $12y$ thousand. The objective function to be maximized is

$$V = 20x + 12y$$

Find the feasibility region   Since there are limits on rehearsal time and budget, there is a constraint for each. Note that neither x nor y can be negative.

$$\begin{cases} 2x + y \le 250 & \text{\color{red}The total rehearsal time must be less than or equal to 250 hours.} \\ 3x + 6y \le 600 & \text{\color{red}The total cost must be less than or equal to \$600 thousand.} \\ x \ge 0, y \ge 0 & \text{\color{red}The numbers of skits and musical numbers must be greater than or equal to 0.} \end{cases}$$

We graph the inequalities to find the feasibility region shown in Figure 4-40 and find the coordinates of each corner point.

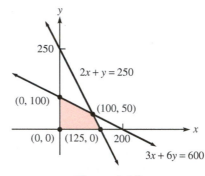

Figure 4-40

Find the maximum income   The coordinates of the corner points of the feasible region are $(0, 0)$, $(0, 100)$, $(100, 50)$, and $(125, 0)$. To find the maximum income, we substitute each pair of coordinates into the objective function.

Corner point	$V = 20x + 12y$
$(0, 0)$	$V = 20(0) + 12(0) = 0$
$(0, 100)$	$V = 20(0) + 12(100) = 1{,}200$
$(100, 50)$	$V = 20(100) + 12(50) = 2{,}600$
$(125, 0)$	$V = 20(125) + 12(0) = 2{,}500$

Maximum income will occur if 100 comedy skits and 50 musical numbers are scheduled. The maximum income will be 2,600 thousand dollars, or $2,600,000.

SELF CHECK 5   If a survey reveals the public wants only to watch comedy skits, what is the maximum income the station will make?

Everyday connections
Fuel Oil Production

Crude Oil, Gasoline, and Natural Gas Futures

Heating Oil NY Harbor	$3.0962
Gasoline NY Harbor	$3.1129

Suppose that the refining process at United States refineries requires the production of at least three gallons of gasoline for each gallon of fuel oil. To meet the anticipated demands of winter, at least 2.10 million gallons of fuel oil will need to be produced per day. The demand for gasoline is not more than 9 million gallons a day. The wholesale price of gasoline is $3.1129 per gallon and the wholesale price of fuel oil is $3.0962/gal.

1. How much of each should be produced in order to maximize revenue?

2. What is the maximum revenue?

Source: http://www.wtrg.com/#Crude

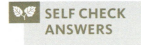
NOW TRY THIS

The FDA set the standard that fat calories should be limited to between 20% and 35% of the total daily intake, carbohydrates between 45% and 65%, and protein between 10% and 35%. Due to diabetes, Dale must limit his carbohydrate intake to the minimum (45%) and consume between 1,600 and 2,000 calories per day. Let x represent the number of fat calories and y the number of protein calories.

1. Between what numbers of calories can Dale consume daily in carbohydrates?

2. Between what numbers of calories does this leave for fats (x) and protein (y) combined?

3. Between what numbers of calories can he allot to fats?

4. Between what numbers of calories can be allotted to proteins?

5. What are the corner points in this graph?

6. Which one will minimize fat but maximize protein?

4.5 Exercises

WARM-UPS *Evaluate $P = 2x + 5y$ when*

1. $x = 0, y = 4$ **2.** $x = 2, y = 0$

Solve each system of linear equations.

3. $\begin{cases} x + y = 5 \\ 3x + 2y = 12 \end{cases}$ **4.** $\begin{cases} x + 3y = 6 \\ 4x + 3y = 12 \end{cases}$

REVIEW *Consider the line passing through $(3, 5)$ and $(-1, -4)$.*

5. Find the slope of the line.
6. Write the equation of the line in general form.
7. Write the equation of the line in slope-intercept form.

8. Write the equation of the line that passes through the origin and is parallel to the line.

VOCABULARY AND CONCEPTS *Fill in the blanks.*

9. In a linear program, the inequalities are called _____.
10. Ordered pairs that satisfy the constraints of a linear program are called _____ solutions.
11. The function to be maximized (or minimized) in a linear program is called the _____ function.
12. The objective function of a linear program attains a maximum (or minimum), subject to the constraints, at a _____ or along an ____ of the feasibility region.

GUIDED PRACTICE *Maximize P subject to the following constraints. SEE EXAMPLE 1. (OBJECTIVE 1)*

13. $P = 2x + 3y$
$\begin{cases} x \geq 0 \\ y \geq 0 \\ x + y \leq 4 \end{cases}$

14. $P = 3x + 2y$
$\begin{cases} x \geq 0 \\ y \geq 0 \\ x + y \leq 4 \end{cases}$

15. $P = y + \dfrac{1}{2}x$
$\begin{cases} x \geq 0 \\ y \geq 0 \\ 2y - x \leq 1 \\ y - 2x \geq -2 \end{cases}$

16. $P = 4y - x$
$\begin{cases} x \leq 2 \\ y \geq 0 \\ x + y \geq 1 \\ 2y - x \leq 1 \end{cases}$

Minimize P subject to the following constraints. SEE EXAMPLE 2. (OBJECTIVE 1)

17. $P = 5x + 12y$
$\begin{cases} x \geq 0 \\ y \geq 0 \\ x + y \leq 4 \end{cases}$

18. $P = 3x + 6y$
$\begin{cases} x \geq 0 \\ y \geq 0 \\ x + y \leq 4 \end{cases}$

19. $P = 3y + x$
$$\begin{cases} x \geq 0 \\ y \geq 0 \\ 2y - x \leq 1 \\ y - 2x \geq -2 \end{cases}$$

20. $P = 5y + x$
$$\begin{cases} x \leq 2 \\ y \geq 0 \\ x + y \geq 1 \\ 2y - x \leq 1 \end{cases}$$

ADDITIONAL PRACTICE *Maximize P subject to the following constraints.*

21. $P = 2x + y$
$$\begin{cases} y \geq 0 \\ y - x \leq 2 \\ 2x + 3y \leq 6 \\ 3x + y \leq 3 \end{cases}$$

22. $P = x - 2y$
$$\begin{cases} x + y \leq 5 \\ y \leq 3 \\ x \leq 2 \\ x \geq 0 \\ y \geq 0 \end{cases}$$

23. $P = 3x - 2y$
$$\begin{cases} x \leq 1 \\ x \geq -1 \\ y - x \leq 1 \\ x - y \leq 1 \end{cases}$$

24. $P = x - y$
$$\begin{cases} 5x + 4y \leq 20 \\ y \leq 5 \\ x \geq 0 \\ y \geq 0 \end{cases}$$

Minimize P subject to the following constraints.

25. $P = 6x + 2y$
$$\begin{cases} y \geq 0 \\ y - x \leq 2 \\ 2x + 3y \leq 6 \\ 3x + y \leq 3 \end{cases}$$

26. $P = 2y - x$
$$\begin{cases} x \geq 0 \\ y \geq 0 \\ x + y \leq 5 \\ x + 2y \geq 2 \end{cases}$$

27. $P = 2x - 2y$
$$\begin{cases} x \leq 1 \\ x \geq -1 \\ y - x \leq 1 \\ x - y \leq 1 \end{cases}$$

28. $P = y - 2x$
$$\begin{cases} x + 2y \leq 4 \\ 2x + y \leq 4 \\ x + 2y \geq 2 \\ 2x + y \geq 2 \end{cases}$$

APPLICATIONS *Use an objective function and a system of inequalities that describe the constraints in each problem. Use a graph of the feasibility region, showing the corner points to find the maximum or minimum value of the objective function.* **SEE EXAMPLES 3–5. (OBJECTIVE 2)**

29. Furniture Two woodworkers, Tom and Carlos, bring in $100 for making a table and $80 for making a chair. On average, Tom must work 3 hours and Carlos 2 hours to make a chair. Tom must work 2 hours and Carlos 6 hours to make a table. If neither wants to work more than 42 hours per week, how many tables and how many chairs should they make each week to maximize their income? Find the maximum income.

	Table	Chair	Time available
Income ($)	100	80	
Tom's time (hr)	2	3	42
Carlos's time (hr)	6	2	42

30. Crafts Two artists, Nina and Rob, make yard ornaments. They bring in $80 for each wooden snowman they make and $64 for each wooden Santa Claus. On average, Nina must work 4 hours and Rob 2 hours to make a snowman. Nina must work 3 hours and Rob 4 hours to make a Santa Claus. If neither wants to work more than 20 hours per week, how many of each ornament should they make each week to maximize their income? Find the maximum income.

	Snowman	Santa Claus	Time available
Income ($)	80	64	
Nina's time (hr)	4	3	20
Rob's time (hr)	2	4	20

31. Inventories An electronics store manager stocks from 20 to 30 IBM-compatible computers and from 30 to 50 Macintosh computers. There is room in the store to stock up to 60 computers. The manager receives a commission of $50 on the sale of each IBM-compatible computer and $40 on the sale of each Macintosh computer. If the manager can sell all of the computers, how many should she stock to maximize her commissions? Find the maximum commission.

Inventory	IBM	Macintosh
Minimum	20	30
Maximum	30	50
Commission	$50	$40

32. Diet A diet requires at least 16 units of Vitamin C and at least 34 units of Vitamin B complex. Two food supplements are available that provide these nutrients in the amounts and costs shown in the table. How much of each should be used to minimize the cost?

Supplement	Vitamin C	Vitamin B	Cost
A	3 units/g	2 units/g	3¢/g
B	2 units/g	6 units/g	4¢/g

33. Production Manufacturing DVD players and TVs requires the use of the electronics, assembly, and finishing departments of a factory, according to the following schedule:

	Hours for DVD player	Hours for TV	Hours available per week
Electronics	3	4	180
Assembly	2	3	120
Finishing	2	1	60

Each DVD player has a profit of $40, and each TV has a profit of $32. How many DVD players and TVs should be manufactured weekly to maximize profit? Find the maximum profit.

34. Production   A company manufactures one type of computer chip that runs at 1.66 GHz and another that runs at 2.66 GHz. The company can make a maximum of 50 fast chips per day and a maximum of 100 slow chips per day. It takes 6 hours to make a fast chip and 3 hours to make a slow chip, and the company's employees can provide up to 360 hours of labor per day. If the company makes a profit of $20 on each 2.66-GHz chip and $27 on each 1.66-GHz chip, how many of each type should be manufactured to earn the maximum profit?

35. Financial planning   A stockbroker has $200,000 to invest in stocks and bonds. She wants to invest at least $100,000 in stocks and at least $50,000 in bonds. If stocks have an annual yield of 9% and bonds have an annual yield of 7%, how much should she invest in each to maximize her income? Find the maximum return.

36. Production   A small country exports soybeans and flowers. Soybeans require 8 workers per acre, flowers require 12 workers per acre, and 100,000 workers are available. Government contracts require that there be at least 3 times as many acres of soybeans as flowers planted. It costs $250 per acre to plant soybeans and $300 per acre to plant flowers, and there is a budget of $3 million. If the profit from soybeans is $1,600 per acre and the profit from flowers is $2,000 per acre, how many acres of each crop should be planted to maximize profit? Find the maximum profit.

WRITING ABOUT MATH

37. What is meant by the constraints of a linear program?

38. What is meant by a feasible solution of a linear program?

SOMETHING TO THINK ABOUT

39. Try to construct a linear programming problem. What difficulties do you encounter?

40. Try to construct a linear programming problem that will have a maximum at every point along an edge of the feasibility region.

Projects

PROJECT 1

A farmer is building a machine shed onto his barn, as shown in the illustration. It is to be 12 feet wide, 20 feet long, and h_2 must be no more than 20 feet. For all of the shed to be useful for storing machinery, h_1 must be at least 6 feet. For the roof to shed rain and melting snow adequately, the slope of the roof must be at least $\frac{1}{2}$, but to be easily shingled, it must have a slope that is no greater than 1.

a. Represent on a graph all of the possible values for h_1 and h_2, subject to the constraints listed above.

b. The farmer wants to minimize the construction costs while still making sure that the shed is large enough for his purposes. He does this by setting a lower bound on the volume of the shed (3,000 cubic feet) and then minimizing the surface area of the walls that must be built. The volume of the shed can be expressed in a formula that contains h_1 and h_2. Derive this formula, and include the volume restriction in your design constraints. Then find the dimensions that will minimize the total area of the two ends of the shed and the outside wall. (The inner wall is already present as a wall of the barn and therefore involves no new cost.)

PROJECT 2

Knowing any three points on the graph of a parabolic function is enough to determine the equation of that parabola. It follows that for any three points that could possibly lie on a parabola, there is exactly one parabola that passes through those points, and this parabola will have the equation $y = ax^2 + bx + c$ for appropriate a, b, and c.

a. For a set of three points to lie on the graph of a parabolic function, no two of the points can have the same x-coordinate, and not all three can have the same y-coordinate. Explain why we need these restrictions.

b. Suppose that the points $(1, 3)$ and $(2, 6)$ are on the graph of a parabola. What restrictions would have to be placed on a and b to guarantee that the y-intercept of the parabola has an absolute value of 4 or less? Can $(-1, 8)$ be a third point on such a parabola?

Reach for Success

EXTENSION OF STUDY STRATEGIES

Let's consider how you can increase test preparedness *prior* to an exam. What would you do differently if you only knew then what you know now?

There is a known learning pattern called the Forgetting Curve, first recognized in 1885 by Hermann Ebbinghaus. The basic premise is that by reviewing material from a one-hour lecture for ten minutes the next day you are firing the synapses that make connections in your brain. Then, reviewing the material just a few minutes every day after that significantly increases your ability to recall that information. When you repeat this information, you send a message to your brain, "Here it is again; it must be important! I need to know this."

According to the graph, by Day 7, how much time would you need to spend to remember the material?_____

By Day 30, how much time would you need to review the material?_____

Can you think of factors that might influence the speed of forgetting?_____
_____

Do you feel overwhelmed by this amount of review? Does it seem to take too much time?_____

All-night study sessions rarely help with exam performance. Would it surprise you to know that it actually takes *less* time to review daily than it does to relearn the material the night before a test?_____

Give this a try and share your experience with your instructor.

You might not be surprised to learn that difficulty of the material, its connectedness to a previously learned topic, stress, and lack of sleep could all influence how well you can remember the material. While studying mathematics, work toward understanding the material. That will help increase the connections to other material and make it easier to remember.

4   *Review*

SECTION 4.1   Solving Linear Inequalities in One Variable

DEFINITIONS AND CONCEPTS	EXAMPLES
Trichotomy property: For any real numbers a and b, exactly one of the following is true: $\qquad a < b,\ a = b,\ \text{or } a > b$	Exactly one of these statements is true. $\qquad 2 < 3, 2 = 3, \text{ or } 2 > 3.$
Transitive properties (a, b, and c are real numbers): If $a < b$ and $b < c$, then $a < c$. If $a > b$ and $b > c$, then $a > c$.	If $-2 < 5$ and $5 < 10$, then $-2 < 10$. If $20 > 7$ and $7 > -5$, then $20 > -5$.
Properties of inequality: If a and b are real numbers and $a < b$, then $\qquad a + c < b + c$ $\qquad a - c < b - c$ $\qquad ac < bc \quad (c > 0)$ $\qquad ac > bc \quad (c < 0)$ $\qquad \dfrac{a}{c} < \dfrac{b}{c} \quad (c > 0)$ $\qquad \dfrac{a}{c} > \dfrac{b}{c} \quad (c < 0)$	To solve the linear inequality $3(2x + 6) < 18$, we use the same steps as for solving equations. $\qquad 3(2x + 6) < 18$ $\qquad 6x + 18 < 18 \qquad$ Use the distributive property to remove parentheses. $\qquad\qquad 6x < 0 \qquad$ Subtract 18 from both sides. $\qquad\qquad\ x < 0 \qquad$ Divide both sides by 6. The solution set is $\{x \mid x < 0\}$. The graph is shown below. The parenthesis at 0 indicates that 0 is not included in the solution set. In interval notation the solution is $(-\infty, 0)$. $\qquad\qquad\qquad\quad$ 0
If both sides of an inequality are multiplied (or divided) by a negative number, another inequality results, but with the opposite relationship from the original inequality and the symbol must be reversed.	To solve $-4(3x - 4) \leq -8$, we proceed as follows: $\qquad -4(3x - 4) \leq -8$ $\qquad -12x + 16 \leq -8 \qquad$ Use the distributive property to remove parentheses. $\qquad\qquad -12x \leq -24 \qquad$ Subtract 16 from both sides. $\qquad\qquad\quad x \geq 2 \qquad$ Divide both sides by -12 and reverse inequality symbol. The solution set is $\{x \mid x \geq 2\}$. The graph is shown below. The bracket at 2 indicates that 2 is included in the solution set. In interval notation the solution is $[2, \infty)$. $\qquad\qquad\qquad\quad$ 2
Compound inequalities: $c < x < d$ is equivalent to $c < x$ and $x < d$.	To solve the inequality $-7 \leq 2x - 5 < 3$, we isolate x between the inequality symbols. $\qquad -7 \leq 2x - 5 < 3$ $\qquad\ -2 \leq 2x < 8 \qquad$ Add 5 to all three parts. $\qquad\ -1 \leq x < 4 \qquad$ Divide all three parts by 2.

The solution set is $\{x \mid -1 \le x < 4\}$, which is the interval $[-1, 4)$. The graph is shown below.

The word *or* between two inequalities indicates that only one of the inequalities needs to be true to make the entire statement true.

To solve $x \le 2$ or $x > 7$, we must find all numbers that make either inequality true. The following graph shows those numbers. The solution set is $\{x \mid x \le 2 \text{ or } x > 7\}$, which is the union of two intervals $(-\infty, 2] \cup (7, \infty)$.

REVIEW EXERCISES

Graph each inequality and give each solution set in interval notation.

1. $4(x - 2) \le 4$

2. $3x + 4 > 10$

3. $\dfrac{1}{3}x - 2 \ge \dfrac{1}{2}x + 2$

4. $\dfrac{7}{4}(x + 3) < \dfrac{3}{8}(x - 3)$

5. $3 < 3x + 4 < 10$

6. $4x > 3x + 2 > x - 3$

7. $2x - 3 \ge -5$ or $2x - 3 > 5$

8. Investing   A woman invests \$25,000 at 3% annual interest. How much more must she invest at 5% so that her annual income is at least \$3,000?

SECTION 4.2   Solving Absolute Value Equations and Inequalities in One Variable

DEFINITIONS AND CONCEPTS	EXAMPLES
If $x \ge 0$, $\lvert x \rvert = x$. If $x < 0$, $\lvert x \rvert = -x$.	$\lvert 4 \rvert = 4$ $\lvert -4 \rvert = -(-4) = 4$
If $k > 0$, $\lvert x \rvert = k$ is equivalent to $x = k$ or $x = -k$. $\overset{\textstyle -k \quad 0 \quad k}{\longleftrightarrow}$	To solve the equation $\lvert 2x - 3 \rvert = 1$, we can write it as $$2x - 3 = 1 \quad \text{or} \quad -(2x - 3) = 1$$ and solve each equation for x: $$\begin{aligned} 2x - 3 &= 1 &\text{or}& & -(2x - 3) &= 1 \\ 2x &= 4 & & & -2x + 3 &= 1 \\ x &= 2 & & & -2x &= -2 \\ & & & & x &= 1 \end{aligned}$$ Verify that both solutions check.
$\lvert a \rvert = \lvert b \rvert$ is equivalent to $a = b$ or $a = -b$.	To solve the equation $\lvert 4x - 3 \rvert = \lvert 2x + 15 \rvert$, we note that the equation is true when $4x - 3 = 2x + 15$ or when $4x - 3 = -(2x + 15)$. We then solve each equation for x. $$\begin{aligned} 4x - 3 &= 2x + 15 &\text{or}& & 4x - 3 &= -(2x + 15) \\ 4x &= 2x + 18 & & & 4x - 3 &= -2x - 15 \\ 2x &= 18 & & & 6x &= -12 \\ x &= 9 & & & x &= -2 \end{aligned}$$ Verify that both solutions check.

If $k > 0$, $|x| < k$ is equivalent to $-k < x < k$.

To solve the inequality $|2x - 3| < 1$, we note that $|2x - 3| < 1$ is equivalent to $-1 < 2x - 3 < 1$, an inequality that we can solve.

$$-1 < 2x - 3 < 1$$
$$2 < 2x < 4 \qquad \text{Add 3 to each part.}$$
$$1 < x < 2 \qquad \text{Divide each part by 2.}$$

The solution set contains all numbers between 1 and 2, not including either 1 or 2. This is the interval $(1, 2)$, whose graph is shown below.

$|x| > k$ is equivalent to $x < -k$ or $x > k$.

To solve the inequality $|5x - 5| > 15$, we write the inequality as two separate inequalities and solve each one for x. Since $|5x - 5| > 15$ is equivalent to

$$5x - 5 < -15 \quad \text{or} \quad 5x - 5 > 15$$

we have

$$5x - 5 < -15 \quad \text{or} \quad 5x - 5 > 15$$
$$5x < -10 \qquad\qquad 5x > 20 \qquad \text{Add 5 to both sides.}$$
$$x < -2 \qquad\qquad x > 4 \qquad \text{Divide both sides by 5.}$$

Thus,

$$x < -2 \quad \text{or} \quad x > 4$$

This is the interval $(-\infty, -2) \cup (4, \infty)$, whose graph appears below.

REVIEW EXERCISES

Find each absolute value.

9. $|-9|$

10. $|8|$

11. $-|7|$

12. $-|-6|$

Graph each function.

13. $f(x) = |x + 1| - 3$

14. $f(x) = |x - 2| + 1$

Solve and check each equation.

15. $|4x + 3| = 7$

16. $\left| \dfrac{3}{2}x - 4 \right| + 2 = 11$

17. $\left| \dfrac{2 - x}{3} \right| = 4$

18. $|6x + 1| = |4x - 3|$

19. $|5x - 4| = |4x - 5|$

20. $\left| \dfrac{3 - 2x}{2} \right| = \left| \dfrac{3x - 2}{3} \right|$

Solve each inequality. Give the solution in interval notation and graph it.

21. $|2x + 7| < 3$

22. $|5 - 3x| \leq 14$

23. $\left| \dfrac{2}{3}x + 14 \right| < 0$

24. $\left| \dfrac{1 - 5x}{3} \right| > 7$

25. $|3x - 8| - 3 \geq 1$

26. $\left| \dfrac{3}{2}x - 14 \right| \geq 0$

SECTION 4.3   Solving Linear Inequalities in Two Variables

DEFINITIONS AND CONCEPTS	EXAMPLES
To graph a linear inequality in x and y, graph the boundary line, and then use a test point to determine which side of the boundary should be shaded.	To graph the inequality $3x - 2y > 6$, we start by graphing $3x - 2y = 6$ to find the boundary line. We draw a dashed line because equality is not included. To determine which half-plane represents $3x - 2y > 6$, we select a test point not on the line and determine whether the coordinates satisfy the inequality. A convenient test point is the origin, $(0, 0)$.

$$3x - 2y > 6$$
$$3(0) - 2(0) \overset{?}{>} 6 \qquad \text{Substitute 0 for } x \text{ and 0 for } y.$$
$$0 > 6$$

If the original inequality is $<$ or $>$, the boundary line will be dashed because the points that lie on the line are not included in the solution. If the original inequality is $\leq$ or $\geq$, the boundary line will be solid.

Since the coordinates do not satisfy the inequality, the origin is not in the half-plane that is the graph of $3x - 2y > 6$. The graph is shown below.

REVIEW EXERCISES

Graph each inequality on the coordinate plane.

27. $2x + 3y > 6$

28. $y \leq 4 - x$

29. $-2 < x < 4$

30. $y \leq -2$ or $y > 1$

SECTION 4.4   Solving Systems of Linear and Quadratic Inequalities in Two Variables

DEFINITIONS AND CONCEPTS	EXAMPLES
To graph a system of linear inequalities in x and y, graph each boundary and determine whether each line will be solid or dashed based on the inequality symbol. Use a test point to determine which side of each of the boundaries should be shaded. The solution will be the region shaded by both inequalities. Graph the solution set, if there is one, on a new set of coordinate axes.	To graph the solution set of $\begin{cases} x - 2y \leq 2 \\ 2x + y > 4 \end{cases}$, we graph each inequality on one set of coordinate axes as shown.

The graph of $x - 2y \leq 2$ includes the solid line graph of the equation $x - 2y = 2$ and all points above it.

The graph of $2x + y > 4$ contains the points to the right of the graph of the equation $2x + y = 4$. Since the edge is not included, we draw it as a dashed line.

The area where the half-planes intersect represents the solution of the system of inequalities, because any point in that region has coordinates that will satisfy both inequalities.

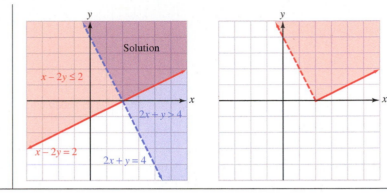

REVIEW EXERCISES

Graph the solution set of each system of inequalities.

31. $\begin{cases} y \geq x + 1 \\ 3x + 2y < 6 \end{cases}$ **32.** $\begin{cases} y \geq x^2 - 4 \\ y < x + 3 \end{cases}$

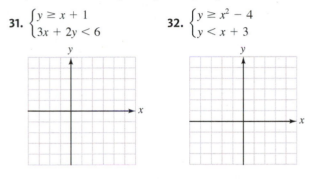

SECTION 4.5 Solving Systems Using Linear Programming

DEFINITIONS AND CONCEPTS	EXAMPLES
If a linear function, subject to the constraints of a system of linear inequalities in two variables, attains a maximum or a minimum value, that value will occur at a corner or along an entire edge of the region R that represents the solution of the system.	To solve a linear program, we maximize (or minimize) a function (called the objective function) subject to given conditions on its variables, called constraints.

To find the maximum value P of the function $P = 3x + 2y$ subject to the constraints

$$\begin{cases} x - y \leq 4 \\ x + y \leq 6 \\ x \geq 0 \\ y \geq 0 \end{cases}$$

we solve the system of inequalities to find the feasibility region R shown in the figure below. The coordinates of its corner points are $(0, 0)$, $(4, 0)$, $(0, 6)$, and $(5, 1)$.

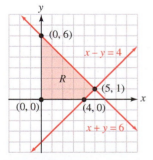

Since the maximum value of P will occur at a corner of R, we substitute the coordinates of each corner point into the objective function $P = 3x + 2y$ and find the one that gives the maximum value of P.

Point	$P = 3x + 2y$
$(0,0)$	$P = 3(0) + 2(0) = 0$
$(4,0)$	$P = 3(4) + 2(0) = 12$
$(5,1)$	$P = 3(5) + 2(1) = 17$
$(0,6)$	$P = 3(0) + 2(6) = 12$

The maximum value $P = 17$ occurs when $x = 5$ and $y = 1$.

REVIEW EXERCISES

33. Maximize $P = 3x + y$ subject to $\begin{cases} x \geq 0 \\ y \geq 0 \\ x + y \leq 4 \end{cases}$

34. Fertilizer   A company manufactures fertilizers X and Y. Each 50-pound bag requires three ingredients, which are available in the limited quantities shown in the table.

Ingredient	Number of pounds in fertilizer X	Number of pounds in fertilizer Y	Total number of pounds available
Nitrogen	6	10	20,000
Phosphorus	8	6	16,400
Potash	6	4	12,000

The profit on each bag of fertilizer X is \$6, and on each bag of Y, \$5. How many bags of each should be produced to maximize profit?

---

4   Test

Graph the solution of each inequality and state the solution in interval notation.

1. $-2(2x + 3) \geq 14$

2. $-2 < \dfrac{x - 4}{3} < 4$

Write each expression without absolute value symbols.

3. $|6 - 10|$

4. $|4\pi - 4|$

Graph each function.

5. $f(x) = |x + 1| - 4$

6. $f(x) = |x - 2| + 3$

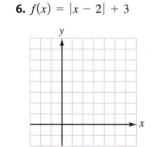

Solve each equation.

7. $|5x - 6| = 9$

8. $|4 - 3x| - 2 = 17$

9. $|3x + 7| + 8 = 5$

10. $|2 - 4x| = |6x - 8|$

Graph the solution of each inequality and state the solution in interval notation.

11. $|x + 3| \leq 4$

12. $|2x - 4| > 22$

13. $|4 - 2x| > 2$

14. $|2x - 4| \leq -2$

Graph each inequality.

15. $3x + 2y \geq 6$

16. $-2 \leq y < 5$

Solve each system.

17. $\begin{cases} 2x - 3y \geq 6 \\ y \leq -x + 1 \end{cases}$

18. $\begin{cases} y \geq x^2 \\ y < x + 3 \end{cases}$

19. Maximize $P = 3x - y$ subject to $\begin{cases} y \geq 1 \\ y \leq 2 \\ y \leq 3x + 1 \\ x \leq 1 \end{cases}$

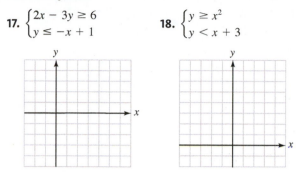

Cumulative Review

1. Draw a number line and graph the prime numbers from 50 to 60.

2. Find the additive inverse of -7.

Evaluate each expression when $x = 2$ and $y = -4$.

3. $x^2 - y$

4. $\dfrac{x^2 - y^2}{3x + y}$

Simplify each expression. Assume no variable is zero.

5. $(x^3 x^5)^3$

6. $(x^2)^3 (x^4)^2$

7. $\left(\dfrac{x^3}{x^5}\right)^{-2}$

8. $\dfrac{a^2 b^n}{a^n b^2}$

9. Write 455,000,000 in scientific notation.

10. Write 0.000012 in scientific notation.

Solve each equation.

11. $4x - 10 = 12$

12. $6(x - 1) = 2(x + 3)$

13. $\dfrac{5b}{2} - 10 = \dfrac{b}{3} + 3$

14. $2a - 5 = -2a + 4(a - 2) + 1$

Determine whether the lines represented by the equations are parallel, perpendicular, or neither.

15. $5x + 3y = 9$, $3x - 5y = 10$

16. $3x = y + 4$, $y = 3(x - 4) - 1$

17. Write the equation of the line passing through $(-2, 3)$ and perpendicular to the graph of $3x + y = 8$.

18. Solve the formula $A = \frac{1}{2} h(b_1 + b_2)$ for h.

Find each value, given that $f(x) = 3x^2 - x$.

19. $f(-1)$

20. $f(-3)$

21. Use graphing to solve: $\begin{cases} 2x + y = 5 \\ x - 2y = 0 \end{cases}$

22. Use substitution to solve: $\begin{cases} 3x + y = 4 \\ 2x - 3y = -1 \end{cases}$

23. Use elimination to solve: $\begin{cases} x + 2y = -2 \\ 2x - y = 6 \end{cases}$

24. Solve: $\begin{cases} \dfrac{x}{10} + \dfrac{y}{5} = \dfrac{1}{2} \\ \dfrac{x}{2} - \dfrac{y}{5} = \dfrac{13}{10} \end{cases}$

25. Solve: $\begin{cases} x + y + z = 1 \\ 2x - y - z = -4 \\ x - 2y + z = 4 \end{cases}$

26. Solve: $\begin{cases} x + 2y + 3z = 1 \\ 3x - 2y + z = -1 \\ 2y + 2z = 1 \end{cases}$

27. Evaluate: $\begin{vmatrix} 3 & -2 \\ 1 & -1 \end{vmatrix}$

28. Evaluate: $\begin{vmatrix} 2 & 3 & -1 \\ -1 & -1 & 2 \\ 4 & 1 & -1 \end{vmatrix}$

Use Cramer's rule to solve each system.

29. $\begin{cases} 4x - 3y = -1 \\ 3x + 4y = -7 \end{cases}$

30. $\begin{cases} x - 2y - z = -2 \\ 3x + y - z = 6 \\ 2x - y + z = -1 \end{cases}$

Solve each inequality and state the answer using interval notation.

31. $-3(x - 4) \geq x - 32$ **32.** $-8 < -3x + 1 < 10$

Solve each equation.

33. $|5x - 7| = 3$ **34.** $|2x - 1| = |3x + 4|$

Solve each inequality and state the answer using interval notation.

35. $|3x - 2| \leq 4$ **36.** $|2x + 3| - 1 > 4$

Solve each inequality.

37. $2x - 3y \leq 12$ **38.** $3 > x \geq -2$

     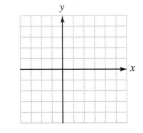

Solve each system.

39. $\begin{cases} 3x - 2y < 6 \\ y < -x + 2 \end{cases}$ **40.** $\begin{cases} y < x + 2 \\ 3x + y \leq 6 \end{cases}$

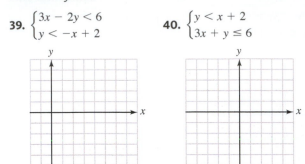

41. Research   To conduct an experiment with mice and rats, a researcher will place the animals into one of two mazes for the number of minutes shown in the table. Find the greatest number of animals that can be used in this experiment.

	Time per mouse	Time per rat	Time available
Maze 1	12 min	8 min	240 min
Maze 2	10 min	15 min	300 min

Polynomials, Polynomial Functions, and Equations

5

Careers and Mathematics

CRAFT AND FINE ARTISTS

Artists create art to communicate ideas, thoughts, or feelings. They use a variety of methods—painting, sculpting, or illustrating. Craft artists create objects typically designed to be functional, while fine artists typically create works for aesthetic value.

Training requirements for artists vary by specialty. However, it is very difficult to become skilled enough to make a living without some formal training.

Job Outlook:
Employment of craft and fine artists and related workers is expected to grow 5% through 2020. Competition for salaried jobs and freelance work in some areas is expected to be intense.

Annual Earnings:
$18,928–$91,603

For More Information:
http://www.bls.gov/ooh/arts-and-design/craft-and-fine-artists.htm

For a Sample Application:
See Problem 87 in Section 5.8.

REACH FOR SUCCESS

5.1 Polynomials and Polynomial Functions
5.2 Adding and Subtracting Polynomials
5.3 Multiplying Polynomials
5.4 The Greatest Common Factor and Factoring by Grouping
5.5 The Difference of Two Squares; the Sum and Difference of Two Cubes
5.6 Factoring Trinomials
5.7 Summary of Factoring Techniques
5.8 Solving Equations by Factoring
■ *Projects*
REACH FOR SUCCESS EXTENSION
CHAPTER REVIEW
CHAPTER TEST

In this chapter

Polynomials can be considered to be the numbers of algebra. In this chapter, we will learn how to add, subtract, and multiply them. Then we will learn how to reverse the operation of multiplication and find the factors of a known product. We conclude by solving equations using factoring.

289

Reach for Success Organizing Your Time

Life during your college years should be fun but it also requires time-management skills if you want to be academically successful. Although you previously examined a typical 24-hour day and a week, looking at a longer period will help you identify potential time conflicts.

The popularity of electronic calendars has increased significantly. Regardless of the format, it is important to find a calendar that works for you!

© iStockphoto.com/Ryan Balderas

Populate a two-week calendar with all the graded assignments from your syllabus in each of your classes.

Begin by filling in the dates in the chart. Then

1. Record all exams in the boxes.
2. Include due dates for any papers.
3. Add any labs/homework assignments.
4. Record employment hours including any scheduled overtime.

Sun ____	Mon ____	Tue ____	Wed ____	Thu ____	Fri ____	Sat ____
Sun ____	Mon ____	Tue ____	Wed ____	Thu ____	Fri ____	Sat ____

Do you know of any other time requirements for this month? If so, record them in your chart. Be sure to include any personal appointments and life events (e.g., birthdays, weddings).

Your calendar might look like this:

Sun 1	Mon 2	Tue 3	Wed 4	Thu 5	Fri 6	Sat 7
	English outline due	Dentist 1 p.m. Work	Work	Mathematics test	RJ's school play	Volunteer work Wedding anniversary
Sun 8 Sociology group project meeting 2 p.m.	Mon 9 Work Kenley's Dr. appt. 2:30 p.m.	Tue 10 Turn in history paper Math lab due	Wed 11 Read history chapters 13–15	Thu 12 Sociology group project due Work	Fri 13 English paper due Fred's birthday	Sat 14 Work

A Successful Study Strategy . . .

Use a planner to schedule your entire semester. By seeing even a week at a time, you will be able to avoid staying up all night to finish a paper or prepare for an exam. All-nighters rarely benefit anyone!

At the end of the chapter you will find an additional exercise to guide you in planning for a successful semester.

Section
5.1

Polynomials and Polynomial Functions

Objectives

1. Classify a polynomial as a monomial, binomial, trinomial, or none of these.
2. Determine the degree of a polynomial.
3. Evaluate a polynomial or a polynomial function.
4. Graph a polynomial function.

Vocabulary

algebraic term	binomial	descending order
polynomial	trinomial	ascending order
monomial	degree of a polynomial	polynomial function

Getting Ready

Write each expression using exponents.

1. $3aabb$
2. $-5xxxy$
3. $4pp + 7qq$
4. $aaa - bb$

In this section, we discuss important algebraic expressions called *polynomials*. We then examine several polynomial functions and their graphs.

1 Classify a polynomial as a monomial, binomial, trinomial, or none of these.

Algebraic terms are expressions that contain constants and/or variables. Some examples are

$$17, \qquad 9x, \qquad 15y^2, \qquad \text{and} \qquad -24x^4y^5$$

The *numerical coefficient* of 17 is 17. The numerical coefficients of the remaining terms are 9, 15, and -24, respectively.

A **polynomial** is a term or the sum of two or more algebraic terms whose variables have whole-number exponents.

POLYNOMIAL IN ONE VARIABLE

A **polynomial in one variable** (say, x) is a term or the sum of two or more terms of the form ax^n, where a is a real number and n is a whole number.

The following expressions are polynomials in x. Note that $-17 = -17x^0$.

$$3x^2 + 2x, \qquad \frac{3}{2}x^5 - \frac{7}{3}x^4 - \frac{8}{3}x^3, \qquad \text{and} \qquad 19x^{20} + \sqrt{3}x^{14} + 4.5x^{11} - 17$$

The following expressions are not polynomials.

$$\frac{2x}{x^2 + 1}, \qquad x^{1/2} - 1, \qquad \text{and} \qquad x^{-3} + 2x$$

The first expression is the quotient of two polynomials, and the last two expressions have exponents that are not whole numbers.

POLYNOMIAL IN MORE THAN ONE VARIABLE	A **polynomial in several variables** (say x, y, and z) is a term or the sum of two or more terms of the form $ax^m y^n z^p$; where a is a real number and m, n, and p are whole numbers.

The following expressions are polynomials in more than one variable.

$$3xy, \qquad 5x^2y + 2yz^3 - 3xz, \qquad \text{and} \qquad u^2v^2w^2 + x^3y^3 + 1$$

A polynomial with one term is called a **monomial**, a polynomial with two terms is called a **binomial**, and a polynomial with three terms is called a **trinomial**:

Monomials *(one term)*	*Binomials* *(two terms)*	*Trinomials* *(three terms)*
$2x^3$	$2x^4 + 5$	$2x^3 + 4x^2 + 3$
a^2b	$-17t^{45} - 3xy$	$3mn^3 - m^2n^3 + 7n$
$3x^3y^5z^2$	$32x^{13}y^5 + 47x^3yz$	$-12x^5y^2 + 13x^4y^3 - 7x^3y^3$

2 Determine the degree of a polynomial.

Because the variable x occurs three times as a factor in the monomial $2x^3$, the monomial is called a *third-degree monomial* or a *monomial of degree 3*. The monomial $3x^3y^5z^2$ is called a *monomial of degree 10*, because the variables x, y, and z occur as factors a total of ten times. These examples illustrate the following definition.

DEGREE OF A MONOMIAL	If $a \neq 0$, the **degree of ax^n** is n. The degree of a monomial containing several variables is the sum of the exponents on those variables.

COMMENT   Since $a \neq 0$ in the previous definition, 0 has no defined degree.

EXAMPLE 1   Find the degree of   **a.** $3x^4$ **b.** $-18x^3y^2z^{12}$ **c.** 4^7xy^4 **d.** 3.

Solution   **a.** $3x^4$ is a monomial of degree 4, because the variable x occurs as a factor four times.

b. $-18x^3y^2z^{12}$ is a monomial of degree 17, because the sum of the exponents on the variables is 17.

c. 4^7xy^4 is a monomial of degree 5, because the sum of the exponents on the variables is 5.

d. 3 is a monomial of degree 0, because $3 = 3x^0$.

SELF CHECK 1   Find the degree of   **a.** $-12a^5$ **b.** $8a^3b^2$ **c.** x^5yz^3 **d.** 15.

We define the degree of a polynomial by considering the degrees of each of its terms.

DEGREE OF A POLYNOMIAL	The **degree of a polynomial** is the same as the degree of the term in the polynomial with largest degree.

EXAMPLE 2   Find the degree of each polynomial: **a.** $3x^5 + 4x^2 + 7$ **b.** $7x^2y^8 - 3xy$ **c.** $3x + 2y - xy$ **d.** $18x^2y^3 - 12x^7y^2 + 3x^9y^3 - 3$

Solution **a.** $3x^5 + 4x^2 + 7$ is a trinomial of degree 5, because the largest degree of the three terms is 5.

b. $7x^2y^8 - 3xy$ is a binomial of degree 10 because the largest sum of the exponents in any term is 10.

c. $3x + 2y - x^1y^1$ is a trinomial of degree 2.

d. $18x^2y^3 - 12x^7y^2 + 3x^9y^3 - 3$ is a polynomial of degree 12.

SELF CHECK 2   Find the degree of $7a^3b^2 - 14a^2b^4$.

If the terms of a polynomial in one variable are written so that the exponents decrease as we move from left to right, we say that the terms are written with their exponents in **descending order**. If the terms are written so that the exponents increase as we move from left to right, we say that the terms are written with their exponents in **ascending order**.

EXAMPLE 3   Write the terms of $7x^2 - 5x^4 + 3x + 2x^3 - 1$ in **a.** descending order   **b.** ascending order.

Solution **a.** $-5x^4 + 2x^3 + 7x^2 + 3x - 1$
b. $-1 + 3x + 7x^2 + 2x^3 - 5x^4$

SELF CHECK 3   Write the terms of $3x^2 + 2x^4 - 5x - 3$ in descending order.

In the following polynomial, the exponents on x are in descending order, and the exponents on y are in ascending order.

$$7x^4 - 2x^3y + 4x^2y^2 - 8xy^3 + 12y^4$$

3   Evaluate a polynomial or a polynomial function.

A function of the form $y = P(x)$, where $P(x)$ is a polynomial, is called a **polynomial function**. Some examples are

$$P(x) = -5x - 10 \quad \text{and} \quad Q(t) = 2t^2 + 5t - 2 \qquad \text{Read } P(x) \text{ as ``}P\text{ of } x\text{'' and } Q(t) \text{ as ``}Q \text{ of } t\text{.''}$$

POLYNOMIAL FUNCTIONS	A **polynomial function in one variable** (say, x) is defined by an equation of the form $y = P(x)$, where $P(x)$ is a polynomial in the variable x. The **degree of the polynomial function** $y = P(x)$ is the same as the degree of the polynomial.

To evaluate a polynomial function at a specific value of its variable, we substitute the value of the variable and simplify. For example, to evaluate $P(x)$ at $x = 1$, we substitute 1 for x and simplify.

$$P(x) = x^6 + 4x^5 - 3x^2 + x - 2$$
$$P(1) = (1)^6 + 4(1)^5 - 3(1)^2 + 1 - 2 \quad \text{Substitute 1 for } x.$$
$$= 1 + 4 - 3 + 1 - 2$$
$$= 1$$

Thus, $P(1) = 1$.

EXAMPLE 4   Given the polynomial function $P(x) = 3x^2 - 2x + 7$, find

    a. $P(-2)$ **b.** $P(a)$ **c.** $P(-2x)$.

Solution   **a.** $P(x) = 3x^2 - 2x + 7$ **b.** $P(x) = 3x^2 - 2x + 7$

    $P(-2) = 3(-2)^2 - 2(-2) + 7$ $P(a) = 3(a)^2 - 2(a) + 7$

        $= 3(4) + 4 + 7$ $= 3a^2 - 2a + 7$

        $= 23$

    c. $P(x) = 3x^2 - 2x + 7$

    $P(-2x) = 3(-2x)^2 - 2(-2x) + 7$

        $= 12x^2 + 4x + 7$

SELF CHECK 4   Given $Q(x) = 12x^2 - 4x + 7$, find   **a.** $Q(-3)$ **b.** $Q(2a)$.

To evaluate polynomials with more than one variable, we substitute values for the variables into the polynomial and simplify.

EXAMPLE 5   Evaluate $4x^2y - 5xy^3$ at $x = 3$ and $y = -2$.

Solution   We substitute 3 for x and -2 for y and simplify.

$$4x^2y - 5xy^3 = 4(3)^2(-2) - 5(3)(-2)^3$$
$$= 4(9)(-2) - 5(3)(-8)$$
$$= -72 + 120$$
$$= 48$$

SELF CHECK 5   Evaluate $3ab^2 - 2a^2b$ at $a = 2$ and $b = -3$.

EXAMPLE 6   **HEIGHT OF A ROCKET**   If a toy rocket is launched straight up with an initial velocity of 128 feet per second, its height h (in feet) above the ground after t seconds is given by the polynomial function

$$P(t) = -16t^2 + 128t \quad \text{The height } h \text{ is the value } P(t).$$

Find the height of the rocket at   **a.** 0 second   **b.** 3 seconds   **c.** 7.9 seconds.

Solution   **a.** To find the height at 0 second, we find $P(0)$ by substituting 0 for t and simplifying.

$$P(t) = -16t^2 + 128t$$
$$P(0) = -16(0)^2 + 128(0)$$
$$= 0$$

At 0 second, the rocket is on the ground waiting to be launched.

b. To find the height at 3 seconds, we find $P(3)$ by substituting 3 for t and simplifying.

$$P(\mathbf{3}) = -16(\mathbf{3})^2 + 128(\mathbf{3})$$
$$= -16(9) + 384$$
$$= -144 + 384$$
$$= 240$$

At 3 seconds, the height of the rocket is 240 feet.

c. To find the height at 7.9 seconds, we find $P(7.9)$ by substituting 7.9 for t and simplifying.

$$P(\mathbf{7.9}) = -16(\mathbf{7.9})^2 + 128(\mathbf{7.9})$$
$$= -16(62.41) + 1{,}011.2$$
$$= -998.56 + 1{,}011.2$$
$$= 12.64$$

At 7.9 seconds, the height is 12.64 feet. It has fallen nearly back to Earth.

SELF CHECK 6   Find the height of the rocket at 4 seconds.

4   Graph a polynomial function.

In earlier sections we discussed three basic polynomial functions. The graph of the linear function $f(x) = 3x + 1$, the graph of the quadratic function $f(x) = x^2$, and the graph of the cubic function $f(x) = x^3$ are shown in Figure 5-1.

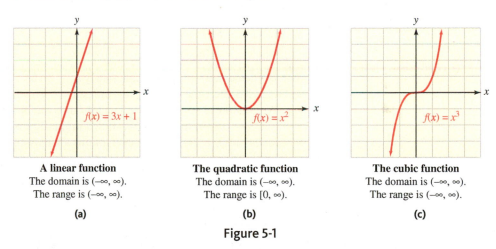

A linear function	The quadratic function	The cubic function
The domain is $(-\infty, \infty)$.	The domain is $(-\infty, \infty)$.	The domain is $(-\infty, \infty)$.
The range is $(-\infty, \infty)$.	The range is $[0, \infty)$.	The range is $(-\infty, \infty)$.
(a)	**(b)**	**(c)**

Figure 5-1

We have seen that the graphs of many polynomial functions are translations or reflections of these graphs. For example, the graph of $f(x) = x^2 - 2$ is the graph of $f(x) = x^2$ translated 2 units downward, as shown in Figure 5-2(a) on the next page. The graph of $f(x) = -x^2$ is the graph of $f(x) = x^2$ reflected about the x-axis, as shown in Figure 5-2(b).

Srinivasa Ramanujan
1887–1920

Ramanujan was one of India's most prominent mathematicians. When in high school, he read *Synopsis of Elementary Results in Pure Mathematics* and from this book taught himself mathematics. He entered college but failed several times because he would study only mathematics. He went on to teach at Cambridge University in England.

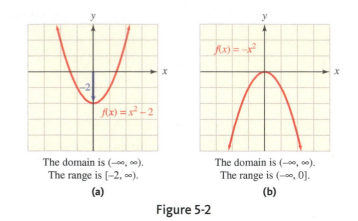

The domain is $(-\infty, \infty)$.
The range is $[-2, \infty)$.

(a)

The domain is $(-\infty, \infty)$.
The range is $(-\infty, 0]$.

(b)

Figure 5-2

In Example 6, we saw that a polynomial function can describe (or model) the height of a rocket. Since the height (h) of the rocket depends on time (t), we say that the height is a function of time, and we can write $h = f(t)$. To graph this function, we can make a table of values, plot the points, and join them with a smooth curve, as shown in Example 7.

EXAMPLE 7   Graph $h = f(t) = -16t^2 + 128t$.

Solution   From Example 6, we saw that

when $t = 3$, then $h = 240$.

when $t = 0$, then $h = 0$.

These ordered pairs and others that satisfy $h = -16t^2 + 128t$ are given in the table shown in Figure 5-3. Here the ordered pairs are pairs (t, h), where t is the time and h is the height.

We plot these pairs on a horizontal t-axis and a vertical h-axis and join the resulting points to obtain the parabola shown in the figure. From the graph, we can see that 4 seconds into the flight, the rocket attains a maximum height of 256 feet.

$$h = f(t) = -16t^2 + 128t$$

t	$f(t)$	$(t, f(t))$
0	0	$(0, 0)$
1	112	$(1, 112)$
2	192	$(2, 192)$
3	240	$(3, 240)$
4	256	$(4, 256)$
5	240	$(5, 240)$
6	192	$(6, 192)$
7	112	$(7, 112)$
8	0	$(8, 0)$

Figure 5-3

SELF CHECK 7   Use Figure 5-3 to estimate the height of the rocket at 6.5 seconds.

COMMENT   The parabola shown in Figure 5-3 describes the height of the rocket in relation to time. It does not show the path of the rocket. The rocket goes straight up and then comes back down.

Everyday connections
Stretching Rubber Bands

$f(x) = 1,071x^3 - 619.6x^2 + 149x - 8.059$

The polynomial function shown in the figure models the stretching force on a rubber band when a mass is attached to one end.

Use the graph of the function to answer the following questions.

1. What mass stretched the rubber band with a force of approximately 5.5 newtons?

2. What was the stretching force caused by a mass of 0.28 kg?

3. What is the predicted stretching force caused by a mass of 0.32 kg?

4. At what point would the model break down?

Source: http://www.madphysics.com/exp/hysteresis _and_rubber_bands.htm

Accent
on technology

▸ Graphing Polynomial Functions

We can graph polynomial functions with a TI-84 graphing calculator. For example, to graph $P(t) = -16t^2 + 128t$, we can use window settings of $[0, 8]$ for x and $[0, 260]$ for y to see the parabola shown in Figure 5-4(a).

We can use the VALUE feature in the CALC menu to estimate the height of the rocket for any number of seconds into the flight. From the graph, press **2ⁿᵈ TRACE** (CALC) **1** (VALUE) and enter a value. We will use 1.6. Figure 5-4(b) shows that the height of the rocket 1.6 seconds into the flight is 163.84 ft. To find other heights, just type in a new value.

$P(t) = -16t^2 + 128t$

$Y_1 = -16X^2 + 128X$

X = 1.6   Y = 163.84

(a)   (b)

Figure 5-4

For instruction regarding the use of a Casio graphing calculator, please refer to the Casio Keystroke Guide in the back of the book.

NOW TRY THIS

1. Graph the polynomial function $f(x) = x^3 - 4x$ and find the x-intercepts.

2. Given $P(x, y) = 3x^2y - 4xy^2$, find $P(-1, 5)$.

5.1   Exercises

WARM-UPS *Determine whether each algebraic expression is a polynomial.*

1. $\dfrac{x^2}{5}$

2. $3x^2 + 7 + 5$

3. $\dfrac{5}{x^2}$

4. $4x^{-1} + 3x$

5. 20

6. $\dfrac{2}{3}x^2 - \dfrac{4}{5}x$

7. $x^{1/2} + 3$

8. $5\sqrt[3]{x}$

REVIEW *Write each expression with a single exponent.*

9. $(y^4y^3)^2$

10. $\dfrac{a^5b^7}{a^7b^5}$

11. $\dfrac{3(y^3)^{10}}{y^3y^4}$

12. $\dfrac{4x^{-4}x^5}{2x^{-6}}$

13. **Astronomy**   The distance from Mars to the Sun is about 114,000,000 miles. Express this number in scientific notation.

14. **Science**   One angstrom is about 0.0000001 millimeter. Express this number in scientific notation.

VOCABULARY AND CONCEPTS *Fill in the blanks.*

15. A polynomial is a term or the ____ of two or more algebraic terms whose variables have _____ number exponents.

16. A monomial is a polynomial with ___ term.

17. A _____ is a polynomial with two terms.

18. A _____ is a polynomial with three terms.

19. The polynomial $3x^5 + 3x^3 - x^2 - 7x$ is written in _____ ____ of powers of x.

20. The polynomial $-2 + 3y^2 + y^3 + 8y^4$ is written in _____ ____ of powers of y.

21. The equation $y = P(x)$, where $P(x)$ is a polynomial, defines a function, because each input value x determines ___ output value y.

22. The degree of a polynomial is the same as the degree of the term in the polynomial with _____ degree.

GUIDED PRACTICE *Classify each polynomial as a monomial, a binomial, a trinomial, or not a polynomial. (OBJECTIVE 1)*

23. $-5x^3$

24. $3y^2 + 2y$

25. $3x^2y - 2x + 3y$

26. $a^2 + b^2$

27. $x^2 - y^2$

28. $\dfrac{17}{2}x^3 + 3x^{-2} - x - 4$

29. 11

30. $6a^2b^3$

Find the degree of each monomial. SEE EXAMPLE 1. (OBJECTIVE 2)

31. $-5x^7$

32. $-10x$

33. 121

34. $3^2x^2y^3$

Find the degree of each polynomial. SEE EXAMPLE 2. (OBJECTIVE 2)

35. $3x^2 + 2$

36. $6x^3 + 4x^2$

37. $4x^8 + 3x^2y^4$

38. $19x^2y^4 - y^{10}$

39. $4x^2y^3 - 5y^3zw^6$

40. $x^7y^4z + w^{10}$

Write each polynomial with the exponents on x in descending order and y in ascending order. SEE EXAMPLE 3. (OBJECTIVE 2)

41. $3x - 2x^4 + 7 - 5x^2$

42. $-x^2 + 3x^5 - 7x + 3x^3$

43. $4y^2 - 2y^5 + 7y - 5y^3$

44. $y^3 + 3y^2 + 8y^4 - 2$

Consider the polynomial function $P(x) = 2x^2 + x + 2$ and find each value. SEE EXAMPLE 4. (OBJECTIVE 3)

45. $P(0)$

46. $P(1)$

47. $P(-2)$

48. $P(-3)$

49. $P(5a)$

50. $P(2t)$

51. $P(-4x)$

52. $P(-2x)$

Find each value when $x = 2$ and $y = -3$. SEE EXAMPLE 5. (OBJECTIVE 3)

53. $x^2 + y^2$

54. $x^3 + y^3$

55. $x^3 - y^3$

56. $x^2 - y^2$

57. $4x^2y^2 + 2xy$

58. $8xy - xy^2$

59. $-2xy^2 + x^2y$

60. $-x^3y - x^2y^2$

The height h, in feet, of a ball shot straight up with an initial velocity of 64 feet per second is given by the polynomial function $h = f(t) = -16t^2 + 64t$. Find the height of the ball after the given number of seconds. SEE EXAMPLE 6. (OBJECTIVE 3)

61. 0 second

62. 3 seconds

63. 2 seconds

64. 4 seconds

 Graph each polynomial function. Check your work with a graphing calculator. SEE EXAMPLE 7. (OBJECTIVE 4)

65. $f(x) = x^2 + 2$

66. $f(x) = x^3 - 2$

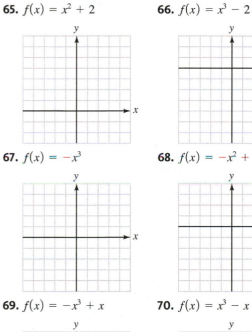

67. $f(x) = -x^3$

68. $f(x) = -x^2 + 1$

69. $f(x) = -x^3 + x$

70. $f(x) = x^3 - x$

71. $f(x) = x^2 - 2x + 1$

72. $f(x) = x^2 - 2x - 3$

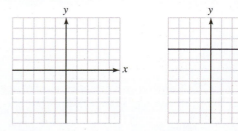

ADDITIONAL PRACTICE *If $P(x) = -3x + 2$, find each value.*

73. $P(0)$

74. $P(3)$

75. $P(-1)$

76. $P(-3)$

Write each polynomial with the exponents on x in descending order. Then state the degree of each polynomial.

77. $ax - ax^3 + 2a^4x^5 - 4a^3x^2$

78. $4x^2y^7 - 3x^5y^2 + 4x^3y^3 - 2x^4y^6 + 5x^6$

Write each polynomial with the exponents on y in ascending order. Then state the degree of each polynomial.

79. $5x^3y^6 + 2x^4y - 5x^3y^3 + x^5y^7 - 2y^4$

80. $-x^3y^2 + x^2y^3 - 2x^3y + x^7y^6 - 3x^6$

 Use a calculator to find each value when $x = 3.7$, $y = -2.5$, and $z = 8.9$.

81. x^2y

82. xyz^2

83. $\dfrac{x^2}{z^2}$

84. $\dfrac{z^3}{y^2}$

85. $\dfrac{x + y + z}{xyz}$

86. $\dfrac{x + yz}{xy + z}$

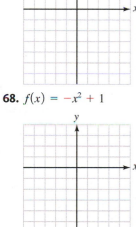 *Use a graphing calculator to graph each polynomial function. Use window settings of $[-4, 6]$ for x and $[-5, 5]$ for y.*

87. $f(x) = 2.75x^2 - 4.7x + 1.5$

88. $f(x) = -2.5x^2 + 1.7x + 3.2$

89. $f(x) = 0.25x^2 - 0.5x - 2.5$

90. $f(x) = 0.37x^2 - 1.4x - 1.5$

APPLICATIONS

91. Juggling During a performance, a juggler tosses one ball straight upward while continuing to juggle three others. The height $f(t)$, in feet, of the ball is given by the polynomial function $f(t) = -16t^2 + 32t + 4$, where t is the time in seconds since the ball was thrown. Find the height of the ball 1 second after it is tossed upward.

f(t)

92. Juggling Find the height of the ball discussed in Exercise 91, 2 seconds after it is tossed upward.

Braking distance *The number of feet that a car travels before stopping depends on the driver's reaction time and the braking distance. See the illustration. For one driver, the stopping distance d is given by the polynomial function* $d(v) = 0.04v^2 + 0.9v$ *where v is the speed of the car. Find the stopping distance for each of the following speeds.*

93. 30 mph   94. 50 mph
95. 60 mph   96. 70 mph

Making gutters *A rectangular sheet of metal will be used to make a rain gutter by bending up its sides, as shown in the illustration. Since a cross section is a rectangle, the cross-sectional area is the product of its length and width, and the capacity c of the gutter is a polynomial function of x:* $f(x) = -2x^2 + 12x$. *Find the capacity for each value of x.*

97. 1 inch   98. 2 inches
99. 3 inches   100. 4 inches

101. **Roller coasters** The polynomial function
$$f(x) = 0.001x^3 - 0.12x^2 + 3.6x + 10$$
models the path of a portion of the track of a roller coaster. Find the height of the track for $x = 0, 20, 40$, and 60 meters.

102. **Customer service** A technical help service of a computer software company has found that on Mondays, the polynomial function $C(t) = -0.0625t^4 + t^3 + 16t$ approximates the number of calls received at any one time. If t represents the time (in hours) since the service opened at 8:00 A.M., how many calls are expected at noon?

WRITING ABOUT MATH

103. Explain how to find the degree of a polynomial.
104. Explain why $P(x) = x^6 + 4x^5 - 3x^2 + x - 2$ defines a function.

SOMETHING TO THINK ABOUT

105. If $P(x) = x^2 - 5x$, is $P(2) + P(3) = P(2 + 3)$?
106. If $P(x) = x^3 - 3x$, is $P(2) - P(3) = P(2 - 3)$?
107. If $P(x) = x^2 - 2x - 3$, find $P(P(0))$.
108. If $P(x) = 2x^2 - x - 5$, find $P(P(-1))$.
109. Graph $f(x) = x^2$, $f(x) = 2x^2$, and $f(x) = 4x^2$. What do you discover?
110. Graph $f(x) = x^2$, $f(x) = \frac{1}{2}x^2$, and $f(x) = \frac{1}{4}x^2$. What do you discover?

Section 5.2
Adding and Subtracting Polynomials

Objectives

1 Combine like terms.
2 Add two or more polynomials.
3 Subtract two polynomials.

Use the distributive property to remove parentheses.

1. $2(x + 3)$ **2.** $-4(2x - 5)$

3. $\frac{5}{2}(2x + 6)$ **4.** $-\frac{2}{3}(3x - 9)$

In this section, we will add and subtract polynomials. To do so, we first will review how to combine like terms.

1 Combine like terms.

Recall that when terms have the same variables with the same exponents, they are called *like* or *similar terms*.

- $3x^2$, $5x^2$, and $7x^2$ are like terms, because they have the same variables with the same exponents.
- $5x^3y^2$, $17x^3y^2$, and $103x^3y^2$ are like terms, because they have the same variables with the same exponents.
- $4x^4y^2$, $12xy^5$, and $98x^7y^9$ are unlike terms. They have the same variables, but with different exponents.
- $3x^4y$ and $5x^4z^2$ are unlike terms. They have different variables.

The distributive property enables us to combine like terms. For example,

Hypatia
A.D. 370–A.D. 415

Hypatia is the earliest known woman in the history of mathematics. She was a professor at the University of Alexandria. Because of her scientific beliefs, she was considered to be a heretic. At the age of 45, she was attacked by a mob and murdered for her beliefs.

$$3x + 7x = (3 + 7)x \qquad \text{Use the distributive property.}$$
$$= 10x \qquad 3 + 7 = 10$$
$$5x^2y^3 + 22x^2y^3 = (5 + 22)x^2y^3 \qquad \text{Use the distributive property.}$$
$$= 27x^2y^3 \qquad 5 + 22 = 27$$
$$9xy^4 + 6xy^4 + xy^4 = 9xy^4 + 6xy^4 + 1xy^4 \qquad xy^4 = 1xy^4$$
$$= (9 + 6 + 1)xy^4 \qquad \text{Use the distributive property.}$$
$$= 16xy^4 \qquad 9 + 6 + 1 = 16$$

The results of the previous examples suggest that to add like terms, *we add their numerical coefficients and keep the same variables with the same exponents.*

COMMENT The terms in the following binomials cannot be combined, because they are not like terms.

$$3x^2 - 5y^2, \quad -2a^2 + 3a^3, \quad \text{and} \quad 5y^2 + 17xy$$

2 Add two or more polynomials.

To add polynomials, we use the distributive property to remove parentheses and combine like terms, whenever possible.

EXAMPLE 1 Add $(3x^2 - 2x + 4)$ and $(2x^2 + 4x - 3)$.

Solution $(3x^2 - 2x + 4) + (2x^2 + 4x - 3)$

$= 3x^2 - 2x + 4 + 2x^2 + 4x - 3$ Use the distributive property to remove parentheses.

$= 3x^2 + 2x^2 - 2x + 4x + 4 - 3$ Use the commutative property of addition to get the terms involving x^2 together, the terms involving x together, and the constants together.

$= 5x^2 + 2x + 1$ Combine like terms.

SELF CHECK 1 Add $(2a^2 - 3a + 5)$ and $(5a^2 - 4a - 2)$.

EXAMPLE 2 Add $(5x^3y^2 + 4x^2y^3)$ and $(-2x^3y^2 + 5x^2y^3)$.

Solution $(5x^3y^2 + 4x^2y^3) + (-2x^3y^2 + 5x^2y^3)$

$= 5x^3y^2 + 4x^2y^3 - 2x^3y^2 + 5x^2y^3$

$= 5x^3y^2 - 2x^3y^2 + 4x^2y^3 + 5x^2y^3$

$= 3x^3y^2 + 9x^2y^3$

SELF CHECK 2 Add $(6a^2b^3 + 5a^3b^2)$ and $(-3a^2b^3 - 2a^3b^2)$.

The additions in Examples 1 and 2 also can be done by aligning the terms vertically.

$$\begin{array}{r} 3x^2 - 2x + 4 \\ + \ 2x^2 + 4x - 3 \\ \hline 5x^2 + 2x + 1 \end{array} \qquad \begin{array}{r} 5x^3y^2 + 4x^2y^3 \\ + \ -2x^3y^2 + 5x^2y^3 \\ \hline 3x^3y^2 + 9x^2y^3 \end{array}$$

3 Subtract two polynomials.

To subtract one monomial from another, we use the relationship $a - b = a + (-b)$. In words, we add the negative (or opposite) of the monomial that is to be subtracted. We will then apply this principle to subtract two polynomials.

EXAMPLE 3 Perform each subtraction.

a. $8x^2 - 3x^2 = 8x^2 + (-3x^2)$ **b.** $3x^2y - 9x^2y = 3x^2y + (-9x^2y)$

$= 5x^2$ $= -6x^2y$

c. $-5x^5y^3z^2 - 3x^5y^3z^2 = -5x^5y^3z^2 + (-3x^5y^3z^2)$

$= -8x^5y^3z^2$

SELF CHECK 3 Perform each subtraction: **a.** $-2a^2b^3 - 5a^2b^3$ **b.** $-2a^2b^3 - (-5a^2b^3)$

To subtract polynomials, we use the distributive property to remove parentheses and combine like terms, whenever possible.

EXAMPLE 4 Perform each subtraction.

a. $(8x^3y + 2x^2y) - (2x^3y - 3x^2y)$

$= (8x^3y + 2x^2y) + \mathbf{-1}(2x^3y - 3x^2y)$ $a - b = a + (-b)$.

$= 8x^3y + 2x^2y - 2x^3y + 3x^2y$ Use the distributive property to remove parentheses.

$= 6x^3y + 5x^2y$ Combine like terms.

b. $(3rt^2 + 4r^2t^2) - (8rt^2 - 4r^2t^2 + r^3t^2)$

$$= (3rt^2 + 4r^2t^2) + \mathbf{-1}(8rt^2 - 4r^2t^2 + r^3t^2) \qquad a - b = a + (-b).$$

$$= 3rt^2 + 4r^2t^2 - 8rt^2 + 4r^2t^2 - r^3t^2 \qquad \text{Use the distributive property to remove parentheses.}$$

$$= -5rt^2 + 8r^2t^2 - r^3t^2 \qquad \text{Combine like terms.}$$

SELF CHECK 4   Perform the subtraction: $(6a^2b^3 - 2a^2b^2) - (-2a^2b^3 + a^2b^2)$

To subtract polynomials in vertical form, we add the negative (or opposite) of the polynomial that is being subtracted.

$$\begin{array}{r} 8x^3y + 2x^2y \\ - \underline{2x^3y - 3x^2y} \end{array} \Rightarrow \begin{array}{r} 8x^3y + 2x^2y \\ + \underline{-2x^3y + 3x^2y} \\ 6x^3y + 5x^2y \end{array}$$

To add or subtract multiples of one polynomial and another, we use the distributive property to remove parentheses and then combine like terms.

EXAMPLE 5   Simplify: $3(2x^2 + 4x - 7) - 2(3x^2 - 4x - 5)$

Solution $3(2x^2 + 4x - 7) - 2(3x^2 - 4x - 5)$

$$= 6x^2 + 12x - 21 - 6x^2 + 8x + 10 \qquad \text{Use the distributive property to remove parentheses.}$$

$$= 20x - 11 \qquad \text{Combine like terms.}$$

SELF CHECK 5   Simplify: $-2(3a^3 + a^2) - 5(2a^3 - a^2 + a)$

SELF CHECK ANSWERS
1. $7a^2 - 7a + 3$ 2. $3a^2b^3 + 3a^3b^2$ 3. **a.** $-7a^2b^3$ **b.** $3a^2b^3$ 4. $8a^2b^3 - 3a^2b^2$
5. $-16a^3 + 3a^2 - 5a$

NOW TRY THIS

Simplify.

1. $x^{2n-1}x^{3n-2}$

2. $\dfrac{x^{2n-1}}{x^{3n-2}}$

5.2   Exercises

WARM-UPS *Simplify.*

1. $2 \cdot 3 + 4 \cdot 5$

2. $5 \cdot 2 - 7 \cdot 3$

3. $4(5 + 5) - 6(1 + 2)$

4. $4(2 + 3) + 7(2 + 4)$

5. $-(6 + 2)$

6. $-(3 - 4)$

REVIEW *Solve each inequality. State the result in interval notation.*

7. $5x + 6 \le -9$

8. $\dfrac{2}{3}x + 5 > 11$

9. $|x - 4| < 5$

10. $|2x + 1| \ge 7$

VOCABULARY AND CONCEPTS *Fill in the blanks.*

11. If two algebraic terms have the same variables each with the same _____, they are called like terms.

12. $-6a^3b^2$ and $7a^2b^3$ are _____ terms.

13. To add two like monomials, we add their numerical _____ and keep the same variables with the same exponents.

14. To subtract one like monomial from another, we add the _____ of the monomial that is to be subtracted.

Determine whether the terms are like or unlike terms. If they are like terms, combine them.

15. $5x, 20x$

16. $-8x, 3y$

17. $7x, 7y$

18. $3mn, 5mn$

19. $3r^2t^3, -8r^2t^3$

20. $3u^2v, -10u^2v$

21. $9x^2y^3, 3x^2y^2$

22. $27x^6y^4z, 9x^6y^4z^2$

GUIDED PRACTICE *Simplify each expression. (OBJECTIVE 1)*

23. $9y + 7y$

24. $-2y + 16y$

25. $5x^3y^2z - 3x^3y^2z$

26. $8wxy - 12wxy$

27. $4xy^3 + 6x^3y - 5xy^3$

28. $3ab^4 - 4a^2b^2 - 2ab^4 + 2a^2b^2$

29. $(3x^2y)^2 + 2x^4y^2 - x^4y^2$

30. $(5x^2y^4)^3 - (5x^3y^6)^2$

Perform each operation. SEE EXAMPLE 1. (OBJECTIVE 2)

31. $(5a + 8) + (2a + 3)$

32. $(5y - 2) + (-2y + 3)$

33. $(3x^2 + 2x + 1) + (-2x^2 - 7x + 5)$

34. $(-2a^2 - 5a - 7) + (-3a^2 + 7a + 1)$

Perform each operation. SEE EXAMPLE 2. (OBJECTIVE 2)

35. $(10ab^2 + 6) + (3ab^2 - 11)$

36. $(-3p^3q^2 - 2pq) + (7p^3q^2 - 3pq)$

37. $(-4a^2b^4 + 2a^2b^2) + (8a^2b^4 - 2a^2b^2)$

38. $(8m^3n^3 - 3mn) + (10m^3n^3 - 2mn)$

Perform each operation. SEE EXAMPLE 3. (OBJECTIVE 3)

39. $14x^2 - 5x^2$

40. $3y^2 - 5y^2$

41. $7x^3y^2 - 4x^3y^2$

42. $-2x^5y^3 - 8x^5y^3$

Perform each operation. SEE EXAMPLE 4. (OBJECTIVE 3)

43. $(5t + 7) - (3t + 6)$

44. $(6z - 5) - (-2z + 1)$

45. $(-a^2 + 2a + 3) - (4a^2 - 2a - 1)$

46. $(4x^2 - 6x + 5) - (2x^2 + x + 7)$

47. $(7pq^2 + 3pq) - (2pq^2 - 2pq)$

48. $(-2a^2b^2 - 2ab) - (5a^2b^2 + 3ab)$

49. $(10xy - 3x^2y^2) - (-3xy - 5x^2y^2)$

50. $(5c^2d + 3cd) - (7c^2d + 5cd)$

Simplify each expression. SEE EXAMPLE 5. (OBJECTIVES 2–3)

51. $7(x + 1) - 4(x + 4)$

52. $-2(x - 4) + 5(x + 1)$

53. $-3(t - 5) - 6(t - 2)$

54. $4(a + 5) - 3(a - 1)$

55. $2(x^3 + x^2) + 3(2x^3 - x^2)$

56. $-3(4x^3 - 2x^2 + 4) - 4(3x^3 + 4x^2 + 3x) + 5(3x - 4)$

57. $5(2a^2 + 4a - 2) - 2(-3a^2 - a + 12) - 2(a^2 + 3a - 5)$

58. $-2(2b^2 - 3b + 3) + 3(3b^2 + 2b - 8) - (3b^2 - b + 4)$

ADDITIONAL PRACTICE *Simplify each expression.*

59. $(x^2 + 2x + 1) + (2x^2 - 2x + 1)$

60. $(x^2 + 2x + 1) - (2x^2 - 2x + 1)$

61. $(2x^2 - x - 3) + (x^2 - 3x - 1)$

62. $(2x^2 - x - 3) - (x^2 - 3x - 1)$

63. $(7y^3 + 4y^2 + y + 3) + (-8y^3 - y + 3)$

64. $(3x^2 + 4x - 3) + (2x^2 - 3x - 1) + (x^2 + x + 7)$

65. $(3x^3 - 2x + 3) + (4x^3 + 3x^2 - 2) + (-4x^3 - 3x^2 + x + 12)$

66. $(x^4 - 3x^2 + 4) + (-2x^4 - x^3 + 3x^2) + (3x^2 + 2x + 1)$

67. $(4x^3 + 2x - 3) - (5x^3 + 2x^2 + 1)$

68. $(-2x^2 + 6x + 5) - (-4x^2 - 7x + 2) - (4x^2 + 10x + 5)$

69. $(3y^2 - 2y + 4) + [(2y^2 - 3y + 2) - (y^2 + 4y + 3)]$

70. $(-t^2 - t - 1) - [(t^2 + 3t - 1) - (-2t^2 + 4)]$

71. $2(y^2 + 5y) - 7(y^2 - 3)$

72. $-3(2m - n) + 2(m - 3n)$

73. $5(p - 2q) - 4(2p + q)$

74. $-5(2x^3 + 7x^2 + 4x) - 2(3x^3 - 4x^2 - 4x)$

75. $-3(3a^2 + 4b^3 + 7) + 4(5a^2 - 2b^3 + 3)$

76. $4(3z^2 - 4z + 5) + 6(-2z^2 - 3z + 4) - 2(4z^2 + 3z - 5)$

Add the polynomials.

77.
$$3x^3 - 2x^2 + 4x - 3$$
$$-2x^3 + 3x^2 + 3x - 2$$
$$\underline{5x^3 - 7x^2 + 7x - 12}$$

78.
$$7a^3 \qquad\; + 3a + 7$$
$$-2a^3 + 4a^2 \qquad\; - 13$$
$$\underline{3a^3 - 3a^2 + 4a + 5}$$

79.
$$-2y^4 - 2y^3 + 4y^2 - 3y + 10$$
$$-3y^4 + 7y^3 - \; y^2 + 14y - 3$$
$$\qquad\quad -3y^3 - 5y^2 - 5y + 7$$
$$\underline{-4y^4 + \; y^3 - 13y^2 + 14y - 2}$$

80.
$$17t^4 + 3t^3 - 2t^2 - 3t + 4$$
$$-12t^4 - 2t^3 + 3t^2 - 5t - 17$$
$$-2t^4 - 7t^3 + 4t^2 + 12t - 5$$
$$\underline{5t^4 + \; t^3 + 5t^2 - 13t + 12}$$

Subtract the bottom polynomial from the top polynomial.

81.
$$\begin{array}{r} 3x^2 - 4x + 17 \\ \underline{2x^2 + 4x - 5} \end{array}$$

82.
$$\begin{array}{r} -2y^2 - 4y + 3 \\ \underline{3y^2 + 10y - 5} \end{array}$$

83.
$$\begin{array}{r} -5y^3 + 4y^2 - 11y + 3 \\ \underline{-2y^3 - 14y^2 + 17y - 32} \end{array}$$

84.
$$\begin{array}{r} 17x^4 - 3x^2 - 65x - 12 \\ \underline{23x^4 + 14x^2 + 3x - 23} \end{array}$$

APPLICATIONS

85. Geometry   Write a polynomial that represents the perimeter of the triangle.

86. Geometry   Write a polynomial that represents the perimeter of the rectangle.

87. Carpentry   The wooden plank is $(2x^2 + 5x + 1)$ meters long. If the shorter end is cut off, write a polynomial to express the length of the remaining piece.

88. Gardening   Write a polynomial that expresses how long the hose will be if the shorter pieces are joined together.

Consider the following information. If a house is purchased for $125,000 and is expected to appreciate $1,100 per year, its value y after x years is given by the polynomial function $f(x) = 1{,}100x + \$125{,}000$.

89. Value of a house   Find the expected value of the house in 10 years.

90. Value of a house   A second house is purchased for $150,000 and is expected to appreciate $1,400 per year.
  a. Find a polynomial function that will give the value y of the house in x years.
  b. Find the value of the second house after 12 years.

91. Value of two houses   Find one polynomial function that will give the combined value y of both houses after x years.

92. Value of two houses   Find the value of the two houses after 20 years by
  a. substituting 20 into the polynomial functions
$$f(x) = 1{,}100x + 125{,}000 \text{ and}$$
$$f(y) = 1{,}400x + 150{,}000$$
  and adding.
  b. substituting 20 into the result of Exercise 91.

Consider the following information. A business bought two cars, one for $16,600 and the other for $19,200. The first car is expected to depreciate $2,100 per year and the second $2,700 per year.

93. Value of a car   Find a polynomial function that will give the value of the first car after x years.

94. Value of a car   Find a polynomial function that will give the value of the second car after x years.

95. Value of two cars   Find one polynomial function that will give the value of both cars after x years.

96. Value of two cars   In two ways, find the total value of the cars after 3 years.

WRITING ABOUT MATH

97. Explain why the terms x^2y and xy^2 are not like terms.
98. Explain how to recognize like terms, and how to add them.

SOMETHING TO THINK ABOUT

99. Find the difference when $3x^2 + 4x - 3$ is subtracted from the sum of $-2x^2 - x + 7$ and $5x^2 + 3x - 1$.
100. Find the difference when $8x^3 + 2x^2 - 1$ is subtracted from the sum of $x^2 + x + 2$ and $2x^3 - x + 9$.
101. Find the sum when $2x^2 - 4x + 3$ minus $8x^2 + 5x - 3$ is added to $-2x^2 + 7x - 4$.
102. Find the sum when $7x^3 - 4x$ minus $x^2 + 2$ is added to $5 + 3x$.

Section 5.3

Multiplying Polynomials

Objectives

1. Multiply two monomials.
2. Multiply a polynomial by a monomial.
3. Multiply a polynomial by a polynomial.
4. Multiply two binomials using the distributive property or the "FOIL method."
5. Square a binomial and multiply two conjugate binomials.
6. Solve an application requiring multiplication of polynomials.
7. Multiply two expressions that are not polynomials.

Vocabulary

conjugate binomials   difference of two squares   perfect square trinomials

Getting Ready

Simplify.

1. $(3a)(4)$

2. $4aaa(a)$

3. $-7a^3 \cdot a$

4. $12a^3a^2$

Use the distributive property to remove parentheses.

5. $2(a + 4)$

6. $a(a - 3)$

7. $-4(a - 3)$

8. $-2b(b + 2)$

We now discuss how to multiply polynomials.

1 Multiply two monomials.

In an earlier section, we saw that to multiply one monomial by another, *we multiply the numerical factors (coefficients) and then multiply the variable factors.*

EXAMPLE 1   Perform each multiplication.

a. $(3x^2)(6x^3)$ **b.** $(-8x)(2y)(xy)$ **c.** $(2a^3b)(-7b^2c)(-12ac^4)$

Solution   We can use the commutative and associative properties of multiplication to rearrange the factors and regroup the numbers.

a. $(3x^2)(6x^3) = 3 \cdot x^2 \cdot 6 \cdot x^3$

$= (3 \cdot 6)(x^2 \cdot x^3)$

$= 18x^5$

b. $(-8x)(2y)(xy) = -8 \cdot x \cdot 2 \cdot y \cdot x \cdot y$

$= (-8 \cdot 2) \cdot x \cdot x \cdot y \cdot y$

$= -16x^2y^2$

c. $(2a^3b)(-7b^2c)(-12ac^4) = 2 \cdot a^3 \cdot b \cdot (-7) \cdot b^2 \cdot c \cdot (-12) \cdot a \cdot c^4$

$= 2(-7)(-12) \cdot a^3 \cdot a \cdot b \cdot b^2 \cdot c \cdot c^4$

$= 168a^4b^3c^5$

SELF CHECK 1 Multiply: **a.** $(-2a^3)(4a^2)$ **b.** $(-5b^3)(-3a)(a^2b)$

2 **Multiply a polynomial by a monomial.**

To multiply a polynomial by a monomial, *we use the distributive property and multiply each term of the polynomial by the monomial.*

EXAMPLE 2 Perform each multiplication. **a.** $3x^2(6xy + 3y^2)$ **b.** $5x^3y^2(xy^3 - 2x^2y)$
c. $-2ab^2(3bz - 2az + 4z^3)$

Solution We can use the distributive property to remove parentheses.

a. $3x^2(6xy + 3y^2) = 3x^2 \cdot 6xy + 3x^2 \cdot 3y^2$

$= 18x^3y + 9x^2y^2$

b. $5x^3y^2(xy^3 - 2x^2y) = 5x^3y^2 \cdot xy^3 + 5x^3y^2 \cdot (-2x^2y)$

$= 5x^4y^5 - 10x^5y^3$

c. $-2ab^2(3bz - 2az + 4z^3) = -2ab^2 \cdot 3bz + (-2ab^2) \cdot (-2az) + (-2ab^2) \cdot 4z^3$

$= -6ab^3z + 4a^2b^2z - 8ab^2z^3$

SELF CHECK 2 Multiply: $-2a^2(a^2 - a + 3)$

3 **Multiply a polynomial by a polynomial.**

To multiply a polynomial by a polynomial, we use the distributive property more than once.

EXAMPLE 3 Perform each multiplication. **a.** $(3x + 2)(4x + 9)$ **b.** $(2a - b)(3a^2 - 4ab + b^2)$

Solution We can use the distributive property twice to remove parentheses.

a. $(3x + 2)(4x + 9) = (3x + 2) \cdot 4x + (3x + 2) \cdot 9$

$= 12x^2 + 8x + 27x + 18$

$= 12x^2 + 35x + 18$ Combine like terms.

b. $(2a - b)(3a^2 - 4ab + b^2)$

$= (2a - b)3a^2 + (2a - b)(-4ab) + (2a - b)b^2$

$= 6a^3 - 3a^2b - 8a^2b + 4ab^2 + 2ab^2 - b^3$

$= 6a^3 - 11a^2b + 6ab^2 - b^3$

SELF CHECK 3 Multiply: **a.** $(3a + 2)(5a - 4)$ **b.** $(2a + b)(3a + 2ab - b)$

The results of Example 3 suggest that to multiply one polynomial by another, *we multiply each term of one polynomial by each term of the other polynomial and combine like terms, when possible.*

In the next example, we organize the work done in Example 3 vertically.

EXAMPLE 4　Perform each multiplication.

a.
$$
\begin{array}{r}
3x + 2 \\
\times\quad 4x + 9 \\
\hline
27x + 18 \qquad \longleftarrow\; 9(3x + 2)\\
12x^2 + 8x \qquad\quad \longleftarrow\; 4x(3x + 2)\\
\hline
12x^2 + 35x + 18 \qquad \text{Add.}
\end{array}
$$

b.
$$
\begin{array}{r}
3a^2 - 4ab + b^2 \\
\times\qquad\qquad 2a - b \\
\hline
-3a^2b + 4ab^2 - b^3 \qquad \longleftarrow\; -b(3a^2 - 4ab + b^2)\\
6a^3 - 8a^2b + 2ab^2 \qquad\quad \longleftarrow\; 2a(3a^2 - 4ab + b^2)\\
\hline
6a^3 - 11a^2b + 6ab^2 - b^3 \qquad \text{Add.}
\end{array}
$$

SELF CHECK 4　Perform each multiplication.

a.
$$
\begin{array}{r}
2x + 3 \\
\times\quad x - 9
\end{array}
$$

b.
$$
\begin{array}{r}
3x^2 + 2x - 5 \\
\times\qquad\quad 2x + 1
\end{array}
$$

4　Multiply two binomials using the distributive property or the "FOIL method."

When multiplying two binomials, the distributive property requires that each term of one binomial be multiplied by each term of the other binomial. This fact can be emphasized by drawing lines to show the indicated products. For example, to multiply $(3x + 2)$ and $(x + 4)$ we can write

First terms　　　Last terms

$$(3x + 2)(x + 4) = 3x \cdot x + 3x \cdot 4 + 2 \cdot x + 2 \cdot 4$$

Inner terms

$$= 3x^2 + 12x + 2x + 8$$

Outer terms

$$= 3x^2 + 14x + 8 \qquad \text{Combine like terms.}$$

COMMENT　It is important to recognize that the acronym FOIL only applies when multiplying two binomials. The distributive property can be applied to any number of terms.

We note that

- the product of the **First** terms is $3x^2$,
- the product of the **Outer** terms is $12x$,
- the product of the **Inner** terms is $2x$, and
- the product of the **Last** terms is 8.

This scheme is called the FOIL method of multiplying two binomials. FOIL is an acronym for **First** terms, **Outer** terms, **Inner** terms, and **Last** terms. Of course, the resulting terms of the products must be combined, if possible.

To multiply binomials by sight, we find the product of the first terms, then find the products of the outer terms and the inner terms and add them (when possible), and then find the product of the last terms.

EXAMPLE 5　Find each product.

a. $(2x - 3)(3x + 2) = 6x^2 - 5x - 6$

The middle term of $-5x$ in the result comes from combining the outer and inner products of $4x$ and $-9x$.

$$4x + (-9x) = -5x$$

b. $(3x + 1)(3x + 4) = 9x^2 + 15x + 4$

The middle term of $+15x$ in the result comes from combining the products $12x$ and $3x$.

$$12x + 3x = 15x$$

c. $(4x - y)(2x + 3y) = 8x^2 + 10xy - 3y^2$

The middle term of $+10xy$ in the result comes from combining the products $12xy$ and $-2xy$.

$$12xy - 2xy = 10xy$$

SELF CHECK 5   Multiply: $(3a + 4b)(2a - b)$

5 Square a binomial and multiply two conjugate binomials.

We can square a binomial by using it as a factor in a multiplication two times.

EXAMPLE 6   Find each square: **a.** $(x + y)^2$ **b.** $(x - y)^2$

Solution   We note that by definition $(x + y)^2 = (x + y)(x + y)$ and find the product $(x + y)(x + y)$ by multiplying each term of the first binomial by each term of the second binomial. Then we combine like terms.

a. $(x + y)^2 = (x + y)(x + y)$

$$\begin{aligned} &= x^2 + xy + xy + y^2 \quad \text{Use the distributive property.} \\ &= x^2 + 2xy + y^2 \quad \text{Combine like terms.} \end{aligned}$$

b. $(x - y)^2 = (x - y)(x - y)$

$$\begin{aligned} &= x^2 - xy - xy + y^2 \quad \text{Use the distributive property.} \\ &= x^2 - 2xy + y^2 \quad \text{Combine like terms.} \end{aligned}$$

SELF CHECK 6   Find each square: **a.** $(a + 2)^2$ **b.** $(a - 4)^2$

We see that the square of the binomial, $(x + y)^2$, is the square of the first term, plus twice the product of the terms, plus the square of the last term. This product can be illustrated graphically as shown in Figure 5-5 on the next page.

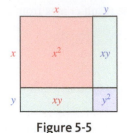

Figure 5-5

The area of the large square is the product of its length and width: $(x + y)(x + y) = (x + y)^2$.

The area of the large square is also the sum of its four pieces: $x^2 + xy + xy + y^2 = x^2 + 2xy + y^2$.

Thus, $(x + y)^2 = x^2 + 2xy + y^2$.

For a geometric interpretation of $(x - y)^2$, see Exercise 125.

Binomials with the same terms, but with opposite signs between their terms, are called **conjugate binomials**. In the next example, we will show that the product of two conjugate binomials is also a binomial.

EXAMPLE 7 Multiply: $(x + y)(x - y)$

Solution

$$(x + y)(x - y) = x^2 - xy + xy - y^2 \qquad \text{Use the distributive property.}$$
$$= x^2 - y^2 \qquad \text{Combine like terms.}$$

SELF CHECK 7 Multiply: $(a + 3)(a - 3)$

From this example, we see that the product of the sum of two quantities and the difference of the same two quantities is the square of the first quantity minus the square of the second quantity. Such a product is called the **difference of two squares**. For a geometric interpretation, see Exercise 126.

The products discussed in Examples 6 and 7 are called *special products*. Since they occur so often, it is useful to recognize their forms.

SPECIAL PRODUCT FORMULAS	$(x + y)^2 = (x + y)(x + y) = x^2 + 2xy + y^2$
	$(x - y)^2 = (x - y)(x - y) = x^2 - 2xy + y^2$
	$(x + y)(x - y) = x^2 - y^2$

Because $x^2 + 2xy + y^2 = (x + y)^2$ and $x^2 - 2xy + y^2 = (x - y)^2$, the two trinomials are called **perfect square trinomials**.

COMMENT The squares $(x + y)^2$ and $(x - y)^2$ have trinomials for their products. Do not forget to write the middle terms in these products.

Also remember that the product $(x + y)(x - y)$ is the binomial $x^2 - y^2$.

At first, the expression $3[x^2 - 2(x + 3)]$ does not look like a polynomial. However, if we remove the parentheses and the brackets, it takes on the form of a polynomial.

$$3[x^2 - 2(x + 3)] = 3[x^2 - 2x - 6]$$
$$= 3x^2 - 6x - 18$$

If an expression has one set of grouping symbols that is enclosed within another set, we always eliminate the inner set first (see Exercise 105).

EXAMPLE 8   Multiply: $(p + q)(2p - q)(p + 2q)$

Solution   First we find the product of $p + q$ and $2p - q$. Then we multiply that product by $p + 2q$.

$$(p + q)(2p - q)(p + 2q) = (2p^2 - pq + 2pq - q^2)(p + 2q)$$

$$= (2p^2 + pq - q^2)(p + 2q) \qquad \text{Combine like terms.}$$

$$= (2p^2 + pq - q^2)(p) + (2p^2 + pq - q^2)(2q) \qquad \text{Distribute the multiplication by } 2p^2 + pq - q^2.$$

$$= 2p^3 + p^2q - pq^2 + 4p^2q + 2pq^2 - 2q^3 \qquad \text{Use the distributive property to remove parentheses.}$$

$$= 2p^3 + 5p^2q + pq^2 - 2q^3 \qquad \text{Combine like terms.}$$

COMMENT   Because of the associative property, we can find the product $(2p - q)(p + 2q)$ first and then multiply the result by $(p + q)$. We will get the same result.

SELF CHECK 8   Multiply: $(x + y)(x + y)(2x + 3y)$

EXAMPLE 9   Multiply: $(2x - 3)(x + 8)(2x + 3)$

Solution   By inspection, we see that we have conjugate binomials, $2x - 3$ and $2x + 3$. We will multiply these first and then multiply by $x + 8$.

$$(2x + 3)(2x - 3)(x + 8) = (4x^2 - 6x + 6x - 9)(x + 8)$$

$$= (4x^2 - 9)(x + 8)$$

$$= 4x^3 + 32x^2 - 9x - 72$$

SELF CHECK 9   Multiply: $(x + y)(x - y)(2x + 3y)$

Perspective

François Vieta (1540–1603) was one of the first to use a mathematical notation that is close to the notation we use today. Trained as a lawyer, Vieta served in the parliament of Brittany and as the personal lawyer of Henry of Navarre. If he had continued as a successful lawyer, Vieta might now be forgotten. However, he lost his job.

When political opposition forced him out of office in 1584, Vieta had time to devote himself entirely to his hobby, mathematics. He studied the work of earlier mathematicians and adapted and improved their ideas. Vieta was the first to use letters to represent unknown numbers, but he did not use modern notation for exponents. To us, his notation seems awkward. For example, what we would write as

$$(x + 1)^3 = x^3 + 3x^2 + 3x + 1$$

Vieta would have written as

$\overline{x + 1}$ cubus aequalis x cubus $+ x$ quad. $3 + x$ in $3 + 1$

---

6   Solve an application requiring multiplication of polynomials.

The profit p earned on the sale of one or more items is given by the formula

$$p = r - c$$

where r is the *revenue* (amount taken in) and c is the *wholesale cost* (amount going out).

If a salesperson has 12 vacuum cleaners and sells them for $225 each, the revenue will be $r = \$(12 \cdot 225) = \$2{,}700$. This illustrates the following formula for finding the revenue r:

$$r = \boxed{\begin{array}{c} \text{number of} \\ \text{items sold } (x) \end{array}} \cdot \boxed{\begin{array}{c} \text{selling price of} \\ \text{each item } (p) \end{array}} = xp = px$$

EXAMPLE 10   **SELLING VACUUM CLEANERS**   A saleswoman has determined that the number of vacuums (x) that she can sell at a price (p) is related by the equation $x = -\frac{2}{25}p + 28$.

a. Find a formula for the revenue r.
b. How much revenue will be taken in if the vacuums are priced at $250?

Solution   a. To find a formula for revenue, we substitute $-\frac{2}{25}p + 28$ for x in the formula $r = px$ and solve for r.

$\quad r = p\boldsymbol{x}$ This is the formula for revenue.

$\quad r = p\left(-\dfrac{2}{25}p + 28\right)$ Substitute $-\frac{2}{25}p + 28$ for x.

$\quad r = -\dfrac{2}{25}p^2 + 28p$ Multiply the polynomials.

b. To find how much revenue will be taken in if the vacuums are priced at $250, we substitute 250 for p in the formula for revenue.

$\quad r = -\dfrac{2}{25}\boldsymbol{p}^2 + 28\boldsymbol{p}$ This is the formula for revenue.

$\quad r = -\dfrac{2}{25}(\mathbf{250})^2 + 28(\mathbf{250})$ Substitute 250 for p.

$\quad r = -5{,}000 + 7{,}000$

$\quad r = 2{,}000$

The revenue will be $2,000.

SELF CHECK 10   How much revenue will be taken in if the vacuums are priced at $325?

7   Multiply two expressions that are not polynomials.

The following examples show that the process of multiplying two polynomials can be applied to expressions that are not polynomials.

EXAMPLE 11   Find the product of $(x^{-2} + y)$ and $(x^2 - y^{-2})$.

Solution   We multiply each term of the second expression by each term of the first expression and then simplify.

$$\begin{aligned}
(x^{-2} + y)(x^2 - y^{-2}) &= x^{-2}x^2 - x^{-2}y^{-2} + yx^2 - yy^{-2} \\
&= x^{-2+2} - x^{-2}y^{-2} + yx^2 - y^{1+(-2)} \\
&= x^0 - \frac{1}{x^2y^2} + x^2y - y^{-1} \quad \text{Recall that } x^{-2} = \frac{1}{x^2} \text{ and } y^{-2} = \frac{1}{y^2}. \\
&= 1 - \frac{1}{x^2y^2} + x^2y - \frac{1}{y} \quad \text{Recall that } x^0 = 1.
\end{aligned}$$

SELF CHECK 11 Multiply: $(a^{-3} + b)(a^2 - b^{-1})$

EXAMPLE 12 Find the product of $(x^n + 2x)$ and $(x^n + 3x^{-n})$.

Solution We multiply each term of the second expression by each term of the first expression and simplify:

$$(x^n + 2x)(x^n + 3x^{-n}) = x^n x^n + x^n(3x^{-n}) + 2x(x^n) + 2x(3x^{-n})$$
$$= x^{n+n} + 3x^{n+(-n)} + 2x^{1+n} + 6x^{1+(-n)}$$
$$= x^{2n} + 3x^0 + 2x^{n+1} + 6x^{1+(-n)}$$
$$= x^{2n} + 3 + 2x^{n+1} + 6x^{1-n}$$

SELF CHECK 12 Multiply: $(a^n + b)(a^n + b^n)$

SELF CHECK ANSWERS

1. a. $-8a^5$ **b.** $15a^3b^4$ **2.** $-2a^4 + 2a^3 - 6a^2$ **3. a.** $15a^2 - 2a - 8$ **b.** $6a^2 + 4a^2b + ab + 2ab^2 - b^2$
4. a. $2x^2 - 15x - 27$ **b.** $6x^3 + 7x^2 - 8x - 5$ **5.** $6a^2 + 5ab - 4b^2$ **6. a.** $a^2 + 4a + 4$ **b.** $a^2 - 8a + 16$
7. $a^2 - 9$ **8.** $2x^3 + 7x^2y + 8xy^2 + 3y^2$ **9.** $2x^3 + 3x^2y - 2xy^2 - 3y^3$ **10.** The revenue will be $650.
11. $\frac{1}{a} - \frac{1}{a^3b} + a^2b - 1$ **12.** $a^{2n} + a^nb^n + a^nb + b^{n+1}$

NOW TRY THIS

1. Given $f(x) = x^2 - 5x + 1$, find $f(x + 2)$.

2. Given the figure of the rectangular solid below, find the volume.

$(2x - 1)$ ft

$(x + 3)$ ft

$(3x + 4)$ ft

3. $[(x + 2) - y][(x + 2) + y]$

5.3 Exercises

WARM-UPS *Determine which property is illustrated by each of the following statements.*

1. $a(x + y) = ax + ay$

2. $4 \cdot x \cdot 3 \cdot x = 4 \cdot 3 \cdot x \cdot x$

3. $(5a^4) a = 5 (a^4 \cdot a)$

4. $3x(x - 2) = 3x \cdot x - 3x \cdot 2$

5. $(x + 3)(x + 1) = (x + 1)(x + 3)$

6. $(x + 4)(x + 5) = (x + 4) \cdot x + (x + 4) \cdot 5$

REVIEW *Let $a = -2$ and $b = 4$ and evaluate each expression.*

7. $|3a - b|$

8. $|ab - b^2|$

9. $-|a^2b - b^0|$

10. $\left| \dfrac{a^3b^2 + ab}{2(ab)^2 - a^3} \right|$

11. Investing A woman owns 200 shares of ABC Company, valued at $125 per share, and 350 shares of WD Company, valued at $75 per share. One day, ABC rose 1.5 points and WD fell 1.5 points. Find the current value of her portfolio.

12. Astronomy One light year is approximately 5,870,000,000,000 miles. Write this number in scientific notation.

VOCABULARY AND CONCEPTS *Fill in the blanks.*

13. To multiply a monomial by a monomial, we multiply the numerical factors and then multiply the _____ factors.

14. To multiply a polynomial by a monomial, we multiply each term of the polynomial by the _____.

15. To multiply a polynomial by a polynomial, we multiply each _____ of one polynomial by each term of the other polynomial.

16. FOIL is an acronym for ____ terms, _____ terms, _____ terms, and ____ terms.

17. $(x + y)^2 = (x + y)(x + y) = $ _____

18. $(x - y)^2 = (x - y)(x - y) = $ _____

19. $(x + y)(x - y) = $ _____. The binomials $x + y$ and $x - y$ are called _____ binomials, and the product $x^2 - y^2$ is called the _____ of two squares.

20. $x^2 + 2xy + y^2$ and $x^2 - 2xy + y^2$ are called _____ trinomials.

GUIDED PRACTICE *Find each product. SEE EXAMPLE 1.*
(OBJECTIVE 1)

21. $(-6a^2b)(4ab^2)$

22. $(-5xy^2)(2xy)$

23. $(-3ab^2c)(5ac^2)$

24. $(-2m^2n)(-4mn^3)$

25. $(4a^2b)(-5a^3b^2)(6a^4)$

26. $(2x^2y^3)(4xy^5)(-5y^6)$

27. $(5x^3y^2)^4\left(\frac{1}{5}x^2\right)^2$

28. $(2ab^2)^3(3a^3b^2)^2$

Find each product. SEE EXAMPLE 2. (OBJECTIVE 2)

29. $7(x + 4)$

30. $-5(a + b)$

31. $-a(a - b)$

32. $y^2(y^2 - 4)$

33. $5x(3x^2 + 4)$

34. $-2x(3x^2 - 2)$

35. $-2x(3x^2 - 3x + 2)$

36. $6a(2a^2 + 5a - 3)$

37. $5a^2b^3(2a^4b - 5b^3)$

38. $-2a^3b(3b^4 - 2a^2b^3)$

39. $7rst(r^2 + s^2 - t^2)$

40. $3x^2yz(x^2 - 2y + 3z^2)$

Find each product. Use either the horizontal or vertical method.
SEE EXAMPLES 3–4. (OBJECTIVE 3)

41. $(x + 4)(x + 2)$

42. $(y + 5)(y - 6)$

43. $(z - 3)(z - 8)$

44. $(x + 3)(x - 5)$

45. $(3y + 1)(2y^2 + 3y + 2)$

46. $(a + 2)(3a^2 + 4a - 2)$

47. $(3x + 4)(2x^2 - 3x + 5)$ **48.** $(4a - 7)(3a^2 + 5a - 2)$

Find each product. SEE EXAMPLE 5. (OBJECTIVE 4)

49. $(2a + 1)(a - 2)$ **50.** $(3b - 1)(2b - 1)$

51. $(2x - 3y)(x + 2y)$ **52.** $(3y + 2z)(y - 3z)$

53. $(3x + y)(3x - 3y)$ **54.** $(2x - y)(3x + 2y)$

55. $(4a - 3b)(2a + 5b)$ **56.** $(3a + 2b)(2a - 7b)$

Find each product. SEE EXAMPLE 6. (OBJECTIVE 5)

57. $(x + 2)^2$ **58.** $(x - 3)^2$

59. $(a - 4)^2$ **60.** $(y + 5)^2$

61. $(2a + b)^2$ **62.** $(a - 2b)^2$

63. $(2x - y)^2$ **64.** $(3m + 4n)^2$

Find each product. SEE EXAMPLE 7. (OBJECTIVE 5)

65. $(x + 2)(x - 2)$ **66.** $(z + 3)(z - 3)$

67. $(a + b)(a - b)$ **68.** $(p + q)(p - q)$

Find each product. SEE EXAMPLE 8. (OBJECTIVE 3)

69. $(2p + q)(p - 2q)^2$
70. $(a - b)(2a + b)^2$
71. $(a + b)^3$
72. $(2m - n)^3$

Find each product. SEE EXAMPLE 9. (OBJECTIVE 3)

73. $(a + b)(a - b)(a - 3b)$
74. $(x - y)(x + 2y)(x - 2y)$
75. $(x + 3)(2x + 1)(x - 3)$
76. $(2x - 3)(2x + 3)(x - 4)$

Find each product. Write all answers without using negative exponents. Assume no division by zero. SEE EXAMPLE 11. (OBJECTIVE 7)

77. $x^3(2x^2 + x^{-2})$
78. $x^{-4}(2x^{-3} - 5x^2)$
79. $x^3y^{-6}z^{-2}(3x^{-2}y^2z - x^3y^{-4})$
80. $ab^{-2}c^{-3}(a^{-4}bc^3 + a^{-3}b^4c^3)$
81. $(x^{-1} + y)(x^{-1} - y)$
82. $(x^{-1} - y)(x^{-1} - y)$
83. $(2x^{-3} + y^3)(2x^3 - y^{-3})$
84. $(5x^{-4} - 4y^2)(5x^2 - 4y^{-4})$

Find each product. SEE EXAMPLE 12. (OBJECTIVE 7)

85. $x^n(x^{2n} - x^n)$ **86.** $(x^n + 1)(x^n - 1)$

87. $a^{2n}(a^n + a^{2n})$ **88.** $(x^n - a^n)(x^n + a^n)$

ADDITIONAL PRACTICE *Simplify each expression. Write any expression with positive exponents only.*

89. $3x(2x + 4) - 3x^2$ **90.** $2y - 3y(y^2 + 4)$

91. $(3y + 1)^2 + (2y - 4)^2$
92. $3(x - 3y)^2 + 2(3x + y)^2$
93. $(2x + 3y)(2x - 3y)$ **94.** $(3a + 4b)(3a - 4b)$

95. $(x - y)(x^2 + xy + y^2)$
96. $(x + y)(x^2 - xy + y^2)$
97. $(-5xx^2)(-3xy)^4$ **98.** $(-2a^2ab^2)^3(-3ab^2b^2)$

99. $4m^2n(-3mn)(m + n)$ **100.** $-3a^2b^3(2b)(3a + b)$

101. $(2x - y)(5x^2 + 3xy + 4y^2)$
102. $(5a + b)(4a^2 - 2ab - b^2)$
103. $(2a - b)(4a^2 + 2ab + b^2)$
104. $(x - 3y)(x^2 + 3xy + 9y^2)$
105. $(2x - 1)[2x^2 - 3(x + 2)]$
106. $(x + 1)^2[x^2 - 2(x + 2)]$
107. $(x + 2)(x^2 - 2x + 4)$
108. $(x - 3)(x^2 + 3x + 9)$
109. $2m(m - n) - (m + n)(m - 2n)$
110. $(2b + 3)(b - 1) - (b + 2)(3b - 1)$
111. $(3x - 4)^2 - (2x + 3)^2$
112. $2(x - y^2)^2 - 3(y^2 + 2x)^2$
113. $5(2y - z)^2 + 4(y + 2z)^2$
114. $3(x + 2z)^2 - 2(2x - z)^2$
115. $(x^n - y^n)(x^n - y^{-n})$ **116.** $(x^n + y^n)(x^n + y^{-n})$

117. $(x^{2n} + y^{2n})(x^{2n} - y^{2n})$ **118.** $(a^{3n} - b^{3n})(a^{3n} + b^{3n})$

119. $(x^n + y^n)(x^n + 1)$ **120.** $(1 - x^n)(x^{-n} - 1)$

Use a calculator to find each product.

121. $(3.21x - 7.85)(2.87x + 4.59)$

122. $(7.44y + 56.7)(-2.1y - 67.3)$

123. $(-17.3y + 4.35)^2$

124. $(-0.31x + 29.3)(-81x - 0.2)$

125. Refer to the illustration and answer the following questions.
   a. Find the area of the large square.
   b. Find the area of square I.
   c. Find the area of rectangle II.
   d. Find the area of rectangle III.
   e. Find the area of square IV.

Use the answers to the preceding questions to show that
$(x - y)^2 = x^2 - 2xy + y^2$.

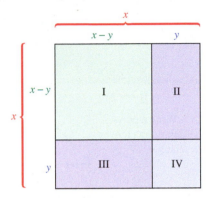

126. Refer to the illustration and answer the following questions.
   a. Find the area of rectangle $ABCD$.
   b. Find the area of rectangle I.
   c. Find the area of rectangle II.

Use the answers to the preceding questions to show that
$(x + y)(x - y) = x^2 - y^2$.

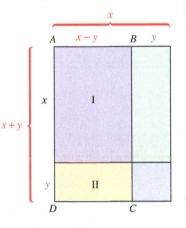

APPLICATIONS *Solve each application.*

Selling TVs *Assume that the number (x) of TVs a salesman can sell at a certain price (p) is given by the equation $x = -\frac{1}{5}p + 90$.* SEE EXAMPLE 10. (OBJECTIVE 6)

127. Find the number of TVs he will sell if the price is $375.
128. If he sold 20 TVs, what was the price?
129. Write a formula for the revenue when x TVs are sold.

130. Find the revenue generated by the sales of the TVs if they are priced at $400 each.
131. **Geometry** Write a polynomial that represents the area of the square garden.

$(2x - 3)$ ft

132. Geometry   Write a polynomial that represents the area of the rectangular field.

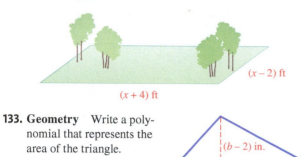

$(x - 2)$ ft

$(x + 4)$ ft

133. Geometry   Write a polynomial that represents the area of the triangle.

$(b - 2)$ in.

$(b + 5)$ in.

134. Geometry   Write a polynomial that represents the volume of the cube.

$(x + 2)$ ft

WRITING ABOUT MATH

135. Explain how to use the distributive property or "FOIL method" to multiply two binomials.

136. Explain how to multiply two trinomials.

SOMETHING TO THINK ABOUT

137. The numbers 0.35×10^7 and 1.96×10^7 both involve the same power of 10. Find their sum.

138. Without converting to standard notation, find the sum: $1.435 \times 10^8 + 2.11 \times 10^7$.   (*Hint*: The first number in the previous exercise is not in scientific notation.)

Section 5.4
The Greatest Common Factor and Factoring by Grouping

Objectives

1. Find the prime-factored form of a natural number.
2. Find the greatest common factor (GCF) of two or more monomials.
3. Factor the greatest common factor (GCF) from a polynomial.
4. Factor a polynomial with four or six terms by grouping.
5. Solve a formula for a specified variable by factoring.

Vocabulary

factoring   greatest common factor (GCF)   prime polynomial
prime-factored form

Getting Ready

Simplify each expression by using the distributive property to remove parentheses.

1. $4(a + 2)$ **2.** $-5(b - 5)$

3. $a(a + 5)$ **4.** $-b(b - 3)$

5. $6a(a + 2b)$ **6.** $-2p^2(3p + 5)$

COMMENT   Use a factoring tree in Example 1:

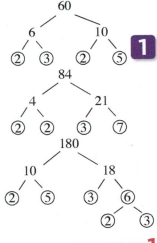

Previously, we used the distributive property to write a product as an addition problem. In this section, we will reverse this process to write an addition problem as a multiplication problem. This reverse process is called **factoring**.

 Find the prime-factored form of a natural number.

If one number a divides a second number b, then a is called a factor of b. For example, because 3 divides 24, it is a factor of 24. Each number in the following list is a factor of 24, because each number divides 24.

$$1, \quad 2, \quad 3, \quad 4, \quad 6, \quad 8, \quad 12, \quad \text{and} \quad 24$$

To factor a natural number means to write it as a product of other natural numbers. If each factor is a prime number, the natural number is said to be written in **prime-factored form**. Example 1 shows how to find the prime-factored forms of 60, 84, and 180, respectively.

EXAMPLE 1   Find the prime factorization of each number.

a. $60 = 6 \cdot 10$
$= 2 \cdot 3 \cdot 2 \cdot 5$
$= 2^2 \cdot 3 \cdot 5$

b. $84 = 4 \cdot 21$
$= 2 \cdot 2 \cdot 3 \cdot 7$
$= 2^2 \cdot 3 \cdot 7$

c. $180 = 10 \cdot 18$
$= 2 \cdot 5 \cdot 3 \cdot 6$
$= 2 \cdot 5 \cdot 3 \cdot 3 \cdot 2$
$= 2^2 \cdot 3^2 \cdot 5$

SELF CHECK 1   Find the prime factorization of 120.

COMMENT   You may begin with any two factors of a number to find the prime factorization. Regardless of the choice of factors, the prime factorization of a number is unique. This is referred to as the Fundamental Theorem of Arithmetic.

2   **Find the greatest common factor (GCF) of two or more monomials.**

The largest natural number that divides 60, 84, and 180 is called the **greatest common factor (GCF)** of the numbers. Because 60, 84, and 180 all have two factors of 2 and one factor of 3, the GCF of these three numbers is $2^2 \cdot 3 = 12$. We note that

$$\frac{60}{12} = 5, \qquad \frac{84}{12} = 7, \qquad \text{and} \qquad \frac{180}{12} = 15$$

There is no natural number greater than 12 that divides 60, 84, and 180 exactly.

Algebraic monomials also can have greatest common factors. The GCF for algebraic monomials is the product of the GCF of the numerical coefficients and the GCF of the variable factors.

EXAMPLE 2   Find the GCF of $6a^2b^3c$, $9a^3b^2c$, and $18a^4c^3$.

Solution   We begin by finding the prime factorization of each monomial coefficient and expand the variables.

$6a^2b^3c = 3 \cdot 2 \cdot a \cdot a \cdot b \cdot b \cdot b \cdot c$
$9a^3b^2c = 3 \cdot 3 \cdot a \cdot a \cdot a \cdot b \cdot b \cdot c$
$18a^4c^3 = 2 \cdot 3 \cdot 3 \cdot a \cdot a \cdot a \cdot a \cdot c \cdot c \cdot c$

Since each monomial has one factor of 3, two factors of a, and one factor of c in common, their GCF is

$$3^1 \cdot a^2 \cdot c^1 = 3a^2c$$

SELF CHECK 2   Find the GCF of $-24x^3y^3$, $3x^3y$, and $18x^2y^2$.

To find the GCF of several monomials, we follow these steps.

FINDING THE GCF

1. Find the prime-factored form of each monomial coefficient and expand the variables, if possible.
2. Identify the prime factors and variable factors that are common to each monomial.
3. Find the product of the factors found in Step 2 with each factor raised to the smallest power that occurs in any one monomial.

3 **Factor the greatest common factor (GCF) from a polynomial.**

We have seen that the distributive property provides a method for multiplying a polynomial by a monomial. For example,

$$2x^3y^3(3x^2 - 4y^3) = 2x^3y^3 \cdot 3x^2 + 2x^3y^3 \cdot (-4y^3)$$
$$= 6x^5y^3 - 8x^3y^6$$

If the product of a multiplication is $6x^5y^3 - 8x^3y^6$, we can use the distributive property to find the individual factors by reversing this process.

$$6x^5y^3 - 8x^3y^6 = 2x^3y^3 \cdot 3x^2 + 2x^3y^3 \cdot (-4y^3)$$
$$= 2x^3y^3(3x^2 - 4y^3)$$

Since $2x^3y^3$ is the GCF of the terms of $6x^5y^3 - 8x^3y^6$, this process is called *factoring out the greatest common factor*.

COMMENT   Although it is true that

$$6x^5y^3 - 8x^3y^6 = 2x^3(3x^2y^3 - 4y^6)$$

the polynomial $6x^5y^3 - 8x^3y^6$ is not completely factored, because y^3 can be factored from $3x^2y^3 - 4y^6$. Whenever the directions state *factor*, be sure to factor it completely.

EXAMPLE 3   Factor: $25a^3b + 15ab^3$

Solution   We begin by finding the prime factorization of each monomial coefficient and expand the variables.

$$25a^3b = 5 \cdot 5 \cdot a \cdot a \cdot a \cdot b$$
$$15ab^3 = 5 \cdot 3 \cdot a \cdot b \cdot b \cdot b$$

Since each term has at least one factor of 5, one factor of a, and one factor of b in common and there are no other common factors, $5ab$ is the GCF of the two terms. We can use the distributive property to factor it out.

$$25a^3b + 15ab^3 = 5ab \cdot 5a^2 + 5ab \cdot 3b^2$$
$$= 5ab(5a^2 + 3b^2)$$

SELF CHECK 3   Factor: $9x^4y^2 - 12x^3y^3$

EXAMPLE 4   Factor: $3xy^2z^3 + 6xyz^3 - 3xz^2$

Solution   We begin by finding the prime factorization of each monomial coefficient and expand the variables.

$$3xy^2z^3 = \mathbf{3} \cdot \mathbf{x} \cdot y \cdot y \cdot \mathbf{z} \cdot \mathbf{z} \cdot z$$
$$6xyz^3 = \mathbf{3} \cdot 2 \cdot \mathbf{x} \cdot y \cdot \mathbf{z} \cdot \mathbf{z} \cdot z$$
$$-3xz^2 = -\mathbf{3} \cdot \mathbf{x} \cdot \mathbf{z} \cdot \mathbf{z}$$

COMMENT   The last term $-3xz^2$ of the given trinomial has a coefficient of -3. When the $3xz^2$ is factored out, remember to write the -1.

Since each term has one factor of 3, one factor of x, and two factors of z in common, and because there are no other common factors, $3xz^2$ is the GCF of the three terms. We can use the distributive property to factor it out.

$$3xy^2z^3 + 6xyz^3 - 3xz^2 = \mathbf{3xz^2} \cdot y^2z + \mathbf{3xz^2} \cdot 2yz + \mathbf{3xz^2} \cdot (-1)$$
$$= \mathbf{3xz^2}(y^2z + 2yz - 1)$$

🌿 **SELF CHECK 4**   Factor: $2a^4b^2 + 6a^3b^2 - 2a^2b$

A polynomial that cannot be factored using only rational numbers is called a **prime polynomial** or an irreducible polynomial over the rational numbers.

EXAMPLE 5   Factor $3x^2 + 4y + 7$, if possible.

Solution   We find the prime factorization of each monomial coefficient and expand the variables.

$$3x^2 = 3 \cdot x \cdot x \qquad 4y = 2 \cdot 2 \cdot y \qquad 7 = 7$$

There are no common factors other than 1. This polynomial is an example of a prime polynomial.

🌿 **SELF CHECK 5**   Factor $6a^3 + 7b^2 + 5$, if possible.

EXAMPLE 6   Factor the negative of the GCF from $-6u^2v^3 + 8u^3v^2$.

Solution   Because the GCF of the two terms is $2u^2v^2$, the negative of the GCF is $-2u^2v^2$. To factor out $-2u^2v^2$, we proceed as follows:

$$-6u^2v^3 + 8u^3v^2 = \mathbf{-2u^2v^2} \cdot 3v + (\mathbf{-2u^2v^2}) \cdot (-4u)$$
$$= \mathbf{-2u^2v^2}(3v - 4u)$$

🌿 **SELF CHECK 6**   Factor the negative of the GCF from $-8a^2b^2 - 12ab^3$.

We also can factor out common factors with variable exponents.

EXAMPLE 7   Factor x^{2n} from $x^{4n} + x^{3n} + x^{2n}$, if n is a natural number.

Solution   We write the trinomial as $x^{2n} \cdot x^{2n} + x^{2n} \cdot x^n + x^{2n} \cdot 1$ and factor out x^{2n}.

$$x^{4n} + x^{3n} + x^{2n} = \mathbf{x^{2n}} \cdot x^{2n} + \mathbf{x^{2n}} \cdot x^n + \mathbf{x^{2n}} \cdot 1$$
$$= \mathbf{x^{2n}}(x^{2n} + x^n + 1)$$

🌿 **SELF CHECK 7**   Factor a^n from $a^{6n} + a^{3n} - a^n$.

A common factor can have more than one term. For example, in the expression

$$x(a + b) + y(a + b)$$

the binomial $(a + b)$ is a factor of both terms. We can factor it out to obtain

$$x(a + b) + y(a + b) = (a + b)x + (a + b)y \qquad \text{Use the commutative property of multiplication.}$$

$$= (a + b)(x + y) \qquad \text{Factor out } a + b.$$

EXAMPLE 8 Factor: $a(x - y + z) - b(x - y + z)$

Solution We can factor out the GCF of $(x - y + z)$.

$$a(x - y + z) - b(x - y + z)$$
$$= (x - y + z)\,a + (x - y + z)(-b)$$
$$= (x - y + z)(a - b)$$

 SELF CHECK 8 Factor: $x(a + b - c) - y(a + b - c)$

4 Factor a polynomial with four or six terms by grouping.

Suppose that we want to factor

$$ac + ad + bc + bd$$

Although no factor is common to all four terms, there is a common factor of a in the first two terms and a common factor of b in the last two terms. We can factor out these common factors to obtain

$$ac + ad + bc + bd = a(c + d) + b(c + d)$$

We can now factor out the common factor of $(c + d)$ on the right side:

$$ac + ad + bc + bd = (c + d)(a + b) \qquad \text{This is factored form.}$$

COMMENT Because of the commutative property of multiplication, $(a + b)(c + d)$ can be written in the equivalent form $(c + d)(a + b)$.

The grouping in this type of problem is not always unique, but the factorization is. For example, if we write the expression $ac + ad + bc + bd$ in the form

$$ac + bc + ad + bd$$

and factor c from the first two terms and d from the last two terms, we obtain

$$ac + bc + ad + bd = c(a + b) + d(a + b)$$
$$= (a + b)(c + d)$$

The method used in the previous explanation is called *factoring by grouping*.

EXAMPLE 9 Factor: $3ax^2 + 3bx^2 + a + 5bx + 5ax + b$

Solution Although no factor is common to all six terms, $3x^2$ can be factored out of the first two terms, and $5x$ can be factored out of the fourth and fifth terms to obtain

$$3ax^2 + 3bx^2 + a + 5bx + 5ax + b = 3x^2(a + b) + a + 5x(b + a) + b$$

Applying the commutative property of addition, this result can be written in the form

$$3ax^2 + 3bx^2 + a + 5bx + 5ax + b = 3x^2(a + b) + 5x(a + b) + (a + b)$$

Since $(a + b)$ is common to all three terms, it can be factored out to obtain

$$3ax^2 + 3bx^2 + a + 5bx + 5ax + b = (a + b)(3x^2 + 5x + 1)$$

 SELF CHECK 9 Factor: $4x^3 + 3x^2 + x + 4x^2y + 3xy + y$

To factor an expression, it is often necessary to factor more than once, as the following example illustrates.

EXAMPLE 10   Factor: $3x^3y - 4x^2y^2 - 6x^2y + 8xy^2$

Solution   We begin by factoring out the common factor of xy.

$$3x^3y - 4x^2y^2 - 6x^2y + 8xy^2 = \boldsymbol{xy}(3x^2 - 4xy - 6x + 8y)$$

We can now factor $3x^2 - 4xy - 6x + 8y$ by grouping:

COMMENT   Whenever you factor an expression, always factor it completely. Each factor of a completely factored expression will be prime over the rational numbers except possibly for a monomial GCF.

$3x^3y - 4x^2y^2 - 6x^2y + 8xy^2$
$= xy(3x^2 - 4xy - 6x + 8y)$
$= xy[x(\boldsymbol{3x - 4y}) - 2(\boldsymbol{3x - 4y})]$ Factor x from $3x^2 - 4xy$ and -2 from $-6x + 8y$.
$= xy(\boldsymbol{3x - 4y})(x - 2)$ Factor out $3x - 4y$, the GCF.

Because each factor is prime, the factorization is complete.

SELF CHECK 10   Factor: $3a^3b + 3a^2b - 2a^2b^2 - 2ab^2$

---

5   Solve a formula for a specified variable by factoring.

Factoring often is required to solve a literal equation for one of its variables.

EXAMPLE 11   **ELECTRONICS**   The formula $r_1r_2 = rr_2 + rr_1$ is used in electronics to relate the combined resistance r of two resistors wired in parallel. The variable r_1 represents the resistance of the first resistor, and the variable r_2 represents the resistance of the second. Solve for r_2.

Solution   To isolate r_2 on one side of the equation, we move all terms involving r_2 on the left side and all terms not involving r_2 on the right side. We then proceed as follows:

$$r_1r_2 = rr_2 + rr_1$$
$$r_1r_2 - rr_2 = rr_1 \qquad \text{Subtract } rr_2 \text{ from both sides.}$$
$$r_2(r_1 - r) = rr_1 \qquad \text{Factor out } r_2 \text{ on the left side.}$$
$$r_2 = \frac{rr_1}{r_1 - r} \qquad \text{Divide both sides by } r_1 - r.$$

SELF CHECK 11   Solve $A = p + prt$ for p.

SELF CHECK ANSWERS

1. $2^3 \cdot 3 \cdot 5$ **2.** $3x^2y$ **3.** $3x^3y^2(3x - 4y)$ **4.** $2a^2b(a^2b + 3ab - 1)$ **5.** a prime polynomial
6. $-4ab^2(2a + 3b)$ **7.** $a^n(a^{5n} + a^{2n} - 1)$ **8.** $(a + b - c)(x - y)$ **9.** $(x + y)(4x^2 + 3x + 1)$
10. $ab(3a - 2b)(a + 1)$ **11.** $p = \frac{A}{1 + rt}$

NOW TRY THIS

Factor out the smallest power of x in each expression.

1. $x^{5n} + x^{2n} - 3x^n$

2. $x^{4/3} + x - 3x^{1/3}$

3. $x^{-2} + x^{-1} - 3x$

Solve for y.

4. $3xy + 2x = 5y - 4$

5.4   Exercises

WARM-UPS *State the number of terms in each expression.*

1. $5x^2 + 3x$
2. $6y^2 - 7y + 1$
3. $x^3 - x^2 + 6x - 8$
4. $a^2b^3 + ab^2 + a^3b^2 - 5a^2b - 5a^2b^2 + b^3$

Identify each number as prime or composite.

5. 13
6. 91

REVIEW *Perform each multiplication.*

7. $(a + 9)(a - 9)$
8. $(3y + 7)(3y - 7)$
9. $(4r^2 + 3s)(4r^2 - 3s)$
10. $(5a + 2b^3)(5a - 2b^3)$
11. $(n - 5)(n^2 + 5n + 25)$
12. $(p - q)(p^2 + pq + q^2)$

VOCABULARY AND CONCEPTS *Fill in the blanks.*

13. The process of finding the individual factors of a known product is called _____.
14. If a natural number is written as the product of prime factors, we say that it is written in _____ form.
15. The abbreviation GCF stands for _____.
16. If a polynomial cannot be factored, it is called a _____ polynomial or an _____ polynomial.

GUIDED PRACTICE *Find the prime-factored form of each number. SEE EXAMPLE 1. (OBJECTIVE 1)*

17. 24
18. 36
19. 135
20. 98
21. 128
22. 357
23. 325
24. 288

Find the GCF of each set of monomials. SEE EXAMPLE 2. (OBJECTIVE 2)

25. 36, 48
26. 45, 75
27. 42, 36, 98
28. 16, 40, 60
29. $4a^2b, 8a^3c$
30. $6x^3y^2z, 9xyz^2$
31. $18x^4y^3z^2, -12xy^2z^3$
32. $6x^2y^3, 24xy^3, 40x^2y^2z^3$

Complete each factorization. (OBJECTIVE 3)

33. $6a - 24 = 6(a - \boxed{})$
34. $5t + 25 = 5(t + \boxed{})$
35. $8z^2 + 2z = 2z(4z + \boxed{})$
36. $9t^3 - 3t^2 = 3t^2(3t - \boxed{})$

Factor each expression. SEE EXAMPLE 3. (OBJECTIVE 3)

37. $2x + 8$
38. $3y - 9$
39. $2x^2 - 6x$
40. $3y^3 + 3y^2$
41. $7x^2 + 14x$
42. $15x^2y - 10x^2y^2$
43. $63x^3y^2 + 81x^2y^4$
44. $28a^3b^2c - 70a^3bc^3$

Factor. SEE EXAMPLE 4. (OBJECTIVE 3)

45. $9a^3b^3 - 15a^2b^2 + 3ab$
46. $15x^2y + 45x^3y^2 - 60x^4y^3$
47. $45x^{10}y^3 - 63x^7y^7 + 81x^{10}y^{10}$
48. $48u^6v^6 - 16u^4v^4 - 3u^6v^3$

Factor each expression, if possible. SEE EXAMPLE 5. (OBJECTIVE 3)

49. $16x^2y^2 + 9ab$
50. $14r^2s^3 + 15t^6$
51. $11m^3n^2 - 12x^2y$
52. $33a^3b^4c - 16xyz$

Factor out the negative of the greatest common factor. SEE EXAMPLE 6. (OBJECTIVE 3)

53. $-5a - 10$
54. $-6b + 12$
55. $-2x^4 + x^3$
56. $-7a^3 - a^2$
57. $-6x^2 - 3xy$
58. $-18y^4 - 27y^2$
59. $-36a^2b^2 + 48ab^3$
60. $-21t^5 + 28t^3$

Factor out the designated factor. Assume no division by 0. SEE EXAMPLE 7. (OBJECTIVE 3)

61. x^{3n} from $x^{9n} + x^{6n} + x^{3n}$
62. y^{2n} from $y^{5n} + y^{4n} - y^{2n}$
63. y^n from $2y^{n+2} - 3y^{n+3}$
64. x^n from $4x^{n+3} - 5x^{n-5}$

Factor each expression. SEE EXAMPLE 8. (OBJECTIVE 3)

65. $4(x + y) + t(x + y)$
66. $5(a - b) - t(a - b)$
67. $(a - b)r - (a - b)s$
68. $(x + y)u + (x + y)v$
69. $(u + v)^2 - (u + v)$
70. $a(x - y) - (x - y)^2$
71. $-a(x + y) + b(x + y)$
72. $-bx(a - b) - cx(a - b)$
73. $(a + b)x - (a + b)y + (a + b)z$
74. $(p - q)a + (p - q)b - (p - q)c$
75. $x(a - b + c) - y(a - b + c) + z(a - b + c)$
76. $-ab(x + y - z) - bc(x + y - z) - ac(x + y - z)$

Factor each expression. SEE EXAMPLE 9. (OBJECTIVE 4)

77. $3c - cd + 3d - c^2$
78. $x^2 + 4y - xy - 4x$
79. $a^2 - 4b + ab - 4a$
80. $7u + v^2 - 7v - uv$

81. $ax + bx - a - b$

82. $x^2y - ax - xy + a$

83. $x^2 + xy + xz + ax + ay + az$

84. $ab - b^2 - bc + ad - bd - cd$

Factor each expression. SEE EXAMPLE 10. (OBJECTIVE 4)

85. $mpx + mqx + npx + nqx$

86. $abd - abe + acd - ace$

87. $x^2y + xy^2 + 2xyz + xy^2 + y^3 + 2y^2z$

88. $a^3 - 2a^2b + a^2c - a^2b + 2ab^2 - abc$

89. $2n^4p - 2n^2 - n^3p^2 + np + 2mn^3p - 2mn$

90. $a^2c^3 + ac^2 + a^3c^2 - 2a^2bc^2 - 2bc^2 + c^3$

91. $2a^3b - 5a^2b^2 - 6a^2b + 15ab^2$

92. $3x^4y + 4x^3y^2 - 6x^3y - 8x^2y^2$

Solve for the indicated variable. SEE EXAMPLE 11. (OBJECTIVE 5)

93. $r_1r_2 = rr_2 + rr_1$ for r_1

94. $r_1r_2 = rr_2 + rr_1$ for r

95. $d_1d_2 = fd_2 + fd_1$ for f

96. $d_1d_2 = fd_2 + fd_1$ for d_1

97. $b^2x^2 + a^2y^2 = a^2b^2$ for a^2

98. $b^2x^2 + a^2y^2 = a^2b^2$ for b^2

99. $S(1 - r) = a - lr$ for r

100. $Sn = (n - 2)180°$ for n

ADDITIONAL PRACTICE *Factor completely. Assume no division by 0.*

101. $27z^3 + 12z^2 + 3z$ **102.** $25t^6 - 10t^3 + 5t^2$

103. $25x^3 - 14y^3 + 36x^3y^3$

104. $9m^4n^3p^2 + 18m^2n^3p^4 - 27m^3n^4p$

105. $-63u^3v^6z^9 + 28u^2v^7z^2 - 21u^3v^3z^4$

106. $-56x^4y^3z^2 - 72x^3y^4z^5 + 80xy^2z^3$

107. $4y^{-2n}$ from $8y^{2n} + 12 + 16y^{-2n}$

108. $7x^{-3n}$ from $21x^{6n} + 7x^{3n} + 14$

109. $3(m + n + p) + x(m + n + p)$

110. $x(x - y - z) + y(x - y - z)$

111. $(x + y)(x + y) + z(x + y)$

112. $8(p - q)^2 + 2(p - q)$

113. $ax + ay + bx + by$

114. $ar + as - br - bs$

115. $x^2 + 2x + xy + 2y$

116. $2c + 2d - cd - d^2$

117. x^{-2} from $x^4 - 5x^6$

118. y^{-4} from $7y^4 + y$

119. $a^{-3}b^{-1}$ from $a^{-3}b - 5a^{-6}b^{-1}$

120. $p^{-4}q^{-2}$ from $4p^3q^{-2} + p^{-7}q^{-5}$

Solve for the indicated variable.

121. $H(a + b) = 2ab$ for a

122. $H(a + b) = 2ab$ for b

123. $3xy - x = 2y + 3$ for y

124. $x(5y + 3) = y - 1$ for y

APPLICATIONS

125. Framing a picture The dimensions of a family portrait and the frame in which it is mounted are given in the illustration. Write an algebraic expression that describes
 a. the area of the picture frame.
 b. the area of the portrait.
 c. the area of the mat used in the framing. Express the result in factored form.

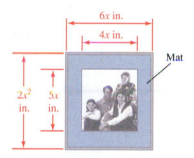

126. Aircraft carriers The illustration shows the deck of the aircraft carrier *Enterprise*. The rectangular landing area of $(x^3 + 4x^2 + 5x + 20)$ ft^2 is shaded. If the dimensions can be determined by factoring the expression, write an expression that represents the width of the landing area.

WRITING ABOUT MATH

127. Explain how to find the greatest common factor of two natural numbers.

128. Explain how to recognize when a number is prime.

SOMETHING TO THINK ABOUT

129. Pick two natural numbers. Divide their product by their greatest common factor. The result is called the **least common multiple** of the two numbers you picked. Why?

130. The number 6 is called a **perfect number**, because the sum of all the divisors of 6 is twice 6: $1 + 2 + 3 + 6 = 12$. Verify that 28 is also a perfect number.

If the greatest common factor of several terms is 1, the terms are called relatively prime. Determine whether the terms in each set are relatively prime.

131. $14, 45$ **132.** $24, 63, 112$

133. $60, 28, 36$ **134.** $55, 49, 78$

135. $12x^2y, 5ab^3, 35x^2b^3$ **136.** $18uv, 25rs, 12rsuv$

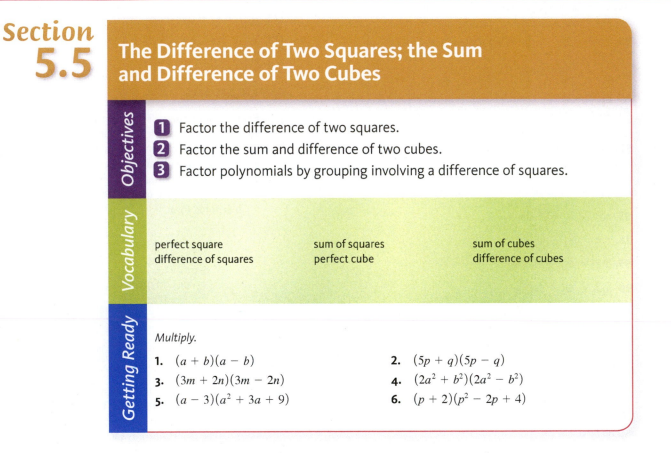

Section 5.5

The Difference of Two Squares; the Sum and Difference of Two Cubes

Objectives

1 Factor the difference of two squares.

2 Factor the sum and difference of two cubes.

3 Factor polynomials by grouping involving a difference of squares.

Vocabulary

perfect square   sum of squares   sum of cubes
difference of squares   perfect cube   difference of cubes

Getting Ready

Multiply.

1. $(a + b)(a - b)$ **2.** $(5p + q)(5p - q)$

3. $(3m + 2n)(3m - 2n)$ **4.** $(2a^2 + b^2)(2a^2 - b^2)$

5. $(a - 3)(a^2 + 3a + 9)$ **6.** $(p + 2)(p^2 - 2p + 4)$

In this section, we will discuss three special types of factoring problems. These types can be factored using the appropriate factoring formula.

1 Factor the difference of two squares.

To factor the difference of two squares, it is helpful to know the first 20 integers that are **perfect squares**.

1, 4, 9, 16, 25, 36, 49, 64, 81, 100, 121, 144, 169, 196, 225, 256, 289, 324, 361, 400

Expressions like $x^6y^4z^2$ are also perfect squares, because they can be written as the square of another quantity:

$$x^6y^4z^2 = (x^3y^2z)^2$$

All exponential expressions that have even-numbered exponents are perfect squares.

In Section 5.3, we developed the special product formula

$$(x + y)(x - y) = x^2 - y^2$$

The binomial $x^2 - y^2$ is called the **difference of two squares**, because x^2 represents the square of x, y^2 represents the square of y, and $x^2 - y^2$ represents the difference of these squares.

The equation $(x + y)(x - y) = (x^2 - y^2)$ can be written in reverse order to give a pattern for factoring the difference of two squares.

FACTORING THE DIFFERENCE OF TWO SQUARES	$x^2 - y^2 = (x + y)(x - y)$

The pattern above shows that the difference of the squares of two quantities factors into a product of two conjugate binomials.

If we think of the difference of two squares as the square of a **F**irst quantity minus the square of a **L**ast quantity, we have the formula

$$F^2 - L^2 = (F + L)(F - L)$$

and we say: *To factor the square of a **F**irst quantity minus the square of a **L**ast quantity, we multiply the **F**irst plus the **L**ast by the **F**irst minus the **L**ast.*

EXAMPLE 1   Factor: $49x^2 - 16$

Solution   We can write $49x^2 - 16$ in the form $(7x)^2 - (4)^2$ and use the pattern for factoring the difference of two squares:

$$\begin{array}{cccccc} \mathbf{F}^2 & - \mathbf{L}^2 = & (\mathbf{F} & + \mathbf{L})(\mathbf{F} & - \mathbf{L}) \\ \downarrow & \downarrow & \downarrow & \downarrow \ \downarrow & \downarrow \end{array}$$ Substitute $7x$ for F and 4 for L.
$$(\mathbf{7x})^2 - \mathbf{4}^2 = (\mathbf{7x + 4})(\mathbf{7x - 4})$$

We can verify this result by multiplication.

$$(7x + 4)(7x - 4) = 49x^2 - 28x + 28x - 16$$
$$= 49x^2 - 16$$

🌿 **SELF CHECK 1**   Factor: $81p^2 - 25$

COMMENT   Expressions such as $x^2 + y^2$ or $(7x)^2 + 4^2$, which represent the **sum of two squares**, cannot be factored in the real-number system unless there are common factors. Thus, the binomials $x^2 + y^2$ and $49x^2 + 16$ are prime binomials.

EXAMPLE 2   Factor: $64a^4 - 25b^2$

Solution   We can write $64a^4 - 25b^2$ in the form $(8a^2)^2 - (5b)^2$ and use the formula for factoring the difference of two squares.

$$(\mathbf{8a^2})^2 - (\mathbf{5b})^2 = (\mathbf{8a^2 + 5b})(\mathbf{8a^2 - 5b})$$

Verify this result by multiplication.

🌿 **SELF CHECK 2**   Factor: $36r^4 - s^2$

EXAMPLE 3 Factor: $x^4 - 1$

Solution Because the binomial is the difference of the squares of x^2 and 1, it factors into the sum of x^2 and 1, and the difference of x^2 and 1.

$$x^4 - 1 = (x^2)^2 - (1)^2$$
$$= (x^2 + 1)(x^2 - 1)$$

The factor $x^2 + 1$ is the sum of two squares and is prime. However, the factor $x^2 - 1$ is the difference of two squares and can be factored as $(x + 1)(x - 1)$. Thus,

$$x^4 - 1 = (x^2 + 1)(x^2 - 1)$$
$$= (x^2 + 1)(x + 1)(x - 1)$$

SELF CHECK 3 Factor: $a^4 - 16$

EXAMPLE 4 Factor: $(x + y)^4 - z^4$

Solution This expression is the difference of two squares that can be factored:

$$(x + y)^4 - z^4 = [(x + y)^2]^2 - (z^2)^2$$
$$= [(x + y)^2 + z^2][(x + y)^2 - z^2]$$

The factor $(x + y)^2 + z^2$ is the sum of two squares and is prime. However, the factor $(x + y)^2 - z^2$ is the difference of two squares and can be factored as $(x + y + z)(x + y - z)$. Thus,

$$(x + y)^4 - z^4 = [(x + y)^2 + z^2][(x + y)^2 - z^2]$$
$$= [(x + y)^2 + z^2](x + y + z)(x + y - z)$$

SELF CHECK 4 Factor: $(a - b)^4 - c^4$

Sometimes we do not know there is a difference of squares until we factor out any common factors. The factoring process is easier when all common factors are factored out first. Be sure to include the GCF as part of the factored form of the answer.

EXAMPLE 5 Factor: $2x^4y - 32y$

Solution
$$2x^4y - 32y = 2y(x^4 - 16) \qquad \text{Factor out } 2y, \text{ the GCF.}$$
$$= 2y(x^2 + 4)(x^2 - 4) \qquad \text{Factor } x^4 - 16, \text{ the difference of two squares.}$$
$$= 2y(x^2 + 4)(x + 2)(x - 2) \qquad \text{Factor } x^2 - 4, \text{ the difference of two squares.}$$

SELF CHECK 5 Factor: $3a^4 - 3$

2 Factor the sum and difference of two cubes.

The number 64 is called a **perfect cube**, because $4^3 = 64$. To factor the sum or difference of two cubes, it is helpful to know the first ten positive perfect cubes:

1, 8, 27, 64, 125, 216, 343, 512, 729, 1,000

Expressions like $x^9y^6z^3$ are also perfect cubes, because they can be written as the cube of another quantity:

$$x^9y^6z^3 = (x^3y^2z)^3$$

To find patterns for factoring the **sum or difference of two cubes**, we use the following product formulas:

$$(1) \qquad (x + y)(x^2 - xy + y^2) = x^3 + y^3$$
$$(2) \qquad (x - y)(x^2 + xy + y^2) = x^3 - y^3$$

To verify Equation 1, we multiply $x^2 - xy + y^2$ by $x + y$ and see that the product is $x^3 + y^3$.

$$(x + y)(x^2 - xy + y^2) = (x + y)x^2 + (x + y)(-xy) + (x + y)y^2$$
$$= x^3 + x^2y - x^2y - xy^2 + xy^2 + y^3$$
$$= x^3 + y^3 \qquad \text{Combine like terms.}$$

Equation 2 also can be verified by multiplication.

If we write Equations 1 and 2 in reverse order, we have the formulas for factoring the sum and difference of two cubes.

FACTORING THE SUM AND DIFFERENCE OF TWO CUBES	$x^3 + y^3 = (x + y)(x^2 - xy + y^2)$ $x^3 - y^3 = (x - y)(x^2 + xy + y^2)$

If we think of the sum of two cubes as the sum of the cube of a **F**irst quantity plus the cube of a **L**ast quantity, we have the relationship

$$F^3 + L^3 = (F + L)(F^2 - FL + L^2)$$

and we say to factor the cube of a First quantity plus the cube of a Last quantity, we multiply the sum of the First and Last by

- *the First squared*
- *minus the First times the Last*
- *plus the Last squared.*

The pattern for factoring the difference of two cubes is

$$F^3 - L^3 = (F - L)(F^2 + FL + L^2)$$

and we say to factor the cube of a First quantity minus the cube of a Last quantity, we multiply the difference of the First and Last by

- *the First squared*
- *plus the First times the Last*
- *plus the Last squared.*

EXAMPLE 6   Factor: $a^3 + 8$

Solution   Since $a^3 + 8$ can be written as $a^3 + 2^3$, we have the sum of two cubes, which factors as follows:

$$F^3 + L^3 = (F + L)(F^2 - F \cdot L + L^2)$$
$$\downarrow \quad \downarrow \qquad \downarrow \quad \downarrow \quad \downarrow \qquad \downarrow \quad \downarrow \qquad \downarrow$$
$$a^3 + 2^3 = (a + 2)(a^2 - a \cdot 2 + 2^2)$$
$$= (a + 2)(a^2 - 2a + 4)$$

Thus, $a^3 + 8 = (a + 2)(a^2 - 2a + 4)$. Check this result by multiplication.

SELF CHECK 6   Factor: $p^3 + 27$

EXAMPLE 7   Factor: $27a^3 - 64b^3$

Solution   We recognize 27 as 3^3 and 64 as 4^3 and both variables also are cubes. Since $27a^3 - 64b^3$ can be written as $(3a)^3 - (4b)^3$, we have the difference of two cubes, which factors as follows:

$$\mathbf{F^3} - \mathbf{L^3} = (\mathbf{F} - \mathbf{L})(\mathbf{F^2} + \mathbf{F}\ \mathbf{L} + \mathbf{L^2})$$
$$(3a)^3 - (4b)^3 = (3a - 4b)[(3a)^2 + (3a)(4b) + (4b)^2]$$
$$= (3a - 4b)(9a^2 + 12ab + 16b^2)$$

Thus, $27a^3 - 64b^3 = (3a - 4b)(9a^2 + 12ab + 16b^2)$. Check this result by multiplication.

SELF CHECK 7   Factor: $8p^3 - 27q^3$

Sometimes we do not know we have a sum or difference of two cubes until we factor out the GCF.

EXAMPLE 8   Factor: $2a^5 + 128a^2$

Solution   This does not fit the pattern of the sum of two cubes; however, the binomial has a common factor of $2a^2$. We factor out the GCF to obtain

$$2a^5 + 128a^2 = 2a^2(a^3 + 64)$$

Then we factor $a^3 + 64$ as the sum of two cubes to obtain

$$2a^5 + 128a^2 = 2a^2(a + 4)(a^2 - 4a + 16)$$

SELF CHECK 8   Factor: $3x^5 + 24x^2$

EXAMPLE 9   Factor: $x^6 - 64$

Solution   This expression can be considered as the difference of two squares or the difference of two cubes. If we treat it as the difference of two squares, it factors into the product of a sum and a difference.

$$x^6 - 64 = (x^3)^2 - 8^2$$
$$= (x^3 + 8)(x^3 - 8)$$

Each of these factors further, however, for one is the sum of two cubes and the other is the difference of two cubes:

$$x^6 - 64 = (x + 2)(x^2 - 2x + 4)(x - 2)(x^2 + 2x + 4)$$

SELF CHECK 9   Factor: $x^6 - 1$

EXAMPLE 10   Factor: $a^3 - (c + d)^3$

We recognize this expression as the difference of two cubes.

Solution $$a^3 - (c + d)^3 = [a - (c + d)][a^2 + a(c + d) + (c + d)^2]$$
$$= (a - c - d)(a^2 + ac + ad + c^2 + 2cd + d^2)$$

SELF CHECK 10   Factor: $(p + q)^3 - r^3$

3 Factor polynomials by grouping involving a difference of squares.

EXAMPLE 11 Factor: $x^2 - y^2 + x - y$

Solution If we group the first two terms and factor the difference of two squares, we have

$$x^2 - y^2 + x - y = (x + y)(x - y) + (x - y) \qquad \text{Factor } x^2 - y^2, \text{ the difference of two squares.}$$

$$= (x - y)(x + y + 1) \qquad \text{Factor out } x - y, \text{ the GCF.}$$

SELF CHECK 11 Factor: $a^2 - b^2 + a + b$

SELF CHECK ANSWERS

1. $(9p + 5)(9p - 5)$ 2. $(6r^2 + s)(6r^2 - s)$ 3. $(a^2 + 4)(a + 2)(a - 2)$
4. $[(a - b)^2 + c^2](a - b + c)(a - b - c)$ 5. $3(a^2 + 1)(a + 1)(a - 1)$ 6. $(p + 3)(p^2 - 3p + 9)$
7. $(2p - 3q)(4p^2 + 6pq + 9q^2)$ 8. $3x^2(x + 2)(x^2 - 2x + 4)$
9. $(x + 1)(x^2 - x + 1)(x - 1)(x^2 + x + 1)$ 10. $(p + q - r)(p^2 + 2pq + q^2 + pr + qr + r^2)$
11. $(a + b)(a - b + 1)$

NOW TRY THIS

Factor.

1. $\frac{1}{9}x^2 - \frac{1}{4}$

2. $x^{2/3} - 1$

3. $x^{3n} - 1$

5.5 Exercises

WARM-UPS *Write any perfect squares as the square of a quantity or perfect cubes as the cube of a quantity.*

1. $36a^4$ 2. $144x^4$
3. y^6 4. $64y^6$
5. $8b^6$ 6. $81m^6$

REVIEW *Perform each multiplication.*

7. $(x + 1)(x + 1)$
8. $(2m - 3)(m - 2)$
9. $(2m + n)(2m + n)$
10. $(3m - 2n)(3m - 2n)$
11. $(a + 4)(a + 3)$
12. $(3b + 2)(2b - 5)$
13. $(4r - 3s)(2r - s)$
14. $(5a - 2b)(3a + 4b)$

VOCABULARY AND CONCEPTS *Fill in the blanks.*

15. Write the first ten perfect-square natural numbers. _____ _____

16. The difference of two squares is factored as
    $p^2 - q^2 = (p + q)$ _____

17. The sum of two squares _____ be factored over the real numbers except for possible common factors.

18. Write the first ten perfect-cube natural numbers. _____ _____

19. The sum of two cubes is factored as $p^3 + q^3 = (p + q)$ _____ .

20. The difference of two cubes is factored as $p^3 - q^3 = (p - q)$ _____ .

GUIDED PRACTICE *The directions "Factor" will always imply factor completely. Factor.* SEE EXAMPLE 1. *(OBJECTIVE 1)*

21. $x^2 - 4$

22. $y^2 - 9$

23. $t^2 - 225$

24. $p^2 - 400$

25. $b^2 - 81$

26. $a^2 - 64$

Factor. SEE EXAMPLE 2. *(OBJECTIVE 1)*

27. $16x^4 - 81y^2$

28. $144a^2 - b^4$

29. $625a^2 - 169b^4$

30. $81a^4 - 49b^2$

Factor. SEE EXAMPLE 3. *(OBJECTIVE 1)*

31. $x^4 - y^4$

32. $81a^4 - 16b^4$

33. $16a^4 - 81b^4$

34. $256x^4y^4 - z^8$

Factor. SEE EXAMPLE 4. *(OBJECTIVE 1)*

35. $(x + y)^2 - z^2$

36. $a^2 - (b - c)^2$

37. $(a - b)^2 - c^2$

38. $(m + n)^2 - p^4$

Factor. SEE EXAMPLE 5. *(OBJECTIVE 1)*

39. $2x^2 - 288$

40. $8x^2 - 72$

41. $2x^3 - 32x$

42. $3x^3 - 243x$

43. $5x^3 - 125x$

44. $6x^4 - 216x^2$

45. $r^2s^2t^2 - t^2x^4y^2$

46. $16a^4b^3c^4 - 64a^2bc^6$

Factor. SEE EXAMPLE 6. *(OBJECTIVE 2)*

47. $x^3 + 27$

48. $y^3 + 1$

49. $p^3 + 64$

50. $a^3 + 125$

Factor. SEE EXAMPLE 7. *(OBJECTIVE 2)*

51. $t^3 - v^3$

52. $p^3 - q^3$

53. $x^3 - 8y^3$

54. $64a^3 - 125b^6$

Factor. SEE EXAMPLE 8. *(OBJECTIVE 2)*

55. $5x^3 + 625$

56. $2x^3 - 128$

57. $4x^5 - 256x^2$

58. $2x^6 + 54x^3$

59. $128u^2v^3 - 2t^3u^2$

60. $56rs^2t^3 + 7rs^2v^6$

61. $(a + b)x^3 + 27(a + b)$

62. $(c - d)r^3 - (c - d)s^3$

Factor. SEE EXAMPLE 9. *(OBJECTIVE 2)*

63. $x^6 - y^6$

64. $64 - a^6$

65. $1 - a^6$

66. $b^6 - a^6$

Factor. SEE EXAMPLE 10. *(OBJECTIVE 2)*

67. $x^3 - (y + z)^3$

68. $a^3 - (b + c)^3$

69. $(a - b)^3 - c^3$

70. $(x + y)^3 - 1$

Factor. SEE EXAMPLE 11. *(OBJECTIVE 3)*

71. $a^2 - b^2 + a + b$

72. $x^2 - y^2 - x - y$

73. $a^2 - b^2 + 2a - 2b$

74. $m^2 - n^2 + 3m + 3n$

ADDITIONAL PRACTICE *Factor. Assume variable exponents are natural numbers.*

75. $x^2 + 25$

76. $y^2 + 49$

77. $4y^4 - 1$

78. $4y^2 + 9z^4$

79. $m^3 + n^3$

80. $27a^3 + b^3$

81. $81x^3 + y^3$

82. $x^9 + y^9$

83. $x^3 + (y - z)^3$

84. $x^3 - (y + 2)^3$

85. $4a^2b^4c^6 - 9d^8$

86. $36x^4y^2 - 49z^4$

87. $8x^6 + 125y^3$

88. $125x^3y^6 + 216z^9$

89. $1{,}000a^6 - 343b^3c^6$

90. $225a^4 - 16b^8c^{12}$

91. $2x + y + 4x^2 - y^2$

92. $m - 2n + m^2 - 4n^2$

93. $x^{3n} - 8$

94. $a^{3m} + 64$

95. $a^{3m} + b^{3n}$

96. $x^{6m} - y^{3n}$

97. $2x^{6m} + 16y^{3m}$

98. $24 + 3c^{3m}$

APPLICATIONS

99. Physics The illustration shows a time-sequence picture of a falling apple. Factor the expression, which gives the difference in the distance fallen by the apple during the time interval from t_1 to t_2 seconds.

This distance is
$0.5gt_1^2 - 0.5gt_2^2$

100. Darts A circular dart board has a series of rings around a solid center, called the bullseye. To find the area of the outer white ring, we can use the formula

$$A = \pi R^2 - \pi r^2$$

Factor the expression on the right side of the equation.

101. Candy To find the amount of chocolate in the outer coating of a malted-milk ball, we can use the formula for the volume V of the chocolate shell:

$$V = \frac{4}{3}\pi r_1^3 - \frac{4}{3}\pi r_2^3$$

Factor the expression on the right side.

102. Movie stunts The distance that a stuntman is above the ground t seconds after he jumps from the top of the building shown in the illustration is given by the formula

$$h = 144 - 16t^2$$

Factor the right side of the equation.

144 ft

WRITING ABOUT MATH

103. Describe the pattern used to factor the difference of two squares.

104. Describe the patterns used to factor the sum and the difference of two cubes.

SOMETHING TO THINK ABOUT

105. Factor $x^{32} - y^{32}$.

106. Find the error in this proof that $2 = 1$.

$$x = y$$
$$x^2 = xy$$
$$x^2 - y^2 = xy - y^2$$
$$(x + y)(x - y) = y(x - y)$$
$$\frac{(x + y)(x - y)}{x - y} = \frac{y(x - y)}{x - y}$$
$$x + y = y$$
$$y + y = y$$
$$2y = y$$
$$\frac{2y}{y} = \frac{y}{y}$$
$$2 = 1$$

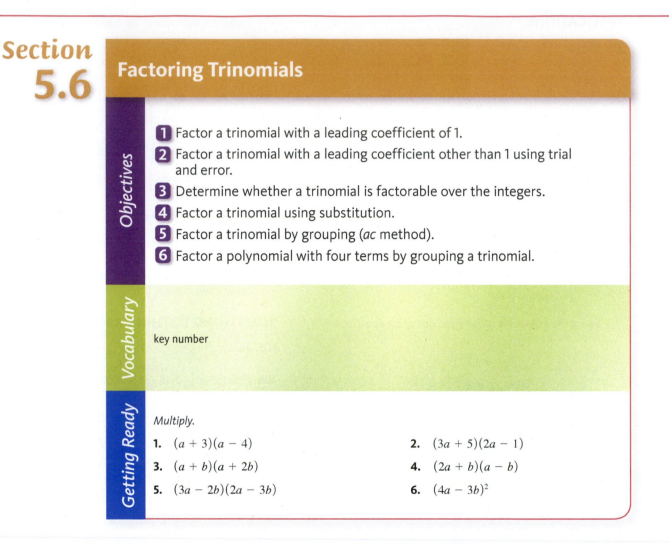

In this section, we will learn how to factor trinomials. We will consider two methods: a trial-and-error method and a factoring-by-grouping (*ac*) method.

1 Factor a trinomial with a leading coefficient of 1.

Recall the product of two binomials is often a trinomial.

$$(x + a)(x + b) = x^2 + bx + ax + ab$$
$$= x^2 + ax + bx + ab$$
$$= x^2 + (a + b)x + ab$$

We will reverse this process to factor a trinomial with a leading coefficient of 1 as the product of two binomials. For example, to factor $x^2 + 7x + 12$, we must find two binomials $(x + a)$ and $(x + b)$ such that

$$x^2 + 7x + 12 = (x + a)(x + b)$$

where $ab = 12$ and $a + b = +7$.

To find the numbers a and b, we list the possible factorizations of $+12$ and find the one where the sum of the factors is $+7$.

COMMENT   A systematic method for finding all the factors could be to create a list of values beginning with 1.

The one to choose
↓

$$\underbrace{1, 12 \qquad 2, 6 \qquad 3, 4}_{+, + \text{ combinations}} \qquad \underbrace{-1, -12 \qquad -2, -6 \qquad -3, -4}_{-, - \text{ combinations}}$$

Thus, $a = 4$, $b = 3$, and

$$x^2 + 7x + 12 = (x + \textbf{\textit{a}})(x + \textbf{\textit{b}})$$
$$x^2 + 7x + 12 = (x + \textbf{4})(x + \textbf{3}) \qquad \text{Substitute 4 for } a \text{ and 3 for } b.$$

This factorization can be verified by multiplying $(x + 4)$ and $(x + 3)$ and observing that the product is $x^2 + 7x + 12$.

Because of the commutative property of multiplication, the order of the factors is not important.

To factor trinomials with leading coefficients of 1, we follow these steps.

FACTORING A TRINOMIAL WITH A LEADING COEFFICIENT OF 1

1. Write the trinomial in descending powers of one variable.
2. List the factorizations of the third term of the trinomial.
3. Select the factorization where the sum of the factors is the coefficient of the middle term.

EXAMPLE 1   Factor: $x^2 - 6x + 8$

Solution   Since the trinomial is written in descending powers of x, we can move to Step 2 and list the possible factorizations of the third term, which is $+8$.

The one to choose
↓

$$\underbrace{1, 8 \qquad 2, 4}_{+, + \text{ combinations}} \qquad \underbrace{-1, -8 \qquad -2, -4}_{-, - \text{ combinations}}$$

In the trinomial, the coefficient of the middle term is -6. The only factorization where the sum of the factors is -6 is $-2(-4)$. Thus, $a = -2$, $b = -4$, and

$$x^2 - 6x + 8 = (x + \textbf{\textit{a}})(x + \textbf{\textit{b}})$$
$$= (x - \textbf{2})(x - \textbf{4}) \qquad \text{Substitute } -2 \text{ for } a \text{ and } -4 \text{ for } b.$$

We can verify this result by multiplication:

$$(x - 2)(x - 4) = x^2 - 4x - 2x + 8$$
$$= x^2 - 6x + 8$$

SELF CHECK 1   Factor: $a^2 - 7a + 12$

EXAMPLE 2   Factor: $x^2 + 6x + 9$

Solution   Since the trinomial is written in descending powers of x, we can move to Step 2 and list the possible factors of the third term, which is $+9$.

The one to choose
↓

$$\underbrace{1, 9 \qquad 3, 3}_{+, + \text{ combinations}} \qquad \underbrace{-1, -9 \qquad -3, -3}_{-, - \text{ combinations}}$$

In the trinomial, the coefficient of the middle term is $+6$. The only factorization where the sum of the factors is $+6$ is $3(3)$. Thus $a = 3$, $b = 3$ and

$$x^2 + 6x + 9 = (x + \boldsymbol{a})(x + \boldsymbol{b})$$
$$= (x + \boldsymbol{3})(x + \boldsymbol{3})$$

which is generally written as $(x + 3)^2$. This result can be verified by multiplication:

$$(x + 3)(x + 3) = x^2 + 3x + 3x + 9$$
$$= x^2 + 6x + 9$$

🌿 **SELF CHECK 2**   Factor: $x^2 + 10x + 25$

Perfect-square trinomials can be factored by using the following special product formulas.

(1) $(x + y)(x + y) = x^2 + 2xy + y^2$
(2) $(x - y)(x - y) = x^2 - 2xy + y^2$

We note that $x^2 + 6x + 9$ can be written in the form $x^2 + 2(3)x + 3^2$. If $y = 3$, this form matches the right side of Equation 1. Thus, $x^2 + 6x + 9$ factors as

$$x^2 + 6x + 9 = x^2 + 2(3)x + 3^2$$
$$= (x + 3)(x + 3)$$

EXAMPLE 3   Factor: $-x + x^2 - 12$

Solution   We begin by writing the trinomial in descending powers of x:

$$-x + x^2 - 12 = x^2 - x - 12$$

The possible factorizations of the third term, which is -12, are

The one to choose
↓

$$1, -12 \quad 2, -6 \quad 3, -4 \quad\quad -1, 12 \quad -2, 6 \quad -3, 4$$

$+, -$ combinations $-, +$ combinations

In the trinomial, the coefficient of the middle term is -1. The only factorization where the sum of the factors is -1 is $3(-4)$. Thus, $a = 3$, $b = -4$, and

$$-x + x^2 - 12 = x^2 - x - 12$$
$$= (x + \boldsymbol{a})(x + \boldsymbol{b})$$
$$= (x + \boldsymbol{3})(x - \boldsymbol{4}) \quad \text{Substitute 3 for } a \text{ and } -4 \text{ for } b.$$

🌿 **SELF CHECK 3**   Factor: $-3a + a^2 - 10$

EXAMPLE 4   Factor: $30x - 4xy - 2xy^2$

Solution   First write the trinomial in descending powers of y.

$$30x - 4xy - 2xy^2 = -2xy^2 - 4xy + 30x$$
$$= \boldsymbol{-2x}(y^2 + 2y - 15) \quad \text{Factor out } -2x, \text{ the GCF.}$$

COMMENT Be sure to include all factors in the final result! Remember that you must be able to multiply all factors and obtain the original polynominal.

To factor $y^2 + 2y - 15$, we list the factors of -15 and find the pair whose sum is 2.

The one to choose
↓

$$1, -15 \quad\quad 3, -5 \quad\quad -1, 15 \quad\quad -3, 5$$

$+, -$ combinations $-, +$ combinations

The only factorization where the sum of the factors is 2 (the coefficient of the middle term of $y^2 + 2y - 15$) is $5(-3)$. Thus, $a = 5$, $b = -3$, and

$$30x - 4xy - 2xy^2 = -2x(y^2 + 2y - 15)$$
$$= -2x(y + 5)(y - 3)$$

SELF CHECK 4   Factor: $18a + 3ab - 3ab^2$

---

2 **Factor a trinomial with a leading coefficient other than 1 using trial and error.**

There are more combinations of factors to consider when factoring trinomials with a leading coefficient other than 1.

EXAMPLE 5   Factor completely: $5x^2 + 7x + 2$

Solution   We must find two binomials of the form $(ax + b)$ and $(cx + d)$ such that

$$5x^2 + 7x + 2 = (ax + b)(cx + d)$$

Since the first term of the trinomial $5x^2 + 7x + 2$ is $5x^2$, the first terms of the binomial factors must be $5x$ and x.

$$5x^2 + 7x + 2 = (5x + b)(x + d)$$

The product of the last terms must be $+2$ and the sum of the products of the outer and inner terms must be $+7x$. We must find two numbers whose product is $+2$ that will give a middle term of $+7x$. However, the factors of 2 must be multiplied by factors of 5 to produce this middle term of $7x$ as illustrated.

$$5x^2 + 7x + 2 = (5x + b)(x + d)$$

$O + I = 7x$

Since $1(2)$ and $-1(-2)$ give a product of 2, there are four possible combinations to consider, because the position of b and d affects the results.

$$(5x + 1)(x + 2) \quad\quad (5x - 1)(x - 2)$$
$$\mathbf{(5x + 2)(x + 1)} \quad\quad (5x - 2)(x - 1)$$

Of these possibilities, only the one printed in red gives the correct middle term of $7x$. Thus,

$$5x^2 + 7x + 2 = (5x + 2)(x + 1)$$

We can verify this result by multiplication:

$$(5x + 2)(x + 1) = 5x^2 + 5x + 2x + 2$$
$$= 5x^2 + 7x + 2$$

SELF CHECK 5   Factor completely: $15x^2 + 22x + 8$

3   Determine whether a trinomial is factorable over the integers.

If a trinomial has the form $ax^2 + bx + c$, with integer coefficients and $a \neq 0$, we can test to see whether it is factorable into two binomials over the integers.

COMMENT   Prior to testing a trinomial for factorability into two binomials with integer coefficients, factor out any GCF.

- If the value of $b^2 - 4ac$ is a perfect square, the trinomial can be factored using only integers.
- If the value is not a perfect square, the trinomial cannot be factored using only integers.

For example, $5x^2 + 7x + 2$ is a trinomial in the form $ax^2 + bx + c$ with

$$a = 5, \quad b = 7, \quad \text{and} \quad c = 2$$

For this trinomial, the value of $b^2 - 4ac$ is

$$b^2 - 4ac = 7^2 - 4(5)(2) \qquad \text{Substitute 7 for } b, 5 \text{ for } a, \text{ and } 2 \text{ for } c.$$
$$= 49 - 40$$
$$= 9$$

Since 9 is a perfect square, the trinomial is factorable. (Its factorization is $(5x + 2)(x + 1)$, as shown in Example 5.)

TEST FOR FACTORABILITY   A trinomial of the form $ax^2 + bx + c$, with integer coefficients and $a \neq 0$, will factor into two binomials with integer coefficients if the value of $b^2 - 4ac$ is a perfect square. If $b^2 - 4ac = 0$, the factors will be the same.

EXAMPLE 6   Determine whether the polynomial $3p^2 - 4p - 4$ is factorable. If it is, factor it.

Solution   In the trinomial, $a = 3$, $b = -4$, and $c = -4$. To see whether it factors, we evaluate $b^2 - 4ac$.

$$b^2 - 4ac = (-4)^2 - 4(3)(-4) \qquad \text{Substitute } -4 \text{ for } b, 3 \text{ for } a, \text{ and } -4 \text{ for } c.$$
$$= 16 + 48$$
$$= 64$$

Since 64 is a perfect square, it is factorable.

To factor the trinomial, we note that the first terms of the binomial factors must be $3p$ and p to give the first term of $3p^2$.

$$3p^2$$
$$3p^2 - 4p - 4 = (3p + ?)(p + ?)$$

The product of the last terms must be -4, and the sum of the products of the outer terms and the inner terms must be $-4p$.

$$-4$$
$$3p^2 - 4p - 4 = (3p + ?)(p + ?)$$
$$O + I = -4p$$

Because $1(-4)$, $-1(4)$, and $2(-2)$ all give a product of -4, there are six possible combinations to consider:

$$(3p + 1)(p - 4) \qquad (3p - 4)(p + 1)$$
$$(3p - 1)(p + 4) \qquad (3p + 4)(p - 1)$$
$$(3p + 2)(p - 2) \qquad (3p - 2)(p + 2)$$

Of these possibilities, only the one printed in red gives the correct middle term of $-4p$. Thus,

$$3p^2 - 4p - 4 = (3p + 2)(p - 2)$$

SELF CHECK 6   Determine whether the polynomial $4q^2 - 9q - 9$ is factorable. If it is, factor it.

EXAMPLE 7   Factor $4y^2 - 3y - 5$, if possible.

Solution   In the trinomial, $a = 4$, $b = -3$, and $c = -5$. To see whether it is factorable, we evaluate $b^2 - 4ac$ by substituting the values of a, b, and c.

$$\begin{aligned} b^2 - 4ac &= (-3)^2 - 4(4)(-5) \\ &= 9 + 80 \\ &= 89 \end{aligned}$$

Since 89 is not a perfect square, the trinomial is not factorable using only integer coefficients. It is prime.

SELF CHECK 7   Factor $5a^2 - 8a + 2$, if possible.

To factor trinomials, the following hints are helpful.

FACTORING A TRINOMIAL

1. Write the trinomial in descending powers of one variable.
2. Factor out any greatest common factor (including -1 if that is necessary to make the coefficient of the first term positive).
3. Test the trinomial for factorability. If it factors, the form will be $(ax + b)(cx + d)$.
4. When the sign of the first term of a trinomial is $+$ and the sign of the third term is $+$, the signs between the terms of each binomial factor are the same as the sign of the middle term of the trinomial.
   When the sign of the first term is $+$ and the sign of the third term is $-$, the signs between the terms of the binomial factors are opposites.
5. Construct a list of binomial factors including all combinations of first and last terms. Multiply each until you find the one that works.
6. Check the factorization by multiplication, including any GCF found in Step 2.

EXAMPLE 8   Factor: $24y + 10xy - 6x^2y$

Solution   We write the trinomial in descending powers of x and factor out the common factor of $-2y$:

$$\begin{aligned} 24y + 10xy - 6x^2y &= -6x^2y + 10xy + 24y \\ &= -2y(3x^2 - 5x - 12) \end{aligned}$$

In the trinomial $3x^2 - 5x - 12$, $a = 3$, $b = -5$, and $c = -12$.

$$\begin{aligned} b^2 - 4ac &= (-5)^2 - 4(3)(-12) \\ &= 25 + 144 \\ &= 169 \end{aligned}$$

Since 169 is a perfect square, the trinomial will factor.

Since the sign of the third term of $3x^2 - 5x - 12$ is $-$, the signs between the binomial factors are opposites. Because the first term is $3x^2$, the first terms of the binomial factors must be $3x$ and x.

$$-2y(3x^2 - 5x - 12) = -2y(3x \qquad)(x \qquad)$$

The product of the last terms must be -12, and the sum of the product of the outer terms and the product of the inner terms must be $-5x$.

$$-2y(3x^2 - 5x - 12) = -2y(3x \quad ?)(x \quad ?)$$
$$O + I = -5x$$

Since $1(-12)$, $2(-6)$, $3(-4)$, $-1(12)$, $-2(6)$, and $-3(4)$ all give a product of -12, there are 12 possible combinations to consider.

Column 1	Column 2
$(3x + 1)(x - 12)$	$(3x - 12)(x + 1)$
$(3x + 2)(x - 6)$	$(3x - 6)(x + 2)$
$(3x + 3)(x - 4)$	$(3x - 4)(x + 3)$
$(3x + 12)(x - 1)$	$(3x - 1)(x + 12)$
$(3x + 6)(x - 2)$	$(3x - 2)(x + 6)$
The one to choose → $(3x + 4)(x - 3)$	$(3x - 3)(x + 4)$

- Column 1 pairs the positive value with $3x$.
- Column 2 pairs the negative value with $3x$.

The combinations marked in blue cannot work, because one of the binomial factors has a common factor. This implies that $3x^2 - 5x - 12$ would have a common factor, which it does not.

After mentally multiplying the remaining combinations, we find that only $(3x + 4)(x - 3)$ gives the correct middle term of $-5x$.

$$24y + 10xy - 6x^2y = -2y(3x^2 - 5x - 12)$$
$$= -2y(3x + 4)(x - 3)$$

Verify this result by multiplication.

SELF CHECK 8   Factor: $9b - 6a^2b - 3ab$

EXAMPLE 9   Factor: $6y + 13x^2y + 6x^4y$

Solution   We write the trinomial in descending powers of x and factor out the common factor of y to obtain

$$6y + 13x^2y + 6x^4y = 6x^4y + 13x^2y + 6y$$
$$= y(6x^4 + 13x^2 + 6)$$

A test of the trinomial $6x^4 + 13x^2 + 6$ will show that it does factor. Since the coefficients of the first and last terms are positive, the signs between the terms in each binomial will be $+$.

The first term of the trinomial is $6x^4$. Thus, the first terms of the binomial factors must be either $2x^2$ and $3x^2$ or x^2 and $6x^2$.

The product of the last terms of the binomial factors must be 6. We must find two numbers whose product is 6 that will lead to a middle term of $13x^2$. Remember these factors of 6 must be paired with the leading factors of $6x$ to produce a middle term of $13x^2$. After trying some combinations, we find the one that works.

$$6y + 13x^2y + 6x^4y = y(6x^4 + 13x^2 + 6)$$
$$= y(2x^2 + 3)(3x^2 + 2)$$

Verify this result by multiplication.

SELF CHECK 9   Factor: $4b + 11a^2b + 6a^4b$

EXAMPLE 10   Factor $x^{2n} + x^n - 2$, where n is a natural number.

Solution   A test of the trinomial $x^{2n} + x^n - 2$ $(a = 1, b = 1, c = -2)$ will show that it does factor. Since the first term is x^{2n}, the first terms of the binomial factors must be x^n and x^n.

$$\overset{x^{2n}}{\overbrace{}}$$
$$x^{2n} + x^n - 2 = (x^n \quad)(x^n \quad)$$

Since the third term of the trinomial is -2, the last terms of the binomial factors must have opposite signs, have a product of -2, and lead to a middle term of x^n. The only combination that works is

$$x^{2n} + x^n - 2 = (x^n + 2)(x^n - 1)$$

Verify this result by multiplication.

SELF CHECK 10   Factor: $a^{2n} + a^n - 6$

4   Factor a trinomial using substitution.

Sometimes factoring a trinomial is easier if we use a substitution.

EXAMPLE 11   Factor: $(x + y)^2 + 7(x + y) + 12$

Solution   If we let $z = (x + y)$, we can write the trinomial $(x + y)^2 + 7(x + y) + 12$ as $z^2 + 7z + 12$. The trinomial $z^2 + 7z + 12$ factors as $(z + 4)(z + 3)$.

COMMENT   Remember to write the final factorization using the variables in the original expression.

To find the factorization of $(x + y)^2 + 7(x + y) + 12$, we substitute $(x + y)$ for z in the expression $(z + 4)(z + 3)$ to obtain

$$z^2 + 7z + 12 = (z + 4)(z + 3)$$
$$(x + y)^2 + 7(x + y) + 12 = (x + y + 4)(x + y + 3)$$

SELF CHECK 11   Factor: $(a + b)^2 - 3(a + b) - 10$

5   Factor a trinomial by grouping (*ac* method).

Factoring by grouping can be used to help factor trinomials of the form $ax^2 + bx + c$. For example, to factor $6x^2 + 7x - 3$ $(a = 6, b = 7, c = -3)$, we proceed as follows:

1. We find the product ac: $6(-3) = -18$. This is called the **key number**.
2. We list the factors of -18.

$$1(-18) \qquad -1(18)$$
$$2(-9) \qquad -2(9)$$
$$3(-6) \qquad -3(6)$$

3. We find the two factors of the key number -18 whose sum is $b = 7$:

$$-2(9) = -18 \qquad \text{and} \qquad 9 + (-2) = 7$$

4. We use the factors 9 and -2 as coefficients of two terms, the sum of which is $+7x$. We replace $+7x$ with $+9x$ and $-2x$.

$$6x^2 + 7x - 3 = 6x^2 + 9x - 2x - 3$$

5. We factor by grouping:

$$6x^2 + 9x - 2x - 3 = 3x(2x + 3) - 1(2x + 3)$$
$$= (2x + 3)(3x - 1) \qquad \text{Factor out } (2x + 3), \text{ the GCF.}$$

We can verify this factorization by multiplication.

EXAMPLE 12   Factor: $10x^2 - 13x - 3$

Solution   Since $a = 10$ and $c = -3$ in the trinomial, $ac = -30$. We list the factors of -30.

$$1, -30 \quad 2, -15 \quad 3, -10 \quad 5, -6 \quad -1, 30 \quad -2, 15 \quad -3, 10 \quad -5, 6$$

We must find two factors of -30 whose sum is -13. The two factors are 2 and -15. We use these factors as coefficients of two terms, the sum of which is $-13x$. We replace $-13x$ with $+2x$ and $-15x$.

$$10x^2 - 13x - 3 = 10x^2 + 2x - 15x - 3$$

Finally, we factor by grouping.

$$10x^2 + 2x - 15x - 3 = 2x(5x + 1) - 3(5x + 1)$$
$$= (5x + 1)(2x - 3) \qquad \text{Factor out } (5x + 1), \text{ the GCF.}$$

Thus, $10x^2 - 13x - 3 = (5x + 1)(2x - 3)$.

SELF CHECK 12   Factor: $15a^2 + 17a - 4$

COMMENT   The box method is the *ac* method presented in a different format.

Another strategy sometimes used for factoring a trinomial is the "box method." To factor $6x^2 + 5x - 4$ using this technique, follow the steps:

First, create a 2×2 box.

In the top left corner, put the first term, $6x^2$.

In the bottom right corner, put the last term, -4.

Multiply these two terms to obtain $-24x^2$. Find two factors of $-24x^2$ that will add to give the middle term of the trinomial, $5x$. These will be $8x$ and $-3x$.

Place these factors in the other two boxes.

Factor the greatest common factor from each row and each column.

	$6x^2$	$-3x$
	$8x$	-4

		$2x$	-1
$3x$		$6x^2$	$-3x$
4		$8x$	-4

The sum of the factors of the rows, $3x + 4$, and the sum of the factors of the columns, $2x - 1$, are the factors of the original trinomial, $6x^2 + 5x - 4 = (3x + 4)(2x - 1)$.

6 Factor a polynomial with four terms by grouping a trinomial.

In the next example, we will group three terms together and factor a trinomial.

EXAMPLE 13   Factor: $x^2 + 6x + 9 - z^2$

Solution   By inspection, the first three terms form a perfect trinomial square. We group the first three terms together and factor the trinomial to obtain

$$x^2 + 6x + 9 - z^2 = (x + 3)(x + 3) - z^2$$
$$= (x + 3)^2 - z^2$$

Notice $(x + 3)^2 - z^2$ is the difference of two squares. We now can factor the difference of two squares to obtain

$$x^2 + 6x + 9 - z^2 = (x + 3 + z)(z + 3 - z)$$

 SELF CHECK 13   Factor: $a^2 + 4a + 4 - b^2$

 SELF CHECK ANSWERS

1. $(a - 4)(a - 3)$ **2.** $(x + 5)^2$ **3.** $(a + 2)(a - 5)$ **4.** $-3a(b + 2)(b - 3)$
5. $(5x + 4)(3x + 2)$ **6.** $(4q + 3)(q - 3)$ **7.** prime **8.** $-3b(2a + 3)(a - 1)$
9. $b(2a^2 + 1)(3a^2 + 4)$ **10.** $(a^n + 3)(a^n - 2)$ **11.** $(a + b + 2)(a + b - 5)$
12. $(3a + 4)(5a - 1)$ **13.** $(a + 2 + b)(a + 2 - b)$

NOW TRY THIS

Factor.

1. $x^2 - x + \frac{1}{4}$

2. $3x^{2/3} - x^{1/3} - 2$

3. $3(\tan x)^2 - \tan x \cos x - 2(\cos x)^2$

5.6   Exercises

WARM-UPS *Find two numbers with the given product and sum.*

1. product of 60; sum of 23

2. product of -60; sum of 11

3. product of -20; sum of -8

4. product of 72; sum of -17

5. product of -30; sum of 13

6. product of 84; sum of -25

REVIEW *Solve each equation.*

7. $\dfrac{2 + x}{11} = 3$

8. $\dfrac{3y - 12}{2} = 9$

9. $\dfrac{2}{3}(5t - 3) = 38$

10. $3(p + 2) = 4p$

11. $11r + 6(3 - r) = 3$ **12.** $2q^2 - 9 = q(q + 3) + q^2$

VOCABULARY AND CONCEPTS *Fill in the blanks.*

13. $x^2 + \underline{\hspace{1.5cm}} = (x + y)(x + y)$
   $x^2 - \underline{\hspace{1.5cm}} = (x - y)(x - y)$

14. $(x + y)(x - y) = \underline{\hspace{1.5cm}}$

15. To factor a trinomial of the form $ax^2 + bx + c$ using the ac method, we must find the product ac. This product is called the _____.

16. If _____ is a perfect square, the trinomial $ax^2 + bx + c$ will factor using _____ coefficients.

Complete each factorization.

17. $x^2 + 5x + 6 = (x + 3)(\underline{\hspace{1cm}})$
18. $x^2 - 6x + 8 = (x - 4)(\underline{\hspace{1cm}})$
19. $x^2 + 2x - 15 = (x + 5)(\underline{\hspace{1cm}})$
20. $x^2 - 3x - 18 = (x - 6)(\underline{\hspace{1cm}})$
21. $2a^2 + 9a + 4 = (\underline{\hspace{1cm}})(a + 4)$
22. $6p^2 - 5p - 4 = (\underline{\hspace{1cm}})(2p + 1)$
23. $4m^2 + 8mn + 3n^2 = (\underline{\hspace{1cm}})(2m + n)$
24. $12r^2 + 5rs - 2s^2 = (\underline{\hspace{1cm}})(3r + 2s)$

GUIDED PRACTICE *The directions "Factor" will always imply factor completely. Factor, if possible.* **SEE EXAMPLE 1.** *(OBJECTIVE 1)*

25. $x^2 + 9x + 8$ **26.** $y^2 + 7y + 6$

27. $x^2 - 7x + 10$ **28.** $c^2 - 7c + 12$

29. $b^2 + 8b + 18$ **30.** $x^2 - 12x + 35$

31. $x^2 - x - 30$ **32.** $a^2 + 4a - 45$

Factor. **SEE EXAMPLE 2.** *(OBJECTIVE 1)*

33. $x^2 + 2x + 1$ **34.** $y^2 - 2y + 1$

35. $a^2 - 18a + 81$ **36.** $b^2 + 12b + 36$

37. $4y^2 + 4y + 1$ **38.** $9x^2 + 6x + 1$

39. $9b^2 - 12b + 4$ **40.** $4a^2 - 12a + 9$

Factor. **SEE EXAMPLE 3.** *(OBJECTIVE 1)*

41. $-4a + a^2 - 32$ **42.** $2x + x^2 + 15$

43. $-5x + x^2 + 6$ **44.** $-3x + x^2 - 4$

Factor out all common monomials first and then factor the remaining trinomial. **SEE EXAMPLE 4.** *(OBJECTIVE 1)*

45. $a^2b^2 - 13ab^2 + 22b^2$
46. $a^2b^2x^2 - 18a^2b^2x + 81a^2b^2$
47. $b^2x^2 - 12bx^2 + 35x^2$
48. $c^3x^2 + 11c^3x - 42c^3$

Factor. **SEE EXAMPLE 5.** *(OBJECTIVE 2)*

49. $6y^2 + 7y + 2$ **50.** $6x^2 - 11x + 3$

51. $6x^2 - 5x - 4$ **52.** $18y^2 - 3y - 10$

Test each trinomial for factorability and factor, if possible. **SEE EXAMPLE 6.** *(OBJECTIVE 3)*

53. $8a^2 + 6a - 9$ **54.** $15b^2 + 4b - 4$

55. $6z^2 + 17z + 12$ **56.** $12x^2 - 7x - 10$

Test each trinomial for factorability and factor, if possible. **SEE EXAMPLE 7.** *(OBJECTIVE 3)*

57. $5x^2 + 4x + 1$ **58.** $6x + 4 + 9x^2$

59. $9x^2 + 12x + 16$ **60.** $2x^2 - 17x + 29$

Test each trinomial for factorability and factor, if possible. **SEE EXAMPLE 8.** *(OBJECTIVE 3)*

61. $-12z - 14yz - 4y^2z$ **62.** $-15y + 70xy - 40x^2y$

63. $-5z + 27y^2z - 10y^4z$ **64.** $-36x^4y - 24x^2y - 16y$

Test each trinomial for factorability and factor each polynomial completely, if possible. **SEE EXAMPLE 9.** *(OBJECTIVE 3)*

65. $x^3 - 60xy^2 + 7x^2y$ **66.** $3x^3 - 10x^2 + 3x$

67. $14x + 9x^3 + x^5$ **68.** $12ax - 45a + 12ax^2$

Factor each expression. Assume any variable exponent is a natural number. **SEE EXAMPLE 10.** *(OBJECTIVE 3)*

69. $x^{2n} + 2x^n + 1$ **70.** $b^{2n} - b^n - 6$

71. $2a^{6n} - 3a^{3n} - 2$ **72.** $x^{4n} - 2x^{2n} + 1$

Factor each expression by substitution. Assume any variable exponent is a natural number. **SEE EXAMPLE 11.** *(OBJECTIVE 4)*

73. $(x + 1)^2 + 2(x + 1) + 1$
74. $(a + b)^2 - 2(a + b) + 1$
75. $(a + b)^2 - 2(a + b) - 24$
76. $(x - y)^2 + 3(x - y) - 10$

Use grouping (ac method) to factor each trinomial. **SEE EXAMPLE 12.** *(OBJECTIVE 5)*

77. $a^2 - 17a + 16$ **78.** $b^2 - 4b - 21$

79. $2u^2 + 5u + 3$ **80.** $6y^2 + 5y - 6$

81. $20r^2 - 7rs - 6s^2$ **82.** $6s^2 + st - 12t^2$

83. $20u^2 + 19uv + 3v^2$ **84.** $12m^2 + mn - 6n^2$

Factor each expression by first grouping a trinomial. **SEE EXAMPLE 13.** *(OBJECTIVE 6)*

85. $x^2 + 4x + 4 - y^2$
86. $x^2 - 6x + 9 - 4y^2$
87. $x^2 + 2x + 1 - 9z^2$
88. $x^2 + 10x + 25 - 16z^2$
89. $c^2 - 4a^2 + 4ab - b^2$

90. $4c^2 - a^2 - 6ab - 9b^2$

91. $a^2 - b^2 + 8a + 16$

92. $a^2 + 14a - 25b^2 + 49$

ADDITIONAL PRACTICE *Factor, if possible.*

93. $a^2 - 3ab - 4b^2$ **94.** $b^2 + 2bc - 80c^2$

95. $9z^2 + 24z + 16$ **96.** $16z^2 - 24z + 9$

97. $a^2 + 5a - 50$ **98.** $b^2 + 9b - 36$

99. $y^2 - 4y - 21$ **100.** $8x^2 - 10x + 3$

101. $x^2 + 4x - 28$ **102.** $4a^2 + 20a + 3$

103. $-3x^2 + 15x - 18$ **104.** $-2y^2 - 16y + 40$

105. $3x^2 + 12x - 63$ **106.** $2y^2 + 4y - 48$

107. $-4x^2 + 4x + 80$ **108.** $-5a^2 + 40a - 75$

109. $-4x^2 - 9 + 12x$ **110.** $5a^2 + 45b^2 - 30ab$

111. $-3a^2 + ab + 2b^2$ **112.** $-2x^2 + 3xy + 5y^2$

113. $2y^2 + yt - 6t^2$ **114.** $3x^2 - 10xy - 8y^2$

115. $3t^3 - 3t^2 + t$ **116.** $8x^2z + 6xyz + 9y^2z$

117. $-90x^2 + 2 - 8x$ **118.** $20x - 4x^2 - 25$

119. $x^{4n} + 2x^{2n}y^{2n} + y^{4n}$ **120.** $y^{6n} + 2y^{3n}z + z^2$

121. $6x^{2n} + 7x^n - 3$ **122.** $12y^{4n} + 10y^{2n} + 2$

123. $21x^4 - 10x^3 - 16x^2$ **124.** $16x^3 - 50x^2 + 36x$

125. $x^4 + 8x^2 + 15$ **126.** $x^4 + 11x^2 + 24$

127. $y^4 - 13y^2 + 30$ **128.** $y^4 - 13y^2 + 42$

129. $a^4 - 13a^2 + 36$

130. $b^4 - 17b^2 + 16$

131. $z^4 - z^2 - 12$

132. $c^4 - 8c^2 - 9$

133. $4x^3 + x^6 + 3$

134. $a^6 - 2 + a^3$

135. $6(x + y)^2 - 7(x + y) - 20$

136. $2(x - z)^2 + 9(x - z) + 4$

137. $4x^2 - z^2 + 4xy + y^2$

138. $x^2 - 4xy - 4z^2 + 4y^2$

APPLICATIONS

139. Ice cubes The surface area of one face of an ice cube is given by $x^2 + 6x + 9$. Write a binomial that expresses the length of an edge of the cube.

140. Checkers The area of a square checkerboard is represented by $25x^2 - 40x + 16$. Write a binomial that expresses the length of one side of the board.

141. Office furniture The area of the desktop is given by the expression $(4x^2 + 20x - 11)$ in.². Factor this expression to find the expressions that can represent its length and width. Then determine the difference in the length and width of the desktop.

142. Storage The volume of the 8-foot-wide portable storage container is given by the expression $(72x^2 + 120x - 400)$ ft³. If its dimensions can be determined by factoring the expression, find the height and the length of the container.

WRITING ABOUT MATH

143. Explain how you would factor -1 from a trinomial.

144. Explain how you would test the polynomial $ax^2 + bx + c$ for factorability.

SOMETHING TO THINK ABOUT

145. Because it is the difference of two squares, $x^2 - q^2$ always factors. Does the test for factorability predict this?

146. The polynomial $ax^2 + ax + a$ factors, because a is a common factor. Does the test for factorability predict this? If not, is there something wrong with the test? Explain.

Section 5.7

Summary of Factoring Techniques

Objectives

1 Factor a polynomial by applying an appropriate technique.

Getting Ready

Multiply.

1. $3pq(p - q)$ **2.** $(2p + 3q)(2p - 3q)$

3. $(p + 6)(p - 1)$ **4.** $(2p - 3q)(3p - 2q)$

5. $2(p - 1)(p + 4)$ **6.** $(p + 2)(p^2 - 2p + 4)$

In this section, we will discuss ways to approach any polynomial that requires factoring.

1 Factor a polynomial by applying an appropriate technique.

Suppose we want to factor the trinomial

$$x^2y^2z^3 + 7xy^2z^3 + 6y^2z^3$$

We begin by checking for any common factors. Because the trinomial has a common monomial factor of y^2z^3, we factor it out:

$$x^2y^2z^3 + 7xy^2z^3 + 6y^2z^3 = \mathbf{\color{red}y^2z^3}(x^2 + 7x + 6)$$

We note that $x^2 + 7x + 6$ is a trinomial that can be factored as $(x + 6)(x + 1)$. Thus,

$$x^2y^2z^3 + 7xy^2z^3 + 6y^2z^3 = y^2z^3(\mathbf{\color{red}x^2 + 7x + 6})$$
$$= y^2z^3(\mathbf{\color{red}x + 6})(\mathbf{\color{red}x + 1})$$

Verify this result by multiplying the factors.

To identify the type of factoring required, we follow these steps.

RECOGNIZING FACTORING TYPES

1. Arrange the polynomial in either descending or ascending order.

2. Factor out the greatest common factor.

3. If an expression has two terms, check to see if the factoring pattern is

    a. the **difference of two squares:** $x^2 - y^2 = (x + y)(x - y)$

    b. the **sum of two cubes:** $x^3 + y^3 = (x + y)(x^2 - xy + y^2)$

    c. the **difference of two cubes:** $x^3 - y^3 = (x - y)(x^2 + xy + y^2)$

4. If an expression has three terms, attempt to factor it as a **trinomial**.

5. If an expression has four or more terms, try factoring by **grouping**.

6. Continue until each individual factor is prime, except possibly a monomial GCF.

7. Check the results by multiplying.

EXAMPLE 1   Factor: $48a^4c^3 - 3b^4c^3$

Solution   We begin by factoring out the common monomial factor of $3c^3$:

$$48a^4c^3 - 3b^4c^3 = \mathbf{3c^3}(16a^4 - b^4)$$

Since the expression $16a^4 - b^4$ has two terms, we check to see whether it is the difference of two squares, which it is. As the difference of two squares, it factors as $(4a^2 + b^2)(4a^2 - b^2)$.

$$48a^4c^3 - 3b^4c^3 = 3c^3(\mathbf{16a^4 - b^4}) \qquad 16a^4 - b^4 = (4a^2)^2 - (b^2)^2$$
$$= 3c^3(\mathbf{4a^2 + b^2})(\mathbf{4a^2 - b^2})$$

The binomial $4a^2 + b^2$ is the sum of two squares and is prime. However, $4a^2 - b^2$ is the difference of two squares and factors as $(2a + b)(2a - b)$.

$$48a^4c^3 - 3b^4c^3 = 3c^3(16a^4 - b^4)$$
$$= 3c^3(4a^2 + b^2)(\mathbf{4a^2 - b^2}) \qquad 4a^2 - b^2 = (2a)^2 - b^2$$
$$= 3c^3(4a^2 + b^2)(\mathbf{2a + b})(\mathbf{2a - b})$$

Since each of the individual factors is prime, the factorization is complete. Verify this result by multiplying the factors.

SELF CHECK 1   Factor: $3p^4r^3 - 3q^4r^3$

EXAMPLE 2   Factor: $x^5y + x^2y^4 - x^3y^3 - y^6$

Solution   We begin by factoring out the common monomial factor of y:

$$x^5y + x^2y^4 - x^3y^3 - y^6 = \mathbf{y}(x^5 + x^2y^3 - x^3y^2 - y^5)$$

Because the expression $x^5 + x^2y^3 - x^3y^2 - y^5$ has four terms, we try factoring by grouping to obtain

$$x^5y + x^2y^4 - x^3y^3 - y^6 = y(\mathbf{x^5 + x^2y^3 - x^3y^2 - y^5}) \qquad \text{Factor out } y, \text{ the GCF.}$$
$$= y[\mathbf{x^2(x^3 + y^3) - y^2(x^3 + y^3)}] \qquad \text{Factor by grouping.}$$
$$= y(\mathbf{x^3 + y^3})(\mathbf{x^2 - y^2}) \qquad \text{Factor out } (x^3 + y^3), \text{ the GCF.}$$

Finally, we factor $x^3 + y^3$ (the sum of two cubes) and $x^2 - y^2$ (the difference of two squares) to obtain

$$x^5y + x^2y^4 - x^3y^3 - y^6 = \mathbf{y(x + y)(x^2 - xy + y^2)(x + y)(x - y)}$$

Because each of the individual factors is prime, the factorization is complete. Verify this result by multiplying the factors.

SELF CHECK 2   Factor: $a^5p - a^3b^2p + a^2b^3p - b^5p$

EXAMPLE 3   Factor: $x^3 + 5x^2 + 6x + x^2y + 5xy + 6y$

Solution   Since there are four or more terms, we try factoring by grouping. We can factor x from the first three terms and y from the last three terms and proceed as follows:

$$x^3 + 5x^2 + 6x + x^2y + 5xy + 6y$$
$$= x(\mathbf{x^2 + 5x + 6}) + y(\mathbf{x^2 + 5x + 6})$$
$$= (\mathbf{x^2 + 5x + 6})(x + y) \qquad \text{Factor out } (x^2 + 5x + 6), \text{ the GCF.}$$
$$= (\mathbf{x + 3})(\mathbf{x + 2})(x + y) \qquad \text{Factor } x^2 + 5x + 6.$$

Verify this result by multiplying the factors.

SELF CHECK 3   Factor: $a^3 - 5a^2 + 6a + a^2b - 5ab + 6b$

NOW TRY THIS

Factor.

1. $64x^2 + 64$ **2.** $64x^2 - 1$

3. $64x^3 - 1$ **4.** $64x^2 + 1$

5.7   Exercises

WARM-UPS *State the number of terms in each expression and identify any common monomial factor other than 1 (GCF).*

1. $3a^2 - 3b^2$ **2.** $x^2 + b^2$

3. $2x^2 + 4x + 4$ **4.** $2x^2 + x + 1$

5. $27a^3 + 64$ **6.** $3x^3 - 24$

REVIEW *Perform the operations.*

7. $(7a^2 + 6a - 3) + (5a^2 - 2a - 4)$

8. $(-4b^2 - 3b - 2) - (3b^2 - 2b + 5)$

9. $5(2y^2 - 3y + 3) - 2(3y^2 - 2y + 6)$

10. $4(3x^2 + 3x + 3) + 3(x^2 - 3x - 4)$

11. $(n + 5)(n - 9)$

12. $(3p + 4q)(2p - 3q)$

VOCABULARY AND CONCEPTS *Fill in the blanks.*

13. In any factoring problem, always factor out the greatest _____ first.

14. If an expression has two terms, check to see whether the problem type is the _____ of two squares, the sum of two _____, or the _____ of two cubes.

15. If an expression has three terms, try to factor it as a _____.

16. If an expression has four or more terms, try factoring it by _____.

PRACTICE *Factor each polynomial completely, if possible.*

17. $x^2 - 10x + 25$

18. $10 - 11x - 6x^2$

19. $64x^3 - 125y^3$

20. $3x^2y + 6xy^2 - 12xy$

21. $xy - ty + xs - ts$

22. $bc + b + cd + d$

23. $36a^2 - 49b^2$

24. $27x^9 - y^3$

25. $12x^2 + 52x + 35$

26. $12x^2 + 14x - 6$

27. $6x^2 - 14x + 8$

28. $12x^2 - 12$

29. $16y^2 + 81$

30. $56x^2 - 15x + 1$

31. $7x^2 - 57x + 8$

32. $4x^2y^2 + 4xy^2 + y^2$

33. $100z^2 - 81t^2$

34. $x^3 + a^6y^3$

35. $4x^2y^2z^2 - 26x^2y^2z^3$

36. $3y^3 - 24$

37. $12x^2 - 9x + 8$

38. $4x^3y^3 + 256$

39. $ae + bf + af + be$

40. $a^2x^2 + b^2y^2 + b^2x^2 + a^2y^2$

41. $5(x + y)^2 + (x + y) - 4$

42. $(x - y)^3 + 125$

43. $625x^4 - 256y^4$

44. $2(a - b)^2 + 5(a - b) + 3$

45. $8x^4 - 6x^3 + 12x^2 + 24x$

46. $36x^4 - 36$

47. $6x^2 - 63 - 13x$

48. $2x^6 + 2y^6$

49. $x^4 - x^4y^4$

50. $a^4 - 13a^2 + 36$

51. $x^4 - 17x^2 + 16$

52. $x^2 + 6x + 9 - y^2$

53. $x^2 + 10x + 25 - y^8$

54. $4x^2 + 4x + 1 - 4y^2$

55. $9x^2 - 6x + 1 - 25y^2$

56. $x^2 - y^2 - 2y - 1$

57. $64x^3 + 81$

58. $a^2 - b^2 + 4b - 4$

59. $x^5 + x^2 - x^3 - 1$

60. $x^5 - x^2 - 4x^3 + 4$

61. $x^5 - 9x^3 + 8x^2 - 72$

62. $x^5 - 4x^3 - 8x^2 + 32$

63. $2x^5z - 2x^2y^3z - 2x^3y^2z + 2y^5z$

64. $x^2y^3 - 4x^2y - 9y^3 + 36y$

65. $x^{2m} - x^m - 6$

66. $a^{2n} - b^{2n}$

67. $a^{3n} - b^{3n}$

68. $x^{3m} + y^{3m}$

69. $x^3 - 3x^2 + 2x - x^2y + 3xy - 2y$

70. $a^3 + 3a^2 + 2a + a^2b + 3ab + 2b$

WRITING ABOUT MATH

71. If you have the choice of factoring a polynomial as the difference of two squares or as the difference of two cubes, which do you do first? Why?

72. Explain how you can know that your factorization is correct.

SOMETHING TO THINK ABOUT

73. Factor: $6x^{-2} - 5x^{-1} - 6$ (*Hint*: Assume no division by zero.)

74. Factor: $x^{-4} - y^{-4}$ (Assume no division by zero.)

75. Factor: $x^4 + x^2 + 1$ (*Hint*: Add and subtract x^2.)

76. Factor: $x^4 + 7x^2 + 16$ (*Hint*: Add and subtract x^2.)

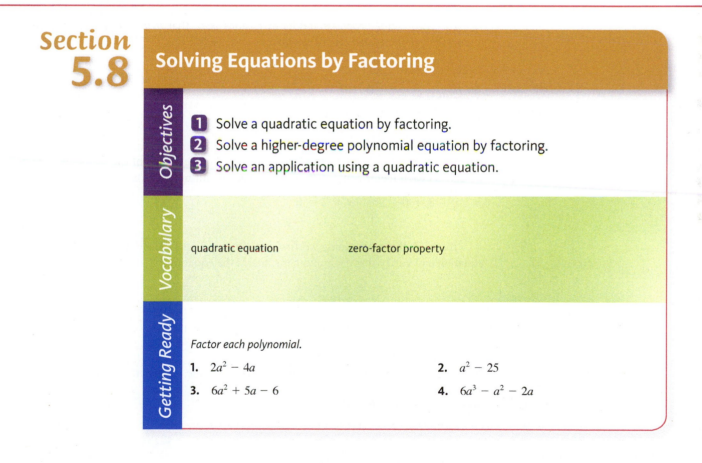

Section 5.8

Solving Equations by Factoring

Objectives

1. Solve a quadratic equation by factoring.
2. Solve a higher-degree polynomial equation by factoring.
3. Solve an application using a quadratic equation.

Vocabulary

quadratic equation   zero-factor property

Getting Ready

Factor each polynomial.

1. $2a^2 - 4a$

2. $a^2 - 25$

3. $6a^2 + 5a - 6$

4. $6a^3 - a^2 - 2a$

In this section, we will learn how to solve quadratic equations by factoring. Then we will use this skill to solve applications.

1   Solve a quadratic equation by factoring.

An equation such as $3x^2 + 4x - 7 = 0$ or $-5y^2 + 3y + 8 = 0$ is called a *quadratic* (or *second-degree*) equation.

QUADRATIC EQUATIONS

A **quadratic equation** is any equation that can be written in the form

$$ax^2 + bx + c = 0$$

where a, b, and c are real numbers and $a \neq 0$.

Many quadratic equations can be solved by factoring and then using the **zero-factor property**.

ZERO-FACTOR PROPERTY

Assume a and b are real numbers.

If $ab = 0$, then $a = 0$ or $b = 0$.

The zero-factor property and its extensions state that *if the product of two or more factors is* 0, *then at least one of the factors must be* 0.

EXAMPLE 1   Solve: $x^2 + 5x + 6 = 0$

Solution   We factor the left side of the equation.

$$(x + 3)(x + 2) = 0$$

Since the product of $(x + 3)$ and $(x + 2)$ is 0, then at least one of the factors is 0. Thus, we can set each factor equal to 0 and solve each resulting linear equation for x:

$$x + 3 = 0 \quad \text{or} \quad x + 2 = 0$$
$$x = -3 \qquad\qquad x = -2$$

To check these solutions, we first substitute -3 and then -2 for x in the original equation and verify that each number satisfies the equation.

$$x^2 + 5x + 6 = 0 \qquad\qquad x^2 + 5x + 6 = 0$$
$$(-3)^2 + 5(-3) + 6 \stackrel{?}{=} 0 \qquad\qquad (-2)^2 + 5(-2) + 6 \stackrel{?}{=} 0$$
$$9 - 15 + 6 \stackrel{?}{=} 0 \qquad\qquad 4 - 10 + 6 \stackrel{?}{=} 0$$
$$0 = 0 \qquad\qquad 0 = 0$$

Both -3 and -2 are solutions, because both satisfy the equation.

SELF CHECK 1   Solve: $x^2 - 5x + 6 = 0$

EXAMPLE 2   Solve: $3x^2 + 6x = 0$

Solution   We factor the left side, set each factor equal to 0, and solve each resulting equation for x.

$$3x^2 + 6x = 0$$
$$3x(x + 2) = 0 \qquad\qquad \text{Factor out the greatest common factor of } 3x.$$
$$3x = 0 \quad \text{or} \quad x + 2 = 0 \qquad \text{Set each factor equal to 0.}$$
$$x = 0 \qquad\qquad x = -2$$

Verify that both solutions check.

SELF CHECK 2   Solve: $4p^2 - 12p = 0$

EXAMPLE 3   Solve: $x^2 - 16 = 0$

Solution   We factor the difference of two squares on the left side, set each factor equal to 0, and solve each resulting equation.

$$x^2 - 16 = 0$$
$$(x + 4)(x - 4) = 0$$
$$x + 4 = 0 \quad \text{or} \quad x - 4 = 0$$
$$x = -4 \quad \quad \quad x = 4$$

Verify that both solutions check.

SELF CHECK 3   Solve: $a^2 - 81 = 0$

Many equations that do not appear to be quadratic can be put into quadratic form $(ax^2 + bx + c = 0)$ and then solved by factoring.

EXAMPLE 4   Solve: $x = \dfrac{6}{5} - \dfrac{6}{5}x^2$

Solution   We write the equation in quadratic form and then solve by factoring.

$$x = \frac{6}{5} - \frac{6}{5}x^2$$

$$5x = 6 - 6x^2 \qquad \text{Multiply both sides by 5 to clear fractions.}$$

$$6x^2 + 5x - 6 = 0 \qquad \text{Add } 6x^2 \text{ to both sides and subtract 6 from both sides.}$$

$$(3x - 2)(2x + 3) = 0 \qquad \text{Factor the trinomial.}$$

$$3x - 2 = 0 \quad \text{or} \quad 2x + 3 = 0 \qquad \text{Set each factor equal to 0.}$$

$$3x = 2 \quad \quad \quad 2x = -3$$

$$x = \frac{2}{3} \quad \quad \quad x = -\frac{3}{2}$$

Verify that both solutions check.

SELF CHECK 4   Solve: $x = \dfrac{6}{7}x^2 - \dfrac{3}{7}$

COMMENT   To solve a quadratic equation by factoring, be sure to set the quadratic polynomial equal to 0 before factoring and using the zero-factor property.

▸Solving Quadratic
Equations

To solve a quadratic equation such as $x^2 + 4x - 5 = 0$ with a TI-84 graphing calculator, we can use standard window settings of $[-10, 10]$ for x and $[-10, 10]$ for y and graph $y = x^2 + 4x - 5$, as shown in Figure 5-6(a).

We can find the solutions by using the ZERO feature found in the CALC menu. We must use the process twice to find the two solutions.

(a)   (b)   (c)

Figure 5-6

For instruction regarding the use of a Casio graphing calculator, please refer to the Casio Keystroke Guide in the back of the book.

2   Solve a higher-degree polynomial equation by factoring.

We can solve many polynomial equations with degrees greater than 2 by factoring.

EXAMPLE 5   Solve: $6x^3 - x^2 - 2x = 0$

Solution   We factor x from the polynomial on the left side and proceed as follows:

$$6x^3 - x^2 - 2x = 0$$
$$x(6x^2 - x - 2) = 0 \qquad \text{Factor out } x, \text{ the GCF.}$$
$$x(3x - 2)(2x + 1) = 0 \qquad \text{Factor } 6x^2 - x - 2.$$
$$x = 0 \quad \text{or} \quad 3x - 2 = 0 \quad \text{or} \quad 2x + 1 = 0 \qquad \text{Set each factor equal to 0.}$$
$$x = \frac{2}{3} \qquad \qquad x = -\frac{1}{2}$$

Verify that each of the three solutions checks.

SELF CHECK 5   Solve: $5x^3 + 13x^2 - 6x = 0$

EXAMPLE 6   Solve: $x^4 - 5x^2 + 4 = 0$

Solution   We factor the trinomial on the left side and proceed as follows:

$$x^4 - 5x^2 + 4 = 0$$
$$(x^2 - 1)(x^2 - 4) = 0$$
$$(x + 1)(x - 1)(x + 2)(x - 2) = 0 \qquad \text{Factor } x^2 - 1 \text{ and } x^2 - 4.$$
$$x + 1 = 0 \quad \text{or} \quad x - 1 = 0 \quad \text{or} \quad x + 2 = 0 \quad \text{or} \quad x - 2 = 0$$
$$x = -1 \qquad x = 1 \qquad x = -2 \qquad x = 2$$

Verify that each of the four solutions checks.

SELF CHECK 6   Solve: $a^4 - 13a^2 + 36 = 0$

EXAMPLE 7 Given that $f(x) = 6x^3 + x^2 - 2x$, find all x such that $f(x) = 0$.

Solution We first set $f(x)$ equal to 0.

$$f(x) = 6x^3 + x^2 - 2x$$
$$0 = 6x^3 + x^2 - 2x \qquad \text{Substitute 0 for } f(x).$$

Then we solve for x.

$$6x^3 + x^2 - 2x = 0$$
$$x(6x^2 + x - 2) = 0 \qquad \text{Factor out } x, \text{ the GCF.}$$
$$x(2x - 1)(3x + 2) = 0 \qquad \text{Factor } 6x^2 + x - 2.$$
$$x = 0 \quad \text{or} \quad 2x - 1 = 0 \quad \text{or} \quad 3x + 2 = 0 \qquad \text{Set each factor equal to 0.}$$
$$x = \frac{1}{2} \qquad x = -\frac{2}{3}$$

Verify that $f(0) = 0$, $f\left(\frac{1}{2}\right) = 0$, and $f\left(-\frac{2}{3}\right) = 0$.

SELF CHECK 7 Given that $f(x) = x^4 - 4x^2$, find all x such that $f(x) = 0$.

Accent on technology

▶Solving Equations

To solve the equation $x^4 - 5x^2 + 4 = 0$ with a TI-84 graphing calculator, we can use window settings of $[-6, 6]$ for x and $[-5, 10]$ for y and graph $y = x^4 - 5x^2 + 4$ as shown in Figure 5-7. Then we can identify the values of x that make $y = 0$. They are $x = -2, -1, 1$, and 2. If the x-coordinates of the x-intercepts were not obvious, we could obtain their values by using the ZERO or INTERSECT feature.

$y = x^4 - 5x^2 + 4$

Figure 5-7

For instruction regarding the use of a Casio graphing calculator, please refer to the Casio Keystroke Guide in the back of the book.

---

3 Solve an application using a quadratic equation.

EXAMPLE 8 **FINDING THE DIMENSIONS OF A TRUSS** The width of the truss shown in Figure 5-8 on the next page is 3 times its height. The area of the triangle is 96 square feet. Find its width and height.

Analyze the problem We can let x represent the height of the truss in feet. Then $3x$ represents its width.

Form and solve an equation We can substitute x for h, $3x$ for b, and 96 for A in the formula for the area of a triangle $A = \frac{1}{2}bh$ and solve for x.

Figure 5-8

$$A = \frac{1}{2}bh$$

$$96 = \frac{1}{2}(3x)x$$

$$192 = 3x^2 \qquad \text{Multiply both sides by 2 to clear fractions.}$$

$$64 = x^2 \qquad \text{Divide both sides by 3.}$$

$$0 = x^2 - 64 \qquad \text{Subtract 64 from both sides.}$$

$$0 = (x + 8)(x - 8) \qquad \text{Factor the difference of two squares.}$$

$$x + 8 = 0 \quad \text{or} \quad x - 8 = 0 \qquad \text{Set each factor equal to zero.}$$

$$x = -8 \qquad \qquad x = 8$$

State the conclusion Since the height of a triangle cannot be negative, we must discard the negative solution. Thus, the height of the truss is 8 feet, and its width is $3(8)$, or 24 feet.

Check the result The area of a triangle with a width (base) of 24 feet and a height of 8 feet is 96 square feet:

$$A = \frac{1}{2}bh = \frac{1}{2}(24)(8) = 12(8) = 96$$

🌿 **SELF CHECK 8** Find the width and height of the truss if the area is 150 square feet.

EXAMPLE 9 **BALLISTICS** If the initial velocity of an object shot straight up into the air is 176 feet per second, when will it hit the ground?

Analyze the problem Recall the formula for the height of an object thrown or shot straight up into the air is $h = vt - 16t^2$, where h represents height in feet, v represents initial velocity in feet per second, and t represents the number of seconds since the object was released.

Form and solve an equation When the object hits the ground, its height will be 0. Thus, we set h equal to 0, set v equal to 176, and solve for t.

$$h = vt - 16t^2$$

$$0 = 176t - 16t^2$$

$$0 = 16t(11 - t) \qquad \text{Factor out } 16t, \text{ the GCF.}$$

$$16t = 0 \quad \text{or} \quad 11 - t = 0 \qquad \text{Set each factor equal to 0.}$$

$$t = 0 \qquad \qquad t = 11$$

State the conclusion When $t = 0$, the object's height above the ground is 0 feet, because it has not been released. When $t = 11$, the height is again 0 feet, because the object has hit the ground. The solution is 11 seconds.

Check the result Verify that $h = 0$ when $t = 11$.

🌿 **SELF CHECK 9** If the intial velocity of an object shot straight up into the air is 128 feet per second, when will it hit the ground?

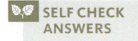
NOW TRY THIS

Factor or solve as indicated.

1. $6x^2 + 5x - 6$

2. $25x^2 = 25x$

3. $6x^2 - 6 = 5x$

4. $25x^3 - 25x$

5.8 Exercises

WARM-UPS *Solve each equation.*

1. $(x - 2)(x - 3) = 0$

2. $(x + 4)(x - 2) = 0$

3. $(x - 2)(x - 3)(x + 1) = 0$

4. $(x + 3)(x + 2)(x - 5)(x - 6) = 0$

REVIEW

5. List the prime numbers less than 10.

6. List the composite numbers between 7 and 17.

7. The formula for the volume of a sphere is $V = \frac{4}{3}\pi r^3$. Find the volume when $r = 21.23$ centimeters. Round the answer to the nearest hundredth.

8. The formula for the volume of a cone is $V = \frac{1}{3}\pi r^2 h$. Find the volume when $r = 12.33$ meters and $h = 14.7$ meters. Round the answer to the nearest hundredth.

VOCABULARY AND CONCEPTS *Fill in the blanks.*

9. A quadratic equation is any equation that can be written in the form _____ $(a \neq 0)$.

10. **The zero-factor property:** If a and b are real numbers, and if _____, then $a = 0$ or $b = 0$.

GUIDED PRACTICE *Solve. SEE EXAMPLE 1. (OBJECTIVE 1)*

11. $z^2 + 9z + 20 = 0$

12. $w^2 + 7w + 12 = 0$

13. $y^2 - 7y + 12 = 0$

14. $x^2 - 3x + 2 = 0$

Solve. SEE EXAMPLE 2. (OBJECTIVE 1)

15. $6m^2 + 24m = 0$

16. $5y^2 - 10y = 0$

17. $x^2 + x = 0$

18. $x^2 - 3x = 0$

Solve. SEE EXAMPLE 3. (OBJECTIVE 1)

19. $x^2 - 9 = 0$

20. $y^2 - 36 = 0$

21. $4a^2 - 25 = 0$

22. $25x^2 - 36 = 0$

Solve. SEE EXAMPLE 4. (OBJECTIVE 1)

23. $\dfrac{3a^2}{2} = \dfrac{1}{2} - a$

24. $x^2 = \dfrac{1}{6}(5x + 4)$

25. $\dfrac{1}{2}x^2 - \dfrac{5}{4}x = -\dfrac{1}{2}$

26. $x^2 + \dfrac{7}{3}x = -\dfrac{5}{4}$

Solve. SEE EXAMPLE 5. (OBJECTIVE 2)

27. $x^3 - 4x^2 - 21x = 0$

28. $x^3 + 8x^2 - 9x = 0$

29. $x^3 + x^2 = 0$

30. $2x^4 + 8x^3 = 0$

Solve. SEE EXAMPLE 6. (OBJECTIVE 2)

31. $x^4 - 26x^2 + 25 = 0$

32. $y^4 - 10y^2 + 9 = 0$

33. $x^4 - 20x^2 + 64 = 0$

34. $a^4 - 29a^2 + 100 = 0$

Find all values of x that will make $f(x) = 0$. SEE EXAMPLE 7. (OBJECTIVES 1–2)

35. $f(x) = x^2 - 49$

36. $f(x) = x^2 + 11x$

37. $f(x) = 2x^2 + 5x - 3$

38. $f(x) = 3x^2 - x - 2$

ADDITIONAL PRACTICE *Solve.*

39. $y^2 - 7y + 6 = 0$

40. $n^2 - 5n + 6 = 0$

41. $3m^2 + 10m + 3 = 0$

42. $2r^2 + 5r + 3 = 0$

43. $b(6b - 7) = 10$ **44.** $2y(4y + 3) = 9$

45. $3a(a^2 + 5a) = -18a$ **46.** $7t^3 = 2t\left(t + \dfrac{5}{2}\right)$

47. $x^3 + 3x^2 - x - 3 = 0$ **48.** $x^3 - x^2 - 4x + 4 = 0$

49. $2r^3 + 3r^2 - 18r - 27 = 0$

50. $3s^3 - 2s^2 - 3s + 2 = 0$

51. $5y^3 - 45y = 0$ **52.** $2z^3 - 200z = 0$

53. $2y^2 - 5y + 2 = 0$ **54.** $2x^2 - 3x + 1 = 0$

55. $2x^2 - x - 1 = 0$ **56.** $2x^2 - 3x - 5 = 0$

57. $3s^2 - 5s - 2 = 0$ **58.** $8t^2 + 10t - 3 = 0$

59. $10x^2 + 9x = 7$ **60.** $5z^2 = 6 - 13z$

61. $5y^2 - 25y = 0$ **62.** $3y^2 - 48 = 0$

63. $x(x - 6) + 9 = 0$ **64.** $x^2 + 8(x + 2) = 0$

65. $x^2 + 6x + 8 = 0$ **66.** $x^2 + 9x + 20 = 0$

67. $x\left(3x + \dfrac{22}{5}\right) = 1$ **68.** $x\left(\dfrac{x}{11} - \dfrac{1}{7}\right) = \dfrac{6}{77}$

69. $\dfrac{x^2(6x + 37)}{35} = x$ **70.** $x^2 = -\dfrac{4x^3(3x + 5)}{3}$

71. $3y^3 + y^2 = 4(3y + 1)$ **72.** $w^3 + 16 = w(w + 16)$

Find all values of x that will make f(x) = 0.

73. $f(x) = 5x^3 + 3x^2 - 2x$ **74.** $f(x) = 3x^4 - 12x^2$

Use a graphing calculator to find the real solutions of each equation, if any exist. If an answer is not exact, give the answer to the nearest hundredth.

75. $x^2 - 4x + 7 = 0$ **76.** $2x^2 - 7x + 4 = 0$

77. $-3x^3 - 2x^2 + 5 = 0$ **78.** $-2x^3 - 3x - 5 = 0$

79. Integers   The product of two consecutive even integers is 224. Find the integers.

80. Integers   The product of two consecutive odd integers is 143. Find the integers.

81. Integers   The sum of the squares of two consecutive positive integers is 85. Find the integers.

82. Integers   The sum of the squares of three consecutive positive integers is 77. Find the integers.

APPLICATIONS *Use an equation to solve.* **SEE EXAMPLES 8–9.** *(OBJECTIVE 3)*

83. Geometry   One side of a rectangle is three times as long as another. If its area is 147 square centimeters, find its dimensions.

84. Geometry   Find the dimensions of the rectangle, given that its area is 375 square feet.

85. Geometry   Find the height of the triangle, given that its area is 162 square centimeters.

86. Geometry   Find the perimeter of the rectangle.

87. Fine arts   An artist intends to paint a 60-square-foot mural on a large wall as shown in the illustration. Find the dimensions of the mural if the artist leaves a border of uniform width around it.

88. Gardening   A woman plans to use one-fourth of her 48-foot-by-100-foot rectangular backyard to plant a garden. Find the perimeter of the garden if the length is to be 40 feet greater than the width.

89. Architecture   The rectangular room shown in the illustration is twice as long as it is wide. It is divided into two rectangular parts by a partition, positioned as shown. If the larger part of the room contains 560 square feet, find the dimensions of the entire room.

90. Perimeter of a square If the length of each side of a square is increased by 4 inches, the area of the square becomes 9 times as much. Find the perimeter of the original square.

91. Time of flight After how many seconds will an object hit the ground if it is thrown upward with an initial velocity of 160 feet per second?

92. Time of flight After how many seconds will an object hit the ground if it is thrown upward with an initial velocity of 208 feet per second?

93. Ballistics The muzzle velocity of a cannon is 480 feet per second. If a cannonball is fired vertically, at what times will it be at a height of 3,344 feet?

94. Ballistics A slingshot can provide an initial velocity of 128 feet per second. At what times will a stone, shot vertically upward, be 192 feet above the ground?

95. Winter recreation The length of the rectangular ice-skating rink is 20 meters greater than twice its width. Find the width.

96. Carpentry A 285-square-foot room is 4 feet longer than it is wide. What length of crown molding is needed to trim the perimeter of the ceiling?

97. Designing a swimming pool Building codes require that the rectangular swimming pool in the illustration be surrounded by a uniform-width walkway of at least 516 square feet. If the length of the pool is 10 feet less than twice the width, how wide should the border be?

98. House construction The formula for the area of a trapezoid is $A = \frac{h(B + b)}{2}$. The area of the truss is 44 square feet. Find the height of the truss if the shorter base is the same as the height.

WRITING ABOUT MATH

99. Describe the steps for solving an application problem.

100. Explain how to check the solution of an application problem.

SOMETHING TO THINK ABOUT *Find a quadratic equation with the given solutions.*

101. 3, 5   **102.** −2, 6

103. 0, −5   **104.** $\frac{1}{2}, \frac{1}{3}$

Projects

PROJECT 1

Mac operates a bait shop located between two sharp bends on the Snarly River. The area around Mac's shop has recently become a popular hiking and camping area. Mac has decided to produce some maps of the area for the use of the visitors. Although he knows the region well, he has very little idea of the actual distances from one location to another. What he knows is:

1. Big Falls, a beautiful waterfall on Snarly River, is due east of Mac's.

2. Grandview Heights, a fabulous rock climbing site, is due west of Mac's, right on the river.

3. Foster's General Store, the only sizable camping and climbing outfitter in the area, is located on the river some distance west and north of Mac's.

Mac hires an aerial photographer to take pictures of the area, with some surprising results. If Mac's bait shop
(continued)

is treated as the origin of a coordinate system, with the y-axis running north–south and the x-axis running east–west, then on the domain $-4 \leq x \leq 4$ (where the units are miles), the river follows the curve

$$P(x) = \frac{1}{4}(x^3 - x^2 - 6x)$$

The aerial photograph is shown in the illustration.

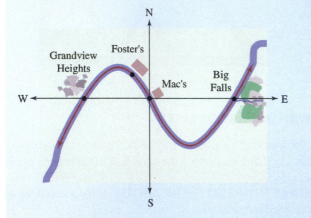

a. Mac would like to include on his maps the exact locations (relative to Mac's) of Big Falls and Grandview Heights. Find these for him, and explain to Mac why your answers must be correct.

b. Mac and Foster have determined that Foster's General Store is 0.7 miles west of the bait shop. Since it is on the river, it is a bit north as well. They decide that to promote business, they will join together to clear a few campsites in the region bordered by the straight-line paths that run between Mac's and Foster's, Mac's and Grandview Heights, and Foster's and Grandview Heights. If they clear 1 campsite for each 40 full acres of area, how many campsites can they put in?   (*Hint:* A square mile contains 640 acres.)

c. A path runs in a straight line directly southeast (along the line $y = -x$) from Mac's to the river. How far east and how far south has a hiker on this trail traveled when he or she reaches the river?

PROJECT 2

The rate at which fluid flows through a cylindrical pipe, or any cylinder-shaped tube (an artery, for instance), is

$$\text{Velocity of fluid flow} = V = \frac{p}{nL}(R^2 - r^2)$$

Cross section of tube

Fluid that is a distance r from center of tube

Wall of tube

where p is the difference in pressure between the two ends of the tube, L is the length of the tube, R is the radius of the tube, and n is the *viscosity constant*, a measure of the thickness of a fluid. Since the variable r represents the distance from the center of the tube, $0 \leq r \leq R$. In most situations, p, L, R, and n are constants, so V is a function of r.

$$V(r) = \frac{p}{nL}(R^2 - r^2)$$

It can be shown that the velocity of a fluid moving in the tube depends on how far from the center (or how close to the wall of the tube) the fluid is.

a. Consider a pipe with radius 5 centimeters and length 60 centimeters. Suppose that $p = 15$ and $n = 0.001$ (water has a viscosity of approximately 0.001). Find the velocity of the fluid at the center of the pipe. The units of measurement for V will be centimeters per second.

   Find the velocity of the fluid when it is halfway between the center and the wall of the pipe. What percent of the velocity at the center of the pipe does this represent? Where in the pipe is the fluid flowing with a velocity of 4,000 centimeters per second?

b. Suppose that the situation is the same, but the fluid is now machinery oil, with a viscosity of 0.15. Answer the same questions as in part **a**, except find where in the pipe the oil flows with a velocity of 15 cm per second. Note that the oil travels at a much slower speed than water.

c. Medical doctors use various methods to increase the rate of blood flow through an artery. The patient may take a drug that "thins the blood" (lowers its viscosity) or a drug that dilates the artery, or the patient may undergo angioplasty, a surgical procedure that widens the canal through which the blood passes. Explain why each of these increases the velocity V at a given distance r from the center of the artery.

Reach for Success

EXTRA PRACTICE EXERCISE **EXTENSION OF ORGANIZING YOUR TIME**

Now that you have created a calendar of all your responsibilities, let's look at how to establish *priorities*. This will allow you to avoid potential conflicts that could create unnecessary stress. Planning is the key to academic success!

Establish *your* criteria for dealing with scheduling conflicts. You cannot make a decision without knowing what criteria to use.

You might consider:

- Financial: Is it going to cost a lot to do it?
- Time: Do you have the time to do it?
- Importance: Are there reasons to do it?
- Risk: What happens if you do not do it or if you are late?
- Ability to complete: Do you have the skills needed?

What criteria do you use to establish priorities?

1. _____
2. _____
3. _____
4. _____

Now, look at the two-week calendar below. Do you see any potential scheduling conflicts?

_____

Sun 1	Mon 2	Tue 3	Wed 4	Thu 5	Fri 6	Sat 7
	English outline due	Dentist 1 p.m. Work	Work	Mathematics test	RJ's school play	Volunteer work Wedding anniversary
Sun 8	Mon 9	Tue 10	Wed 11	Thu 12	Fri 13	Sat 14
Sociology group project meeting 2 p.m.	Work Kenley's Dr. appt. 2:30 p.m.	Turn in history paper Math lab due	Read history chapters 13–15	Sociology group project due Work	English paper due Fred's birthday	Work

If you have to work on Tuesday and Wednesday the 3rd and 4th, when do you study for the mathematics test due on the 5th?

Your plan: _____

_____

As you can see, the second week is busy with many assignments. To avoid time conflicts, consider completing some of the assignments the first week rather than waiting until the second week. What other strategies could you use?

_____

_____

_____

Most instructors are willing to help you with study skills in addition to mathematical skills. Do not hesitate to ask them!

5   Review

SECTION 5.1   Polynomials and Polynomial Functions

DEFINITIONS AND CONCEPTS	EXAMPLES
The **degree of a polynomial** is the degree of the term with highest degree contained within the polynomial.	The degree of $2x^2 - 14x^5 + 8x + 5x^3 - 7$ is 5 because 5 is the highest degree of any term.
If the terms of a polynomial in one variable are written so that the exponents decrease as we move from left to right, the terms are written with their exponents in descending order.	The polynomial in descending order is $-14x^5 + 5x^3 + 2x^2 + 8x - 7$.
If $P(x)$ is a polynomial in x then $P(r)$ is the value of the polynomial at $x = r$.	Given $P(x) = 2x^2 - 6x + 1$, find $P(-1)$. $$P(-1) = 2(-1)^2 - 6(-1) + 1$$ $$= 2(1) + 6 + 1$$ $$= 9$$

REVIEW EXERCISES

1. Find the degree of $P(x) = 3x^5 + 4x^3 + 2$.

2. Find the degree of $9x^2y + 13x^3y^2 + 8x^4y^4$ and write the polynomial in descending powers of x.

Find each value if $P(x) = -x^2 + 4x + 6$.

3. $P(-1)$

4. $P(-2)$

5. $P(-t)$

6. $P(z)$

Graph each function.

7. $f(x) = x^3 - 1$

8. $f(x) = x^2 - 2x$

SECTION 5.2   Adding and Subtracting Polynomials

DEFINITIONS AND CONCEPTS	EXAMPLES
To *add* polynomials, add their like terms.	To add $(2x^2 - 2x - 3)$ and $(3x^2 + 4x + 4)$, remove parentheses and combine like terms. $$(2x^2 - 2x - 3) + (3x^2 + 4x + 4)$$ $$= 2x^2 - 2x - 3 + 3x^2 + 4x + 4$$ $$= 5x^2 + 2x + 1$$
To *subtract* polynomials, add the opposite of the polynomial that is to be subtracted from the other polynomial.	To subtract $(3x^2 + 4x - 4)$ from $(2x^2 - 2x - 3)$, write the expression as a subtraction and use the distributive property to remove the parentheses. This will change the sign of each term in the second polynomial. Combine like terms. $$(2x^2 - 2x - 3) - (3x^2 + 4x - 4)$$ $$= 2x^2 - 2x - 3 - 3x^2 - 4x + 4$$ $$= -x^2 - 6x + 1$$

REVIEW EXERCISES

Simplify each expression.

9. $(5x^2 + 7x - 2) + (4x^2 - 10x - 3)$

10. $(4x^3 + 4x^2 + 7) - (-2x^3 - x - 2)$

11. $(2x^2 - 5x + 9) - (x^2 - 3) - (-3x^2 + 4x - 7)$

12. $2(7x^3 - 6x^2 + 4x - 3) - 3(7x^3 + 6x^2 + 4x - 3)$

SECTION 5.3   Multiplying Polynomials

DEFINITIONS AND CONCEPTS	EXAMPLES
To multiply monomials, multiply their numerical factors and multiply their variable factors.	$(-5a^2b)(6ab^3) = -5 \cdot 6 \cdot a^2 \cdot a \cdot b \cdot b^3$ $= -30a^3b^4$
To multiply a polynomial by a monomial, multiply each term of the polynomial by the monomial.	$4x^2(3x^2 + 6x - 1) = 4x^2 \cdot 3x^2 + 4x^2 \cdot 6x + 4x^2 \cdot (-1)$ $= 12x^4 + 24x^3 - 4x^2$
To multiply polynomials, multiply each term of one polynomial by each term of the other polynomial. Then combine like terms, if possible.	$(6x - 2)(5x^3 - 8x + 9)$ $= 30x^4 - 48x^2 + 54x - 10x^3 + 16x - 18$ $= 30x^4 - 10x^3 - 48x^2 + 70x - 18$
To square a binomial, use it as a factor in a multiplication two times.	$(2x + 7)^2$ $(2x + 7)(2x + 7)$ $4x^2 + 14x + 14x + 49$ $4x^2 + 28x + 49$
The product of two conjugate binomials is also a binomial, the difference of two squares.	$(5x + 2)(5x - 2)$ $25x^2 - 10x + 10x - 4$ $25x^2 - 4$

REVIEW EXERCISES

Find each product.

13. $(8a^2b^2)(-2abc)$ **14.** $(-3xy^2z)(2xz^3)$

Find each product.

15. $2xy^2(x^3y - 4xy^5)$ **16.** $a^2b(a^2 + 2ab + b^2)$

Find each product.

17. $(6x + 7)(3x - 2)$ **18.** $(3x + 2)(2x - 4)$

19. $(3x - 8y)^2$ **20.** $(3x + 2)(3x - 2)$

21. $(3x^2 - 2)(x^2 - x + 2)$

22. $(2a - b)(a + b)(a - 2b)$

SECTION 5.4   The Greatest Common Factor and Factoring by Grouping

DEFINITIONS AND CONCEPTS	EXAMPLES
Always factor out common monomial factors as the first step in factoring a polynomial.	To factor $36x^3 - 6x^2 + 12x$, use the distributive property to factor out the common factor of $6x$. $6x(6x^2 - x + 2)$ Factor the common term.
If an expression has four or more terms, try to factor the expression by grouping.	$25x^3 - 10x^2 + 20x - 8$ $= (25x^3 - 10x^2) + (20x - 8)$ Group the first two terms and the last two terms. $= 5x^2(5x - 2) + 4(5x - 2)$ Factor each grouping. $= (5x - 2)(5x^2 + 4)$ Factor out $(5x - 2)$, the GCF.
Factoring is often required to solve a literal equation.	Solve $A = bc + bd$ for b. (Assume no division by 0.) $A = b(c + d)$ $\dfrac{A}{(c + d)} = \dfrac{b(c + d)}{(c + d)}$ $\dfrac{A}{c + d} = b$

REVIEW EXERCISES

Factor each expression.

23. $-6x - 18$

24. $8x^3 + 32x^2$

25. $5x^2y^3 - 10xy^2$

26. $7a^4b^2 + 49a^3b$

27. $-8x^2y^3z^4 - 12x^4y^3z^2$

28. $12a^6b^4c^2 + 15a^2b^4c^6$

29. $27x^3y^3z^3 + 81x^4y^5z^2 - 90x^2y^3z^7$

30. $-36a^5b^4c^2 + 60a^7b^5c^3 - 24a^2b^3c^7$

31. Factor x^n from $x^{2n} + x^n$.

32. Factor y^{2n} from $y^{2n} - y^{4n}$.

33. Factor x^{-2} from $x^{-4} - x^{-2}$.

34. Factor a^{-3} from $a^6 + 1$.

35. $5x^2(x + y)^3 - 15x^3(x + y)^4$

36. $-49a^3b^2(a - b)^4 + 63a^2b^4(a - b)^3$

Factor each expression.

37. $mn + 5n + 6m + 30$

38. $ac + bc - 3a - 3b$

39. $x^4 + 4y + 4x^2 + x^2y$

40. $a^5 + b^2c + a^2c + a^3b^2$

Solve for the indicated variable.

41. $S = 2wh + 2wl + 2lh$ for h

42. $S = 2wh + 2wl + 2lh$ for l

SECTION 5.5   The Difference of Two Squares; the Sum and Difference of Two Cubes

DEFINITIONS AND CONCEPTS	EXAMPLES
Difference of two squares: $$x^2 - y^2 = (x + y)(x - y)$$	$64x^2 - 25$ $\quad = (8x)^2 - (5)^2$ Write each factor as a square. $\quad = (8x + 5)(8x - 5)$ Factor.
Sum of two cubes: $$x^3 + y^3 = (x + y)(x^2 - xy + y^2)$$	$64x^3 + 125$ $\quad = (4x)^3 + (5)^2$ Write each factor as a cube. $\quad = (4x + 5)[(4x)^2 - (4x)(5) + (5)^2]$ Factor. $\quad = (4x + 5)(16x^2 - 20x + 25)$ Simplify.
Difference of two cubes: $$x^3 - y^3 = (x - y)(x^2 + xy + y^2)$$	$27a^3 - 1$ $\quad = (3a)^3 - (1)^3$ Write each factor as a cube. $\quad = (3a - 1)[(3a)^2 + (3a)(1) + (1)^2]$ Factor. $\quad = (3a - 1)(9a^2 + 3a + 1)$ Simplify.

REVIEW EXERCISES

Factor.

43. $z^2 - 36$

44. $y^4 - 16$

45. $x^2y^4 - 64z^6$

46. $a^2b^2 + c^2$

47. $(x + z)^2 - t^2$

48. $c^2 - (a + b)^2$

49. $2x^4 - 98$

50. $3x^6 - 300x^2$

51. $x^3 + 343$

52. $3p^3 + 81$

53. $8y^3 - 512$

54. $4x^3y + 108yz^3$

SECTION 5.6   Factoring Trinomials

DEFINITIONS AND CONCEPTS	EXAMPLES
Factoring a trinomial with a leading coefficient of 1: **1.** Write the trinomial in descending powers of one variable. **2.** List the factorizations of the third term of the trinomial. **3.** Select the factorization in which the sum of the factors is the coefficient of the middle term.	To factor $x^2 - 8x + 12$, first note that the factors of 12 are 1 and 12 -1 and -12 2 and 6 -2 and -6 3 and 4 -3 and -4 Since -2 and -6 are the only pair whose sum is -8, the factorization is $(x - 2)(x - 6)$.

Special-product formulas: $$x^2 + 2xy + y^2 = (x + y)(x + y)$$ $$x^2 - 2xy + y^2 = (x - y)(x - y)$$	The trinomial $9a^2 + 30a + 25$ can be written in the form $(3a)^2 - 2(3a)(5) + (5)^2$. This can be factored as $$(3a + 5)(3a + 5) \quad \text{or} \quad (3a + 5)^2$$ The trinomial $4a^2 - 12a + 9$ can be written in the form $(2a)^2 - 2(2a)(3) + (3)^2$. This can be factored as $$(2a - 3)(2a - 3) \quad \text{or} \quad (2a - 3)^2$$
Test for factorability: A trinomial the form $ax^2 + bx + c \quad (a \neq 0)$ will factor into two binomials with integer coefficients if $b^2 - 4ac$ is a perfect square. If $b^2 - 4ac = 0$, the factors will be the same.	To determine whether $5x^2 - 8x + 3$ is factorable with integer coefficients, note that $a = 5$, $b = -8$, and $c = 3$. Then calculate $b^2 - 4ac$. $$(-8)^2 - 4(5)(3) = 64 - 60 = 4$$ Since 4 is a perfect square, $5x^2 - 8x + 3$ is factorable with integer coefficients.
Factoring a trinomial with a leading coefficient other than 1 by trial and error. **1.** Write the trinomial in descending powers of one variable. **2.** Factor out any greatest common factor (including -1 if that is necessary to make the coefficient of the first term positive). **3.** Test the trinomial for factorability. **4.** When the sign of the first term of a trinomial is $+$ and the sign of the third term is $+$, the signs between the terms of each binomial factor are the same as the sign of the middle term of the trinomial. When the sign of the first term is $+$ and the sign of the third term is $-$, the signs between the terms of the binomials are opposites. **5.** Try various combinations as first terms and last terms until you find the one that works. **6.** Check the factorization by multiplication, including the GCF found in Step 2.	To factor $5x^2 - 8x + 3$, follow the steps shown on the left. The polynomial is already written in descending powers of x. In this polynomial, there are no common factors. We previously have determined that $5x^2 - 8x + 3$ is factorable. Since the last term is $+$ and the middle term $-$, the signs in both sets of parentheses will be $-$. The factors of the first term 5 are 5 and 1. The factors of the last term 3 are 3 and 1. The possible combinations are $$(5x - 3)(x - 1) \quad \text{and} \quad (5x - 1)(x - 3)$$ The pair that will give a middle term of $-8x$ is $(5x - 3)(x - 1)$. Verify by multiplying.
Factoring a trinomial by grouping (ac method). **1.** Write the trinomial in $ax^2 + bx + c$ order. **2.** Find the product ac, the key number. **3.** List all the factors of ac. **4.** Find the two factors of the key number whose sum is b. **5.** Use those factors as coefficients of two terms, the sum of which is bx, to replace the middle term. **6.** Factor by grouping. **7.** Check by multiplication.	Factor $5x^2 - 8x + 3$. $$a = 5, b = -8, c = 3$$ $$a \cdot c = 5 \cdot 3 = 15$$ $$1, 15 \quad 3, 5 \quad -1, -15 \quad \mathbf{-3, -5}$$ $$5x^2 - 3x - 5x + 3$$ $$x(5x - 3) - 1(5x - 3)$$ $$(5x - 3)(x - 1)$$ Verify by multiplying.

REVIEW EXERCISES

Factor, if possible.

55. $x^2 + 18x + 81$

56. $a^2 - 14a + 49$

57. $y^2 + 21y + 20$

58. $z^2 - 11z + 30$

59. $-x^2 - 3x + 28$

60. $y^2 - 5y - 36$

61. $4a^2 - 5a + 1$

62. $3b^2 + 2b + 1$

63. $7x^2 + x + 2$

64. $-15x^2 + 14x + 8$

65. $y^3 + y^2 - 2y$

66. $2a^4 + 4a^3 - 6a^2$

67. $-8x^2 + 22x - 12$

68. $8x^2 - 4x - 24$
69. $15x^2 - 57xy - 12y^2$
70. $30x^2 + 65xy + 10y^2$

71. $24x^2 - 23xy - 12y^2$
72. $14x^2 + 13xy - 12y^2$

SECTION 5.7   Summary of Factoring Techniques

DEFINITIONS AND CONCEPTS	EXAMPLES

Identifying factoring patterns:
 	
1. Arrange the polynomial in either descending or ascending order.
2. Factor out the greatest common factor.
3. If an expression has two terms, check to see whether it is
   a. the difference of two squares:
$$x^2 - y^2 = (x + y)(x - y)$$
   b. the sum of two cubes:
$$x^3 + y^3 = (x + y)(x^2 - xy + y^2)$$
   c. the difference of two cubes:
$$x^3 - y^3 = (x - y)(x^2 + xy + y^2)$$
4. If an expression has three terms, attempt to factor it as a **trinomial.**
5. If an expression has four or more terms, try factoring by **grouping.**
6. Continue until each individual factor is prime (except possibly a monomial GCF).
7. Check the results by multiplying.
	To factor $ax^3 - bx^3 - axy^2 + bxy^2$, first factor out the common factor of x to get
	$$x(ax^2 - bx^2 - ay^2 + by^2)$$
	Because the expression in the parentheses has four terms, factor it by grouping.
	$$x(ax^2 - bx^2 - ay^2 + by^2) = x[x^2(a - b) - y^2(a - b)]$$
	Then factor out $a - b$ to obtain
	$$= x(a - b)(x^2 - y^2)$$
	and finally factor the difference of two squares to obtain
	$$= x(a - b)(x + y)(x - y)$$
	Thus,
	$$ax^3 - bx^3 - axy^2 + bxy^2 = x(a - b)(x + y)(x - y)$$
	To check, verify that
	$$x(a - b)(x + y)(x - y) = ax^3 - bx^3 - axy^2 + bxy^2$$

REVIEW EXERCISES

Factor each expression, if possible.

73. $10x^3 + 5x^2 - 15x$
74. $3x^2y - 12xy - 63y$
75. $z^2 - 4 + zx - 2x$
76. $x^2 + 4x + 4 - y^2$

77. $x^2 + 4x + 4 - 4p^4$
78. $8y^3 - 125$
79. $x^{2m} + 2x^m - 3$
80. $x^4 - 5x^2 - 36$

SECTION 5.8   Solving Equations by Factoring

DEFINITIONS AND CONCEPTS	EXAMPLES
A quadratic equation is any equation that can be written in the form $ax^2 + bx + c = 0$, where a, b, and c are real numbers and $a \neq 0$. **Zero-factor property:** If $xy = 0$, then $x = 0$ or $y = 0$.	To solve $7x^2 = 2\left(x + \frac{5}{2}\right)$, proceed as follows: $$7x^2 = 2x + 5 \qquad \text{Use the distributive property to remove parentheses.}$$ $$7x^2 - 2x - 5 = 0 \qquad \text{Subtract } 2x \text{ and } 5 \text{ from both sides to write the equation in quadratic form.}$$ $$(7x + 5)(x - 1) = 0 \qquad \text{Factor the trinomial.}$$ $$7x + 5 = 0 \quad \text{or} \quad x - 1 = 0 \qquad \text{Use the zero-factor property.}$$ $$x = -\frac{5}{7} \qquad \qquad x = 1 \qquad \text{Solve each equation.}$$ Verify that the solutions are $-\frac{5}{7}$ and 1.

REVIEW EXERCISES

Solve each equation.

81. $5x^2 - 7x = 0$

82. $2x^2 - 18 = 0$

83. $12x^2 + 4x - 5 = 0$

84. $7y^2 - 37y + 10 = 0$

85. $t^2(15t - 2) = 8t$

86. $3u^3 = u(19u + 14)$

87. Volume   The volume V of the rectangular solid is given by the formula $V = lwh$, where l is its length, w is its width, and h is its height. If the volume is 840 cubic centimeters, the length is 12 centimeters, and the width exceeds the height by 3 centimeters, find the height.

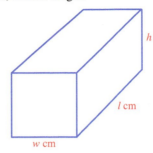

88. Volume   The volume of the pyramid is given by the formula $V = \frac{Bh}{3}$, where B is the area of its base and h is its height. The volume of the pyramid is 1,020 cubic meters. Find the dimensions of its rectangular base if one edge of the base is 3 meters longer than the other and the height of the pyramid is 9 meters.

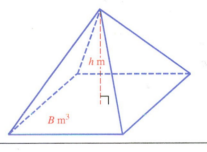

5   Test

Find the degree of each polynomial.

1. $4x^5 - 3x^3 - 7x^6 + 5x$

2. $3x^5y^3 - x^8y^2 + 2x^9y^4 - 3x^2y^5 + 4$

Let $P(x) = -3x^2 + 2x - 1$ and find each value.

3. $P(-3)$ 　　　　　　　　　**4.** $P(-1)$

5. Graph the function $f(x) = x^2 + 2x$.

Perform each operation.

6. $(2y^2 + 4y + 3) + (3y^2 - 3y - 4)$

7. $(-3u^2 + 2u - 7) - (u^2 + 7)$

8. $3(2a^2 - 4a + 2) - 4(-a^2 - 3a - 4)$

9. $-2(2x^2 - 2) + 3(x^2 + 5x - 2)$

10. $(3x^3y^2z)(-2xy^{-1}z^3)$

11. $-5a^2b(3ab^3 - 2ab^4)$

12. $(z + 9)(z - 9)$

13. $(5x + 3)(6x - 7)$

14. $(u - v)^2(2u + v)$

Factor or solve as indicated.

15. $4x^2y - 12xy^2$

16. $12a^3b^2c - 3a^2b^2c^2 + 6abc^3$

17. Factor y^n from $x^2y^{n+2} + y^n$.

18. $4x^2 - 12x + 9$

19. $(u - v)r + (u - v)s$

20. $ax - xy + ay - y^2$

21. $x^2 - 64$

22. $2x^2 - 32$

23. $4y^4 - 64$

24. $b^3 + 64$

25. $5x^2 - 8x - 6$

26. $3u^3 - 24$

27. $a^2 - 9a - 10$

28. $6b^2 + b - 2$

29. $6u^2 + 9u - 6$

30. $20r^2 - 15r - 5$

31. $x^{2n} + 2x^n + 1$

32. $x^2 + 6x + 9 - y^2$

33. $x^2 + x - 6 = 0$

34. $5x^2 + 9x - 18 = 0$

35. $x^2 = 9x + 10$

36. $4x^3 + 24x^2 = -35x$

37. Solve $r_1 r_2 - r_2 r = r_1 r$ for r.

38. Solve $x^2 - 5x - 6 = 0$ for x.

39. Integers The product of two consecutive positive integers is 156. Find their sum.

40. Preformed concrete The slab of concrete in the illustration is twice as long as it is wide. The area in which it is placed includes a 1-foot-wide border of 70 square feet. Find the dimensions of the slab.

Rational Expressions

© Erik Isakson/Tetra images/Jupiterimages

Careers and Mathematics

ENVIRONMENTAL ENGINEER

Using the principles of biology and chemistry, environmental engineers develop solutions to environmental problems. They are involved in water-and air-pollution control, recycling, waste disposal, and public health issues. Almost half of all environmental engineers work in professional, scientific, and technical services and about one third are employed in federal, state, and local government agencies. A bachelor's degree is required for almost all entry-level jobs.

Job Outlook:
Employment of environmental engineers is expected to increase about 22% through 2020, faster than the average for all occupations.

Annual Earnings:
$48,980-$119,060

For More Information:
http://www.bls.gov/ooh/architecture-and-engineering/environmental-engineers.htm

For a Sample Application:
See Problem 107 in Section 6.1.

REACH FOR SUCCESS

6.1 Finding the Domain of Rational Functions and Simplifying Rational Expressions

6.2 Multiplying and Dividing Rational Expressions

6.3 Adding and Subtracting Rational Expressions

6.4 Simplifying Complex Fractions

6.5 Solving Equations Containing Rational Expressions

6.6 Dividing Polynomials

6.7 Synthetic Division

6.8 Proportion and Variation

  ▪ *Projects*
  REACH FOR SUCCESS EXTENSION
  CHAPTER REVIEW
  CHAPTER TEST
  CUMULATIVE REVIEW

In this chapter

Rational expressions are the fractions of algebra. In this chapter, we will learn how to simplify, add, subtract, multiply, and divide them. Then we will solve equations containing rational expressions.

Reach for Success Preparing for a Test

Preparation is necessary to be successful in any event. You have practiced and worked out the kinks. You are well rested, and you are motivated. Now is the time. Are you ready?

When preparing for class tests, what can you do to increase your confidence and feel less anxious?

© bikeriderlondon/Shutterstock.com

To know what you need to spend time reviewing, you might want to organize and categorize your notes according to:

- what you clearly understand.
- what is a little unclear.
- what you do not understand at all.

If you have note cards, move the cards that you understand to the back of the deck. If you do not have note cards, highlight the topics in your notebook that you do not understand. Continue to work on these. Seek help.

List the objectives that are giving you the most difficulty.

1. _____
2. _____
3. _____
4. _____
5. _____
6. _____
7. _____
8. _____

How many questions will be on the exam?
_____

Will vocabulary be tested? _____

Will the formulas be provided? _____

Determine how much time you should allow yourself for each question on a test if you have 60 minutes to take a 20-question test. _____

Write down the formulas that you will need for the exam.

_____

Use the Chapter Review at the end of the chapter to prepare for the exam. After completing the review, take the Chapter Test without notes, grade it, and mark the objectives you missed.

If the test includes completion, multiple choice, true/false, essay, or matching, be certain to practice those formats.

Did your instructor provide a written or computerized review for the test? _____ If so, work it as if it were the test—in a single sitting and without notes. How did you do? _____

Note any objectives you missed.

_____

If your instructor did not have a review, make up your own practice test, alone or with a group. Then, take it as described above.

How will you further review any objectives you missed?

_____

A Successful Study Strategy . . .

 Begin your review early and practice every day to increase your confidence and thus reduce anxiety.

At the end of the chapter you will find an additional exercise to guide you in planning for a successful semester.

Section 6.1

Finding the Domain of Rational Functions and Simplifying Rational Expressions

Objectives

1. Find all values of a variable for which a rational expression is undefined.
2. Find the domain of a rational function.
3. Simplify a rational expression.
4. Simplify a rational expression containing factors that are negatives.
5. Evaluate a rational function that models an application.

Vocabulary

rational expressions   rational functions   asymptotes

Getting Ready

Simplify each fraction.

1. $\dfrac{6}{8}$ **2.** $\dfrac{12}{15}$ **3.** $\dfrac{-25}{65}$ **4.** $\dfrac{-49}{-63}$

We begin our discussion of rational expressions by finding all values for which a rational expression is undefined. We also consider the domain of a rational function. Then, we will discuss how to simplify rational expressions.

1 **Find all values of a variable for which a rational expression is undefined.**

Rational expressions are fractions that indicate the quotient of two polynomials.

$$\frac{3x}{x-7}, \qquad \frac{5m+n}{8m+16}, \qquad \text{and} \qquad \frac{a^3+2a^2+7}{2a^2-5a+4}$$

To evaluate a rational expression for a given value, we replace the variable in the expression with the given value and simplify. For example, to evaluate $\frac{x+1}{x-3}$ for $x = -1$, $x = 0$, and $x = 3$, we proceed as follows:

$$\frac{x+1}{x-3} = \frac{-1+1}{-1-3} \qquad \frac{x+1}{x-3} = \frac{0+1}{0-3} \qquad \frac{x+1}{x-3} = \frac{3+1}{3-3}$$

$$= \frac{0}{-4} \qquad\qquad\qquad = \frac{1}{-3} \qquad\qquad\qquad = \frac{4}{0} \quad \text{(undefined)}$$

$$= 0 \qquad\qquad\qquad\qquad = -\frac{1}{3}$$

Thus, in the expression $\frac{x+1}{x-3}$, x cannot be 3 because 3 would make the denominator 0.

EXAMPLE 1   Find all values of x for which each expression is undefined.

a. $\dfrac{9x}{5x-4}$ b. $\dfrac{3x+2}{x^2+x-6}$

Solution   a. Because the denominator of the rational expression cannot be 0, the denominator $5x-4$ cannot be 0. So we must exclude all values of x such that $5x-4=0$. To find these values we will solve the linear equation.

$$5x-4=0$$

$$5x=4 \qquad \text{Add 4 to both sides.}$$

$$x=\frac{4}{5} \qquad \text{Divide both sides by 5.}$$

Since substituting $\frac{4}{5}$ for x will make the denominator 0, the expression will be undefined when $x=\frac{4}{5}$.

b. Because the denominator of the rational expression cannot be 0, the denominator x^2+x-6 cannot be 0. So we must exclude all values of x such that $x^2+x-6=0$. To find these values, we will solve the quadratic equation.

$$x^2+x-6=0$$

$$(x+3)(x-2)=0 \qquad \text{Factor the trinomial.}$$

$$x+3=0 \quad \text{or} \quad x-2=0 \qquad \text{Set each factor equal to 0.}$$

$$x=-3 \qquad\qquad x=2 \qquad \text{Solve each linear equation.}$$

Since substituting 2 or -3 for x will make the denominator 0, the expression will be undefined when $x=2$ or $x=-3$.

 SELF CHECK 1   Find all values of x for which each expression is undefined.

a. $\dfrac{5x+2}{6x-9}$ b. $\dfrac{3x}{5x^2+3x-2}$

2   Find the domain of a rational function.

Rational expressions often define **rational functions**. For example, the equation $f(x)=\dfrac{9x}{5x-4}$ defines a function because it sets up a correspondence in which every number x determines exactly one value of y. Since division by 0 is undefined, we must exclude any values that make the denominator 0 from the domain of the function. From Example 1(a), we saw that x cannot be $\frac{4}{5}$. However, the domain of a function is defined to be all values that can replace the variable. Thus, the domain of $f(x)=\dfrac{9x}{5x-4}$ is the set of all real numbers except $\frac{4}{5}$. In interval notation, the domain is $\left(-\infty,\frac{4}{5}\right)\cup\left(\frac{4}{5},\infty\right)$.

EXAMPLE 2   Find the domain of $f(x)=\dfrac{3x+2}{x^2+x-6}$.

Solution   From Example 1(b), we saw that $x=2$ and $x=-3$ make the denominator of the fraction $\frac{3x+2}{x^2+x-6}$ equal to 0. Thus, the domain of the function is the set of all real numbers except -3 and 2. In interval notation, the domain is $(-\infty,-3)\cup(-3,2)\cup(2,\infty)$.

 SELF CHECK 2   Find the domain of $f(x)=\dfrac{x^2+1}{x-2}$.

If we graph the function in Example 2 as illustrated in the Accent on Technology, we notice lines referred to as **asymptotes**.

Accent
on technology

▶ Finding the Domain
   and Range of a Function

We can find the domain and range of the function in Example 2 by looking at its graph using a TI-84 graphing calculator. If we use window settings of $[-10, 10]$ for x and $[-10, 10]$ for y and graph the function

$$f(x) = \frac{3x + 2}{x^2 + x - 6}$$

we will obtain the graph in Figure 6-1(a).

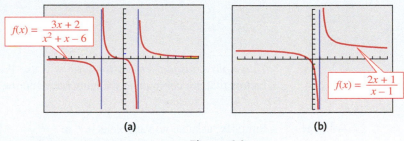

$$f(x) = \frac{3x + 2}{x^2 + x - 6}$$

$$f(x) = \frac{2x + 1}{x - 1}$$

(a)   (b)

Figure 6-1

COMMENT   Due to the variations in calculators, it is recommended you consult the owner's manual. For example, the TI-84 Plus does not display asymptotes.

From the figure, we can see that

- As x approaches -3 from the left, the values of y decrease, and the graph approaches the vertical line $x = -3$.
- As x approaches -3 from the right, the values of y increase, and the graph approaches the vertical line $x = -3$.

We also can see that

- As x approaches 2 from the left, the values of y decrease, and the graph approaches the vertical line $x = 2$.
- As x approaches 2 from the right, the values of y increase, and the graph approaches the vertical line $x = 2$.

The lines $x = -3$ and $x = 2$ are *vertical asymptotes.* Although the vertical lines in the graph appear to be the graphs of $x = -3$ and $x = 2$, they are not. Graphing calculators draw graphs by connecting dots whose x-coordinates are close together. Often, when two such points straddle a vertical asymptote and their y-coordinates are far apart, the calculator draws a line between them anyway, producing what appears to be a vertical asymptote. If you set your calculator to dot mode instead of connected mode, the vertical lines will not appear.
  From Figure 6-1(a), we also can see that

- As x increases to the right of 2, the values of y decrease and approach the value $y = 0$.
- As x decreases to the left of -3, the values of y increase and approach the value $y = 0$.

The line $y = 0$ (the x-axis) is a *horizontal asymptote.* Graphing calculators do not draw lines that appear to be horizontal asymptotes.
  From the graph, we can see that all real numbers x, except -3 and 2, give a value of y. This confirms that the domain of the function is $(-\infty, -3) \cup (-3, 2) \cup (2, \infty)$. We also can see that y can be any value. Thus, the range is $(-\infty, \infty)$.
  To find the domain and range of the function $f(x) = \frac{2x + 1}{x - 1}$, we use a calculator to draw the graph shown in Figure 6-1(b). From this graph, we can see that the line $x = 1$ is a vertical asymptote and that the line $y = 2$ is a horizontal asymptote. Since x can be any real number except 1, the domain is the interval $(-\infty, 1) \cup (1, \infty)$. Since y can be any value except 2, the range is $(-\infty, 2) \cup (2, \infty)$.
  For instruction regarding the use of a Casio graphing calculator, please refer to the Casio Keystroke Guide in the back of the book.

3 Simplify a rational expression.

Since rational expressions are the fractions of algebra, the rules for arithmetic fractions apply.

PROPERTIES OF RATIONAL EXPRESSIONS

If there are no divisions by 0, then

1. $\dfrac{a}{b} = \dfrac{c}{d}$ if and only if $ad = bc$ **2.** $\dfrac{a}{1} = a$ and $\dfrac{a}{a} = 1$

3. $\dfrac{ak}{bk} = \dfrac{a}{b}$ **4.** $-\dfrac{a}{b} = \dfrac{-a}{b} = \dfrac{a}{-b}$

To simplify rational expressions, we will use Property 3, which enables us to divide out factors that are common to the numerator and the denominator.

EXAMPLE 3   Simplify; assume no variable is 0.   **a.** $\dfrac{10k}{25k^2}$ **b.** $\dfrac{-8y^3z^5}{6y^4z^3}$

Solution   Factor each numerator and denominator and divide out the common factors.

a. $\dfrac{10k}{25k^2} = \dfrac{5 \cdot 2 \cdot k}{5 \cdot 5 \cdot k \cdot k}$

$= \dfrac{\overset{1}{\cancel{5}} \cdot 2 \cdot \overset{1}{\cancel{k}}}{\underset{1}{\cancel{5}} \cdot 5 \cdot \underset{1}{\cancel{k}} \cdot k}$

$= \dfrac{2}{5k}$

b. $\dfrac{-8y^3z^5}{6y^4z^3} = \dfrac{-2 \cdot 4 \cdot y \cdot y \cdot y \cdot z \cdot z \cdot z \cdot z \cdot z}{2 \cdot 3 \cdot y \cdot y \cdot y \cdot y \cdot z \cdot z \cdot z}$

$= \dfrac{-2 \cdot 4 \cdot \cancel{y} \cdot \cancel{y} \cdot \cancel{y} \cdot \cancel{z} \cdot \cancel{z} \cdot \cancel{z} \cdot z \cdot z}{2 \cdot 3 \cdot \cancel{y} \cdot \cancel{y} \cdot \cancel{y} \cdot y \cdot \cancel{z} \cdot \cancel{z} \cdot \cancel{z}}$

$= -\dfrac{4z^2}{3y}$

SELF CHECK 3   Simplify: $\dfrac{-12a^4b^2}{-3ab^4}$ Assume no variable is 0.

The rational expressions in Example 3 also can be simplified by using the rules of exponents:

$\dfrac{10k}{25k^2} = \dfrac{5 \cdot 2}{5 \cdot 5} k^{1-2}$

$= \dfrac{2}{5} \cdot k^{-1}$

$= \dfrac{2}{5} \cdot \dfrac{1}{k}$

$= \dfrac{2}{5k}$

$\dfrac{-8y^3z^5}{6y^4z^3} = \dfrac{-2 \cdot 4}{2 \cdot 3} y^{3-4} z^{5-3}$

$= \dfrac{-4}{3} \cdot y^{-1} z^2$

$= -\dfrac{4}{3} \cdot \dfrac{1}{y} \cdot \dfrac{z^2}{1}$

$= -\dfrac{4z^2}{3y}$

EXAMPLE 4   Simplify: $\dfrac{x^2 - 16}{x + 4}$ Assume no denominator is 0.

Solution   We factor $x^2 - 16$ and use the fact that $\dfrac{x+4}{x+4} = 1$.

$\dfrac{x^2 - 16}{x + 4} = \dfrac{\overset{1}{\cancel{(x + 4)}}(x - 4)}{1\underset{1}{\cancel{(x + 4)}}}$

$= \dfrac{x - 4}{1}$ Divide out the common factor $(x + 4)$.

$= x - 4$

🌿 **SELF CHECK 4**   Simplify: $\dfrac{x^2 - 9}{x - 3}$ Assume no denominator is 0.

Accent on technology

▸ Checking an Algebraic Simplification

To show that the simplification in Example 4 is correct, we can graph the following functions with a TI-84 graphing calculator and show that the graphs look the same.

$$f(x) = \frac{x^2 - 16}{x + 4} \quad \text{and} \quad g(x) = x - 4$$

(See Figure 6-2.) Except for the point where $x = -4$, the graphs are the same.

The point where $x = -4$ is excluded from the graph of $f(x) = \dfrac{x^2 - 16}{x + 4}$, because -4 is not in the domain of f. However, graphing calculators do not show that this point is excluded. The point where $x = -4$ is included in the graph of $g(x) = x - 4$, because -4 is in the domain of g.

(a)   (b)

Figure 6-2

For instruction regarding the use of a Casio graphing calculator, please refer to the Casio Keystroke Guide in the back of the book.

EXAMPLE 5   Simplify: $\dfrac{2x^2 + 11x + 12}{3x^2 + 11x - 4}$ Assume no denominator is 0.

Solution   We factor the numerator and denominator and use the fact that $\dfrac{x + 4}{x + 4} = 1$.

$$\frac{2x^2 + 11x + 12}{3x^2 + 11x - 4} = \frac{(2x + 3)(x + 4)}{(3x - 1)(x + 4)} \qquad \text{Divide out the common factor } (x + 4).$$

$$= \frac{2x + 3}{3x - 1}$$

COMMENT   Only *factors* that are common to the entire numerator and the entire denominator can be divided out. *Terms* common to both the numerator and denominator cannot be divided out. For example, it is incorrect to divide out the common term of 3 in the following simplification, because doing so gives a wrong answer.

$$\frac{3 + 7}{3} = \frac{\cancel{3} + 7}{\cancel{3}} = \frac{1 + 7}{1} = 8 \qquad \text{The correct simplification is } \frac{3 + 7}{3} = \frac{10}{3}.$$

The 3s in the fraction $\dfrac{2x + 3}{3x - 1}$ cannot be divided out either, because the 3 in the numerator is a term and the 3 in the denominator is a factor of the first term only. To be divided out, the 3 must be a factor of the entire numerator and the entire denominator.

🌿 **SELF CHECK 5**   Simplify: $\dfrac{2x^2 + 7x - 15}{2x^2 + 13x + 15}$ Assume no denominator is 0.

EXAMPLE 6 Simplify: $\dfrac{(x^2 + 2x)(x^2 + 2x - 3)}{(x^2 + x - 2)(x^2 + 3x)}$ Assume no denominator is 0.

Solution We factor the numerator and denominator and divide out all common factors:

$$\frac{(x^2 + 2x)(x^2 + 2x - 3)}{(x^2 + x - 2)(x^2 + 3x)} = \frac{\overset{1}{\cancel{x}}(\overset{1}{\cancel{x+2}})(\overset{1}{\cancel{x+3}})(\overset{1}{\cancel{x-1}})}{\underset{1}{\cancel{(x+2)}}\underset{1}{\cancel{(x-1)}}\underset{1}{\cancel{x}}\underset{1}{\cancel{(x+3)}}}$$ Divide out common factors.

$$= 1$$

SELF CHECK 6 Simplify: $\dfrac{(a^2 + 4a)(a^2 - a - 2)}{a(a^2 + 2a - 8)}$ Assume no denominator is 0.

Many rational expressions we shall encounter are already in simplified form. For example, to attempt to simplify

$$\frac{x^2 + xa + 2x + 2a}{x^2 + x - 6}$$

we factor the numerator and denominator and divide out any common factors:

$$\frac{x^2 + xa + 2x + 2a}{x^2 + x - 6} = \frac{x(x + a) + 2(x + a)}{(x - 2)(x + 3)} = \frac{(x + a)(x + 2)}{(x - 2)(x + 3)}$$

Since there are no common factors in the numerator and denominator, the result cannot be simplified.

4 ## Simplify a rational expression containing factors that are negatives.

To simplify $\frac{b - a}{a - b}$ $(a \neq b)$, we factor -1 from the numerator and divide out any factors common to both the numerator and the denominator:

$$\frac{b - a}{a - b} = \frac{-a + b}{a - b} \quad (a \neq b) \quad \text{Use the commutative property of addition.}$$

$$= \frac{-1(a - b)}{a - b} \qquad \text{Factor out } -1.$$

$$= \frac{-1(\overset{1}{\cancel{a - b}})}{\underset{1}{\cancel{(a - b)}}} \qquad \text{Divide out the common factor } (a - b).$$

$$= -1$$

In general, we have the following principle.

QUOTIENT OF A QUANTITY AND ITS OPPOSITE The quotient of any nonzero quantity and its negative (or opposite) is -1.

EXAMPLE 7 Simplify: $\dfrac{3x^2 - 10xy - 8y^2}{4y^2 - xy}$ Assume no denominator is 0.

Solution We factor the numerator and denominator and note that because $x - 4y$ and $4y - x$ are negatives, their quotient is -1.

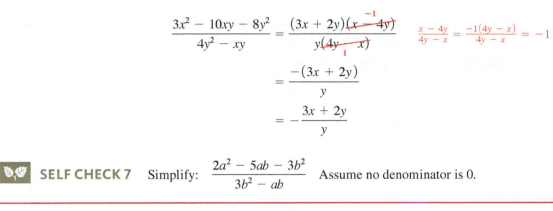

$$\frac{3x^2 - 10xy - 8y^2}{4y^2 - xy} = \frac{(3x + 2y)\overset{-1}{(x - 4y)}}{y\underset{1}{(4y - x)}} \qquad \frac{x - 4y}{4y - x} = \frac{-1(4y - x)}{4y - x} = -1$$

$$= \frac{-(3x + 2y)}{y}$$

$$= -\frac{3x + 2y}{y}$$

SELF CHECK 7 Simplify: $\dfrac{2a^2 - 5ab - 3b^2}{3b^2 - ab}$ Assume no denominator is 0.

5 ### Evaluate a rational function that models an application.

Rational expressions often define functions. For example, if the cost of subscribing to an online research service is \$6 per month plus \$1.50 per hour of access time, the average (mean) hourly cost of the service is the total monthly cost, divided by the number of hours of access time.

$$\bar{c} = \frac{c}{n} = \frac{1.50n + 6}{n} \qquad \text{$\bar{c}$ is the mean hourly cost, c is the total monthly cost, and n is the number of hours the service is used.}$$

The function

(1) $$\bar{c} = f(n) = \frac{1.50n + 6}{n} \quad (n > 0)$$

Sonya Kovalevskaya
1850–1891
This talented young Russian woman hoped to study mathematics at the University of Berlin, but strict rules prohibited women from attending lectures. Undaunted, Sonya studied privately with the great mathematician Karl Weierstrauss and published several important papers.

gives the mean hourly cost of using the research service for n hours per month.

Figure 6-3 shows the graph of the rational function $\bar{c} = f(n) = \frac{1.50n + 6}{n}$ $(n > 0)$. Since $n > 0$, the domain of this function is the interval $(0, \infty)$.

Figure 6-3

From the graph, we can see that the mean hourly cost decreases as the number of hours of access time increases. Since the cost of each extra hour of access time is \$1.50, the mean hourly cost can approach \$1.50 but never drop below it. Thus, the graph of the function approaches the line $y = 1.5$ as n increases without bound. When a graph approaches a line as the independent variable gets large, we call the line an asymptote. The line $y = 1.5$ is a *horizontal asymptote* of this graph.

From Figure 6-3 we see that as n gets smaller and approaches 0, the graph approaches the y-axis but never touches it. The y-axis is a *vertical asymptote* of this graph.

EXAMPLE 8   Find the mean hourly cost when the service described in Figure 6-3 is used for
a. 3 hours.   **b.** 70.4 hours.

Solution   **a.** To find the mean hourly cost for 3 hours of access time, we substitute 3 for n in Equation 1 and simplify:

$$\bar{c} = f(\mathbf{3}) = \frac{1.50(\mathbf{3}) + 6}{\mathbf{3}} = 3.5$$

The mean hourly cost for 3 hours of access time is $3.50.

b. To find the mean hourly cost for 70.4 hours of access time, we substitute 70.4 for n in Equation 1 and simplify:

$$\bar{c} = f(\mathbf{70.4}) = \frac{1.50(\mathbf{70.4}) + 6}{\mathbf{70.4}} = 1.585227273$$

The mean hourly cost for 70.4 hours of access time is approximately $1.59.

SELF CHECK 8   Find the mean hourly cost when the service is used for 5 hours.

SELF CHECK ANSWERS

1. a. $\frac{3}{2}$ b. $\frac{2}{5}, -1$ 2. $(-\infty, 2) \cup (2, \infty)$ 3. $\frac{4a^3}{b^2}$ 4. $x + 3$ 5. $\frac{2x - 3}{2x + 3}$ 6. $a + 1$
7. $-\frac{2a + b}{b}$ 8. $2.70

NOW TRY THIS

A vertical asymptote appears at any value for which a simplified rational function is undefined. A hole in the graph of a function occurs at any value resulting from a factor common to the numerator and denominator of the original function.

For each of the functions below, determine where the vertical asymptote(s) occurs (if any) and where the hole(s) occurs (if any). Graph the original function and any simplified function and compare their graphs. What do you notice?

1. $f(x) = \dfrac{x^2 + 4}{x^2 - 9}$

2. $g(x) = \dfrac{x^3 - 3x^2}{15x - 5x^2}$

3. $h(x) = \dfrac{x^2 - 6x + 9}{x^2 - 5x + 6}$

Note: Graph the functions using the window settings of $x = [-4.7, 4.7]$ and $y = [-5, 5]$ in order to see the holes. Use the TRACE feature to verify the location of the holes. (There will be no value for y.)

6.1 Exercises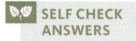

WARM-UPS *Evaluate.*

1. $\dfrac{x}{x - 2}$ when $x = 0$

2. $\dfrac{x}{x - 2}$ when $x = 2$

3. $\dfrac{y + 1}{y^2 - 9}$ when $y = -3$

4. $\dfrac{y + 1}{y^2 - 9}$ when $y = -1$

5. $\dfrac{x - 1}{1 - x}$ when $x = 2$

6. $\dfrac{3 - y}{y - 3}$ when $y = 1$

REVIEW *Factor each expression.*

7. $3x^2 - 9x$

8. $6t^2 - 5t - 6$

9. $27x^6 + 64y^3$

10. $x^2 + ax + 2x + 2a$

VOCABULARY AND CONCEPTS *Fill in the blanks.*

11. If a fraction is the quotient of two polynomials, it is called a _____ expression.

12. The denominator of a fraction can never be _.

13. If a graph approaches a line, the line is called an _____.

14. $\dfrac{a}{b} = \dfrac{c}{d}$ if and only if _____.

15. $\dfrac{a}{1} = $ _ and $\dfrac{a}{a} = $ _, provided that $a \neq $ _.

16. Rational expressions can often define a _____ function.

17. $\dfrac{ak}{bk} = $ ___ , provided that $b \neq $ _ and $k \neq 0$.

18. The quotient of any nonzero quantity and its negative is __.

GUIDED PRACTICE *Find all values of the variable for which the rational expression is undefined.* **SEE EXAMPLE 1. (OBJECTIVE 1)**

19. $\dfrac{3x - 8}{x + 6}$

20. $\dfrac{x + 7}{x - 9}$

21. $\dfrac{5x + 1}{3x + 4}$

22. $\dfrac{3x - 13}{x^2 + x - 2}$

State the domain of each function in interval notation.
SEE EXAMPLE 2. (OBJECTIVE 2)

23. $f(x) = \dfrac{3x - 8}{x + 6}$

24. $f(x) = \dfrac{x + 7}{x - 9}$

25. $f(x) = \dfrac{5x + 1}{3x + 4}$

26. $f(x) = \dfrac{6x - 11}{x^2 - x - 6}$

Use a graphing calculator to graph each function. From the graph, determine its domain. **(OBJECTIVE 2)**

27. $f(x) = \dfrac{x}{x - 2}$

28. $f(x) = \dfrac{x + 2}{x}$

29. $f(x) = \dfrac{x + 1}{x^2 - 4}$

30. $f(x) = \dfrac{x - 2}{x^2 - 3x - 4}$

Simplify each expression when possible. Assume no denominator is zero. **SEE EXAMPLE 3. (OBJECTIVE 3)**

31. $-\dfrac{18}{24}$

32. $\dfrac{69}{15}$

33. $-\dfrac{112}{36}$

34. $-\dfrac{49}{21}$

35. $\dfrac{12x^3}{3x}$

36. $\dfrac{26y^5}{12y^3}$

37. $\dfrac{-24x^3y^4}{18x^4y^3}$

38. $\dfrac{15a^5b^4}{21b^3c^2}$

Simplify each expression when possible. Assume no denominator is zero. **SEE EXAMPLE 4. (OBJECTIVE 3)**

39. $\dfrac{x^2 - 81}{x + 9}$

40. $\dfrac{y - 8}{y^2 - 64}$

41. $\dfrac{x + y}{x^2 - y^2}$

42. $\dfrac{x - y}{x^2 - y^2}$

Simplify each expression when possible. Assume no denominator is zero. **SEE EXAMPLE 5. (OBJECTIVE 3)**

43. $\dfrac{x^2 + 2x - 15}{x^2 - 25}$

44. $\dfrac{5x - 10}{x^2 - 4x + 4}$

45. $\dfrac{2a^2 + 5a + 3}{6a + 9}$

46. $\dfrac{2x^2 + 2x - 12}{x^3 + 3x^2 - 4x - 12}$

47. $\dfrac{x^2 + 2x + 1}{x^2 + 4x + 3}$

48. $\dfrac{6x^2 + x - 2}{8x^2 + 2x - 3}$

49. $\dfrac{4x^2 + 24x + 32}{16x^2 + 8x - 48}$

50. $\dfrac{-4x - 4 + 3x^2}{4x^2 - 2 - 7x}$

Simplify each expression when possible. Assume no denominator is zero. **SEE EXAMPLE 6. (OBJECTIVE 3)**

51. $\dfrac{(x^2 - 1)(x + 1)}{(x^2 - 2x + 1)^2}$

52. $\dfrac{(x^2 + 2x + 1)(x^2 - 2x + 1)}{(x^2 - 1)^2}$

53. $\dfrac{(2x^2 + 3xy + y^2)(3a + b)}{(x + y)(2xy + 2bx + y^2 + by)}$

54. $\dfrac{(x - 1)(6ax + 9x + 4a + 6)}{(3x + 2)(2ax - 2a + 3x - 3)}$

Simplify each expression when possible. Assume no denominator is zero. **SEE EXAMPLE 7. (OBJECTIVE 4)**

55. $\dfrac{(p + q)(p - r)}{(r - p)(p + q)}$

56. $\dfrac{a - b}{b^2 - a^2}$

57. $\dfrac{-3ab^2(a - b)}{9ab(b - a)}$

58. $\dfrac{12 - 3x^2}{x^2 - x - 2}$

Each of the following applications models a rational function. Evaluate each one for the given values. **SEE EXAMPLE 8. (OBJECTIVE 5)**

The time, *t*, it takes to travel 600 miles is a function of the mean rate of speed, r: $t = f(r) = \dfrac{600}{r}$. Find t for each value of r.

59. 30 mph

60. 40 mph

61. 50 mph

62. 60 mph

Suppose the cost (in dollars) of removing p% of the pollution in a river is given by the function $c = f(p) = \frac{50{,}000p}{100 - p}$ $(0 \le p < 100)$. Find the cost of removing each percent of pollution.

63. 10%   **64.** 30%

65. 50%   **66.** 80%

ADDITIONAL PRACTICE *Simplify each expression when possible. Assume no denominator is zero.*

67. $\dfrac{288}{312}$

68. $\dfrac{144}{72}$

69. $-\dfrac{96a^5}{112a^3}$

70. $-\dfrac{147x^2y^4}{84x^3y}$

71. $\dfrac{(5x^4)^2}{15x^5}$

72. $\dfrac{(2x^3y^4)^3}{16(x^4y)^4}$

73. $\dfrac{-14x^2(x + y)^2}{49x^3(x + y)}$

74. $\dfrac{35a^4(a - 5)}{28a(a - 5)^3}$

75. $\dfrac{9y^2(y - z)}{21y(y - z)^2}$

76. $\dfrac{(a - b)(c - d)}{(c - d)(a - b)}$

77. $\dfrac{4x^2 + 8x + 3}{6 + x - 2x^2}$

78. $\dfrac{6x^2 + 13x + 6}{6 - 5x - 6x^2}$

79. $\dfrac{y^3 - 8}{20 - 5y^2}$

80. $\dfrac{x^3 - 27}{3x^2 - 8x - 3}$

81. $\dfrac{x^2 + x - 30}{x^2 - x - 20}$

82. $\dfrac{6x^2 - 7x - 5}{2x^2 + 5x + 2}$

83. $\dfrac{x^2 + y^2}{x + y}$

84. $\dfrac{2x^2 - 3x - 9}{2x^2 + 3x - 9}$

85. $\dfrac{3m - 6n}{3n - 6m}$

86. $\dfrac{x^4 - y^4}{(x^2 + 2xy + y^2)(x^2 + y^2)}$

87. $\dfrac{x^3 + 8}{x^2 - 2x + 4}$

88. $\dfrac{a^2 - 4}{a^3 - 8}$

89. $\dfrac{x^2 + 3x + 9}{x^3 - 27}$

90. $\dfrac{x^2 - x - 2}{2x + 4}$

91. $\dfrac{3x + 6y}{x + 2y}$

92. $\dfrac{y - xy}{xy - x}$

93. $\dfrac{4a^2 - 9b^2}{2a^2 - ab - 6b^2}$

94. $\dfrac{x^2 + 2xy}{x + 2y + x^2 - 4y^2}$

95. $\dfrac{x - y}{x^3 - y^3 - x + y}$

96. $\dfrac{(m + n)^3}{m^2 + 2mn + n^2}$

97. $\dfrac{px - py + qx - qy}{px + qx + py + qy}$

98. $\dfrac{6xy - 4x - 9y + 6}{6y^2 - 13y + 6}$

99. $\dfrac{m^3 - mn^2}{mn^2 + m^2n - 2m^3}$

100. $\dfrac{p^3 + p^2q - 2pq^2}{pq^2 + p^2q - 2p^3}$

101. $\dfrac{3x^2 - 3y^2}{x^2 + 2y + 2x + yx}$

102. $\dfrac{ax + by + ay + bx}{a^2 - b^2}$

The function $f(t) = \frac{t^2 + 2t}{2t + 2}$ gives the number of days it would take two construction crews, working together, to frame a house that crew 1 (working alone) could complete in t days and crew 2 (working alone) could complete in t + 2 days.

103. If crew 1 can frame a certain house in 15 days, how long would it take both crews working together?

104. If crew 2 can frame a certain house in 20 days, how long would it take both crews working together?

The function $f(t) = \frac{t^2 + 3t}{2t + 3}$ gives the number of hours it would take two pipes to fill a pool that the larger pipe (working alone) could fill in t hours and the smaller pipe (working alone) could fill in t + 3 hours.

105. If the smaller pipe can fill a pool in 7 hours, how long would it take both pipes to fill the pool?

106. If the larger pipe can fill a pool in 8 hours, how long would it take both pipes to fill the pool?

APPLICATIONS

107. Environmental cleanup   The cost (in dollars) of removing p% of the pollution in a river is approximated by the rational function

$$f(p) = \frac{50{,}000p}{100 - p} \quad (0 \le p < 100)$$

Find the cost of removing each percent of pollution.

a. 40%   **b.** 70%

108. Directory costs   The average (mean) cost for a service club to publish a directory of its members is given by the rational function

$$f(x) = \frac{1.25x + 800}{x}$$

where x is the number of directories printed. Find the average cost per directory if

a. 700 directories are printed.

b. 2,500 directories are printed.

A service club wants to publish a directory of its members. Some investigation shows that the cost of typesetting and photography will be \$700 and the cost of printing each directory will be \$1.25.

109. Find a function that gives the total cost c of printing x directories.

110. Find a function that gives the mean cost per directory, $\bar{c}$, of printing x directories.

111. Find the total cost of printing 500 directories.

112. Find the mean cost per directory if 500 directories are printed.

113. Find the mean cost per directory if 1,000 directories are printed.

114. Find the mean cost per directory if 2,000 directories are printed.

An electric company charges $7.50 per month plus 9¢ for each kilowatt hour (kWh) of electricity used.

115. Find a function that gives the total cost c of n kWh of electricity.

116. Find a function that gives the mean cost per kWh, $\bar{c}$, when using n kWh.

117. Find the total cost for using 775 kWh.

118. Find the mean cost per kWh when 775 kWh are used.

119. Find the mean cost per kWh when 1,000 kWh are used.

120. Find the mean cost per kWh when 1,200 kWh are used.

WRITING ABOUT MATH

121. Explain how to simplify a rational expression.

122. Explain how to recognize that a rational expression is in lowest terms.

SOMETHING TO THINK ABOUT

123. A student compares an answer of $\frac{a-3b}{2b-a}$ to an answer of $\frac{3b-a}{a-2b}$. Are the two answers the same?

124. Is this work correct? Explain.
$$\frac{3x^2+6}{3y} = \frac{\cancel{3}x^2+6}{\cancel{3}y} = \frac{x^2+6}{y}$$

125. In which parts can you divide out the 4s?

  a. $\dfrac{4x}{4y}$ **b.** $\dfrac{4x}{x+4}$

  c. $\dfrac{4+x}{4+y}$ **d.** $\dfrac{4x}{4+4y}$

126. In which parts can you divide out the 3s?

  a. $\dfrac{3x+3y}{3z}$ **b.** $\dfrac{3(x+y)}{3x+y}$

  c. $\dfrac{3x+3y}{3a-3b}$ **d.** $\dfrac{x+3}{3y}$

Section 6.2

Multiplying and Dividing Rational Expressions

Objectives

1. Multiply two rational expressions and write the result in simplest form.
2. Find the square of a rational expression.
3. Divide two rational expressions and write the result in simplest form.
4. Perform operations on three or more rational expressions.

Getting Ready

Simplify each fraction.

1. $\dfrac{45}{30}$ **2.** $\dfrac{72}{180}$ **3.** $\dfrac{600}{450}$ **4.** $\dfrac{210}{45}$

In this section, we will learn how to multiply and divide rational expressions. After mastering these skills, we will consider mixed operations.

1 Multiply two rational expressions and write the result in simplest form.

In Section 6.1, we reviewed four basic properties of fractions. We now review the process for multiplying fractions.

MULTIPLYING FRACTIONS

If no denominators are 0, then

$$\frac{a}{b} \cdot \frac{c}{d} = \frac{a \cdot c}{b \cdot d} = \frac{ac}{bd}$$

Thus, *to multiply two fractions, we multiply the numerators and multiply the denominators.*

$$\frac{3}{5} \cdot \frac{2}{7} = \frac{3 \cdot 2}{5 \cdot 7}$$

$$= \frac{6}{35}$$

$$\frac{4}{7} \cdot \frac{5}{8} = \frac{4 \cdot 5}{7 \cdot 8}$$

$$= \frac{\overset{1}{2} \cdot \overset{1}{2} \cdot 5}{7 \cdot \underset{1}{2} \cdot \underset{1}{2} \cdot 2} \qquad \frac{2}{2} = 1$$

$$= \frac{5}{14}$$

The same rule applies to rational expressions. If $t \neq 0$, then

$$\frac{x^2 y}{t} \cdot \frac{xy^3}{t^3} = \frac{x^2 y \cdot xy^3}{tt^3}$$

$$= \frac{x^2 x \cdot yy^3}{t^4}$$

$$= \frac{x^3 y^4}{t^4}$$

EXAMPLE 1   Find the product of $\dfrac{x^2 - 6x + 9}{x}$ and $\dfrac{x^2}{x - 3}$.   Assume no denominators are 0.

Solution   We multiply the numerators and multiply the denominators and then simplify.

$$\frac{x^2 - 6x + 9}{x} \cdot \frac{x^2}{x - 3} = \frac{(x^2 - 6x + 9)x^2}{x(x - 3)} \qquad \text{Multiply the numerators and multiply the denominators.}$$

$$= \frac{(x - 3)(x - 3)xx}{x(x - 3)} \qquad \text{Factor the numerator.}$$

$$= \frac{\overset{1}{(x - 3)}(x - 3)\overset{1}{x}x}{\underset{1}{x(x - 3)}} \qquad \frac{x - 3}{x - 3} = 1 \text{ and } \frac{x}{x} = 1$$

$$= x(x - 3)$$

🌿 **SELF CHECK 1**   Multiply: $\dfrac{a^2 + 6a + 9}{a} \cdot \dfrac{a^3}{a + 3}$ Assume no denominators are 0.

Accent on technology

▶ Checking an Algebraic Simplification

We can check the simplification in Example 1 by graphing the rational functions $f(x) = \left(\frac{x^2 - 6x + 9}{x}\right)\left(\frac{x^2}{x - 3}\right)$ and $g(x) = x(x - 3)$ as in Figure 6-4 and observing that the graphs look the same. Note that 0 and 3 are not included in the domain of the first function.

(a)   (b)

Figure 6-4

For instruction regarding the use of a Casio graphing calculator, please refer to the Casio Keystroke Guide in the back of the book.

EXAMPLE 2   Multiply: $\dfrac{x^2 - x - 6}{x^2 - 4} \cdot \dfrac{x^2 + x - 6}{x^2 - 9}$ Assume no denominators are 0.

Solution $\dfrac{x^2 - x - 6}{x^2 - 4} \cdot \dfrac{x^2 + x - 6}{x^2 - 9}$

$= \dfrac{(x^2 - x - 6)(x^2 + x - 6)}{(x^2 - 4)(x^2 - 9)}$ Multiply the numerators and multiply the denominators.

$= \dfrac{(x - 3)(x + 2)(x + 3)(x - 2)}{(x + 2)(x - 2)(x + 3)(x - 3)}$ Factor the polynomials.

$= \dfrac{\overset{1}{\cancel{(x - 3)}}\overset{1}{\cancel{(x + 2)}}\overset{1}{\cancel{(x + 3)}}\overset{1}{\cancel{(x - 2)}}}{\underset{1}{\cancel{(x + 2)}}\,\underset{1}{\cancel{(x - 2)}}\,\underset{1}{\cancel{(x + 3)}}\,\underset{1}{\cancel{(x - 3)}}}$ $\frac{x - 3}{x - 3} = 1, \frac{x + 2}{x + 2} = 1, \frac{x + 3}{x + 3} = 1,$ and $\frac{x - 2}{x - 2} = 1$

COMMENT   Note that when all factors divide out, the result is 1 and not 0.

$= 1$

SELF CHECK 2   Multiply: $\dfrac{a^2 + a - 6}{a^2 - 9} \cdot \dfrac{a^2 - a - 6}{a^2 - 4}$ Assume no denominators are 0.

EXAMPLE 3   Multiply: $\dfrac{6x^2 + 5x - 4}{2x^2 + 5x + 3} \cdot \dfrac{8x^2 + 6x - 9}{12x^2 + 7x - 12}$ Assume no denominators are 0.

Solution $\dfrac{6x^2 + 5x - 4}{2x^2 + 5x + 3} \cdot \dfrac{8x^2 + 6x - 9}{12x^2 + 7x - 12}$

$= \dfrac{(6x^2 + 5x - 4)(8x^2 + 6x - 9)}{(2x^2 + 5x + 3)(12x^2 + 7x - 12)}$ Multiply the numerators and multiply the denominators.

$= \dfrac{(3x + 4)(2x - 1)(4x - 3)(2x + 3)}{(2x + 3)(x + 1)(3x + 4)(4x - 3)}$ Factor the polynomials.

$= \dfrac{\overset{1}{\cancel{(3x + 4)}}(2x - 1)\overset{1}{\cancel{(4x - 3)}}\overset{1}{\cancel{(2x + 3)}}}{\underset{1}{\cancel{(2x + 3)}}(x + 1)\underset{1}{\cancel{(3x + 4)}}\underset{1}{\cancel{(4x - 3)}}}$ $\frac{3x + 4}{3x + 4} = 1, \frac{4x - 3}{4x - 3} = 1,$ and $\frac{2x + 3}{2x + 3} = 1$

$= \dfrac{2x - 1}{x + 1}$

SELF CHECK 3   Multiply: $\dfrac{2a^2 + 5a - 12}{2a^2 + 11a + 12} \cdot \dfrac{2a^2 - 3a - 9}{2a^2 - a - 3}$ Assume no denominators are 0.

EXAMPLE 4   Multiply: $(2x - x^2) \cdot \dfrac{x}{x^2 - 5x + 6}$ Assume no denominators are 0.

Solution

$(2x - x^2) \cdot \dfrac{x}{x^2 - 5x + 6}$

COMMENT In Examples 1–4, we can obtain the same answers if we factor first and divide out the common factors before we multiply.

$= \dfrac{2x - x^2}{1} \cdot \dfrac{x}{x^2 - 5x + 6}$ Write $2x - x^2$ as $\dfrac{2x - x^2}{1}$.

$= \dfrac{(2x - x^2)x}{1(x^2 - 5x + 6)}$ Multiply the fractions.

$= \dfrac{x(\overset{-1}{\cancel{2 - x}})x}{(\cancel{x - 2})(x - 3)}$ Factor out x and note that the quotient of any nonzero quantity and its negative is -1.

$= \dfrac{-x^2}{x - 3}$

Since $\dfrac{-a}{b} = -\dfrac{a}{b}$, the $-$ sign can be written in front of the fraction. For this reason, the final result can be written as

$-\dfrac{x^2}{x - 3}$

SELF CHECK 4   Multiply: $(x^2 + 5x + 6) \cdot \dfrac{2x^2}{2x^2 + 4x}$ Assume no denominators are 0.

2 Find the square of a rational expression.

To square a rational expression, we can write the expression twice and then multiply as we have in the previous examples.

EXAMPLE 5   Find: $\left(\dfrac{x^2 + x - 1}{2x + 3}\right)^2$ Assume no denominators are 0.

Solution   To square the expression, we write it as a factor twice and perform the multiplication.

$\left(\dfrac{x^2 + x - 1}{2x + 3}\right)^2 = \left(\dfrac{x^2 + x - 1}{2x + 3}\right)\left(\dfrac{x^2 + x - 1}{2x + 3}\right)$

$= \dfrac{(x^2 + x - 1)(x^2 + x - 1)}{(2x + 3)(2x + 3)}$ Multiply the numerators and multiply the denominators.

$= \dfrac{x^4 + 2x^3 - x^2 - 2x + 1}{4x^2 + 12x + 9}$ Do the multiplications.

SELF CHECK 5   Find: $\left(\dfrac{x + 5}{x^2 - 6x}\right)^2$ Assume no denominators are 0.

Divide two rational expressions and write the result in simplest form.

Recall the process for dividing fractions.

DIVIDING FRACTIONS

If no denominators are 0, then

$$\frac{a}{b} \div \frac{c}{d} = \frac{a}{b} \cdot \frac{d}{c} = \frac{ad}{bc}$$

Thus, *to divide two fractions, we can invert the divisor and multiply.*

$$\frac{3}{5} \div \frac{2}{7} = \frac{3}{5} \cdot \frac{7}{2}$$

$$= \frac{3 \cdot 7}{5 \cdot 2}$$

$$= \frac{21}{10}$$

$$\frac{4}{7} \div \frac{2}{21} = \frac{4}{7} \cdot \frac{21}{2}$$

$$= \frac{4 \cdot 21}{7 \cdot 2}$$

$$= \frac{\overset{1}{2} \cdot 2 \cdot 3 \cdot \overset{1}{7}}{\underset{1}{7} \cdot \underset{1}{2}}$$

$$= 6$$

The same rule applies to rational expressions.

$$\frac{x^2}{y^3 z^2} \div \frac{x^2}{yz^3} = \frac{x^2}{y^3 z^2} \cdot \frac{yz^3}{x^2} \qquad \text{Invert the divisor and multiply.}$$

$$= \frac{x^2 yz^3}{x^2 y^3 z^2} \qquad \text{Multiply the numerators and the denominators.}$$

$$= x^{2-2} y^{1-3} z^{3-2} \qquad \text{To divide exponential expressions with the same base, keep the base and subtract the exponents.}$$

$$= x^0 y^{-2} z^1 \qquad \text{Simplify the exponents.}$$

$$= 1 \cdot y^{-2} \cdot z \qquad x^0 = 1$$

$$= \frac{z}{y^2} \qquad y^{-2} = \frac{1}{y^2}$$

EXAMPLE 6 Divide: $\dfrac{x^3 + 8}{x + 1} \div \dfrac{x^2 - 2x + 4}{2x^2 - 2}$ Assume no denominators are 0.

Solution We invert the divisor and multiply.

$$\frac{x^3 + 8}{x + 1} \div \frac{x^2 - 2x + 4}{2x^2 - 2}$$

$$= \frac{x^3 + 8}{x + 1} \cdot \frac{2x^2 - 2}{x^2 - 2x + 4} \qquad \text{Invert the divisor and multiply.}$$

$$= \frac{(x^3 + 8)(2x^2 - 2)}{(x + 1)(x^2 - 2x + 4)} \qquad \text{Multiply the numerators and the denominators.}$$

$$= \frac{(x + 2)\overset{1}{\cancel{(x^2 - 2x + 4)}} 2 \overset{1}{\cancel{(x + 1)}}(x - 1)}{\underset{1}{\cancel{(x + 1)}} \underset{1}{\cancel{(x^2 - 2x + 4)}}} \qquad \frac{x^2 - 2x + 4}{x^2 - 2x + 4} = 1, \frac{x + 1}{x + 1} = 1$$

$$= 2(x + 2)(x - 1)$$

 SELF CHECK 6 Divide: $\dfrac{x^3 - 8}{x - 1} \div \dfrac{x^2 + 2x + 4}{3x^2 - 3x}$ Assume no denominators are 0.

EXAMPLE 7 Divide: $\dfrac{x^2 - 4}{x - 1} \div (x - 2)$ Assume no denominators are 0.

Solution $\dfrac{x^2 - 4}{x - 1} \div (x - 2)$

$= \dfrac{x^2 - 4}{x - 1} \div \dfrac{x - 2}{1}$ Write $(x - 2)$ as a fraction with a denominator of 1.

$= \dfrac{x^2 - 4}{x - 1} \cdot \dfrac{1}{x - 2}$ Invert the divisor and multiply.

$= \dfrac{x^2 - 4}{(x - 1)(x - 2)}$ Multiply the numerators and the denominators.

$= \dfrac{(x + 2)\cancel{(x - 2)}}{(x - 1)\cancel{(x - 2)}}$ Factor $(x^2 - 4)$ and divide out the common factor, $(x - 2)$.

$= \dfrac{x + 2}{x - 1}$

SELF CHECK 7 Divide: $\dfrac{a^3 - 9a}{a - 3} \div (a^2 + 3a)$ Assume no denominators are 0.

---

4 Perform operations on three or more rational expressions.

The following example involves both a division and a multiplication.

EXAMPLE 8 Simplify: $\dfrac{x^2 + 2x - 3}{6x^2 + 5x + 1} \div \dfrac{2x^2 - 2}{2x^2 - 5x - 3} \cdot \dfrac{6x^2 + 4x - 2}{x^2 - 2x - 3}$ Assume no denominators are 0.

Solution Since multiplications and divisions are done in order from left to right, we write the division as a multiplication by inverting the divisor

$\left(\dfrac{x^2 + 2x - 3}{6x^2 + 5x + 1} \div \dfrac{2x^2 - 2}{2x^2 - 5x - 3} \right) \dfrac{6x^2 + 4x - 2}{x^2 - 2x - 3}$

$= \left(\dfrac{x^2 + 2x - 3}{6x^2 + 5x + 1} \cdot \dfrac{2x^2 - 5x - 3}{2x^2 - 2} \right) \dfrac{6x^2 + 4x - 2}{x^2 - 2x - 3}$

and then multiply the rational expressions and simplify the result.

$= \dfrac{(x^2 + 2x - 3)(2x^2 - 5x - 3)(6x^2 + 4x - 2)}{(6x^2 + 5x + 1)(2x^2 - 2)(x^2 - 2x - 3)}$

$= \dfrac{(x + 3)\cancel{(x - 1)}\cancel{(2x + 1)}\cancel{(x - 3)}2(3x - 1)\cancel{(x + 1)}}{(3x + 1)\cancel{(2x + 1)}2(x + 1)\cancel{(x - 1)}\cancel{(x - 3)}\cancel{(x + 1)}}$ Divide out common factors.

$= \dfrac{(x + 3)(3x - 1)}{(3x + 1)(x + 1)}$

SELF CHECK 8 Simplify: $\dfrac{2x^2 - x - 1}{x^2 + 5x + 4} \div \dfrac{x^2 + x - 2}{2x^2 + 5x + 2} \cdot \dfrac{x^2 + 5x + 4}{4x^2 + 4x + 1}$ Assume no denominators are 0.

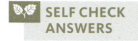
NOW TRY THIS

Simplify.

1. $\dfrac{6x+11}{6} \div \dfrac{5x}{7}$

2. $\dfrac{3x^2-x-2}{x^2-1} \div \dfrac{x^2-3x+2}{3x^2-4x-4}$

3. $\dfrac{3-2x}{5x-2} \cdot \dfrac{x}{2x^2-3x}$

6.2 Exercises

WARM-UPS *Simplify.*

1. $\dfrac{2 \cdot 7}{7 \cdot 2^2}$

2. $\dfrac{3 \cdot 5^2}{5 \cdot 2 \cdot 3}$

3. $\dfrac{3(5-1)}{3^2(1-5)}$

4. $\dfrac{3^2 \cdot 11}{2 \cdot 3 \cdot 11}$

5. $\dfrac{2^2 \cdot 3^3}{2 \cdot 3^2}$

6. $\dfrac{5^4(7-3) \cdot 2^4}{2^3 \cdot 5^6(3-7)}$

REVIEW *Perform each operation.*

7. $-7x^3(4x^4 - x^2)$

8. $(2x-1)^2$

9. $(m^n + 2)(m^n - 2)$

10. $(3b^{-n} + c)(b^{-n} - c)$

VOCABULARY AND CONCEPTS *Fill in the blanks.*

11. $\dfrac{a}{b} \cdot \dfrac{c}{d} = $ ____

12. $\dfrac{a}{b} \div \dfrac{c}{d} = $ ____

13. The denominator of a fraction cannot be __.

14. $\dfrac{a+1}{a+1} = $ __

GUIDED PRACTICE *Perform the operations and simplify. Assume no division by 0.* **(OBJECTIVE 1)**

15. $\dfrac{10}{9} \cdot \dfrac{3}{5} \cdot \dfrac{7}{8}$

16. $-\dfrac{5}{6} \cdot \dfrac{3}{7} \cdot \dfrac{14}{25}$

17. $-\dfrac{8}{13} \div \dfrac{12}{26}$

18. $\dfrac{17}{12} \div \dfrac{34}{3}$

19. $\dfrac{x^2y^2}{cd} \cdot \dfrac{c^{-2}d^2}{x}$

20. $\dfrac{a^{-2}b^2}{x^{-1}y} \cdot \dfrac{a^4b^4}{x^2y^3}$

21. $\dfrac{-x^2y^{-2}}{x^{-1}y^{-3}} \div \dfrac{x^{-3}y^2}{x^4y^{-1}}$

22. $\dfrac{(a^3)^2}{b^{-1}} \div \dfrac{(a^3)^{-2}}{b^{-1}}$

Perform the operations and simplify. **SEE EXAMPLES 1–2. (OBJECTIVE 1)**

23. $\dfrac{x^2-4}{x^3} \cdot \dfrac{x^2}{x+2}$

24. $\dfrac{a+6}{a^2-16} \cdot \dfrac{3a-12}{3a+18}$

25. $\dfrac{x^2-x-6}{x^2-4} \cdot \dfrac{x^2-x-2}{9-x^2}$

26. $\dfrac{x^4-3x^2-4}{x^4-1} \cdot \dfrac{x^2+3x+2}{x^2+4x+4}$

27. $\dfrac{x^3+2x^2+4x+8}{y^2-1} \cdot \dfrac{y^2+2y+1}{x^4-16}$

28. $\dfrac{x^2+3x+yx+3y}{x^2-9} \cdot \dfrac{x-3}{x+3}$

29. $\dfrac{m^2-n^2}{2x^2+3x-2} \cdot \dfrac{2x^2+5x-3}{n^2-m^2}$

30. $\dfrac{p^3-q^3}{q^2-p^2} \cdot \dfrac{q^2+pq}{p^3+p^2q+pq^2}$

Perform the operations and simplify. **SEE EXAMPLE 3. (OBJECTIVE 1)**

31. $\dfrac{2x^2-x-3}{x^2-1} \cdot \dfrac{x^2+x-2}{2x^2+x-6}$

32. $\dfrac{3t^2-t-2}{6t^2-5t-6} \cdot \dfrac{4t^2-9}{2t^2+5t+3}$

33. $\dfrac{2p^2-5p-3}{p^2-9} \cdot \dfrac{2p^2+5p-3}{2p^2+5p+2}$

34. $\dfrac{9x^2+3x-20}{3x^2-7x+4} \cdot \dfrac{3x^2-5x+2}{9x^2+18x+5}$

Perform the operations and simplify. SEE EXAMPLE 4. (OBJECTIVE 1)

35. $(9 - x^2) \cdot \dfrac{x}{x^3 - 4x^2 + 3x}$

36. $(x^2 - x - 2) \cdot \dfrac{x^2 + 3x + 2}{x^2 - 4}$

37. $\dfrac{x}{2x^2 + x} \cdot (2x^2 - 9x - 5)$

38. $\dfrac{2x^2 - x - 1}{x - 2} \cdot (2x^2 - 3x - 2)$

Find the square of each rational expression. SEE EXAMPLE 5. (OBJECTIVE 2)

39. $\left(\dfrac{x - 7}{x^2 + 3}\right)^2$

40. $\left(\dfrac{4y^3 - y}{y + 2}\right)^2$

41. $\left(\dfrac{2m^2 - m - 3}{x^2 - 1}\right)^2$

42. $\left(\dfrac{-k - 3}{x^2 - x + 1}\right)^2$

Perform the operations and simplify. SEE EXAMPLE 6. (OBJECTIVE 3)

43. $\dfrac{x^2 - 36}{x^2 - 81} \div \dfrac{x + 6}{x - 9}$

44. $\dfrac{a^2 - 9}{a^2 - 49} \div \dfrac{a + 3}{a + 7}$

45. $\dfrac{3n^2 + 5n - 2}{12n^2 - 13n + 3} \div \dfrac{n^2 + 3n + 2}{4n^2 + 5n - 6}$

46. $\dfrac{8y^2 - 14y - 15}{6y^2 - 11y - 10} \div \dfrac{4y^2 - 9y - 9}{3y^2 - 7y - 6}$

47. $\dfrac{a^2 + 2a - 35}{12x} \div \dfrac{ax - 3x}{a^2 + 4a - 21}$

48. $\dfrac{x^2 - 4}{2b - bx} \div \dfrac{x^2 + 4x + 4}{2b + bx}$

49. $\dfrac{x^2 - 6x + 9}{4 - x^2} \div \dfrac{x^2 - 9}{x^2 - 8x + 12}$

50. $\dfrac{2x^2 - 7x - 4}{20 - x - x^2} \div \dfrac{2x^2 - 9x - 5}{x^2 - 25}$

Perform the operations and simplify. SEE EXAMPLE 7. (OBJECTIVE 3)

51. $\dfrac{x^2 - 4}{x} \div (x + 2)$

52. $(2x^2 - 15x + 25) \div \dfrac{2x^2 - 3x - 5}{x + 1}$

53. $\dfrac{x^3 - 25x}{x^2} \div (x^2 - 10x + 25)$

54. $\dfrac{3x^2 - 2x - 1}{x^2 - 1} \div (x + 1)$

Perform the operations and simplify. SEE EXAMPLE 8. (OBJECTIVE 4)

55. $\dfrac{3x^2y^2}{6x^3y} \cdot \dfrac{-4x^7y^{-2}}{18x^{-2}y} \div \dfrac{36x}{18y^{-2}}$

56. $\dfrac{9ab^3}{7xy} \cdot \dfrac{14xy^2}{27z^3} \div \dfrac{18a^2b^2x}{3z^2}$

57. $(5x + 10) \cdot \dfrac{x^3}{10x - 20} \div \dfrac{2x + 4}{4}$

58. $(4x^2 - 9) \div \dfrac{2x^2 + 5x + 3}{x + 2} \div (2x - 3)$

59. $\dfrac{2x^2 - 2x - 4}{x^2 + 2x - 8} \cdot \dfrac{3x^2 + 15x}{x + 1} \div \dfrac{4x^2 - 100}{x^2 - x - 20}$

60. $\dfrac{6a^2 - 7a - 3}{a^2 - 1} \div \dfrac{4a^2 - 12a + 9}{a^2 - 1} \cdot \dfrac{2a^2 - a - 3}{3a^2 - 2a - 1}$

61. $\dfrac{2t^2 + 5t + 2}{t^2 - 4t + 16} \div \dfrac{t + 2}{t^3 + 64} \div \dfrac{2t^3 + 9t^2 + 4t}{t + 1}$

62. $\dfrac{a^6 - b^6}{a^4 - a^3b} \cdot \dfrac{a^3}{a^4 + a^2b^2 + b^4} \div \dfrac{1}{a}$

ADDITIONAL PRACTICE *Simplify each expression. Assume no division by 0.*

63. $\dfrac{x^2 - y^2}{2x^2 + 2xy + x + y} \cdot \dfrac{2x^2 - 5x - 3}{yx - 3y - x^2 + 3x}$

64. $\dfrac{ax + ay + bx + by}{x^3 - 27} \cdot \dfrac{x^2 + 3x + 9}{xc + xd + yc + yd}$

65. $\dfrac{x^3 + y^3}{x^3 - y^3} \div \dfrac{x^2 - xy + y^2}{x^2 + xy + y^2}$

66. $\dfrac{2x^2 + 3xy + y^2}{y^2 - x^2} \div \dfrac{6x^2 + 5xy + y^2}{2x^2 - xy - y^2}$

67. $(x^2 - 7x + 10) \div (x - 2) \div (x + 5)$

68. $(x^2 - x - 6) \div [(x - 3) \div (x - 2)]$

69. $\dfrac{3x^2 - 2x}{3x + 2} \div (3x - 2) \div \dfrac{3x}{3x - 3}$

70. $(2x^2 - 3x - 2) \div \dfrac{2x^2 - x - 1}{x - 2} \div (x - 1)$

71. $\dfrac{2x^2 + 5x - 3}{x^2 + 2x - 3} \div \left(\dfrac{x^2 + 2x - 35}{x^2 - 6x + 5} \div \dfrac{x^2 - 9x + 14}{2x^2 - 5x + 2}\right)$

72. $\dfrac{x^2 - 4}{x^2 - x - 6} \div \left(\dfrac{x^2 - x - 2}{x^2 - 8x + 15} \cdot \dfrac{x^2 - 3x - 10}{x^2 + 3x + 2}\right)$

73. $\dfrac{x^2 - x - 12}{x^2 + x - 2} \div \dfrac{x^2 - 6x + 8}{x^2 - 3x - 10} \cdot \dfrac{x^2 - 3x + 2}{x^2 - 2x - 15}$

74. $\dfrac{4x^2 - 10x + 6}{x^4 - 3x^3} \div \dfrac{2x - 3}{2x^3} \cdot \dfrac{x - 3}{2x - 2}$

APPLICATIONS

75. **Area of a triangle**
Find an expression that represents the area of the triangle.

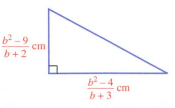

$\dfrac{b^2 - 9}{b + 2}$ cm

$\dfrac{b^2 - 4}{b + 3}$ cm

76. Storage capacity Find the capacity of the trunk.

$\dfrac{x+2}{x^2+3x}$ in.

$\dfrac{2x+8}{x^2+4x+4}$ in.

$\dfrac{x^2+3x+2}{x+4}$ in.

77. Distance, rate, and time Complete the following table.

Rate (mph)	Time (hr)	Distance (m)
$\dfrac{k^2-k-6}{k-4}$	$\dfrac{k^2-16}{k^2-2k-3}$	

78. Lab experiments The following table shows data obtained from a physics experiment in which k and k_1 are constants. Complete the table.

Trial	Rate (m/sec)	Time (sec)	Distance (m)
1	$\dfrac{k_1^2+3k_1+2}{k_1-3}$	$\dfrac{k_1^2-3k_1}{k_1+1}$	
2	$\dfrac{k_2^2+6k_2+5}{k_2+1}$	k_2+6	$k_2^2+11k_2+30$

WRITING ABOUT MATH

79. Explain how to multiply two rational expressions.

80. Explain how to divide one rational expression by another.

SOMETHING TO THINK ABOUT
Insert either a multiplication or a division symbol in each box to make a true statement.

81. $\dfrac{x^2}{y}\ \square\ \dfrac{x}{y^2}\ \square\ \dfrac{x^2}{y^3}=\dfrac{x^3}{y^2}$

82. $\dfrac{x^2}{y}\ \square\ \dfrac{x}{y^2}\ \square\ \dfrac{x^2}{y^2}=\dfrac{y^3}{x}$

Section 6.3

Adding and Subtracting Rational Expressions

Objectives

1. Add and subtract two rational expressions with like denominators.
2. Find the least common denominator of two or more rational expressions.
3. Add and subtract two rational expressions with unlike denominators.
4. Perform operations on three or more rational expressions.

Getting Ready

Determine whether the fractions are equal.

1. $\dfrac{3}{5}, \dfrac{3}{5}$

2. $\dfrac{3}{7}, \dfrac{18}{40}$

3. $\dfrac{8}{13}, \dfrac{40}{65}$

4. $\dfrac{7}{25}, \dfrac{42}{150}$

5. $\dfrac{23}{32}, \dfrac{46}{66}$

6. $\dfrac{8}{9}, \dfrac{64}{72}$

We now discuss how to add and subtract rational expressions. After mastering these skills, we will simplify expressions that involve more than one operation.

1 Add and subtract two rational expressions with like denominators.

Rational expressions with like denominators are added and subtracted according to the following procedures.

ADDING AND SUBTRACTING RATIONAL EXPRESSIONS

If no denominators are 0, then

$$\frac{a}{b} + \frac{c}{b} = \frac{a+c}{b} \quad \text{and} \quad \frac{a}{b} - \frac{c}{b} = \frac{a-c}{b}$$

In words, *we add (or subtract) rational expressions with like denominators by adding (or subtracting) the numerators and keeping the common denominator.* Whenever possible, we should simplify the result.

EXAMPLE 1   Simplify; assume no denominators are 0.   **a.** $\frac{17}{22} + \frac{13}{22}$ **b.** $\frac{3}{2x} + \frac{7}{2x}$

c. $\frac{4x}{x+2} - \frac{7x}{x+2}$

Solution   **a.** $\frac{17}{22} + \frac{13}{22} = \frac{17+13}{22}$ **b.** $\frac{3}{2x} + \frac{7}{2x} = \frac{3+7}{2x}$

$$= \frac{30}{22} \qquad\qquad\qquad = \frac{10}{2x}$$

$$= \frac{15 \cdot 2}{11 \cdot 2} \qquad\qquad\quad = \frac{2 \cdot 5}{2 \cdot x}$$

$$= \frac{15}{11} \qquad\qquad\qquad = \frac{5}{x}$$

c. $\frac{4x}{x+2} - \frac{7x}{x+2} = \frac{4x-7x}{x+2}$

$$= \frac{-3x}{x+2}$$

SELF CHECK 1   Simplify; assume no denominators are 0.   **a.** $\frac{5}{3a} - \frac{2}{3a}$ **b.** $\frac{3a}{a-2} + \frac{2a}{a-2}$

Accent on technology

▸Checking Algebra

We can check the subtraction in Example 1(c) by graphing the rational functions $f(x) = \frac{4x}{x+2} - \frac{7x}{x+2}$, shown in Figure 6-5(a), and $g(x) = \frac{-3x}{x+2}$, shown in Figure 6-5(b), and observing that the graphs appear the same. Note that there is a vertical asymptote at $x = -2$ because -2 is not in the domain of either function.

(a)   (b)

Figure 6-5

For instruction regarding the use of a Casio graphing calculator, please refer to the Casio Keystroke Guide in the back of the book.

2 **Find the least common denominator of two or more rational expressions.**

When adding fractions with unlike denominators, we write the fractions as fractions having the smallest common denominator possible, called the *least* (or lowest) *common denominator (LCD)*.

Suppose we have the fractions $\frac{1}{12}$, $\frac{1}{20}$, and $\frac{1}{35}$. To find the LCD of these fractions, we first find the prime factorizations of each denominator.

$$12 = 4 \cdot 3 = 2^2 \cdot 3$$
$$20 = 4 \cdot 5 = 2^2 \cdot 5$$
$$35 = 5 \cdot 7$$

Since the LCD is the smallest number that can be exactly divided by 12, 20, and 35, it must contain factors of $2^2, 3, 5,$ and 7, and we have

$$\text{LCD} = 2^2 \cdot 3 \cdot 5 \cdot 7 = 420$$

We can find the LCD of several rational expressions in a similar way by following these steps.

Charles Babbage
1792–1871
In 1823, Babbage built a steam-powered digital calculator, which he called a difference engine. Thought to be a crackpot by his London neighbors, Babbage was a visionary. His machine embodied principles still used in modern computers.

FINDING THE LCD

1. Factor the denominator of each rational expression into its prime factors.
2. List the different factors of each denominator.
3. Write each factor found in Step 2 to the highest power that occurs in any one factorization.
4. The LCD is the product of the factors to their highest powers found in Step 3.

EXAMPLE 2   Find the LCD of $\dfrac{1}{x^2 + 7x + 6}$, $\dfrac{3}{x^2 - 36}$, and $\dfrac{5}{x^2 + 12x + 36}$. Assume no denominators are 0.

Solution   We factor each denominator:

$$x^2 + 7x + 6 = (x + 6)(x + 1)$$
$$x^2 - 36 = (x + 6)(x - 6)$$
$$x^2 + 12x + 36 = (x + 6)(x + 6) = (x + 6)^2$$

and list the individual factors:

$$x + 6, \qquad x + 1, \qquad \text{and} \qquad x - 6$$

To find the LCD, we use the highest power of each of these factors:

$$\text{LCD} = (x + 6)^2(x + 1)(x - 6)$$

❧ SELF CHECK 2   Find the LCD of $\dfrac{1}{a^2 - 25}$ and $\dfrac{7}{a^2 + 7a + 10}$.   Assume no denominators are 0.

3 Add and subtract two rational expressions with unlike denominators.

To add or subtract rational expressions with unlike denominators, we write them as expressions with a common denominator. When the denominators are negatives (opposites), we can multiply one of them by 1, written in the form $\frac{-1}{-1}$, to obtain a common denominator.

EXAMPLE 3   Add: $\dfrac{x}{x-y} + \dfrac{y}{y-x}$ Assume no denominator is 0.

Solution

$$\frac{x}{x-y} + \frac{y}{y-x} = \frac{x}{x-y} + \left(\frac{-1}{-1}\right)\frac{y}{y-x} \qquad \frac{-1}{-1} = 1; \text{ multiplying a fraction by}$$
$$\text{1 does not change its value.}$$

$$= \frac{x}{x-y} + \frac{-y}{-y+x} \qquad \text{Multiply.}$$

$$= \frac{x}{x-y} + \frac{-y}{x-y} \qquad \text{Apply the commutative property of addition.}$$

$$= \frac{x-y}{x-y} \qquad \text{Add the numerators and keep the common denominator.}$$

$$= 1 \qquad \text{Simplify.}$$

SELF CHECK 3   Add: $\dfrac{2a}{a-b} + \dfrac{b}{b-a}$ Assume no denominators are 0.

If the operation is addition or subtraction of two or more rational expressions and the denominators are different, we often have to multiply one or more of them by 1, written in some appropriate form, to obtain a common denominator. Each fraction must be written using this new denominator, the LCD, prior to adding or subtracting.

EXAMPLE 4   Simplify; assume no denominators are 0.   **a.** $\dfrac{2}{3} - \dfrac{3}{2}$ **b.** $\dfrac{3}{x} + \dfrac{4}{y}$

Solution   **a.** $\dfrac{2}{3} - \dfrac{3}{2} = \dfrac{2}{3} \cdot \dfrac{2}{2} - \dfrac{3}{2} \cdot \dfrac{3}{3}$ $\frac{2}{2} = 1$ and $\frac{3}{3} = 1$. The LCD is 6.

$$= \frac{2 \cdot 2}{3 \cdot 2} - \frac{3 \cdot 3}{2 \cdot 3} \qquad \text{Multiply the fractions by multiplying their numerators and their denominators.}$$

$$= \frac{4}{6} - \frac{9}{6} \qquad \text{Simplify.}$$

$$= \frac{-5}{6} \qquad \text{Subtract the numerators and keep the common denominator.}$$

b. $\dfrac{3}{x} + \dfrac{4}{y} = \dfrac{3}{x} \cdot \dfrac{y}{y} + \dfrac{4}{y} \cdot \dfrac{x}{x}$ $\frac{y}{y} = 1$ and $\frac{x}{x} = 1$. The LCD is xy.

$$= \frac{3y}{xy} + \frac{4x}{xy} \qquad \text{Multiply the fractions by multiplying their numerators and their denominators.}$$

$$= \frac{3y + 4x}{xy} \qquad \text{Add the numerators and keep the common denominator.}$$

SELF CHECK 4   Simplify: $\dfrac{5}{a} - \dfrac{7}{b}$ Assume no denominators are 0.

EXAMPLE 5 Simplify; assume no denominators are 0. **a.** $3 + \dfrac{7}{x-2}$ **b.** $\dfrac{4x}{x+2} - \dfrac{7x}{x-2}$

Solution **a.** $3 + \dfrac{7}{x-2} = \dfrac{3}{1} + \dfrac{7}{x-2}$ $\qquad 3 = \dfrac{3}{1}$

$\qquad\qquad = \dfrac{3(x-2)}{1(x-2)} + \dfrac{7}{x-2}$ $\qquad \dfrac{x-2}{x-2} = 1.$ The LCD is $x-2$.

$\qquad\qquad = \dfrac{3x-6}{x-2} + \dfrac{7}{x-2}$ $\qquad$ Use the distributive property to remove parentheses.

$\qquad\qquad = \dfrac{3x-6+7}{x-2}$ $\qquad$ Add the numerators and keep the common denominator.

$\qquad\qquad = \dfrac{3x+1}{x-2}$ $\qquad$ Simplify.

b. $\dfrac{4x}{x+2} - \dfrac{7x}{x-2}$

$\qquad = \dfrac{4x(x-2)}{(x+2)(x-2)} - \dfrac{(x+2)7x}{(x+2)(x-2)}$ $\qquad$ Write each fraction with the LCD, $(x-2)(x+2)$.

$\qquad = \dfrac{(4x^2-8x)-(7x^2+14x)}{(x+2)(x-2)}$ $\qquad$ Multiply in the numerators, subtract the numerators, and keep the common denominator.

$\qquad = \dfrac{4x^2-8x-7x^2-14x}{(x+2)(x-2)}$ $\qquad$ Use the distributive property to remove parentheses in the numerator.

$\qquad = \dfrac{-3x^2-22x}{(x+2)(x-2)}$ $\qquad$ Simplify the numerator.

$\qquad = \dfrac{-x(3x+22)}{(x+2)(x-2)}$ $\qquad$ Factor the numerator to see whether the fraction can be simplified.

COMMENT The $-$ sign between the fractions in Step 1 of part b applies to both terms of $7x^2 + 14x$.

SELF CHECK 5 Simplify; assume no denominators are 0. **a.** $4 + \dfrac{5}{x-1}$

b. $\dfrac{3a}{a+3} - \dfrac{5a}{a-3}$

EXAMPLE 6 Simplify: $\dfrac{x}{x^2-2x+1} + \dfrac{3}{x^2-1}$ Assume no denominators are 0.

Solution We factor each denominator to find the LCD:

$$x^2 - 2x + 1 = (x-1)(x-1) = (x-1)^2$$
$$x^2 - 1 = (x+1)(x-1)$$

The LCD is $(x-1)^2(x+1)$.

We now write the rational expressions with their denominators in factored form and write them as expressions with an LCD of $(x-1)^2(x+1)$. Then we add.

$$\dfrac{x}{x^2-2x+1} + \dfrac{3}{x^2-1}$$

$$= \dfrac{x}{(x-1)(x-1)} + \dfrac{3}{(x+1)(x-1)}$$ $\qquad$ Factor each denominator.

$$= \dfrac{x(x+1)}{(x-1)(x-1)(x+1)} + \dfrac{3(x-1)}{(x+1)(x-1)(x-1)}$$ $\qquad$ Write each expression with the LCD, $(x-1)^2(x+1)$.

$$= \frac{x^2 + x + 3x - 3}{(x - 1)(x - 1)(x + 1)}$$ Add the numerators and keep the denominators.

$$= \frac{x^2 + 4x - 3}{(x - 1)^2(x + 1)}$$ Combine like terms.

This result does not simplify.

SELF CHECK 6 Simplify: $\dfrac{a}{a^2 - 4a + 4} - \dfrac{2}{a^2 - 4}$ Assume no denominators are 0.

EXAMPLE 7 Simplify: $\dfrac{3x}{x - 1} - \dfrac{2x^2 + 3x - 2}{(x + 1)(x - 1)}$ Assume no denominators are 0.

Solution We write each rational expression in a form having the LCD of $(x + 1)(x - 1)$, remove the resulting parentheses in the first numerator, do the subtraction, and simplify.

$$\frac{3x}{x - 1} - \frac{2x^2 + 3x - 2}{(x + 1)(x - 1)} = \frac{(x + 1)3x}{(x + 1)(x - 1)} - \frac{2x^2 + 3x - 2}{(x + 1)(x - 1)}$$ Write the first expression with the LCD, $(x + 1)(x - 1)$.

$$= \frac{3x^2 + 3x}{(x + 1)(x - 1)} - \frac{2x^2 + 3x - 2}{(x + 1)(x - 1)}$$ Multiply.

$$= \frac{3x^2 + 3x - (2x^2 + 3x - 2)}{(x + 1)(x - 1)}$$ Subtract the numerators and keep the denominators.

$$= \frac{3x^2 + 3x - 2x^2 - 3x + 2}{(x + 1)(x - 1)}$$ Use the distributive property to remove parentheses in the numerator.

$$= \frac{x^2 + 2}{(x + 1)(x - 1)}$$ Combine like terms.

SELF CHECK 7 Simplify: $\dfrac{2a}{a + 2} - \dfrac{a^2 - 4a + 4}{a^2 + a - 2}$ Assume no denominators are 0.

COMMENT Whenever we subtract one rational expression from another, we must remember to subtract each term of the numerator in the second expression.

4 Perform operations on three or more rational expressions.

We can use the previous techniques to simplify expressions that contain both an addition and a subtraction.

EXAMPLE 8 Simplify: $\dfrac{2x}{x^2 - 4} - \dfrac{1}{x^2 - 3x + 2} + \dfrac{x + 1}{x^2 + x - 2}$ Assume no denominators are 0.

Solution We factor each denominator $x^2 - 4 = (x + 2)(x - 2)$, $x^2 - 3x + 2 = (x - 2)(x - 1)$, and $x^2 + x - 2 = (x + 2)(x - 1)$ to find the LCD:

$$\text{LCD} = (x + 2)(x - 2)(x - 1)$$

We then write each rational expression with the LCD as its denominator and do the subtraction and addition.

$$\frac{2x}{x^2 - 4} - \frac{1}{x^2 - 3x + 2} + \frac{x + 1}{x^2 + x - 2}$$

$$= \frac{2x}{(x - 2)(x + 2)} - \frac{1}{(x - 2)(x - 1)} + \frac{x + 1}{(x - 1)(x + 2)}$$

$$= \frac{2x(x - 1)}{(x - 2)(x + 2)(x - 1)} - \frac{1(x + 2)}{(x - 2)(x - 1)(x + 2)} + \frac{(x + 1)(x - 2)}{(x - 1)(x + 2)(x - 2)}$$

$$= \frac{2x(x - 1) - 1(x + 2) + (x + 1)(x - 2)}{(x + 2)(x - 2)(x - 1)}$$ Add the numerators and keep the denominators.

$$= \frac{2x^2 - 2x - x - 2 + x^2 - x - 2}{(x + 2)(x - 2)(x - 1)}$$ Use the distributive property to remove parentheses in the numerator.

$$= \frac{3x^2 - 4x - 4}{(x + 2)(x - 2)(x - 1)}$$ Combine like terms.

$$= \frac{(3x + 2)\cancel{(x - 2)}}{(x + 2)\cancel{(x - 2)}(x - 1)}$$ Factor the numerator and divide out the common factor $(x - 2)$.

$$= \frac{3x + 2}{(x + 2)(x - 1)}$$

SELF CHECK 8   Simplify: $\dfrac{y}{y^2 + y} - \dfrac{1}{y^2 - 1} + \dfrac{2}{y}$ Assume no denominators are 0.

EXAMPLE 9   Simplify: $\left(\dfrac{x^2}{x - 2} + \dfrac{4}{2 - x}\right)^2$ Assume no denominators are 0.

Solution   We do the addition within the parentheses first. Since the denominators are negatives (opposites) of each other, we can write the rational expressions with a common denominator by multiplying both the numerator and denominator of $\dfrac{4}{2 - x}$ by -1. Then we can add, simplify, and square the result.

$$\left(\frac{x^2}{x - 2} + \frac{4}{2 - x}\right)^2 = \left[\frac{x^2}{x - 2} + \frac{(-1)4}{(-1)(2 - x)}\right]^2 \quad \frac{-1}{-1} = 1$$

$$= \left[\frac{x^2}{x - 2} + \frac{-4}{x - 2}\right]^2$$ Use the distributive property to remove parentheses and apply the commutative property of addition.

$$= \left[\frac{x^2 - 4}{x - 2}\right]^2$$ Add.

$$= \left[\frac{(x + 2)(x - 2)}{x - 2}\right]^2$$ Factor the numerator.

$$= (x + 2)^2$$ Divide out the common factor $(x - 2)$.

$$= x^2 + 4x + 4$$ Square the binomial.

SELF CHECK 9   Simplify: $\left(\dfrac{a^2}{a - 3} - \dfrac{9 - 6a}{3 - a}\right)^2$ Assume no denominators are 0.

SELF CHECK ANSWERS

1. a. $\frac{1}{a}$ **b.** $\frac{5a}{a - 2}$ **2.** $(a + 5)(a + 2)(a - 5)$ **3.** $\frac{2a - b}{a - b}$ **4.** $\frac{5b - 7a}{ab}$ **5. a.** $\frac{4x + 1}{x - 1}$ **b.** $\frac{-2a(a + 12)}{(a + 3)(a - 3)}$

6. $\frac{a^2 + 4}{(a - 2)^2(a + 2)}$ **7.** $\frac{a^2 + 2a - 4}{(a + 2)(a - 1)}$ **8.** $\frac{3y^2 - 2y - 2}{y(y + 1)(y - 1)}$ **9.** $a^2 - 6a + 9$

NOW TRY THIS

1. $5x(x - 2)^{-1} - 6(x + 3)^{-1}$

2. $8(x - 2)^{-2} - 30(x - 2)^{-1} + 7$

3. $\left(\dfrac{2}{5} + \dfrac{2}{x}\right) \div \left(\dfrac{x - 2}{x} - \dfrac{3}{5}\right)$

6.3   Exercises

WARM-UPS *Simplify.*

1. $x + x$

2. $3a - a$

3. $5x + (4x + 3)$

4. $2a - (a - 4)$

5. $(3y + 1) - (y + 3)$

6. $(x - 5) + (4x - 5)$

REVIEW *Graph each interval on a number line.*

7. $(-1, 4]$

8. $(-\infty, -5] \cup [4, \infty)$

Solve each formula for the indicated variable.

9. $V = \pi r^2 h$ for h

10. $S = \dfrac{a - lr}{1 - r}$ for a

VOCABULARY AND CONCEPTS *Fill in the blanks.*

11. $\dfrac{a}{b} + \dfrac{c}{b} = $ _____ $(b \neq 0)$

12. $\dfrac{a}{b} - \dfrac{c}{b} = $ _____ $(b \neq 0)$

13. To subtract fractions with like denominators, we _____ the numerators and ____ the common denominator.

14. To add fractions with like denominators, we ___ the numerators and keep the _____ denominator.

15. The abbreviation for the least common denominator is ____.

16. To find the LCD of several fractions, we _____ each denominator and use each factor to the _____ power that it appears in any factorization.

GUIDED PRACTICE *Perform the operations and simplify. Assume no denominators are 0. SEE EXAMPLE 1. (OBJECTIVE 1)*

17. $\dfrac{4}{9} + \dfrac{2}{9}$

18. $\dfrac{5}{11} + \dfrac{2}{11}$

19. $\dfrac{10}{33} - \dfrac{21}{33}$

20. $\dfrac{9}{20} - \dfrac{13}{20}$

21. $\dfrac{5}{6x} + \dfrac{3}{6x}$

22. $\dfrac{5}{3z^2} - \dfrac{6}{3z^2}$

23. $\dfrac{a}{a + b} - \dfrac{7}{a + b}$

24. $\dfrac{x}{x + 4} + \dfrac{5}{x + 4}$

25. $\dfrac{3x}{2x + 2} + \dfrac{x + 4}{2x + 2}$

26. $\dfrac{4y}{y - 4} - \dfrac{16}{y - 4}$

27. $\dfrac{2x}{x - 5} - \dfrac{10}{x - 5}$

28. $\dfrac{9x}{x - y} - \dfrac{9y}{x - y}$

29. $\dfrac{5x}{x + 1} + \dfrac{3}{x + 1} - \dfrac{2x}{x + 1}$

30. $\dfrac{4}{a + 4} - \dfrac{2a}{a + 4} + \dfrac{3a}{a + 4}$

31. $\dfrac{3(x^2 + x)}{x^2 - 5x + 6} + \dfrac{-3(x^2 - x)}{x^2 - 5x + 6}$

32. $\dfrac{2x + 4}{x^2 + 13x + 12} - \dfrac{x + 3}{x^2 + 13x + 12}$

The denominators of several fractions are given. Find the LCD. SEE EXAMPLE 2. (OBJECTIVE 2)

33. $8, 12, 18$

34. $10, 15, 28$

35. $x^2 - 4, x^2 + 2x$

36. $3y^2 - 6y, 3y(y - 4)$

37. $x^3 + 27, x^2 + 6x + 9$

38. $x^3 - 1, x^2 - 2x + 1$

39. $2x^2 + 5x + 3, 4x^2 + 12x + 9, x^2 + 2x + 1$

40. $2x^2 + 5x + 3, 4x^2 + 12x + 9, 4x + 6$

Perform the operations and simplify. Assume no denominators are 0. SEE EXAMPLE 3. (OBJECTIVE 3)

41. $\dfrac{x + 3}{x - 4} - \dfrac{x - 11}{4 - x}$

42. $\dfrac{3 - x}{2 - x} + \dfrac{x - 1}{x - 2}$

43. $\dfrac{2a + 1}{3a - 2} - \dfrac{a - 4}{2 - 3a}$

44. $\dfrac{4}{x - 2} - \dfrac{5}{2 - x}$

Perform the operations and simplify. Assume no denominators are 0. SEE EXAMPLE 4. (OBJECTIVE 3)

45. $\dfrac{1}{5} + \dfrac{1}{4}$

46. $\dfrac{5}{6} + \dfrac{2}{7}$

47. $\dfrac{a}{2} + \dfrac{2a}{5}$

48. $\dfrac{b}{6} + \dfrac{3a}{4}$

49. $\dfrac{2}{7x} + \dfrac{5}{2y}$

50. $\dfrac{3}{8a} + \dfrac{4}{5b}$

51. $\dfrac{3a}{2b} - \dfrac{2b}{3a}$

52. $\dfrac{5m}{2n} - \dfrac{3n}{4m}$

Perform the operations and simplify. Assume no denominators are 0.
SEE EXAMPLE 5. (OBJECTIVE 3)

53. $4 + \dfrac{7}{x - 3}$

54. $\dfrac{4}{b - 2} - 3$

55. $\dfrac{9}{y - 3} + 6$

56. $5 - \dfrac{6}{x + 1}$

57. $\dfrac{3}{x + 2} + \dfrac{5}{x - 4}$

58. $\dfrac{7}{a + 5} - \dfrac{4}{a + 2}$

59. $\dfrac{x + 2}{x + 5} - \dfrac{x - 3}{x + 7}$

60. $\dfrac{7}{x + 3} + \dfrac{4x}{x + 6}$

Perform the operations and simplify. Assume no denominators are 0.
SEE EXAMPLE 6. (OBJECTIVE 3)

61. $\dfrac{x}{x^2 + 5x + 6} + \dfrac{x}{x^2 - 4}$

62. $\dfrac{x}{3x^2 - 2x - 1} + \dfrac{4}{3x^2 + 10x + 3}$

63. $\dfrac{4}{x^2 - 2x - 3} - \dfrac{x}{3x^2 - 7x - 6}$

64. $\dfrac{2a}{a^2 - 2a - 8} + \dfrac{3}{a^2 - 5a + 4}$

Perform the operations and simplify. Assume no denominators are 0.
SEE EXAMPLE 7. (OBJECTIVE 3)

65. $\dfrac{2x + y}{y - 2x} + \dfrac{3x^2 + y}{y^2 - 4xy + 4x^2}$

66. $\dfrac{2x}{x^2 - 1} - \dfrac{x}{x^2 + 2x + 1}$

67. $\dfrac{x}{1 - x} - \dfrac{2x^2 - 1}{x^2 - 2x + 1}$

68. $\dfrac{5}{x} + \dfrac{x + 1}{x^3 - 5x^2}$

Perform the operations and simplify. Assume no denominators are 0.
SEE EXAMPLE 8. (OBJECTIVE 4)

69. $\dfrac{8}{x^2 - 9} + \dfrac{2}{x - 3} - \dfrac{6}{x}$

70. $\dfrac{x}{x^2 - 4} - \dfrac{x}{x + 2} + \dfrac{2}{x}$

71. $\dfrac{x}{x + 1} - \dfrac{x}{1 - x^2} + \dfrac{1}{x}$

72. $\dfrac{y}{y - 2} - \dfrac{2}{y + 2} - \dfrac{-8}{4 - y^2}$

73. $\dfrac{y + 4}{y^2 + 7y + 12} - \dfrac{y - 4}{y + 3} + \dfrac{47}{y + 4}$

74. $\dfrac{3}{x + 1} - \dfrac{2}{x - 1} + \dfrac{x + 3}{x^2 - 1}$

75. $\dfrac{2}{x - 2} + \dfrac{3}{x + 2} - \dfrac{x - 1}{x^2 - 4}$

76. $\dfrac{x - 2}{x^2 - 3x} + \dfrac{2x - 1}{x^2 + 3x} - \dfrac{2}{x^2 - 9}$

Raise each rational expression to the stated power. Assume no denominators are 0. SEE EXAMPLE 9. (OBJECTIVE 3)

77. $\left(\dfrac{1}{x - 1} + \dfrac{1}{1 - x}\right)^2$

78. $\left(\dfrac{1}{a - 1} - \dfrac{1}{1 - a}\right)^2$

79. $\left(\dfrac{x}{x - 3} + \dfrac{3}{3 - x}\right)^3$

80. $\left(\dfrac{2y}{y + 4} + \dfrac{8}{y + 4}\right)^3$

ADDITIONAL PRACTICE *Perform the operations and simplify. Assume no denominators are 0.*

81. $\dfrac{2}{x} - \dfrac{5}{y}$

82. $\dfrac{9}{a} + \dfrac{3}{b}$

83. $2x + 3 + \dfrac{1}{x + 1}$

84. $x + 2 + \dfrac{4}{x - 2}$

85. $\dfrac{2}{x - 1} - \dfrac{2x}{x^2 - 1} - \dfrac{x}{x^2 + 2x + 1}$

86. $\dfrac{5}{x^2 - 25} - \dfrac{3}{2x^2 - 9x - 5} + 1$

87. $a + \dfrac{1}{a + 3}$

88. $5 - \dfrac{1}{y + 2}$

89. $\dfrac{x + 3}{2x^2 - 5x + 2} - \dfrac{3x - 1}{x^2 - x - 2}$

90. $\dfrac{a + b}{3} + \dfrac{a - b}{7}$

91. $\dfrac{x - y}{2} + \dfrac{x + y}{3}$

92. $1 + x - \dfrac{x}{x - 5}$

93. $2 - x + \dfrac{3}{x - 9}$

94. $\dfrac{3x}{x - 1} - 2x - x^2$

95. $\dfrac{23}{x - 1} + 4x - 5x^2$

96. $\dfrac{3x}{2x - 1} + \dfrac{x + 1}{3x + 2} + \dfrac{2x}{6x^3 + x^2 - 2x}$

97. $\dfrac{3x}{x - 3} + \dfrac{4}{x - 2} - \dfrac{5x}{x^3 - 5x^2 + 6x}$

98. $\dfrac{2x - 1}{x^2 + x - 6} - \dfrac{3x - 5}{x^2 - 2x - 15} + \dfrac{2x - 3}{x^2 - 7x + 10}$

99. $2 + \dfrac{4a}{a^2 - 1} - \dfrac{2}{a + 1}$

100. $\dfrac{a}{a - 1} - \dfrac{a + 1}{2a - 2} + a$

101. $\dfrac{x + 5}{2x^2 - 2} + \dfrac{x}{2x + 2} - \dfrac{3}{x - 1}$

102. $\dfrac{a}{2-a} + \dfrac{3}{a-2} - \dfrac{3a-2}{a^2-4}$

103. $\dfrac{a}{a-b} + \dfrac{b}{a+b} + \dfrac{a^2+b^2}{b^2-a^2}$

104. $\dfrac{1}{x+y} - \dfrac{1}{x-y} - \dfrac{2y}{y^2-x^2}$

105. $\dfrac{7n^2}{m-n} + \dfrac{3m}{n-m} - \dfrac{3m^2-n}{m^2-2mn+n^2}$

106. $\dfrac{3b}{2a-b} + \dfrac{2a-1}{b-2a} - \dfrac{3a^2+b}{b^2-4ab+4a^2}$

107. $\dfrac{m+1}{m^2+2m+1} + \dfrac{m-1}{m^2-2m+1} + \dfrac{2}{m^2-1}$

108. $\dfrac{a+2}{a^2+3a+2} + \dfrac{a-1}{a^2-1} + \dfrac{3}{a+1}$

109. Show that $\dfrac{a}{b} + \dfrac{c}{d} = \dfrac{ad+bc}{bd}$.

110. Show that $\dfrac{a}{b} - \dfrac{c}{d} = \dfrac{ad-bc}{bd}$.

APPLICATIONS

111. Monuments   Find an expression that represents the total height of the monument.

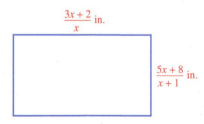

$\dfrac{3x+6}{x}$ ft

x ft

$\dfrac{9x-2}{2}$ ft

112. Perimeter of a rectangle   Write an expression that represents the perimeter of the rectangle.

$\dfrac{3x+2}{x}$ in.

$\dfrac{5x+8}{x+1}$ in.

113. Width of a ring   Find an expression that represents the width of the ring.

$\dfrac{8x-6}{x}$ cm

$\dfrac{7x+5}{2}$ cm

114. Work rates   After working together on a common job, one worker completes $\dfrac{x+2}{4}$ of the work and another completes $\dfrac{x+3}{6}$ of the work. How much more did the first worker accomplish than the second?

WRITING ABOUT MATH

115. Explain how to find the least common denominator.

116. Explain how to add two fractions.

SOMETHING TO THINK ABOUT

117. Find the error:

$$\dfrac{8x+2}{5} - \dfrac{3x+8}{5}$$
$$= \dfrac{8x+2-3x+8}{5}$$
$$= \dfrac{5x+10}{5}$$
$$= x+2$$

118. Find the error:

$$\dfrac{(x+y)^2}{2} + \dfrac{(x-y)^2}{3}$$
$$= \dfrac{3\cdot(x+y)^2}{3\cdot 2} + \dfrac{2\cdot(x-y)^2}{2\cdot 3}$$
$$= \dfrac{3x^2+3y^2+2x^2-2y^2}{6}$$
$$= \dfrac{5x^2+y^2}{6}$$

Section
6.4

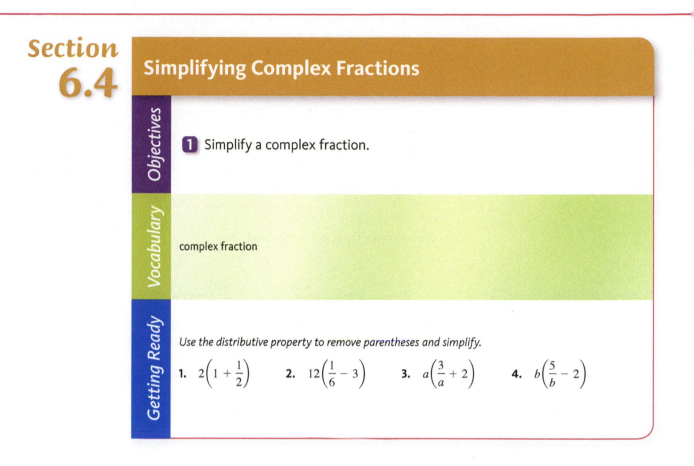

Simplifying Complex Fractions

Objectives

1 Simplify a complex fraction.

Vocabulary

complex fraction

Getting Ready

Use the distributive property to remove parentheses and simplify.

1. $2\left(1 + \dfrac{1}{2}\right)$ **2.** $12\left(\dfrac{1}{6} - 3\right)$ **3.** $a\left(\dfrac{3}{a} + 2\right)$ **4.** $b\left(\dfrac{5}{b} - 2\right)$

In this section, we will consider **complex fractions**, fractions that have a fraction in their numerator and/or their denominator. Examples of complex fractions are

$$\dfrac{\dfrac{3a}{b}}{\dfrac{6ac}{b^2}}, \qquad \dfrac{\dfrac{2}{x} + 1}{3 + x}, \qquad \text{and} \qquad \dfrac{\dfrac{1}{x} + \dfrac{1}{y}}{\dfrac{1}{x} - \dfrac{1}{y}}$$

 Simplify a complex fraction.

We can use two methods to simplify the complex fraction

$$\dfrac{\dfrac{3a}{b}}{\dfrac{6ac}{b^2}}$$

In **Method 1**, we eliminate the rational expressions in the numerator and denominator by writing the complex fraction as a division and using the division rule for fractions:

$$\dfrac{\dfrac{3a}{b}}{\dfrac{6ac}{b^2}} = \dfrac{3a}{b} \div \dfrac{6ac}{b^2}$$

$$= \dfrac{3a}{b} \cdot \dfrac{b^2}{6ac} \qquad \text{Invert the divisor and multiply.}$$

$$= \dfrac{b}{2c} \qquad \text{Multiply and simplify.}$$

In **Method 2**, we eliminate the rational expressions in the numerator and denominator by multiplying the expression by 1, written in the form $\frac{b^2}{b^2}$. We use $\frac{b^2}{b^2}$, because b^2 is the LCD of $\frac{3a}{b}$ and $\frac{6ac}{b^2}$.

$$\frac{\dfrac{3a}{b}}{\dfrac{6ac}{b^2}} = \frac{\boldsymbol{b^2} \cdot \dfrac{3a}{b}}{\boldsymbol{b^2} \cdot \dfrac{6ac}{b^2}} \qquad \frac{b^2}{b^2} = 1$$

$$= \frac{3ab}{6ac} \qquad \text{Simplify the fractions in the numerator and denominator.}$$

$$= \frac{b}{2c} \qquad \text{Divide out the common factor of } 3a.$$

With either method, the result is the same.

EXAMPLE 1   Simplify: $\dfrac{\dfrac{2}{x} + 1}{3 + x}$ Assume no denominators are 0.

Solution   **Method 1:**   We can add in the numerator and proceed as follows:

$$\frac{\dfrac{2}{x} + \boldsymbol{1}}{3 + x} = \frac{\dfrac{2}{x} + \dfrac{\boldsymbol{x}}{\boldsymbol{x}}}{\dfrac{3 + x}{1}} \qquad \text{Write 1 as } \tfrac{x}{x} \text{ and } 3 + x \text{ as } \tfrac{3 + x}{1}.$$

$$= \frac{\dfrac{2 + x}{x}}{\dfrac{3 + x}{1}} \qquad \text{Add } \tfrac{2}{x} \text{ and } \tfrac{x}{x} \text{ to obtain } \tfrac{2 + x}{x}.$$

$$= \frac{2 + x}{x} \div \frac{3 + x}{1} \qquad \text{Write the complex fraction as a division.}$$

$$= \frac{2 + x}{x} \cdot \frac{1}{3 + x} \qquad \text{Invert the divisor and multiply.}$$

$$= \frac{2 + x}{x^2 + 3x} \qquad \text{Multiply the numerators and multiply the denominators.}$$

COMMENT   In Method 1 we are trying to write the numerator as a single fraction and the denominator as a single fraction so that we can invert the divisor and multiply.

Method 2:   To eliminate the denominator of x, we can multiply the numerator and the denominator by x, the LCD of the rational expressions in the complex fraction.

$$\frac{\dfrac{2}{x} + 1}{3 + x} = \frac{\boldsymbol{x}\left(\dfrac{2}{x} + 1\right)}{\boldsymbol{x}(3 + x)} \qquad \frac{x}{x} = 1$$

$$= \frac{\boldsymbol{x} \cdot \dfrac{2}{x} + \boldsymbol{x} \cdot 1}{\boldsymbol{x} \cdot 3 + \boldsymbol{x} \cdot x} \qquad \text{Use the distributive property to remove parentheses.}$$

$$= \frac{2 + x}{x^2 + 3x} \qquad \text{Simplify.}$$

SELF CHECK 1   Simplify: $\dfrac{\dfrac{3}{a} + 2}{2 + a}$ Assume no denominator is 0.

Accent
on technology

▸ Checking Algebra

We can check the simplification in Example 1 by graphing the functions

$$f(x) = \dfrac{\frac{2}{x} + 1}{3 + x}$$

shown in Figure 6-6(a), and

$$g(x) = \dfrac{2 + x}{x^2 + 3x}$$

shown in Figure 6-6(b), and observing that the graphs look the same. Each graph has window settings of $[-5, 3]$ for x and $[-6, 6]$ for y.

(a)   (b)

Figure 6-6

For instruction regarding the use of a Casio graphing calculator, please refer to the Casio Keystroke Guide in the back of the book.

EXAMPLE 2   Simplify: $\dfrac{\frac{1}{x} + \frac{1}{y}}{\frac{1}{x} - \frac{1}{y}}$ Assume no denominators are 0.

Solution   **Method 1:**   We can add in the numerator and in the denominator and proceed as follows:

$$\dfrac{\frac{1}{x} + \frac{1}{y}}{\frac{1}{x} - \frac{1}{y}} = \dfrac{\frac{1 \cdot y}{x \cdot y} + \frac{x \cdot 1}{x \cdot y}}{\frac{1 \cdot y}{x \cdot y} - \frac{x \cdot 1}{x \cdot y}}$$

Write each fraction as an expression with the common denominator of xy.

$$= \dfrac{\frac{y + x}{xy}}{\frac{y - x}{xy}}$$

Add the fractions in the numerator and subtract the fractions in the denominator.

$$= \dfrac{y + x}{xy} \div \dfrac{y - x}{xy}$$

Write the complex fraction as a division.

$$= \dfrac{y + x}{xy} \cdot \dfrac{xy}{y - x}$$

Invert the divisor and multiply.

$$= \dfrac{y + x}{y - x}$$

Multiply the numerators and multiply the denominators and simplify.

Method 2:   We can multiply the numerator and denominator by xy (the LCD of the rational expressions appearing in the complex fraction) and simplify.

$$\frac{\dfrac{1}{x} + \dfrac{1}{y}}{\dfrac{1}{x} - \dfrac{1}{y}} = \frac{xy\left(\dfrac{1}{x} + \dfrac{1}{y}\right)}{xy\left(\dfrac{1}{x} - \dfrac{1}{y}\right)}$$

$$= \frac{xy \cdot \dfrac{1}{x} + xy \cdot \dfrac{1}{y}}{xy \cdot \dfrac{1}{x} - xy \cdot \dfrac{1}{y}} \qquad \text{Use the distributive property to remove parentheses.}$$

$$= \frac{y + x}{y - x} \qquad\qquad\qquad \text{Simplify.}$$

SELF CHECK 2   Simplify: $\dfrac{\dfrac{1}{a} - \dfrac{1}{b}}{\dfrac{1}{a} + \dfrac{1}{b}}$ Assume no denominators are 0.

EXAMPLE 3   Simplify: $\dfrac{x^{-1} + y^{-1}}{x^{-2} - y^{-2}}$ Assume no denominators are 0.

Solution   **Method 1:**   We can use the rule $n^{-1} = \dfrac{1}{n}$ and proceed as follows:

$$\frac{x^{-1} + y^{-1}}{x^{-2} - y^{-2}} = \frac{\dfrac{1}{x} + \dfrac{1}{y}}{\dfrac{1}{x^2} - \dfrac{1}{y^2}} \qquad \text{Write the fraction using positive exponents.}$$

$$= \frac{\dfrac{y}{xy} + \dfrac{x}{xy}}{\dfrac{y^2}{x^2y^2} - \dfrac{x^2}{x^2y^2}} \qquad \begin{array}{l}\text{Write each fraction in the numerator as an} \\ \text{expression with an LCD of } xy, \text{ and the} \\ \text{fractions in the denominator as expressions} \\ \text{with an LCD of } x^2y^2.\end{array}$$

$$= \frac{\dfrac{y + x}{xy}}{\dfrac{y^2 - x^2}{x^2y^2}} \qquad \begin{array}{l}\text{Add the fractions in the numerator and} \\ \text{denominator.}\end{array}$$

$$= \frac{y + x}{xy} \div \frac{y^2 - x^2}{x^2y^2} \qquad \text{Write the fraction as a division.}$$

$$= \frac{y + x}{xy} \cdot \frac{x^2y^2}{(y - x)(y + x)} \qquad \text{Invert the divisor, multiply, and factor } (y^2 - x^2).$$

$$= \frac{\overset{1}{\cancel{(y + x)}}\,\overset{x\ \ y}{\cancel{x^2y^2}}}{\underset{1\ \ 1}{\cancel{xy}}\,(y - x)\underset{1}{\cancel{(y + x)}}} \qquad \begin{array}{l}\text{Multiply the numerators and the denominators} \\ \text{and divide out the common factors of } x, y, \\ \text{and } (y + x).\end{array}$$

$$= \frac{xy}{y - x} \qquad\qquad \text{Simplify.}$$

Method 2: We can use the rule $n^{-1} = \dfrac{1}{n}$ and then multiply both numerator and denominator by $x^2 y^2$, the LCD of the rational expressions, and proceed as follows:

$$\frac{x^{-1} + y^{-1}}{x^{-2} - y^{-2}} = \frac{\dfrac{1}{x} + \dfrac{1}{y}}{\dfrac{1}{x^2} - \dfrac{1}{y^2}}$$

Write the fraction without negative exponents.

$$= \frac{x^2 y^2 \left(\dfrac{1}{x} + \dfrac{1}{y}\right)}{x^2 y^2 \left(\dfrac{1}{x^2} - \dfrac{1}{y^2}\right)}$$

Multiply numerator and denominator by $x^2 y^2$, the LCD.

$$= \frac{x^2 y^2 \cdot \dfrac{1}{x} + x^2 y^2 \cdot \dfrac{1}{y}}{x^2 y^2 \cdot \dfrac{1}{x^2} - x^2 y^2 \cdot \dfrac{1}{y^2}}$$

Use the distributive property to remove parentheses.

$$= \frac{xy^2 + x^2 y}{y^2 - x^2}$$

Simplify.

$$= \frac{xy \overset{1}{\cancel{(y + x)}}}{\cancel{(y + x)}(y - x)}$$

Factor the numerator and denominator.

$$= \frac{xy}{y - x}$$

Divide out the common factor $(y + x)$.

COMMENT $x^{-1} + y^{-1}$ means $\dfrac{1}{x} + \dfrac{1}{y}$, and $(x + y)^{-1}$ means $\dfrac{1}{x + y}$.

SELF CHECK 3 Simplify: $\dfrac{a^{-2} + b^{-2}}{a^{-1} - b^{-1}}$ Assume no denominators are 0.

Perspective

Each of the complex fractions in the list

$$1 + \frac{1}{2}, \quad 1 + \frac{1}{1 + \dfrac{1}{2}}, \quad 1 + \frac{1}{1 + \dfrac{1}{1 + \dfrac{1}{2}}}, \ldots$$

can be simplified by using the value of the expression preceding it. For example, to simplify the second expression in the list, replace $1 + \frac{1}{2}$ with $\frac{3}{2}$:

$$1 + \frac{1}{1 + \dfrac{1}{2}} = 1 + \frac{1}{\dfrac{3}{2}} = 1 + \frac{2}{3} = \frac{5}{3}$$

To simplify the third expression, we replace $1 + \dfrac{1}{1 + \dfrac{1}{2}}$ with $\dfrac{5}{3}$:

$$1 + \frac{1}{1 + \dfrac{1}{1 + \dfrac{1}{2}}} = 1 + \frac{1}{\dfrac{5}{3}} = 1 + \frac{3}{5} = \frac{8}{5}$$

The complex fractions in the list simplify to the fractions $\frac{3}{2}, \frac{5}{3},$ and $\frac{8}{5}$. The decimal values of these fractions get closer and closer to the irrational number $1.61803398875\ldots$, known as the *golden ratio*. This number often appears in the architecture of the ancient Greeks and Egyptians. For example, the width of the stairs in front of the Greek Parthenon (Illustration 1), divided by the building's height, is

(continued)

the golden ratio. The height of the triangular face of the Great Pyramid of Cheops (Illustration 2), divided by the pyramid's width, is also the golden ratio.

Illustration 1

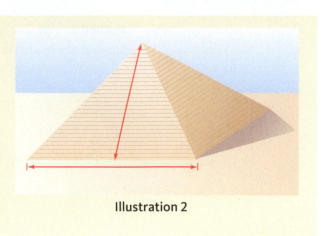

Illustration 2

EXAMPLE 4   Simplify: $\dfrac{\dfrac{2x}{1-\dfrac{1}{x}}+3}{3-\dfrac{2}{x}}$ Assume no denominators are 0.

Solution   In more complicated complex fractions such as this one, it is usually easiest to use Method 2. To do so, we begin by multiplying the numerator and denominator of

$$\frac{2x}{1-\dfrac{1}{x}}$$

by x to eliminate the complex fraction in the numerator.

$$\frac{\dfrac{2x}{1-\dfrac{1}{x}}+3}{3-\dfrac{2}{x}} = \frac{\dfrac{x\cdot 2x}{x\left(1-\dfrac{1}{x}\right)}+3}{3-\dfrac{2}{x}} \qquad \frac{x}{x}=1$$

$$= \frac{\dfrac{2x^2}{x\cdot 1-x\cdot\dfrac{1}{x}}+3}{3-\dfrac{2}{x}} \qquad \text{Use the distributive property to remove parentheses.}$$

$$= \frac{\dfrac{2x^2}{x-1}+3}{3-\dfrac{2}{x}} \qquad \text{Simplify.}$$

We then multiply the numerator and denominator of the previous fraction by $x(x - 1)$, the LCD of $\frac{2x^2}{x-1}$, 3, and $\frac{2}{x}$, and simplify:

$$\frac{\dfrac{2x}{1 - \dfrac{1}{x}} + 3}{3 - \dfrac{2}{x}} = \frac{x(x-1)\left(\dfrac{2x^2}{x-1} + 3\right)}{x(x-1)\left(3 - \dfrac{2}{x}\right)}$$

Multiply the numerator and denominator by $x(x - 1)$, the LCD.

$$= \frac{x(x-1) \cdot \dfrac{2x^2}{x-1} + x(x-1) \cdot 3}{x(x-1) \cdot 3 - x(x-1) \cdot \dfrac{2}{x}}$$

Use the distributive property.

$$= \frac{2x^3 + 3x(x-1)}{3x(x-1) - 2(x-1)}$$

Simplify.

$$= \frac{2x^3 + 3x^2 - 3x}{3x^2 - 3x - 2x + 2}$$

Use the distributive property to remove parentheses.

$$= \frac{2x^3 + 3x^2 - 3x}{3x^2 - 5x + 2}$$

This result does not simplify.

 SELF CHECK 4   Simplify: $\dfrac{3 + \dfrac{2}{a}}{\dfrac{2a}{1 + \dfrac{1}{a}} + 3}$ Assume no denominators are 0.

 SELF CHECK ANSWERS   **1.** $\frac{2a+3}{a^2+2a}$ **2.** $\frac{b-a}{b+a}$ **3.** $\frac{b^2+a^2}{ab(b-a)}$ **4.** $\frac{(3a+2)(a+1)}{a(2a^2+3a+3)}$

NOW TRY THIS

Assume no denominators are **0.**

1. Simplify: $\left(\dfrac{x+11}{3x-3} - \dfrac{2}{x}\right) \div \left(\dfrac{3}{2x} + \dfrac{x-9}{6x-6}\right)$

2. Simplify: $\dfrac{\dfrac{2}{x} - \dfrac{x+11}{3x-3}}{\dfrac{9-x}{6x-6} - \dfrac{3}{2x}}$

3. Did the form of the problem influence your choice of method for simplifying? Do you find one method easier than the other? Discuss your choice with a classmate.

6.4   Exercises

WARM-UPS *Divide. Assume no denominators are 0.*

1. $\dfrac{2}{5} \div \dfrac{3}{5}$

2. $\dfrac{x}{y} \div \dfrac{z}{y}$

3. $\dfrac{a+b}{a} \div \dfrac{a-b}{a}$

4. $\dfrac{x-7}{x} \div \dfrac{x+7}{x}$

5. $\dfrac{6x}{y} \div \dfrac{4x^2}{y^2}$

6. $\dfrac{10a^4}{7b} \div \dfrac{15a^2}{14b^2}$

REVIEW *Solve each equation.*

7. $\dfrac{8(a-5)}{3} = 2(a-4)$

8. $\dfrac{3t^2}{5} + \dfrac{7t}{10} = \dfrac{3t+6}{5}$

9. $a^4 - 20a^2 + 64 = 0$

10. $|2x-1| = 9$

VOCABULARY AND CONCEPTS *Fill in the blanks.*

11. A _____ fraction is a fraction that has fractions in its numerator or denominator.

12. The fraction $\dfrac{\frac{a}{b}}{\frac{c}{d}}$ is equivalent to $\dfrac{a}{b} \ \_\_ \ \dfrac{c}{d}$.

GUIDED PRACTICE *Simplify using any method. Assume no denominators are 0. (OBJECTIVE 1)*

13. $\dfrac{\frac{1}{3}}{\frac{4}{9}}$

14. $-\dfrac{\frac{3}{4}}{\frac{1}{2}}$

15. $\dfrac{\frac{8y}{x^2}}{\frac{12yz}{x}}$

16. $\dfrac{\frac{5t^4}{9x}}{\frac{2t}{18x}}$

17. $\dfrac{\frac{x^2+5x+6}{3xy}}{\frac{x^2-9}{6xy}}$

18. $\dfrac{\frac{x^2-4x-5}{4xy}}{\frac{x^2-25}{8xy}}$

19. $\dfrac{\frac{1}{2}-\frac{2}{3}}{\frac{2}{3}+\frac{1}{2}}$

20. $\dfrac{\frac{2}{3}+\frac{4}{5}}{\frac{2}{5}-\frac{1}{3}}$

Simplify using any method. Assume no denominators are 0.
SEE EXAMPLE 1. (OBJECTIVE 1)

21. $\dfrac{\frac{3}{x}+1}{x+2}$

22. $\dfrac{2+\frac{5}{y}}{y-3}$

23. $\dfrac{\frac{1}{x}-\frac{1}{y}}{xy}$

24. $\dfrac{pq}{\frac{2}{p}+\frac{3}{q}}$

25. $\dfrac{\frac{5}{a}+\frac{2}{b}}{ab}$

26. $\dfrac{ab}{\frac{3}{a}-\frac{2}{b}}$

27. $\dfrac{5xy}{1+\frac{1}{xy}}$

28. $\dfrac{3a}{a+\frac{1}{a}}$

Simplify using any method. Assume no denominators are 0.
SEE EXAMPLE 2. (OBJECTIVE 1)

29. $\dfrac{\frac{p}{q}-\frac{q}{p}}{\frac{1}{q}-\frac{1}{p}}$

30. $\dfrac{\frac{y}{x}-\frac{x}{y}}{\frac{1}{y}-\frac{1}{x}}$

31. $\dfrac{\frac{1}{a}-\frac{1}{b}}{\frac{a}{b}-\frac{b}{a}}$

32. $\dfrac{\frac{1}{a}+\frac{1}{b}}{\frac{a}{b}-\frac{b}{a}}$

Simplify using any method. Assume no denominators are 0.
SEE EXAMPLE 3. (OBJECTIVE 1)

33. $\dfrac{x^{-1}+y^{-1}}{x^{-1}-y^{-1}}$

34. $\dfrac{(x+y)^{-1}}{x^{-1}+y^{-1}}$

35. $\dfrac{x-y^{-2}}{y-x^{-2}}$

36. $\dfrac{x^{-2}-y^{-2}}{x^{-1}-y^{-1}}$

Simplify using any method. Assume no denominators are 0.
SEE EXAMPLE 4. (OBJECTIVE 1)

37. $\dfrac{1+\frac{a}{b}}{1-\frac{a}{1-\frac{a}{b}}}$

38. $\dfrac{1+\frac{2}{1+\frac{a}{b}}}{1-\frac{a}{b}}$

39. $\dfrac{x-\frac{1}{x}}{1+\frac{1}{\frac{1}{x}}}$

40. $\dfrac{\frac{a^2+3a+4}{ab}}{2+\frac{3+a}{\frac{2}{a}}}$

ADDITIONAL PRACTICE *Simplify each complex fraction. Assume no denominators are 0.*

41. $\dfrac{x+1-\frac{6}{x}}{\frac{1}{x}}$

42. $\dfrac{x-1-\frac{2}{x}}{\frac{x}{3}}$

43. $\dfrac{1+\frac{x}{y}}{1-\frac{x}{y}}$

44. $\dfrac{\frac{x}{y}+1}{1-\frac{x}{y}}$

45. $\dfrac{1+\frac{6}{x}+\frac{8}{x^2}}{1+\frac{1}{x}-\frac{12}{x^2}}$

46. $\dfrac{1-x-\frac{2}{x}}{\frac{6}{x^2}+\frac{1}{x}-1}$

47. $\dfrac{\frac{1}{a+1}+1}{\frac{3}{a-1}+1}$

48. $\dfrac{2+\frac{3}{x+1}}{\frac{1}{x}+x+x^2}$

49. $\dfrac{x^{-1}+y^{-1}}{x}$

50. $\dfrac{x^{-1}-y^{-1}}{y}$

51. $\dfrac{y}{x^{-1}-y^{-1}}$

52. $\dfrac{x^{-1}+y^{-1}}{(x+y)^{-1}}$

53. $\dfrac{b}{b + \dfrac{2}{2 + \frac{1}{2}}}$

54. $\dfrac{2y}{y - \dfrac{y}{3 - \frac{1}{2}}}$

55. $a + \dfrac{a}{1 + \dfrac{a}{a + 1}}$

56. $b + \dfrac{b}{1 - \dfrac{b + 1}{b}}$

57. $\dfrac{-\dfrac{2}{3}}{\dfrac{6}{9}}$

58. $\dfrac{\dfrac{11}{18}}{\dfrac{22}{27}}$

59. $\dfrac{\dfrac{1}{2} + \dfrac{1}{3}}{\dfrac{1}{4}}$

60. $\dfrac{\dfrac{1}{4} - \dfrac{1}{5}}{\dfrac{1}{3}}$

61. $\dfrac{x + y}{x^{-1} + y^{-1}}$

62. $\dfrac{x - y}{x^{-1} - y^{-1}}$

63. $\dfrac{x - \dfrac{1}{1 - \frac{x}{2}}}{\dfrac{3}{x + \frac{2}{3}} - x}$

64. $\dfrac{\dfrac{2x}{x - \frac{1}{x}} - \dfrac{1}{x}}{2x + \dfrac{2x}{1 - \frac{1}{x}}}$

65. $\dfrac{2x + \dfrac{1}{2 - \frac{x}{2}}}{\dfrac{4}{\frac{x}{2} - 2} - x}$

66. $\dfrac{3x - \dfrac{1}{3 - \frac{x}{2}}}{\dfrac{3}{\frac{x}{2} - 3} + x}$

67. $\dfrac{\dfrac{1}{x^2 + 3x + 2} + \dfrac{1}{x^2 + x - 2}}{\dfrac{3x}{x^2 - 1} - \dfrac{x}{x + 2}}$

68. $\dfrac{\dfrac{1}{x^2 - 1} - \dfrac{2}{x^2 + 4x + 3}}{\dfrac{2}{x^2 + 2x - 3} + \dfrac{1}{x + 3}}$

APPLICATIONS

69. Engineering The stiffness k of the shaft is given by the formula

$$k = \dfrac{1}{\dfrac{1}{k_1} + \dfrac{1}{k_2}}$$

where k_1 and k_2 are the individual stiffnesses of each section. Simplify the complex fraction.

70. Electronics In electronic circuits, resistors oppose the flow of an electric current. To find the total resistance of a parallel combination of the two resistors shown in the illustration, we can use the formula

$$\text{Total resistance} = \dfrac{1}{\dfrac{1}{R_1} + \dfrac{1}{R_2}}$$

where R_1 is the resistance of the first resistor and R_2 is the resistance of the second. Simplify the complex fraction.

71. Data analysis Use the data in the table to find the average measurement for the three-trial experiment.

	Trial 1	Trial 2	Trial 3
Measurement	$\dfrac{k}{2}$	$\dfrac{k}{3}$	$\dfrac{k}{2}$

72. Transportation If a car travels a distance d_1 at a speed s_1, and then travels a distance d_2 at a speed s_2, the average (mean) speed is given by the formula

$$\bar{s} = \dfrac{d_1 + d_2}{\dfrac{d_1}{s_1} + \dfrac{d_2}{s_2}}$$

Simplify the complex fraction.

WRITING ABOUT MATH

73. There are two methods used to simplify a complex fraction. Explain one of them.

74. Explain the other method of simplifying a complex fraction.

SOMETHING TO THINK ABOUT *Assume no denominators are 0.*

75. Simplify: $(x^{-1}y^{-1})(x^{-1} + y^{-1})^{-1}$

76. Simplify: $[(x^{-1} + 1)^{-1} + 1]^{-1}$

Section 6.5

Solving Equations Containing Rational Expressions

Objectives

1. Solve a rational equation.
2. Solve a formula involving a rational equation for a specified variable.
3. Solve an application involving a rational equation.

Vocabulary

rational equation   extraneous solution

Getting Ready

Solve each equation.

1. $\dfrac{3}{x} = \dfrac{6}{9}$ 2. $\dfrac{x}{8} = \dfrac{2}{x}$

We now show how to solve equations that contain rational expressions. Then we will use these skills to solve applications.

1 Solve a rational equation.

If an equation contains one or more rational expressions, it is called a **rational equation**. Some examples are

$$\frac{3}{5} + \frac{7}{x+2} = 2, \qquad \frac{x+3}{x-3} = \frac{2}{x^2-4}, \qquad \text{and} \qquad \frac{-x^2+10}{x^2-1} + \frac{3x}{x-1} = \frac{2x}{x+1}$$

To solve a rational equation, we can multiply both sides of the equation by a nonzero expression *to clear the equation of fractions*.

EXAMPLE 1   Solve: $\dfrac{x+1}{5} - 2 = -\dfrac{4}{x}$

Solution   We note that x cannot be 0, because this would give a 0 in a denominator. If $x \neq 0$, we can clear the equation of fractions by multiplying both sides by $5x$, the LCD.

$$\frac{x+1}{5} - 2 = -\frac{4}{x}$$

$$5x\left(\frac{x+1}{5} - 2\right) = 5x\left(-\frac{4}{x}\right) \qquad \text{Multiply both sides by } 5x \text{, the LCD.}$$

$$x(x+1) - 10x = -20 \qquad \text{Use the distributive property to remove parentheses and simplify.}$$

$$x^2 + x - 10x = -20 \qquad \text{Use the distributive property to remove parentheses.}$$

$$x^2 - 9x + 20 = 0 \qquad \text{Combine like terms and add 20 to both sides.}$$
$$(x - 5)(x - 4) = 0 \qquad \text{Factor } (x^2 - 9x + 20).$$
$$x - 5 = 0 \quad \text{or} \quad x - 4 = 0 \qquad \text{Set each factor equal to 0.}$$
$$x = 5 \qquad\quad x = 4$$

Since 4 and 5 both satisfy the original equation, the solutions are 4 and 5.

SELF CHECK 1   Solve: $a + \dfrac{2}{3} = \dfrac{2a - 12}{3(a - 3)}$

EXAMPLE 2   Solve: $\dfrac{3}{5} + \dfrac{7}{x + 2} = 2$

Solution   We first note that x cannot be -2, because this would give a 0 in the denominator of $\dfrac{7}{x+2}$. If $x \neq -2$, we can multiply both sides of the equation by the LCD of the fractions, which is $5(x + 2)$.

$$5(x + 2)\left(\dfrac{3}{5} + \dfrac{7}{x + 2}\right) = 5(x + 2)2$$

$$5(x + 2)\left(\dfrac{3}{5}\right) + 5(x + 2)\left(\dfrac{7}{x + 2}\right) = 5(x + 2)2 \qquad \text{Use the distributive property.}$$

$$3(x + 2) + 5(7) = 10(x + 2) \qquad \text{Simplify.}$$

$$3x + 6 + 35 = 10x + 20 \qquad \begin{array}{l}\text{Use the distributive property to}\\\text{remove parentheses and simplify.}\end{array}$$

$$3x + 41 = 10x + 20 \qquad \text{Add.}$$

$$-7x = -21 \qquad \begin{array}{l}\text{Add } -10x \text{ and } -41 \text{ to both}\\\text{sides to write in quadratic form.}\end{array}$$

$$x = 3 \qquad \text{Divide both sides by } -7.$$

The solution is 3.

Check: To check, we substitute 3 for x in the original equation and simplify.

$$\dfrac{3}{5} + \dfrac{7}{x + 2} = 2$$

$$\dfrac{3}{5} + \dfrac{7}{3 + 2} \stackrel{?}{=} 2$$

$$\dfrac{3}{5} + \dfrac{7}{5} \stackrel{?}{=} 2$$

$$2 = 2 \qquad \text{The result checks.}$$

SELF CHECK 2   Solve: $\dfrac{2}{5} + \dfrac{5}{x - 2} = \dfrac{29}{10}$

Accent
on technology

▶ Solving Equations

To use a graphing calculator to approximate the solution of $\frac{3}{5} + \frac{7}{x + 2} = 2$, we graph the functions $f(x) = \frac{3}{5} + \frac{7}{x + 2}$ and $g(x) = 2$. If we use window settings of $[-10, 10]$ for x and $[-10, 10]$ for y, we will obtain the graph shown in Figure 6-7(a).

We can find the intersection point by using the INTERSECT feature found in the CALC menu. The intersection point is $(3, 2)$ but since the original equation, $\frac{3}{5} + \frac{7}{x + 2} = 2$, is only in terms of x, the solution is 3.

Figure 6-7

For instruction regarding the use of a Casio graphing calculator, please refer to the Casio Keystroke Guide in the back of the book.

When we multiply both sides of an equation by a quantity that contains a variable, we can obtain false solutions, called **extraneous solutions**. We must exclude extraneous solutions from the solution set of an equation.

EXAMPLE 3 Solve: $\dfrac{2(x + 1)}{x - 3} = \dfrac{x + 5}{x - 3}$

Solution We start by noting that x cannot be 3, because this would give a 0 in a denominator. If $x \neq 3$, we can clear the equation of fractions by multiplying both sides by $(x - 3)$.

$$\frac{2(x + 1)}{x - 3} = \frac{x + 5}{x - 3}$$

$$(x - 3)\frac{2(x + 1)}{x - 3} = (x - 3)\frac{x + 5}{x - 3} \qquad \text{Multiply both sides by } (x - 3).$$

$$2(x + 1) = x + 5 \qquad \text{Simplify.}$$

$$2x + 2 = x + 5 \qquad \text{Use the distributive property to remove parentheses.}$$

$$x + 2 = 5 \qquad \text{Subtract } x \text{ from both sides.}$$

$$x = 3 \qquad \text{Subtract 2 from both sides.}$$

Since x cannot be 3, the 3 is extraneous and must be discarded. This equation has no solution, denoted by $\varnothing$.

SELF CHECK 3 Solve: $\dfrac{6}{x - 6} + 3 = \dfrac{x}{x - 6}$

EXAMPLE 4   Solve: $\dfrac{-x^2 + 10}{x^2 - 1} + \dfrac{3x}{x - 1} = \dfrac{2x}{x + 1}$

Solution   We first note that x cannot be 1 or -1, because this would give a 0 in a denominator. If $x \neq 1$ and $x \neq -1$, we can clear the equation of fractions by multiplying both sides by the LCD of the three rational expressions, $(x + 1)(x - 1)$, and proceed as follows:

$$\frac{-x^2 + 10}{x^2 - 1} + \frac{3x}{x - 1} = \frac{2x}{x + 1}$$

$$\frac{-x^2 + 10}{(x + 1)(x - 1)} + \frac{3x}{x - 1} = \frac{2x}{x + 1} \qquad \text{Factor } x^2 - 1.$$

$$(x + 1)(x - 1)\left(\frac{-x^2 + 10}{(x + 1)(x - 1)} + \frac{3x}{x - 1}\right) = (x + 1)(x - 1)\left(\frac{2x}{x + 1}\right) \qquad \begin{array}{l}\text{Multiply both sides by the LCD} \\ (x + 1)(x - 1) \text{ to clear fractions.}\end{array}$$

$$\frac{(x + 1)(x - 1)(-x^2 + 10)}{(x + 1)(x - 1)} + \frac{3x(x + 1)(x - 1)}{x - 1} = \frac{2x(x + 1)(x - 1)}{x + 1} \qquad \text{Use the distributive property.}$$

$$-x^2 + 10 + 3x(x + 1) = 2x(x - 1) \qquad \text{Divide out common factors.}$$

$$-x^2 + 10 + 3x^2 + 3x = 2x^2 - 2x \qquad \begin{array}{l}\text{Use the distributive property to} \\ \text{remove parentheses.}\end{array}$$

$$2x^2 + 10 + 3x = 2x^2 - 2x \qquad \text{Combine like terms.}$$

$$10 + 3x = -2x \qquad \text{Subtract } 2x^2 \text{ from both sides.}$$

$$10 + 5x = 0 \qquad \text{Add } 2x \text{ to both sides.}$$

$$5x = -10 \qquad \text{Subtract 10 from both sides.}$$

$$x = -2 \qquad \text{Divide both sides by 5.}$$

The solution is -2. Verify that -2 is a solution of the original equation.

SELF CHECK 4   Solve: $\dfrac{2x^2}{x^2 - 4} = \dfrac{3}{x + 2} + \dfrac{2x}{x - 2}$

---

2   **Solve a formula involving a rational equation for a specified variable.**

Many formulas must be cleared of fractions before we can solve them for a specific variable.

EXAMPLE 5   Solve: $\dfrac{1}{r} = \dfrac{1}{r_1} + \dfrac{1}{r_2}$ for r

Solution   This formula is from electronics. Since r, r_1, and r_2 represent resistances, they will not be zero.

$$\frac{1}{r} = \frac{1}{r_1} + \frac{1}{r_2}$$

$$\frac{rr_1r_2}{r} = \frac{rr_1r_2}{r_1} + \frac{rr_1r_2}{r_2} \qquad \text{Multiply both sides by } rr_1r_2, \text{ the LCD.}$$

$$r_1r_2 = rr_2 + rr_1 \qquad \text{Simplify each fraction.}$$

$$r_1r_2 = r(r_2 + r_1) \qquad \text{Factor out } r \text{ on the right side.}$$

$$\frac{r_1r_2}{r_2 + r_1} = r \qquad \text{Divide both sides by } (r_2 + r_1).$$

SELF CHECK 5   Solve: $\dfrac{1}{a} - \dfrac{1}{b} = 1$ for b $(a, b \neq 0)$

Solve an application involving a rational equation.

EXAMPLE 6   **DRYWALLING A HOUSE**   A contractor knows that one crew can drywall a house in 4 days and that another crew can drywall the same house in 5 days. One day must be allowed for the plaster coat to dry. If the contractor uses both crews, can the house be ready for painting in 4 days?

Analyze the problem   We will let x represent the number of days it takes for both crews to drywall the house.

Form an equation   Since the first crew can drywall the house in 4 days, it can complete $\frac{1}{4}$ of the job in 1 day. Since the second crew can drywall the house in 5 days, it can complete $\frac{1}{5}$ of the job in 1 day. If it takes x days for both crews to finish the house, together they can complete $\frac{1}{x}$ of the job in 1 day. The amount of work the first crew can do in 1 day plus the amount of work the second crew can do in 1 day equals the amount of work both crews can do in 1 day. This gives the following equation:

What crew 1 can complete in one day	plus	what crew 2 can complete in one day	equals	what they can complete together in one day.
$\dfrac{1}{4}$	$+$	$\dfrac{1}{5}$	$=$	$\dfrac{1}{x}$

Solve the equation   We can solve this equation as follows.

$$20x\left(\frac{1}{4} + \frac{1}{5}\right) = 20x\left(\frac{1}{x}\right) \qquad \text{Multiply both sides by } 20x, \text{ the LCD of } \tfrac{1}{4}, \tfrac{1}{5}, \text{ and } \tfrac{1}{x}.$$

$$5x + 4x = 20 \qquad \text{Use the distributive property to remove parentheses and simplify.}$$

$$9x = 20 \qquad \text{Combine like terms.}$$

$$x = \frac{20}{9} \qquad \text{Divide both sides by 9.}$$

State the conclusion   Because 1 day is necessary for drying, the drywallers must complete their work in 3 days. Since it will take $2\frac{2}{9}$ days for both crews to drywall the house, it will be ready for painting in less than 4 days.

Check the result
$$\frac{1}{4} + \frac{1}{5} \overset{?}{=} \frac{1}{\frac{20}{9}}$$

$$\frac{1}{4} + \frac{1}{5} \overset{?}{=} \frac{9}{20}$$

$$\frac{9}{20} = \frac{9}{20} \qquad \text{The result checks.}$$

SELF CHECK 6   If one crew can drywall the house in 5 days and another in 6 days, can the house be ready for painting in 4 days with 1 day allowed for the drywall to dry?

EXAMPLE 7   **DRIVING TO A CONVENTION**   A man drove 200 miles to a convention. Because of road construction, his average speed on the return trip was 10 mph less than his average speed going to the convention. If the return trip took 1 hour longer, how fast did he drive in each direction?

Analyze the problem   Let r represent the average rate of speed in mph going to the meeting. Then $(r - 10)$ represents the average rate of speed on the return trip. We will need the formula $d = rt$ for this problem. (d is distance, r is the rate of speed, and t is time.)

Form an equation   We know the distance and the rate for each part of the trip, and the formula for time is $t = \dfrac{d}{r}$. We can organize the given information in the table shown in Figure 6-8.

	Rate	·	Time	=	Distance
Going	r		$\dfrac{200}{r}$		200
Returning	$r - 10$		$\dfrac{200}{r - 10}$		200

Figure 6-8

Because the return trip took 1 hour longer, we can form the following equation:

The time it took to travel to the convention	plus	1	equals	the time it took to return.
$\dfrac{200}{r}$	$+$	1	$=$	$\dfrac{200}{r - 10}$

Solve the equation   We can solve the equation as follows:

$$r(r - 10)\left(\frac{200}{r} + 1\right) = r(r - 10)\left(\frac{200}{r - 10}\right) \qquad \text{Multiply both sides by } r(r - 10), \text{ the LCD.}$$

$$200(r - 10) + r(r - 10) = 200r \qquad \text{Use the distributive property to remove parentheses and simplify.}$$

$$200r - 2{,}000 + r^2 - 10r = 200r \qquad \text{Use the distributive property to remove parentheses.}$$

$$r^2 - 10r - 2{,}000 = 0 \qquad \text{Subtract } 200r \text{ from both sides.}$$

$$(r - 50)(r + 40) = 0 \qquad \text{Factor } (r^2 - 10r - 2{,}000).$$

$$r - 50 = 0 \quad \text{or} \quad r + 40 = 0 \qquad \text{Set each factor equal to 0.}$$

$$r = 50 \quad \mid \quad r = -40$$

State the conclusion   We must exclude the solution of -40, because a speed cannot be negative. Thus, the man averaged 50 mph going to the convention, and he averaged $50 - 10$, or 40 mph, returning.

Check the result   At 50 mph, the 200-mile trip took 4 hours. At 40 mph, the return trip took 5 hours, which is 1 hour longer.

SELF CHECK 7   If the trip was 300 miles, how fast did he drive in each direction?

EXAMPLE 8   **A RIVER CRUISE**   The Forest City Queen can make a 9-mile trip down the Rock River and return in a total of 1.6 hours. If the riverboat travels 12 mph in still water, find the speed of the current in the Rock River.

Analyze the problem   We can let c represent the speed of the current in mph. Since the boat travels 12 mph and a current of c mph pushes the boat while it is going downstream, the speed of the boat going downstream is $(12 + c)$ mph. On the return trip, the current pushes against the boat, and its speed is $(12 - c)$ mph.

Form an equation   Since $t = \dfrac{d}{r}$, the time required for the downstream leg of the trip is $\dfrac{9}{12 + c}$ hours, and the time required for the upstream leg of the trip is $\dfrac{9}{12 - c}$ hours.

We can organize this information in the table shown in Figure 6-9 on the next page. Furthermore, we know that the total time required for the round trip is 1.6 or $\dfrac{8}{5}$ hours.

	Rate	·	Time	=	Distance
Going downstream	$12 + c$		$\dfrac{9}{12 + c}$		9
Going upstream	$12 - c$		$\dfrac{9}{12 - c}$		9

Figure 6-9

This gives the following equation.

The time it takes to travel downstream	plus	the time it takes to travel upstream	equals	the total time for the round trip.
$\dfrac{9}{12 + c}$	$+$	$\dfrac{9}{12 - c}$	$=$	$\dfrac{8}{5}$

Solve the equation　We can multiply both sides of this equation by $5(12 + c)(12 - c)$, the LCD, to clear it of fractions and proceed as follows:

$$\frac{5(12 + c)(12 - c)9}{12 + c} + \frac{5(12 + c)(12 - c)9}{12 - c} = \frac{5(12 + c)(12 - c)8}{5}$$

$$45(12 - c) + 45(12 + c) = 8(12 + c)(12 - c) \qquad \tfrac{12 + c}{12 + c} = 1, \tfrac{12 - c}{12 - c} = 1, \text{ and } \tfrac{5}{5} = 1$$

$$540 - 45c + 540 + 45c = 8(144 - c^2) \qquad \text{Multiply.}$$

$$1{,}080 = 1{,}152 - 8c^2 \qquad \text{Combine like terms and multiply.}$$

$$8c^2 - 72 = 0 \qquad \text{Add } 8c^2 \text{ and } -1{,}152 \text{ to both sides.}$$

$$c^2 - 9 = 0 \qquad \text{Divide both sides by 8.}$$

$$(c + 3)(c - 3) = 0 \qquad \text{Factor } c^2 - 9.$$

$$c + 3 = 0 \quad \text{or} \quad c - 3 = 0 \qquad \text{Set each factor equal to 0.}$$

$$c = -3 \quad | \quad c = 3$$

State the conclusion　Since the current cannot be a negative rate, the apparent solution of -3 must be discarded. Thus, the speed of the current in the Rock River is 3 mph.

Check the result　The boat travels 9 miles downstream at a speed of 15 mph ($12 + 3$) and 9 miles back upstream at a speed of 9 mph ($12 - 3$). It takes $\frac{3}{5}$ hr to go downstream and 1 hour to return for a total trip of $1\frac{3}{5}$ or 1.6 hours.

SELF CHECK 8　If the Forest City Queen can make a 12-mile trip down the Rock River and return in a total of 2.25 hours, what is the speed of the current?

SELF CHECK ANSWERS　 **1.** 1, 2　**2.** 4　**3.** 6 is extraneous, $\varnothing$　**4.** $\frac{6}{7}$　**5.** $b = \frac{a}{1 - a}$　**6.** It will take $2\frac{8}{11}$ days to drywall the house. Yes, it will be ready to paint in 4 days.　**7.** He drove 60 mph to the convention and 50 mph back.　**8.** The current is 4 mph.

NOW TRY THIS

Solve:

1. $\dfrac{3x-5}{x^2-x-6}=0$

2. $\dfrac{2}{(x-2)^2}-\dfrac{5}{x-2}+3=0$

3. $\dfrac{1}{x^2-5x+6}+\dfrac{x-2}{x^2-x-6}=\dfrac{x}{x^2-4}$

6.5 Exercises

WARM-UPS *Find the LCD. Assume no denominators are 0.*

1. $\dfrac{1}{4},\dfrac{5}{x+3},\dfrac{7}{12}$

2. $\dfrac{6}{x-2},\dfrac{5}{6},\dfrac{1}{2}$

3. $\dfrac{1}{a},\dfrac{3}{a-5}$

4. $\dfrac{2}{x+7},\dfrac{7}{x+2}$

5. $\dfrac{1}{x^2-9},\dfrac{5}{x-3}$

6. $\dfrac{x}{x+5},\dfrac{2x+1}{x^2+6x+5}$

REVIEW *Simplify each expression. Write each answer without negative exponents. Assume no variables or denominators are 0.*

7. $(m^4n^{-2})^{-3}$

8. $\dfrac{a^{-1}}{a^{-1}+1}$

9. $\dfrac{a^0+2a^0-3a^0}{(a-b)^0}$

10. $(4x^{-2}+3)(2x-4)$

VOCABULARY AND CONCEPTS *Fill in the blanks.*

11. If an equation contains one or more rational expressions, it is called a _____ equation.

12. A false solution to an equation is called an _____ solution.

GUIDED PRACTICE *Solve each equation. SEE EXAMPLE 1. (OBJECTIVE 1)*

13. $\dfrac{1}{5}+\dfrac{8}{x}=1$

14. $\dfrac{1}{3}-\dfrac{10}{x}=-3$

15. $\dfrac{34}{x}-\dfrac{3}{2}=-\dfrac{13}{20}$

16. $\dfrac{1}{2}+\dfrac{7}{x}=2+\dfrac{1}{x}$

17. $\dfrac{3}{4y}+\dfrac{4}{y}=9$

18. $\dfrac{2}{x}+\dfrac{1}{2}=\dfrac{7}{2x}$

19. $\dfrac{7}{5x}-\dfrac{1}{2}=\dfrac{5}{6x}+\dfrac{1}{3}$

20. $\dfrac{2}{x}+\dfrac{1}{2}=\dfrac{9}{4x}-\dfrac{1}{2x}$

Solve each equation. SEE EXAMPLE 2. (OBJECTIVE 1)

21. $\dfrac{5}{x+4}+\dfrac{1}{x+4}=x-1$

22. $\dfrac{2}{x-1}+\dfrac{x-2}{3}=\dfrac{4}{x-1}$

23. $\dfrac{3}{x+1}-\dfrac{x-2}{2}=\dfrac{x-2}{x+1}$

24. $\dfrac{2}{x-3}+\dfrac{3}{4}=\dfrac{17}{2x}$

Solve each equation. If a solution is extraneous, so indicate. SEE EXAMPLE 3. (OBJECTIVE 1)

25. $\dfrac{x+1}{x}-\dfrac{x-1}{x}=0$

26. $\dfrac{x-3}{x-1}-\dfrac{2x-4}{x-1}=0$

27. $\dfrac{2x+1}{x-3}-\dfrac{x+4}{x-3}=0$

28. $\dfrac{4x+5}{x+1}-\dfrac{3x+4}{x+1}=0$

Solve each equation. SEE EXAMPLE 4. (OBJECTIVE 1)

29. $\dfrac{x}{x+2}=1-\dfrac{3x+2}{x^2+4x+4}$

30. $\dfrac{3+2a}{a^2+6+5a}+\dfrac{2-5a}{a^2-4}=\dfrac{2-3a}{a^2-6+a}$

31. $\dfrac{5}{x+3}+\dfrac{1}{x-1}=\dfrac{1}{x^2+2x-3}$

32. $\dfrac{5}{2z^2+z-3}-\dfrac{2}{2z+3}=\dfrac{z+1}{z-1}-1$

Solve each formula for the indicated variable. SEE EXAMPLE 5. (OBJECTIVE 2)

33. $\dfrac{1}{R}=\dfrac{1}{r_1}+\dfrac{1}{r_2}$ for r_1

34. $S=\dfrac{a}{1-r}$ for r

35. $\dfrac{1}{p} + \dfrac{1}{q} = \dfrac{1}{f}$ for f

36. $\dfrac{1}{p} + \dfrac{1}{q} = \dfrac{1}{f}$ for p

ADDITIONAL PRACTICE *Solve each equation. Assume no denominators are 0. If a solution is extraneous, so indicate.*

37. $\dfrac{3 - 5y}{2 + y} = \dfrac{3 + 5y}{2 - y}$

38. $\dfrac{b - 4}{b - 3} = \dfrac{b + 1}{b + 3}$

39. $\dfrac{30}{y - 2} + \dfrac{24}{y - 5} = 13$

40. $\dfrac{x + 4}{x + 7} - \dfrac{x}{x + 3} = \dfrac{3}{8}$

41. $\dfrac{x - 4}{x - 3} + \dfrac{x - 2}{x - 3} = x - 3$

42. $1 = \dfrac{3}{x - 2} - \dfrac{12}{x^2 - 4}$

43. $\dfrac{4 - 3a}{5 + a} = \dfrac{4 - 3a}{5 - a}$

44. $\dfrac{y + 2}{y + 4} = \dfrac{y - 2}{y - 1}$

45. $\dfrac{y - 3}{y + 2} = 3 - \dfrac{1 - 2y}{y + 2}$

46. $\dfrac{3a - 5}{a - 1} - 2 = \dfrac{2a}{1 - a}$

47. $\dfrac{x + 2}{x + 3} - 1 = \dfrac{1}{3 - 2x - x^2}$

48. $\dfrac{x - 3}{x - 2} - \dfrac{1}{x} = \dfrac{x - 3}{x}$

49. $\dfrac{y^2}{y + 2} + 4 = \dfrac{y + 6}{y + 2} + 3$

50. $\dfrac{5}{y - 1} + \dfrac{3}{y - 3} = \dfrac{8}{y - 2}$

51. $\dfrac{5}{x + 4} - \dfrac{1}{3} = \dfrac{x - 1}{x}$

52. $\dfrac{3}{a - 2} - \dfrac{1}{a - 1} = \dfrac{7}{(a - 2)(a - 1)}$

53. $\dfrac{5}{x + 6} - \dfrac{3}{x - 4} = \dfrac{2}{x + 3}$

54. $\dfrac{a - 1}{a + 3} - \dfrac{1 - 2a}{3 - a} = \dfrac{2 - a}{a - 3}$

Solve each formula for the indicated variable.

55. $S = \dfrac{a - lr}{1 - r}$ for r

56. $H = \dfrac{2ab}{a + b}$ for b

57. $\dfrac{1}{R} = \dfrac{1}{r_1} + \dfrac{1}{r_2} + \dfrac{1}{r_3}$ for R

58. $\dfrac{1}{R} = \dfrac{1}{r_1} + \dfrac{1}{r_2} + \dfrac{1}{r_3}$ for r_1

APPLICATIONS *Use a rational equation to solve each application.* SEE EXAMPLES 6–8. (OBJECTIVE 3)

59. House painting If one painter can paint a house in 5 days and another painter can paint the same house in 3 days, how long will it take them to paint the house working together?

60. Reading proof A proofreader can read 200 pages in 6 hours, and a second proofreader can read 200 pages in 9 hours. If they both work on a 200-page book, can they meet a 3-hour deadline?

61. Storing corn In 10 minutes, a conveyor belt can move 1,000 bushels of corn into the storage bin shown in the illustration. A smaller belt can move 1,000 bushels to the storage bin in 14 minutes. If both belts are used, how long will it take to move 1,000 bushels to the storage bin?

62. Roofing a house One roofing crew can finish a 2,800-square-foot roof in 12 hours, and another crew can do the job in 10 hours. If they work together, can they finish before a predicted rain in 5 hours?

63. Draining a pool A drain can empty the swimming pool shown in the illustration in 3 days. A second drain can empty the pool in 2 days. How long will it take to empty the pool if both drains are used?

64. Filling a pool A pipe can fill a pool in 9 hours. If a second pipe is also used, the pool can be filled in 3 hours. How long would it take the second pipe alone to fill the pool?

65. Filling a pond One pipe can fill a pond in 3 weeks, and a second pipe can fill the pond in 5 weeks. However, evaporation and seepage can empty the pond in 10 weeks. If both pipes are used, how long will it take to fill the pond?

66. Housecleaning Sally can clean the house in 6 hours, and her father can clean the house in 4 hours. Sally's younger brother, Dennis, can completely mess up the house in 8 hours. If Sally and her father clean and Dennis plays, how long will it take to clean the house?

67. Walking for charity A man bicycles 5 mph faster than he can walk. He bicycles 24 miles and walks back along the same route for a charity race in 11 hours. How fast does he walk?

68. Finding rates Two trains made the same 315-mile run. Since one train traveled 10 mph faster than the other, it arrived 2 hours earlier. Find the speed of each train.

69. Train travel A train traveled 120 miles from Freeport to Chicago and returned the same distance in a total time of 5 hours. If the train traveled 20 mph slower on the return trip, how fast did the train travel in each direction?

70. Time on the road A car traveled from Rockford to Chicago in 3 hours less time than it took a second car to travel from Rockford to St. Louis. If the cars traveled at the same average speed, how long was the first driver on the road?

71. Boating A man can drive a motorboat 45 miles down the Rock River in the same amount of time that he can drive it 27 miles upstream. Find the speed of the current if the speed of the boat is 12 mph in still water.

72. Rowing a boat A woman who can row 3 mph in still water rows 10 miles downstream on the Eagle River and returns upstream in a total of 12 hours. Find the speed of the current.

73. Aviation A plane that can fly 340 mph in still air can fly 200 miles with the wind in the same amount of time that it can fly 140 miles against the wind. Find the velocity of the wind.

74. Aviation An airplane can fly 650 miles with the wind in the same amount of time as it can fly 475 miles against the wind. If the wind speed is 40 mph, find the speed of the plane in still air.

75. Discount buying A repairman purchased some washing-machine motors for a total of $224. When the unit cost decreased by $4, he was able to buy one extra motor for the same total price. How many motors did he buy originally?

76. Price increases An appliance store manager bought several microwave ovens for a total of $1,800. When her unit cost increased by $25, she was able to buy one less oven for the same total price. How many ovens did she buy originally?

77. Stretching a vacation A student saved $1,200 for a trip to Europe. By cutting $20 from her daily expenses, she was able to stay three extra days. How long had she originally planned to be gone?

78. Focal length The design of a camera lens uses the equation

$$\frac{1}{f} = \frac{1}{s_1} + \frac{1}{s_2}$$

which relates the focal length f of a lens to the image distance s_1 and the object distance s_2. Find the focal length of the lens shown in the illustration. (*Hint*: Convert feet to inches.)

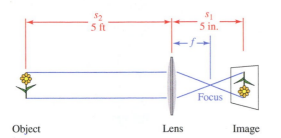

79. Lens maker's formula The focal length f of a lens is given by the lens maker's formula

$$\frac{1}{f} = 0.6\left(\frac{1}{r_1} + \frac{1}{r_2}\right)$$

where f is the focal length of the lens and r_1 and r_2 are the radii of the two circular surfaces. Find the focal length of the lens shown in the illustration.

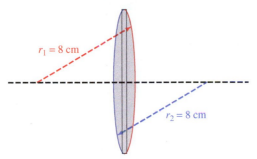

80. Engineering The stiffness of the shaft shown is given by the formula

$$k = \frac{1}{\dfrac{1}{k_1} + \dfrac{1}{k_2}}$$

where k_1 and k_2 represent the respective stiffness of each section. If the stiffness k_2 of Section 2 is 4,200,000 in. lb/rad, and the design specifications require that the stiffness k of the entire shaft be 1,900,000 in. lb/rad, what must the stiffness k_1 of Section 1 be?

WRITING ABOUT MATH

81. Why is it necessary to check the solutions of a rational equation?

82. Explain the steps you would use to solve a rational equation.

SOMETHING TO THINK ABOUT

83. Invent a rational equation that has an extraneous solution of 3.

84. Solve: $(x - 1)^{-1} - x^{-1} = 6^{-1}$

Section 6.6

Dividing Polynomials

Objectives

1. Divide a monomial by a monomial.
2. Divide a polynomial by a monomial.
3. Divide a polynomial by a polynomial.
4. Divide a polynomial with one or more missing terms by a polynomial.

Vocabulary

algorithm	dividend	remainder
divisor	quotient	

Getting Ready

Simplify the fraction or perform the division.

1. $\dfrac{16}{24}$ **2.** $-\dfrac{45}{81}$ **3.** $23\overline{)115}$ **4.** $32\overline{)512}$

We have previously seen how to add, subtract, and multiply polynomials. We now consider how to divide them.

1 Divide a monomial by a monomial.

In Example 1, we review how to divide a monomial by a monomial.

EXAMPLE 1 Simplify: $(3a^2b^3) \div (2a^3b)$ Assume no division by 0.

Solution We can use the rules of exponents (Method 1) or divide out all common factors (Method 2).

Method 1	*Method 2*
$\dfrac{3a^2b^3}{2a^3b} = \dfrac{3}{2}a^{2-3}b^{3-1}$	$\dfrac{3a^2b^3}{2a^3b} = \dfrac{3aabbb}{2aaab}$
$= \dfrac{3}{2}a^{-1}b^2$	$= \dfrac{3\cancel{aa}b\cancel{bb}}{2\cancel{aa}a\cancel{b}}$
$= \dfrac{3}{2}\left(\dfrac{1}{a}\right)\dfrac{b^2}{1}$	$= \dfrac{3b^2}{2a}$
$= \dfrac{3b^2}{2a}$	

SELF CHECK 1 Simplify: $\dfrac{6x^3y^2}{8x^2y^3}$ Assume no division by 0.

Divide a polynomial by a monomial.

In Example 2, we use the fact that $\frac{a+b}{c} = \frac{a}{c} + \frac{b}{c}$ $(c \neq 0)$ to divide a polynomial by a monomial.

EXAMPLE 2   Divide: $(4x^3y^2 + 3xy^5 - 12xy)$ by $3x^2y^3$ Assume no division by 0.

Solution   We can write the division in the form $\frac{4x^3y^2 + 3xy^5 - 12xy}{3x^2y^3}$ and proceed as follows:

$$\frac{4x^3y^2 + 3xy^5 - 12xy}{3x^2y^3} = \frac{4x^3y^2}{3x^2y^3} + \frac{3xy^5}{3x^2y^3} - \frac{12xy}{3x^2y^3} \qquad \text{Write the fraction as three separate fractions.}$$

$$= \frac{4x}{3y} + \frac{y^2}{x} - \frac{4}{xy^2} \qquad \text{Simplify each fraction.}$$

SELF CHECK 2   Simplify: $\dfrac{8a^3b^4 - 4a^4b^2 + a^2b^2}{4a^2b^2}$ Assume no division by 0.

3 **Divide a polynomial by a polynomial.**

COMMENT   The process of dividing polynomials is the same as that for dividing numbers. Divide $\dfrac{3568}{67}$.

There is an **algorithm** (a repeated series of steps) to use when the divisor is not a monomial. It is very similar to the division algorithm we use in arithmetic. To use the division algorithm to divide $(x^2 + 7x + 12)$ by $(x + 4)$, we write the division in long division form and proceed as follows:

$$\begin{array}{r} 5 \\ 67\overline{)3568} \end{array}$$

$$\begin{array}{r} x \\ x+4\overline{)x^2 + 7x + 12} \end{array} \qquad \text{How many times does } x \text{ divide } x^2? \frac{x^2}{x} = x. \text{ Place the } x \text{ in the quotient.}$$

$$\begin{array}{r} 5 \\ 67\overline{)3568} \\ -335 \\ \hline 21 \end{array}$$

$$\begin{array}{r} x \\ x+4\overline{)x^2 + 7x + 12} \\ -\ x^2 + 4x \\ \hline 3x + 12 \end{array} \qquad \begin{array}{l} \text{Multiply each term in the divisor by } x \text{ to obtain } (x^2 + 4x), \\ \text{subtract } (x^2 + 4x) \text{ from } (x^2 + 7x), \text{ and bring down the 12.} \end{array}$$

$$\begin{array}{r} 53 \\ 67\overline{)3568} \\ -335 \\ \hline 218 \end{array}$$

$$\begin{array}{r} x + 3 \\ x+4\overline{)x^2 + 7x + 12} \\ -\ x^2 + 4x \\ \hline 3x + 12 \end{array} \qquad \begin{array}{l} \text{How many times does } x \text{ divide } 3x? \frac{3x}{x} = 3. \text{ Place the } +3 \text{ in} \\ \text{the quotient.} \end{array}$$

$$\begin{array}{r} 53 \\ 67\overline{)3568} \\ -335 \\ \hline 218 \\ -201 \\ \hline 17 \end{array}$$

$$\begin{array}{r} x + 3 \\ x+4\overline{)x^2 + 7x + 12} \\ -\ x^2 + 4x \\ \hline 3x + 12 \\ -\ 3x + 12 \\ \hline 0 \end{array} \qquad \begin{array}{l} \text{Multiply each term in the divisor by 3 to obtain } (3x + 12), \text{ and} \\ \text{subtract } (3x + 12) \text{ from } (3x + 12) \text{ to obtain 0.} \end{array}$$

$$53\frac{17}{67}$$

We can check the quotient by multiplying the divisor by the quotient. The product should be the dividend.

Divisor	·	**quotient**	+	**remainder**	=	**dividend**
$(x + 4)$		$(x + 3)$	+	0	=	$x^2 + 7x + 12$ The quotient checks.

The division process stops when the result of the subtraction is a constant or a polynomial with degree less than the degree of the divisor. Here, the quotient is $(x + 3)$ and the remainder is 0.

EXAMPLE 3   Divide: $(2a^3 + 9a^2 + 5a - 6)$ by $(2a + 3)$ Assume no division by 0.

Solution

$$\begin{array}{r} a^2 \\ 2a + 3 \overline{) 2a^3 + 9a^2 + 5a - 6} \end{array}$$

How many times does $2a$ divide $2a^3$? $\frac{2a^3}{2a} = a^2$. Place a^2 in the quotient.

$$\begin{array}{r} a^2 \\ 2a + 3 \overline{) 2a^3 + 9a^2 + 5a - 6} \\ -\ \underline{2a^3 + 3a^2} \\ 6a^2 + 5a \end{array}$$

Multiply each term in the divisor by a^2 to obtain $(2a^3 + 3a^2)$, subtract $(2a^3 + 3a^2)$ from $(2a^3 + 9a^2)$, and bring down the $5a$.

$$\begin{array}{r} a^2 + 3a \\ 2a + 3 \overline{) 2a^3 + 9a^2 + 5a - 6} \\ -\ \underline{2a^3 + 3a^2} \\ 6a^2 + 5a \end{array}$$

How many times does $2a$ divide $6a^2$? $\frac{6a^2}{2a} = 3a$. Place the $+3a$ in the quotient.

$$\begin{array}{r} a^2 + 3a \\ 2a + 3 \overline{) 2a^3 + 9a^2 + 5a - 6} \\ -\ \underline{2a^3 + 3a^2} \\ 6a^2 + 5a \\ -\ \underline{6a^2 + 9a} \\ -4a - 6 \end{array}$$

Multiply each term in the divisor by $3a$ to obtain $(6a^2 + 9a)$, subtract $(6a^2 + 9a)$ from $(6a^2 + 5a)$, and bring down -6.

$$\begin{array}{r} a^2 + 3a - 2 \\ 2a + 3 \overline{) 2a^3 + 9a^2 + 5a - 6} \\ -\ \underline{2a^3 + 3a^2} \\ 6a^2 + 5a \\ -\ \underline{6a^2 + 9a} \\ -4a - 6 \end{array}$$

How many times does $2a$ divide $-4a$? $-\frac{4a}{2a} = -2$. Place the -2 in the quotient.

$$\begin{array}{r} a^2 + 3a - 2 \\ 2a + 3 \overline{) 2a^3 + 9a^2 + 5a - 6} \\ -\ \underline{2a^3 + 3a^2} \\ 6a^2 + 5a \\ -\ \underline{6a^2 + 9a} \\ -4a - 6 \\ -\ \underline{-4a - 6} \\ 0 \end{array}$$

Multiply each term in the divisor by -2 to obtain $(-4a - 6)$, and subtract $(-4a - 6)$ from $(-4a - 6)$ to obtain 0.

Since the remainder is 0, the quotient is $a^2 + 3a - 2$. We can check the quotient by verifying that

$$\underbrace{(2a + 3)}_{\text{Divisor}} \cdot \underbrace{(a^2 + 3a - 2)}_{\text{quotient}} + \underbrace{0}_{\text{remainder}} = \underbrace{2a^3 + 9a^2 + 5a - 6}_{\text{dividend}}$$

SELF CHECK 3   Divide: $2p + 1 \overline{) 2p^3 - 5p^2 - p + 1}$ Assume no division by 0.

EXAMPLE 4   Divide: $\dfrac{3x^3 + 2x^2 - 3x + 8}{x - 2}$ Assume no division by 0.

Solution

$$
\begin{array}{r}
3x^2 + 8x + 13 \\
x - 2 \overline{)3x^3 + 2x^2 - 3x + 8} \\
\underline{-\ 3x^3 - 6x^2} \\
8x^2 - 3x \\
\underline{-\ 8x^2 - 16x} \\
13x + 8 \\
\underline{-\ 13x - 26} \\
34
\end{array}
$$

This division results in a quotient of $(3x^2 + 8x + 13)$ and a remainder of 34. It is common to form a fraction with the remainder as the numerator and the divisor as the denominator and to write the result as

$$3x^2 + 8x + 13 + \frac{34}{x - 2}.$$

To check, we verify that

$$
\underbrace{(x - 2)}_{\text{Divisor}} \cdot \underbrace{(3x^2 + 8x + 13)}_{\text{quotient}} + \underbrace{34}_{\text{remainder}} = \underbrace{3x^3 + 2x^2 - 3x + 8}_{\text{dividend}}
$$

🌿 **SELF CHECK 4**   Divide: $a - 3 \overline{)2a^3 + 3a^2 - a + 2}$ Assume no division by 0.

EXAMPLE 5   Divide: $\dfrac{-9x + 8x^3 + 10x^2 - 9}{3 + 2x}$ Assume no division by 0.

Solution   The division algorithm works best when the polynomials in the dividend and the divisor are written in descending powers of the variable. We use the commutative property of addition to rearrange the terms and proceed with the division.

$$
\begin{array}{r}
4x^2 - x - 3 \\
2x + 3 \overline{)8x^3 + 10x^2 - 9x - 9} \\
\underline{-\ 8x^3 + 12x^2} \\
-2x^2 - 9x \\
-\ \underline{-2x^2 - 3x} \\
-6x - 9 \\
-\ \underline{-6x - 9} \\
0
\end{array}
$$

Thus,

$$\frac{-9x + 8x^3 + 10x^2 - 9}{3 + 2x} = 4x^2 - x - 3$$

🌿 **SELF CHECK 5**   Divide: $\dfrac{-4a + 15a^2 + 18a^3 - 4}{2 + 3a}$ Assume no division by 0.

4   Divide a polynomial with one or more missing terms by a polynomial.

Sometimes the divisor or the dividend will have a missing term.

EXAMPLE 6   Divide: $\dfrac{8x^3 + 1}{2x + 1}$ Assume no division by 0.

Solution   When we write the terms in the dividend in descending powers of x, we see that the terms involving x^2 and x are missing in the numerator. We must include the terms $0x^2$ and $0x$ in the dividend or leave spaces for them.

$$
\begin{array}{r}
4x^2 - 2x + 1 \\
2x + 1 \overline{)8x^3 + 0x^2 + 0x + 1} \\
-\underline{8x^3 + 4x^2} \\
-4x^2 + 0x \\
-\underline{-4x^2 - 2x} \\
2x + 1 \\
-\underline{2x + 1} \\
0
\end{array}
$$

Thus,

$$\frac{8x^3 + 1}{2x + 1} = 4x^2 - 2x + 1$$

SELF CHECK 6   Divide: $3a - 1\overline{)27a^3 - 1}$ Assume no division by 0.

EXAMPLE 7   Divide: $\dfrac{-17x^2 + 5x + x^4 + 2}{x^2 - 1 + 4x}$ Assume no division by 0.

Solution   We write the problem with the divisor and the dividend in descending powers of x. After leaving space for the missing term in the dividend, we proceed as follows:

$$
\begin{array}{r}
x^2 - 4x \\
x^2 + 4x - 1 \overline{)x^4 - 17x^2 + 5x + 2} \\
-\underline{x^4 + 4x^3 - x^2} \\
-4x^3 - 16x^2 + 5x \\
-\underline{-4x^3 - 16x^2 + 4x} \\
x + 2
\end{array}
$$

This division gives a quotient of $(x^2 - 4x)$ and a remainder of $(x + 2)$.

$$\frac{-17x^2 + 5x + x^4 + 2}{x^2 - 1 + 4x} = x^2 - 4x + \frac{x + 2}{x^2 + 4x - 1}$$

SELF CHECK 7   Divide: $\dfrac{2a^2 + 3a^3 + a^4 + a}{a^2 + 1 - 2a}$ Assume no division by 0.

SELF CHECK ANSWERS   1. $\frac{3x}{4y}$ 2. $2ab^2 - a^2 + \frac{1}{4}$ 3. $p^2 - 3p + 1$ 4. $2a^2 + 9a + 26 + \frac{80}{a - 3}$ 5. $6a^2 + a - 2$
6. $9a^2 + 3a + 1$ 7. $a^2 + 5a + 11 + \frac{18a - 11}{a^2 - 2a + 1}$

> **NOW TRY THIS**
>
> *Assume no division by* **0.**
>
> **1.** Divide: $\dfrac{10x^3 + 25x^2 + 5x}{5x}$
>
> **2.** The product of three binomials with integer coefficients is $(12x^3 - 32x^2 - 19x + 60)$. If one of the binomials is $(2x - 3)$, find the other two.
>
> **3.** Divide: $\dfrac{x^4 - 3x^3 + 2x^2 - 7}{x^2 - 2}$

6.6   Exercises

WARM-UPS   *Write the polynomial in descending powers of x.*

1. $x - 5 + x^2$ **2.** $4 + 3x^2 + 7x$

3. $5x^2 - x + 2x^3 - 6$

4. $6x + 4 - 2x^2 + 11x^3$

REVIEW   *Use the distributive property to remove parentheses and simplify.*

5. $2(x^2 + 4x - 1) + 3(2x^2 - 2x + 2)$

6. $3(2a^2 - 3a + 2) - 4(2a^2 + 4a - 7)$

7. $-2(3y^3 - 2y + 7) - 3(y^2 + 2y - 4) + 4(y^3 + 2y - 1)$

8. $3(4y^3 + 3y - 2) + 2(3y^2 - y + 3) - 5(2y^3 - y^2 - 2)$

VOCABULARY AND CONCEPTS   *Fill in the blanks.*

9. $\dfrac{a}{b} = \underline{} \cdot a \quad (b \neq 0)$

10. The division _____ is a repeated series of steps used to do a long division.

11. To check a division problem, the divisor $\cdot$ _____ $+$ remainder = dividend.

12. If a polynomial is divided by $3a - 2$ and the quotient is $3a^2 + 5$ with a remainder of 6, we usually write the result as

$3a^2 + 5 + $ _____ .

GUIDED PRACTICE   *Perform each division. Write each answer using positive exponents and assume no division by 0.*
SEE EXAMPLE 1. (OBJECTIVE 1)

13. $(2x^4y^5) \div (12x^2y^8)$

14. $(36a^{10}b) \div (6a^4b^4)$

15. $\dfrac{33a^{-2}b^2}{44a^2b^{-2}}$

16. $\dfrac{-63a^4b^{-3}}{81a^{-3}b^3}$

17. $\dfrac{45x^{-2}y^{-3}t^0}{-63x^{-1}y^4t^2}$

18. $\dfrac{112a^0b^2c^{-3}}{48a^4b^0c^4}$

19. $\dfrac{-65a^{2n}b^nc^{3n}}{-15a^nb^{-n}c}$

20. $\dfrac{-32x^{-3n}y^{-2n}z}{40x^{-2}y^{-n}z^{n+1}}$

Perform each division. Write each answer using positive exponents and assume no division by 0. SEE EXAMPLE 2. (OBJECTIVE 2)

21. $(4x^2 - x^3) \div (6x)$

22. $(5y^4 + 45y^3) \div (15y^2)$

23. $\dfrac{6x^5y^2 + 16x^3y}{4x^2y^2}$

24. $\dfrac{3a^3y^2 - 18a^4y^3}{27a^2y^2}$

25. $\dfrac{28x^2y^3 + 4xy - 20x^3y^2}{4xy}$

26. $\dfrac{45a^5b^3 - 63a^4b^4 - 9a^2b^3}{-9a^2b^3}$

27. $\dfrac{24x^6y^7 - 12x^5y^{12} + 36xy}{48x^2y^3}$

28. $\dfrac{9x^4y^3 + 18x^2y - 27xy^4}{9x^3y^3}$

Divide. Assume no division by 0. SEE EXAMPLE 3. (OBJECTIVE 3)

29. $(x^2 + 7x + 10) \div (x + 5)$

30. $(x^2 - 5x + 6) \div (x - 3)$

31. $x - 2\overline{)x^2 - 8x + 12}$

32. $x + 7\overline{)x^2 + 10x + 21}$

33. $\dfrac{6x^2 - x - 12}{2x - 3}$

34. $\dfrac{6x^2 - 7x + 2}{3x - 2}$

35. $\dfrac{15x^3 - 4x^2 - 13x + 6}{3x - 2}$

36. $\dfrac{16x^3 + 16x^2 - 9x - 5}{4x + 5}$

Divide. Assume no division by 0. SEE EXAMPLE 4. (OBJECTIVE 3)

37. $\dfrac{4a^3 + a^2 - 3a + 7}{a + 1}$

38. $\dfrac{3x^3 - 2x^2 + x + 6}{x - 1}$

39. $\dfrac{6x^3 + 11x^2 - x + 10}{2x + 3}$

40. $\dfrac{4x^3 - 12x^2 + 17x + 12}{2x - 3}$

Divide. Assume no division by 0. SEE EXAMPLE 5. (OBJECTIVE 3)

41. $\dfrac{-5x - 14 + x^2}{x + 2}$ **42.** $\dfrac{6x + 5x^2 - 8}{x + 2}$

43. $\dfrac{-9x^2 + 8x + 9x^3 - 4}{3x - 2}$ **44.** $\dfrac{6x^2 + 8x^3 - 13x + 3}{4x - 3}$

Divide. Assume no division by 0. SEE EXAMPLES 6–7. (OBJECTIVE 4)

45. $\dfrac{27a^3 - 8b^3}{3a - 2b}$ **46.** $\dfrac{64a^3 - 27b^3}{4a - 3b}$

47. $\dfrac{a^3 + 1}{a - 1}$ **48.** $\dfrac{8a^3 - 28}{2a - 3}$

ADDITIONAL PRACTICE *Divide. Assume no division by 0.*

49. $\dfrac{6x^2 - x - 12}{2x + 3}$

50. $\dfrac{6x^3 + 11x^2 - x - 2}{3x - 2}$

51. $\dfrac{x^4 + 2x^3 + 4x^2 + 3x + 2}{x^2 + x + 2}$

52. $\dfrac{2x^4 + 3x^3 + 3x^2 - 5x - 3}{2x^2 - x - 1}$

53. $\dfrac{6y - 4 + 10y^2}{5y - 2}$ **54.** $\dfrac{14x - 15 + 8x^2}{4x - 3}$

55. $\dfrac{x^3 + 3x + 5x^2 + 6 + x^4}{x^2 + 3}$ **56.** $\dfrac{13x + 16x^4 + 3x^2 + 3}{4x + 3}$

57. $x^2 - 2\overline{)x^6 - x^4 + 2x^2 - 8}$

58. $x^2 + 3\overline{)x^6 + 2x^4 - 6x^2 - 9}$

59. $\dfrac{5x^n y^n - 4x^{5n} y^{5n} - 3x^{2n} y^{2n}}{x^n y^n}$

60. $\dfrac{2a^n - 3a^n b^{2n} - 6b^{4n}}{a^n b^{n-1}}$

61. $\dfrac{3a^{-2} b^3 - 6a^2 b^{-3} + 9a^{-2}}{12a^{-1} b}$

62. $\dfrac{4x^3 y^{-2} + 8x^{-2} y^2 - 12y^4}{12x^{-1} y^{-1}}$

63. $\dfrac{3x^2 + 9x^3 + 4x + 4}{3x + 2}$

64. $\dfrac{27x + 23x^2 + 6x^3}{2x + 3}$

65. $\dfrac{15a^3 - 29a^2 + 16}{3a - 4}$

66. $\dfrac{x^5 + 3x + 2}{x^3 + 1 + 2x}$

67. $\dfrac{-10xy + x^2 + 16y^2}{x - 2y}$

68. $\dfrac{24y + 24 + 6y^2}{y - 2}$

69. $2x + y\overline{)32x^5 + y^5}$

70. $3x - y\overline{)81x^4 - y^4}$

APPLICATIONS

71. Geometry Find an expression for the length of the longer sides of the rectangle.

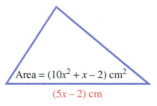

$(3x + 2)$ ft | Area $= (9x^2 + 21x + 10)$ ft^2

72. Geometry Find an expression for the height of the triangle.

Area $= (10x^2 + x - 2)$ cm^2

$(5x - 2)$ cm

73. Winter travel Complete the following table, which lists the rate (mph), time traveled (hr), and distance traveled (mi) by an Alaskan trail guide using two different means of travel.

	r	$\cdot$	t	$=$	d
Dog sled			$4x + 7$		$12x^2 + 13x - 14$
Snowshoes	$3x + 4$				$3x^2 + 19x + 20$

74. Pricing Complete the table for two items sold at a produce store.

	Price per lb	$\cdot$	Number of lb	$=$	Value
Cashews	$x^2 + 2x + 4$				$x^4 + 4x^2 + 16$
Sunflower seeds			$x^2 + 6$		$x^4 - x^2 - 42$

WRITING ABOUT MATH

75. Explain how to divide a monomial by a monomial.

76. Explain how to check the result of a division problem.

SOMETHING TO THINK ABOUT

77. Since 6 is a factor of 24, 6 divides 24 with no remainder. Decide whether $2x - 3$ is a factor of $10x^2 - x - 21$.

78. Is $x - 1$ a factor of $x^5 - 1$?

Section 6.7

Synthetic Division

Objectives

1. Divide a polynomial by a binomial of the form $(x - r)$ using synthetic division.
2. Apply the remainder theorem to evaluate a polynomial.
3. Apply the factor theorem to determine whether a specific value is a zero of a polynomial.

Vocabulary

synthetic division · factor theorem · zero of a polynomial
remainder theorem

Getting Ready

Divide each polynomial P(x) by x − 2 and find P(2).

1. $P(x) = x^2 - x - 1$ **2.** $P(x) = x^2 + x + 3$

---

1. ## Divide a polynomial by a binomial of the form $(x - r)$ using synthetic division.

There is a method, called **synthetic division**, that we can use to divide a polynomial by a binomial of the form $(x - r)$. To see how it works, we consider the division of $(4x^3 - 5x^2 - 11x + 20)$ by $(x - 2)$.

On the left is the long division, and on the right is the same division with the variables and their exponents removed and the numbers in the quotient moved to the left. The various powers of x can be remembered without actually writing them, because the exponents of the terms in the divisor, dividend, and quotient were written in descending order.

We can further shorten the version on the right. The numbers printed in color on the previous page need not be written, because they are duplicates of the numbers above them. Thus, we can write the division in the following form:

$$
\begin{array}{r}
4 \quad 3 - \ \ 5 \quad\quad\quad \\
1-2 \overline{)4-5-11\ \ \ 20} \\
-\ \ \underline{-\ 8\quad\quad\quad} \\
3\quad\quad\quad\ \\
-\ \ \underline{-\ \ 6\quad} \\
-\ \ 5\quad \\
-\ \ \underline{\ \ 10} \\
10
\end{array}
$$

We can shorten the process further by compressing the work vertically and eliminating the 1 (the coefficient of x in the divisor):

$$
\begin{array}{r}
4\quad 3\ \ -5\quad\quad\ \\
-2\overline{)4-5\ -11\quad 20} \\
-\ \ \underline{-8\ \ -6\quad 10} \\
3\ \ -5\quad 10
\end{array}
$$

If we write the 4 in the quotient on the bottom line, the bottom line gives the coefficients of the quotient and the remainder. If we eliminate the top line, the division appears as follows:

$$
\begin{array}{r}
-2\underline{|}\ \ 4\ \ -5\ \ -11\ \ \ \ 20 \\
-\ \ \underline{\quad\quad\quad -8\ \ \ -6\ \ \ 10} \\
4\ \ \ \ 3\ \ \ \ -5\ \ \ 10
\end{array}
$$

The bottom line was obtained by subtracting the middle line from the top line. If we replace the -2 in the divisor by $+2$, the division process will reverse the signs of every entry in the middle line, and then the bottom line can be obtained by addition. This gives the final form of the synthetic division.

$$
\begin{array}{r}
+2\underline{|}\ \ 4\ \ -5\ \ -11\ \ \ \ 20 \\
+\ \ \underline{\quad\quad\quad\ \ 8\ \ \ \ \ 6\ \ -10} \\
4\ \ \ \ 3\ \ \ -5\ \ \ \ 10
\end{array}
$$

The coefficients of the dividend

The coefficients of the quotient and the remainder

Thus,

$$
\frac{4x^3 - 5x^2 - 11x + 20}{x - 2} = 4x^2 + 3x - 5 + \frac{10}{x - 2}
$$

EXAMPLE 1   Use synthetic division to divide $(6x^2 + 5x - 2)$ by $(x - 5)$. Assume no division by 0.

Solution   We write the coefficients in the dividend and the 5 in the divisor in the following form:

$$
5\underline{|}\ \ 6\ \ 5\ \ -2
$$

Then we follow these steps:

$$
5\underline{|}\ \ 6\ \ 5\ \ -2 \qquad \text{\color{red}Begin by bringing down the 6.}
$$
$$
6
$$

$$
5\underline{|}\ \ 6\ \ 5\ \ -2 \qquad \text{\color{red}Multiply 5 by 6 to obtain 30.}
$$
$$
\underline{\quad\ \ 30\quad}
$$
$$
6
$$

$$\begin{array}{r|rrr} 5 & 6 & 5 & -2 \\ & & 30 & \\ \hline & 6 & 35 & \end{array}$$ Add 5 and 30 to obtain 35.

$$\begin{array}{r|rrr} 5 & 6 & 5 & -2 \\ & & 30 & 175 \\ \hline & 6 & 35 & \end{array}$$ Multiply 35 by 5 to obtain 175.

$$\begin{array}{r|rrr} 5 & 6 & 5 & -2 \\ & & 30 & 175 \\ \hline & 6 & 35 & 173 \end{array}$$ Add −2 and 175 to obtain 173.

COMMENT When using synthetic division the quotient will be one degree less than the dividend.

The numbers 6 and 35 represent the quotient $(6x + 35)$, and 173 is the remainder. Thus,

$$\frac{6x^2 + 5x - 2}{x - 5} = 6x + 35 + \frac{173}{x - 5}$$

SELF CHECK 1 Use synthetic division to divide $(6x^2 - 5x + 2)$ by $(x - 5)$. Assume no division by 0.

EXAMPLE 2 Use synthetic division to divide $(5x^3 + x^2 - 3)$ by $(x - 2)$. Assume no division by 0.

Solution We begin by writing

$$\begin{array}{r|rrrr} 2 & 5 & 1 & \mathbf{0} & -3 \end{array}$$ Write 0 for the coefficient of x, the missing term.

and complete the division as follows:

$$\begin{array}{r|rrrr} 2 & 5 & 1 & \mathbf{0} & -3 \\ & & 10 & & \\ \hline & 5 & 11 & & \end{array} \qquad \begin{array}{r|rrrr} 2 & 5 & 1 & \mathbf{0} & -3 \\ & & 10 & 22 & \\ \hline & 5 & 11 & 22 & \end{array} \qquad \begin{array}{r|rrrr} 2 & 5 & 1 & \mathbf{0} & -3 \\ & & 10 & 22 & 44 \\ \hline & 5 & 11 & 22 & 41 \end{array}$$

The numbers 5, 11, and 22 represent the quotient $5x^2 + 11x + 22$ and 41 is the remainder. Thus,

$$\frac{5x^3 + x^2 - 3}{x - 2} = 5x^2 + 11x + 22 + \frac{41}{x - 2}$$

SELF CHECK 2 Use synthetic division to divide $(5x^3 - x^2 + 3)$ by $(x - 2)$. Assume no division by 0.

EXAMPLE 3 Use synthetic division to divide $(5x^2 + 6x^3 + 2 - 4x)$ by $(x + 2)$. Assume no division by 0.

Solution First, we write the dividend with the exponents in descending order.

$$6x^3 + 5x^2 - 4x + 2$$

Then we write the divisor in $(x - r)$ form: $x - (-2)$. Using synthetic division, we begin by writing

$$\begin{array}{r|rrrr} -2 & 6 & 5 & -4 & 2 \\ & & & & \\ \hline \end{array}$$

and complete the division.

$$\begin{array}{r|rrrr} -2 & 6 & 5 & -4 & 2 \\ & & -12 & 14 & -20 \\ \hline & 6 & -7 & 10 & -18 \end{array}$$

Thus,

$$\frac{5x^2 + 6x^3 + 2 - 4x}{x + 2} = 6x^2 - 7x + 10 + \frac{-18}{x + 2}$$

SELF CHECK 3 Divide $(2x - 4x^2 + 3x^3 - 3)$ by $(x - 1)$. Assume no division by 0.

2 **Apply the remainder theorem to evaluate a polynomial.**

Synthetic division not only provides a method for simplifying some divisions, it can also be used to evaluate a polynomial by applying the **remainder theorem**.

REMAINDER THEOREM If a polynomial $P(x)$ is divided by $(x - r)$, the remainder is $P(r)$.

We illustrate the remainder theorem in the next example.

EXAMPLE 4 Let $P(x) = 2x^3 - 3x^2 - 2x + 1$. Find

a. $P(3)$. **b.** the remainder when $P(x)$ is divided by $(x - 3)$.

Solution **a.** $P(3) = 2(3)^3 - 3(3)^2 - 2(3) + 1$ Substitute 3 for x.

$= 2(27) - 3(9) - 6 + 1$

$= 54 - 27 - 6 + 1$

$= 22$

b. We can use the following synthetic division to find the remainder when $P(x) = 2x^3 - 3x^2 - 2x + 1$ is divided by $x - 3$.

$$\begin{array}{r|rrrr} 3 & 2 & -3 & -2 & 1 \\ & & 6 & 9 & 21 \\ \hline & 2 & 3 & 7 & 22 \end{array}$$ The remainder is 22.

The results of parts **a** and **b** illustrate that when $P(x)$ is divided by $(x - 3)$, the remainder is $P(3)$.

SELF CHECK 4 Use the polynomial of Example 4 to find:
a. $P(2)$. **b.** the remainder when the polynomial is divided by $(x - 2)$.

COMMENT It is often more efficient to find $P(r)$ by using synthetic division than by substituting r for x in $P(x)$. This is especially true if r is a fraction or decimal.

3 **Apply the factor theorem to determine whether a specific value is a zero of a polynomial.**

Recall that if two quantities are multiplied, each is called a *factor* of the product. Thus, $(x - 2)$ is one factor of $6x - 12$, because $6(x - 2) = 6x - 12$. The **factor theorem** enables us to find one factor of a polynomial if the remainder of a certain division is 0.

FACTOR THEOREM If $P(x)$ is a polynomial in x, then

$$P(r) = 0 \quad \text{if and only if} \quad (x - r) \text{ is a factor of } P(x).$$

If $P(x)$ is a polynomial in x and if $P(r) = 0$, r is called a **zero of the polynomial**.

EXAMPLE 5 Let $P(x) = 3x^3 - 5x^2 + 3x - 10$. Show that

a. $P(2) = 0$.   **b.** $(x - 2)$ is a factor of $P(x)$.

Solution **a.** $P(2) = 3(2)^3 - 5(2)^2 + 3(2) - 10$

$= 3(8) - 5(4) + 6 - 10$

$= 24 - 20 + 6 - 10$

$= 0$

$P(2) = 0$

Therefore, 2 is a zero of the polynomial.

b. Divide $P(x)$ by $(x - 2)$.

$$\begin{array}{r|rrrr} 2 & 3 & -5 & 3 & -10 \\ & & 6 & 2 & 10 \\ \hline & 3 & 1 & 5 & 0 \end{array}$$

Because the remainder is 0, the numbers 3, 1, and 5 in the synthetic division represent the quotient $(3x^2 + x + 5)$. Thus,

$$\underbrace{(x-2)}_{\text{Divisor}} \cdot \underbrace{(3x^2 + x + 5)}_{\text{quotient}} + \underbrace{0}_{\text{remainder}} = \underbrace{3x^3 - 5x^2 + 3x - 10}_{\text{the dividend, } P(x)}$$

or

$$(x - 2)(3x^2 + x + 5) = 3x^3 - 5x^2 + 3x - 10$$

Thus, $(x - 2)$ is a factor of $P(x)$.

SELF CHECK 5 Use the polynomial $P(x) = x^2 - 5x + 6$ and show that $P(2) = 0$ and that $(x - 2)$ is a factor of $P(x)$.

The result in Example 5 is true because the remainder, $P(2)$, is 0. If the remainder had not been 0, then $(x - 2)$ would not have been a factor of $P(x)$.

Accent on technology

▸ Approximating Zeros of Polynomials

COMMENT The x-coordinates of the x-intercepts of the graph represent what we call the real zeros of a polynomial.

We can use a graphing calculator to approximate the real zeros of a polynomial function $f(x)$. For example, to find the real zeros of $f(x) = 2x^3 - 6x^2 + 7x - 21$, we graph the function as in Figure 6-10 on the next page.

It appears from the figure that the function f has a zero at $x = 3$. To be certain, find $f(3)$.

$f(3) = 2(3)^3 - 6(3)^2 + 7(3) - 21$ Substitute 3 for x.

$= 2(27) - 6(9) + 21 - 21$

$= 0$

From the factor theorem, we know that if $P(3) = 0$, then $(x - 3)$ is a factor of the polynomial. To find the other factors, we can synthetically divide by 3.

$$\begin{array}{r|rrrr} 3 & 2 & -6 & 7 & -21 \\ & & 6 & 0 & 21 \\ \hline & 2 & 0 & 7 & 0 \end{array}$$

(continued)

Thus, $f(x) = (x - 3)(2x^2 + 7)$. Since $(2x^2 + 7)$ cannot be factored over the real numbers, we can conclude that 3 is the only real zero of the polynomial function.

$f(x) = 2x^3 - 6x^2 + 7x - 21$

Figure 6-10

For instructions regarding the use of a Casio graphing calculator, please refer to the Casio Keystroke Guide in the back of the book.

SELF CHECK ANSWERS

1. $6x + 25 + \frac{127}{x - 5}$ **2.** $5x^2 + 9x + 18 + \frac{39}{x - 2}$ **3.** $3x^2 - x + 1 + \frac{-2}{x - 1}$ **4.** both are 1
5. $(x - 2)(x - 3) = x^2 - 5x + 6$

NOW TRY THIS

1. Given that $(x + 2)$ is a factor of $(2x^3 - 3x^2 - 11x + 6)$, find the remaining factors.

2. Solve $2x^3 - 3x^2 - 11x + 6 = 0$ without a calculator.

3. Find the x-intercepts for $f(x) = 2x^3 - 3x^2 - 11x + 6$.

6.7   Exercises

For Exercise Set 6.7, assume no division by 0.

WARM-UPS *Use long division to find each remainder.*

1. $(2x^2 + 3x - 18) \div (x + 4)$
2. $(4x^2 + 15x + 8) \div (x + 3)$

Evaluate.

3. $2x^2 + 3x - 18$ when $x = -4$
4. $4x^2 + 15x + 8$ when $x = -3$

REVIEW *Let $f(x) = 3x^2 + 2x - 1$ and find each value.*

5. $f(-1)$ **6.** $f(3)$
7. $f(2a)$ **8.** $f(-2t)$

Remove parentheses and simplify.

9. $5(x^2 - 3x + 2) + 4(3x^2 - x + 1)$
10. $-2(3y^3 - 2y + 7) - 3(y^2 + 2y - 4) + 4(y^3 + 2y - 1)$

VOCABULARY AND CONCEPTS *Fill in the blanks.*

11. In order to use synthetic division, the divisor must be in the form ____.

12. If the degree of the leading term of the dividend is 5 and the divisor is $(x - r)$, the degree of the leading term of the quotient will be __.

13. The remainder theorem states that if a polynomial $P(x)$ is divided by $(x - r)$, the remainder is ____.

14. The factor theorem states that if $P(x)$ is a polynomial in x, then $P(r) = 0$ if and only if ____ is a factor of $P(x)$.

GUIDED PRACTICE *Use synthetic division to perform each division. SEE EXAMPLE 1. (OBJECTIVE 1)*

15. $(x^2 - 13x + 30) \div (x - 9)$
16. $(x^2 - 6x + 5) \div (x - 5)$
17. $(x^2 + 4x - 10) \div (x + 5)$
18. $(x^2 + x - 6) \div (x - 2)$

Use synthetic division to perform each division. SEE EXAMPLE 2. (OBJECTIVE 1)

19. $(2x^3 - 5x - 6) \div (x - 2)$

20. $(x^3 - x + 6) \div (x - 2)$

21. $(4x^3 - 2x + 36) \div (x + 2)$

22. $(2x^3 - 5x^2 - 9) \div (x - 3)$

Use synthetic division to perform each division. SEE EXAMPLE 3. (OBJECTIVE 1)

23. $(5x^2 + 6x^3 + 4) \div (x + 1)$

24. $(-13x^2 + 2x^3 - 5 + 23x) \div (x - 4)$

25. $(4x^2 + 3x^3 + 8) \div (x + 2)$

26. $(4 - 3x^2 + x) \div (x - 4)$

Let $P(x) = 2x^3 - 4x^2 + 2x - 1$. Evaluate $P(x)$ for the given value by using the remainder theorem. Then evaluate by substituting the value of x into the polynomial and simplifying. SEE EXAMPLE 4. (OBJECTIVE 2)

27. $P(1)$

28. $P(2)$

29. $P(-3)$

30. $P(-1)$

31. $P(3)$

32. $P(-4)$

33. $P(-5)$

34. $P(4)$

Use the remainder theorem and synthetic division to find $P(r)$. (OBJECTIVE 2)

35. $P(x) = x^3 - 4x^2 + x - 2; r = 2$

36. $P(x) = x^3 - 3x^2 + x + 1; r = 1$

37. $P(x) = x^4 - 2x^3 + x^2 - 3x + 2; r = -2$

38. $P(x) = 2x^3 + x + 2; r = 3$

Use the factor theorem and determine whether the first expression is a factor of $P(x)$. Factor, if possible. SEE EXAMPLE 5. (OBJECTIVE 3)

39. $x - 3; P(x) = x^3 - 3x^2 + 5x - 15$

40. $x + 1; P(x) = x^3 + 2x^2 - 2x - 3$

41. $x - 6; P(x) = 2x^3 - 8x^2 + x - 36$

42. $x + 2; P(x) = 3x^2 - 7x + 4$

ADDITIONAL PRACTICE *Divide using synthetic division.*

43. $(3x^3 - 10x^2 + 5x - 6) \div (x - 3)$

44. $(2x^3 - 9x^2 + 10x - 3) \div (x - 3)$

45. $(x^2 - 5x + 14) \div (x + 2)$

46. $(5x^2 - 8x - 2) \div (x - 2)$

47. $(x^2 + 8 + 6x) \div (x + 4)$

48. $(x^2 - 15 - 2x) \div (x + 3)$

49. $(x^2 + 13x + 42) \div (x + 6)$

50. $(4x^3 + 5x^2 - 1) \div (x + 2)$

Use the remainder theorem and synthetic division to find $P(r)$.

51. $P(x) = x^3 + x^2 + 1; r = -2$

52. $P(x) = x^5 + 3x^4 - x^2 + 1; r = -1$

53. $P(x) = 2x^3 - x^2 + 4x - 5; r = \dfrac{1}{2}$

54. $P(x) = 3x^3 + x^2 - 8x + 4; r = \dfrac{2}{3}$

Use the factor theorem and determine whether the first expression is a factor of $P(x)$. Factor, if possible.

55. $x; P(x) = 7x^3 - 5x^2 - 8x$

56. $x; P(x) = 5x^3 + 6x + 2$

57. $x - 5; P(x) = x^3 - 125$

58. $x - 2; P(x) = x^3 + 8$

Evaluate $P(x)$ for the given value using the remainder theorem.

59. $P(x) = 3x^5 + 1; c = -\dfrac{1}{2}$

60. $P(x) = 5x^7 - 7x^4 + x^2 + 1; c = 2$

Let $Q(x) = x^4 - 3x^3 + 2x^2 + x - 3$. Evaluate $Q(x)$ for the given value by using the remainder theorem.

61. $Q(-1)$

62. $Q(0)$

63. $Q(2)$

64. $Q(-2)$

Use a calculator to work each problem.

65. Find 2^6 by using synthetic division to evaluate the polynomial $P(x) = x^6$ at $x = 2$. Then check the answer by evaluating 2^6 with a calculator.

66. Find $(-3)^5$ by using synthetic division to evaluate the polynomial $P(x) = x^5$ at $x = -3$. Then check the answer by evaluating $(-3)^5$ with a calculator.

WRITING ABOUT MATH

67. If you are given $P(x)$, explain how to use synthetic division to calculate $P(a)$.

68. Explain the factor theorem.

SOMETHING TO THINK ABOUT *Suppose that $P(x) = x^{100} - x^{99} + x^{98} - x^{97} + \cdots + x^2 - x + 1$.*

69. Find the remainder when $P(x)$ is divided by $(x - 1)$.

70. Find the remainder when $P(x)$ is divided by $(x + 1)$.

Proportion and Variation

Objectives

1. Solve a proportion.
2. Solve an application involving a proportion.
3. Use similar triangles to determine a missing length of one of the sides.
4. Solve an application involving direct variation.
5. Solve an application involving inverse variation.
6. Solve an application involving joint variation.
7. Solve an application involving combined variation.

Vocabulary

ratio	means	inverse variation
unit cost	similar triangles	joint variation
rates	direct variation	combined variation
proportion	constant of proportionality	
extremes	(constant of variation)	

Getting Ready

Solve each equation.

1. $30k = 70$
2. $\dfrac{k}{4{,}000^2} = 90$

Classify each function as a linear function or a rational function.

3. $f(x) = 3x$
4. $f(x) = \dfrac{3}{x} \quad (x > 0)$

In this section, we will discuss ratio and proportion. Then we will use these skills to solve variation problems.

1 Solve a proportion.

The comparison of two numbers is often called a **ratio**. For example, the fraction $\frac{2}{3}$ can be read as "the ratio of 2 to 3." Some more examples of ratios are

$$\frac{4x}{7y} \quad \text{(the ratio of } 4x \text{ to } 7y\text{)} \qquad \text{and} \qquad \frac{x-2}{3x} \quad \text{(the ratio of } (x-2) \text{ to } 3x\text{)}$$

Ratios often are used to express **unit costs**, such as the cost per pound of ground round steak.

$$\text{The cost of a package of ground round} \rightarrow \frac{\$18.75}{5 \text{ lb}} = \$3.75 \text{ per lb} \leftarrow \text{The cost per pound}$$
$$\text{The weight of the package} \rightarrow$$

Ratios also are used to express **rates**, such as an average rate of speed.

$$\text{A distance traveled} \rightarrow \frac{372 \text{ miles}}{6 \text{ hours}} = 62 \text{ mph} \leftarrow \text{The average rate of speed}$$
$$\text{in a period of time} \rightarrow$$

An equation indicating that two ratios are equal is called a **proportion**. Two examples of proportions are

$$\frac{1}{4} = \frac{2}{8} \quad \text{and} \quad \frac{4}{7} = \frac{12}{21}$$

In the proportion $\frac{a}{b} = \frac{c}{d}$, a and d are called the **extremes** and b and c are called the **means**. To develop a fundamental property of proportions, we suppose that

$$\frac{a}{b} = \frac{c}{d}$$

is a proportion and multiply both sides of the equation by bd, the LCD, to obtain

$$bd\left(\frac{a}{b}\right) = bd\left(\frac{c}{d}\right)$$

$$\frac{\cancel{b}da}{\cancel{b}} = \frac{b\cancel{d}c}{\cancel{d}}$$

$$ad = bc$$

Thus, if $\frac{a}{b} = \frac{c}{d}$, then $ad = bc$. This shows that in a proportion, the *product of the extremes equals the product of the means.*

EXAMPLE 1   Solve: $\dfrac{x+1}{x} = \dfrac{x}{x+2}$ Assume no denominators are 0.

Solution   We will use the fact that *in a proportion, the product of the extremes is equal to the product of the means.*

$$\frac{x+1}{x} = \frac{x}{x+2}$$

$(x+1)(x+2) = x \cdot x$ In a proportion, the product of the extremes equals the product of the means.

$x^2 + 3x + 2 = x^2$ Multiply.

$3x + 2 = 0$ Subtract x^2 from both sides.

$x = -\dfrac{2}{3}$ Subtract 2 from both sides and divide by 3.

Thus, the solution is $-\frac{2}{3}$.

 SELF CHECK 1   Solve: $\dfrac{x-1}{x} = \dfrac{x}{x+3}$ Assume no denominators are 0.

EXAMPLE 2   Solve: $\dfrac{5a+2}{2a} = \dfrac{18}{a+4}$ Assume no denominators are 0.

Solution   We will use the fact that *in a proportion, the product of the extremes is equal to the product of the means.*

$$\frac{5a+2}{2a} = \frac{18}{a+4}$$

$(5a+2)(a+4) = 2a(18)$ In a proportion, the product of the extremes equals the product of the means.

$5a^2 + 22a + 8 = 36a$ Multiply.

$5a^2 - 14a + 8 = 0$ Subtract $36a$ from both sides because this is a quadratic equation.

$(5a - 4)(a - 2) = 0$ Factor.

$$5a - 4 = 0 \quad \text{or} \quad a - 2 = 0 \qquad \text{\color{red}{Set each factor equal to 0.}}$$
$$5a = 4 \qquad\qquad\qquad a = 2 \qquad \text{\color{red}{Solve each linear equation.}}$$
$$a = \frac{4}{5}$$

Thus, the solutions are $\frac{4}{5}$ and 2.

 SELF CHECK 2   Solve: $\dfrac{3x + 1}{12} = \dfrac{x}{x + 2}$ Assume no denominators are 0.

2   Solve an application involving a proportion.

EXAMPLE 3   **GROCERY SHOPPING**   If 7 pears cost $2.73, how much will 11 pears cost?

Solution   We can let c represent the cost in dollars of 11 pears. The price per pear of 7 pears is $\frac{\$2.73}{7}$, and the price per pear of 11 pears is $\frac{\$c}{11}$. Since these costs are equal, we can set up and solve the following proportion.

$$\frac{2.73}{7} = \frac{c}{11}$$

$$11(2.73) = 7c \qquad \text{\color{red}{In a proportion, the product of the extremes is equal to the product of the means.}}$$

$$30.03 = 7c \qquad \text{\color{red}{Multiply.}}$$

$$\frac{30.03}{7} = c \qquad \text{\color{red}{Divide both sides by 7.}}$$

$$c = 4.29 \qquad \text{\color{red}{Simplify.}}$$

Eleven pears will cost $4.29.

 SELF CHECK 3   How much will 28 pears cost?

Everyday connections
Highest and Longest Jumpers

A kangaroo rat can jump 45 times its body length, the longest jumper of all mammals. Its body length averages 4.5 inches. The longest jump by a human is 29 feet 4 inches (Mike Powell (6 feet 2 inches) of USA in 1991).

A flea measures 0.029 inch high and can jump up 4.06 inches. The highest jump by a human is 8 feet 5 inches (Javier Sotomayor (6 feet 5 inches) of Cuba in 1991).

Sources: http://www.sports-reference.com/olympics/ athletes/po/mike-powell-1.html

http://www.iaaf.org/athletes/biographies/ country=cubathcode=2718/index.html

1. If the height of a 6 foot 2 inch man were proportional to that of a kangaroo rat, how far could the man jump?

2. If the distance the kangaroo rat could jump were proportional to the distance the man could jump, how far would the kangaroo rat be able to jump?

3. If the height of a 6 foot 5 inch man were proportional to that of a flea, how high could the man jump?

4. If the height a flea could jump were proportional to the height a man could jump, how high would the flea be able to jump?

3 Use similar triangles to determine a missing length of one of the sides.

If two angles of one triangle have the same measure as two angles of a second triangle, the triangles will have the same shape. In this case, we call the triangles **similar triangles**. Here are some facts about similar triangles.

SIMILAR TRIANGLES

If two triangles are similar, then

1. the three angles of the first triangle have the same measures, respectively, as the three angles of the second triangle.

2. the lengths of all corresponding sides are in proportion.

The triangles shown in Figure 6-11 are similar because the measures of their corresponding angles are the same. Therefore, the measures of their corresponding sides are in proportion.

$$\frac{2}{4} = \frac{x}{2x}, \qquad \frac{x}{2x} = \frac{1}{2}, \qquad \frac{1}{2} = \frac{2}{4}$$

COMMENT It is also true that if the measures of the corresponding sides of two triangles are in proportion, the triangles are similar.

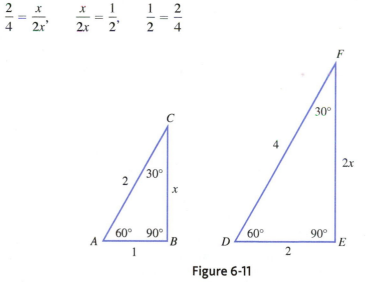

Figure 6-11

The properties of similar triangles often enable us to find the lengths of the sides of a triangle indirectly. For example, on a sunny day we can find the height of a tree and stay safely on the ground.

EXAMPLE 4 **HEIGHT OF A TREE** A tree casts a shadow of 29 feet at the same time as a vertical yardstick casts a shadow of 2.5 feet. Find the height of the tree.

Solution We will let h represent the height of the tree in feet. Refer to Figure 6-12 on the next page, which illustrates the triangles determined by the tree and its shadow, and the yardstick and its shadow. Because the triangles have the same shape, they are similar, and the measures of their corresponding sides are in proportion. We can find the value of h by setting up and solving the following proportion.

$$\frac{h}{3} = \frac{29}{2.5}$$

$2.5h = 3(29)$ In a proportion, the product of the extremes is equal to the product of the means.

$2.5h = 87$ Simplify.

$h = 34.8$ Divide both sides by 2.5.

The tree is about 35 feet tall.

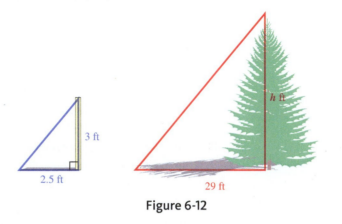

Figure 6-12

SELF CHECK 4   Find the height of the tree if its shadow is 25 feet long.

---

4   **Solve an application involving direct variation.**

To introduce **direct variation**, we consider the formula

$$C = \pi D$$

for the circumference of a circle, where C is the circumference, D is the diameter, and $\pi \approx 3.14159$. If we double the diameter of a circle, we determine another circle with a larger circumference C_1 such that

$$C_1 = \pi(2D) = 2\pi D = 2C$$

Thus, doubling the diameter results in doubling the circumference. Likewise, if we triple the diameter, we triple the circumference.

COMMENT In this section, we will assume that k is a positive number.

In this formula, we say that the variables C and D *vary directly*, or that they are *directly proportional*. This is because as one variable gets larger, so does the other, and in a predictable way. In this example, the constant π is called the **constant of variation** or the *constant of proportionality*.

DIRECT VARIATION	The words "y varies directly with x" or "y is directly proportional to x" mean that $y = kx$ for some nonzero constant k. The constant k is called the constant of variation or the constant of proportionality.

Since the relationship for direct variation ($y = kx$) defines a linear function, its graph is always a line with a y-intercept at the origin. The graph of $y = kx$ appears in Figure 6-13 for three positive values of k.

One real-world example of direct variation is Hooke's law from physics. Hooke's law states that the distance a spring will stretch varies directly with the force that is applied to it.

If d represents a distance and f represents a force, Hooke's law is expressed mathematically as

$$d = kf$$

where k is the constant of variation. If the spring stretches 10 inches when a weight of 6 pounds is attached, the value of k can be found as follows:

$d = k\mathbf{f}$

$\mathbf{10} = k(\mathbf{6})$ Substitute 10 for d and 6 for f.

$\dfrac{5}{3} = k$ Divide both sides by 6 and simplify.

Figure 6-13

To find the force required to stretch the spring a distance of 35 inches, we solve the equation $d = kf$ for f, with $d = 35$ and $k = \frac{5}{3}$.

$d = \mathbf{k}f$

$\mathbf{35} = \dfrac{\mathbf{5}}{\mathbf{3}}f$ Substitute 35 for d and $\frac{5}{3}$ for k.

$105 = 5f$ Multiply both sides by 3.

$21 = f$ Divide both sides by 5.

The force required to stretch the spring a distance of 35 inches is 21 pounds.

EXAMPLE 5   **DIRECT VARIATION**   The distance traveled in a given time is directly proportional to the speed. If a car travels 70 miles at 30 mph, how far will it travel in the same time at 45 mph?

Solution   The words *distance is directly proportional to speed* can be expressed by the equation

(1) $d = ks$

where d is distance, k is the constant of variation, and s is the speed. To find the value of k, we substitute 70 for d and 30 for s, and solve.

$d = \mathbf{k}s$

$\mathbf{70} = k(\mathbf{30})$

$k = \dfrac{7}{3}$

To find the distance traveled at 45 mph, we substitute $\frac{7}{3}$ for k and 45 for s in Equation 1 and simplify.

$d = \mathbf{k}s$

$d = \dfrac{7}{3}(\mathbf{45})$

$= 105$

In the time it took to travel 70 miles at 30 mph, the car could travel 105 miles at 45 mph.

SELF CHECK 5   How far will the car travel in the same time at 60 mph?

5 Solve an application involving inverse variation.

In the formula $w = \frac{12}{l}$, w gets smaller as l gets larger, and w gets larger as l gets smaller. This is an example of **inverse variation**. Since these variables vary in opposite directions in a predictable way, we say that the variables *vary inversely*, or that they are *inversely proportional*. The constant 12 is the constant of variation.

INVERSE VARIATION

The words "y varies inversely with x" or "y is inversely proportional to x" mean that $y = \frac{k}{x}$ for some nonzero constant k. The constant k is called the constant of variation.

The formula for inverse variation $\left(y = \frac{k}{x}\right)$ defines a rational function. The graph of $y = \frac{k}{x}$ appears in Figure 6-14 for three positive values of k.

Because of gravity, an object in space is attracted to Earth. The force of this attraction varies inversely with the square of the distance from the object to Earth's center.

If f represents the force and d represents the distance, this information can be expressed by the equation

$$f = \frac{k}{d^2}$$

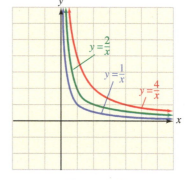

Figure 6-14

If we know that an object 4,000 miles from Earth's center is attracted to Earth with a force of 90 pounds, we can find the value of k.

$$f = \frac{k}{d^2}$$

$$90 = \frac{k}{4{,}000^2} \qquad \text{Substitute 90 for } f \text{ and 4,000 for } d.$$

$$k = 90(4{,}000^2)$$

$$= 1.44 \times 10^9$$

To find the force of attraction when the object is 5,000 miles from Earth's center, we proceed as follows:

$$f = \frac{k}{d^2}$$

$$f = \frac{1.44 \times 10^9}{5{,}000^2} \qquad \text{Substitute } 1.44 \times 10^9 \text{ for } k \text{ and 5,000 for } d.$$

$$= 57.6$$

The object will be attracted to Earth with a force of 57.6 pounds when it is 5,000 miles from Earth's center.

EXAMPLE 6 **LIGHT INTENSITY** The intensity I of light received from a light source varies inversely with the square of the distance from the source. If the intensity of a light source 4 feet from an object is 8 candelas, find the intensity at a distance of 2 feet.

Solution The words *intensity varies inversely with the square of the distance d* can be expressed by the equation

$$I = \frac{k}{d^2}$$

To find the value of k, we substitute 8 for I and 4 for d and solve.

$$I = \frac{k}{d^2}$$

$$8 = \frac{k}{4^2}$$

$$128 = k$$

To find the intensity when the object is 2 feet from the light source, we substitute 2 for d and 128 for k and simplify.

$$I = \frac{k}{d^2}$$

$$I = \frac{128}{2^2}$$

$$= 32$$

The intensity at 2 feet is 32 candelas.

 SELF CHECK 6   Find the intensity at a distance of 8 feet.

6   Solve an application involving joint variation.

There are times when one variable varies with the product of several variables. For example, the area of a triangle varies directly with the product of its base and height:

$$A = \frac{1}{2} bh$$

Such variation is called **joint variation**.

JOINT VARIATION	If one variable varies directly with the product of two or more variables, the relationship is called joint variation. If y varies jointly with x and z, then $y = kxz$. The nonzero constant k is called the constant of variation.

EXAMPLE 7   **VOLUME OF A CONE**   The volume V of a cone varies jointly with its height h and the area of its base b. If $V = 6 \text{ cm}^3$ when $h = 3$ cm and $B = 6 \text{ cm}^2$, find V when $h = 2$ cm and $B = 8 \text{ cm}^2$.

Solution   The words V varies jointly with h and B can be expressed by the equation

$$V = khB \qquad \text{The relationship also can be read as "V is directly proportional to the product of h and B."}$$

We can find the value of k by substituting 6 for V, 3 for h, and 6 for B.

$$V = khB$$

$$6 = k(3)(6)$$

$$6 = k(18)$$

$$\frac{1}{3} = k \qquad \text{Divide both sides by 18; } \frac{6}{18} = \frac{1}{3}.$$

To find V when $h = 2$ and $B = 8$, we substitute these values into the formula $V = \frac{1}{3}hB$.

$$V = \frac{1}{3}\,hB$$

$$V = \left(\frac{1}{3}\right)(2)(8)$$

$$= \frac{16}{3}$$

The volume is $5\frac{1}{3}$ cm^3.

 SELF CHECK 7   Find V when $B = 10$ cm.

---

7   Solve an application involving combined variation.

Many applied problems involve a combination of direct and inverse variation. Such variation is called **combined variation**.

EXAMPLE 8   **BUILDING HIGHWAYS**   The time it takes to build a highway varies directly with the length of the road, but inversely with the number of workers. If it takes 100 workers 4 weeks to build 2 miles of highway, how long will it take 80 workers to build 10 miles of highway?

Solution   We can let t represent the time in weeks, l represent the length in miles, and w represent the number of workers. The relationship among these variables can be expressed by the equation

$$t = \frac{kl}{w}$$

We substitute 4 for t, 100 for w, and 2 for l to find the value of k:

$$4 = \frac{k(2)}{100}$$

$400 = 2k$ Multiply both sides by 100.

$200 = k$ Divide both sides by 2.

We now substitute 80 for w, 10 for l, and 200 for k in the equation $t = \frac{kl}{w}$ and simplify:

$$t = \frac{kl}{w}$$

$$t = \frac{200(10)}{80}$$

$$= 25$$

It will take 25 weeks for 80 workers to build 10 miles of highway.

 SELF CHECK 8   How long will it take 60 workers to build 6 miles of highway?

 SELF CHECK ANSWERS   **1.** $\frac{3}{2}$ **2.** $\frac{2}{3}$, 1   **3.** \$10.92   **4.** 30 ft   **5.** 140 mi   **6.** 2 candelas   **7.** $6\frac{2}{3}$ cm^3 **8.** 20 weeks

NOW TRY THIS

1. Solve $\dfrac{5x + 2}{6x + 3} = \dfrac{2x}{2x + 1}$ as a proportion.

2. Solve $\dfrac{5x + 2}{6x + 3} = \dfrac{2x}{2x + 1}$ by multiplying by the LCD.

Explain to a classmate why you did not get the same answer as in Problem 1.

The time T, in hours, required for a satellite to complete an orbit around Earth varies directly as the radius r of the orbit measured from the center of Earth and inversely as the velocity in miles per hour. The radius of Earth is approximately 4,000 miles.

3. The space shuttle made one orbit around Earth in 1.5 hours at a rate of 17,000 mph. Its altitude above Earth's surface was about 200 miles. Find the constant of variation rounded to 2 decimal places.

4. The Pentagon's Defense Satellite Communications System (DSCS) orbits at 23,500 miles above Earth's surface at a speed of approximately 6,955 mph. Using the same constant of variation, how long is its orbit? What can you conclude about the satellite?

6.8   Exercises

Fill in each box with "=" or "≠."

1. $\dfrac{15}{4} \ \square\ \dfrac{105}{28}$

2. $\dfrac{16}{17} \ \square\ \dfrac{48}{51}$

3. $\dfrac{30}{49} \ \square\ \dfrac{120}{245}$

4. $\dfrac{36}{39} \ \square\ \dfrac{124}{146}$

Find the value of k.

5. $t = \dfrac{kl}{w}$ when $t = 20$, $l = 4$, and $w = 40$

6. $a = kbc$ when $a = 12$, $b = 2$, and $c = 3$

REVIEW *Simplify each expression.*

7. $(x^4 \cdot x^{-5})^2$

8. $\left(\dfrac{a^3 a^5}{a^{-2}}\right)^3$

9. $\dfrac{3y^0 - 5y^0}{y^0}$

10. $\left(\dfrac{2r^{-2}r^{-3}}{4r^{-5}}\right)^{-3}$

11. Write 470,000 in scientific notation.

12. Write 0.000047 in scientific notation.

13. Write 2.5×10^{-3} in standard notation.

14. Write 2.5×10^4 in standard notation.

VOCABULARY AND CONCEPTS *Fill in the blanks.*

15. Ratios are used to express _____ and ____.

16. An equation stating that two ratios are equal is called a _____.

17. In a proportion, the product of the _____ is equal to the product of the ____.

18. If two angles of one triangle have the same measures as two angles of a second triangle, the triangles are ____.

19. The equation $y = kx$ indicates ____ variation.

20. The equation $y = \dfrac{k}{x}$ indicates ____ variation.

21. Inverse variation is represented by a ____ function.

22. The relationship for direct variation ($y = kx$) is represented by a ____ function through the origin.

23. The equation $y = kxz$ indicates ____ variation and k represents the _____.

24. The equation $y = \dfrac{kx}{z}$ indicates _____ variation.

Determine whether the graph represents direct variation, inverse variation, or neither.

25.   26.

27.   28.

GUIDED PRACTICE *Solve each proportion for the variable.*
SEE EXAMPLE 1. (OBJECTIVE 1)

29. $\dfrac{x}{8} = \dfrac{28}{32}$

30. $\dfrac{4}{y} = \dfrac{6}{27}$

31. $\dfrac{r-2}{3} = \dfrac{r}{5}$

32. $\dfrac{5}{n} = \dfrac{7}{n+2}$

33. $\dfrac{x+2}{x-2} = \dfrac{7}{3}$

34. $\dfrac{4}{x+3} = \dfrac{3}{5}$

35. $\dfrac{5}{5z+3} = \dfrac{2z}{2z^2+6}$

36. $\dfrac{z+2}{z+6} = \dfrac{z-4}{z-2}$

Solve each proportion. **SEE EXAMPLE 2. (OBJECTIVE 1)**

37. $\dfrac{1}{x+3} = \dfrac{-2x}{x+5}$

38. $\dfrac{x-1}{x+1} = \dfrac{2}{3x}$

39. $\dfrac{a-4}{a+2} = \dfrac{a-5}{a+1}$

40. $\dfrac{9t+6}{t(t+3)} = \dfrac{7}{t+3}$

Express each sentence as a formula. **SEE EXAMPLES 5–8.**
(OBJECTIVES 4–7)

41. *A* varies directly with the square of *p*.

42. *p* varies directly with *q*.

43. *a* varies inversely with the square of *b*.

44. *v* varies inversely with the cube of *r*.

45. *B* varies jointly with *m* and *n*.

46. *C* varies jointly with *x*, *y*, and *z*.

47. *X* varies directly with *w* and inversely with *q*.

48. *P* varies directly with *x* and inversely with the square of *q*.

ADDITIONAL PRACTICE *Solve each proportion.*

49. $\dfrac{2}{c} = \dfrac{c-3}{2}$

50. $\dfrac{x}{7} = \dfrac{7}{x}$

51. $\dfrac{9}{5x} = \dfrac{3x}{15}$

52. $\dfrac{2}{x+6} = \dfrac{-2x}{5}$

53. $\dfrac{a+6}{a+4} = \dfrac{a-3}{a-4}$

54. $\dfrac{x+4}{5} = \dfrac{3(x-2)}{3}$

Express each sentence as a formula.

55. *P* varies directly with the square of *a*, and inversely with the cube of *j*.

56. *M* varies inversely with the cube of *n*, and jointly with *x* and the square of *z*.

Express each formula in words. In each formula, k is the constant of variation.

57. $L = kmn$

58. $P = \dfrac{km}{n}$

59. $E = kab^2$

60. $U = krs^2t$

61. $X = \dfrac{kx^2}{y^2}$

62. $Z = \dfrac{kw}{xy}$

63. $R = \dfrac{kL}{d^2}$

64. $e = \dfrac{kPL}{A}$

APPLICATIONS *Use an equation to solve.* **SEE EXAMPLE 3.**
(OBJECTIVE 2)

65. Selling shirts Consider the following ad. How much will 5 shirts cost?

66. Cooking A recipe requires four 16-ounce bottles of catsup to make two gallons of spaghetti sauce. How many bottles are needed to make ten gallons of sauce?

67. Gas consumption A car gets 38 mpg. How much gas will be needed to go 323 miles?

68. Model railroading An HO-scale model railroad engine is 9 inches long. The HO scale is 87 feet to 1 foot. How long is a real engine?

69. Hobbies Standard dollhouse scale is 1 inch to 1 foot. Heidi's dollhouse is 32 inches wide. How wide would it be if it were a real house?

70. Staffing A school board has determined that there should be 3 teachers for every 50 students. How many teachers are needed for an enrollment of 2,700 students?

71. Drafting In a scale drawing, a 280-foot antenna tower is drawn 7 inches high. The building next to it is drawn 2 inches high. How tall is the actual building?

72. Mixing fuel The instructions on a can of oil intended to be added to lawnmower gasoline read:

Recommended	Gasoline	Oil
50 to 1	6 gal	16 oz

Are these instructions correct? (*Hint:* There are 128 ounces in 1 gallon.)

73. Recommended dosage The recommended child's dose of the sedative hydroxine is 0.006 gram per kilogram of body mass. Find the dosage for a 30-kg child.

74. Body mass The proper dose of the antibiotic cephalexin in children is 0.025 gram per kilogram of body mass. Find the mass of a child receiving a $1\frac{1}{8}$-gram dose.

Use similar triangles to solve. SEE EXAMPLE 4. (OBJECTIVE 3)

75. Height of a tree   A tree casts a shadow of 28 feet at the same time as a 6-foot man casts a shadow of 4 feet. Find the height of the tree.

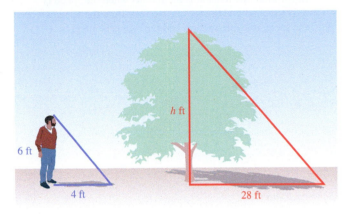

76. Height of a flagpole   A man places a mirror on the ground and sees the reflection of the top of a flagpole. The two triangles in the illustration are similar. Find the height, *h*, of the flagpole.

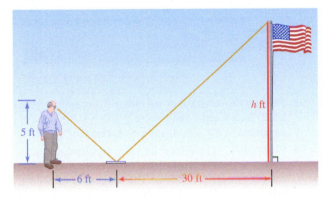

77. Width of a river   Use the dimensions in the illustration to find *w*, the width of the river. The two triangles in the illustration are similar.

78. Flight paths   An airplane ascends 150 feet at a constant rate as it flies a horizontal distance of 1,000 feet. How much altitude will it gain as it flies a horizontal distance of 1 mile? (*Hint*: 5,280 feet = 1 mile.)

79. Flight paths   An airplane descends 1,350 feet at a constant rate as it flies a horizontal distance of 1 mile. How much altitude is lost as it flies a horizontal distance of 5 miles?

80. Ski runs   A ski course with $\frac{1}{2}$ mile of horizontal run falls 100 feet in every 300 feet of run. Find the height of the hill.

Solve each variation. SEE EXAMPLE 5. (OBJECTIVE 4)

81. Area of a circle   The area of a circle varies directly with the square of its radius. The constant of variation is π. Find the area of a circle with a radius of 8 inches.

82. Falling objects   An object in free fall travels a distance *s* that is directly proportional to the square of the time *t*. If an object falls 1,024 feet in 8 seconds, how far will it fall in 10 seconds?

83. Finding distance   The distance that a car can travel is directly proportional to the number of gallons of gasoline it consumes. If a car can travel 288 miles on 12 gallons of gasoline, how far can it travel on a full tank of 18 gallons?

84. Farming   A farmer's harvest in bushels varies directly with the number of acres planted. If 8 acres can produce 144 bushels, how many acres are required to produce 1,152 bushes?

Solve each variation. SEE EXAMPLE 6. (OBJECTIVE 5)

85. Farming   The length of time that a given number of bushels of corn will last when feeding cattle varies inversely with the number of animals. If *x* bushels will feed 25 cows for 10 days, how long will the feed last for 10 cows?

86. Geometry   For a fixed area, the length of a rectangle is inversely proportional to its width. A rectangle has a width of 18 feet and a length of 12 feet. If the length is increased to 16 feet, find the width.

87. Gas pressure   Under constant temperature, the volume occupied by a gas is inversely proportional to the pressure applied. If the gas occupies a volume of 20 cubic inches under a pressure of 6 pounds per square inch, find the volume when the gas is subjected to a pressure of 10 pounds per square inch.

88. Value of a boat   The value of a boat usually varies inversely with its age. If a boat is worth $7,000 when it is 3 years old, how much will it be worth when it is 7 years old?

Solve each variation. SEE EXAMPLE 7. (OBJECTIVE 6)

89. Geometry   The area of a rectangle varies jointly with its length and width. If both the length and the width are tripled, by what factor is the area multiplied?

90. Geometry The volume of a rectangular solid varies jointly with its length, width, and height. If the length is doubled, the width is tripled, and the height is doubled, by what factor is the volume multiplied?

91. Costs of a trucking company The costs incurred by a trucking company vary jointly with the number of trucks in service and the number of hours each is used. When 4 trucks are used for 6 hours each, the costs are $1,800. Find the costs of using 10 trucks, each for 12 hours.

92. Storing oil The number of gallons of oil that can be stored in a cylindrical tank varies jointly with the height of the tank and the square of the radius of its base. The constant of proportionality is 23.5. Find the number of gallons that can be stored in the cylindrical tank shown in the illustration.

Solve each variation. SEE EXAMPLE 8. (OBJECTIVE 7)

93. Building construction The deflection of a beam is inversely proportional to its width and the cube of its depth. If the deflection of a 4-inch-by-4-inch beam is 1.1 inches, find the deflection of a 2-inch-by-8-inch beam positioned as in the illustration.

94. Building construction Find the deflection of the beam in Exercise 93 when the beam is positioned as in the illustration.

95. Gas pressure The pressure of a certain amount of gas is directly proportional to the temperature (measured on the Kelvin scale) and inversely proportional to the volume. A sample of gas at a pressure of 1 atmosphere occupies a volume of 1 cubic meter at a temperature of 273 Kelvin. When heated, the gas expands to twice its volume, but the pressure remains constant. To what temperature is it heated?

96. Tension A yo-yo, twirled at the end of a string, is kept in its circular path by the tension of the string. The tension T is directly proportional to the square of the speed s and inversely proportional to the radius r of the circle. In the illustration, the tension is 32 pounds when the speed is 8 feet/second and the radius is 6 feet. Find the tension when the speed is 4 feet/second and the radius is 3 feet.

Solve each variation.

97. Organ pipes The frequency of vibration of air in an organ pipe is inversely proportional to the length l of the pipe. If a pipe 2 feet long vibrates 256 times per second, how many times per second will a 6-foot pipe vibrate?

98. Finding the constant of variation A quantity l varies jointly with x and y and inversely with z. If the value of l is 24 when $x = 12$, $y = 4$, and $z = 8$, find k.

99. Electronics The voltage (in volts) measured across a resistor is directly proportional to the current (in amperes) flowing through the resistor. The constant of variation is the **resistance** (in ohms). If 6 volts is measured across a resistor carrying a current of 2 amperes, find the resistance.

100. Electronics The power (in watts) lost in a resistor (in the form of heat) is directly proportional to the square of the current (in amperes) passing through it. The constant of proportionality is the resistance (in ohms). What power is lost in a 5-ohm resistor carrying a 3-ampere current?

WRITING ABOUT MATH

101. Explain the terms *means* and *extremes*.

102. Distinguish between a *ratio* and a *proportion*.

103. Explain the term *joint variation*.

104. Explain why the equation $\frac{y}{x} = k$ indicates that y varies directly with x.

SOMETHING TO THINK ABOUT

105. As temperature increases on the Fahrenheit scale, it also increases on the Celsius scale. Is this direct variation? Explain.

106. As the cost of a purchase (less than $5) increases, the amount of change received from a five-dollar bill decreases. Is this inverse variation? Explain.

Projects

PROJECT 1

Suppose that a motorist usually makes an 80-mile trip at an average speed of 50 mph.

a. How much time does the driver save by increasing her rate by 5 mph? By 10 mph? By 15 mph? Give all answers to the nearest minute.

b. Consider your work in part **a**, and find a formula that will tell how much time is saved if the motorist travels x mph faster than 50 mph. That is, find an expression involving x that will give the time saved by traveling $(50 + x)$ mph instead of 50 mph.

c. Find a formula that will give the time saved on a trip of d miles by traveling at an average rate that is x mph faster than a usual speed of y mph. When you have this formula, simplify it into a nice, compact form.

d. Test your formula by repeating part **a** using the formula. The answers you obtain should agree with those found earlier.

e. Use your formula to solve the following problem. Every holiday season, Kurt and Ellen travel to visit their relatives, a distance of 980 miles. Under normal circumstances, they can average 60 mph during the trip. However, improved roads will enable them to travel 4 mph faster this year. How much time (to the nearest minute) will they save on this year's trip? How much faster than normal (to the nearest tenth of a mph) would Kurt and Ellen have to travel to save two hours?

PROJECT 2

By cross-fertilization among a number of species of corn, Professor Greenthumb has succeeded in creating a miracle hybrid corn plant. Under ideal conditions, the plant will produce twice as much corn as an ordinary hybrid. However, the new hybrid is very sensitive to the amount of sun and water it receives. If conditions are not ideal, the corn yield diminishes.

To determine what yield the plants will have, daily measurements record the amount of sun the plants receive (the sun index, S) and the amount of water they receive (W, measured in millimeters). Then the daily yield diminution is calculated using the formula

$$\text{Daily yield diminution} = \frac{S^3 + W^3 - (S + W)^2}{(S + W)^3}$$

At the end of a 100-day growing season, the daily yield diminutions are added together to make the total diminution (D). The yield for the year is then determined using the formula

$$\text{Annual yield} = 1 - 0.01D$$

The resulting decimal is the percent of maximum yield that the corn plants will achieve. Remember that maximum yield for these plants is twice the yield of an ordinary corn plant.

As Greenthumb's research assistant, you have been asked to handle a few questions that have come up regarding her research.

a. First, show that if $S = W = 2$, the daily yield diminution is 0. These would be the optimal conditions for Greenthumb's plants.

b. Now suppose that for each day of the 100-day growing season, the sun index is 8, and 6 millimeters of rain fall on the plants. Find the daily yield diminution. To the nearest tenth of a percent, what percentage of the yield of normal plants will the special hybrid plants produce?

c. Show that when the daily yield diminution is 0.5 for each of the 100 days in the growing season, the annual yield will be 0.5 (exactly the yield of ordinary corn plants). Now suppose that through the use of an irrigation system, you arrange for the corn to receive 10 millimeters of rain each day. What would the sun index have to be each day to give an annual yield of 0.5? To answer this question, first simplify the daily yield diminution formula. Do this *before* you substitute any numbers into the formula.

d. Another assistant is also working with Greenthumb's formula, but he is getting different results. Something is wrong; for the situation given in part **b**, he finds that the daily yield diminution for one day is 34. He simplified Greenthumb's formula first, then substituted in the values for S and W. He shows you the following work. Find all of his mistakes and explain what he did wrong in making each of them.

$$\text{Daily yield diminution} = \frac{S^3 + W^3 - (S + W)^2}{(S + W)^3}$$
$$= \frac{S^3 + W^3 - S^2 + W^2}{(S + W)^3}$$
$$= \frac{(S^3 + W^3) - (S^2 - W^2)}{(S + W)^3}$$
$$= \frac{(S + W) \cdot (S^2 + SW + W^2) - (S + W) \cdot (S - W)}{(S + W)^3}$$

$$= \frac{(S^2 + SW + W^2) - S - W}{(S + W)^2}$$

$$= \frac{S^2 + SW + W^2 - S - W}{S^2 + W^2}$$

$$= SW - S - W$$

So for $S = 8$ and $W = 6$, he gets 34 as the daily yield diminution.

Reach for Success
EXTENSION OF PREPARING FOR A TEST

Now that you have a few strategies for preparing for a test, consider how to be emotionally and physically ready. *Emotional intelligence* is defined as the ability to identify one's emotions and then manage those emotions. A good start to physical fitness is a healthy lifestyle that includes proper nutrition, exercise, and sleep. If you feel good physically you will more likely feel good emotionally. Feeling emotionally and physically strong should contribute to improved test performance.

On a scale of 1 to 10, how would you rate your emotional readiness for the next test if 10 represents emotionally strong?

1 2 3 4 5 6 7 8 9 10

What strategies might you employ to improve your emotional readiness?

1. _____

2. _____

3. _____

On a scale of 1 to 10, how would you rate your physical health if 10 represents healthy and well?

1 2 3 4 5 6 7 8 9 10

What strategies might you employ to improve your physical health?

1. _____

2. _____

3. _____

Talk with your professor or talk with your advisor for additional strategies for successful test preparation.

6 Review

SECTION 6.1 Finding the Domain of Rational Functions and Simplifying Rational Expressions

DEFINITIONS AND CONCEPTS	EXAMPLES
Division by 0 is undefined. Restricted values are any values in a rational expression that will cause the denominator to be 0.	To find any restricted values for the variable in the expression $\dfrac{6x + 7}{x^2 - 5x - 6}$, factor the denominator to obtain $$\dfrac{6x + 7}{(x - 6)(x + 1)}$$ The restricted values are 6 and -1 because both of these will create a 0 in the denominator.
Finding the domain of a rational function: The **domain** of a rational function is all values of the variable for which the function is defined. On the graph of a simplified rational expression, a vertical asymptote occurs at any value for which the expression is undefined.	Write the domain of $f(x) = \dfrac{6x + 7}{x^2 - 5x - 6}$ in interval notation. From the work above, we see that 6 and -1 are not permissible replacements for x in the expression $\dfrac{6x + 7}{x^2 - 5x - 6}$. Therefore, the domain of $f(x)$ is the set of real numbers except 6 and -1. In interval notation, the domain is $(-\infty, -1) \cup (-1, 6) \cup (6, \infty)$. There will be vertical asymptotes at $x = -1$ and $x = 6$ on the graph of the function.
Simplifying a rational expression: To simplify a rational expression, factor the numerator and denominator and divide out all factors common to the numerator and denominator. $$\dfrac{ak}{bk} = \dfrac{a}{b}$$ Assume no division by zero.	To simplify $\dfrac{x^2 - 5x + 6}{x^2 - 9}$ (assume no division by zero), factor the numerator and the denominator and divide out the resulting common factor: $\dfrac{(x - 3)(x - 2)}{(x + 3)(x - 3)}$ Factor the numerator and the denominator. $\dfrac{x - 2}{x + 3}$ Divide out $(x - 3)$.

REVIEW EXERCISES

1. Evaluate $f(x) = \dfrac{4x + 3}{x}$ when $x = 3$.

2. Find all values of the variable for which $\dfrac{5x}{2x^2 - x - 21}$ is undefined.

3. Graph the rational function $f(x) = \dfrac{3x + 2}{x}$ $(x > 0)$. Find the equation of the vertical asymptote.

4. Write the domain of $f(x) = \dfrac{x + 2}{x - 1}$ in interval notation.

Simplify each rational expression. Assume no division by 0.

5. $\dfrac{212m^3n}{588m^2n^3}$

6. $\dfrac{x^2 - 81}{x^2 + 18x + 81}$

7. $\dfrac{x^2 + 6x + 36}{x^3 - 216}$

8. $\dfrac{x^2 - 2x + 4}{2x^3 + 16}$

9. $\dfrac{5x - 5y}{4y - 4x}$

10. $\dfrac{2m - 2n}{n - m}$

11. $\dfrac{m^3 + m^2n - 2mn^2}{2m^3 - mn^2 - m^2n}$

12. $\dfrac{ac - ad + bc - bd}{d^2 - c^2}$

SECTION 6.2   Multiplying and Dividing Rational Expressions

DEFINITIONS AND CONCEPTS	EXAMPLES

Multiplying rational expressions:

To multiply two rational expressions, follow the same procedure as multiplying two numerical fractions.

$$\frac{a}{b} \cdot \frac{c}{d} = \frac{ac}{bd} \quad \text{Assume no division by 0.}$$

$$\frac{a+6}{(ax+4a-4x-16)} \cdot \frac{3a-12}{3a+18}$$

$$= \frac{(a+6)(3a-12)}{(ax+4a-4x-16)(3a+18)} \quad \text{Multiply the numerators and multiply the denominators.}$$

$$= \frac{(a+6)3(a-4)}{(a-4)(x+4)3(a+6)} \quad \text{Factor.}$$

$$= \frac{1}{x+4} \quad \text{Divide out all common factors.}$$

Dividing rational expressions:

To divide two rational expressions, follow the same procedure as dividing two numerical fractions.

$$\frac{a}{b} \div \frac{c}{d} = \frac{a}{b} \cdot \frac{d}{c} \quad \text{Assume no division by 0.}$$

$$\frac{x^2-4}{2b-bx} \div \frac{x^2+4x+4}{2b+bx}$$

$$= \frac{x^2-4}{2b-bx} \cdot \frac{2b+bx}{x^2+4x+4} \quad \text{Invert the divisor and multiply.}$$

$$= \frac{(x^2-4)(2b+bx)}{(2b-bx)(x^2+4x+4)} \quad \text{Multiply the numerators and multiply the denominators.}$$

$$= \frac{(x-2)(x+2)b(2+x)}{b(2-x)(x+2)(x+2)} \quad \text{Factor.}$$

$$= \frac{x-2}{2-x} \quad \text{Divide out all common factors.}$$

$$= -1 \quad (x-2) \text{ and } (2-x) \text{ are opposites.}$$

REVIEW EXERCISES

Perform the operations and simplify. Assume no division by 0.

13. $\dfrac{x^2+4x+4}{x^2-x-6} \cdot \dfrac{x^2-9}{x^2+5x+6}$

14. $\dfrac{x^3-8}{x^2+2x+4} \div \dfrac{x^2-4}{x+2}$

15. $\dfrac{x^2+3x+2}{x^2-x-6} \cdot \dfrac{3x^2-3x}{x^2-3x-4} \div \dfrac{x^2+3x+2}{x^2-2x-8}$

16. $\dfrac{x^2-x-6}{x^2-3x-10} \div \dfrac{x^2-x}{x^2-5x} \cdot \dfrac{x^2-4x+3}{x^2-6x+9}$

SECTION 6.3   Adding and Subtracting Rational Expressions

DEFINITIONS AND CONCEPTS	EXAMPLES

Adding and subtracting rational expressions with the same denominator:

To add or subtract two rational expressions with like denominators, combine the numerators and keep the denominator. Simplify, if possible.

$$\frac{a}{b} + \frac{c}{b} = \frac{a+c}{b} \quad \text{Assume no division by 0.}$$

$$\frac{a}{b} - \frac{c}{b} = \frac{a-c}{b} \quad (b \neq 0)$$

$$\frac{2x+4}{x^2+13x+12} - \frac{x+3}{x^2+13x+12}$$

$$= \frac{2x+4-x-3}{x^2+13x+12} \quad \text{Subtract the numerators and keep the common denominator.}$$

$$= \frac{x+1}{x^2+13x+12} \quad \text{Combine like terms.}$$

$$= \frac{x+1}{(x+12)(x+1)} \quad \text{Factor the denominator.}$$

$$= \frac{1}{x+12} \quad \text{Divide out the common factor, } (x+1).$$

Finding the LCD of two rational expressions:

To find the LCD of two fractions, factor each denominator and use each factor the greatest number of times that it appears in any one denominator. The product of these factors is the LCD.

Adding and subtracting rational expressions with different denominators:

To add or subtract two rational expressions with different denominators, find the LCD of the two expressions and write each fraction as an equivalent fraction with the new denominator. Combine the numerators and keep the denominator. Simplify if possible.

To find the LCD of $\dfrac{2a}{a^2 - 2a - 8}$ and $\dfrac{3}{a^2 - 5a + 4}$, factor each denominator to obtain

$$\frac{2a}{(a - 4)(a + 2)} \qquad \frac{3}{(a - 4)(a - 1)}$$

The LCD is $(a - 4)(a + 2)(a - 1)$.

Add: $\dfrac{2a}{a^2 - 2a - 8} + \dfrac{3}{a^2 - 5a + 4}$

$$= \frac{2a}{(a - 4)(a + 2)} + \frac{3}{(a - 4)(a - 1)} \qquad \text{\textcolor{red}{Factor each denominator.}}$$

Write each fraction with the LCD of $(a - 4)(a + 2)(a - 1)$, found above.

$$= \frac{2a\textcolor{red}{(a - 1)}}{(a - 4)(a + 2)\textcolor{red}{(a - 1)}} + \frac{3\textcolor{red}{(a + 2)}}{(a - 4)(a - 1)\textcolor{red}{(a + 2)}}$$

$$= \frac{2a^2 - 2a}{(a - 4)(a + 2)(a - 1)} + \frac{3a + 6}{(a - 4)(a - 1)(a + 2)}$$

$$= \frac{2a^2 + a + 6}{(a - 4)(a + 2)(a - 1)}$$

REVIEW EXERCISES

Perform the operations and simplify.

17. $\dfrac{3x}{x - y} + \dfrac{7}{x - y}$

18. $\dfrac{5y - 3}{y^2 + 3} - \dfrac{4(y - 3)}{y^2 + 3}$

19. $\dfrac{3}{x + 2} + \dfrac{2}{x + 3}$

20. $\dfrac{4x}{x - 4} - \dfrac{3}{x + 3}$

21. $\dfrac{2x}{x + 1} + \dfrac{3x}{x + 2} + \dfrac{4x}{x^2 + 3x + 2}$

22. $\dfrac{5x}{x - 3} + \dfrac{5}{x^2 - 5x + 6} + \dfrac{x + 3}{x - 2}$

23. $\dfrac{3(x + 2)}{x^2 - 1} - \dfrac{2}{x + 1} + \dfrac{4(x + 3)}{x^2 - 2x + 1}$

24. $\dfrac{-2(3 + x)}{x^2 + 6x + 9} + \dfrac{3(x + 2)}{x^2 - 6x + 9} - \dfrac{1}{x^2 - 9}$

SECTION 6.4 Simplifying Complex Fractions

DEFINITIONS AND CONCEPTS

EXAMPLES

Simplifying complex fractions:

A fraction that has a fraction in its numerator and/or its denominator is called a **complex fraction**.

There are two ways to simplify a complex fraction.

Method 1: Eliminate the rational expressions in the numerator and denominator by writing the complex fraction as a division and using the division process for fractions.

Method 2: Find the LCD of all denominators in the complex fraction. Eliminate the rational expressions in the numerator and denominator by multiplying the expression by 1, written in the form $\dfrac{\text{LCD}}{\text{LCD}}$.

To use Method 2 to simplify $\dfrac{\dfrac{1}{a} + \dfrac{1}{b}}{\dfrac{a}{b} - \dfrac{b}{a}}$, proceed as follows:

$$\frac{\dfrac{1}{a} + \dfrac{1}{b}}{\dfrac{a}{b} - \dfrac{b}{a}} \cdot \frac{\textcolor{red}{ab}}{\textcolor{red}{ab}} = \frac{\dfrac{1}{a}(ab) + \dfrac{1}{b}(ab)}{\dfrac{a}{b}(ab) - \dfrac{b}{a}(ab)} \qquad \text{\textcolor{red}{Multiply by } } \frac{ab}{ab} = 1.$$

$$= \frac{b + a}{a^2 - b^2} \qquad \text{\textcolor{red}{Simplify.}}$$

$$= \frac{b + a}{(a + b)(a - b)} \qquad \text{\textcolor{red}{Factor.}}$$

$$= \frac{1}{a - b} \qquad \text{\textcolor{red}{Simplify.}}$$

REVIEW EXERCISES

Simplify each complex fraction. Assume no denominators are zero.

25. $\dfrac{\dfrac{5}{a} - \dfrac{4}{b}}{ab}$

26. $\dfrac{\dfrac{3}{p} + \dfrac{4}{q}}{\dfrac{4}{p} - \dfrac{3}{q}}$

29. $\dfrac{1 - \dfrac{1}{x} - \dfrac{2}{x^2}}{1 + \dfrac{4}{x} + \dfrac{3}{x^2}}$

30. $\dfrac{x^{-1} + 1}{x + 1}$

27. $\dfrac{2x + 3 + \dfrac{1}{x}}{x + 2 + \dfrac{1}{x}}$

28. $\dfrac{6x + 13 + \dfrac{6}{x}}{6x + 5 - \dfrac{6}{x}}$

31. $\dfrac{x^{-1} - y^{-1}}{x^{-1} + y^{-1}}$

32. $\dfrac{(x - y)^{-2}}{x^{-2} - y^{-2}}$

SECTION 6.5   Solving Equations Containing Rational Expressions

DEFINITIONS AND CONCEPTS	EXAMPLES
Solving rational equations: To solve a rational equation, we multiply both sides of the equation by the LCD of the rational expressions in the equation to clear it of fractions. Because this method can produce *extraneous solutions,* it is necessary to check all possible solutions.	To solve $1 = \dfrac{3}{x - 2} - \dfrac{12}{x^2 - 4}$, first determine that x cannot be 2 or -2. Then multiply both sides by the LCD of $(x + 2)(x - 2)$. $$(x + 2)(x - 2)(1) = \left(\dfrac{3}{x - 2} - \dfrac{12}{(x + 2)(x - 2)} \right)(x + 2)(x - 2)$$ $(x + 2)(x - 2) = 3(x + 2) - 12$ $x^2 - 4 = 3x + 6 - 12$ Use the distributive property to remove parentheses. $x^2 - 4 = 3x - 6$ Combine like terms. $x^2 - 3x + 2 = 0$ Add $-3x$ and 6 to both sides to write in quadratic form. $(x - 2)(x - 1) = 0$ Factor. $x - 2 = 0$ or $x - 1 = 0$ Set each factor equal to 0. $x = 2$ $\quad$ $x = 1$ Solve each linear equation. Since 2 is a restricted value, it is extraneous. The only solution is 1.
Solving a formula for a specified variable: Clear the formula of any fractions and isolate the identified variable.	To solve $S = \dfrac{a}{1 - r}$ for r, proceed as follows: $(1 - r)\, S = \dfrac{a}{1 - r}\, (1 - r)$ Multiply both sides by $(1 - r)$. $S - rS = a$ Use the distributive property to remove parentheses. $S = a + rS$ Add rS to both sides. $S - a = rS$ Subtract a from both sides. $\dfrac{S - a}{S} = r$ Divide both sides by S. The result can also be written as $1 - \dfrac{a}{S} = r$.

REVIEW EXERCISES

Solve each equation, if possible.

33. $\dfrac{3}{y} - \dfrac{1}{12} = \dfrac{7}{4y}$

34. $\dfrac{2}{x + 5} - \dfrac{1}{6} = \dfrac{1}{x + 4}$

35. $\dfrac{2(x - 5)}{x - 2} = \dfrac{6x + 12}{4 - x^2}$

36. $\dfrac{7}{x + 9} - \dfrac{x + 2}{2} = \dfrac{x + 4}{x + 9}$

Solve each formula for the indicated variable.

37. $\dfrac{x^2}{a^2} - \dfrac{y^2}{b^2} = 1$ for a^2

38. $H = \dfrac{2ab}{a + b}$ for b

39. Trip length Traffic reduced Jim's usual speed by 10 mph, which lengthened his 200-mile trip by 1 hour. Find his usual average speed.

40. Flying speed On a 600-mile trip, a pilot can save 30 minutes by increasing her usual speed by 40 mph. Find her usual speed.

41. Draining a tank If one outlet pipe can drain a tank in 24 hours and another pipe can drain the tank in 36 hours, how long will it take for both pipes to drain the tank?

42. Siding a house Two men have estimated that they can side a house in 8 days. If one of them, who could have sided the house alone in 14 days, gets sick, how long will it take the other man to side the house alone?

SECTION 6.6   Dividing Polynomials

DEFINITIONS AND CONCEPTS	EXAMPLES
Dividing a monomial by a monomial: To find the quotient of two monomials, express the quotient as a rational expression and apply the rules of exponents to simplify.	To simplify $\dfrac{25x^4y^7}{5xy^9}$ by using the rule of exponents, proceed as follows: $\dfrac{25x^4y^7}{5xy^9} = 5 \cdot x^{4-1} \cdot y^{7-9}$ $\quad \dfrac{x^4}{x} = x^{4-1}, \dfrac{y^7}{y^9} = y^{7-9}$ $\quad\quad\quad = 5 \cdot x^3 \cdot y^{-2}$ Simplify. $\quad\quad\quad = 5x^3 \cdot \dfrac{1}{y^2}$ Write with positive exponents. $\quad\quad\quad = \dfrac{5x^3}{y^2}$ Write as a single fraction.
Dividing a polynomial by a monomial: To find the quotient of a polynomial and a monomial, divide each term of the polynomial by the monomial and simplify each resulting rational expression.	To divide $\dfrac{45a^5b^3 - 63a^4b^4 - 9a^2b^3}{-9a^2b^3}$, write the fraction as three separate fractions and simplify each one. $\dfrac{45a^5b^3}{-9a^2b^3} - \dfrac{63a^4b^4}{-9a^2b^3} - \dfrac{9a^2b^3}{-9a^2b^3} = -5a^3 + 7a^2b + 1$
Dividing a polynomial by a polynomial: To find the quotient of two polynomials, write the polynomials in descending order and then use long division.	To divide $\dfrac{4a^3 + a^2 - 3a + 7}{a + 1}$, use long division. $\begin{array}{r} 4a^2 - 3a \\ a + 1 \overline{\smash{)}4a^3 + a^2 - 3a + 7} \\ \underline{-\ 4a^3 + 4a^2\phantom{{}-3a+7}} \\ -3a^2 - 3a \\ \underline{-\ -3a^2 - 3a} \\ +7 \end{array}$ Thus, $\dfrac{4a^3 + a^2 - 3a + 7}{a + 1} = 4a^2 - 3a + \dfrac{7}{a + 1}$.

REVIEW EXERCISES

Perform each division. Assume no division by zero.

43. $\dfrac{-8x^7y^5}{24x^5y^8}$

44. $\dfrac{30x^3y^2 - 15x^2y - 10xy^2}{-10xy}$

45. $(3x^2 + 13xy - 10y^2) \div (3x - 2y)$

46. $(2x^3 + 7x^2 + 3 + 4x) \div (2x + 3)$

SECTION 6.7   Synthetic Division

DEFINITIONS AND CONCEPTS	EXAMPLES
Dividing a polynomial by a binomial of the form $(x - r)$ using synthetic division: Write the coefficients of the polynomial in the dividend and r in the divisor. Use multiplication and addition to complete the division.	Use synthetic division to perform the division $$(5x^2 - 8x - 2) \div (x - 2)$$ $$\begin{array}{r|rrr} 2 & 5 & -8 & -2 \\ & & 10 & 4 \\ \hline & \mathbf{5} & \mathbf{2} & \mathbf{2} \end{array}$$ Thus, $(5x^2 - 8x - 2) \div (x - 2) = \mathbf{5}x + \mathbf{2} + \dfrac{\mathbf{2}}{x - 2}$.
Remainder theorem: If a polynomial $P(x)$ is divided by $(x - r)$, then the remainder is $P(r)$.	Given $P(x) = 8x^3 - 2x^2 - 9$, find the remainder when $P(x)$ is divided by $x + 2$. $$\begin{array}{r|rrrr} -2 & 8 & -2 & 0 & -9 \\ & & -16 & 36 & -72 \\ \hline & 8 & -18 & 36 & \mathbf{-81} \end{array}$$ Insert a 0 for the missing term. The remainder is -81. Thus, $P(-2) = -81$.
Factor theorem: If $P(x)$ is divided by $(x - r)$, then $P(r) = 0$, if and only if $(x - r)$ is a factor of $P(x)$.	Determine whether $(x + 4)$ is a factor of $(x^4 + 4x^3 + 9x^2 + 37x + 4)$. $$\begin{array}{r|rrrrr} -4 & 1 & 4 & 9 & 37 & 4 \\ & & -4 & 0 & -36 & -4 \\ \hline & 1 & 0 & 9 & 1 & 0 \end{array}$$ Since the remainder is 0, $(x + 4)$ is a factor of $(x^4 + 4x^3 + 9x^2 + 37x + 4)$.

REVIEW EXERCISES

Use the factor theorem and synthetic division to determine whether the first expression is a factor of P(x). Factor, if possible.

47. $x - 3$; $P(x) = x^3 - 7x^2 + 9x + 9$

48. $x + 5$; $P(x) = x^3 + 4x^2 - 5x + 5$
(*Hint:* Write $(x + 5)$ as $x - (-5)$.)

SECTION 6.8   Proportion and Variation

DEFINITIONS AND CONCEPTS	EXAMPLES
Solving a proportion: In a proportion, the product of the extremes is equal to the product of the means.	To solve $\dfrac{x - 1}{x + 1} = \dfrac{2}{3x}$, proceed as follows: $3x(x - 1) = 2(x + 1)$ The product of the extremes is equal to the product of the means. $3x^2 - 3x = 2x + 2$ Use the distributive property to remove parentheses. $3x^2 - 5x - 2 = 0$ Subtract $2x$ and 2 from both sides. $(3x + 1)(x - 2) = 0$ Factor. $3x + 1 = 0$ or $x - 2 = 0$ Set each factor equal to 0. $x = -\dfrac{1}{3}$ $\Big|$ $x = 2$ Solve each linear equation. The solutions are $-\dfrac{1}{3}, 2$.

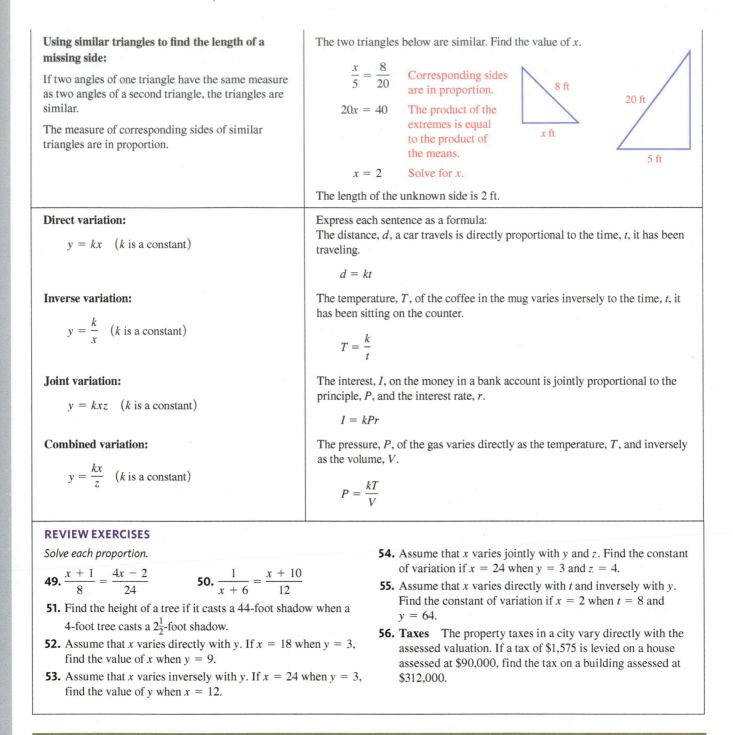

Using similar triangles to find the length of a missing side:

If two angles of one triangle have the same measure as two angles of a second triangle, the triangles are similar.

The measure of corresponding sides of similar triangles are in proportion.

The two triangles below are similar. Find the value of x.

$\dfrac{x}{5} = \dfrac{8}{20}$ Corresponding sides are in proportion.

$20x = 40$ The product of the extremes is equal to the product of the means.

$x = 2$ Solve for x.

The length of the unknown side is 2 ft.

8 ft

20 ft

x ft

5 ft

Direct variation:

$y = kx$ (k is a constant)

Inverse variation:

$y = \dfrac{k}{x}$ (k is a constant)

Joint variation:

$y = kxz$ (k is a constant)

Combined variation:

$y = \dfrac{kx}{z}$ (k is a constant)

Express each sentence as a formula:
The distance, d, a car travels is directly proportional to the time, t, it has been traveling.

$$d = kt$$

The temperature, T, of the coffee in the mug varies inversely to the time, t, it has been sitting on the counter.

$$T = \dfrac{k}{t}$$

The interest, I, on the money in a bank account is jointly proportional to the principle, P, and the interest rate, r.

$$I = kPr$$

The pressure, P, of the gas varies directly as the temperature, T, and inversely as the volume, V.

$$P = \dfrac{kT}{V}$$

REVIEW EXERCISES

Solve each proportion.

49. $\dfrac{x+1}{8} = \dfrac{4x-2}{24}$

50. $\dfrac{1}{x+6} = \dfrac{x+10}{12}$

51. Find the height of a tree if it casts a 44-foot shadow when a 4-foot tree casts a $2\frac{1}{2}$-foot shadow.

52. Assume that x varies directly with y. If $x = 18$ when $y = 3$, find the value of x when $y = 9$.

53. Assume that x varies inversely with y. If $x = 24$ when $y = 3$, find the value of y when $x = 12$.

54. Assume that x varies jointly with y and z. Find the constant of variation if $x = 24$ when $y = 3$ and $z = 4$.

55. Assume that x varies directly with t and inversely with y. Find the constant of variation if $x = 2$ when $t = 8$ and $y = 64$.

56. Taxes   The property taxes in a city vary directly with the assessed valuation. If a tax of $1,575 is levied on a house assessed at $90,000, find the tax on a building assessed at $312,000.

---

6   Test ✿

Simplify each rational expression. Assume no division by 0.

1. $\dfrac{-12x^2y^3z^2}{18x^3y^4z^2}$

2. $\dfrac{5x+15}{x^2-9}$

3. $\dfrac{7a-28b}{4b-a}$

4. Given $f(x) = \dfrac{x+6}{x-3}$

a. State all values of the variable for which $f(x)$ is undefined.

b. Write the domain of $f(x)$ in interval notation.

5. Evaluate $f(x) = \dfrac{7x - 4}{x}$ when $x = 6$.

Perform the operations and simplify, if possible. Write all answers using positive exponents. Assume no division by 0.

6. $\dfrac{x^2 y^{-2}}{x^3 z^2} \cdot \dfrac{x^2 z^4}{y^2 z}$

7. $\dfrac{x^2 - 3x - 4}{12x - 3} \cdot \dfrac{3}{(x - 4)}$

8. $\dfrac{u^2 + 5u + 6}{u^2 - 4} \cdot \dfrac{u^2 - 5u + 6}{u^2 - 9}$

9. $\dfrac{x^3 + y^3}{4} \div \dfrac{x^2 - xy + y^2}{2x + 2y}$

10. $\dfrac{x^2}{x + 1} - \dfrac{x^2 - 2}{x + 1}$

11. $\dfrac{a^2 + 7a + 12}{a + 3} \div \dfrac{16 - a^2}{a - 4}$

12. $\dfrac{6y}{y + 6} + \dfrac{36}{y + 6}$

13. $\dfrac{3w}{w - 5} + \dfrac{w + 10}{5 - w}$

14. $\dfrac{2}{r} + \dfrac{r}{s}$

15. $\dfrac{x + 2}{x + 1} - \dfrac{x + 1}{x + 2}$

Simplify each complex fraction. Assume no division by 0.

16. $\dfrac{\dfrac{2u^2 w^3}{v^2}}{\dfrac{4uw^4}{uv}}$

17. $\dfrac{\dfrac{x}{y} + \dfrac{1}{2}}{\dfrac{x}{2} - \dfrac{1}{y}}$

Solve each equation.

18. $\dfrac{2}{x - 1} + \dfrac{5}{x + 2} = \dfrac{11}{x + 2}$

19. $\dfrac{u - 2}{u - 3} + 3 = u + \dfrac{u - 4}{3 - u}$

Solve each formula for the indicated variable.

20. $\dfrac{x^2}{a^2} + \dfrac{y^2}{b^2} = 1$ for x^2

21. $\dfrac{1}{r} = \dfrac{1}{r_1} + \dfrac{1}{r_2}$ for r_2

22. Sailing time A boat sails a distance of 440 nautical miles. If the boat had averaged 11 nautical miles more each day, the trip would have required 2 fewer days. How long did the trip take?

23. Investing A student can earn $300 interest annually by investing in a certificate of deposit at a certain interest rate. If she were to receive an annual interest rate that is 4% higher, she could receive the same annual interest by investing $2,000 less. How much would she invest at each rate?

24. Divide: $\dfrac{18x^2 y^3 - 12x^3 y^2 + 9xy}{-3xy^4}$

25. Divide: $(6x^3 + 5x^2 - 2) \div (2x - 1)$

26. Find the remainder: $\dfrac{x^3 - 4x^2 + 5x + 3}{x + 1}$

27. Use synthetic division to find the remainder when $(4x^3 + 3x^2 + 2x - 1)$ is divided by $(x - 2)$.

28. Find the height of a tree that casts a shadow of 12 feet when a vertical yardstick casts a shadow of 2 feet.

29. Solve the proportion: $\dfrac{3}{x - 2} = \dfrac{x + 3}{2x}$

30. V varies inversely with t. If $V = 55$ when $t = 20$, find the value of t when $V = 75$.

❧ Cumulative Review ❧

Simplify each expression. Assume no division by 0.

1. $x^4 y^3 x^7 y^{-5}$

2. $\dfrac{a^3 b^6}{a^7 b^2}$

3. $\left(\dfrac{2a^2}{3b^4}\right)^{-4}$

4. $\left(\dfrac{x^{-2} y^3}{x^2 x^3 y^4}\right)^{-3}$

Write each number in standard notation.

5. 7.23×10^6

6. 7.12×10^{-4}

Solve each equation.

7. $\dfrac{a + 2}{5} - \dfrac{8}{5} = 4a - \dfrac{a + 9}{2}$

8. $\dfrac{3x - 4}{6} - \dfrac{x - 2}{2} = \dfrac{-2x - 3}{3}$

Find the slope of the line with the given properties.

9. passing through $(-2, 5)$ and $(4, 10)$

10. has an equation of $3x + 4y = 13$

11. parallel to a line with equation of $y = -5x + 2$

12. perpendicular to a line with equation of $y = 3x + 2$

Let $f(x) = x^2 - 2x$ and find each value.

13. $f(3)$

14. $f(-2)$

15. $f\left(\dfrac{2}{5}\right)$

16. $f(t - 1)$

17. Express as a formula: *y varies directly with the product of x and z, and inversely with r.*

18. Does the graph below represent a function?

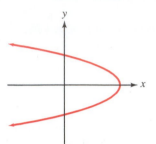

Solve each inequality, graph the solution set, and write the solution in interval notation.

19. $x - 2 \leq 3x + 1 \leq 5x - 4$

20. $\left| \dfrac{3a}{5} - 2 \right| + 1 \geq \dfrac{6}{5}$

21. Is $4x - 7x^2$ a monomial, a binomial, or a trinomial?

22. Find the degree of $3 + x^2y + 17x^3y^4$.

23. If $f(x) = -x^3 + 7x - 2$, find $f(-2)$.

24. Graph: $y = 2x^2 - 3$

Perform the operations and simplify.

25. $(3x^2 - 2x + 7) + (-2x^2 + 2x + 5) + (3x^2 - 4x + 2)$

26. $(-5x^2 + 3x + 4) - (-2x^2 + 3x + 7)$

27. $(4x - 3)(2x + 7)$

28. $(2x^n - 1)(x^n + 2)$

Factor each expression.

29. $3r^2s^3 - 6rs^4$

30. $x(a - b) - 4(a - b)$

31. $xu + yv + xv + yu$

32. $81x^4 - 16y^4$

33. $27y^3 - 64x^3$

34. $6x^2 + 5x - 6$

35. $9x^2 - 30x + 25$

36. $15x^2 - x - 6$

37. $27a^3 + 8b^3$

38. $6x^2 + x - 35$

39. $x^2 + 10x + 25 - y^4$

40. $y^2 - x^2 + 4x - 4$

Solve each equation.

41. $x^3 - 9x = 0$ **42.** $6x^2 + 7 = -23x$

Simplify each expression. Assume no divisions by zero.

43. $\dfrac{2x^2y + xy - 6y}{3x^2y + 5xy - 2y}$

44. $\dfrac{x^2 - 4}{x^2 + 9x + 20} \div \dfrac{x^2 + 5x + 6}{x^2 + 4x - 5} \cdot \dfrac{x^2 + 3x - 4}{(x - 1)^2}$

45. $\dfrac{2}{x + y} + \dfrac{3}{x - y} - \dfrac{x - 3y}{x^2 - y^2}$

46. $\dfrac{\dfrac{a}{b} + b}{a - \dfrac{b}{a}}$

Solve each equation.

47. $\dfrac{5x - 3}{x + 2} = \dfrac{5x + 3}{x - 2}$

48. $\dfrac{3}{x - 2} + \dfrac{x^2}{(x + 3)(x - 2)} = \dfrac{x + 4}{x + 3}$

Perform the operation.

49. Divide using any method: $(x^2 + 9x + 20) \div (x + 5)$

50. Divide using any method: $(2x^2 + 4x - x^3 + 3) \div (x - 1)$

Radical Expressions and Equations

© Mircea BEZERGHEANU/Shutterstock.com

Careers and Mathematics

PHOTOGRAPHERS

Photographers produce and preserve images that paint a picture, tell a story, or record an event. They use either a traditional camera that records images on silver halide film that is developed into prints or a digital camera that electronically records images.

More than half of all photographers are self-employed. Employers usually seek applicants with a "good eye," imagination, and creativity. Entry-level positions in photojournalism generally require a college degree in journalism or photography.

Job Outlook:
Employment of photographers is expected to increase about as fast as the average for all occupations through 2020. However, photographers can expect intense competition for job openings because the work is attractive to many people.

Annual Earnings:
$18,380–$37,370

For More Information:
http://www.bls.gov/oco/ocos264.htm

For a Sample Application:
See Problem 109 in Section 7.5.

REACH FOR SUCCESS

7.1 Radical Expressions

7.2 Applications of the Pythagorean Theorem and the Distance Formula

7.3 Rational Exponents

7.4 Simplifying and Combining Radical Expressions

7.5 Multiplying Radical Expressions and Rationalizing

7.6 Radical Equations

7.7 Complex Numbers

■ *Projects*

REACH FOR SUCCESS EXTENSION

CHAPTER REVIEW

CHAPTER TEST

In this chapter

In this chapter, we will reverse the squaring process and learn how to find square roots of numbers. We also will learn how to find other roots of numbers, simplify radical expressions, solve radical equations, and simplify and perform operations on complex numbers.

Reach for Success Examining Test Results

Now that you have your exam score, what are you going to do with it?

Regardless of whether your score is high or low, it is important that you take the time to find out *why* you missed any of the questions so that you might do better the next time.

Look at different types of errors that can be made on a test. First, consider the *careless errors*. These are generally sign errors, arithmetic errors, miscopying from one line to another, and misreading the directions.

Go over your test and identify any problems for which you lost points for careless errors.

How many points did you lose? _____

By reviewing your work prior to submitting your test, you may be able to avoid losing points to careless errors.

Now look at *conceptual errors*. These errors reflect a lack of understanding the material. (These will require you to answer some tough questions and to be totally truthful with yourself!)

Review your test and identify any problems for which you lost points because you did not understand the material.

How many points did you lose?_____

Now, think back to the days prior to the test.

Were you in class the day these topics were covered?_____

Did you do *all* the homework assigned for the topics?_____

Did you ask questions in class?_____

Did you go to your professor or the mathematics/tutor center for additional help?_____

Next, look at *test-taking strategies*. These include writing down formulas at the start of the test, using all the allotted time, and previewing questions to work from easier to more difficult.

Did you write down formulas at the start of the test? _____

How much of the allotted time did you use? _____

Did you preview to work questions you understood first? _____

Set up an appointment with your instructor to discuss your exam errors. This is especially critical if your instructor does not allow you to keep the exam.

Describe the result of the consultation with your instructor. _____

_____

To separate anxiety interference from conceptual errors, rework the problems by covering up your test and sliding a blank sheet of paper down as you progress. What was the result of covering the problems and reworking? _____

_____

If you now get them correct, math anxiety might be interfering and you should consult with your instructor.

If you still miss them, it might be a lack of understanding and you should seek tutoring.

A Successful Study Strategy . . .

 Review each test in detail. Consider test-taking as an opportunity to show what you know. This positive attitude can improve performance.

At the end of the chapter you will find an additional exercise to guide you in planning for a successful semester.

Section 7.1

Radical Expressions

Objectives

1. Simplify a perfect-square root.
2. Simplify a perfect-square root expression.
3. Simplify a perfect-cube root.
4. Simplify a perfect nth root.
5. Find the domain of a square-root function and a cube-root function.
6. Use a square root to solve an application.

Vocabulary

square root	integer squares	index
radical sign	cube root	square-root function
radicand	odd root	cube-root function
principal square root	even root	standard deviation

Getting Ready

Find each power.

1. 0^2 **2.** 4^2 **3.** $(-4)^2$ **4.** -4^2

5. $\left(\dfrac{2}{5}\right)^3$ **6.** $\left(-\dfrac{3}{4}\right)^4$ **7.** $(7xy)^2$ **8.** $(7xy)^3$

In this section, we will discuss perfect square roots and other perfect roots of algebraic expressions. We also will consider their related functions.

1   Simplify a perfect-square root.

When solving an equation, we often must find what number must be squared to obtain a second number a. If such a number can be found, it is called a **square root** of a. For example,

- 0 is a square root of 0, because $0^2 = 0$.
- 4 is a square root of 16, because $4^2 = 16$.
- -4 is a square root of 16, because $(-4)^2 = 16$.
- $7xy$ is a square root of $49x^2y^2$, because $(7xy)^2 = 49x^2y^2$.
- $-7xy$ is a square root of $49x^2y^2$, because $(-7xy)^2 = 49x^2y^2$.

All positive numbers have two real-number square roots: one that is positive and one that is negative.

EXAMPLE 1 Find the two square roots of 121.

Solution The two square roots of 121 are 11 and -11, because

$$11^2 = 121 \qquad \text{and} \qquad (-11)^2 = 121$$

 SELF CHECK 1 Find the two square roots of 144.

To express square roots, we use the symbol $\sqrt{}$, called a **radical sign**. For example,

$$\sqrt{121} = 11 \qquad \text{Read as "The positive square root of 121 is 11."}$$
$$-\sqrt{121} = -11 \qquad \text{Read as "The negative square root of 121 is } -11\text{."}$$

The number under the radical sign is called a **radicand**.

COMMENT   The **principal square root** of a positive number is always positive. Although 5 and -5 are both square roots of 25, only 5 is the principal square root. The radical expression $\sqrt{25}$ represents 5. The radical expression $-\sqrt{25}$ represents -5.

SQUARE ROOT OF a	If $a > 0$, $\sqrt{a}$ is the positive number, the principal square root of a, whose square is a.

In symbols,

$$\left(\sqrt{a}\right)^2 = a$$

If $a = 0$, $\sqrt{a} = \sqrt{0} = 0$. The principal square root of 0 is 0.

If $a < 0$, $\sqrt{a}$ is not a real number.

Because of the previous definition, the square root of any number squared is that number. For example,

$$\left(\sqrt{10}\right)^2 = \sqrt{10} \cdot \sqrt{10} = 10 \qquad \left(\sqrt{a}\right)^2 = \sqrt{a} \cdot \sqrt{a} = a$$

EXAMPLE 2   Simplify each radical.

a. $\sqrt{1} = 1$ **b.** $\sqrt{81} = 9$

c. $-\sqrt{81} = -9$ **d.** $-\sqrt{225} = -15$

e. $\sqrt{\dfrac{1}{4}} = \dfrac{1}{2}$ **f.** $-\sqrt{\dfrac{16}{121}} = -\dfrac{4}{11}$

g. $\sqrt{0.04} = 0.2$ **h.** $-\sqrt{0.0009} = -0.03$

SELF CHECK 2   Simplify: **a.** $-\sqrt{49}$ **b.** $\sqrt{\dfrac{25}{49}}$ **c.** $\sqrt{0.0036}$

Numbers such as $1, 4, 9, 16, 49,$ and $1,600$ are called **integer squares**, because each one is the square of an integer. The square root of every integer square is an integer.

$$\sqrt{1} = 1 \qquad \sqrt{4} = 2 \qquad \sqrt{9} = 3 \qquad \sqrt{16} = 4 \qquad \sqrt{49} = 7 \qquad \sqrt{1,600} = 40$$

Perspective

CALCULATING SQUARE ROOTS

The Bakhshali manuscript is an early mathematical manuscript that was discovered in India in the late 19th century. Mathematical historians estimate that the manuscript was written sometime around A.D. 400. One section of the manuscript presents a procedure for calculating square roots using arithmetic. Specifically, we can use the formula

$$\sqrt{Q} = A + \frac{b}{2A} - \left(\frac{b^2}{4A(2A^2 + b)}\right)$$

where $A^2 = $ a perfect square close to the number Q, and $b = Q - A^2$.

For example, if we want to compute an approximation of $\sqrt{21}$, we can choose $A^2 = 16$. Thus, $A = 4$ and $b = 21 - 16 = 5$. So we obtain

$$\sqrt{21} = 4 + \frac{5}{(2)(4)} - \left(\frac{5^2}{(4)(4)((2)(4)^2 + 5)}\right)$$
$$= 4 + \frac{5}{8} - \left(\frac{25}{(16)(37)}\right)$$

$$= 4 + \frac{5}{8} - \frac{25}{592} \approx 4.58277027$$

Using the square root key on a calculator, we see that, to nine decimal places, $\sqrt{21} = 4.582575695$. Therefore, the formula gives an answer that is correct to three decimal places.

1. Use the formula to approximate $\sqrt{105}$. How accurate is your answer?

2. Use the formula to approximate $\sqrt{627}$. How accurate is your answer?

Source: http://www.gap-system.org/~history/ HistTopics/Bakhshali_manuscript.html

---

Accent on technology

▸ Approximating Square Roots

The square roots of many positive integers are not rational numbers. For example, $\sqrt{11}$ is an *irrational number*. To find an approximate value of $\sqrt{11}$ with a calculator, we enter these numbers and press these keys.

11 **2ND** $\sqrt{}$ Using a scientific calculator

2ND X² ($\sqrt{}$) 11 **ENTER**   Using a graphing calculator

Either way, we will see that

$$\sqrt{11} \approx 3.31662479$$

For instructions regarding the use of a Casio graphing calculator, please refer to the Casio Keystroke Guide in the back of the book.

Square roots of negative numbers are not real numbers. For example, $\sqrt{-9}$ is not a real number, because no real number squared equals -9. Square roots of negative numbers come from a set called *imaginary numbers*, which we will discuss later in this chapter.

2 Simplify a perfect-square root expression.

If $x \neq 0$, the positive number x^2 has x and $-x$ for its two square roots. To denote the principal square root of x^2, we must know whether x is positive or negative.

If $x > 0$, we can write

$$\sqrt{x^2} = x \quad \sqrt{x^2} \text{ represents the positive square root of } x^2, \text{ which is } x.$$

If x is negative, then $-x > 0$, and we can write

$$\sqrt{x^2} = -x \quad \sqrt{x^2} \text{ represents the positive square root of } x^2, \text{ which is } -x.$$

If we do not know whether x is positive or negative, we must use absolute value symbols to guarantee that $\sqrt{x^2}$ is positive.

DEFINITION OF $\sqrt{x^2}$

If x can be any real number, then

$$\sqrt{x^2} = |x|$$

EXAMPLE 3 Simplify each expression. Assume that x can be any real number.

a. $\sqrt{16x^2} = \sqrt{(4x)^2}$ Write $16x^2$ as $(4x)^2$.

$\quad = |4x|$ Because $16x^2 = (|4x|)^2$. Since x could be negative, absolute value symbols are needed.

$\quad = 4|x|$ Since 4 is a positive constant in the product $4x$, we write it outside the absolute value symbols.

b. $\sqrt{x^2 + 2x + 1}$

$= \sqrt{(x + 1)^2}$ Factor $x^2 + 2x + 1$.

$= |x + 1|$ Because $(x + 1)$ can be negative, absolute value symbols are needed.

c. $\sqrt{x^4} = x^2$ Because $x^4 = (x^2)^2$, and $x^2 \geq 0$, no absolute value symbols are needed.

SELF CHECK 3   Simplify: **a.** $\sqrt{25a^2}$ **b.** $\sqrt{x^2 + 4x + 4}$ **c.** $\sqrt{16a^4}$

---

3   Simplify a perfect-cube root.

The **cube root of x** is any number whose cube is x. For example,

4 is a cube root of 64, because $4^3 = 64$.

$3x^2y$ is a cube root of $27x^6y^3$, because $(3x^2y)^3 = 27x^6y^3$.

$-2y$ is a cube root of $-8y^3$, because $(-2y)^3 = -8y^3$.

CUBE ROOT OF a   The cube root of a is denoted as $\sqrt[3]{a}$ and is the number whose cube is a. In symbols,

$$\left(\sqrt[3]{a}\right)^3 = a$$

If a is any real number, then

$$\sqrt[3]{a^3} = a$$

We note that 64 has two real-number square roots, 8 and -8. However, 64 has only one real-number cube root, 4, because 4 is the only real number whose cube is 64. *Since every real number has exactly one real cube root, it is unnecessary to use absolute value symbols when simplifying cube roots.*

EXAMPLE 4   Simplify each radical.

a. $\sqrt[3]{125} = 5$ Because $5^3 = 5 \cdot 5 \cdot 5 = 125$

b. $\sqrt[3]{\dfrac{1}{8}} = \dfrac{1}{2}$ Because $\left(\dfrac{1}{2}\right)^3 = \dfrac{1}{2} \cdot \dfrac{1}{2} \cdot \dfrac{1}{2} = \dfrac{1}{8}$

c. $\sqrt[3]{-27x^3} = -3x$ Because $(-3x)^3 = (-3x)(-3x)(-3x) = -27x^3$

d. $\sqrt[3]{-\dfrac{8a^3}{27b^3}} = -\dfrac{2a}{3b}$ $\left(-\dfrac{2a}{3b}\right)^3 = \left(-\dfrac{2a}{3b}\right)\left(-\dfrac{2a}{3b}\right)\left(-\dfrac{2a}{3b}\right) = -\dfrac{8a^3}{27b^3}$

e. $\sqrt[3]{0.216x^3y^6} = 0.6xy^2$ $(0.6xy^2)^3 = (0.6xy^2)(0.6xy^2)(0.6xy^2) = 0.216x^3y^6$

SELF CHECK 4   Simplify: **a.** $\sqrt[3]{1{,}000}$ **b.** $\sqrt[3]{\dfrac{1}{27}}$ **c.** $\sqrt[3]{125a^3}$ **d.** $\sqrt[3]{-64x^3}$

COMMENT   The previous examples suggest that if a can be factored into three equal factors, any one of those factors is a cube root of a.

4   Simplify a perfect nth root.

Just as there are square roots and cube roots, there are fourth roots, fifth roots, sixth roots, and so on.

In the radical $\sqrt[n]{x}$, n is called the **index** (or **order**) of the radical. When the index is 2, the radical is a square root, and we usually do not write the index.

$$\sqrt[2]{x} = \sqrt{x}$$

COMMENT When n is an even number greater than 1 and $x < 0$, $\sqrt[n]{x}$ is not a real number. For example, $\sqrt[4]{-81}$ is not a real number, because no real number raised to the 4th power is -81. However, when n is odd, $\sqrt[n]{x}$ is a real number.

When n is an odd natural number greater than 1, $\sqrt[n]{x}$ represents an **odd root**. Since every real number has only one real nth root when n is odd, we do not need to use absolute value symbols when finding odd roots. For example,

$$\sqrt[5]{243} = 3 \qquad \text{because } 3^5 = 243$$
$$\sqrt[7]{-128x^7} = -2x \qquad \text{because } (-2x)^7 = -128x^7$$

When n is an even natural number greater than 1, $\sqrt[n]{x}$ represents an **even root**. In this case, there will be one positive and one negative real nth root. For example, the two real sixth roots of 729 are 3 and -3, because $3^6 = 729$ and $(-3)^6 = 729$. When finding even roots, we use absolute value symbols to guarantee that the principal nth root is positive.

$$\sqrt[4]{(-3)^4} = |-3| = 3 \qquad \textcolor{red}{3^4 = (-3)^4. \text{ We also could simplify this as follows:}}$$
$$\textcolor{red}{\sqrt[4]{(-3)^4} = \sqrt[4]{81} = 3}$$
$$\sqrt[6]{729x^6} = |3x| = 3|x| \qquad \textcolor{red}{(3|x|)^6 = 729x^6. \text{ The absolute value symbols guarantee}}$$
$$\textcolor{red}{\text{that the sixth root is positive.}}$$

EXAMPLE 5 Simplify each radical.

a. $\sqrt[4]{625} = 5$, because $5^4 = 625$ $\qquad$ $\textcolor{red}{\text{Read } \sqrt[4]{625} \text{ as "the fourth root of 625."}}$

b. $\sqrt[5]{-32} = -2$, because $(-2)^5 = -32$ $\qquad$ $\textcolor{red}{\text{Read } \sqrt[5]{-32} \text{ as "the fifth root of } -32\text{."}}$

c. $\sqrt[6]{\dfrac{1}{64}} = \dfrac{1}{2}$, because $\left(\dfrac{1}{2}\right)^6 = \dfrac{1}{64}$ $\qquad$ $\textcolor{red}{\text{Read } \sqrt[6]{\frac{1}{64}} \text{ as "the sixth root of } \frac{1}{64}\text{."}}$

d. $\sqrt[7]{10^7} = 10$, because $10^7 = 10^7$ $\qquad$ $\textcolor{red}{\text{Read } \sqrt[7]{10^7} \text{ as "the seventh root of } 10^7\text{."}}$

SELF CHECK 5 Simplify: **a.** $\sqrt[4]{\dfrac{1}{81}}$ $\qquad$ **b.** $\sqrt[5]{10^5}$

When finding the nth root of an nth power, we can use the following rules.

DEFINITION OF $\sqrt[n]{a^n}$

If n is an odd natural number greater than 1, then $\sqrt[n]{a^n} = a$.

If n is an even natural number, then $\sqrt[n]{a^n} = |a|$.

EXAMPLE 6 Simplify each radical. Assume that x can be any real number.

Solution **a.** $\sqrt[5]{x^5} = x$ $\qquad$ $\textcolor{red}{\text{Since } n \text{ is odd, absolute value symbols are not needed.}}$

b. $\sqrt[4]{16x^4} = |2x| = 2|x|$ $\qquad$ $\textcolor{red}{\text{Since } n \text{ is even and } x \text{ can be negative, absolute}}$
$\textcolor{red}{\text{value symbols are needed to guarantee that the}}$
$\textcolor{red}{\text{result is positive.}}$

c. $\sqrt[6]{(x + 4)^6} = |x + 4|$ $\qquad$ $\textcolor{red}{\text{Absolute value symbols are needed to guarantee}}$
$\textcolor{red}{\text{that the result is positive.}}$

d. $\sqrt[3]{(x + 1)^3} = x + 1$ $\qquad$ $\textcolor{red}{\text{Since } n \text{ is odd, absolute value symbols are not needed.}}$

e. $\sqrt{(x^2 + 6x + 9)^2} = \sqrt{[(x + 3)^2]^2}$ $\qquad$ $\textcolor{red}{\text{Factor } x^2 + 6x + 9.}$
$\qquad\qquad\qquad\quad = \sqrt{(x + 3)^4}$
$\qquad\qquad\qquad\quad = (x + 3)^2$ $\qquad$ $\textcolor{red}{\text{Since } (x + 3)^2 \text{ is always positive, absolute value}}$
$\textcolor{red}{\text{symbols are not needed.}}$

SELF CHECK 6   Simplify. Assume that a can be any real number.   **a.** $\sqrt[4]{16a^4}$
b. $\sqrt[5]{(a + 5)^5}$ **c.** $\sqrt{(x^2 + 10x + 25)^2}$

We summarize the possibilities for $\sqrt[n]{x}$ as follows:

DEFINITION FOR $\sqrt[n]{x}$

Assume n is a natural number greater than 1 and x is a real number.

If $x > 0$, then $\sqrt[n]{x}$ is the positive number such that $\left(\sqrt[n]{x}\right)^n = x$.

If $x = 0$, then $\sqrt[n]{x} = 0$.

If $x < 0$ $\begin{cases} \text{and n is odd, then } \sqrt[n]{x} \text{ is the real number such that } \left(\sqrt[n]{x}\right)^n = x. \\ \text{and n is even, then } \sqrt[n]{x} \text{ is not a real number.} \end{cases}$

5 Find the domain of a square-root function
and a cube-root function.

Since there is one principal square root for every nonnegative real number x, the equation $f(x) = \sqrt{x}$ determines a function, called the **square-root function**.

EXAMPLE 7   Consider the function $f(x) = \sqrt{x}$.
a. Find the domain.   **b.** Graph the function.   **c.** Find the range.

Solution   **a.** To find the domain, we note that $x \geq 0$ in the function because the radicand must be nonnegative. Thus, the domain is the set of nonnegative real numbers. In interval notation, the domain is $[0, \infty)$.

b. We can create a table of values and plot points to obtain the graph shown in Figure 7-1(a). If we use a graphing calculator, we can choose window settings of $[-1, 9]$ for x and $[-2, 5]$ for y to see the graph shown in Figure 7-1(b).

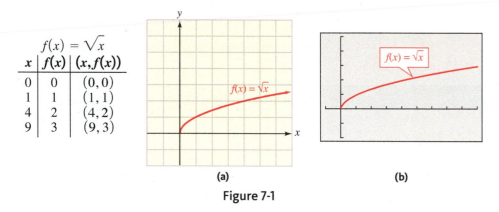

$f(x) = \sqrt{x}$

x	$f(x)$	$(x, f(x))$
0	0	$(0, 0)$
1	1	$(1, 1)$
4	2	$(4, 2)$
9	3	$(9, 3)$

(a)   (b)

Figure 7-1

c. From either graph, we can conclude that the range of the function is the set of nonnegative real numbers, which is the interval $[0, \infty)$. The graph also confirms that the domain is the interval $[0, \infty)$.

SELF CHECK 7   Graph $f(x) = \sqrt{x} + 2$ and compare it to the graph of $f(x) = \sqrt{x}$. Find the domain and the range.

The graphs of many functions are translations or reflections of the square-root function. For example, if $k > 0$,

- The graph of $f(x) = \sqrt{x} + k$ is the graph of $f(x) = \sqrt{x}$ translated k units upward.
- The graph of $f(x) = \sqrt{x} - k$ is the graph of $f(x) = \sqrt{x}$ translated k units downward.
- The graph of $f(x) = \sqrt{x + k}$ is the graph of $f(x) = \sqrt{x}$ translated k units to the left.
- The graph of $f(x) = \sqrt{x - k}$ is the graph of $f(x) = \sqrt{x}$ translated k units to the right.
- The graph of $f(x) = -\sqrt{x}$ is the graph of $f(x) = \sqrt{x}$ reflected about the x-axis.

The equation $f(x) = \sqrt[3]{x}$ defines the **cube-root function**. From the graph shown in Figure 7-2(a), we can see that the domain and range of the function $f(x) = \sqrt[3]{x}$ are the set of real numbers, $\mathbb{R}$, and in interval notation $(-\infty, \infty)$. Note that the graph of $f(x) = \sqrt[3]{x}$ passes the vertical line test. Figures 7-2(b) and 7-2(c) show several translations of the cube-root function.

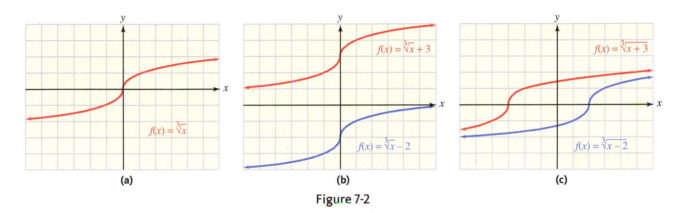

(a)   **(b)**   **(c)**

Figure 7-2

EXAMPLE 8   Consider the function $f(x) = \sqrt[3]{x + 3} + 1$.
   a. Find the domain.   **b.** Graph the function.   **c.** Find the range.

Solution   **a.** To find the domain, we note that the radicand is not restricted because the function is a cube root. Therefore, the domain is all real numbers, $\mathbb{R}$, and represented by the interval $(-\infty, \infty)$.

b. This graph will be the same shape as $f(x) = \sqrt[3]{x}$, translated 3 units to the left and 1 unit upward. (See Figure 7-3(a).) We can confirm this graph by using a graphing calculator. If we select window settings of $[-10, 5]$ for x and $[-5, 5]$ for y, we will obtain the graph shown in Figure 7-3(b).

(a)   **(b)**

Figure 7-3

c. From either graph, we can conclude that the range is the interval $(-\infty, \infty)$. The graph also confirms that the domain is the interval $(-\infty, \infty)$.

SELF CHECK 8   Graph $f(x) = -\sqrt[3]{x} - 2$ and find the domain and range.

6   Use a square root to solve an application.

In the real world, there are many models that involve square root relationships.

EXAMPLE 9   **PERIOD OF A PENDULUM**   The *period of a pendulum* is the time required for the pendulum to swing back and forth to complete one cycle. (See Figure 7-4.) The period t (in seconds) is a function of the pendulum's length l in feet, which is defined by the formula

$$t = f(l) = 2\pi\sqrt{\frac{l}{32}}$$

Find the period of a pendulum that is 5 feet long.

Solution   We substitute 5 for l in the formula and simplify.

$$t = 2\pi\sqrt{\frac{l}{32}}$$

$$t = 2\pi\sqrt{\frac{5}{32}} \qquad \text{Substitute.}$$

$$\approx 2.483647066 \qquad \text{Use a calculator.}$$

To the nearest tenth, the period is 2.5 seconds.

Figure 7-4

SELF CHECK 9   To the nearest hundredth, find the period of a pendulum that is 3 feet long.

Accent on technology

▸ Finding the Period of a Pendulum

To solve Example 9 with a graphing calculator with window settings of $[-2, 10]$ for x and $[-2, 10]$ for y, we graph the function $f(x) = 2\pi\sqrt{\frac{x}{32}}$, as in Figure 7-5(a). We use the value feature of the calculator to find the value of y when $x = 5$. From the graph, press **2ⁿᵈ** **TRACE** (CALC) **ENTER**. Enter 5 **ENTER**. The period is given by the y-value shown on the screen.

(a)   (b)

Figure 7-5

For instructions regarding the use of a Casio graphing calculator, please refer to the Casio Keystroke Guide in the back of the book.

In statistics, the **standard deviation** of a data set is a measure of how tightly the data points are grouped around the mean (average) of the data set. The smaller the value, the more tightly the data points are grouped around the mean.

To see how to compute the standard deviation of a distribution, we consider the distribution 4, 5, 5, 8, 13 and construct the following table.

Original terms	Mean of the distribution	Differences (original term minus mean)	Squares of the differences from the mean
4	7	−3	9
5	7	−2	4
5	7	−2	4
8	7	1	1
13	7	6	36

The population *standard deviation* of the distribution is the positive square root of the mean of the numbers shown in column 4 of the table.

$$\text{Standard deviation} = \sqrt{\frac{\text{sum of the squares of the differences from the mean}}{\text{number of differences}}}$$

$$= \sqrt{\frac{9 + 4 + 4 + 1 + 36}{5}}$$

$$= \sqrt{\frac{54}{5}}$$

$$\approx 3.286335345 \quad \text{Use a calculator.}$$

To the nearest hundredth, the standard deviation of the given distribution is 3.29.

The symbol for the population standard deviation is σ, the lowercase Greek letter *sigma*.

EXAMPLE 10   Which of the following distributions is more tightly grouped around the mean?
a. $3, 5, 7, 8, 12$ **b.** $1, 4, 6, 11$

Solution   We compute the standard deviation of each distribution.

a.

Original terms	Mean of the distribution	Differences (original term minus mean)	Squares of the differences from the mean
3	7	−4	16
5	7	−2	4
7	7	0	0
8	7	1	1
12	7	5	25

$$\sigma = \sqrt{\frac{16 + 4 + 0 + 1 + 25}{5}} = \sqrt{\frac{46}{5}} \approx 3.03$$

b.

Original terms	Mean of the distribution	Differences (original term minus mean)	Squares of the differences from the mean
1	5.5	−4.5	20.25
4	5.5	−1.5	2.25
6	5.5	0.5	0.25
11	5.5	5.5	30.25

$$\sigma = \sqrt{\frac{20.25 + 2.25 + 0.25 + 30.25}{4}} = \sqrt{\frac{53}{4}} \approx 3.64$$

Since the standard deviation for the first distribution is less than the standard deviation for the second, the first distribution is more tightly grouped around the mean.

🌿 **SELF CHECK 10**   Which of the following distributions is more tightly grouped around the mean?
a. 4, 6, 8, 10, 11   **b.** 3, 5, 7, 9

🌿 **SELF CHECK ANSWERS**

1. 12, −12   **2. a.** −7   **b.** $\frac{5}{7}$ **c.** 0.06   **3. a.** $5|a|$ **b.** $|x + 2|$ **c.** $4a^2$
4. a. 10   **b.** $\frac{1}{3}$ **c.** $5a$ **d.** $-4x$ **5. a.** $\frac{1}{3}$ **b.** 10   **6. a.** $2|a|$ **b.** $a + 5$ **c.** $(x + 5)^2$
7. It is 2 units higher.   D: $[0, \infty)$ **8.**   D: $(-\infty, \infty)$
R: $[2, \infty)$ R: $(-\infty, \infty)$

$f(x) = \sqrt{x} + 2$ $f(x) = -\sqrt[3]{x} - 2$

9. 1.92 sec   **10.** Distribution **b** is more tightly grouped around the mean.

NOW TRY THIS

1. Simplify: $\sqrt{100x^{100}}$

2. Without using a calculator, between which two integers will the value of each expression be found?
a. $\sqrt{8}$ **b.** $\sqrt{54}$ **c.** $\sqrt[3]{54}$

3. Given that $2^{-1} = \frac{1}{2}$, $2^0 = 1$, $2^1 = 2$, $2^2 = 4$, and $2^3 = 8$, between which two integers would you expect the value of each expression to be found?
a. $2^{1/2}$ **b.** $2^{3/2}$ **c.** $2^{5/2}$

7.1 Exercises 🌿

WARM-UPS *Fill in the blanks.*

1. $(\underline{})^2 = 25$ **2.** $(\underline{})^2 = 81$

3. $(\underline{})^2 = 4x^2$ **4.** $(\underline{})^2 = 36a^2$

5. $(\underline{})^3 = 8$ **6.** $(\underline{})^3 = -27$

7. $(\underline{})^4 = 81$ **8.** $(\underline{})^4 = 16$

REVIEW *Simplify each rational expression. Assume no denominators are zero.*

9. $\dfrac{x^2 + 7x + 10}{x^2 - 4}$ **10.** $\dfrac{a^3 - b^3}{b^2 - a^2}$

Perform the operations. Assume no denominators are zero.

11. $\dfrac{x^2 - x - 6}{x^2 - 2x - 3} \cdot \dfrac{x^2 - 1}{x^2 + x - 2}$

12. $\dfrac{x^2 - 3x - 4}{x^2 - 5x + 6} \div \dfrac{x^2 - 2x - 3}{x^2 - x - 2}$

13. $\dfrac{3}{m + 1} + \dfrac{3m}{m - 1}$ **14.** $\dfrac{2x + 3}{3x - 1} - \dfrac{x - 4}{2x + 1}$

VOCABULARY AND CONCEPTS *Fill in the blanks.*

15. $5x^2$ is the square root of $25x^4$, because _____ $= 25x^4$ and 6 is a square root of 36 because _____.

16. The numbers $1, 4, 9, 16, 25, \ldots$ are called _____.

17. The principal square root of x $(x > 0)$ is the _____ square root of x.

18. The graph of $f(x) = \sqrt{x} + 3$ is the graph of $f(x) = \sqrt{x}$ translated _ units __.

19. The graph of $f(x) = \sqrt{x + 5}$ is the graph of $f(x) = \sqrt{x}$ translated _ units to the ___.

20. When n is an odd natural number greater than 1, $\sqrt[n]{x}$ represents an ___ root.

21. Given the radical $\sqrt[a]{b}$, the symbol $\sqrt{}$ is the _____ sign, a is the _____, and b is the _____.

22. The square-root function $f(x) = \sqrt{x}$ has the domain _____, while the cube-root function $f(x) = \sqrt[3]{x}$ has the domain _____.

23. $\sqrt{x^2} = $ ___   **24.** $\left(\sqrt[3]{x}\right)^3 = $ ___

25. $\sqrt[3]{x^3} = $ ___   **26.** $\sqrt{0} = $ ___

27. When n is a positive ____ number, $\sqrt[n]{x}$ represents an even root.

28. The _____ deviation of a set of numbers is a measure of how tightly the data points are grouped around the mean of the data set.

Identify the radicand in each expression.

29. $-\sqrt{7y^2}$

30. $6\sqrt{ab}$

31. $ab^2\sqrt{a^2 + b^3}$

32. $\dfrac{1}{2}x\sqrt{\dfrac{x}{y}}$

GUIDED PRACTICE *Find the two square roots of each number. SEE EXAMPLE 1. (OBJECTIVE 1)*

33. 100

34. 25

35. 49

36. 169

Find each square root, if possible. SEE EXAMPLE 2. (OBJECTIVE 1)

37. $\sqrt{121}$

38. $\sqrt{144}$

39. $-\sqrt{64}$

40. $-\sqrt{1}$

41. $-\sqrt{\dfrac{25}{49}}$

42. $\sqrt{\dfrac{49}{81}}$

43. $\sqrt{-4}$

44. $\sqrt{-121}$

45. $\sqrt{0.16}$

46. $\sqrt{0.25}$

Find each square root. Assume that all variables are unrestricted, and use absolute value symbols when necessary. SEE EXAMPLE 3. (OBJECTIVE 2)

47. $\sqrt{25x^2}$

48. $\sqrt{16b^2}$

49. $\sqrt{64y^4}$

50. $\sqrt{16y^4}$

51. $\sqrt{(x + 3)^2}$

52. $\sqrt{(a + 6)^2}$

53. $\sqrt{a^2 + 6a + 9}$

54. $\sqrt{x^2 + 14x + 49}$

Simplify each cube root. SEE EXAMPLE 4. (OBJECTIVE 3)

55. $\sqrt[3]{1}$

56. $\sqrt[3]{512}$

57. $\sqrt[3]{-27}$

58. $\sqrt[3]{-8}$

59. $\sqrt[3]{-\dfrac{64}{27}}$

60. $\sqrt[3]{\dfrac{125}{216}}$

61. $\sqrt[3]{0.064}$

62. $\sqrt[3]{0.001}$

63. $\sqrt[3]{125y^3}$

64. $\sqrt[3]{-27x^6}$

65. $\sqrt[3]{-1,000p^3q^3}$

66. $\sqrt[3]{343a^6b^3}$

Simplify each radical, if possible. SEE EXAMPLE 5. (OBJECTIVE 4)

67. $-\sqrt[5]{243}$

68. $\sqrt[5]{-32}$

69. $-\sqrt[4]{16}$

70. $-\sqrt[4]{625}$

71. $\sqrt[6]{64}$

72. $\sqrt[6]{729}$

73. $\sqrt[4]{\dfrac{16}{625}}$

74. $\sqrt[5]{-\dfrac{243}{32}}$

75. $\sqrt[4]{-256}$

76. $\sqrt[6]{-729}$

77. $-\sqrt[5]{-\dfrac{1}{32}}$

78. $-\sqrt[4]{\dfrac{81}{256}}$

Simplify each radical. Assume that all variables are unrestricted, and use absolute value symbols where necessary. SEE EXAMPLE 6. (OBJECTIVE 4)

79. $\sqrt[4]{81x^4}$

80. $\sqrt[6]{64x^6}$

81. $\sqrt[3]{8a^3}$

82. $\sqrt[5]{32a^5}$

83. $\sqrt[4]{\dfrac{1}{16}x^4}$

84. $\sqrt[4]{\dfrac{1}{81}x^8}$

85. $\sqrt[4]{x^{12}}$

86. $\sqrt[8]{x^{24}}$

87. $\sqrt[5]{-x^5}$

88. $\sqrt[3]{-x^6}$

89. $\sqrt[3]{-27a^6}$

90. $\sqrt[5]{-32x^5}$

Graph each function and state the domain and range. SEE EXAMPLE 7. (OBJECTIVE 5)

91. $f(x) = \sqrt{x + 4}$

92. $f(x) = -\sqrt{x - 2}$

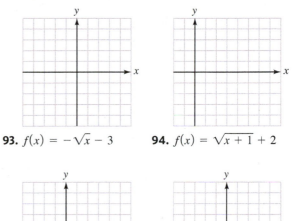

93. $f(x) = -\sqrt{x} - 3$

94. $f(x) = \sqrt{x + 1} + 2$

Graph each function and state the domain and range. SEE EXAMPLE 8. (OBJECTIVE 5)

95. $f(x) = \sqrt[3]{x} - 1$

96. $f(x) = \sqrt[3]{x + 1} + 2$

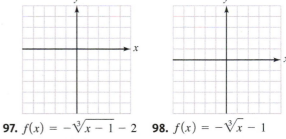

97. $f(x) = -\sqrt[3]{x - 1} - 2$

98. $f(x) = -\sqrt[3]{x} - 1$

ADDITIONAL PRACTICE *Simplify each radical. Assume that all variables are unrestricted, and use absolute value symbols where necessary.*

99. $\sqrt{(-4)^2}$ **100.** $\sqrt{(-9)^2}$

101. $\sqrt{-36}$ **102.** $-\sqrt{-4}$

103. $\sqrt{(-5b)^2}$ **104.** $\sqrt{(-8c)^2}$

105. $\sqrt{t^2 + 24t + 144}$ **106.** $\sqrt{m^2 + 30m + 225}$

107. $\sqrt[3]{-\dfrac{1}{8}m^6n^3}$ **108.** $\sqrt[3]{\dfrac{27}{1,000}a^6b^6}$

109. $\sqrt[3]{0.008z^9}$ **110.** $\sqrt[3]{0.064s^9t^6}$

111. $\sqrt[25]{(x + 2)^{25}}$ **112.** $\sqrt[44]{(x + 4)^{44}}$

113. $\sqrt[8]{0.00000001x^{16}y^8}$ **114.** $\sqrt[5]{0.00032x^{10}y^5}$

Use a calculator to find each square root. Give the answer to four decimal places.

115. $\sqrt{12}$ **116.** $\sqrt{340}$

117. $\sqrt{679.25}$ **118.** $\sqrt{0.0063}$

APPLICATIONS *For Exercises 119–122, use a calculator to solve each application.* **SEE EXAMPLES 9–10. (OBJECTIVE 6)**

119. Period of a pendulum The longest pendulum in a working clock is 13 feet. Find the period of the pendulum.

120. Period of a pendulum A small clock has a pendulum that is two inches in length. Find the period of the pendulum.

121. Statistics Find the standard deviation of the following distribution to the nearest hundredth: $2, 5, 5, 6, 7$

122. Statistics Find the standard deviation of the following distribution to the nearest hundredth: $3, 6, 7, 9, 11, 12$

123. Statistics In statistics, the formula

$$s_{\bar{x}} = \frac{s}{\sqrt{N}}$$

gives an estimate of the standard error of the mean. Find $s_{\bar{x}}$ to four decimal places when $s = 65$ and $N = 30$.

124. Statistics In statistics, the formula

$$\sigma_{\bar{x}} = \frac{\sigma}{\sqrt{N}}$$

gives the standard deviation of means of samples of size N. Find $\sigma_{\bar{x}}$ to four decimal places when $\sigma = 12.7$ and $N = 32$.

125. Radius of a circle The radius r of a circle is given by the formula $r = \sqrt{\dfrac{A}{\pi}}$, where A is its area. Find the radius of a circle whose area is 9π square units.

126. Diagonal of a baseball diamond The diagonal d of a square is given by the formula $d = \sqrt{2s^2}$, where s is the length of each side. Find the diagonal of the baseball diamond.

127. Falling objects The time t (in seconds) that it will take for an object to fall a distance of s feet is given by the formula

$$t = \frac{\sqrt{s}}{4}$$

If a stone is dropped down a 256-foot well, how long will it take it to hit bottom?

128. Law enforcement Police sometimes use the formula $s = k\sqrt{l}$ to estimate the speed s (in mph) of a car involved in an accident. In this formula, l is the length of the skid in feet, and k is a constant depending on the condition of the pavement. For wet pavement, $k \approx 3.24$. How fast was a car going if its skid was 400 feet on wet pavement?

129. Electronics When the resistance in a circuit is 18 ohms, the current I (measured in amperes) and the power P (measured in watts) are related by the formula

$$I = \sqrt{\frac{P}{18}}$$

Find the current used by an electrical appliance that is rated at 980 watts.

130. Medicine The approximate pulse rate p (in beats per minute) of an adult who is t inches tall is given by the formula

$$p = \frac{590}{\sqrt{t}}$$

Find the approximate pulse rate of an adult who is 71 inches tall.

WRITING ABOUT MATH

131. If x is any real number, then $\sqrt{x^2} = x$ is not correct. Explain.

132. If x is any real number, then $\sqrt[3]{x^3} = |x|$ is not correct. Explain.

SOMETHING TO THINK ABOUT

133. Is $\sqrt{x^2 - 4x + 4} = x - 2$? What are the exceptions?

134. When is $\sqrt{x^2} \neq x$?

Section 7.2

Applications of the Pythagorean Theorem and the Distance Formula

Objectives

1. Apply the Pythagorean theorem to find the length of one side of a right triangle.
2. Find the distance between two points on the coordinate plane.

Vocabulary

leg
hypotenuse

Pythagorean theorem

distance formula

Getting Ready

Evaluate each expression.

1. $3^2 + 4^2$ **2.** $5^2 + 12^2$

3. $(5 - 2)^2 + (2 + 1)^2$ **4.** $(111 - 21)^2 + (60 - 4)^2$

In this section, we will discuss the Pythagorean theorem, a theorem that shows the relationship of the sides of a right triangle. We will then use this theorem to develop a formula to calculate the distance between two points on the coordinate plane.

1 Apply the Pythagorean theorem to find the length of one side of a right triangle.

If we know the lengths of the two **legs** of a right triangle, we can find the length of the **hypotenuse** (the side opposite the 90° angle) by using the **Pythagorean theorem**. In fact, if we know the lengths of any two sides of a right triangle, we can find the length of the third side.

PYTHAGOREAN THEOREM

If a and b are the lengths of two legs of a right triangle and c is the length of the hypotenuse, then

$$a^2 + b^2 = c^2$$

In any right triangle, the square of the length of the hypotenuse is equal to the sum of the squares of the lengths of the two legs.

Suppose the right triangle shown in Figure 7-6 has legs of length 3 and 4 units. To find the length of the hypotenuse, we can use the Pythagorean theorem.

$$a^2 + b^2 = c^2$$
$$3^2 + 4^2 = c^2 \quad \text{Substitute.}$$
$$9 + 16 = c^2 \quad \text{Simplify.}$$
$$25 = c^2 \quad \text{Add.}$$

Figure 7-6

To find the value of c, we ask "what number when squared is equal to 25?" There are two such numbers: the positive square root of 25 and the negative square root of 25. Since c represents the length of the hypotenuse and cannot be negative, it follows that c is the positive square root of 25.

$$\sqrt{25} = c \quad \text{Recall that the radical symbol } \sqrt{} \text{ represents the positive, or principal,}$$
$$\text{square root of a number.}$$
$$5 = c$$

The length of the hypotenuse is 5 units.

EXAMPLE 1   **FIGHTING FIRES**   To fight a forest fire, the forestry department plans to clear a rectangular fire break around the fire, as shown in Figure 7-7. Crews are equipped with mobile communications with a 3,000-yard range. Can crews at points A and B remain in radio contact?

Solution   Points A, B, and C form a right triangle. The lengths of its sides are represented as a, b, and c in yards, where a is opposite point A, b is opposite point B, and c is opposite point C. To find the distance c, we can use the Pythagorean theorem, substituting 2,400 for a and 1,000 for b and solving for c.

$$a^2 + b^2 = c^2$$
$$2,400^2 + 1,000^2 = c^2 \quad \text{Substitute.}$$
$$5,760,000 + 1,000,000 = c^2 \quad \text{Square each value.}$$
$$6,760,000 = c^2 \quad \text{Add.}$$
$$\sqrt{6,760,000} = c \quad \text{Since } c \text{ represents a length, it must be the positive square}$$
$$\text{root of 6,760,000.}$$
$$2,600 = c \quad \text{Use a calculator to find the square root.}$$

Figure 7-7

The two crews are 2,600 yards apart. Because this distance is less than the 3,000-yard range of the radios, they can communicate.

SELF CHECK 1   Can the crews communicate if $b = 1,850$ yards?

Perspective

Pythagoras was a teacher. Although it was unusual at that time, his classes were coeducational. He and his followers formed a secret society with two rules: Membership was for life, and members could not reveal the secrets they knew.

Much of their teaching was good mathematics, but some ideas were strange. To them, numbers were sacred. Because beans were used as counters to represent numbers, Pythagoreans refused to eat beans. They also believed that the *only* numbers were the whole numbers.

To them, fractions were not numbers; $\frac{2}{3}$ was just a way of comparing the whole numbers 2 and 3. They believed that whole numbers were the building blocks of the universe. The basic Pythagorean doctrine was, "All things are numbers," and they meant *whole* numbers.

The Pythagorean theorem was an important discovery of the Pythagorean school, yet it caused some controversy. The right triangle in the illustration has two legs of length 1. By the Pythagorean theorem, the length of the hypotenuse is $\sqrt{2}$. One of their own group, Hippasus of Metapontum, discovered that $\sqrt{2}$ is an irrational number: There are *no* whole numbers a and b that make the fraction $\frac{a}{b}$ exactly equal to $\sqrt{2}$. This discovery was not appreciated by the other Pythagoreans. How could everything in the universe be described with whole numbers, when the side of this simple triangle couldn't? The Pythagoreans had a choice. Either expand their beliefs, or cling to the old. According to legend, the group was at sea at the time of the discovery. Rather than upset the system, they threw Hippasus overboard.

---

2 ## Find the distance between two points on the coordinate plane.

We can use the Pythagorean theorem to develop a formula to find the distance between any two points that are graphed on a rectangular coordinate system.

To find the distance d between points P and Q shown in Figure 7-8, we construct the right triangle PRQ. Because line segment RQ is vertical, point R will have the same x-coordinate as point Q. Because line segment PR is horizontal, point R will have the same y-coordinate as point P. The distance between P and R is $|x_2 - x_1|$, and the distance between R and Q is $|y_2 - y_1|$. We apply the Pythagorean theorem to the right triangle PRQ to get

$$(PQ)^2 = (PR)^2 + (RQ)^2 \qquad \text{Read } PQ \text{ as "the length of segment } PQ."$$
$$d^2 = |x_2 - x_1|^2 + |y_2 - y_1|^2 \qquad \text{Substitute the value of each expression.}$$
$$d^2 = (x_2 - x_1)^2 + (y_2 - y_1)^2 \qquad |x_2 - x_1|^2 = (x_2 - x_1)^2 \text{ and } |y_2 - y_1|^2 = (y_2 - y_1)^2$$

(1) $$d = \sqrt{(x_2 - x_1)^2 + (y_2 - y_1)^2} \qquad \text{Since } d \text{ represents a length, it must be the positive square root of } (x_2 - x_1)^2 + (y_2 - y_1)^2.$$

Equation 1 is called the **distance formula**.

Pythagoras of Samos
569?–475? B.C.

Pythagoras is thought to be the world's first pure mathematician. Although he is famous for the theorem that bears his name, he is often called "the father of music," because a society he led discovered some of the fundamentals of musical harmony. This secret society had numerology as its religion. The society is also credited with the discovery of irrational numbers.

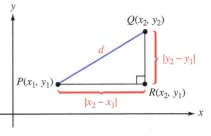

Figure 7-8

DISTANCE FORMULA

The distance d between two points (x_1, y_1) and (x_2, y_2) is given by the formula

$$d = \sqrt{(x_2 - x_1)^2 + (y_2 - y_1)^2}$$

EXAMPLE 2 Find the distance between the points $(-2, 3)$ and $(4, -5)$.

Solution To find the distance, we can use the distance formula by substituting 4 for x_2, -2 for x_1, -5 for y_2, and 3 for y_1.

COMMENT Recall that a square root symbol is a grouping symbol and that any radicand must be simplified first.

$$\sqrt{a^2 + b^2} \neq \sqrt{a} + \sqrt{b}$$

$$
\begin{aligned}
d &= \sqrt{(x_2 - x_1)^2 + (y_2 - y_1)^2} \\
&= \sqrt{[4 - (-2)]^2 + (-5 - 3)^2} && \text{Substitute.} \\
&= \sqrt{(4 + 2)^2 + (-5 - 3)^2} && \text{Simplify.} \\
&= \sqrt{6^2 + (-8)^2} && \text{Simplify.} \\
&= \sqrt{36 + 64} && \text{Square each value.} \\
&= \sqrt{100} && \text{Add.} \\
&= 10 && \text{Take the square root.}
\end{aligned}
$$

The distance between the two points is 10 units.

SELF CHECK 2 Find the distance between $P(-2, -2)$ and $Q(3, 10)$.

EXAMPLE 3 **BUILDING A FREEWAY** In a city, streets run north and south, and avenues run east and west. Streets are 850 feet apart and avenues are 850 feet apart. The city plans to construct a straight freeway from the intersection of 25th Street and 8th Avenue to the intersection of 115th Street and 64th Avenue. How long will the freeway be?

Solution We can represent the roads by the coordinate system in Figure 7-9, where the units on each axis represent 850 feet. We represent the end of the freeway at 25th Street and 8th Avenue by the point $(x_1, y_1) = (25, 8)$. The other end is $(x_2, y_2) = (115, 64)$.

Figure 7-9

We can use the distance formula to find the number of units between the two designated points.

$$
\begin{aligned}
d &= \sqrt{(x_2 - x_1)^2 + (y_2 - y_1)^2} \\
d &= \sqrt{(115 - 25)^2 + (64 - 8)^2} && \text{Substitute.} \\
&= \sqrt{90^2 + 56^2} && \text{Subtract.} \\
&= \sqrt{8{,}100 + 3{,}136} && \text{Square each value.} \\
&= \sqrt{11{,}236} && \text{Add.} \\
&= 106 && \text{Use a calculator to find the square root.}
\end{aligned}
$$

There are approximately 106 units between the two designated points, and because each unit is 850 feet, the length of the freeway is $106(850) = 90{,}100$ feet. Since 5,280 feet $= 1$ mile, we can divide 90,100 by 5,280 to convert 90,100 feet to 17.064394 miles. Thus, the freeway will be about 17 miles long.

SELF CHECK 3 Find the diagonal distance in feet from the intersection of 25th Street and 8th Avenue to the intersection of 28th Street and 12th Avenue.

SELF CHECK ANSWERS 1. no 2. 13 units 3. 4,250 ft

NOW TRY THIS

1. Determine whether the points $(4, -2), (-2, -4)$, and $(-4, 2)$ are the vertices of an *isosceles triangle* (two equal sides), an *equilateral triangle* (three equal sides), or neither. Is the triangle a right triangle?
2. Find the distance between points with coordinates of $(2x + 1, x + 1)$ and $(2x - 3, x - 2)$.

7.2 Exercises

WARM-UPS *Evaluate each expression.*

1. $\sqrt{625}$
2. $\sqrt{289}$
3. $\sqrt{7^2 + 24^2}$
4. $\sqrt{15^2 + 8^2}$
5. $\sqrt{5^2 - 3^2}$
6. $\sqrt{5^2 - 4^2}$

REVIEW *Find each product.*

7. $(2x + 5)(3x - 4)$
8. $(6y - 7)(4y + 5)$
9. $(4a - 3b)(5a - 2b)$
10. $(4r - 3)(2r^2 + 3r - 4)$

VOCABULARY AND CONCEPTS *Fill in the blanks.*

11. In a right triangle, the side opposite the 90° angle is called the _____.
12. In a right triangle, the two shorter sides are called ___.
13. If a and b are the lengths of two legs of a right triangle and c is the length of the hypotenuse, then _____.
14. In any right triangle, the square of the length of the hypotenuse is equal to the ___ of the squares of the lengths of the two ___. This fact is known as the _____.
15. If $x^2 = 25$ and x is positive, we can conclude that x is the _____ square root of 25. Thus, $x = 5$.
16. The formula for finding the distance between two points on a rectangular coordinate system is $d =$ _____.

GUIDED PRACTICE *The lengths of two sides of the right triangle ABC shown in the illustration are given. Find the length of the third side. SEE EXAMPLE 1. (OBJECTIVE 1)*

17. $a = 6$ ft and $b = 8$ ft
18. $a = 10$ cm and $c = 26$ cm

19. $b = 18$ m and $c = 82$ m
20. $b = 7$ ft and $c = 25$ ft
21. $a = 14$ in. and $c = 50$ in.
22. $a = 8$ cm and $b = 15$ cm
23. $a = \sqrt{8}$ mi and $b = \sqrt{8}$ mi
24. $a = \sqrt{11}$ ft and $b = \sqrt{38}$ ft

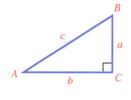

Find the distance between the given points. If an answer is not exact, use a calculator and give an approximation to the nearest tenth.
SEE EXAMPLE 2. (OBJECTIVE 2)

25. $(0, 0), (3, -4)$
26. $(0, 0), (-6, 8)$
27. $(10, 8), (-2, 3)$
28. $(5, 9), (8, 13)$
29. $(-7, 12), (-1, 4)$
30. $(-5, -2), (7, 3)$
31. $(12, -6), (-3, 2)$
32. $(-14, -9), (10, -2)$
33. $(-3, 5), (-5, -5)$
34. $(2, -3), (4, -8)$
35. $(-9, 3), (4, 7)$
36. $(-1, -3), (-5, 8)$

ADDITIONAL PRACTICE *In Exercises 37–40, use a calculator to approximate each value to the nearest foot. The baseball diamond is a square, 90 feet on a side.*

37. Baseball   How far must a catcher throw the ball to throw out a runner stealing second base?

38. Baseball   In baseball, the pitcher's mound is 60 feet, 6 inches from home plate. How far from the mound is second base?

39. Baseball   If the third baseman fields a ground ball 10 feet directly behind third base along the third base line, how far must he throw the ball to throw a runner out at first base?

40. Baseball   The shortstop fields a grounder at a point one-third of the way from second base to third base. How far will he have to throw the ball to make an out at first base?

For Exercises 41–44, approximate each answer to the nearest tenth.

41. Geometry   Find the length of the diagonal of one of the faces of the cube.

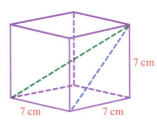

7 cm   7 cm   7 cm

42. Geometry   Find the length of the diagonal of the cube shown in the illustration in Exercise 41.

43. Supporting a weight   A weight placed on the tight wire pulls the center down 1 foot. By how much is the wire stretched? Round the answer to the nearest hundredth of a foot.

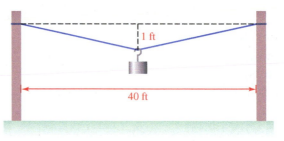

1 ft

40 ft

44. Supporting a weight   If the weight in Exercise 43 pulls the center down 2 feet, by how much would the wire stretch? Round the answer to the nearest tenth of a foot.

APPLICATIONS   *Solve each application.* **SEE EXAMPLE 1.** *(OBJECTIVE 1)*

45. Sailing   Refer to the sailboat in the illustration. How long must a rope be to fasten the top of the mast to the bow?

d

12 ft

5 ft

46. Carpentry   The gable end of the roof shown is divided in half by a vertical brace. Find the distance from an eave to the peak.

17 ft

h

30 ft

47. Reach of a ladder   The base of the 37-foot ladder in the illustration is 9 feet from the wall. Will the top reach a window ledge that is 35 feet above the ground?

37 ft

h ft

9 ft

48. Geometry   The side, s, of a square with area A square feet is given by the formula $s = \sqrt{A}$. Find the perimeter of a square with an area of 49 square feet.

Solve each application. **SEE EXAMPLE 3.** *(OBJECTIVE 2)*

49. Telephone service   The telephone cable in the illustration currently runs from A to B to C to D. How much cable is required to run from A to D directly?

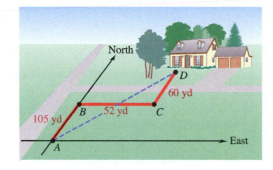

North

D

60 yd

105 yd   B   52 yd   C

A   East

50. Electric service   The power company routes its lines as shown in the illustration. How much wire could be saved by going directly from A to E?

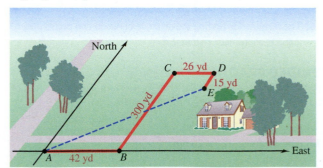

North

C   26 yd   D

15 yd

300 yd   E

A   42 yd   B   East

51. Surface area of a cube The total surface area, A, of a cube is related to its volume, V, by the formula $A = 6\sqrt[3]{V^2}$. Find the surface area of a cube with a volume of 8 cubic centimeters.

52. Area of many cubes A grain of table salt is a cube with a volume of approximately 6×10^{-6} cubic in., and there are about 1.5 million grains of salt in one cup. Find the total surface area of the salt in one cup. (See Exercise 51.)

Use a calculator to help answer each question.

53. Packing a tennis racket The diagonal d of a rectangular box with dimensions $a \times b \times c$ is given by

$$d = \sqrt{a^2 + b^2 + c^2}$$

Will the racket fit in the shipping carton?

54. Packing a tennis racket Will the racket in Exercise 53 fit in a carton with dimensions that are 15 in. · 15 in. · 17 in.?

55. Shipping packages A delivery service won't accept a package for shipping if any dimension exceeds 21 inches. An archaeologist wants to ship a 36-inch femur bone. Will it fit in a 3-inch-tall box that has a 21-inch-square base?

56. Shipping packages Can the archaeologist in Exercise 55 ship the femur bone in a cubical box 21 inches on an edge?

57. Geometry Show that the point $(5, 1)$ is equidistant from points $(7, 0)$ and $(3, 0)$.

58. Geometry Show that a triangle with vertices at $(2, 3)$, $(-3, 4)$, and $(1, -2)$ is a right triangle. (*Hint*: If the Pythagorean theorem holds, the triangle is a right triangle.)

59. Geometry Show that a triangle with vertices at $(-2, 4)$, $(2, 8)$, and $(6, 4)$ is isosceles.

60. Geometry Show that a triangle with vertices at $(-2, 13)$, $(-8, 9)$, and $(-2, 5)$ is isosceles.

WRITING ABOUT MATH

61. State the Pythagorean theorem.

62. Explain the distance formula.

SOMETHING TO THINK ABOUT

63. Body mass The formula

$$I = \frac{703w}{h^2}$$

(where w is weight in pounds and h is height in inches) can be used to estimate body mass index, I. The scale shown in the table can be used to judge a person's risk of heart attack. A girl weighing 104 pounds is 54.1 inches tall. Find her estimated body mass index.

20–26	normal
27–29	higher risk
30 and above	very high risk

64. What is the risk of a heart attack for a man who is 6 feet tall and weighs 210 pounds?

65. Bowling The velocity, v, of an object after it has fallen d feet is given by the equation $v^2 = 64d$. If an inexperienced bowler lofts the ball 4 feet, with what velocity does it strike the alley?

66. Bowling Using the formula from Exercise 65, find the velocity if the bowler lofts the ball 3 feet. Round to the nearest tenth.

Section 7.3

Rational Exponents

Objectives

1. Simplify an expression that contains a positive rational exponent with a numerator of 1.
2. Simplify an expression that contains a positive rational exponent with a numerator other than 1.
3. Simplify an expression that contains a negative rational exponent.
4. Simplify an expression that contains rational exponents by applying the properties of exponents.
5. Simplify a radical expression by first writing it as an expression with a rational exponent.

Getting Ready

Simplify each expression. Assume no variable is zero.

1. x^3x^4 2. $(a^3)^4$ 3. $\dfrac{a^8}{a^4}$ 4. a^0

5. x^{-4} 6. $\dfrac{1}{x^{-3}}$ 7. $\left(\dfrac{b^2}{c^3}\right)^3$ 8. $(a^2a^3)^2$

---

1 Simplify an expression that contains a positive rational exponent with a numerator of 1.

We have seen that positive integer exponents indicate the number of times that a base is to be used as a factor in a product. For example, x^5 means that x is to be used as a factor five times.

$$x^5 = \overbrace{x \cdot x \cdot x \cdot x \cdot x}^{5 \text{ factors of } x}$$

Furthermore, we recall the following properties of exponents.

RULES OF EXPONENTS

If there are no divisions by 0, then for all integers m and n,

1. $x^m x^n = x^{m+n}$ 2. $(x^m)^n = x^{mn}$ 3. $(xy)^n = x^n y^n$

4. $\left(\dfrac{x}{y}\right)^n = \dfrac{x^n}{y^n}$ 5. $x^0 = 1 \quad (x \neq 0)$ 6. $x^{-n} = \dfrac{1}{x^n}$

7. $\dfrac{x^m}{x^n} = x^{m-n}$ 8. $\left(\dfrac{x}{y}\right)^{-n} = \left(\dfrac{y}{x}\right)^n$ 9. $\dfrac{1}{x^{-n}} = x^n$

To show how to raise bases to rational powers, we consider the expression $10^{1/2}$. Since rational exponents must obey the same rules as integer exponents, the square of $10^{1/2}$ is equal to 10.

$$(10^{1/2})^2 = 10^{(1/2)2} \quad \text{Keep the base and multiply the exponents.}$$
$$= 10^1 \quad \tfrac{1}{2} \cdot 2 = 1$$
$$= 10 \quad 10^1 = 10$$

However, we have seen that

$$\left(\sqrt{10}\right)^2 = 10$$

Since $(10^{1/2})^2$ and $\left(\sqrt{10}\right)^2$ both equal 10, we define $10^{1/2}$ to be $\sqrt{10}$. Likewise, we define

$$10^{1/3} \text{ to be } \sqrt[3]{10} \qquad \text{and} \qquad 10^{1/4} \text{ to be } \sqrt[4]{10}$$

RATIONAL EXPONENTS If n is a natural number greater than 1, and $\sqrt[n]{x}$ is a real number, then

$$x^{1/n} = \sqrt[n]{x}$$

EXAMPLE 1 Simplify each expression. Assume all variables represent nonnegative values.

a. $9^{1/2}$ **b.** $-\left(\dfrac{16}{9}\right)^{1/2}$ **c.** $(-64)^{1/3}$ **d.** $16^{1/4}$ **e.** $\left(\dfrac{1}{32}\right)^{1/5}$ **f.** $0^{1/8}$

g. $-(32x^5)^{1/5}$ **h.** $(xyz)^{1/4}$

Solution **a.** $9^{1/2} = \sqrt{9} = 3$ **b.** $-\left(\dfrac{16}{9}\right)^{1/2} = -\sqrt{\dfrac{16}{9}} = -\dfrac{4}{3}$

c. $(-64)^{1/3} = \sqrt[3]{-64} = -4$ **d.** $16^{1/4} = \sqrt[4]{16} = 2$

e. $\left(\dfrac{1}{32}\right)^{1/5} = \sqrt[5]{\dfrac{1}{32}} = \dfrac{1}{2}$ **f.** $0^{1/8} = \sqrt[8]{0} = 0$

g. $-(32x^5)^{1/5} = -\sqrt[5]{32x^5} = -2x$ **h.** $(xyz)^{1/4} = \sqrt[4]{xyz}$

SELF CHECK 1 Simplify each expression. Assume that all variables represent nonnegative values.

a. $16^{1/2}$ **b.** $\left(\dfrac{27}{8}\right)^{1/3}$ **c.** $-(16x^4)^{1/4}$

EXAMPLE 2 Write each radical using a rational exponent. **a.** $\sqrt[4]{5xyz}$ **b.** $\sqrt[5]{xy^2}$

Solution **a.** $\sqrt[4]{5xyz} = (5xyz)^{1/4}$ **b.** $\sqrt[5]{xy^2} = (xy^2)^{1/5}$

SELF CHECK 2 Write each radical using a rational exponent.

a. $\sqrt[6]{4ab}$ **b.** $\sqrt[3]{a^2b^4}$

As with radicals, when n is even in the expression $x^{1/n}(n > 1)$, there are two real nth roots and we must use absolute value symbols to guarantee that the simplified result is positive.

When n is odd, there is only one real nth root, and we do not need to use absolute value symbols.

When n is even and x is negative, the expression $x^{1/n}$ is not a real number.

EXAMPLE 3   Assume that all variables can be any real number, and simplify each expression using absolute value symbols when necessary.

a. $(-27x^3)^{1/3} = -3x$

$-27x^3 = (-3x)^3$. Since n is odd, no absolute value symbols are needed.

b. $(49x^2)^{1/2} = |7x| = 7|x|$

$49x^2 = (|7x|)^2$. Since $7x$ can be negative, absolute value symbols are needed.

c. $(256a^8)^{1/8} = 2|a|$

$256a^8 = (2|a|)^8$. Since a can be any real number, $2a$ can be negative. Thus, absolute value symbols are needed.

d. $[(y + 1)^2]^{1/2} = |y + 1|$

$(y + 1)^2 = |y + 1|^2$. Since y can be any real number, $y + 1$ can be negative, and the absolute value symbols are needed.

e. $(25b^4)^{1/2} = 5b^2$

$25b^4 = (5b^2)^2$. Since $b^2 \geq 0$, no absolute value symbols are needed.

f. $(-256x^4)^{1/4}$ is not a real number.

No real number raised to the 4th power is $-256x^4$.

SELF CHECK 3   Simplify each expression using absolute value symbols when necessary.

a. $(625a^4)^{1/4}$ b. $(b^4)^{1/2}$ c. $(-8x^3)^{1/3}$ d. $(64x^2)^{1/2}$
e. $(-9x^4)^{1/2}$

We summarize the cases as follows.

SUMMARY OF THE DEFINITIONS OF $x^{1/n}$

Assume n is a natural number greater than 1 and x is a real number.

If $x > 0$, then $x^{1/n}$ is the positive number such that $(x^{1/n})^n = x$.

If $x = 0$, then $x^{1/n} = 0$.

If $x < 0$ $\begin{cases} \text{and } n \text{ is odd, then } x^{1/n} \text{ is the real number such that } (x^{1/n})^n = x. \\ \text{and } n \text{ is even, then } x^{1/n} \text{ is not a real number.} \end{cases}$

2   Simplify an expression that contains a positive rational exponent with a numerator other than 1.

We can extend the definition of $x^{1/n}$ to include rational exponents with numerators other than 1. For example, since $4^{3/2}$ can be written as $(4^{1/2})^3$, we have

$$4^{3/2} = (4^{1/2})^3 = (\sqrt{4})^3 = 2^3 = 8$$

Thus, we can simplify $4^{3/2}$ by cubing the square root of 4. We can also simplify $4^{3/2}$ by taking the square root of 4 cubed.

$$4^{3/2} = (4^3)^{1/2} = 64^{1/2} = \sqrt{64} = 8$$

In general, we have the following strategy.

CHANGING FROM RATIONAL EXPONENTS TO RADICALS

If m and n are positive integers, $x \geq 0$, and $\frac{m}{n}$ is in simplified form, then

$$x^{m/n} = (\sqrt[n]{x})^m = \sqrt[n]{x^m}$$

We can interpret $x^{m/n}$ in two ways:

1. $x^{m/n}$ means the *m*th power of the *n*th root of *x*.

2. $x^{m/n}$ means the *n*th root of the *m*th power of *x*.

EXAMPLE 4   Simplify each expression.   **a.** $27^{2/3}$ **b.** $\left(\dfrac{1}{16}\right)^{3/4}$ **c.** $(-8x^3)^{4/3}$

Solution   **a.** $27^{2/3} = \left(\sqrt[3]{27}\right)^2$ or $27^{2/3} = \sqrt[3]{27^2}$
$\qquad\qquad = 3^2 \qquad\qquad\qquad\qquad = \sqrt[3]{729}$
$\qquad\qquad = 9 \qquad\qquad\qquad\qquad\ = 9$

b. $\left(\dfrac{1}{16}\right)^{3/4} = \left(\sqrt[4]{\dfrac{1}{16}}\right)^3$ or $\left(\dfrac{1}{16}\right)^{3/4} = \sqrt[4]{\left(\dfrac{1}{16}\right)^3}$
$\qquad\qquad\quad = \left(\dfrac{1}{2}\right)^3 \qquad\qquad\qquad\qquad = \sqrt[4]{\dfrac{1}{4{,}096}}$
$\qquad\qquad\quad = \dfrac{1}{8} \qquad\qquad\qquad\qquad\quad = \dfrac{1}{8}$

COMMENT   To avoid large numbers, it is usually better to find the root of the base first, as shown in Example 4.

c. $(-8x^3)^{4/3} = \left(\sqrt[3]{-8x^3}\right)^4$ or $(-8x^3)^{3/4} = \sqrt[3]{(-8x^3)^4}$
$\qquad\qquad\quad = (-2x)^4 \qquad\qquad\qquad\qquad = \sqrt[3]{4{,}096x^{12}}$
$\qquad\qquad\quad = 16x^4 \qquad\qquad\qquad\qquad\ = 16x^4$

 SELF CHECK 4   Simplify: **a.** $16^{3/2}$ **b.** $(-27x^6)^{2/3}$

Accent on technology

▶ Rational Exponents

We can evaluate expressions containing rational exponents using the exponential key **y^x** or **x^y** on a scientific calculator. For example, to evaluate $10^{2/3}$, we enter

10 **y^x** (2 ÷ 3) =   4.641588834

Note that parentheses were used when entering the power. Without them, the calculator would interpret the entry as $10^2 \div 3$.

To evaluate the exponential expression using a graphing calculator, we use the **∧** key, which raises a base to a power. Again, we use parentheses when entering the power.

10 **∧** (2 ÷ 3) **ENTER**   10∧(2/3)
                                         4.641588834

To the nearest hundredth, $10^{2/3} \approx 4.64$.

For instructions regarding the use of a Casio graphing calculator, please refer to the Casio Keystroke Guide in the back of the book.

---

3   **Simplify an expression that contains a negative rational exponent.**

To be consistent with the definition of negative integer exponents, we define $x^{-m/n}$ as follows.

DEFINITION OF $x^{-m/n}$ If *m* and *n* are natural numbers, $\dfrac{m}{n}$ is in simplified form, and $x^{1/n}$ is a real number $(x \neq 0)$, then

$$x^{-m/n} = \dfrac{1}{x^{m/n}} \qquad \text{and} \qquad \dfrac{1}{x^{-m/n}} = x^{m/n}$$

EXAMPLE 5   Write each expression without negative exponents.

a. $64^{-1/2}$ **b.** $(16x^{-4})^{-1/4}$ **c.** $(-32x^5)^{-2/5}$ **d.** $(-16)^{-3/4}$

Solution   **a.** $64^{-1/2} = \dfrac{1}{64^{1/2}}$

$= \dfrac{1}{8}$

b. $(16x^{-4})^{-1/4} = 16^{-1/4}(x^{-4})^{-1/4}$

$= \dfrac{1}{16^{1/4}}x$

$= \dfrac{1}{2}x \quad \text{or} \quad \dfrac{x}{2}$

c. $(-32x^5)^{-2/5} = \dfrac{1}{(-32x^5)^{2/5}} \quad (x \neq 0)$

$= \dfrac{1}{[(-32x^5)^{1/5}]^2}$

$= \dfrac{1}{(-2x)^2}$

$= \dfrac{1}{4x^2}$

d. $(-16)^{-3/4}$ is not a real number, because $(-16)^{1/4}$ is not a real number.

SELF CHECK 5   Write each expression without negative exponents.

a. $25^{-3/2}$ **b.** $(-27a^{-3})^{-2/3}$ **c.** $(-81)^{-3/4}$

COMMENT   A base of 0 raised to a negative power is undefined, because 0^{-2} would equal $\dfrac{1}{0^2}$, which is undefined since we cannot divide by 0.

4   Simplify an expression that contains rational exponents by applying the properties of exponents.

We can use the properties of exponents to simplify many expressions with rational exponents.

EXAMPLE 6   Write all answers without negative exponents. Assume that all variables represent positive numbers. Thus, no absolute value symbols are necessary.

a. $5^{2/7}5^{3/7} = 5^{2/7+3/7}$ Apply the rule $x^m x^n = x^{m+n}$.

$= 5^{5/7}$ Add: $\dfrac{2}{7} + \dfrac{3}{7} = \dfrac{5}{7}$

b. $(5^{2/7})^3 = 5^{(2/7)(3)}$ Apply the rule $(x^m)^n = x^{mn}$.

$= 5^{6/7}$ Multiply: $\dfrac{2}{7}(3) = \dfrac{2}{7}\left(\dfrac{3}{1}\right) = \dfrac{6}{7}$

c. $(a^{2/3}b^{1/2})^6 = (a^{2/3})^6(b^{1/2})^6$ Apply the rule $(xy)^n = x^n y^n$.

$= a^{12/3}b^{6/2}$ Apply the rule $(x^m)^n = x^{mn}$ twice.

$= a^4 b^3$ Simplify the exponents.

d. $\dfrac{a^{8/3}a^{1/3}}{a^2} = a^{8/3+1/3-2}$ Apply the rules $x^m x^n = x^{m-n}$ and $\dfrac{x^m}{x^n} = x^{m-n}$.

$= a^{8/3+1/3-6/3}$ $2 = \dfrac{6}{3}$

$= a^{3/3}$ $\dfrac{8}{3} + \dfrac{1}{3} - \dfrac{6}{3} = \dfrac{3}{3}$

$= a$ $\dfrac{3}{3} = 1$

SELF CHECK 6   Simplify: **a.** $(x^{1/3}y^{3/2})^6$ **b.** $\dfrac{x^{5/3}x^{2/3}}{x^{1/3}}$

EXAMPLE 7 Assume that all variables represent positive numbers and perform the operations.

COMMENT Note that $a + a^{7/5} \neq a^{1+7/5}$. The expression $a + a^{7/5}$ cannot be simplified, because a and $a^{7/5}$ are not like terms.

a. $a^{4/5}(a^{1/5} + a^{3/5}) = a^{4/5}a^{1/5} + a^{4/5}a^{3/5}$ Use the distributive property to remove parentheses.

$= a^{4/5+1/5} + a^{4/5+3/5}$ Apply the rule $x^m x^n = x^{m+n}$.

$= a^{5/5} + a^{7/5}$ Simplify the exponents.

$= a + a^{7/5}$

b. $x^{1/2}(x^{-1/2} + x^{1/2}) = x^{1/2}x^{-1/2} + x^{1/2}x^{1/2}$ Use the distributive property to remove parentheses.

$= x^{1/2-1/2} + x^{1/2+1/2}$ Apply the rule $x^m x^n = x^{m+n}$.

$= x^0 + x^1$ Simplify.

$= 1 + x$ $x^0 = 1$

c. $(x^{2/3} + 1)(x^{2/3} - 1) = x^{4/3} - x^{2/3} + x^{2/3} - 1$ Use the distributive property to remove parentheses.

$= x^{4/3} - 1$ Combine like terms.

d. $(x^{1/2} + y^{1/2})^2 = (x^{1/2} + y^{1/2})(x^{1/2} + y^{1/2})$ $(x + y)^2 = (x + y)(x + y)$

$= x + 2x^{1/2}y^{1/2} + y$ Use the distributive property to remove parentheses.

SELF CHECK 7 Assume that all variables represent positive numbers and perform the operations.
a. $p^{1/5}(p^{4/5} + p^{2/5})$ **b.** $(p^{2/3} + q^{1/3})(p^{2/3} - q^{1/3})$

---

5 Simplify a radical expression by first writing it as an expression with a rational exponent.

We can simplify many radical expressions by using the following steps.

USING RATIONAL EXPONENTS TO SIMPLIFY RADICALS

1. Write the radical expression as an exponential expression with rational exponents.
2. Simplify the rational exponents.
3. Write the exponential expression as a radical.

EXAMPLE 8 Simplify. Assume variables represent positive values.
a. $\sqrt[4]{3^2}$ **b.** $\sqrt[8]{x^6}$ **c.** $\sqrt[9]{27x^6y^3}$

Solution **a.** $\sqrt[4]{3^2} = 3^{2/4}$ Apply the rule $\sqrt[n]{x^m} = x^{m/n}$.

$= 3^{1/2}$ $\frac{2}{4} = \frac{1}{2}$

$= \sqrt{3}$ Write using radical notation.

b. $\sqrt[8]{x^6} = x^{6/8}$ Apply the rule $\sqrt[n]{x^m} = x^{m/n}$.

$= x^{3/4}$ $\frac{6}{8} = \frac{3}{4}$

$= \sqrt[4]{x^3}$ Write using radical notation.

c. $\sqrt[9]{27x^6y^3} = (3^3x^6y^3)^{1/9}$ Write 27 as 3^3 and write the radical as an exponential expression.

$= 3^{3/9}x^{6/9}y^{3/9}$ Raise each factor to the $\frac{1}{9}$ power by multiplying the fractional exponents.

$= 3^{1/3}x^{2/3}y^{1/3}$ Simplify each fractional exponent.

$= (3x^2y)^{1/3}$ Apply the rule $(xy)^n = x^ny^n$.

$= \sqrt[3]{3x^2y}$ Write using radical notation.

SELF CHECK 8   Simplify. Assume variables represent positive values.

a. $\sqrt[6]{3^3}$ **b.** $\sqrt[10]{x^8}$ **c.** $\sqrt[4]{49x^2y^2}$

SELF CHECK ANSWERS

1. a. 4   **b.** $\frac{3}{2}$ **c.** $-2x$ **2. a.** $(4ab)^{1/6}$ **b.** $(a^2b^4)^{1/3}$ **3. a.** $5|a|$ **b.** b^2 **c.** $-2x$ **d.** $8|x|$

e. not a real number   **4. a.** 64   **b.** $9x^4$ **5. a.** $\frac{1}{125}$ **b.** $\frac{1}{9}a^2$ or $\frac{a^2}{9}$ **c.** not a real number   **6. a.** x^2y^9

b. x^2 **7. a.** $p + p^{3/5}$ **b.** $p^{4/3} - q^{2/3}$ **8. a.** $\sqrt{3}$ **b.** $\sqrt[5]{x^4}$ **c.** $\sqrt{7xy}$

NOW TRY THIS

1. Evaluate each exponential expression, if possible.

a. $-64^{-2/3}$ **b.** $-64^{-3/2}$ **c.** $(-64)^{-2/3}$

d. $(-64)^{-3/2}$

2. Simplify each expression and then write the answer using radical notation. Assume all variables represent positive numbers.

a. $x^{a/4} \cdot x^{a/2}$ **b.** $\dfrac{x^{n/m}}{x^{(n-1)/m}}$

3. Simplify $\dfrac{\sqrt{5}}{\sqrt[3]{5}}$ and write the answer using radical notation.

7.3   Exercises

Assume no division by zero.

WARM-UPS *Simplify.*

1. -7^2 **2.** -9^2

3. 5^3 **4.** 2^3

5. 4^{-2} **6.** 10^{-3}

7. $\sqrt{4^3}$ **8.** $\sqrt{2^4}$

9. $\sqrt[3]{8^2}$ **10.** $\sqrt[3]{3^2 \cdot 3}$

REVIEW *Solve each inequality.*

11. $8x - 7 \le 1$ **12.** $4(6 - 2y) < -16$

13. $\dfrac{4}{5}(r - 3) > \dfrac{2}{3}(r + 2)$ **14.** $-4 < 2x - 4 \le 8$

15. Mixing solutions How much water must be added to 5 pints of a 20% alcohol solution to dilute it to a 15% solution?

16. Selling apples A grocer bought some boxes of apples for $70. However, 4 boxes were spoiled. The grocer sold the remaining boxes at a profit of $2 each. How many boxes did the grocer sell if she managed to break even?

VOCABULARY AND CONCEPTS *Fill in the blanks.*

17. $a^4 = $ _____

18. $a^ma^n = $ ____

19. $(a^m)^n = $ ___

20. $(ab)^n = $ ____

21. $\left(\dfrac{a}{b}\right)^n = $ ___

22. $a^0 = $ _, provided $a \ne$ _.

23. $a^{-n} = $ ___ , provided $a \ne$ _.

24. $\dfrac{a^m}{a^n} = $ ____, provided $a \ne 0$.

25. $\left(\dfrac{a}{b}\right)^{-n} = $ ___

26. $x^{1/n} = $ ___

27. $(x^n)^{1/n} = $ ___, provided n is even.

28. $x^{m/n} = \sqrt[n]{x^m} = $ _____

GUIDED PRACTICE *Change each expression into radical notation.* (OBJECTIVE 1)

29. $5^{1/4}$ **30.** $26^{1/2}$

31. $8^{1/5}$ **32.** $29^{1/8}$

33. $(13a)^{1/7}$ **34.** $(4ab)^{1/6}$

35. $\left(\dfrac{1}{2}x^3y\right)^{1/4}$ **36.** $\left(\dfrac{3}{4}a^2b^2\right)^{1/5}$

37. $(6a^3b)^{1/4}$ **38.** $(5pq^2)^{1/3}$

39. $(x^2+y^2)^{1/2}$ **40.** $(a^4+b^4)^{1/4}$

Simplify each expression, if possible. SEE EXAMPLE 1. (OBJECTIVE 1)

41. $49^{1/2}$ **42.** $64^{1/2}$

43. $27^{1/3}$ **44.** $125^{1/3}$

45. $\left(\dfrac{1}{9}\right)^{1/2}$ **46.** $\left(\dfrac{1}{16}\right)^{1/2}$

47. $\left(\dfrac{1}{8}\right)^{1/3}$ **48.** $\left(\dfrac{1}{16}\right)^{1/4}$

49. $-81^{1/4}$ **50.** $-125^{1/3}$

51. $(-25)^{1/2}$ **52.** $(-216)^{1/2}$

Change each radical to an exponential expression. SEE EXAMPLE 2. (OBJECTIVE 1)

53. $\sqrt{7}$ **54.** $\sqrt[3]{12}$

55. $\sqrt[4]{3a}$ **56.** $\sqrt[5]{8xz}$

57. $5\sqrt[7]{b}$ **58.** $4\sqrt[3]{p}$

59. $\sqrt[6]{\dfrac{1}{7}abc}$ **60.** $\sqrt[7]{\dfrac{3}{8}p^2q}$

61. $\sqrt[5]{\dfrac{1}{2}mn}$ **62.** $\sqrt[8]{\dfrac{2}{7}p^2q}$

63. $\sqrt[3]{x^2+y^2}$ **64.** $\sqrt[4]{a^4-b^4}$

Simplify each expression. Assume that all variables can be any real number, and use absolute value symbols if necessary. SEE EXAMPLE 3. (OBJECTIVE 1)

65. $(25y^2)^{1/2}$ **66.** $(16x^4)^{1/4}$

67. $(243x^5)^{1/5}$ **68.** $(-27x^3)^{1/3}$

69. $[(x+1)^4]^{1/4}$ **70.** $[(x+5)^3]^{1/3}$

71. $(-81x^{12})^{1/4}$ **72.** $(-16x^4)^{1/2}$

Simplify each expression, if possible. SEE EXAMPLE 4. (OBJECTIVE 2)

73. $25^{3/2}$ **74.** $27^{2/3}$

75. $81^{3/4}$ **76.** $100^{3/2}$

77. $(-27x^3)^{4/3}$ **78.** $(-1,000\,x^{12})^{2/3}$

79. $\left(\dfrac{1}{8}\right)^{2/3}$ **80.** $\left(\dfrac{4}{9}\right)^{3/2}$

Write each expression without using negative exponents. Assume that all variables represent positive numbers. SEE EXAMPLE 5. (OBJECTIVE 3)

81. $4^{-1/2}$ **82.** $8^{-1/3}$

83. $4^{-3/2}$ **84.** $25^{-5/2}$

85. $(16x^2)^{-3/2}$ **86.** $(81c^4)^{-3/2}$

87. $(-27y^3)^{-2/3}$ **88.** $(-8z^9)^{-2/3}$

89. $\left(\dfrac{1}{4}\right)^{-3/2}$ **90.** $\left(\dfrac{4}{25}\right)^{-3/2}$

91. $\left(\dfrac{27}{8}\right)^{-4/3}$ **92.** $\left(\dfrac{25}{49}\right)^{-3/2}$

Perform the operations. Write answers without negative exponents. Assume that all variables represent positive numbers. SEE EXAMPLE 6. (OBJECTIVE 4)

93. $5^{4/9}5^{4/9}$ **94.** $4^{2/5}4^{2/5}$

95. $(4^{1/5})^3$ **96.** $(3^{1/3})^5$

97. $6^{-2/3}6^{-4/3}$ **98.** $5^{1/3}5^{-5/3}$

99. $\dfrac{9^{4/5}}{9^{3/5}}$ **100.** $\dfrac{7^{2/3}}{7^{1/2}}$

101. $\dfrac{7^{1/2}}{7^0}$ **102.** $\dfrac{3^{4/3}3^{1/3}}{3^{2/3}}$

103. $\dfrac{2^{5/6}2^{1/3}}{2^{1/2}}$ **104.** $\dfrac{5^{1/3}5^{1/2}}{5^{1/3}}$

105. $(x^{1/4}\cdot x^{3/2})^8$ **106.** $(a^{1/3}\cdot a^{3/4})^{12}$

107. $(a^{2/3})^{1/3}$ **108.** $(t^{4/5})^{10}$

Perform the operations. Write answers without negative exponents. Assume that all variables represent positive numbers. SEE EXAMPLE 7. (OBJECTIVE 4)

109. $y^{1/3}(y^{2/3}+y^{5/3})$ **110.** $y^{2/5}(y^{-2/5}+y^{3/5})$

111. $x^{3/5}(x^{7/5}-x^{2/5}+1)$ **112.** $(x^{1/2}+2)(x^{1/2}-2)$

113. $(x^{1/2}+y^{1/2})(x^{1/2}-y^{1/2})$ **114.** $(x^{2/3}-x)(x^{2/3}+x)$

115. $(x^{2/3}+y^{2/3})^2$ **116.** $(a^{3/2}-b^{3/2})^2$

Use rational exponents to simplify each radical. Assume that all variables represent positive numbers. SEE EXAMPLE 8. (OBJECTIVE 5)

117. $\sqrt[6]{p^3}$ **118.** $\sqrt[8]{q^2}$

119. $\sqrt[4]{25b^2}$ **120.** $\sqrt[9]{-8x^6}$

ADDITIONAL PRACTICE *Simplify each expression, if possible. Assume all variables represent positive numbers. Write answers without negative exponents.*

121. $16^{1/4}$ **122.** $625^{1/4}$

123. $32^{1/5}$ **124.** $0^{1/5}$

125. $0^{1/3}$ **126.** $(-243)^{1/5}$

127. $(-27)^{1/3}$ **128.** $(-125)^{1/3}$

129. $(25x^4)^{3/2}$ **130.** $(27a^3b^3)^{2/3}$

131. $\left(\dfrac{8x^3}{27}\right)^{2/3}$ **132.** $\left(\dfrac{27}{64y^6}\right)^{2/3}$

133. $(-32p^5)^{-2/5}$ **134.** $(16q^6)^{-5/2}$

135. $\left(-\dfrac{8x^3}{27}\right)^{-1/3}$ **136.** $\left(\dfrac{16}{81y^4}\right)^{-3/4}$

137. $(a^{1/2}b^{1/3})^{3/2}$ **138.** $(a^{3/5}b^{3/2})^{2/3}$

139. $(mn^{-2/3})^{-3/5}$

140. $(r^{-2}s^3)^{1/3}$

141. $\dfrac{(4x^3y)^{1/2}}{(9xy)^{1/2}}$

142. $\dfrac{(27x^3y)^{1/3}}{(8xy^2)^{2/3}}$

143. $(27x^{-3})^{-1/3}$

144. $(16a^{-2})^{-1/2}$

145. $x^{4/3}(x^{2/3} + 3x^{5/3} - 4)$

146. $(x^{1/3} + x^2)(x^{1/3} - x^2)$

147. $(x^{-1/2} - x^{1/2})^2$

148. $(a^{1/2} - b^{2/3})^2$

Use a calculator to evaluate each expression. Round to the nearest hundredth.

149. $15^{1/3}$

150. $50.5^{1/4}$

151. $1.045^{1/5}$

152. $(-1,000)^{2/5}$

Use a calculator to evaluate each expression. Round to the nearest hundredth.

153. $17^{-1/2}$

154. $2.45^{-2/3}$

155. $(-0.25)^{-1/5}$

156. $(-17.1)^{-3/7}$

WRITING ABOUT MATH

157. Explain how you would decide whether $a^{1/n}$ is a real number.

158. The expression $(a^{1/2} + b^{1/2})^2$ is not equal to $a + b$. Explain.

SOMETHING TO THINK ABOUT

159. The fraction $\frac{2}{4}$ is equal to $\frac{1}{2}$. Is $16^{2/4}$ equal to $16^{1/2}$? Explain.

160. How would you evaluate an expression with a mixed-number exponent? For example, what is $8^{1\frac{1}{3}}$? What is $25^{2\frac{1}{2}}$? Explain.

Section 7.4

Simplifying and Combining Radical Expressions

Objectives

1. Simplify a radical expression by applying the properties of radicals.
2. Add and subtract two or more radical expressions.
3. Find the length of a side of a 30°–60°–90° triangle and a 45°–45°–90° triangle.

Vocabulary

like (similar) radicals   altitude

Getting Ready

Simplify each radical. Assume that all variables represent positive numbers.

1. $\sqrt{225}$ **2.** $\sqrt{576}$ **3.** $\sqrt[3]{125}$ **4.** $\sqrt[3]{343}$

5. $\sqrt{16x^4}$ **6.** $\sqrt{\dfrac{64}{121}x^6}$ **7.** $\sqrt[3]{27a^3b^9}$ **8.** $\sqrt[3]{-8a^{12}}$

In this section, we will introduce the multiplication and division properties of radicals and use them to simplify radical expressions. Then we will add and subtract radical expressions and apply these techniques to find the lengths of the sides of special right triangles.

1   Simplify a radical expression by applying the properties of radicals.

Many properties of exponents have counterparts in radical notation. Because $a^{1/n}b^{1/n} = (ab)^{1/n}$, we have

(1) $\sqrt[n]{a}\sqrt[n]{b} = \sqrt[n]{ab}$

For example,

$$\sqrt{5}\,\sqrt{5} = \sqrt{5 \cdot 5} = \sqrt{5^2} = 5$$

$$\sqrt[3]{7x}\,\sqrt[3]{49x^2} = \sqrt[3]{7x \cdot 7^2x^2} = \sqrt[3]{7^3 \cdot x^3} = 7x$$

$$\sqrt[4]{2x^3}\,\sqrt[4]{8x} = \sqrt[4]{2x^3 \cdot 2^3x} = \sqrt[4]{2^4 \cdot x^4} = 2x \quad (x > 0)$$

If we write Equation 1 in a different order, we have the following rule.

MULTIPLICATION PROPERTY OF RADICALS	If $\sqrt[n]{a}$ and $\sqrt[n]{b}$ are real numbers, then $$\sqrt[n]{ab} = \sqrt[n]{a}\sqrt[n]{b}$$

Carl Friedrich Gauss
1777–1855

Many people consider Gauss to be the greatest mathematician of all time. He made contributions in the areas of number theory, solutions of equations, geometry of curved surfaces, and statistics. For his efforts, he earned the title "Prince of the Mathematicians."

If all radicals represent real numbers, *the nth root of the product of two numbers is equal to the product of their nth roots.*

COMMENT   The multiplication property of radicals applies to the nth root of the product of two numbers. There is no such property for sums or differences. A radical symbol is a grouping symbol. Thus, any addition or subtraction within the radicand must be completed first.

A second property of radicals involves quotients. Because

$$\frac{a^{1/n}}{b^{1/n}} = \left(\frac{a}{b}\right)^{1/n} \quad (b \neq 0)$$

it follows that

(2) $\dfrac{\sqrt[n]{a}}{\sqrt[n]{b}} = \sqrt[n]{\dfrac{a}{b}} \quad (b \neq 0)$

For example,

$$\frac{\sqrt{8x^3}}{\sqrt{2x}} = \sqrt{\frac{8x^3}{2x}} = \sqrt{4x^2} = 2x \quad (x > 0)$$

$$\frac{\sqrt[3]{54x^5}}{\sqrt[3]{2x^2}} = \sqrt[3]{\frac{54x^5}{2x^2}} = \sqrt[3]{27x^3} = 3x \quad (x \neq 0)$$

If we write Equation 2 in a different order, we have the following rule.

DIVISION PROPERTY OF RADICALS	If $\sqrt[n]{a}$ and $\sqrt[n]{b}$ are real numbers, then $$\sqrt[n]{\frac{a}{b}} = \frac{\sqrt[n]{a}}{\sqrt[n]{b}} \quad (b \neq 0)$$

As long as all radicals represent real numbers, *the nth root of the quotient of two numbers is equal to the quotient of their nth roots.*

A radical expression is said to be in simplified form when each of the following statements is true.

SIMPLIFIED FORM OF A RADICAL EXPRESSION	A radical expression is in simplified form when

1. Each numerical factor in the radicand, in prime factored form, appears to a power that is less than the index of the radical.

2. Each variable factor in the radicand appears to a power that is less than the index of the radical.

3. The radicand contains no fractions or negative numbers.

4. No radicals appear in the denominator of a fraction.

EXAMPLE 1 Simplify: **a.** $\sqrt{12}$ **b.** $\sqrt{98}$ **c.** $\sqrt[3]{54}$

Solution **a.** Recall that squares of integers, such as $1, 4, 9, 16, 25,$ and 36, are *perfect squares.* To simplify $\sqrt{12}$, we factor 12 so that one factor is the largest perfect square that divides 12. Since 4 is the largest perfect-square factor of 12, we write 12 as $4 \cdot 3$, use the multiplication property of radicals, and simplify.

COMMENT Use a factoring tree.

a.

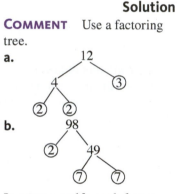

$$\sqrt{12} = \sqrt{\mathbf{4 \cdot 3}} \qquad \text{Write 12 as } 4 \cdot 3.$$
$$= \sqrt{4}\sqrt{3} \qquad \sqrt{4 \cdot 3} = \sqrt{4}\sqrt{3}$$
$$= 2\sqrt{3} \qquad \sqrt{4} = 2$$

b. Since the largest perfect-square factor of 98 is 49, we have

$$\sqrt{98} = \sqrt{\mathbf{49 \cdot 2}} \qquad \text{Write 98 as } 49 \cdot 2.$$
$$= \sqrt{49}\sqrt{2} \qquad \sqrt{49 \cdot 2} = \sqrt{49}\sqrt{2}$$
$$= 7\sqrt{2} \qquad \sqrt{49} = 7$$

In parts **a** and **b**, each factor pair produces a single number outside the radical.

c.

In part **c**, each factor triple produces a single number outside the radical.

c. Numbers that are cubes of integers, such as $1, 8, 27, 64, 125,$ and 216, are called *perfect cubes.* Since the largest perfect-cube factor of 54 is 27, we have

$$\sqrt[3]{54} = \sqrt[3]{\mathbf{27 \cdot 2}} \qquad \text{Write 54 as } 27 \cdot 2.$$
$$= \sqrt[3]{27}\sqrt[3]{2} \qquad \sqrt[3]{27 \cdot 2} = \sqrt[3]{27}\sqrt[3]{2}$$
$$= 3\sqrt[3]{2} \qquad \sqrt[3]{27} = 3$$

SELF CHECK 1 Simplify: **a.** $\sqrt{20}$ **b.** $\sqrt[3]{24}$

EXAMPLE 2 Simplify: **a.** $\sqrt{\dfrac{15}{49x^2}}$ $(x > 0)$ **b.** $\sqrt[3]{\dfrac{10x^2}{27y^6}}$ $(y \neq 0)$

Solution **a.** We can write the square root of the quotient as the quotient of the square roots and simplify the denominator. Since $x > 0$, we have

$$\sqrt{\frac{15}{49x^2}} = \frac{\sqrt{15}}{\sqrt{49x^2}}$$
$$= \frac{\sqrt{15}}{7x}$$

b. We can write the cube root of the quotient as the quotient of two cube roots. Since $y \neq 0$, we have

$$\sqrt[3]{\frac{10x^2}{27y^6}} = \frac{\sqrt[3]{10x^2}}{\sqrt[3]{27y^6}}$$

$$= \frac{\sqrt[3]{10x^2}}{3y^2}$$

SELF CHECK 2   Simplify: **a.** $\sqrt{\dfrac{11}{36a^2}}$ $(a > 0)$ **b.** $\sqrt[3]{\dfrac{8a^2}{125y^3}}$ $(y \neq 0)$

EXAMPLE 3   Simplify each expression. Assume that all variables represent positive numbers.

   a. $\sqrt{128a^5}$ **b.** $\sqrt[3]{24x^5}$ **c.** $\dfrac{\sqrt{45xy^2}}{\sqrt{5x}}$ **d.** $\dfrac{\sqrt[3]{-432x^5}}{\sqrt[3]{8x}}$

Solution   **a.** We can write $128a^5$ as $64a^4 \cdot 2a$ and use the multiplication property of radicals.

$$\sqrt{128a^5} = \sqrt{64a^4 \cdot 2a} \qquad \text{\color{red}{$64a^4$ is the largest perfect square that divides $128a^5$.}}$$

$$= \sqrt{64a^4}\sqrt{2a} \qquad \text{\color{red}{Use the multiplication property of radicals.}}$$

$$= 8a^2\sqrt{2a} \qquad \text{\color{red}{$\sqrt{64a^4} = 8a^2$}}$$

COMMENT Be sure to include the index of radicals in each step to avoid errors.

b. We can write $24x^5$ as $8x^3 \cdot 3x^2$ and use the multiplication property of radicals.

$$\sqrt[3]{24x^5} = \sqrt[3]{8x^3 \cdot 3x^2} \qquad \text{\color{red}{$8x^3$ is the largest perfect cube that divides $24x^5$.}}$$

$$= \sqrt[3]{8x^3}\sqrt[3]{3x^2} \qquad \text{\color{red}{Use the multiplication property of radicals.}}$$

$$= 2x\sqrt[3]{3x^2} \qquad \text{\color{red}{$\sqrt[3]{8x^3} = 2x$}}$$

c. We can write the quotient of the square roots as the square root of a quotient.

$$\frac{\sqrt{45xy^2}}{\sqrt{5x}} = \sqrt{\frac{45xy^2}{5x}} \qquad \text{\color{red}{Use the quotient property of radicals.}}$$

$$= \sqrt{9y^2} \qquad \text{\color{red}{Simplify the fraction.}}$$

$$= 3y$$

d. We can write the quotient of the cube roots as the cube root of a quotient.

$$\frac{\sqrt[3]{-432x^5}}{\sqrt[3]{8x}} = \sqrt[3]{\frac{-432x^5}{8x}} \qquad \text{\color{red}{Use the quotient property of radicals.}}$$

$$= \sqrt[3]{-54x^4} \qquad \text{\color{red}{Simplify the fraction.}}$$

$$= \sqrt[3]{-27x^3 \cdot 2x} \qquad \text{\color{red}{$-27x^3$ is the largest perfect cube that divides $-54x^4$.}}$$

$$= \sqrt[3]{-27x^3}\sqrt[3]{2x} \qquad \text{\color{red}{Use the multiplication property of radicals.}}$$

$$= -3x\sqrt[3]{2x}$$

SELF CHECK 3   Simplify each expression. Assume that all variables represent positive numbers.

   a. $\sqrt{98b^3}$ **b.** $\sqrt[3]{54y^5}$ **c.** $\dfrac{\sqrt{50ab^2}}{\sqrt{2a}}$

To simplify more complicated radicals, we can use the prime factorization of the radicand to find its perfect-square factors. For example, to simplify $\sqrt{3{,}168x^5y^7}$, we first find the prime factorization of $3{,}168x^5y^7$.

$$3{,}168x^5y^7 = 2^5 \cdot 3^2 \cdot 11 \cdot x^5 \cdot y^7$$

Then we have

$$\sqrt{3{,}168x^5y^7} = \sqrt{2^4 \cdot 3^2 \cdot x^4 \cdot y^6 \cdot 2 \cdot 11 \cdot x \cdot y}$$

$$= \sqrt{2^4 \cdot 3^2 \cdot x^4 \cdot y^6}\sqrt{2 \cdot 11 \cdot x \cdot y}$$ Write each perfect square under the left radical and each nonperfect square under the right radical.

$$= 2^2 \cdot 3x^2y^3\sqrt{22xy}$$

$$= 12x^2y^3\sqrt{22xy}$$

2 Add and subtract two or more radical expressions.

Radical expressions with the same index and the same radicand are called **like** or **similar radicals**. For example, $3\sqrt{2}$ and $5\sqrt{2}$ are like radicals. However,

$3\sqrt{5}$ and $5\sqrt{2}$ are not like radicals, because the radicands are different.

$3\sqrt{5}$ and $2\sqrt[3]{5}$ are not like radicals, because the indexes are different.

We often can combine like terms. For example, to simplify the expression $3\sqrt{2} + 2\sqrt{2}$, we use the distributive property to factor out $\sqrt{2}$ and simplify.

$$3\sqrt{2} + 2\sqrt{2} = (3 + 2)\sqrt{2}$$

$$= 5\sqrt{2}$$

Radicals with the same index but different radicands can often be written as like radicals. For example, to simplify the expression $\sqrt{27} - \sqrt{12}$, we simplify both radicals and combine the resulting like radicals.

$$\sqrt{27} - \sqrt{12} = \sqrt{9 \cdot 3} - \sqrt{4 \cdot 3}$$

$$= \sqrt{9}\sqrt{3} - \sqrt{4}\sqrt{3}$$ Apply the multiplication property of radicals.

$$= 3\sqrt{3} - 2\sqrt{3}$$ Simplify.

$$= (3 - 2)\sqrt{3}$$ Factor out $\sqrt{3}$, the GCF.

$$= \sqrt{3}$$

As the previous examples suggest, we can use the following procedure to add or subtract radicals.

| **ADDING AND SUBTRACTING RADICALS** | To add or subtract radicals, simplify each radical and combine all like radicals by adding the coefficients and keeping the common radical. |

EXAMPLE 4   Simplify: $2\sqrt{12} - 3\sqrt{48} + 3\sqrt{3}$

Solution   We simplify each radical separately and combine like radicals.

$$2\sqrt{12} - 3\sqrt{48} + 3\sqrt{3} = 2\sqrt{4 \cdot 3} - 3\sqrt{16 \cdot 3} + 3\sqrt{3}$$

$$= 2\sqrt{4}\sqrt{3} - 3\sqrt{16}\sqrt{3} + 3\sqrt{3}$$

$$= 2(2)\sqrt{3} - 3(4)\sqrt{3} + 3\sqrt{3}$$

$$= 4\sqrt{3} - 12\sqrt{3} + 3\sqrt{3}$$

$$= (4 - 12 + 3)\sqrt{3}$$

$$= -5\sqrt{3}$$

SELF CHECK 4   Simplify: $3\sqrt{75} - 2\sqrt{12} + 2\sqrt{48}$

EXAMPLE 5   Simplify: $\sqrt[3]{16} - \sqrt[3]{54} + \sqrt[3]{24}$

Solution   We simplify each radical separately and combine like radicals.

COMMENT   We cannot combine $-\sqrt[3]{2}$ and $2\sqrt[3]{3}$, because the radicals have different radicands.

$$
\begin{aligned}
\sqrt[3]{16} - \sqrt[3]{54} + \sqrt[3]{24} &= \sqrt[3]{8 \cdot 2} - \sqrt[3]{27 \cdot 2} + \sqrt[3]{8 \cdot 3} \\
&= \sqrt[3]{8}\sqrt[3]{2} - \sqrt[3]{27}\sqrt[3]{2} + \sqrt[3]{8}\sqrt[3]{3} \\
&= 2\sqrt[3]{2} - 3\sqrt[3]{2} + 2\sqrt[3]{3} \\
&= -\sqrt[3]{2} + 2\sqrt[3]{3}
\end{aligned}
$$

SELF CHECK 5   Simplify: $\sqrt[3]{24} - \sqrt[3]{16} + \sqrt[3]{54}$

Remember to include the index of radicals in each step.

EXAMPLE 6   Simplify: $\sqrt[3]{16x^4} + \sqrt[3]{54x^4} - \sqrt[3]{-128x^4}$

Solution   We simplify each radical separately, factor out $\sqrt[3]{2x}$, and combine like radicals.

$$
\begin{aligned}
\sqrt[3]{16x^4} &+ \sqrt[3]{54x^4} - \sqrt[3]{-128x^4} \\
&= \sqrt[3]{8x^3 \cdot 2x} + \sqrt[3]{27x^3 \cdot 2x} - \sqrt[3]{-64x^3 \cdot 2x} \\
&= \sqrt[3]{8x^3}\sqrt[3]{2x} + \sqrt[3]{27x^3}\sqrt[3]{2x} - \sqrt[3]{-64x^3}\sqrt[3]{2x} \\
&= 2x\sqrt[3]{2x} + 3x\sqrt[3]{2x} + 4x\sqrt[3]{2x} \\
&= (2x + 3x + 4x)\sqrt[3]{2x} \\
&= 9x\sqrt[3]{2x}
\end{aligned}
$$

SELF CHECK 6   Simplify: $\sqrt{32x^3} + \sqrt{50x^3} - \sqrt{18x^3}$ $(x > 0)$

3   Find the length of a side of a 30°−60°−90° triangle and a 45°−45°−90° triangle.

An isosceles right triangle is a right triangle with two legs of equal length. If we know the length of one leg of an isosceles right triangle, we can use the Pythagorean theorem to find the length of the hypotenuse. Since the triangle shown in Figure 7-10 is a right triangle, we have

Figure 7-10

$$
\begin{aligned}
c^2 &= a^2 + a^2 && \text{Use the Pythagorean theorem.} \\
c^2 &= 2a^2 && \text{Combine like terms.} \\
c &= \sqrt{2a^2} && \text{Since } c \text{ represents a length, take the positive square root.} \\
c &= a\sqrt{2} && \sqrt{2a^2} = \sqrt{2}\sqrt{a^2} = \sqrt{2}a = a\sqrt{2}. \text{ No absolute value symbols are needed,} \\
& && \text{because } a \text{ is positive.}
\end{aligned}
$$

Thus, *in an isosceles right triangle, the length of the hypotenuse is the length of one leg times* $\sqrt{2}$.

EXAMPLE 7   **GEOMETRY**   If one leg of the isosceles right triangle shown in Figure 7-10 is 10 feet long, find the length of the hypotenuse.

Solution   Since the length of the hypotenuse is the length of a leg times $\sqrt{2}$, we have

$$c = 10\sqrt{2}$$

The length of the hypotenuse is $10\sqrt{2}$ feet. To two decimal places, the length is 14.14 feet.

SELF CHECK 7   Find the length of the hypotenuse of an isosceles right triangle if one leg is 12 meters long.

If the length of the hypotenuse of an isosceles right triangle is known, we can use the Pythagorean theorem to find the length of each leg.

EXAMPLE 8   **GEOMETRY**   Approximate the length of each leg of the isosceles right triangle shown in Figure 7-11 to two decimal places.

Solution   We use the Pythagorean theorem.

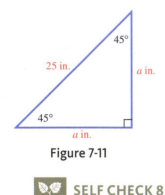

Figure 7-11

$$c^2 = a^2 + a^2$$

$$25^2 = 2a^2 \qquad \text{Substitute 25 for } c \text{ and combine like terms.}$$

$$\frac{625}{2} = a^2 \qquad \text{Square 25 and divide both sides by 2.}$$

$$\sqrt{\frac{625}{2}} = a \qquad \text{Since } a \text{ represents a length, take the positive square root.}$$

$$a \approx 17.67766953 \qquad \text{Use a calculator.}$$

To two decimal places, the length is 17.68 inches.

SELF CHECK 8   If the length of the hypotenuse of an isosceles right triangle is 48 inches, approximate the length of each leg to two decimal places.

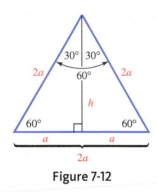

Figure 7-12

From geometry, we know that an *equilateral triangle* is a triangle with three sides of equal length and three 60° angles. An **altitude** is the perpendicular distance from a vertex of a triangle to the opposite side. If an altitude is drawn upon the base of an equilateral triangle, as shown in Figure 7-12, it bisects the base and divides the triangle into two 30°–60°–90° triangles. We can see that the shortest leg of each 30°–60°–90° triangle is a units long. Recall that in any triangle, the side opposite the smallest angle is the shortest side and the side opposite the largest angle is the longest. Thus,

The length of the shorter leg of a 30°–60°–90° *triangle is half as long as its hypotenuse.*

We can find the length of the altitude, h, by using the Pythagorean theorem.

$$a^2 + h^2 = (2a)^2$$

$$a^2 + h^2 = 4a^2 \qquad (2a)^2 = (2a)(2a) = 4a^2$$

$$h^2 = 3a^2 \qquad \text{Subtract } a^2 \text{ from both sides.}$$

$$h = \sqrt{3a^2} \qquad \text{Since } h \text{ represents a length, take the positive square root.}$$

$$h = a\sqrt{3} \qquad \sqrt{3a^2} = \sqrt{3}\sqrt{a^2} = a\sqrt{3}. \text{ No absolute value symbols are needed, because } a \text{ is positive.}$$

COMMENT   In summary, the lengths of the sides are as follows:

30°–60°–90°; $s, s\sqrt{3}, 2s$

45°–45°–90°; $s, s, s\sqrt{2}$

Thus,

The length of the longer leg is the length of the shorter leg times $\sqrt{3}$ *and the length of the hypotenuse is twice the length of the shorter leg.*

EXAMPLE 9   **GEOMETRY**   Find the length of the hypotenuse and the longer leg of the right triangle shown in Figure 7-13.

Solution   The known length of 6 cm is opposite the 30° angle. Since the shorter leg of the triangle is half as long as its hypotenuse, the hypotenuse is 12 centimeters long.

Since the length of the longer leg is the length of the shorter leg times $\sqrt{3}$, the longer leg is $6\sqrt{3}$ (about 10.39) centimeters long.

Figure 7-13

🍃 **SELF CHECK 9** Find the length of the hypotenuse and the longer leg of a 30°–60°–90° right triangle if the shorter leg is 8 centimeters long.

EXAMPLE 10 **GEOMETRY** Find the length of each leg of the triangle shown in Figure 7-14.

Solution Since the shorter leg of a 30°–60°–90° triangle is half as long as its hypotenuse, the shorter leg is $\frac{9}{2}$ centimeters long.

Since the length of the longer leg is the length of the shorter leg times $\sqrt{3}$, the longer leg is $\frac{9}{2}\sqrt{3}$ (or about 7.79) centimeters long.

Figure 7-14

🍃 **SELF CHECK 10** Find the length of the longer leg of a right triangle if its hypotenuse is 10 cm long.

🍃 **SELF CHECK ANSWERS**

1. a. $2\sqrt{5}$ b. $2\sqrt[3]{3}$ 2. a. $\frac{\sqrt{11}}{6a}$ b. $\frac{2\sqrt[3]{a^2}}{5y}$ 3. a. $7b\sqrt{2b}$ b. $3y\sqrt[3]{2y^2}$ c. $5b$ 4. $19\sqrt{3}$ 5. $2\sqrt[3]{3}+\sqrt[3]{2}$
6. $6x\sqrt{2x}$ 7. $12\sqrt{2}$ m 8. To two decimal places, the length is 33.94 inches. 9. 16 cm, $8\sqrt{3}$ cm
10. $5\sqrt{3}$ cm

NOW TRY THIS

Simplify.

1. $\sqrt[3]{8x-8}+\sqrt[3]{x-1}$

2. $\dfrac{5\sqrt[3]{32}}{2}$

7.4 Exercises 🍃

WARM-UPS *Factor such that one factor is the largest perfect square.*

1. 28
2. 24
3. 18
4. 50

Factor such that one factor is the largest perfect cube.

5. 16
6. 40
7. 54
8. 108

Combine like terms.

9. $12x^2+12x^2$
10. $20y^3-17y^3$

REVIEW *Perform each operation.*

11. $7x^3y(-3x^4y^{-5})$

12. $-2a^2b^{-2}(4a^{-2}b^4-2a^2b+3a^3b^2)$
13. $(3t+2)^2$
14. $(5r-3s)(5r+2s)$
15. $2p-5\overline{)6p^2-7p-25}$
16. $3m+n\overline{)6m^3-m^2n+2mn^2+n^3}$

VOCABULARY AND CONCEPTS *Fill in the blanks. Assume a and b are positive numbers and n is a natural number greater than 1.*

17. $\sqrt[n]{ab}=$ _____

18. $\sqrt[n]{\dfrac{a}{b}}=$ _____

19. If two radicals have the same index and the same radicand, they are called ___ radicals.

20. The perpendicular distance from a vertex of a triangle to the opposite side is called an _____.

GUIDED PRACTICE *Simplify each expression. Assume that all variables represent positive numbers.* (OBJECTIVE 1)

21. $\sqrt{6}\sqrt{6}$

22. $\sqrt{11}\sqrt{11}$

23. $\sqrt{t}\sqrt{t}$

24. $-\sqrt{z}\sqrt{z}$

25. $\sqrt[3]{5x^2}\sqrt[3]{25x}$

26. $\sqrt[4]{25a}\sqrt[4]{25a^3}$

27. $\dfrac{\sqrt{500}}{\sqrt{5}}$

28. $\dfrac{\sqrt{128}}{\sqrt{2}}$

29. $\dfrac{\sqrt{98x^3}}{\sqrt{2x}}$

30. $\dfrac{\sqrt{125a^5}}{\sqrt{5a}}$

31. $\dfrac{\sqrt{180ab^4}}{\sqrt{5ab^2}}$

32. $\dfrac{\sqrt{112ab^3}}{\sqrt{7ab}}$

33. $\dfrac{\sqrt[3]{48}}{\sqrt[3]{6}}$

34. $\dfrac{\sqrt[3]{64}}{\sqrt[3]{8}}$

35. $\dfrac{\sqrt[3]{189a^4}}{\sqrt[3]{7a}}$

36. $\dfrac{\sqrt[3]{243x^7}}{\sqrt[3]{9x}}$

Simplify. SEE EXAMPLE 1. (OBJECTIVE 1)

37. $\sqrt{20}$

38. $\sqrt{8}$

39. $-\sqrt{200}$

40. $-\sqrt{250}$

41. $\sqrt[3]{80}$

42. $\sqrt[3]{270}$

43. $\sqrt[3]{-81}$

44. $\sqrt[3]{-72}$

45. $\sqrt[4]{32}$

46. $\sqrt[4]{48}$

47. $\sqrt[5]{96}$

48. $\sqrt[7]{256}$

Simplify. Assume no denominators are 0. SEE EXAMPLE 2. (OBJECTIVE 1)

49. $\sqrt{\dfrac{5}{49a^2}}$

50. $\sqrt{\dfrac{11}{36x^2}}$

51. $\sqrt[3]{\dfrac{7a^3}{64}}$

52. $\sqrt[3]{\dfrac{4b^3}{125}}$

53. $\sqrt[4]{\dfrac{3p^4}{10,000q^4}}$

54. $\sqrt[5]{\dfrac{4r^5}{243s^{10}}}$

55. $\sqrt[5]{\dfrac{3m^{15}}{32n^{10}}}$

56. $\sqrt[6]{\dfrac{5a^6}{64b^{12}}}$

Simplify. Assume that all variables represent positive numbers. SEE EXAMPLE 3. (OBJECTIVE 1)

57. $\sqrt{63y^3}$

58. $\sqrt{125x^3}$

59. $\sqrt{48a^5}$

60. $\sqrt{160b^5}$

61. $-\sqrt{112a^3}$

62. $\sqrt{147a^5}$

63. $\sqrt{175a^2b^3}$

64. $\sqrt{128a^3b^5}$

65. $\dfrac{-\sqrt{150ab^4}}{\sqrt{2a}}$

66. $\dfrac{\sqrt{360xy^3}}{\sqrt{3xy}}$

67. $\dfrac{\sqrt[3]{-108x^6y}}{\sqrt[3]{2y}}$

68. $\dfrac{-\sqrt[3]{-243a^4b}}{\sqrt[3]{3ab}}$

69. $\sqrt[3]{16x^{12}y^3}$

70. $\sqrt[3]{40a^3b^6}$

71. $\sqrt{\dfrac{z^2}{16x^2}}$

72. $\sqrt{\dfrac{b^4}{64a^8}}$

Simplify. SEE EXAMPLE 4. (OBJECTIVE 2)

73. $\sqrt{3} + \sqrt{27}$

74. $\sqrt{8} + \sqrt{32}$

75. $\sqrt{2} - \sqrt{8}$

76. $\sqrt{20} - \sqrt{125}$

77. $\sqrt{98} - \sqrt{50}$

78. $\sqrt{72} - \sqrt{200}$

79. $3\sqrt{125} + 4\sqrt{45}$

80. $2\sqrt{72} + 5\sqrt{32}$

81. $3\sqrt{54} + 2\sqrt{28} - 4\sqrt{96}$

82. $4\sqrt{300} - 2\sqrt{108} + 3\sqrt{200}$

Simplify. SEE EXAMPLE 5. (OBJECTIVE 2)

83. $\sqrt[3]{24} + \sqrt[3]{81}$

84. $\sqrt[3]{16} + \sqrt[3]{128}$

85. $\sqrt[3]{32} - \sqrt[3]{108}$

86. $\sqrt[3]{80} - \sqrt[3]{10,000}$

87. $2\sqrt[3]{125} - 5\sqrt[3]{64}$

88. $\sqrt[3]{81} - \sqrt[3]{24}$

89. $2\sqrt[3]{16} - \sqrt[3]{54} - 3\sqrt[3]{128}$

90. $\sqrt[3]{250} - 4\sqrt[3]{5} + \sqrt[3]{16}$

Simplify. All variables represent positive numbers. SEE EXAMPLE 6. (OBJECTIVE 2)

91. $\sqrt[3]{81x^5} - \sqrt[3]{24x^5}$

92. $\sqrt[3]{16x^4} - \sqrt[3]{54x^4}$

93. $\sqrt{25yz^2} + \sqrt{9yz^2}$

94. $\sqrt{36xy^2} + \sqrt{49xy^2}$

95. $\sqrt{y^5} - \sqrt{9y^5} - \sqrt{25y^5}$

96. $\sqrt{8y^7} + \sqrt{32y^7} - \sqrt{2y^7}$

97. $3\sqrt[3]{-2x^4} - \sqrt[3]{54x^4}$

98. $2\sqrt[3]{320a^5} - 2\sqrt[3]{-135a^5}$

Use the figure for Exercises 99–106. Find the lengths of the remaining sides of the triangle. SEE EXAMPLE 7. (OBJECTIVE 3)

99. $b = \dfrac{2}{3}$

100. $b = \dfrac{3}{10}$

101. $a = 5\sqrt{2}$

102. $a = 12\sqrt{2}$

Find the lengths of the remaining sides of the triangle. SEE EXAMPLE 8. (OBJECTIVE 3)

103. $c = 5\sqrt{2}$

104. $c = 8\sqrt{2}$

105. $c = 7\sqrt{2}$

106. $c = 16\sqrt{2}$

Use the figure for Exercises 107–114. Find the lengths of the remaining sides of the triangle. **SEE EXAMPLE 9. (OBJECTIVE 3)**

107. $a = 5$

108. $a = 8$

109. $b = 9\sqrt{3}$

110. $b = 18\sqrt{3}$

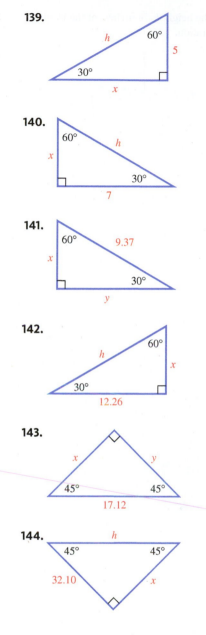

Find the lengths of the remaining sides of the triangle. **SEE EXAMPLE 10. (OBJECTIVE 3)**

111. $c = 24$

112. $c = 8$

113. $c = 15$

114. $c = 25$

ADDITIONAL PRACTICE *Simplify. Assume that all variables represent positive numbers.*

115. $4\sqrt{2x} + 6\sqrt{2x}$

116. $\sqrt{25y^2z} - \sqrt{16y^2z}$

117. $\sqrt[4]{32x^{12}y^4}$

118. $\sqrt[5]{64x^{10}y^5}$

119. $\sqrt[4]{\dfrac{5x}{16z^4}}$

120. $\sqrt[3]{\dfrac{11a^2}{125b^6}}$

121. $\sqrt{98} - \sqrt{50} - \sqrt{72}$

122. $\sqrt{20} + \sqrt{125} - \sqrt{80}$

123. $3\sqrt[3]{27} + 12\sqrt[3]{216}$

124. $14\sqrt[4]{32} - 15\sqrt[4]{162}$

125. $23\sqrt[4]{768} + \sqrt[4]{48}$

126. $3\sqrt[4]{512} + 2\sqrt[4]{32}$

127. $4\sqrt[4]{243} - \sqrt[4]{48}$

128. $\sqrt[4]{48} - \sqrt[4]{243} - \sqrt[4]{768}$

129. $6\sqrt[3]{5y} + 3\sqrt[3]{5y}$

130. $8\sqrt[5]{7a^2} - 7\sqrt[5]{7a^2}$

131. $10\sqrt[6]{12xyz} - \sqrt[6]{12xyz}$

132. $3\sqrt[4]{x^4y} - 2\sqrt[4]{x^4y}$

133. $\sqrt[5]{x^6y^2} + \sqrt[5]{32x^6y^2} + \sqrt[5]{x^6y^2}$

134. $\sqrt[3]{xy^4} + \sqrt[3]{8xy^4} - \sqrt[3]{27xy^4}$

135. $\sqrt{x^2 + 2x + 1} + \sqrt{x^2 + 2x + 1}$

136. $\sqrt{4x^2 + 12x + 9} + \sqrt{9x^2 + 6x + 1}$

Find the lengths of the remaining sides in each triangle. Approximate each answer to two decimal places.

137.

138.

139.

140.

141.

142.

143.

144.

APPLICATIONS *Find the exact answer and then approximate to the nearest hundredth.*

145. Hardware The sides of a regular hexagonal nut are 10 millimeters long. Find the height h in millimeters of the nut.

146. Ironing boards Find the height h, in inches, of the ironing board shown in the illustration.

WRITING ABOUT MATH

147. Explain how to recognize like radicals.

148. Explain how to combine like radicals.

SOMETHING TO THINK ABOUT

149. Find the sum.

$$\sqrt{3} + \sqrt{3^2} + \sqrt{3^3} + \sqrt{3^4} + \sqrt{3^5}$$

150. Can you find any numbers a and b such that

$$\sqrt{a + b} = \sqrt{a} + \sqrt{b}$$

Section 7.5

Multiplying Radical Expressions and Rationalizing

Objectives

1. Multiply two radical expressions.
2. Rationalize the denominator of a fraction that contains a radical expression.
3. Rationalize the numerator of a fraction that contains a radical expression.
4. Solve an application containing a radical expression.

Vocabulary

rationalize denominators   conjugates   rationalize numerators

Getting Ready

Simplify.

1. $a^3 a^4$ **2.** $\dfrac{b^5}{b^2}$ **3.** $a(a - 2)$ **4.** $3b^2(2b + 3)$

5. $(a + 2)(a - 5)$ **6.** $(2a + 3b)(2a - 3b)$

We now learn how to multiply radical expressions and rationalize the denominators or numerators of radical expressions. Then we will use these skills to solve applications.

1   Multiply two radical expressions.

Radical expressions with the same index can be multiplied and divided.

EXAMPLE 1   Multiply: **a.** $\left(3\sqrt{6}\right)\left(2\sqrt{3}\right)$ **b.** $\left(\sqrt[3]{3a}\right)\left(\sqrt[3]{9a^4}\right)$

Solution   We use the commutative and associative properties of multiplication to multiply the coefficients and the radicals separately. Then we simplify any radicals in the product, if possible.

a. $\left(3\sqrt{6}\right)\left(2\sqrt{3}\right) = 3(2)\sqrt{6}\sqrt{3}$ Multiply the coefficients and multiply the radicals.

$$= 6\sqrt{18}$$ $3(2) = 6$ and $\sqrt{6}\sqrt{3} = \sqrt{18}$

$$= 6\sqrt{9}\sqrt{2}$$ $\sqrt{18} = \sqrt{9\cdot 2} = \sqrt{9}\sqrt{2}$

$$= 6(3)\sqrt{2}$$ Simplify.

$$= 18\sqrt{2}$$

b. $\left(\sqrt[3]{3a}\right)\left(\sqrt[3]{9a^4}\right) = \sqrt[3]{27a^5}$ Multiply the radicals.

$$= \sqrt[3]{27a^3 \cdot a^2}$$ Factor $27a^5$.

$$= \sqrt[3]{27a^3}\sqrt[3]{a^2}$$ $\sqrt[3]{ab} = \sqrt[3]{a}\sqrt[3]{b}$

$$= 3a\sqrt[3]{a^2}$$ $\sqrt[3]{27a^3} = 3a$

SELF CHECK 1   Multiply: **a.** $\left(-2\sqrt{8}\right)\left(5\sqrt{6}\right)$ **b.** $\left(\sqrt[3]{4a^2}\right)\left(\sqrt[3]{6a^2}\right)$

To multiply a radical expression with two or more terms by a radical expression, we use the distributive property to remove parentheses and then simplify each resulting term, if possible.

EXAMPLE 2   Multiply: $3\sqrt{3}\left(4\sqrt{8} - 5\sqrt{10}\right)$

Solution $3\sqrt{3}\left(4\sqrt{8} - 5\sqrt{10}\right)$

$$= 3\sqrt{3}\cdot 4\sqrt{8} - 3\sqrt{3}\cdot 5\sqrt{10}$$ Use the distributive property to remove parentheses.

$$= 12\sqrt{24} - 15\sqrt{30}$$ Multiply the coefficients and multiply the radicals.

$$= 12\sqrt{4}\sqrt{6} - 15\sqrt{30}$$ Use the multiplication property of radicals.

$$= 12(2)\sqrt{6} - 15\sqrt{30}$$ Simplify.

$$= 24\sqrt{6} - 15\sqrt{30}$$

SELF CHECK 2   Multiply: $4\sqrt{2}\left(3\sqrt{5} - 2\sqrt{8}\right)$

To multiply two radical expressions, each with two or more terms, we use the distributive property as we did when we multiplied two polynomials. Then we simplify each resulting term, if possible.

EXAMPLE 3   Multiply: $\left(\sqrt{7} + \sqrt{2}\right)\left(\sqrt{7} - 3\sqrt{2}\right)$

Solution $\left(\sqrt{7} + \sqrt{2}\right)\left(\sqrt{7} - 3\sqrt{2}\right)$

$$= \sqrt{7}\sqrt{7} - 3\sqrt{7}\sqrt{2} + \sqrt{2}\sqrt{7} - 3\sqrt{2}\sqrt{2}$$

$$= 7 - 3\sqrt{14} + \sqrt{14} - 3(2)$$

$$= 7 - 2\sqrt{14} - 6$$

$$= 1 - 2\sqrt{14}$$

SELF CHECK 3   Multiply: $\left(\sqrt{5} + 2\sqrt{3}\right)\left(\sqrt{5} - \sqrt{3}\right)$

EXAMPLE 4 Multiply: $\left(\sqrt{3x} - \sqrt{5}\right)\left(\sqrt{2x} + \sqrt{10}\right)$

Solution $\left(\sqrt{3x} - \sqrt{5}\right)\left(\sqrt{2x} + \sqrt{10}\right)$

COMMENT Note that x in the answer is not under the radical in the first term, but it is under the radical in the second and third terms.

$$= \sqrt{3x}\sqrt{2x} + \sqrt{3x}\sqrt{10} - \sqrt{5}\sqrt{2x} - \sqrt{5}\sqrt{10}$$
$$= \sqrt{6x^2} + \sqrt{30x} - \sqrt{10x} - \sqrt{50}$$
$$= \sqrt{6}\sqrt{x^2} + \sqrt{30x} - \sqrt{10x} - \sqrt{25}\sqrt{2}$$
$$= \sqrt{6}x + \sqrt{30x} - \sqrt{10x} - 5\sqrt{2}$$

🌿 **SELF CHECK 4** Multiply: $\left(\sqrt{x} + 1\right)\left(\sqrt{x} - 3\right)$

COMMENT It is important to draw radical signs so they completely cover the radicand, but no more than the radicand. For example $\sqrt{6x}$ and $\sqrt{6}x$ are not the same expressions. To avoid confusion, we can use the commutative property of multiplication and write an expression such as $\sqrt{6}x$ in the form $x\sqrt{6}$.

2 Rationalize the denominator of a fraction that contains a radical expression.

To divide radical expressions, we **rationalize the denominator** of a fraction to write the denominator with a rational number. For example, to divide $\sqrt{70}$ by $\sqrt{3}$ we write the division as the fraction

$$\frac{\sqrt{70}}{\sqrt{3}}$$

To rationalize the radical in the denominator, we multiply the numerator and the denominator by a number that will result in a perfect square under the radical. Because $3 \cdot 3 = 9$ and 9 is a perfect square, $\sqrt{3}$ is such a number.

$$\frac{\sqrt{70}}{\sqrt{3}} = \frac{\sqrt{70} \cdot \sqrt{3}}{\sqrt{3} \cdot \sqrt{3}} \qquad \text{Multiply numerator and denominator by } \sqrt{3}.$$
$$= \frac{\sqrt{210}}{3} \qquad \text{Multiply the radicals.}$$

Since there is no radical in the denominator and $\sqrt{210}$ cannot be simplified, the expression $\dfrac{\sqrt{210}}{3}$ is in simplest form, and the division is complete.

EXAMPLE 5 Rationalize each denominator.

a. $\sqrt{\dfrac{20}{7}}$ **b.** $\dfrac{4}{\sqrt[3]{2}}$

Solution **a.** We first write the square root of the quotient as the quotient of two square roots.

$$\sqrt{\frac{20}{7}} = \frac{\sqrt{20}}{\sqrt{7}} \qquad \text{Recall, } \sqrt[n]{\frac{a}{b}} = \frac{\sqrt[n]{a}}{\sqrt[n]{b}} \quad b \neq 0.$$

Because the denominator is a square root, we must then multiply the numerator and the denominator by a number that will result in a rational number in the denominator.

Such a number is $\sqrt{7}$.

$$\frac{\sqrt{20}}{\sqrt{7}} = \frac{\sqrt{20} \cdot \sqrt{7}}{\sqrt{7} \cdot \sqrt{7}}$$ Multiply numerator and denominator by $\sqrt{7}$.

$$= \frac{\sqrt{140}}{7}$$ Multiply the radicals.

$$= \frac{2\sqrt{35}}{7}$$ Simplify $\sqrt{140}$: $\sqrt{140} = \sqrt{4 \cdot 35} = \sqrt{4}\sqrt{35} = 2\sqrt{35}$

Another approach is to simplify as follows:

$$\sqrt{\frac{20}{7}} = \sqrt{\frac{20 \cdot 7}{7 \cdot 7}}$$

$$= \sqrt{\frac{4 \cdot 5 \cdot 7}{49}}$$

$$= \frac{\sqrt{4}\sqrt{35}}{\sqrt{49}}$$

$$= \frac{2\sqrt{35}}{7}$$

b. Since the denominator is a cube root, we multiply the numerator and the denominator by a cube root of a number that will result in a perfect cube under the radical sign. Since 8 is a perfect cube, and $8 = 2 \cdot 4$, then $\sqrt[3]{4}$ is such a number.

$$\frac{4}{\sqrt[3]{2}} = \frac{4 \cdot \sqrt[3]{4}}{\sqrt[3]{2} \cdot \sqrt[3]{4}}$$ Multiply numerator and denominator by $\sqrt[3]{4}$.

$$= \frac{4\sqrt[3]{4}}{\sqrt[3]{8}}$$ Multiply the radicals in the denominator.

$$= \frac{4\sqrt[3]{4}}{2}$$ Simplify.

$$= 2\sqrt[3]{4}$$

Here is another approach:

$$\frac{4}{\sqrt[3]{2}} = \frac{4}{\sqrt[3]{2}} \cdot \frac{\sqrt[3]{2^2}}{\sqrt[3]{2^2}}$$

$$= \frac{4\sqrt[3]{4}}{\sqrt[3]{2^3}}$$

$$= \frac{4\sqrt[3]{4}}{2}$$

$$= 2\sqrt[3]{4}$$

SELF CHECK 5 Rationalize each denominator. **a.** $\sqrt{\dfrac{8}{5}}$ **b.** $\dfrac{5}{\sqrt[4]{3}}$

EXAMPLE 6 Rationalize the denominator: $\dfrac{\sqrt[3]{5}}{\sqrt[3]{18}}$

Solution We multiply the numerator and the denominator by a number that will result in a perfect cube under the radical sign in the denominator.

Since 216 is the smallest perfect cube that is divisible by 18 (216 ÷ 18 = 12), multiplying the numerator and the denominator by $\sqrt[3]{12}$ will result in the smallest possible perfect cube under the radical in the denominator.

$$\dfrac{\sqrt[3]{5}}{\sqrt[3]{18}} = \dfrac{\sqrt[3]{5} \cdot \sqrt[3]{12}}{\sqrt[3]{18} \cdot \sqrt[3]{12}} \qquad \text{Multiply numerator and denominator by } \sqrt[3]{12}.$$

$$= \dfrac{\sqrt[3]{60}}{\sqrt[3]{216}} \qquad \text{Multiply the radicals.}$$

$$= \dfrac{\sqrt[3]{60}}{6} \qquad \text{Simplify.}$$

There is another approach. Since $18 = 2 \cdot 3^2$, we need factors of $2^2 \cdot 3$ to obtain a perfect cube $2^3 \cdot 3^3$.

$$\dfrac{\sqrt[3]{5}}{\sqrt[3]{18}} = \dfrac{\sqrt[3]{5}\sqrt[3]{2^2 \cdot 3}}{\sqrt[3]{2 \cdot 3^2}\sqrt[3]{2^2 \cdot 3}}$$

$$= \dfrac{\sqrt[3]{5}\sqrt[3]{12}}{\sqrt[3]{2^3 \cdot 3^3}}$$

$$= \dfrac{\sqrt[3]{60}}{2 \cdot 3}$$

$$= \dfrac{\sqrt[3]{60}}{6}$$

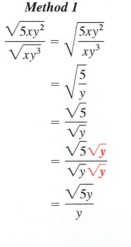 **SELF CHECK 6** Rationalize the denominator. $\dfrac{\sqrt[3]{2}}{\sqrt[3]{9}}$

EXAMPLE 7 Rationalize the denominator of $\dfrac{\sqrt{5xy^2}}{\sqrt{xy^3}}$ (*x* and *y* are positive numbers).

Solution

Method 1

$$\dfrac{\sqrt{5xy^2}}{\sqrt{xy^3}} = \sqrt{\dfrac{5xy^2}{xy^3}}$$

$$= \sqrt{\dfrac{5}{y}}$$

$$= \dfrac{\sqrt{5}}{\sqrt{y}}$$

$$= \dfrac{\sqrt{5}\sqrt{y}}{\sqrt{y}\sqrt{y}}$$

$$= \dfrac{\sqrt{5y}}{y}$$

Method 2

$$\dfrac{\sqrt{5xy^2}}{\sqrt{xy^3}} = \sqrt{\dfrac{5xy^2}{xy^3}}$$

$$= \sqrt{\dfrac{5}{y}}$$

$$= \sqrt{\dfrac{5 \cdot y}{y \cdot y}}$$

$$= \dfrac{\sqrt{5y}}{\sqrt{y^2}}$$

$$= \dfrac{\sqrt{5y}}{y}$$

COMMENT When possible, simplify prior to rationalizing.

SELF CHECK 7 Rationalize the denominator. $\dfrac{\sqrt{4ab^3}}{\sqrt{2a^2b^2}}$ $(a > 0, b > 0)$

To rationalize the denominator of a fraction with square roots in a binomial denominator, we can multiply the numerator and denominator by the **conjugate** of the denominator.

CONJUGATES	The conjugate of $(a + b)$ is $(a - b)$, and the conjugate of $(a - b)$ is $(a + b)$.

If we multiply an expression such as $(5 + \sqrt{2})$ by its conjugate $(5 - \sqrt{2})$, we will obtain an expression without any radical terms.

$$(5 + \sqrt{2})(5 - \sqrt{2}) = 25 - 5\sqrt{2} + 5\sqrt{2} - 2$$
$$= 23$$

EXAMPLE 8   Rationalize the denominator. $\dfrac{1}{\sqrt{2} + 1}$

Solution   We multiply the numerator and denominator of the fraction by $(\sqrt{2} - 1)$, which is the conjugate of the denominator.

$$\frac{1}{\sqrt{2} + 1} = \frac{1(\sqrt{2} - 1)}{(\sqrt{2} + 1)(\sqrt{2} - 1)} \qquad \frac{\sqrt{2} - 1}{\sqrt{2} - 1} = 1$$

$$= \frac{\sqrt{2} - 1}{(\sqrt{2})^2 - \sqrt{2} + \sqrt{2} - 1} \qquad \text{Multiply.}$$

$$= \frac{\sqrt{2} - 1}{2 - 1} \qquad \text{Simplify.}$$

$$= \sqrt{2} - 1 \qquad \frac{\sqrt{2} - 1}{2 - 1} = \frac{\sqrt{2} - 1}{1} = \sqrt{2} - 1$$

SELF CHECK 8   Rationalize the denominator. $\dfrac{2}{\sqrt{3} + 1}$

EXAMPLE 9   Rationalize the denominator. $\dfrac{\sqrt{x} + \sqrt{2}}{\sqrt{x} - \sqrt{2}}$ $(x > 0, x \neq 2)$

Solution   We multiply the numerator and denominator by $(\sqrt{x} + \sqrt{2})$, which is the conjugate of the denominator, and simplify.

$$\frac{\sqrt{x} + \sqrt{2}}{\sqrt{x} - \sqrt{2}} = \frac{(\sqrt{x} + \sqrt{2})(\sqrt{x} + \sqrt{2})}{(\sqrt{x} - \sqrt{2})(\sqrt{x} + \sqrt{2})}$$

$$= \frac{x + \sqrt{2x} + \sqrt{2x} + 2}{(\sqrt{x})^2 + \sqrt{2x} - \sqrt{2x} - (\sqrt{2})^2} \qquad \text{Use the distributive property.}$$

$$= \frac{x + 2\sqrt{2x} + 2}{x - 2} \qquad \text{Combine like terms.}$$

SELF CHECK 9   Rationalize the denominator. $\dfrac{\sqrt{x} - \sqrt{2}}{\sqrt{x} + \sqrt{2}}$ $(x > 0)$

3   Rationalize the numerator of a fraction that contains a radical expression.

In calculus, we sometimes have to **rationalize a numerator** by multiplying the numerator and denominator of the fraction by the conjugate of the numerator. The process is similar to rationalizing the denominator, but the denominator may still contain a radical expression.

EXAMPLE 10   Rationalize the numerator. $\dfrac{\sqrt{x} - 3}{\sqrt{x}}$ $(x > 0)$

Solution   We multiply the numerator and denominator by $\left(\sqrt{x} + 3\right)$, which is the conjugate of the numerator.

$$\frac{\sqrt{x} - 3}{\sqrt{x}} = \frac{\left(\sqrt{x} - 3\right)\left(\sqrt{x} + 3\right)}{\sqrt{x}\left(\sqrt{x} + 3\right)}$$

$$= \frac{x + 3\sqrt{x} - 3\sqrt{x} - 9}{x + 3\sqrt{x}}$$

$$= \frac{x - 9}{x + 3\sqrt{x}}$$

 SELF CHECK 10   Rationalize the numerator. $\dfrac{\sqrt{x} + 3}{\sqrt{x}}$ $(x > 0)$

---

4   Solve an application containing a radical expression.

© Brocorwin/Shutterstock.com

Figure 7-15

EXAMPLE 11   **PHOTOGRAPHY**   Many camera lenses (see Figure 7-15) have an adjustable opening called the *aperture*, which controls the amount of light passing through the lens. The *f-number* of a lens is its *focal length* divided by the diameter of its circular aperture.

$$f\text{-number} = \frac{f}{d} \qquad f \text{ is the focal length, and } d \text{ is the diameter of the aperture.}$$

A lens with a focal length of 12 centimeters and an aperture with a diameter of 6 centimeters has an *f*-number of $\frac{12}{6}$ and is an *f*/2 lens. If the area of the aperture is reduced to admit half as much light, the *f*-number of the lens will change. Find the new *f*-number.

Solution   We first find the area of the aperture when its diameter is 6 centimeters.

$A = \pi r^2$ This is the formula for the area of a circle.

$A = \pi(\mathbf{3})^2$ Since a radius is half the diameter, substitute 3 for r.

$A = 9\pi$

When the size of the aperture is reduced to admit half as much light, the area of the aperture will be $\frac{9\pi}{2}$ square centimeters. To find the diameter of a circle with this area, we proceed as follows:

$A = \pi r^2$ This is the formula for the area of a circle.

$\dfrac{9\pi}{2} = \pi\left(\dfrac{d}{2}\right)^2$ Substitute $\frac{9\pi}{2}$ for A and $\frac{d}{2}$ for r.

$\dfrac{9\pi}{2} = \dfrac{\pi d^2}{4}$ $\left(\frac{d}{2}\right)^2 = \frac{d^2}{4}$

$\dfrac{4}{\pi}\left(\dfrac{9\pi}{2}\right) = \dfrac{4}{\pi}\left(\dfrac{\pi d^2}{4}\right)$ Multiply both sides by $\frac{4}{\pi}$, the reciprocal of the coefficient of d^2.

$18 = d^2$ Simplify.

$d = 3\sqrt{2}$ $\sqrt{18} = \sqrt{9}\sqrt{2} = 3\sqrt{2}$

Since the focal length of the lens is still 12 centimeters and the diameter is now $3\sqrt{2}$ centimeters, the new *f*-number of the lens is

$$f\text{-number} = \frac{f}{d} = \frac{\mathbf{12}}{\mathbf{3\sqrt{2}}} \qquad \text{Substitute 12 for } f \text{ and } 3\sqrt{2} \text{ for } d.$$

$$= \frac{4}{\sqrt{2}} \qquad \text{Simplify.}$$

$$= \frac{4\sqrt{2}}{2} \qquad \text{Rationalize the denominator.}$$

$$= 2\sqrt{2} \qquad \text{Simplify.}$$

$$\approx 2.828427125 \qquad \text{Use a calculator.}$$

The lens is now an $f/2.8$ lens.

SELF CHECK 11 If the area of the aperature is enlarged to admit twice as much light, find the f-number.

SELF CHECK ANSWERS

1. a. $-40\sqrt{3}$ b. $2a\sqrt[3]{3a}$ 2. $12\sqrt{10} - 32$ 3. $-1 + \sqrt{15}$ 4. $x - 2\sqrt{x} - 3$ 5. a. $\frac{2\sqrt{10}}{5}$ b. $\frac{5\sqrt[3]{27}}{3}$

6. $\frac{\sqrt[3]{6}}{3}$ 7. $\frac{\sqrt{2ab}}{a}$ 8. $\sqrt{3} - 1$ 9. $\frac{x - 2\sqrt{2x} + 2}{x - 2}$ 10. $\frac{x - 9}{x - 3\sqrt{x}}$ 11. The lens is now an $f/1.4$ lens.

NOW TRY THIS

Find the domain of each of the following. State your answer in interval notation.

1. $f(x) = \dfrac{3x - 2}{\sqrt{x} + 1}$ 2. $g(x) = \dfrac{3x - 2}{\sqrt{x + 1}}$

3. $h(x) = \dfrac{\sqrt{x - 2}}{\sqrt{x} + 1}$

7.5   Exercises

WARM-UPS *Simplify each expression. Assume that all variables represent positive numbers.*

1. $\sqrt{5}\sqrt{5}$
2. $\sqrt{11}\sqrt{11}$
3. $\sqrt{xy}\sqrt{xy}$
4. $\sqrt{ab}\sqrt{ab}$
5. $\sqrt[3]{5}\sqrt[3]{5}\sqrt[3]{5}$
6. $\sqrt[3]{7}\sqrt[3]{7}\sqrt[3]{7}$
7. $\sqrt[3]{a^2b}\sqrt[3]{ab^2}$
8. $\sqrt[3]{xy^2}\sqrt[3]{x^2y}$

REVIEW *Solve.*

9. $\dfrac{2}{3 - a} = 1$
10. $5(s - 4) = -5(s - 4)$
11. $\dfrac{8}{b - 2} + \dfrac{3}{2 - b} = -\dfrac{1}{b}$
12. $\dfrac{2}{x - 2} + \dfrac{1}{x + 1} = \dfrac{1}{(x + 1)(x - 2)}$

VOCABULARY AND CONCEPTS *Fill in the blanks.*

13. To multiply $2\sqrt{7}$ by $3\sqrt{5}$, we multiply _ by 3 and then multiply ____ by ____.
14. To multiply $2\sqrt{5}(3\sqrt{8} + \sqrt{3})$, we use the _____ property to remove parentheses and simplify each resulting term.

15. The conjugate of $(\sqrt{x} + 1)$ is _____.
16. To multiply $(\sqrt{3} + \sqrt{2})(\sqrt{3} - 2\sqrt{2})$, we use the _____.
17. To rationalize the denominator of $\dfrac{1}{\sqrt{3} - 1}$, multiply both the numerator and denominator by the _____ of the denominator, $\sqrt{3} + 1$.
18. To rationalize the numerator of $\dfrac{\sqrt{5} + 2}{\sqrt{5} - 2}$, multiply both the numerator and denominator by _____.

GUIDED PRACTICE *Simplify. All variables represent positive values. SEE EXAMPLE 1. (OBJECTIVE 1)*

19. $\sqrt{2}\sqrt{8}$
20. $\sqrt{3}\sqrt{27}$
21. $\sqrt{5}\sqrt{10}$
22. $\sqrt{7}\sqrt{35}$
23. $2\sqrt{3}\sqrt{6}$
24. $3\sqrt{11}\sqrt{33}$
25. $\sqrt[3]{5}\sqrt[3]{25}$
26. $\sqrt[3]{7}\sqrt[3]{49}$
27. $\sqrt[3]{5r^2s}\sqrt[3]{2r}$
28. $\sqrt[3]{3xy^2}\sqrt[3]{9x^3}$
29. $\sqrt{ab^3}\sqrt{ab}$
30. $\sqrt{8x}\sqrt{2x^3y}$
31. $(2\sqrt{3x^3y})(5\sqrt{6xy^2})$
32. $(7\sqrt{18a})(3\sqrt{3a^3b})$

33. $\sqrt{x(x+3)}\sqrt{x^3(x+3)}$

34. $\sqrt{y^2(x+y)}\sqrt{(x+y)^3}$

Simplify. *SEE EXAMPLE 2. (OBJECTIVE 1)*

35. $3\sqrt{5}\left(4-\sqrt{5}\right)$ **36.** $2\sqrt{7}\left(3\sqrt{7}-1\right)$

37. $3\sqrt{2}\left(4\sqrt{3}+2\sqrt{7}\right)$ **38.** $-2\sqrt{3}\left(3\sqrt{5}-6\sqrt{7}\right)$

Simplify. *SEE EXAMPLE 3. (OBJECTIVE 1)*

39. $\left(\sqrt{2}+1\right)\left(\sqrt{2}-3\right)$

40. $\left(2\sqrt{3}+1\right)\left(\sqrt{3}-1\right)$

41. $\left(3\sqrt{2}+2\right)\left(\sqrt{2}+1\right)$

42. $\left(4\sqrt{3}-3\right)\left(\sqrt{3}+2\right)$

Simplify. All variables represent positive values. *SEE EXAMPLE 4. (OBJECTIVE 1)*

43. $\left(4\sqrt{x}+3\right)\left(2\sqrt{x}-5\right)$

44. $\left(5\sqrt{y}-4\right)\left(2\sqrt{y}-5\right)$

45. $\left(\sqrt{3a}+\sqrt{6}\right)\left(\sqrt{2a}-\sqrt{3}\right)$

46. $\left(\sqrt{7y}-\sqrt{5}\right)\left(\sqrt{14y}+\sqrt{10}\right)$

Rationalize each denominator. *SEE EXAMPLE 5. (OBJECTIVE 2)*

47. $\sqrt{\dfrac{1}{7}}$ **48.** $\sqrt{\dfrac{5}{3}}$

49. $\sqrt{\dfrac{2}{3}}$ **50.** $\sqrt{\dfrac{3}{2}}$

51. $\dfrac{\sqrt{24}}{\sqrt{5}}$ **52.** $\dfrac{\sqrt{8}}{\sqrt{3}}$

53. $\dfrac{\sqrt{7}}{\sqrt{12}}$ **54.** $\dfrac{\sqrt{5}}{\sqrt{32}}$

Rationalize each denominator. *SEE EXAMPLE 6. (OBJECTIVE 2)*

55. $\dfrac{3}{\sqrt[3]{16}}$ **56.** $\dfrac{2}{\sqrt[3]{6}}$

57. $\dfrac{\sqrt[3]{7}}{\sqrt[3]{72}}$ **58.** $\dfrac{\sqrt[3]{9}}{\sqrt[3]{54}}$

Rationalize each denominator. All variables represent positive values. *SEE EXAMPLE 7. (OBJECTIVE 2)*

59. $\dfrac{\sqrt{8x^2y}}{\sqrt{xy}}$ **60.** $\dfrac{\sqrt{9xy}}{\sqrt{3x^2y}}$

61. $\dfrac{\sqrt{10xy^2}}{\sqrt{2xy^3}}$ **62.** $\dfrac{\sqrt{5ab^2c}}{\sqrt{10abc}}$

Rationalize each denominator. *SEE EXAMPLE 8. (OBJECTIVE 2)*

63. $\dfrac{1}{\sqrt{2}-1}$ **64.** $\dfrac{3}{\sqrt{3}-1}$

65. $\dfrac{\sqrt{2}}{\sqrt{5}+3}$ **66.** $\dfrac{\sqrt{3}}{\sqrt{3}-2}$

67. $\dfrac{\sqrt{3}+1}{\sqrt{3}-1}$ **68.** $\dfrac{\sqrt{2}-1}{\sqrt{2}+1}$

69. $\dfrac{\sqrt{7}-\sqrt{2}}{\sqrt{2}+\sqrt{7}}$ **70.** $\dfrac{\sqrt{3}+\sqrt{2}}{\sqrt{3}-\sqrt{2}}$

Rationalize each denominator. All variables represent positive values. *SEE EXAMPLE 9. (OBJECTIVE 2)*

71. $\dfrac{2}{\sqrt{x}+1}$ **72.** $\dfrac{3}{\sqrt{x}-2}$

73. $\dfrac{x}{\sqrt{x}-4}$ **74.** $\dfrac{2x}{\sqrt{x}+1}$

75. $\dfrac{\sqrt{x}-\sqrt{y}}{\sqrt{x}+\sqrt{y}}$ **76.** $\dfrac{\sqrt{x}+\sqrt{y}}{\sqrt{x}-\sqrt{y}}$

Rationalize each numerator. All variables represent positive values. *SEE EXAMPLE 10. (OBJECTIVE 3)*

77. $\dfrac{\sqrt{3}+1}{2}$ **78.** $\dfrac{\sqrt{5}-1}{2}$

79. $\dfrac{\sqrt{x}+3}{x}$ **80.** $\dfrac{2+\sqrt{x}}{5x}$

ADDITIONAL PRACTICE   Simplify. All variables represent positive values.

81. $\left(3\sqrt[3]{9}\right)\left(2\sqrt[3]{3}\right)$ **82.** $\left(2\sqrt[3]{16}\right)\left(-\sqrt[3]{4}\right)$

83. $\sqrt{5ab}\sqrt{5a}$

84. $\sqrt{15rs^2}\sqrt{10r}$

85. $\sqrt[3]{a^5b}\sqrt[3]{16ab^5}$

86. $\sqrt[3]{3x^4y}\sqrt[3]{18x}$

87. $\sqrt[3]{6x^2(y+z)^2}\sqrt[3]{18x(y+z)}$

88. $\sqrt[3]{9x^2y(z+1)^2}\sqrt[3]{6xy^2(z+1)}$

89. $-2\sqrt{5x}\left(4\sqrt{2x}-3\sqrt{3}\right)$

90. $3\sqrt{7t}\left(2\sqrt{7t}+3\sqrt{3t^2}\right)$

91. $\left(\sqrt{5z}+\sqrt{3}\right)\left(\sqrt{5z}+\sqrt{3}\right)$

92. $\left(\sqrt{5x}+\sqrt{3}\right)\left(\sqrt{5x}-\sqrt{3}\right)$

93. $\left(3\sqrt{2r}-2\right)^2$

94. $\left(2\sqrt{3t}+5\right)^2$

95. $-2\left(\sqrt{3x}+\sqrt{3}\right)^2$

96. $3\left(\sqrt{5x}-\sqrt{3}\right)^2$

Rationalize each denominator. All variables represent positive values.

97. $\dfrac{3}{\sqrt[3]{9}}$ **98.** $\dfrac{2}{\sqrt[3]{a}}$

99. $\dfrac{1}{\sqrt[4]{4}}$ **100.** $\dfrac{4}{\sqrt[4]{32}}$

101. $\dfrac{1}{\sqrt[5]{16}}$ **102.** $\dfrac{1}{\sqrt[5]{2}}$

103. $\dfrac{\sqrt[3]{4a^2}}{\sqrt[3]{2ab}}$

104. $\dfrac{\sqrt[3]{9x}}{\sqrt[3]{3xy}}$

105. $\dfrac{2z-1}{\sqrt{2z-1}}$

106. $\dfrac{3t-1}{\sqrt{3t+1}}$

Rationalize each numerator. All variables represent positive values.

107. $\dfrac{\sqrt{x}+\sqrt{y}}{\sqrt{x}}$

108. $\dfrac{\sqrt{x}-\sqrt{y}}{\sqrt{x}+\sqrt{y}}$

APPLICATIONS *Solve each application.* **SEE EXAMPLE 11.**
(OBJECTIVE 4)

109. Photography We have seen that a lens with a focal length of 12 centimeters and an aperture $3\sqrt{2}$ centimeters in diameter is an $f/2.8$ lens. Find the f-number if the area of the aperture is again cut in half.

110. Photography A lens with a focal length of 12 centimeters and an aperture 3 centimeters in diameter is an $f/4$ lens. Find the f-number if the area of the aperture is cut in half.

111. Targets The radius r of the target is given by the formula

$$r = \sqrt{\dfrac{A}{\pi}}$$

where A is the area. Write the formula in a form in which the denominator is not part of the radicand.

112. Pulse rates The approximate pulse rate (in beats per minute) of an adult who is t inches tall is given by the function

$$p(t) = \dfrac{590}{\sqrt{t}}$$

Write the formula in a form in which the denominator is a rational expression.

113. If the hypotenuse of an isosceles right triangle is 8 cm, find the length of each leg.

114. The hypotenuse of a 45°–45°–90° triangle is 14 m. Find the length of each leg.

115. The longer leg of a 30°–60°–90° triangle is 6 ft. Find the length of remaining sides.

116. The altitude of an equilateral triangle is 24 mm. Find the lengths of the sides of the triangle.

WRITING ABOUT MATH

117. Explain how to simplify a fraction with the monomial denominator $\sqrt[3]{3}$.

118. Explain how to simplify a fraction with the monomial denominator $\sqrt[3]{9}$.

SOMETHING TO THINK ABOUT *Assume that x is a rational number.*

119. Write the numerator of $\dfrac{\sqrt{x}-3}{4}$ as a rational number.

120. Rationalize the numerator: $\dfrac{2\sqrt{3x}+4}{\sqrt{3x}-1}$

Section 7.6

Radical Equations

Objectives

1. Solve a radical equation containing one radical.
2. Solve a radical equation containing two radicals.
3. Solve a radical equation containing three radicals.
4. Solve a formula containing a radical for a specified variable.

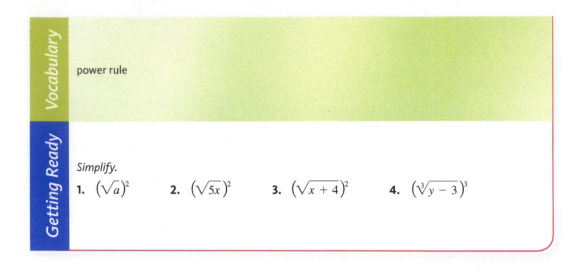

Vocabulary

power rule

Getting Ready

Simplify.

1. $\left(\sqrt{a}\right)^2$ 2. $\left(\sqrt{5x}\right)^2$ 3. $\left(\sqrt{x+4}\right)^2$ 4. $\left(\sqrt[3]{y-3}\right)^3$

In this section, we will solve equations that contain radicals. To do so, we will use the **power rule**. Then, we will use this rule to solve an application.

1 **Solve a radical equation containing one radical.**

THE POWER RULE   If x, y, and n are real numbers and $x = y$, then

$$x^n = y^n$$

When we raise both sides of an equation to the same power, the resulting equation might not be equivalent to the original equation. For example, if we square both sides of the equation

(1) $x = 3$ With a solution set of $\{3\}$

we obtain the equation

(2) $x^2 = 9$ With a solution set of $\{3, -3\}$

Equations 1 and 2 are not equivalent, because they have different solution sets, and the solution -3 of Equation 2 does not satisfy Equation 1. Since raising both sides of an equation to the same power can produce an equation with roots (extraneous) that do not satisfy the original equation, we must check each possible solution in the original equation.

EXAMPLE 1   Solve: $\sqrt{x+3} = 4$

Solution   Because the radical term is isolated, to eliminate the radical, we apply the power rule by squaring both sides of the equation, and proceed as follows:

$$\sqrt{x+3} = 4$$
$$\left(\sqrt{x+3}\right)^2 = (4)^2 \quad \text{Square both sides of the equation.}$$
$$x + 3 = 16 \quad \text{Simplify.}$$
$$x = 13 \quad \text{Subtract 3 from both sides.}$$

To check the apparent solution of 13, we can substitute 13 for x and determine whether it satisfies the original equation.

$$\sqrt{x + 3} = 4$$
$$\sqrt{13 + 3} \stackrel{?}{=} 4 \qquad \text{Substitute 13 for } x.$$
$$\sqrt{16} \stackrel{?}{=} 4$$
$$4 = 4$$

Since 13 satisfies the original equation, it is a solution.

SELF CHECK 1   Solve: $\sqrt{a - 2} = 3$

To solve an equation containing radicals, we follow these steps.

SOLVING AN EQUATION CONTAINING RADICALS

1. Isolate one radical expression on one side of the equation.
2. Raise both sides of the equation to the power that is the same as the index of the radical.
3. Solve the resulting equation. If it still contains a radical, go back to Step 1.
4. Check the possible solutions to eliminate the extraneous ones.

EXAMPLE 2   **HEIGHT OF A BRIDGE**   The distance d (in feet) that an object will fall in t seconds is given by the formula

$$t = \sqrt{\frac{d}{16}}$$

To find the height of a bridge, a man drops a stone into the water. (See Figure 7-16.) If it takes the stone 3 seconds to hit the water, how far above the river is the bridge?

Figure 7-16

Solution   We substitute 3 for t in the formula and solve for d.

$$t = \sqrt{\frac{d}{16}}$$
$$3 = \sqrt{\frac{d}{16}}$$
$$9 = \frac{d}{16} \qquad \text{Apply the power rule (square both sides) to eliminate the square root.}$$
$$144 = d \qquad \text{Multiply both sides by 16.}$$

The bridge is 144 feet above the river.

SELF CHECK 2   How high is the bridge if it takes 4 seconds for the stone to hit the water?

COMMENT   Be certain to isolate the radical term prior to applying the power rule.

EXAMPLE 3   Solve: $\sqrt{3x + 1} + 1 = x$

Solution   We first subtract 1 from both sides to isolate the radical. Then, to eliminate the radical, we square both sides of the equation and proceed as follows:

$$\sqrt{3x + 1} + 1 = x$$

$$\sqrt{3x + 1} = x - 1 \qquad \text{Subtract 1 from both sides.}$$

$$\left(\sqrt{3x + 1}\right)^2 = (x - 1)^2 \qquad \text{Apply the power rule.}$$

$$3x + 1 = x^2 - 2x + 1 \qquad \text{Simplify.}$$

$$0 = x^2 - 5x \qquad \text{Subtract } 3x \text{ and 1 from both sides to write the equation in quadratic form.}$$

$$0 = x(x - 5) \qquad \text{Factor.}$$

$$x = 0 \quad \text{or} \quad x - 5 = 0 \qquad \text{Set each factor equal to 0.}$$

$$x = 0 \quad \Big| \qquad x = 5$$

We must check each possible solution to see whether it satisfies the original equation.

Check: $\sqrt{3x + 1} + 1 = x$ $\sqrt{3x + 1} + 1 = x$

$\sqrt{3(0) + 1} + 1 \overset{?}{=} 0$ $\sqrt{3(5) + 1} + 1 \overset{?}{=} 5$

$\sqrt{1} + 1 \overset{?}{=} 0$ $\sqrt{16} + 1 \overset{?}{=} 5$

$2 \neq 0$ $5 = 5$

Since 0 does not check, it is extraneous and must be discarded. The only solution of the original equation is 5.

SELF CHECK 3   Solve: $\sqrt{4x + 1} + 1 = x$

Accent on technology

▸Solving Equations Containing Radicals

To find solutions for $\sqrt{3x + 1} + 1 = x$ with a graphing calculator, we graph the functions $f(x) = \sqrt{3x + 1} + 1$ and $g(x) = x$, and then adjust the window settings to $[-5, 10]$ for x and $[-2, 8]$ for y as in Figure 7-17(a).

We can find the exact x-coordinate of the intersection point by using the INTERSECT command found in the CALC menu. We see that $x = 5$, as in Figure 7-17(b).

(a)   (b)

Figure 7-17

For instructions regarding the use of a Casio graphing calculator, please refer to the Casio Keystroke Guide in the back of the book.

EXAMPLE 4   Solve: $\sqrt[3]{x^3 + 7} = x + 1$

Solution   To eliminate the radical, we cube both sides of the equation and proceed as follows:

$$\sqrt[3]{x^3 + 7} = x + 1$$

$$\left(\sqrt[3]{x^3 + 7}\right)^3 = (x + 1)^3 \qquad \text{Cube both sides to eliminate the cube root.}$$

$$x^3 + 7 = x^3 + 3x^2 + 3x + 1 \qquad \text{Simplify.}$$

$$0 = 3x^2 + 3x - 6 \qquad \text{Subtract } x^3 \text{ and 7 from both sides to write the equation in quadratic form.}$$

$$0 = x^2 + x - 2 \qquad \text{Divide both sides by 3.}$$

$$0 = (x + 2)(x - 1) \quad \text{Factor the trinomial.}$$

$$x + 2 = 0 \quad \text{or} \quad x - 1 = 0 \qquad \text{Set each factor equal to 0.}$$
$$x = -2 \qquad\qquad x = 1$$

We check each possible solution to see whether it satisfies the original equation.

Check:
$$\sqrt[3]{x^3 + 7} = x + 1 \qquad\qquad \sqrt[3]{x^3 + 7} = x + 1$$
$$\sqrt[3]{(-2)^3 + 7} \overset{?}{=} -2 + 1 \qquad\qquad \sqrt[3]{1 + 7} \overset{?}{=} 1 + 1$$
$$\sqrt[3]{-8 + 7} \overset{?}{=} -1 \qquad\qquad \sqrt[3]{8} \overset{?}{=} 2$$
$$\sqrt[3]{-1} \overset{?}{=} -1 \qquad\qquad 2 = 2$$
$$-1 = -1$$

Both solutions satisfy the original equation; thus the solutions are -2 and 1.

 SELF CHECK 4   Solve: $\sqrt[3]{x^3 + 8} = x + 2$

---

2 **Solve a radical equation containing two radicals.**

When more than one radical appears in an equation, it is often necessary to apply the power rule more than once.

EXAMPLE 5   Solve: $\sqrt{x} + \sqrt{x + 2} = 2$

Solution   We subtract $\sqrt{x}$ from both sides to isolate one radical on one side of the equation. To remove one of the radicals, we square both sides of the equation.

$$\sqrt{x} + \sqrt{x + 2} = 2$$

$$\sqrt{x + 2} = 2 - \sqrt{x} \qquad \text{Subtract } \sqrt{x} \text{ from both sides.}$$

$$\left(\sqrt{x + 2}\right)^2 = \left(2 - \sqrt{x}\right)^2 \qquad \text{Square both sides to eliminate the square root on the left.}$$

$$x + 2 = 4 - 4\sqrt{x} + x \qquad \begin{array}{l}\left(2 - \sqrt{x}\right)\left(2 - \sqrt{x}\right) = \\ 4 - 2\sqrt{x} - 2\sqrt{x} + x = 4 - 4\sqrt{x} + x\end{array}$$

$$2 = 4 - 4\sqrt{x} \qquad \text{Subtract } x \text{ from both sides.}$$

$$-2 = -4\sqrt{x} \qquad \text{Subtract 4 from both sides.}$$

$$\frac{1}{2} = \sqrt{x} \qquad \text{Divide both sides by } -4.$$

$$\frac{1}{4} = x \qquad \text{Square both sides to eliminate the square root on the right side.}$$

Check:
$$\sqrt{x} + \sqrt{x + 2} = 2$$
$$\sqrt{\frac{1}{4}} + \sqrt{\frac{1}{4} + 2} \overset{?}{=} 2$$
$$\frac{1}{2} + \sqrt{\frac{9}{4}} \overset{?}{=} 2$$
$$\frac{1}{2} + \frac{3}{2} \overset{?}{=} 2$$
$$2 = 2$$

The solution checks.

 SELF CHECK 5   Solve: $\sqrt{a} + \sqrt{a + 3} = 3$

3 Solve a radical equation containing three radicals.

EXAMPLE 6 Solve: $\sqrt{x+2} + \sqrt{2x} = \sqrt{18-x}$

Solution In this case, one radical is already isolated on one side of the equation, so we begin by squaring both sides. Then we proceed as follows.

$$\sqrt{x+2} + \sqrt{2x} = \sqrt{18-x}$$

$$\left(\sqrt{x+2} + \sqrt{2x}\right)^2 = \left(\sqrt{18-x}\right)^2 \qquad \text{Square both sides to eliminate the square root on the right side.}$$

$$\left(\sqrt{x+2} + \sqrt{2x}\right)\left(\sqrt{x+2} + \sqrt{2x}\right) = 18-x \qquad \text{Simplify.}$$

$$\left(\sqrt{x+2}\right)^2 + \left(\sqrt{x+2}\right)\left(\sqrt{2x}\right) + \left(\sqrt{2x}\right)\left(\sqrt{x+2}\right) + \left(\sqrt{2x}\right)^2 = 18-x \qquad \text{Multiply.}$$

$$x+2+\sqrt{2x(x+2)} + \sqrt{2x(x+2)} + 2x = 18-x \qquad \text{Simplify.}$$

$$3x+2+2\sqrt{2x(x+2)} = 18-x \qquad \text{Combine like terms.}$$

$$2\sqrt{2x(x+2)} = 16-4x \qquad \text{Subtract } 3x \text{ and 2 from both sides.}$$

$$\sqrt{2x(x+2)} = 8-2x \qquad \text{Divide both sides by 2.}$$

$$\left(\sqrt{(2x)(x+2)}\right)^2 = (8-2x)^2 \qquad \text{Square both sides to eliminate the square root on the left side.}$$

$$2x(x+2) = 64-32x+4x^2$$

$$2x^2+4x = 64-32x+4x^2 \qquad \text{Use the distributive property to remove parentheses.}$$

$$0 = 2x^2-36x+64 \qquad \text{Write the equation in quadratic form.}$$

$$0 = x^2-18x+32 \qquad \text{Divide both sides by 2.}$$

$$0 = (x-16)(x-2) \qquad \text{Factor the trinomial.}$$

$$x-16 = 0 \quad \text{or} \quad x-2 = 0 \qquad \text{Set each factor equal to 0.}$$

$$x = 16 \qquad \qquad x = 2$$

Verify that 2 satisfies the equation, but 16 does not. Thus, the only solution is 2.

SELF CHECK 6 Solve: $\sqrt{3x+4} + \sqrt{x+9} = \sqrt{x+25}$

---

4 Solve a formula containing a radical for a specified variable.

To *solve a formula for a variable* means to isolate that variable on one side of the equation, with all other quantities on the other side.

EXAMPLE 7 **DEPRECIATION RATES** Some office equipment that is now worth V dollars originally cost C dollars 3 years ago. The rate r at which it has depreciated is given by

$$r = 1 - \sqrt[3]{\frac{V}{C}}$$

Solve the formula for C.

Solution We begin by isolating the cube root on the right side of the equation.

$$r = 1 - \sqrt[3]{\frac{V}{C}}$$

$$r-1 = -\sqrt[3]{\frac{V}{C}} \qquad \text{Subtract 1 from both sides.}$$

$$(r-1)^3 = \left(-\sqrt[3]{\frac{V}{C}}\right)^3 \qquad \text{To eliminate the radical, cube both sides.}$$

$$(r - 1)^3 = -\frac{V}{C}$$ Simplify the right side.

$$C(r - 1)^3 = -V$$ Multiply both sides by C.

$$C = -\frac{V}{(r - 1)^3}$$ Divide both sides by $(r - 1)^3$.

 SELF CHECK 7   A formula used in statistics to determine the size of a sample to obtain a desired degree of accuracy is

$$E = z_0\sqrt{\frac{pq}{n}}$$

Solve the formula for n.

SELF CHECK ANSWERS   **1.** 11 **2.** 256 ft **3.** 6; 0 is extraneous **4.** 0, −2 **5.** 1 **6.** 0; $-\frac{196}{3}$ is extraneous **7.** $n = \frac{z_0^2 pq}{E^2}$

NOW TRY THIS

Solve.

1. $x^{1/3} = 2$

2. $x^{2/3} = 4$

3. $(x + 1)^{-1/2} = 3$

7.6 Exercises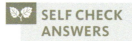

WARM-UPS *Simplify.*

1. $(x^{1/2})^2$ **2.** $(a^{1/3})^3$

3. $[(x + 3)^{1/2}]^2$ **4.** $[(x - 4)^{1/3}]^3$

5. $(x + 7)^2$ **6.** $(x - 6)^2$

REVIEW *If $f(x) = 3x^2 - 4x + 2$, find each quantity.*

7. $f(-2)$ **8.** $f(-3)$

9. $f(3)$ **10.** $f\left(\frac{1}{2}\right)$

VOCABULARY AND CONCEPTS *Fill in the blanks.*

11. If x, y, and n are real numbers and $x = y$, then _____, called the _____.

12. When solving equations containing radicals, first _____ one radical on one side of the equation.

13. To solve the equation $\sqrt{x + 4} = 5$, we first _____ both sides.

14. To solve the equation $\sqrt[3]{x + 4} = 2$, we first ____ both sides.

15. Squaring both sides of an equation can introduce _____ solutions.

16. Always remember to _____ the solutions of an equation containing radicals to eliminate any _____ solutions.

GUIDED PRACTICE *Solve each equation. SEE EXAMPLE 1. (OBJECTIVE 1)*

17. $\sqrt{4x + 5} = 3$ **18.** $\sqrt{7x - 10} = 12$

19. $\sqrt{4x - 8} + 4 = 10$ **20.** $\sqrt{6x + 13} - 2 = 5$

21. $\sqrt[3]{7n - 1} = 3$ **22.** $\sqrt[3]{9p + 10} = 4$

23. $x = \frac{\sqrt{12x - 5}}{2}$ **24.** $x = \frac{\sqrt{16x - 12}}{2}$

Solve each equation. Identify any extraneous solution. SEE EXAMPLE 3. (OBJECTIVE 1)

25. $\sqrt{2y - 5} + 4 = y$ **26.** $\sqrt{3x + 13} - 3 = x$

27. $\sqrt{-5x + 24} = 6 - x$ **28.** $\sqrt{-x + 2} = x - 2$

Solve each equation. SEE EXAMPLE 4. (OBJECTIVE 1)

29. $\sqrt[3]{x^3 - 1} = x - 1$ **30.** $\sqrt[3]{x^3 - 7} = x - 1$

31. $\sqrt[3]{x^3 + 56} - 2 = x$ **32.** $\sqrt[3]{x^3 - 19} = x - 3$

Solve each equation. Identify any extraneous solution.
SEE EXAMPLE 5. (OBJECTIVE 2)

33. $2\sqrt{4x + 1} = \sqrt{x + 4}$ **34.** $\sqrt{3(x + 4)} = \sqrt{5x - 12}$

35. $\sqrt{x + 5} = \sqrt{7 - x}$ **36.** $\sqrt{10 - x} = \sqrt{2x + 4}$

37. $1 + \sqrt{z} = \sqrt{z + 3}$ **38.** $\sqrt{x + 5} = \sqrt{x + 35}$

39. $\sqrt{4s + 1} - \sqrt{6s} = -1$

40. $\sqrt{y + 7} + 3 = \sqrt{y + 4}$

Solve each equation. Identify any extraneous solution.
SEE EXAMPLE 6. (OBJECTIVE 3)

41. $\sqrt{y} + \sqrt{5} = \sqrt{y + 15}$

42. $\sqrt{x + 1} + \sqrt{3x} = \sqrt{5x + 1}$

43. $\sqrt{3x} - \sqrt{x + 1} = \sqrt{x - 2}$

44. $\sqrt{x + 2} + \sqrt{2x - 3} = \sqrt{11 - x}$

Solve each formula for the indicated variable. SEE EXAMPLE 7.
(OBJECTIVE 4)

45. $v = \sqrt{2gh}$ for g **46.** $d = 1.4\sqrt{h}$ for h

47. $T = 2\pi\sqrt{\dfrac{l}{32}}$ for l **48.** $d = \sqrt[3]{\dfrac{12V}{\pi}}$ for V

49. $r = \sqrt[3]{\dfrac{A}{P}} - 1$ for A **50.** $r = \sqrt[3]{\dfrac{A}{P}} - 1$ for P

51. $L_A = L_B\sqrt{1 - \dfrac{v^2}{c^2}}$ for v^2 **52.** $R_1 = \sqrt{\dfrac{A}{\pi} - R_2^{\,2}}$ for A

ADDITIONAL PRACTICE   *Solve each equation. Identify any extraneous solution.*

53. $\sqrt{y + 2} = 4 - y$

54. $\sqrt{22y + 86} = y + 9$

55. $2\sqrt{x} = \sqrt{5x - 16}$

56. $3\sqrt{x} = \sqrt{3x + 12}$

57. $\sqrt{2y + 1} = 1 - 2\sqrt{y}$

58. $\sqrt{u} + 3 = \sqrt{u - 3}$

59. $\sqrt{2x + 5} + \sqrt{x + 2} = 5$

60. $\sqrt{2x + 5} + \sqrt{2x + 1} + 4 = 0$

61. $5r + 4 = \sqrt{5r + 20} + 4r$

62. $\sqrt{x}\sqrt{x + 16} = 15$

63. $\sqrt{x}\sqrt{x + 6} = 4$

64. $\dfrac{6}{\sqrt{x + 5}} = \sqrt{x}$

65. $\sqrt[4]{x^4 + 4x^2 - 4} = -x$

66. $\sqrt[4]{8x - 8} + 2 = 0$

67. $2 + \sqrt{u} = \sqrt{2u + 7}$

68. $\sqrt[4]{12t + 4} + 2 = 0$

69. $u = \sqrt[4]{u^4 - 6u^2 + 24}$

70. $\sqrt{6t + 1} - 3\sqrt{t} = -1$

71. $\sqrt{x - 5} - \sqrt{x + 3} = 4$

72. $\sqrt[4]{10p + 1} = \sqrt[4]{11p - 7}$

73. $\sqrt{x + 8} - \sqrt{x - 4} = -2$

74. $\sqrt[4]{10y + 2} = 2\sqrt[4]{2}$

75. $\sqrt{z - 1} + \sqrt{z + 2} = 3$

76. $\sqrt{16v + 1} + \sqrt{8v + 1} = 12$

77. $\sqrt{\sqrt{a} + \sqrt{a + 8}} = 2$

78. $\sqrt{\sqrt{2y} - \sqrt{y - 1}} = 1$

79. $\dfrac{\sqrt{2x}}{\sqrt{x + 2}} = \sqrt{x - 1}$

80. $\sqrt{8 - x} - \sqrt{3x - 8} = \sqrt{x - 4}$

APPLICATIONS   *Solve each application. SEE EXAMPLE 2.*
(OBJECTIVE 1)

81. Highway design   A curve banked at 8° will accommodate traffic traveling s mph if the radius of the curve is r feet, according to the formula $s = 1.45\sqrt{r}$. If engineers expect 65-mph traffic, what radius should they specify?

82. Horizon distance   The higher a lookout tower is built, the farther an observer can see. That distance d (called the *horizon distance*, measured in miles) is related to the height h of the observer (measured in feet) by the formula $d = 1.4\sqrt{h}$. How tall must a lookout tower be to see the edge of the forest, 25 miles away?

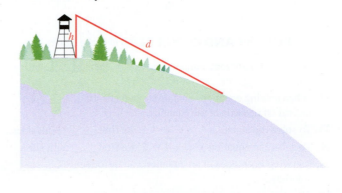

83. Generating power The power generated by a windmill is related to the velocity of the wind by the formula

$$v = \sqrt[3]{\frac{P}{0.02}}$$

where P is the power (in watts) and v is the velocity of the wind (in mph). Find the speed of the wind when the windmill is generating 500 watts of power.

84. Carpentry During construction, carpenters often brace walls as shown in the illustration, where the length of the brace is given by the formula

$$l = \sqrt{f^2 + h^2}$$

If a carpenter nails a 10-ft brace to the wall 6 feet above the floor, how far from the base of the wall should he nail the brace to the floor?

Use a graphing calculator.

85. Depreciation The formula

$$r = 1 - \sqrt[n]{\frac{S}{C}}$$

gives the annual depreciation rate r of a car that had an original cost of C dollars, a useful life of n years, and a trade-in value of S dollars. Find the annual depreciation rate of a car that cost $22,000 and was sold 15 years later as salvage for $900. Give the result to the nearest percent.

86. Savings accounts The interest rate r earned by a savings account after n compoundings is given by the formula

$$\sqrt[n]{\frac{V}{P}} - 1 = r$$

where V is the current value and P is the original principal. What interest rate r was paid on an account in which a deposit of $1,000 grew to $1,338.23 after 5 compoundings?

87. Marketing The number of wrenches that will be produced at a given price can be predicted by the formula $s = \sqrt{5x}$, where s is the supply (in thousands) and x is the price (in dollars). If the demand, d, for wrenches can be predicted by the formula $d = \sqrt{100 - 3x^2}$, find the equilibrium price.

88. Marketing The number of footballs that will be produced at a given price can be predicted by the formula $s = \sqrt{23x}$, where s is the supply (in thousands) and x is the price (in dollars). If the demand, d, for footballs can be predicted by the formula $d = \sqrt{312 - 2x^2}$, find the equilibrium price.

89. Medicine The resistance R to blood flow through an artery can be found using the formula

$$r = \sqrt[4]{\frac{8kl}{\pi R}}$$

where r is the radius of the artery, k is the viscosity of blood, and l is the length of the artery. Solve the formula for R.

90. Generating power The power P generated by a windmill is given by the formula

$$s = \sqrt[3]{\frac{P}{0.02}}$$

where s is the speed of the wind. Solve the formula for P.

WRITING ABOUT MATH

91. If both sides of an equation are raised to the same power, the resulting equation might not be equivalent to the original equation. Explain.

92. Explain why you must check each apparent solution of a radical equation.

SOMETHING TO THINK ABOUT

93. Solve: $\sqrt[3]{2x} = \sqrt{x}$ **94.** Solve: $\sqrt[4]{x} = \sqrt{\frac{x}{4}}$

Section 7.7

Complex Numbers

Objectives

1. Simplify an imaginary number.
2. Write a complex number in $a + bi$ form.
3. Determine whether two complex numbers are equal.
4. Simplify an expression containing complex numbers.
5. Rationalize the denominator of a fraction that contains a complex number.
6. Find a specified power of i and apply this pattern to simplifying an expression.
7. Find the absolute value of a complex number.

Vocabulary

imaginary number   complex conjugates   absolute value of a complex
complex number   number

Getting Ready

Perform the following operations.

1. $(3x + 5) + (4x - 5)$
2. $(3x + 5) - (4x - 5)$
3. $(3x + 5)(4x - 5)$
4. $(3x + 5)(3x - 5)$

We have seen that square roots of negative numbers are not real numbers. However, there is a set of numbers, called the *complex numbers*, in which negative numbers do have square roots. In this section, we will discuss this broader set of numbers.

1 Simplify an imaginary number.

Consider the number $\sqrt{-3}$. Since no real number squared is -3, $\sqrt{-3}$ is not a real number. For years, people believed that numbers such as

$$\sqrt{-1}, \quad \sqrt{-3}, \quad \sqrt{-4}, \quad \text{and} \quad \sqrt{-9}$$

were nonsense. In the 17th century, René Descartes (1596–1650) called them **imaginary numbers**. Today, imaginary numbers have many important uses, such as describing the behavior of alternating current in electronics.

The imaginary number $\sqrt{-1}$ often is denoted by the letter i:

$$i = \sqrt{-1}$$

Because i represents the square root of -1, it follows that

$$i^2 = -1$$

Perspective

The Pythagoreans (ca. 500 B.C.) understood the universe as a harmony of whole numbers. They did not classify fractions as numbers, and were upset that $\sqrt{2}$ was not the ratio of whole numbers. For 2,000 years, little progress was made in the understanding of the various kinds of numbers.

The father of algebra, François Vieta (1540–1603), understood the whole numbers, fractions, and certain irrational numbers. But he was unable to accept negative numbers, and certainly not imaginary numbers.

René Descartes (1596–1650) thought these numbers to be nothing more than figments of his imagination, so he called them *imaginary numbers*. Leonhard Euler (1707–1783) used the letter i for $\sqrt{-1}$; Augustin Cauchy (1789–1857) used the term *conjugate*; and Carl Gauss (1777–1855) first used the word *complex*.

Today, we accept complex numbers without question, but it took many centuries and the work of many mathematicians to make them respectable.

If we assume that multiplication of imaginary numbers is commutative and associative, then

$$(2i)^2 = 2^2 i^2$$
$$= 4(-1) \quad i^2 = -1$$
$$= -4$$

Since $(2i)^2 = -4$, $2i$ is a square root of -4, and we can write

$$\sqrt{-4} = 2i$$

This result also can be obtained by using the multiplication property of radicals:

$$\sqrt{-4} = \sqrt{4(-1)} = \sqrt{4}\sqrt{-1} = 2i$$

EXAMPLE 1 Simplify:

a. $\sqrt{-25}$ **b.** $\sqrt{\dfrac{-100}{49}}$

Solution We can use the multiplication property of radicals to simplify any imaginary number. For example,

a. $\sqrt{-25} = \sqrt{25(-1)} = \sqrt{25}\sqrt{-1} = 5i$

b. $\sqrt{\dfrac{-100}{49}} = \sqrt{\dfrac{100}{49}(-1)} = \dfrac{\sqrt{100}}{\sqrt{49}}\sqrt{-1} = \dfrac{10}{7}i$

SELF CHECK 1 Simplify: **a.** $\sqrt{-36}$ **b.** $\sqrt{\dfrac{-4}{9}}$

These examples illustrate the following rule.

PROPERTIES OF RADICALS

If at least one of a and b is a nonnegative real number, then

$$\sqrt{ab} = \sqrt{a}\sqrt{b} \quad \text{and} \quad \sqrt{\dfrac{a}{b}} = \dfrac{\sqrt{a}}{\sqrt{b}} \quad (b \neq 0)$$

COMMENT   If a and b are both negative, then $\sqrt{ab} \neq \sqrt{a}\sqrt{b}$. Recall, the simplified form of a radical expression does not contain a negative radicand. Therefore the $\sqrt{-16}$ and the $\sqrt{-4}$ must be simplified prior to multiplying. For example, if $a = -16$ and $b = -4$, we have

$$\sqrt{(-16)}\sqrt{(-4)} = (4i)(2i) = 8i^2 = 8(-1) = -8$$

The imaginary numbers are a subset of a set of numbers called the *complex numbers*.

COMPLEX NUMBERS

A **complex number** is any number that can be written in the standard form $a + bi$, where a and b are real numbers and $i = \sqrt{-1}$.

In the complex number $a + bi$, a is called the *real part*, and b is called the *imaginary part*.

If $b = 0$, the complex number $a + bi$ is a real number. If $b \neq 0$ and $a = 0$, the complex number $0 + bi$ (or just bi) is an imaginary number.

Any imaginary number can be expressed in bi form. For example,

COMMENT   The expression $\sqrt{3}i$ is sometimes written as $i\sqrt{3}$ to make it clear that i is not part of the radicand.

$$\sqrt{-1} = i$$
$$\sqrt{-9} = \sqrt{9(-1)} = \sqrt{9}\sqrt{-1} = 3i$$
$$\sqrt{-3} = \sqrt{3(-1)} = \sqrt{3}\sqrt{-1} = \sqrt{3}i$$

The relationship among the real numbers, the imaginary numbers, and the complex numbers is shown in Figure 7-18.

Figure 7-18

---

2   Write a complex number in $a + bi$ form.

The next example shows how to write complex numbers in $(a + bi)$ form. It is common to use $(a - bi)$ as a substitute for $a + (-b)i$.

EXAMPLE 2   Write each number in $a + bi$ form.

a. $7 = 7 + 0i$ **b.** $3i = 0 + 3i$

c. $4 - \sqrt{-16} = 4 - \sqrt{-1(16)}$ **d.** $5 + \sqrt{-11} = 5 + \sqrt{-1(11)}$
$\phantom{4 - \sqrt{-16}} = 4 - \sqrt{16}\sqrt{-1}$ $\phantom{5 + \sqrt{-11}} = 5 + \sqrt{11}\sqrt{-1}$
$\phantom{4 - \sqrt{-16}} = 4 - 4i$ $\phantom{5 + \sqrt{-11}} = 5 + \sqrt{11}i$

 SELF CHECK 2   Write $3 - \sqrt{-25}$ in $a + bi$ form.

3 Determine whether two complex numbers are equal.

EQUALITY OF COMPLEX NUMBERS	The complex numbers $a + bi$ and $c + di$ are equal if and only if $$a = c \quad \text{and} \quad b = d$$

Because of the preceding definition, complex numbers are equal when their real parts are equal and their imaginary parts are equal.

EXAMPLE 3

a. $2 + 3i = \sqrt{4} + \dfrac{6}{2}i$ because $2 = \sqrt{4}$ and $3 = \dfrac{6}{2}$.

b. $4 - 5i = \dfrac{12}{3} - \sqrt{25}\,i$ because $4 = \dfrac{12}{3}$ and $-5 = -\sqrt{25}$.

c. $x + yi = 4 + 7i$ if and only if $x = 4$ and $y = 7$.

SELF CHECK 3   Is $2 + 3i = \sqrt{4} - \dfrac{6}{2}i$?

4 Simplify an expression containing complex numbers.

Now that we can simplify a complex number, we will move to adding, subtracting, and multiplying them.

ADDITION AND SUBTRACTION OF COMPLEX NUMBERS	Complex numbers are added and subtracted as if they were binomials: $$(a + bi) + (c + di) = (a + c) + (b + d)i$$ $$(a + bi) - (c + di) = (a + bi) + (-c - di) = (a - c) + (b - d)i$$

The preceding definition suggests that when adding or subtracting two complex numbers, we add or subtract the real parts and then add or subtract the imaginary parts.

EXAMPLE 4   Perform the operations.

a. $(8 + 4i) + (12 + 8i) = 8 + 4i + 12 + 8i$
$$= 20 + 12i$$

b. $(7 - 4i) + (9 + 2i) = 7 - 4i + 9 + 2i$
$$= 16 - 2i$$

c. $(-6 + i) - (3 - 4i) = -6 + i - 3 + 4i$
$$= -9 + 5i$$

d. $(2 - 4i) - (-4 + 3i) = 2 - 4i + 4 - 3i$
$$= 6 - 7i$$

SELF CHECK 4   Perform the operations.
a. $(3 - 5i) + (-2 + 7i)$ **b.** $(3 - 5i) - (-2 + 7i)$

To multiply a complex number by an imaginary number, we use the distributive property to remove parentheses and simplify. For example,

$$\overset{\frown}{-5i(4 - 8i)} = -5i(4) - (-5i)8i \qquad \text{Use the distributive property to remove parentheses.}$$
$$= -20i + 40i^2 \qquad \text{Simplify.}$$
$$= -20i + 40(-1) \qquad \text{Remember that } i^2 = -1.$$
$$= -40 - 20i$$

To multiply two complex numbers, we use the following definition.

MULTIPLYING COMPLEX NUMBERS

Complex numbers are multiplied as if they were binomials, with $i^2 = -1$:

$$(a + bi)(c + di) = ac + adi + bci + bdi^2$$
$$= ac + adi + bci + bd(-1)$$
$$= (ac - bd) + (ad + bc)i$$

EXAMPLE 5 Multiply the complex numbers.

a. $2i(7 - 3i) = 14i - 6i^2$ Use the distributive property to remove parentheses.
$$= 14i + 6 \qquad i^2 = -1$$
$$= 6 + 14i$$

b. $(2 + 3i)(3 - 2i) = 6 - 4i + 9i - 6i^2$ Use the distributive property to remove parentheses.
$$= 6 + 5i + 6 \qquad i^2 = -1, \text{ combine } -4i \text{ and } 9i.$$
$$= 12 + 5i$$

c. $(3 + i)(1 + 2i) = 3 + 6i + i + 2i^2$ Use the distributive property to remove parentheses.
$$= 3 + 7i - 2 \qquad i^2 = -1, \text{ combine } 6i \text{ and } i.$$
$$= 1 + 7i$$

d. $(-4 + 2i)(2 + i) = -8 - 4i + 4i + 2i^2$ Use the distributive property to remove parentheses.
$$= -8 - 2 \qquad i^2 = -1, \text{ combine } -4i \text{ and } 4i.$$
$$= -10$$

SELF CHECK 5   Multiply: **a.** $3(4 - 5i)$ **b.** $(-2 + 3i)(3 - 2i)$

COMPLEX CONJUGATES   The complex numbers $(a + bi)$ and $(a - bi)$ are called **complex conjugates**.

For example,

$(3 + 4i)$ and $(3 - 4i)$ are complex conjugates.
$(5 - 7i)$ and $(5 + 7i)$ are complex conjugates.

EXAMPLE 6   Find the product of $(3 + i)$ and its complex conjugate.

Solution   The complex conjugate of $(3 + i)$ is $(3 - i)$. We can find the product as follows:

$$(3 + i)(3 - i) = 9 - 3i + 3i - i^2 \quad \text{Use the distributive property.}$$
$$= 9 - i^2 \qquad\qquad\quad \text{Combine like terms.}$$
$$= 9 - (-1) \qquad\qquad i^2 = -1$$
$$= 10$$

 SELF CHECK 6   Multiply: $(2 + 3i)(2 - 3i)$

The product of the complex number $a + bi$ and its complex conjugate $a - bi$ is the real number $a^2 + b^2$, as the following work shows:

$$(a + bi)(a - bi) = a^2 - abi + abi - b^2i^2 \quad \begin{array}{l}\text{Use the distributive property to remove}\\ \text{parentheses.}\end{array}$$
$$= a^2 - b^2(-1) \qquad\qquad\quad i^2 = -1$$
$$= a^2 + b^2$$

5   ## Rationalize the denominator of a fraction that contains a complex number.

If $b \neq 0$, the complex number $a + bi$ contains the square root $i = \sqrt{-1}$. Since a square root cannot remain in the denominator of a fraction, we often have to rationalize a denominator when dividing complex numbers.

EXAMPLE 7   Divide and write the result in $a + bi$ form: $\dfrac{1}{3 + i}$

Solution   We can rationalize the denominator by multiplying the numerator and the denominator by the complex conjugate of the denominator.

$$\frac{1}{3 + i} = \frac{1}{3 + i} \cdot \frac{3 - i}{3 - i} \qquad \begin{array}{l}\text{Multiply the numerator and denominator by the complex}\\ \text{conjugate of the denominator.}\end{array}$$

$$= \frac{3 - i}{9 - 3i + 3i - i^2} \qquad \text{Multiply.}$$

$$= \frac{3 - i}{9 - (-1)} \qquad\qquad i^2 = -1$$

$$= \frac{3 - i}{10} \qquad\qquad\quad \text{Subtract.}$$

$$= \frac{3}{10} - \frac{1}{10}i \qquad\qquad \text{Write the result in } a + bi \text{ form.}$$

 SELF CHECK 7   Divide and write the result in $a + bi$ form. $\dfrac{1}{5 - i}$

EXAMPLE 8 Write $\dfrac{3-i}{2+i}$ in $a+bi$ form.

Solution We multiply the numerator and the denominator of the fraction by the complex conjugate of the denominator.

$$\frac{3-i}{2+i} = \frac{3-i}{2+i} \cdot \frac{\mathbf{2-i}}{\mathbf{2-i}} \qquad \tfrac{2-i}{2-i}=1$$

$$= \frac{6-3i-2i+i^2}{4-2i+2i-i^2} \qquad \text{Multiply the numerators and multiply the denominators.}$$

$$= \frac{6-5i+(-1)}{4-(-1)} \qquad \text{Combine like terms and substitute } i^2 \text{ with } -1.$$

$$= \frac{5-5i}{4-(\mathbf{-1})} \qquad \text{Subtract.}$$

$$= \frac{5(1-i)}{5} \qquad \text{Factor out 5, the GCF.}$$

$$= 1-i \qquad \text{Simplify.}$$

🌱 **SELF CHECK 8** Rationalize the denominator: $\dfrac{2+i}{5-i}$

EXAMPLE 9 Write $\dfrac{4+\sqrt{-16}}{2+\sqrt{-4}}$ in $a+bi$ form.

Solution $$\frac{4+\sqrt{-16}}{2+\sqrt{-4}} = \frac{4+4i}{2+2i} \qquad \text{Write each number in } a+bi \text{ form.}$$

$$= \frac{\overset{1}{\cancel{2(2+2i)}}}{\underset{1}{\cancel{2+2i}}} \qquad \text{Factor out 2, the GCF, and simplify.}$$

$$= 2+0i$$

🌱 **SELF CHECK 9** Divide: $\dfrac{6+\sqrt{-81}}{2+\sqrt{-9}}$

COMMENT To avoid mistakes, always write complex numbers in $a+bi$ form prior to performing any operations with complex numbers.

6 Find a specified power of i and apply this pattern to simplifying an expression.

The powers of i produce an interesting pattern:

$$i = \sqrt{-1} = \mathbf{i} \qquad\qquad i^5 = i^4 i = 1i = \mathbf{i}$$
$$i^2 = \left(\sqrt{-1}\right)^2 = \mathbf{-1} \qquad i^6 = i^4 i^2 = 1(-1) = \mathbf{-1}$$
$$i^3 = i^2 i = -1i = \mathbf{-i} \qquad i^7 = i^4 i^3 = 1(-i) = \mathbf{-i}$$
$$i^4 = i^2 i^2 = (-1)(-1) = \mathbf{1} \qquad i^8 = i^4 i^4 = (1)(1) = \mathbf{1}$$

The pattern continues: $i, -1, -i, 1, \ldots$. When simplifying a specified power of i the calculation is easier if we can write the expression in terms of i^4 since $i^4 = 1$. If there is a remainder when the exponent is divided by 4, use the pattern $i, -1, -i, 1$ to simplify the expression as illustrated in the next example.

EXAMPLE 10   Simplify: i^{29}

Solution   We note that 29 divided by 4 gives a quotient of 7 and a remainder of 1. Thus, $29 = 4 \cdot 7 + 1$, and

$$i^{29} = i^{4 \cdot 7 + 1} \qquad \textcolor{red}{29 = 4 \cdot 7 + 1}$$
$$= (i^4)^7 \cdot i \qquad \textcolor{red}{i^{4 \cdot 7 + 1} = i^{4 \cdot 7} \cdot i^1 = (i^4)^7 \cdot i}$$
$$= 1^7 \cdot i \qquad \textcolor{red}{i^4 = 1}$$
$$= i$$

SELF CHECK 10   Simplify: i^{31}

The results of Example 10 illustrate the following process.

POWERS OF i

If n is a natural number that has a remainder of r when divided by 4, then

$$i^n = i^r.$$

When n is divisible by 4, the remainder r is 0 and $i^0 = 1$.

EXAMPLE 11   Simplify: i^{55}

Solution   We divide 55 by 4 and get a remainder of 3. Therefore,

$$i^{55} = i^3 = -i$$

SELF CHECK 11   Simplify: i^{62}

EXAMPLE 12   Simplify each expression. If a denominator has a factor of i, rationalize the denominator. Write the result in $a + bi$ form.

a. $2i^2 + 4i^3 = 2(\textcolor{red}{-1}) + 4(\textcolor{red}{-i})$
$$= -2 - 4i$$

b. $\dfrac{3}{2i} = \dfrac{3}{2i} \cdot \dfrac{\textcolor{red}{i}}{\textcolor{red}{i}} \qquad \textcolor{red}{\frac{i}{i} = 1}$
$$= \dfrac{3i}{2i^2}$$
$$= \dfrac{3i}{2(\textcolor{red}{-1})}$$
$$= \dfrac{3i}{-2}$$
$$= 0 - \dfrac{3}{2}i$$

c. $-\dfrac{5}{i} = -\dfrac{5}{i} \cdot \dfrac{\textcolor{red}{i}}{\textcolor{red}{i}} \qquad \textcolor{red}{\frac{i}{i} = 1}$
$$= -\dfrac{5(i)}{i^2}$$
$$= -\dfrac{5i}{-1}$$
$$= 5i$$
$$= 0 + 5i$$

d. $\dfrac{6}{i^3} = \dfrac{6\textcolor{red}{i}}{i^3 \textcolor{red}{i}} \qquad \textcolor{red}{\frac{i}{i} = 1}$
$$= \dfrac{6i}{i^4}$$
$$= \dfrac{6i}{1}$$
$$= 6i$$
$$= 0 + 6i$$

🌿 **SELF CHECK 12**   Simplify and write the result in $a + bi$ form.   **a.** $3i^3 - 2i^2$ **b.** $\dfrac{2}{3i}$

---

7   Find the absolute value of a complex number.

ABSOLUTE VALUE OF A COMPLEX NUMBER	The **absolute value** of the complex number $a + bi$ is $\sqrt{a^2 + b^2}$. In symbols, $$|a + bi| = \sqrt{a^2 + b^2}$$

EXAMPLE 13   Find each absolute value.

a. $\begin{aligned} |3 + 4i| &= \sqrt{3^2 + 4^2} \\ &= \sqrt{9 + 16} \\ &= \sqrt{25} \\ &= 5 \end{aligned}$

b. $\begin{aligned} |3 - 4i| &= \sqrt{3^2 + (-4)^2} \\ &= \sqrt{9 + 16} \\ &= \sqrt{25} \\ &= 5 \end{aligned}$

c. $\begin{aligned} |-5 - 12i| &= \sqrt{(-5)^2 + (-12)^2} \\ &= \sqrt{25 + 144} \\ &= \sqrt{169} \\ &= 13 \end{aligned}$

d. $\begin{aligned} |a + 0i| &= \sqrt{a^2 + 0^2} \\ &= \sqrt{a^2} \\ &= |a| \end{aligned}$

🌿 **SELF CHECK 13**   Evaluate: $|5 + 12i|$

🌿 **SELF CHECK ANSWERS**

1. a. $6i$ **b.** $\frac{2}{3}i$ **2.** $3 - 5i$ **3.** no   **4. a.** $1 + 2i$ **b.** $5 - 12i$ **5. a.** $12 - 15i$ **b.** $13i$ **6.** 13 **7.** $\frac{5}{26} + \frac{1}{26}i$
8. $\frac{9}{26} + \frac{7}{26}i$ **9.** $3 + 0i$ **10.** $-i$ **11.** -1 **12. a.** $2 - 3i$ **b.** $0 - \frac{2}{3}i$ **13.** 13

NOW TRY THIS

1. Simplify: $-\sqrt{-8}\sqrt{-2}$
2. Evaluate $3x^2 - 2x - 4$ for $x = 2 - 3i$.

7.7   Exercises 🌿

WARM-UPS   *Identify each number as a real number or not a real number.*

1. $\sqrt{-9}$

2. $-\sqrt{25}$

3. $\sqrt{a}$ if $a \geq 0$

4. $\sqrt{x}$ if $x < 0$

5. $-\sqrt{-1}$

6. $\sqrt{-1}$

Find the conjugate for each expression.

7. $3 + \sqrt{5}$ **8.** $-7 - \sqrt{2}$

Combine like terms.

9. $5 + 3x + 7 + 4x$

10. $-3 - 4x + 9 - 6x$

REVIEW *Perform each operation.*

11. $\dfrac{x^2 - x - 6}{9 - x^2} \cdot \dfrac{x^2 + x - 6}{x^2 - 4}$ **12.** $\dfrac{3x + 4}{x - 2} + \dfrac{x - 4}{x + 2}$

13. Wind speed A plane that can fly 200 mph in still air makes a 330-mile flight with a tail wind and returns, flying into the same wind. Find the speed of the wind if the total flying time is $3\frac{1}{3}$ hours.

14. Finding rates A student drove a distance of 135 miles at an average speed of 50 mph. How much faster would he have to drive on the return trip to save 30 minutes of driving time?

VOCABULARY AND CONCEPTS *Fill in the blanks.*

15. $\sqrt{-1} = \_$ **16.** $i^6 = \_\_\_$

17. $i^7 = \_\_\_$ **18.** $i^8 = \_$

19. $\sqrt{-1}, \sqrt{-3}, \sqrt{-4}$ are examples of _____ numbers.

20. $\sqrt{ab} = \_\_\_\_$, provided a and b are not both negative.

21. $\sqrt{\dfrac{a}{b}} = \_\_\_\_$ $(b \neq 0)$, provided a and b are not both negative.

22. $3 + 5i, 2 - 7i$, and $5 - \frac{1}{2}i$ are examples of _____ numbers.

23. The real part of $5 + 7i$ is __. The imaginary part is __.

24. $a + bi = c + di$ if and only if $a = \_$ and $b = \_$.

25. $a + bi$ and $a - bi$ are called complex _____.

26. $|a + bi| = _____$

GUIDED PRACTICE *Write each imaginary number in simplified form.* **SEE EXAMPLE 1. (OBJECTIVE 1)**

27. $\sqrt{-25}$ **28.** $\sqrt{-16}$

29. $\sqrt{-121}$ **30.** $\sqrt{-81}$

31. $\sqrt{-7}$ **32.** $\sqrt{-29}$

33. $\sqrt{-8}$ **34.** $\sqrt{-24}$

Write each number in a + bi form. **SEE EXAMPLE 2. (OBJECTIVE 2)**

35. 9 **36.** $4i$

37. $9 + \sqrt{-9}$ **38.** $16 + \sqrt{-36}$

Determine whether the complex numbers are equal. **SEE EXAMPLE 3. (OBJECTIVE 3)**

39. $3 + 7i, \sqrt{9} + (5 + 2)i$ **40.** $\sqrt{4} + \sqrt{25}i, 2 - (-5)i$

41. $\sqrt{4} + \sqrt{-4}, 2 - 2i$ **42.** $\sqrt{-9} - i, 4i$

Simplify each expression. Write all answers in standard form. **SEE EXAMPLE 4. (OBJECTIVE 4)**

43. $(4 + 7i) + (3 - 5i)$ **44.** $(5 + 3i) - (6 - 9i)$

45. $(7 - 3i) - (4 + 2i)$ **46.** $(8 + 3i) + (-7 - 2i)$

47. $(8 + 5i) + (7 + 2i)$ **48.** $(-6 - 8i) - (-9 - 3i)$

49. $(1 + i) - 2i + (5 - 7i)$ **50.** $(-9 + i) - 5i + (2 + 7i)$

Simplify each expression. Write all answers in standard form. **SEE EXAMPLE 5. (OBJECTIVE 4)**

51. $6i(3 - 4i)$ **52.** $-4i(3 + 4i)$

53. $-5i(5 - 5i)$ **54.** $2i(7 + 2i)$

55. $(2 + i)(3 - i)$ **56.** $(4 - i)(2 + i)$

57. $(5 + 2i)(4 - 6i)$ **58.** $(3 - 2i)(4 - 3i)$

59. $\left(4 + \sqrt{3}i\right)\left(3 - \sqrt{3}i\right)$ **60.** $\left(5 + \sqrt{3}i\right)\left(2 - \sqrt{3}i\right)$

61. $(4 + 3i)^2$ **62.** $(3 - 2i)^2$

Find the product of the complex number and its complex conjugate. **SEE EXAMPLE 6. (OBJECTIVE 4)**

63. $6 - 5i$ **64.** $7 - 2i$

65. $3 + 4i$ **66.** $5 + 2i$

Divide and write each expression in standard form. **SEE EXAMPLE 7. (OBJECTIVE 5)**

67. $\dfrac{5}{2 - i}$ **68.** $\dfrac{26}{3 - 2i}$

69. $\dfrac{10}{5 - i}$ **70.** $\dfrac{4}{2 + i}$

Divide and write each expression in standard form. **SEE EXAMPLE 8. (OBJECTIVE 5)**

71. $\dfrac{3 - 2i}{3 + 2i}$ **72.** $\dfrac{2 + 3i}{2 - 3i}$

73. $\dfrac{3 + 2i}{3 + i}$ **74.** $\dfrac{2 - 5i}{2 + 5i}$

Divide and write each expression in standard form. **SEE EXAMPLE 9. (OBJECTIVE 5)**

75. $\dfrac{-12}{7 - \sqrt{-1}}$ **76.** $\dfrac{4}{3 + \sqrt{-1}}$

77. $\dfrac{7}{5 - \sqrt{-9}}$ **78.** $\dfrac{3}{2 + \sqrt{-25}}$

Simplify each expression. **SEE EXAMPLES 10–11. (OBJECTIVE 6)**

79. i^5 **80.** i^7

81. i^{10} **82.** i^{12}

83. i^{37} **84.** i^{19}

85. i^{27} **86.** i^{22}

87. i^{64} **88.** i^{82}

89. i^{97} **90.** i^{200}

Simplify each expression. **SEE EXAMPLE 12. (OBJECTIVES 5–6)**

91. $5i^3 + 2i^2$ **92.** $4i^2 - 3i^3$

93. $\dfrac{1}{i}$ **94.** $\dfrac{1}{i^3}$

95. $\dfrac{4}{5i^3}$

96. $\dfrac{2}{3i}$

97. $\dfrac{3i}{8\sqrt{-9}}$

98. $\dfrac{5i^3}{2\sqrt{-4}}$

99. $\dfrac{-3}{5i^5}$

100. $\dfrac{-4}{6i^7}$

Find each value. SEE EXAMPLE 13. (OBJECTIVE 7)

101. $|6 + 8i|$

102. $|12 + 5i|$

103. $|12 - 5i|$

104. $|3 - 4i|$

105. $|5 + 7i|$

106. $|6 - 5i|$

107. $\left| \dfrac{3}{5} - \dfrac{4}{5}i \right|$

108. $\left| \dfrac{5}{13} + \dfrac{12}{13}i \right|$

ADDITIONAL PRACTICE Are the two numbers equal?

109. $8 + 5i,\ 2^3 + \sqrt{25}i^3$

110. $4 - 7i,\ -4i^2 + 7i^3$

Simplify each expression. Write the answer in standard form.

111. $\left(8 - \sqrt{-1}\right)\left(-2 - \sqrt{-16}\right)$

112. $\left(-1 + \sqrt{-4}\right)\left(2 + \sqrt{-9}\right)$

113. $(2 + 3i)^2$

114. $(1 - 3i)^2$

115. $\dfrac{5i}{6 + 2i}$

116. $\dfrac{-4i}{2 - 6i}$

117. $\dfrac{\sqrt{5} - \sqrt{3}i}{\sqrt{5} + \sqrt{3}i}$

118. $\dfrac{\sqrt{3} + \sqrt{2}i}{\sqrt{3} - \sqrt{2}i}$

119. $\left(\dfrac{i}{3 + 2i}\right)^2$

120. $\left(\dfrac{5 + i}{2 + i}\right)^2$

121. $(5 + 3i) - (3 - 5i) + \sqrt{-1}$

122. $(8 + 7i) - \left(-7 - \sqrt{-64}\right) + (3 - i)$

123. $\left(-8 - \sqrt{3}i\right) - \left(7 - 3\sqrt{3}i\right)$

124. $\left(4 + 3\sqrt{5}i\right) + \left(-6 - 4\sqrt{5}i\right)$

125. $(2 + i)(2 - i)(1 + i)$

126. $(3 + 2i)(3 - 2i)(i + 1)$

127. $(3 + i)[(3 - 2i) + (2 + i)]$

128. $(2 - 3i)[(5 - 2i) - (2i + 1)]$

APPLICATIONS *In electronics, the formula V = IR is called Ohm's law. It gives the relationship in a circuit between the voltage V (in volts), the current I (in amperes), and the resistance R (in ohms).*

129. Electronics Find V when $I = 2 - 3i$ amperes and $R = 2 + i$ ohms.

130. Electronics Find R when $I = 3 - 2i$ amperes and $V = 18 + i$ volts.

In electronics, the formula $Z = \dfrac{V}{I}$ is used to find the impedance Z of a circuit, where V is the voltage and I is the current.

131. Electronics Find the impedance of a circuit when the voltage is $1.7 + 0.5i$ and the current is $0.5i$.

132. Electronics Find the impedance of a circuit when the voltage is $1.6 - 0.4i$ and the current is $-0.2i$.

WRITING ABOUT MATH

133. Determine how to decide whether two complex numbers are equal.

134. Define the complex conjugate of a complex number.

135. Show that $1 - 5i$ is a solution of $x^2 - 2x + 26 = 0$.

136. Show that $3 - 2i$ is a solution of $x^2 - 6x + 13 = 0$.

137. Show that i is a solution of $x^4 - 3x^2 - 4 = 0$.

138. Show that $2 + i$ is *not* a solution of $x^2 + x + 1 = 0$.

SOMETHING TO THINK ABOUT

139. Rationalize the numerator: $\dfrac{3 - i}{2}$

140. Rationalize the numerator: $\dfrac{2 + 3i}{2 - 3i}$

Projects

PROJECT 1

The size of a TV screen is measured along the diagonal of its screen, as shown in the illustrations. The screen of a traditional TV has an aspect ratio of 4:3. This means that the ratio of the width of the screen to its height is $\dfrac{4}{3}$. The screen of a wide-screen set has an aspect ratio of 16:9. This means that the ratio of the width of the screen to its height is $\dfrac{16}{9}$.

50 in.

a. Find the width and height of the traditional-screen set shown in the illustration on the previous page. $\left(\textit{Hint: } \frac{4}{3} = \frac{4x}{3x}.\right)$

b. Find the viewing area of the traditional-screen set in square inches.

c. Find the width and height of the wide-screen set shown in the illustration above.

d. Find the viewing area of the wide-screen set in square inches.

e. Which set has the larger viewing area? Give the answer as a percent.

PROJECT 2

Tom and Brian arrange to have a bicycle race. Each leaves his own house at the same time and rides to the other's house, whereupon the winner of the race calls his own house and leaves a message for the loser. A map of the race is shown in the illustration. Brian stays on the highway, averaging 21 mph. Tom knows that he and Brian are evenly matched when biking on the highway, so he cuts across country for the first part of his trip, averaging 15 mph. When Tom reaches the highway at point A, he turns right and follows the highway, averaging 21 mph.

Tom and Brian never meet during the race and, amazingly, the race is a tie. Each of them calls the other at exactly the same moment!

a. How long (to the nearest second) did it take each person to complete the race?

b. How far from the intersection of the two highways is point A? (*Hint:* Set the travel times for Brian and Tom equal to each other. You may find two answers, but only one of them matches all of the information.)

c. Show that if Tom had started straight across country for Brian's house (in order to minimize the distance he had to travel), he would have lost the race. By how much time (to the nearest second) would he have lost? Then show that if Tom had biked across country to a point 9 miles from the intersection of the two highways, he would have won the race. By how much time (to the nearest second) would he have won?

Reach for Success REVIEWING YOUR GAME PLAN

In an activity in a previous chapter, you set goals for this course. It is now time to check your progress toward achieving them.

A. Fill in the blanks from your *original goals* for this course.

The grade I am willing to work to achieve is a/an _____.

Considering other course commitments as well as any work and family commitments, state the number of hours *outside of class* you realistically believe you can devote each week to this one course. _____

List at least three things you are willing to add to your game plan that will support your success in this course.

1. _____

2. _____

3. _____

What else can you do to improve your performance in this class? _____

_____

B. Considering your performance in this course to date, please answer these questions.

What is your approximate grade in this class now? _____ Is this the grade you want to earn? _____ If this is your goal grade, or a higher one, congratulations! It appears you stayed with your game plan and should continue with this plan for the remainder of the semester.

If the grade is lower than what you want, let's look at some factors that may have contributed to this.

List the items on your game plan that you have done to be successful in this course.

_____

Congratulations on following through with these items!

If there are items on your game plan that you have not done, list these and consider why you have not been able to follow through with your plan. Some reasons might be work schedule, unreliable child care, other course demands, content more difficult than anticipated, motivational issues, or personal issues.

_____

C. Now, let's determine if there are changes you can make now to turn the semester around.

Is it necessary to modify the grade you would be working toward? _____ What is that new grade? _____

Of the items listed in Part B above, consider and list any changes you can make now to improve your chance of success in this course before the end of the semester.

_____

_____

D. Is your success in this course a high enough priority to make the changes you listed above? _____

7   Review

SECTION 7.1   Radical Expressions

DEFINITIONS AND CONCEPTS	EXAMPLES
Simplifying radicals: Assume n is a natural number greater than 1 and x is a real number. If $x > 0$, then $\sqrt[n]{x}$ is the positive number such that $\left(\sqrt[n]{x}\right)^n = x$. If $x = 0$, then $\sqrt[n]{x} = 0$. If $x < 0$, and n is odd, $\sqrt[n]{x}$ is the real number such that $\left(\sqrt[n]{x}\right)^n = x$. If $x < 0$, and n is even, $\sqrt[n]{x}$ is not a real number.	$\sqrt{25} = 5$ because $5^2 = 25$ (the principal square root). $\sqrt[4]{0} = 0$ because $0^4 = 0$. $\sqrt[3]{-64} = -4$ because $(-4)^3 = -64$. $\sqrt{-4}$ is not a real number because there is no real number when squared that equals -4.
If n is an even natural number, $\sqrt[n]{a^n} = |a|$ If n is an odd natural number, greater than 1, $\sqrt[n]{a^n} = a$	Simplify. $\sqrt{64x^2} = 8|x|$ Absolute value bars are necessary because x could be a negative number. $\sqrt[3]{8x^3} = 2x$ Absolute value bars are not necessary because the index is odd.
Finding the domain of a radical function: If $f(x) = \sqrt[n]{x}$, then the domain of $f(x)$ is $(-\infty, \infty)$ when n is odd. If n is even, find the domain by setting the radicand greater than or equal to 0.	To find the domain of $g(x) = \sqrt[3]{x - 9}$, we note that in a cube root the radicand can be any real number. Therefore, x can be any real number, and the domain is $(-\infty, \infty)$. To find the domain of $f(x) = \sqrt{x + 4}$, set the radicand to be greater than or equal to 0, and solve for x. $x + 4 \geq 0$ The radicand must be ≥ 0. $x \geq -4$ Subtract 4 from each side. The domain is $[-4, \infty)$.
Standard deviation of a data set: Standard deviation $= \sqrt{\dfrac{\text{sum of the squares of the differences from the mean}}{\text{number of differences}}}$	Find the standard deviation of the data set $1, 3, 4, 8$. <table><tr><th>Original terms</th><th>Mean</th><th>Difference</th><th>Square of the differences</th></tr><tr><td>1</td><td>4</td><td>-3</td><td>9</td></tr><tr><td>3</td><td>4</td><td>-1</td><td>1</td></tr><tr><td>4</td><td>4</td><td>0</td><td>0</td></tr><tr><td>8</td><td>4</td><td>4</td><td>16</td></tr></table> The sum of the squares of the differences is 26 and there are 4 values in the data set. The standard deviation is $\sqrt{\dfrac{26}{4}} \approx 2.549509757$. To the nearest hundredth, the standard deviation is 2.55.

REVIEW EXERCISES

Simplify each radical. Assume that x can be any number.

1. $\sqrt{81}$

2. $-\sqrt{169}$

3. $-\sqrt{36}$

4. $\sqrt{225}$

5. $\sqrt[3]{-27}$

6. $-\sqrt[3]{216}$

7. $\sqrt[4]{625}$

8. $\sqrt[5]{-32}$

9. $\sqrt{25x^2}$

10. $\sqrt{x^2 + 6x + 9}$

11. $\sqrt[3]{27a^6b^3}$

12. $\sqrt[4]{256x^8y^4}$

Graph each function and state the domain and range.

13. $f(x) = \sqrt{x} + 2$

14. $f(x) = -\sqrt{x} - 1$

15. $f(x) = -\sqrt{x} + 2$

16. $f(x) = -\sqrt[3]{x} + 3$

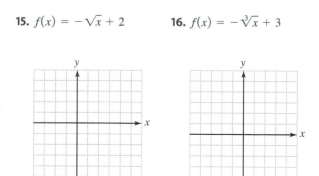

Consider the distribution 4, 8, 12, 16, 20.

17. Find the mean of the distribution.

18. Find the standard deviation.

SECTION 7.2   Applications of the Pythagorean Theorem and the Distance Formula

DEFINITIONS AND CONCEPTS	EXAMPLES
The Pythagorean theorem: If a and b are the lengths of the legs of a right triangle and c is the length of the hypotenuse, then $$a^2 + b^2 = c^2$$	To find the length of the hypotenuse of a right triangle with legs of length 9 ft and 12 ft, proceed as follows: $a^2 + b^2 = c^2$ Use the Pythagorean theorem. $(9)^2 + (12)^2 = c^2$ Substitute the values. $81 + 144 = c^2$ Square each value. $225 = c^2$ Add. $\sqrt{225} = c$ Since c is a length, take the positive square root. $15 = c$ $\sqrt{225} = 15$ The hypotenuse is 15 ft.
The distance formula: The distance between two points, (x_1, y_1) and (x_2, y_2), on a coordinate plane is $$d = \sqrt{(x_2 - x_1)^2 + (y_2 - y_1)^2}$$	To find the distance between $(8, -1)$ and $(5, 3)$, use the distance formula: $d = \sqrt{(x_2 - x_1)^2 + (y_2 - y_1)^2}$ Use the distance formula. $= \sqrt{(5 - 8)^2 + [3 - (-1)]^2}$ Substitute the values. $= \sqrt{(-3)^2 + 4^2}$ Subtract. $= \sqrt{9 + 16}$ Square each value. $= \sqrt{25}$ Add. $= 5$ Simplify. The distance between the points is 5 units.

REVIEW EXERCISES

In Exercises 19–20, the horizon distance d (measured in miles) is related to the height h (measured in feet) of the observer by the formula $d = 1.4\sqrt{h}$.

19. View from a submarine A submarine's periscope extends 4.7 feet above the surface. About how far away is the horizon?

20. View from a submarine About how far out of the water must a submarine periscope extend to provide a 4-mile horizon?

21. Sailing A technique called *tacking* allows a sailboat to make progress into the wind. A sailboat follows the course in the illustration. Find *d*, the distance the boat advances into the wind.

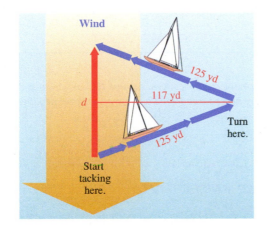

22. Communications Some campers 3,900 yards from a highway are talking to truckers on a citizen's band radio with an 8,900-yard range. Over what length of highway can these conversations take place?

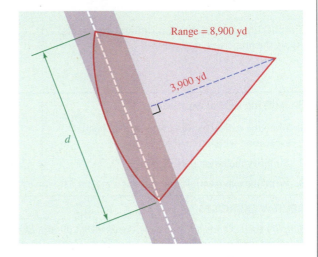

23. Find the distance between points $(1, 5)$ and $(9, -1)$.

24. Find the distance between points $(-4, 6)$ and $(-2, 8)$. Give the result to the nearest hundredth.

SECTION 7.3   Rational Exponents

DEFINITIONS AND CONCEPTS	EXAMPLES
Rational exponents with a numerator of 1: If n is a natural number $(n > 1)$ and $\sqrt[n]{x}$ is a real number, then $x^{1/n} = \sqrt[n]{x}$. If n is an even natural number, $(x^n)^{1/n} = \lvert x \rvert$. Assume n is a natural number $(n > 1)$ and x is a real number.	Simplify: $16^{1/2} = \sqrt{16} = 4$ $(25x^2)^{1/2} = \sqrt{25x^2} = 5\lvert x \rvert$
If $x > 0$, then $x^{1/n}$ is the positive number such that $(x^{1/n})^n = x$. If $x = 0$, then $x^{1/n} = 0$. If $x < 0$ and n is odd, then $x^{1/n}$ is the real number such that $(x^{1/n})^n = x$. If $x < 0$ and n is even, then $x^{1/n}$ is not a real number.	$64^{1/2} = \sqrt{64} = 8$ $0^{1/4} = \sqrt[4]{0} = 0$ $(-27)^{1/3} = \sqrt[3]{-27} = -3$ $(-64)^{1/2} = \sqrt{-64}$ is not a real number.
Assume m and n are positive integers and $x > 0$. **Rational exponents with a numerator other than 1:** $x^{m/n} = \sqrt[n]{x^m} = (\sqrt[n]{x})^m$	Simplify: $64^{2/3} = \sqrt[3]{(64)^2} = \sqrt[3]{4{,}096} = 16$ $64^{2/3} = (\sqrt[3]{64})^2 = 4^2 = 16$

Negative rational exponents: $$x^{-m/n} = \frac{1}{x^{m/n}}$$ $$\frac{1}{x^{-m/n}} = x^{m/n}$$	Simplify: $$25^{-1/2} = \frac{1}{25^{1/2}} = \frac{1}{\sqrt{25}} = \frac{1}{5}$$ $$\frac{1}{16^{-3/2}} = 16^{3/2} = \left(\sqrt{16}\right)^3 = 4^3 = 64$$
Simplifying expressions with rational exponents: Apply the properties of exponents.	$\dfrac{x^{3/4} \cdot x^{2/3}}{x^{7/6}} = x^{3/4 + 2/3 - 7/6}$ Use the rules $x^m \cdot x^n = x^{m-n}$ and $\frac{x^m}{x^n} = x^{m-n}$. $\qquad = x^{1/4}$ $\frac{3}{4} + \frac{2}{3} - \frac{7}{6} = \frac{1}{4}$ $\qquad = \sqrt[4]{x}$ Write in radical notation.
Simplifying radical expressions: 1. Write the radical expression as an exponential expression with rational exponents. 2. Simplify the rational exponents. 3. Write the exponential expression as a radical.	$\sqrt[4]{81x^2} = (81x^2)^{1/4}$ Use the rule $\sqrt[n]{x} = x^{1/n}$. $\qquad = (3^4 x^2)^{1/4}$ $81 = 3^4$ $\qquad = 3x^{1/2}$ Use the rule $(xy)^m = x^m \cdot y^m$. $\qquad = 3\sqrt{x}$ Write in radical notation.

REVIEW EXERCISES

Simplify each expression, if possible. Assume that all variables represent positive numbers.

25. $81^{1/2}$ **26.** $-49^{1/2}$

27. $9^{3/2}$ **28.** $16^{3/2}$

29. $(-27)^{1/3}$ **30.** $-8^{2/3}$

31. $8^{-2/3}$ **32.** $64^{-1/3}$

33. $-49^{5/2}$ **34.** $\dfrac{1}{16^{5/2}}$

35. $\left(\dfrac{27}{125}\right)^{-2/3}$ **36.** $\left(\dfrac{4}{9}\right)^{-3/2}$

37. $(27x^3y)^{1/3}$ **38.** $(16x^2y^4)^{1/4}$

39. $(25x^3y^4)^{3/2}$ **40.** $(8u^2v^3)^{-2/3}$

Perform the multiplications. Assume that all variables represent positive numbers and write all answers without negative exponents.

41. $5^{1/4}5^{1/2}$ **42.** $a^{5/9}a^{2/9}$

43. $u^{1/2}(u^{1/2} - u^{-1/2})$ **44.** $v^{2/3}(v^{1/3} + v^{4/3})$

45. $(x^{1/2} + y^{1/2})^2$

46. $(a^{2/3} + b^{2/3})(a^{2/3} - b^{2/3})$

Simplify each expression. Assume that all variables are positive.

47. $\sqrt[6]{5^2}$ **48.** $\sqrt[8]{x^4}$

49. $\sqrt[9]{27a^3b^6}$ **50.** $\sqrt[4]{25a^2b^2}$

SECTION 7.4   Simplifying and Combining Radical Expressions

DEFINITIONS AND CONCEPTS	**EXAMPLES**
Properties of radicals: Assume n is a natural number greater than 1. $\sqrt[n]{ab} = \sqrt[n]{a}\,\sqrt[n]{b}$ if a and b are not both negative $\sqrt[n]{\dfrac{a}{b}} = \dfrac{\sqrt[n]{a}}{\sqrt[n]{b}}$ $(b \neq 0)$	Simplify: $\sqrt{24} = \sqrt{4 \cdot 6}$ $\sqrt[3]{24} = \sqrt[3]{8 \cdot 3}$ $\qquad = \sqrt{4}\sqrt{6}$ $\qquad = \sqrt[3]{8}\sqrt[3]{3}$ $\qquad = 2\sqrt{6}$ $\qquad = 2\sqrt[3]{3}$ $\sqrt{\dfrac{21}{64x^6}} = \dfrac{\sqrt{21}}{\sqrt{64x^6}}$ $\sqrt[3]{\dfrac{21}{64x^6}} = \dfrac{\sqrt[3]{21}}{\sqrt[3]{64x^6}}$ $\qquad = \dfrac{\sqrt{21}}{8x^3}$ $\qquad = \dfrac{\sqrt[3]{21}}{4x^2}$
Adding and subtracting radical expressions: Like radicals, radical expressions with the same index and the same radicand, can be combined by addition and subtraction. Radicals that are not alike often can be simplified to radicals that are alike and then combined.	Add: $9\sqrt{2} + 4\sqrt{2} = (9 + 4)\sqrt{2} = 13\sqrt{2}$ $\sqrt{2} + \sqrt{18} = \sqrt{2} + \sqrt{9}\sqrt{2}$ $\qquad = \sqrt{2} + 3\sqrt{2}$ $\qquad = 4\sqrt{2}$

Special right triangles:
In a 45°–45°–90° triangle, the length of the hypotenuse is the length of one leg times $\sqrt{2}$.

The shorter leg of a 30°–60°–90° triangle is half as long as the hypotenuse. The longer leg is the length of the shorter leg times $\sqrt{3}$.

If each leg of an isosceles triangle is 17 cm, the hypotenuse measures $17\sqrt{2}$ cm.

If the shorter leg of a 30°–60°–90° triangle is 12 units, the hypotenuse is 24 cm and the longer leg is $12\sqrt{3}$ units.

REVIEW EXERCISES

Simplify each expression. Assume that all variables represent positive numbers.

51. $\sqrt{176}$ **52.** $\sqrt[3]{250}$

53. $\sqrt[4]{32}$ **54.** $\sqrt[5]{96}$

55. $\sqrt{8x^3}$ **56.** $\sqrt{24x^6y^5}$

57. $\sqrt[3]{16x^5y^4}$ **58.** $\sqrt[3]{54x^7y^3}$

59. $\dfrac{\sqrt{32x^3}}{\sqrt{2x}}$ **60.** $\dfrac{\sqrt[3]{16x^5}}{\sqrt[3]{2x^2}}$

61. $\sqrt[3]{\dfrac{2a^2b}{27x^3}}$ **62.** $\sqrt{\dfrac{17xy}{64a^4}}$

Simplify and combine like radicals. Assume that all variables represent positive numbers.

63. $\sqrt{12} + \sqrt{27}$ **64.** $\sqrt{18} - \sqrt{32}$

65. $2\sqrt[3]{3} - \sqrt[3]{24}$ **66.** $\sqrt[4]{32} + 2\sqrt[4]{162}$

67. $2x\sqrt{8} + 2\sqrt{200x^2} + \sqrt{50x^2}$

68. $3\sqrt{27a^3} - 2a\sqrt{3a} + 5\sqrt{75a^3}$

69. $\sqrt[3]{54} - 3\sqrt[3]{16} + 4\sqrt[3]{128}$

70. $2\sqrt[4]{32x^5} + 4\sqrt[4]{162x^5} - 5x\sqrt[4]{512x}$

71. Geometry Find the length of the hypotenuse of an isosceles right triangle whose legs measure 7 meters.

72. Geometry The hypotenuse of a 30°–60°–90° triangle measures $12\sqrt{3}$ centimeters. Find the length of each leg.

Find the value of x to two decimal places.

73.   **74.**

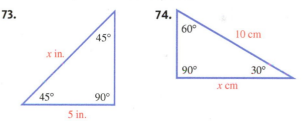

SECTION 7.5 Multiplying Radical Expressions and Rationalizing

DEFINITIONS AND CONCEPTS	EXAMPLES
Multiplying radical expressions: If two radicals have the same index, they can be multiplied.	Multiply: $\sqrt{3x}\,\sqrt{6x} = \sqrt{18x^2}$ $(x > 0)$ $\qquad\qquad = \sqrt{9x^2}\,\sqrt{2}$ $\qquad\qquad = 3x\sqrt{2}$ $(5 - 2\sqrt{3})(7 + 6\sqrt{3})$ $= 35 + 30\sqrt{3} - 14\sqrt{3} - 12\sqrt{9}$ Distribute. $= 35 + 16\sqrt{3} - 12(3)$ Combine like terms. $= 35 + 16\sqrt{3} - 36$ Multiply. $= -1 + 16\sqrt{3}$ Simplify.
Rationalizing the denominator: To eliminate a single radical in the denominator, we multiply the numerator and the denominator by a number that will result in a perfect square (or cube, or fourth power, etc.) under the radical in the denominator determined by the index.	Rationalize: $\sqrt{\dfrac{5}{8}} = \dfrac{\sqrt{5}}{\sqrt{8}}$ $\qquad \sqrt{\dfrac{a}{b}} = \dfrac{\sqrt{a}}{\sqrt{b}}$ $\qquad = \dfrac{\sqrt{5}}{\sqrt{8}} \cdot \dfrac{\sqrt{2}}{\sqrt{2}}$ Multiply by 1: $\dfrac{\sqrt{2}}{\sqrt{2}} = 1$ $\qquad = \dfrac{\sqrt{10}}{\sqrt{16}}$ Multiply radicals. $\qquad = \dfrac{\sqrt{10}}{4}$ Simplify.

To rationalize a fraction whose denominator has two terms with one or both containing square roots, we multiply its numerator and denominator by the *conjugate* of its denominator.	Rationalize: $$\frac{5}{4 - \sqrt{2}} \cdot \frac{\textbf{4} + \sqrt{\textbf{2}}}{\textbf{4} + \sqrt{\textbf{2}}}$$ Multiply the numerator and the denominator by the conjugate of the denominator. $$= \frac{20 + 5\sqrt{2}}{16 - 2}$$ Multiply. $$= \frac{20 + 5\sqrt{2}}{14}$$ Simplify.

REVIEW EXERCISES

Simplify each expression. Assume that all variables represent positive numbers.

75. $\left(5\sqrt{3}\right)\left(2\sqrt{5}\right)$ **76.** $2\sqrt{6}\sqrt{216}$

77. $\sqrt{5x}\sqrt{4x}$ **78.** $\sqrt[3]{3x^2}\sqrt[3]{9x^4}$

79. $-\sqrt[3]{2x^2}\sqrt[3]{4x}$ **80.** $-\sqrt[4]{256x^5y^{11}}\sqrt[4]{625x^9y^3}$

81. $\sqrt{3}\left(\sqrt{27} - 4\right)$ **82.** $\sqrt{5}\left(\sqrt{5} - 6\right)$

83. $\sqrt{5}\left(\sqrt{2} - 1\right)$ **84.** $\sqrt{3}\left(\sqrt{3} + \sqrt{2}\right)$

85. $\left(\sqrt{5} + 1\right)\left(\sqrt{5} - 1\right)$
86. $\left(\sqrt{3} + \sqrt{2}\right)\left(\sqrt{3} + \sqrt{2}\right)$
87. $\left(\sqrt{x} + \sqrt{y}\right)\left(\sqrt{x} - \sqrt{y}\right)$
88. $\left(2\sqrt{u} + 3\right)\left(3\sqrt{u} - 4\right)$

Rationalize each denominator.

89. $\dfrac{1}{\sqrt{3}}$ **90.** $\dfrac{\sqrt{3}}{\sqrt{5}}$

91. $\dfrac{x}{\sqrt{xy}}$ **92.** $\dfrac{\sqrt[3]{uv}}{\sqrt[3]{u^5v^7}}$

93. $\dfrac{2}{\sqrt{2} - 1}$ **94.** $\dfrac{\sqrt{2}}{\sqrt{3} - 1}$

95. $\dfrac{2x - 32}{\sqrt{x} + 4}$ **96.** $\dfrac{\sqrt{a} + 1}{\sqrt{a} - 1}$

Rationalize each numerator. All variables represent positive numbers.

97. $\dfrac{\sqrt{3}}{5}$ **98.** $\dfrac{\sqrt[3]{9}}{3}$

99. $\dfrac{3 - \sqrt{x}}{2}$ **100.** $\dfrac{\sqrt{a} - \sqrt{b}}{\sqrt{a}}$

SECTION 7.6   Radical Equations

DEFINITIONS AND CONCEPTS	EXAMPLES
The power rule: Assume x, y, and n are real numbers. If $x = y$, then $x^n = y^n$. When we raise both sides of an equation to the same power it can lead to extraneous solutions. Be sure to check all possible solutions. Prior to applying the power rule be certain to isolate one radical.	To solve $\sqrt{x + 4} = x - 2$, proceed as follows: $$\left(\sqrt{x + 4}\right)^2 = \left(x - 2\right)^2$$ Square both sides of the equation to eliminate the square root. $$x + 4 = x^2 - 4x + 4$$ Square the binomial. $$0 = x^2 - 5x$$ Subtract x and 4 from both sides to write the equation in quadratic form. $$0 = x(x - 5)$$ Factor $x^2 - 5x$. $$x = 0 \quad \text{or} \quad x - 5 = 0$$ Set each factor equal to 0. $$x = 0 \qquad\qquad x = 5$$ **Check:** $$\sqrt{x + 4} = x - 2 \qquad \sqrt{x + 4} = x - 2$$ $$\sqrt{\textbf{0} + 4} \stackrel{?}{=} \textbf{0} - 2 \qquad \sqrt{\textbf{5} + 4} \stackrel{?}{=} \textbf{5} - 2$$ $$\sqrt{4} \stackrel{?}{=} -2 \qquad\qquad \sqrt{9} \stackrel{?}{=} 3$$ $$2 \neq -2 \qquad\qquad\qquad 3 = 3$$ Since 0 does not check, it is extraneous and must be discarded. The only solution of the original equation is 5.

REVIEW EXERCISES

Solve each equation.

105. $\sqrt{2x + 5} - \sqrt{2x} = 1$ **106.** $\sqrt[3]{x^3 + 8} = x + 2$

101. $\sqrt{3y - 11} = \sqrt{y + 7}$ **102.** $u = \sqrt{25u - 144}$

103. $\sqrt{2x + 6} + 1 = x$ **104.** $\sqrt{z + 1} + \sqrt{z} = 2$

SECTION 7.7   Complex Numbers

DEFINITIONS AND CONCEPTS	EXAMPLES
Simplifying imaginary numbers: $\sqrt{-1}$ is defined as the **imaginary number** i.	Simplify: $\sqrt{-12} = \sqrt{-4}\sqrt{3}$ Write -12 as $-4(3)$. $\phantom{\sqrt{-12}} = \sqrt{-1}\sqrt{4}\sqrt{3}$ Write -4 as the product of -1 and 4. $\phantom{\sqrt{-12}} = i(2)\sqrt{3}$ $\sqrt{-1} = i$ $\phantom{\sqrt{-12}} = 2\sqrt{3}i$
Operations with complex numbers: If a, b, c, and d are real numbers and $i^2 = -1$, $a + bi = c + di$ if and only if $a = c$ and $b = d$ $(a + bi) + (c + di) = (a + c) + (b + d)i$ $(a + bi) - (c + di) = (a - c) + (b - d)i$ $(a + bi)(c + di) = (ac - bd) + (ad + bc)i$	$3 + 5i = \sqrt{9} + \frac{10}{2}i$ because $\sqrt{9} = 3$ and $\frac{10}{2} = 5$. Perform the indicated operations: $(6 + 5i) + (2 + 9i) = (6 + 2) + (5 + 9)i = 8 + 14i$ $(6 + 5i) - (2 + 9i) = (6 - 2) + (5 - 9)i = 4 - 4i$ $(6 - 5i)(2 + 9i) = 12 + 54i - 10i - 45i^2$ $ = 12 + 44i - 45(-1)$ $ = 12 + 44i + 45$ $ = 57 + 44i$
Dividing complex numbers: To divide complex numbers, write the division as a fraction, and rationalize the denominator.	$\dfrac{6}{3 + i} \cdot \dfrac{3 - i}{3 - i} = \dfrac{18 - 6i}{9 - i^2}$ Multiply the numerator and the denominator by the conjugate of the denominator. Multiply the fractions. $ = \dfrac{18 - 6i}{9 + 1}$ $i^2 = -1$ $ = \dfrac{18 - 6i}{10}$ Add. $ = \dfrac{18}{10} - \dfrac{6}{10}i$ Write in $a + bi$ form. $ = \dfrac{9}{5} - \dfrac{3}{5}i$ Simplify each fraction.
Powers of i $i = \sqrt{-1}$ $i^2 = -1$ $i^3 = -i$ $i^4 = 1$ If n is a natural number that has a remainder of r when divided by 4, then $i^n = i^r$. When n is divisible by 4, the remainder r is 0 and $i^0 = 1$.	Simplify: $i^{81} = (i^4)^{20} \cdot i$ $\phantom{i^{81}} = (i^1)^{20} \cdot i$ $\phantom{i^{81}} = i$

Absolute value of a complex number: $$\lvert a + bi \rvert = \sqrt{a^2 + b^2}$$	Find the absolute value:
	$\lvert 4 + 8i \rvert = \sqrt{4^2 + 8^2}$ $\lvert a + bi \rvert = \sqrt{a^2 + b^2}$
	$= \sqrt{16 + 64}$ Square each value.
	$= \sqrt{80}$ Add.
	$= \sqrt{16 \cdot 5}$ Factor: $80 = 16(5)$
	$= 4\sqrt{5}$ Simplify: $\sqrt{16 \cdot 5} = \sqrt{16}\,\sqrt{5} = 4\sqrt{5}$

REVIEW EXERCISES

Perform the operations and give all answers in a + bi form.

107. $(6 + 3i) + (4 - 15i)$

108. $(-5 - 22i) - (-7 + 28i)$

109. $\left(-32 + \sqrt{-144}\right) - \left(64 + \sqrt{-81}\right)$

110. $\left(-8 + \sqrt{-8}\right) + \left(6 - \sqrt{-32}\right)$

111. $(2 - 7i)(-3 + 4i)$

112. $(-4 + 5i)(3 + 2i)$

113. $\left(5 - \sqrt{-27}\right)\left(-6 + \sqrt{-12}\right)$

114. $\left(2 + \sqrt{-128}\right)\left(3 - \sqrt{-98}\right)$

115. $\dfrac{3}{4i}$

116. $\dfrac{-2}{5i^3}$

117. $\dfrac{6}{2 + i}$

118. $\dfrac{7}{3 - i}$

119. $\dfrac{4 + i}{4 - i}$

120. $\dfrac{3 - i}{3 + i}$

121. $\dfrac{3}{5 + \sqrt{-4}}$

122. $\dfrac{2}{3 - \sqrt{-9}}$

Simplify.

123. $\lvert 9 + 12i \rvert$

124. $\lvert 24 - 10i \rvert$

125. i^{39}

126. i^{52}

---

7   Test

Find each root.

1. $\sqrt{36}$

2. $\sqrt{125}$

3. $\sqrt{4x^2}$

4. $\sqrt[3]{8x^3}$

Graph each function and find the domain and range.

5. $f(x) = \sqrt{x - 2}$

6. $f(x) = \sqrt[3]{x} + 3$

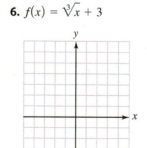

Use a calculator.

7. Shipping crates   The diagonal brace on the shipping crate shown in the illustration is 53 inches. Find the height, h, of the crate.

h in.   53 in.

45 in.

8. Pendulums   The 2-meter pendulum rises 0.1 meter at the extremes of its swing. Find the width w of the swing in meters.

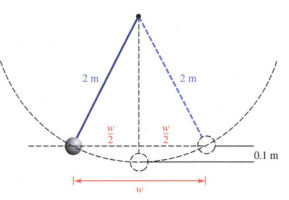

Find the distance between the points.

9. $(6, 8), (0, 0)$ **10.** $(-2, 5), (22, 12)$

Simplify each expression. Assume that all variables represent positive numbers, and write answers without using negative exponents.

11. $81^{1/4}$ **12.** $-64^{2/3}$

13. $36^{-3/2}$ **14.** $\left(-\dfrac{8}{27}\right)^{-2/3}$

15. $\dfrac{2^{5/3}2^{1/6}}{2^{1/2}}$ **16.** $\dfrac{(8x^3y)^{1/2}(8xy^5)^{1/2}}{(x^3y^6)^{1/3}}$

Simplify each expression. Assume that all variables represent positive numbers.

17. $\sqrt{108}$ **18.** $\sqrt{250x^3y^5}$

19. $\dfrac{\sqrt[3]{240x^{10}y^4}}{\sqrt[3]{6xy}}$ **20.** $\sqrt{\dfrac{3a^5}{48a^7}}$

Simplify each expression. Assume that the variables are unrestricted.

21. $\sqrt{45x^2}$ **22.** $\sqrt{48x^6}$

23. $\sqrt[3]{16x^9}$ **24.** $\sqrt{18x^4y^9}$

Simplify. Assume that all variables represent positive numbers.

25. $\sqrt{12} - \sqrt{27}$

26. $2\sqrt[3]{40} - \sqrt[3]{5,000} + 4\sqrt[3]{625}$

27. $2\sqrt{48y^5} - 3y\sqrt{12y^3}$

28. $\sqrt[4]{768z^5} + z\sqrt[4]{48z}$

Perform each operation and simplify, if possible. All variables represent positive numbers.

29. $-2\sqrt{xy}\left(3\sqrt{x} + \sqrt{xy^3}\right)$

30. $\left(3\sqrt{2} + \sqrt{3}\right)\left(2\sqrt{2} - 3\sqrt{3}\right)$

Rationalize each denominator.

31. $\dfrac{1}{\sqrt{5}}$ **32.** $\dfrac{3t - 1}{\sqrt{3t} - 1}$

Rationalize each numerator.

33. $\dfrac{\sqrt{3}}{\sqrt{7}}$ **34.** $\dfrac{\sqrt{a} + \sqrt{b}}{\sqrt{a} - \sqrt{b}}$

Solve.

35. $\sqrt[3]{6n + 4} - 4 = 0$

36. $1 - \sqrt{u} = \sqrt{u - 3}$

Perform the operations. Give all answers in a + bi form.

37. $(2 + 4i) + (-3 + 7i)$

38. $\left(3 - \sqrt{-9}\right) - \left(-1 + \sqrt{-16}\right)$

39. $2i(3 - 4i)$ **40.** $(3 + 2i)(-4 - i)$

41. $\dfrac{1}{\sqrt{2i}}$ **42.** $\dfrac{2 + i}{3 - i}$

Quadratic Equations and Functions; Inequalities; and Algebra, Composition, and Inverses of Functions

8

Careers and Mathematics

POLICE OFFICERS AND DETECTIVES

People depend on police officers and detectives to protect their lives and property. Law enforcement officers, some of whom are state or federal special agents, perform these duties in a variety of ways. Uniformed police officers maintain regular patrols and respond to calls for service, and some serve as undercover officers. Police work can be very dangerous and stressful. Detectives collect evidence for criminal cases.

Job Outlook:
Employment of police officers and detectives is expected to grow by 7%, slower than average for all occupations through 2020.

Annual Earnings:
$35,600–$88,870

For More Information:
http://www.bls.gov/oco/ocos160.htm

For a Sample Application:
See Problem 89 in Section 8.1.

REACH FOR SUCCESS

8.1 Solving Quadratic Equations Using the Square-Root Property and by Completing the Square

8.2 Solving Quadratic Equations Using the Quadratic Formula

8.3 The Discriminant and Equations That Can Be Written in Quadratic Form

8.4 Graphs of Quadratic Functions

8.5 Quadratic and Other Nonlinear Inequalities

8.6 Algebra and Composition of Functions

8.7 Inverses of Functions

■ *Projects*
  REACH FOR SUCCESS EXTENSION
  CHAPTER REVIEW
  CHAPTER TEST
  CUMULATIVE REVIEW

In this chapter

In this chapter, we will discuss general methods for solving quadratic equations and quadratic inequalities, and we will consider the graphs of quadratic functions. Finally, we will discuss algebraic operations, composition, and inverses of functions.

Reach for Success Accessing Academic Resources

When a baseball pitcher starts to struggle with a certain type of pitch, he may fall behind in the ball/strike count. Often, a suggestion from his manager or pitching coach can give him the guidance he needs to correct it.

If you do not feel you are performing as well as you should, or would like, then perhaps it is time to seek help. Using the varied resources available to you can improve your chances of a successful semester.

© Dennis Ku/Shutterstock.com

Obtain a hard copy or view an online version of your college catalog. Where did you find it?

_____

Search for Academic Resources. List the resources that could help with your mathematics class.

_____

_____

Does your college have a mathematics lab or tutoring center? _____

If so, where is it located? _____

> What are the hours of operation?
> M: _____ T: _____ W: _____
> Th: _____ F: _____ S/Su: _____
>
> Write down at least two days and times that you could go to the center to get help for an hour or more.
>
> _____

If not, what other resources are available? _____

In addition to, or in place of, on-campus services, some colleges offer online tutoring.

Does your college offer online tutoring services for mathematics? _____

If so, what is the process to get enrolled? _____

_____

If not, have you tried the software that accompanies your textbook? _____

Colleges usually offer a wide variety of resources to help students succeed.

Are there other services at your college that might support your performance in this class?

If so, what are they? _____

Does your college offer a College Success or Study Skills course? _____

If so, would you consider enrolling? _____ Explain.

_____

A Successful Study Strategy . . .

 Know where to go to get help with your mathematics. In addition to those listed above, remember that your instructor may be your best resource.

At the end of the chapter you will find an additional exercise to guide you in planning for a successful semester.

Solving Quadratic Equations Using the Square-Root Property and by Completing the Square

Objectives

1 Solve a quadratic equation by factoring.
2 Solve a quadratic equation by applying the square-root property.
3 Solve a quadratic equation by completing the square.
4 Solve an application requiring the use of the square-root property.

Vocabulary

square-root property   completing the square   compound interest
extracting the roots

Getting Ready

Factor each expression.

1. $x^2 - 25$ **2.** $b^2 - 81$

3. $6x^2 + x - 2$ **4.** $4x^2 - 4x - 3$

We begin this section by reviewing how to solve quadratic equations by factoring. We will then discuss how to solve these equations by applying the **square-root property**, sometimes called ***extracting the roots***, by completing the square, and then use these skills to solve applications.

1 ### Solve a quadratic equation by factoring.

Recall that a *quadratic equation* is an equation of the form $ax^2 + bx + c = 0$ $(a \neq 0)$, where a, b, and c are real numbers. As a review, we will solve the first two examples by factoring.

EXAMPLE 1   Solve: $x^2 = 9$

Solution   To solve this quadratic equation by factoring, we proceed as follows:

$$x^2 = 9$$
$$x^2 - 9 = 0 \qquad \text{Subtract 9 from both sides.}$$
$$(x + 3)(x - 3) = 0 \qquad \text{Factor the binomial.}$$
$$x + 3 = 0 \quad \text{or} \quad x - 3 = 0 \qquad \text{Set each factor equal to 0.}$$
$$x = -3 \quad \bigg| \quad x = 3 \qquad \text{Solve each linear equation.}$$

Check: *For x = −3*   *For x = 3*

$$x^2 = 9 \qquad\qquad x^2 = 9$$

$$(-3)^2 \stackrel{?}{=} 9 \qquad (3)^2 \stackrel{?}{=} 9$$

$$9 = 9 \qquad\qquad 9 = 9$$

Since both results check, the solutions are 3 and −3.

SELF CHECK 1   Solve: $p^2 = 64$

EXAMPLE 2   Solve: $6x^2 - 7x - 3 = 0$

Solution   To solve this quadratic equation by factoring, we proceed as follows:

$$6x^2 - 7x - 3 = 0$$

$$(2x - 3)(3x + 1) = 0 \qquad \text{Factor.}$$

$$2x - 3 = 0 \quad \text{or} \quad 3x + 1 = 0 \qquad \text{Set each factor equal to 0.}$$

$$x = \frac{3}{2} \qquad\qquad x = -\frac{1}{3} \qquad \text{Solve each linear equation.}$$

Check: *For x = $\frac{3}{2}$*   *For x = $-\frac{1}{3}$*

$$6x^2 - 7x - 3 = 0 \qquad\qquad 6x^2 - 7x - 3 = 0$$

$$6\left(\frac{3}{2}\right)^2 - 7\left(\frac{3}{2}\right) - 3 \stackrel{?}{=} 0 \qquad 6\left(-\frac{1}{3}\right)^2 - 7\left(-\frac{1}{3}\right) - 3 \stackrel{?}{=} 0$$

$$6\left(\frac{9}{4}\right) - 7\left(\frac{3}{2}\right) - 3 \stackrel{?}{=} 0 \qquad 6\left(\frac{1}{9}\right) - 7\left(-\frac{1}{3}\right) - 3 \stackrel{?}{=} 0$$

$$\frac{27}{2} - \frac{21}{2} - \frac{6}{2} \stackrel{?}{=} 0 \qquad\qquad \frac{2}{3} + \frac{7}{3} - \frac{9}{3} \stackrel{?}{=} 0$$

$$0 = 0 \qquad\qquad\qquad 0 = 0$$

Since both results check, the solutions are $\frac{3}{2}$ and $-\frac{1}{3}$.

SELF CHECK 2   Solve: $6m^2 - 5m + 1 = 0$

Unfortunately, many quadratic expressions do not factor easily. For example, it would be difficult to solve $2x^2 + 4x + 1 = 0$ by factoring, because $2x^2 + 4x + 1$ cannot be factored by using only integers.

2   Solve a quadratic equation by applying the square-root property.

To develop general methods for solving all quadratic equations, we first solve $x^2 = c$ by a method similar to the one used in Example 1.

$$x^2 = c$$

$$x^2 - c = 0 \qquad\qquad \text{Subtract } c \text{ from both sides.}$$

$$x^2 - \left(\sqrt{c}\right)^2 = 0 \qquad\qquad c = \left(\sqrt{c}\right)^2$$

$$\left(x + \sqrt{c}\right)\left(x - \sqrt{c}\right) = 0 \qquad\qquad \text{Factor the difference of two squares.}$$

$$x + \sqrt{c} = 0 \quad \text{or} \quad x - \sqrt{c} = 0 \qquad \text{Set each factor equal to 0.}$$

$$x = -\sqrt{c} \qquad\qquad x = \sqrt{c} \qquad \text{Solve each linear equation.}$$

The two solutions of $x^2 = c$ are $x = \sqrt{c}$ and $x = -\sqrt{c}$.

| THE SQUARE-ROOT PROPERTY (EXTRACTING THE ROOTS) | The equation $x^2 = c$ has two solutions. They are $$x = \sqrt{c} \quad \text{or} \quad x = -\sqrt{c}$$ |

We often use the symbol $\pm\sqrt{c}$ to represent the two solutions $\sqrt{c}$ and $-\sqrt{c}$. The symbol $\pm\sqrt{c}$ is read as "the positive or negative square root of c."

EXAMPLE 3 Use the square-root property to solve $x^2 - 12 = 0$.

Solution We can write the equation as $x^2 = 12$ and use the square-root property.

$$x^2 - 12 = 0$$
$$x^2 = 12 \qquad \text{Add 12 to both sides.}$$
$$x = \sqrt{12} \quad \text{or} \quad x = -\sqrt{12} \qquad \text{Use the square-root property.}$$
$$x = 2\sqrt{3} \quad \mid \quad x = -2\sqrt{3} \qquad \sqrt{12} = \sqrt{4}\sqrt{3} = 2\sqrt{3}$$

The solutions can be written as $\pm 2\sqrt{3}$. Verify that each one satisfies the equation.

SELF CHECK 3 Use the square-root property to solve $x^2 - 18 = 0$.

EXAMPLE 4 Use the square-root property to solve $(x - 3)^2 = 16$.

Solution We can use the square-root property.

$$(x - 3)^2 = 16$$
$$x - 3 = \sqrt{16} \quad \text{or} \quad x - 3 = -\sqrt{16} \qquad \text{Use the square-root property.}$$
$$x - 3 = 4 \quad \mid \quad x - 3 = -4 \qquad \sqrt{16} = 4 \text{ and } -\sqrt{16} = -4$$
$$x = 3 + 4 \quad \mid \quad x = 3 - 4 \qquad \text{Add 3 to both sides.}$$
$$x = 7 \quad \mid \quad x = -1 \qquad \text{Simplify.}$$

Verify that each solution satisfies the equation.

SELF CHECK 4 Use the square-root property to solve $(x + 2)^2 = 9$.

Because we have learned to work with radicals and complex numbers, we can now solve any quadratic equation.

EXAMPLE 5 Use the square-root property to solve $9x^2 + 25 = 0$.

Solution We can write the equation as $x^2 = -\dfrac{25}{9}$ and use the square-root property.

$$9x^2 + 25 = 0$$
$$x^2 = -\frac{25}{9} \qquad \begin{array}{l}\text{Subtract 25 from both sides}\\ \text{and divide both sides by 9.}\end{array}$$
$$x = \sqrt{-\frac{25}{9}} \quad \text{or} \quad x = -\sqrt{-\frac{25}{9}} \qquad \text{Use the square-root property.}$$
$$x = \sqrt{\frac{25}{9}}\sqrt{-1} \quad \mid \quad x = -\sqrt{\frac{25}{9}}\sqrt{-1} \qquad \sqrt{-\frac{25}{9}} = \sqrt{\frac{25}{9}(-1)} = \sqrt{\frac{25}{9}}\sqrt{-1}$$
$$x = \frac{5}{3}i \quad \mid \quad x = -\frac{5}{3}i \qquad \sqrt{\frac{25}{9}} = \frac{5}{3}; \sqrt{-1} = i$$

Check:

$$9x^2 + 25 = 0$$
$$9\left(\frac{5}{3}i\right)^2 + 25 \stackrel{?}{=} 0$$
$$9\left(\frac{25}{9}\right)i^2 + 25 \stackrel{?}{=} 0$$
$$25(-1) + 25 \stackrel{?}{=} 0$$
$$0 = 0$$

$$9x^2 + 25 = 0$$
$$9\left(-\frac{5}{3}i\right)^2 + 25 \stackrel{?}{=} 0$$
$$9\left(\frac{25}{9}\right)i^2 + 25 \stackrel{?}{=} 0$$
$$25(-1) + 25 \stackrel{?}{=} 0$$
$$0 = 0$$

Since both results check, the solutions are $\pm\frac{5}{3}i$.

SELF CHECK 5 Use the square-root property to solve $4x^2 + 36 = 0$.

---

3 Solve a quadratic equation by completing the square.

All quadratic equations can be solved by a method called **completing the square**. This method is based on the special products

$$x^2 + 2ax + a^2 = (x + a)^2 \qquad \text{and} \qquad x^2 - 2ax + a^2 = (x - a)^2$$

Recall that the trinomials $x^2 + 2ax + a^2$ and $x^2 - 2ax + a^2$ are both *perfect-square trinomials*, because both factor as the square of a binomial. In each case, the coefficient of the first term is 1 and if we take one-half of the coefficient of x in the middle term and square it, we obtain the third term.

$$\left[\frac{1}{2}(2a)\right]^2 = a^2 \qquad\qquad \tfrac{1}{2}(2a) = \left(\tfrac{1}{2}\cdot 2\right)a = a$$

$$\left[\frac{1}{2}(-2a)\right]^2 = (-a)^2 = a^2 \qquad\qquad \tfrac{1}{2}(-2a) = \left(\tfrac{1}{2}\cdot -2\right)a = -a$$

EXAMPLE 6 Find the number that when added to each binomial results in a perfect-square trinomial: **a.** $x^2 + 10x$ **b.** $x^2 - 6x$ **c.** $x^2 - 11x$

Solution **a.** To make $x^2 + 10x$ a perfect-square trinomial, we first find one-half of 10 to obtain 5 and square 5 to get 25.

$$\left[\frac{1}{2}(10)\right]^2 = (5)^2 = 25$$

If we add 25 to $x^2 + 10x$ we obtain $x^2 + 10x + 25$. This is a perfect-square trinomial because $x^2 + 10x + 25 = (x + 5)^2$.

b. To make $x^2 - 6x$ a perfect-square trinomial, we first find one-half of -6 to obtain -3 and square -3 to get 9.

$$\left[\frac{1}{2}(-6)\right]^2 = (-3)^2 = 9$$

If we add 9 to $x^2 - 6x$ we obtain $x^2 - 6x + 9$. This is a perfect-square trinomial because $x^2 - 6x + 9 = (x - 3)^2$.

c. To make $x^2 - 11x$ a perfect-square trinomial, we first find one-half of -11 to obtain $-\frac{11}{2}$ and square $-\frac{11}{2}$ to get $\frac{121}{4}$.

$$\left[\frac{1}{2}(-11)\right]^2 = \left(-\frac{11}{2}\right)^2 = \frac{121}{4}$$

If we add $\frac{121}{4}$ to $x^2 - 11x$ we obtain $x^2 - 11x + \frac{121}{4}$. This is a perfect-square trinomial because $x^2 - 11x + \frac{121}{4} = \left(x - \frac{11}{2}\right)^2$.

SELF CHECK 6 Find the number that when added to $a^2 - 5a$ results in a perfect-square trinomial.

To see geometrically why completing the square works on $x^2 + 10x$, we refer to Figure 8-1(a), which shows a polygon with an area of $x^2 + 10x$. To turn the polygon into a square, we can divide the area of $10x$ into two areas of $5x$ and then reassemble the polygon as shown in Figure 8-1(b). To fill in the missing corner, we must add a square with an area of $5^2 = 25$. Thus, we complete the square.

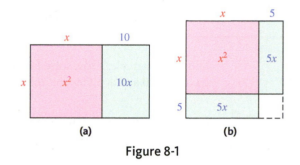

Figure 8-1

To solve an equation of the form $ax^2 + bx + c = 0$ $(a \neq 0)$ by completing the square, we use the following steps.

COMPLETING THE SQUARE

1. Make sure that the coefficient of x^2 is 1. If not, divide both sides of the equation by the coefficient of x^2.

2. If necessary, add a number to both sides of the equation to place the constant term on the right side of the equal sign.

3. Complete the square:

   a. Find one-half of the coefficient of x and square it.

   b. Add the square to both sides of the equation.

4. Factor the trinomial square on one side of the equation and combine like terms on the other side.

5. Solve the resulting equation by using the square-root property.

EXAMPLE 7 Solve $x^2 + 8x + 7 = 0$ by completing the square.

Solution **Step 1** In this example, the coefficient of x^2 is already 1.

Step 2 We add -7 to both sides to place the constant on the right side of the equal sign.

$$x^2 + 8x + 7 = 0$$
$$x^2 + 8x \quad\;\; = -7$$

Step 3 The coefficient of x is 8, one-half of 8 is 4, and $4^2 = 16$. To complete the square, we add 16 to both sides.

$$x^2 + 8x + 16 = -7 + 16$$

Step 4   Since the left side of the equation is a perfect-square trinomial, we can factor it as $(x + 4)^2$ and simplify on the right side to obtain

$$(x + 4)^2 = 9$$

Step 5   We can then solve the resulting equation by using the square-root property.

$$(x + 4)^2 = 9$$

$$x + 4 = \sqrt{9} \quad \text{or} \quad x + 4 = -\sqrt{9}$$

$$x + 4 = 3 \quad\quad\quad\quad x + 4 = -3$$

$$x = -1 \quad\quad\quad\quad\quad x = -7$$

After checking both results, we see that the solutions are -1 and -7.

🌿 **SELF CHECK 7**   Solve $a^2 + 5a + 4 = 0$ by completing the square.

Although Example 7 was solved by completing the square, we could solve it by factoring the trinomial and applying the zero factor property. The results would be the same.

EXAMPLE 8   Solve $6x^2 + 5x - 6 = 0$ by completing the square.

Solution   **Step 1**   To make the coefficient of x^2 equal to 1, we divide both sides by 6.

$$6x^2 + 5x - 6 = 0$$

$$\frac{6x^2}{6} + \frac{5}{6}x - \frac{6}{6} = \frac{0}{6} \quad \textcolor{red}{\text{Divide both sides by 6.}}$$

$$x^2 + \frac{5}{6}x - 1 = 0 \quad \textcolor{red}{\text{Simplify.}}$$

Step 2   We add 1 to both sides to place the constant on the right side.

$$x^2 + \frac{5}{6}x \quad\quad = 1$$

Step 3   The coefficient of x is $\frac{5}{6}$, one-half of $\frac{5}{6}$ is $\frac{5}{12}$, and $\left(\frac{5}{12}\right)^2 = \frac{25}{144}$. To complete the square, we add $\frac{25}{144}$ to both sides.

$$x^2 + \frac{5}{6}x + \textcolor{red}{\frac{25}{144}} = 1 + \textcolor{red}{\frac{25}{144}}$$

Step 4   Since the left side of the equation is a perfect-square trinomial, we can factor it as $\left(x + \frac{5}{12}\right)^2$ and simplify the right side to obtain

$$\left(x + \frac{5}{12}\right)^2 = \frac{169}{144} \quad \textcolor{red}{1 + \frac{25}{144} = \frac{144}{144} + \frac{25}{144} = \frac{169}{144}}$$

Step 5   We can solve this equation by using the square-root property.

$$x + \frac{5}{12} = \sqrt{\frac{169}{144}} \quad \text{or} \quad x + \frac{5}{12} = -\sqrt{\frac{169}{144}} \quad \textcolor{red}{\text{Apply the square-root property.}}$$

$$x + \frac{5}{12} = \frac{13}{12} \quad\quad\quad x + \frac{5}{12} = -\frac{13}{12} \quad \textcolor{red}{\sqrt{\frac{169}{144}} = \frac{13}{12}}$$

$$x = -\frac{5}{12} + \frac{13}{12} \quad\quad x = -\frac{5}{12} - \frac{13}{12} \quad \textcolor{red}{\text{Subtract } \frac{5}{12} \text{ from both sides.}}$$

$$x = \frac{2}{3} \quad\quad\quad\quad\quad x = -\frac{3}{2} \quad \textcolor{red}{\text{Add and simplify each fraction.}}$$

After checking both results, we see that the solutions are $\frac{2}{3}$ and $-\frac{3}{2}$. Note that this equation could be solved by factoring.

SELF CHECK 8   Solve $6p^2 - 5p - 6 = 0$ by completing the square.

EXAMPLE 9   Solve $2x^2 + 4x + 1 = 0$ by completing the square.

Solution

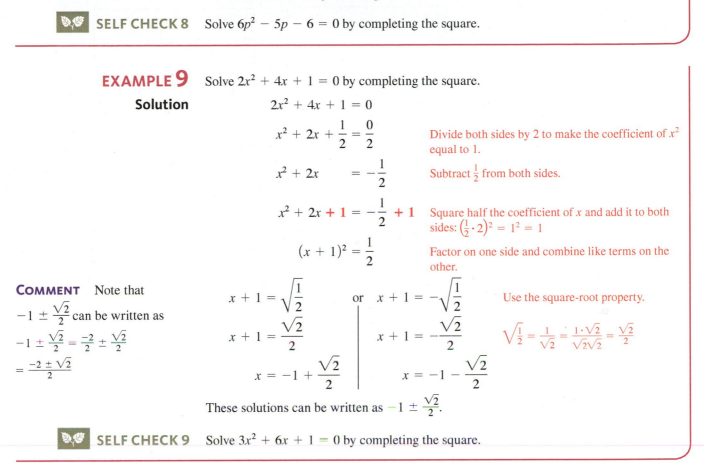

$$2x^2 + 4x + 1 = 0$$

$$x^2 + 2x + \frac{1}{2} = \frac{0}{2}$$ Divide both sides by 2 to make the coefficient of x^2 equal to 1.

$$x^2 + 2x = -\frac{1}{2}$$ Subtract $\frac{1}{2}$ from both sides.

$$x^2 + 2x + 1 = -\frac{1}{2} + 1$$ Square half the coefficient of x and add it to both sides: $\left(\frac{1}{2} \cdot 2\right)^2 = 1^2 = 1$

$$(x + 1)^2 = \frac{1}{2}$$ Factor on one side and combine like terms on the other.

COMMENT   Note that $-1 \pm \frac{\sqrt{2}}{2}$ can be written as
$$-1 \pm \frac{\sqrt{2}}{2} = \frac{-2}{2} \pm \frac{\sqrt{2}}{2}$$
$$= \frac{-2 \pm \sqrt{2}}{2}$$

$$x + 1 = \sqrt{\frac{1}{2}} \quad \text{or} \quad x + 1 = -\sqrt{\frac{1}{2}}$$ Use the square-root property.

$$x + 1 = \frac{\sqrt{2}}{2} \qquad\qquad x + 1 = -\frac{\sqrt{2}}{2}$$ $\sqrt{\frac{1}{2}} = \frac{1}{\sqrt{2}} = \frac{1 \cdot \sqrt{2}}{\sqrt{2}\sqrt{2}} = \frac{\sqrt{2}}{2}$

$$x = -1 + \frac{\sqrt{2}}{2} \qquad\qquad x = -1 - \frac{\sqrt{2}}{2}$$

These solutions can be written as $-1 \pm \frac{\sqrt{2}}{2}$.

SELF CHECK 9   Solve $3x^2 + 6x + 1 = 0$ by completing the square.

Accent on technology

▸ Checking Solutions of Quadratic Equations

We can use a graphing calculator to check the solutions of the equation $2x^2 + 4x + 1 = 0$ found in Example 9. Using a TI-84 Plus calculator, we first find an approximation of the decimal value of $-1 + \frac{\sqrt{2}}{2}$ by pressing these keys:

-1 **+ 2ND X²($\surd$)2)** ÷ **2 ENTER**

We will obtain the screen shown in Figure 8-2(a) on the next page. We can now store this decimal value in the calculator by pressing these keys:

STO *X, T, θ, n* **ENTER**

Finally, we enter $2x^2 + 4x + 1$ by pressing

2 *X, T, θ, n* **X² (²) + 4** *X, T, θ, n* **+ 1**

After pressing **ENTER** one more time, we will obtain the screen shown in Figure 8-2(b). The 0 on the screen confirms that $-1 + \frac{\sqrt{2}}{2}$ satisfies the equation $2x^2 + 4x + 1 = 0$ and is a solution.

(continued)

We can check the other solution in a similar way.

(a)   (b)

Figure 8-2

For instructions regarding the use of a Casio graphing calculator, please refer to the Casio Keystroke Guide in the back of the book.

In the next example, the solutions are complex numbers.

EXAMPLE 10   Solve $3x^2 + 2x + 2 = 0$ by completing the square.

Solution
$$3x^2 + 2x + 2 = 0$$

$$x^2 + \frac{2}{3}x + \frac{2}{3} = \frac{0}{3}$$ Divide both sides by 3 to make the coefficient of x^2 equal to 1.

$$x^2 + \frac{2}{3}x \quad\quad = -\frac{2}{3}$$ Subtract $\frac{2}{3}$ from both sides.

$$x^2 + \frac{2}{3}x + \frac{1}{9} = -\frac{2}{3} + \frac{1}{9}$$ Square half the coefficient of x and add it to both sides. $\left(\frac{1}{2}\cdot\frac{2}{3}\right)^2 = \left(\frac{1}{3}\right)^2 = \frac{1}{9}$

$$\left(x + \frac{1}{3}\right)^2 = -\frac{5}{9}$$ Factor on one side and combine terms on the other: $\frac{1}{9} - \frac{2}{3} = \frac{1}{9} - \frac{6}{9} = -\frac{5}{9}.$

$$x + \frac{1}{3} = \sqrt{-\frac{5}{9}} \quad \text{or} \quad x + \frac{1}{3} = -\sqrt{-\frac{5}{9}}$$ Use the square-root property.

$$x + \frac{1}{3} = \sqrt{\frac{5}{9}}\sqrt{-1} \quad\quad x + \frac{1}{3} = -\sqrt{\frac{5}{9}}\sqrt{-1}$$ $\sqrt{-\frac{5}{9}} = \sqrt{\frac{5}{9}(-1)} = \sqrt{\frac{5}{9}}\sqrt{-1}$

$$x + \frac{1}{3} = \frac{\sqrt{5}}{3}i \quad\quad x + \frac{1}{3} = -\frac{\sqrt{5}}{3}i$$ $\sqrt{\frac{5}{9}} = \frac{\sqrt{5}}{\sqrt{9}} = \frac{\sqrt{5}}{3}$

$$x = -\frac{1}{3} + \frac{\sqrt{5}}{3}i \quad\quad x = -\frac{1}{3} - \frac{\sqrt{5}}{3}i$$ Subtract $\frac{1}{3}$ from both sides.

These solutions are written as $x = -\frac{1}{3} \pm \frac{\sqrt{5}}{3}i$.

SELF CHECK 10   Solve $x^2 + 4x + 6 = 0$ by completing the square.

4   **Solve an application requiring the use of the square-root property.**

Many applications involving equations containing squared terms can be solved using the square-root property.

EXAMPLE 11   **DVDs**   A DVD used for recording movies has a surface area of 17.72 square inches on one side. Find the radius of a disc.

Solution   The formula for the area of a circular disc is $A = \pi r^2$. We can find the radius of a disc by substituting 17.72 for A and solving for r.

$$A = \pi r^2$$

$$17.72 = \pi r^2 \qquad \text{Substitute 17.72 for } A.$$

$$\frac{17.72}{\pi} = r^2 \qquad \text{Divide both sides by } \pi.$$

$$r = \sqrt{\frac{17.72}{\pi}} \quad \text{or} \quad r = -\sqrt{\frac{17.72}{\pi}} \qquad \text{Use the square-root property.}$$

Since the radius of a disc cannot be negative, we will discard the negative result. Thus, the radius of a disc is $\sqrt{\frac{17.72}{\pi}}$ inches or, to the nearest hundredth, 2.37 inches.

SELF CHECK 11   The surface area of the DVD is 114.31 square centimeters. Find the radius of the disc.

When you deposit money in a bank account, it earns interest. If you leave the money in the account, the earned interest is deposited back into the account and also earns interest. When this is the case, the account is earning **compound interest**. There is a formula we can use to compute the amount in an account at any time t.

FORMULA FOR COMPOUND INTEREST

If P dollars is deposited in an account and interest is paid once a year at an annual rate r, the amount A in the account after t years is given by the formula

$$A = P(1 + r)^t$$

EXAMPLE 12   **SAVING MONEY**   A woman invests $10,000 in an account. Find the annual interest rate if the account is worth $11,025 in 2 years.

Solution   We substitute 11,025 for A, 10,000 for P, and 2 for t in the compound interest formula and solve for r.

$$A = P(1 + r)^t$$

$$11{,}025 = 10{,}000(1 + r)^2 \qquad \text{Substitute.}$$

$$\frac{11{,}025}{10{,}000} = (1 + r)^2 \qquad \text{Divide both sides by 10,000.}$$

$$1.1025 = (1 + r)^2 \qquad \tfrac{11{,}025}{10{,}000} = 1.1025$$

$$1 + r = 1.05 \quad \text{or} \quad 1 + r = -1.05 \qquad \text{Use the square-root property:}$$
$$\sqrt{1.1025} = 1.05$$

$$r = 0.05 \quad \Big| \quad r = -2.05 \qquad \text{Subtract 1 from both sides.}$$

Since an interest rate cannot be negative, we must discard the result of -2.05. Thus, the annual interest rate is 0.05, or 5%.

We can check this result by substituting 0.05 for r, 10,000 for P, and 2 for t in the formula and confirming that the deposit of $10,000 will grow to $11,025 in 2 years.

$$A = P(1 + r)^t = 10{,}000(1 + 0.05)^2 = 10{,}000(1.1025) = 11{,}025$$

SELF CHECK 12   Find the annual interest rate if the account is worth $10,920.25 in 2 years.

NOW TRY THIS

1. Solve using the square-root property.

   a. $(3x + 5)^2 = 18$

   b. $(x + 6)^2 = 0$

2. Solve $x^2 - 2\sqrt{2}x + 1 = 0$ by completing the square.

8.1 Exercises

WARM-UPS *Multiply.*

1. $(x + 5)(x - 5)$ **2.** $(x - 10)(x + 10)$
3. $(x + 2)^2$ **4.** $(x - 5)^2$
5. $(x - 7)^2$ **6.** $(x + 4)^2$

REVIEW *Solve each equation or inequality.*

7. $\dfrac{t + 9}{2} + \dfrac{t + 2}{5} = \dfrac{8}{5} + 4t$

8. $\dfrac{1 - 5x}{2x} + 4 = \dfrac{x + 3}{x}$

9. $3(t - 3) + 3t \le 2(t + 1) + t + 1$

10. $-2(y + 4) - 3y + 8 \ge 3(2y - 3) - y$

VOCABULARY AND CONCEPTS *Fill in the blanks.*

11. The square-root property, also called extracting the roots, states that the solutions of $x^2 = c$ are _____ and _____.

12. To complete the square on x in $x^2 + 6x = 17$, find one-half of _, square it to obtain _, and add _ to both sides of the equation.

13. The symbol $\pm$ is read as _____.

14. The formula for annual compound interest is _____.

GUIDED PRACTICE *Use factoring to solve each equation. SEE EXAMPLE 1. (OBJECTIVE 1)*

15. $x^2 - 81 = 0$ **16.** $y^2 - 121 = 0$
17. $2y^2 - 72 = 0$ **18.** $4y^2 - 64 = 0$

Use factoring to solve each equation. SEE EXAMPLE 2. (OBJECTIVE 1)

19. $y^2 + 7y + 12 = 0$ **20.** $x^2 + 9x + 20 = 0$

21. $6s^2 + 11s - 10 = 0$ **22.** $3x^2 + 10x - 8 = 0$

Use the square-root property to solve each equation. SEE EXAMPLE 3. (OBJECTIVE 2)

23. $x^2 = 36$ **24.** $x^2 = 144$
25. $z^2 = 5$ **26.** $u^2 = 24$

Use the square-root property to solve each equation. SEE EXAMPLE 4. (OBJECTIVE 2)

27. $(y + 3)^2 = 9$ **28.** $(y - 1)^2 = 4$
29. $(x - 2)^2 - 5 = 0$ **30.** $(x - 4)^2 - 7 = 0$

Use the square-root property to solve each equation. SEE EXAMPLE 5. (OBJECTIVE 2)

31. $16p^2 + 49 = 0$ **32.** $q^2 + 25 = 0$
33. $4m^2 + 81 = 0$ **34.** $9n^2 + 64 = 0$

Find the number that when added to the binomial will make it a perfect-square trinomial. SEE EXAMPLE 6. (OBJECTIVE 3)

35. $x^2 + 4x$ **36.** $x^2 - 6x$
37. $x^2 - 3x$ **38.** $x^2 + 7x$

Use completing the square to solve each equation. SEE EXAMPLE 7. (OBJECTIVE 3)

39. $x^2 + 4x - 12 = 0$ **40.** $x^2 + 6x + 5 = 0$
41. $x^2 - 6x + 8 = 0$ **42.** $x^2 + 8x + 15 = 0$

Use completing the square to solve each equation. SEE EXAMPLE 8. (OBJECTIVE 3)

43. $6x^2 + 11x + 3 = 0$ **44.** $6x^2 + x - 2 = 0$

45. $6x^2 - 7x - 5 = 0$ **46.** $4x^2 - x - 3 = 0$

47. $9 - 6r = 8r^2$ **48.** $11m - 10 = 3m^2$

49. $x + 1 = 2x^2$ **50.** $-2 = 2x^2 - 5x$

Use completing the square to solve each equation. SEE EXAMPLE 9. (OBJECTIVE 3)

51. $\dfrac{7x + 1}{5} = -x^2$ **52.** $\dfrac{3x^2}{8} = \dfrac{1}{8} - x$

53. $3x^2 - 6x + 1 = 0$ **54.** $3x^2 + 9x + 5 = 0$

Use completing the square to solve each equation. SEE EXAMPLE 10.
(OBJECTIVE 3)

55. $p^2 + 2p + 2 = 0$ **56.** $x^2 - 8x + 17 = 0$

57. $y^2 + 8y + 18 = 0$ **58.** $t^2 + t + 3 = 0$

ADDITIONAL PRACTICE   *Solve using any method.*

59. $9x - 8 = x^2$ **60.** $5t - 6 = t^2$

61. $3x^2 - 16 = 0$ **62.** $5x^2 - 49 = 0$

63. $(s - 7)^2 - 9 = 0$ **64.** $(t + 4)^2 = 16$

65. $(x + 5)^2 - 3 = 0$ **66.** $(x + 3)^2 - 7 = 0$

67. $2z^2 - 5z + 2 = 0$ **68.** $2x^2 - x - 1 = 0$

69. $3m^2 - 2m - 3 = 0$ **70.** $4p^2 + 2p + 3 = 0$

71. $5x^2 + 15x = 0$ **72.** $5x^2 + 11x = 0$

73. $x^2 + 5x + 4 = 0$ **74.** $x^2 - 11x + 30 = 0$

75. $x^2 - 9x - 10 = 0$ **76.** $x^2 - 7x + 10 = 0$

77. $2x^2 - x + 8 = 0$ **78.** $4x^2 + 2x + 5 = 0$

Solve for the indicated variable. Assume that all variables represent positive numbers. Express all radicals in simplified form.

79. $2d^2 = 3h$ for d **80.** $2x^2 = d^2$ for d

81. $E = mc^2$ for c **82.** $S = \frac{1}{2}gt^2$ for t

Find all values of x that will make f(x) = 0.

83. $f(x) = 2x^2 + x - 5$ **84.** $f(x) = 3x^2 - 2x - 4$

85. $f(x) = x^2 + x - 3$ **86.** $f(x) = x^2 + 2x - 4$

APPLICATIONS   *Solve each application. SEE EXAMPLES 11–12.*
(OBJECTIVE 4)

87. Falling object   The distance s (in feet) that an object will fall in t seconds is given by the formula $s = 16t^2$. How long will it take an object to fall 256 feet?

88. Pendulum   The time (in seconds) it takes a pendulum to swing back and forth to complete one cycle is related to its length l (in feet) by the formula:

$$l = \frac{32t^2}{4\pi^2}$$

How long will it take a 5-foot pendulum to swing through one cycle? Give the result to the nearest hundredth.

89. Law enforcement   To estimate the speed s (in mph) of a car involved in an accident, police often use the formula $s^2 = 10.5l$, where l is the length of any skid mark. Approximately how fast was a car going that was involved in an accident and left skid marks of 495 feet?

90. Medicine   The approximate pulse rate (in beats per minute) of an adult who is t inches tall is given by the formula

$$p^2 = \frac{348,100}{t}$$

Find the pulse rate of an adult who is 64 inches tall.

91. Saving money   A student invests $8,500 in a savings account drawing interest that is compounded annually. Find the annual rate if the money grows to $9,193.60 in 2 years.

92. Saving money   A woman invests $12,500 in a savings account drawing interest that is compounded annually. Find the annual rate if the money grows to $14,045 in 2 years.

93. Flags   In 1912, an order by President Taft fixed the width and length of the U.S. flag in the ratio 1 to 1.9. If 100 square feet of cloth are to be used to make a U.S. flag, estimate its dimensions to the nearest $\frac{1}{4}$ foot.

94. Accidents   The height h (in feet) of an object that is dropped from a height of s feet is given by the formula $h = s - 16t^2$, where t is the time the object has been falling. A 5-foot-tall woman on a sidewalk looks directly overhead and sees a window washer drop a bottle from 4 stories up. How long does she have to get out of the way? (A story is 10 feet.)

WRITING ABOUT MATH

95. Explain why we did not round to the nearest tenth in Exercise 94.

96. Explain why a cannot be 0 in the quadratic equation $ax^2 + bx + c = 0$.

SOMETHING TO THINK ABOUT

97. What number must be added to $x^2 + \sqrt{3}x$ to make it a perfect-square trinomial?

98. Solve $x^2 + \sqrt{3}x - \frac{1}{4} = 0$ by completing the square.

Section 8.2

Solving Quadratic Equations Using the Quadratic Formula

Objectives

1. Solve a quadratic equation using the quadratic formula.
2. Solve a formula for a specified variable using the quadratic formula.
3. Solve an application involving a quadratic equation.

Vocabulary

quadratic formula

Getting Ready

Add a number to each binomial to complete the square. Then write the resulting trinomial as the square of a binomial.

1. $x^2 + 12x$ **2.** $x^2 - 7x$

Evaluate $\sqrt{b^2 - 4ac}$ for the following values.

3. $a = 6, b = 1, c = -2$ **4.** $a = 4, b = -4, c = -3$

Solving quadratic equations by completing the square is often tedious. Fortunately, there is a more direct way. In this section, we will develop a formula, called the **quadratic formula**, that we can use to solve quadratic equations primarily with arithmetic. To develop this formula, we will use the skills we learned in the last section and complete the square.

1 ## Solve a quadratic equation using the quadratic formula.

To develop a formula to solve quadratic equations, we will solve the general quadratic equation $ax^2 + bx + c = 0 \ (a \neq 0)$ by completing the square.

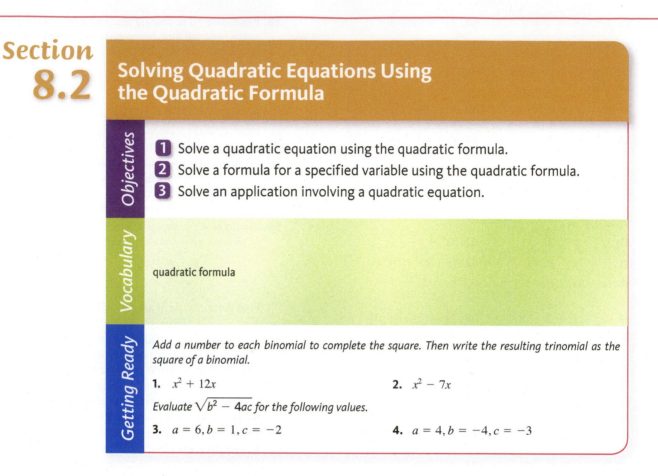

$$ax^2 + bx + c = 0$$

$$\frac{ax^2}{a} + \frac{bx}{a} + \frac{c}{a} = \frac{0}{a}$$ To make the coefficient of x^2 equal to 1, we divide both sides by a.

$$x^2 + \frac{bx}{a} = -\frac{c}{a}$$ $\frac{0}{a} = 0$; subtract $\frac{c}{a}$ from both sides.

$$x^2 + \frac{bx}{a} + \left(\frac{b}{2a}\right)^2 = \left(\frac{b}{2a}\right)^2 - \frac{c}{a}$$ Complete the square on x by adding $\left(\frac{b}{2a}\right)^2$ to both sides.

$$x^2 + \frac{b}{a}x + \frac{b^2}{4a^2} = \frac{b^2}{4a^2} - \frac{4ac}{4aa}$$ Simplify and obtain a common denominator on the right side.

$$\left(x + \frac{b}{2a}\right)^2 = \frac{b^2 - 4ac}{4a^2}$$ Factor the left side and add the fractions on the right side.

$$(1) \qquad x + \frac{b}{2a} = \pm\sqrt{\frac{b^2 - 4ac}{4a^2}}$$ Use the square-root property.

We can solve Equation 1 by writing it as two separate equations.

$$x + \frac{b}{2a} = \sqrt{\frac{b^2 - 4ac}{4a^2}} \qquad \text{or} \qquad x + \frac{b}{2a} = -\sqrt{\frac{b^2 - 4ac}{4a^2}}$$

$$x + \frac{b}{2a} = \frac{\sqrt{b^2 - 4ac}}{2a} \qquad\qquad x + \frac{b}{2a} = -\frac{\sqrt{b^2 - 4ac}}{2a}$$

$$x = -\frac{b}{2a} + \frac{\sqrt{b^2 - 4ac}}{2a} \qquad\qquad x = -\frac{b}{2a} - \frac{\sqrt{b^2 - 4ac}}{2a}$$

$$x = \frac{-b + \sqrt{b^2 - 4ac}}{2a} \qquad\qquad x = \frac{-b - \sqrt{b^2 - 4ac}}{2a}$$

These two solutions represent the *quadratic formula*.

THE QUADRATIC FORMULA

The solutions of $ax^2 + bx + c = 0$ $(a \neq 0)$ are obtained by the formula

$$x = \frac{-b \pm \sqrt{b^2 - 4ac}}{2a}$$

COMMENT Be sure to draw the fraction bar under both parts of the numerator, and be sure to draw the radical sign exactly over $b^2 - 4ac$.

EXAMPLE 1 Use the quadratic formula to solve $2x^2 - 3x - 5 = 0$.

Solution In this equation $a = 2$, $b = -3$, and $c = -5$.

$$x = \frac{-b \pm \sqrt{b^2 - 4ac}}{2a}$$

$$= \frac{-(-3) \pm \sqrt{(-3)^2 - 4(2)(-5)}}{2(2)} \qquad \text{Substitute 2 for } a, -3 \text{ for } b, \text{ and } -5 \text{ for } c.$$

$$= \frac{3 \pm \sqrt{9 + 40}}{4} \qquad \text{Simplify.}$$

$$= \frac{3 \pm \sqrt{49}}{4} \qquad \text{Add.}$$

$$= \frac{3 \pm 7}{4} \qquad \text{Simplify the radical.}$$

$$x = \frac{3 + 7}{4} \quad \text{or} \quad x = \frac{3 - 7}{4}$$

$$x = \frac{10}{4} \qquad\qquad x = \frac{-4}{4}$$

$$x = \frac{5}{2} \qquad\qquad x = -1$$

After checking the results, we see that the solutions are $\frac{5}{2}$ and -1. Note that this equation can be solved by factoring.

SELF CHECK 1 Use the quadratic formula to solve $3x^2 - 5x - 2 = 0$.

It is important to write the quadratic equation in quadratic form as $ax^2 + bx + c = 0$, prior to identifying the values of a, b and c.

EXAMPLE 2   Use the quadratic formula to solve $2x^2 + 1 = -4x$.

Solution   We begin by writing the equation in $ax^2 + bx + c = 0$ form (called *standard form*) before identifying a, b, and c.

$$2x^2 + 4x + 1 = 0$$

In this equation, $a = 2$, $b = 4$, and $c = 1$.

$$x = \frac{-b \pm \sqrt{b^2 - 4ac}}{2a}$$

$$= \frac{-4 \pm \sqrt{4^2 - 4(2)(1)}}{2(2)} \qquad \text{Substitute 2 for } a, 4 \text{ for } b, \text{ and 1 for } c.$$

$$= \frac{-4 \pm \sqrt{16 - 8}}{4} \qquad \text{Simplify.}$$

$$= \frac{-4 \pm \sqrt{8}}{4} \qquad \text{Subtract.}$$

$$= \frac{-4 \pm 2\sqrt{2}}{4} \qquad \sqrt{8} = \sqrt{4 \cdot 2} = \sqrt{4}\sqrt{2} = 2\sqrt{2}$$

$$= \frac{-2 \pm \sqrt{2}}{2} \qquad \frac{-4 \pm 2\sqrt{2}}{4} = \frac{2(-2 \pm \sqrt{2})}{4} = \frac{-2 \pm \sqrt{2}}{2}$$

Note that these solutions can be written as $-1 \pm \frac{\sqrt{2}}{2}$.

SELF CHECK 2   Use the quadratic formula to solve $3x^2 - 2x - 3 = 0$.

In the next example, the solutions are complex numbers.

EXAMPLE 3   Use the quadratic formula to solve $x^2 + x = -1$.

Solution   We begin by writing the equation in standard form before identifying a, b, and c.

$$x^2 + x + 1 = 0$$

In this equation, $a = 1$, $b = 1$, and $c = 1$.

$$x = \frac{-b \pm \sqrt{b^2 - 4ac}}{2a}$$

$$= \frac{-1 \pm \sqrt{1^2 - 4(1)(1)}}{2(1)} \qquad \text{Substitute 1 for } a, 1 \text{ for } b, \text{ and 1 for } c.$$

$$= \frac{-1 \pm \sqrt{1 - 4}}{2} \qquad \text{Simplify the expression under the radical.}$$

$$= \frac{-1 \pm \sqrt{-3}}{2} \qquad \text{Subtract.}$$

$$= \frac{-1 \pm \sqrt{3}i}{2} \qquad \text{Simplify the radical expression.}$$

Note that these solutions should be written in $a + bi$ form as $-\frac{1}{2} \pm \frac{\sqrt{3}}{2}i$.

SELF CHECK 3   Use the quadratic formula to solve $a^2 + 2a = -3$.

Perspective

THE FIBONACCI SEQUENCE AND THE GOLDEN RATIO

Perhaps one of the most intriguing examples of how a mathematical idea can represent natural phenomena is the *Fibonacci Sequence*, a list of whole numbers that is generated by a very simple rule. This sequence was first developed by the Italian mathematician Leonardo da Pisa, more commonly known as Fibonacci. The Fibonacci Sequence is the following list of numbers

$$1, 1, 2, 3, 5, 8, 13, 21, \ldots$$

where each successive number in the list is obtained by adding the two preceding numbers. Although Fibonacci originally developed this sequence to solve a mathematical puzzle, subsequent study of the numbers in this sequence has uncovered many examples in the natural world in which this sequence emerges. For example, the arrangement of the seeds on the face of a sunflower, the hibernation periods of certain insects, and the branching patterns of many plants all give rise to Fibonacci numbers.

Among the many special properties of these numbers is the fact that, as we generate more and more numbers in the list, the ratio of successive numbers approaches a constant value. This value is designated by the symbol ϕ and often is referred to as the "Golden Ratio." One way to calculate the value of ϕ is to solve the quadratic equation $\phi^2 - \phi - 1 = 0$.

Leonardo Fibonacci

1. Using the quadratic formula, find the exact value of ϕ.

2. Using a calculator, find a decimal approximation of ϕ, correct to three decimal places.

Note: Negative solutions are not applicable in this context.

2 Solve a formula for a specified variable using the quadratic formula.

EXAMPLE 4 An object thrown straight up with an initial velocity of v_0 feet per second will reach a height of s feet in t seconds according to the formula $s = -16t^2 + v_0 t$. Solve the formula for t.

Solution We begin by writing the equation in standard form:

$$s = -16t^2 + v_0 t$$
$$16t^2 - v_0 t + s = 0$$

In this equation, $a = 16$, $b = -v_0$, and $c = s$. We can use the quadratic formula to solve for t.

$$t = \frac{-b \pm \sqrt{b^2 - 4ac}}{2a}$$

$$t = \frac{-(-v_0) \pm \sqrt{(-v_0)^2 - 4(16)(s)}}{2(16)} \qquad \text{Substitute into the quadratic formula.}$$

$$t = \frac{v_0 \pm \sqrt{v_0^2 - 64s}}{32} \qquad \text{Simplify.}$$

SELF CHECK 4 Solve the formula $s = -16t^2 + 128t$ for t.

3 Solve an application involving a quadratic equation.

EXAMPLE 5 **DIMENSIONS OF A RECTANGLE** Find the dimensions of the rectangle shown in Figure 8-3, given that its area is 253 cm².

Figure 8-3

Solution If we let w represent the width of the rectangle in cm, then $(w + 12)$ represents its length in cm. Since the area of the rectangle is 253 square centimeters, we can form the equation

$$w(w + 12) = 253 \quad \text{Area of a rectangle = width · length}$$

and solve it as follows:

$$w(w + 12) = 253$$
$$w^2 + 12w = 253 \quad \text{Use the distributive property to remove parentheses.}$$
$$w^2 + 12w - 253 = 0 \quad \text{Write the equation in quadratic form.}$$

Solution by factoring

$$(w - 11)(w + 23) = 0$$
$$w - 11 = 0 \quad \text{or} \quad w + 23 = 0$$
$$w = 11 \qquad\qquad w = -23$$

Solution by quadratic formula

$$w = \frac{-12 \pm \sqrt{12^2 - 4(1)(-253)}}{2(1)}$$
$$= \frac{-12 \pm \sqrt{144 + 1{,}012}}{2}$$
$$= \frac{-12 \pm \sqrt{1{,}156}}{2}$$
$$= \frac{-12 \pm 34}{2}$$
$$w = 11 \quad \text{or} \quad w = -23$$

Since the rectangle cannot have a negative width, we discard the solution of -23. Thus, the only solution is $w = 11$. Since the rectangle is 11 centimeters wide and $(11 + 12)$ centimeters long, its dimensions are 11 centimeters by 23 centimeters.

Check: 23 is 12 more than 11, and the area of a rectangle with dimensions of 23 centimeters by 11 centimeters is 253 square centimeters.

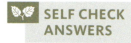 **SELF CHECK 5** Find the dimensions of the rectangle shown given that its area is 189 square centimeters.

SELF CHECK ANSWERS 1. $2, -\frac{1}{3}$ 2. $\frac{1}{3} \pm \frac{\sqrt{10}}{3}$ 3. $-1 \pm \sqrt{2}i$ 4. $t = 4 \pm \frac{\sqrt{256 - s}}{4}$ 5. The dimensions are 9 cm by 21 cm.

NOW TRY THIS

1. The length of a rectangular garden is 1 ft less than 3 times the width. If the area is 44 square feet, find the length of the garden.
2. The product of 2 consecutive integers is 90. Find the numbers.
3. Solve $x^3 - 8 = 0$. (*Hint*: Recall how to factor the difference of cubes.)
4. Graph $y = x^3 - 8$ and identify the x-intercept(s).

8.2   Exercises

WARM-UPS *Identify a, b, and c in each quadratic equation.*

1. $5x^2 - 3x = 4$

2. $-2x^2 + x = 5$

REVIEW *Solve for the indicated variable.*

3. $Ax + By = C$ for y

4. $R = \dfrac{kL}{d^2}$ for L

Simplify each radical.

5. $\sqrt{45}$

6. $\sqrt{288}$

7. $\dfrac{5}{\sqrt{5}}$

8. $\dfrac{1}{2 - \sqrt{3}}$

VOCABULARY AND CONCEPTS *Fill in the blanks.*

9. In the quadratic equation $7x^2 - 4x - 9 = 0$, $a = \_\_$, $b = \_\_$, and $c = \_\_$.

10. The solutions of $ax^2 + bx + c = 0$ $(a \neq 0)$ are given by

the quadratic formula, which is $x = $ _____.

GUIDED PRACTICE *Solve each equation using the quadratic formula.* **SEE EXAMPLE 1. (OBJECTIVE 1)**

11. $x^2 + 6x + 5 = 0$

12. $x^2 - 3x + 2 = 0$

13. $x^2 - 5x - 14 = 0$

14. $x^2 - 2x - 35 = 0$

15. $x^2 + 9x = -20$

16. $y^2 - 18y = -81$

17. $2x^2 - x - 3 = 0$

18. $3x^2 - 10x + 8 = 0$

Solve each equation using the quadratic formula. **SEE EXAMPLE 2. (OBJECTIVE 1)**

19. $15x^2 - 14x = 8$

20. $4x^2 = -5x + 6$

21. $8u = -4u^2 - 3$

22. $4t + 3 = 4t^2$

23. $5x^2 + 5x + 1 = 0$

24. $4w^2 + 6w + 1 = 0$

25. $5x^2 + 2x - 1 = 0$

26. $7x^2 + 4x - 1 = 0$

Solve each equation using the quadratic formula. **SEE EXAMPLE 3. (OBJECTIVE 1)**

27. $x^2 + 2x + 2 = 0$

28. $5x^2 = 2x - 1$

29. $x^2 + 5x + 7 = 0$

30. $x^2 + 3x + 3 = 0$

31. $3x^2 - 4x = -2$

32. $2x^2 + 3x = -3$

33. $3x^2 - 2x = -3$

34. $3x^2 + 2x + 1 = 0$

Solve each formula for the indicated variable. **SEE EXAMPLE 4. (OBJECTIVE 2)**

35. $C = \dfrac{N^2 - N}{2}$, for N

(The formula for a selection sort in data processing)

36. $A = 2\pi r^2 + 2\pi hr$, for r

(The formula for the surface area of a right circular cylinder)

37. $x^2 - kx = -ay$ for x

38. $xy^2 + 3xy + 7 = 0$ for y

ADDITIONAL PRACTICE *Solve each equation using any method.*

39. $6x^2 - 5x - 4 = 0$

40. $2x^2 + 5x - 3 = 0$

41. $\dfrac{x^2}{2} + \dfrac{5}{2}x = -1$ **42.** $-3x = \dfrac{x^2}{2} + 2$

43. $3x^2 - 2 = 2x$ **44.** $-9x = 2 - 3x^2$

45. $16y^2 + 8y - 3 = 0$ **46.** $16x^2 + 16x + 3 = 0$

Find all x-values that will make f(x) = 0.

47. $f(x) = 4x^2 + 4x - 19$ **48.** $f(x) = 9x^2 + 12x - 8$

49. $f(x) = 3x^2 + 2x + 2$ **50.** $f(x) = 5x^2 + 2x + 1$

Use the quadratic formula and a calculator to solve each equation. Give all answers to the nearest hundredth.

51. $0.7x^2 - 3.5x - 25 = 0$
52. $-4.5x^2 + 0.2x + 3.75 = 0$

Note that a and b are the solutions to the equation
$(x - a)(x - b) = 0.$

53. Find a quadratic equation that has a solution set of $\{3, 5\}$.

54. Find a quadratic equation that has a solution set of $\{-2, 7\}$.

55. Find a third-degree equation that has a solution set of $\{2, 3, -4\}$.

56. Find a fourth-degree equation that has a solution set of $\{2, -2, 5, -5\}$.

APPLICATIONS *Solve. SEE EXAMPLE 5. (OBJECTIVE 3)*

57. Dimensions of a rectangle
The rectangle has an area of 96 square feet. Find its dimensions.

$(x + 4)$ ft

x ft

58. Dimensions of a window The area of the window is 77 square feet. Find its dimensions.

$(2x - 3)$ ft

x ft

59. Side of a square The area of a square is numerically equal to its perimeter. Find the length of each side of the square.

60. Perimeter of a rectangle A rectangle is 2 inches longer than it is wide. Numerically, its area exceeds its perimeter by 11. Find the perimeter.

Solve.

61. Base of a triangle The height of a triangle is 5 centimeters longer than three times its base. Find the base of the triangle if its area is 6 square centimeters.

62. Height of a triangle The height of a triangle is 4 meters longer than twice its base. Find the height if the area of the triangle is 15 square meters.

63. Integer The product of two consecutive even integers is 168. Find the integers.   (*Hint*: If one even integer is x, the next consecutive even integer is $x + 2$.)

64. Integer The product of two consecutive odd integers is 143. Find the integers.   (*Hint*: If one odd integer is x, the next consecutive odd integer is $x + 2$.)

65. Integer The sum of the squares of two consecutive integers is 85. Find the integers.   (*Hint*: If one integer is x, the next consecutive positive integer is $x + 1$.)

66. Integer The sum of the squares of three consecutive integers is 77. Find the integers.   (*Hint*: If one integer is x, the next consecutive positive integer is $x + 1$, and the third is $x + 2$.)

67. Rates A woman drives her snowmobile 150 miles at the rate of r mph. She could have gone the same distance in 2 hours less time if she had increased her speed by 20 mph. Find r.

68. Rates Jeff bicycles 160 miles at the rate of r mph. The same trip would have taken 2 hours longer if he had decreased his speed by 4 mph. Find r.

69. Concert tickets Tickets to a concert cost $4, and the projected attendance is 300 people. It is further projected that for every 10¢ increase in ticket price, the average attendance will decrease by 5. At what ticket price will the nightly receipts be $1,248?

70. Bus fares A bus company has 3,000 passengers daily, paying a $1.25 fare. For each 25¢ increase in fare, the company estimates that it will lose 80 passengers. What is the smallest increase in fare that will produce a $4,970 daily revenue?

71. Profit The *Gazette's* profit is $20 per year for each of its 3,000 subscribers. Management estimates that the profit per subscriber will increase by 1¢ for each additional subscriber over the current 3,000. How many subscribers will bring a total profit of $120,000?

72. Interest rates A woman invests $1,000 in a mutual fund for which interest is compounded annually at a rate r. After one year, she deposits an additional $2,000. After two years, the balance A in the account is

$$A = 1{,}000(1 + r)^2 + 2{,}000(1 + r)$$

If this amount is $3,368.10, find r.

73. Framing a picture The frame around the picture in the illustration has a constant width. To the nearest hundredth, how wide is the frame if its area equals the area of the picture?

12 in.

10 in.

74. Metal fabrication A box with no top is to be made by cutting a 2-inch square from each corner of the square sheet of metal shown in the illustration. After bending up the sides, the volume of the box is to be 200 cubic inches. How large should the piece of metal be?

🔳 *Use a calculator.*

75. Labor force The labor force participation rate P (in percent) for workers ages 16 and older from 1966 to 2008 is approximated by the quadratic equation

$$P = -0.0072x^2 + 0.4904x + 58.2714$$

where $x = 0$ corresponds to the year 1966, $x = 1$ corresponds to 1967, and so on. (Thus, $0 \leq x \leq 42$.) In what year in this range were 65% of the workers ages 16 and older part of the workforce?

76. Space program The yearly budget B (in billions of dollars) for the National Aeronautics and Space Administration (NASA) is approximated by the quadratic equation

$$B = 0.0518x^2 - 0.2122x + 14.1112$$

where x is the number of years since 1997 and $0 \leq x \leq 12$. In what year does the model indicate that NASA's budget was about $17 billion?

77. Chemistry A weak acid (0.1 M concentration) breaks down into free cations (the hydrogen ion, H^+) and anions (A^-). When this acid dissociates, the following equilibrium equation is established:

$$\frac{[H^+][A^-]}{[HA]} = 4 \times 10^{-4}$$

where $[H^+]$, the hydrogen ion concentration, is equal to $[A^-]$, the anion concentration. $[HA]$ is the concentration of the undissociated acid itself. Find $[H^+]$ at equilibrium. (*Hint:* If $[H^+] = x$, then $[HA] = 0.1 - x$.)

78. Chemistry A saturated solution of hydrogen sulfide (0.1 M concentration) dissociates into cation $[H^+]$ and anion $[HS^-]$, where $[H^+] = [HS^-]$. When this solution dissociates, the following equilibrium equation is established:

$$\frac{[H^+][HS^-]}{[HHS]} = 1.0 \times 10^{-7}$$

Find $[H^+]$. (*Hint:* If $[H^+] = x$, then $[HHS] = 0.1 - x$.)

WRITING ABOUT MATH

79. Explain why $x = -b \pm \dfrac{\sqrt{b^2 - 4ac}}{2a}$ is not a correct statement of the quadratic formula.

80. Explain why $x = \dfrac{b \pm \sqrt{b^2 - 4ac}}{2a}$ is not a correct statement of the quadratic formula.

SOMETHING TO THINK ABOUT *All of the equations we have solved so far have had rational-number coefficients. However, the quadratic formula can be used to solve quadratic equations with irrational or even imaginary coefficients. Try solving each of the following equations.*

81. $x^2 + 2\sqrt{2}x - 6 = 0$ **82.** $\sqrt{2}x^2 + x - \sqrt{2} = 0$

83. $x^2 - 3ix - 2 = 0$ **84.** $ix^2 + 3x - 2i = 0$

Section 8.3

The Discriminant and Equations That Can Be Written in Quadratic Form

Objectives

1 Use the discriminant to determine the type of solutions to a given quadratic equation.

2 Solve an equation that can be written in quadratic form.

3 Verify the solutions of a quadratic equation by showing that the sum of the solutions is $-\dfrac{b}{a}$ and the product is $\dfrac{c}{a}$.

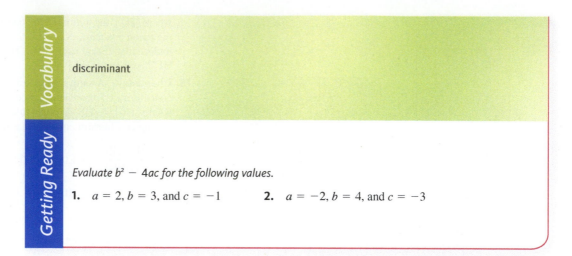

Vocabulary

discriminant

Getting Ready

Evaluate $b^2 - 4ac$ for the following values.

1. $a = 2, b = 3$, and $c = -1$ **2.** $a = -2, b = 4$, and $c = -3$

Recall that we can test a trinomial of the form $ax^2 + bx + c$ $(a \neq 0)$ for factorability by evaluating $b^2 - 4ac$. If $b^2 - 4ac$ is a perfect square, the trinomial is factorable. We recognize that $b^2 - 4ac$ is the radicand in the quadratic formula. We can use that part of the quadratic formula to determine the type of solutions that a quadratic equation will have without solving it first.

1 Use the discriminant to determine the type of solutions to a given quadratic equation.

Suppose that the coefficients a, b, and c in the equation $ax^2 + bx + c = 0$ $(a \neq 0)$ are real numbers. Then the solutions of the equation are given by the quadratic formula

$$x = \frac{-b \pm \sqrt{b^2 - 4ac}}{2a} \quad (a \neq 0)$$

The value of $b^2 - 4ac$, called the **discriminant**, determines the type of solutions for any quadratic equation. Note that $b^2 - 4ac$ is a radicand; thus if $b^2 - 4ac \geq 0$, the solutions are real numbers and if $b^2 - 4ac < 0$, the solutions are nonreal complex numbers.

THE DISCRIMINANT

Given $ax^2 + bx + c = 0$ $(a \neq 0)$ and a, b, and c are real numbers,

if $b^2 - 4ac > 0$, there are two unequal real solutions.
if $b^2 - 4ac = 0$, there are two equal real solutions (called a *double root*).
if $b^2 - 4ac < 0$, the two solutions are complex conjugates.

If we further restrict a, b, and c to the rational numbers and the discriminant

is a perfect square greater than 0, there are two unequal rational solutions.
is positive but not a perfect square, the solutions are irrational and unequal.

EXAMPLE 1 Determine the type of solutions for the equations.

a. $x^2 + x + 1 = 0$ **b.** $3x^2 + 5x + 2 = 0$

Solution   **a.** We calculate the discriminant for $x^2 + x + 1 = 0$.

$$b^2 - 4ac = 1^2 - 4(1)(1) \qquad a = 1, b = 1, \text{ and } c = 1$$
$$= -3$$

Since $b^2 - 4ac < 0$, the solutions will be two complex conjugates.

b. We calculate the discriminant for $3x^2 + 5x + 2 = 0$.

$$b^2 - 4ac = 5^2 - 4(3)(2) \qquad a = 3, b = 5, \text{ and } c = 2$$
$$= 25 - 24$$
$$= 1$$

Since $b^2 - 4ac > 0$ and $b^2 - 4ac$ is a perfect square, there will be two unequal rational solutions.

 SELF CHECK 1   Determine the type of solutions.
a. $x^2 + x - 1 = 0$
b. $4x^2 - 10x + 25 = 0$

EXAMPLE 2   What value of k will make the solutions of the equation $kx^2 - 12x + 9 = 0$ equal?

Solution   We calculate the discriminant:

$$b^2 - 4ac = (-12)^2 - 4(k)(9) \qquad a = k, b = -12, \text{ and } c = 9$$
$$= 144 - 36k$$
$$= -36k + 144$$

Since the solutions are to be equal, we let $-36k + 144 = 0$ and solve for k.

$$-36k + 144 = 0$$
$$-36k = -144 \qquad \text{Subtract 144 from both sides.}$$
$$k = 4 \qquad \text{Divide both sides by } -36.$$

If $k = 4$, the solutions will be equal. Verify this by solving $4x^2 - 12x + 9 = 0$ and showing that the solutions are equal.

SELF CHECK 2   Find the value of k that will make the solutions of $kx^2 - 20x + 25 = 0$ equal.

---

2   Solve an equation that can be written in quadratic form.

Many equations that are not quadratic can be written in quadratic form $(ax^2 + bx + c = 0)$ and then solved using the techniques discussed in previous sections. For example, an inspection of the equation $x^4 - 5x^2 + 4 = 0$ shows that

The leading term x^4 is the square of x^2, the variable part of the middle term:
$$x^4 = (x^2)^2$$

$$x^4 - 5x^2 + 4 = 0$$

The last term is a constant.

To solve the equation $x^4 - 5x^2 + 4 = 0$, we can write the equation in a different form and proceed as follows:

$$x^4 - 5x^2 + 4 = 0$$
$$(x^2)^2 - 5(x^2) + 4 = 0$$

If we substitute u for each x^2, we will obtain a quadratic equation with the variable u that we can solve by factoring.

$$u^2 - 5u + 4 = 0 \qquad \text{Let } x^2 = u.$$
$$(u - 4)(u - 1) = 0 \qquad \text{Factor } u^2 - 5u + 4.$$
$$u - 4 = 0 \quad \text{or} \quad u - 1 = 0 \qquad \text{Set each factor equal to 0.}$$
$$u = 4 \qquad \qquad u = 1$$

Since $u = x^2$, it follows that $x^2 = 4$ or $x^2 = 1$. Thus,

$$x^2 = 4 \qquad \text{or} \qquad x^2 = 1$$
$$x = 2 \quad \text{or} \quad x = -2 \quad \bigg| \quad x = 1 \quad \text{or} \quad x = -1$$

This equation has four solutions: $1, -1, 2$, and -2. Verify that each one satisfies the original equation. Note that this equation can be solved by factoring.

EXAMPLE 3   Solve: $x - 7\sqrt{x} + 12 = 0$

Solution   We examine the leading term and middle term.

The leading term x is the
square of $\sqrt{x}$, the variable $x - 7\sqrt{x} + 12 = 0$
part of the middle term:
$$x = \left(\sqrt{x}\right)^2$$

If we write x as $\left(\sqrt{x}\right)^2$, the equation takes the form

$$\left(\sqrt{x}\right)^2 - 7\sqrt{x} + 12 = 0$$

and it is said to be *quadratic in* $\sqrt{x}$. We can solve this equation by letting $u = \sqrt{x}$ and factoring.

$$u^2 - 7u + 12 = 0 \qquad\qquad \text{Replace each } \sqrt{x} \text{ with } u.$$
$$(u - 3)(u - 4) = 0 \qquad\qquad \text{Factor } u^2 - 7u + 12.$$
$$u - 3 = 0 \quad \text{or} \quad u - 4 = 0 \qquad \text{Set each factor equal to 0.}$$
$$u = 3 \quad \bigg| \qquad u = 4$$

To find the value of x, we undo the substitutions by replacing each u with $\sqrt{x}$. Then we solve the radical equations by squaring both sides.

$$\sqrt{x} = 3 \quad \text{or} \quad \sqrt{x} = 4$$
$$x = 9 \quad \bigg| \qquad x = 16$$

The solutions are 9 and 16. Verify that both satisfy the original equation.

SELF CHECK 3   Solve: $x - 5\sqrt{x} + 6 = 0$

EXAMPLE 4   Solve: $2m^{2/3} - 2 = 3m^{1/3}$

Solution   First we subtract $3m^{1/3}$ from both sides to write the equation in quadratic form with descending powers of m.

$$2m^{2/3} - 3m^{1/3} - 2 = 0$$

Then we write the equation in the form

$$2(m^{1/3})^2 - 3m^{1/3} - 2 = 0 \quad (m^{1/3})^2 = m^{2/3}$$

If we substitute u for $m^{1/3}$, this equation can be written in a form that can be solved by factoring.

$$2u^2 - 3u - 2 = 0 \qquad\qquad \text{Replace each } m^{1/3} \text{ with } u.$$
$$(2u + 1)(u - 2) = 0 \qquad\qquad \text{Factor } 2u^2 - 3u - 2.$$
$$2u + 1 = 0 \quad \text{or} \quad u - 2 = 0 \qquad \text{Set each factor equal to 0.}$$
$$u = -\frac{1}{2} \quad \bigg| \qquad u = 2$$

To find the value of m, we undo the substitutions by replacing each u with $m^{1/3}$ and solve each resulting equation by cubing both sides.

$$m^{1/3} = -\frac{1}{2} \quad \text{or} \quad m^{1/3} = 2$$

$$(m^{1/3})^3 = \left(-\frac{1}{2}\right)^3 \qquad (m^{1/3})^3 = (2)^3 \quad \text{Cube both sides.}$$

$$m = -\frac{1}{8} \qquad\qquad m = 8 \qquad \text{Simplify.}$$

The solutions are $-\frac{1}{8}$ and 8. Verify that both satisfy the original equation.

🍂 **SELF CHECK 4**   Solve: $a^{2/3} = -3a^{1/3} + 10$

EXAMPLE 5   Solve: $\dfrac{24}{x} + \dfrac{12}{x+1} = 11$

Solution   Since the denominator cannot be 0, x cannot be 0 or -1. If either 0 or -1 appears as a possible solution, it is extraneous and must be discarded.

$$\frac{24}{x} + \frac{12}{x+1} = 11$$

$$x(x+1)\left(\frac{24}{x} + \frac{12}{x+1}\right) = x(x+1)11 \qquad \text{Multiply both sides by } x(x+1), \text{ the LCD.}$$

$$24(x+1) + 12x = (x^2 + x)11 \qquad \text{Simplify.}$$

$$24x + 24 + 12x = 11x^2 + 11x \qquad \text{Use the distributive property to remove parentheses.}$$

$$36x + 24 = 11x^2 + 11x \qquad \text{Combine like terms.}$$

$$0 = 11x^2 - 25x - 24 \qquad \text{Write the equation in quadratic form.}$$

$$0 = (11x + 8)(x - 3) \qquad \text{Factor } 11x^2 - 25x - 24.$$

$$11x + 8 = 0 \quad \text{or} \quad x - 3 = 0 \qquad \text{Set each factor equal to 0.}$$

$$x = -\frac{8}{11} \qquad\qquad x = 3$$

Verify that $-\frac{8}{11}$ and 3 satisfy the original equation.

🍂 **SELF CHECK 5**   Solve: $\dfrac{12}{x} + \dfrac{6}{x+3} = 5$

EXAMPLE 6   Solve: $15a^{-2} - 8a^{-1} + 1 = 0$

Solution   First we write the equation in the form

$$15(a^{-1})^2 - 8a^{-1} + 1 = 0 \quad (a^{-1})^2 = a^{-2}$$

If we substitute u for a^{-1}, this equation can be written in a form that can be solved by factoring.

$$15u^2 - 8u + 1 = 0 \qquad \text{Replace each } a^{-1} \text{ with } u.$$

$$(5u - 1)(3u - 1) = 0 \qquad \text{Factor } 15u^2 - 8u + 1.$$

$$5u - 1 = 0 \quad \text{or} \quad 3u - 1 = 0 \qquad \text{Set each factor equal to 0.}$$

$$u = \frac{1}{5} \qquad\qquad u = \frac{1}{3}$$

To find the value of a, we undo the substitutions by replacing each u with a^{-1} and solve each resulting equation.

$$a^{-1} = \frac{1}{5} \quad \bigg| \quad a^{-1} = \frac{1}{3}$$

$$\frac{1}{a} = \frac{1}{5} \quad \bigg| \quad \frac{1}{a} = \frac{1}{3} \qquad a^{-1} = \frac{1}{a}$$

$$5 = a \quad \bigg| \quad 3 = a \qquad \text{Solve the proportions.}$$

The solutions are 5 and 3. Verify that both satisfy the original equation.

SELF CHECK 6   Solve: $28c^{-2} - 3c^{-1} - 1 = 0$

EXAMPLE 7   Solve the formula $s = 16t^2 - 32$ for t.

Solution   We proceed as follows:

$$s = 16t^2 - 32$$

$$s + 32 = 16t^2 \qquad \text{Add 32 to both sides.}$$

$$\frac{s + 32}{16} = t^2 \qquad \text{Divide both sides by 16.}$$

$$t^2 = \frac{s + 32}{16} \qquad \text{Write } t^2 \text{ on the left side.}$$

$$t = \pm\sqrt{\frac{s + 32}{16}} \qquad \text{Apply the square-root property.}$$

$$t = \pm\frac{\sqrt{s + 32}}{\sqrt{16}} \qquad \sqrt{\frac{a}{b}} = \frac{\sqrt{a}}{\sqrt{b}}$$

$$t = \pm\frac{\sqrt{s + 32}}{4}$$

SELF CHECK 7   Solve $a^2 + b^2 = c^2$ for a.

3   Verify the solutions of a quadratic equation by showing that the sum of the solutions is $-\frac{b}{a}$ and the product is $\frac{c}{a}$.

SOLUTIONS OF A QUADRATIC EQUATION   If r_1 and r_2 are the solutions of the quadratic equation $ax^2 + bx + c = 0$, with $a \neq 0$, then

$$r_1 + r_2 = -\frac{b}{a} \quad \text{and} \quad r_1 r_2 = \frac{c}{a}$$

Proof   We note that the solutions to the equation are given by the quadratic formula

$$r_1 = \frac{-b + \sqrt{b^2 - 4ac}}{2a} \quad \text{and} \quad r_2 = \frac{-b - \sqrt{b^2 - 4ac}}{2a}$$

Thus,

$$r_1 + r_2 = \frac{-b + \sqrt{b^2 - 4ac}}{2a} + \frac{-b - \sqrt{b^2 - 4ac}}{2a}$$

$$= \frac{-b + \sqrt{b^2 - 4ac} - b - \sqrt{b^2 - 4ac}}{2a}$$ Keep the denominator and add the numerators.

$$= -\frac{2b}{2a}$$ Combine like terms.

$$= -\frac{b}{a}$$ Simplify.

and

$$r_1 r_2 = \frac{-b + \sqrt{b^2 - 4ac}}{2a} \cdot \frac{-b - \sqrt{b^2 - 4ac}}{2a}$$

$$= \frac{b^2 - (b^2 - 4ac)}{4a^2}$$ Multiply the numerators and multiply the denominators.

$$= \frac{b^2 - b^2 + 4ac}{4a^2}$$ Use the distributive property to remove parentheses.

$$= \frac{4ac}{4a^2}$$ Combine like terms.

$$= \frac{c}{a}$$ Simplify.

It can also be shown that if

$$r_1 + r_2 = -\frac{b}{a} \quad \text{and} \quad r_1 r_2 = \frac{c}{a}$$

then r_1 and r_2 are solutions of $ax^2 + bx + c = 0$. We can use this fact to check the solutions of quadratic equations.

EXAMPLE 8 Show that $\frac{3}{2}$ and $-\frac{1}{3}$ are solutions of $6x^2 - 7x - 3 = 0$.

Solution Since $a = 6$, $b = -7$, and $c = -3$, we have

$$-\frac{b}{a} = -\frac{-7}{6} = \frac{7}{6} \quad \text{and} \quad \frac{c}{a} = \frac{-3}{6} = -\frac{1}{2}$$

Since $\frac{3}{2} + \left(-\frac{1}{3}\right) = \frac{7}{6}$ and $\left(\frac{3}{2}\right)\left(-\frac{1}{3}\right) = -\frac{1}{2}$, these numbers are solutions. Solve the equation to verify that the roots are $\frac{3}{2}$ and $-\frac{1}{3}$.

SELF CHECK 8 Are $-\frac{3}{2}$ and $\frac{1}{3}$ solutions of $6x^2 + 7x - 3 = 0$?

SELF CHECK ANSWERS **1. a.** two irrational and unequal **b.** two complex conjugates **2.** 4 **3.** 4, 9 **4.** 8, −125 **5.** 3, $-\frac{12}{5}$ **6.** −7, 4 **7.** $a = \pm\sqrt{c^2 - b^2}$ **8.** yes

NOW TRY THIS

1. Solve: $x - 3\sqrt{x} - 4 = 0$

2. Without substituting, show that $(3 + 5i)$ and $(3 - 5i)$ are solutions of $x^2 - 6x + 34 = 0$.

3. Find the discriminant of $\sqrt{2}x^2 - \sqrt{65}x - 2\sqrt{2} = 0$.

8.3   Exercises

WARM-UPS *Find $b^2 - 4ac$ when*

1. $a = 1, b = 1, c = 1$ **2.** $a = 2, b = 1, c = 1$

Are the following numbers solutions of $x^2 - 2x + 2 = 0$?

3. $1 - i$ **4.** $-1 - i$
5. $-1 + i$ **6.** $1 + i$

REVIEW *Solve each equation. Assume no division by 0.*

7. $\dfrac{1}{4} + \dfrac{1}{t} = \dfrac{1}{2t}$ **8.** $\dfrac{p - 3}{3p} + \dfrac{1}{2p} = \dfrac{1}{4}$

9. Find the slope of the line passing through $(-3, -2)$ and $(4, 1)$.
10. Write an equation of the line passing through $(-3, -2)$ and $(4, 1)$ in general form.

VOCABULARY AND CONCEPTS *Consider the equation $ax^2 + bx + c = 0$ $(a \neq 0)$, and fill in the blanks.*

11. The discriminant is _____.
12. If $b^2 - 4ac < 0$, the solutions of the equation are two complex _____.
13. If $b^2 - 4ac$ is a nonzero perfect square, the solutions are _____ numbers and _____.
14. If r_1 and r_2 are the solutions of the equation, then $r_1 + r_2 =$ ____ and $r_1 r_2 =$ ___.

GUIDED PRACTICE *Use the discriminant to determine what type of solutions exist for each quadratic equation. Do not solve the equation. SEE EXAMPLE 1. (OBJECTIVE 1)*

15. $9x^2 + 6x + 1 = 0$ **16.** $6x^2 - 5x - 6 = 0$

17. $10x^2 + 2x + 3 = 0$ **18.** $3x^2 + 10x - 2 = 0$

19. $6x^2 = 8x - 1$ **20.** $9x^2 = 12x - 4$

21. $x(2x - 3) = 20$ **22.** $x(x - 3) = -10$

Find the values of k that will make the solutions of each given quadratic equation equal. SEE EXAMPLE 2. (OBJECTIVE 1)

23. $x^2 + kx + 16 = 0$
24. $kx^2 - 12x + 4 = 0$
25. $4x^2 + 9 = kx$
26. $9x^2 - kx + 25 = 0$

Solve each equation. (OBJECTIVE 2)

27. $x^4 - 17x^2 + 16 = 0$ **28.** $x^4 - 10x^2 + 9 = 0$

29. $x^4 - 3x^2 = -2$ **30.** $x^4 - 19x^2 = -48$

31. $x^4 = 6x^2 - 5$ **32.** $x^4 = 8x^2 - 7$

33. $2x^4 - 10x^2 = -8$ **34.** $3x^4 + 12 = 15x^2$

Solve each equation. SEE EXAMPLE 3. (OBJECTIVE 2)

35. $x - 7\sqrt{x} + 10 = 0$ **36.** $x - 5\sqrt{x} + 4 = 0$

37. $2x - \sqrt{x} = 3$ **38.** $3x - 4 = -4\sqrt{x}$

39. $2x + x^{1/2} - 3 = 0$ **40.** $2x - x^{1/2} - 1 = 0$

41. $3x + 5x^{1/2} + 2 = 0$ **42.** $3x - 4x^{1/2} + 1 = 0$

Solve each equation. SEE EXAMPLE 4. (OBJECTIVE 2)

43. $x^{2/3} + 5x^{1/3} + 6 = 0$
44. $x^{2/3} - 2x^{1/3} - 3 = 0$
45. $3m^{2/3} - m^{1/3} - 2 = 0$
46. $4m^{2/3} - 8m^{1/3} + 3 = 0$

Solve each equation. SEE EXAMPLE 5. (OBJECTIVE 2)

47. $x + 10 + \dfrac{9}{x} = 0$ **48.** $x - 4 + \dfrac{3}{x} = 0$

49. $x + 3 = \dfrac{28}{x}$ **50.** $x + \dfrac{15}{x} = 8$

51. $\dfrac{1}{x - 1} + \dfrac{3}{x + 1} = 2$ **52.** $\dfrac{6}{x - 2} - \dfrac{12}{x - 1} = -1$

53. $\dfrac{1}{x + 2} + \dfrac{24}{x + 3} = 13$ **54.** $\dfrac{3}{x} + \dfrac{4}{x + 1} = 2$

Solve each equation. SEE EXAMPLE 6. (OBJECTIVE 2)

55. $x^{-4} - 2x^{-2} + 1 = 0$
56. $4x^{-4} + 1 = 5x^{-2}$
57. $8a^{-2} - 10a^{-1} - 3 = 0$
58. $2y^{-2} - 5y^{-1} = 3$

Solve each equation for the indicated variable. SEE EXAMPLE 7. (OBJECTIVE 2)

59. $x^2 + y^2 = r^2$ for x
60. $x^2 + y^2 = r^2$ for y

61. $xy^2 + 5xy + 3 = 0$ for y

62. $kx = ay - x^2$ for x

Solve each equation and verify that the sum of the solutions is $-\frac{b}{a}$ and that the product of the solutions is $\frac{c}{a}$. SEE EXAMPLE 8. (OBJECTIVE 3)

63. $12x^2 - 5x - 2 = 0$ **64.** $8x^2 - 2x - 3 = 0$

65. $2x^2 + 5x + 1 = 0$ **66.** $3x^2 + 9x + 1 = 0$

67. $3x^2 - 2x + 4 = 0$ **68.** $2x^2 - x + 4 = 0$

69. $x^2 + 2x + 5 = 0$ **70.** $x^2 - 4x + 13 = 0$

ADDITIONAL PRACTICE

71. Use the discriminant to determine whether the solutions of $1{,}492x^2 + 1{,}776x - 1{,}984 = 0$ are real numbers.

72. Use the discriminant to determine whether the solutions of $1{,}776x^2 - 1{,}492x + 1{,}984 = 0$ are real numbers.

Solve using any method.

73. $4x - 5\sqrt{x} - 9 = 0$ **74.** $9x - 5\sqrt{x} = 4$

75. $3x^{2/3} - x^{1/3} - 2 = 0$ **76.** $4x^{2/3} + 4x^{1/3} + 1 = 0$

77. $2x^4 + 24 = 26x^2$ **78.** $4x^4 = -9 + 13x^2$

79. $x^4 - 24 = -2x^2$ **80.** $x^4 - 7x^2 = 18$

81. $4(2x - 1)^2 - 3(2x - 1) - 1 = 0$

82. $4(x^2 - 1)^2 + 13(x^2 - 1) + 9 = 0$

83. $x^{-2/3} - 2x^{-1/3} - 3 = 0$ **84.** $4x^{-1} - 5x^{-1/2} - 9 = 0$

85. $x + \dfrac{2}{x - 2} = 0$ **86.** $x + \dfrac{x + 5}{x - 3} = 0$

87. $8(m + 1)^{-2} - 30(m + 1)^{-1} + 7 = 0$

88. $2(p - 2)^{-2} + 3(p - 2)^{-1} - 5 = 0$

89. $I = \dfrac{k}{d^2}$ for d

90. $V = \dfrac{1}{3}\pi r^2 h$ for r

91. $\sigma = \sqrt{\dfrac{\Sigma x^2}{N} - \mu^2}$ for μ^2

92. $\sigma = \sqrt{\dfrac{\Sigma x^2}{N} - \mu^2}$ for N

Find the values of k that will make the solutions of each given quadratic equation equal.

93. $(k - 1)x^2 + (k - 1)x + 1 = 0$

94. $(k + 3)x^2 + 2kx + 4 = 0$

95. $(k + 4)x^2 + 2kx + 9 = 0$

96. $(k + 15)x^2 + (k - 30)x + 4 = 0$

97. Determine k such that the solutions of $3x^2 + 4x = k$ are complex numbers.

98. Determine k such that the solutions of $kx^2 - 4x = 7$ are complex numbers.

WRITING ABOUT MATH

99. Describe how to predict what type of solutions the equation $3x^2 - 4x + 5 = 0$ will have.

100. How is the discriminant related to the quadratic formula?

SOMETHING TO THINK ABOUT

101. Can a quadratic equation with integer coefficients have one real and one complex solution? Why?

102. Can a quadratic equation with complex coefficients have one real and one complex solution? Why?

Section 8.4

Graphs of Quadratic Functions

Objectives

1 Graph a quadratic function of the form $f(x) = ax^2$.
2 Use a vertical translation of $f(x) = ax^2$ to graph $f(x) = ax^2 + k$.
3 Use a horizontal translation of $f(x) = ax^2$ to graph $f(x) = a(x - h)^2$.
4 Use both a vertical and horizontal translation of $f(x) = ax^2$ to graph $f(x) = a(x - h)^2 + k$.
5 Graph a quadratic function written in standard form by writing it in the form $f(x) = a(x - h)^2 + k$.
6 Find the vertex of a parabola using $\left(-\frac{b}{2a}, f\left(-\frac{b}{2a}\right)\right)$.
7 Graph a quadratic function written in standard form by finding the vertex, the axis of symmetry, and the x- and y-intercepts.
8 Solve an application using a quadratic function.

Vocabulary

quadratic functions   vertex   variance
parabola   axis of symmetry

Getting Ready

If $f(x) = 3x^2 + x - 2$, find each value.

1. $f(0)$ 2. $f(1)$ 3. $f(-1)$ 4. $f(-2)$

If $x = -\frac{b}{2a}$, find the value of x when a and b have the following values.

5. $a = 3$ and $b = -6$ 6. $a = 5$ and $b = -40$

In this section, we consider graphs of second-degree polynomial functions, called **quadratic functions**.

The graph shown in Figure 8-4 shows the height (in relation to time) of a toy rocket launched straight up into the air.

From the graph, we can see that the height of the rocket 2 seconds after it was launched is about 128 feet and that the height of the rocket 5 seconds after it was launched is 80 feet.

The parabola shown in Figure 8-4 is the graph of a quadratic function.

QUADRATIC FUNCTION

A quadratic function is a second-degree polynomial function of the form

$$f(x) = ax^2 + bx + c \quad (a \neq 0)$$

where a, b, and c are real numbers.

We begin the discussion of graphing quadratic functions by considering the graph of $f(x) = ax^2 + bx + c$, where $b = 0$ and $c = 0$.

COMMENT   Note that the graph describes the height of the rocket, not the path of the rocket. The rocket goes straight up and comes straight down.

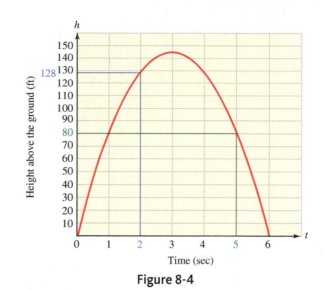

Figure 8-4

1   Graph a quadratic function of the form $f(x) = ax^2$.

EXAMPLE 1   Graph: **a.** $f(x) = x^2$ **b.** $g(x) = 3x^2$ **c.** $h(x) = \frac{1}{3}x^2$

Solution   We create a table of ordered pairs that satisfy each equation, plot each point, and join them with a smooth curve, as in Figure 8-5. We note that the graph of $h(x) = \frac{1}{3}x^2$ is wider than the graph of $f(x) = x^2$, and that the graph of $g(x) = 3x^2$ is narrower than the graph of $f(x) = x^2$. In the function $f(x) = ax^2$, the smaller the value of $|a|$, the wider the graph.

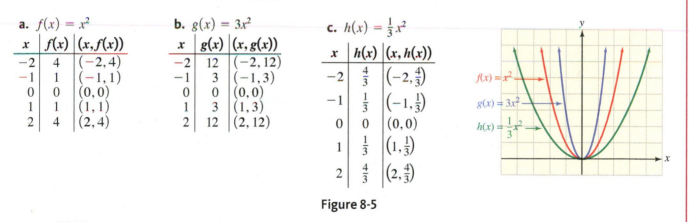

a. $f(x) = x^2$

x	$f(x)$	$(x, f(x))$
-2	4	$(-2, 4)$
-1	1	$(-1, 1)$
0	0	$(0, 0)$
1	1	$(1, 1)$
2	4	$(2, 4)$

b. $g(x) = 3x^2$

x	$g(x)$	$(x, g(x))$
-2	12	$(-2, 12)$
-1	3	$(-1, 3)$
0	0	$(0, 0)$
1	3	$(1, 3)$
2	12	$(2, 12)$

c. $h(x) = \frac{1}{3}x^2$

x	$h(x)$	$(x, h(x))$
-2	$\frac{4}{3}$	$(-2, \frac{4}{3})$
-1	$\frac{1}{3}$	$(-1, \frac{1}{3})$
0	0	$(0, 0)$
1	$\frac{1}{3}$	$(1, \frac{1}{3})$
2	$\frac{4}{3}$	$(2, \frac{4}{3})$

Figure 8-5

SELF CHECK 1   On the same set of coordinate axes, graph each function.
a. $f(x) = 2x^2$ **b.** $f(x) = \frac{1}{2}x^2$

If we consider the graph of $f(x) = -3x^2$, we will see that it opens downward and has the same shape as the graph of $g(x) = 3x^2$. This is called a *reflection*.

EXAMPLE 2   Graph: $f(x) = -3x^2$

Solution   We create a table of ordered pairs that satisfy the equation, plot each point, and join them with a smooth curve, as in Figure 8-6 on the next page.

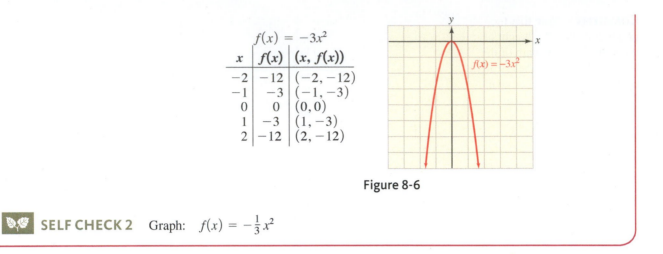

$f(x) = -3x^2$

x	$f(x)$	$(x, f(x))$
-2	-12	$(-2, -12)$
-1	-3	$(-1, -3)$
0	0	$(0, 0)$
1	-3	$(1, -3)$
2	-12	$(2, -12)$

Figure 8-6

 SELF CHECK 2   Graph: $f(x) = -\frac{1}{3}x^2$

The graphs of quadratic functions are called **parabolas**. They open upward when $a > 0$ and downward when $a < 0$. The lowest point (*minimum*) of a parabola that opens upward, or the highest point (*maximum*) of a parabola that opens downward, is called the **vertex** of the parabola. The vertex of the parabola shown in Figure 8-6 is the point $(0, 0)$.

The vertical line, called an **axis of symmetry**, that passes through the vertex divides the parabola into two congruent halves. The axis of symmetry of the parabola shown in Figure 8-6 is the y-axis, written as the equation $x = 0$.

2   Use a vertical translation of $f(x) = ax^2$ to graph $f(x) = ax^2 + k$.

Now let us consider a vertical translation of $f(x) = ax^2$ to graph $f(x) = ax^2 + k$. This means the graph of $f(x) = ax^2$ will retain its shape, but will be shifted k units upward or downward. In this case the vertex will shift to the point $(0, k)$. The axis of symmetry of the parabola is a vertical line written as the equation $x = 0$.

EXAMPLE 3   Graph: **a.** $f(x) = 2x^2$ **b.** $g(x) = 2x^2 + 3$ **c.** $h(x) = 2x^2 - 3$

Solution   We create a table of ordered pairs that satisfy each equation, plot each point, and join them with a smooth curve, as in Figure 8-7. We note that the graph of $g(x) = 2x^2 + 3$ is identical to the graph of $f(x) = 2x^2$, except that it has been translated 3 units upward. The graph of $h(x) = 2x^2 - 3$ is identical to the graph of $f(x) = 2x^2$, except that it has been translated 3 units downward.

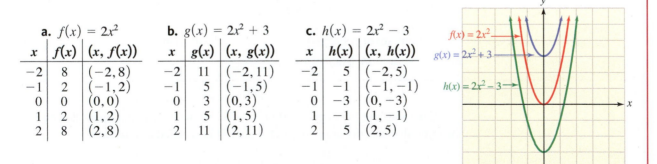

a. $f(x) = 2x^2$

x	$f(x)$	$(x, f(x))$
-2	8	$(-2, 8)$
-1	2	$(-1, 2)$
0	0	$(0, 0)$
1	2	$(1, 2)$
2	8	$(2, 8)$

b. $g(x) = 2x^2 + 3$

x	$g(x)$	$(x, g(x))$
-2	11	$(-2, 11)$
-1	5	$(-1, 5)$
0	3	$(0, 3)$
1	5	$(1, 5)$
2	11	$(2, 11)$

c. $h(x) = 2x^2 - 3$

x	$h(x)$	$(x, h(x))$
-2	5	$(-2, 5)$
-1	-1	$(-1, -1)$
0	-3	$(0, -3)$
1	-1	$(1, -1)$
2	5	$(2, 5)$

Figure 8-7

SELF CHECK 3   On the same set of coordinate axes, graph each function and tell how it differs from the graph of $f(x) = x^2$.
a. $f(x) = x^2 + 1$
b. $f(x) = x^2 - 5$

The results of Example 3 confirm the following facts.

VERTICAL TRANSLATIONS OF GRAPHS

If $y = f(x)$ is a function and k is a positive number, then

• The graph of $y = f(x) + k$ is identical to the graph of $y = f(x)$, except that it is translated k units upward.

• The graph of $y = f(x) - k$ is identical to the graph of $y = f(x)$, except that it is translated k units downward.

3   Use a horizontal translation of $f(x) = ax^2$ to graph $f(x) = a(x - h)^2$.

Now let us consider horizontal translations of $f(x) = ax^2$ to graph $f(x) = a(x - h)^2$. This means the graph of $f(x) = ax^2$ will retain its shape but will be shifted h units to the right or left. In this case the vertex will be shifted to the point $(h, 0)$. The axis of symmetry of the parabola is a vertical line written as the equation $x = h$.

EXAMPLE 4   Graph: **a.** $f(x) = 2x^2$ **b.** $g(x) = 2(x - 3)^2$ **c.** $h(x) = 2(x + 3)^2$

Solution   We create a table of ordered pairs that satisfy each equation, plot each point, and join them with a smooth curve, as in Figure 8-8. We note that the graph of $g(x) = 2(x - 3)^2$ is identical to the graph of $f(x) = 2x^2$, except that it has been translated 3 units to the right. The graph of $h(x) = 2(x + 3)^2$ is identical to the graph of $f(x) = 2x^2$, except that it has been translated 3 units to the left.

a. $f(x) = 2x^2$

x	$f(x)$	$(x, f(x))$
-2	8	$(-2, 8)$
-1	2	$(-1, 2)$
0	0	$(0, 0)$
1	2	$(1, 2)$
2	8	$(2, 8)$

b. $g(x) = 2(x - 3)^2$

x	$g(x)$	$(x, g(x))$
1	8	$(1, 8)$
2	2	$(2, 2)$
3	0	$(3, 0)$
4	2	$(4, 2)$
5	8	$(5, 8)$

c. $h(x) = 2(x + 3)^2$

x	$h(x)$	$(x, h(x))$
-5	8	$(-5, 8)$
-4	2	$(-4, 2)$
-3	0	$(-3, 0)$
-2	2	$(-2, 2)$
-1	8	$(-1, 8)$

Figure 8-8

SELF CHECK 4   On the same set of coordinate axes, graph each function and tell how it differs from the graph of $f(x) = x^2$.
a. $f(x) = (x - 2)^2$
b. $f(x) = (x + 5)^2$

The results of Example 4 confirm the following facts.

HORIZONTAL TRANSLATIONS OF GRAPHS	If $y = f(x)$ is a function and h is a positive number, then
	• The graph of $y = f(x - h)$ is identical to the graph of $y = f(x)$, except that it is translated h units to the right.
	• The graph of $y = f(x + h)$ is identical to the graph of $y = f(x)$, except that it is translated h units to the left.

4 Use both a vertical and horizontal translation of $f(x) = ax^2$ to graph $f(x) = a(x - h)^2 + k$.

We will now consider both the vertical and horizontal translation of $f(x) = ax^2$ to graph $f(x) = a(x - h)^2 + k$. This means the graph of $f(x) = ax^2$ will retain its shape but will be shifted h units to the right or left and k units upward or downward. In this case the vertex will be shifted to the point (h, k) and the axis of symmetry of the parabola will be a vertical line written as the equation $x = h$.

EXAMPLE 5   Graph: $f(x) = 2(x - 3)^2 - 4$

Solution   The graph of $f(x) = 2(x - 3)^2 - 4$ is identical to the graph of $g(x) = 2(x - 3)^2$, except that it has been translated 4 units downward. The graph of $g(x) = 2(x - 3)^2$ is identical to the graph of $h(x) = 2x^2$, except that it has been translated 3 units to the right. Thus, to graph $f(x) = 2(x - 3)^2 - 4$, we can graph $h(x) = 2x^2$ and shift it 3 units to the right and then 4 units downward, as shown in Figure 8-9.

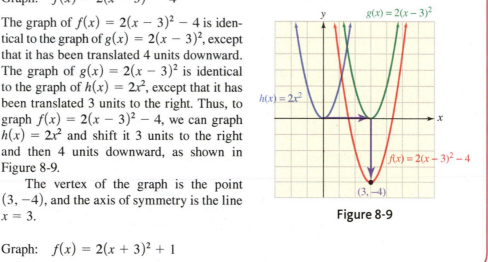

The vertex of the graph is the point $(3, -4)$, and the axis of symmetry is the line $x = 3$.

Figure 8-9

SELF CHECK 5   Graph: $f(x) = 2(x + 3)^2 + 1$

The results of Example 5 confirm the following facts.

VERTEX AND AXIS OF SYMMETRY OF A PARABOLA	The graph of the function
	$$f(x) = a(x - h)^2 + k \quad (a \neq 0)$$
	is a parabola with vertex at (h, k). (See Figure 8-10.)
	The parabola opens upward when $a > 0$ and downward when $a < 0$. The axis of symmetry is the vertical line $x = h$.

Figure 8-10

5   Graph a quadratic function written in standard form by writing it in the form $f(x) = a(x - h)^2 + k$.

To graph functions of the form $f(x) = ax^2 + bx + c$, we can complete the square to write the function in the form $f(x) = a(x - h)^2 + k$.

EXAMPLE 6   Graph: $f(x) = 2x^2 - 4x - 1$

Solution   We complete the square on x to write the function in the form $f(x) = a(x - h)^2 + k$.

$$f(x) = 2x^2 - 4x - 1$$

$$f(x) = 2(x^2 - 2x) - 1 \qquad \text{Factor 2 from } (2x^2 - 4x).$$

$$f(x) = \mathbf{2}(x^2 - 2x \mathbf{+ 1}) - 1 \mathbf{- 2} \qquad \text{Complete the square on } x. \text{ Since this adds 2 to the}$$
right side, we also subtract 2 from the right side.

(1) $\qquad f(x) = 2(x - 1)^2 - 3 \qquad \text{Factor } (x^2 - 2x + 1) \text{ and combine like terms.}$

From Equation 1, we can see that the vertex will be at the point $(1, -3)$. We can plot the vertex and a few points on either side of the vertex and draw the graph, which appears in Figure 8-11.

COMMENT   Note that this is the graph of $f(x) = 2x^2$ translated one unit to the right and three units downward.

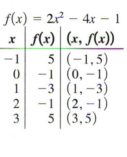

$f(x) = 2x^2 - 4x - 1$

x	$f(x)$	$(x, f(x))$
-1	5	$(-1, 5)$
0	-1	$(0, -1)$
1	-3	$(1, -3)$
2	-1	$(2, -1)$
3	5	$(3, 5)$

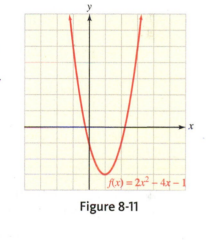

$f(x) = 2x^2 - 4x - 1$

Figure 8-11

SELF CHECK 6   Graph: $f(x) = 2x^2 - 4x + 1$

6   Find the vertex of a parabola using $\left(-\dfrac{b}{2a}, f\left(-\dfrac{b}{2a}\right)\right)$.

We can derive a formula for the vertex of the graph of $f(x) = ax^2 + bx + c$ by completing the square in the same manner as we did in Example 6.

$$f(x) = ax^2 + bx + c$$

$$f(x) = a\left(x^2 + \frac{b}{a}x\right) + c \qquad \text{Factor } a \text{ from the first two terms.}$$

$$f(x) = a\left(x^2 + \frac{b}{a}x + \frac{b^2}{4a^2}\right) + c - \frac{b^2}{4a} \qquad \text{Complete the square on } x. \text{ This adds to } \frac{b^2}{4a^2}.$$
Since $\frac{a \cdot b^2}{4a^2} = \frac{b^2}{4a}$, we also subtract $\frac{b^2}{4a}$ from the right side.

$$f(x) = a\left[\left(x + \frac{b}{2a}\right)^2\right] + \frac{4ac - b^2}{4a} \qquad \text{Factor } x^2 + \frac{b}{a}x + \frac{b^2}{4a^2} \text{ and combine like terms.}$$

$$f(x) = a\left[x - \left(-\frac{b}{2a}\right)\right]^2 + \frac{4ac - b^2}{4a} \qquad \text{Compare to } f(x) = a(x - h)^2 + k.$$

$\uparrow$ $\qquad\qquad\qquad\qquad$ $\uparrow$
h $\qquad\qquad\qquad\qquad\quad$ k

The x-coordinate of the vertex is $-\frac{b}{2a}$. The y-coordinate of the vertex is $\frac{4ac - b^2}{4a}$. We can also find the y-coordinate of the vertex by substituting the x-coordinate, $-\frac{b}{2a}$, for x in the quadratic function and simplifying.

| **FORMULA FOR THE VERTEX OF A PARABOLA** | The vertex of the graph of the quadratic function $f(x) = ax^2 + bx + c$ is $$\left(-\frac{b}{2a}, f\left(-\frac{b}{2a}\right)\right)$$ and the axis of symmetry of the parabola is the vertical line $x = -\frac{b}{2a}$. |

EXAMPLE 7   Find the vertex of the graph of $f(x) = 2x^2 - 4x - 1$.

Solution   The function is written in $f(x) = ax^2 + bx + c$ form, where $a = 2$, $b = -4$, and $c = -1$. We can find the x-coordinate of the vertex by evaluating $-\frac{b}{2a}$.

$$-\frac{b}{2a} = -\frac{-4}{2(2)} = -\frac{-4}{4} = 1$$

COMMENT This is the same as finding $f(1)$.

We can find the y-coordinate by evaluating $f\left(-\frac{b}{2a}\right)$.

$$f\left(-\frac{b}{2a}\right) = f(1) = 2(1)^2 - 4(1) - 1 = -3$$

The vertex is the point $(1, -3)$. This agrees with the result we obtained in Example 6 by completing the square.

SELF CHECK 7   Find the vertex of the graph of $f(x) = 3x^2 - 12x + 8$.

7   Graph a quadratic function written in standard form by finding the vertex, the axis of symmetry, and the x- and y-intercepts.

Much can be determined about the graph of $f(x) = ax^2 + bx + c$ from the coefficients a, b, and c. This information is summarized as follows:

| **GRAPHING A QUADRATIC FUNCTION** $f(x) = ax^2 + bx + c$ | **1.** Determine whether the parabola opens upward or downward.

  If $a > 0$, the parabola opens upward.
  If $a < 0$, the parabola opens downward.
2. Find the vertex and axis of symmetry.
  a. The x-coordinate of the vertex of the parabola is $-\frac{b}{2a}$.
  b. To find the y-coordinate of the vertex, substitute $-\frac{b}{2a}$ for x and find $f\left(-\frac{b}{2a}\right)$.
  c. The axis of symmetry is the vertical line passing through the vertex. The axis of symmetry is the equation $x = -\frac{b}{2a}$.
3. Find the intercepts.
  a. The y-intercept is determined by the value of $f(x)$ when $x = 0$. The y-intercept is $(0, c)$.
  b. The x-intercepts (if any) are determined by the values of x that make $f(x) = 0$. To find them, solve the quadratic equation $ax^2 + bx + c = 0$.
4. Plot another point using the axis of symmetry.
5. Draw a smooth curve through the points. |

EXAMPLE 8 Graph: $f(x) = -2x^2 - 8x - 8$

Solution **Step 1** *Determine whether the parabola opens upward or downward.* The function is in the form $f(x) = ax^2 + bx + c$, with $a = -2$, $b = -8$, and $c = -8$. Since $a < 0$, the parabola opens downward.

Step 2 *Find the vertex and draw the axis of symmetry.* To find the coordinates of the vertex, we evaluate $-\frac{b}{2a}$ by substituting -2 for a and -8 for b.

$$x = -\frac{b}{2a} = -\frac{-8}{2(-2)} = -2$$

We then find $f(-2)$.

$$f\left(-\frac{b}{2a}\right) = f(-2) = -2(-2)^2 - 8(-2) - 8 = -8 + 16 - 8 = 0$$

The vertex of the parabola is the point $(-2, 0)$. The axis of symmetry is the line $x = -2$.

Step 3 *Find the x- and y-intercepts.* Since $c = -8$, the y-intercept of the parabola is $(0, -8)$. The point $(-4, -8)$, two units to the left of the axis of symmetry, must also be on the graph. We plot both points in black on the graph.

To find the x-intercepts, we set $f(x)$ equal to 0 and solve the resulting quadratic equation.

$$f(x) = -2x^2 - 8x - 8$$
$$0 = -2x^2 - 8x - 8 \qquad \text{Set } f(x) = 0.$$
$$0 = x^2 + 4x + 4 \qquad \text{Divide both sides by } -2.$$
$$0 = (x + 2)(x + 2) \qquad \text{Factor the trinomial.}$$
$$x + 2 = 0 \quad \text{or} \quad x + 2 = 0 \qquad \text{Set each factor equal to 0.}$$
$$x = -2 \qquad \qquad x = -2$$

Since the solutions are the same, the graph has only one x-intercept: $(-2, 0)$. This point is the vertex of the parabola and has already been plotted.

Step 4 *Plot another point.* Finally, we find another point on the parabola. If $x = -3$, then $f(-3) = -2$. We plot $(-3, -2)$ and use symmetry to determine that $(-1, -2)$ is also on the graph. Both points are in black.

Step 5 *Draw a smooth curve through the points,* as shown in Figure 8.12.

$$f(x) = -2x^2 - 8x - 8$$

x	$f(x)$	$(x, f(x))$
-3	-2	$(-3, -2)$

Figure 8-12

 SELF CHECK 8 Graph: $f(x) = 2x^2 - 8x + 8$

Everyday connections
Astronaut Training

Some astronauts train for weightlessness in space travel by flying in C9B aircraft that follows a series of maneuvers to produce a "parabolic flight." A typical training mission may last up to 3 hours with the astronaut experiencing 40 to 50 weightlessness sessions. The aircraft altitude is modeled by the equation

$$f(x) = -\frac{1,600}{169}x^2 + \frac{8,000}{13}x + 24,000$$

where x represents time and $f(x)$ represents altitude.

The flight never flies below 24,000 feet during the training.

1. Find the vertex of the parabola and explain its meaning in the context of this problem.

2. How long does one flight last?

3. Weightlessness occurs for the middle 25 seconds. At what altitude will weightlessness first occur? Round your answer to the nearest foot.

http://jsc-aircraft-ops.jsc.nasa.gov/Reduced_Gravity/about.html

Accent on technology

▸Graphing Quadratic Functions

To find the coordinates of the vertex of a graph we use the MINIMUM (or MAXIMUM) command found in the CALC menu. We first graph the function $f(x) = 0.7x^2 + 2x - 3.5$ as in Figure 8-13(a). We then select 3 in the CALC menu, enter -3 for a left guess, and press **ENTER**. We then enter 0 for a right guess and press **ENTER**. After pressing **ENTER** again, we will obtain the minimum value $(-1.42857, -4.928571)$, as shown in Figure 8-13(b).

(a)   (b)

Figure 8-13

The solutions of the quadratic equation $0.7x^2 + 2x - 3.5 = 0$ are the values of x that will make $f(x) = 0$ in the function $f(x) = 0.7x^2 + 2x - 3.5$. To approximate these values, we solve the equation by using the ZERO command found in the CALC menu. From the graph in Figure 8.14(a), we select 2 in the CALC menu to obtain Figure 8-14(b). We enter -5 for a left guess and press **ENTER**. We then enter -2 for a right guess and press **ENTER**. After pressing **ENTER** again, we will obtain Figure 8-14(c). We can find the second solution from the graph as Figure 8-15(a) by selecting 2 in the CALC

(a)   (b)   (c)

Figure 8-14

(continued)

menu and enter 0 for a left guess and press **ENTER** to obtain Figure 8-15(b). By entering 3 and then **ENTER** we obtain Figure 8-15(c). The solutions, to three decimal points, are -4.082 and 1.225.

(a)   (b)   (c)

Figure 8-15

For instructions regarding the use of a Casio graphing calculator, please refer to the Casio Keystroke Guide in the back of the book.

8   Solve an application using a quadratic function.

EXAMPLE 9   **BALLISTICS**   The ball shown in Figure 8-16(a) is thrown straight up by a Major League Baseball pitcher with a velocity of 128 feet per second. The function $s = h(t) = -16t^2 + 128t$ gives the relation between t (the time measured in seconds) and s (the number of feet the ball is above the ground). How long will it take the ball to reach its maximum height, and what is that height?

(a)   (b)

Figure 8-16

Solution   The graph of $s = h(t) = -16t^2 + 128t$ is a parabola. Since the coefficient of t^2 is negative, it opens downward. The time it takes the ball to reach its maximum height is given by the t-coordinate of its vertex, and the maximum height of the ball is given by the s-coordinate of the vertex. To find the vertex, we find its t-coordinate and s-coordinate. To find the t-coordinate, we compute

$$-\frac{b}{2a} = -\frac{128}{2(-16)} \qquad b = 128 \text{ and } a = -16$$

$$= -\frac{128}{-32}$$

$$= 4$$

To find the *s*-coordinate, we substitute 4 for *t* in $h(t) = -16t^2 + 128t$.

$$h(4) = -16(4)^2 + 128(4)$$
$$= -256 + 512$$
$$= 256$$

Since $t = 4$ and $s = 256$ are the coordinates of the vertex, the ball will reach a maximum height in 4 seconds and that maximum height will be 256 feet.

To solve this with a graphing calculator with window settings of $[0, 10]$ for *x* and $[0, 300]$ for *y*, we graph the function $h(t) = -16t^2 + 128t$ to obtain the graph in Figure 8-16(b). By using the MAXIMUM command found under the CALC menu in the same way as the MINIMUM command, we can determine that the ball reaches a height of 256 feet in 4 seconds.

SELF CHECK 9 If the ball shown is thrown straight up with a velocity of 96 feet per second, how long will it take the ball to reach its maximum height and what is that height?

EXAMPLE 10 **MAXIMIZING AREA** A man wants to build the rectangular pen shown in Figure 8-17(a) to house his dog. If he uses one side of his barn, find the maximum area that he can enclose with 80 feet of fencing.

Solution If we use *w* to represent the width of the pen in feet, the length is represented by $(80 - 2w)$ feet. Since the area *A* of the pen is the product of its length and width, we have

$$A = (80 - 2w)w$$
$$= 80w - 2w^2$$
$$= -2w^2 + 80w$$

Since the graph of $A = -2w^2 + 80w$ is a parabola opening downward, the maximum area will be given by the second-coordinate of the vertex of the graph. To find the vertex, we find its first coordinate by letting $b = 80$ and $a = -2$ and computing $-\frac{b}{2a}$.

$$-\frac{b}{2a} = -\frac{80}{2(-2)} = 20$$

We can then find the second coordinate of the vertex by substituting 20 into the function $A = -2w^2 + 80w$.

$$A = -2w^2 + 80w$$
$$= -2(20)^2 + 80(20)$$
$$= -2(400) + 1{,}600$$
$$= -800 + 1{,}600$$
$$= 800$$

Thus, the coordinates of the vertex of the graph of the quadratic function are $(20, 800)$, and the maximum area is 800 square feet. This occurs when the width is 20 feet.

To solve this using a graphing calculator with window settings of $[0, 50]$ for *x* and $[0, 1{,}000]$ for *y*, we graph the function $A = -2w^2 + 80w$ to obtain the graph in Figure 8-17(b). By using the MAXIMUM command, we can determine that the maximum area is 800 square feet when the width is 20 feet.

(a)

(b)

Figure 8-17

🌿 **SELF CHECK 10**   Find the maximum area the man can enclose with 130 feet of fencing.

In statistics, the square of the standard deviation is called the **variance**, a way of looking at the variability of scores.

EXAMPLE 11   **VARIANCE**   If p is the chance that a person selected at random has the H1N1 virus, then $(1 - p)$ is the chance that the person does not have the H1N1 virus. If 100 people are randomly sampled, we know from statistics that the variance of this type of sample distribution will be $100p(1 - p)$. Find the value of p that will maximize the variance.

Solution   The variance is given by the function

$$v(p) = 100p(1 - p) \qquad \text{or} \qquad v(p) = -100p^2 + 100p$$

$$a = -100, b = 100, c = 0$$

$$\frac{-b}{2a} = \frac{-100}{2(-100)} = \frac{1}{2}$$

A value of $\frac{1}{2}$ or 0.5 will give maximum variance.

In this setting, all values of p are between 0 and 1, including 0 and 1.

We can use window settings of $[0, 1]$ for x when graphing the function $v(p) = -100p^2 + 100p$ on a graphing calculator. If we also use window settings of $[0, 30]$ for y, we will obtain the graph shown in Figure 8-18(a). After using the MAXIMUM command to obtain Figure 8-18(b), we can see that a value of 0.5 will yield the maximum variance.

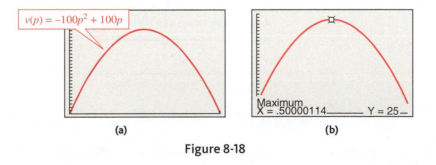

(a)   (b)

Figure 8-18

🌿 **SELF CHECK 11**   What is the maximum variance in this example?

SELF CHECK ANSWERS

1. $f(x) = 2x^2$ $f(x) = \frac{1}{2}x^2$

2. $f(x) = -\frac{1}{3}x^2$

3. $y \; f(x) = x^2 + 1$ $f(x) = x^2 - 5$
   a. shifted 1 unit upward
   b. shifted 5 units downward

4. $f(x) = (x + 5)^2$ $f(x) = (x - 2)^2$
   a. shifted 2 units to the right
   b. shifted 5 units to the left

5. $f(x) = 2(x + 3)^2 + 1$

6. $f(x) = 2x^2 - 4x + 1$

7. $(2, -4)$

8. $f(x) = 2x^2 - 8x + 8$

9. It will take 3 sec for the ball to reach its maximum height of 144 ft.   10. The maximum area he can enclose is 2,112.5 ft². 11. The maximum variance is 25.

NOW TRY THIS

Answer the following questions about the graph of $f(x)$.

1. What is the vertex?

2. What is the axis of symmetry?

3. What is the domain?

4. What is the range?

5. For what values of x will $f(x) = 0$?

6. For what values of x will y be positive? ($f(x) > 0$)

7. For what values of x will y be negative? ($f(x) < 0$)

8.4   Exercises

WARM-UPS *Find the equation of a vertical line passing through the given point.*

1. $(3, 5)$ 2. $(-2, 7)$
3. $(-1, 0)$ 4. $(6, 4)$

Solve.

5. $f(x) = x^2 - 9$ 6. $f(x) = x^2 - 2x - 3$

REVIEW *Find the value of x.*

7.

$(3x + 5)°$ $(5x - 15)°$

8. Lines *r* and *s* are parallel.

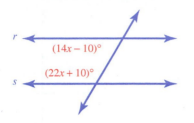

9. Travel Madison and St. Louis are 385 miles apart. One train leaves Madison and heads toward St. Louis at the rate of 30 mph. Three hours later, a second train leaves Madison, bound for St. Louis. If the second train travels at the rate of 55 mph, in how many hours will the faster train overtake the slower train?

10. Investing A woman invests $25,000, some at 7% annual interest and the rest at 8%. If the annual income from both investments is $1,900, how much is invested at the higher rate?

VOCABULARY AND CONCEPTS *Fill in the blanks.*

11. A quadratic function is a second-degree polynomial function that can be written in the form _____, where _____.

12. The graphs of quadratic functions are called _____.

13. The highest (_____) or lowest (_____) point on a parabola is called the _____.

14. A vertical line that divides a parabola into two congruent halves is called an ____ of symmetry.

15. The graph of $y = f(x) + k$ $(k > 0)$ is identical to the graph of $y = f(x)$, except that it is translated k units _____.

16. The graph of $y = f(x) - k$ $(k > 0)$ is a vertical translation of the graph of $y = f(x)$, k units _____.

17. The graph of $y = f(x - h)$ $(h > 0)$ is a horizontal translation of the graph of $y = f(x)$, h units _____.

18. The graph of $y = f(x + h)$ $(h > 0)$ is identical to the graph of $y = f(x)$, except that it is translated h units _____.

19. The graph of $y = f(x) = ax^2 + bx + c$ $(a \neq 0)$ opens _____ when $a > 0$.

20. In statistics, the square of the standard deviation is called the _____.

GUIDED PRACTICE *Graph each function. SEE EXAMPLES 1–2. (OBJECTIVE 1)*

21. $f(x) = x^2$

22. $f(x) = -x^2$

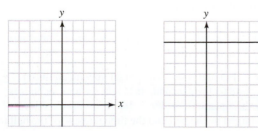

23. $f(x) = -2x^2$

24. $f(x) = 3x^2$

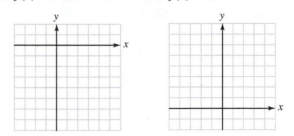

Graph each function. SEE EXAMPLE 3. (OBJECTIVE 2)

25. $f(x) = x^2 + 2$

26. $f(x) = x^2 - 3$

27. $f(x) = x^2 - 1$

28. $f(x) = x^2 + 4$

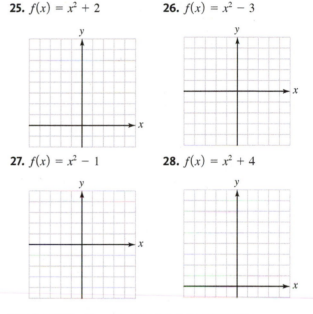

Graph each function. SEE EXAMPLE 4. (OBJECTIVE 3)

29. $f(x) = -(x - 2)^2$

30. $f(x) = (x + 2)^2$

31. $f(x) = -(x - 3)^2$

32. $f(x) = (x + 3)^2$

Graph each function. SEE EXAMPLE 5. (OBJECTIVE 4)

33. $f(x) = (x - 3)^2 + 2$ **34.** $f(x) = (x + 1)^2 - 2$

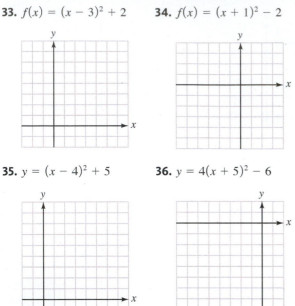

35. $y = (x - 4)^2 + 5$ **36.** $y = 4(x + 5)^2 - 6$

*Find the coordinates of the vertex and the axis of symmetry of the graph of each equation. If necessary, complete the square on x to write the equation in the form $y = a(x - h)^2 + k$. **Do not graph the equation.** SEE EXAMPLE 6. (OBJECTIVE 5)*

37. $y = (x - 1)^2 + 2$ **38.** $y = 2(x - 2)^2 - 1$

39. $y = 2(x + 3)^2 - 4$ **40.** $y = -3(x + 1)^2 + 3$

41. $y = -3x^2$ **42.** $y = 3x^2 - 3$

43. $y = 2x^2 - 4x$ **44.** $y = 3x^2 + 6x$

Use $\left(-\frac{b}{2a}, f\left(-\frac{b}{2a}\right)\right)$ to find the vertex of each function. SEE EXAMPLE 7. (OBJECTIVE 6)

45. $f(x) = 3x^2 + 6x + 1$ **46.** $f(x) = 5x^2 - 10x$

47. $f(x) = -3x^2 + 12x + 4$ **48.** $f(x) = 3x^2 - 8x + 5$

Find the coordinates of the vertex and the axis of symmetry of the graph of each equation. Use $\left(-\frac{b}{2a}, f\left(-\frac{b}{2a}\right)\right)$ to find the vertex and graph the equation. SEE EXAMPLE 8. (OBJECTIVE 7)

49. $f(x) = -2x^2 + 4x + 1$ **50.** $f(x) = 3x^2 - 12x + 9$

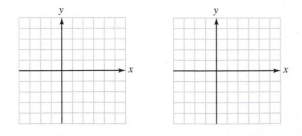

51. $f(x) = 3x^2 - 12x + 10$ **52.** $f(x) = -2x^2 + 4x + 3$

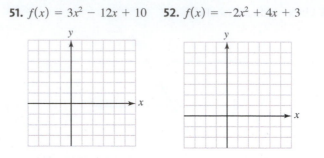

ADDITIONAL PRACTICE *Find the vertex and axis of symmetry using any method. Do not graph.*

53. $y = -4x^2 + 16x + 5$ **54.** $y = 5x^2 + 20x + 25$

55. $y - 7 = 6x^2 - 5x$ **56.** $y - 7 = 4x^2 + 10x$

Graph each function.

57. $f(x) = x^2 + x - 6$ **58.** $f(x) = x^2 - x - 6$

 Use a graphing calculator to find the coordinates of the vertex of the graph of each quadratic function. Give results to the nearest hundredth.

59. $y = 2x^2 - x + 1$ **60.** $y = x^2 + 5x - 6$

61. $y = 7 + x - x^2$ **62.** $y = 2x^2 - 3x + 2$

Use a graphing calculator to solve each equation. If a result is not exact, give the result to the nearest hundredth.

63. $x^2 + 9x - 10 = 0$ **64.** $2x^2 - 5x - 3 = 0$

65. $0.5x^2 - 0.7x - 3 = 0$ **66.** $2x^2 - 0.5x - 2 = 0$

67. The equation $y - 3 = (x + 7)^2$ represents a quadratic function whose graph is a parabola. Find its vertex.

68. Show that $y = ax^2$, where $a \neq 0$, represents a quadratic function whose vertex is at the origin.

APPLICATIONS *Solve. Use a graphing calculator if necessary. SEE EXAMPLES 9–11. (OBJECTIVE 8)*

69. Ballistics If a ball is thrown straight up with an initial velocity of 48 feet per second, its height s after t seconds is given by the equation $s = 48t - 16t^2$. Find the maximum height attained by the ball and the time it takes for the ball to reach that height.

70. Ballistics From the top of the building, a ball is thrown straight up with an initial velocity of 32 feet per second. The equation $s = -16t^2 + 32t + 48$ gives the height s of the ball t seconds after it is thrown. Find the maximum height reached by the ball and the time it takes for the ball to hit the ground. (*Hint*: Let $s = 0$ and solve for t.)

48 ft

919

s

71. Maximizing area Find the dimensions of the rectangle of maximum area that can be constructed with 400 feet of fencing. Find the maximum area.

72. Fencing a field A farmer wants to fence in three sides of a rectangular field with 1,000 feet of fencing. The other side of the rectangle will be a river. If the enclosed area is to be maximum, find the dimensions of the field.

1,000 ft

73. Finding the variance If p is the chance that a person sampled at random has high blood pressure, $1 - p$ is the chance that the person does not. If 50 people are sampled at random, the variance of the sample will be $50p(1 - p)$. What two values of p will give a variance of 9.375?

74. Finding the variance If p is the chance that a person sampled at random smokes, then $1 - p$ is the chance that the person doesn't. If 75 people are sampled at random, the variance of the sample will be $75p(1 - p)$. What two values of p will give a variance of 12?

75. Police investigations A police officer seals off the scene of a car collision using a roll of yellow police tape that is 300 feet long. What dimensions should be used to seal off the maximum rectangular area around the collision? What is the maximum area?

POLICE LINE   DO NOT CROSS

76. Operating costs The cost C in dollars of operating a certain concrete-cutting machine is related to the number of minutes n the machine is run by the function

$$C(n) = 2.2n^2 - 66n + 655$$

For what number of minutes is the cost of running the machine a minimum? What is the minimum cost?

77. Water usage The height (in feet) of the water level in a reservoir over a 1-year period is modeled by the function

$$H(t) = 3.3t^2 - 59.4t + 281.3$$

How low did the water level get that year?

78. School enrollment The total annual enrollment (in millions) in U.S. elementary and secondary schools for the years 1975–1996 is given by the function

$$E(x) = 0.058x^2 - 1.162x + 50.604$$

For this period, what was the lowest enrollment?

79. Maximizing revenue The revenue R received for selling x stereos is given by the equation

$$R = -\frac{x^2}{1,000} + 10x$$

Find the number of stereos that must be sold to obtain the maximum revenue.

80. Maximizing revenue In Exercise 79, find the maximum revenue.

81. Maximizing revenue The revenue received for selling x radios is given by the formula

$$R = -\frac{x^2}{728} + 9x$$

How many radios must be sold to obtain the maximum revenue? Find the maximum revenue.

82. Maximizing revenue The revenue received for selling x stereos is given by the formula

$$R = -\frac{x^2}{5} + 80x - 1,000$$

How many stereos must be sold to obtain the maximum revenue? Find the maximum revenue.

83. Maximizing revenue When priced at $30 each, a toy has annual sales of 4,000 units. The manufacturer estimates that each $1 increase in cost will decrease sales by 100 units. Find the unit price that will maximize total revenue. (*Hint*: Total revenue = price · the number of units sold.)

84. Maximizing revenue When priced at $57, one type of camera has annual sales of 525 units. For each $1 the camera is reduced in price, management expects to sell an additional 75 cameras. Find the unit price that will maximize total revenue. (*Hint*: Total revenue = price · the number of units sold.)

WRITING ABOUT MATH

85. The graph of $y = ax^2 + bx + c$ $(a \neq 0)$ passes the vertical line test. Explain why this shows that the equation defines a function.

86. The graph of $x = y^2 - 2y$ is a parabola. Explain why its graph does not represent a function.

SOMETHING TO THINK ABOUT

87. Can you use a graphing calculator to find solutions of the equation $x^2 + x + 1 = 0$? What is the problem? How do you interpret the result?

88. Complete the square on x in the equation $y = ax^2 + bx + c$ and show that the vertex of the parabolic graph is the point with coordinates of

$$\left(-\frac{b}{2a}, \frac{4ac - b^2}{4a}\right)$$

Section 8.5

Quadratic and Other Nonlinear Inequalities

Objectives

1. Solve a quadratic inequality in one variable.
2. Solve a rational inequality in one variable.
3. Graph a nonlinear inequality in two variables.

Vocabulary

quadratic inequality   critical values   critical points

Getting Ready

Factor each trinomial.

1. $x^2 + 2x - 15$ **2.** $x^2 - 3x + 2$

We have previously solved linear inequalities in one variable. We will now discuss how to solve quadratic, rational, and other nonlinear inequalities.

1 Solve a quadratic inequality in one variable.

Quadratic inequalities in one variable, say x, are inequalities that can be written in one of the following forms, where $a \neq 0$:

$$ax^2 + bx + c < 0 \qquad ax^2 + bx + c > 0$$
$$ax^2 + bx + c \leq 0 \qquad ax^2 + bx + c \geq 0$$

To solve any of these inequalities, we must find the values of x that make the inequality true. There are two algebraic methods we may use to find these values.

Method 1 To solve

$$x^2 + x - 6 < 0$$

we can factor the trinomial to obtain

$$(x + 3)(x - 2) < 0$$

Since the product of $(x + 3)$ and $(x - 2)$ is to be less than 0, the values of the factors must be opposite in sign. This will happen when one of the factors is positive and the other is negative.

To keep track of the sign of $(x + 3)$, we can construct the following graph.

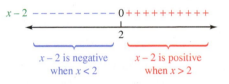

To keep track of the sign of $(x - 2)$, we can construct the following graph.

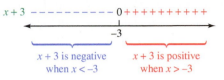

We can merge these graphs as shown in Figure 8-19 and note where the signs of the factors are opposite. The product $(x + 3)(x - 2)$ will be less than 0 when

$$-3 < x < 2$$

The graph of the solution set is shown on the number line in the figure.

Figure 8-19

The solution in interval notation is $(-3, 2)$.

Method 2   To solve the inequality $x^2 + x - 6 < 0$ is to first solve its related quadratic equation $x^2 + x - 6 = 0$. The solutions to this equation are called **critical values** and they establish points on the number line, called **critical points**.

$$x^2 + x - 6 = 0$$
$$(x + 3)(x - 2) = 0$$
$$x + 3 = 0 \quad \text{or} \quad x - 2 = 0$$
$$x = -3 \quad | \quad x = 2$$

The graphs of these critical values establish the three intervals shown on the number line. To determine which intervals are solutions, we create a table of intervals, test a number in each interval, and determine if it satisfies the inequality, as shown in Table 8-1.

TABLE 8-1			
Interval	**Test value**	**Inequality $x^2 + x - 6 < 0$**	**Result**
$(-\infty, -3)$	-6	$(-6)^2 - (-6) - 6 \overset{?}{<} 0$ $36 < 0$ false	The numbers in this interval are not solutions.
$(-3, 2)$	0	$(0)^2 - (0) - 6 \overset{?}{<} 0$ $-6 < 0$ true	The numbers in this interval are solutions.
$(2, \infty)$	5	$(5)^2 - (5) - 6 \overset{?}{<} 0$ $14 < 0$ false	The numbers in this interval are not solutions.

Figure 8-20

The solution set is the same result as using Method 1, the interval $(-3, 2)$, as shown in Figure 8-20.

COMMENT   Be sure to use a bracket rather than a parenthesis if an end point is to be included.

EXAMPLE 1   Solve $x^2 + 2x - 3 \geq 0$, graph the solution set, and state the solution in interval notation.

Solution   **Method 1**   We can factor the trinomial to obtain $(x - 1)(x + 3)$ and construct the sign chart shown in Figure 8-21.

Figure 8-21

- $(x - 1)$ is 0 when $x = 1$, is positive when $x > 1$, and is negative when $x < 1$.
- $(x + 3)$ is 0 when $x = -3$, is positive when $x > -3$, and is negative when $x < -3$.

The product of $(x - 1)$ and $(x + 3)$ will be greater than 0, a positive value, when the signs of the binomial factors are the same. The numbers -3 and 1 are also included, because equality is included in the original inequality. The graph of the solution set is shown on the number line in Figure 8-21. Thus, the solution set is the union of two intervals

$$(-\infty, -3] \cup [1, \infty)$$

Method 2   We can obtain the same result by first solving the related quadratic equation $x^2 + 2x - 3 = 0$ and establishing critical points on the number line.

$$x^2 + 2x - 3 = 0$$
$$(x + 3)(x - 1) = 0$$
$$x + 3 = 0 \quad \text{or} \quad x - 1 = 0$$
$$x = -3 \quad \bigg| \quad x = 1$$

The graphs of these critical values establish the three intervals shown on the number line. To determine which intervals are solutions, we create a table of intervals and test a number in each interval to see whether it satisfies the inequality, as shown in Table 8-2.

		TABLE 8-2	
Interval	**Test value**	**Inequality $x^2 + 2x - 3 \geq 0$**	**Result**
$(-\infty, -3]$	-5	$(-5)^2 + 2(-5) - 3 \overset{?}{\geq} 0$	The numbers in this interval
		$12 \geq 0$ true	are solutions.
$[-3, 1]$	0	$(0)^2 + 2(0) - 3 \overset{?}{\geq} 0$	The numbers in this interval
		$-3 \geq 0$ false	are not solutions.
$[1, \infty)$	4	$(4)^2 + 2(4) - 3 \overset{?}{\geq} 0$	The numbers in this interval
		$21 \geq 0$ true	are solutions.

From the table, we see that numbers in the intervals $(-\infty, -3]$ and $[1, \infty)$ satisfy the inequality. Because the quadratic inequality contains equality, the critical values of $x = -3$ and $x = 1$ also satisfy the inequality. Thus, the solution set is the union of two intervals: $(-\infty, -3] \cup [1, \infty)$, as shown in Figure 8-22.

Figure 8-22

SELF CHECK 1   Solve $x^2 + 2x - 15 > 0$, graph the solution set, and state the solution in interval notation.

2   Solve a rational inequality in one variable.

Making a sign chart is useful for solving many inequalities that are neither linear nor quadratic.

EXAMPLE 2   Solve $\dfrac{1}{x} < 6$ and graph the solution set.

Solution   We subtract 6 from both sides to make the right side equal to 0, find a common denominator, and add the fractions:

$$\frac{1}{x} < 6$$

$$\frac{1}{x} - 6 < 0 \qquad \text{Subtract 6 from both sides.}$$

$$\frac{1}{x} - \frac{6x}{x} < 0 \qquad \text{Write each fraction with the same denominator.}$$

$$\frac{1 - 6x}{x} < 0 \qquad \text{Subtract the numerators and keep the common denominator.}$$

The fraction $\dfrac{1 - 6x}{x}$ will be less than 0 when the numerator and denominator are opposite in sign.

To use Method 1, we make a sign chart, as in Figure 8-23.

- The denominator x is undefined when $x = 0$, is positive when $x > 0$, and is negative when $x < 0$.
- The numerator $1 - 6x$ is 0 when $x = \dfrac{1}{6}$, is positive when $x < \dfrac{1}{6}$, and is negative when $x > \dfrac{1}{6}$.

The graph of this union is shown in Figure 8-23.
The solution is the union of two intervals.

$$(-\infty, 0) \cup \left(\frac{1}{6}, \infty\right)$$

Figure 8-23

To solve this inequality by Method 2, find the critical values by finding the values of x that make the numerator of $\dfrac{1 - 6x}{x}$ equal to 0 and the values of x that make the denominator undefined. The critical values are the numbers $\dfrac{1}{6}$ and 0. Use a test value to determine which of the intervals satisfy the inequality, as shown in Table 8-3.

TABLE 8-3			
Interval	**Test value**	**Inequality $\frac{1}{x} < 6$**	**Result**
$(-\infty, -0)$	-1	$\dfrac{1}{(-1)} \overset{?}{<} 6$ $-1 < 6$ true	The numbers in this interval are solutions.
$\left(0, \frac{1}{6}\right)$	$\dfrac{1}{12}$	$\dfrac{1}{\left(\frac{1}{12}\right)} \overset{?}{<} 6$ $12 < 6$ false	The numbers in this interval are not solutions.
$\left(\frac{1}{6}, \infty\right)$	1	$\dfrac{1}{(1)} \overset{?}{<} 6$ $1 < 6$ true	The numbers in this interval are solutions.

The solution is $(-\infty, 0) \cup \left(\frac{1}{6}, \infty\right)$.

COMMENT   Since we do not know whether x is positive, 0, or negative, multiplying both sides of the inequality $\frac{1}{x} < 6$ by x can be misleading and introduces a three-case situation:

- If $x > 0$, then $1 < 6x$.
- If $x = 0$, then the fraction $\frac{1}{x}$ is undefined.
- If $x < 0$, then $1 > 6x$.

If you multiply both sides by x and solve $1 < 6x$, you are considering only one case and will obtain only part of the answer.

SELF CHECK 2   Solve $\frac{3}{x} > 5$ and graph the solution set.

EXAMPLE 3   Solve $\dfrac{x^2 - 3x + 2}{x - 3} \geq 0$ and graph the solution set.

Solution   We write the fraction with the numerator in factored form.

$$\frac{(x - 2)(x - 1)}{x - 3} \geq 0$$

To keep track of the signs of the binomials, we use Method 1 and construct the sign chart shown in Figure 8-24. The fraction will be positive in the intervals where all factors are positive, or where two factors are negative. The numbers 1 and 2 are included, because they make the numerator (and thus the fraction) equal to 0. The number 3 is not included, because the denominator is undefined for this value.

The solution is the union of two intervals $[1, 2] \cup (3, \infty)$. The graph appears in Figure 8-24.

To solve this inequality by Method 2, we can find the critical values by finding the values of x that make the numerator of $\dfrac{(x - 2)(x - 1)}{x - 3}$ equal to 0 and the values of x that make the denominator undefined. The critical values are the numbers 1, 2, and 3. Use a test value to determine which of the intervals satisfy the inequality, as shown in Table 8-4.

Figure 8-24

		TABLE 8-4	
Interval	**Test value**	**Inequality** $\dfrac{x^2 - 3x + 2}{x - 3} \geq 0$	**Result**
$(-\infty, 1]$	0	$\dfrac{(0)^2 - 3(0) + 2}{(0) - 3} \overset{?}{\geq} 0$ $-\dfrac{2}{3} \geq 0$ false	The numbers in this interval are not solutions.
$[1, 2]$	1.5	$\dfrac{(1.5)^2 - 3(1.5) + 2}{(1.5) - 3} \overset{?}{\geq} 0$ $\dfrac{1}{6} \geq 0$ true	The numbers in this interval are solutions.
$[2, 3)$	2.5	$\dfrac{(2.5)^2 - 3(2.5) + 2}{(2.5 - 3)} \overset{?}{\geq} 0$ $-\dfrac{3}{2} \geq 0$ false	The numbers in this interval are not solutions.
$(3, \infty)$	4	$\dfrac{(4)^2 - 3(4) + 2}{(4 - 1)} \geq 0$ $6 \geq 0$ true	The numbers in this interval are solutions.

The solution set is $[1, 2] \cup (3, \infty)$.

SELF CHECK 3   Solve $\dfrac{x + 2}{x^2 - 2x - 3} > 0$ and graph the solution set.

EXAMPLE 4   Solve $\dfrac{3}{x-1} < \dfrac{2}{x}$ and graph the solution set.

Solution   We subtract $\dfrac{2}{x}$ from both sides to get 0 on the right side and proceed as follows:

$$\frac{3}{x-1} < \frac{2}{x}$$

$$\frac{3}{x-1} - \frac{2}{x} < 0 \qquad \text{Subtract } \tfrac{2}{x} \text{ from both sides.}$$

$$\frac{3x}{(x-1)x} - \frac{2(x-1)}{x(x-1)} < 0 \qquad \text{Write each fraction with the same denominator.}$$

$$\frac{3x - 2x + 2}{x(x-1)} < 0 \qquad \text{Keep the denominator and subtract the numerators.}$$

$$\frac{x+2}{x(x-1)} < 0 \qquad \text{Combine like terms.}$$

To use Method 1, we can keep track of the signs of the three factors with the sign chart shown in Figure 8-25. The fraction will be negative in the intervals with either one or three negative factors. The numbers 0 and 1 are not included, because the denominator will be undefined, and the number -2 is not included, because it does not satisfy the strict inequality.

The solution is the union of two intervals $(-\infty, -2) \cup (0, 1)$, as shown in Figure 8-25.

Figure 8-25

To solve this inequality by Method 2, we can find the critical values by finding the values of x that make the numerator of $\dfrac{x+2}{x(x-1)}$ equal to 0 and the values of x that make the denominator undefined. The critical values are the numbers $-2, 0,$ and 1. They will form the intervals $(-\infty, -2), (-2, 0), (0, 1),$ and $(1, \infty)$. Use test values to determine which of the intervals satisfy the inequality.

The solution set is $(-\infty, -2) \cup (0, 1)$.

SELF CHECK 4   Solve $\dfrac{2}{x+2} > \dfrac{1}{x}$ and graph the solution set.

Accent
on technology

▸ Solving Inequalities

To find the solutions of $x^2 + 2x - 3 \geq 0$ (Example 1) by graphing with a TI-84 calculator, we can use window settings of $[-10, 10]$ for x and $[-10, 10]$ for y and graph the quadratic function $y = x^2 + 2x - 3$, as in Figure 8-26 on the next page. The solutions of the inequality will be those values of x for which the graph of $y = x^2 + 2x - 3$ lies above or on the x-axis. We can use the zero feature to find the critical values and interpret the graph to determine that the solution is the interval $(-\infty, -3] \cup [1, \infty)$.

(continued)

Figure 8-26

To find the solutions of $\frac{3}{x-1} < \frac{2}{x}$ (Example 4), we first write the inequality in the form

$$\frac{3}{x-1} - \frac{2}{x} < 0$$

Then we use window settings of $[-5, 5]$ for x and $[-3, 3]$ for y and graph the function $y = \frac{3}{x-1} - \frac{2}{x}$, as in Figure 8-27(a). The solutions of the inequality will be those numbers x for which the graph lies below the x-axis.

We can trace to see that the graph is below the x-axis when x is less than -2. Since we cannot see the graph in the interval $0 < x < 1$, we redraw the graph using window settings of $[-1, 2]$ for x and $[-25, 10]$ for y. See Figure 8-27(b).

We can now see that the graph is below the x-axis in the interval $(0, 1)$. Thus, the solution of the inequality is the union of two intervals:

$$(-\infty, -2) \cup (0, 1)$$

COMMENT Graphing calculators cannot determine whether a critical value is included in a solution set. You must make that determination yourself.

(a)   (b)

Figure 8-27

For instructions regarding the use of a Casio graphing calculator, please refer to the Casio Keystroke Guide in the back of the book.

3   Graph a nonlinear inequality in two variables.

We now consider the graphs of nonlinear inequalities in two variables.

EXAMPLE 5   Graph: $y < -x^2 + 4$

Solution   The graph of $y = -x^2 + 4$ is the parabolic boundary separating the region representing $y < -x^2 + 4$ and the region representing $y > -x^2 + 4$.

We graph $y = -x^2 + 4$ as a dashed parabola, because equality is not included in the original inequality. We may use $(0, 0)$ as a test point to determine shading. Since the coordinates of the origin satisfy the original inequality $y < -x^2 + 4$, the point $(0, 0)$ is in the graph and we shade below the parabolic boundary. The complete graph is shown in Figure 8-28.

Figure 8-28

🌿 **SELF CHECK 5**   Graph: $y \geq -x^2 + 4$

EXAMPLE 6   Graph: $x \leq |y|$

Solution   We first graph $x = |y|$ as in Figure 8-29(a), using a solid line because equality is included in the original inequality. Since the origin is on the graph, we cannot use it as a test point. However, another point, such as $(1, 0)$, will do. We substitute 1 for x and 0 for y into the inequality to obtain

$$x \leq |y|$$
$$1 \leq |0|$$
$$1 \leq 0$$

Since $1 \leq 0$ is a false statement, the point $(1, 0)$ does not satisfy the original inequality and is not part of the graph. Thus, the graph of $x \leq |y|$ is to the left of the boundary.
The complete graph is shown in Figure 8-29(b).

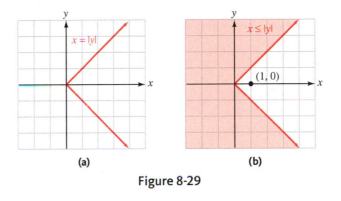

(a)   (b)

Figure 8-29

🌿 **SELF CHECK 6**   Graph: $x \geq -|y|$

🌿 **SELF CHECK ANSWERS**

1. $(-\infty, -5) \cup (3, \infty)$ 2. $\left(0, \frac{3}{5}\right)$
3. $(-2, -1) \cup (3, \infty)$
4. $(-2, 0) \cup (2, \infty)$ 5.   6.

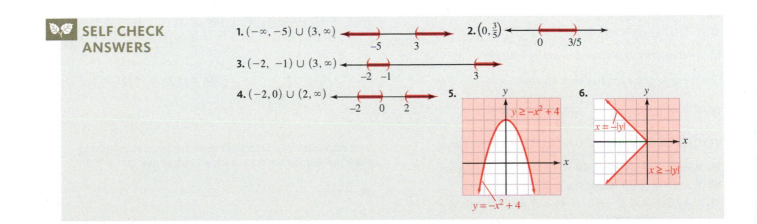

NOW TRY THIS

Find the domain of each of the following.

1. $f(x) = \sqrt{x - 6}$

2. $g(x) \geq x^2 - 5x - 6$

3. $h(x) = \sqrt{x^2 - 5x - 6}$

4. $k(x) = \dfrac{3x - 5}{\sqrt{x^2 - 5x - 6}}$

8.5 Exercises

WARM-UPS *Determine the value of x when (x − 2) is*

1. 0.
2. positive.
3. negative.

Determine the value of x when (x + 3) is

4. 0.
5. positive.
6. negative.

Find the undefined values.

7. $\dfrac{x - 2}{x + 3}$
8. $\dfrac{x}{2x - 1}$

REVIEW *Write each expression as an equation.*

9. y varies directly with the square of x.

10. p varies inversely with q.

11. y varies jointly with a and b.

12. d varies directly with t but inversely with u^2.

Find the slope of the graph of each equation.

13. $y = -7x + 2$
14. $\dfrac{2x - y}{5} = 8$

VOCABULARY AND CONCEPTS *Fill in the blanks.*

15. When $x > 3$, the binomial $x - 3$ is _____ than zero.

16. When $x < 3$, the binomial $x - 3$ is ___ than zero.

17. The expression $x^2 + 10x + 16 \leq 0$ is an example of a _____ inequality.

18. The expression $\frac{x + 5}{x - 7} > 0$ is an example of a _____ inequality.

19. If $x = 3$, the fraction $\frac{1}{x - 3}$ is _____.

20. To solve $x^2 + 2x - 3 < 0$, we first find the solutions of the related equation $x^2 + 2x - 3 = 0$. The solutions are called _____. They establish points on a number line that separate the line into _____.

21. To keep track of the signs of factors in a product or quotient, we can use a ____ chart.

22. The inequality $|x - 5| < 0$ will be graphed with a _____ line.

GUIDED PRACTICE *Solve each inequality. State each result in interval notation and graph the solution set.* SEE EXAMPLE 1. (*OBJECTIVE 1*)

23. $x^2 - 5x + 4 < 0$
24. $x^2 - 3x - 4 > 0$

25. $x^2 - 8x + 15 > 0$
26. $x^2 + 2x - 8 < 0$

27. $x^2 + x - 12 \leq 0$
28. $x^2 + 7x + 12 \geq 0$

29. $x^2 + 2x \geq 15$
30. $x^2 - 8x \leq -15$

Solve each inequality. State each result in interval notation and graph the solution set. SEE EXAMPLE 2. (*OBJECTIVE 2*)

31. $\dfrac{3}{x} < 6$
32. $\dfrac{4}{x} > 12$

33. $\dfrac{10}{x} \geq 5$
34. $-\dfrac{6}{x} < 12$

Solve each inequality. State each result in interval notation and graph the solution set. SEE EXAMPLE 3. (OBJECTIVE 2)

35. $\dfrac{x^2 - 5x + 4}{x + 3} < 0$

36. $\dfrac{x^2 + x - 6}{x - 4} \geq 0$

37. $\dfrac{x^2 + x - 20}{x + 2} \geq 0$

38. $\dfrac{x^2 + x - 20}{x - 4} < 0$

Solve each inequality. State each result in interval notation and graph the solution set. SEE EXAMPLE 4. (OBJECTIVE 2)

39. $\dfrac{3}{x - 2} < \dfrac{4}{x}$

40. $\dfrac{-6}{x + 1} \geq \dfrac{1}{x}$

41. $\dfrac{-5}{x + 2} \geq \dfrac{4}{2 - x}$

42. $\dfrac{-6}{x - 3} < \dfrac{5}{3 - x}$

Graph each inequality. SEE EXAMPLE 5. (OBJECTIVE 3)

43. $y < x^2 + 1$

44. $y > x^2 - 3$

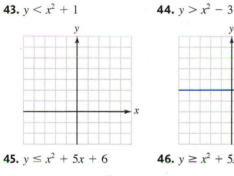

45. $y \leq x^2 + 5x + 6$

46. $y \geq x^2 + 5x + 4$

Graph each inequality. SEE EXAMPLE 6. (OBJECTIVE 3)

47. $y < |x + 4|$

48. $y \geq |x - 3|$

49. $y \leq -|x| + 2$

50. $y > |x| - 2$

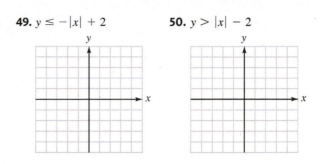

ADDITIONAL PRACTICE Solve each inequality. State each result in interval notation and graph the solution set.

51. $x^2 - 10x < -25$

52. $x^2 + 6x \geq -9$

53. $x^2 \geq 9$

54. $x^2 \geq 16$

55. $\dfrac{x^2 - 6x + 9}{x + 4} < 0$

56. $\dfrac{2x^2 - 5x + 2}{x + 2} > 0$

57. $\dfrac{6x^2 - 5x + 1}{2x + 1} > 0$

58. $\dfrac{6x^2 + 11x + 3}{3x - 1} < 0$

59. $\dfrac{7}{x - 3} \geq \dfrac{2}{x + 4}$

60. $\dfrac{-5}{x - 4} < \dfrac{3}{x + 1}$

61. $(x + 2)^2 > 0$

62. $(x - 5)^2 < 0$

63. $3x^2 - 75 < 0$

64. $3x^2 - 243 < 0$

65. $-\dfrac{5}{x} < 3$

66. $\dfrac{4}{x} \geq 8$

67. $\dfrac{x}{x + 4} \leq \dfrac{1}{x + 1}$

68. $\dfrac{x}{x + 9} \geq \dfrac{1}{x + 1}$

69. $\dfrac{x}{x + 16} > \dfrac{1}{x + 1}$

70. $\dfrac{x}{x + 25} < \dfrac{1}{x + 1}$

Graph each inequality.

71. $y \geq (x - 1)^2$

72. $y \leq (x + 2)^2$

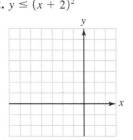

73. $-x^2 - y + 6 > -x$

74. $y > (x + 3)(x - 2)$

Use a graphing calculator to solve each inequality. Give the answer in interval notation.

75. $x^2 - 2x - 3 < 0$

76. $x^2 + x - 6 > 0$

77. $\dfrac{x + 3}{x - 2} > 0$

78. $\dfrac{3}{x} < 2$

WRITING ABOUT MATH

79. Explain why $(x - 4)(x + 5)$ will be positive only when the signs of $(x - 4)$ and $(x + 5)$ are the same.

80. Explain how to find the graph of $y \geq x^2$.

SOMETHING TO THINK ABOUT

81. Under what conditions will the fraction $\dfrac{(x - 1)(x + 4)}{(x + 2)(x + 1)}$ be positive?

82. Under what conditions will the fraction $\dfrac{(x - 1)(x + 4)}{(x + 2)(x + 1)}$ be negative?

Section 8.6

Algebra and Composition of Functions

Objectives

1. Find the sum, difference, product, and quotient of two functions.
2. Find the composition of two functions.
3. Find the difference quotient of a function.
4. Solve an application requiring the composition of two functions.

Vocabulary

composition   composite functions   difference quotient

Getting Ready

Simplify.

1. $2x + 1 + x - 2$

2. $2x + 1 - (x - 2)$

3. $(2x + 1)(x - 2)$

Throughout the text, we have talked about functions. In this section, we will show how to add, subtract, multiply, and divide them, processes that are sometimes referred to as the algebra of functions. We also will show how to find the composition of two functions.

 Find the sum, difference, product, and quotient of two functions.

We now consider how functions can be added, subtracted, multiplied, and divided.

OPERATIONS ON FUNCTIONS

If the domains and ranges of functions f and g are subsets of the real numbers,

the *sum* of f and g, denoted as $f + g$, is defined by

$$(f + g)(x) = f(x) + g(x)$$

the *difference* of f and g, denoted as $f - g$, is defined by

$$(f - g)(x) = f(x) - g(x)$$

the *product* of f and g, denoted as $f \cdot g$, is defined by

$$(f \cdot g)(x) = f(x)g(x)$$

the *quotient* of f and g, denoted as f/g, is defined by

$$(f/g)(x) = \frac{f(x)}{g(x)} \quad (g(x) \neq 0)$$

The domain of each of these functions is the set of real numbers x that are in the domain of both f and g. In the case of the quotient, there is the further restriction that $g(x) \neq 0$.

EXAMPLE 1   Let $f(x) = 2x^2 + 1$ and $g(x) = 5x - 3$. Find each function and its domain.

a. $f + g$ **b.** $f - g$

Solution   **a.** $(f + g)(x) = f(x) + g(x)$
$$= (2x^2 + 1) + (5x - 3)$$
$$= 2x^2 + 5x - 2$$

The domain of $f + g$ is the set of real numbers that are in the domain of both f and g. Since the domain of both f and g is the interval $(-\infty, \infty)$, the domain of $f + g$ is also the interval $(-\infty, \infty)$.

b. $(f - g)(x) = f(x) - g(x)$
$$= (2x^2 + 1) - (5x - 3)$$
$$= 2x^2 + 1 - 5x + 3 \qquad \text{Use the distributive property to remove parentheses.}$$
$$= 2x^2 - 5x + 4 \qquad \text{Combine like terms.}$$

Since the domain of both f and g is $(-\infty, \infty)$, the domain of $f - g$ is also the interval $(-\infty, \infty)$.

 SELF CHECK 1   Let $f(x) = 3x - 2$ and $g(x) = 2x^2 + 3x$. Find each function and its domain.

a. $f + g$ **b.** $f - g$

EXAMPLE 2   Let $f(x) = 2x^2 + 1$ and $g(x) = 5x - 3$. Find each function and its domain.

a. $f \cdot g$ **b.** f/g

Solution   **a.** $(f \cdot g)(x) = f(x)g(x)$

$$= (2x^2 + 1)(5x - 3)$$

$$= 10x^3 - 6x^2 + 5x - 3 \quad \text{Multiply.}$$

The domain of $f \cdot g$ is the set of real numbers that are in the domain of both f and g. Since the domain of both f and g is the interval $(-\infty, \infty)$, the domain of $f \cdot g$ is also the interval $(-\infty, \infty)$.

b. $(f/g)(x) = \dfrac{f(x)}{g(x)}$

$$= \dfrac{2x^2 + 1}{5x - 3}$$

Since the denominator of the fraction cannot be $0, x \neq \frac{3}{5}$. The domain of f/g is the union of two intervals $\left(-\infty, \frac{3}{5}\right) \cup \left(\frac{3}{5}, \infty\right)$.

SELF CHECK 2   Let $f(x) = 2x^2 - 3$ and $g(x) = x - 1$. Find each function and its domain.

a. $f \cdot g$ **b.** f/g

---

2   Find the composition of two functions.

We have seen that a function can be represented by a machine: We put in a number from the domain, and a number from the range comes out. For example, if we put the number 2 into the machine shown in Figure 8-30(a), the number $f(2) = 5(2) - 2 = 8$ comes out. In general, if we put x into the machine shown in Figure 8-30(b), the value $f(x)$ comes out.

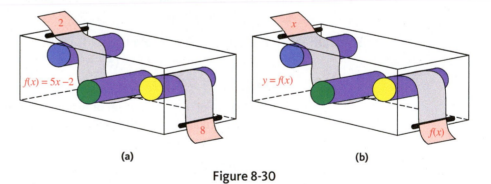

(a)   (b)

Figure 8-30

Often one quantity is a function of a second quantity that depends, in turn, on a third quantity. For example, the cost of a car trip is a function of the gasoline consumed. The amount of gasoline consumed, in turn, is a function of the number of miles driven. Such chains of dependence can be analyzed mathematically as **compositions of functions**.

The function machines shown in Figure 8-31 illustrate the composition of functions f and g. When we put a number x into the function g, $g(x)$ comes out. The value $g(x)$ goes into function f, which transforms $g(x)$ into $f(g(x))$. This two-step process defines a new function, called a **composite function**. If the function machines for g and f were connected to make a single machine, that machine would be named $f \circ g$, read as "f composition g."

To be in the domain of the composite function $f \circ g$, a number x has to be in the domain of g. Also, the output of g must be in the domain of f. Thus, the domain of $f \circ g$ consists of those numbers x that are in the domain of g, and for which $g(x)$ is in the domain of f.

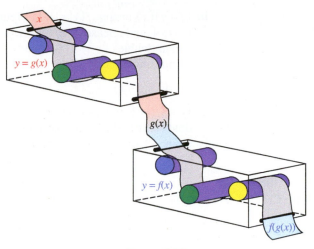

Figure 8-31

COMPOSITE FUNCTIONS   The **composite function** $f \circ g$ is defined by

$$(f \circ g)(x) = f(g(x))$$

COMMENT   Note that in this example, $(f \circ g)(x) \neq (g \circ f)(x)$. This shows that the composition of functions is not commutative.

For example, if $f(x) = 4x - 5$ and $g(x) = 3x + 2$, then

$$
\begin{aligned}
(f \circ g)(x) &= f(g(x)) \\
&= f(3x + 2) \\
&= 4(3x + 2) - 5 \\
&= 12x + 8 - 5 \\
&= 12x + 3
\end{aligned}
\qquad
\begin{aligned}
(g \circ f)(x) &= g(f(x)) \\
&= g(4x - 5) \\
&= 3(4x - 5) + 2 \\
&= 12x - 15 + 2 \\
&= 12x - 13
\end{aligned}
$$

EXAMPLE 3   Let $f(x) = 2x + 1$ and $g(x) = x - 4$. Find

a. $(f \circ g)(9)$ **b.** $(f \circ g)(x)$ **c.** $(g \circ f)(-2)$

Solution   **a.** $(f \circ g)(9)$ means $f(g(9))$. In Figure 8-32(a), function g receives the number 9, subtracts 4, and releases the number 5. The 5 then goes into the f function, which doubles 5 and adds 1. The final result, 11, is the output of the composite function $f \circ g$:

$$(f \circ g)(9) = f(g(9)) = f(5) = 2(5) + 1 = 11$$

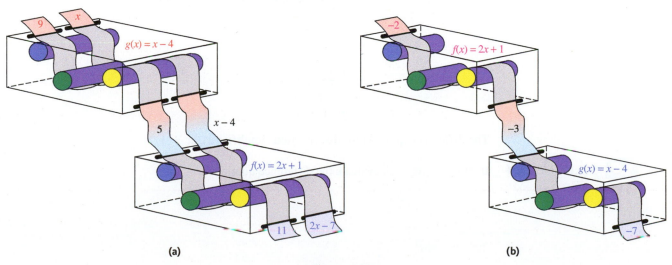

(a)   (b)

Figure 8-32

b. $(f \circ g)(x)$ means $f(g(x))$. In Figure 8-32(a), function g receives the number x, subtracts 4, and releases the number $(x - 4)$. The $(x - 4)$ then goes into the f function, which doubles $(x - 4)$ and adds 1. The final result, $(2x - 7)$, is the output of the composite function $f \circ g$.

$$(f \circ g)(x) = f(g(x)) = f(x - 4) = 2(x - 4) + 1 = 2x - 7$$

c. $(g \circ f)(-2)$ means $g(f(-2))$. In Figure 8-32(b), function f receives the number -2, doubles it and adds 1, and releases -3 into the g function. Function g subtracts 4 from -3 and releases a final result of -7. Thus,

$$(g \circ f)(-2) = g(f(-2)) = g(-3) = -3 - 4 = -7$$

 SELF CHECK 3   Let $f(x) = 3x + 2$ and $g(x) = 9x - 5$. Find

a. $(f \circ g)(2)$ **b.** $(g \circ f)(x)$ **c.** $(g \circ f)(-4)$

---

3   **Find the difference quotient of a function.**

An important function in calculus, called the **difference quotient**, represents the slope of a line at a given point on the graph of a function. The difference quotient is defined as follows:

$$\frac{f(x + h) - f(x)}{h}, \quad h \neq 0$$

EXAMPLE 4   If $f(x) = x^2 - 4$, evaluate the difference quotient.

Solution   First, we evaluate $f(x + h)$.

$$f(x) = x^2 - 4$$
$$f(x + h) = (x + h)^2 - 4 \qquad \text{Substitute } x + h \text{ for } h.$$
$$= x^2 + 2xh + h^2 - 4 \quad (x + h)^2 = x^2 + 2hx + h^2$$

Then we note that $f(x) = x^2 - 4$. We can now substitute the values of $f(x + h)$ and $f(x)$ into the difference quotient and simplify.

$$\frac{f(x + h) - f(x)}{h} = \frac{(x^2 + 2xh + h^2 - 4) - (x^2 - 4)}{h}$$

$$= \frac{x^2 + 2xh + h^2 - 4 - x^2 + 4}{h} \qquad \text{Use the distributive property to remove parentheses.}$$

$$= \frac{2xh + h^2}{h} \qquad \text{Combine like terms.}$$

$$= \frac{h(2x + h)}{h} \qquad \text{Factor out } h, \text{ the GCF, in the numerator.}$$

$$= 2x + h \qquad \text{Divide out } h; \frac{h}{h} = 1.$$

The difference quotient for this function simplifies as $2x + h$.

 SELF CHECK 4   If $f(x) = 3x^2 - 4$, evaluate the difference quotient.

4 Solve an application requiring the composition of two functions.

EXAMPLE 5 **TEMPERATURE CHANGE** A laboratory sample is removed from a cooler at a temperature of 15° Fahrenheit. Technicians are warming the sample at a controlled rate of 3° F per hour. Express the sample's Celsius temperature as a function of the time, t (in hours), since it was removed from refrigeration.

Solution The temperature of the sample is 15° F when $t = 0$. Because it warms at 3° F per hour, it warms $(3t)°$ after t hours. The Fahrenheit temperature after t hours is given by the function

$$F(t) = 3t + 15$$

The Celsius temperature is a function of the Fahrenheit temperature, given by the formula

$$C(F) = \frac{5}{9}(F - 32)$$

To express the sample's Celsius temperature as a function of time, we find the composition function $C \circ F$.

$$(C \circ F)(t) = C(F(t))$$

$$= \frac{5}{9}(\boldsymbol{F(t)} - 32)$$

$$= \frac{5}{9}[(\boldsymbol{3t + 15}) - 32] \quad \text{Substitute } 3t + 15 \text{ for } F(t).$$

$$= \frac{5}{9}(3t - 17) \quad \text{Combine like terms.}$$

$$= \frac{15}{9}t - \frac{85}{9} \quad \text{Use the distributive property to remove parentheses.}$$

$$= \frac{5}{3}t - \frac{85}{9} \quad \text{Simplify.}$$

 SELF CHECK 5 If the sample temperature is 51° Fahrenheit and technicians are cooling the sample at 3° F per hour, express the sample's Celsius temperature as a function of the time since it was put into refrigeration.

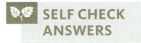 **SELF CHECK ANSWERS** 1. a. $2x^2 + 6x - 2, (-\infty, \infty)$ b. $-2x^2 - 2, (-\infty, \infty)$ 2. a. $2x^3 - 2x^2 - 3x + 3, (-\infty, \infty)$ b. $\frac{2x^2 - 3}{x - 1}, (-\infty, 1) \cup (1, \infty)$ 3. a. 41 b. $27x + 13$ c. -95 4. $6x + 3h$ 5. $(C \circ F)(t) = \frac{95}{9} - \frac{5}{3}t$

NOW TRY THIS

Given $f(x) = 3x - 2$ and $g(x) = x^2 - 5x + 1$, find

1. $(g \circ f)(-1)$
2. $(g \circ f)(x)$
3. $(f \circ f)(x)$

8.6 Exercises ✿

Assume no denominators are 0.

WARM-UPS *If $f(x) = 2x + 3$ and $g(x) = 4x^2 - 2$, find the following.*

1. $f(-3)$
2. $g(-2)$
3. $g(x + 1)$
4. $f(x + 1)$

Simplify.

5. $(2x + 3) + (4x^2 - 2)$
6. $(4x^2 - 2) - (2x + 3)$
7. $(2x + 3) - (4x^2 - 2)$
8. $(2x + 3)(4x^2 - 2)$

REVIEW *Simplify each expression.*

9. $\dfrac{5x^2 - 13x - 6}{9 - x^2}$

10. $\dfrac{2x^3 + 14x^2}{3 + 2x - x^2} \cdot \dfrac{x^2 - 3x}{x}$

11. $\dfrac{8 + 2x - x^2}{12 + x - 3x^2} \div \dfrac{3x^2 + 5x - 2}{3x - 1}$

12. $\dfrac{x - 1}{1 + \dfrac{x}{x - 2}}$

VOCABULARY AND CONCEPTS *Fill in the blanks.*

13. $(f + g)(x) = $ _____
14. $(f - g)(x) = $ _____
15. $(f \cdot g)(x) = $ _____

16. $(f/g)(x) = $ _____ $\quad (g(x) \neq 0)$

17. In Exercises 13–15, the domain of each function is the set of real numbers x that are in the _____ of both f and g.
18. The _____ of functions f and g is denoted by $(f \circ g)(x)$ or $f \circ g$.
19. $(f \circ g)(x) = $ _____
20. The difference quotient is defined as _____.
21. In calculus, the difference quotient represents the slope of a line at a _____ on a graph.
22. In the difference quotient, $h \neq $ _.

GUIDED PRACTICE *Let $f(x) = 3x$ and $g(x) = 4x$. Find each function and its domain.* **SEE EXAMPLE 1. (OBJECTIVE 1)**

23. $f + g$
24. $f - g$
25. $g - f$
26. $g + f$

Let $f(x) = 3x$ and $g(x) = 4x$. Find each function and its domain. **SEE EXAMPLE 2. (OBJECTIVE 1)**

27. $f \cdot g$
28. f/g
29. g/f
30. $g \cdot f$

Let $f(x) = 2x + 1$ and $g(x) = x - 3$. Find each function and its domain. **SEE EXAMPLES 1–2. (OBJECTIVE 1)**

31. $f + g$
32. $f - g$
33. $g - f$
34. $g + f$
35. $f \cdot g$
36. f/g
37. g/f
38. $g \cdot f$

Let $f(x) = 2x + 1$ and $g(x) = x^2 - 1$. Find each value. **SEE EXAMPLE 3. (OBJECTIVE 2)**

39. $(f \circ g)(4)$
40. $(g \circ f)(4)$
41. $(g \circ f)(-3)$
42. $(f \circ g)(-3)$
43. $(f \circ g)(0)$
44. $(g \circ f)(0)$
45. $(f \circ g)\left(\dfrac{1}{4}\right)$
46. $(g \circ f)\left(\dfrac{1}{4}\right)$
47. $(f \circ g)(x)$
48. $(g \circ f)(x)$
49. $(g \circ f)(3x)$
50. $(f \circ g)(3x)$

Find $\dfrac{f(x + h) - f(x)}{h}$. **SEE EXAMPLE 4. (OBJECTIVE 3)**

51. $f(x) = 4x + 5$
52. $f(x) = 5x - 1$
53. $f(x) = x^2$
54. $f(x) = x^2 - 1$
55. $f(x) = 2x^2 - 1$
56. $f(x) = 3x^2$
57. $f(x) = x^2 + x$
58. $f(x) = x^2 - x$
59. $f(x) = x^2 + 3x - 4$
60. $f(x) = x^2 - 4x + 3$
61. $f(x) = 2x^2 + 3x - 7$
62. $f(x) = 3x^2 - 2x + 4$

ADDITIONAL PRACTICE *Let $f(x) = 3x - 2$ and $g(x) = 2x^2 + 1$. Find each function and its domain.*

63. $f - g$
64. $f + g$
65. f/g
66. $f \cdot g$

Let $f(x) = x^2 - 1$ and $g(x) = x^2 - 4$. Find each function and its domain.

67. $f - g$
68. $f + g$
69. g/f
70. $g \cdot f$

Let $f(x) = 3x - 2$ and $g(x) = x^2 + x$. Find each value.

71. $(f \circ g)(4)$ **72.** $(g \circ f)(4)$
73. $(g \circ f)(-2)$ **74.** $(f \circ g)(-2)$
75. $(g \circ f)(0)$ **76.** $(f \circ g)(0)$
77. $(g \circ f)(x)$ **78.** $(f \circ g)(x)$

For problems 79–90, find $\dfrac{f(x) - f(a)}{x - a}$ $(x \neq a)$.

79. $f(x) = 4x - 7$ **80.** $f(x) = 6x + 5$
81. $f(x) = x^2$ **82.** $f(x) = x^2 - 1$
83. $f(x) = 2x^2 - 1$ **84.** $f(x) = 3x^2$

85. $f(x) = x^2 + x$ **86.** $f(x) = x^2 - x$

87. $f(x) = x^2 + 3x - 4$ **88.** $f(x) = x^2 - 4x + 3$

89. $f(x) = 2x^2 + 3x - 7$ **90.** $f(x) = 3x^2 - 2x + 4$

91. If $f(x) = x + 1$ and $g(x) = 2x - 5$, show that $(f \circ g)(x) \neq (g \circ f)(x)$.
92. If $f(x) = x^2 + 1$ and $g(x) = 3x^2 - 2$, show that $(f \circ g)(x) \neq (g \circ f)(x)$.
93. If $f(x) = x^2 + 2x - 3$, find $f(a)$, $f(h)$, and $f(a + h)$. Then show that $f(a + h) \neq f(a) + f(h)$.
94. If $g(x) = 2x^2 + 10$, find $g(a)$, $g(h)$, and $g(a + h)$. Then show that $g(a + h) \neq g(a) + g(h)$.
95. If $f(x) = x^3 - 1$, find $\dfrac{f(x + h) - f(x)}{h}$.
96. If $f(x) = x^3 + 2$, find $\dfrac{f(x + h) - f(x)}{h}$.

APPLICATIONS *Solve each application. SEE EXAMPLE 5. (OBJECTIVE 4)*

97. Alloys A molten alloy must be cooled slowly to control crystallization. When removed from the furnace, its temperature is 2,700° F, and it will be cooled at 200° F per hour. Express the Celsius temperature as a function of the number of hours t since cooling began.

98. Weather forecasting A high-pressure area promises increasingly warmer weather for the next 48 hours. The temperature is now 34° Celsius and will rise 1° C every 6 hours. Express the Fahrenheit temperature as a function of the number of hours from now.

WRITING ABOUT MATH

99. Explain how to find the domain of f/g.
100. Explain why the difference quotient represents the slope of a line passing through $(x, f(x))$ and $(x + h, f(x + h))$.

SOMETHING TO THINK ABOUT

101. Is composition of functions associative? Choose functions f, g, and h and determine whether $[f \circ (g \circ h)](x) = [(f \circ g) \circ h](x)$.
102. Choose functions f, g, and h and determine whether $f \circ (g + h) = f \circ g + f \circ h$.

Section 8.7

Inverses of Functions

Objectives

1. Determine whether a function is one-to-one.
2. Apply the horizontal line test to determine whether the graph of a function is one-to-one.
3. Find the inverse of a function.

Vocabulary

one-to-one function   inverse of a function   identity function
horizontal line test

Getting Ready

Solve each equation for y.

1. $x = 3y + 2$ **2.** $x = \dfrac{3}{2}y + 5$

We already know that real numbers have inverses. For example, the additive inverse of 3 is -3, because $3 + (-3) = 0$. The multiplicative inverse of 3 is $\frac{1}{3}$, because $3\left(\frac{1}{3}\right) = 1$. In a similar way, functions have inverses. After discussing one-to-one functions, we will learn how to find the inverse of a function.

1   Determine whether a function is one-to-one.

Recall that for each input into a function, there is a single output. For some functions, different inputs have the same output, as shown in Figure 8-33(a). For other functions, different inputs have different outputs, as shown in Figure 8-33(b).

Not a one-to-one function
(a)

A one-to-one function
(b)

Figure 8-33

When every output of a function corresponds to exactly one input, we say that the function is *one-to-one*.

ONE-TO-ONE FUNCTIONS	A function is called **one-to-one** if each input value of x in the domain determines a different output value of y in the range.

EXAMPLE 1   Determine whether the following are one-to-one.   **a.** $f(x) = x^2$ **b.** $f(x) = x^3$

Solution   **a.** The function $f(x) = x^2$ is not one-to-one, because different input values x can determine the same output value y. For example, inputs of 3 and -3 produce the same output value of 9.

$$f(3) = 3^2 = 9 \quad \text{and} \quad f(-3) = (-3)^2 = 9$$

b. The function $f(x) = x^3$ is one-to-one, because different input values x determine different output values of y for all x. This is because different numbers have different cubes.

SELF CHECK 1   Determine whether $f(x) = 2x + 3$ is one-to-one.

2   Apply the horizontal line test to determine whether
the graph of a function is one-to-one.

Similar to the vertical line test used to determine if a graph represents a function, a **horizontal
line test** can be used to determine whether the graph of a function represents a one-to-one
function. If every horizontal line that intersects the graph of a function does so only once, the
function is one-to-one. Otherwise, the function is not one-to-one. See Figure 8-34.

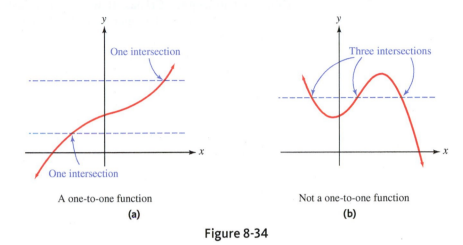

A one-to-one function
(a)

Not a one-to-one function
(b)

Figure 8-34

EXAMPLE 2   The graphs in Figure 8-35 represent functions. Use the horizontal line test to determine
whether the graphs represent one-to-one functions.

Solution   **a.** Because many horizontal lines intersect the graph shown in Figure 8-35(a) twice, the
graph does not represent a one-to-one function.

b. Because each horizontal line that intersects the graph in Figure 8-35(b) does so
exactly once, the graph does represent a one-to-one function.

(a)   **(b)**

Figure 8-35

 SELF CHECK 2   Does the following graph represent a one-to-one function?

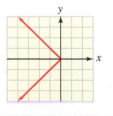

COMMENT   Use the vertical line test to determine whether a graph represents a function. If it does, use the horizontal line test to determine whether the function is one to-one.

3   Find the inverse of a function.

The function defined by $C = \frac{5}{9}(F - 32)$ is the formula that we use to convert degrees Fahrenheit to degrees Celsius. If we substitute a Fahrenheit reading into the formula, we obtain a Celsius reading. For example, if we substitute 41° for F we obtain a Celsius reading of 5°.

$$C = \frac{5}{9}(\mathbf{F} - 32)$$

$$= \frac{5}{9}(\mathbf{41} - 32) \quad \text{\textcolor{red}{Substitute 41 for }} F.$$

$$= \frac{5}{9}(9)$$

$$= 5$$

If we want to find a Fahrenheit reading from a Celsius reading, we need a formula into which we can substitute a Celsius reading and have a Fahrenheit reading result. Such a formula is $F = \frac{9}{5}C + 32$, which takes the Celsius reading of 5° and turns it back into a Fahrenheit reading of 41°.

$$F = \frac{9}{5}\mathbf{C} + 32$$

$$= \frac{9}{5}(\mathbf{5}) + 32 \quad \text{\textcolor{red}{Substitute 5 for }} C.$$

$$= 41$$

The functions defined by these two formulas do opposite things. The first turns 41° F into 5° C, and the second turns 5° C back into 41° F. For this reason, we say that the functions are *inverses* of each other.

If f is the function determined by the table shown in Figure 8-36(a), it turns the number 1 into 10, 2 into 20, and 3 into 30. Since the inverse of f must turn 10 back into 1, 20 back into 2, and 30 back into 3, it consists of the ordered pairs shown in Figure 8-36(b).

Figure 8-36

We note that the domain of f and the range of its inverse is {1, 2, 3}. The range of f and the domain of its inverse is {10, 20, 30}.

This example suggests that to form the **inverse of a function** f, we simply interchange the coordinates of each ordered pair that determines f. When the inverse of a function is also a function, we call it f *inverse* and denote it with the symbol f^{-1}.

COMMENT   The symbol $f^{-1}(x)$ is read as "the inverse of $f(x)$" or just "f inverse." The -1 in the notation $f^{-1}(x)$ is not an exponent.

FINDING THE INVERSE OF A ONE-TO-ONE FUNCTION	We find the inverse of a function as follows:

We find the inverse of a function as follows:

1. Replace $f(x)$ with y.

2. Interchange the variables x and y.

3. Solve the resulting equation for y. If the inverse is also a function, use the notation $f^{-1}(x)$.

EXAMPLE 3 If $f(x) = 4x + 2$, find the inverse of f and determine whether it is a function.

Solution Because $f(x) = 4x + 2$ is a linear equation, its graph is a line. The horizontal line test indicates that the function is one-to-one. Therefore, its inverse is also a function. To find the inverse, we replace $f(x)$ with y and interchange the positions of x and y.

$$f(x) = 4x + 2$$
$$y = 4x + 2 \quad \text{Replace } f(x) \text{ with } y.$$
$$x = 4y + 2 \quad \text{Interchange the variables } x \text{ and } y.$$

Then we solve the equation for y.

$$x = 4y + 2$$
$$x - 2 = 4y \quad \text{Subtract 2 from both sides.}$$
$$y = \frac{x - 2}{4} \quad \text{Divide both sides by 4 and write } y \text{ on the left side.}$$

Expressing the inverse in function notation, we have

$$f^{-1}(x) = \frac{x - 2}{4}$$

SELF CHECK 3 If $f(x) = -5x - 3$, find the inverse of f and determine whether it is a function.

COMMENT If the inverse of a function is also a function, we write the inverse function using function notation.

To emphasize an important relationship between a function and its inverse, we substitute some number x, such as $x = 3$, into the function $f(x) = 4x + 2$ of Example 3. The corresponding value of y produced is

$$f(3) = 4(3) + 2 = 14$$

If we substitute 14 into the inverse function, f^{-1}, the corresponding value of y that is produced is

$$f^{-1}(14) = \frac{14 - 2}{4} = 3$$

Thus, the function f turns 3 into 14, and the inverse function f^{-1} turns 14 back into 3. In general, *the composition of a function and its inverse is the **identity function**.*

Recall that in the real number system, 0 is called the *additive identity* because $0 + x = x$ and 1 is called the *multiplicative identity* because $1 \cdot x = x$. There is an identity for functions as well. The identity function is defined by the equation $I(x) = x$. Under this function, the value that corresponds to any real number x is x itself. If f is any function, the composition of f with the identity function is the function f:

$$(f \circ I)(x) = (I \circ f)(x) = f(x)$$

We can show this as follows:

$(f \circ I)(x)$ means $f(I(x))$. Because $I(x) = x$, we have

$$(f \circ I)(x) = f(\mathbf{I(x)}) = f(\mathbf{x})$$

$(I \circ f)(x)$ means $I(f(x))$. Because I passes any number through unchanged, we have $I(f(x)) = f(x)$ and

$$(I \circ f)(x) = I(\mathbf{f(x)}) = \mathbf{f(x)}$$

To prove that $f(x) = 4x + 2$ and $f^{-1}(x) = \dfrac{x - 2}{4}$ are inverse functions, we must show that their composition (in both directions) is the identity function:

$$
\begin{array}{ll}
(f \circ f^{-1})(x) = f(\mathbf{f^{-1}(x)}) & \qquad (f^{-1} \circ f)(x) = f^{-1}(\mathbf{f(x)}) \\[2mm]
\qquad = f\left(\dfrac{\mathbf{x-2}}{\mathbf{4}}\right) & \qquad\qquad = f^{-1}(\mathbf{4x+2}) \\[2mm]
\qquad = 4\left(\dfrac{\mathbf{x-2}}{\mathbf{4}}\right) + 2 & \qquad\qquad = \dfrac{\mathbf{4x+2}-2}{4} \\[2mm]
\qquad = x - 2 + 2 & \qquad\qquad = \dfrac{4x}{4} \\[2mm]
\qquad = x & \qquad\qquad = x
\end{array}
$$

Thus, $(f \circ f^{-1})(x) = (f^{-1} \circ f)(x) = x$, which is the identity function $I(x)$.

EXAMPLE 4 The set of all pairs (x, y) determined by $3x + 2y = 6$ is a function. Find its inverse function, and graph the function and its inverse on one coordinate system.

Solution To find the inverse function of $3x + 2y = 6$, we interchange x and y to obtain

$$3y + 2x = 6$$

and then solve the equation for y.

$$
\begin{array}{ll}
3y + 2x = 6 & \\
3y = -2x + 6 & \text{Subtract } 2x \text{ from both sides.} \\[2mm]
y = -\dfrac{2}{3}x + 2 & \text{Divide both sides by 3.}
\end{array}
$$

Since the resulting equation represents a function, we write it in inverse function notation as

$$f^{-1}(x) = -\frac{2}{3}x + 2$$

The graphs of $3x + 2y = 6$ and $f^{-1}(x) = -\frac{2}{3}x + 2$ appear in Figure 8-37.

Figure 8-37

COMMENT The dashed line $y = x$ shown in Figure 8-37 is included to illustrate the symmetry of the function and its inverse.

SELF CHECK 4 Find the inverse of the function defined by $2x - 3y = 6$. Graph the function and its inverse on one coordinate system.

In Example 4, the graph of $3x + 2y = 6$ and $f^{-1}(x) = -\frac{2}{3}x + 2$ are symmetric about the line $y = x$. In general, *any function and its inverse are symmetric about the line $y = x$*, because when the coordinates (a, b) satisfy an equation, the coordinates (b, a) will satisfy its inverse.

In each example so far, the inverse of a function has been another function. This is not always true, as the following example will show.

EXAMPLE 5   Find the inverse of the function determined by $f(x) = x^2$.

Solution

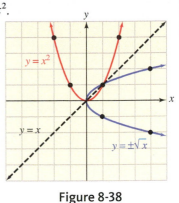

$y = x^2$ Replace $f(x)$ with y.

$x = y^2$ Interchange x and y.

$y = \pm\sqrt{x}$ Use the square-root property and write y on the left side.

When the inverse $y = \pm\sqrt{x}$ is graphed as in Figure 8-38, we see that the graph does not pass the vertical line test. Thus, it is not a function.

The graph of $y = x^2$ is also shown in the figure. As expected, the graphs of $y = x^2$ and $y = \pm\sqrt{x}$ are symmetric about the line $y = x$.

Figure 8-38

🍃 **SELF CHECK 5**   Find the inverse of the function determined by $f(x) = 4x^2$.

EXAMPLE 6   Find the inverse of $f(x) = x^3$.

Solution   To find the inverse, we proceed as follows:

$y = x^3$ Replace $f(x)$ with y.

$x = y^3$ Interchange the variables x and y.

$\sqrt[3]{x} = y$ Take the cube root of both sides.

We note that to each number x there corresponds one real cube root. Thus, $y = \sqrt[3]{x}$ represents a function. In function notation, we have

$$f^{-1}(x) = \sqrt[3]{x}$$

🍃 **SELF CHECK 6**   Find the inverse of $f(x) = x^5$.

If a function is not one-to-one, we may be able to restrict its domain to create a one-to-one function.

EXAMPLE 7   Find the inverse of $f(x) = x^2$ $(x \geq 0)$. Then determine whether the inverse is a function. Graph the function and its inverse.

Solution   The domain is restricted to $x \geq 0$ so that $f(x)$ is one-to-one. The inverse of the function $f(x) = x^2$ with $x \geq 0$ is

$y = x^2$ with $x \geq 0$ Replace $f(x)$ with y.

$x = y^2$ with $y \geq 0$ Interchange the variables x and y.

$y = \pm\sqrt{x}$ with $y \geq 0$ Solve for y and write y on the left side.

Considering the restriction $y \geq 0$, the equation can be written more simply as

$y = \sqrt{x}$ This is the inverse of $f(x) = x^2$.

In this equation, each number x gives only one value of y. Thus, the inverse is a function, which we can write as

$$f^{-1}(x) = \sqrt{x}$$

The graphs of the two functions appear in Figure 8-39 on the next page. The line $y = x$ is included so that we can see that the graphs are symmetric about the line $y = x$.

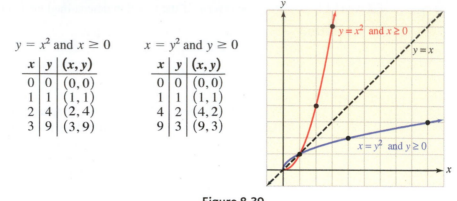

$y = x^2$ and $x \geq 0$			$x = y^2$ and $y \geq 0$		
x	y	(x, y)	x	y	(x, y)
0	0	$(0, 0)$	0	0	$(0, 0)$
1	1	$(1, 1)$	1	1	$(1, 1)$
2	4	$(2, 4)$	4	2	$(4, 2)$
3	9	$(3, 9)$	9	3	$(9, 3)$

Figure 8-39

SELF CHECK 7   Find the inverse of $f(x) = x^2 - 8$ $(x \geq 0)$. Graph the function and its inverse.

SELF CHECK ANSWERS

1. yes   **2.** no   **3.** $f^{-1}(x) = -\frac{1}{5}x - \frac{3}{5}$; yes   **4.** $f^{-1}(x) = \frac{3}{2}x + 3$

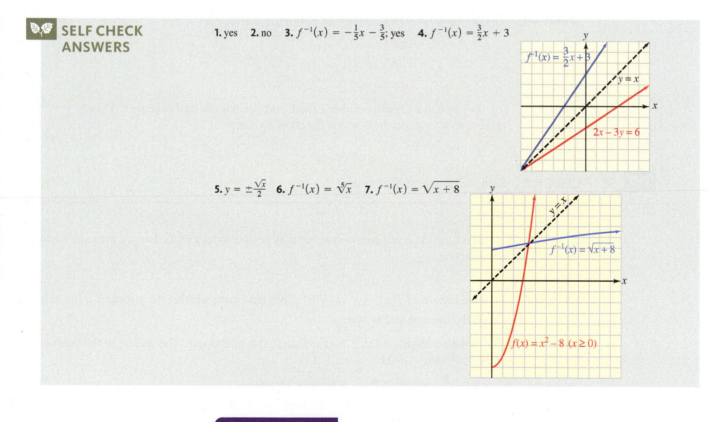

5. $y = \pm\frac{\sqrt{x}}{2}$ **6.** $f^{-1}(x) = \sqrt[5]{x}$ **7.** $f^{-1}(x) = \sqrt{x + 8}$

NOW TRY THIS

1. Find the inverse of $f(x) = \frac{1}{x}$.

2. Given $f(x) = \frac{5x + 1}{3x - 2}$, find $f^{-1}(x)$.

8.7 Exercises

WARM-UPS *Given* $f(x) = 2x + 1$ *and* $g(x) = \frac{1}{2}x - \frac{1}{2}$, *find the following and write the results as ordered pairs.*

1. $f(0)$
2. $g(1)$
3. $f(3)$
4. $g(7)$
5. $f(-1)$
6. $g(-1)$

REVIEW *Write each complex number in* $a + bi$ *form or find each value.*

7. $7 - \sqrt{-49}$
8. $(6 - 4i) + (2 + 7i)$

9. $(3 + 4i)(2 - 3i)$
10. $\dfrac{6 + 7i}{3 - 4i}$

11. $|6 - 8i|$
12. $\left| \dfrac{2 + i}{3 - i} \right|$

VOCABULARY AND CONCEPTS *Fill in the blanks.*

13. A function is called _____ if each input determines a different output.
14. If every _____ line that intersects the graph of a function does so only once, the function is one-to-one.
15. If a one-to-one function turns an input of 2 into an output of 5, the inverse function will turn 5 into __.
16. The symbol $f^{-1}(x)$ is read as _____ or _____.
17. $(f \circ f^{-1})(x) = (f^{-1} \circ f)(x) = $ __.
18. The graphs of a function and its inverse are symmetrical about the line _____.

GUIDED PRACTICE *Determine whether each function is one-to-one.* **SEE EXAMPLE 1. (OBJECTIVE 1)**

19. $f(x) = 5x$
20. $f(x) = |x|$
21. $f(x) = x^4$
22. $f(x) = x^3 - 2$

Each graph represents a function. Use the horizontal line test to determine whether the function is one-to-one. **SEE EXAMPLE 2. (OBJECTIVE 2)**

23.
24.

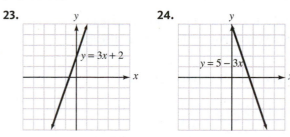

25. $y = \dfrac{x + 5}{2}$
26. $y = \dfrac{5 - x}{2}$
27. $y = 3x^2 + 2$
28. $y = 5 - x^2$
29. $y = \sqrt[3]{x}$
30. $y = \sqrt{x}$

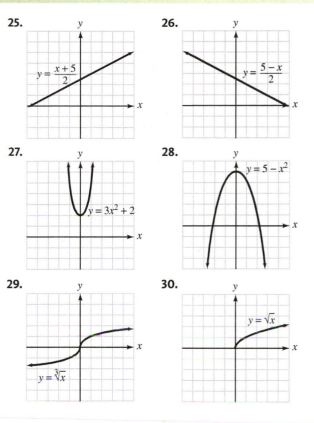

Find the inverse of each set of ordered pairs (x, y) *and determine whether the inverse is a function.* **(OBJECTIVE 3)**

31. $\{(4, 3), (3, 2), (2, 1)\}$

32. $\{(4, 1), (5, 1), (6, 1), (7, 1)\}$

33. $\{(1, 2), (2, 3), (1, 3), (1, 5)\}$

34. $\{(-1, -1), (0, 0), (1, 1), (2, 2)\}$

Find the inverse of each function and express it in the form $y = f^{-1}(x)$. *Verify each result by showing that* $(f \circ f^{-1})(x) = (f^{-1} \circ f)(x) = (x)$. **SEE EXAMPLE 3. (OBJECTIVE 3)**

35. $f(x) = 4x + 1$
36. $y + 1 = 5x$
37. $x + 4 = 5y$
38. $x = 3y + 1$

Find the inverse of each function. Then graph the function and its inverse on one coordinate system. Draw the line of symmetry on the graph. **SEE EXAMPLE 4. (OBJECTIVE 3)**

39. $y = 4x + 3$

40. $x = 3y - 1$

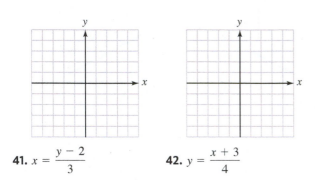

41. $x = \dfrac{y - 2}{3}$

42. $y = \dfrac{x + 3}{4}$

Find the inverse of each function and determine whether it is a function. If it is a function, express it in function notation. **SEE EXAMPLES 5–6. (OBJECTIVE 3)**

43. $y = x^2 + 9$

44. $y = x^2 + 5$

45. $y = x^3$

46. $xy = 4$

Graph each equation and its inverse on one set of coordinate axes. Draw the line of symmetry. **SEE EXAMPLE 7. (OBJECTIVE 3)**

47. $y = x^2 + 1, x \geq 0$

48. $y = \dfrac{1}{4}x^2 - 3, x \geq 0$

49. $y = \sqrt{x}, x \geq 0$

50. $y = -\sqrt{x}, x \geq 0$

ADDITIONAL PRACTICE *Find the inverse of each function. Do not rationalize denominators.*

51. $\{(1, 1), (2, 4), (3, 9), (4, 16)\}$

52. $\{(1, 1), (2, 1), (3, 1), (4, 1)\}$

53. $y = -\sqrt[3]{x}$

54. $y = \sqrt[3]{x}$

55. $f(x) = 2x^3 - 3$

56. $f(x) = \dfrac{3}{x^3} - 1$

57. $f(x) = \dfrac{x - 7}{3}$

58. $f(x) = \dfrac{2x + 6}{3}$

59. $4x - 5y = 20$

60. $3x + 5y = 15$

Graph each equation and its inverse on one set of coordinate axes. Draw the line of symmetry.

61. $3x - y = 5$

62. $2x + 3y = 9$

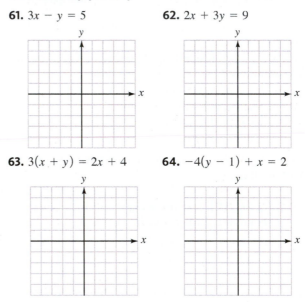

63. $3(x + y) = 2x + 4$

64. $-4(y - 1) + x = 2$

WRITING ABOUT MATH

65. Explain the purpose of the vertical line test.

66. Explain the purpose of the horizontal line test.

SOMETHING TO THINK ABOUT

67. Find the inverse of $y = \dfrac{x + 1}{x - 1}$.

68. Using the functions of Exercise 67, show that $(f \circ f^{-1})(x) = x$.

Projects

PROJECT 1

Ballistics is the study of how projectiles fly. The general formula for the height above the ground of an object thrown straight up or down is given by the function

$$h(t) = -16t^2 + v_0 t + h_0$$

where h is the object's height (in feet) above the ground t seconds after it is thrown. The initial velocity v_0 is the velocity with which the object is thrown, measured in feet per second. The initial height h_0 is the object's height (in feet) above the ground when it is thrown. (If $v_0 > 0$, the object is thrown upward; if $v_0 < 0$, the object is thrown downward.)

This formula takes into account the force of gravity, but disregards the force of air resistance. It is much more accurate for a smooth, dense ball than for a crumpled piece of paper.

One act in the Bungling Brothers Circus is Amazing Glendo's cannonball-catching act. A cannon fires a ball vertically into the air; Glendo, standing on a platform above the cannon, uses his catlike reflexes to catch the ball as it passes by on its way toward the roof of the big top. As the balls fly past, they are within Glendo's reach only during a two-foot interval of their upward path.

As an investigator for the company that insures the circus, you have been asked to find answers to the following questions. The answers will determine whether or not Bungling Brothers' insurance policy will be renewed.

a. In the first part of the act, cannonballs are fired from the end of a six-foot cannon with an initial velocity of 80 feet per second. Glendo catches one ball between 40 and 42 feet above the ground. Then he lowers his platform and catches another ball between 25 and 27 feet above the ground.

  1. Show that if Glendo missed a cannonball, it would hit the roof of the 56-foot-tall big top. How long would it take for a ball to hit the big top? To prevent this from happening, a special net near the roof catches and holds any missed cannonballs.

  2. Find (to the nearest thousandth of a second) how long the cannonballs are within Glendo's reach for each of his catches. Which catch is easier? Why does your answer make sense? Your company is willing to insure against injuries to Glendo if he has at least 0.025 second to make each catch. Should the insurance be offered?

b. For Glendo's grand finale, the special net at the roof of the big top is removed, making Glendo's catch more significant to the people in the audience, who worry that if Glendo misses, the tent will collapse around them. To make it even more dramatic, Glendo's arms are tied to restrict his reach to a one-foot interval of the ball's flight, and he stands on a platform just under the peak of the big top, so that his catch is made at the very last instant (between 54 and 55 feet above the ground). For this part of the act, however, Glendo has the cannon charged with less gunpowder, so that the muzzle velocity of the cannon is 56 feet per second. Show work to prove that Glendo's big finale is in fact his easiest catch, and that even if he misses, the big top is never in any danger of collapsing, so insurance should be offered against injury to the audience.

PROJECT 2

The center of Sterlington is the intersection of Main Street (running east–west) and Due North Road (running north–south). The recreation area for the townspeople is Robin Park, a few blocks from there. The park is bounded on the south by Main Street and on every other side by Parabolic Boulevard, named for its distinctive shape. In fact, if Main Street and Due North Road were used as the axes of a rectangular coordinate system, Parabolic Boulevard would have the equation $y = -(x - 4)^2 + 5$, where each unit on the axes is 100 yards.

The city council has recently begun to consider whether or not to put two walkways through the park. (See Illustration 1.) The walkways would run from two points on Main Street and converge at the northernmost point of the park, dividing the area of the park exactly into thirds.

The city council is pleased with the esthetics of this arrangement but needs to know two important facts.

Illustration 1

(continued)

a. For planning purposes, they need to know exactly where on Main Street the walkways would begin.
b. To budget for the construction, they need to know how long the walkways will be.

Provide answers for the city council, along with explanations and work to show that your answers are correct. Use the formula shown in Illustration 2, credited to Archimedes (287–212 B.C.), for the area under a parabola but above a line perpendicular to the axis of symmetry of the parabola.

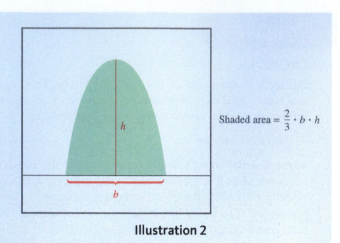

Shaded area $= \frac{2}{3} \cdot b \cdot h$

Illustration 2

Reach for Success

EXTENSION OF ACCESSING ACADEMIC RESOURCES

Now that you know where to go to get mathematics help, let's move on and look at some other resources.

An advisor can discuss your workload and course load to help you find a good balance. An advisor can also explain the consequences of failing a course or falling below a 2.0 GPA.	Do you have any issues for which an advisor would be good resource? If so, list them. _____ _____ _____
Some colleges assign students to a specific academic advisor while others offer "drop in" service to students.	Were you assigned an academic advisor?_____ If so, what is your advisor's name?_____ What is his/her phone number?_____ Where is his/her office? _____ Have you met with your advisor this semester?_____ If not, where is the advising office located?_____ What is the telephone number?_____ Have you met with any advisor this semester?_____

The following may not apply to all students.	
Financial aid officers can provide information on obtaining money for college expenses through scholarships, grants, and loans.	Could you benefit from speaking with a financial aid officer? _____ Write the office phone number here. _____ Where is the office located? _____
The Office of Veteran Affairs is a valuable resource for all students who have served in the armed forces at any time.	Could you benefit from speaking with a Veterans' Affairs officer? _____ Write the office phone number here. _____ Where is the office located? _____
Most colleges offer personal counseling on a short-term basis.	Could you benefit from speaking with a personal counselor? _____ Write the phone number of the counseling office. _____ Where is it located?_____

Taking advantage of one or more of these resources may help guide you on the path to success.

8 Review

SECTION 8.1 Solving Quadratic Equations Using the Square-Root Property and by Completing the Square

DEFINITIONS AND CONCEPTS	EXAMPLES
Square-root property: The equation $x^2 = c$ has two solutions: $x = \sqrt{c}$ and $x = -\sqrt{c}$	To solve $x^2 - 28 = 0$, proceed as follows: $x^2 = 28$ Add 28 to both sides. $x = \sqrt{28}$ or $x = -\sqrt{28}$ Use the square-root property. $x = 2\sqrt{7}$ | $x = -2\sqrt{7}$ $\sqrt{28} = \sqrt{4}\sqrt{7} = 2\sqrt{7}$ The solutions are $2\sqrt{7}$ and $-2\sqrt{7}$, or $\pm 2\sqrt{7}$.

Completing the square:
 	
1. Be certain that the coefficient of x^2 is 1. If not, divide both sides of the equation by the coefficient of x^2.
 	
2. If necessary, move the constant term to one side of the equal sign.
 	
3. Complete the square:
  a. Find one-half of the coefficient of x and square it.
  b. Add the square to both sides of the equation.
 	
4. Factor the trinomial square on one side of the equation and combine like terms on the other side.
 	
5. Solve the resulting equation by using the square-root property.
	To solve $2x^2 - 12x + 24 = 0$, we divide both sides of the equation by 2 to obtain 1 as the coefficient of x^2.
	
	1. $\dfrac{2x^2}{2} - \dfrac{12x}{2} + \dfrac{24}{2} = \dfrac{0}{2}$ Divide both sides by 2.
	
	$x^2 - 6x + 12 = 0$ Simplify.
	
	2. $x^2 - 6x \qquad = -12$ Subtract 12 from both sides.
	
	3. Since $\left[\frac{1}{2}(-6)\right]^2 = (-3)^2 = 9$, we complete the square by adding 9 to both sides.
	
	$x^2 - 6x + 9 = -12 + 9$
	
	4. $(x - 3)^2 = -3$ Factor the left side and combine terms on the right side.
	
	5. $x - 3 = \sqrt{-3}$ or $x - 3 = -\sqrt{-3}$ Use the square-root property.
	$x - 3 = \sqrt{3}i$ | $x - 3 = -\sqrt{3}i$ Simplify $\sqrt{-3}$.
	$x = 3 + \sqrt{3}i$ | $x = 3 - \sqrt{3}i$ Add 3 to both sides.
	
	The solutions are $3 \pm \sqrt{3}i$.

REVIEW EXERCISES

Solve each equation by factoring or by using the square-root property.

1. $12x^2 + x - 6 = 0$ **2.** $6x^2 + 17x + 5 = 0$

3. $15x^2 + 2x - 8 = 0$ **4.** $(x + 3)^2 = 16$

Solve each equation by completing the square.

5. $x^2 + 8x + 12 = 0$

6. $2x^2 - 9x + 7 = 0$

7. $2x^2 - x - 5 = 0$

SECTION 8.2 Solving Quadratic Equations Using the Quadratic Formula

DEFINITIONS AND CONCEPTS	EXAMPLES
Quadratic formula: The solutions of $ax^2 + bx + c = 0$ $(a \neq 0)$ are represented by the formula	To solve $2x^2 - 5x + 4 = 0$, note that in the equation $a = 2$, $b = -5$, and $c = 4$, and substitute these values into the quadratic formula. $x = \dfrac{-b \pm \sqrt{b^2 - 4ac}}{2a}$

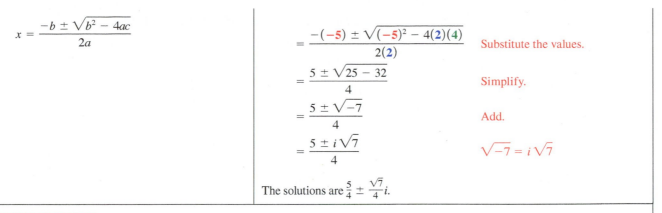

$$x = \frac{-b \pm \sqrt{b^2 - 4ac}}{2a}$$

$$= \frac{-(-5) \pm \sqrt{(-5)^2 - 4(2)(4)}}{2(2)}$$ Substitute the values.

$$= \frac{5 \pm \sqrt{25 - 32}}{4}$$ Simplify.

$$= \frac{5 \pm \sqrt{-7}}{4}$$ Add.

$$= \frac{5 \pm i\sqrt{7}}{4}$$ $\sqrt{-7} = i\sqrt{7}$

The solutions are $\frac{5}{4} \pm \frac{\sqrt{7}}{4}i$.

REVIEW EXERCISES

Solve each equation by using the quadratic formula.

8. $x^2 - 5x - 6 = 0$ **9.** $x^2 = 7x$

10. $2x^2 + 13x - 7 = 0$ **11.** $5x^2 - 2x - 3 = 0$

12. $2x^2 = x + 2$ **13.** $x^2 + x + 2 = 0$

14. Dimensions of a rectangle A rectangle is 2 centimeters longer than it is wide. If both the length and width are doubled, its area is increased by 72 square centimeters. Find the dimensions of the original rectangle.

15. Dimensions of a rectangle A rectangle is 1 foot longer than it is wide. If the length is tripled and the width is doubled, its area is increased by 30 square feet. Find the dimensions of the original rectangle.

16. Ballistics If a rocket is launched straight up into the air with an initial velocity of 112 feet per second, its height after t seconds is given by the formula $h = 112t - 16t^2$, where h represents the height of the rocket in feet. After launch, how long will it be before it hits the ground?

17. Ballistics What is the maximum height of the rocket discussed in Exercise 16?

SECTION 8.3   The Discriminant and Equations That Can Be Written in Quadratic Form

DEFINITIONS AND CONCEPTS	**EXAMPLES**

The discriminant is $b^2 - 4ac$, where a, b, and c are real numbers and $a \neq 0$.
 	
If $b^2 - 4ac > 0$, the equation $ax^2 + bx + c = 0$ has two real unequal solutions.
 	
If $b^2 - 4ac = 0$, the equation $ax^2 + bx + c = 0$ has two real equal solutions (called a *double root*).
 	
If $b^2 - 4ac < 0$, the equation $ax^2 + bx + c = 0$ has two solutions that are complex conjugates.
 	
Further, if a, b, and c are rational numbers and the discriminant:
 	
    is a perfect square greater than 0, there are two unequal rational solutions.
 	
    is positive but not a perfect square, the solutions are irrational and unequal.
	To determine the type of solutions for the equation $x^2 - x + 9 = 0$, we calculate the discriminant.
	
	$b^2 - 4ac = (-1)^2 - 4(1)(9)$ $a = 1, b = -1$, and $c = 9$.
	$\qquad\quad = 1 - 36$
	$\qquad\quad = -35$
	
	Since $b^2 - 4ac < 0$, the two solutions are complex conjugates.
	
	To determine the type of solutions for the equation $2x^2 + 4x - 5 = 0$, we calculate the discriminant.
	
	$b^2 - 4ac = 4^2 - 4(2)(-5)$ $a = 2, b = 4$, and $c = -5$.
	$\qquad\quad = 16 + 40$
	$\qquad\quad = 56$
	
	Since $b^2 - 4ac > 0$, the solutions are unequal and irrational.

Solving equations quadratic in form: Use u substitution to write a quadratic equation.	To solve the equation $x^4 - 3x^2 - 4 = 0$, we can proceed as follows: $x^4 - 3x^2 - 4 = 0$ $(x^2)^2 - 3(x^2) - 4 = 0$ $u^2 - 3u - 4 = 0$ Substitute u for x^2. $(u - 4)(u + 1) = 0$ Factor $u^2 - 5u + 4$. $u - 4 = 0$ or $u + 1 = 0$ Set each factor equal to 0. $u = 4 \qquad\qquad u = -1$

Since $u = x^2$, it follows that $x^2 = 4$ or $x^2 = -1$. Thus,

$$x^2 = 4 \qquad \text{or} \qquad x^2 = -1$$
$$x = 2 \quad \text{or} \quad x = -2 \quad | \quad x = i \quad \text{or} \quad x = -i$$

| **Verifying solutions:** If r_1 and r_2 are solutions of $ax^2 + bx + c = 0$ $(a \neq 0)$, then $$r_1 + r_2 = -\frac{b}{a} \quad \text{and} \quad r_1 r_2 = \frac{c}{a}$$ | To verify that $\frac{5}{3}$ and 2 are the solutions of $3x^2 - 11x + 10 = 0$, make the following calculations: $$-\frac{b}{a} = -\frac{-11}{3} = \frac{11}{3} \quad \text{and} \quad \frac{c}{a} = \frac{10}{3}$$ Since $\frac{5}{3} + 2 = \frac{5}{3} + \frac{6}{3} = \frac{11}{3}$ and $\frac{5}{3}(2) = \frac{5}{3} \cdot \frac{2}{1} = \frac{10}{3}$, $\frac{5}{3}$ and 2 are the solutions of $3x^2 - 11x + 10 = 0$. |

REVIEW EXERCISES

Use the discriminant to determine the types of solutions that exist for each equation.

18. $2x^2 + 5x - 4 = 0$

19. $4x^2 - 5x + 7 = 0$

20. Find the values of k that will make the solutions of $(k - 8)x^2 + (k + 16)x = -49$ equal.

21. Find the values of k such that the solutions of $3x^2 + 4x = k + 1$ will be real numbers.

Solve each equation.

22. $x - 8x^{1/2} + 7 = 0$

23. $a^{2/3} + a^{1/3} - 6 = 0$

24. $\dfrac{1}{x + 1} - \dfrac{1}{x} = -\dfrac{1}{x + 1}$

25. $\dfrac{6}{x + 2} + \dfrac{6}{x + 1} = 5$

26. Find the sum of the solutions of the equation $3x^2 - 14x + 3 = 0$.

27. Find the product of the solutions of the equation $3x^2 - 14x + 3 = 0$.

SECTION 8.4   Graphs of Quadratic Functions

DEFINITIONS AND CONCEPTS	**EXAMPLES**
Graphing quadratic functions: If f is a function and k and h positive numbers, then: The graph of $f(x) + k$ is identical to the graph of $f(x)$, except that it is translated k units upward. The graph of $f(x) - k$ is identical to the graph of $f(x)$, except that it is translated k units downward. The graph of $f(x - h)$ is identical to the graph of $f(x)$, except that it is translated h units to the right. The graph of $f(x + h)$ is identical to the graph of $f(x)$, except that it is translated h units to the left.	Graph each of the following. **a.** $f(x) = 4x^2$ **b.** $f(x) = 4x^2 - 3$ **c.** $f(x) = 4(x - 3)^2$
Finding the vertex of a parabola: If $a \neq 0$, the graph of $y = a(x - h)^2 + k$ is a parabola with vertex at (h, k). It opens upward when $a > 0$ and downward when $a < 0$.	The vertex of the graph of $y = 2(x - 3)^2 - 5$ is $(3, -5)$. Since $2 > 0$, the graph will open upward.

The coordinates of the vertex of the graph of	Find the axis of symmetry and the vertex of the graph of $f(x) = 2x^2 - 4x - 5$.

The coordinates of the vertex of the graph of

$$f(x) = ax^2 + bc + c \quad (a \neq 0)$$

are $\left(-\frac{b}{2a}, f\left(-\frac{b}{2a}\right)\right)$.

The axis of symmetry is the vertical line $x = -\frac{b}{2a}$.

Find the axis of symmetry and the vertex of the graph of $f(x) = 2x^2 - 4x - 5$.

$$x\text{-coordinate} = -\frac{b}{2a} = -\frac{-4}{2(2)} = -\frac{-4}{4} = \frac{4}{4} = 1$$

The axis of symmetry is $x = 1$.

To find the y-coordinate of the vertex, substitute 1 for x in $f(x) = 2x^2 - 4x - 5$.

$$f(x) = 2x^2 - 4x - 5$$
$$f(1) = 2(1)^2 - 4(1) - 5$$
$$= 2 - 4 - 5$$
$$= -7$$

The vertex is $(1, -7)$.

REVIEW EXERCISES

Graph each function and state the coordinates of the vertex of the resulting parabola.

28. $y = 2x^2 - 3$

29. $y = -2x^2 - 1$

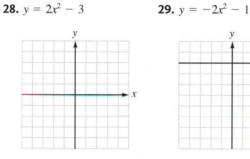

30. $y = -4(x - 2)^2 + 1$

31. $y = 5x^2 + 10x - 1$

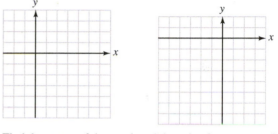

32. Find the vertex of the graph and the axis of symmetry of $f(x) = 4x^2 - 16x - 3$.

SECTION 8.5   Quadratic and Other Nonlinear Inequalities

DEFINITIONS AND CONCEPTS	EXAMPLES

Solving quadratic inequalities:

Method 1 To solve a quadratic inequality in one variable, make a sign chart and determine which intervals are solutions.

Method 1 To solve $x^2 - 2x - 8 > 0$, factor the trinomial to obtain $(x - 4)(x + 2)$ and construct the chart shown. The critical values will occur at 4 and -2.

$x - 4$ is positive when $x > 4$, and is negative when $x < 4$.

$x + 2$ is positive when $x > -2$, and is negative when $x < -2$.

The product of $x - 4$ and $x + 2$ will be greater than 0 when the signs of the binomial factors are the same. This occurs in the intervals $(-\infty, -2)$ and $(4, \infty)$. The numbers -2 and 4 are not included, because equality is not indicated in the original inequality. Thus, the solution set in interval notation is

$$(-\infty, -2) \cup (4, \infty)$$

The graph of the solution set is shown on the number line below.

Method 2 Solve the related quadratic equation and establish critical points on the number line. Use a test number in each interval to determine whether it satisfies the inequality.

Method 2 To solve $x^2 - 2x - 8 > 0$, solve the related quadratic equation to determine the critical values.

$$x^2 - 2x - 8 = 0$$
$$(x - 4)(x + 2) = 0$$
$$x = 4 \quad \text{or} \quad x = -2$$

The critical values are -2 and 4. These create the intervals $(-\infty, -2)$, $(-2, 4)$, and $(4, \infty)$. Use a test value to determine which of the intervals satisfy the inequality.

Interval	Test value	Inequality $x^2 - 2x - 8 > 0$	Result
$(-\infty, -2)$	-3	$(-3)^2 - 2(-3) - 8 > 0$ $7 > 0$ true	The numbers in this interval satisfy the inequality.
$(-2, 4)$	0	$(0)^2 - 2(0) - 8 > 0$ $-8 > 0$ false	The numbers in this interval do not satisfy the inequality.
$(4, \infty)$	5	$(5)^2 - 2(5) - 8 > 0$ $7 > 0$ true	The numbers in this interval satisfy the inequality.

The solution is $(-\infty, -2) \cup (4, \infty)$.

Graphing rational inequalities:

To solve inequalities with rational expressions, set the inequality to 0. Combine the fractions, and then factor the numerator and denominator. Use a sign chart to determine the solution.

For an alternative method, see the section text.

To solve $\frac{1}{x} \geq -4$, add 4 to both sides to set the right side equal to 0 and proceed as follows:

$$\frac{1}{x} + 4 \geq 0$$

$$\frac{1}{x} + \frac{4x}{x} \geq 0 \quad \text{\color{red}Write each fraction with a common denominator.}$$

$$\frac{1 + 4x}{x} \geq 0 \quad \text{\color{red}Add.}$$

Finally, make a sign chart, as shown.

The denominator x is undefined when $x = 0$, positive when $x > 0$, and negative when $x < 0$. The numerator $1 + 4x$ is 0 when $x = -\frac{1}{4}$, positive when $x > -\frac{1}{4}$, and negative when $x < -\frac{1}{4}$. The critical values occur when the numerator is 0 or the denominator is undefined: 0 and $-\frac{1}{4}$.

The fraction $\frac{1 + 4x}{x}$ will be greater than or equal to 0 when the numerator and denominator are the same sign and when the numerator is 0. This occurs in the interval

$$\left(-\infty, -\frac{1}{4}\right] \cup (0, \infty)$$

The graph of this interval is shown.

Graphing nonlinear inequalities in two variables:

To graph a nonlinear inequality, first graph the equation. Then determine which region represents the graph of the inequality and whether to include the boundary.

To graph $x > |y|$, first graph $x = |y|$ as in the illustration below using a dashed line, because equality is not indicated in the original inequality.

Since the origin is on the graph, we cannot use it as a test point. We select another point, such as $(1, 0)$. We substitute 1 for x and 0 for y into the inequality to get

$$\color{blue}x > |y|$$
$$\color{blue}1 > |0|$$
$$1 > 0$$

Since $1 > 0$ is a true statement, the point $(1, 0)$ satisfies the inequality and is part of the graph. Thus, the graph of $x > |y|$ is to the right of the boundary. The complete graph is shown.

REVIEW EXERCISES

Solve each inequality. State each result in interval notation and graph the solution set.

33. $x^2 + 2x - 35 > 0$

34. $x^2 + 7x - 18 < 0$

35. $\frac{3}{x} \leq 5$

36. $\frac{2x^2 + 5x - 7}{x - 4} > 0$

Use a graphing calculator to solve each inequality. Compare the results with Review Exercises 33–36.

Graph each inequality.

37. $x^2 + 2x - 35 > 0$ **38.** $x^2 + 7x - 18 < 0$

41. $y < \frac{1}{2}x^2 - 1$ **42.** $y \geq -|x|$

39. $\frac{3}{x} \leq 5$ **40.** $\frac{2x^2 - x - 28}{x - 1} > 0$

SECTION 8.6   Algebra and Composition of Functions

DEFINITIONS AND CONCEPTS	EXAMPLES
Operations with functions: $(f + g)(x) = f(x) + g(x)$ $(f - g)(x) = f(x) - g(x)$ $(f \cdot g)(x) = f(x)g(x)$ $(f/g)(x) = \dfrac{f(x)}{g(x)} \quad (g(x) \neq 0)$ The domain of each of these functions is the set of real numbers x that are in the domain of both f and g. In the case of the quotient, there is the further restriction that $g(x) \neq 0$.	Given $f(x) = 6x - 2$ and $g(x) = x^2 + 4$, find each function and its domain. **a.** $f + g$ **b.** $f - g$ **c.** $f \cdot g$ **d.** f/g **a.** $f + g = f(x) + g(x)$ $\quad = (6x - 2) + (x^2 + 4)$ $\quad = x^2 + 6x + 2$ Combine like terms. D: $(-\infty, \infty)$ **b.** $f - g = (6x - 2) - (x^2 + 4)$ $\quad = 6x - 2 - x^2 - 4$ Use the distributive property to remove parentheses. $\quad = -x^2 + 6x - 6$ Combine like terms. D: $(-\infty, \infty)$ **c.** $f \cdot g = (6x - 2)(x^2 + 4)$ $\quad = 6x^3 + 24x - 2x^2 - 8$ Multiply. $\quad = 6x^3 - 2x^2 + 24x - 8$ Write exponents in descending order. D: $(-\infty, \infty)$ **d.** $\dfrac{g}{f} = \dfrac{x^2 + 4}{6x - 2}$ D: $\left(-\infty, \frac{1}{3}\right) \cup \left(\frac{1}{3}, \infty\right)$
Composition of functions: $(f \circ g)(x) = f(g(x))$	Let $f(x) = 5x - 4$ and $g(x) = x^2 + 2$. Find **a.** $(f \circ g)(x)$ **b.** $(g \circ f)(x)$ **a.** $(f \circ g)(x)$ means $f(g(x))$. $\quad f(g(x)) = f(x^2 + 2)$ $\qquad = 5(x^2 + 2) - 4$ Substitute $(x^2 + 2)$ for x. $\qquad = 5x^2 + 10 - 4$ Multiply. $\qquad = 5x^2 + 6$ Add.

b. $(g \circ f)(x)$ means $g(f(x))$.

$$g(f(x)) = g(5x - 4)$$

$$= (5x - 4)^2 + 2 \qquad \text{Substitute } (5x - 4) \text{ for } x.$$

$$= 25x^2 - 40x + 16 + 2 \qquad \text{Square the binomial.}$$

$$= 25x^2 - 40x + 18 \qquad \text{Add.}$$

The difference quotient:

The difference quotient is defined as follows:

$$\frac{f(x + h) - f(x)}{h}, \quad h \neq 0$$

To evaluate the difference quotient for $f(x) = x^2 + 2x - 6$, first evaluate $f(x + h)$.

$$f(x) = x^2 + 2x - 6$$

$$f(x + h) = (x + h)^2 + 2(x + h) - 6 \qquad \text{Substitute } (x + h) \text{ for } h.$$

$$= x^2 + 2xh + h^2 + 2x + 2h - 6 \qquad (x + h)^2 = x^2 + 2hx + h^2$$

Now substitute the values of $f(x + h)$ and $f(x)$ into the difference quotient and simplify.

$$\frac{f(x + h) - f(x)}{h}$$

$$= \frac{(x^2 + 2xh + h^2 + 2x + 2h - 6) - (x^2 + 2x - 6)}{h}$$

$$= \frac{x^2 + 2xh + h^2 + 2x + 2h - 6 - x^2 - 2x + 6}{h} \qquad \text{Use the distributive property to remove parentheses.}$$

$$= \frac{2xh + h^2 + 2h}{h} \qquad \text{Combine like terms.}$$

$$= \frac{h(2x + h + 2)}{h} \qquad \text{Factor out } h, \text{ the GCF, in the numerator.}$$

$$= 2x + h + 2 \qquad \text{Divide out } h.$$

The difference quotient for this function is $2x + h + 2$.

REVIEW EXERCISES

Let $f(x) = 2x$ and $g(x) = x + 1$. Find each function or value.

For 43–46 only, find the domain.

43. $f + g$

44. $f - g$

45. $f \cdot g$

46. f/g

47. $(f \circ g)(3)$

48. $(g \circ f)(-2)$

49. $(f \circ g)(x)$

50. $(g \circ f)(x)$

SECTION 8.7   Inverses of Functions

DEFINITIONS AND CONCEPTS

EXAMPLES

Horizontal line test:

If every horizontal line that intersects the graph of a function does so only once, the function is one-to-one.

Determine whether the graph of each function is one-to-one.

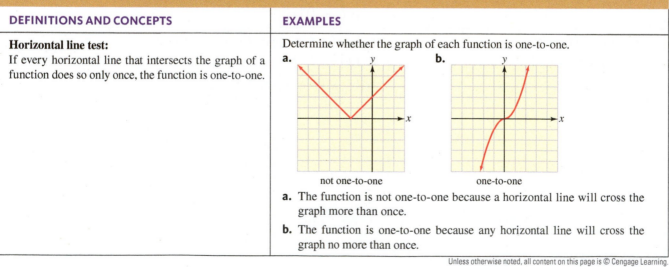

a.   not one-to-one   b.   one-to-one

a. The function is not one-to-one because a horizontal line will cross the graph more than once.

b. The function is one-to-one because any horizontal line will cross the graph no more than once.

<table>
<tr><td>

Finding the inverse of a one-to-one function:
We find the inverse of a function as follows:

1. Replace $f(x)$ with y.

2. Interchange the variables x and y.

3. Solve the resulting equation for y. If the inverse is also a function, use the notation $f^{-1}(x)$.

</td><td>

To find the inverse of $f(x) = x^3 - 5$, we proceed as follows:

$$y = x^3 - 5 \quad \text{Replace } f(x) \text{ with } y.$$
$$x = y^3 - 5 \quad \text{Interchange the variables } x \text{ and } y.$$
$$x + 5 = y^3 \quad \text{Add 5 to both sides.}$$
$$\sqrt[3]{x + 5} = y \quad \text{Take the cube root of both sides.}$$

Because there corresponds one real cube root for each x, $y = \sqrt[3]{x + 5}$ represents a function. In function notation, we describe the inverse as

$$f^{-1}(x) = \sqrt[3]{x + 5}$$

</td></tr>
</table>

REVIEW EXERCISES

Graph each function and use the horizontal line test to decide whether the function is one-to-one.

51. $f(x) = 2(x - 3)$ **52.** $f(x) = x(2x - 3)$

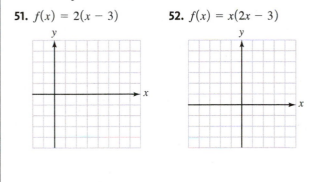

53. $f(x) = -3(x - 2)^2 + 5$ **54.** $f(x) = |x|$

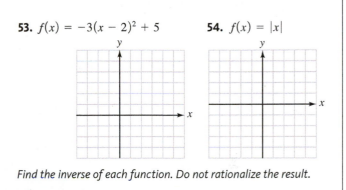

Find the inverse of each function. Do not rationalize the result.

55. $f(x) = 7x - 2$

56. $f(x) = 4x + 5$

57. $y = 2x^2 - 1 \quad (x \geq 0)$

---

8   Test

Solve each equation by factoring.

1. $x^2 - 3x - 28 = 0$ **2.** $x(6x + 19) = -15$

Solve each equation using the square root method.

3. $x^2 - 144 = 0$ **4.** $(x - 5)^2 = -6$

Solve each equation by completing the square.

5. $x^2 + 6x + 7 = 0$ **6.** $x^2 - 5x - 3 = 0$

Solve each equation by the quadratic formula.

7. $2x^2 + 5x + 1 = 0$ **8.** $x^2 - 2x + 6 = 0$

9. Determine whether the solutions of $4x^2 + 3x + 10 = 0$ are real or nonreal numbers.

10. For what value(s) of k are the solutions of $4x^2 - 2kx + k - 1 = 0$ equal?

11. One leg of a right triangle is 14 inches longer than the other, and the hypotenuse is 26 inches. Find the length of the shorter leg.

12. Solve: $2y - 3y^{1/2} + 1 = 0$

13. Graph $f(x) = \frac{1}{2}x^2 - 4$ and state the coordinates of its vertex.

14. Find the vertex and axis of symmetry of the graph of $f(x) = -3x^2 + 12x - 5$.

15. Graph: $y \le -x^2 + 3$

Solve each inequality and graph the solution set.

16. $x^2 - 2x - 8 > 0$ **17.** $\dfrac{x - 2}{x + 3} \le 0$

Let $f(x) = 4x$ and $g(x) = x - 1$. Find each function and its domain.

18. $g + f$

19. $f - g$

20. $g \cdot f$

21. g/f

Let $f(x) = 4x$ and $g(x) = x - 1$. Find each value.

22. $(g \circ f)(1)$ **23.** $(f \circ g)(0)$

24. $(f \circ g)(-1)$ **25.** $(g \circ f)(-2)$

Let $f(x) = 4x$ and $g(x) = x - 1$. Find each function.

26. $(f \circ g)(x)$ **27.** $(g \circ f)(x)$

Find the inverse of the function.

28. $3x + 2y = 12$

∽ Cumulative Review ∾

Find the domain and range of each function.

1. $f(x) = 5x^2 - 2$

2. $f(x) = -|x - 4|$

Write the equation of the line with the given properties.

3. $m = 4$, passing through $(-3, -5)$

4. parallel to the graph of $2x + 3y = 6$ and passing through $(0, -2)$

Perform each operation.

5. $(2a^2 + 4a - 7) - 2(3a^2 - 4a)$

6. $(4x - 3)(5x + 2)$

Factor each expression completely using only integers.

7. $x^4 - 16y^4$

8. $15x^2 - 2x - 8$

Solve each equation.

9. $x^2 - 8x - 9 = 0$ **10.** $6a^3 - 2a = a^2$

Simplify each expression. Assume that all variables represent positive numbers.

11. $\sqrt{36a^2b^4}$ **12.** $\sqrt{48t^3}$

13. $\sqrt[3]{-64y^6}$ **14.** $\sqrt[3]{\dfrac{128x^4}{2x}}$

15. $8^{-1/3}$ **16.** $27^{2/3}$

17. $\dfrac{y^{2/3} \, y^{5/3}}{y^{1/3}}$ **18.** $\dfrac{x^{5/3} \, x^{1/2}}{x^{3/4}}$

Graph each function and state the domain and the range.

19. $f(x) = \sqrt{x - 2}$ **20.** $f(x) = -\sqrt{x + 2}$

Perform the operations.

21. $(x^{2/3} - x^{1/3})(x^{2/3} + x^{1/3})$

22. $(x^{-1/2} + x^{1/2})^2$

Simplify each statement. Assume no division by 0.

23. $\sqrt{50} - \sqrt{8} + \sqrt{32}$

24. $-3\sqrt[4]{32} - 2\sqrt[4]{162} + 5\sqrt[4]{48}$

25. $3\sqrt{2}(2\sqrt{3} - 4\sqrt{12})$

26. $\dfrac{5}{\sqrt[3]{x}}$

27. $\dfrac{\sqrt{x} + 2}{\sqrt{x} - 1}$ **28.** $\sqrt[6]{x^3 \, y^3}$

Solve each equation.

29. $5\sqrt{x + 2} = x + 8$ **30.** $\sqrt{x} + \sqrt{x + 2} = 2$

31. Find the length of the hypotenuse of the right triangle shown in Illustration 1.

32. Find the length of the hypotenuse of the right triangle shown in Illustration 2.

Illustration 1   **Illustration 2**

33. Find the distance between $(-2, 6)$ and $(4, 14)$.

34. Solve $x^2 + 25 = 0$ using the square root method.

35. Use the method of completing the square to solve $2x^2 + x - 3 = 0$.

36. Use the quadratic formula to solve $3x^2 + 4x - 1 = 0$.

37. Graph $f(x) = \frac{1}{2}x^2 + 5$. State the coordinates of its vertex and identify the axis of symmetry.

38. Graph $f(x) \leq -x^2 + 3$ and find the coordinates of its vertex.

Write each expression as a real number or as a complex number in a + bi form.

39. $(5 + 7i) + (8 - 10i)$ **40.** $(7 - 4i) - (12 + 3i)$

41. $(6 + 5i)(6 - 5i)$ **42.** $(3 + i)(3 - 3i)$

43. $(3 - 2i) - (4 + i)^2$ **44.** $\dfrac{5}{3 - i}$

45. $|3 + 2i|$ **46.** $|5 - 6i|$

47. For what values of k will the solutions of $2x^2 + 4x = k$ be equal?

48. Solve: $a - 7a^{1/2} + 12 = 0$

Solve each inequality and graph the solution set on the number line.

49. $x^2 - x - 6 > 0$ **50.** $\dfrac{x - 3}{x + 2} \leq 0$

Let $f(x) = 3x^2 + 2$ and $g(x) = 2x - 1$. Find each value or composite function.

51. $f(-2)$ **52.** $(g \circ f)(4)$

53. $(f \circ g)(x)$ **54.** $(g \circ f)(x)$

Find the inverse of each function.

55. $f(x) = 3x + 2$ **56.** $f(x) = x^3 + 4$

Exponential and Logarithmic Functions

© Fotokost:c/Shutterstock.com

Careers and Mathematics

PEST-CONTROL WORKERS

Few people welcome roaches, rats, mice, spiders, termites, fleas, ants, and bees into their homes. It is the job of pest-control workers to locate, identify, destroy, control, and repel these pests.

Roughly 86 percent of pest control workers are employed in services related to the building industry, primarily in states with warmer climates. About 7 percent are self-employed.

Both federal and state laws require pest-control workers to be certified.

Job Outlook:
Employment of pest control workers is expected to grow 26% between now and 2020, which is faster than the average for all occupations.

Yearly Earnings:
$20,340–$46,930

For More Information:
http://www.bls.gov/oco/ocos254.htm

For a Sample Application:
See Problem 89 in Section 9.6.

REACH FOR SUCCESS

9.1 Exponential Functions
9.2 Base-*e* Exponential Functions
9.3 Logarithmic Functions
9.4 Natural Logarithms
9.5 Properties of Logarithms
9.6 Exponential and Logarithmic Equations
■ *Projects*
REACH FOR SUCCESS EXTENSION
CHAPTER REVIEW
CHAPTER TEST

In this chapter

In this chapter, we will discuss two functions that are important in many applications of mathematics. Exponential functions are used to compute compound interest, find radioactive decay, and model population growth. Logarithmic functions are used to measure acidity of solutions, drug dosage, gain of an amplifier, magnitude of earthquakes, and safe noise levels in factories.

Reach for Success   Taking Notes in Class

You're trying to take notes in class but your instructor is talking faster than you can write. You need to know what is important and how you can reflect this in your notes.

When reading, most students use a highlighter to identify what is important. In a mathematics class, consider using two different colored inks to clarify the procedures.

A professor speaks an average of 150 words per minute but you can only write at a rate of 35 words per minute and can think at a much faster rate of 350 words per minute.

What strategies can you use to take effective notes and help you stay focused during class?

1. _____

2. _____

In your reading-intensive courses, notes are generally presented from left to right, whereas in mathematics, problems can be worked vertically. This can pose a challenge in identifying what is important in each problem.

We use a red color change to emphasize any processes we use to complete the exercise. We use black to show the results of the processes.

In each blank at the right, write why you think a different color is used.

Solve the equation:

$$5x - 9 = 3(x + 3)$$
$$5x - 9 = 3(x + 3)$$
$$5x - 9 = 3x + 9$$
$$\underline{\quad -3x \quad -3x \quad}$$
$$2x - 9 = 9$$
$$\underline{\quad +9 \quad +9 \quad}$$
$$\frac{2x}{2} = \frac{18}{2}$$
$$x = 9$$

Read the Instructions
Original example

_____
(Why the red arrows?)

_____
(Why the red − 3x?)

_____
(Why the red + 9?)

_____
(Why the red division by 2?)

Try this problem. Solve the equation using a pencil and a red pen appropriately.

The steps have been identified to help you.

$$6x + 3 = -5(x - 2)$$
$$6x + 3 = -5(x - 2)$$

_____

_____

_____

_____

_____

Original example

Insert red arrows where necessary.

Write the result of distributing.

Add 5x to each side of the equation. (Show your work in red.)

Write the result of the addition.

Subtract 3 from each side of the equation. (Show your work in red.)

Write the result of the subtraction.

Divide each term by 11. (Show your work in red.)

Write the result of the division.

A Successful Study Strategy . . .

Use at least two colors of pens when taking notes in your mathematics class. Reviewing your notes is more effective because they have been "highlighted."

At the end of the chapter you will find an additional exercise to guide you in planning for a successful college experience.

Section
9.1

Exponential Functions

Objectives

1. Graph an exponential function.
2. Graph a translation of an exponential function.
3. Evaluate an application containing an exponential function.

Vocabulary

exponential function
increasing function
decreasing function

compound interest
present value
future value

periodic interest rate
compounding period

Getting Ready

Find each value.

1. 2^3 2. $25^{1/2}$ 3. 5^{-2} 4. $\left(\dfrac{3}{2}\right)^{-3}$

The graph in Figure 9-1 shows the balance in an investment account in which $10,000 was invested in 2000 at 9% annual interest, compounded monthly. The graph shows that in the year 2010, the value of the account was $25,000, and in the year 2030, the value will be approximately $147,000. The curve shown in Figure 9-1 is the graph of a function called an *exponential function*, the topic of this section.

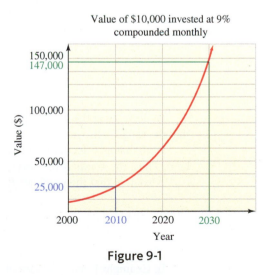

Value of $10,000 invested at 9% compounded monthly

Figure 9-1

1   Graph an exponential function.

If $b > 0$ and $b \neq 1$, the function $f(x) = b^x$ is called an **exponential function**. Since x can be any real number, its domain is the set of real numbers. This is the interval $(-\infty, \infty)$. Since b is positive, the value of $f(x)$ is positive and the range is the set of positive numbers. This is the interval $(0, \infty)$.

Since $b \neq 1$, an exponential function cannot be the constant function $f(x) = 1^x$, in which $f(x) = 1$ for every real number x.

EXPONENTIAL FUNCTIONS	An exponential function with base b is defined by the equation
	$f(x) = b^x \ (b > 0, b \neq 1, \text{and } x \text{ is a real number})$

The domain of any exponential function is the interval $(-\infty, \infty)$. The range is the interval $(0, \infty)$.

Since the domain and range of $f(x) = b^x$ are subsets of the real numbers, we can graph exponential functions on a rectangular coordinate system.

EXAMPLE 1   Graph: $f(x) = 2^x$

Solution   To graph $f(x) = 2^x$, we find several points (x, y) whose coordinates satisfy the equation, plot the points, and join them with a smooth curve, as shown in Figure 9-2.

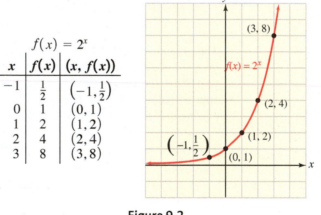

$f(x) = 2^x$

x	$f(x)$	$(x, f(x))$
-1	$\frac{1}{2}$	$\left(-1, \frac{1}{2}\right)$
0	1	$(0, 1)$
1	2	$(1, 2)$
2	4	$(2, 4)$
3	8	$(3, 8)$

Figure 9-2

By looking at the graph, we can verify that the domain is the interval $(-\infty, \infty)$ and that the range is the interval $(0, \infty)$.

Note that as x decreases, the values of $f(x)$ decrease and approach 0, but will never be 0. Thus, the x-axis is the horizontal asymptote of the graph.

Also note that the graph of $f(x) = 2^x$ passes through the points $(0, 1)$ and $(1, 2)$.

SELF CHECK 1   Graph: $f(x) = 4^x$

In Example 1, as the values of x increase the values of $f(x)$ increase. When the graph of a function rises as we move to the right, we call the function an **increasing function**. When $b > 1$, the larger the value of b, the steeper the curve.

EXAMPLE 2   Graph: $f(x) = \left(\frac{1}{2}\right)^x$

Solution   We find and plot pairs (x, y) that satisfy the equation. The graph of $f(x) = \left(\frac{1}{2}\right)^x$ appears in Figure 9-3.

$f(x) = \left(\frac{1}{2}\right)^x$

x	$f(x)$	$(x, f(x))$
-2	4	$(-2, 4)$
-1	2	$(-1, 2)$
0	1	$(0, 1)$
1	$\frac{1}{2}$	$\left(1, \frac{1}{2}\right)$

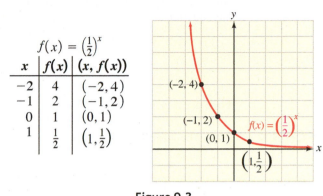

Figure 9-3

By looking at the graph, we can see that the domain is the interval $(-\infty, \infty)$ and that the range is the interval $(0, \infty)$.

In this case, as x increases, the values of $f(x)$ decrease and approach 0. The x-axis is a horizontal asymptote. Note that the graph of $f(x) = \left(\frac{1}{2}\right)^x$ passes through the points $(0, 1)$ and $\left(1, \frac{1}{2}\right)$.

SELF CHECK 2   Graph: $f(x) = \left(\frac{1}{4}\right)^x$

In Example 2, as the values of x increase the values of $f(x)$ decrease. When the graph of a function drops as we move to the right, we call the function a **decreasing function**. When $0 < b < 1$, the smaller the value of b, the steeper the curve.

Examples 1 and 2 illustrate the following properties of exponential functions.

PROPERTIES OF EXPONENTIAL FUNCTIONS

The domain of the exponential function $f(x) = b^x$ is the interval $(-\infty, \infty)$.

The range is the interval $(0, \infty)$.

The graph has a y-intercept of $(0, 1)$.

The x-axis is an asymptote of the graph.

The graph of $f(x) = b^x$ passes through the point $(1, b)$.

If $b > 1$, then $f(x) = b^x$ is an increasing function.

Increasing function

If $0 < b < 1$, then $f(x) = b^x$ is a decreasing function.

Decreasing function

EXAMPLE 3 From the graph of $f(x) = b^x$ shown in Figure 9-4, find the value of b.

Solution We first note that the graph passes through $(0, 1)$. Since the point $(2, 9)$ is on the graph, we substitute 9 for y and 2 for x in the equation $y = b^x$ to obtain

$$y = b^x$$
$$9 = b^2$$
$$3 = b \qquad \text{Because 3 is the positive number whose square is 9}$$

The base b is 3.

Figure 9-4

SELF CHECK 3 From the graph of $f(x) = b^x$ shown to the right, find the value of b.

$(0, 1)$ $\left(1, \dfrac{1}{4}\right)$

$f(x) = \left(\dfrac{1}{4}\right)^x$

Accent
on technology

▶ Graphing One-To-One Exponential Functions

To use a graphing calculator to graph $f(x) = \left(\dfrac{2}{3}\right)^x$ and $f(x) = \left(\dfrac{3}{2}\right)^x$, we enter the right sides of the equations. The screen will show the following equations.

$$\backslash Y_1 = (2/3)^\wedge X$$
$$\backslash Y_2 = (3/2)^\wedge X$$

If we use window settings of $[-10, 10]$ for x and $[-2, 10]$ for y and press **GRAPH**, we will obtain the graph shown in Figure 9-5.

$f(x) = \left(\dfrac{3}{2}\right)^x$ $f(x) = \left(\dfrac{2}{3}\right)^x$

Figure 9-5

The graph of $f(x) = \left(\dfrac{2}{3}\right)^x$ passes through the points $(0, 1)$ and $\left(1, \dfrac{2}{3}\right)$. Since $\dfrac{2}{3} < 1$, the function is decreasing.

The graph of $f(x) = \left(\dfrac{3}{2}\right)^x$ passes through the points $(0, 1)$ and $\left(1, \dfrac{3}{2}\right)$. Since $\dfrac{3}{2} > 1$, the function is increasing.

Since both graphs pass the horizontal line test, each function is one-to-one.

For instructions regarding the use of a Casio graphing calculator, please refer to the Casio Keystroke Guide in the back of the book.

2 Graph a translation of an exponential function.

We have seen that when $k > 0$ the graph of

$y = f(x) + k$ is the graph of $y = f(x)$ translated k units upward.

$y = f(x) - k$ is the graph of $y = f(x)$ translated k units downward.

$y = f(x - k)$ is the graph of $y = f(x)$ translated k units to the right.

$y = f(x + k)$ is the graph of $y = f(x)$ translated k units to the left.

EXAMPLE 4 On one set of axes, graph $f(x) = 2^x$ and $f(x) = 2^x + 3$.

Solution The graph of $f(x) = 2^x + 3$ is identical to the graph of $f(x) = 2^x$, except that it is translated 3 units upward. (See Figure 9-6.)

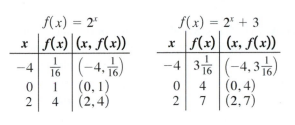

$f(x) = 2^x$

x	$f(x)$	$(x, f(x))$
-4	$\frac{1}{16}$	$\left(-4, \frac{1}{16}\right)$
0	1	$(0, 1)$
2	4	$(2, 4)$

$f(x) = 2^x + 3$

x	$f(x)$	$(x, f(x))$
-4	$3\frac{1}{16}$	$\left(-4, 3\frac{1}{16}\right)$
0	4	$(0, 4)$
2	7	$(2, 7)$

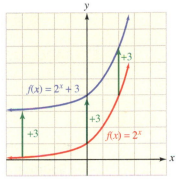

Figure 9-6

SELF CHECK 4 On one set of axes, graph $f(x) = 4^x$ and $f(x) = 4^x - 3$.

EXAMPLE 5 On one set of axes, graph $f(x) = 2^x$ and $f(x) = 2^{(x+3)}$.

Solution The graph of $f(x) = 2^{(x+3)}$ is identical to the graph of $f(x) = 2^x$, except that it is translated 3 units to the left. (See Figure 9-7.)

$f(x) = 2^x$

x	$f(x)$	$(x, f(x))$
0	1	$(0, 1)$
2	4	$(2, 4)$
4	16	$(4, 16)$

$f(x) = 2^{(x+3)}$

x	$f(x)$	$(x, f(x))$
0	8	$(0, 8)$
-1	4	$(-1, 4)$
-2	2	$(-2, 2)$

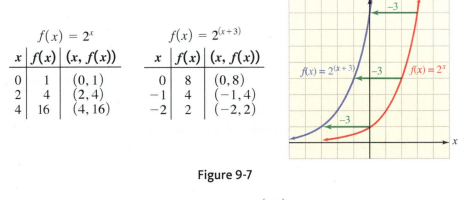

Figure 9-7

SELF CHECK 5 On one set of axes, graph $f(x) = 4^x$ and $f(x) = 4^{(x-3)}$.

The graphs of $f(x) = kb^x$ and $f(x) = b^{kx}$ are vertical and horizontal stretchings, respectively, of the graph of $f(x) = b^x$. To graph these functions, we can plot several points and join them with a smooth curve or use a graphing calculator.

To use a TI-84 graphing calculator to graph the exponential function $f(x) = 3(2^{x/3})$, we enter the right side of the equation. The display will show the equation

$\backslash Y_1 = 3(2^\wedge(X/3))$

If we use window settings of $[-10, 10]$ for x and $[-2, 18]$ for y and press **GRAPH**, we will obtain the graph shown in Figure 9-8.

Figure 9-8

For instructions regarding the use of a Casio graphing calculator, please refer to the Casio Keystroke Guide in the back of the book.

3   Evaluate an application containing an exponential function.

EXAMPLE 6   **CELL PHONE GROWTH**   The worldwide cellular telephone industry continues to experience exponential growth. The exponential function $S(n) = 1.6(1.173)^n$ approximates the number of cellular telephone subscribers in billions from 2004 to 2011, where n is the number of years since 2004.

a. How many subscribers were there in 2004?

b. How many subscribers were there in 2011?

Solution   **a.** To find the number of subscribers in 2004, we substitute 0 for n in the function and find $S(0)$.

$$S(\mathbf{n}) = 1.6(1.173)^n$$
$$S(\mathbf{0}) = 1.6(1.173)^{\mathbf{0}} \qquad \text{Substitute 0 for } n.$$
$$= 1.6 \cdot 1 \qquad\qquad (1.173)^0 = 1$$
$$= 1.6$$

In 2004, there were approximately 1.6 billion cellular telephone subscribers in the world.

b. To find the number of subscribers in 2011, we substitute 7 for n in the function and find $S(7)$.

$$S(\mathbf{n}) = 1.6(1.173)^n$$
$$S(\mathbf{7}) = 1.6(1.173)^{\mathbf{7}} \qquad \text{Substitute 7 for } n.$$
$$\approx 4.88884274 \qquad \text{Use a calculator to find an approximation.}$$

In 2011, there were approximately 4.89 billion cellular telephone subscribers.

 SELF CHECK 6   If the cellular telephone industry continues to grow at the same rate, approximate the number of subscribers in 2020.

Calculating compound interest is another application that involves an exponential function.

If we deposit P in an account paying an annual simple interest rate r, we can find the amount A in the account at the end of t years by using the formula $A = P + Prt$ or $A = P(1 + rt)$.

Suppose that we deposit $500 in an account that pays interest every six months. Then $P = 500$, and after six months $\left(\frac{1}{2} \text{ year}\right)$, the amount in the account will be

$$A = 500(1 + rt)$$

$$= 500\left(1 + r \cdot \frac{1}{2}\right) \quad \text{Substitute } \tfrac{1}{2} \text{ for } t.$$

$$= 500\left(1 + \frac{r}{2}\right)$$

The account will begin the second six-month period with a value of $\$500\left(1 + \frac{r}{2}\right)$. After the second six-month period, the amount will be

$$A = P(1 + rt)$$

$$A = \left[500\left(1 + \frac{r}{2}\right)\right]\left(1 + r \cdot \frac{1}{2}\right) \quad \text{Substitute } 500\left(1 + \tfrac{r}{2}\right) \text{ for } P \text{ and } \tfrac{1}{2} \text{ for } t.$$

$$= 500\left(1 + \frac{r}{2}\right)\left(1 + \frac{r}{2}\right)$$

$$= 500\left(1 + \frac{r}{2}\right)^2$$

At the end of the third six-month period, the amount in the account will be

$$A = 500\left(1 + \frac{r}{2}\right)^3$$

In this discussion, the earned interest is deposited back in the account and also earns interest. When this is the case, we say that the account is earning **compound interest**.

FORMULA FOR COMPOUND INTEREST

If $\$P$ is deposited in an account and interest is paid k times a year at an annual rate r, the amount A in the account after t years is given by

$$A = P\left(1 + \frac{r}{k}\right)^{kt}$$

EXAMPLE 7   **SAVING FOR COLLEGE**   To save for college, parents invest $12,000 for their newborn child in a mutual fund that should average a 10% annual return. If the quarterly interest is reinvested, how much will be available in 18 years?

Solution   We substitute 12,000 for P, 0.10 for r, and 18 for t into the formula for compound interest and calculate A. Since interest is paid quarterly, $k = 4$.

$$A = P\left(1 + \frac{r}{k}\right)^{kt}$$

$$A = 12{,}000\left(1 + \frac{0.10}{4}\right)^{4(18)}$$

$$= 12{,}000(1 + 0.025)^{72}$$

$$= 12{,}000(1.025)^{72}$$

$$\approx 71{,}006.74 \qquad \text{Use a calculator.}$$

In 18 years, the account should be worth $71,006.74.

SELF CHECK 7   How much would be available if the parents invest $20,000?

In business applications, the initial amount of money deposited is the **present value** (PV). The amount to which the money will grow is called the **future value** (FV).

The interest rate used for each compounding period is the **periodic interest rate** (i), and the number of times interest is compounded is the number of **compounding periods** (n). Using these definitions, we have an alternative formula for compound interest.

FORMULA FOR FUTURE VALUE	$FV = PV(1 + i)^n$ where $i = \dfrac{r}{k}$ and $n = kt$

This formula appears on business calculators. To use this formula to solve Example 7, we proceed as follows:

$$FV = \mathbf{PV}(1 + \mathbf{i})^n$$
$$FV = \mathbf{12{,}000}(1 + \mathbf{0.025})^{72} \qquad i = \tfrac{0.10}{4} = 0.025 \text{ and } n = 4(18) = 72$$
$$\approx \$71{,}006.74$$

Accent on technology

▸ Solving Investment Problems

Suppose \$1 is deposited in an account earning 6% annual interest, compounded monthly. To use a graphing calculator to estimate how much money will be in the account in 100 years, we can substitute 1 for P, 0.06 for r, and 12 for k into the formula

$$A = \mathbf{P}\left(1 + \frac{\mathbf{r}}{\mathbf{k}}\right)^{kt}$$

$$A = \mathbf{1}\left(1 + \frac{\mathbf{0.06}}{\mathbf{12}}\right)^{12t}$$

and simplify to obtain

$$A = (1.005)^{12t}$$

We now graph $A = (1.005)^{12t}$ using a TI-84 calculator with window settings of $[0, 120]$ for t and $[0, 400]$ for A to obtain the graph shown in Figure 9-9. From the graph, we can see that the money grows slowly in the early years and rapidly in the later years.

We can use the VALUE feature under the CALC menu to determine that exactly \$397.44 will be in the account in 100 years.

For instructions regarding the use of a Casio graphing calculator, please refer to the Casio Keystroke Guide in the back of the book.

Figure 9-9

SELF CHECK ANSWERS

4.   5.   **6.** ≈ 20.55 billion   **7.** $118,344.56

NOW TRY THIS

If you were given $1 on May 1, $2 on May 2, $4 on May 3, $8 on May 4, and double the previous day's earnings each day after, how much would you earn on

a. May 10?
b. May 15?
c. May 21?

Try to find an equation to model this situation.

9.1 Exercises

WARM-UPS *If x = 2, evaluate each expression.*

1. 3^x

2. 7^x

3. $3(6^x)$

4. $5^{(x-1)}$

If x = −2, evaluate each expression.

5. 3^x

6. 7^x

7. $3(6^x)$

8. $5^{(x-1)}$

REVIEW *In the illustration, lines r and s are parallel.*

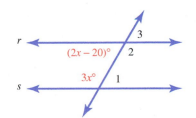

9. Find the value of x.
10. Find the measure of ∠1.
11. Find the measure of ∠2.
12. Find the measure of ∠3.

VOCABULARY AND CONCEPTS *Fill in the blanks.*

13. If $b > 0$ and $b \neq 1$, $f(x) = b^x$ is called an _____ function.
14. The _____ of an exponential function is $(-\infty, \infty)$.
15. The range of an exponential function is the interval _____.

16. The graph of $f(x) = 3^x$ passes through the points $(0, \_)$ and $(1, \_)$.
17. If $b > 1$, then $f(x) = b^x$ is an _____ function.
18. If $0 < b < 1$, then $f(x) = b^x$ is a _____ function.
19. A formula for compound interest is $A = $ _____.
20. An alternative formula for compound interest is $FV = $ _____, where PV stands for present value, i stands for the _____, and n stands for the number of _____.
21. The graph of $f(x) = b^x$ passes through the point $(1, \_)$.
22. The _ axis is an asymptote of the graph $f(x) = b^x$.
23. $f(x) = b^{(x+1)}$ is the graph of $f(x) = b^x$ translated 1 unit to the ___.
24. $f(x) = b^x + 1$ is the graph of $f(x) = b^x$ translated 1 unit _____.

GUIDED PRACTICE *Graph each exponential function. Check your work with a graphing calculator.* SEE EXAMPLE 1. (OBJECTIVE 1)

25. $f(x) = 3^x$

26. $f(x) = 5^x$

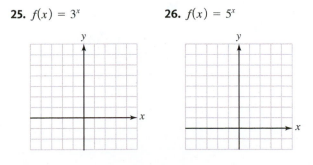

27. $f(x) = \left(\frac{5}{2}\right)^x$ **28.** $f(x) = \left(\frac{4}{3}\right)^x$

 Graph each exponential function. Check your work with a graphing calculator. *SEE EXAMPLE 2. (OBJECTIVE 1)*

29. $f(x) = \left(\frac{1}{3}\right)^x$ **30.** $f(x) = \left(\frac{1}{5}\right)^x$

31. $f(x) = \left(\frac{2}{5}\right)^x$ **32.** $f(x) = \left(\frac{3}{4}\right)^x$

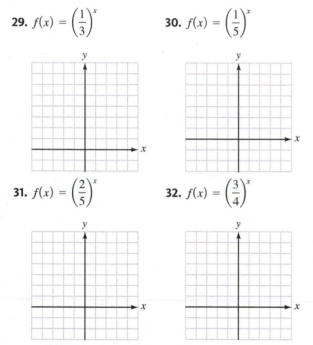

Find the value of b that would cause the graph of $f(x) = b^x$ to look like the graph indicated. *SEE EXAMPLE 3. (OBJECTIVE 1)*

33.   **34.**

35.   **36.**

 Graph each exponential function. Check your work with a graphing calculator. *SEE EXAMPLES 4–5. (OBJECTIVE 2)*

37. $f(x) = 3^x - 2$ **38.** $f(x) = 2^x + 1$

39. $f(x) = 3^{(x-1)}$ **40.** $f(x) = 2^{(x+1)}$

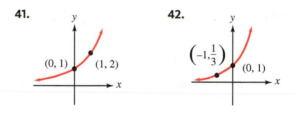

ADDITIONAL PRACTICE *Find the value of b that would cause the graph of $f(x) = b^x$ to look like the graph indicated.*

41.   **42.**

Use a graphing calculator to graph each function. Determine whether the function is an increasing or a decreasing function.

43. $f(x) = \frac{1}{2}\left(3^{x/2}\right)$ **44.** $f(x) = -3\left(2^{x/3}\right)$

45. $f(x) = 2\left(3^{-x/2}\right)$ **46.** $f(x) = -\frac{1}{4}\left(2^{-x/2}\right)$

APPLICATIONS *Evaluate each application.* SEE EXAMPLES 6–7. (OBJECTIVE 3)

47. Cell phone usage   Refer to Example 6 and state the number of cell phone users in 2005.

48. Cell phone usage   Refer to Example 6 and state the number of cell phone users in 2010.

49. Radioactive decay   A radioactive material decays according to the formula $A = A_0\left(\frac{2}{3}\right)^t$, where A_0 is the initial amount present and t is measured in years. Find an expression for the amount present in 8 years.

50. Bacteria cultures   A colony of 6 million bacteria is growing in a culture medium. The population P after t hours is given by the formula $P = (6 \times 10^6)(2.3)^t$. Find the population after 4 hours.

Assume that there are no deposits or withdrawals.

51. Compound interest   An initial deposit of $10,000 earns 8% interest, compounded quarterly. How much will be in the account after 10 years?

52. Compound interest   An initial deposit of $10,000 earns 8% interest, compounded monthly. How much will be in the account after 10 years?

53. Comparing interest rates   How much more interest could $100,000 earn in 10 years, compounded quarterly, if the annual interest rate were $4\frac{1}{2}$% instead of 4%?

54. Comparing savings plans   Which institution in the two ads provides the better investment?

> *Fidelity Savings & Loan*
> Earn 5.25%
> compounded monthly

> **Union Trust**
> Money Market Account
> paying 5.35%
> compounded annually

55. Compound interest   If $1 had been invested on July 4, 1776, at 5% interest, compounded annually, what would it be worth on July 4, 2076?

56. Frequency of compounding $10,000 is invested in each of two accounts, both paying 6% annual interest. In the first account, interest compounds quarterly, and in the second account, interest compounds daily. Find the difference between the accounts after 20 years.

57. Discharging a battery   The charge remaining in a battery decreases as the battery discharges. The charge C (in coulombs) after t days is given by the formula $C = (3 \times 10^{-4})(0.7)^t$. Find the charge after 5 days.

58. Town population   The population of North Rivers is decreasing exponentially according to the formula $P = 3,745(0.93)^t$, where t is measured in years from the present date. Find the population in 6 years, 9 months.

59. Salvage value   A small business purchases a computer for $4,700. It is expected that its value each year will be 75% of its value in the preceding year. If the business disposes of the computer after 5 years, find its salvage value (the value after 5 years).

60. Louisiana Purchase   In 1803, the United States acquired territory from France in the Louisiana Purchase. The country doubled its territory by adding 827,000 square miles of land for $15 million. If the value of the land has appreciated at the rate of 6% each year, approximate what one square mile of land would be worth in 2023?

WRITING ABOUT MATH

61. If the world population is increasing exponentially, why is there cause for concern?

62. How do the graphs of $y = b^x$ differ when $b > 1$ and $0 < b < 1$?

SOMETHING TO THINK ABOUT

63. In the definition of the exponential function, b could not equal 0. Why not?

64. In the definition of the exponential function, b could not be negative. Why not?

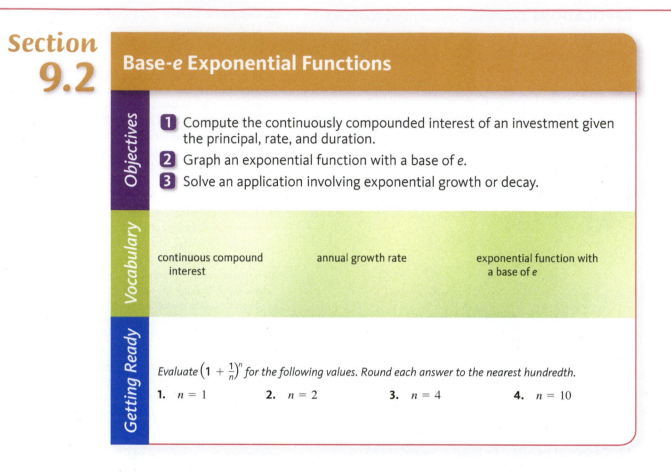

Section 9.2

Base-*e* Exponential Functions

Objectives

1. Compute the continuously compounded interest of an investment given the principal, rate, and duration.
2. Graph an exponential function with a base of *e*.
3. Solve an application involving exponential growth or decay.

Vocabulary

continuous compound interest   annual growth rate   exponential function with a base of *e*

Getting Ready

Evaluate $\left(1 + \frac{1}{n}\right)^n$ for the following values. Round each answer to the nearest hundredth.

1. $n = 1$ 2. $n = 2$ 3. $n = 4$ 4. $n = 10$

In this section, we will discuss special exponential functions with a base of *e*, an important irrational number.

1 Compute the continuously compounded interest of an investment given the principal, rate, and duration.

If a bank pays interest twice a year, we say that interest is *compounded semiannually*. If it pays interest four times a year, we say that interest is *compounded quarterly*. If it pays interest continuously (infinitely many times in a year), we say that interest is *compounded continuously*.

To develop the formula for **continuous compound interest**, we start with the formula

$$A = P\left(1 + \frac{r}{k}\right)^{kt} \qquad \text{This is the formula for compound interest.}$$

and substitute rn for k. Since r and k are positive numbers, so is n.

$$A = P\left(1 + \frac{r}{rn}\right)^{rnt}$$

We can then simplify the fraction $\frac{r}{rn}$ and use the commutative property of multiplication to change the order of the exponents.

$$A = P\left(1 + \frac{1}{n}\right)^{nrt}$$

Finally, we can use a property of exponents to write this formula as

(1) $A = P\left[\left(1 + \dfrac{1}{n}\right)^{n}\right]^{rt}$ Use the property $a^{mn} = (a^{m})^{n}$.

To find the value of $\left(1 + \frac{1}{n}\right)^{n}$, we use a calculator to evaluate it for several values of n, as shown in Table 9-1.

Leonhard Euler
1707–1783
Euler first used the letter *i* to represent $\sqrt{-1}$, the letter *e* for the base of natural logarithms, and the symbol Σ for summation. Euler was one of the most prolific mathematicians of all time, contributing to almost all areas of mathematics. Much of his work was accomplished after he became blind.

TABLE 9-1	
n	$\left(1 + \dfrac{1}{n}\right)^{n}$
1	2
2	2.25
4	2.44140625 . . .
12	2.61303529 . . .
365	2.71456748 . . .
1,000	2.71692393 . . .
100,000	2.71826823 . . .
1,000,000	2.71828046 . . .

The results suggest that as n gets larger, the value of $\left(1 + \frac{1}{n}\right)^{n}$ approaches an irrational number with a value of 2.71828. . . . This number is called *e*, which has the following approximate value.

$e \approx \mathbf{2.718281828459}$

In continuous compound interest, k (the number of compoundings) is infinitely large. Since k, r, and n are all positive and $k = rn$, as k gets very large (approaches infinity), then so does n. Therefore, we can replace $\left(1 + \frac{1}{n}\right)^{n}$ in Equation 1 with *e* to obtain

$A = Pe^{rt}$

FORMULA FOR EXPONENTIAL GROWTH

If a quantity P increases or decreases at an annual rate r, compounded continuously, then the amount A after t years is given by

$A = Pe^{rt}$

If time is measured in years, then r is called the **annual growth rate**. If r is negative, the "growth" represents a decrease, commonly referred to as *decay*.

To compute the amount to which $12,000 will grow if invested for 18 years at 10% annual interest, compounded continuously, we substitute 12,000 for P, 0.10 for r, and 18 for t in the formula for exponential growth:

$A = \boldsymbol{Pe^{rt}}$
$\quad = \mathbf{12{,}000}e^{0.10(18)}$
$\quad = 12{,}000e^{1.8}$
$\quad \approx 72{,}595.76957$ Use a calculator.

After 18 years, the account will contain $72,595.77. This is $1,589.03 more than the result in Example 7 in the previous section, where interest was compounded quarterly.

EXAMPLE 1   **CONTINUOUS COMPOUND INTEREST**   If $25,000 accumulates interest at an annual rate of 8%, compounded continuously, find the balance in the account in 50 years.

Solution   We substitute 25,000 for P, 0.08 for r, and 50 for t.

$$A = Pe^{rt}$$
$$A = 25,000e^{(0.08)(50)}$$
$$= 25,000e^4$$
$$\approx 1,364,953.751 \quad \text{Use a calculator.}$$

In 50 years, the balance will be \$1,364,953.75—over one million dollars.

SELF CHECK 1   Find the balance in 60 years.

2   Graph an exponential function with a base of e.

The function $f(x) = e^x$ is called the **exponential function** with a base of e. To graph the function $f(x) = e^x$, we plot several points and join them with a smooth curve, as shown in Figure 9-10.

EXAMPLE 2   Graph: $f(x) = e^x$

Solution

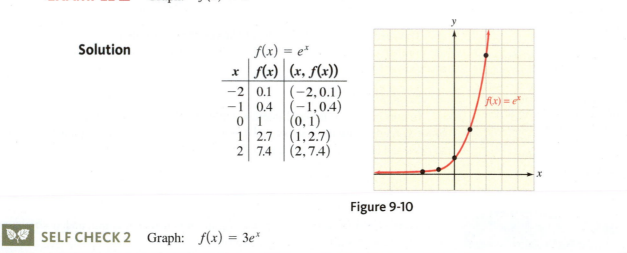

$$f(x) = e^x$$

x	$f(x)$	$(x, f(x))$
-2	0.1	$(-2, 0.1)$
-1	0.4	$(-1, 0.4)$
0	1	$(0, 1)$
1	2.7	$(1, 2.7)$
2	7.4	$(2, 7.4)$

Figure 9-10

SELF CHECK 2   Graph: $f(x) = 3e^x$

Accent on technology

▶Translations of the Exponential Function

Figure 9-11(a) shows the graphs of $f(x) = e^x$, $f(x) = e^x + 5$, and $f(x) = e^x - 3$. To graph these functions on a TI-84 graphing calculator with window settings of $[-3, 6]$ for x and $[-5, 15]$ for y, we enter the right sides of the equations after the symbols $Y_1 =$, $Y_2 =$, and $Y_3 =$. To enter e^x, press the keys **2nd LN** (e^x) **X, t, θ, n**). The display will show

$$Y_1 = e^{\wedge}(x)$$
$$Y_2 = e^{\wedge}(x) + 5$$
$$Y_3 = e^{\wedge}(x) - 3$$

After graphing these functions, we can see that the graph of $f(x) = e^x + 5$ is 5 units above the graph of $f(x) = e^x$, and that the graph of $f(x) = e^x - 3$ is 3 units below the graph of $f(x) = e^x$.

Figure 9-11(b) shows the calculator graphs of $f(x) = e^x$, $f(x) = e^{x+5}$, and $f(x) = e^{x-3}$. To graph these functions with window settings of $[-7, 10]$ for x and $[-5, 15]$ for y, we enter the right sides of the equations after the symbols $Y_1 =$, $Y_2 =$, $Y_3 =$. The display will show

$$Y_1 = e^{\wedge}(x)$$
$$Y_2 = e^{\wedge}(x + 5)$$
$$Y_3 = e^{\wedge}(x - 3)$$

After graphing these functions, we can see that the graph of $f(x) = e^{x+5}$ is 5 units to the left of the graph of $f(x) = e^x$, and that the graph of $f(x) = e^{x-3}$ is 3 units to the right of the graph of $f(x) = e^x$.

(a)   (b)

Figure 9-11

For instructions regarding the use of a Casio graphing calculator, please refer to the Casio Keystroke Guide in the back of the book.

Accent
on technology

▶ Graphing Exponential
  Functions with a Base of *e*

Figure 9-12 shows the calculator graph of $f(x) = 3e^{-x/2}$. To graph this function with window settings of $[-7, 10]$ for x and $[-5, 15]$ for y, we enter the right side of the equation after the symbol $Y_1 =$. The display will show the equation

$$Y_1 = 3(e^{\wedge}(-x/2))$$

The graph has a *y*-intercept of $(0, 3)$.

Figure 9-12

For instructions regarding the use of a Casio graphing calculator, please refer to the Casio Keystroke Guide in the back of the book.

3 **Solve an application involving exponential growth or decay.**

An equation based on the exponential function provides a model for population growth. In the Malthusian model for population growth, the future or past population of a colony is related to the present population by the formula $A = Pe^{rt}$.

EXAMPLE 3   **CITY PLANNING**   The population of a city is currently 15,000, but changing economic conditions are causing the population to decrease by 2% each year. If this trend continues, estimate the population in 30 years.

Solution   Since the population is decreasing by 2% each year, the annual growth rate is -2%, or -0.02. We can substitute -0.02 for r, 30 for t, and 15,000 for P in the formula for exponential growth and find the value of A.

$$A = \boldsymbol{P}e^{rt}$$

$$A = \boldsymbol{15{,}000}e^{-0.02(30)}$$

$$= 15{,}000e^{-0.6}$$

$$\approx 8{,}232.174541$$

In 30 years, city planners expect a population of approximately 8,232.

SELF CHECK 3 Estimate the population in 50 years.

The English economist Thomas Robert Malthus (1766–1834) pioneered in population study. He believed that poverty and starvation were unavoidable, because the human population tends to grow exponentially, but the food supply tends to grow linearly.

EXAMPLE 4 **POPULATION GROWTH** Suppose that a country with a population of 1,000 people is growing exponentially according to the formula

$$P = 1{,}000e^{0.02t}$$

where t is in years and 0.02 represents a 2% growth rate. Furthermore, assume that the food supply measured in adequate food per day per person is growing linearly according to the formula

$$y = 30.625x + 2{,}000$$

where x is in years. In how many years will the population outstrip the food supply?

Solution We can use a graphing calculator, with window settings of $[0, 100]$ for x and $[0, 10{,}000]$ for y. After graphing the functions, we obtain Figure 9-13(a).

Figure 9-13

If we use the INTERSECT feature in the CALC menu, we obtain Figure 9-13(b). The food supply will be adequate for about 71.7 years when the population will be about 4,196.

SELF CHECK 4 Suppose that the population grows at a 3% rate and the food supply growth remains the same. For how many years will the food supply be adequate?

The atomic structure of a radioactive material changes as the material emits radiation. The amount of radioactive material that is present decays exponentially according to the following formula.

RADIOACTIVE DECAY FORMULA

The amount A of radioactive material present at a time t is given by the formula

$$A = A_0 e^{kt}$$

where A_0 is the amount that was present at $t = 0$ and k is a negative number.

EXAMPLE 5   **RADIOACTIVE DECAY**   The radioactive material radon-22 decays according to the formula $A = A_0 e^{-0.181t}$, where t is expressed in days. How much radon-22 will be left if a sample of 100 grams decays for 20 days?

Solution   To find the number of grams of radon-22 that will be left, we substitute 100 for A_0 and 20 for t and simplify.

COMMENT   Note that the radioactive decay formula is the same as the formula for exponential growth, except for the variables.

$$A = \mathbf{A_0}e^{-0.181t}$$
$$A = \mathbf{100}e^{-0.181(\mathbf{20})}$$
$$= 100e^{-3.62} \qquad \textcolor{red}{-0.181 \cdot 20 = -3.62}$$
$$= 100(0.0267826765) \qquad \textcolor{red}{e^{-3.62} \approx 0.0267826765}$$
$$\approx 2.678267649 \qquad \textcolor{red}{\text{Multiply.}}$$

To the nearest hundredth, 2.68 grams of radon-22 will be left in 20 days.

SELF CHECK 5   To the nearest hundredth, how much radon-22 will be left in 30 days?

SELF CHECK ANSWERS   **1.** $3,037,760.44   **2.**   **3.** approximately 5,518   **4.** about 38.6 years   **5.** 0.44 gram

$f(x) = 3e^x$

NOW TRY THIS

Give all answers to the nearest milligram (mg).

The half-life (how much time has passed when half of the drug remains) of ibuprofen is approximately 1.8 hours. This translates to $k \approx -0.39$. (We will learn how to compute this value in Section 9.6.)

1. If a woman takes two 200 mg tablets, how many mg of ibuprofen are in her system 4 hours after taking the medication?

2. If she takes 2 more tablets 4 hours after taking the first dose, how many mg are in her system 6 hours after the first dose?

3. If she doesn't take any more medication, how many mg are in her system 12 hours after the first dose?   18 hours?

9.2   Exercises

WARM-UPS   *Use a calculator to find each value to the nearest hundredth.*

1. e^0

2. e^1

3. $2e^3$

4. $-3e^2$

5. $e^{-1.5}$

6. $e^{-3.7}$

REVIEW   *Simplify each expression. Assume that all variables represent positive numbers.*

7. $\sqrt{320x^7}$

8. $\sqrt[3]{-81x^4y^5}$

9. $5\sqrt{98y^3} - 4y\sqrt{18y}$

10. $\sqrt[4]{48z^5} + \sqrt[4]{768z^5}$

VOCABULARY AND CONCEPTS *Fill in the blanks.*

11. To two decimal places, the value of e is ____.

12. The formula for continuous compound interest is $A =$ ____.

13. Since $e > 1$, the base-e exponential function is a(n) _____ function.

14. The graph of the exponential function $y = e^x$ passes through the points $(0, 1)$ and ____.

15. The Malthusian population growth formula is _____.

16. The Malthusian theory is pessimistic, because _____ grows exponentially, but food supplies grow _____.

GUIDED PRACTICE *Graph each function. Check your work with a graphing calculator. Compare each graph to the graph of $f(x) = e^x$.* **SEE EXAMPLE 2.** *(OBJECTIVE 2)*

17. $f(x) = e^x + 1$ **18.** $f(x) = e^x - 2$

19. $f(x) = e^{x+3}$ **20.** $f(x) = e^{x-5}$

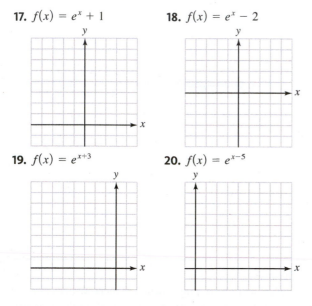

Determine whether the graph of $f(x) = e^x$, where x is a natural number, could look like the graph shown here. **(OBJECTIVE 2)**

21.   **22.**

23.   **24.**

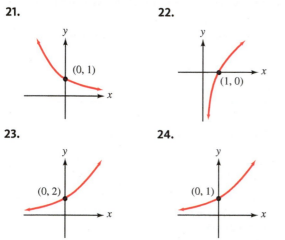

ADDITIONAL PRACTICE *Graph each function.*

25. $f(x) = -e^x$ **26.** $f(x) = -e^x + 1$

27. $f(x) = 2e^x$ **28.** $f(x) = \dfrac{1}{2}e^x$

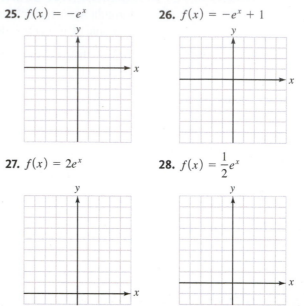

APPLICATIONS *Use a calculator to help solve. Assume that there are no deposits or withdrawals.* **SEE EXAMPLE 1.** *(OBJECTIVE 1)*

29. Continuous compound interest An investment of $9,000 earns 4% interest, compounded continuously. What will the investment be worth in 20 years?

30. Continuous compound interest An investment of $6,000 earns 7% interest, compounded continuously. What will the investment be worth in 35 years?

31. Determining the initial deposit An account now contains $25,000 and has been accumulating 5% annual interest, compounded continuously, for 15 years. Find the initial deposit.

32. Determining a previous balance An account now contains $8,000 and has been accumulating 8% annual interest, compounded continuously. How much was in the account 6 years ago?

Use a calculator to help solve. **SEE EXAMPLE 3.** *(OBJECTIVE 3)*

33. World population growth Earth's population is approximately 6 billion people and is growing at an annual rate of 1.9%. Assuming a Malthusian growth model, find the world population in 30 years.

34. World population growth Earth's population is approximately 6 billion people and is growing at an annual rate of 1.9%. Assuming a Malthusian growth model, find the world population in 40 years.

35. World population growth Assuming a Malthusian growth model and an annual growth rate of 1.9%, by what factor will Earth's current population increase in 50 years? (See Exercise 33.)

36. Population growth The growth of a population is modeled by

$$P = 173e^{0.03t}$$

How large will the population be when $t = 30$?

Use a calculator to help solve. SEE EXAMPLE 4. (OBJECTIVE 3)

37. In Example 4, suppose that better farming methods change the formula for food growth to $y = 31x + 2,000$. How long will the food supply be adequate?

38. In Example 4, suppose that a birth-control program changed the formula for population growth to $P = 1,000e^{0.01t}$. How long will the food supply be adequate?

Use a calculator to help solve. Round each answer to the nearest hundredth. SEE EXAMPLE 5. (OBJECTIVE 3)

39. Radioactive decay The radioactive material iodine-131 decays according to the formula $A = A_0e^{-0.087t}$, where t is expressed in days. To the nearest hundredth, how much iodine-131 will be left if a sample of 75 grams decays for 40 days?

40. Radioactive decay The radioactive material strontium-89 decays according to the formula $A = A_0e^{-0.013t}$, where t is expressed in days. To the nearest hundredth, how much strontium-89 will be left if a sample of 25 grams decays for 45 days?

41. Radioactive decay The radioactive material tin-126 decays according to the formula $A = A_0e^{-0.00000693t}$, where t is expressed in years. To the nearest hundredth, how much tin-126 will be left if a sample of 2,500 grams decays for 100 years?

42. Radioactive decay The radioactive material plutonium-239 decays according to the formula $A = A_0e^{-0.0000284t}$, where t is expressed in years. To the nearest hundredth, how much plutonium-239 will be left if a sample of 75 grams decays for 50,000 years?

Use a graphing calculator to solve.

43. Compounding methods An initial deposit of $5,000 grows at an annual rate of 8.5% for 5 years. Compute the final balances resulting from continuous compounding and annual compounding.

44. Compounding methods An initial deposit of $30,000 grows at an annual rate of 8% for 20 years. Compute the final balances resulting from continuous compounding and annual compounding.

45. Population decline The decline of a population is modeled by

$$P = 8,000e^{-0.008t}$$

How large will the population be when $t = 50$?

46. Epidemics The spread of foot and mouth disease through a herd of cattle can be modeled by the formula

$$P = P_0e^{0.27t} \quad (t \text{ is in days})$$

If a rancher does not act quickly to treat two cases, how many cattle will have the disease in 10 days?

47. Medicine The concentration, x, of a certain drug in an organ after t minutes is given by $x = 0.08(1 - e^{-0.1t})$. Find the concentration of the drug after 30 minutes.

48. Medicine Refer to Exercise 47. Find the initial concentration of the drug.

49. Skydiving Before her parachute opens, a skydiver's velocity v in meters per second is given by $v = 50(1 - e^{-0.2t})$. Find the initial velocity.

50. Skydiving Refer to Exercise 49 and find the velocity after 20 seconds.

51. Free-falling objects After t seconds, a certain falling object has a velocity v given by $v = 50(1 - e^{-0.3t})$. Which is falling faster after 2 seconds, this object or the skydiver in Exercise 49?

52. Alcohol absorption In one individual, the blood alcohol level t minutes after drinking two shots of whiskey is given by $P = 0.3(1 - e^{-0.05t})$. Find the blood alcohol level after 15 minutes.

53. Depreciation A camping trailer originally purchased for $4,570 is continuously losing value at the rate of 6% per year. Find its value when it is $6\frac{1}{2}$ years old.

54. Depreciation A boat purchased for $7,500 has been continuously decreasing in value at the rate of 2% each year. It is now 8 years, 3 months old. Find its value.

WRITING ABOUT MATH

55. Explain why the graph of $f(x) = e^x - 5$ is 5 units below the graph of $f(x) = e^x$.

56. Explain why the graph of $f(x) = e^{(x+5)}$ is 5 units to the left of the graph of $f(x) = e^x$.

SOMETHING TO THINK ABOUT

57. The value of e can be calculated to any degree of accuracy by adding the first several terms of the following list:

$$1, 1, \frac{1}{2}, \frac{1}{2 \cdot 3}, \frac{1}{2 \cdot 3 \cdot 4}, \frac{1}{2 \cdot 3 \cdot 4 \cdot 5}, \cdots$$

The more terms that are added, the closer the sum will be to e. Add the first six numbers in the preceding list. To how many decimal places is the sum accurate?

58. Graph the function defined by the equation

$$y = f(x) = \frac{e^x + e^{-x}}{2}$$

from $x = -2$ to $x = 2$. The graph will look like a parabola, but it is not. The graph, called a **catenary**, is important in the design of power distribution networks, because it represents the shape of a uniform flexible cable whose ends are suspended from the same height. The function is called the **hyperbolic cosine function**.

59. If $e^{t+10} = ke^t$, find k.

60. If $e^{5t} = k^t$, find k.

Section 9.3

Logarithmic Functions

Objectives

1. Write a logarithmic function as an exponential function and write an exponential function as a logarithmic function.
2. Graph a logarithmic function.
3. Graph a vertical and horizontal translation of a logarithmic function.
4. Evaluate a common logarithm.
5. Solve an application involving a logarithm.

Vocabulary

logarithmic function   common logarithm   Richter scale
logarithm   decibel

Getting Ready

Find each value.

1. 7^0 2. 5^2 3. 5^{-2} 4. $16^{1/2}$

In this section, we consider the inverse function of an exponential function $f(x) = b^x$. The inverse function is called a *logarithmic function*. These functions can be used to solve applications from fields such as electronics, seismology, and business.

1 Write a logarithmic function as an exponential function and write an exponential function as a logarithmic function.

Since the exponential function $y = b^x$ is one-to-one, it has an inverse function defined by the equation $x = b^y$. To express this inverse function in the form $y = f^{-1}(x)$, we must solve the equation $x = b^y$ for y. To do this, we need the following definition.

LOGARITHMIC FUNCTIONS

If $b > 0$ and $b \neq 1$, the **logarithmic function with base b** is defined by

$$y = \log_b x \quad \text{if and only if} \quad x = b^y$$

COMMENT Since the domain of the logarithmic function is the set of positive numbers, the logarithm of 0 or the logarithm of a negative number is not defined in the real-number system.

The domain of the logarithmic function is the interval $(0, \infty)$. The range is the interval $(-\infty, \infty)$.

Since the function $y = \log_b x$ is the inverse of the one-to-one exponential function $y = b^x$, the logarithmic function is also one-to-one.

The previous definition guarantees that any pair (x, y) that satisfies the equation $y = \log_b x$ also satisfies the equation $x = b^y$.

$$\log_4 1 = \mathbf{0} \qquad \text{because} \qquad 1 = 4^{\mathbf{0}}$$

$$\log_5 25 = \mathbf{2} \qquad \text{because} \qquad 25 = 5^{\mathbf{2}}$$

$$\log_5 \frac{1}{25} = \mathbf{-2} \qquad \text{because} \qquad \frac{1}{25} = 5^{-2}$$

$$\log_{16} 4 = \mathbf{\frac{1}{2}} \qquad \text{because} \qquad 4 = 16^{1/2}$$

COMMENT   Since $b^y = x$ is equivalent to $y = \log_b x$, then $b^{\log_b x} = x$ by substitution.

$$\log_2 \frac{1}{8} = \mathbf{-3} \qquad \text{because} \qquad \frac{1}{8} = 2^{-3}$$

$$\log_b x = \mathbf{y} \qquad \text{because} \qquad x = b^y$$

In each of these examples, the **logarithm** of a number is an exponent. In fact,

$\log_b x$ is the exponent to which b is raised to get x.

In equation form, we write

$$\boldsymbol{b^{\log_b x} = x}$$

EXAMPLE 1   Find the value of y in each equation.   **a.** $\log_6 1 = y$ **b.** $\log_3 27 = y$ **c.** $\log_5 \frac{1}{5} = y$
d. $\log_3 81 = x$

Solution   **a.** We can write the equation $\log_6 1 = y$ as the equivalent exponential equation $6^y = 1$. Since $6^{\mathbf{0}} = 1$, it follows that $y = 0$. Thus,

$$\log_6 1 = 0$$

b. $\log_3 27 = y$ is equivalent to $3^y = 27$. Since $3^3 = 27$, it follows that $3^y = 3^{\mathbf{3}}$, and $y = 3$. Thus,

$$\log_3 27 = 3$$

c. $\log_5 \frac{1}{5} = y$ is equivalent to $5^y = \frac{1}{5}$. Since $5^{-1} = \frac{1}{5}$, it follows that $5^y = 5^{-1}$, and $y = -1$. Thus,

$$\log_5 \frac{1}{5} = -1$$

d. $\log_3 81 = x$ is equivalent to $3^x = 81$. Because $3^4 = 81$, it follows that $3^x = 3^{\mathbf{4}}$. Thus, $x = 4$.

SELF CHECK 1   Find the value of y in each equation.   **a.** $\log_3 9 = y$ **b.** $\log_2 64 = y$
c. $\log_5 \frac{1}{125} = y$ **d.** $\log_2 32 = x$

EXAMPLE 2   Find the value of x in each equation.   **a.** $\log_x 125 = 3$ **b.** $\log_4 x = 3$

Solution   **a.** $\log_x 125 = 3$ is equivalent to $x^3 = 125$. Because $5^3 = 125$, it follows that $x^3 = \mathbf{5^3}$. Thus, $x = 5$.

b. $\log_4 x = 3$ is equivalent to $4^3 = \mathbf{x}$. Because $4^3 = \mathbf{64}$, it follows that $x = 64$.

SELF CHECK 2   Find the value of x in each equation.   **a.** $\log_x 8 = 3$ **b.** $\log_5 x = 2$

EXAMPLE 3   Find the value of x in each equation.
a. $\log_{1/3} x = 2$ **b.** $\log_{1/3} x = -2$ **c.** $\log_{1/3} \frac{1}{27} = x$

Solution   **a.** $\log_{1/3} x = 2$ is equivalent to $\left(\frac{1}{3}\right)^2 = x$. Thus, $x = \frac{1}{9}$.

b. $\log_{1/3} x = -2$ is equivalent to $\left(\frac{1}{3}\right)^{-2} = x$. Thus,

$$x = \left(\frac{1}{3}\right)^{-2} = 3^2 = 9$$

c. $\log_{1/3} \frac{1}{27} = x$ is equivalent to $\left(\frac{1}{3}\right)^x = \frac{1}{27}$. Because $\left(\frac{1}{3}\right)^3 = \frac{1}{27}$, it follows that $x = 3$.

SELF CHECK 3   Find the value of x in each equation.   **a.** $\log_{1/4} x = 3$ **b.** $\log_{1/4} x = -2$

2 Graph a logarithmic function.

To graph the logarithmic function $f(x) = \log_2 x$, we calculate and plot several points with coordinates (x, y) that satisfy the equation $x = 2^y$. After joining these points with a smooth curve, we have the graph shown in Figure 9-14(a).

COMMENT   Substituting values for y and solving for x may be easier.

To graph $f(x) = \log_{1/2} x$, we calculate and plot several points with coordinates (x, y) that satisfy the equation $x = \left(\frac{1}{2}\right)^y$. After joining these points with a smooth curve, we have the graph shown in Figure 9-14(b).

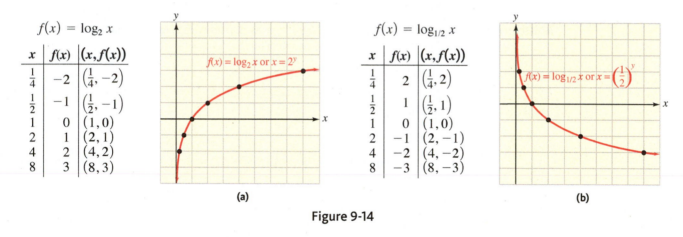

$f(x) = \log_2 x$

x	$f(x)$	$(x, f(x))$
$\frac{1}{4}$	-2	$\left(\frac{1}{4}, -2\right)$
$\frac{1}{2}$	-1	$\left(\frac{1}{2}, -1\right)$
1	0	$(1, 0)$
2	1	$(2, 1)$
4	2	$(4, 2)$
8	3	$(8, 3)$

$f(x) = \log_{1/2} x$

x	$f(x)$	$(x, f(x))$
$\frac{1}{4}$	2	$\left(\frac{1}{4}, 2\right)$
$\frac{1}{2}$	1	$\left(\frac{1}{2}, 1\right)$
1	0	$(1, 0)$
2	-1	$(2, -1)$
4	-2	$(4, -2)$
8	-3	$(8, -3)$

(a)   (b)

Figure 9-14

EXAMPLE 4   Graph the function defined by $f(x) = \log_3 x$.

Solution   To graph the logarithmic function, we calculate and plot several points with coordinates (x, y) that satisfy the equation $x = 3^y$. (See Figure 9-15.)

x	$f(x)$	$(x, f(x))$
1	0	$(1, 0)$
3	1	$(3, 1)$
9	2	$(9, 2)$

Figure 9-15

SELF CHECK 4   Graph the function defined by $f(x) = \log_4 x$.

The graphs of all logarithmic functions $(y = \log_b x)$ are similar to those in Figure 9-16. If $b > 1$, the logarithmic function is increasing, as in Figure 9-16(a). If $0 < b < 1$, the logarithmic function is decreasing, as in Figure 9-16(b).

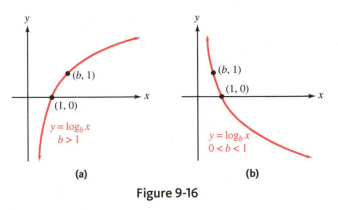

Figure 9-16

The graph of $y = \log_b x$ has the following properties.

1. It passes through the point $(1, 0)$.

2. It passes through the point $(b, 1)$.

3. The y-axis is an asymptote.

4. The domain is $(0, \infty)$ and the range is $(-\infty, \infty)$.

The exponential and logarithmic functions are inverses of each other and, therefore, have symmetry about the line $y = x$. The graphs $y = \log_b x$ and $y = b^x$ are shown in Figure 9-17(a) when $b > 1$, and in Figure 9-17(b) when $0 < b < 1$.

Figure 9-17

 Graph a vertical and horizontal translation of a logarithmic function.

The graphs of many functions involving logarithms are translations of the basic logarithmic graphs.

EXAMPLE 5 Graph the function defined by $f(x) = 3 + \log_2 x$.

Solution The graph of $f(x) = 3 + \log_2 x$ is identical to the graph of $f(x) = \log_2 x$, except that it is translated 3 units upward. (See Figure 9-18.)

Figure 9-18

SELF CHECK 5 Graph: $f(x) = \log_3 x - 2$

EXAMPLE 6 Graph: $f(x) = \log_{1/2}(x - 1)$

Solution The graph of $f(x) = \log_{1/2}(x - 1)$ is identical to the graph of $f(x) = \log_{1/2} x$, except that it is translated 1 unit to the right. (See Figure 9-19.)

Figure 9-19

 SELF CHECK 6 Graph: $f(x) = \log_{1/3}(x + 2)$

Accent on technology

▸ Graphing Logarithmic Functions

Graphing calculators can graph logarithmic functions directly only if the base of the logarithmic function is 10 or e. To use a TI-84 calculator to graph $f(x) = -2 + \log_{10}\left(\frac{1}{2}x\right)$, we enter the right side of the equation after the symbol Y$_1$=. The display will show the equation

$$Y_1 = -2 + \log(1/2*x)$$

If we use window settings of $[-1, 5]$ for x and $[-4, 1]$ for y and press **GRAPH**, we will obtain the graph shown in Figure 9-20.

Figure 9-20

For instructions regarding the use of a Casio graphing calculator, please refer to the Casio Keystroke Guide in the back of the book.

---

4 Evaluate a common logarithm.

For computational purposes and in many applications, we will use base-10 logarithms (also called **common logarithms**). When the base b is not indicated in the notation $\log x$, we assume that $b = 10$:

> **$\log x$** means **$\log_{10} x$**

Because base-10 logarithms appear so often, it is a good idea to become familiar with the following base-10 logarithms:

$$\log_{10} \frac{1}{100} = -2 \quad \text{because} \quad 10^{-2} = \frac{1}{100}$$

$$\log_{10} \frac{1}{10} = -1 \quad \text{because} \quad 10^{-1} = \frac{1}{10}$$

$$\log_{10} 1 = 0 \quad \text{because} \quad 10^0 = 1$$

$$\log_{10} 10 = 1 \quad \text{because} \quad 10^1 = 10$$

$$\log_{10} 100 = 2 \quad \text{because} \quad 10^2 = 100$$

$$\log_{10} 1{,}000 = 3 \quad \text{because} \quad 10^3 = 1{,}000$$

In general, we have

> **$\log_{10} 10^x = x$**

Before calculators, extensive tables provided logarithms of numbers. Today, logarithms can be computed quickly with a calculator. For example, to find log 32.58 with a scientific calculator, we enter these numbers and press these keys:

32.58 **LOG**

The display will read **1.51295108** . To four decimal places, log 32.58 = 1.5130.
To use a TI-84 graphing calculator, we enter these numbers and press these keys:

LOG 32.58 **)** **ENTER**

The display will read **LOG (32.58)** .
1.51295108

To four decimal places, log 32.58 = 1.5130.
For instructions regarding the use of a Casio graphing calculator, please refer to the Casio Keystroke Guide in the back of the book.

EXAMPLE 7 Find the value of x in the equation $\log x = 0.3568$. Round to four decimal places.

Solution The equation $\log x = 0.3568$ is equivalent to $10^{0.3568} = x$. To find the value of x with a calculator, we can enter these numbers and press these keys:

Scientific Calculator	*Graphing Calculator*
10 **y^x** 0.3568 **=**	10 **∧** 0.3568 **ENTER**

Either way, the display will read **2.274049951**. To four decimal places,

$$x = 2.2740$$

 SELF CHECK 7 Find the value of x: $\log x = 2.7$. Round to four decimal places.

5 Solve an application involving a logarithm.

Common logarithms are used in electrical engineering to express the voltage gain (or loss) of an electronic device such as an amplifier. The unit of gain (or loss), called the **decibel**, is defined by a logarithmic relation.

DECIBEL VOLTAGE GAIN If E_O is the output voltage of a device and E_I is the input voltage, the decibel voltage gain is given by

$$\text{dB gain} = 20 \log \frac{E_O}{E_I}$$

If the input to an amplifier is 0.4 volt and the output is 50 volts, we can find the decibel voltage gain by substituting 0.4 for E_I and 50 for E_O into the formula for dB gain:

$$\text{dB gain} = 20 \log \frac{E_O}{E_I}$$

$$\text{dB gain} = 20 \log \frac{50}{0.4}$$

$$= 20 \log 125$$

$$\approx 42 \qquad \text{Use a calculator.}$$

The amplifier provides a 42-decibel voltage gain.

In seismology, the study of earthquakes, common logarithms are used to measure the magnitude (ground motion) of earthquakes on the **Richter scale**. The magnitude of an earthquake is given by the following logarithmic function.

RICHTER SCALE

If R is the magnitude of an earthquake, A is the amplitude (measured in micrometers), and P is the period (the time of one oscillation of Earth's surface, measured in seconds), then

$$R = \log \frac{A}{P}$$

EXAMPLE 8   **MEASURING EARTHQUAKES**   Find the measure on the Richter scale of an earthquake with an amplitude of 10,000 micrometers (1 centimeter) and a period of 0.1 second.

Solution   We substitute 10,000 for A and 0.1 for P in the Richter scale formula and simplify.

$$R = \log \frac{A}{P}$$

$$R = \log \frac{10,000}{0.1}$$

$$= \log 100,000$$

$$= 5$$

The earthquake measures 5 on the Richter scale.

SELF CHECK 8   Find the measure on the Richter scale of an earthquake with an amplitude of 3,162,277 micrometers and a period of 0.1 second.

SELF CHECK ANSWERS

1. a. 2   b. 6   c. −3   d. 5   2. a. 2   b. 25   3. a. $\frac{1}{64}$ b. 16   4.

5.   6.   7. 501.1872   8. It measures 7.5 on the Richter scale.

NOW TRY THIS

In 1989, an earthquake measuring 7.1 on the Richter scale rocked San Francisco.

1. Write a logarithmic equation to describe the earthquake.

2. Write the equation in exponential form.

In 1964, an earthquake measuring 9.1 on the Richter scale devastated Juneau, Alaska.

3. Write the equation in exponential form.

4. Given that each earthquake had the same period, the amplitude in Juneau was how many times greater than that in San Francisco?

9.3   Exercises

WARM-UPS *Fill in the blanks.*

1. $2^{\square} = 8$

2. $3^{\square} = 9$

3. $\square^3 = 125$

4. $\square^3 = 27$

5. $5^{\square} = 25$

6. $\square^4 = 81$

7. $2^{\square} = 32$

8. $\square^{-3} = \dfrac{1}{27}$

9. $\square^{-2} = \dfrac{1}{4}$

REVIEW *Solve each equation.*

10. $\sqrt[3]{6x + 4} = 4$

11. $\sqrt{3x - 4} = \sqrt{-7x + 2}$

12. $\sqrt{a + 1} - 1 = 3a$

13. $3 - \sqrt{t - 3} = \sqrt{t}$

VOCABULARY AND CONCEPTS *Fill in the blanks.*

14. The equation $y = \log_b x$ is equivalent to _____ .

15. The domain of the logarithmic function is the interval _____.

16. The _____ of the logarithmic function is the interval $(-\infty, \infty)$.

17. $b^{\log_b x} = $ __.

18. Because an exponential function is one-to-one, it has an _____ function that is called a _____ function.

19. $\log_b x$ is the _____ to which b is raised to obtain x.

20. The y-axis is an _____ to the graph of $f(x) = \log_b x$.

21. The graph of $f(x) = \log_b x$ passes through the points _____ and _____.

22. A logarithm with a base of 10 is called a _____ logarithm and $\log_{10} 10^x = $ __.

23. The decibel voltage gain is found using the equation dB gain = _____ .

24. The magnitude of an earthquake is measured by the formula $R = $ ____ .

GUIDED PRACTICE *Write each equation in exponential form.* (OBJECTIVE 1)

25. $\log_4 64 = 3$

26. $\log_5 5 = 1$

27. $\log_{1/2} \dfrac{1}{8} = 3$

28. $\log_{1/3} 1 = 0$

29. $\log_4 \dfrac{1}{64} = -3$

30. $\log_6 \dfrac{1}{36} = -2$

31. $\log_{1/2} \dfrac{1}{8} = 3$

32. $\log_{1/5} 25 = -2$

Write each equation in logarithmic form. (OBJECTIVE 1)

33. $7^2 = 49$

34. $10^4 = 10,000$

35. $6^{-2} = \dfrac{1}{36}$

36. $5^{-3} = \dfrac{1}{125}$

37. $\left(\dfrac{1}{2}\right)^{-5} = 32$

38. $\left(\dfrac{1}{3}\right)^{-3} = 27$

39. $x^y = z$

40. $m^n = p$

Find each value of x. SEE EXAMPLE 1. (OBJECTIVE 1)

41. $\log_2 32 = x$

42. $\log_3 9 = x$

43. $\log_5 125 = x$

44. $\log_6 216 = x$

Find each value of x. SEE EXAMPLE 2. (OBJECTIVE 1)

45. $\log_7 x = 2$

46. $\log_5 x = 0$

47. $\log_6 x = 1$

48. $\log_2 x = 4$

49. $\log_{25} x = \dfrac{1}{2}$

50. $\log_4 x = \dfrac{1}{2}$

51. $\log_5 x = -2$

52. $\log_{27} x = -\dfrac{1}{3}$

53. $\log_x 5^3 = 3$

54. $\log_x 5 = 1$

55. $\log_x \dfrac{9}{4} = 2$

56. $\log_x \dfrac{\sqrt{3}}{3} = \dfrac{1}{2}$

Find each value of x. SEE EXAMPLE 3. (OBJECTIVE 1)

57. $\log_{1/2} \dfrac{1}{8} = x$

58. $\log_{1/3} \dfrac{1}{81} = x$

59. $\log_{1/4} x = 3$

60. $\log_{1/3} x = 4$

Graph each function. Determine whether each function is an increasing or decreasing function. SEE EXAMPLE 4. (OBJECTIVE 2)

61. $f(x) = \log_3 x$

62. $f(x) = \log_{1/3} x$

63. $f(x) = \log_{1/2} x$

64. $f(x) = \log_4 x$

Graph each pair of inverse functions on a single coordinate system. (OBJECTIVE 2)

65. $f(x) = 2^x$
$g(x) = \log_2 x$

66. $f(x) = \left(\dfrac{1}{2}\right)^x$
$g(x) = \log_{1/2} x$

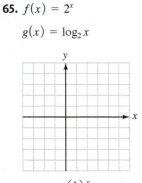

67. $f(x) = \left(\dfrac{1}{4}\right)^x$
$g(x) = \log_{1/4} x$

68. $f(x) = 4^x$
$g(x) = \log_4 x$

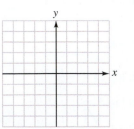

Graph each function. SEE EXAMPLES 5–6. (OBJECTIVE 3)

69. $f(x) = 3 + \log_3 x$

70. $f(x) = \log_{1/3} x - 1$

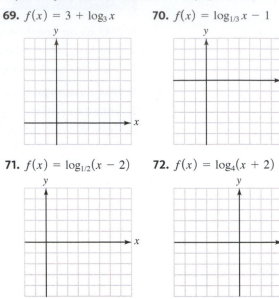

71. $f(x) = \log_{1/2}(x - 2)$

72. $f(x) = \log_4(x + 2)$

Use a calculator to find each value. Give answers to four decimal places. (OBJECTIVE 4)

73. $\log 8.25$

74. $\log 0.77$

75. $\log 0.00867$

76. $\log 375.876$

Use a calculator to find each value of y. If an answer is not exact, give the answer to two decimal places. SEE EXAMPLE 7. (OBJECTIVE 4)

77. $\log y = 4.24$

78. $\log y = 0.926$

79. $\log y = -3.71$

80. $\log y = -0.28$

ADDITIONAL PRACTICE *Find each value of x.*

81. $\log_{36} x = -\dfrac{1}{2}$

82. $\log_{27} x = -\dfrac{1}{3}$

83. $\log_{1/2} 8 = x$

84. $\log_{1/2} 16 = x$

85. $\log_{1/2} x = -3$

86. $\log_{1/4} x = -2$

87. $\log_4 2 = x$

88. $\log_{125} 5 = x$

89. $\log_{100} \dfrac{1}{1,000} = x$

90. $\log_{5/2} \dfrac{4}{25} = x$

91. $\log_{27} 9 = x$

92. $\log_{12} x = 0$

93. $\log_{2\sqrt{2}} x = 2$

94. $\log_4 8 = x$

95. $\log_x \dfrac{1}{64} = -3$

96. $\log_x \dfrac{1}{100} = -2$

97. $2^{\log_2 4} = x$

98. $3^{\log_3 5} = x$

99. $x^{\log_4 6} = 6$

100. $x^{\log_3 8} = 8$

101. $\log 10^3 = x$

102. $\log 10^{-2} = x$

103. $10^{\log x} = 100$

104. $10^{\log x} = \dfrac{1}{10}$

Use a calculator to find each value of y. If an answer is not exact, give the answer to two decimal places.

105. $\log y = 1.4023$ **106.** $\log y = 2.6490$

107. $\log y = \log 8$ **108.** $\log y = \log 7$

Find the value of b, if any, that would cause the graph of $f(x) = \log_b x$ to look like the graph indicated.

109.   **110.**

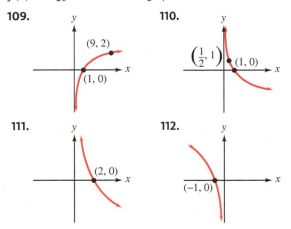

111.   **112.**

Use a calculator to help solve. If an answer is not exact, round to the nearest tenth. SEE EXAMPLE 8. (OBJECTIVE 5)

113. Earthquakes An earthquake has an amplitude of 5,000 micrometers and a period of 0.2 second. Find its measure on the Richter scale.

114. Earthquakes The period of an earthquake with amplitude of 80,000 micrometers is 0.08 second. Find its measure on the Richter scale.

115. Earthquakes An earthquake has a period of $\frac{1}{4}$ second and an amplitude of 2,500 micrometers. Find its measure on the Richter scale.

116. Earthquakes By what factor must the amplitude of an earthquake change to increase its magnitude by 1 point on the Richter scale? Assume that the period remains constant.

Use a calculator to help solve. If an answer is not exact, round to the nearest tenth.

117. Depreciation Business equipment is often depreciated using the double declining-balance method. In this method, a piece of equipment with a life expectancy of N years, costing C, will depreciate to a value of V in n years, where n is given by the formula

$$n = \frac{\log V - \log C}{\log\left(1 - \dfrac{2}{N}\right)}$$

A computer that cost $17,000 has a life expectancy of 5 years. If it has depreciated to a value of $2,000, how old is it?

118. Depreciation See Exercise 117. A printer worth $470 when new had a life expectancy of 12 years. If it is now worth $189, how old is it?

119. Time for money to grow If P is invested at the end of each year in an annuity earning annual interest at a rate r, the amount in the account will be A after n years, where

$$n = \frac{\log\left(\dfrac{Ar}{P} + 1\right)}{\log(1 + r)}$$

If $1,000 is invested each year in an annuity earning 12% annual interest, how long will it take for the account to be worth $20,000?

120. Time for money to grow If $5,000 is invested each year in an annuity earning 8% annual interest, how long will it take for the account to be worth $50,000? (See Exercise 119.)

APPLICATIONS Use a calculator to help solve. If an answer is not exact, round to the nearest tenth. (OBJECTIVE 5)

121. Finding the gain of an amplifier Find the dB gain of an amplifier if the input voltage is 0.71 volt when the output voltage is 20 volts.

122. Finding the gain of an amplifier Find the dB gain of an amplifier if the output voltage is 2.8 volts when the input voltage is 0.05 volt.

123. dB gain of an amplifier Find the dB gain of the amplifier.

124. dB gain of an amplifier An amplifier produces an output of 80 volts when driven by an input of 0.12 volts. Find the amplifier's dB gain.

WRITING ABOUT MATH

125. Describe the appearance of the graph of $y = f(x) = \log_b x$ when $0 < b < 1$ and when $b > 1$.

126. Explain why it is impossible to find the logarithm of a negative number.

SOMETHING TO THINK ABOUT

127. Graph $f(x) = -\log_3 x$. How does the graph compare to the graph of $f(x) = \log_3 x$?

128. Find a logarithmic function that passes through the points $(1, 0)$ and $(5, 1)$.

129. Explain why an earthquake measuring 7 on the Richter scale is much worse than an earthquake measuring 6.

Section 9.4

Natural Logarithms

Objectives

1. Evaluate a natural logarithm.
2. Solve a logarithmic equation with a calculator.
3. Graph a natural logarithmic function.
4. Solve an application involving a natural logarithm.

Vocabulary

natural logarithm

Getting Ready

Evaluate each expression.

1. $\log_4 16$ **2.** $\log_2 \frac{1}{8}$ **3.** $\log_5 5$ **4.** $\log_7 1$

In this section, we will discuss special logarithmic functions with a base of e. They play an important role in advanced mathematics and applications in other scientific fields.

 Evaluate a natural logarithm.

We have seen the importance of base-e exponential functions in mathematical models of events in nature. Base-e logarithms are just as important. They are called **natural logarithms** or Napierian logarithms, after John Napier (1550–1617), and usually are written as $\ln x$, rather than $\log_e x$:

> $\ln x$ **means** $\log_e x$

As with all logarithmic functions, the domain of $f(x) = \ln x$ is the interval $(0, \infty)$, and the range is the interval $(-\infty, \infty)$.

We have seen that the logarithm of a number is an exponent. For natural logarithms,

> **$\ln x$ is the exponent to which e is raised to obtain x.**

In equation form, we write

> $e^{\ln x} = x$

To find the base-e logarithms of numbers, we can use a calculator.

John Napier
1550–1617

Napier is famous for his work with natural logarithms. In fact, natural logarithms are often called *Napierian logarithms*. He also invented a device, called *Napier's rods*, that did multiplications mechanically. His device was a forerunner of modern-day computers.

Accent
on technology

▸ Evaluating Logarithms

To use a scientific calculator to find the value of ln 9.87, we enter these numbers and press these keys:

9.87 **LN**

The display will read **2.289499853**. To four decimal places, ln 9.87 = 2.2895.
To use a TI-84 graphing calculator, we enter

LN 9.87 **)** **ENTER**

The display will read **LN (9.87)**
  2.289499853

For instructions regarding the use of a Casio graphing calculator, please refer to the Casio Keystroke Guide in the back of the book.

EXAMPLE 1 Use a calculator to find each value. **a.** ln 17.32 **b.** ln(log 0.05)

Solution **a.** We can enter these numbers and press these keys:

Scientific Calculator *Graphing Calculator*

17.32 **LN** **LN** 17.32 **)** **ENTER**

Either way, the result is 2.851861903.

b. We can enter these numbers and press these keys:

Scientific Calculator *Graphing Calculator*

0.05 **LOG LN** **LN (LOG** 0.05 **))** **ENTER**

Either way, we obtain an error, because log 0.05 is a negative number. Because the domain of ln x is $(0, \infty)$, we cannot take the logarithm of a negative number.

SELF CHECK 1 Find each value to four decimal places.
a. ln π **b.** $\ln\left(\log \frac{1}{2}\right)$

2 Solve a logarithmic equation with a calculator.

EXAMPLE 2 Find the value of x to four decimal places.
a. ln x = 1.335 **b.** ln x = log 5.5

Solution **a.** The equation ln x = 1.335 is equivalent to $e^{1.335} = x$. Enter $e^{1.335}$ into the calculator and the display will read 3.799995946. To four decimal places,

$x = 3.8000$

b. The equation ln x = log 5.5 is equivalent to $e^{\log 5.5} = x$. Enter $e^{\log 5.5}$ into the calculator and the display will read 2.096695826. To four decimal places,

$x = 2.0967$

SELF CHECK 2 Find the value of x to four decimal places.
a. ln x = 2.5437 **b.** log x = ln 5

3   Graph a natural logarithmic function.

The equation $y = \ln x$ is equivalent to the equation $x = e^y$. To graph $f(x) = \ln x$, we can plot points that satisfy the equation $x = e^y$ and join them with a smooth curve, as shown in Figure 9-21(a). Figure 9-21(b) shows the calculator graph.

$$f(x) = \ln x$$

x	$f(x)$	$(x, f(x))$
$\frac{1}{e} \approx 0.4$	-1	$(0.4, -1)$
1	0	$(1, 0)$
$e \approx 2.7$	1	$(2.7, 1)$
$e^2 \approx 7.4$	2	$(7.4, 2)$

(a)   (b)

Figure 9-21

EXAMPLE 3   Graph: $f(x) = \ln x - 1$

Solution   The graph of $f(x) = \ln x - 1$ is 1 unit below the graph of $f(x) = \ln x$. (See Figure 9-22.)

Figure 9-22

SELF CHECK 3   Graph: $f(x) = \ln x + 1$

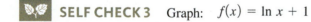

Accent
on technology

▶ Graphing Logarithmic
Functions

Many graphs of logarithmic functions involve translations of the graph of $f(x) = \ln x$. For example, Figure 9-23 shows calculator graphs of the functions $f(x) = \ln x$, $f(x) = \ln x + 2$, and $f(x) = \ln x - 3$.

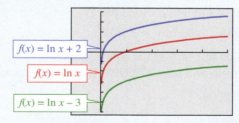

The graph of $f(x) = \ln x + 2$ is 2 units above the graph of $f(x) = \ln x$.

The graph of $f(x) = \ln x - 3$ is 3 units below the graph of $f(x) = \ln x$.

Figure 9-23

Figure 9-24 shows the calculator graphs of the functions $f(x) = \ln x$, $f(x) = \ln(x - 2)$, and $f(x) = \ln(x + 2)$.

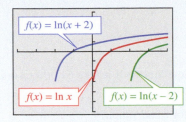

The graph of $f(x) = \ln(x - 2)$ is 2 units to the right of the graph of $f(x) = \ln x$.

The graph of $f(x) = \ln(x + 2)$ is 2 units to the left of the graph of $f(x) = \ln x$.

Figure 9-24

For instructions regarding the use of a Casio graphing calculator, please refer to the Casio Keystroke Guide in the back of the book.

4   Solve an application involving a natural logarithm.

If a population grows exponentially at a certain annual rate, the time required for the population to double is called the *doubling time*.

EXAMPLE 4   **DOUBLING TIME**   If the Earth's population continues to grow at the approximate rate of 2% per year, how long will it take for its population to double?

Solution   In Section 9.2, we learned that the formula for population growth is $A = Pe^{rt}$, where P is the population at time $t = 0$, A is the population after t years, and r is the annual rate of growth, compounded continuously. Since we want to find out how long it takes for the population to double, we can substitute $2P$ for A and 0.02 for r in the formula and proceed as follows.

$$A = Pe^{rt}$$
$$2P = Pe^{0.02t} \qquad \text{Substitute } 2P \text{ for } A \text{ and } 0.02 \text{ for } r.$$
$$2 = e^{0.02t} \qquad \text{Divide both sides by } P.$$
$$\ln 2 = 0.02t \qquad \ln 2 \text{ is the exponent to which } e \text{ is raised to obtain 2.}$$
$$34.65735903 \approx t \qquad \text{Divide both sides by } 0.02 \text{ and simplify.}$$

The population will double in about 35 years.

SELF CHECK 4   If the world population's annual growth rate could be reduced to 1.5% per year, what would be the doubling time? Give the result to the nearest year.

By solving the formula $A = Pe^{rt}$ for t, we can obtain a simpler formula for finding the doubling time.

$$A = Pe^{rt}$$
$$2P = Pe^{rt} \qquad \text{Substitute } 2P \text{ for } A.$$
$$2 = e^{rt} \qquad \text{Divide both sides by } P.$$
$$\ln 2 = rt \qquad \ln 2 \text{ is the exponent to which } e \text{ is raised to get 2.}$$
$$\frac{\ln 2}{r} = t \qquad \text{Divide both sides by } r.$$

This result gives a specific formula for finding the doubling time.

FORMULA FOR DOUBLING TIME	If r is the annual rate (compounded continuously) and t is the time required for a population to double, then
	$$t = \frac{\ln 2}{r}$$

EXAMPLE 5   **DOUBLING TIME**   How long will it take $1,000 to double at an annual rate of 8%, compounded continuously?

Solution   We can substitute 0.08 for r and simplify:

$$t = \frac{\ln 2}{\textcolor{red}{r}}$$

COMMENT To find the doubling time, you can use either the method in Example 4 or the method in Example 5.

$$t = \frac{\ln 2}{\textcolor{red}{0.08}}$$

$$\approx 8.664339757$$

It will take about $8\frac{2}{3}$ years for the money to double.

SELF CHECK 5   How long will it take at 9%, compounded continuously?

SELF CHECK ANSWERS

1. a. 1.1447 **b.** no value **2. a.** 12.7267 **b.** 40.6853 **3.** **4.** 46 years
5. about 7.7 years

$f(x) = \ln(x + 1)$

NOW TRY THIS

Between 2000 and 2007, McKinney, TX, was ranked overall as the fastest growing city in the United States.

1. The population more than doubled from 54,369 to 115,620 during this period. To the nearest tenth, what was the average growth rate?

2. If the growth rate slowed to half that above and was estimated to remain steady for the next 10 years, project the population in 2017.

3. Discuss with another student (or in a group) the difficulties a city might face with such a growth rate.

9.4   Exercises

WARM- UPS *Write each logarithmic equation as an exponential equation.*

1. $\log_3 9 = 2$ **2.** $\log_a b = c$

Write each exponential equation as a logarithmic equation.

3. $2^3 = 8$ **4.** $x^y = z$

REVIEW *Write the equation of the required line.*

5. having a slope of -7 and a y-intercept of $(0, 2)$

6. passing through the point $(3, 2)$ and perpendicular to the line $y = \frac{2}{3}x - 12$

7. parallel to the line $3x + 2y = 9$ and passing through the point $(-3, 5)$

8. vertical line through the point $(4, 1)$

9. horizontal line through the point $(-3, 6)$

Simplify each expression. Assume no denominators are 0.

10. $\dfrac{3x - 5}{9x^2 - 25}$

11. $\dfrac{x + 1}{x} + \dfrac{x - 1}{x + 1}$

12. $\dfrac{x^2 + 3x + 2}{3x + 12} \cdot \dfrac{x + 4}{x^2 - 4}$

13. $\dfrac{1 + \frac{y}{x}}{\frac{y}{x} - 1}$

VOCABULARY AND CONCEPTS *Fill in the blanks.*

14. The expression _____ means ___ and is called a _____ logarithm.

15. The domain of the function $f(x) = \ln x$ is the interval _____ and the range is the interval _____.

16. The graph of $f(x) = \ln x$ has the _____ as an asymptote.

17. In the expression $\log x$, the base is understood to be ___.

18. In the expression $\ln x$, the base is understood to be ___.

19. If a population grows exponentially at a rate r, the time it will take the population to double is given by the formula

$t = \dfrac{\quad}{\quad}$.

20. The logarithm of a negative number is _____.

 GUIDED PRACTICE *Use a calculator to find each value, if possible. Express all answers to four decimal places.* SEE EXAMPLE 1. (OBJECTIVE 1)

21. $\ln 25.25$

22. $\ln 0.523$

23. $\ln 9.89$

24. $\ln 0.00725$

Use a calculator to find the values of y, if possible. Express all answers to four decimal places. SEE EXAMPLE 2. (OBJECTIVE 2)

25. $\ln y = 2.3015$

26. $\ln y = 1.548$

27. $\ln y = 3.17$

28. $\ln y = 0.837$

Use a graphing calculator to graph each function. SEE EXAMPLE 3. (OBJECTIVE 3)

29. $y = -\ln x$

30. $y = \ln x^2$

31. $y = \ln(-x)$

32. $y = \ln\left(\dfrac{1}{2}x\right)$

Determine whether the graph could represent the graph of $y = \ln x$.

33.

34.

35.

36.

 ADDITIONAL PRACTICE *Use a calculator to find each value, if possible. Express all answers to four decimal places.*

37. $\log(\ln 3)$

38. $\ln(\log 28.8)$

39. $\ln(\log 0.7)$

40. $\log(\ln 0.2)$

Use a calculator to find each value of y, if possible. Express all answers to four decimal places.

41. $\ln y = -4.72$

42. $\ln y = -0.48$

43. $\log y = \ln 4$

44. $\ln y = \log 5$

 APPLICATIONS *Use a calculator to solve. Round each answer to the nearest tenth.* SEE EXAMPLES 4–5. (OBJECTIVE 4)

45. Population growth See the ad. How long will it take the population of River City to double?

River City
A growing community
- 6 parks
- 10 churches
- 12% annual growth
- Low crime rate

46. Population growth A population growing at an annual rate r will triple in a time t given by the formula

$t = \dfrac{\ln 3}{r}$

How long will it take the population of a town growing at the rate of 8% per year to triple?

47. Doubling money How long will it take $1,000 to double if it is invested at an annual rate of 5%, compounded continuously?

48. Tripling money Find the length of time for $25,000 to triple if invested at 6% annual interest, compounded continuously. (See Exercise 46.)

49. Making Jell-O   After the contents of a package of Jell-O are combined with boiling water, the mixture is placed in a refrigerator whose temperature remains a constant 38° F. Estimate the number of hours t that it will take for the Jell-O to cool to 50° F using the formula

$$t = -\frac{1}{0.9} \ln \frac{50 - T_r}{212 - T_r}$$

where T_r is the temperature of the refrigerator.

50. Forensic medicine   To estimate the number of hours t that a murder victim had been dead, a coroner used the formula

$$t = \frac{1}{0.25} \ln \frac{98.6 - T_s}{82 - T_s}$$

where T_s is the temperature of the surroundings where the body was found. If the crime took place in an apartment where the thermostat was set at 72° F, approximately how long ago did the murder occur?

WRITING ABOUT MATH

51. The time it takes money to double at an annual rate r, compounded continuously, is given by the formula $t = (\ln 2)/r$. Explain why money doubles more quickly as the rate increases.

52. The time it takes money to triple at an annual rate r, compounded continuously, is given by the formula $t = (\ln 3)/r$. Explain why money triples less quickly as the rate decreases.

SOMETHING TO THINK ABOUT

53. Use the formula $P = P_0 e^{rt}$ to verify that P will be three times as large as P_0 when $t = \frac{\ln 3}{r}$.

54. Use the formula $P = P_0 e^{rt}$ to verify that P will be four times as large as P_0 when $t = \frac{\ln 4}{r}$.

55. Find a formula to find how long it will take a sum of money to become five times as large.

56. Use a graphing calculator to graph

$$f(x) = \frac{1}{1 + e^{-2x}}$$

and discuss the graph.

Section 9.5

Properties of Logarithms

Objectives

1. Simplify a logarithmic expression by applying a property or properties of logarithms.
2. Expand a logarithmic expression.
3. Write a logarithmic expression as a single logarithm.
4. Evaluate a logarithm by applying properties of logarithms.
5. Apply the change-of-base formula.
6. Solve an application using one or more logarithmic properties.

Vocabulary

change-of-base formula   pH of a solution

Getting Ready

Simplify each expression. Assume $x \neq 0$.

1. $x^m x^n$ **2.** x^0 **3.** $(x^m)^n$ **4.** $\dfrac{x^m}{x^n}$

In this section, we will consider many properties of logarithms. We will then use these properties to solve applications.

 Simplify a logarithmic expression by applying a property or properties of logarithms.

Since logarithms are exponents, the properties of exponents have counterparts in the theory of logarithms. We begin with four basic properties.

PROPERTIES OF LOGARITHMS

If b is a positive number and $b \neq 1$, then

1. $\log_b 1 = 0$ 2. $\log_b b = 1$
3. $\log_b b^x = x$ 4. $b^{\log_b x} = x$ $(x > 0)$

Properties 1 through 4 follow directly from the definition of a logarithm.

1. $\log_b 1 = 0$, because $b^0 = 1$.
2. $\log_b b = 1$, because $b^1 = b$.
3. $\log_b b^x = x$, because $b^x = b^x$.
4. $b^{\log_b x} = x$, because $\log_b x$ is the exponent to which b is raised to get x.

Properties 3 and 4 also indicate that the composition of the exponential and logarithmic functions with the same base (in both directions) is the identity function. This is expected, because the exponential and logarithmic functions with the same base are inverse functions.

EXAMPLE 1   Simplify each expression.   **a.** $\log_5 1$ **b.** $\log_3 3$ **c.** $\log_7 7^3$ **d.** $b^{\log_b 7}$

Solution   **a.** By Property 1, $\log_5 1 = 0$, because $5^0 = 1$.
   b. By Property 2, $\log_3 3 = 1$, because $3^1 = 3$.
   c. By Property 3, $\log_7 7^3 = 3$, because $7^3 = 7^3$.
   d. By Property 4, $b^{\log_b 7} = 7$, because $\log_b 7$ is the power to which b is raised to get 7.

SELF CHECK 1   Simplify.   **a.** $\log_4 1$ **b.** $\log_5 5$ **c.** $\log_2 2^4$ **d.** $5^{\log_5 2}$

The next two properties state that

   The logarithm of a product is the sum of the logarithms.

   The logarithm of a quotient is the difference of the logarithms.

THE PRODUCT AND QUOTIENT PROPERTIES OF LOGARITHMS

If M, N, and b are positive numbers and $b \neq 1$, then

5. $\log_b MN = \log_b M + \log_b N$ 6. $\log_b \dfrac{M}{N} = \log_b M - \log_b N$

Proof   To prove the product property of logarithms, we let $x = \log_b M$ and $y = \log_b N$. We use the definition of logarithms to write each equation in exponential form.

$$M = b^x \quad \text{and} \quad N = b^y$$

Then $MN = b^x b^y$ and a property of exponents gives

$$MN = b^{x+y} \quad \textcolor{red}{b^x b^y = b^{x+y}\text{: Keep the base and add the exponents.}}$$

We write this exponential equation in logarithmic form as

$$\log_b MN = x + y$$

Substituting the values of x and y completes the proof.

$$\log_b MN = \log_b M + \log_b N$$

The proof of the quotient property of logarithms is similar.

COMMENT　By the product property of logarithms, the logarithm of a *product* is equal to the *sum* of the logarithms. The logarithm of a sum or a difference usually does not simplify. In general,

$$\log_b (M + N) \ne \log_b M + \log_b N$$
$$\log_b (M - N) \ne \log_b M - \log_b N$$

By the quotient property of logarithms, the logarithm of a *quotient* is equal to the *difference* of the logarithms. The logarithm of a quotient is not the quotient of the logarithms:

$$\log_b \frac{M}{N} \ne \frac{\log_b M}{\log_b N}$$

Accent on technology

▸ Verifying Properties of Logarithms

We can use a calculator to illustrate the product property of logarithms by showing that

$$\ln[(3.7)(15.9)] = \ln 3.7 + \ln 15.9$$

We calculate the left and right sides of the equation separately and compare the results. To use a calculator to find $\ln[(3.7)(15.9)]$, we enter these numbers and press these keys:

3.7 **×** 15.9 **=** **LN**　　　Using a scientific calculator

LN 3.7 **×** 15.9 **)** **ENTER**　　Using a TI-84 graphing calculator

The display will read **4.074651929**.

To find $\ln 3.7 + \ln 15.9$, we enter these numbers and press these keys:

3.7 **LN** **+** 15.9 **LN** **=**　　　Using a scientific calculator

LN 3.7 **)** **+** **LN** 15.9 **)** **ENTER**　　Using a TI-84 graphing calculator

The display will read **4.074651929**. Since the left and right sides are equal, the equation is true.

For instructions regarding the use of a Casio graphing calculator, please refer to the Casio Keystroke Guide in the back of the book.

The power rule of logarithms states that

The logarithm of an expression to a power is the power times the logarithm of the expression.

THE POWER RULE OF LOGARITHMS

If M, p, and b are positive numbers and $b \ne 1$, then

7. $\log_b M^p = p \log_b M$

Proof To prove the power rule, we let $x = \log_b M$, write the expression in exponential form, and raise both sides to the *p*th power:

$$M = b^x$$

$(M)^p = (b^x)^p$ Raise both sides to the *p*th power.

$M^p = b^{px}$ Keep the base and multiply the exponents.

Using the definition of logarithms gives

$$\log_b M^p = p\boldsymbol{x}$$

Substituting the value for *x* completes the proof.

$$\log_b M^p = p \, \boldsymbol{\log_b M}$$

The logarithmic property of equality states that

If the logarithms of two numbers are equal, the numbers are equal.

THE LOGARITHMIC PROPERTY OF EQUALITY

If *x*, *y*, and *b* are positive numbers and $b \neq 1$, then

8. If $\log_b x = \log_b y$, then $x = y$.

The logarithmic property of equality follows from the fact that the logarithmic function is a one-to-one function. It will be important in the next section when we solve logarithmic equations.

2 Expand a logarithmic expression.

We can use the properties of logarithms to write a logarithm as the sum or difference of several logarithms.

EXAMPLE 2 Assume that *b*, *x*, *y*, and *z* are positive numbers and $b \neq 1$. Write each expression in terms of the logarithms of *x*, *y*, and *z*.

a. $\log_b (xyz)$ **b.** $\log_b \dfrac{xy}{z}$

Solution **a.** $\log_b (xyz) = \log_b x + \log_b y + \log_b z$ The log of a product is the sum of the logs.

b. $\log_b \dfrac{xy}{z} = \log_b(xy) - \log_b z$ The log of a quotient is the difference of the logs.

$= (\log_b x + \log_b y) - \log_b z$ The log of a product is the sum of the logs.

$= \log_b x + \log_b y - \log_b z$ Remove parentheses.

SELF CHECK 2 Write $\log_b \dfrac{x}{yz}$ in terms of the logarithms of *x*, *y*, and *z*.

EXAMPLE 3   Assume that b, x, y, and z are positive numbers and $b \neq 1$. Write each expression in terms of the logarithms of x, y, and z.

    a. $\log_b(x^2 y^3 z)$ **b.** $\log_b \dfrac{\sqrt{x}}{y^3 z}$

Solution   **a.** $\log_b(x^2 y^3 z) = \log_b x^2 + \log_b y^3 + \log_b z$ The log of a product is the sum of the logs.

                                         $= 2\log_b x + 3\log_b y + \log_b z$ The log of an expression to a power is the power times the log of the expression.

    b. $\log_b \dfrac{\sqrt{x}}{y^3 z} = \log_b \sqrt{x} - \log_b(y^3 z)$ The log of a quotient is the difference of the logs.

                       $= \log_b x^{1/2} - (\log_b y^3 + \log_b z)$ $\sqrt{x} = x^{1/2}$. The log of a product is the sum of the logs.

                       $= \dfrac{1}{2}\log_b x - (3\log_b y + \log_b z)$ The log of a power is the power times the log.

                       $= \dfrac{1}{2}\log_b x - 3\log_b y - \log_b z$ Use the distributive property to remove parentheses.

SELF CHECK 3   Write $\log_b \sqrt[4]{\dfrac{x^3 y}{z}}$ in terms of the logarithms of x, y, and z.

---

3   **Write a logarithmic expression as a single logarithm.**

We can use the properties of logarithms to combine several logarithms into one logarithm.

EXAMPLE 4   Assume that b, x, y, and z are positive numbers and $b \neq 1$. Write each expression as one logarithm.

    a. $3\log_b x + \dfrac{1}{2}\log_b y$ **b.** $\dfrac{1}{2}\log_b(x-2) - \log_b y + 3\log_b z$

Solution   **a.** $3\log_b x + \dfrac{1}{2}\log_b y = \log_b x^3 + \log_b y^{1/2}$ A power times a log is the log of the power.

                                  $= \log_b(x^3 y^{1/2})$ The sum of two logs is the log of a product.

                                  $= \log_b(x^3 \sqrt{y})$ $y^{1/2} = \sqrt{y}$

    b. $\dfrac{1}{2}\log_b(x-2) - \log_b y + 3\log_b z$

         $= \log_b(x-2)^{1/2} - \log_b y + \log_b z^3$ A power times a log is the log of the power.

         $= \log_b \dfrac{(x-2)^{1/2}}{y} + \log_b z^3$ The difference of two logs is the log of the quotient.

         $= \log_b \dfrac{z^3 \sqrt{x-2}}{y}$ The sum of two logs is the log of a product.

SELF CHECK 4   Write the expression as one logarithm.

    $2\log_b x + \dfrac{1}{2}\log_b y - 2\log_b(x-y)$

We summarize the properties of logarithms.

PROPERTIES OF LOGARITHMS	If b, M, and N are positive numbers and $b \neq 1$, then

1. $\log_b 1 = 0$ **2.** $\log_b b = 1$

3. $\log_b b^x = x$ **4.** $b^{\log_b x} = x$

5. $\log_b MN = \log_b M + \log_b N$ **6.** $\log_b \dfrac{M}{N} = \log_b M - \log_b N$

7. $\log_b M^p = p \log_b M$ **8.** If $\log_b x = \log_b y$, then $x = y$.

4 Evaluate a logarithm by applying properties of logarithms.

EXAMPLE 5 Given that $\log 2 \approx 0.3010$ and $\log 3 \approx 0.4771$, find approximations without using a calculator for the following:
a. $\log 6$ **b.** $\log 9$ **c.** $\log 18$ **d.** $\log 2.5$

Solution

a. $\log 6 = \log(2 \cdot 3)$

$= \log 2 + \log 3$ The log of a product is the sum of the logs.

$\approx 0.3010 + 0.4771$ Substitute the approximate value of each logarithm.

≈ 0.7781

b. $\log 9 = \log(3^2)$

$= 2 \log 3$ The log of a power is the power times the log.

$\approx 2(0.4771)$ Substitute the approximate value of log 3.

≈ 0.9542

c. $\log 18 = \log(2 \cdot 3^2)$

$= \log 2 + \log 3^2$ The log of a product is the sum of the logs.

$= \log 2 + 2 \log 3$ The log of a power is the power times the log.

$\approx 0.3010 + 2(0.4771)$ Substitute the approximate value of each logarithm.

≈ 1.2552

d. $\log 2.5 = \log\left(\dfrac{5}{2}\right)$

$= \log 5 - \log 2$ The log of a quotient is the difference of the logs.

$= \log \dfrac{10}{2} - \log 2$ Write 5 as $\frac{10}{2}$.

$= \log 10 - \log 2 - \log 2$ The log of a quotient is the difference of the logs.

$= 1 - 2 \log 2$ $\log_{10} 10 = 1$

$\approx 1 - 2(0.3010)$ Substitute the approximate value of log 2.

≈ 0.3980

SELF CHECK 5 Use the values given in Example 5 and approximate:
a. $\log 1.5$ **b.** $\log 0.2$

5 Apply the change-of-base formula.

If we know the base-a logarithm of a number, we can find its logarithm to some other base b with a formula called the **change-of-base formula**.

CHANGE-OF-BASE FORMULA	If a, b, and x are real numbers, $b > 0$, and $b \neq 1$, then $$\log_b x = \frac{\log_a x}{\log_a b}$$

Proof To prove this formula, we begin with the equation $\log_b x = y$.

$$y = \log_b x$$

$$x = b^y \qquad \text{Write the equation in exponential form.}$$

$$\log_a x = \log_a b^y \qquad \text{Take the base-}a\text{ logarithm of both sides.}$$

$$\log_a x = y \log_a b \qquad \text{The log of a power is the power times the log.}$$

$$y = \frac{\log_a x}{\log_a b} \qquad \text{Divide both sides by } \log_a b.$$

$$\log_b x = \frac{\log_a x}{\log_a b} \qquad \text{Refer to the first equation and substitute } \log_b x \text{ for } y.$$

If we know logarithms to base a (for example, $a = 10$), we can find the logarithm of x to a new base b. We simply divide the base-a logarithm of x by the base-a logarithm of b.

EXAMPLE 6 Approximate to 4 decimal places $\log_4 9$ using base-10 logarithms.

Solution We substitute 4 for b, 10 for a, and 9 for x into the change-of-base formula.

COMMENT $\frac{\log_a x}{\log_a b}$ means that one logarithm is to be divided by the other.

$$\log_b x = \frac{\log_a x}{\log_a b}$$

$$\log_4 9 = \frac{\log_{10} 9}{\log_{10} 4}$$

$$\approx 1.584962501$$

To four decimal places, $\log_4 9 = 1.5850$.

SELF CHECK 6 Approximate $\log_5 3$ to four decimal places using base-10 logarithms.

COMMENT It does not matter what base you choose when applying the change-of-base formula. You could use base-e (natural logarithm) and obtain the same result. In the example above, $\log_4 9 = \frac{\ln 9}{\ln 4} \approx 1.584962501$.

6 Solve an application using one or more logarithmic properties.

Common logarithms are used to express the acidity of solutions. The more acidic a solution, the greater the concentration of hydrogen ions. This concentration is indicated by the pH scale, or hydrogen ion index. The **pH of a solution** is defined by the following equation.

pH OF A SOLUTION	If $[\text{H}^+]$ is the hydrogen ion concentration in gram-ions per liter, then $$\text{pH} = -\log[\text{H}^+]$$

EXAMPLE 7 **FINDING THE pH OF A SOLUTION** Find the pH of pure water, which has a hydrogen ion concentration of 10^{-7} gram-ions per liter.

Solution Since pure water has approximately 10^{-7} gram-ions per liter, its pH is

$$pH = -\log[\mathbf{H^+}]$$
$$pH = -\log \mathbf{10^{-7}}$$
$$= -(-7)\log 10 \qquad \text{The log of a power is the power times the log.}$$
$$= -(-7)(1) \qquad \log 10 = 1$$
$$= 7$$

SELF CHECK 7 Find the pH of black coffee, which has a hydrogen ion concentration of 10^{-5} gram-ions per liter.

EXAMPLE 8 **FINDING THE HYDROGEN ION CONCENTRATION** Find the hydrogen ion concentration of seawater if its pH is 8.5.

Solution To find its hydrogen ion concentration, we substitute 8.5 for the pH and find $[H^+]$.

$$8.5 = -\log[H^+]$$
$$-8.5 = \log[H^+] \qquad \text{Multiply both sides by } -1.$$
$$[H^+] = 10^{-8.5} \qquad \text{Write the equation in exponential form.}$$

We can use a calculator to find that

$$[H^+] \approx 3.2 \times 10^{-9} \text{ gram-ions per liter}$$

SELF CHECK 8 Find the hydrogen ion concentration of healthy skin if its pH is 4.5.

In physiology, experiments suggest that the relationship between the loudness and the intensity of sound is logarithmic and has been named the Weber–Fechner law.

WEBER–FECHNER LAW If L is the apparent loudness of a sound, I is the actual intensity, and k is a constant, then

$$L = k \ln I$$

EXAMPLE 9 **WEBER–FECHNER LAW** Find the increase in intensity that will cause the apparent loudness of a sound to double.

Solution If the original loudness L_O is caused by an actual intensity I_O, then

$$(1) \qquad L_O = k \ln I_O$$

To double the apparent loudness, we multiply both sides of Equation 1 by 2 and use the power rule of logarithms:

$$\mathbf{2}\,L_O = \mathbf{2}\,k \ln I_O$$
$$= k \ln(I_O)^2$$

To double the loudness of a sound, the intensity must be squared.

SELF CHECK 9 What decrease in intensity will cause a sound to be half as loud?

NOW TRY THIS

Evaluate.

1. $\log 5 + \log 20$ **2.** $\log_3 24 - \log_3 8$ **3.** $7\log_2 \frac{1}{2} - \log_2 \frac{1}{8}$

9.5 Exercises

WARM-UPS *Find the value of x in each equation.*

1. $\log_5 25 = x$ **2.** $\log_x 5 = 1$

3. $\log_7 x = 3$ **4.** $\log_2 x = -2$

5. $\log_9 x = \frac{1}{2}$ **6.** $\log_x 4 = 2$

7. $\log_{1/3} x = 2$ **8.** $\log_{16} 4 = x$

9. $\log_x \frac{1}{4} = -2$

REVIEW *Consider the line that passes through* $(-2, 3)$ *and* $(4, -4)$.

10. Find the slope of the line.

11. Find the distance between the points.

12. Find the midpoint of the segment.

13. Write an equation of the line.

VOCABULARY AND CONCEPTS *Fill in the blanks. Assume all variables are positive numbers and* $b \neq 1$.

14. $\log_b 1 = \_$

15. $\log_b b = \_$

16. $\log_b MN = \log_b \_ + \log_b \_$

17. $b^{\log_b x} = \_$

18. If $\log_b x = \log_b y$, then $\_ = \_$

19. $\log_b \frac{M}{N} = \log_b M \_ \log_b N$

20. $\log_b x^p = p \cdot \log_b \_$

21. $\log_b b^x = \_$

22. $\log_b(A + B) \_\_\_ \log_b A + \log_b B$

23. $\log_b A + \log_b B \_\_\_ \log_b AB$

24. The change-of-base formula states that $\log_b x = \_\_\_\_$.

GUIDED PRACTICE *Simplify each expression. SEE EXAMPLE 1. (OBJECTIVE 1)*

25. $\log_3 1 = \_$ **26.** $\log_7 7 = \_$

27. $\log_6 6^2 = \_$ **28.** $7^{\log_7 9} = \_$

29. $5^{\log_5 10} = \_\_$ **30.** $\log_4 4^3 = \_$

31. $\log_2 2 = \_$ **32.** $\log_4 1 = \_$

Use a calculator to verify each equation. (OBJECTIVE 1)

33. $\log[(2.5)(3.7)] = \log 2.5 + \log 3.7$

34. $\ln \frac{11.3}{6.1} = \ln 11.3 - \ln 6.1$

35. $\ln(2.25)^4 = 4\ln 2.25$

36. $\log 45.37 = \frac{\ln 45.37}{\ln 10}$

Assume that x, y, z, and b are positive numbers (b ≠ 1). Use the properties of logarithms to write each expression in terms of the logarithms of x, y, and z. SEE EXAMPLE 2. (OBJECTIVE 2)

37. $\log_b 7xy$ **38.** $\log_b 4xz$

39. $\log_b \frac{5x}{y}$ **40.** $\log_b \frac{x}{yz}$

Assume that x, y, z, and b are positive numbers (b ≠ 1). Use the properties of logarithms to write each expression in terms of the logarithms of x, y, and z. SEE EXAMPLE 3. (OBJECTIVE 2)

41. $\log_b x^3 y^2$ **42.** $\log_b xy^2 z^3$

43. $\log_b \frac{x^3\sqrt{z}}{y^5}$

44. $\log_b \sqrt{xy}$

Assume that x, y, z, and b are positive numbers (b ≠ 1). Use the properties of logarithms to write each expression as the logarithm of a single quantity. SEE EXAMPLE 4. (OBJECTIVE 3)

45. $\log_b(x - 3) - 5 \log_b x$

46. $\log_b x + \log_b(x + 2) - \log_b 8$

47. $7 \log_b x + \dfrac{1}{2} \log_b z$

48. $-2 \log_b x - 3 \log_b y + \log_b z$

Assume that log 4 ≈ 0.6021, log 7 ≈ 0.8451, and log 9 ≈ 0.9542. Use these values and the properties of logarithms to approximate each value. Do not use a calculator. SEE EXAMPLE 5. (OBJECTIVE 4)

49. $\log 28$

50. $\log \dfrac{7}{9}$

51. $\log 2.25$

52. $\log 1.75$

53. $\log \dfrac{49}{36}$

54. $\log \dfrac{4}{63}$

55. $\log 252$

56. $\log 49$

Use a calculator and the change-of-base formula to find each logarithm to four decimal places. SEE EXAMPLE 6. (OBJECTIVE 5)

57. $\log_3 7$

58. $\log_7 3$

59. $\log_{1/3} 3$

60. $\log_{1/2} 6$

61. $\log_3 8$

62. $\log_5 10$

63. $\log_{\sqrt{2}} \sqrt{5}$

64. $\log_\pi e$

ADDITIONAL PRACTICE *Simplify each expression.*

65. $\log_7 1 = \_$

66. $\log_9 9 = \_$

67. $\log_3 3^7 = \_$

68. $5^{\log_5 8} = \_$

69. $8^{\log_8 10} = \_\_$

70. $\log_4 4^2 = \_$

71. $\log_9 9 = \_$

72. $\log_3 1 = \_$

Assume that x, y, z, and b are positive numbers (b ≠ 1). Use the properties of logarithms to write each expression in terms of the logarithms of x, y, and z.

73. $\log_b\left(\dfrac{xy}{z}\right)^{1/3}$

74. $\log_b \dfrac{x^4 \sqrt{y}}{z^2}$

75. $\log_b \dfrac{\sqrt[3]{x}}{\sqrt[4]{yz}}$

76. $\log_b \sqrt[4]{\dfrac{x^3 y^2}{z^4}}$

Assume that x, y, z, and b are positive numbers (b ≠ 1). Use the properties of logarithms to write each expression as the logarithm of a single quantity. Simplify, if possible.

77. $3 \log_b(x + 2) - 4 \log_b y + \dfrac{1}{2} \log_b(x + 1)$

78. $3 \log_b(x + 1) - 2 \log_b(x + 2) + \log_b x$

79. $\log_b\left(\dfrac{x}{z} + x\right) - \log_b\left(\dfrac{y}{z} + y\right)$

80. $\log_b(xy + y^2) - \log_b(xz + yz) + \log_b z$

Assume that log 4 ≈ 0.6021, log 7 ≈ 0.8451, and log 9 ≈ 0.9542. Use these values and the properties of logarithms to approximate each value. Do not use a calculator.

81. $\log 112$

82. $\log 324$

83. $\log \dfrac{144}{49}$

84. $\log \dfrac{324}{63}$

Use a calculator to verify each equation.

85. $\log \sqrt{24.3} = \dfrac{1}{2} \log 24.3$ **86.** $\ln 8.75 = \dfrac{\log 8.75}{\log e}$

Determine whether each statement is true. If a statement is false, explain why.

87. $\log_b 0 = 1$

88. $\log_b(x + y) \neq \log_b x + \log_b y$

89. $\log_b xy = (\log_b x)(\log_b y)$

90. $\log_b ab = \log_b a + 1$

91. $\log_7 7^7 = 7$

92. $7^{\log_7 7} = 7$

93. $\dfrac{\log_b A}{\log_b B} = \log_b A - \log_b B$

94. $\log_b(A - B) = \dfrac{\log_b A}{\log_b B}$

95. $3 \log_b \sqrt[3]{a} = \log_b a$

96. $\dfrac{1}{3} \log_b a^3 = \log_b a$

97. $\log_b \dfrac{1}{a} = -\log_b a$

98. $\log_b 2 = \log_2 b$

APPLICATIONS *Use a calculator to find each value.*
SEE EXAMPLES 7–9. (OBJECTIVE 6)

99. pH of a solution   Find the pH of a solution with a hydrogen ion concentration of 1.7×10^{-5} gram-ions per liter.

100. Hydrogen ion concentration   Find the hydrogen ion concentration of a saturated solution of calcium hydroxide whose pH is 13.2.

101. Aquariums   To test for safe pH levels in a fresh-water aquarium, a test strip is compared with the scale shown in the illustration. Find the corresponding range in the hydrogen ion concentration.

102. pH of pickles   The hydrogen ion concentration of sour pickles is 6.31×10^{-4}. Find the pH.

103. Change in loudness   If the intensity of a sound is doubled, find the apparent change in loudness.

104. Change in loudness   If the intensity of a sound is tripled, find the apparent change in loudness.

105. Change in intensity   What change in intensity of sound will cause an apparent tripling of the loudness?

106. Change in intensity   What increase in the intensity of a sound will cause the apparent loudness to be multiplied by 4?

WRITING ABOUT MATH

107. Explain why $\ln(\log 0.9)$ is undefined in the real-number system.

108. Explain why $\log_b(\ln 1)$ is undefined in the real-number system.

SOMETHING TO THINK ABOUT

109. Show that $\ln(e^x) = x$.

110. If $\log_b 3x = 1 + \log_b x$, find the value of x.

111. Show that $\log_{b^2} x = \frac{1}{2} \log_b x$.

112. Show that $e^{x \ln a} = a^x$.

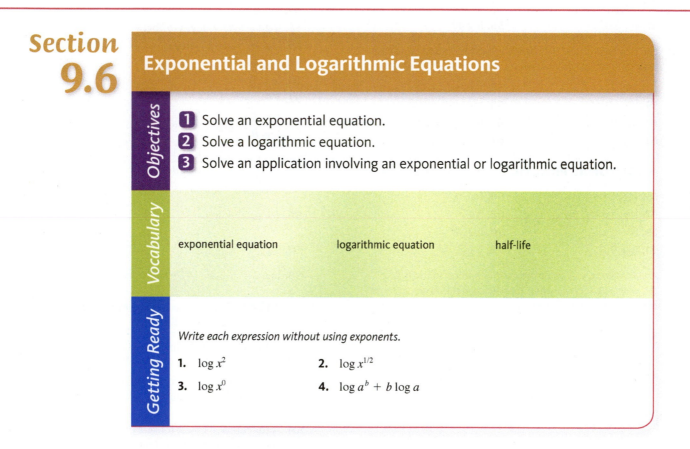

Section 9.6

Exponential and Logarithmic Equations

Objectives

1. Solve an exponential equation.
2. Solve a logarithmic equation.
3. Solve an application involving an exponential or logarithmic equation.

Vocabulary

exponential equation   logarithmic equation   half-life

Getting Ready

Write each expression without using exponents.

1. $\log x^2$
2. $\log x^{1/2}$
3. $\log x^0$
4. $\log a^b + b \log a$

An **exponential equation** is an equation that contains a variable in one of its exponents. Some examples of exponential equations are

$$3^x = 5, \qquad 6^{x-3} = 2^x, \qquad \text{and} \qquad 3^{2x+1} - 10(3^x) + 3 = 0$$

A **logarithmic equation** is an equation with logarithmic expressions that contain a variable. Some examples of logarithmic equations are

$$\log(2x) = 25, \qquad \ln x - \ln(x - 12) = 24, \qquad \text{and} \qquad \log x = \log \frac{1}{x} + 4$$

In this section, we will learn how to solve many of these equations.

1 Solve an exponential equation.

One technique we use to solve exponential equations is to apply one of the properties:

if $a = b$, then $\log a = \log b$, or

if $a = b$, then $\ln a = \ln b$.

EXAMPLE 1 Solve: $4^x = 7$

Solution Since logarithms of equal numbers are equal, we can take the common or natural logarithm of both sides of the equation and obtain a new equation. We will use the common logarithm. The power rule of logarithms then provides a process of writing the variable x as a coefficient instead of an exponent.

$$4^x = 7$$

$$\log(4^x) = \log(7) \qquad \text{Take the common logarithm of both sides.}$$

$$x \log 4 = \log 7 \qquad \text{The log of a power is the power times the log.}$$

$$(1) \qquad x = \frac{\log 7}{\log 4} \qquad \text{Divide both sides by } (\log 4).$$

$$\approx 1.403677461 \qquad \text{Use a calculator.}$$

To four decimal places, $x = 1.4037$.

SELF CHECK 1 Solve $5^x = 4$. Give the result to four decimal places.

COMMENT The right side of Equation 1 calls for a division, not a subtraction.

$$\frac{\log 7}{\log 4} \quad \text{means} \quad (\log 7) \div (\log 4)$$

It is the expression $\log\left(\frac{7}{4}\right)$ that means $\log 7 - \log 4$.

EXAMPLE 2 Solve: $73 = 1.6(1.03)^t$

Solution We divide both sides by 1.6 to isolate the exponential expression. Thus, we obtain

$$\frac{73}{1.6} = 1.03^t$$

and solve the equation. If $a = b$ then $b = a$.

$$1.03^t = \frac{73}{1.6}$$

$$\log(1.03^t) = \log\left(\frac{73}{1.6}\right) \qquad \text{Take the common logarithm of both sides.}$$

$$t \log 1.03 = \log \frac{73}{1.6} \qquad \text{The logarithm of a power is the power times the logarithm.}$$

$$t = \frac{\log \frac{73}{1.6}}{\log 1.03} \qquad \text{Divide both sides by } \log 1.03.$$

$$t \approx 129.2493444 \qquad \text{Use a calculator.}$$

To four decimal places, $t = 129.2493$.

SELF CHECK 2 Solve $47 = 2.5(1.05)^t$. Give the result to 4 decimal places.

EXAMPLE 3 Solve: $6^{x-3} = 2^x$

Solution

$$6^{x-3} = 2^x$$

$\log(6^{x-3}) = \log(2^x)$ Take the common logarithm of both sides.

$(x - 3)\log 6 = x \log 2$ The log of a power is the power times the log.

$x \log 6 - 3 \log 6 = x \log 2$ Use the distributive property to remove parentheses.

$x \log 6 - x \log 2 = 3 \log 6$ Add (3 log 6) and subtract (x log 2) from both sides.

$x(\log 6 - \log 2) = 3 \log 6$ Factor out x, the GCF, on the left side.

$$x = \frac{3 \log 6}{\log 6 - \log 2}$$ Divide both sides by (log 6 − log 2).

$x \approx 4.892789261$ Use a calculator.

To four decimal places, $x = 4.8928$.

SELF CHECK 3 Solve to four decimal places: $5^{x-2} = 3^x$

Another technique we use to solve exponential equations is to write the expressions with the same base and then set their corresponding exponents equal.

If $a^n = a^m$, then $n = m$.

EXAMPLE 4 Solve: $2^{x^2+2x} = \frac{1}{2}$

Solution We can use the fact that both expressions involve a base of 2 to write the equation in a different form. Since $\frac{1}{2} = 2^{-1}$, we can write the equation in the form

$$2^{x^2+2x} = 2^{-1}$$

Equal quantities with equal bases have equal exponents. Therefore,

$x^2 + 2x = -1$

$x^2 + 2x + 1 = 0$ Add 1 to both sides.

$(x + 1)(x + 1) = 0$ Factor the trinomial.

$x + 1 = 0$ or $x + 1 = 0$ Set each factor equal to 0.

$x = -1$ | $x = -1$ −1 is a double root.

Verify that −1 satisfies the equation.

SELF CHECK 4 Solve: $3^{x^2-2x} = \frac{1}{3}$

Accent on technology

▶ Solving Exponential Equations

To use a TI-84 graphing calculator to approximate the solutions of $2^{x^2+2x} = \frac{1}{2}$ (see Example 4), we can graph the two corresponding functions

$$f(x) = 2^{x^2+2x} \quad \text{and} \quad f(x) = \frac{1}{2}$$

and find the intersection.

If we use window settings of $[-4, 4]$ for x and $[-2, 6]$ for y, we obtain the graph shown in Figure 9-25(a).

We use the intersect feature to find the point(s) of intersection as illustrated in Figure 9.25(b). Since $x = -1$ is the only intersection point, −1 is the only solution.

Figure 9-25

For instructions regarding the use of a Casio graphing calculator, please refer to the Casio Keystroke Guide in the back of the book.

2 Solve a logarithmic equation.

In each of the following examples, we use the properties of logarithms to change a logarithmic equation into an algebraic equation. Recall that the logarithmic property of equality states that if $\log_b x = \log_b y$, then $x = y$. Be certain to check any apparent solutions.

EXAMPLE 5 Solve: $\log_b(3x + 2) - \log_b(2x - 3) = 0$

Solution

$$\log_b(3x + 2) - \log_b(2x - 3) = 0$$

$$\log_b(3x + 2) = \log_b(2x - 3) \qquad \text{Add } \log_b(2x - 3) \text{ to both sides.}$$

$$3x + 2 = 2x - 3 \qquad \text{If } \log_b r = \log_b s, \text{ then } r = s.$$

$$x = -5 \qquad \text{Subtract } 2x \text{ and } 2 \text{ from both sides.}$$

Check:

$$\log_b(3x + 2) - \log_b(2x - 3) = 0$$

$$\log_b[3(-5) + 2] - \log_b[2(-5) - 3] \stackrel{?}{=} 0$$

$$\log_b(-13) - \log_b(-13) \stackrel{?}{=} 0$$

COMMENT Example 5 illustrates that you must check the apparent solutions of all logarithmic equations.

Since we cannot take the logarithm of a negative number, the apparent solution of -5 must be discarded. Since this equation has no solution, its solution set is $\varnothing$.

SELF CHECK 5 Solve: $\log_b(5x + 2) - \log_b(7x - 2) = 0$

COMMENT To solve a logarithmic equation, consider whether writing the equation in exponential form would help you solve for the variable.

EXAMPLE 6 Solve: $\log x + \log(x - 3) = 1$

Solution

$$\log x + \log(x - 3) = 1$$

$$\log[x(x - 3)] = 1 \qquad \text{The sum of two logs is the log of a product.}$$

$$x(x - 3) = 10^1 \qquad \text{Use the definition of logarithms to write the equation in exponential form.}$$

$$x^2 - 3x - 10 = 0 \qquad \text{Use the distributive property to remove parentheses and write in quadratic form.}$$

$$(x + 2)(x - 5) = 0 \qquad \text{Factor the trinomial.}$$

$$x + 2 = 0 \quad \text{or} \quad x - 5 = 0 \qquad \text{Set each factor equal to 0.}$$

$$x = -2 \qquad \qquad x = 5$$

Check: The number -2 is not a solution, because we cannot take the logarithm of a negative number. It is extraneous. We will check the remaining number, 5.

$$\log x + \log(x - 3) = 1$$

$$\log 5 + \log(5 - 3) \overset{?}{=} 1 \qquad \text{Substitute 5 for } x.$$

$$\log 5 + \log 2 \overset{?}{=} 1$$

$$\log 10 \overset{?}{=} 1 \qquad \text{The sum of two logs is the log of a product.}$$

$$1 = 1 \qquad \log 10 = 1$$

Since 5 satisfies the equation, it is a solution.

SELF CHECK 6   Solve: $\log x + \log(x + 3) = 1$

EXAMPLE 7   Solve: $\dfrac{\log(5x - 6)}{\log x} = 2$

Solution   We know that the $\log x$ cannot equal 0, thus, we can multiply both sides of the equation by $\log x$ to obtain

$$\log(5x - 6) = 2 \log x$$

and apply the power rule of logarithms to obtain

$$\log(5x - 6) = \log x^2$$

By Property 8 of logarithms, $5x - 6 = x^2$. Thus,

$$5x - 6 = x^2$$

$$0 = x^2 - 5x + 6 \qquad \text{Write in quadratic form.}$$

$$0 = (x - 3)(x - 2) \qquad \text{Factor the trinomial.}$$

$$x - 3 = 0 \quad \text{or} \quad x - 2 = 0 \qquad \text{Set each factor equal to 0.}$$

$$x = 3 \qquad\qquad x = 2 \qquad \text{Solve each equation.}$$

Verify that both 2 and 3 satisfy the equation.

SELF CHECK 7   Solve: $\dfrac{\log(5x + 6)}{\log x} = 2$

Accent on technology

▶ Solving Logarithmic Equations

To use a graphing calculator to approximate the solutions of $\log x + \log(x - 3) = 1$ (see Example 6), we can graph the corresponding functions

$$f(x) = \log(x) + \log(x - 3) \quad \text{and} \quad f(x) = 1$$

If we use window settings of $[0, 10]$ for x and $[-2, 2]$ for y, we obtain the graph shown in Figure 9-26. Since the solution of the equation is the intersection, we can find the solution by using the INTERCEPT command. The solution is $x = 5$.

Figure 9-26

For instructions regarding the use of a Casio graphing calculator, please refer to the Casio Keystroke Guide in the back of the book.

3 Solve an application involving an exponential or logarithmic equation.

We have previously discussed that the amount A of radiation present in a radioactive material decays exponentially according to the formula $A = A_0 e^{kt}$, where A_0 is the amount of radioactive material present at time $t = 0$, and k is a negative number.

Experiments have determined the time it takes for one-half of a sample of a radioactive element to decompose. That time is a constant, called the given material's **half-life**.

EXAMPLE 8 **HALF-LIFE OF RADON-22** In Example 5 of Section 9.2, we learned that radon-22 decays according to the formula $A = A_0 e^{-0.181t}$, where t is expressed in days. Find the material's half-life to the nearest hundredth.

Solution At time $t = 0$, the amount of radioactive material is A_0. At the end of one half-life, the amount present will be $\frac{1}{2}A_0$. To find the material's half-life, we can substitute $\frac{1}{2}A_0$ for A in the formula and solve for t.

$$A = A_0 e^{-0.181t}$$

$$\frac{1}{2}A_0 = A_0 e^{-0.181t} \qquad \text{Substitute } \tfrac{1}{2}A_0 \text{ for } A.$$

$$\frac{1}{2} = e^{-0.181t} \qquad \text{Divide both sides by } A_0.$$

$$\ln\left(\frac{1}{2}\right) = \ln(e^{-0.181t}) \qquad \text{Take the natural logarithm of both sides.}$$

COMMENT We can divide both sides of the equation by A_0 because we know it is not zero.

$$-0.6931471806 \approx -0.181t \ln e \qquad \text{Find } \ln\tfrac{1}{2}, \text{ and use the property } \ln M^p = p \ln M.$$

$$3.829542434 \approx t \qquad \text{Divide both sides by } -0.181 \text{ and note that } \ln e = 1.$$

To the nearest hundredth, the half-life of radon-22 is 3.83 days.

 SELF CHECK 8 To the nearest hundredth, find the half-life of iodine-131 given that it decays according to the formula $A = A_0 e^{-0.087t}$ where t is expressed in days.

When a living organism dies, the oxygen/carbon dioxide cycle common to all living things stops and carbon-14, a radioactive isotope with a half-life of 5,730 years, is no longer absorbed. By measuring the amount of carbon-14 present in an ancient object, archaeologists can estimate the object's age. The formula for radioactive decay is $A = A_0 e^{kt}$, where A_0 is the original amount of carbon-14 present, t is the age of the object, and k is a negative number.

EXAMPLE 9   **CARBON-14 DATING**   To the nearest hundred years, how old is a wooden statue that retains one-third of its original carbon-14 content?

Solution   We can find the value of k in the formula $A = A_0e^{kt}$ by using the fact that after 5,730 years, half of the original amount of carbon-14 will remain.

$$A = A_0e^{kt}$$

$$\frac{1}{2}A_0 = A_0e^{k(5,730)}$$ Substitute $\frac{1}{2}A_0$ for A and 5,730 for t.

$$\frac{1}{2} = e^{5,730k}$$ Divide both sides by A_0.

$$\ln\left(\frac{1}{2}\right) = \ln(e^{5,730k})$$ Take the natural logarithm of both sides.

COMMENT   Do not round any values until the last step.

$$-0.6931471806 \approx 5,730k \ln e$$ Find $\ln\frac{1}{2}$, and use the property $\ln M^p = p \ln M$.

$$-0.000120968094 \approx k$$ Divide both sides by 5,730 and note that $\ln e = 1$.

Thus, the formula for radioactive decay for carbon-14 can be written as

$$A \approx A_0e^{-0.000120968094t}$$

Since $\frac{1}{3}$ of the original carbon-14 still remains, we can proceed as follows:

$$A \approx A_0e^{-0.000120968094t}$$

$$\frac{1}{3}A_0 \approx A_0e^{-0.000120968094t}$$ Substitute $\frac{1}{3}A_0$ for A.

$$\frac{1}{3} \approx e^{-0.000120968094t}$$ Divide both sides by A_0.

$$\ln\left(\frac{1}{3}\right) \approx \ln(e^{-0.000120968094t})$$ Take the natural logarithm of both sides.

$$-1.098612289 \approx (-0.000120968094t) \ln e$$ Find $\ln\frac{1}{3}$, and use the property $\ln M^p = p \ln M$.

$$9081.835155 \approx t$$ Divide both sides by -0.000120968094 and note that $\ln e = 1$.

To the nearest one hundred years, the statue is 9,100 years old.

SELF CHECK 9   To the nearest hundred years, find the age of a statue that retains 25% of its original carbon-14 content.

Recall that when there is sufficient food and space, populations of living organisms tend to increase exponentially according to the Malthusian growth model.

MALTHUSIAN GROWTH MODEL   If P is the population at some time t, P_0 is the initial population at $t = 0$, and k is the rate of growth, then

$$P = P_0e^{kt}$$

COMMENT   Note that this formula is the same as all exponential growth formulas, except for the variables.

EXAMPLE 10 **POPULATION GROWTH** The bacteria in a laboratory culture increased from an initial population of 500 to 1,500 in 3 hours. How long will it take for the population to reach 10,000?

Solution We substitute 500 for P_0, 1,500 for P, and 3 for t and simplify to find k:

$$P = P_0 e^{kt}$$
$$1{,}500 = 500(e^{k3}) \qquad \text{Substitute 1,500 for } P, 500 \text{ for } P_0, \text{ and 3 for } t.$$
$$3 = e^{3k} \qquad \text{Divide both sides by 500.}$$
$$3k = \ln 3 \qquad \text{Write the equation in logarithmic form.}$$
$$k = \frac{\ln 3}{3} \qquad \text{Divide both sides by 3.}$$

To find when the population will reach 10,000, we substitute 10,000 for P, 500 for P_0, and $\frac{\ln 3}{3}$ for k in the equation $P = P_0 e^{kt}$ and solve for t:

$$P = P_0 e^{kt}$$
$$10{,}000 = 500 e^{[(\ln 3)/3]t}$$
$$20 = e^{[(\ln 3)/3]t} \qquad \text{Divide both sides by 500.}$$
$$\left(\frac{\ln 3}{3}\right)t = \ln 20 \qquad \text{Write the equation in logarithmic form.}$$
$$t = \frac{3 \ln 20}{\ln 3} \qquad \text{Multiply both sides by } \frac{3}{\ln 3}.$$
$$\approx 8.180499084 \qquad \text{Use a calculator.}$$

The culture will reach 10,000 bacteria in about 8 hours.

 SELF CHECK 10 How long will it take to reach 20,000?

EXAMPLE 11 **GENERATION TIME** If a medium is inoculated with a bacterial culture that contains 1,000 cells per milliliter, how many generations will pass by the time the culture has grown to a population of 1 million cells per milliliter?

Solution During bacterial reproduction, the time required for a population to double is called the *generation time*. If b bacteria are introduced into a medium, then after the generation time of the organism has elapsed, there are $2b$ cells. After another generation, there are $2(2b)$, or $4b$ cells, and so on. After n generations, the number of cells present will be

(1) $\qquad B = b \cdot 2^n$

To find the number of generations that have passed while the population grows from b bacteria to B bacteria, we solve Equation 1 for n.

$$\log B = \log(b \cdot 2^n) \qquad \text{Take the common logarithm of both sides.}$$
$$\log B = \log b + n \log 2 \qquad \text{Apply the product and power rules of logarithms.}$$
$$\log B - \log b = n \log 2 \qquad \text{Subtract } \log b \text{ from both sides.}$$
$$n = \frac{1}{\log 2}(\log B - \log b) \qquad \text{Multiply both sides by } \frac{1}{\log 2}.$$

(2) $\qquad n = \dfrac{1}{\log 2}\left(\log \dfrac{B}{b}\right) \qquad \text{Use the quotient rule of logarithms.}$

Equation 2 is a formula that gives the number of generations that will pass as the population grows from b bacteria to B bacteria.

To find the number of generations that have passed while a population of 1,000 cells per milliliter has grown to a population of 1 million cells per milliliter, we substitute 1,000 for b and 1,000,000 for B in Equation 2 on the previous page and solve for n.

$$n = \frac{1}{\log 2}\log\frac{1,000,000}{1,000}$$

$$= \frac{1}{\log 2}\log 1,000 \qquad \text{Simplify.}$$

$$\approx 3.321928095(3) \qquad \tfrac{1}{\log 2} \approx 3.321928095 \text{ and } \log 1,000 = 3.$$

$$\approx 9.965784285$$

Approximately 10 generations will have passed.

SELF CHECK 11 How many generations will pass by the time the culture has grown to a population of 10 million cells per milliliter?

Everyday connections
U.S. Population Growth

Historical and Projected U.S. Population

Population growth in the United States can be modeled by an exponential function of the form $P(t) = P_0 \cdot e^{rt}$, where $P_0 =$ the initial population during a given time interval, and t represents the number of years in that time interval.

1. Given that the United States population was approximately 200 million in 1970 and approximately 280 million in 2000, determine the growth rate of the population during this time period.

2. Given that the United States population was 280 million in 2000 and approximately 300 million in 2010, determine the growth rate of the population during this time period.

3. Using the population growth rate from Question 2, to the nearest year how long would it take for the United States population to double?

Source: http://www.npg.org/popfacts.htm

NOW TRY THIS

Given $f(x) = \log_3(x + 2) + \log_3 x$,

1. solve $f(x) = 1$.

2. solve $f(x) = 2$.

9.6   Exercises

WARM-UPS *Write each logarithmic equation as an exponential equation.*

1. $\log x = 2$
2. $\log y = 1$
3. $\log(x + 1) = 1$
4. $\log(x - 2) = 3$

🖩 *Simplify, using a calculator, and approximate to four decimal places.*

5. $\dfrac{\log 4}{\log 3}$
6. $\log\left(\dfrac{4}{3}\right)$
7. $-\log\left(\dfrac{5}{7}\right)$
8. $-\dfrac{\log 5}{\log 7}$

REVIEW *Solve each equation.*

9. $4x^2 - 64x = 0$
10. $9y^2 - 49 = 0$
11. $3p^2 + 10p = 8$
12. $4t^2 + 1 = -6t$

VOCABULARY AND CONCEPTS *Fill in the blanks.*

13. An equation with a variable as an exponent is called a(n) _____ equation.

14. An equation with a logarithmic expression that contains a variable is a(n) _____ equation.

15. The formula for radioactive decay is $A =$ _____.

16. The _____ of a radioactive element is determined by how long it takes for half of a sample to decompose.

GUIDED PRACTICE *Solve each exponential equation. If an answer is not exact, give the answer to four decimal places. SEE EXAMPLE 1. (OBJECTIVE 1)*

17. $3^x = 8$
18. $7^x = 12$
19. $e^t = 50$
20. $e^{-t} = 0.25$

Solve each exponential equation. If an answer is not exact, give the answer to four decimal places. SEE EXAMPLE 2. (OBJECTIVE 1)

21. $7 = 4.3(1.01)^t$
22. $52 = 5.1(1.03)^t$
23. $5 = 2.1(1.04)^t$
24. $61 = 1.5(1.02)^t$

Solve each exponential equation. Give the answer to four decimal places. SEE EXAMPLE 3. (OBJECTIVE 1)

25. $6^{x-2} = 4$
26. $5^{x+1} = 3$
27. $4^{x+1} = 7^x$
28. $5^{x-3} = 3^{2x}$

Solve each exponential equation. If an answer is not exact, give the answer to four decimal places. SEE EXAMPLE 4. (OBJECTIVE 1)

29. $5^{x^2 + 2x} = 125$
30. $3^{x^2 - 3x} = 81$
31. $3^{x^2 + 4x} = \dfrac{1}{81}$
32. $7^{x^2 + 3x} = \dfrac{1}{49}$

🖩 *Use a calculator to solve each equation, if possible. Give all answers to the nearest tenth. (OBJECTIVE 1)*

33. $2^{x+1} = 7$
34. $3^{x-1} = 2^x$
35. $2^{x^2 - 2x} - 8 = 0$
36. $3^x - 10 = 3^{-x}$

Solve each logarithmic equation. Check all solutions. SEE EXAMPLE 5. (OBJECTIVE 2)

37. $\log 9x = \log 27$
38. $\log 5x = \log 45$
39. $\log(4x + 3) = \log(x + 9)$
40. $\log(x^2 + 4x) = \log(x^2 + 16)$

Solve each logarithmic equation. Check all solutions. SEE EXAMPLE 6. (OBJECTIVE 2)

41. $\log x^2 = 2$
42. $\log x^3 = 3$
43. $\log x + \log(x - 3) = 1$
44. $\log x + \log(x + 9) = 1$
45. $\log x + \log(x - 15) = 2$
46. $\log x + \log(x + 21) = 2$
47. $\log(x + 90) = 3 - \log x$
48. $\log(x - 90) = 3 - \log x$

Solve each logarithmic equation. Check all solutions. SEE EXAMPLE 7. (OBJECTIVE 2)

49. $\dfrac{\log(2x + 1)}{\log(x - 1)} = 2$

50. $\dfrac{\log(4x + 9)}{\log(2x - 3)} = 2$

51. $\dfrac{\log(3x + 4)}{\log x} = 2$

52. $\dfrac{\log(8x - 7)}{\log x} = 2$

Use a graphing calculator to solve each equation. If an answer is not exact, give all answers to the nearest tenth. (OBJECTIVE 2)

53. $\log x + \log(x - 15) = 2$

54. $\log x + \log(x + 3) = 1$

55. $\ln(2x + 5) - \ln 3 = \ln(x - 1)$

56. $\log(x^2 + 4x)^2 = 1$

ADDITIONAL PRACTICE *Solve each exponential equation. If an answer is not exact, give the answer to four decimal places.*

57. $5^x = 4^x$

58. $3^{2x} = 4^x$

59. $7^{x^2} = 10$

60. $8^{x^2} = 11$

61. $8^{x^2} = 9^x$

62. $5^{x^2} = 2^{5x}$

Solve each logarithmic equation. Check all solutions.

63. $\log(5 - 4x) - \log(x + 25) = 0$

64. $\log(3x + 5) - \log(2x + 6) = 0$

65. $\log(x - 6) - \log(x - 2) = \log\dfrac{5}{x}$

66. $\log(3 - 2x) - \log(x + 9) = 0$

Solve each equation. If the answer is not exact, round to four decimal places.

67. $4^{x+2} - 4^x = 15$ (*Hint:* $4^{x+2} = 4^x 4^2$.)

68. $3^{x+3} + 3^x = 84$ (*Hint:* $3^{x+3} = 3^x 3^3$.)

69. $\dfrac{\log(5x + 6)}{2} = \log x$

70. $\dfrac{1}{2}\log(4x + 5) = \log x$

71. $\log_3 x = \log_3\left(\dfrac{1}{x}\right) + 4$

72. $\log_5(7 + x) + \log_5(8 - x) - \log_5 2 = 2$

73. $2(3^x) = 6^{2x}$

74. $2(3^{x+1}) = 3(2^{x-1})$

75. $\log x^2 = (\log x)^2$

76. $\log(\log x) = 1$

77. $2\log_2 x = 3 + \log_2(x - 2)$

78. $2\log_3 x - \log_3(x - 4) = 2 + \log_3 2$

79. $\log(7y + 1) = 2\log(y + 3) - \log 2$

80. $2\log(y + 2) = \log(y + 2) - \log 12$

81. $\log\dfrac{4x + 1}{2x + 9} = 0$

82. $\log\dfrac{2 - 5x}{2(x + 8)} = 0$

APPLICATIONS *Solve each application.* SEE EXAMPLE 8. (OBJECTIVE 3)

83. **Half-life**   To the nearest day, find the half-life of strontium-89, given that it decays according to the formula $A = A_0 e^{-0.013t}$.

84. **Half-life**   To the nearest thousand years, find the half-life of plutonium-239, given that it decays according to the formula $A = A_0 e^{-0.0000284t}$.

85. **Radioactive decay**   In two years, 20% of a radioactive element decays. Find its half-life.

86. **Tritium decay**   The half-life of tritium is 12.4 years. How long will it take for 25% of a sample of tritium to decompose?

Solve each application. SEE EXAMPLE 9. (OBJECTIVE 3)

87. **Carbon-14 dating**   The bone fragment shown in the illustration contains 60% of the carbon-14 that it is assumed to have had initially. How old is it?

© Falk Kienas/Shutterstock.com

88. **Carbon-14 dating**   Only 10% of the carbon-14 in a small wooden bowl remains. How old is the bowl?

Solve each application. SEE EXAMPLE 10. (OBJECTIVE 3)

89. **Rodent control**   The rodent population in a city is currently estimated at 30,000. If it is expected to double every 5 years, when will the population reach 1 million?

90. **Population growth**   The population of a city is expected to triple every 15 years. When can the city planners expect the present population of 140 persons to double?

91. **Bacterial culture**   A bacterial culture doubles in size every 24 hours. By how much will it have increased in 36 hours?

92. **Bacterial growth**   A bacterial culture grows according to the formula

$$P = P_0 a^t$$

If it takes 5 days for the culture to triple in size, how long will it take to double in size?

Solve each application. SEE EXAMPLE 11. (OBJECTIVE 3)

93. **Medicine**   If a medium is inoculated with a bacterial culture containing 500 cells per milliliter, how many generations will have passed by the time the culture contains 5×10^6 cells per milliliter?

94. Medicine   If a medium is inoculated with a bacterial culture containing 800 cells per milliliter, how many generations will have passed by the time the culture contains 6×10^7 cells per milliliter?

Solve each application.

95. Thorium decay   An isotope of thorium, ^{227}Th, has a half-life of 18.4 days. How long will it take for 80% of the sample to decompose?

96. Lead decay   An isotope of lead, ^{201}Pb, has a half-life of 8.4 hours. How many hours ago was there 30% more of the substance?

97. Compound interest   If $500 is deposited in an account paying 8.5% annual interest, compounded semiannually, how long will it take for the account to increase to $800?

98. Continuous compound interest   In Exercise 97, how long will it take if the interest is compounded continuously?

99. Compound interest   If $1,300 is deposited in a savings account paying 9% interest, compounded quarterly, how long will it take the account to increase to $2,100?

100. Compound interest   A sum of $5,000 deposited in an account grows to $7,000 in 5 years. Assuming annual compounding, what interest rate is being paid?

101. Rule of seventy   A rule of thumb for finding how long it takes an investment to double is called the **rule of seventy**. To apply the rule, divide 70 by the interest rate written as a percent. At 5%, it takes $\frac{70}{5} = 14$ years to double an investment. At 7%, it takes $\frac{70}{7} = 10$ years. Explain why this formula works.

102. Oceanography   The intensity I of a light a distance x meters beneath the surface of a lake decreases exponentially. From the illustration, find the depth at which the intensity will be 20%.

WRITING ABOUT MATH

103. Explain how to solve the equation $2^x = 7$.

104. Explain how to solve the equation $x^2 = 7$.

SOMETHING TO THINK ABOUT

105. Without solving the following equation, find the values of x that cannot be a solution:

$$\log(x - 3) - \log(x^2 + 2) = 0$$

106. Solve the equation $x^{\log x} = 10,000$.

Projects

PROJECT 1

When an object moves through air, it encounters air resistance. So far, all ballistics problems in this text have ignored air resistance. We now consider the case where an object's fall is affected by air resistance.

At relatively low velocities ($v < 200$ feet per second), the force resisting an object's motion is a constant multiple of the object's velocity:

Resisting force $= f_r = bv$

where b is a constant that depends on the size, shape, and texture of the object, and has units of kilograms per second. This is known as **Stokes' law of resistance**.

In a vacuum, the downward velocity of an object dropped with an initial velocity of 0 feet per second is

$v(t) = 32t$ (no air resistance)

t seconds after it is released. However, with air resistance, the velocity is given by the formula

$$v(t) = \frac{32m}{b}\left(1 - e^{-(b/m)t}\right)$$

where m is the object's mass (in kilograms). There is also a formula for the distance an object falls (in feet) during the first t seconds after release, taking into account air resistance:

$$d(t) = \frac{32m}{b}t - \frac{32m^2}{b^2}\left(1 - e^{-(b/m)t}\right)$$

Without air resistance, the formula would be

$$d(t) = 16t^2$$

a. Fearless Freda, a renowned skydiving daredevil, performs a practice dive from a hot-air balloon with an altitude of 5,000 feet. With her parachute on, Freda has a mass of 75 kg, so that $b = 15$ kg/sec. How far (to the nearest foot) will Freda fall in 5 seconds? Compare this with the answer you get by disregarding air resistance.

b. What downward velocity (to the nearest ft/sec) does Freda have after she has fallen for 2 seconds? For

(continued)

5 seconds? Compare these answers with the answers you get by disregarding air resistance.

c. Find Freda's downward velocity after falling for 20, 22, and 25 seconds. (Without air resistance, Freda would hit the ground in less than 18 seconds.) Note that Freda's velocity increases only slightly. This is because for a large enough velocity, the force of air resistance almost counteracts the force of gravity; after Freda has been falling for a few seconds, her velocity becomes nearly constant. The constant velocity that a falling object approaches is called the *terminal velocity*.

$$\text{Terminal velocity} = \frac{32m}{b}$$

Find Freda's terminal velocity for her practice dive.

d. In Freda's show, she dives from a hot-air balloon with an altitude of only 550 feet, and pulls her ripcord when her velocity is 100 feet per second. (She can't tell her speed, but she knows how long it takes to reach that speed.) It takes a fall of 80 more feet for the chute to open fully, but then the chute increases the force of air resistance, making $b = 80$. After that, Freda's velocity approaches the terminal velocity of an object with this new b-value.

To the nearest hundredth of a second, how long should Freda fall before she pulls the ripcord? To the nearest foot, how close is she to the ground when she pulls the ripcord? How close to the ground is she when the chute takes full effect? At what velocity will Freda hit the ground?

PROJECT 2

If an object at temperature T_0 is surrounded by a constant temperature T_s (for instance, an oven or a large amount of fluid that has a constant temperature), the temperature of the object will change with time t according to the formula

$$T(t) = T_s + (T_0 - T_s)e^{-kt}$$

This is **Newton's law of cooling and warming**. The number k is a constant that depends on how well the object absorbs and dispels heat.

In the course of brewing "yo ho! grog," the dread pirates of Hancock Isle have learned that it is important that their rather disgusting, soupy mash be heated slowly to allow all of the ingredients a chance to add their particular offensiveness to the mixture. However, after the mixture has simmered for several hours, it is equally important that the grog be cooled very quickly, so that it retains its potency. The kegs of grog are then stored in a cool spring.

By trial and error, the pirates have learned that by placing the mash pot into a tub of boiling water (100° C), they can heat the mash in the correct amount of time. They have also learned that they can cool the grog to the temperature of the spring by placing it in ice caves for 1 hour.

With a thermometer, you find that the pirates heat the mash from 20° C to 95° C and then cool the grog from 95° C to 7° C. Calculate how long the pirates cook the mash, and how cold the ice caves are. Assume that $k = 0.5$, and t is measured in hours.

Reach for Success

EXTENSION OF TAKING NOTES IN CLASS

You've learned to use two colors to take notes in class. In the previous exercise we listed *what* we did in each step. In this excercise, we will list both *what* steps we took and *why*.

To make an example easier to understand, we explain the steps we take, and why we take them, to find the answer. This serves two purposes: • to use the vocabulary necessary for success in your mathematics course; and • to help you understand the mathematical processes.	Find the slope and the *y*-intercept for the line with equation $5x + 3y = 12$ $5x + 3y = 12x$ $\underline{-5x \quad\quad -5x}$ $\dfrac{3y}{3} = -\dfrac{5x}{3} + \dfrac{12}{3}$ $y = -\frac{5}{3}x + 4$ The slope is $-\frac{5}{3}$ The y-intercept is (0, 4)	In the blanks below list the steps taken to complete the exercise. The explanation of *why* has already been listed for you. Original exercise. _____ (to isolate the *y*-term) _____ (to obtain a leading coefficient of 1 on *y*) _____ This is in *y* = *mx* + *b* form. (slope/*y*-intercept form) _____ The slope is the coefficient of the *x*-term, and the *y*-intercept is a point that must be expressed as an ordered pair.
For each step taken in answering the question, state • what steps were taken, • explain why we took those steps. Fill in the last two blanks of the exercise in the second column and explain how you arrived at those answers in the third column.	Find the slope of the line perpendicular to the line $3x + y = 15$ $3x + y = 15$ $\underline{-3x \quad\quad -3x}$ $\quad\quad y = -3x + 15$ The slope of the line is _____. The slope of a line perpendicular to this line is _____.	Step: _____ (Why?) _____ (Why?) _____ (Why?) _____

You can use the techniques to reinforce your understanding of the material, even if your instructor does not use these methods when explaining problems in class.

9 Review 🌿

SECTION 9.1   Exponential Functions

DEFINITIONS AND CONCEPTS	EXAMPLES
An exponential function with base b is defined by the equation $f(x) = b^x$ ($b > 0, b \neq 1$ and x is a real number) The domain is $(-\infty, \infty)$ and the range is $(0, \infty)$.	Graph $f(x) = 4^x$ and $f(x) = \left(\frac{1}{4}\right)^x$.
Increasing/decreasing functions: The graph of $f(x) = b^x$ is an increasing function if $b > 1$ and decreasing if $0 < b < 1$.	The graph of $f(x) = 4^x$ is an increasing function because as the values of x increase the values of $f(x)$ also increase. The graph of $f(x) = \left(\frac{1}{4}\right)^x$ is a decreasing function because as the values of x increase the values of $f(x)$ decrease.
Graphing translations: Graphing a translation of an exponential function follows the same rules as for polynomial functions.	Graph $f(x) = 4^x - 2$ and $f(x) = 4^{x-2}$. The graph of $f(x) = 4^x - 2$ is the same as the graph of $f(x) = 4^x$ but shifted 2 units downward. The graph of $f(x) = 4^{x-2}$ is the same as the graph of $f(x) = 4^x$ but shifted 2 units to the right.
Compound interest: The amount of money in an account with an initial deposit of $\$P$ at an annual interest rate of $r\%$, compounded k times a year for t years, can be found using the formula $A = P\left(1 + \frac{r}{k}\right)^{kt}$.	To find the balance in an account after 3 years when $5,000 is deposited at 6% interest, compounded quarterly, we note that $P = 5{,}000 \qquad r = 0.06 \qquad k = 4 \qquad t = 3$ We can substitute these values into the formula for compound interest and proceed as follows: $A = P\left(1 + \dfrac{r}{k}\right)^{kt}$ $A = 5{,}000\left(1 + \dfrac{0.06}{4}\right)^{4(3)}$ $A = 5{,}000(1 + 0.015)^{12}$ $A = 5{,}000(1.015)^{12}$ $A \approx 5{,}978.090857 \qquad$ Use a calculator. There will be $5,978.09 in the account after 3 years.

REVIEW EXERCISES

1. Sketch the graph of an increasing exponential function.

2. Sketch the graph of a decreasing exponential function.

Graph the function defined by each equation.

3. $f(x) = 3^x$ **4.** $f(x) = \left(\dfrac{1}{3}\right)^x$

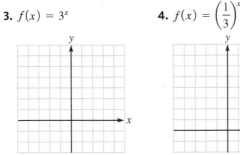

5. The graph of $f(x) = 6^x$ will pass through the points $(0, x)$ and $(1, y)$. Find the values of x and y.

6. Give the domain and range of the function $f(x) = 7^x$.

Graph each function by using a translation.

7. $f(x) = \left(\dfrac{1}{2}\right)^x - 2$ **8.** $f(x) = \left(\dfrac{1}{2}\right)^{x+2}$

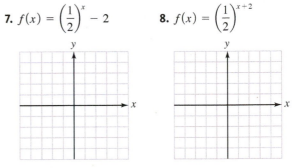

9. Savings How much will $25,000 become if it earns 5% per year for 35 years, compounded quarterly?

SECTION 9.2   Base-*e* Exponential Functions

DEFINITIONS AND CONCEPTS	EXAMPLES
$e \approx 2.71828182845904$ **Continuous compound interest:** The amount of money A in an account with an initial deposit of $\$P$ at an annual interest rate of $r\%$, compounded continuously for t years, can be found using the formula $A = Pe^{rt}$	To find the balance in an account after 3 years when $5,000 is deposited at 6% interest, compounded continuously, we note that $P = 5{,}000 \quad r = 0.06 \quad t = 3$ We can substitute these values into the formula $A = Pe^{rt}$ and simplify to obtain $A = Pe^{rt}$ $A = 5{,}000e^{0.06(3)}$ $A \approx 5{,}986.09$ Use a calculator. There will be $5,986.09 in the account.
Malthusian population growth: The same formula for continuous compound interest is used for population growth, where A is the new population, P is the previous population, r is the growth rate, and t is the number of years. $A = Pe^{rt}$	To find the population of a town in 5 years with a current population of 4,000 and a growth rate of 1.5%, we note that $P = 4{,}000 \quad r = 0.015 \quad t = 5$ We can substitute these values into the formula $A = Pe^{rt}$ to obtain $A = Pe^{rt}$ $A = 4{,}000e^{0.015(5)}$ $A \approx 4{,}311.54$ Use a calculator. There will be about 4,312 people in 5 years.
Radioactive decay: The formula $A = A_0e^{kt}$ can be used to determine the amount of material left (A) when an initial amount (A_0) has been decaying for t years at a rate of k percent per year. The value of k in this type of problem will be negative.	The radioactive material radon-22 decays with a rate of $k = -0.181$ for t days. To find how much radon-22 will be left if a sample of 100 grams decays for 30 days, we can substitute the values into the radioactive decay formula. $A = A_0e^{kt}$ $A = 100e^{-0.181(30)}$ $A \approx .4383095803$ Use a calculator. There will be approximately 0.44 grams remaining.

REVIEW EXERCISES

10. If $25,000 accumulates interest at an annual rate of 5%, compounded continuously, how much will be in the account in 35 years?

Graph each function.

11. $f(x) = e^x + 1$ **12.** $f(x) = e^{x-3}$

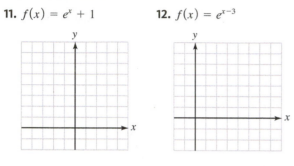

13. U.S. population The population of the United States is approximately 275,000,000 people. Find the population in 50 years if $k = 0.015$.

14. Radioactive decay The radioactive material strontium-90 decays according to the formula $A = A_0 e^{-0.0244t}$, where t is expressed in years. To the nearest hundredth, how much of the material will remain if a sample of 50 grams decays for 20 years?

SECTION 9.3   Logarithmic Functions

DEFINITIONS AND CONCEPTS	EXAMPLES
If $b > 0$ and $b \neq 1$, $$y = \log_b x \quad \text{if and only if} \quad x = b^y$$ The graph of $f(x) = \log_b x$ has the following properties. **1.** It passes through the point $(1, 0)$. **2.** It passes through the point $(b, 1)$. **3.** The y-axis is an asymptote. **4.** The domain is $(0, \infty)$ and the range is $(-\infty, \infty)$. If $b > 1$, $y = \log_b x$ is an increasing function. If $0 < b < 1$, $y = \log_b x$ is a decreasing function.	$y = \log_3 27$ is equivalent to $27 = 3^y$. Since $\log_4 x = 2$ is equivalent to $x = 4^2$, we have $x = $ **16**. Since $\log_2 32 = x$ is equivalent to $32 = 2^x$ or $2^5 = 2^x$, we have $x = 5$. Since $\log_x 64 = 3$ is equivalent to $64 = x^3$ or $4^3 = x^3$, we have $x = 4$. Graph $f(x) = \log_2 x$ *graph of* $f(x) = \log_2 x$
If E_O is the output voltage of a device and E_I is the input voltage, the decibel voltage gain is given by $$\text{dB gain} = 20 \log \frac{E_O}{E_I}$$	If the input to an amplifier is 0.6 volt and the output is 40 volts, find the decibel voltage gain. $$\text{dB gain} = 20 \log \frac{40}{0.6}$$ $$\approx 36.5\text{-decibel voltage gain}$$
If R is the magnitude of an earthquake, A is the amplitude (measured in micrometers), and P is the period (the time of one oscillation of Earth's surface, measured in seconds), then $$R = \log \frac{A}{P}$$	An earthquake has an amplitude of 8,000 micrometers and a period of 0.02 second. Find the earthquake's measure on the Richter scale. $$R = \log \frac{8,000}{0.02}$$ $$\approx 5.6 \text{ on the Richter scale}$$

REVIEW EXERCISES

15. State the domain and range of the logarithmic function $y = \log_b x$.

16. Explain why the functions $y = b^x$ and $y = \log_b x$ are called inverse functions.

Find each value.

17. $\log_4 16$ **18.** $\log_{16} \dfrac{1}{4}$

19. $\log_\pi 1$ **20.** $\log_5 0.04$

21. $\log_b \sqrt{b}$ **22.** $\log_a \sqrt[3]{a}$

Find the value of x.

23. $\log_3 x = 4$

24. $\log_{\sqrt{3}} x = 4$

25. $\log_{\sqrt{3}} x = 6$

26. $\log_{0.1} 10 = x$

27. $\log_x 2 = -\dfrac{1}{3}$

28. $\log_x 16 = 4$

29. $\log_{0.25} x = -1$

30. $\log_{0.125} x = -\dfrac{1}{3}$

31. $\log_{\sqrt{2}} 32 = x$

32. $\log_{\sqrt{5}} x = -4$

33. $\log_{\sqrt{3}} 9\sqrt{3} = x$

34. $\log_{\sqrt{5}} 5\sqrt{5} = x$

Graph each function.

35. $f(x) = \log(x - 2)$

36. $f(x) = 3 + \log x$

Graph each pair of equations on one set of coordinate axes.

37. $y = 4^x$ and $y = \log_4 x$

38. $y = \left(\dfrac{1}{3}\right)^x$ and $y = \log_{1/3} x$

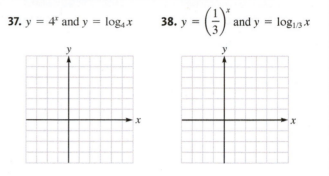

39. dB gain An amplifier has an output of 18 volts when the input is 0.04 volt. Find the dB gain.

40. Earthquakes An earthquake had a period of 0.3 second and an amplitude of 7,500 micrometers. Find its measure on the Richter scale.

SECTION 9.4   **Natural Logarithms**

DEFINITIONS AND CONCEPTS	EXAMPLES
$\ln x$ means $\log_e x$.	Use a calculator to approximate $\ln 5.89$ to 4 decimal places. $$\ln 5.89 \approx 1.7733$$
Population doubling time: $$A = A_0 e^{kt} \quad \text{or} \quad t = \frac{\ln 2}{r}$$	To find the time it takes for the population of a colony to double if the growth rate is 1.5% per year, substitute 0.015 into the formula for doubling time: $$t = \frac{\ln 2}{r}$$ $$t = \frac{\ln 2}{0.015}$$ $$t \approx 46.2098 \quad \text{Use a calculator.}$$ The colony will double in population in about 46 years.

REVIEW EXERCISES

Use a calculator to find each value to four decimal places.

41. $\ln 362$

42. $\ln(\log 7.85)$

Find the value of x.

43. $\ln x = 2.336$

44. $\ln x = \log 8.8$

Graph each function.

45. $f(x) = 1 + \ln x$

46. $f(x) = \ln(x + 1)$

47. U.S. population How long will it take the population of the United States to double if the growth rate is 3% per year?

SECTION 9.5   Properties of Logarithms

DEFINITIONS AND CONCEPTS	EXAMPLES
Properties of logarithms: If all variables are positive numbers and $b \neq 1$, **1.** $\log_b 1 = 0$ **2.** $\log_b b = 1$ **3.** $\log_b b^x = x$ **4.** $b^{\log_b x} = x$ **5.** $\log_b MN = \log_b M + \log_b N$ **6.** $\log_b \dfrac{M}{N} = \log_b M - \log_b N$ **7.** $\log_b M^p = p \log_b M$ **8.** If $\log_b x = \log_b y$, then $x = y$.	 **1.** $\log_b 1 = 0$ because $b^0 = 1$. **2.** $\log_b b = 1$ because $b^1 = b$. **3.** $\log_b b^x = x$ because $b^x = b^x$. **4.** $b^{\log_b x} = x$ because $b^y = x$ and $y = \log_b x$. **5.** $\log_5 xy = \log_5 x + \log_5 y$ **6.** $\log_5 \dfrac{x}{y} = \log_5 x - \log_5 y$ **7.** $\log_5 x^3 = 3 \log_5 x$ **8.** If $\log_3 x = \log_3 9$, then $x = 9$.
Change-of-base formula: If a, b, and x are real numbers, $b > 0$, and $b \neq 0$, then $$\log_b x = \frac{\log_a x}{\log_a b}$$	To evaluate $\log_7 16$, substitute 7 and 16 into the change-of-base formula. $$\log_7 16 = \frac{\log_{10} 16}{\log_{10} 7} \approx 1.4248 \quad \text{Use a calculator.}$$ or $$\log_7 16 = \frac{\ln 16}{\ln 7} \approx 1.4248 \quad \text{Use a calculator.}$$
If $[H^+]$ is the hydrogen ion concentration in gram-ions per liter, then $$pH = -\log [H^+]$$	Find the pH of a solution that has a hydrogen concentration of 10^{-4} grams per liter $$pH = -\log(10^{-4})$$ $$= 4$$
If L is the apparent loudness of a sound, I is the actual intensity, and k is a constant, then $$L = k \ln I$$	Find the decrease in intensity that will cause the apparent loudness of a sound to be a fourth as loud. $$\frac{1}{4} L_0 = \frac{1}{4} k \ln I_0$$ $$= k \ln (I_0)^{1/4}$$ To decrease the loudness of a sound by a fourth, the intensity is the fourth root.

REVIEW EXERCISES

Simplify each expression.

48. $\log_5 1$ **49.** $\log_9 9$

50. $\log_7 7^3$ **51.** $8^{\log_8 5}$

52. $\log 10^5$ **53.** $\ln 1$

54. $10^{\log_{10} 7}$ **55.** $e^{\ln 2}$

56. $\log_a a^6$ **57.** $\ln e^3$

Write each expression in terms of the logarithms of x, y, and z. Simplify, if possible.

58. $\log_b \dfrac{x^5 y^4}{z^2}$

59. $\log_b \sqrt{\dfrac{x}{yz^2}}$

Write each expression as the logarithm of one quantity.

60. $3 \log_b x - 5 \log_b y + 7 \log_b z$

61. $\dfrac{1}{2} \log_b x + 3 \log_b y - 7 \log_b z$

Assume that $\log a = 0.6$, $\log b = 0.36$, and $\log c = 2.4$. Find the value of each expression.

62. $\log abc$ **63.** $\log a^2 b$

64. $\log \dfrac{ac}{b}$ **65.** $\log \dfrac{a^2}{c^3 b^2}$

66. To four decimal places, approximate $\log_5 17$.

67. pH of grapefruit   The pH of grapefruit juice is about 3.1. Find its hydrogen ion concentration.

68. Change in loudness   Find the decrease in loudness if the intensity is cut in half.

SECTION 9.6 Exponential and Logarithmic Equations

DEFINITIONS AND CONCEPTS	EXAMPLES
Solving exponential and logarithmic equations: If $a^n = a^m$, then $n = m$	**1.** Solve each equation. **a.** $\quad 9^x = 3^{x-1}$ $\quad\quad (3^2)^x = 3^{x-1}$ $\quad\quad 3^{2x} = 3^{x-1}$ $\quad\quad 2x = x - 1$ $\quad\quad x = -1$
If the bases of an exponential equation are not the same, take the logarithm of both sides. Then use the properties of logarithms to solve for the variable. Use the properties of logarithms to combine multiple logarithms into a single logarithm.	**b.** $\quad 9^x = 5$ $\quad \ln(9^x) = \ln 5 \quad$ Take the natural logarithm of both sides. $\quad x \ln 9 = \ln 5 \quad$ The logarithm of a power is the power times the logarithm. $\quad x = \dfrac{\ln 5}{\ln 9} \quad$ Divide both sides by $\ln 9$. $\quad x \approx .7325 \quad$ Use a calculator. **c.** $\log_2(x + 1) + \log_2(x - 1) = 3$ $\quad \log_2\left[(x+1)(x-1)\right] = 3 \quad$ Write as a single logarithm. $\quad\quad \log_2(x^2 - 1) = 3 \quad$ Multiply inside the brackets. $\quad\quad\quad 2^3 = x^2 - 1 \quad$ Write as an exponential expression. $\quad\quad\quad\quad 8 = x^2 - 1 \quad$ Simplify. $\quad\quad\quad\quad 9 = x^2 \quad$ Add 1 to both sides. $\quad\quad\quad\quad x = 3, -3 \quad$ Take the square root of both sides. Since 3 checks, it is a solution. Since -3 does not check, it is extraneous.

REVIEW EXERCISES

Solve each equation for x. If an answer is not exact, round to four decimal places.

69. $3^x = 7$

70. $5^{x+2} = 625$

71. $25 = 5.5(1.05)^t$

72. $4^{2t-1} = 64$

73. $2^x = 3^{x-1}$

74. $3^{x^2+4x} = \dfrac{1}{27}$

75. $\log x + \log(29 - x) = 2$

76. $\log_2 x + \log_2(x - 2) = 3$

77. $\log_2(x + 2) + \log_2(x - 1) = 2$

78. $\dfrac{\log(7x - 12)}{\log x} = 2$

79. $\log x + \log(x - 5) = \log 6$

80. $\log 3 - \log(x - 1) = -1$

81. $e^{x \ln 2} = 9$

82. $\ln x = \ln(x - 1)$

83. $\ln x = \ln(x - 1) + 1$

84. $\ln x = \log_{10} x$ (*Hint:* Use the change-of-base formula.)

85. Carbon-14 dating A wooden statue found in Egypt has a carbon-14 content that is two-thirds of that found in living wood. If the half-life of carbon-14 is 5,730 years, how old is the statue?

9 Test

Graph each function.

1. $f(x) = 2^x + 1$

2. $f(x) = 2^{-x}$

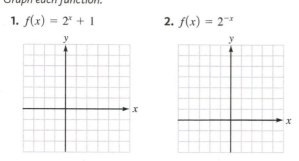

Solve each equation.

3. Radioactive decay A radioactive material decays according to the formula $A = A_0(2)^{-t}$. How much of a 3-gram sample will be left in 6 years?

4. Investing An initial deposit of \$1,000 earns 6% interest, compounded twice a year. How much will be in the account in one year?

5. Graph the function $f(x) = e^x$.

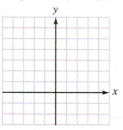

6. Investing An account contains \$2,000 and has been earning 8% interest, compounded continuously. How much will be in the account in 10 years?

Find the value of x.

7. $\log_9 81 = x$

8. $\log_x 81 = 4$

9. $\log_3 x = -3$

10. $\log_x 49 = 2$

11. $\log_{3/2} \dfrac{9}{4} = x$

12. $\log_{2/3} x = -3$

Graph each function.

13. $f(x) = -\log_3 x$

14. $f(x) = \ln x$

Write each expression in terms of the logarithms of a, b, and c.

15. $\log xy^3z^4$

16. $\ln\sqrt{\dfrac{a}{b^2c}}$

Write each expression as a logarithm of a single quantity.

17. $\dfrac{1}{2}\log(a + 2) + \log b - 3\log c$

18. $\dfrac{1}{3}(\log a - 2\log b) - \log c$

Assume that $\log 2 \approx 0.3010$ *and* $\log 3 \approx 0.4771$. *Find each value. Do not use a calculator.*

19. $\log 18$

20. $\log \dfrac{8}{3}$

Use the change-of-base formula to find each logarithm. Do not simplify the answer.

21. $\log_3 2$

22. $\log_\pi e$

Determine whether each statement is true. If a statement is not true, explain why.

23. $\log_a ab = 1 + \log_a b$

24. $\dfrac{\log a}{\log b} = \log a - \log b$

25. $\log a^{-3} = \dfrac{1}{3\log a}$

26. $\ln(-x) = -\ln x$

27. Find the pH of a solution with a hydrogen ion concentration of 3.7×10^{-7}. (*Hint*: pH $= -\log[\text{H}^+]$.)

28. Find the dB gain of an amplifier when $E_O = 60$ volts and $E_I = 0.3$ volt. (*Hint*: dB gain $= 20\log(E_O/E_I)$.)

Solve each equation.

29. $7^x = 4$

30. $3^{x-1} = 100^x$

Solve each equation.

31. $\log(3x + 1) = \log(5x - 7)$

32. $\log x + \log(x - 9) = 1$

Conic Sections and More Graphing

Careers and Mathematics

WATER TRANSPORTATION OCCUPATIONS

The movement of cargo and passengers between nations depends on workers in the water transportation occupations, also known on commercial ships as merchant mariners. They operate and maintain deep-sea merchant ships, tugboats, ferries, and excursion vessels.

About 17 percent of water transportation employees work in inland waters, primarily the Mississippi River system; 23 percent on the deep seas and the Great Lakes; and another 24 percent in harbor operations.

Job Outlook:
Employment in this industry is projected to grow 20 percent through 2020. This is faster than average for all occupations.

Annual Earnings:
$24,890–$99,690

For More Information:
http://www.bls.gov/oco/ocos247.htm

For a Sample Application:
See Problem 38 in Section 10.3.

REACH FOR SUCCESS

10.1 The Circle and the Parabola

10.2 The Ellipse

10.3 The Hyperbola

10.4 Systems Containing Second-Degree Equations

10.5 Piecewise-Defined Functions and the Greatest Integer Function

■ *Projects*

REACH FOR SUCCESS EXTENSION

CHAPTER REVIEW

CHAPTER TEST

CUMULATIVE REVIEW

In this chapter

We have seen that the graphs of linear functions are straight lines, and that the graphs of quadratic functions are parabolas. In this chapter, we will discuss some special curves called conic sections. Then we will discuss piecewise-defined functions and step functions.

Reach for Success   Learning Beyond Remembering

Can you name all five of the Great Lakes shown in the photo? Being able to do so means that you have learned and stored this information in your brain in such a way that you can access it any time you need it.

In this exercise, you will explore Bloom's Taxonomy of learning at the first three levels.

Provided by the SeaWiFS Project, NASA/Goddard Space Flight Center, and ORBIMAGE

If you couldn't name all five, does this list help you select the correct names?

Cumberland	*Erie*	*Huron*	*Michigan*
Ontario	*Victoria*	*Salt Lake*	*Superior*

If you could not name the five lakes without this list, you have learned only to *recognize* their names. The information is stored in your brain, but in a way that requires a "prompt" to be retrieved.

The triangle is an updated illustration of learning levels as inspired by psychologist Benjamin Bloom in 1956.

Creating
Evaluating
Analyzing
Applying
Understanding
Remembering

Ideally, you would like to learn material beyond recognition and remembering, to understanding, applying, analyzing, evaluating, and ultimately creating.

To help you *recall* the names of the Great Lakes, you can use the mnemonic device HOMES (Huron, Ontario, Michigan, Erie, Superior). Vocabulary and formulas in mathematics are necessary and creating your own mnemonic devices may make it easier to remember them.

Recalling the names of the Great Lakes is still at the *Remembering* level of Bloom's Taxonomy. "Explain why they are called the Great Lakes" is at the *Understanding* level. "Demonstrate the size of the Great Lakes in terms of one of the states" is at the *Applying* level.

Below are three questions you might see on a test covering the topic of slope. For each, identify the level of Bloom's Taxonomy corresponding to the degree of learning required: Remembering, Understanding, or Applying.

1. Given a table of data, find the slope of the line it represents. _____

2. Write the formula used to calculate the slope of a line between two points. _____

3. Given a graph of the national debt between 2008 and 2012, interpret how steeply it grew. _____

Do an Internet search for the levels of Bloom's Taxonomy. Check your answers to the previous three statements against the descriptions you find.

In your own words, explain the difference between

1. the Remembering level and the Understanding level.

_____

2. the Understanding level and the Applying level.

_____

It is impossible for a textbook to provide an example of every variation of a mathematics problem. Therefore it is important that you *learn* the skills in such a way that you can recall and understand what you've learned and can recall and apply it to a new situation. Many of the *Now Try This* exercises in this book are designed to help you transition to higher levels of learning.

A Successful Study Strategy . . .

Practicing your mathematics every day will help you learn at levels beyond remembering.

At the end of the chapter you will find an additional exercise to guide you to a successful semester.

Section 10.1

The Circle and the Parabola

The Circle and the Parabola

Objectives

1. Find the center and radius of a circle given an equation in standard form.
2. Find the center and radius of a circle given an equation in general form.
3. Write an equation of a circle in both standard and general form given the center and the radius.
4. Solve an application involving a circle.
5. Graph a parabola of the form $x = (y - k)^2 + h$.
6. Solve an application involving a parabola.

Vocabulary

conic section
circle
center of a circle
radius of a circle

standard form of the equation
  of a circle
point circle

general form of the equation
  of a circle
paraboloid

Getting Ready

Square each binomial.

1. $(x - 2)^2$ **2.** $(x + 4)^2$

What number must be added to each binomial to make it a perfect square trinomial?

3. $x^2 + 9x$ **4.** $x^2 - 12x$

In this chapter, we will introduce a group of curves called *conic sections*.

The graphs of second-degree equations in *x* and *y* represent figures that were investigated in the 17th century by René Descartes (1596–1650) and Blaise Pascal (1623–1662). Descartes discovered that graphs of second-degree equations fall into one of several categories: a pair of lines, a point, a circle, a parabola, an ellipse, a hyperbola, or no graph at all. Because all of these graphs can be formed by the intersection of a plane and a right-circular cone, they are called **conic sections**. See Figure 10-1 on the next page.

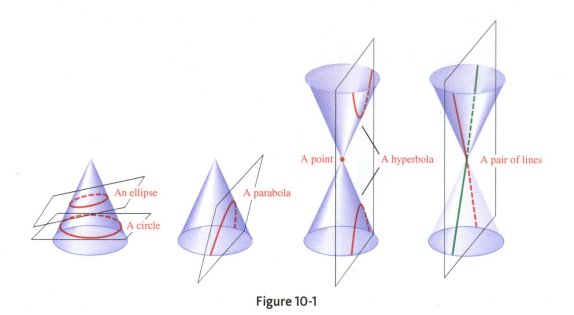

Figure 10-1

1   Find the center and radius of a circle given an equation in standard form.

A *circle* is one of the most common of the conic sections, generally seen in circular gears, pizza cutters, and Ferris wheels. Because of the circle's importance, we will begin the study of conic sections with the circle.

Every conic section can be represented by a second-degree equation in *x* and *y*. To find the form of an equation of a circle, we use the following definition.

THE CIRCLE	A **circle** is the set of all points in a plane that are a fixed distance from a point, called its **center**. The fixed distance is the **radius** of the circle.

Figure 10-2

To develop the general equation of a circle, we must write the equation of a circle with a radius of *r* and with center at some point $C(h, k)$, as in Figure 10-2. This task is equivalent to finding all points $P(x, y)$ such that the length of line segment *CP* is *r*. We can use the distance formula to find *r*.

$$r = \sqrt{(x - h)^2 + (y - k)^2}$$

We then square both sides to obtain

(1) $r^2 = (x - h)^2 + (y - k)^2$ or $(x - h)^2 + (y - k)^2 = r^2$

This equation is called the **standard form of the equation of a circle** with radius *r* and center at the point with coordinates (h, k).

STANDARD FORM OF THE EQUATION OF A CIRCLE WITH CENTER AT (h, k)	Any equation that can be written in the form $(x - h)^2 + (y - k)^2 = r^2$ has a graph that is a circle with radius *r* and center at point (h, k).

If $r = 0$, the graph reduces to a single point called a **point circle**. If $r^2 < 0$, a circle does not exist. If both coordinates of the center are 0, the center of the circle is the origin.

| STANDARD FORM OF THE EQUATION OF A CIRCLE WITH CENTER AT $(0, 0)$ | Any equation that can be written in the form $$x^2 + y^2 = r^2$$ has a graph that is a circle with radius r and center at the origin. |

EXAMPLE 1   Find the center and the radius of each circle and then graph it.
a. $(x - 4)^2 + (y - 1)^2 = 9$ **b.** $x^2 + y^2 = 25$ **c.** $(x + 3)^2 + y^2 = 12$

Solution   **a.** We can determine the center and the radius of a circle when its equation is written in standard form.

$$(x - \mathbf{4})^2 + (y - \mathbf{1})^2 = \mathbf{9}$$

$h = 4, k = 1$, and $r^2 = 9$. Since the radius of a circle must be positive, $r = 3$.

$$(x - \mathbf{h})^2 + (y - \mathbf{k})^2 = \mathbf{r^2}$$

The center of the circle is $(h, k) = (4, 1)$ and the radius is 3.

To plot four points on the circle, we move up, down, left, and right 3 units from the center, as shown in Figure 10-3(a). Then we draw a circle through the points to obtain the graph of $(x - 4)^2 + (y - 1)^2 = 9$, as shown in Figure 10-3(b).

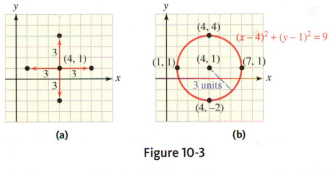

(a)   (b)

Figure 10-3

b. To verify that the center of the circle is the origin, we can write $x^2 + y^2 = 25$ in the following way:

$$(x - \mathbf{0})^2 + (y - \mathbf{0})^2 = \mathbf{25}$$

$h = 0, k = 0$, and $r^2 = 25$. Since the radius of a circle must be positive, $r = 5$.

$$\mathbf{h} \qquad \mathbf{k} \quad \mathbf{r^2}$$

This illustrates that the center of the circle is at $(0, 0)$ and the radius is 5.

To plot four points on the circle, we move up, down, left, and right 5 units from the center. Then we draw a circle through the points to obtain the graph of $x^2 + y^2 = 25$, as shown in Figure 10-4.

c. To determine h in the equation $(x + 3)^2 + y^2 = 12$, it is helpful to write $x + 3$ as $x - (-3)$.

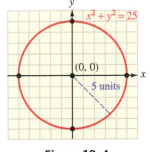

Figure 10-4

Standard form requires a minus symbol here.
↓
$$[x - (\mathbf{-3})]^2 + (y - \mathbf{0})^2 = \mathbf{12}$$

$h = -3, k = 0$, and $r^2 = 12$

$$\mathbf{h} \qquad\qquad \mathbf{k} \quad \mathbf{r^2}$$

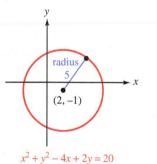

Figure 10-5

If $r^2 = 12$, by the square root property

$$r = \pm\sqrt{12} = \pm 2\sqrt{3}$$

Because the radius must be positive, $r = 2\sqrt{3}$. The center of the circle is at $(-3, 0)$ and the radius is $2\sqrt{3}$.

To plot four points on the circle, we move upward, downward, left, and right $2\sqrt{3} \approx 3.5$ units from the center. Then we draw a circle through the points to obtain the graph of $(x + 3)^2 + y^2 = 12$, as shown in Figure 10-5.

SELF CHECK 1   Find the center and the radius of each circle and then graph it.
a. $(x - 3)^2 + (y + 4)^2 = 4$ **b.** $x^2 + y^2 = 8$

2 Find the center and radius of a circle given an equation in general form.

Another important form of the equation of a circle is called the **general form of the equation of a circle**.

GENERAL FORM OF THE EQUATION OF A CIRCLE

The equation of any circle can be written in the form

$$x^2 + y^2 + Dx + Ey + F = 0$$

EXAMPLE 2   Graph: $x^2 + y^2 - 4x + 2y - 20 = 0$

Solution   Since the equation matches the general form of the equation of a circle, we know that its graph will be a circle. To find its center and radius, we must complete the square on both x and y and write the equation in standard form.

$$x^2 + y^2 - 4x + 2y = 20 \qquad \text{Add 20 to both sides.}$$
$$x^2 - 4x + y^2 + 2y = 20 \qquad \text{Rearrange terms using the commutative property of addition.}$$

To complete the square on x and y, add 4 and 1 to both sides.

$$x^2 - 4x \mathbin{\color{blue}+ 4} + y^2 + 2y \mathbin{\color{blue}+ 1} = 20 \mathbin{\color{red}+ 4 + 1}$$
$$(x - 2)^2 + (y + 1)^2 = 25 \qquad \text{Factor } x^2 - 4x + 4 \text{ and } y^2 + 2y + 1.$$

We can now see that this result is the standard equation of a circle with a radius of 5 and center at $h = 2$ and $k = -1$. If we plot the center and draw a circle with a radius of 5 units, we will obtain the circle shown in Figure 10-6.

$x^2 + y^2 - 4x + 2y = 20$

Figure 10-6

SELF CHECK 2   Write the equation $x^2 + y^2 + 2x - 4y - 11 = 0$ in standard form and graph it.

3 Write an equation of a circle in both standard and general form given the center and the radius.

EXAMPLE 3   Find the general form of the equation of the circle with radius 5 and center at $(3, 2)$.

Solution   We substitute 5 for r, 3 for h, and 2 for k in the standard form of a circle and proceed as follows:

$$(x - \mathbf{\color{green}h})^2 + (y - \mathbf{\color{green}k})^2 = \mathbf{\color{red}r}^2$$
$$(x - \mathbf{\color{green}3})^2 + (y - \mathbf{\color{green}2})^2 = \mathbf{\color{red}5}^2$$

$$x^2 - 6x + 9 + y^2 - 4y + 4 = 25 \quad \color{red}{(x - 3)^2 = x^2 - 6x + 9; (y - 2)^2 = y^2 - 4y + 4}$$
$$x^2 + y^2 - 6x - 4y - 12 = 0 \quad \color{red}{\text{Subtract 25 from both sides and simplify.}}$$

The general form of the equation is $x^2 + y^2 - 6x - 4y - 12 = 0$.

SELF CHECK 3   Find the general form of the equation of the circle with radius 6 and center at $(2, 3)$.

Accent on technology

▶ Graphing Circles

Since the graphs of circles fail the vertical line test, their equations do not represent functions. It is somewhat more difficult to use a graphing calculator to graph equations that are not functions. For example, to graph the circle described by $(x - 1)^2 + (y - 2)^2 = 4$, we must split the equation into two functions and graph each one separately. We begin by solving the equation for y.

$$(x - 1)^2 + (y - 2)^2 = 4$$
$$(y - 2)^2 = 4 - (x - 1)^2 \quad \color{red}{\text{Subtract } (x - 1)^2 \text{ from both sides.}}$$
$$y - 2 = \pm\sqrt{4 - (x - 1)^2} \quad \color{red}{\text{Use the square-root property.}}$$
$$y = 2 \pm \sqrt{4 - (x - 1)^2} \quad \color{red}{\text{Add 2 to both sides.}}$$

This equation defines two functions. If we use window settings of $[-3, 5]$ for x and $[-3, 5]$ for y and graph the functions

$$y = 2 + \sqrt{4 - (x - 1)^2} \quad \text{and} \quad y = 2 - \sqrt{4 - (x - 1)^2}$$

Using a TI-84 calculator, we see the distorted circle shown in Figure 10-7(a). To see a better representation, graphing calculators have a square feature, ZSquare, that displays an equal unit distance on both the x- and y-axes. After applying this feature (zoom 5), we obtain the circle shown in Figure 10-7(b).

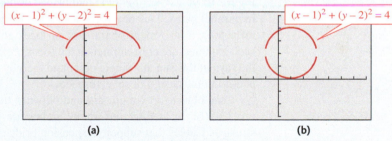

(a)   (b)

Figure 10-7

For instructions regarding the use of a Casio graphing calculator, please refer to the Casio Keystroke Guide in the back of the book.

---

4   Solve an application involving a circle.

EXAMPLE 4   **TV TRANSLATORS**   The broadcast area of a TV station is bounded by the circle $x^2 + y^2 = 3,600$, where x and y are measured in miles. A translator station picks up the signal and retransmits it from the center of a circular area bounded by $(x + 30)^2 + (y - 40)^2 = 1,600$. Find the location of the translator and the greatest distance from the main transmitter that the signal can be received.

Solution   The coverage of the TV station is bounded by $x^2 + y^2 = 60^2$, a circle centered at the origin with a radius of 60 miles, as shown in Figure 10-8 on the next page. Because the translator is at the center of the circle $(x + 30)^2 + (y - 40)^2 = 1,600$, it is located at $(-30, 40)$, a point 30 miles west and 40 miles north of the TV station. The radius of the translator's coverage is $\sqrt{1,600}$, or 40 miles.

As shown in the figure, the greatest distance of reception is the sum of A, the distance from the translator to the TV station, and 40 miles, the radius of the translator's coverage.

To find A, we use the distance formula to find the distance between $(x_1, y_1) = (-30, 40)$ and the origin, $(x_2, y_2) = (0, 0)$.

$$A = \sqrt{(x_1 - x_2)^2 + (y_1 - y_2)^2}$$
$$A = \sqrt{(-30 - 0)^2 + (40 - 0)^2}$$
$$= \sqrt{(-30)^2 + 40^2}$$
$$= \sqrt{2{,}500}$$
$$= 50$$

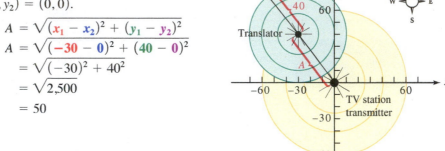

Figure 10-8

Since the translator is 50 miles from the TV station and it broadcasts the signal an additional 40 miles, the greatest reception distance is $50 + 40$, or 90 miles.

SELF CHECK 4   If the circular area is bounded by $(x - 60)^2 + (y + 80)^2 = 1{,}600$, find the location of the translator.

---

5   ## Graph a parabola of the form $x = (y - k)^2 + h$.

Parabolas can be rotated to generate dish-shaped surfaces called **paraboloids**, as shown in Figure 10-9(a). Any light or sound placed at the *focus* of a paraboloid is reflected outward in parallel paths. This property makes parabolic surfaces ideal for flashlight and headlight reflectors. It also makes parabolic surfaces good antennas, because signals captured by such antennas are concentrated at the focus. Parabolic mirrors are capable of concentrating the rays of the Sun at a single point and thereby generating tremendous heat. This property is used in the design of solar furnaces.

Any object thrown upward and outward travels in a parabolic path, as shown in Figure 10-9(b). In architecture, many arches are parabolic in shape, because this provides strength. Cables that support suspension bridges hang in a form similar to that of a parabola. (See Figure 10-9(c).)

Parabolas

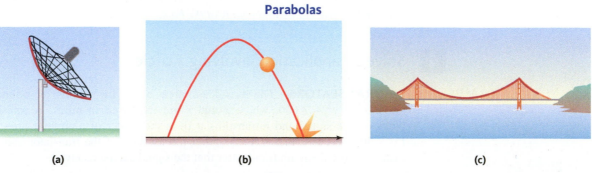

(a)   (b)   (c)

Figure 10-9

We have seen that equations of the form $y = a(x - h)^2 + k$, with $a \neq 0$, represent parabolas with the vertex at the point (h, k). They open upward when $a > 0$ and downward when $a < 0$.

Equations of the form $x = a(y - k)^2 + h$ $(a \neq 0)$, also represent parabolas with vertex at point (h, k). However, they open to the right when $a > 0$ and to the left when $a < 0$. Parabolas that open to the right or left do not represent functions, because their graphs fail the vertical line test.

Standard equations of many parabolas are summarized in the following table.

EQUATIONS OF PARABOLAS	Parabola opening	Vertex at origin	Vertex at (h, k)
	Upward	$y = ax^2$ $(a > 0)$	$y = a(x - h)^2 + k$ $(a > 0)$
	Downward	$y = ax^2$ $(a < 0)$	$y = a(x - h)^2 + k$ $(a < 0)$
	Right	$x = ay^2$ $(a > 0)$	$x = a(y - k)^2 + h$ $(a > 0)$
	Left	$x = ay^2$ $(a < 0)$	$x = a(y - k)^2 + h$ $(a < 0)$

EXAMPLE 5   Graph each equation: **a.** $x = \dfrac{1}{2} y^2$ **b.** $x = -2(y - 2)^2 + 3$

Solution   **a.** We can create a table of ordered pairs that satisfy the equation, plot each pair, and draw the parabola, as in Figure 10-10(a). Because the equation is of the form $x = ay^2$ with $a > 0$, the parabola opens to the right and has its vertex at the origin.

b. We can create a table of ordered pairs that satisfy the equation, plot each pair, and draw the parabola, as in Figure 10-10(b). Because the equation is of the form $x = a(y - k)^2 + h$ $(a < 0)$, the parabola opens to the left and has its vertex at the point with coordinates $(3, 2)$.

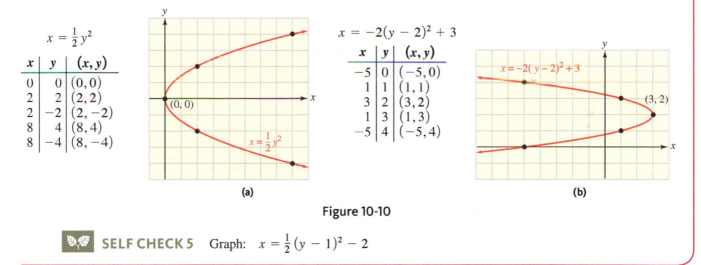

$x = \dfrac{1}{2} y^2$

x	y	(x, y)
0	0	$(0, 0)$
2	2	$(2, 2)$
2	-2	$(2, -2)$
8	4	$(8, 4)$
8	-4	$(8, -4)$

$x = -2(y - 2)^2 + 3$

x	y	(x, y)
-5	0	$(-5, 0)$
1	1	$(1, 1)$
3	2	$(3, 2)$
1	3	$(1, 3)$
-5	4	$(-5, 4)$

(a)   (b)

Figure 10-10

SELF CHECK 5   Graph: $x = \dfrac{1}{2}(y - 1)^2 - 2$

The general forms of the equations of a parabola are as follows:

GENERAL FORM OF THE EQUATION OF A PARABOLA THAT OPENS UPWARD OR DOWNWARD	The general form of the equation of a parabola that opens upward or downward is $$y = ax^2 + bx + c \quad (a \neq 0)$$ If $a > 0$, the parabola opens upward. If $a < 0$, the parabola opens downward.
GENERAL FORM OF THE EQUATION OF A PARABOLA THAT OPENS LEFT OR RIGHT	The general form of the equation of a parabola that opens to the left or to the right is $$x = ay^2 + by + c \quad (a \neq 0)$$ If $a > 0$, the parabola opens to the right. If $a < 0$, the parabola opens to the left.

EXAMPLE 6 Graph: $x = -2y^2 + 12y - 15$

Solution This equation is in the general form of a parabola that opens left or right. Since $a = -2$, the parabola opens to the left. To find the coordinates of its vertex, we write the equation in standard form by completing the square on y.

COMMENT Notice the similarity to writing an equation in $y = a(x - h)^2 + k$ form.

$$x = -2y^2 + 12y - 15$$
$$= -2(y^2 - 6y) - 15 \quad \text{Factor out } -2 \text{ from } -2y^2 + 12y.$$
$$= -2(y^2 - 6y + 9) - 15 + 18 \quad \text{Subtract and add 18; } -2(9) = -18.$$
$$= -2(y - 3)^2 + 3$$

Because the equation is written in the form $x = a(y - k)^2 + h$, we can determine that the parabola has its vertex at $(3, 3)$. The graph is shown in Figure 10-11.

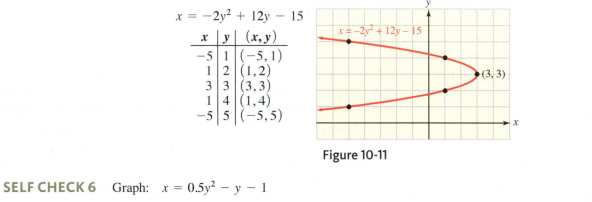

$x = -2y^2 + 12y - 15$

x	y	(x, y)
-5	1	$(-5, 1)$
1	2	$(1, 2)$
3	3	$(3, 3)$
1	4	$(1, 4)$
-5	5	$(-5, 5)$

Figure 10-11

 SELF CHECK 6 Graph: $x = 0.5y^2 - y - 1$

COMMENT To find the y-coordinate of the vertex for any horizontal parabola $(x = ay^2 + by + c)$, we could compute $-\frac{b}{2a}$. To find the x-coordinate, we could substitute the value of $-\frac{b}{2a}$ for y and find the value of x.

6 Solve an application involving a parabola.

EXAMPLE 7 **GATEWAY ARCH** The shape of the Gateway Arch in St. Louis is approximately a parabola 630 feet high and 630 feet wide, as shown in Figure 10-12(a). How high is the arch 100 feet from its foundation?

Solution We place the parabola in a coordinate system as in Figure 10-12(b), with ground level on the x-axis and the vertex of the parabola at the point $(h, k) = (0, 630)$.

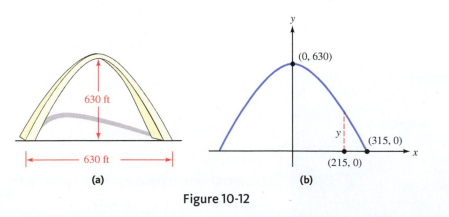

Figure 10-12

The equation of this downward-opening parabola has the form

$$y = a(x - \boldsymbol{h})^2 + \boldsymbol{k} \qquad \text{With } a < 0$$
$$= a(x - \boldsymbol{0})^2 + \boldsymbol{630} \qquad \text{Substitute } h = 0 \text{ and } k = 630.$$
$$= ax^2 + 630 \qquad\qquad \text{Simplify.}$$

Because the Gateway Arch is 630 feet wide at its base, the parabola passes through the point $\left(\frac{630}{2}, 0\right)$, or $(315, 0)$. To find the value of a in the equation of the parabola, we proceed as follows:

$$\boldsymbol{y} = a\boldsymbol{x}^2 + 630$$
$$\boldsymbol{0} = a(\boldsymbol{315})^2 + 630 \qquad \text{Substitute 315 for } x \text{ and 0 for } y.$$
$$\frac{-630}{315^2} = a \qquad\qquad\qquad \text{Subtract 630 from both sides and divide both sides by } 315^2.$$
$$-\frac{2}{315} = a \qquad\qquad\qquad \text{Simplify: } \frac{-630}{315^2} = \frac{-2}{315}$$

The equation of the parabola that approximates the shape of the Gateway Arch is

$$y = -\frac{2}{315}x^2 + 630$$

To find the height of the arch at a point 100 feet from its foundation, we substitute $315 - 100$, or 215, for x in the equation of the parabola and solve for y.

$$y = -\frac{2}{315}\boldsymbol{x}^2 + 630$$
$$= -\frac{2}{315}(\boldsymbol{215})^2 + 630$$
$$= 336.5079365$$

At a point 100 feet from the foundation, the height of the arch is about 337 feet.

SELF CHECK 7   How high is the arch 150 feet from its foundation?

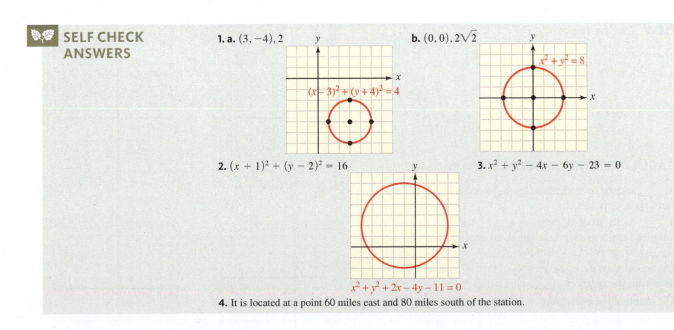

SELF CHECK ANSWERS

1. a. $(3, -4), 2$ b. $(0, 0), 2\sqrt{2}$

$(x - 3)^2 + (y + 4)^2 = 4$

$x^2 + y^2 = 8$

2. $(x + 1)^2 + (y - 2)^2 = 16$ 3. $x^2 + y^2 - 4x - 6y - 23 = 0$

$x^2 + y^2 + 2x - 4y - 11 = 0$

4. It is located at a point 60 miles east and 80 miles south of the station.

5.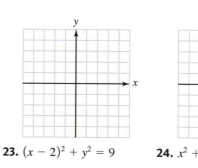
$x = \frac{1}{2}(y-1)^2 - 2$

6.
$x = 0.5y^2 - y - 1$

7. From a point 150 feet from the foundation, the height of the arch is about 457 feet.

NOW TRY THIS

The equation $4y^2 + (4x - 1)y + x^2 - 5x - 3 = 0$ is an equation for a parabola that is neither vertical nor horizontal. Use the quadratic formula to solve for y and then graph the equation with a calculator.

10.1 Exercises

WARM-UPS *Determine whether the graph of the parabola opens upward or downward.*

1. $y = -5x^2 + 4x + 1$

2. $y = 8x^2 - 7x - 1$

3. $y = x^2 - 6x - 7$

4. $y = -4x^2 - 7x + 1$

Factor.

5. $x^2 + 14x + 49$

6. $x^2 - 18x + 81$

7. $x^2 - 20x + 100$

8. $x^2 + 12x + 36$

REVIEW *Solve each equation.*

9. $|2x - 5| = 7$

10. $\left| \dfrac{4 - 3x}{5} \right| = 12$

11. $|3x + 4| = |5x - 2|$

12. $|6 - 4x| = |x + 2|$

VOCABULARY AND CONCEPTS *Fill in the blanks.*

13. A _____ section is determined by the intersection of a plane and a right-circular cone.

14. A _____ is the set of all points in a _____ that are a fixed distance from a given point. The fixed distance is called the _____ and the point is called the _____.

15. The equation of the circle $x^2 + (y - 3)^2 = 16$ is in _____ form with the center at _____ and a radius _.

16. The graph of the equation $x^2 + y^2 = 0$ is a _____.

17. The equation $x^2 + y^2 - 10x - 8y - 8 = 0$ is an equation of a _____ written in _____ form.

18. The graph of $y = ax^2$ $(a > 0)$ is a _____ with vertex at the _____ that opens _____.

19. The graph of $x = a(y - 2)^2 + 3$ $(a > 0)$ is a _____ with vertex at _____ that opens to the _____.

20. The graph of $x = a(y - 1)^2 - 3$ $(a < 0)$ is a _____ with vertex at _____ that opens to the ___.

GUIDED PRACTICE *Graph each equation and state the center and radius of the circle.* *SEE EXAMPLE 1. (OBJECTIVE 1)*

21. $x^2 + y^2 = 9$

22. $x^2 + y^2 = 16$

23. $(x - 2)^2 + y^2 = 9$

24. $x^2 + (y - 3)^2 = 4$

25. $(x - 2)^2 + (y - 4)^2 = 4$ **26.** $(x - 3)^2 + (y - 2)^2 = 4$

33. $x^2 + y^2 - 2x + 4y = -1$ **34.** $x^2 + y^2 + 4x + 2y = 4$

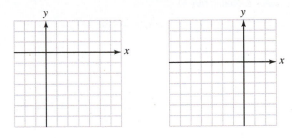

27. $(x + 3)^2 + (y - 1)^2 = 16$

35. $x^2 + y^2 + 6x - 4y = -12$

28. $(x - 1)^2 + (y + 4)^2 = 9$

36. $x^2 + y^2 + 8x + 2y = -13$

Write the equation of the circle with the following properties in standard form and in general form. **SEE EXAMPLE 3. (OBJECTIVE 3)**

37. center at origin; radius 3

38. center at origin; radius 4

Graph each circle. Give the coordinates of the center and state the radius. **SEE EXAMPLE 2. (OBJECTIVE 2)**

39. center at $(2, 4)$; radius 7

29. $x^2 + y^2 + 2x - 8 = 0$ **30.** $x^2 + y^2 - 4y = 12$

40. center at $(5, 3)$; radius 2

41. center at $(-4, 7)$; radius 10

42. center at $(5, -4)$; radius 6

43. center at the origin; diameter $8\sqrt{5}$

31. $9x^2 + 9y^2 - 12y = 5$ **32.** $4x^2 + 4y^2 + 4y = 15$

44. center at the origin; diameter $4\sqrt{3}$

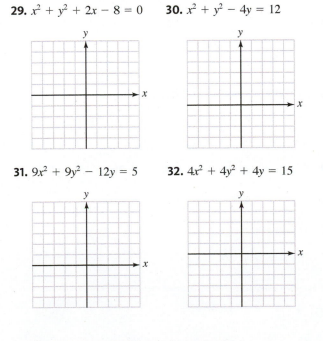

State the vertex of each parabola and graph the equation.
SEE EXAMPLE 5. (OBJECTIVE 5)

45. $x = y^2$ **46.** $x = -y^2 + 1$

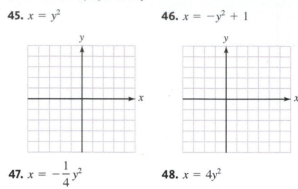

47. $x = -\dfrac{1}{4} y^2$ **48.** $x = 4y^2$

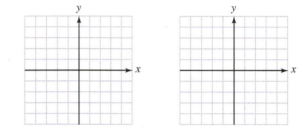

State the vertex of each parabola and graph the equation.
SEE EXAMPLE 6. (OBJECTIVE 5)

49. $y^2 + 4x - 6y = -1$ **50.** $x = \tfrac{1}{2} y^2 + 2y$

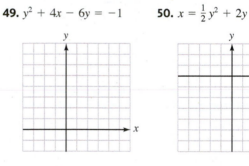

51. $y = -x^2 - x + 1$ **52.** $x^2 - 2y - 2x = -7$

ADDITIONAL PRACTICE *Graph each equation.*

53. $x^2 + (y + 3)^2 = 1$ **54.** $(x + 4)^2 + y^2 = 1$

55. $y = 2(x - 1)^2 + 3$ **56.** $y = -2(x + 1)^2 + 2$

57. $y = x^2 + 4x + 5$ **58.** $y = -x^2 - 2x + 3$

59. $y = 3(x + 1)^2 - 2$ **60.** $y = -3(x - 1)^2 - 2$

Use a graphing calculator to graph each equation.

61. $3x^2 + 3y^2 = 16$ **62.** $2x^2 + 2y^2 = 9$

63. $(x + 1)^2 + y^2 = 16$ **64.** $x^2 + (y - 2)^2 = 4$

65. $x = 2y^2$ **66.** $x = y^2 - 4$

67. $x^2 - 2x + y = 6$ **68.** $x = -2(y - 1)^2 + 2$

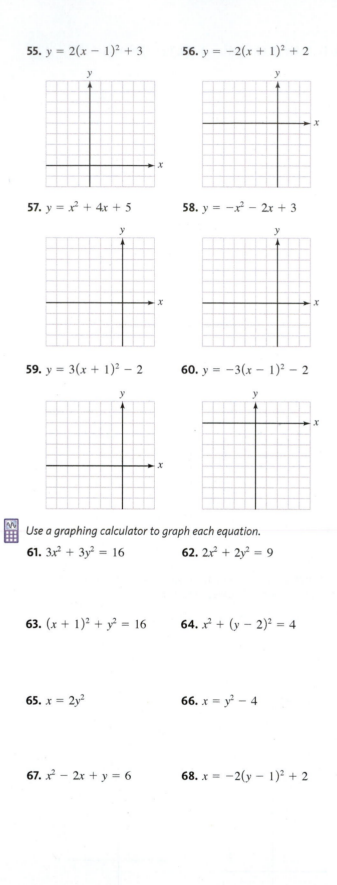

APPLICATIONS *Solve each application.* **SEE EXAMPLE 4.**
(OBJECTIVE 4)

69. Meshing gears For design purposes, the large gear is the circle $x^2 + y^2 = 16$. The smaller gear is a circle centered at $(7, 0)$ and tangent (touching at 1 point) to the larger circle. Find the equation of the smaller gear.

(7, 0)

70. Width of a walkway The following walkway is bounded by the two circles $x^2 + y^2 = 2,500$ and $(x - 10)^2 + y^2 = 900$, measured in feet. Find the largest and the smallest width of the walkway.

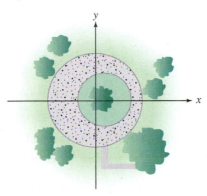

71. Broadcast ranges Radio stations applying for licensing may not use the same frequency if their broadcast areas overlap. One station's coverage is bounded by $x^2 + y^2 - 8x - 20y + 16 = 0$, and the other's by $x^2 + y^2 + 2x + 4y - 11 = 0$. May they be licensed for the same frequency?

72. Highway design Engineers want to join two sections of highway with a curve that is one-quarter of a circle as shown in the illustration. The equation of the circle is $x^2 + y^2 - 16x - 20y + 155 = 0$, where distances are measured in kilometers. Find the locations (relative to the center of town) of the intersections of the highway with State and with Main.

New Highway

Main Street

City Hall

State Street

Solve each application. **SEE EXAMPLE 7. (OBJECTIVE 6)**

73. Projectiles The cannonball in the illustration follows the parabolic trajectory $y = 30x - x^2$. Where does it land?

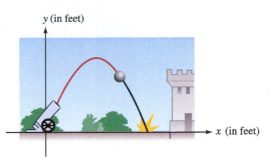

y (in feet)

x (in feet)

74. Projectiles In Exercise 73, how high does the cannonball rise?

75. Path of a comet If the path of a comet is given by the equation $2y^2 - 9x = 18$, how far is it from the Sun at the vertex of the orbit? Distances are measured in astronomical units (AU).

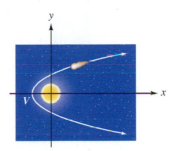

y

V

x

76. Satellite antennas The cross section of the satellite antenna is a parabola given by the equation $y = \frac{1}{16} x^2$, with distances measured in feet. If the dish is 8 feet wide, how deep is it?

8 ft

y

$y = \frac{1}{16} x^2$

x

WRITING ABOUT MATH

77. Explain how to decide from its equation whether the graph of a parabola opens upward, downward, right, or left.

78. From the equation of a circle, explain how to determine the radius and the coordinates of the center.

SOMETHING TO THINK ABOUT

79. From the values of a, h, and k, explain how to determine the number of x-intercepts of the graph of $y = a(x - h)^2 + k$.

80. Under what conditions will the graph of $x = a(y - k)^2 + h$ have no y-intercepts?

Section 10.2

The Ellipse

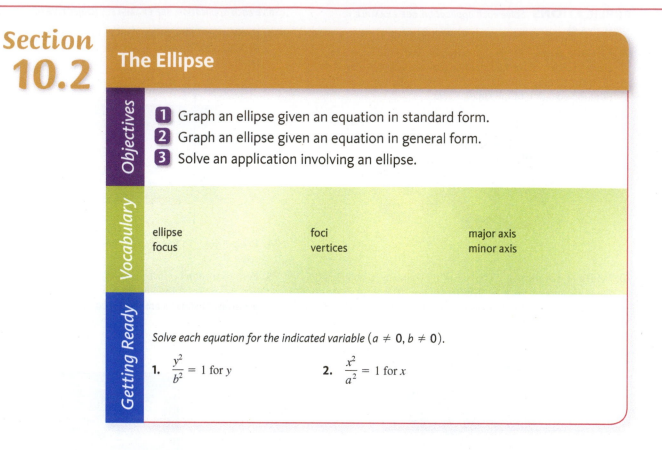

Objectives

1. Graph an ellipse given an equation in standard form.
2. Graph an ellipse given an equation in general form.
3. Solve an application involving an ellipse.

Vocabulary

ellipse   foci   major axis
focus   vertices   minor axis

Getting Ready

Solve each equation for the indicated variable ($a \neq 0, b \neq 0$).

1. $\dfrac{y^2}{b^2} = 1$ for y 2. $\dfrac{x^2}{a^2} = 1$ for x

A third conic section is an oval-shaped curve called an *ellipse*. Ellipses can be nearly round or they can be long and narrow. In this section, we will learn how to construct ellipses and how to graph equations that represent them.

Ellipses have optical and acoustical properties that are useful in architecture and engineering. For example, many arches are portions of an ellipse, because the shape is pleasing to the eye. (See Figure 10-13(a).) The planets and many comets have elliptical orbits. (See Figure 10-13(b).) Gears are often cut into elliptical shapes to provide nonuniform motion. (See Figure 10-13(c).)

Ellipses

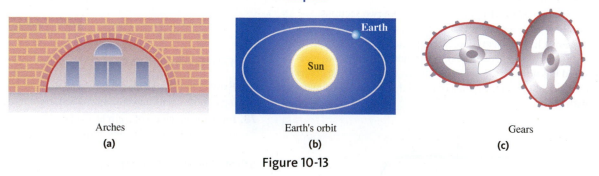

Arches   Earth's orbit   Gears
(a)   (b)   (c)

Figure 10-13

1   Graph an ellipse given an equation in standard form.

THE ELLIPSE

An **ellipse** is the set of all points P in the plane the sum of whose distances from two fixed points is a constant. See Figure 10-14, in which $d_1 + d_2$ is a constant.

Each of the two fixed points is called a **focus**. Midway between the **foci** is the *center* of the ellipse.

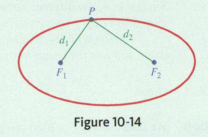

Figure 10-14

We can construct an ellipse by placing two thumbtacks fairly close together, as in Figure 10-15. We then tie each end of a piece of string to a thumbtack, catch the loop with the point of a pencil, and, while keeping the string taut, draw the ellipse.

Figure 10-15

Using this method, we can construct an ellipse of any specific size. For example, to construct an ellipse that is 10 inches wide and 6 inches high, we must find the length of string to use and the distance between thumbtacks.

To do this, we will let a represent the distance between the center and vertex V, as shown in Figure 10-16(a). We will also let c represent the distance between the center of the ellipse and either focus. When the pencil is at vertex V, the length of the string is $c + a + (a - c)$, or just $2a$. Because $2a$ is the 10-inch width of the ellipse, the string needs to be 10 inches long. The distance $2a$ is constant for any point on the ellipse, including point B shown in Figure 10-16(b).

Sir Isaac Newton

1642–1727

Newton was an English scientist and mathematician. Because he was not a good farmer, he went to Cambridge University to become a preacher. When he had to leave Cambridge because of the plague, he made some of his most important discoveries. He is best known in mathematics for developing calculus and in physics for discovering the laws of motion. Newton probably contributed more to science and mathematics than anyone else in history.

(a)   (b)

Figure 10-16

From right triangle BCF' in Figure 10-16(b) and the Pythagorean theorem, we can find the value of c as follows:

$$a^2 = b^2 + c^2 \quad \text{or} \quad c = \sqrt{a^2 - b^2}$$

Since distance b is one-half of the height of the ellipse, $b = 3$. Since $2a = 10$, $a = 5$. We can now substitute $a = 5$ and $b = 3$ into the formula to find the value of c:

$$c = \sqrt{5^2 - 3^2}$$
$$= \sqrt{25 - 9}$$
$$= \sqrt{16}$$
$$= 4$$

Since $c = 4$, the distance between the thumbtacks must be 8 inches. We can construct the ellipse by tying a 10-inch string to thumbtacks that are 8 inches apart.

To graph ellipses, we can create a table of ordered pairs that satisfy the equation, plot them, and join the points with a smooth curve.

EXAMPLE 1   Graph: $\dfrac{x^2}{36} + \dfrac{y^2}{9} = 1$

Solution   We note that the equation can be written in the form

$$\frac{x^2}{6^2} + \frac{y^2}{3^2} = 1 \quad \textcolor{red}{36 = 6^2 \text{ and } 9 = 3^2}$$

After creating a table of ordered pairs that satisfy the equation, plotting each of them, and joining the points with a curve, we obtain the ellipse shown in Figure 10-17.

We note that the center of the ellipse is the origin, the ellipse intersects the x-axis at points $(6, 0)$ and $(-6, 0)$, and the ellipse intersects the y-axis at points $(0, 3)$ and $(0, -3)$.

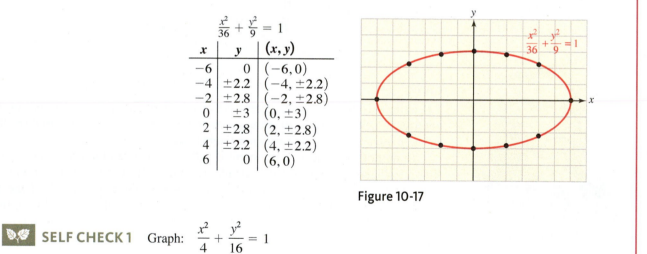

$$\frac{x^2}{36} + \frac{y^2}{9} = 1$$

x	y	(x, y)
-6	0	$(-6, 0)$
-4	± 2.2	$(-4, \pm 2.2)$
-2	± 2.8	$(-2, \pm 2.8)$
0	± 3	$(0, \pm 3)$
2	± 2.8	$(2, \pm 2.8)$
4	± 2.2	$(4, \pm 2.2)$
6	0	$(6, 0)$

Figure 10-17

SELF CHECK 1   Graph: $\dfrac{x^2}{4} + \dfrac{y^2}{16} = 1$

Example 1 illustrates that the graph of

$$\frac{x^2}{a^2} + \frac{y^2}{b^2} = 1$$

is an ellipse centered at the origin. To find the x-intercepts of the graph, we can let $y = 0$ and solve for x.

$$\frac{x^2}{a^2} + \frac{\textcolor{red}{0^2}}{b^2} = 1$$

$$\frac{x^2}{a^2} + 0 = 1$$

$$x^2 = a^2$$

$$x = a \quad \text{or} \quad x = -a$$

The x-intercepts are $(a, 0)$ and $(-a, 0)$.

To find the y-intercepts, we let $x = 0$ and solve for y.

$$\frac{0^2}{a^2} + \frac{y^2}{b^2} = 1$$

$$0 + \frac{y^2}{b^2} = 1$$

$$y^2 = b^2$$

$$y = b \quad \text{or} \quad y = -b$$

The y-intercepts are $(0, b)$ and $(0, -b)$.

In general, we have the following results.

EQUATIONS OF AN ELLIPSE CENTERED AT THE ORIGIN

The equation of an ellipse centered at the origin, with x-intercepts at $V_1(a, 0)$ and $V_2(-a, 0)$ and with y-intercepts of $(0, b)$ and $(0, -b)$, is

$$\frac{x^2}{a^2} + \frac{y^2}{b^2} = 1 \quad (a > b > 0) \quad \text{See Figure 10-18(a).}$$

The equation of an ellipse centered at the origin, with y-intercepts at $V_1(0, a)$ and $V_2(0, -a)$ and x-intercepts of $(b, 0)$ and $(-b, 0)$, is

$$\frac{x^2}{b^2} + \frac{y^2}{a^2} = 1 \quad (a > b > 0) \quad \text{See Figure 10-18(b).}$$

In Figure 10-18, the points V_1 and V_2 are the **vertices** of the ellipse, the midpoint of segment V_1V_2 is the *center* of the ellipse, and the distance between the center and either vertex is a. The segment V_1V_2 is called the **major axis**, and the segment joining either $(0, b)$ and $(0, -b)$ or $(b, 0)$ and $(-b, 0)$ is called the **minor axis**.

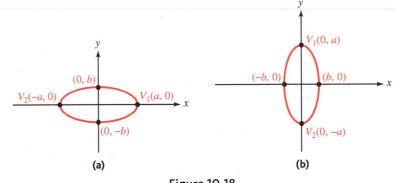

(a)   (b)

Figure 10-18

The equations for ellipses centered at (h, k) are as follows.

STANDARD EQUATION OF A HORIZONTAL ELLIPSE CENTERED AT (h, k)

The equation of a horizontal ellipse centered at (h, k), with major axis parallel to the x-axis, is

$$\frac{(x - h)^2}{a^2} + \frac{(y - k)^2}{b^2} = 1 \quad (a > b > 0)$$

STANDARD EQUATION OF A VERTICAL ELLIPSE CENTERED AT (h, k)

The equation of a vertical ellipse centered at (h, k), with major axis parallel to the y-axis, is

$$\frac{(x - h)^2}{b^2} + \frac{(y - k)^2}{a^2} = 1 \quad (a > b > 0)$$

COMMENT   To determine whether an ellipse is horizontal or vertical, look at the denominators in its standard equation. If the largest denominator is associated with the x-term, the ellipse will be horizontal. If the largest denominator is associated with the y-term, the ellipse will be vertical.

EXAMPLE 2   Graph: $25(x - 2)^2 + 16(y + 3)^2 = 400$

Solution   We first write the equation in standard form.

$$25(x - 2)^2 + 16(y + 3)^2 = 400$$

$$\frac{25(x - 2)^2}{400} + \frac{16(y + 3)^2}{400} = \frac{400}{400} \qquad \text{Divide both sides by 400.}$$

$$\frac{(x - 2)^2}{16} + \frac{(y + 3)^2}{25} = 1 \qquad \text{Simplify each fraction.}$$

$$\frac{(x - 2)^2}{4^2} + \frac{[y - (-3)]^2}{5^2} = 1$$

Because $5 > 4$ this is the equation of a vertical ellipse centered at $(h, k) = (2, -3)$ with major axis parallel to the y-axis and with $b = 4$ and $a = 5$. We first plot the center, as shown in Figure 10-19. Since a is the distance from the center to a vertex, we can locate the vertices by counting 5 units above and 5 units below the center. The vertices are at points $(2, 2)$ and $(2, -8)$.

Since $b = 4$, we can locate two more points on the ellipse by counting 4 units to the left and 4 units to the right of the center. The points $(-2, -3)$ and $(6, -3)$ are also on the graph.

Using these four points as guides, we can draw the ellipse.

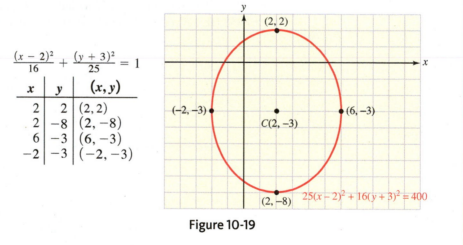

$$\frac{(x - 2)^2}{16} + \frac{(y + 3)^2}{25} = 1$$

x	y	(x, y)
2	2	$(2, 2)$
2	−8	$(2, -8)$
6	−3	$(6, -3)$
−2	−3	$(-2, -3)$

Figure 10-19

SELF CHECK 2   Graph: $16(x - 1)^2 + 9(y + 2)^2 = 144$

Accent on technology

▸ Graphing Ellipses

To use a graphing calculator to graph

$$\frac{(x + 2)^2}{4} + \frac{(y - 1)^2}{25} = 1$$

we first clear the fractions by multiplying both sides by 100 and solving for y.

$$25(x + 2)^2 + 4(y - 1)^2 = 100 \qquad \text{Multiply both sides by 100.}$$

$$4(y - 1)^2 = 100 - 25(x + 2)^2 \qquad \text{Subtract } 25(x + 2)^2 \text{ from both sides.}$$

$$(y - 1)^2 = \frac{100 - 25(x + 2)^2}{4} \qquad \text{Divide both sides by 4.}$$

$$y - 1 = \pm\frac{\sqrt{100 - 25(x + 2)^2}}{2} \qquad \text{Use the square-root property.}$$

$$y = 1 \pm \frac{\sqrt{100 - 25(x + 2)^2}}{2} \qquad \text{Add 1 to both sides.}$$

If we use window settings $[-6, 6]$ for x and $[-6, 6]$ for y and graph the functions

$$y = 1 + \frac{\sqrt{100 - 25(x + 2)^2}}{2} \qquad \text{and} \qquad y = 1 - \frac{\sqrt{100 - 25(x + 2)^2}}{2}$$

On a TI-84 calculator, we will obtain the ellipse shown in Figure 10-20.

$$\frac{(x + 2)^2}{4} + \frac{(y - 1)^2}{25} = 1$$

Figure 10-20

For instructions regarding the use of a Casio graphing calculator, please refer to the Casio Keystroke Guide in the back of the book.

2 Graph an ellipse given an equation in general form.

Another important form of the equation of an ellipse is called the *general form*.

GENERAL FORM OF THE EQUATION OF AN ELLIPSE

The equation of any ellipse can be written in the form

$$Ax^2 + Cy^2 + Dx + Ey + F = 0$$

COMMENT   Notice that there is no B coefficient in the general form. It is reserved for an xy term that would rotate the ellipse off the x- and y-axis. These will not be considered in this course.

We can use completing the square to write the general form of the equation of an ellipse.

EXAMPLE 3   Write $4x^2 + 9y^2 - 16x - 18y - 11 = 0$ in standard form to show that the equation represents an ellipse. Then graph the equation.

Solution   We write the equation in standard form by completing the square on x and y.

$$4x^2 + 9y^2 - 16x - 18y - 11 = 0$$

$$4x^2 + 9y^2 - 16x - 18y = 11 \qquad \text{Add 11 to both sides.}$$

$$4x^2 - 16x + 9y^2 - 18y = 11 \qquad \text{Use the commutative property to rearrange terms.}$$

$$4(x^2 - 4x) + 9(y^2 - 2y) = 11$$

Factor 4 from $4x^2 - 16x$ and factor 9 from $9y^2 - 18y$ to obtain coefficients of 1 for the squared terms.

$$4(x^2 - 4x \mathbin{\color{red}{+}} \color{red}{4}) + 9(y^2 - 2y \mathbin{\color{red}{+}} \color{red}{1}) = 11 \mathbin{\color{red}{+}} \color{red}{16} \mathbin{\color{red}{+}} \color{red}{9}$$

Complete the square to make $x^2 - 4x$ and $y^2 - 2y$ perfect trinomial squares. Since 16 and 9 are added to the left side, add 16 and 9 to the right side.

$$4(x - 2)^2 + 9(y - 1)^2 = 36$$

Factor $x^2 - 4x + 4$ and $y^2 - 2y + 1$.

$$\frac{(x - 2)^2}{9} + \frac{(y - 1)^2}{4} = 1$$

Divide both sides by 36.

Since this equation is of the form $\frac{(x - h)^2}{a^2} + \frac{(y - k)^2}{b^2} = 1$ $(a > b > 0)$, it represents an ellipse with $h = 2$, $k = 1$, $a = 3$, and $b = 2$. Its graph is shown in Figure 10-21.

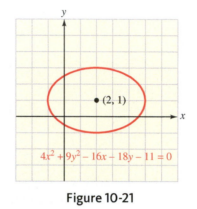

Figure 10-21

SELF CHECK 3   Graph: $4x^2 - 8x + 9y^2 - 36y = -4$

COMMENT   To distinguish between an equation of an ellipse and an equation of a circle, look at the coefficients of the squared terms. If the coefficients are the same, the equation is the equation of a circle. If the coefficients are different, but both positive, the equation is the equation of an ellipse.

3   Solve an application involving an ellipse.

EXAMPLE 4   **LANDSCAPE DESIGN**   A landscape architect is designing an elliptical pool that will fit in the center of a 20-by-30-foot rectangular garden, leaving at least 5 feet of space on all sides. Find the equation of the ellipse.

Solution   We place the rectangular garden in a coordinate system, as in Figure 10-22. To maintain 5 feet of clearance at the ends of the ellipse, the vertices must be the points $V_1(10, 0)$ and $V_2(-10, 0)$. Similarly, the y-intercepts are the points $(0, 5)$ and $(0, -5)$.

The equation of the ellipse has the form

$$\frac{x^2}{a^2} + \frac{y^2}{b^2} = 1$$

with $a = 10$ and $b = 5$. Thus, the equation of the boundary of the pool is

$$\frac{x^2}{100} + \frac{y^2}{25} = 1$$

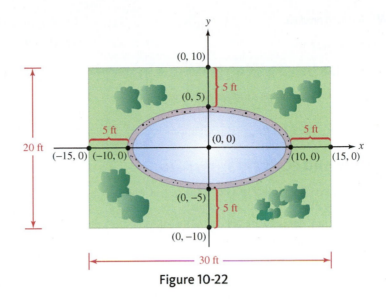

Figure 10-22

🌿 **SELF CHECK 4**   If the orientation of the garden is changed to 30 feet by 20 feet, find the equation of the pool.

🌿 **SELF CHECK ANSWERS**

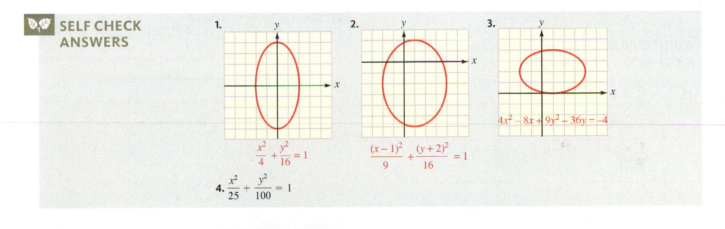

1. $\dfrac{x^2}{4} + \dfrac{y^2}{16} = 1$

2. $\dfrac{(x-1)^2}{9} + \dfrac{(y+2)^2}{16} = 1$

3. $4x^2 - 8x + 9y^2 - 36y = -4$

4. $\dfrac{x^2}{25} + \dfrac{y^2}{100} = 1$

NOW TRY THIS

1. If the vertices of an ellipse are $(5, -1)$ and $(5, 5)$ and the endpoints of the minor axis are $(3, 2)$ and $(7, 2)$, find the equation.

2. The eccentricity of an ellipse is $\frac{c}{a}$. If the center is located at $(4, -1)$ and the eccentricity is $\frac{2}{3}$, $a = 3$, find the equation of the horizontal ellipse.

10.2   Exercises 🌿

WARM-UPS *Evaluate when x = 0. Write as ordered pairs.*   *Evaluate when y = 0. Write as ordered pairs.*

1. $\dfrac{x^2}{4} + \dfrac{y^2}{9} = 1$ 2. $\dfrac{x^2}{36} + \dfrac{y^2}{25} = 1$ 3. $\dfrac{x^2}{4} + \dfrac{y^2}{9} = 1$ 4. $\dfrac{x^2}{36} + \dfrac{y^2}{25} = 1$

REVIEW   *Find each product.*

5. $5x^{-3}y^4(7x^3 + 2y^{-4})$

6. $(3x^{-3} - y^{-3})(3x^{-3} + y^{-3})$

Write each expression without using negative exponents.

7. $\dfrac{x^{-2} + y^{-2}}{x^{-2} - y^{-2}}$

8. $\dfrac{2x^{-3} - 2y^{-3}}{4x^{-3} + 4y^{-3}}$

VOCABULARY AND CONCEPTS   *Fill in the blanks.*

9. An _____ is the set of all points in a plane the ___ of whose distances from two fixed points is a constant.

10. The fixed points in Exercise 9 are the ___ of the ellipse.

11. The midpoint of the line segment joining the foci of an ellipse is called the _____ of the ellipse.

12. The graph of $\frac{x^2}{a^2} + \frac{y^2}{b^2} = 1$ $(a > b > 0)$ has vertices at _____, y-intercepts at _____, and eccentricity of _.

13. The center of the ellipse with an equation of $\frac{x^2}{a^2} + \frac{y^2}{b^2} = 1$ is ____ with the _____ having a length of $2a$ and the minor axis having a length of __.

14. The center of the ellipse with an equation of $\frac{(x - h)^2}{a^2} + \frac{(y - k)^2}{b^2} = 1$ is the point ____ .

GUIDED PRACTICE   *Graph each equation.* SEE EXAMPLE 1. (OBJECTIVE 1)

15. $\dfrac{x^2}{4} + \dfrac{y^2}{9} = 1$

16. $x^2 + \dfrac{y^2}{9} = 1$

17. $\dfrac{(x - 2)^2}{9} + \dfrac{(y - 1)^2}{4} = 1$ **18.** $\dfrac{(x - 1)^2}{9} + \dfrac{(y - 3)^2}{4} = 1$

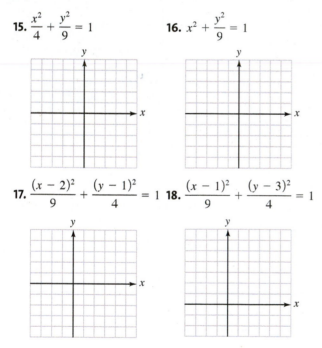

Graph each equation. SEE EXAMPLE 2. (OBJECTIVE 1)

19. $x^2 + 9y^2 = 9$

20. $25x^2 + 9y^2 = 225$

21. $(x + 1)^2 + 4(y + 2)^2 = 4$

22. $9(x - 5)^2 + (y + 2)^2 = 9$

Write each equation in standard form and graph it. SEE EXAMPLE 3. (OBJECTIVE 2)

23. $x^2 + 4y^2 - 4x + 8y + 4 = 0$

24. $x^2 + 4y^2 - 2x - 16y = -13$

25. $9x^2 + 4y^2 - 18x + 16y = 11$

26. $16x^2 + 25y^2 - 160x - 200y + 400 = 0$

ADDITIONAL PRACTICE *Graph each equation.*

27. $\dfrac{x^2}{9} + \dfrac{y^2}{16} = 1$

28. $\dfrac{x^2}{25} + \dfrac{y^2}{36} = 1$

29. $\dfrac{(x-2)^2}{16} + \dfrac{y^2}{25} = 1$

(2, 0)

30. $\dfrac{x^2}{25} + \dfrac{(y+1)^2}{36} = 1$

(0, −1)

31. $16x^2 + 4y^2 = 64$

32. $4x^2 + 9y^2 = 36$

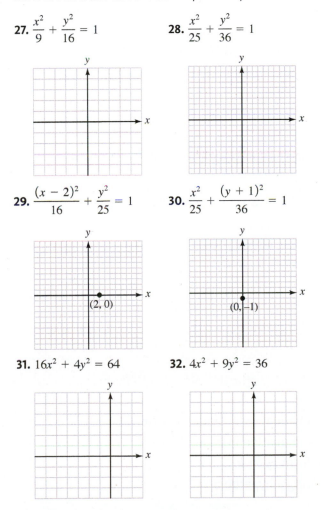

33. $25(x+1)^2 + 9y^2 = 225$

34. $4(x-6)^2 + 25(y-3)^2 = 100$

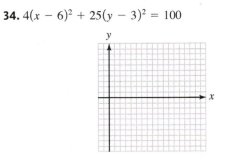

Use a graphing calculator to graph each equation.

35. $\dfrac{x^2}{9} + \dfrac{y^2}{4} = 1$

36. $x^2 + 16y^2 = 16$

37. $\dfrac{x^2}{4} + \dfrac{(y-1)^2}{9} = 1$

38. $\dfrac{(x+1)^2}{9} + \dfrac{(y-2)^2}{4} = 1$

APPLICATIONS *Solve each application.* SEE EXAMPLE 4.
(OBJECTIVE 3)

39. Fitness equipment   With elliptical cross-training equipment, the feet move through the natural elliptical pattern that
one experiences when walking, jogging, or running. Write the
equation of the elliptical pattern shown below.

10 in.

24 in.

40. Pool tables   Find the equation of the outer edge of the elliptical pool table shown below. Assume the red ball is at the focus.

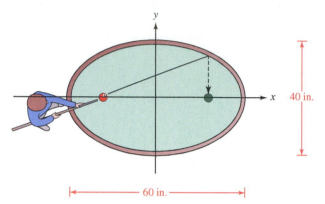

41. Designing an underpass   The arch of the underpass is half of an ellipse. Find the equation of the arch.

42. Calculating clearance   Find the height of the elliptical arch in Exercise 41 at a point 10 feet from the center of the roadway.

43. Area of an ellipse   The area A of the ellipse

$$\frac{x^2}{a^2} + \frac{y^2}{b^2} = 1$$

is given by $A = \pi ab$. Find the area of the ellipse $25x^2 + 4y^2 = 100$.

44. Area of a track   The elliptical track is bounded by the ellipses $4x^2 + 9y^2 = 576$ and $9x^2 + 25y^2 = 900$. Find the area of the track. (See Exercise 43.)

WRITING ABOUT MATH

45. Explain how to find the x- and the y-intercepts of the graph of the ellipse

$$\frac{x^2}{a^2} + \frac{y^2}{b^2} = 1$$

46. Explain the relationship between the center, focus, and vertex of an ellipse.

SOMETHING TO THINK ABOUT

47. What happens to the graph of $\frac{x^2}{a^2} + \frac{y^2}{b^2} = 1$ when $a = b$?

48. Explain why the graph of $x^2 + 4x + y^2 - 8y + 30 = 0$ does not exist.

Section 10.3

The Hyperbola

Objectives

1 Graph a hyperbola given an equation in standard form.
2 Graph a hyperbola given an equation in general form.
3 Graph a hyperbola of the form $xy = k$.
4 Solve an application involving a hyperbola.

Vocabulary

hyperbola   fundamental rectangle   asymptotes

Getting Ready

Find the value of y when $\frac{x^2}{25} - \frac{y^2}{9} = 1$ and x is the given value. Give each result to the nearest tenth.

1. $x = 6$ **2.** $x = -7$

The last conic section, the *hyperbola*, is a curve with two branches. In this section, we will graph equations that represent hyperbolas.

Hyperbolas are the basis of a navigational system known as LORAN (LOng RAnge Navigation). (See Figure 10-23.) They are also used to find the source of a distress signal, are the basis for the design of hypoid gears, and describe the paths of some comets.

Hyperbola

Figure 10-23

1   Graph a hyperbola given an equation in standard form.

THE HYPERBOLA

A **hyperbola** is the set of all points P in the plane for which the difference of the distances of each point from two fixed points, the foci, is a constant.

See Figure 10-24, in which $d_1 - d_2$ is a constant.

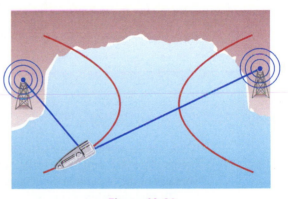

Figure 10-24

The graph of the standard form of the equation

$$\frac{x^2}{25} - \frac{y^2}{9} = 1$$

is a hyperbola. To graph the equation, we create a table of ordered pairs that satisfy the equation, plot each pair, and join the points with a smooth curve as in Figure 10-25.

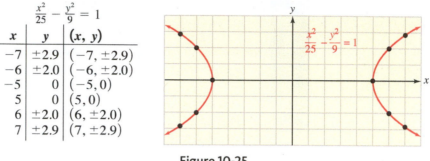

$$\frac{x^2}{25} - \frac{y^2}{9} = 1$$

x	y	(x, y)
-7	± 2.9	$(-7, \pm 2.9)$
-6	± 2.0	$(-6, \pm 2.0)$
-5	0	$(-5, 0)$
5	0	$(5, 0)$
6	± 2.0	$(6, \pm 2.0)$
7	± 2.9	$(7, \pm 2.9)$

Figure 10-25

This graph is centered at the origin and intersects the x-axis at $(5, 0)$ and $(-5, 0)$. We also note that the graph does not intersect the y-axis.

It is possible to draw a hyperbola by plotting only 4 points. For example, if we want to graph the hyperbola with an equation of

$$\frac{x^2}{a^2} - \frac{y^2}{b^2} = 1$$

we first look at the x- and y-intercepts. To find the x-intercepts, we let $y = 0$ and solve for x:

$$\frac{x^2}{a^2} - \frac{\mathbf{0}^2}{b^2} = 1$$

$$x^2 = a^2$$

$$x = \pm a$$

Thus, the hyperbola crosses the x-axis at the points $V_1(a, 0)$ and $V_2(-a, 0)$, called the **vertices** of the hyperbola. See Figure 10-26.

To find the y-intercepts, we let $x = 0$ and solve for y:

$$\frac{\mathbf{0}^2}{a^2} - \frac{y^2}{b^2} = 1$$

$$y^2 = -b^2$$

$$y = \pm\sqrt{-b^2}$$

Figure 10-26

Since b^2 is always positive, $\sqrt{-b^2}$ is an imaginary number. This means that the hyperbola does not cross the y-axis.

We construct a rectangle, called the **fundamental rectangle**, whose sides pass horizontally through $\pm b$ on the y-axis and vertically through $\pm a$ on the x-axis. The extended diagonals of the rectangle will be **asymptotes** of the hyperbola, which can be useful when sketching the graph.

EQUATION OF A HYPERBOLA CENTERED AT THE ORIGIN WITH VERTICES ON THE x-AXIS

Any equation that can be written in the form

$$\frac{x^2}{a^2} - \frac{y^2}{b^2} = 1$$

has a graph that is a hyperbola centered at the origin. The x-intercepts are the vertices $V_1(a, 0)$ and $V_2(-a, 0)$. There are no y-intercepts.

The asymptotes of the hyperbola are the extended diagonals of the fundamental rectangle shown in Figure 10-27.

Figure 10-27

The branches of the hyperbola in previous discussions open to the left and to the right. It is possible for hyperbolas to have different orientations with respect to the x- and y-axes. For example, the branches of a hyperbola can open upward and downward. In that case, the following equation applies.

EQUATION OF A HYPERBOLA CENTERED AT THE ORIGIN WITH VERTICES ON THE y-AXIS

Any equation that can be written in the form

$$\frac{y^2}{a^2} - \frac{x^2}{b^2} = 1$$

has a graph that is a hyperbola centered at the origin, as in Figure 10-28. The y-intercepts are the vertices $V_1(0, a)$ and $V_2(0, -a)$. There are no x-intercepts.

The asymptotes of the hyperbola are the extended diagonals of the rectangle shown in Figure 10-28.

Figure 10-28

COMMENT To determine whether a hyperbola opens horizontally, as in Figure 10-27, or vertically, as in Figure 10-28, we look at the signs of the terms. If the term containing x^2 is positive, the hyperbola will open horizontally. If the term containing y^2 is positive, the hyperbola will open vertically.

EXAMPLE 1 Graph: $9y^2 - 4x^2 = 36$

Solution To write the equation in standard form, we divide both sides by 36 to obtain

$$\frac{9y^2}{36} - \frac{4x^2}{36} = 1$$

$$\frac{y^2}{4} - \frac{x^2}{9} = 1 \quad \text{Simplify each fraction.}$$

Because the term containing y^2 is positive, the hyperbola will open vertically. We can find the y-intercepts of the graph by letting $x = 0$ and solving for y:

$$\frac{y^2}{4} - \frac{\mathbf{0}^2}{9} = 1$$

$$y^2 = 4$$

Thus, $y = \pm 2$, and the vertices of the hyperbola are $V_1(0, 2)$ and $V_2(0, -2)$. (See Figure 10-29.)

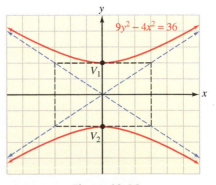

Figure 10-29

Since $\pm\sqrt{9} = \pm 3$, we can use the points $(3, 0)$ and $(-3, 0)$ on the x-axis to help draw the fundamental rectangle. We then draw its extended diagonals as asymptotes to help sketch the hyperbola.

SELF CHECK 1   Graph: $9x^2 - 4y^2 = 36$

Accent on technology

▸ Graphing Hyperbolas

To graph $\frac{x^2}{9} - \frac{y^2}{16} = 1$ using a graphing calculator, we follow the same procedure that we used for circles and ellipses. To write the equation as two functions, we solve for y to obtain $y = \pm\frac{\sqrt{16x^2 - 144}}{3}$. Then we graph the following two functions in a square window setting on a TI-84 calculator to obtain the graph of the hyperbola shown in Figure 10-30.

$$y = \frac{\sqrt{16x^2 - 144}}{3} \qquad \text{and} \qquad y = -\frac{\sqrt{16x^2 - 144}}{3}$$

Figure 10-30

For instructions regarding the use of a Casio graphing calculator, please refer to the Casio Keystroke Guide in the back of the book.

If a hyperbola is centered at a point with coordinates (h, k), the following equations apply.

EQUATIONS OF HYPERBOLAS CENTERED AT (h, k)

Any equation that can be written in the form

$$\frac{(x - h)^2}{a^2} - \frac{(y - k)^2}{b^2} = 1$$

is a hyperbola centered at (h, k) that opens left and right (horizontally).

Any equation of the form

$$\frac{(y - k)^2}{a^2} - \frac{(x - h)^2}{b^2} = 1$$

is a hyperbola centered at (h, k) that opens upward and downward (vertically).

EXAMPLE 2   Graph: $\dfrac{(x - 3)^2}{16} - \dfrac{(y + 1)^2}{4} = 1$

Solution   We write the equation in the form

$$\frac{(x - 3)^2}{4^2} - \frac{[y - (-1)]^2}{2^2} = 1$$

to see that its graph will be a hyperbola centered at the point $(h, k) = (3, -1)$. Its vertices are located at $a = 4$ units to the right and left of center, at $(7, -1)$ and $(-1, -1)$. Since $b = 2$, we can count 2 units above and below center to locate points $(3, 1)$ and $(3, -3)$. With these points, we can draw the fundamental rectangle along with its extended diagonals to locate the asymptotes. We can then sketch the hyperbola, as shown in Figure 10-31.

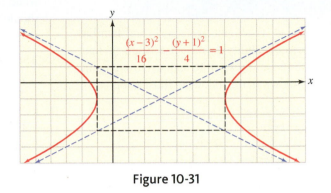

$$\frac{(x-3)^2}{16} - \frac{(y+1)^2}{4} = 1$$

Figure 10-31

🌿 **SELF CHECK 2**   Graph: $\dfrac{(x+2)^2}{9} - \dfrac{(y-1)^2}{4} = 1$

2 Graph a hyperbola given an equation in general form.

Another important form of the equation of a hyperbola is called the *general form*.

GENERAL FORM OF THE EQUATION OF A HYPERBOLA

The equation of any hyperbola can be written in the form

$$Ax^2 - Cy^2 + Dx + Ey + F = 0$$

Notice that for a hyperbola the coefficient for either the x^2 or y^2 term is negative. Recall for an ellipse the coefficients were both positive. As we did with ellipses we will need to complete the square to identify critical information for graphing.

EXAMPLE 3   Write the equation $x^2 - y^2 - 2x + 4y - 12 = 0$ in standard form to show that the equation represents a hyperbola. Then graph it.

Solution   We proceed as follows.

$$x^2 - y^2 - 2x + 4y - 12 = 0$$

$$x^2 - y^2 - 2x + 4y = 12 \qquad \text{Add 12 to both sides.}$$

$$x^2 - 2x - y^2 + 4y = 12 \qquad \text{Use the commutative property to rearrange the } x \text{ and } y \text{ terms.}$$

$$x^2 - 2x - 1(y^2 - 4y) = 12 \qquad \text{Factor } -1 \text{ from } -y^2 + 4y.$$

We then complete the square on x and y to make perfect trinomial squares.

$$x^2 - 2x + 1 - (y^2 - 4y + 4) = 12 + 1 - 4$$

We then factor $x^2 - 2x + 1$ and $y^2 - 4y + 4$ to obtain

$$(x - 1)^2 - (y - 2)^2 = 9$$

$$\frac{(x-1)^2}{9} - \frac{(y-2)^2}{9} = 1 \qquad \text{Divide both sides by 9.}$$

This is the equation of a hyperbola centered at $(1, 2)$. Its graph is shown in Figure 10-32.

$x^2 - y^2 - 2x + 4y = 12$

Figure 10-32

🌿 **SELF CHECK 3**   Graph: $x^2 - 4y^2 + 2x - 8y = 7$

Everyday connections
Focus on Conics

Satellite dishes and flashlight reflectors are familiar examples of a conic's ability to reflect a beam of light or to concentrate incoming satellite signals at one point. That property is shown in the following illustration.

1. Another form of an equation for a parabola with the vertex at $(0, 0)$ is $x^2 = 4py$, where p is the distance from the vertex to the focus. Find the coordinate of the focus for the parabola defined by the equation $x^2 = y$.

An ellipse has two foci, the points marked in the following illustration. Any light or signal that starts at one focus will be reflected to the other. This property is the basis of whispering galleries, where a person standing at one focus can clearly hear another person speaking at the other focus.

2. If the height of an elliptical rotunda is 32 feet and its width is 80 feet, how far from the center of the room should two people stand to witness the "whispering chamber" effect?

The focal property of the ellipse is also used in *lithotripsy*, a medical procedure for treating kidney stones. The patient is placed in an elliptical tank of water with the kidney stone at one focus. Shock waves from a small controlled explosion at the other focus are concentrated on the stone, pulverizing it.

The hyperbola also has two foci, the two points labeled F in the following illustration. As in the ellipse, light aimed at one focus is reflected toward the other. Hyperbolic mirrors are used in some reflecting telescopes.

3   Graph a hyperbola of the form $xy = k$.

There is a special type of hyperbola (also centered at the origin) that does not intersect either the x- or the y-axis. These hyperbolas have equations of the form $xy = k$, where $k \neq 0$.

EXAMPLE 4   Graph: $xy = -8$

Solution   We create a table of ordered pairs, plot each pair, and join the points with a smooth curve to obtain the hyperbola in Figure 10-33.

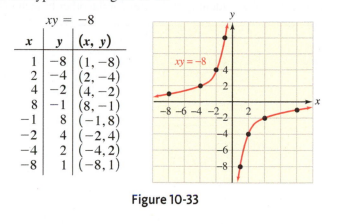

$xy = -8$

x	y	(x, y)
1	-8	$(1, -8)$
2	-4	$(2, -4)$
4	-2	$(4, -2)$
8	-1	$(8, -1)$
-1	8	$(-1, 8)$
-2	4	$(-2, 4)$
-4	2	$(-4, 2)$
-8	1	$(-8, 1)$

Figure 10-33

SELF CHECK 4   Graph: $xy = 6$

The result in Example 4 illustrates the following general equation.

EQUATIONS OF HYPERBOLAS OF THE FORM $xy = k$ Any equation of the form $xy = k$, where $k \neq 0$, has a graph that is a hyperbola that does not intersect either the x-axis or the y-axis.

4   Solve an application involving a hyperbola.

EXAMPLE 5   **ATOMIC STRUCTURE**   In an experiment that led to the discovery of the atomic structure of matter, Lord Rutherford (1871–1937) shot high-energy alpha particles toward a thin sheet of gold. Because many were reflected, Rutherford showed the existence of the nucleus of a gold atom. The alpha particle in Figure 10-34 is repelled by the nucleus at the origin; it travels along the hyperbolic path given by $4x^2 - y^2 = 16$. How close does the particle come to the nucleus?

Solution   To find the distance from the nucleus at the origin, we must find the coordinates of the vertex V. To do so, we write the equation of the particle's path in standard form:

$$4x^2 - y^2 = 16$$

$$\frac{4x^2}{16} - \frac{y^2}{16} = \frac{16}{16}$$ Divide both sides by 16.

$$\frac{x^2}{4} - \frac{y^2}{16} = 1$$ Simplify.

$$\frac{x^2}{2^2} - \frac{y^2}{4^2} = 1$$ Write 4 as 2^2 and 16 as 4^2.

This equation is in the form

$$\frac{x^2}{a^2} - \frac{y^2}{b^2} = 1$$

Figure 10-34

with $a = 2$. Thus, the vertex of the path is $(2, 0)$. The particle is never closer than 2 units from the nucleus.

SELF CHECK 5   Identify the vertex for the other branch of the hyperbola.

SELF CHECK ANSWERS

1. $9x^2 - 4y^2 = 36$

2. $\dfrac{(x+2)^2}{9} - \dfrac{(y-1)^2}{4} = 1$

3. $x^2 - 4y^2 + 2x - 8y = 7$

4. $xy = 6$

5. $(-2, 0)$

NOW TRY THIS

Given the equation $\dfrac{x^2}{a^2} - \dfrac{y^2}{b^2} = 1$, the equations for the asymptotes are $y = \dfrac{b}{a}x$, and $y = -\dfrac{b}{a}x$.

1. Find the equations for the asymptotes for the hyperbola described by the equation $\dfrac{x^2}{25} - \dfrac{y^2}{9} = 1$.

2. Find the equations for the asymptotes for the hyperbola described by the equation $\dfrac{y^2}{4} - \dfrac{x^2}{9} = 1$.

3. Find the equations for the asymptotes for the hyperbola described by the equation $\dfrac{(x+2)^2}{9} - \dfrac{(y-1)^2}{4} = 1$.

10.3 Exercises

WARM-UPS *Evaluate when $y = 0$. Write as ordered pairs.*

1. $\dfrac{x^2}{4} - \dfrac{y^2}{9} = 1$

2. $\dfrac{x^2}{36} - \dfrac{y^2}{16} = 1$

REVIEW *Factor each expression.*

3. $-5x^4 + 10x^3 - 15x^2$

4. $9a^2 - b^2$

5. $14a^2 - 15ab - 9b^2$

6. $27p^3 - 64q^3$

VOCABULARY AND CONCEPTS *Fill in the blanks.*

7. A _____ is the set of all points in a plane for which the _____ of the distances from two fixed points is a constant.

8. The fixed points in Exercise 7 are the ___ of the hyperbola.

9. The midpoint of the line segment joining the foci of a hyperbola is called the _____ of the hyperbola.

10. To graph a hyperbola, we locate its center and vertices, sketch the _____, and sketch the asymptotes.

11. The hyperbolic graph of $\frac{x^2}{a^2} - \frac{y^2}{b^2} = 1$ has x-intercepts of _____. There are no _____.

12. The center of the hyperbola with an equation of $\frac{x^2}{a^2} - \frac{y^2}{b^2} = 1$ is the point _____. The center of the hyperbola with an equation of $\frac{(x-h)^2}{a^2} - \frac{(y-k)^2}{b^2} = 1$ is the point _____.

GUIDED PRACTICE *Graph each hyperbola.* SEE EXAMPLE 1. (OBJECTIVE 1)

13. $\frac{x^2}{9} - \frac{y^2}{4} = 1$

14. $\frac{x^2}{4} - \frac{y^2}{4} = 1$

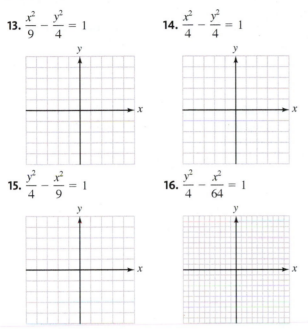

15. $\frac{y^2}{4} - \frac{x^2}{9} = 1$

16. $\frac{y^2}{4} - \frac{x^2}{64} = 1$

Graph each hyperbola. SEE EXAMPLE 2. (OBJECTIVE 1)

17. $\frac{(x-2)^2}{9} - \frac{y^2}{16} = 1$

18. $\frac{(x+2)^2}{16} - \frac{(y-3)^2}{25} = 1$

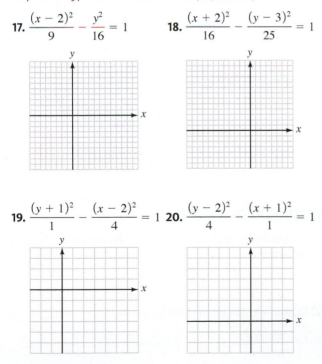

19. $\frac{(y+1)^2}{1} - \frac{(x-2)^2}{4} = 1$ **20.** $\frac{(y-2)^2}{4} - \frac{(x+1)^2}{1} = 1$

Write each equation in standard form and graph it. SEE EXAMPLE 3. (OBJECTIVE 2)

21. $4x^2 - y^2 + 8x - 4y = 4$ **22.** $x^2 - 9y^2 - 4x - 54y = 86$

23. $4y^2 - x^2 + 8y + 4x = 4$ **24.** $y^2 - 4x^2 - 4y - 8x = 4$

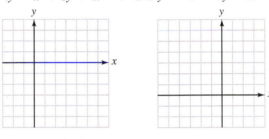

Graph each hyperbola. SEE EXAMPLE 4. (OBJECTIVE 3)

25. $xy = 10$

26. $xy = -10$

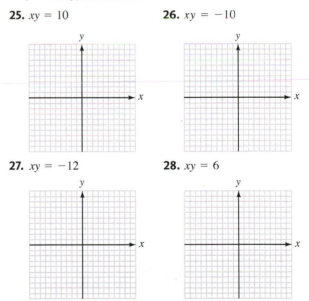

27. $xy = -12$

28. $xy = 6$

ADDITIONAL PRACTICE *Graph each equation.*

29. $25x^2 - y^2 = 25$

30. $9x^2 - 4y^2 = 36$

31. $4(x + 3)^2 - (y - 1)^2 = 4$ **32.** $(x + 5)^2 - 16y^2 = 16$

 Use a graphing calculator to graph each equation.

33. $\dfrac{x^2}{9} - \dfrac{y^2}{4} = 1$ **34.** $y^2 - 16x^2 = 16$

35. $\dfrac{x^2}{4} - \dfrac{(y - 1)^2}{9} = 1$ **36.** $\dfrac{(y + 1)^2}{9} - \dfrac{(x - 2)^2}{4} = 1$

APPLICATIONS *Solve each application. SEE EXAMPLE 5.* **(OBJECTIVE 4)**

37. Alpha particles The particle in the illustration approaches the nucleus at the origin along the path $9y^2 - x^2 = 81$. How close does the particle come to the nucleus?

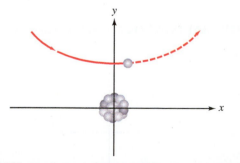

38. LORAN By determining the difference of the distances between the ship and two land-based radio transmitters,

the LORAN system places the ship on the hyperbola $x^2 - 4y^2 = 576$. If the ship is also 5 miles out to sea, find its coordinates.

39. Electrostatic repulsion Two similarly charged particles are shot together for an almost head-on collision, as shown in the illustration. They repel each other and travel the two branches of the hyperbola given by $x^2 - 4y^2 = 4$. How close do they get?

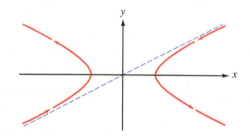

40. Sonic boom The position of the sonic boom caused by faster-than-sound aircraft is the hyperbola $y^2 - x^2 = 25$ in the coordinate system shown below. How wide is the hyperbola 5 units from its vertex?

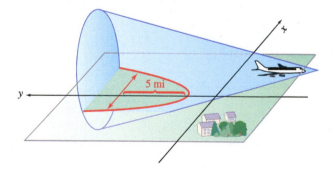

WRITING ABOUT MATH

41. Explain how to find the x- and the y-intercepts of the graph of the hyperbola

$$\frac{x^2}{a^2} - \frac{y^2}{b^2} = 1$$

42. Explain why the graph of the hyperbola

$$\frac{x^2}{a^2} - \frac{y^2}{b^2} = 1$$

has no y-intercept.

SOMETHING TO THINK ABOUT

43. Describe the fundamental rectangle of

$$\frac{x^2}{a^2} - \frac{y^2}{b^2} = 1$$

when $a = b$.

44. The hyperbolas $x^2 - y^2 = 1$ and $y^2 - x^2 = 1$ are called **conjugate hyperbolas**. Graph both on the same axes. What do they have in common?

Section 10.4

Systems Containing Second-Degree Equations

Objectives

1. Solve a system of equations containing a second-degree equation by graphing.

2. Solve a system of equations containing a second-degree equation by substitution.

3. Solve a system of equations containing a second-degree equation by elimination (addition).

Getting Ready

Add the left sides and the right sides of the following equations.

1.
$$3x^2 + 2y^2 = 12$$
$$\underline{-3x^2 - 3y^2 = 32}$$

2.
$$-7x^2 - 5y^2 = -17$$
$$\underline{12x^2 + 5y^2 = 25}$$

We now discuss ways to solve systems of two equations in two variables where at least one of the equations is of second degree.

 Solve a system of equations containing a second-degree equation by graphing.

EXAMPLE 1   Solve by graphing: $\begin{cases} x^2 + y^2 = 25 \\ 2x + y = 10 \end{cases}$

Solution   The graph of $x^2 + y^2 = 25$ is a circle with center at the origin and radius of 5. The graph of $2x + y = 10$ is a line. Depending on whether the line is a secant (intersecting the circle at two points) or a tangent (intersecting the circle at one point) or does not intersect the circle at all, there are two, one, or no solutions to the system, respectively.

After graphing the circle and the line, as shown in Figure 10-35, we see that there are two intersection points $(3, 4)$ and $(5, 0)$. These are the solutions of the system.

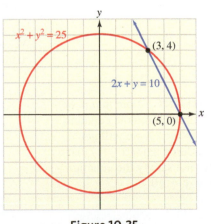

Figure 10-35

> 🌿 **SELF CHECK 1**   Solve by graphing: $\begin{cases} x^2 + y^2 = 13 \\ y = -\frac{1}{5}x + \frac{13}{5} \end{cases}$

Accent on technology

▸ Solving Systems of Equations

To solve Example 1 with a graphing calculator, we solve each equation for y and enter $-2x + 10$ for y_1, $\sqrt{25 - x^2}$ for y_2, and $-\sqrt{25 - x^2}$ for y_3 and graph (see Figure 10-36(a)). We can find the exact answers by using the INTERSECT feature found in the CALC menu. However, we must select the correct two equations to find the intersections. To obtain the solution $(3, 4)$, we use y_1 and y_2. To obtain the solution $(5, 0)$, we use y_1 and y_3 (see Figure 10-36(b) and Figure 10-36(c)).

$x^2 + y^2 = 25$ $2x + y = 10$

(a)

$Y_1 = \sqrt{(25 - X^2)}$ Intersection X = 3   Y = 4

(b)

$Y_3 = -\sqrt{(25 - X^2)}$ X = 5   Y = 0

(c)

Figure 10-36

---

2 Solve a system of equations containing a second-degree equation by substitution.

Algebraic methods also can be used to solve systems of equations. In the next two examples, we will solve systems using the substitution method. In Example 2 we will use the same system we solved by graphing in Example 1.

EXAMPLE 2   Solve using substitution: $\begin{cases} x^2 + y^2 = 25 \\ 2x + y = 10 \end{cases}$

Solution   This system has one second-degree equation and one first-degree equation. The first equation is the equation of a circle and the second equation is the equation of a line. Since a line can intersect a circle in 0, 1, or 2 points, this system can have zero, one, or two solutions.

We can solve the system by substitution. Solving the linear equation for y gives

$$2x + y = 10$$
(1) $$y = -2x + 10$$

We can substitute $-2x + 10$ for y in the second-degree equation and solve the resulting quadratic equation for x:

COMMENT   Note that if we substitute 3 into the equation $x^2 + y^2 = 25$, we will get an ordered pair $(3, -4)$ that does not satisfy the equation $2x + y = 10$.

$$x^2 + y^2 = 25$$
$$x^2 + (-2x + 10)^2 = 25$$
$$x^2 + 4x^2 - 40x + 100 = 25 \qquad \text{Square the binomial.}$$
$$5x^2 - 40x + 75 = 0 \qquad \text{Combine like terms and subtract 25 from both sides.}$$
$$x^2 - 8x + 15 = 0 \qquad \text{Divide both sides by 5.}$$
$$(x - 5)(x - 3) = 0 \qquad \text{Factor } x^2 - 8x + 15.$$

$$x - 5 = 0 \quad \text{or} \quad x - 3 = 0 \qquad \textcolor{red}{\text{Set each factor equal to 0 and solve.}}$$
$$x = 5 \qquad\qquad x = 3$$

If we substitute 5 for x in Equation 1, we get $y = 0$. If we substitute 3 for x in Equation 1, we get $y = 4$. The two solutions of this system are $(5, 0)$ and $(3, 4)$. Check the results by substituting these ordered pairs in both of the original equations.

 SELF CHECK 2   Solve using substitution: $\begin{cases} x^2 + y^2 = 13 \\ y = -\frac{1}{5}x + \frac{13}{5} \end{cases}$

EXAMPLE 3   Solve using substitution: $\begin{cases} 4x^2 + 9y^2 = 5 \\ y = x^2 \end{cases}$

Solution   This system has two second-degree equations. The first equation is the equation of an ellipse and the second equation is the equation of a parabola. Since an ellipse and a parabola can intersect in 0, 1, 2, 3, or 4 points, this system can have zero, one, two, three, or four solutions.

We can solve the system by substitution.

$$4\textcolor{red}{x^2} + 9y^2 = 5$$
$$4\textcolor{red}{y} + 9y^2 = 5 \qquad\qquad \textcolor{red}{\text{Substitute } y \text{ for } x^2.}$$
$$9y^2 + 4y - 5 = 0 \qquad\qquad \textcolor{red}{\text{Write in quadratic form.}}$$
$$(9y - 5)(y + 1) = 0 \qquad\qquad \textcolor{red}{\text{Factor } 9y^2 + 4y - 5.}$$
$$9y - 5 = 0 \quad \text{or} \quad y + 1 = 0 \qquad \textcolor{red}{\text{Set each factor equal to 0 and solve.}}$$
$$y = \frac{5}{9} \qquad\qquad y = -1$$

Since $y = x^2$, the values of x are found by solving the equations

$$x^2 = \frac{5}{9} \quad \text{and} \quad x^2 = -1$$

Because $x^2 = -1$ has no real number solutions, this possibility is discarded. The solutions of $x^2 = \frac{5}{9}$ are

$$x = \frac{\sqrt{5}}{3} \quad \text{or} \quad x = -\frac{\sqrt{5}}{3}$$

The two solutions of the system are

$$\left(\frac{\sqrt{5}}{3}, \frac{5}{9} \right) \quad \text{and} \quad \left(-\frac{\sqrt{5}}{3}, \frac{5}{9} \right)$$

 SELF CHECK 3   Solve using substitution: $\begin{cases} x^2 + y^2 = 20 \\ y = x^2 \end{cases}$

---

3   Solve a system of equations containing a second-degree equation by elimination (addition).

In the next example, we will solve a system by elimination.

EXAMPLE 4   Solve by elimination: $\begin{cases} 3x^2 + 2y^2 = 36 \\ 4x^2 - y^2 = 4 \end{cases}$

Solution   This system has two second-degree equations. The first equation is the equation of an ellipse and the second equation is the equation of a hyperbola. Since an ellipse and a hyperbola can intersect in 0, 1, 2, 3, or 4 points, this system can have 0, 1, 2, 3, or 4 solutions.

Both equations are in the form $ax^2 + by^2 = c$ and we solve the system by elimination.

We keep the first equation and multiply the second equation by 2 to obtain the equivalent system

$$\begin{cases} 3x^2 + 2y^2 = 36 \\ 8x^2 - 2y^2 = 8 \end{cases}$$

We add the equations to eliminate y and solve the resulting equation for x:

$$11x^2 = 44$$
$$x^2 = 4$$
$$x = 2 \quad \text{or} \quad x = -2$$

To find the values of y, we substitute 2 for x and then -2 for x in the first equation and proceed as follows:

For x = 2	*For x = -2*
$3x^2 + 2y^2 = 36$	$3x^2 + 2y^2 = 36$
$3(2)^2 + 2y^2 = 36$	$3(-2)^2 + 2y^2 = 36$
$12 + 2y^2 = 36$	$12 + 2y^2 = 36$
$2y^2 = 24$	$2y^2 = 24$
$y^2 = 12$	$y^2 = 12$
$y = +\sqrt{12}$ or $y = -\sqrt{12}$	$y = +\sqrt{12}$ or $y = -\sqrt{12}$
$y = 2\sqrt{3}$ $\mid$ $y = -2\sqrt{3}$	$y = 2\sqrt{3}$ $\mid$ $y = -2\sqrt{3}$

The four solutions of this system are

$$\left(2, 2\sqrt{3}\right), \left(2, -2\sqrt{3}\right), \left(-2, 2\sqrt{3}\right), \text{and} \left(-2, -2\sqrt{3}\right)$$

SELF CHECK 4   Solve by elimination: $\begin{cases} x^2 + 4y^2 = 16 \\ x^2 - y^2 = 1 \end{cases}$

SELF CHECK ANSWERS

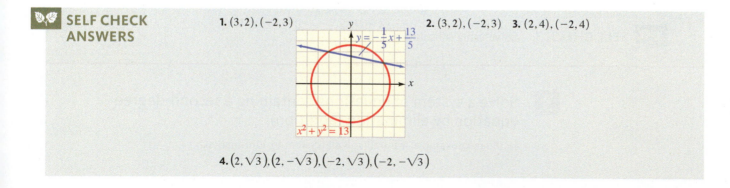

1. $(3, 2), (-2, 3)$ 2. $(3, 2), (-2, 3)$ 3. $(2, 4), (-2, 4)$

4. $\left(2, \sqrt{3}\right), \left(2, -\sqrt{3}\right), \left(-2, \sqrt{3}\right), \left(-2, -\sqrt{3}\right)$

> **NOW TRY THIS**
>
> *Solve by substitution.*
> $$\begin{cases} x^2 - y^2 = 4 \\ 9x^2 + 16y^2 = 144 \end{cases}$$

10.4 Exercises

WARM-UPS *State the possible number of solutions of a system when the graphs of the equations are*

1. a line and a parabola.   **2.** a line and a hyperbola.

3. a circle and a parabola.   **4.** a circle and a hyperbola.

REVIEW *Simplify each radical expression. Assume that all variables represent positive numbers.*

5. $\sqrt{45x^2} - 7\sqrt{80x^2}$

6. $\sqrt{243a^3} - 4a\sqrt{300a}$

7. $\dfrac{\sqrt{18x^3} - x\sqrt{8x}}{\sqrt{32x} - \sqrt{8x}}$

8. $\sqrt[3]{\dfrac{x}{4}} + \sqrt[3]{\dfrac{x}{32}} - \sqrt[3]{\dfrac{x}{500}}$

VOCABULARY AND CONCEPTS *Fill in the blanks.*

9. We can solve systems of equations by _____, elimination (addition), or _____.

10. A line can intersect an ellipse in at most ___ points.

11. A parabola can intersect an ellipse in at most ___ points.

12. An ellipse can intersect a hyperbola in at most ___ points.

GUIDED PRACTICE *Solve each system of equations by graphing.* SEE EXAMPLE 1. (OBJECTIVE 1)

13. $\begin{cases} 8x^2 + 32y^2 = 256 \\ x = 2y \end{cases}$

14. $\begin{cases} x^2 + y^2 = 2 \\ x + y = 2 \end{cases}$

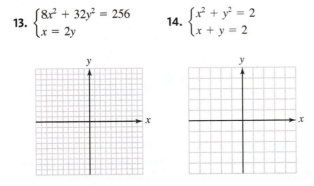

15. $\begin{cases} x^2 + y^2 = 10 \\ y = 3x^2 \end{cases}$

16. $\begin{cases} x^2 + y^2 = 5 \\ x + y = 3 \end{cases}$

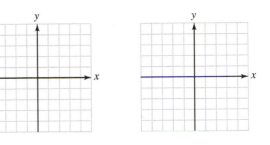

Solve each system by substitution. SEE EXAMPLE 2. (OBJECTIVE 2)

17. $\begin{cases} 25x^2 + 9y^2 = 225 \\ 5x + 3y = 15 \end{cases}$

18. $\begin{cases} x^2 + y^2 = 2 \\ x + y = 2 \end{cases}$

19. $\begin{cases} x^2 + y^2 = 5 \\ x + y = 3 \end{cases}$

20. $\begin{cases} x^2 - x - y = 2 \\ 4x - 3y = 0 \end{cases}$

Solve each system by substitution. SEE EXAMPLE 3. (OBJECTIVE 2)

21. $\begin{cases} x^2 + y^2 = 20 \\ y = x^2 \end{cases}$

22. $\begin{cases} x^2 + y^2 = 73 \\ y = x^2 - 1 \end{cases}$

23. $\begin{cases} x^2 + y^2 = 30 \\ y = x^2 \end{cases}$

24. $\begin{cases} x^2 + y^2 = 90 \\ y = x^2 \end{cases}$

Solve each system by elimination. SEE EXAMPLE 4. (OBJECTIVE 3)

25. $\begin{cases} x^2 + y^2 = 34 \\ x^2 - y^2 = 16 \end{cases}$

26. $\begin{cases} 2x^2 + y^2 = 14 \\ x^2 - y^2 = 7 \end{cases}$

27. $\begin{cases} x^2 + y^2 = 20 \\ x^2 - y^2 = -12 \end{cases}$

28. $\begin{cases} x^2 - 4x - y = -3 \\ x^2 - 4x + y = -3 \end{cases}$

ADDITIONAL PRACTICE *Solve each system using any method.*

29. $\begin{cases} x^2 + y^2 = 25 \\ 12x^2 + 64y^2 = 768 \end{cases}$ **30.** $\begin{cases} x^2 + y^2 = 13 \\ y = x^2 - 1 \end{cases}$

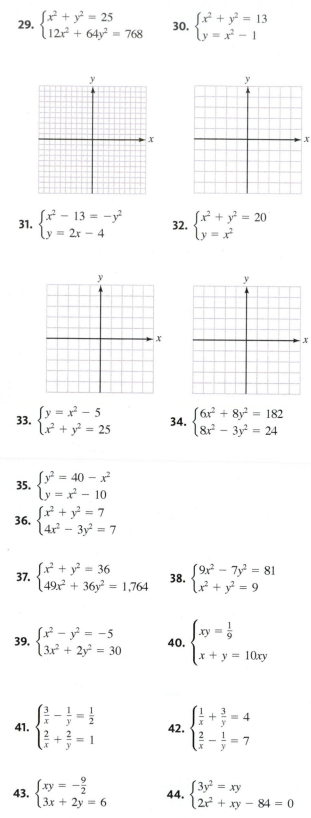

31. $\begin{cases} x^2 - 13 = -y^2 \\ y = 2x - 4 \end{cases}$ **32.** $\begin{cases} x^2 + y^2 = 20 \\ y = x^2 \end{cases}$

33. $\begin{cases} y = x^2 - 5 \\ x^2 + y^2 = 25 \end{cases}$ **34.** $\begin{cases} 6x^2 + 8y^2 = 182 \\ 8x^2 - 3y^2 = 24 \end{cases}$

35. $\begin{cases} y^2 = 40 - x^2 \\ y = x^2 - 10 \end{cases}$

36. $\begin{cases} x^2 + y^2 = 7 \\ 4x^2 - 3y^2 = 7 \end{cases}$

37. $\begin{cases} x^2 + y^2 = 36 \\ 49x^2 + 36y^2 = 1,764 \end{cases}$ **38.** $\begin{cases} 9x^2 - 7y^2 = 81 \\ x^2 + y^2 = 9 \end{cases}$

39. $\begin{cases} x^2 - y^2 = -5 \\ 3x^2 + 2y^2 = 30 \end{cases}$ **40.** $\begin{cases} xy = \frac{1}{9} \\ x + y = 10xy \end{cases}$

41. $\begin{cases} \frac{3}{x} - \frac{1}{y} = \frac{1}{2} \\ \frac{2}{x} + \frac{2}{y} = 1 \end{cases}$ **42.** $\begin{cases} \frac{1}{x} + \frac{3}{y} = 4 \\ \frac{2}{x} - \frac{1}{y} = 7 \end{cases}$

43. $\begin{cases} xy = -\frac{9}{2} \\ 3x + 2y = 6 \end{cases}$ **44.** $\begin{cases} 3y^2 = xy \\ 2x^2 + xy - 84 = 0 \end{cases}$

45. $\begin{cases} xy = \frac{1}{6} \\ y + x = 5xy \end{cases}$ **46.** $\begin{cases} xy = \frac{1}{12} \\ y + x = 7xy \end{cases}$

📈 *Use a graphing calculator to solve each system.*

47. $\begin{cases} x^2 - 6x - y = -5 \\ x^2 - 6x + y = -5 \end{cases}$ **48.** $\begin{cases} x^2 - y^2 = -5 \\ 3x^2 + 2y^2 = 30 \end{cases}$

APPLICATIONS *Solve each application.*

49. Integers   The product of two integers is 32, and their sum is 12. Find the integers.

50. Numbers   The sum of the squares of two numbers is 221, and the sum of the numbers is 9. Find the numbers.

51. Geometry   The area of a rectangle is 63 square centimeters, and its perimeter is 32 centimeters. Find the dimensions of the rectangle.

52. Investing money   Ignacio receives $225 annual income from one investment. Carol invested $500 more than Ignacio, but at an annual rate of 1% less. Carol's annual income is $240. What is the amount and rate of Ignacio's investment?

53. Investing money   Rania receives $67.50 annual income from one investment. Jerome invested $150 more than Rania at an annual rate of $1\frac{1}{2}$% more. Jerome's annual income is $94.50. What is the amount and rate of Rania's investment?   (*Hint*: There are two answers.)

54. Artillery   The shell fired from the base of the hill in the illustration follows the parabolic path $y = -\frac{1}{6}x^2 + 2x$ with distances measured in miles. The hill has a slope of $\frac{1}{3}$. How far from the gun is the point of impact?   (*Hint*: Find the coordinates of the point and then the distance.)

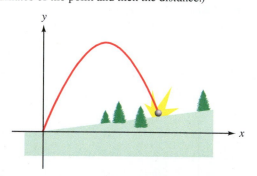

55. Driving rates   Jim drove 306 miles. Jim's brother made the same trip at a speed 17 mph slower than Jim did and required an extra $1\frac{1}{2}$ hours. What was Jim's rate and time?

WRITING ABOUT MATH

56. Describe the benefits of the graphical method for solving a system of equations.
57. Describe the drawbacks of the graphical method.

SOMETHING TO THINK ABOUT

58. The graphs of the two independent equations of a system are parabolas. How many solutions might the system have?
59. The graphs of the two independent equations of a system are hyperbolas. How many solutions might the system have?

Section 10.5

Piecewise-Defined Functions and the Greatest Integer Function

Objectives

1. Graph a piecewise-defined function and determine the open intervals over which the function is increasing, decreasing, and constant.
2. Graph the greatest integer function.
3. Solve an application involving the greatest integer function or a piecewise-defined function.

Vocabulary

increasing   piecewise-defined function   step function
decreasing   greatest integer function

Getting Ready

1. Is $f(x) = x^2$ positive or negative when $x > 0$?
2. Is $f(x) = -x^2$ positive or negative when $x > 0$?
3. What is the largest integer that is less than 98.6?
4. What is the largest integer that is less than -2.7?

If the values of $f(x)$ increase as x increases on an open interval, we say that the function is **increasing** *on the interval* (see Figure 10-37(a) on the next page). If the values of $f(x)$ decrease as x increases on an open interval, we say that the function is **decreasing** *on the interval* (see Figure 10-37(b)). If the values of $f(x)$ remain constant as x increases on an open interval, we say that the function is *constant on the interval* (see Figure 10-37(c)).

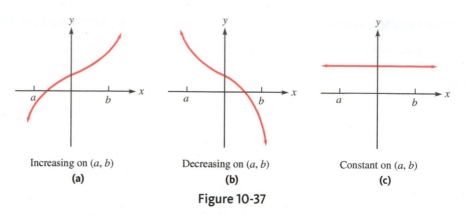

Increasing on (a, b) Decreasing on (a, b) Constant on (a, b)

    (a)   (b)   (c)

Figure 10-37

Some functions are defined by using different equations for different parts of their domains. Such functions are called **piecewise-defined functions**.

1 Graph a piecewise-defined function and determine the open intervals over which the function is increasing, decreasing, and constant.

A basic piecewise-defined function is the definition of $|x|$, which can be written in the form

$$f(x) = \begin{cases} x \text{ when } x \geq 0 \\ -x \text{ when } x < 0 \end{cases}$$

When x is in the interval $[0, \infty)$, we use the function $f(x) = x$ to evaluate. However, when x is in the interval $(-\infty, 0)$, we use the function $f(x) = -x$ to evaluate. The graph of the resulting piecewise function is shown in Figure 10-38.

For $x \geq 0$

x	y	(x, y)
0	0	$(0, 0)$
1	1	$(1, 1)$
2	2	$(2, 2)$
3	3	$(3, 3)$

For $x < 0$

x	y	(x, y)
-4	4	$(-4, 4)$
-3	3	$(-3, 3)$
-2	2	$(-2, 2)$
-1	1	$(-1, 1)$

Figure 10-38

The function, shown in Figure 10-38, is decreasing on the interval $(-\infty, 0)$ and is increasing on the interval $(0, \infty)$.

EXAMPLE 1   Graph the piecewise-defined function given by

$$f(x) = \begin{cases} x^2 \text{ when } x \leq 0 \\ x \text{ when } 0 < x < 2 \\ -1 \text{ when } x \geq 2 \end{cases}$$

and determine where the function is increasing, decreasing, and constant.

Solution   For each number x, we decide which of the three equations will be used to find the corresponding value of y:

- For numbers $x \leq 0$, $f(x)$ is determined by $f(x) = x^2$, and the graph is the left half of a parabola. See Figure 10-39. Since the values of $f(x)$ decrease on this graph as x increases, the function is decreasing on the interval $(-\infty, 0)$.

- For numbers $0 < x < 2$, $f(x)$ is determined by $f(x) = x$, and the graph is part of a line. Since the values of $f(x)$ increase on this graph as x increases, the function is increasing on the interval $(0, 2)$.

- For numbers $x \geq 2$, $f(x)$ is the constant -1, and the graph is part of a horizontal line. Since the values of $f(x)$ remain constant on this line, the function is constant on the interval $(2, \infty)$.

$$f(x) = \begin{cases} x^2 \text{ if } x \leq 0 \\ x \text{ if } 0 < x < 2 \\ -1 \text{ if } x \geq 2 \end{cases}$$

Figure 10-39

The use of solid and open circles on the graph indicates that $f(x) = -1$ when $x = 2$.

Since every number x determines exactly one value y, the domain of this function is the interval $(-\infty, \infty)$. The range is $\{-1\} \cup [0, \infty)$.

SELF CHECK 1   Graph: $f(x) = \begin{cases} 2x \text{ when } x \leq 0 \\ \frac{1}{2}x \text{ when } x > 0 \end{cases}$

2   Graph the greatest integer function.

The **greatest integer function** is important in computer applications. It is a function determined by the equation

$$f(x) = [x] \qquad \text{Read as "} y \text{ equals the greatest integer in } x\text{."}$$

where the value of y that corresponds to x is the greatest integer that is less than or equal to x. For example,

$$[4.7] = 4, \quad \left[2\frac{1}{2}\right] = 2, \quad [\pi] = 3, \quad [-3.7] = -4, \quad [-5.7] = -6$$

COMMENT   One way to help determine the greatest integer for a value that is not an integer is to visualize the number on a number line. The integer directly to the left of the number is the greatest integer.

EXAMPLE 2   Graph: $f(x) = [x]$

Solution   We list several intervals and the corresponding values of the greatest integer function:

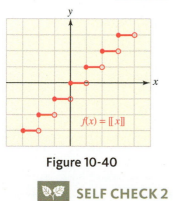

$f(x) = [[x]]$

Figure 10-40

$[0, 1)$ $y = [x] = 0$ For numbers from 0 to 1, not including 1, the greatest integer in the interval is 0.

$[1, 2)$ $y = [x] = 1$ For numbers from 1 to 2, not including 2, the greatest integer in the interval is 1.

$[2, 3)$ $y = [x] = 2$ For numbers from 2 to 3, not including 3, the greatest integer in the interval is 2.

In each interval, the values of y are constant, but they jump by 1 at integer values of x. The graph is shown in Figure 10-40. From the graph, we see that the domain is $(-\infty, \infty)$, and the range is the set of integers $\{\ldots, -3, -2, -1, 0, 1, 2, 3, \ldots\}$.

SELF CHECK 2   Graph: $f(x) = [x] + 1$

Since the greatest integer function contains a series of horizontal line segments, it is an example of a group of functions called **step functions**.

3 Solve an application involving the greatest integer function or a piecewise-defined function.

EXAMPLE 3 **PRINTING STATIONERY** To print stationery, a printer charges $10 for setup charges, plus $20 for each box. Any portion of a box counts as a full box. Graph this step function.

Solution If we order stationery and cancel the order before it is printed, the cost will be $10. Thus, the ordered pair $(0, 10)$ will be on the graph.

If we purchase 1 box, the cost will be $10 for setup plus $20 for printing, for a total cost of $30. Thus, the ordered pair $(1, 30)$ will be on the graph.

The cost of $1\frac{1}{2}$ boxes will be the same as the cost of 2 boxes, or $50. Thus, the ordered pairs $(1.5, 50)$ and $(2, 50)$ will be on the graph. The complete graph is shown in Figure 10-41.

Figure 10-41

SELF CHECK 3 How much will $3\frac{1}{2}$ boxes cost?

EXAMPLE 4 **COMPUTING TAXES** The IRS taxes for a single person earning an adjusted gross income less than $83,500 a year can be computed from the piecewise-defined function

$$f(x) = \begin{cases} 0.10x & \text{if } 0 \le x < 8{,}500 \\ 0.15(x - 8{,}500) + 850 & \text{if } 8{,}500 \le x < 34{,}500 \\ 0.25(x - 34{,}500) + 4{,}750 & \text{if } 34{,}500 \le x < 83{,}500 \end{cases}$$

Compute the taxes owed for an adjusted gross income of $23,257.

Solution Because $23,257 falls between $8,500 and $34,500, we use the function $f(x) = 0.15(x - 8{,}500) + 850$ to compute the taxes owed. We evaluate this function for $x = 23{,}257$.

$$f(x) = 0.15(x - 8{,}500) + 850$$
$$f(\mathbf{23{,}257}) = 0.15(\mathbf{23{,}257} - 8{,}500) + 850$$
$$= 0.15(14{,}757) + 850$$
$$= 2{,}213.55 + 850$$
$$= 3{,}063.55$$

The taxes owed on an adjusted gross income of $23,257 are $3,063.55.

SELF CHECK 4 Compute the taxes owed on an adjusted gross income of $82,368.

SELF CHECK ANSWERS

1.

$$f(x) = \begin{cases} 2x \text{ when } x \le 0 \\ \dfrac{1}{2}x \text{ when } x > 0 \end{cases}$$

2.

$f(x) = [\![x]\!] + 1$

3. $90   4. $16,717.00

NOW TRY THIS

Graph the piecewise-defined function.

$$f(x) = \begin{cases} -x^2 \text{ if } x < 1 \\ [\![x]\!] \text{ if } 1 \le x < 4 \\ x - 1 \text{ if } x \ge 4 \end{cases}$$

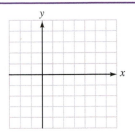

10.5   Exercises

WARM-UPS *For each function, evaluate f(1), f(2), and f(3). Determine if the function values, y, are increasing or decreasing.*

1. $f(x) = x^2$ **2.** $f(x) = |x|$

For each function, find f(−3), f(−2), and f(−1). Determine if the function values, y, are increasing or decreasing.

3. $f(x) = x^2$ **4.** $f(x) = |x|$

REVIEW *Find the value of x. Assume that lines r and s are parallel.*

5.

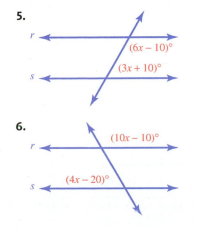

$(6x − 10)°$
$(3x + 10)°$

6.

$(10x − 10)°$

$(4x − 20)°$

VOCABULARY AND CONCEPTS *Fill in the blanks.*

7. Piecewise-defined functions are defined by using different functions for different parts of their _____.

8. When the values $f(x)$ increase as the values of x increase over an interval, we say that the function is _____ over that interval.

9. In a _____ function, as the values of x increase the values of ____ are the same.

10. When the values of $f(x)$ decrease as the values of x _____ over an interval, we say that the function is decreasing over that interval.

11. When the graph of a function contains a series of horizontal line segments, the function is called a ____ function.

12. The function that gives the largest integer that is less than or equal to a number x is called the _____ function.

GUIDED PRACTICE *Give the intervals on which each function is increasing, decreasing, or constant. (OBJECTIVE 1)*

13.

14.

15. 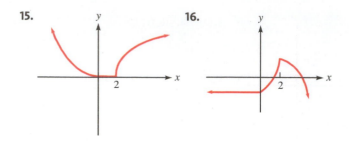   **16.**

Graph each function and state the intervals on which f(x) is increasing, decreasing, or constant. *SEE EXAMPLE 1. (OBJECTIVE 1)*

17. $f(x) = \begin{cases} -1 \text{ when } x \le 0 \\ x \text{ when } x > 0 \end{cases}$ **18.** $f(x) = \begin{cases} -2 \text{ if } x \le 0 \\ x^2 \text{ if } x > 0 \end{cases}$

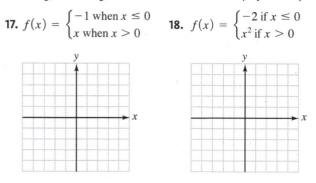

19. $f(x) = \begin{cases} -x \text{ if } x \le 0 \\ x \text{ if } 0 < x < 2 \\ -x \text{ if } x \ge 2 \end{cases}$

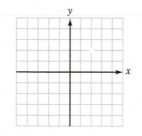

20. $f(x) = \begin{cases} -x \text{ if } x < 0 \\ x^2 \text{ if } 0 \le x \le 1 \\ 1 \text{ if } x > 1 \end{cases}$

Graph each function. *SEE EXAMPLE 2. (OBJECTIVE 2)*

21. $f(x) = -[x]$ **22.** $f(x) = [x] + 2$

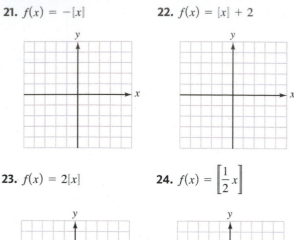

23. $f(x) = 2[x]$ **24.** $f(x) = \left[\!\left[\dfrac{1}{2}x \right]\!\right]$

ADDITIONAL PRACTICE

25. Signum function   Computer programmers use a function denoted by $f(x) = \operatorname{sgn} x$ that is defined in the following way:

$$f(x) = \begin{cases} -1 \text{ if } x < 0 \\ 0 \text{ if } x = 0 \\ 1 \text{ if } x > 0 \end{cases}$$

Graph this function.

26. Heaviside unit step function   This function, used in calculus, is defined by

$$f(x) = \begin{cases} 1 \text{ if } x > 0 \\ 0 \text{ if } x < 0 \end{cases}$$

Graph this function.

APPLICATIONS *Solve each application.* ***SEE EXAMPLES 3 AND 4.***
(OBJECTIVE 3)

27. **Renting a ski jet** A marina charges $20 to rent a ski jet for 1 hour, plus $5 for every extra hour (or portion of an hour). In the illustration, graph the ordered pairs (h, c), where h represents the number of hours and c represents the cost. Find the cost if the ski is used for 2.5 hours.

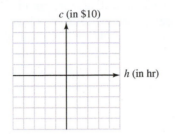

28. **Riding in a taxi** A cab company charges $3 for a trip up to 1 mile, and $2 for every extra mile (or portion of a mile). In the illustration, graph the ordered pairs (m, c), where m represents the number of miles traveled and c represents the cost. Find the cost to ride $10\frac{1}{4}$ miles.

29. **Information access** Computer access to international data network A costs $10 per day plus $8 per hour or fraction of an hour. Network B charges $15 per day, but only $6 per hour or fraction of an hour. For each network, graph the ordered pairs (t, C), where t represents the connect time and C represents the total cost. Find the minimum daily usage at which it would be more economical to use network B.

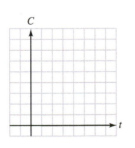

30. **Royalties** A publisher has agreed to pay the author of a novel 7% royalties on sales of the first 50,000 copies and 10% on sales thereafter. If the book sells for $10, express the royalty income I as a function of s, the number of copies sold, and graph the function. (*Hint:* When sales are into the second 50,000 copies, how much was earned on the first 50,000?)

WRITING ABOUT MATH

31. Explain how to decide whether a function is increasing on the interval (a, b).

32. Describe the greatest integer function and why it might be called a step function.

SOMETHING TO THINK ABOUT

33. Find a piecewise-defined function that is increasing on the interval $(-\infty, 6)$ and decreasing on the interval $(6, \infty)$.

34. Find a piecewise-defined function that is constant on the interval $(-\infty, 0)$, increasing on the interval $(0, 3)$, and decreasing on the interval $(3, \infty)$.

Project

The zillionaire G. I. Luvmoney is known for his love of flowers. On his estate, he recently set aside a circular plot of land with a radius of 100 yards to be made into a flower garden. He has hired your landscape design firm to do the job. If Luvmoney is satisfied, he will hire your firm to do more lucrative jobs. Here is Luvmoney's plan.

The center of the circular plot of land is to be the origin of a rectangular coordinate system. You are to make 100 circles, all centered at the origin, with radii of 1 yard, 2 yards, 3 yards, and so on up to the outermost circle, which will have a radius of 100 yards. Inside the innermost circle, he wants a fountain with a circular walkway around it. In the ring between the first and second circle, he wants to plant his favorite kind of flower, in the next ring his second favorite, and so on, until you reach the edge of the circular plot. Luvmoney provides you with a list ranking his 99 favorite flowers.

The first thing he wants to know is the area of each ring, so that he will know how many of each plant to order. Then he wants a simple formula that will give the area of any ring just by substituting in the number of the ring.

He also wants a walkway to go through the garden in the form of a hyperbolic path, following the equation

$$x^2 - \frac{y^2}{9} = 1$$

Luvmoney wants to know the x- and y-coordinates of the points where the path will intersect the circles, so that those points can be marked with stakes to keep gardeners from planting flowers where the walkway will later be built. He wants a formula (or two) that will enable him to put in the number of a circle and get out the intersection points.

Finally, although cost has no importance for Luvmoney, his accountants will want an estimate of the total cost of all of the flowers.

You go back to your office with Luvmoney's list. You find that because the areas of the rings grow from the inside of the garden to the outside, and because of Luvmoney's ranking of flowers, a strange thing happens. The first ring of flowers will cost $360, and the flowers in every ring after that will cost 110% as much as the flowers in the previous ring. That is, the second ring of flowers will cost $360(1.1) = \$396$, the third will cost $435.60, and so on.

Answer all of Luvmoney's questions, and show work that will convince him that you are right.

Reach for Success

EXTENSION OF LEARNING BEYOND REMEMBERING

The brain cells shown in the figure were not connected at one time. No matter your age, when you first hear new information (for example, the formula for finding the slope of a line between two points), your brain starts to grow **dendrites**. Your instructor gives an example and explains the process. When you have to try a problem on your own, your brain connects the dendrites in this cell to other brain cells that have information you need (understanding coordinates of a point, evaluating a variable, and subtracting, dividing, and manipulating signed numbers). This is accomplished by moving the new information through the **axon** to **synapses** at its end. Once in the synapses, the information is "fired" to bridge the gap between this and other cells' dendrites. The more connections that are made (working problems in class), the more deeply you have embedded that information in your brain. We say "you must fire it to wire it," meaning that you must continue to make these connections (do many exercises until you've solidified your understanding) or risk not having enough connections when you need that information (for a test). The phrase "use it or lose it" is more accurate than we ever imagined!

Which of the following are true statements? Circle any/all that apply.

1. To learn new material you must be able to connect it to something you already know.

2. You must practice skills in order to "wire" them.

3. You can forget information if you have not "fired" enough synapses or have not "fired" them recently.

4. If you continue to learn new things, you can grow dendrites at any age.

Your ability to move from the remembering to the higher levels of learning depends on the number of new dendrites and connections you're able to make.

10   Review

SECTION 10.1   The Circle and the Parabola

DEFINITIONS AND CONCEPTS	EXAMPLES

Equations of a circle:

Standard forms:

$$x^2 + y^2 = r^2$$
center $(0, 0)$, radius r

To graph a circle, we need to know the center and the radius. Graph each circle.

a. $x^2 + y^2 = 25$

$C(0, 0)$ Compare to the formula.

$r^2 = 25$ Compare to the formula.

$r = 5$

The center is $(0, 0)$ and the radius is 5 units.

$$(x - h)^2 + (y - k)^2 = r^2$$
center (h, k), radius r

b. $(x - 3)^2 + (y + 4)^2 = 16$

$(x - 3)^2 + [y - (-4)]^2 = 16$ Write in standard form.

$C(3, -4)$ Comparing to the formula, $h = 3$ and $k = -4$.

$r^2 = 16$ Compare to the formula.

$r = 4$ Take the positive square root of 16.

The center is at $(3, -4)$ and the radius is 4 units.

General form:

$$x^2 + y^2 + Dx + Ey + F = 0$$

c. $x^2 + y^2 + 2x + 6y + 6 = 0$

To write the equation in standard form, we can complete the square on both x and y.

$x^2 + y^2 + 2x + 6y = -6$ Subtract 6 from both sides.

$x^2 + 2x + y^2 + 6y = -6$ Rearrange the terms.

$x^2 + 2x + 1 + y^2 + 6y + 9 = -6 + 1 + 9$ Complete the square on x and y.

$(x + 1)^2 + (y + 3)^2 = 4$ Factor and simplify.

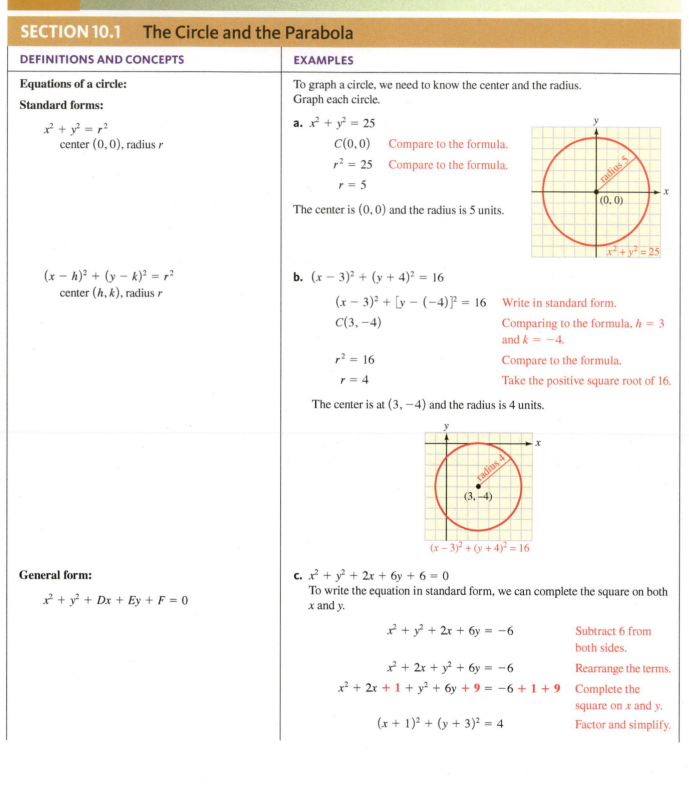

Comparing to the standard form, we see that $C(-1, -3)$, $r^2 = 4$, and $r = 2$. The center is at $(-1, -3)$ with a radius of 2 units.

Equations of parabolas:

Parabola opening	Vertex at origin	
Upward	$y = ax^2$	$(a > 0)$
Downward	$y = ax^2$	$(a < 0)$
Right	$x = ay^2$	$(a > 0)$
Left	$x = ay^2$	$(a < 0)$

Parabola opening	Vertex at (h, k)	
Upward	$y = a(x - h)^2 + k$	$(a > 0)$
Downward	$y = a(x - h)^2 + k$	$(a < 0)$
Right	$x = a(y - k)^2 + h$	$(a > 0)$
Left	$x = a(y - k)^2 + h$	$(a < 0)$

Graph each parabola.

a. $x = y^2$

The parabola is horizontal and opens to the right because $a > 0$. To obtain the graph, we can plot several points and connect them with a smooth curve.

$$x = y^2$$

x	y	(x, y)
0	0	$(0, 0)$
4	2	$(4, 2)$
4	−2	$(4, -2)$
9	3	$(9, 3)$
9	−3	$(9, -3)$

b. $x = -2(y - 1)^2 + 3$

The parabola is horizontal and opens to the left because $a < 0$. To obtain the graph, we can plot several points and connect them with a smooth curve.

$$x = -2(y - 1)^2 + 3$$

x	y	(x, y)
1	0	$(1, 0)$
1	2	$(1, 2)$
3	1	$(3, 1)$
−5	−1	$(-5, -1)$
−5	3	$(-5, 3)$

REVIEW EXERCISES

Graph each equation.

1. $(x - 1)^2 + (y + 2)^2 = 9$ **2.** $x^2 + y^2 = 16$

3. Write the equation in standard form and graph it.

$$x^2 + y^2 + 4x - 2y = 4$$

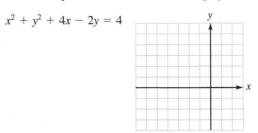

Graph each equation.

4. $x = -3(y - 2)^2 + 5$

5. $x = 2(y + 1)^2 - 2$

6. $y = -2(x - 1)^2 + 3$

7. $y = (x + 2)^2 - 3$

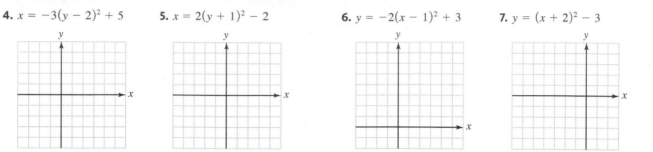

SECTION 10.2 The Ellipse

DEFINITIONS AND CONCEPTS	EXAMPLES

Equations of an ellipse:

Standard forms:

Center at $(0, 0)$

$$\frac{x^2}{a^2} + \frac{y^2}{b^2} = 1 \quad (a > b > 0)$$

$$\frac{x^2}{b^2} + \frac{y^2}{a^2} = 1 \quad (a > b > 0)$$

To graph an ellipse, we need to know the center and the endpoints of the major and minor axes.

Graph each ellipse:

a. $\dfrac{x^2}{9} + \dfrac{y^2}{16} = 1$

Comparing to the formula, the center is $(0, 0)$ and

$$a^2 = 16 \qquad b^2 = 9$$
$$a = 4 \qquad b = 3$$

The ellipse will be vertical because the larger denominator is associated with the y-term. The endpoints of the major axis will be 4 units above and below the center, $(0, 4)$ and $(0, -4)$. The endpoints of the minor axis will be 3 units to the left and right of center, $(3, 0)$ and $(-3, 0)$.

Center at (h, k)

$$\frac{(x - h)^2}{a^2} + \frac{(y - k)^2}{b^2} = 1$$

$$\frac{(x - h)^2}{b^2} + \frac{(y - k)^2}{a^2} = 1$$

In either case,
The length of the major axis is $2a$.
The length of the minor axis is $2b$.

b. $\dfrac{(x - 3)^2}{9} + \dfrac{(y - 4)^2}{4} = 1$

From the formula, the center is $(3, 4)$ and

$$a^2 = 9 \qquad b^2 = 4$$
$$a = 3 \qquad b = 2$$

The ellipse will be horizontal because the larger denominator is associated with the x-term. The endpoints of the major axis will be 3 units to the left and right of the center, $(6, 4)$ and $(0, 4)$. The endpoints of the minor axis will be 2 units above and below the center, $(3, 6)$ and $(3, 2)$.

General form:

$$Ax^2 + Cy^2 + Dx + Ey + F = 0$$

To find the standard equation of the ellipse with equation $4x^2 + y^2 - 8x - 2y - 11 = 0$, proceed as follows:

$$4x^2 + y^2 - 8x - 2y = 11 \qquad \text{Add 11 to both sides.}$$

$$4x^2 - 8x + y^2 - 2y = 11 \qquad \text{Apply the commutative property of addition to rearrange terms.}$$

$$4(x^2 - 2x) + y^2 - 2y = 11$$ Factor to obtain a coefficient of 1 for the term involving *x*-squared.

$$4(x^2 - 2x + \mathbf{1}) + (y^2 - 2y + \mathbf{1}) = 11 + \mathbf{4} + \mathbf{1}$$ Complete the square on both *x* and *y*.

$$4(x - 1)^2 + (y - 1)^2 = 16$$ Factor and simplify.

$$\frac{(x - 1)^2}{4} + \frac{(y - 1)^2}{16} = 1$$ Divide both sides by 16.

REVIEW EXERCISES

Graph each ellipse.

8. $9x^2 + 16y^2 = 144$

9. $\dfrac{(x - 2)^2}{4} + \dfrac{(y - 1)^2}{9} = 1$

10. Write the equation in standard form and graph it.

$$4x^2 + 9y^2 + 8x - 18y = 23$$

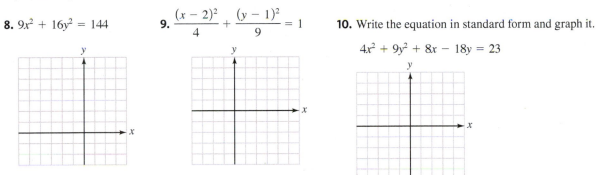

SECTION 10.3 The Hyperbola

DEFINITIONS AND CONCEPTS	**EXAMPLES**

Equations of a hyperbola:

Standard forms:

Center at $(0, 0)$

$\dfrac{x^2}{a^2} - \dfrac{y^2}{b^2} = 1$ opens left or right $V_1(a, 0)$, $V_2(-a, 0)$

$\dfrac{y^2}{a^2} - \dfrac{x^2}{b^2} = 1$ opens upward or downward $V_1(0, a)$, $V_2(0, -a)$

To graph a hyperbola, we need to know where it is centered, the coordinates of the vertices, and the location of the asymptotes.

Graph: $9y^2 - 4x^2 = 36$

$$\frac{9y^2}{36} - \frac{4x^2}{36} = 1$$ Write the equation in standard form.

$$\frac{y^2}{4} - \frac{x^2}{9} = 1$$ Simplify each fraction.

From the previous equation, we can determine that $a = 2$ and $b = 3$. Because the *y*-term is positive, the hyperbola will be vertical and the vertices of the hyperbola are $V_1(0, 2)$ and $V_2(0, -2)$.

Since $b = 3$, we can use the points $(3, 0)$ and $(-3, 0)$ on the *x*-axis to help draw the fundamental rectangle. We then draw its extended diagonals to locate the asymptotes and sketch the hyperbola.

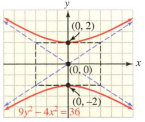

Center at (h, k)

$$\frac{(x - k)^2}{a^2} - \frac{(y - h)^2}{b^2} = 1 \quad \text{opens left or right}$$

$$\frac{(y - k)^2}{a^2} - \frac{(x - h)^2}{b^2} = 1 \quad \begin{array}{l}\text{opens upward or}\\ \text{downward}\end{array}$$

Graph: $\dfrac{(x - 3)^2}{4} - \dfrac{(y + 1)^2}{4} = 1$

From the equation, we see that the hyperbola is centered at $(3, -1)$. Its vertices are located 2 units to the right and left of center, at $(5, -1)$ and $(1, -1)$. Since $b = 2$, we can count 2 units above and below center to locate points $(3, -3)$ and $(3, 1)$. With these points, we can draw the fundamental rectangle along with its extended diagonals to locate the asymptotes. Then we can sketch the hyperbola.

General form:

$$Ax^2 - Cy^2 + Dx + Ey + F = 0$$

Write $x^2 - y^2 + 6x + 4y = 4$ in standard form and graph.

$$x^2 - y^2 + 6x + 4y = 4$$
$$x^2 + 6x - y^2 + 4y - 4 = 0$$
$$x^2 + 6x + 9 - (y^2 - 4y + 4) = 9$$
$$(x + 3)^2 - (y - 2)^2 = 9$$
$$\frac{(x + 3)^2}{9} - \frac{(y - 2)^2}{9} = 1$$

Comparing to the standard equation, this is a horizontal hyperbola with centered at $(-3, 2)$.

$xy = k, k \neq 0$

The equation $xy = 6$ is a hyperbola that does not intersect the x-axis or the y-axis.

REVIEW EXERCISES

Graph each hyperbola.

11. $9x^2 - y^2 = -9$ **12.** $xy = 9$

13. Determine whether the equation $2y^2 - 4x^2 + 8x - 8y = 8$ represents an ellipse or a hyperbola.

14. Write the equation in standard form and graph it.

$$9x^2 - 4y^2 - 18x - 8y = 31$$

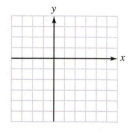

SECTION 10.4   Systems Containing Second-Degree Equations

DEFINITIONS AND CONCEPTS	EXAMPLES
Solve by graphing: To solve a system of equations by graphing, graph each equation. The coordinates of the intersection points of the graphs will be the solutions of the system.	Solve the system by graphing. $$\begin{cases} x^2 + y^2 = 4 \\ x - y = 2 \end{cases}$$ The equation $x^2 + y^2 = 4$ is that of a circle and $x - y = 2$ is that of a line. There is a possibility of 0, 1, or 2 solutions. After graphing the equations, we will find that the two graphs intersect at the points $(2, 0)$ and $(0, -2)$. These are the solutions.
Solve by substitution: Solve one equation for one variable and substitute the result for that variable in the other equation.	Solve the system by substitution. $$\begin{cases} x^2 + 9y^2 = 10 \\ y = x^2 \end{cases}$$ The equation $x^2 + 9y^2 = 10$ is that of an ellipse and $y = x^2$ is that of a parabola. We have the possibility of 0, 1, 2, 3, or 4 solutions. To find the solutions, we notice that one of the equations is already solved for y, and then proceed as follows: $x^2 + 9y^2 = 10$ $y + 9y^2 = 10$ Substitute y for x^2. $9y^2 + y - 10 = 0$ Write in quadratic form. $(9y + 10)(y - 1) = 0$ Factor $9y^2 + y - 10$. $9y + 10 = 0$ or $y - 1 = 0$ Set each factor equal to 0 $y = -\dfrac{10}{9}$ $y = 1$ and solve. Since $y = x^2$, the values of x can be found by solving the equations $x^2 = -\dfrac{10}{9}$ and $x^2 = 1$ Because $x^2 = -\dfrac{10}{9}$ has no real number solutions, this possibility is discarded. The solutions of $x^2 = 1$ are $x = 1$ or $x = -1$. The solutions of the system are $(1, 1)$ and $(-1, 1)$.
Solve by elimination (addition): To solve a system of equations by elimination, add the equations to eliminate one of the variables. Then solve the resulting equation for the other variable. If the original system does not contain opposite coefficients on one of the variables, multiply one or both of the equations by constants that will produce additive inverses.	Solve the system by elimination (addition). $$\begin{cases} 4x^2 - y^2 = 1 \\ 4x^2 + y^2 = 1 \end{cases}$$ The equation $4x^2 - y^2 = 1$ is that of a hyperbola and the equation $4x^2 + y^2 = 1$ is that of an ellipse. We have the possibility of 0, 1, 2, 3, or 4 solutions. If we add the equations $\begin{cases} 4x^2 - y^2 = 1 \\ 4x^2 + y^2 = 1 \end{cases}$, we eliminate y^2. $8x^2 = 2$ $x^2 = \dfrac{1}{4}$ $x = \dfrac{1}{2}, -\dfrac{1}{2}$

After substituting each value of x for y in the first equation, we have

$$4\left(\frac{1}{2}\right)^2 - y^2 = 1 \qquad\qquad 4\left(-\frac{1}{2}\right)^2 - y^2 = 1$$

$$4\left(\frac{1}{4}\right) - y^2 = 1 \qquad\qquad 4\left(\frac{1}{4}\right) - y^2 = 1$$

$$1 - y^2 = 1 \qquad\qquad\qquad 1 - y^2 = 1$$

$$-y^2 = 0 \qquad\qquad\qquad -y^2 = 0$$

$$y = 0 \qquad\qquad\qquad\quad y = 0$$

The solutions are $\left(\frac{1}{2}, 0\right)$ and $\left(-\frac{1}{2}, 0\right)$.

REVIEW EXERCISES

Solve each system.

15. $\begin{cases} x^2 + y^2 = 25 \\ 4x^2 - y^2 = 20 \end{cases}$

16. $\begin{cases} \frac{x^2}{4} - \frac{y^2}{2} = -3 \\ \frac{x^2}{3} + y^2 = 11 \end{cases}$

SECTION 10.5   Piecewise-Defined Functions and the Greatest Integer Function

DEFINITIONS AND CONCEPTS

Piecewise-defined functions:
A piecewise-defined function is a function that has different equations for different intervals of x.

Increasing, decreasing, and constant functions:
A function is increasing on the interval (a, b) if the values of $f(x)$ increase as x increases from a to b.

A function is decreasing on the interval (a, b) if the values of $f(x)$ decrease as x increases from a to b.

A function is constant on the interval (a, b) if the value of $f(x)$ is constant as x increases from a to b.

EXAMPLES

Graph the function

$$f(x) = \begin{cases} -x^2 & \text{if } x < 0 \\ -x & \text{if } 0 \le x < 3 \\ -1 & \text{if } x \ge 3 \end{cases}$$

and determine the intervals where the function is increasing, decreasing, and constant.

For each number x, we decide which of the three equations will be used to find the corresponding value of y:

- For numbers $x < 0$, $f(x)$ is determined by $f(x) = -x^2$, and the graph is the left half of a parabola. Since the values of $f(x)$ increase on this graph as x increases, the function is increasing on the interval $(-\infty, 0)$.
- For numbers $0 < x < 3$, $f(x)$ is determined by $f(x) = -x$, and the graph is part of a line. Since the values of $f(x)$ decrease on this graph as x increases, the function is decreasing on the interval $(0, 3)$.
- For numbers $x \ge 3$, $f(x)$ is the constant -1, and the graph is part of a horizontal line. Since the values of $f(x)$ remain constant on this line, the function is constant on the interval $(3, \infty)$.

Greatest integer function:
The function $f(x) = [x]$ describes the greatest integer function where the value of y that corresponds to x is the greatest integer that is less than or equal to x. To find the greatest integer for a value that is not an integer, visualize the number on a number line and the integer directly to the left is the greatest integer.

Graph: $f(x) = [x] - 1$

We list several intervals and the corresponding values of the greatest integer function:

$[0, 1)$ $f(x) = [x] - 1$ For numbers from 0 to 1, not including 1, the greatest integer in the interval is 0 and then we subtract 1 and graph -1 in the interval.

$[1, 2)$ $f(x) = [\![x]\!] - 1$ For numbers from 1 to 2, not including 2, the greatest integer in the interval is 1 and then we subtract 1 and graph 0 in the interval.

$[2, 3)$ $f(x) = [\![x]\!] - 1$ For numbers from 2 to 3, not including 3, the greatest integer in the interval is 0 and then we subtract 1 and graph 1 in the interval.

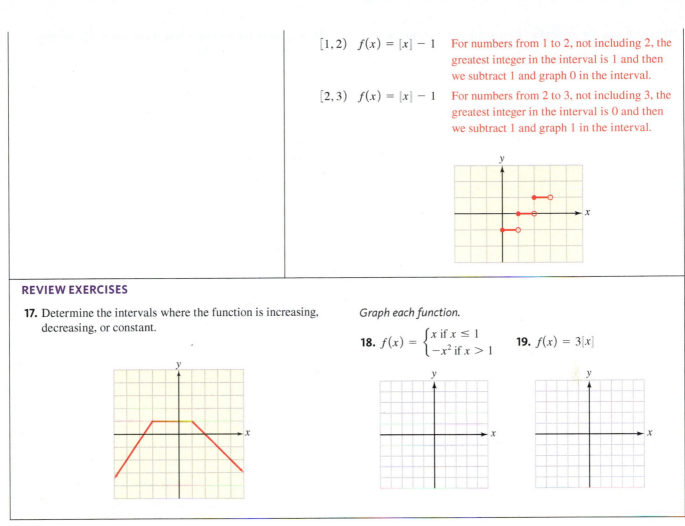

REVIEW EXERCISES

17. Determine the intervals where the function is increasing, decreasing, or constant.

Graph each function.

18. $f(x) = \begin{cases} x & \text{if } x \le 1 \\ -x^2 & \text{if } x > 1 \end{cases}$

19. $f(x) = 3[\![x]\!]$

---

10   Test

1. Find the center and the radius of the circle $(x + 5)^2 + (y - 2)^2 = 9$.

2. Find the center and the radius of the circle $x^2 + y^2 + 8x - 4y = 5$.

Graph each equation.

3. $(x + 1)^2 + (y - 2)^2 = 9$ **4.** $x = (y - 2)^2 - 1$

5. $9x^2 + 4y^2 = 36$ **6.** $\dfrac{(x - 2)^2}{9} - y^2 = 1$

Write each equation in standard form and graph the equation.

7. $4x^2 + y^2 - 24x + 2y = -33$

8. $x^2 + y^2 + 6x - 2y = -1$

9. $x^2 - 9y^2 + 2x + 36y = 44$

Solve each system.

10. $\begin{cases} 3x + 4y = 12 \\ 9x^2 + 16y^2 = 144 \end{cases}$

11. $\begin{cases} x^2 + y^2 = 25 \\ 4x^2 - 9y = 0 \end{cases}$

12. Determine the interval where the function is increasing, decreasing, or constant.

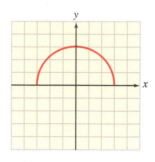

13. Graph: $f(x) = \begin{cases} -x^2, \text{ when } x < 0 \\ -x, \text{ when } x \geq 0 \end{cases}$

14. Graph: $f(x) = 2[x]$

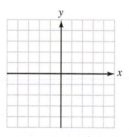

❧ Cumulative Review ❧

Perform the operations.

1. $(2x - 5y)(4x + y)$

2. $(a^n + 1)(a^n - 3)$

Simplify each fraction. Assume no division by 0.

3. $\dfrac{7a - 14}{a^2 - 7a + 10}$

4. $\dfrac{a^4 - 5a^2 + 4}{a^2 + 3a + 2}$

Perform the operations and simplify the result. Assume no division by 0.

5. $\dfrac{a^2 - a - 6}{a^2 - 4} \div \dfrac{a^2 - 9}{a^2 + a - 6}$

6. $\dfrac{2}{a - 2} + \dfrac{3}{a + 2} - \dfrac{a - 1}{a^2 - 4}$

Determine whether the graphs of the linear equations are parallel, perpendicular, or neither.

7. $3x - 4y = 12, \ y = \dfrac{3}{4}x - 5$

8. $y = 3x + 4, \ x = -3y + 4$

Write the equation of each line with the following properties.

9. $m = -6$, passing through $(0, 3)$

10. passing through $(8, -5)$ and $(-5, 4)$

Graph each inequality.

11. $2x - 3y < 6$

12. $y \geq x^2 - 4$

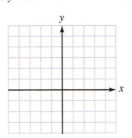

Simplify each expression.

13. $\sqrt{98} + \sqrt{8} - \sqrt{32}$

14. $12\sqrt[3]{648x^4} + 3\sqrt[3]{81x^4}$

Solve each equation.

15. $\sqrt{3a + 1} = a - 1$

16. $x - \dfrac{2x}{x - 5} = 1 - \dfrac{10}{x - 5}$

17. $10a^2 + 21a - 10 = 0$

18. $2x^2 - 2x + 5 = 0$

19. If $f(x) = x^2 - 3$ and $g(x) = 4x + 1$, find $(f \circ g)(x)$.

20. Find the inverse function of $y = 2x^3 - 1$.

21. Graph: $y = \left(\dfrac{1}{2}\right)^x$

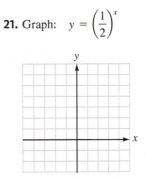

22. Write $y = \log_5 x$ as an exponential equation.

Solve each equation.

23. $2^{x+1} = 8^{x^2 - 1}$

24. $2 \log 5 + \log x - \log 4 = 2$

Graph each equation.

25. $x^2 + (y + 1)^2 = 9$

26. $x^2 - 9(y + 1)^2 = 9$

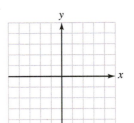

Solve each system.

27. $\begin{cases} x + 2y + 3z = 4 \\ 5x + 6y + 7z = 8 \\ 9x + 10y + 11z = 9 \end{cases}$

28. $\begin{cases} x^2 + y^2 = 5 \\ x^2 - y^2 = 3 \end{cases}$

29. $\begin{cases} x^2 + y^2 = 25 \\ 4x^2 - 9y = 0 \end{cases}$

Miscellaneous Topics

11

© wavebreakmedia ltd./Shutterstock.com

Careers and Mathematics

FINANCIAL ANALYSTS

Financial analysts provide analysis and guidance to businesses to help them with their investment decisions. These specialists gather financial information, analyze it, and make recommendations to their clients. Financial analysts specialize in an industry, product, or geographical area. A bachelor's or graduate degree is required for financial analysts.

Job Outlook:
Employment of financial analysts is expected to grow by 23 percent through 2020. This is faster than the average for all occupations.

Annual Earnings:
$44,130–$141,700

For More Information:
http://www.bls.gov/ooh/business-and-financial/financial-analysts.htm

For a Sample Application:
See Example 8 in Section 11.3.

REACH FOR SUCCESS
11.1 The Binomial Theorem
11.2 Arithmetic Sequences and Series
11.3 Geometric Sequences and Series
11.4 Infinite Geometric Sequences and Series
11.5 Permutations and Combinations
11.6 Probability
■ *Projects*
REACH FOR SUCCESS EXTENSION
CHAPTER REVIEW
CHAPTER TEST
CUMULATIVE REVIEW

In this chapter

In this chapter, we introduce several topics that have applications in advanced mathematics and in many occupational areas. The binomial theorem, permutations, and combinations are used in statistics. Arithmetic and geometric sequences are used in the mathematics of finance.

Reach for Success   Planning for Your Future

The various courses and services that are needed to attain your academic and career goals should fit together like the pieces of a puzzle. Planning for your future involves more than concentrating on a single piece, such as a single course. It involves all of the pieces necessary to create the picture in your particular puzzle.

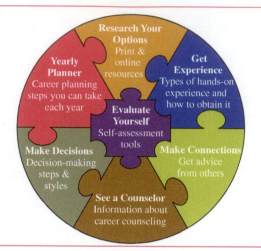

SHORT-TERM EDUCATIONAL PLANNING

This includes ensuring there is sufficient funding to attend college and then selecting the appropriate courses for next semester.

Where can you find your grade point average (GPA)? What factors can affect your GPA?

_____

_____

Check with your financial aid office to find the requirements for both merit-based and need-based financial aid.

_____

_____

List the importance of a high GPA.

1. _____

2. _____

Taking the entire required mathematics sequence consecutively improves your chance of success by more than 50%.

Familiarize yourself with the mathematics sequence required to meet graduation requirements at your college or transfer school for your major. What courses must you take?

_____

_____

To be successful, you need to start in the right course and the right format for you. Assessment tests will typically place you in the right course. The format of the course is usually your choice and varies by college.

Some format possibilities include computer-based lecture, mini-sessions, express, weekend, learning communities, and hybrid, blended, or fully online.

What formats are available at your college for the mathematics courses you need?

_____

_____

_____

Which format is a good fit for you? _____

Explain. _____

A Successful Study Strategy . . .

Take all of your required courses that have prerequisites without interruption in the semester sequence.

At the end of the chapter you will find an additional exercise to guide you in planning for a successful college experience.

Section 11.1

The Binomial Theorem

1. Raise a binomial to a power.
2. Complete a specified number of rows of Pascal's triangle.
3. Simplify an expression involving factorial notation.
4. Apply the binomial theorem to expand a binomial.
5. Find a specified term of a binomial expansion.

Vocabulary

Pascal's triangle   factorial notation   binomial theorem

Getting Ready

Raise each binomial to the indicated power.

1. $(x + 2)^2$ 2. $(x - 3)^2$
3. $(x + 1)^3$ 4. $(x - 2)^3$

We have seen how to square and cube binomials. In this section, we will learn how to raise binomials to higher powers without performing the multiplication and to find any term of a binomial expansion.

1 Raise a binomial to a power.

We have discussed how to raise binomials to positive-integer powers. For example, we know that

$$(a + b)^2 = a^2 + 2ab + b^2$$

and that

$$(a + b)^3 = (a + b)(a + b)^2$$
$$= (a + b)(a^2 + 2ab + b^2)$$
$$= a^3 + 2a^2b + ab^2 + a^2b + 2ab^2 + b^3 \qquad \text{Use the distributive property to remove parentheses}$$
$$= a^3 + 3a^2b + 3ab^2 + b^3 \qquad \text{Combine like terms.}$$

To show how to raise binomials to positive-integer powers without performing the actual multiplication, we consider the following binomial expansions:

$$(a + b)^0 = 1 \qquad \text{1 term}$$
$$(a + b)^1 = a + b \qquad \text{2 terms}$$
$$(a + b)^2 = a^2 + 2ab + b^2 \qquad \text{3 terms}$$

Blaise Pascal

1623–1662

Pascal was torn between the fields of religion and mathematics. Each surfaced at times in his life to dominate his interest. In mathematics, Pascal made contributions to the study of conic sections, probability, and differential calculus. At the age of 19, he invented a calculating machine. He is best known for a triangular array of numbers that bears his name.

$(a + b)^3 = a^3 + 3a^2b + 3ab^2 + b^3$ 4 terms

$(a + b)^4 = a^4 + 4a^3b + 6a^2b^2 + 4ab^3 + b^4$ 5 terms

$(a + b)^5 = a^5 + 5a^4b + 10a^3b^2 + 10a^2b^3 + 5ab^4 + b^5$ 6 terms

$(a + b)^6 = a^6 + 6a^5b + 15a^4b^2 + 20a^3b^3 + 15a^2b^4 + 6ab^5 + b^6$ 7 terms

Several patterns appear in these expansions:

1. Each expansion has one more term than the power of the binomial.

2. The degree of each term in each expansion is equal to the exponent of the binomial that is being expanded. For example, in the expansion of $(a + b)^5$, the sum of the exponents in each term is 5:

$$4 + 1 = 5 \quad 3 + 2 = 5 \quad 2 + 3 = 5 \quad 1 + 4 = 5$$
$$(a + b)^5 = a^5 + 5\,a^4b + 10\,a^3b^2 + 10\,a^2b^3 + 5\,ab^4 + b^5$$

3. The first term in each expansion is a raised to the power of the binomial, and the last term in each expansion is b raised to the power of the binomial.

4. The exponents on a decrease by 1 in each successive term. The exponents of b, beginning with $b^0 = 1$ in the first term, increase by 1 in each successive term. For example, the expansion of $(a + b)^4$ is

$$a^4b^0 + 4a^3b^1 + 6a^2b^2 + 4a^1b^3 + a^0b^4$$

Thus, the variables have the pattern

$$a^n, \quad a^{n-1}b, \quad a^{n-2}b^2, \quad \ldots, \quad ab^{n-1}, \quad b^n$$

2 Complete a specified number of rows of Pascal's triangle.

To see another pattern, we write the coefficients of each expansion in the following triangular array:

					1							Row 0
				1		1						Row 1
			1		2		1					Row 2
		1		3		3		1				Row 3
	1		4		6		4		1			Row 4
1		5		10		10		5		1		Row 5
1	6		15		20		15		6		1	Row 6

In this array, called **Pascal's triangle**, each entry between the 1's is the sum of the closest pair of numbers in the line immediately above it. For example, the first 15 in the bottom row is the sum of the 5 and 10 immediately above it. Pascal's triangle continues with the same pattern forever. The next two lines are

1	7	21	35	35	21	7	1	Row 7
1	8	28	56	70	56	28	8	1	Row 8

EXAMPLE 1 Expand: $(x + y)^5$

Solution The first term in the expansion is x^5, and the exponents of x decrease by 1 in each successive term. A y first appears in the second term, and the exponents on y increase by 1 in each successive term, concluding when the term y^5 is reached. Thus, the variables in the expansion are

$$x^5, \quad x^4y, \quad x^3y^2, \quad x^2y^3, \quad xy^4, \quad y^5$$

The coefficients of these variables are given in row 5 of Pascal's triangle.

1 5 10 10 5 1

Combining this information gives the following expansion:

$$(x + y)^5 = x^5 + 5x^4y + 10x^3y^2 + 10x^2y^3 + 5xy^4 + y^5$$

 SELF CHECK 1   Expand: $(x + y)^4$

EXAMPLE 2   Expand: $(u - v)^4$

Solution   We note that $(u - v)^4$ can be written in the form $[u + (-v)]^4$. The variables in this expansion are

$$u^4, \quad u^3(-v), \quad u^2(-v)^2, \quad u(-v)^3, \quad (-v)^4$$

and the coefficients are given in row 4 of Pascal's triangle.

1   4   6   4   1

Thus, the required expansion is

$$(u - v)^4 = u^4 + 4u^3(-v) + 6u^2(-v)^2 + 4u(-v)^3 + (-v)^4$$
$$= u^4 - 4u^3v + 6u^2v^2 - 4uv^3 + v^4$$

SELF CHECK 2   Expand: $(x - y)^5$

3   Simplify an expression involving factorial notation.

Although Pascal's triangle gives the coefficients of the terms in a binomial expansion, it can be a tedious way to expand a binomial for large powers. To develop a more efficient way, we introduce **factorial notation**.

FACTORIAL NOTATION   If n is a natural number, the symbol $n!$ (read as "n factorial" or as "factorial n") is defined as

$$n! = n(n - 1)(n - 2)(n - 3) \cdots (3)(2)(1)$$

Zero factorial is defined as

$$0! = 1$$

EXAMPLE 3   Write each expression without using factorial notation.
   a. $2!$ **b.** $5!$ **c.** $-9!$ **d.** $(n - 2)!$ **e.** $4! \cdot 0!$

Solution   **a.** $2! = 2 \cdot 1 = 2$

   b. $5! = 5 \cdot 4 \cdot 3 \cdot 2 \cdot 1 = 120$

   c. $-9! = -1 \cdot 9 \cdot 8 \cdot 7 \cdot 6 \cdot 5 \cdot 4 \cdot 3 \cdot 2 \cdot 1 = -362{,}880$

   d. $(n - 2)! = (n - 2)(n - 3)(n - 4) \cdot \cdots \cdot 3 \cdot 2 \cdot 1$

   e. $4! \cdot 0! = (4 \cdot 3 \cdot 2 \cdot 1) \cdot 1 = 24$

COMMENT   According to the previous definition, part **d** is meaningful only if $n - 2$ is a natural number.

SELF CHECK 3   Write each expression without using factorial notation.

a. 6!   **b.** $x!$

Accent
on technology

▶ Factorials

We can find factorials using a calculator. For example, to find 12! with a scientific calculator, we enter

    12 **$x!$** (You may have to use a **2ND** or **SHIFT** key first.)   479001600

To find 12! on a TI-84 graphing calculator, we enter

    12 **MATH ▶ ▶ ▶ 4 ENTER**   12!
                                              479001600

For instructions regarding the use of a Casio graphing calculator, please refer to the Casio Keystroke Guide in the back of the book.

To discover an important property of factorials, we note that

$$5 \cdot 4! = 5 \cdot 4 \cdot 3 \cdot 2 \cdot 1 = 5!$$
$$7 \cdot 6! = 7 \cdot 6 \cdot 5 \cdot 4 \cdot 3 \cdot 2 \cdot 1 = 7!$$
$$10 \cdot 9! = 10 \cdot 9 \cdot 8 \cdot 7 \cdot 6 \cdot 5 \cdot 4 \cdot 3 \cdot 2 \cdot 1 = 10!$$

These examples suggest the following property.

PROPERTY OF FACTORIALS

If n is a positive integer, then

$$n(n - 1)! = n!$$

EXAMPLE 4   Simplify: **a.** $\dfrac{6!}{5!}$ **b.** $\dfrac{10!}{8!(10-8)!}$

Solution   **a.** If we write 6! as $6 \cdot 5!$, we can simplify the fraction by removing the common factor 5! in the numerator and denominator.

$$\frac{6!}{5!} = \frac{6 \cdot 5!}{5!} = \frac{6 \cdot \cancel{5!}}{\cancel{5!}} = 6 \quad \textcolor{red}{\text{Simplify: } \frac{5!}{5!} = 1}$$

b. First, we subtract within the parentheses. Then we write 10! as $10 \cdot 9 \cdot 8!$ and simplify.

$$\frac{10!}{8!(10-8)!} = \frac{10!}{8! \cdot 2!} = \frac{10 \cdot 9 \cdot \cancel{8!}}{\cancel{8!} \cdot 2!} = \frac{5 \cdot 2 \cdot 9}{2 \cdot 1} = 45 \quad \textcolor{red}{\text{Simplify: } \frac{8!}{8!} = 1. \text{ Factor 10 as } 5 \cdot 2 \text{ and simplify: } \frac{2}{2} = 1}$$

 SELF CHECK 4   Simplify: **a.** $\dfrac{4!}{3!}$ **b.** $\dfrac{7!}{5!(7-5)!}$

---

4   Apply the binomial theorem to expand a binomial.

Now that we have developed an understanding of factorial notation, we will use this notation to state the **binomial theorem**, which will provide an alternative method for expanding a binomial for large powers of n.

THE BINOMIAL THEOREM If n is any positive integer, then

$$(a + b)^n = a^n + \frac{n!}{1!(n-1)!} a^{n-1}b + \frac{n!}{2!(n-2)!} a^{n-2}b^2 + \frac{n!}{3!(n-3)!} a^{n-3}b^3$$

$$+ \cdots + \frac{n!}{r!(n-r)!} a^{n-r}b^r + \cdots + b^n$$

In the binomial theorem, the exponents of the variables follow the familiar pattern:

- The sum of the exponents on a and b in each term is n,
- the exponents on a decrease in each subsequent term, and
- the exponents on b increase in each subsequent term.

Only the method of finding the coefficients is different. Except for the first and last terms, the numerator of each coefficient is $n!$. If the exponent of b in a particular term is r, the denominator of the coefficient of that term is $r!(n-r)!$.

EXAMPLE 5 Use the binomial theorem to expand $(a + b)^3$.

Solution We can substitute directly into the binomial theorem and simplify:

$$(a + b)^3 = a^3 + \frac{3!}{1!(3-1)!}a^2b + \frac{3!}{2!(3-2)!}ab^2 + b^3$$

$$= a^3 + \frac{3!}{1! \cdot 2!}a^2b + \frac{3!}{2! \cdot 1!}ab^2 + b^3$$

$$= a^3 + \frac{3 \cdot 2 \cdot 1}{1 \cdot 2 \cdot 1}a^2b + \frac{3 \cdot 2 \cdot 1}{2 \cdot 1 \cdot 1}ab^2 + b^3$$

$$= a^3 + 3a^2b + 3ab^2 + b^3$$

SELF CHECK 5 Use the binomial theorem to expand $(a + b)^4$.

EXAMPLE 6 Use the binomial theorem to expand $(x - y)^4$.

Solution We can write $(x - y)^4$ in the form $[x + (-y)]^4$, substitute directly into the binomial theorem, and simplify:

COMMENT To expand $(x - y)^n$, the signs on the terms will alternate.

$$(x - y)^4 = [x + (-y)]^4$$

$$= x^4 + \frac{4!}{1!(4-1)!}x^3(-y) + \frac{4!}{2!(4-2)!}x^2(-y)^2$$

$$+ \frac{4!}{3!(4-3)!}x(-y)^3 + (-y)^4$$

$$= x^4 - \frac{4 \cdot 3!}{1! \cdot 3!}x^3y + \frac{4 \cdot 3 \cdot 2!}{2! \cdot 2!}x^2y^2 - \frac{4 \cdot 3!}{3! \cdot 1!}xy^3 + y^4$$

$$= x^4 - 4x^3y + 6x^2y^2 - 4xy^3 + y^4 \quad \text{Note the alternating signs.}$$

SELF CHECK 6 Use the binomial theorem to expand $(x - y)^3$.

EXAMPLE 7 Use the binomial theorem to expand $(3u - 2v)^4$.

Solution We write $(3u - 2v)^4$ in the form $[3u + (-2v)]^4$ and let $a = 3u$ and $b = -2v$. Then we can use the binomial theorem to expand $(a + b)^4$.

$$(a + b)^4 = a^4 + \frac{4!}{1!(4 - 1)!}a^3b + \frac{4!}{2!(4 - 2)!}a^2b^2 + \frac{4!}{3!(4 - 3)!}ab^3 + b^4$$
$$= a^4 + 4a^3b + 6a^2b^2 + 4ab^3 + b^4$$

Now we can substitute $3u$ for a and $-2v$ for b and simplify:

$$(3u - 2v)^4 = (3u)^4 + 4(3u)^3(-2v) + 6(3u)^2(-2v)^2 + 4(3u)(-2v)^3 + (-2v)^4$$
$$= 81u^4 - 216u^3v + 216u^2v^2 - 96uv^3 + 16v^4$$

SELF CHECK 7 Use the binomial theorem to expand $(2a - 3b)^3$.

5 Find a specified term of a binomial expansion.

To find the fourth term of the expansion of $(a + b)^9$, we could raise the binomial $a + b$ to the 9th power and look at the fourth term. However, this task would be tedious. By using the binomial theorem, we can construct the fourth term without finding the complete expansion of $(a + b)^9$.

EXAMPLE 8 Find the fourth term in the expansion of $(a + b)^9$.

Solution Since b^1 appears in the second term, b^2 appears in the third term, and so on, the exponent on b in the fourth term is 3. Since the exponent on b added to the exponent on a must equal 9, the exponent on a must be 6. Thus, the variables of the fourth term are

a^6b^3 The sum of the exponents must be 9.

We can find the coefficient of a^6b^3 by using the formula $\frac{n!}{r!(n - r)!}$, where n is the power of the expansion and r is the exponent of the second variable b.

To find the coefficient of the fourth term, we substitute 9 for n and 3 for r and simplify.

$$\frac{n!}{r!(n - r)!} = \frac{9!}{3!(9 - 3)!}$$

The complete fourth term is

$$\frac{9!}{3!(9 - 3)!}a^6b^3 = \frac{9 \cdot 8 \cdot 7 \cdot 6!}{3 \cdot 2 \cdot 1 \cdot 6!}a^6b^3$$
$$= 84a^6b^3$$

SELF CHECK 8 Find the third term of the expansion of $(a + b)^9$.

EXAMPLE 9 Find the sixth term in the expansion of $(x - y)^7$.

Solution We first find the sixth term of $[x + (-y)]^7$. In the sixth term, the exponent on $(-y)$ is 5. Thus, the variables in the sixth term are

$x^2(-y)^5$ The sum of the exponents must be 7.

The coefficient of these variables is

$$\frac{n!}{r!(n - r)!} = \frac{7!}{5!(7 - 5)!}$$

The complete sixth term is

$$\frac{7!}{5!(7-5)!}x^2(-y)^5 = -\frac{7 \cdot 6 \cdot 5!}{5! \cdot 2 \cdot 1}x^2y^5$$
$$= -21x^2y^5$$

🌿 **SELF CHECK 9**   Find the fourth term of the expansion of $(a - b)^7$.

EXAMPLE 10   Find the fourth term of the expansion of $(2x - 3y)^6$.

Solution   We can let $a = 2x$ and $b = -3y$ and find the fourth term of the expansion of $(a + b)^6$:

$$\frac{6!}{3!(6-3)!}a^3b^3 = \frac{6 \cdot 5 \cdot 4 \cdot 3!}{3! \cdot 3 \cdot 2 \cdot 1}a^3b^3$$
$$= 20a^3b^3$$

We can now substitute $2x$ for a and $-3y$ for b and simplify:

$$20\textcolor{blue}{a}^3\textcolor{red}{b}^3 = 20(\textcolor{blue}{2x})^3(\textcolor{red}{-3y})^3$$
$$= -4{,}320x^3y^3$$

The fourth term is $-4{,}320x^3y^3$.

🌿 **SELF CHECK 10**   Find the third term of the expansion of $(2a - 3b)^6$.

🌿 **SELF CHECK ANSWERS**
1. $x^4 + 4x^3y + 6x^2y^2 + 4xy^3 + y^4$ **2.** $x^5 - 5x^4y + 10x^3y^2 - 10x^2y^3 + 5xy^4 - y^5$
3. a. 720 **b.** $x(x - 1)(x - 2) \cdot \cdots \cdot 3 \cdot 2 \cdot 1$ **4. a.** 4 **b.** 21 **5.** $a^4 + 4a^3b + 6a^2b^2 + 4ab^3 + b^4$
6. $x^3 - 3x^2y + 3xy^2 - y^3$ **7.** $8a^3 - 36a^2b + 54ab^2 - 27b^3$ **8.** $36a^7b^2$ **9.** $-35a^4b^3$ **10.** $2{,}160a^4b^2$

NOW TRY THIS

1. Expand $(1 + 2i)^7$ using any method.

Find the specified term of each expansion.

2. fourth term of $\left(5x + \frac{1}{4}\right)^6$

3. fifth term of $(3 - 2i)^8$

11.1   Exercises 🌿

WARM-UPS *Simplify.*

1. $6 \cdot 5 \cdot 4 \cdot 3 \cdot 2 \cdot 1$

2. $4 \cdot 3 \cdot 2 \cdot 1$

3. $\dfrac{6 \cdot 5 \cdot 4 \cdot 3 \cdot 2 \cdot 1}{4 \cdot 3 \cdot 2 \cdot 1}$

4. $\dfrac{7 \cdot 6 \cdot 5 \cdot 4 \cdot 3 \cdot 2 \cdot 1}{5 \cdot 4 \cdot 3 \cdot 2 \cdot 1}$

Square each binomial.

5. $(x + 3y)^2$

6. $(2x + 3y)^2$

7. $(x - 3y)^2$

8. $(2x - 3y)^2$

REVIEW *Find each value of x.*

9. $\log_9 81 = x$

10. $\log_x 100 = 2$

11. $\log_{25} x = \dfrac{1}{2}$

12. $\log_{1/2} \dfrac{1}{8} = x$

VOCABULARY AND CONCEPTS *Fill in the blanks.*

13. Every binomial expansion has ___ more term than the power of the binomial.

14. The first term in the expansion of $(a + b)^{20}$ is ___.

15. The triangular array that can be used to find the coefficients of a binomial expansion is _____ triangle.

16. The symbol 5! is read as "_____."

17. $8 \cdot 7 \cdot 6 \cdot 5 \cdot 4 \cdot 3 \cdot 2 \cdot 1 = \_$ (Write your answer in factorial notation.)

18. $6! = 6 \cdot \_$

19. $0! = \_$

20. According to the binomial theorem, the third term of the expansion of $(a + b)^n$ is _____.

GUIDED PRACTICE   *Expand each expression using Pascal's triangle.* SEE EXAMPLES 1–2. *(OBJECTIVES 1–2)*

21. $(a + b)^3$

22. $(a + b)^4$

23. $(a - b)^4$

24. $(a - b)^3$

Evaluate. SEE EXAMPLE 3. *(OBJECTIVE 3)*

25. $4!$

26. $7!$

27. $-8!$

28. $-6!$

29. $3! + 4!$

30. $2!(3!)$

31. $3!(4!)$

32. $4! + 4!$

Simplify. SEE EXAMPLE 4. *(OBJECTIVE 3)*

33. $\dfrac{9!}{11!}$

34. $\dfrac{13!}{10!}$

35. $\dfrac{49!}{47!}$

36. $\dfrac{101!}{100!}$

37. $\dfrac{9!}{7! \cdot 0!}$

38. $\dfrac{7!}{5! \cdot 0!}$

39. $\dfrac{5!}{3!(5 - 3)!}$

40. $\dfrac{6!}{4!(6 - 4)!}$

Use the binomial theorem to expand each expression. SEE EXAMPLE 5. *(OBJECTIVE 4)*

41. $(x + y)^3$

42. $(x + y)^4$

43. $(a + b)^6$

44. $(a + b)^5$

Use the binomial theorem to expand each expression. SEE EXAMPLE 6. *(OBJECTIVE 4)*

45. $(a - b)^4$

46. $(x - y)^3$

47. $(x - 2y)^3$

48. $(2x - y)^4$

Use the binomial theorem to expand each expression. SEE EXAMPLE 7. *(OBJECTIVE 4)*

49. $(2x + y)^3$

50. $(x + 2y)^3$

51. $(2x + 3y)^3$

52. $(3x - 2y)^3$

Find the specified term of each expansion. SEE EXAMPLE 8. *(OBJECTIVE 5)*

53. $(a + b)^4$; third term

54. $(a + b)^3$; third term

55. $(x + y)^6$; fifth term

56. $(x + y)^7$; fifth term

Find the specified term of each expansion. SEE EXAMPLE 9. *(OBJECTIVE 5)*

57. $(x - y)^4$; fourth term

58. $(x - y)^5$; second term

59. $(x - y)^8$; third term

60. $(x - y)^9$; seventh term

Find the specified term of each expansion. SEE EXAMPLE 10. *(OBJECTIVE 5)*

61. $(4x + y)^5$; third term

62. $(x + 4y)^5$; fourth term

63. $(x - 3y)^4$; second term

64. $(3x - y)^5$; third term

65. $(2x - 5)^7$; fourth term

66. $(2x + 3)^6$; sixth term

67. $(2x - 3y)^5$; fifth term

68. $(3x - 2y)^4$; second term

ADDITIONAL PRACTICE   *Evaluate each expression.*

69. $8(7!)$

70. $4!(5)$

71. $\dfrac{7!}{5!(7 - 5)!}$

72. $\dfrac{8!}{6!(8 - 6)!}$

73. $\dfrac{5!(8 - 5)!}{4!7!}$

74. $\dfrac{6!7!}{(8 - 3)!(7 - 4)!}$

Use a calculator to find each factorial.

75. $11!$

76. $13!$

77. $20!$

78. $55!$

Expand using any method.

79. $(3 + 2y)^4$

80. $(2x + 3)^4$

81. $\left(\dfrac{x}{2} - \dfrac{y}{3}\right)^3$

82. $\left(\dfrac{x}{3} + \dfrac{y}{2}\right)^3$

83. $(x - y)^5$

84. $(a - b)^6$

85. $\left(\dfrac{x}{2} + \dfrac{y}{3}\right)^4$

86. $\left(\dfrac{x}{3} - \dfrac{y}{2}\right)^4$

87. Without referring to the text, write the first ten rows of Pascal's triangle.

88. Find the sum of the numbers in each row of the first ten rows of Pascal's triangle. What is the pattern?

Find the specified term of each expansion.

89. $\left(\sqrt{2}x + \sqrt{3}y\right)^6$; third term

90. $\left(\sqrt{3}x + \sqrt{2}y\right)^5$; second term

91. $\left(\dfrac{x}{2} - \dfrac{y}{3}\right)^4$; second term

92. $\left(\dfrac{x}{3} + \dfrac{y}{2}\right)^5$; fourth term

93. $(x + 5)^6$; third term

94. $(x - 2)^4$; second term

95. $(a + b)^n$; fourth term

96. $(a + b)^n$; third term

97. $(a - b)^n$; fifth term

98. $(a - b)^n$; sixth term

99. $(a + b)^n$; rth term

100. $(a + b)^n$; $(r + 1)$th term

WRITING ABOUT MATH

101. Explain how to construct Pascal's triangle.

102. Explain how to find the variables of the terms in the expansion of $(r + s)^4$.

103. Explain how to find the coefficients in the expansion of $(x + y)^5$.

104. Explain why the signs alternate in the expansion of $(x - y)^9$.

SOMETHING TO THINK ABOUT

105. If we apply the pattern of the coefficients to the coefficient of the first term in a binomial expansion, the coefficient would be $\dfrac{n!}{0!(n - 0)!}$. Show that this expression is 1.

106. If we apply the pattern of the coefficients to the coefficient of the last term in a binomial expansion, the coefficient would be $\dfrac{n!}{n!(n - n)!}$. Show that this expression is 1.

107. Find the sum of the numbers in the designated diagonal rows of Pascal's triangle shown in the illustration. What is the pattern?

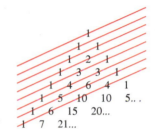

108. Find the constant term in the expansion of $\left(x + \dfrac{1}{x}\right)^{10}$.

109. Find the coefficient of a^5 in the expansion of $\left(a - \dfrac{1}{a}\right)^9$.

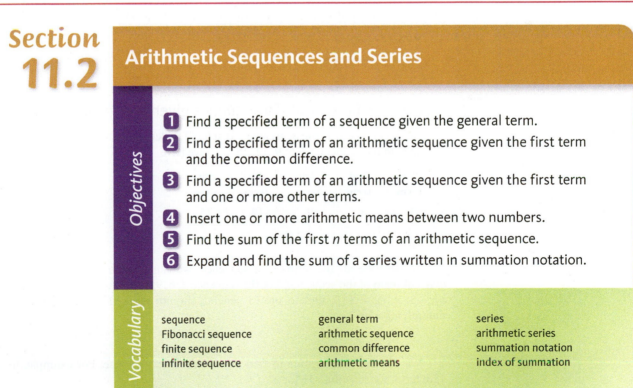

Section 11.2

Arithmetic Sequences and Series

Objectives

1. Find a specified term of a sequence given the general term.
2. Find a specified term of an arithmetic sequence given the first term and the common difference.
3. Find a specified term of an arithmetic sequence given the first term and one or more other terms.
4. Insert one or more arithmetic means between two numbers.
5. Find the sum of the first n terms of an arithmetic sequence.
6. Expand and find the sum of a series written in summation notation.

Vocabulary

sequence	general term	series
Fibonacci sequence	arithmetic sequence	arithmetic series
finite sequence	common difference	summation notation
infinite sequence	arithmetic means	index of summation

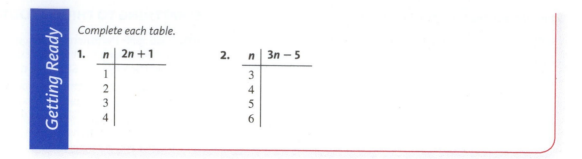

Complete each table.

1.

n	$2n + 1$
1	
2	
3	
4	

2.

n	$3n - 5$
3	
4	
5	
6	

In the next two sections, we will discuss ordered lists of numbers called *sequences*. In this section, we will discuss sequences in general and then examine a special type of sequence called an *arithmetic sequence*.

A **sequence** is a function whose domain is the set of natural numbers. For example, the function $f(n) = 3n + 2$, where n is a natural number, is a sequence. Because a sequence is a function whose domain is the set of natural numbers, we can write its values as a list. If the natural numbers are substituted for n, the function $f(n) = 3n + 2$ generates the list

5, 8, 11, 14, 17, ...

It is common to call the list, as well as the function, a sequence. Each number in the list is called a *term* of the sequence. Other examples of sequences are

$1^3, 2^3, 3^3, 4^3, \ldots$ The ordered list of the cubes of the natural numbers

4, 8, 12, 16, ...   The ordered list of the positive multiples of 4

2, 3, 5, 7, 11, ...   The ordered list of prime numbers

1, 1, 2, 3, 5, 8, 13, 21, ...   The Fibonacci sequence

The **Fibonacci sequence** is named after the 12th-century mathematician Leonardo of Pisa—also known as Fibonacci. Beginning with the 2, each term of the sequence is the sum of the two preceding terms.

Finite sequences contain a finite number of terms and **infinite sequences** contain infinitely many terms. One example of each type of sequence is:

Finite sequence: 1, 5, 9, 13, 17, 21, 25

Infinite sequence: 3, 6, 9, 12, 15, ...   The ellipsis, . . . , indicates that the sequence goes on forever.

1 ## Find a specified term of a sequence given the general term.

In this section, we will use a_n (read as "*a* sub *n*") to denote the *n*th term of a sequence. For example, in the sequence 3, 6, 9, 12, 15, . . . , we have

1st term	2nd term	3rd term	4th term	5th term	
3,	6,	9,	12,	15,	...
↑	↑	↑	↑	↑	
a_1	a_2	a_3	a_4	a_5	

To describe all the terms of a sequence, we can write a formula for a_n, called the **general term** of the sequence. For the sequence 3, 6, 9, 12, 15, . . . , we note that $a_1 = 3 \cdot 1$, $a_2 = 3 \cdot 2$, $a_3 = 3 \cdot 3$, and so on. In general, the *n*th term of the sequence is found by multiplying n by 3.

$$a_n = 3n$$

We can use this formula to find any term of the sequence. For example, to find the 12th term, we substitute 12 for n.

$$a_{12} = 3(12) = 36$$

EXAMPLE 1 Given an infinite sequence with $a_n = 2n - 3$, find:

a. the first four terms   **b.** a_{50}

Solution **a.** To find the first four terms of the sequence, we substitute 1, 2, 3, and 4 for n in $a_n = 2n - 3$ and simplify.

$$a_1 = 2(\mathbf{1}) - 3 = -1 \quad \text{Substitute 1 for } n.$$
$$a_2 = 2(\mathbf{2}) - 3 = 1 \quad \text{Substitute 2 for } n.$$
$$a_3 = 2(\mathbf{3}) - 3 = 3 \quad \text{Substitute 3 for } n.$$
$$a_4 = 2(\mathbf{4}) - 3 = 5 \quad \text{Substitute 4 for } n.$$

The first four terms of the sequence are $-1, 1, 3,$ and 5.

b. To find a_{50}, the 50th term of the sequence, we let $n = 50$:

$$a_{50} = 2(\mathbf{50}) - 3 = 97$$

SELF CHECK 1 Given an infinite sequence with $a_n = 3n + 5$, find:

a. the first three terms   **b.** a_{100}

---

2 Find a specified term of an arithmetic sequence given the first term and the common difference.

One common type of sequence is the arithmetic sequence.

ARITHMETIC SEQUENCE An **arithmetic sequence** is a sequence of the form

$$a_1, \quad a_1 + d, \quad a_1 + 2d, \quad \ldots, \quad a_1 + (n-1)d, \ldots$$

where a_1 is the *first term* and d is the **common difference**. The *nth term* is given by

$$a_n = a_1 + (n-1)d$$

We note that the second term of an arithmetic sequence has an addend of $1d$, the third term has an addend of $2d$, the fourth term has an addend of $3d$, and the nth term has an addend of $(n-1)d$. We also note that the difference between any two consecutive terms in an arithmetic sequence is d.

EXAMPLE 2 An arithmetic sequence has a first term of 5 and a common difference of 4.

a. Write the first six terms of the sequence.

b. Write the 25th term of the sequence.

Solution **a.** Since the first term is $a_1 = 5$ and the common difference is $d = 4$, the first six terms are

$$\underset{a_1}{5,} \quad \underset{a_2}{5+4,} \quad \underset{a_3}{5+2(4),} \quad \underset{a_4}{5+3(4),} \quad \underset{a_5}{5+4(4),} \quad \underset{a_6}{5+5(4)}$$

or

$$5, 9, 13, 17, 21, 25$$

b. The nth term is $a_n = a_1 + (n - 1)d$. Since we want the 25th term, we let $n = 25$:

$$a_n = a_1 + (n - 1)d$$
$$a_{25} = 5 + (25 - 1)4 \qquad \text{Remember that } a_1 = 5 \text{ and } d = 4.$$
$$= 5 + 24(4)$$
$$= 5 + 96$$
$$= 101$$

SELF CHECK 2   **a.** Write the seventh term of the sequence in Example 2.

b. Write the 30th term of the sequence in Example 2.

3   **Find a specified term of an arithmetic sequence given the first term and one or more other terms.**

EXAMPLE 3   The first three terms of an arithmetic sequence are 3, 8, and 13. Find:

a. the 67th term   **b.** the 100th term

Solution   We first find d, the common difference. It is the difference between two successive terms:

$$d = 8 - 3 = 13 - 8 = 5$$

a. We substitute 3 for a_1, 67 for n, and 5 for d in the formula for the nth term and simplify:

$$a_n = a_1 + (n - 1)d$$
$$a_{67} = 3 + (67 - 1)5$$
$$= 3 + 66(5)$$
$$= 333$$

b. We substitute 3 for a_1, 100 for n, and 5 for d in the formula for the nth term, and simplify:

$$a_n = a_1 + (n - 1)d$$
$$a_{100} = 3 + (100 - 1)5$$
$$= 3 + 99(5)$$
$$= 498$$

SELF CHECK 3   Find the 50th term of the sequence in Example 3.

EXAMPLE 4   The first term of an arithmetic sequence is 12, and the 50th term is 3,099. Write the first six terms of the sequence.

Solution   The key is to find the common difference. Because the 50th term of this sequence is 3,099, we can let $n = 50$ and solve the following equation for d:

$$a_n = a_1 + (n - 1)d$$
$$a_{50} = 12 + (50 - 1)d$$
$$3{,}099 = 12 + 49d \qquad \text{Substitute 3,099 for } a_{50} \text{ and simplify.}$$
$$3{,}087 = 49d \qquad \text{Subtract 12 from both sides.}$$
$$63 = d \qquad \text{Divide both sides by 49.}$$

The common difference is 63, and since the first term of the sequence is 12, its first six terms are

12, 75, 138, 201, 264, 327   Add 63 to a term to get the next term.

SELF CHECK 4   The first term of an arithmetic sequence is 15, and the 12th term is 92. Write the first four terms of the sequence.

4 Insert one or more arithmetic means between two numbers.

If numbers are inserted between two numbers a and b to form an arithmetic sequence, the inserted numbers are called **arithmetic means** between a and b.

If a single number is inserted between the numbers a and b to form an arithmetic sequence, that number is called the *arithmetic mean* between a and b.

EXAMPLE 5   Insert two arithmetic means between 6 and 27.

Solution   Because we want to insert two arithmetic means between 6 and 27, this implies that the first term is $a_1 = 6$ and 27 is the fourth term, a_4. We must find the common difference so that the terms

$$\begin{array}{cccc} 6, & 6+d, & 6+2d, & 27 \\ \uparrow & \uparrow & \uparrow & \uparrow \\ a_1 & a_2 & a_3 & a_4 \end{array}$$

form an arithmetic sequence. To find the value of d, we substitute 6 for a_1 and 4 for n into the formula for the nth term:

$$a_n = a_1 + (n-1)d$$
$$a_4 = 6 + (4-1)d$$
$$27 = 6 + 3d \qquad \text{Substitute 27 for } a_4 \text{ and simplify.}$$
$$21 = 3d \qquad \text{Subtract 6 from both sides.}$$
$$7 = d \qquad \text{Divide both sides by 3.}$$

The common difference is 7. The two arithmetic means between 6 and 27 are

$$\begin{array}{ccc} 6 + d = 6 + 7 & \quad \text{or} \quad & 6 + 2d = 6 + 2(7) \\ \qquad = 13 & & \qquad = 6 + 14 \\ & & \qquad = 20 \end{array}$$

The numbers 6, 13, 20, and 27 are the first four terms of an arithmetic sequence.

SELF CHECK 5   Insert two arithmetic means between 8 and 44.

5 Find the sum of the first n terms of an arithmetic sequence.

We now consider a formula that calculates the sum of the first n terms of an arithmetic sequence. To develop this formula, we let S_n represent the sum of the first n terms of an arithmetic sequence:

$$S_n = a_1 + [a_1 + d] + [a_1 + 2d] + \cdots + [a_1 + (n-1)d]$$

We write the same sum again, but in reverse order:

$$S_n = [a_1 + (n-1)d] + [a_1 + (n-2)d] + [a_1 + (n-3)d] + \cdots + a_1$$

We add these two equations together, term by term, to obtain

$$2S_n = [2a_1 + (n-1)d] + [2a_1 + d + dn - 2d] + [2a_1 + 2d + dn - 3d]$$
$$+ \cdots + [2a_1 + (n-1)d]$$
$$= [2a_1 + (n-1)d] + [2a_1 + dn - d] + [2a_1 + dn - d]$$
$$+ \cdots + [2a_1 + (n-1)d]$$
$$2S_n = [2a_1 + (n-1)d] + [2a_1 + (n-1)d] + [2a_1 + (n-1)d]$$
$$+ \cdots + [2a_1 + (n-1)d]$$

Because there are n equal terms on the right side of the preceding equation, we can write

$$2S_n = n[2a_1 + (n-1)d]$$
$$2S_n = n[a_1 + a_1 + (n-1)d] \qquad 2a_1 = a_1 + a_1$$
$$2S_n = n[a_1 + a_n] \qquad \text{Substitute } a_n \text{ for } a_1 + (n-1)d.$$
$$S_n = \frac{n(a_1 + a_n)}{2}$$

This reasoning establishes the following.

SUM OF THE FIRST n TERMS OF AN ARITHMETIC SEQUENCE	The sum of the first n terms of an arithmetic sequence is given by the formula $$S_n = \frac{n(a_1 + a_n)}{2} \quad \text{with} \quad a_n = a_1 + (n-1)d$$ where a_1 is the first term, a_n is the nth term, and n is the number of terms in the sequence.

EXAMPLE 6   Find the sum of the first 40 terms of the arithmetic sequence 4, 10, 16, ...

Solution   In order to find the sum, we will need to identify the common difference, d, and the 40th term, a_{40}. The common difference is $16 - 10 = 10 - 4 = 6$ and $a_{40} = 4 + (40 - 1)6 = 238$. Then we substitute into the formula for S_n:

$$S_n = \frac{n(a_1 + a_{40})}{2}$$
$$S_{40} = \frac{40(4 + 238)}{2}$$
$$= 20(242)$$
$$= 4{,}840$$

The sum of the first 40 terms is 4,840.

SELF CHECK 6   Find the sum of the first 50 terms of the arithmetic sequence 3, 8, 13, ...

When the commas between the terms of a sequence are replaced by $+$ signs, we call the indicated sum a **series**. The sum of the terms of an arithmetic sequence is called an **arithmetic series**. Some examples are

$$4 + 8 + 12 + 16 + 20 + 24 \qquad \text{Since this series has a limited number of terms, it is a finite arithmetic series.}$$

$$5 + 8 + 11 + 14 + 17 + \cdots \qquad \text{Since this series has infinitely many terms, it is an infinite arithmetic series.}$$

Expand and find the sum of a series written in summation notation.

COMMENT This objective introduces summation notation for series, not just arithmetic series.

We can use a shorthand notation for indicating the sum of a finite number of consecutive terms in a series. This notation, called **summation notation**, involves the Greek letter Σ (sigma). The expression

$$\sum_{k=2}^{5} 3k \qquad \textcolor{red}{\text{Read as "the summation of } 3k \text{ from } k = 2 \text{ to } k = 5."}}$$

designates the sum of all terms obtained if we successively substitute the numbers 2, 3, 4, and 5 for k, called the **index of the summation**. Thus, we have

$$\textcolor{red}{k = 2 \quad k = 3 \quad k = 4 \quad k = 5}$$
$$\textcolor{red}{\downarrow \qquad \downarrow \qquad \downarrow \qquad \downarrow}$$
$$\sum_{k=2}^{5} 3k = 3(\textcolor{red}{2}) + 3(\textcolor{red}{3}) + 3(\textcolor{red}{4}) + 3(\textcolor{red}{5})$$
$$= 6 + 9 + 12 + 15$$
$$= 42$$

EXAMPLE 7 Write the series associated with each summation.

   a. $\displaystyle\sum_{k=1}^{4} k$ **b.** $\displaystyle\sum_{k=2}^{5} (k-1)^3$

Solution   **a.** $\displaystyle\sum_{k=1}^{4} k = \textcolor{red}{1} + \textcolor{red}{2} + \textcolor{red}{3} + \textcolor{red}{4}$

     b. $\displaystyle\sum_{k=2}^{5} (k-1)^3 = (\textcolor{red}{2} - 1)^3 + (\textcolor{red}{3} - 1)^3 + (\textcolor{red}{4} - 1)^3 + (\textcolor{red}{5} - 1)^3$
$$= 1^3 + 2^3 + 3^3 + 4^3$$
$$= 1 + 8 + 27 + 64$$

SELF CHECK 7 Write the series associated with the summation $\displaystyle\sum_{t=3}^{5} t^2$.

EXAMPLE 8 Find each sum. **a.** $\displaystyle\sum_{k=3}^{5} (2k+1)$ **b.** $\displaystyle\sum_{k=2}^{5} k^2$ **c.** $\displaystyle\sum_{k=1}^{3} (3k^2 + 3)$

Solution   **a.** $\displaystyle\sum_{k=3}^{5} (2k+1) = [2(\textcolor{red}{3}) + 1] + [2(\textcolor{red}{4}) + 1] + [2(\textcolor{red}{5}) + 1]$
$$= 7 + 9 + 11$$
$$= 27$$

     b. $\displaystyle\sum_{k=2}^{5} k^2 = \textcolor{red}{2}^2 + \textcolor{red}{3}^2 + \textcolor{red}{4}^2 + \textcolor{red}{5}^2$
$$= 4 + 9 + 16 + 25$$
$$= 54$$

c. $\displaystyle\sum_{k=1}^{3}(3k^2 + 3) = [3(\mathbf{1})^2 + 3] + [3(\mathbf{2})^2 + 3] + [3(\mathbf{3})^2 + 3]$

$$= 6 + 15 + 30$$

$$= 51$$

🌿 **SELF CHECK 8**   Evaluate: $\displaystyle\sum_{k=1}^{4}(2k^2 - 3)$

🌿 **SELF CHECK ANSWERS**   **1. a.** 8, 11, 14 **b.** 305 **2. a.** 29 **b.** 121 **3.** 248 **4.** 15, 22, 29, 36 **5.** 20, 32 **6.** 6,275
7. 9 + 16 + 25 **8.** 48

NOW TRY THIS

Given $a_5 = 3a - 2$ and $a_9 = 11a + 10$ in an arithmetic sequence:

1. Find the first 5 terms.

2. Find the sum of the first 10 terms.

11.2   Exercises 🌿

WARM-UPS *Evaluate each algebraic expression for $n = 1$,
$n = 2, n = 3$, and $n = 4$.*

1. $2n$

2. n^2

3. $n + 4$

4. $n - 3$

5. $5 + (n - 1)2$

6. $3 + (n - 1)7$

REVIEW *Perform the operations and simplify, if possible. Assume
no division by 0.*

7. $5(3x^2 - 6x + 4) + 3(4x^2 + 6x - 2)$

8. $(4p + q)(2p^2 - 3pq + 2q^2)$

9. $\dfrac{3a + 4}{a - 2} + \dfrac{3a - 4}{a + 2}$

10. $2t - 3\overline{)8t^4 - 12t^3 + 8t^2 - 16t + 6}$

VOCABULARY AND CONCEPTS *Fill in the blanks.*

11. A _____ is a function whose domain is the set of natural
numbers. If a sequence has a limited number of terms, it is
called a _____ sequence. If it has an unlimited number of
terms, it is an _____ sequence.

12. The sequence 1, 1, 2, 3, 5, 8, 13, 21, . . . is called the _____
sequence.

13. The sequence 3, 9, 15, 21, . . . is an example of an _____
sequence with a common _____ of 6 and a first term of 3.

14. The last term (or nth term) of an arithmetic sequence is given
by the formula _____.

15. If a number c is inserted between two numbers a and b to
form an arithmetic sequence, then c is called the
_____ between a and b.

16. The sum of the first n terms of an arithmetic sequence is
given by the formula $S_n =$ _____ .

17. The indicated sum of the terms of an arithmetic sequence is
called an arithmetic _____.

18. The symbol Σ is the Greek letter _____.

19. $\displaystyle\sum_{k=1}^{7}k$ means _____.

20. In the expression $\displaystyle\sum_{k=1}^{5}(2k - 5)$, k is called the _____ of the
summation.

GUIDED PRACTICE *Given the infinite sequence
$a_n = 3n - 2$, find each value. SEE EXAMPLE 1. (OBJECTIVE 1)*

21. a_2

22. a_3

23. a_{30}

24. a_{50}

*Write the first five terms of each arithmetic sequence with the
given properties. SEE EXAMPLE 2. (OBJECTIVE 2)*

25. $a_1 = 3, d = 2$

26. $a_1 = -2, d = 3$

27. $a_1 = -5, d = -3$

28. $a_1 = 8, d = -5$

*Write (a) the first five terms of each arithmetic sequence with the
given properties; (b) the specified term of the sequence.
SEE EXAMPLE 2. (OBJECTIVE 2)*

29. $a_1 = 4, d = 3$; find a_{15}.

30. $a_1 = -2, d = 3$; find a_{20}.

31. $a_1 = -7, d = -2$; find a_{30}.

32. $a_1 = 10, d = -6$; find a_{12}.

Find the specific term given the first three terms of an arithmetic sequence. SEE EXAMPLE 3. (OBJECTIVE 3)

33. $2, 7, 12, \ldots$; 50th term

34. $10, 7, 4, \ldots$; 23rd term

35. $5, -1, -7, \ldots$; 49th term

36. $-2, 5, 12 \ldots$; 61st term

Write the first five terms of each arithmetic sequence with the given properties. SEE EXAMPLE 4. (OBJECTIVE 3)

37. $a_1 = 5$, fifth term is 29

38. $a_1 = 4$, sixth term is 39

39. $a_1 = -4$, sixth term is -39

40. $a_1 = -5$, fifth term is -37

Insert the specified number of arithmetic means. SEE EXAMPLE 5. (OBJECTIVE 4)

41. Insert three arithmetic means between 2 and 11.

42. Insert four arithmetic means between 5 and 25.

43. Insert four arithmetic means between 10 and 20.

44. Insert three arithmetic means between 20 and 30.

Find the sum of the first n terms of each arithmetic sequence. SEE EXAMPLE 6. (OBJECTIVE 5)

45. $1, 4, 7, \ldots$; $n = 30$

46. $2, 6, 10, \ldots$; $n = 28$

47. $-5, -1, 3, \ldots$; $n = 17$

48. $-7, -1, 5, \ldots$; $n = 15$

Write the series associated with each summation. SEE EXAMPLE 7. (OBJECTIVE 6)

49. $\displaystyle\sum_{k=1}^{4} (3k)$

50. $\displaystyle\sum_{k=1}^{3} (k - 9)^2$

51. $\displaystyle\sum_{k=4}^{6} k^2$

52. $\displaystyle\sum_{k=3}^{5} (-2k)$

Find each sum. SEE EXAMPLE 8. (OBJECTIVE 6)

53. $\displaystyle\sum_{k=1}^{4} 6k$

54. $\displaystyle\sum_{k=2}^{5} 3k$

55. $\displaystyle\sum_{k=3}^{4} (k^2 + 3)$

56. $\displaystyle\sum_{k=2}^{6} (k^2 + 1)$

ADDITIONAL PRACTICE *Write the first five terms of each arithmetic sequence with the given properties.*

57. $d = 7$, sixth term is -83.

58. $d = 3$, seventh term is 12.

59. $d = -3$, seventh term is 16.

60. $d = -5$, seventh term is -12.

61. The 19th term is 131 and the 20th term is 138.

62. The 16th term is 70 and the 18th term is 78.

Find the specified value.

63. Find the 30th term of the arithmetic sequence with $a_1 = 7$ and $d = 12$.

64. Find the 55th term of the arithmetic sequence with $a_1 = -5$ and $d = 4$.

65. Find the 37th term of the arithmetic sequence with a second term of -4 and a third term of -9.

66. Find the 40th term of the arithmetic sequence with a second term of 6 and a fourth term of 16.

67. Find the first term of the arithmetic sequence with a common difference of 11 and whose 27th term is 263.

68. Find the common difference of the arithmetic sequence with a first term of -164 if its 36th term is -24.

69. Find the common difference of the arithmetic sequence with a first term of 40 if its 44th term is 556.

70. Find the first term of the arithmetic sequence with a common difference of -5 and whose 23rd term is -625.

71. Find the arithmetic mean between 10 and 19.

72. Find the arithmetic mean between 5 and 23.

73. Find the arithmetic mean between -4.5 and 7.

74. Find the arithmetic mean between -6.3 and -5.2.

Find the sum of the first n terms.

75. Second term is 7, third term is 12; $n = 12$.

76. Second term is 5, fourth term is 9; $n = 16$.

77. $f(n) = 2n + 1$, nth term is 31; n is a natural number.

78. $f(n) = 4n + 3$, nth term is 23; n is a natural number.

79. Find the sum of the first 50 natural numbers.

80. Find the sum of the first 100 natural numbers.

81. Find the sum of the first 50 odd natural numbers.

82. Find the sum of the first 50 even natural numbers.

Find each sum.

83. $\displaystyle\sum_{k=4}^{4} (2k + 4)$

84. $\displaystyle\sum_{k=3}^{5} (3k^2 - 7)$

APPLICATIONS *Solve each application.*

85. Saving money   Yasmeen puts $60 into a safety deposit box. Each month after that, she puts $50 more in the box. Write the first six terms of an arithmetic sequence that gives the monthly amounts in her savings, and find her savings after 10 years.

86. Installment loans   Maria borrowed $10,000, interest-free, from her mother. She agreed to pay back the loan in monthly installments of $275. Write the first six terms of an arithmetic sequence that shows the balance due after each month, and find the balance due after 17 months.

87. Designing a patio   Each row of bricks in the triangular patio is to have one more brick than the previous row, ending with the longest row of 150 bricks. How many bricks will be needed?

88. Falling objects The equation $s = 16t^2$ represents the distance s in feet that an object will fall in t seconds. After 1 second, the object has fallen 16 feet. After 2 seconds, the object has fallen 64 feet, and so on. Find the distance that the object will fall during the second and third seconds.

89. Falling objects Refer to Exercise 88. How far will the object fall during the 12th second?

90. Interior angles The sums of the angles of several polygons are given in the table. Assuming that the pattern continues, complete the table.

Figure	Number of sides	Sum of angles
Triangle	3	180°
Quadrilateral	4	360°
Pentagon	5	540°
Hexagon	6	720°
Octagon	8	
Dodecagon	12	

WRITING ABOUT MATH

91. Define an arithmetic sequence.

92. Develop the formula for finding the sum of the first n terms of an arithmetic sequence.

SOMETHING TO THINK ABOUT

93. Write the addends of the sum given by

$$\sum_{n=1}^{6}\left(\frac{1}{2}n + 1\right)$$

94. Find the sum of the sequence given in Exercise 93.

95. Show that the arithmetic mean between a and b is the average of a and b: $\frac{a + b}{2}$

96. Show that the sum of the two arithmetic means between a and b is $a + b$.

97. Show that $\sum_{k=1}^{5} 5k = 5\sum_{k=1}^{5} k$.

98. Show that $\sum_{k=3}^{6}(k^2 + 3k) = \sum_{k=3}^{6}k^2 + \sum_{k=3}^{6}3k$.

99. Show that $\sum_{k=1}^{n} 3 = 3n$. (*Hint:* Consider 3 to be $3k^0$.)

100. Show that $\sum_{k=1}^{3}\frac{k^2}{k} \neq \frac{\sum_{k=1}^{3}k^2}{\sum_{k=1}^{3}k}$.

Section 11.3 Geometric Sequences and Series

Objectives

1. Find a specified term of a geometric sequence given the first term and the common ratio.
2. Find a specified term of a geometric sequence given the first term and one or more other terms.
3. Find one or more geometric means given two terms of a sequence.
4. Find the sum of the first n terms of a geometric sequence.
5. Solve an application involving a geometric sequence.

Vocabulary

geometric sequence   common ratio   geometric mean

Complete each table.

1.

n	$5(2^n)$
1	
2	
3	

2.

n	$6(3^n)$
1	
2	
3	

In this section, we consider another common type of sequence called a *geometric sequence*.

1 Find a specified term of a geometric sequence given the first term and the common ratio.

Each term of a geometric sequence is found by multiplying the previous term by the same number.

GEOMETRIC SEQUENCE

A **geometric sequence** is a sequence of the form

$$a_1, \quad a_1 r, \quad a_1 r^2, \quad a_1 r^3, \quad \ldots \quad a_1 r^{n-1}, \quad \ldots$$

where a_1 is the *first term* and r is the **common ratio**. The *nth term* is given by

$$a_n = a_1 r^{n-1}$$

We note that the second term of a geometric sequence has a factor of r^1, the third term has a factor of r^2, the fourth term has a factor of r^3, and the *n*th term has a factor of r^{n-1}. We also note that the quotient obtained when any term is divided by the previous term is r.

EXAMPLE 1   A geometric sequence has a first term of 5 and a common ratio of 3.
a. Write the first five terms of the sequence.
b. Find the ninth term.

Solution   **a.** Since the first term is $a_1 = 5$ and the common ratio is $r = 3$, the first five terms are

$$5, \quad 5(3), \quad 5(3^2), \quad 5(3^3), \quad 5(3^4)$$
$$\uparrow \qquad \uparrow \qquad \uparrow \qquad \uparrow \qquad \uparrow$$
$$a_1 \qquad a_2 \qquad a_3 \qquad a_4 \qquad a_5$$

Each term is found by multiplying the previous term by 3.

or

$$5, 15, 45, 135, 405$$

b. The *n*th term is $a_1 r^{n-1}$ where $a_1 = 5$ and $r = 3$. Because we want the ninth term, we let $n = 9$:

$$a_n = a_1 r^{n-1}$$
$$a_9 = 5(3)^{9-1}$$
$$= 5(3)^8$$
$$= 5(6{,}561)$$
$$= 32{,}805$$

SELF CHECK 1   A geometric sequence has a first term of 3 and a common ratio of 4.
a. Write the first four terms.
b. Find the eighth term.

2 Find a specified term of a geometric sequence given the first term and one or more other terms.

EXAMPLE 2 The first three terms of a geometric sequence are $16, 4,$ and 1. Find the seventh term.

Solution We note that the first term a_1 is 16 and that

$$r = \frac{a_2}{a_1} = \frac{4}{16} = \frac{1}{4} \qquad \text{Also note that } \frac{a_3}{a_2} = \frac{1}{4}.$$

We substitute 16 for a_1, $\frac{1}{4}$ for r, and 7 for n in the formula for the nth term and simplify:

$$a_n = a_1 r^{n-1}$$
$$a_7 = 16\left(\frac{1}{4}\right)^{7-1}$$
$$= 16\left(\frac{1}{4}\right)^6$$
$$= 16\left(\frac{1}{4,096}\right)$$
$$= \frac{1}{256}$$

SELF CHECK 2 Find the tenth term of the sequence in Example 2.

EXAMPLE 3 Write the first five terms of the geometric sequence whose first term is -3 and fourth term is -192.

Solution Here the first term is $a_1 = -3$ and $a_4 = -192$. We substitute -3 for a_1 and -192 for a_4 into the formula for the nth term of a geometric sequence and solve for r.

$$a_n = a_1 r^{n-1}$$
$$-192 = -3r^{4-1}$$
$$64 = r^3 \qquad \text{Divide both sides by } -3.$$
$$4 = r \qquad \text{Take the cube root of both sides.}$$

The first five terms are $-3, -12, -48, -192, -768$.

SELF CHECK 3 Write the first five terms of the geometric sequence whose first term is -64 and sixth term is -2.

3 Find one or more geometric means given two terms of a sequence.

If numbers are inserted between two numbers a and b to form a geometric sequence, the inserted numbers are called **geometric means** between a and b.

If a single number is inserted between the numbers a and b to form a geometric sequence, that number is called the **geometric mean** between a and b.

EXAMPLE 4 Insert two geometric means between 7 and 1,512.

Solution Here the first term is $a_1 = 7$. Because we are inserting two geometric means between 7 and 1,512, the fourth term is $a_4 = 1,512$. To find the common ratio r so that the terms

$$7, \quad \mathbf{7r}, \quad \mathbf{7r^2}, \quad 1,512$$
$$\uparrow \qquad \uparrow \qquad \uparrow \qquad \uparrow$$
$$a_1 \qquad a_2 \qquad a_3 \qquad a_4$$

form a geometric sequence, we substitute 4 for n and 7 for a_1 into the formula for the nth term of a geometric sequence and solve for r.

$$a_n = a_1 r^{n-1}$$
$$a_4 = 7r^{4-1}$$
$$1,512 = 7r^3$$
$$216 = r^3 \qquad \text{Divide both sides by 7.}$$
$$6 = r \qquad \text{Take the cube root of both sides.}$$

The two geometric means between 7 and 1,512 are

$$\mathbf{7r = 7(6) = 42} \qquad \text{and} \qquad \mathbf{7r^2 = 7(6)^2 = 7(36) = 252}$$

The numbers 7, 42, 252, and 1,512 are the first four terms of a geometric sequence.

 SELF CHECK 4 Insert three positive geometric means between 1 and 16.

EXAMPLE 5 Find a geometric mean between 2 and 20.

Solution We want to find the middle term of the three-term geometric sequence

$$2, \quad \mathbf{2r}, \quad 20$$
$$\uparrow \qquad \uparrow \qquad \uparrow$$
$$a_1 \qquad a_2 \qquad a_3$$

with $a_1 = 2$, $a_3 = 20$, and $n = 3$. To find the value of r, we substitute these values into the formula for the nth term of a geometric sequence:

$$a_n = ar^{n-1}$$
$$a_3 = 2r^{3-1}$$
$$20 = 2r^2$$
$$10 = r^2 \qquad \text{Divide both sides by 2.}$$
$$\pm\sqrt{10} = r \qquad \text{Take the square root of both sides.}$$

Because r can be either $\sqrt{10}$ or $-\sqrt{10}$, there are two values for the geometric mean. They are

$$\mathbf{2r = 2\sqrt{10}} \qquad \text{and} \qquad \mathbf{2r = -2\sqrt{10}}$$

The numbers 2, $2\sqrt{10}$, 20 and 2, $-2\sqrt{10}$, 20 both form geometric sequences. The common ratio of the first sequence is $\sqrt{10}$, and the common ratio of the second sequence is $-\sqrt{10}$.

 SELF CHECK 5 Find a geometric mean between 2 and 200.

4 ## Find the sum of the first n terms of a geometric sequence.

There is a formula that gives the sum of the first n terms of a geometric sequence. To develop this formula, we let S_n represent the sum of the first n terms of a geometric sequence.

(1) $S_n = a_1 + a_1 r + a_1 r^2 + a_1 r^3 + \cdots + a_1 r^{n-1}$ This is a geometric series.

We multiply both sides of Equation 1 by r to obtain

(2) $S_n r = a_1 r + a_1 r^2 + a_1 r^3 + \cdots + a_1 r^{n-1} + a_1 r^n$

We now subtract Equation 2 from Equation 1 and solve for S_n:

$$S_n - S_n r = a_1 - a_1 r^n$$
$$S_n(1 - r) = a_1 - a_1 r^n \quad \text{Factor out } S_n \text{ from the left side.}$$
$$S_n = \frac{a_1 - a_1 r^n}{1 - r} \quad \text{Divide both sides by } 1 - r.$$

This reasoning establishes the following formula.

SUM OF THE FIRST n TERMS OF A GEOMETRIC SEQUENCE	The sum of the first n terms of a geometric sequence is given by the formula $$S_n = \frac{a_1 - a_1 r^n}{1 - r} \quad (r \neq 1)$$ where S_n is the sum, a_1 is the first term, r is the common ratio, and n is the number of terms.

EXAMPLE 6   Find the sum of the first six terms of the geometric sequence $250, 50, 10, \ldots$.

Solution   Here $a_1 = 250$, $r = \frac{1}{5}$, and $n = 6$. We substitute these values into the formula for the sum of the first n terms of a geometric sequence and simplify:

$$S_n = \frac{a_1 - a_1 r^n}{1 - r}$$

$$S_n = \frac{250 - 250\left(\dfrac{1}{5}\right)^6}{1 - \dfrac{1}{5}}$$

$$= \frac{250 - 250\left(\dfrac{1}{15,625}\right)}{\dfrac{4}{5}}$$

$$= \frac{5}{4}\left(250 - \frac{250}{15,625}\right)$$

$$= \frac{5}{4}\left(\frac{3,906,000}{15,625}\right)$$

$$= 312.48$$

The sum of the first six terms is 312.48.

SELF CHECK 6   Find the sum of the first five terms of the geometric sequence $100, 20, 4, \ldots$.

5   Solve an application involving a geometric sequence.

EXAMPLE 7   **GROWTH OF A TOWN**   The mayor of Eagle River (population 1,500) predicts a growth rate of 4% each year for the next 10 years. Find the expected population of Eagle River 10 years from now.

Solution   Let P_0 be the initial population of Eagle River. After 1 year, there will be a different population, P_1. The initial population (P_0) plus the growth (the product of P_0 and the rate of growth, r) will equal this new population P_1:

$$P_1 = P_0 + P_0 r = P_0(1 + r)$$

The population after 2 years will be P_2, and

$$
\begin{aligned}
P_2 &= P_1 + P_1 r \\
&= \mathbf{P_1}(1 + r) && \text{Factor out } P_1. \\
&= \mathbf{P_0(1 + r)}(1 + r) && P_1 = P_0(1 + r). \\
&= P_0(1 + r)^2
\end{aligned}
$$

The population after 3 years will be P_3, and

$$
\begin{aligned}
P_3 &= P_2 + P_2 r \\
&= \mathbf{P_2}(1 + r) && \text{Factor out } P_2. \\
&= \mathbf{P_0(1 + r)^2}(1 + r) && P_2 = P_0(1 + r)^2. \\
&= P_0(1 + r)^3
\end{aligned}
$$

The yearly population figures

$$P_0, \quad P_1, \quad P_2, \quad P_3, \quad \dots$$

or

$$P_0, \quad P_0(1 + r), \quad P_0(1 + r)^2, \quad P_0(1 + r)^3, \quad \dots$$

form a geometric sequence with a first term of P_0 and a common ratio of $1 + r$. The population of Eagle River after 10 years is P_{10}, which is the 11th term of this sequence:

$$
\begin{aligned}
a_n &= a r^{n-1} \\
P_{10} = a_{11} &= P_0(1 + r)^{10} \\
&= 1{,}500(1 + 0.04)^{10} \\
&= 1{,}500(1.04)^{10} \\
&\approx 1{,}500(1.480244285) \\
&\approx 2{,}220
\end{aligned}
$$

The expected population 10 years from now is 2,220 people.

SELF CHECK 7   Find the expected population of Eagle River 20 years from now.

EXAMPLE 8   **AMOUNT OF AN ANNUITY**   An *annuity* is a sequence of equal payments made periodically over a length of time. The sum of the payments and the interest earned during the *term* of the annuity is called the *amount* of the annuity.

After a sales clerk works six months, her employer will begin an annuity for her and will contribute $500 every six months to a fund that pays 8% annual interest. After she has been employed for two years, what will be the amount of the annuity?

Solution   Because the payments are to be made semiannually, there will be four payments of $500, each earning a rate of 4% per six-month period. These payments will occur at the end of 6 months, 12 months, 18 months, and 24 months. The first payment, to be made after 6 months, will earn interest for three interest periods. Thus, the amount of the first payment is $500(1.04)^3$. The amounts of each of the four payments after two years are shown in Table 11-1 on the next page.

The amount of the annuity is the sum of the amounts of the individual payments, a sum of $2,123.23.

TABLE 11-1	
Payment (at the end of period)	Amount of payment at the end of 2 years
1	$500(1.04)^3 = 562.43
2	$500(1.04)^2 = 540.80
3	$500(1.04)^1 = 520.00
4	$500 = 500.00
	$A_n = $2,123.23$

SELF CHECK 8 If the interest rate is only 4% annual interest, what will be the amount of the annuity?

SELF CHECK ANSWERS

1. a. 3, 12, 48, 192 **b.** 49,152 **2.** $\frac{1}{16,384}$ **3.** $-64, -32, -16, -8, -4$ **4.** 2, 4, 8 **5.** 20, -20
6. 124.96 **7.** The expected population 20 years from now is 3,286. **8.** $2,060.80

NOW TRY THIS

Given $a_1 = 8x - 12$ and $r = \frac{1}{2}$ in a geometric sequence:

1. Find the first 4 terms.

2. Find the sum of the first 4 terms by adding the terms.

3. Set up the formula for finding the sum of the first 4 terms and simplify it to show that the result is the same as the result obtained in Exercise 2.

11.3 Exercises

WARM-UPS *Simplify.*

1. $2(4^2)$
2. $2(4^3)$
3. $4(3)^{5-1}$
4. $4(3)^{7-1}$
5. $7\left(\frac{1}{3}\right)^{4-1}$
6. $7\left(\frac{1}{3}\right)^{5-1}$

REVIEW *Solve each inequality. Assume no division by 0.*

7. $x^2 - 4x - 5 \le 0$
8. $a^2 - 9a + 20 \ge 0$
9. $\frac{x - 4}{x + 3} \ge 0$
10. $\frac{t^2 + t - 20}{t + 2} < 0$

VOCABULARY AND CONCEPTS *Fill in the blanks.*

11. A sequence of the form $a_1, a_1r, a_1r^2, \ldots$ is called a _____ sequence.

12. The formula for the nth term of a geometric sequence is _____.

13. In a geometric sequence, r is called the _____.

14. A number inserted between two numbers a and b to form a geometric sequence is called a geometric _____ between a and b.

15. The sum of the first n terms of a geometric sequence is given by the formula _____.

16. In the formula for Exercise 15, a_1 is the _____ term of the sequence.

Find the next term in each geometric sequence.

17. 1, 3, 9, . . .

18. $1, \dfrac{1}{3}, \dfrac{1}{9}, \ldots$

Find the common ratio in each geometric sequence.

19. 0.2, 0.5, 1.25, . . .

20. $\sqrt{3}, 3, 3\sqrt{3}, \ldots$

Find the value of x in each geometric sequence.

21. 2, x, 18, 54, . . .

22. $3, x, \dfrac{1}{3}, \dfrac{1}{9}, \ldots$

GUIDED PRACTICE *Write the first five terms of each geometric sequence and find the eighth term.* **SEE EXAMPLE 1. (OBJECTIVE 1)**

23. $a_1 = 4, r = 3$

24. $a_1 = -2, r = 2$

25. $a_1 = -5, r = \dfrac{1}{5}$

26. $a_1 = 8, r = \dfrac{1}{2}$

Given the first three terms of a geometric sequence, find the specified term. **SEE EXAMPLE 2. (OBJECTIVE 2)**

27. 1, 3, 9, . . . ; sixth term

28. 2, 6, 18, . . . ; fifth term

29. $2, 1, \dfrac{1}{2}, \ldots$; seventh term

30. $1, \dfrac{1}{3}, \dfrac{1}{9}, \ldots$; fifth term

Write the first five terms of each geometric sequence. **SEE EXAMPLE 3. (OBJECTIVE 2)**

31. $a_1 = 2, r > 0$, third term is 32.

32. $a_1 = 3$, fourth term is 24.

33. $a_1 = 2, r < 0$, third term is 50.

34. $a_1 = -81$, sixth term is $\dfrac{1}{3}$.

Insert the specified number of geometric means. **SEE EXAMPLE 4. (OBJECTIVE 3)**

35. Insert three positive geometric means between 2 and 162.

36. Insert four geometric means between 3 and 96.

37. Insert four geometric means between −4 and −12,500.

38. Insert three geometric means (two positive and one negative) between −64 and −1,024.

Insert the specified geometric mean between the given numbers. **SEE EXAMPLE 5. (OBJECTIVE 3)**

39. Find the negative geometric mean between 2 and 128.

40. Find the positive geometric mean between 3 and 243.

41. Find the positive geometric mean between 10 and 20.

42. Find the negative geometric mean between 5 and 15.

Find the sum of the first n terms of each geometric sequence. **SEE EXAMPLE 6. (OBJECTIVE 4)**

43. 120, 60, 30, . . . ; $n = 4$

44. 2, −6, 18, . . . ; $n = 6$

45. 2, −6, 18, . . . ; $n = 5$

46. 256, −64, 16, . . . ; $n = 5$

ADDITIONAL PRACTICE *Find the first five terms of the geometric sequence.*

47. The second term is 10, and the third term is 50.

48. The third term is −27, and the fourth term is 81.

49. $a_1 = -64, r < 0$, fifth term is −4.

50. $a_1 = -64, r > 0$, fifth term is −4.

Find the indicated quantity.

51. Find the ninth term of the geometric sequence with $a_1 = 4$ and $r = 3$.

52. Find the 12th term of the geometric sequence with $a_1 = 64$ and $r = \dfrac{1}{2}$.

53. Find the first term of the geometric sequence with a common ratio of −3 and an eighth term of −81.

54. Find the first term of the geometric sequence with a common ratio of 2 and a tenth term of 384.

55. Find the common ratio of the geometric sequence with a first term of −8 and a sixth term of −1,944.

56. Find the common ratio of the geometric sequence with a first term of 12 and a sixth term of $\dfrac{3}{8}$.

57. Find a geometric mean, if possible, between −50 and 10.

58. Find a negative geometric mean, if possible, between −25 and −5.

Find the sum of the first n terms of each geometric sequence.

59. 3, −6, 12, . . . ; $n = 8$

60. 3, 6, 12, . . . ; $n = 8$

61. 3, 6, 12, . . . ; $n = 7$

62. 3, −6, 12, . . . ; $n = 7$

63. The second term is 1, and the third term is $\dfrac{1}{5}$; $n = 4$.

64. The second term is 1, and the third term is 4; $n = 5$.

65. The third term is −2, and the fourth term is 1; $n = 6$.

66. The third term is −3, and the fourth term is 1; $n = 5$.

APPLICATIONS *Use a calculator to help solve.* **SEE EXAMPLES 7–8. (OBJECTIVE 5)**

67. Population growth The population of Union is predicted to increase by 6% each year. What will be the population of Union 5 years from now if its current population is 500?

68. Population decline The population of Bensonville is decreasing by 10% each year. If its current population is 98, what will be the population 8 years from now?

69. Declining savings   John has $10,000 in a safety deposit box. Each year he spends 12% of what is left in the box. How much will be in the box after 15 years?

70. Savings growth   Lu Ling has $5,000 in a savings account earning 12% annual interest. How much will be in her account 10 years from now? (Assume that Lu Ling makes no deposits or withdrawals.)

71. House appreciation   A house appreciates by 6% each year. If the house is worth $70,000 today, how much will it be worth 12 years from now?

72. Motorboat depreciation   A motorboat that cost $5,000 when new depreciates at a rate of 9% per year. How much will the boat be worth in 5 years?

73. Inscribed squares   Each inscribed square in the illustration joins the midpoints of the next larger square. The area of the first square, the largest, is 1 square unit. Find the area of the 12th square.

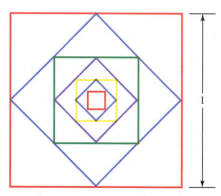

74. Genealogy   The following family tree spans 3 generations and lists 7 people. How many names would be listed in a family tree that spans 10 generations?

75. Annuities   Find the amount of an annuity if $1,000 is paid semiannually for two years at 6% annual interest. Assume that the first of the four payments is made immediately.

76. Annuities   Note that the amounts shown in Table 11-1 form a geometric sequence. Verify the answer for Example 8 by using the formula for the sum of a geometric sequence.

WRITING ABOUT MATH

77. Define a geometric sequence.

78. Develop the formula for finding the sum of the first n terms of a geometric sequence.

SOMETHING TO THINK ABOUT

79. Show that the formula for the sum of the first n terms of a geometric sequence can be found by using the formula
$$S = \frac{a_1 - a_n r}{1 - r}.$$

80. Show that the geometric mean between a and b is $\sqrt{ab}$.

81. If $a > b > 0$, which is larger: the arithmetic mean between a and b or the geometric mean between a and b?

82. Is there a geometric mean between -5 and 5?

83. Show that the formula for the sum of the first n terms of a geometric sequence can be written in the form
$$S_n = \frac{a_1 - a_n r}{1 - r} \quad \text{where } a_n = a_1 r^{n-1}$$

84. Show that the formula for the sum of the first n terms of a geometric sequence can be written in the form
$$S_n = \frac{a_1(1 - r^n)}{1 - r}$$

Section 11.4

Infinite Geometric Sequences and Series

Objectives

1. Find the nth partial sum of an infinite geometric series.
2. Determine whether the sum of an infinite geometric sequence exists.
3. Find the sum of an infinite geometric series, if possible.
4. Write a repeating decimal as a common fraction.

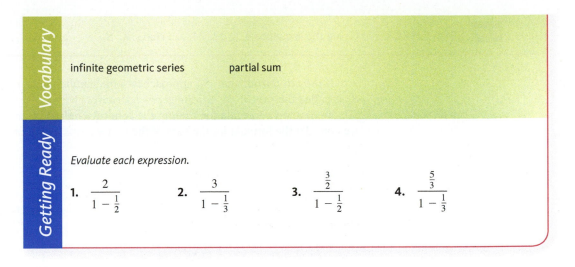

Vocabulary

infinite geometric series   partial sum

Getting Ready

Evaluate each expression.

1. $\dfrac{2}{1-\frac{1}{2}}$ **2.** $\dfrac{3}{1-\frac{1}{3}}$ **3.** $\dfrac{\frac{3}{2}}{1-\frac{1}{2}}$ **4.** $\dfrac{\frac{5}{3}}{1-\frac{1}{3}}$

In this section, we will consider geometric sequences with infinitely many terms.

If we form the sum of the terms of an infinite geometric sequence, we obtain a series called an **infinite geometric series**. For example, if the common ratio r is 3, we have

Infinite geometric sequence	*Infinite geometric series*
2, 6, 18, 54, 162, . . .	$2 + 6 + 18 + 54 + 162 + \cdots$

Under certain conditions, we can find the sum of an infinite geometric series. To define this sum, we consider the geometric series

$$a_1 + a_1r + a_1r^2 + a_1r^3 + \cdots + a_1r^{n-1}$$

- The first **partial sum**, S_1, of the sequence is $S_1 = a_1$.
- The second partial sum, S_2, of the sequence is $S_2 = a_1 + a_1r$.
- The third partial sum, S_3, of the sequence is $S_3 = a_1 + a_1r + a_1r^2$.
- The nth partial sum, S_n, of the sequence is $S_n = a_1 + a_1r + a_1r^2 + \cdots + a_1r^{n-1}$.

1 Find the nth partial sum of an infinite geometric series.

EXAMPLE 1   Given the sequence 2, 6, 18, 54, 162, . . . find
a. S_4 and   **b.** S_7

Solution   We know $a_1 = 2$ and the common ratio is $18 \div 6 = 6 \div 2 = 3$ and we will substitute these values into the formula.

a. $S_n = \dfrac{a_1 - a_1r^n}{1 - r}$

$S_4 = \dfrac{2 - 2(3)^4}{1 - 3}$

$= \dfrac{2 - 2(81)}{-2}$

$= \dfrac{2 - 162}{-2}$

$= \dfrac{-160}{-2}$

$= 80$

The 4th partial sum is 80.

b. $S_n = \dfrac{a_1 - a_1r^n}{1 - r}$

$S_7 = \dfrac{2 - 2(3)^7}{1 - 3}$

$= \dfrac{2 - 2(2{,}187)}{-2}$

$= \dfrac{2 - 4{,}374}{-2}$

$= \dfrac{-4372}{-2}$

$= 2{,}186$

The 7th partial sum is 2,186.

🌿 **SELF CHECK 1** Given the sequence $3, 6, 12, 24, \ldots$ find **a.** S_8 and **b.** S_{15}

If the nth *partial sum*, S_n, approaches some number S_∞ as n approaches infinity, then S_∞ is called the *sum of the infinite geometric series.*

To develop a formula for finding the sum (if it exists) of an infinite geometric series, we consider the formula for the sum of the first n terms of a geometric sequence.

$$S_n = \frac{a_1 - a_1 r^n}{1 - r} \quad (r \neq 1)$$

If $|r| < 1$ and a_1 is constant, then the term $a_1 r^n$ in the above formula approaches 0 as n becomes very large. For example,

$$a_1 \left(\frac{1}{4}\right)^1 = \frac{1}{4} a_1, \qquad a_1 \left(\frac{1}{4}\right)^2 = \frac{1}{16} a_1, \qquad a_1 \left(\frac{1}{4}\right)^3 = \frac{1}{64} a_1$$

and so on. When n is very large, the value of $a_1 r^n$ is negligible, and the term $a_1 r^n$ in the above formula can be ignored. This reasoning justifies the following.

SUM OF AN INFINITE GEOMETRIC SERIES

If a_1 is the first term and r is the common ratio of an infinite geometric sequence, and if $|r| < 1$, the sum of the related geometric series is represented by the formula

$$S_\infty = \frac{a_1}{1 - r}$$

COMMENT Recall that $|r| < 1$ is equivalent to the inequality $-1 < r < 1$. This implies that an infinite geometric series will have a sum if r is between -1 and 1.

2 ## Determine whether the sum of an infinite geometric sequence exists.

To determine whether S_∞ can be calculated, we look at the ratio of the sequence. If $-1 < r < 1$, then the infinite geometric sequence has a sum. If $r < -1$ or $r > 1$, the sum cannot be calculated.

EXAMPLE 2 Determine whether the S_∞ of each infinite geometric sequence can be found.
a. $2, 4, 8, \ldots$ **b.** $3, 1, \frac{1}{3}, \ldots$

Solution We must calculate the common ratio and determine if it falls between -1 and 1.

a. $2, 4, 8, \ldots$
The ratio of the sequence is
$8 \div 4 = 4 \div 2 = 2$. Since 2 does
not fall between -1 and 1, the sum
of the terms of the infinite geometric
sequence cannot be found.

b. $3, 1, \frac{1}{3}, \ldots$
The ratio of the sequence is
$\frac{1}{3} \div 1 = 1 \div 3 = \frac{1}{3}$. Since
$-1 < \frac{1}{3} < 1$, the sum of the terms
of the infinite geometric sequence
can be found.

🌿 **SELF CHECK 2** Determine whether the S_∞ of each infinite geometric sequence can be found.
a. $8, 4, 2, \ldots$ **b.** $-5, 10, -20, \ldots$

In the next example, we will find the S_∞ of an infinite geometric series, if possible.

3   Find the sum of an infinite geometric series, if possible.

EXAMPLE 3   Find the sum of the infinite geometric series $125 + 25 + 5 + \cdots$.

Solution   Here $a_1 = 125$ and $r = \frac{1}{5}$. Since $|r| = \left|\frac{1}{5}\right| = \frac{1}{5} < 1$, we can find the sum of the series.

We substitute 125 for a_1 and $\frac{1}{5}$ for r in the formula $S_\infty = \frac{a_1}{1 - r}$ and simplify:

$$S_\infty = \frac{a_1}{1 - r} = \frac{125}{1 - \frac{1}{5}} = \frac{125}{\frac{4}{5}} = \frac{5}{4}(125) = \frac{625}{4}$$

The sum of the series is $\frac{625}{4}$ or 156.25.

SELF CHECK 3   Find the sum of the infinite geometric series $100 + 20 + 4 + \cdots$.

EXAMPLE 4   Find the sum of the infinite geometric series $64 + (-4) + \frac{1}{4} + \cdots$.

Solution   Here $a_1 = 64$ and $r = -\frac{1}{16}$. Since $|r| = \left|-\frac{1}{16}\right| = \frac{1}{16} < 1$, we can find the sum of the series. We substitute 64 for a_1 and $-\frac{1}{16}$ for r in the formula $S_\infty = \frac{a_1}{1 - r}$ and simplify:

$$S_\infty = \frac{a_1}{1 - r} = \frac{64}{1 - \left(-\frac{1}{16}\right)} = \frac{64}{\frac{17}{16}} = \frac{16}{17}(64) = \frac{1{,}024}{17}$$

The sum of the terms of the infinite geometric sequence $64, -4, \frac{1}{4}, \ldots$ is $\frac{1{,}024}{17}$.

SELF CHECK 4   Find the sum of the infinite geometric series $81 + 27 + 9 + \cdots$.

EXAMPLE 5   Find the sum of the infinite geometric series $\frac{9}{2} + 6 + 8 + \cdots$, if possible.

Solution   Here $a_1 = \frac{9}{2}$ and $r = \frac{4}{3}$. Since $|r| = \left|\frac{4}{3}\right| = \frac{4}{3} > 1$, there is no sum.

SELF CHECK 5   Find the sum of the infinite geometric series $\frac{5}{2} + 5 + 10 + \cdots$, if possible.

4   Write a repeating decimal as a common fraction.

We can use the sum of an infinite geometric series to write a repeating decimal as a common fraction.

EXAMPLE 6   Write $0.\overline{8}$ as a common fraction.

Solution   The decimal $0.\overline{8}$ can be written as an infinite geometric series:

$$0.\overline{8} = 0.888\cdots = \frac{8}{10} + \frac{8}{100} + \frac{8}{1{,}000} + \cdots$$

where $a_1 = \frac{8}{10}$ and $r = \frac{1}{10}$. Because $|r| = \left|\frac{1}{10}\right| = \frac{1}{10} < 1$, we can find the sum as follows:

$$S_\infty = \frac{a_1}{1-r} = \frac{\frac{8}{10}}{1 - \frac{1}{10}} = \frac{\frac{8}{10}}{\frac{9}{10}} = \frac{8}{9}$$

Thus, $0.\overline{8} = \frac{8}{9}$. Long division will verify that $\frac{8}{9} = 0.888. \ldots$

 SELF CHECK 6   Write $0.\overline{5}$ as a common fraction.

EXAMPLE 7   Write $0.\overline{25}$ as a common fraction.

Solution   The decimal $0.\overline{25}$ can be written as an infinite geometric series:

$$0.\overline{25} = 0.252525\ldots = \frac{25}{100} + \frac{25}{10,000} + \frac{25}{1,000,000} + \cdots$$

where $a_1 = \frac{25}{100}$ and $r = \frac{1}{100}$. Since $|r| = \left|\frac{1}{100}\right| = \frac{1}{100} < 1$, we can find the sum as follows:

$$S_\infty = \frac{a_1}{1-r} = \frac{\frac{25}{100}}{1 - \frac{1}{100}} = \frac{\frac{25}{100}}{\frac{99}{100}} = \frac{25}{99}$$

Thus, $0.\overline{25} = \frac{25}{99}$. Long division will verify that this is true.

 SELF CHECK 7   Write $0.\overline{15}$ as a common fraction.

SELF CHECK ANSWERS   **1. a.** 765   **b.** 98,301   **2. a.** yes   **b.** no   **3.** 125   **4.** $\frac{243}{2}$ **5.** no sum because $r > 1$ **6.** $\frac{5}{9}$ **7.** $\frac{5}{33}$

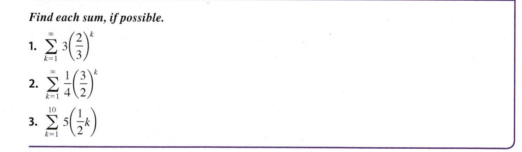

NOW TRY THIS

Find each sum, if possible.

1. $\displaystyle\sum_{k=1}^{\infty} 3\left(\frac{2}{3}\right)^k$

2. $\displaystyle\sum_{k=1}^{\infty} \frac{1}{4}\left(\frac{3}{2}\right)^k$

3. $\displaystyle\sum_{k=1}^{10} 5\left(\frac{1}{2}k\right)$

11.4   Exercises

WARM-UPS *Find the common ratio in each infinite geometric sequence.*

1. $2, 6, 18, \ldots$

2. $100, 50, 25, \ldots$

3. $50, 10, 2, \ldots$

4. $\dfrac{1}{50}, \dfrac{1}{10}, \dfrac{1}{2}, \ldots$

5. $1, \dfrac{1}{5}, \dfrac{1}{25}, \ldots$

6. $\dfrac{2}{5}, \dfrac{1}{5}, \dfrac{1}{10}, \ldots$

Determine whether each equation determines y to be a function of x.

7. $y = -2x^3 + 5$

8. $xy = 12$

9. $3x = y^2 + 4$

10. $x = |y|$

VOCABULARY AND CONCEPTS   *Fill in the blanks.*

11. If a geometric sequence has infinitely many terms, it is called an _____ geometric sequence.

12. The third partial sum of the series $2 + 6 + 18 + 54 + \cdots$ is _____ $= 26$.

13. The formula for the sum of an infinite geometric series with $|r| < 1$ is _____.

14. Write $0.\overline{7}$ as an infinite geometric series.

_____

GUIDED PRACTICE   *Find the sum of each infinite geometric series.* SEE EXAMPLE 3. (OBJECTIVE 3)

15. $48 + 12 + 3 + \cdots$

16. $12 + 6 + 3 + \cdots$

17. $200 + 40 + 8 + \cdots$

18. $45 + 15 + 5 + \cdots$

Find the sum of each infinite geometric series. SEE EXAMPLE 4. (OBJECTIVE 3)

19. $12 + (-6) + 3 + \cdots$

20. $8 + (-4) + 2 + \cdots$

21. $-81 + 27 + (-9) + \cdots$

22. $-45 + 15 + (-5) + \cdots$

Find the sum of each infinite geometric series, if possible. SEE EXAMPLE 5. (OBJECTIVE 3)

23. $-3 + (-6) + (-12) + \cdots$

24. $4 + 12 + 36 + \cdots$

25. $\dfrac{3}{4} + \dfrac{3}{2} + 3 + \cdots$

26. $-\dfrac{5}{6} + \left(-\dfrac{5}{3}\right) + \left(-\dfrac{10}{3}\right) + \cdots$

Write each decimal as a common fraction. Then check the answer by using long division. SEE EXAMPLE 6. (OBJECTIVE 4)

27. $0.\overline{1}$

28. $0.\overline{2}$

29. $-0.\overline{3}$

30. $-0.\overline{4}$

Write each decimal as a common fraction. Then check the answer by using long division. SEE EXAMPLE 7. (OBJECTIVE 4)

31. $0.\overline{12}$

32. $0.\overline{21}$

33. $0.\overline{75}$

34. $0.\overline{57}$

ADDITIONAL PRACTICE   *Find the sum of each infinite geometric series, if possible.*

35. $-54 + 18 + (-6) + \cdots$

36. $-112 + (-28) + (-7) + \cdots$

37. $-\dfrac{27}{2} + (-9) + (-6) + \cdots$

38. $\dfrac{18}{25} + \dfrac{6}{5} + 2 + \cdots$

APPLICATIONS   *Solve each application.*

39. Controlling moths To reduce the population of a destructive moth, biologists release 1,000 sterilized male moths each day into the environment. If 80% of these moths alive one day survive until the next, then the population of sterile males is the sum of the infinite geometric sequence

$$1,000 + 1,000(0.8) + 1,000(0.8)^2 + 1,000(0.8)^3 + \cdots$$

Find the long-term population.

40. Controlling moths If mild weather increases the day-to-day survival rate of the sterile male moths in Exercise 39 to 90%, find the long-term population.

41. Bouncing balls On each bounce, the rubber ball in the illustration rebounds to a height one-half of that from which it fell. Find the total distance the ball travels.

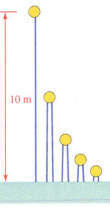

10 m

42. Bouncing balls A golf ball is dropped from a height of 12 feet. On each bounce, it returns to a height two-thirds of that from which it fell. Find the total distance the ball travels.

WRITING ABOUT MATH

43. Why must the absolute value of the common ratio be less than 1 before an infinite geometric series can have a sum?

44. Can an infinite arithmetic series have a sum?

SOMETHING TO THINK ABOUT

45. An infinite geometric series has a sum of 5 and a first term of 1. Find the common ratio.

46. An infinite geometric series has a common ratio of $-\dfrac{2}{3}$ and a sum of 9. Find the first term.

47. Show that $0.\overline{9} = 1$.

48. Show that $1.\overline{9} = 2$.

49. Does $0.999999 = 1$? Explain.

50. If $f(x) = 1 + x + x^2 + x^3 + x^4 + \cdots$, find $f\left(\dfrac{1}{2}\right)$ and $f\left(-\dfrac{1}{2}\right)$.

Section 11.5

Permutations and Combinations

Objectives

1. Use the multiplication principle to determine the number of ways one event can be followed by another.
2. Use permutations to find the number of n things taken r at a time if order is important.
3. Use combinations to find the number of n things taken r at a time.
4. Use combinations to find the coefficients of the terms of a binomial expansion.

Vocabulary

tree diagram   multiplication principle   permutation
event   for events   combination

Getting Ready

Evaluate each expression.

1. $4 \cdot 3 \cdot 2 \cdot 1$

2. $5 \cdot 4 \cdot 3 \cdot 2 \cdot 1$

3. $\dfrac{6 \cdot 5 \cdot 4 \cdot 3 \cdot 2 \cdot 1}{4 \cdot 3 \cdot 2 \cdot 1}$

4. $\dfrac{8 \cdot 7 \cdot 6 \cdot 5 \cdot 4 \cdot 3 \cdot 2 \cdot 1}{2(5 \cdot 4 \cdot 3 \cdot 2 \cdot 1)}$

In this section, we will discuss methods of counting the different ways we can do something like arranging books on a shelf or selecting a committee. These kinds of problems are important in statistics, insurance, telecommunications, and other fields.

1   **Use the multiplication principle to determine the number of ways one event can be followed by another.**

Steven goes to the cafeteria for lunch. He has a choice of three different sandwiches (hamburger, hot dog, or ham and cheese) and four different beverages (cola, root beer, orange, or milk). How many different lunches can he choose?

He has three choices of sandwich, and for any one of these choices, he has four choices of drink. The different options are shown in the **tree diagram** in Figure 11-1.

Figure 11-1

The tree diagram illustrates that he has a total of 12 different lunches from which to choose. One of the possibilities is a hamburger with a cola, and another is a hot dog with milk.

A situation that can have several different outcomes—such as choosing a sandwich—is called an **event**. Choosing a sandwich and choosing a beverage can be thought of as two events. The preceding example illustrates the **multiplication principle for events**.

MULTIPLICATION PRINCIPLE FOR EVENTS	Let E_1 and E_2 be two events. If E_1 can be done in a_1 ways, and if—after E_1 has occurred—E_2 can be done in a_2 ways, the event "E_1 followed by E_2" can be done in $a_1 \cdot a_2$ ways.

EXAMPLE 1   After dinner, Heidi plans to watch the evening news and then a situation comedy on TV. If there are choices of four news broadcasts and two comedies, in how many ways can she choose to watch television?

Solution   Let E_1 be the event "watching the news" and E_2 be the event "watching a comedy." Because there are four ways to accomplish E_1 and two ways to accomplish E_2, the number of choices that she has is $4 \cdot 2 = 8$.

SELF CHECK 1   If Josie has 7 shirts and 5 pairs of pants, how many outfits could be created?

The multiplication principle can be extended to any number of events. In Example 2, we use it to complete the number of ways that we can arrange objects in a row.

EXAMPLE 2   In how many ways can we arrange five books on a shelf?

Solution   We can fill the first space with any of the 5 books, the second space with any of the remaining 4 books, the third space with any of the remaining 3 books, the fourth space with any of the remaining 2 books, and the fifth space with the remaining 1 (or last) book. By the multiplication principle for events, the number of ways in which the books can be arranged is

$$5 \cdot 4 \cdot 3 \cdot 2 \cdot 1 = 120$$

SELF CHECK 2   In how many ways can 4 men line up in a row?

EXAMPLE 3   If Juan has six flags, each of a different color, to hang on a flagpole, how many different arrangements can he display by using four flags?

Solution   Juan must find the number of arrangements of 4 flags when there are 6 flags from which to choose. He can hang any one of the 6 flags in the top position, any one of the remaining 5 flags in the second position, any one of the remaining 4 flags in the third position, and any one of the remaining 3 flags in the lowest position. By the multiplication principle for events, the total number of arrangements is

$$6 \cdot 5 \cdot 4 \cdot 3 = 360$$

SELF CHECK 3   How many different arrangements can he display if each uses three flags?

2   Use permutations to find the number of n things taken r at a time if order is important.

When computing the number of possible arrangements of objects such as books on a shelf or flags on a pole, we are finding the number of **permutations** of those objects. In these cases, *order is important.* A blue flag followed by a yellow flag is different than a yellow flag followed by a blue flag.

In Example 2, we found that the number of permutations of five books, using all five of them, is 120. In Example 3, we found that the number of permutations of six flags, using four of them, is 360.

The symbol $P(n, r)$, read as "the number of permutations of n things taken r at a time," is often used to express permutation situations. In Example 2, we found that $P(5, 5) = 120$. In Example 3, we found that $P(6, 4) = 360$.

EXAMPLE 4   If Sarah has seven flags, each of a different color, to hang on a flagpole, how many different arrangements can she display by using three flags?

Solution   She must find $P(7, 3)$ (the number of permutations of 7 things taken 3 at a time). In the top position Sarah can hang any of the 7 flags, in the middle position any one of the remaining 6 flags, and in the bottom position any one of the remaining 5 flags. According to the multiplication principle for events,

$$P(7, 3) = 7 \cdot 6 \cdot 5 = 210$$

She can display 210 arrangements using only three of the seven flags.

SELF CHECK 4   How many different arrangements can Sarah display using four flags?

COMMENT   Refer to the calculator tear out card for instructions on finding permutations and combinations using your calculator.

Although it is correct to write $P(7, 3) = 7 \cdot 6 \cdot 5$, there is an advantage in changing the form of this answer to obtain a formula for computing $P(7, 3)$:

$$P(7, 3) = 7 \cdot 6 \cdot 5$$

$$= \frac{7 \cdot 6 \cdot 5 \cdot 4 \cdot 3 \cdot 2 \cdot 1}{4 \cdot 3 \cdot 2 \cdot 1} \qquad \text{Multiply both the numerator and denominator by } 4 \cdot 3 \cdot 2 \cdot 1.$$

$$= \frac{7!}{4!}$$

$$= \frac{7!}{(7 - 3)!}$$

The generalization of this process provides the following formula.

COMPUTING $P(n, r)$

The number of permutations of n things taken r at a time is given by the formula

$$P(n, r) = \frac{n!}{(n - r)!}$$

EXAMPLE 5   Compute: **a.** $P(8, 2)$ **b.** $P(7, 5)$ **c.** $P(n, n)$ **d.** $P(n, 0)$

Solution   We substitute into the permutation formula $P(n, r) = \frac{n!}{(n - r)!}$.

a. $P(8, 2) = \dfrac{8!}{(8 - 2)!}$

$\qquad = \dfrac{8 \cdot 7 \cdot 6!}{6!}$

$\qquad = 8 \cdot 7$

$\qquad = 56$

b. $P(7, 5) = \dfrac{7!}{(7 - 5)!}$

$\qquad = \dfrac{7 \cdot 6 \cdot 5 \cdot 4 \cdot 3 \cdot 2!}{2!}$

$\qquad = 7 \cdot 6 \cdot 5 \cdot 4 \cdot 3$

$\qquad = 2,520$

c. $P(n, n) = \dfrac{n!}{(n - n)!}$

$= \dfrac{n!}{0!}$

$= \dfrac{n!}{1}$

$= n!$

d. $P(n, 0) = \dfrac{n!}{(n - 0)!}$

$= \dfrac{n!}{n!}$

$= 1$

 SELF CHECK 5   Compute: **a.** $P(10, 6)$ **b.** $P(10, 0)$

Parts **c** and **d** of Example 5 establish the following formulas.

> **FINDING $P(n, n)$ AND $P(n, 0)$**
>
> The number of permutations of n things taken n at a time and n things taken 0 at a time are given by the formulas
>
> $$P(n, n) = n! \quad \text{and} \quad P(n, 0) = 1$$

EXAMPLE 6   **TV PROGRAMMING   a.** In how many ways can a TV executive arrange the Saturday night lineup of 6 programs if there are 15 programs from which to choose?   **b.** If there are only 6 programs from which to choose?

Solution   **a.** To find the number of permutations of 15 programs taken 6 at a time, we will use the formula $P(n, r) = \dfrac{n!}{(n - r)!}$ with $n = 15$ and $r = 6$.

$$P(15, 6) = \dfrac{15!}{(15 - 6)!}$$

$$= \dfrac{15 \cdot 14 \cdot 13 \cdot 12 \cdot 11 \cdot 10 \cdot 9!}{9!}$$

$$= 15 \cdot 14 \cdot 13 \cdot 12 \cdot 11 \cdot 10$$

$$= 3{,}603{,}600$$

b. To find the number of permutations of 6 programs taken 6 at a time, we use the formula $P(n, n) = n!$ with $n = 6$.

$$P(6, 6) = 6! = 720$$

SELF CHECK 6   How many ways can the executive select 6 programs if there are 20 programs from which to choose?

3   **Use combinations to find the number of n things taken r at a time.**

Suppose that a student must read 4 books from a reading list of 10 books. The order in which he reads them is not important. For the moment, however, let's assume that order is important and find the number of permutations of 10 things taken 4 at a time.

$$P(10, 4) = \dfrac{10!}{(10 - 4)!}$$

$$= \dfrac{10 \cdot 9 \cdot 8 \cdot 7 \cdot 6!}{6!}$$

$$= 10 \cdot 9 \cdot 8 \cdot 7$$

$$= 5{,}040$$

If order is important, there are 5,040 ways of choosing 4 books when there are 10 books from which to choose. However, because the order in which the student reads the books does not matter, the previous result of 5,040 is too large. Since there are 24 (or 4!) ways of ordering the 4 books that are chosen, the result of 5,040 is exactly 24 (or 4!) times too large. Therefore, the number of choices that the student has is the number of permutations of 10 things taken 4 at a time, divided by 24:

$$\frac{P(10,4)}{24} = \frac{5,040}{24} = 210$$

The student has 210 ways of choosing 4 books to read from the list of 10 books.

In situations where *order is not important*, we are interested in **combinations**, not permutations. The symbols $C(n,r)$ and $\binom{n}{r}$ both mean the number of combinations of n things taken r at a time.

If a selection of r books is chosen from a total of n books, the number of possible selections is $C(n,r)$ and there are $r!$ arrangements of the r books in each selection. If we consider the selected books as an ordered grouping, the number of orderings is $P(n,r)$. Therefore, we have

(1) $r! \cdot C(n,r) = P(n,r)$

We can divide both sides of Equation 1 by $r!$ to obtain the formula for finding $C(n,r)$:

$$C(n,r) = \binom{n}{r} = \frac{P(n,r)}{r!} = \frac{n!}{r!(n-r)!}$$

FINDING $C(n,r)$

The number of combinations of n things taken r at a time is given by

$$C(n,r) = \frac{n!}{r!(n-r)!}$$

EXAMPLE 7   Compute: **a.** $C(8,5)$ **b.** $\binom{7}{2}$ **c.** $C(n,n)$ **d.** $C(n,0)$

Solution   We will substitute into the combination formula $C(n,r) = \frac{n!}{r!(n-r)!}$.

a. $C(8,5) = \dfrac{8!}{5!(8-5)!}$

$\qquad\quad = \dfrac{8 \cdot 7 \cdot 6 \cdot 5!}{5! \cdot 3!}$

$\qquad\quad = 8 \cdot 7$

$\qquad\quad = 56$

b. $\binom{7}{2} = \dfrac{7!}{2!(7-2)!}$

$\qquad\quad = \dfrac{7 \cdot 6 \cdot 5!}{2 \cdot 1 \cdot 5!}$

$\qquad\quad = 21$

c. $C(n,n) = \dfrac{n!}{n!(n-n)!}$

$\qquad\quad = \dfrac{n!}{n!(0!)}$

$\qquad\quad = \dfrac{n!}{n!(1)}$

$\qquad\quad = 1$

d. $C(n,0) = \dfrac{n!}{0!(n-0)!}$

$\qquad\quad = \dfrac{n!}{0! \cdot n!}$

$\qquad\quad = \dfrac{1}{0!}$

$\qquad\quad = \dfrac{1}{1}$

$\qquad\quad = 1$

The symbol $C(n,0)$ indicates that we choose 0 things from the available n things.

SELF CHECK 7   Compute: **a.** $C(9,6)$ **b.** $C(10,10)$

Parts **c** and **d** of Example 7 establish the following formulas.

FINDING $C(n, n)$ AND $C(n, 0)$	The number of combinations of n things taken n at a time is 1. The number of combinations of n things taken 0 at a time is 1.

$$C(n,n) = 1 \quad \text{and} \quad C(n,0) = 1$$

EXAMPLE 8 **SELECTING COMMITTEES** If 15 students want to select a committee of 4 students, how many different committees are possible?

Solution Since the ordering of people on each possible committee is not important, we find the number of combinations of 15 people selected 4 at a time:

$$C(15,4) = \frac{15!}{4!(15 - 4)!}$$

$$= \frac{15 \cdot 14 \cdot 13 \cdot 12 \cdot 11!}{4 \cdot 3 \cdot 2 \cdot 1 \cdot 11!}$$

$$= \frac{15 \cdot 14 \cdot 13 \cdot 12}{4 \cdot 3 \cdot 2 \cdot 1}$$

$$= 1,365$$

There are 1,365 possible committees.

 SELF CHECK 8 In how many ways can 20 students select a committee of 5 students?

EXAMPLE 9 **CONGRESS** A committee in Congress consists of ten Democrats and eight Republicans. In how many ways can a subcommittee be chosen if it is to contain five Democrats and four Republicans?

Solution There are $C(10, 5)$ ways of choosing the 5 Democrats and $C(8, 4)$ ways of choosing the 4 Republicans. By the multiplication principle for events, there are $C(10, 5) \cdot C(8, 4)$ ways of choosing the subcommittee:

$$C(10,5) \cdot C(8,4) = \frac{10!}{5!(10 - 5)!} \cdot \frac{8!}{4!(8 - 4)!}$$

$$= \frac{10 \cdot 9 \cdot 8 \cdot 7 \cdot 6 \cdot 5!}{120 \cdot 5!} \cdot \frac{8 \cdot 7 \cdot 6 \cdot 5 \cdot 4!}{24 \cdot 4!}$$

$$= \frac{10 \cdot 9 \cdot 8 \cdot 7 \cdot 6}{120} \cdot \frac{8 \cdot 7 \cdot 6 \cdot 5}{24}$$

$$= 17,640$$

There are 17,640 possible subcommittees.

SELF CHECK 9 In how many ways can a subcommittee be chosen if it is to contain four members from each party?

4 Use combinations to find the coefficients of the terms of a binomial expansion.

We have seen that the expansion of $(x + y)^3$ is

$$(x + y)^3 = \mathbf{1}x^3 + \mathbf{3}x^2y + \mathbf{3}xy^2 + \mathbf{1}y^3$$

and that

$$\binom{3}{0} = 1, \quad \binom{3}{1} = 3, \quad \binom{3}{2} = 3, \quad \text{and} \quad \binom{3}{3} = 1$$

Putting these facts together provides the following way of writing the expansion of $(x + y)^3$.

$$(x + y)^3 = \binom{3}{0}x^3 + \binom{3}{1}x^2y + \binom{3}{2}xy^2 + \binom{3}{3}y^3$$

Likewise, we have

$$(x + y)^4 = \binom{4}{0}x^4 + \binom{4}{1}x^3y + \binom{4}{2}x^2y^2 + \binom{4}{3}xy^3 + \binom{4}{4}y^4$$

The generalization of this idea allows us to state the binomial theorem using combinatorial notation.

THE BINOMIAL THEOREM

If n is any positive integer, then

$$(a + b)^n = \binom{n}{0}a^n + \binom{n}{1}a^{n-1}b + \binom{n}{2}a^{n-2}b^2 + \cdots + \binom{n}{r}a^{n-r}b^r + \cdots + \binom{n}{n}b^n$$

EXAMPLE 10   Use the combinatorial form of the binomial theorem to expand $(x + y)^6$.

Solution $(x + y)^6 = \binom{6}{0}x^6 + \binom{6}{1}x^5y + \binom{6}{2}x^4y^2 + \binom{6}{3}x^3y^3 + \binom{6}{4}x^2y^4 + \binom{6}{5}xy^5 + \binom{6}{6}y^6$

$= x^6 + 6x^5y + 15x^4y^2 + 20x^3y^3 + 15x^2y^4 + 6xy^5 + y^6$

SELF CHECK 10   Use the combinatorial form of the binomial theorem to expand $(a + b)^2$.

EXAMPLE 11   Use the combinatorial form of the binomial theorem to expand $(2x - y)^3$.

Solution $(2x - y)^3 = [2x + (-y)]^3$

$= \binom{3}{0}(2x)^3 + \binom{3}{1}(2x)^2(-y) + \binom{3}{2}(2x)(-y)^2 + \binom{3}{3}(-y)^3$

$= 1(2x)^3 + 3(4x^2)(-y) + 3(2x)(y^2) + 1(-y)^3$

$= 8x^3 - 12x^2y + 6xy^2 - y^3$

SELF CHECK 11   Use the combinatorial form of the binomial theorem to expand $(3a + b)^3$.

SELF CHECK ANSWERS
1. 35   2. 24   3. 120   4. 840   5. a. 151,200   b. 1   6. 27,907,200   7. a. 84   b. 1
8. 15,504   9. 14,700   10. $a^2 + 2ab + b^2$ 11. $27a^3 + 27a^2b + 9ab^2 + b^3$

NOW TRY THIS

1. In Texas, license plate numbering for passenger cars has often been changed to accommodate a growing population and the fact that new plates must be issued every 8 years. Find the greatest number of license plates that could be issued during each of these periods. A or B represents a letter and 0 or 1 represents a number.

1975–1982	AAA-000	(I or O is not used)
1982–1990	000-AAA	(I or O is not used)
1990–1998	BBB-00B	(no vowels or Q used)
1998–2004	B00-BBB	(no vowels or Q used)
2004–2007	000-BBB	(no vowels or Q used)
2007–mid-2009	BBB-000	(no vowels or Q used)

2. Find the greatest number of license plates that could be distributed from 1975–mid-2009.

3. Beginning in mid-2009, Texas began issuing 7-digit license numbers of the form BB0-B000 (no vowels or Q is used). How many license plates can be issued using this form?

4. Personalized license plates can be purchased from an authorized third party with the following specifications. The plate can consist of 1–6 letters or numbers including no more than 2 symbols (dash, space, period, @ for heart, or * for the state icon). A letter or number must be in the first position. How many plates could be created within these parameters?

Source: http://www.licenseplateinfo.com/txchart/pass-tables.html

11.5 Exercises

WARM-UPS *Simplify.*

1. $\dfrac{7!}{3!}$

2. $\dfrac{6!}{4!}$

3. $\dfrac{5!}{1!}$

4. $\dfrac{5!}{0!}$

5. $\dfrac{8!}{3!5!}$

6. $\dfrac{9!}{5!4!}$

REVIEW *Find each value of x. Assume no division by 0.*

7. $|2x - 7| = 17$

8. $2x^2 - x = 15$

9. $\dfrac{3}{x - 5} = \dfrac{8}{x}$

10. $\dfrac{3}{x} = \dfrac{x - 2}{8}$

VOCABULARY AND CONCEPTS *Fill in the blanks.*

11. If an event E_1 can be done in p ways and, after it occurs, a second event E_2 can be done in q ways, the event E_1 followed by E_2 can be done in ____ ways.

12. We can use a ____ diagram to illustrate the multiplication principle.

13. A _____ is an arrangement of objects in which order matters.

14. The symbol _____ means the number of permutations of n things taken r at a time.

15. The formula for the number of permutations of n things taken r at a time is _____.

16. $P(n, n) = $ __ and $P(n, 0) = $ __.

17. A _____ is an arrangement of objects in which order does not matter.

18. The symbol $C(n, r)$ or ___ means the number of _____ of n things taken r at a time.

19. The formula for the number of combinations of n things taken r at a time is _____.

20. $C(n, n) = $ __ and $C(n, 0) = $ __.

GUIDED PRACTICE *Evaluate using permutations.*
SEE EXAMPLE 5. (OBJECTIVE 2)

21. $P(5, 5)$

22. $P(4, 4)$

23. $P(6, 2)$

24. $P(3, 2)$

25. $P(4, 2) \cdot P(5, 2)$

26. $P(n, n) \cdot P(n, 0)$

27. $\dfrac{P(n, n)}{P(n, 0)}$

28. $\dfrac{P(6, 2)}{P(5, 4)}$

Evaluate using combinations. SEE EXAMPLE 7. (OBJECTIVE 3)

29. $C(4, 3)$ **30.** $C(7, 5)$

31. $\binom{8}{5}$ **32.** $\binom{6}{4}$

33. $\binom{7}{4}\binom{7}{5}$ **34.** $\binom{9}{8}\binom{9}{7}$

35. $\dfrac{C(38, 37)}{C(19, 18)}$ **36.** $\dfrac{C(25, 23)}{C(40, 39)}$

Use the combinatorial form of the binomial theorem to expand each binomial. SEE EXAMPLE 10. (OBJECTIVE 4)

37. $(x + y)^4$

38. $(x + y)^8$

39. $(a + b)^5$

40. $(a + b)^7$

Use the combinatorial form of the binomial theorem to expand each binomial. SEE EXAMPLE 11. (OBJECTIVE 4)

41. $(2x + y)^3$

42. $(2x + 1)^4$

43. $(4x - y)^3$

44. $(x - 2y)^5$

ADDITIONAL PRACTICE   *Evaluate.*

45. $C(12, 0) \cdot C(12, 12)$ **46.** $\dfrac{C(8, 0)}{C(8, 1)}$

47. $\dfrac{P(6, 2) \cdot P(7, 3)}{P(5, 1)}$ **48.** $\dfrac{P(8, 3)}{P(5, 3) \cdot P(4, 3)}$

49. $C(n, 2)$ **50.** $C(n, 3)$

Expand the binomial.

51. $(3x - 2)^4$

52. $(3 - x^2)^3$

Find the indicated term of the binomial expansion.

53. $(x - 5y)^5$; fourth term

54. $(2x - y)^5$; third term

55. $(x^2 - y^3)^4$; second term

56. $(x^3 - y^2)^4$; fourth term

APPLICATIONS   *Solve each application. SEE EXAMPLES 1–3. (OBJECTIVE 1)*

57. Arranging an evening   Kyoro plans to go to dinner and see a movie. In how many ways can she arrange her evening if she has a choice of five movies and seven restaurants?

58. Travel choices   Paula has five ways to travel from New York to Chicago, three ways to travel from Chicago to Denver, and four ways to travel from Denver to Los Angeles. How many choices are available to Paula if she travels from New York to Los Angeles?

59. Arranging books   In how many ways can seven books be placed on a shelf?

60. Lining up   In how many ways can the people shown be placed in a line?

Solve each application. SEE EXAMPLES 4 AND 6. (OBJECTIVE 2)

61. Making license plates   How many six-digit license plates can be manufactured? Note that there are ten choices—0, 1, 2, 3, 4, 5, 6, 7, 8, 9—for each digit.

62. Making license plates   How many six-digit license plates can be manufactured if no digit can be repeated?

63. Making license plates   How many six-digit license plates can be manufactured if no license can begin with 0 and if no digit can be repeated?

64. Making license plates   How many license plates can be manufactured with two letters followed by four digits?

65. Phone numbers   How many seven-digit phone numbers are available in area code 815 if no phone number can begin with 0 or 1?

66. Phone numbers   How many ten-digit phone numbers are available if area codes 000 and 911 cannot be used and if no local number can begin with 0 or 1?

67. Arranging books   In how many ways can four novels and five biographies be arranged on a shelf if the novels are placed on the left?

68. Making a ballot   In how many ways can six candidates for mayor and four candidates for the county board be arranged on a ballot if all of the candidates for mayor must be placed on top?

69. Combination locks   How many permutations does a combination lock have if each combination has three numbers, no two numbers of any combination are equal, and the lock has 25 numbers?

70. Combination locks   How many permutations does a combination lock have if each combination has three numbers, no two numbers of any combination are equal, and the lock has 50 numbers?

71. Arranging appointments   The receptionist at a dental office has only three appointment times available before Tuesday, and ten patients have toothaches. In how many ways can the receptionist fill those appointments?

72. Computers   In many computers, a word consists of 32 *bits*—a string of thirty-two 1's and 0's. How many different words are possible?

73. Palindromes   A palindrome is any word, such as *madam* or *radar*, that reads the same backward and forward. How many five-digit numerical palindromes (such as 13531) are there? (*Hint*: A leading 0 would be dropped.)

74. Call letters The call letters of U.S. commercial radio stations have 3 or 4 letters, and the first is always a W or a K. How many radio stations could this system support?

Solve each application. SEE EXAMPLES 8–9. (OBJECTIVE 3)

75. Planning a picnic A class of 14 students wants to pick a committee of 3 students to plan a picnic. How many committees are possible?

76. Choosing books Jeff must read 3 books from a reading list of 15 books. How many choices does he have?

77. Forming committees The number of three-person committees that can be formed from a group of people is ten. How many people are in the group?

78. Forming committees The number of three-person committees that can be formed from a group of people is 20. How many people are in the group?

79. Winning a lottery In one state lottery, anyone who picks the correct 6 numbers (in any order) wins. With the numbers 0 through 99 available, how many choices are possible?

80. Taking a test The instructions on a test read: "Answer any ten of the following fifteen questions. Then choose one of the remaining questions for homework and turn in its solution tomorrow." In how many ways can the questions be chosen?

81. Forming a committee In how many ways can we select a committee of two men and two women from a group containing three men and four women?

82. Forming a committee In how many ways can we select a committee of three women and two men from a group containing five women and three men?

83. Choosing clothes In how many ways can we select 2 shirts and 3 neckties from a group of 12 shirts and 10 neckties?

84. Choosing clothes In how many ways can we select five dresses and two coats from a wardrobe containing nine dresses and three coats?

WRITING ABOUT MATH

85. State the multiplication principle for events.

86. Explain why *permutation lock* would be a better name for a combination lock.

SOMETHING TO THINK ABOUT

87. How many ways could five people stand in line if two people insist on standing together?

88. How many ways could five people stand in line if two people refuse to stand next to each other?

Section 11.6

Probability

Objectives

1. Find a sample space for an experiment.
2. Find the probability of an event.

Vocabulary

probability experiment   sample space   event

Getting Ready

Answer each question.

1. What are the possible outcomes on one roll of one die?

2. What are the possible outcomes after flipping a single coin one time?

The **probability** that an event will occur is a measure of the likelihood of that event. A tossed coin, for example, can land in two ways, either heads or tails. Because one of these two equally likely outcomes is heads, we expect that out of several tosses, about half will be heads. We say that the probability of obtaining heads in a single toss of the coin is $\frac{1}{2}$.

If records show that out of 100 days with weather conditions like today's, 30 have received rain, the weather service will report, "There is a $\frac{30}{100}$ or 30% probability of rain today."

1   Find a sample space for an experiment.

Activities such as tossing a coin, rolling a die, drawing a card, and predicting rain are called **experiments**. For any experiment, a list of all possible outcomes is called a **sample space**. For example, the sample space S for the experiment of tossing two coins is the set

$$S = \{(H, H), (H, T), (T, H), (T, T)\} \quad \text{There are four possible outcomes.}$$

where the ordered pair (H, T) represents the outcome "heads on the first coin and tails on the second coin."

EXAMPLE 1   List the sample space of the experiment "rolling two dice a single time."

Solution   We can list ordered pairs and let the first number be the result on the first die and the second number the result on the second die. The sample space S is the following set of ordered pairs:

$$(1, 1)\ (1, 2)\ (1, 3)\ (1, 4)\ (1, 5)\ (1, 6)$$
$$(2, 1)\ (2, 2)\ (2, 3)\ (2, 4)\ (2, 5)\ (2, 6)$$
$$(3, 1)\ (3, 2)\ (3, 3)\ (3, 4)\ (3, 5)\ (3, 6)$$
$$(4, 1)\ (4, 2)\ (4, 3)\ (4, 4)\ (4, 5)\ (4, 6)$$
$$(5, 1)\ (5, 2)\ (5, 3)\ (5, 4)\ (5, 5)\ (5, 6)$$
$$(6, 1)\ (6, 2)\ (6, 3)\ (6, 4)\ (6, 5)\ (6, 6)$$

By counting, we see that the experiment has 36 equally likely possible outcomes.

SELF CHECK 1   List the sample space of the experiment "flipping one coin followed by rolling 1 die."

2   Find the probability of an event.

An **event** is a subset of the sample space of an experiment. For example, if E is the event "getting at least one heads" in the experiment of tossing two coins, then

$$E = \{(H, H), (H, T), (T, H)\} \quad \text{There are 3 ways of getting at least one heads.}$$

Because the outcome of getting at least one heads can occur in 3 out of 4 possible ways, we say that the *probability* of E is $\frac{3}{4}$, and we write

$$P(E) = P(\text{at least one heads}) = \frac{3}{4}$$

PROBABILITY OF AN EVENT	If a sample space of an experiment has n distinct and equally likely outcomes and E is an event that occurs in s of those ways, the *probability of E* is $$P(E) = \frac{s}{n}$$

Since $0 \le s \le n$, it follows that $0 \le \frac{s}{n} \le 1$. This implies that all probabilities have a value from 0 to 1. If an event cannot happen, its probability is 0. If an event is certain to happen, its probability is 1.

EXAMPLE 2   Find the probability of the event "rolling a sum of 7 on one roll of two dice."

Solution   In the sample space listed in Example 1, the following ordered pairs give a sum of 7:

$$\{(1,6), (2,5), (3,4), (4,3), (5,2), (6,1)\}$$

Since there are 6 ordered pairs whose numbers give a sum of 7 out of a total of 36 equally likely outcomes, we have

$$P(E) = P(\text{rolling a sum of 7}) = \frac{s}{n} = \frac{6}{36} = \frac{1}{6}$$

SELF CHECK 2   Find the probability of rolling a sum of 4.

COMMENT   A standard playing deck of 52 cards has two red suits, hearts and diamonds, and two black suits, clubs and spades. Each suit has 13 cards, including the ace, face cards (king, queen, and jack), and numbers 2 through 10. We will refer to a standard deck of cards in many examples and exercises.

EXAMPLE 3   Find the probability of drawing an ace on one draw from a standard card deck.

Solution   Since there are 4 aces in the deck, the number of favorable outcomes is $s = 4$. Since there are 52 cards in the deck, the total number of possible outcomes is $n = 52$. The probability of drawing an ace is the ratio of the number of favorable outcomes to the number of possible outcomes.

$$P(\text{an ace}) = \frac{s}{n} = \frac{4}{52} = \frac{1}{13}$$

The probability of drawing an ace is $\frac{1}{13}$.

SELF CHECK 3   Find the probability of drawing a red ace on one draw from a standard card deck.

Everyday connections
Winning the Lottery

©iStockphoto.com/David Birkbeck

Suppose a certain state lottery has the following Mega Millions design.

- A spinning machine contains 49 different balls. Each ball is labeled with exactly one number from the list $\{1, 2, 3, \ldots, 49\}$. The spinning machine ensures that each ball has an equal chance of being selected.

- Six different balls are drawn, without replacement, from the spinning machine.

- Each lottery ticket has six different numbers from the list $\{1, 2, 3, \ldots, 49\}$.

- The last number must match the sixth drawn number.

Suppose that a grand-prize-winning ticket consists of all six numbers drawn from the machine. Find the probability associated with the winning ticket.

EXAMPLE 4   Find the probability of drawing 5 cards, all hearts, from a standard card deck.

Solution   The number of ways we can draw 5 hearts from the 13 hearts is $C(13, 5)$, the number of combinations of 13 things taken 5 at a time. The number of ways to draw 5 cards from the deck is $C(52, 5)$, the number of combinations of 52 things taken 5 at a time. The probability of drawing 5 hearts is the ratio of the number of favorable outcomes to the number of possible outcomes.

$$P(5 \text{ hearts}) = \frac{s}{n} = \frac{C(13, 5)}{C(52, 5)}$$

$$P(5 \text{ hearts}) = \frac{\dfrac{13!}{5!8!}}{\dfrac{52!}{5!47!}}$$

$$= \frac{13!}{5!8!} \cdot \frac{5!47!}{52!} \qquad {\color{red}\frac{5!}{5!} = 1}$$

$$= \frac{13 \cdot 12 \cdot 11 \cdot 10 \cdot 9 \cdot 8!}{8!} \cdot \frac{47!}{52 \cdot 51 \cdot 50 \cdot 49 \cdot 48 \cdot 47!}$$

$$= \frac{13 \cdot 12 \cdot 11 \cdot 10 \cdot 9}{52 \cdot 51 \cdot 50 \cdot 49 \cdot 48}$$

$$= \frac{33}{66,640}$$

The probability of drawing 5 hearts is $\frac{33}{66,640}$ or $4.951980792 \times 10^{-4}$.

SELF CHECK 4   Find the probability of drawing 6 cards, all diamonds, from a standard card deck.

SELF CHECK ANSWERS

1. $S = \{(H, 1), (H, 2), (H, 3), (H, 4), (H, 5), (H, 6), (T, 1), (T, 2), (T, 3), (T, 4), (T, 5), (T, 6)\}$ 2. $\frac{1}{12}$
3. $\frac{1}{26}$ 4. $\frac{33}{391,510}$ or $8.428903476 \times 10^{-5}$

NOW TRY THIS

The notation $P(E)$ represents the probability that event E will occur. The notation $P(\overline{E})$ represents the probability that event E will *not* occur. Because either event E or event $\overline{E}$ must occur, $P(E) + P(\overline{E}) = 1$. Thus, $P(E) = 1 - P(\overline{E})$.

Sometimes it is easier to determine $P(E)$ by finding $P(\overline{E})$ and subtracting the result from 1.

1. In a family of 5 children, find the probability that at least one child is a girl.

2. In a clinical trial, the probability of experiencing a serious side effect is 0.15 while the probability of a minor side effect is 0.45, with no overlapping side effects. What is the probability that a person in the trial picked at random experiences no side effects?

11.6   Exercises

WARM-UPS   *Write a combination to represent*

1. the number of ways of drawing a red card from a standard deck.
2. the number of ways of drawing a club card from a standard deck.

REVIEW   *Solve each equation.*

3. $4^{3x} = \dfrac{1}{64}$

4. $6^{-x+2} = \dfrac{1}{36}$

5. $5^{x^2-2x} = 125$

6. $2^{x^2-3x} = 16$

7. $3^{x^2+4x} = \dfrac{1}{81}$

8. $7^{x^2+3x} = \dfrac{1}{49}$

VOCABULARY AND CONCEPTS   *Fill in the blanks.*

9. An _____ is any activity for which the outcome is uncertain.
10. A list of all possible outcomes for an experiment is called a _____.
11. The probability of an event E is defined as $P(E) = $ __.
12. If an event is certain to happen, its probability is _.
13. If an event cannot happen, its probability is _.
14. All probability values are between _ and _, inclusive.
15. Fill in the blanks to find the probability of drawing a red face card from a standard deck.
    a. The number of red face cards is _.
    b. The number of cards in the deck is __.
    c. The probability is __ or __.

16. Fill in the blanks to find the probability of drawing 4 aces from a standard card deck.

    a. The number of ways to draw 4 aces from 4 aces is $C(4, 4) =$ __.

    b. The number of ways to draw 4 cards from 52 cards is $C(52, 4) =$ _____.

    c. The probability is _____.

GUIDED PRACTICE *List the sample space of each experiment.*
SEE EXAMPLE 1. (OBJECTIVE 1)

17. Rolling one die followed by tossing one coin

18. Tossing three coins

19. Selecting a letter of the alphabet

20. Picking a one-digit number

Two dice are rolled. Find the probability of each event.
SEE EXAMPLE 2. (OBJECTIVE 2)

21. Rolling a sum of 5

22. Rolling a sum of 6

23. Rolling a sum less than 6

24. Rolling a sum greater than 9

Find the probability of each event. SEE EXAMPLE 3. (OBJECTIVE 2)

25. Drawing a diamond on one draw from a standard card deck

26. Drawing a face card from a standard deck

27. Drawing a red face card from a standard deck

28. Drawing the ace of spades from a standard deck

Find the probability of each event. SEE EXAMPLE 4. (OBJECTIVE 2)

29. Drawing 6 diamonds from a standard deck without replacing the cards after each draw

30. Drawing 5 aces from a standard deck without replacing the cards after each draw

31. Drawing 5 clubs from the black cards in a standard deck

32. Drawing 5 cards, all red, from a standard deck

ADDITIONAL PRACTICE *An ordinary die is rolled once. Find the probability of each event.*

33. Rolling a 2

34. Rolling a number greater than 4

35. Rolling a number larger than 1 but less than 6

36. Rolling an odd number

Balls numbered from 1 to 42 are placed in a container and stirred. If one is drawn at random, find the probability of each result.

37. The number is less than 20.

38. The number is less than 50.

39. The number is a prime number.

40. The number is less than 10 or greater than 40.

Refer to the following spinner. If the spinner is spun, find the probability of each event. Assume that the spinner never stops on a line.

41. The spinner stops on red.

42. The spinner stops on green.

43. The spinner stops on brown.

44. The spinner stops on yellow.

Find the probability of each event.

45. Drawing a face card from a standard deck followed by a 10 after replacing the first card

46. Drawing an ace from a standard deck followed by a 10 after replacing the first card

47. Rolling a sum of 7 on one roll of two dice

48. Drawing a red egg from a basket containing 5 red eggs and 7 blue eggs

49. Drawing a yellow egg from a basket containing 5 red eggs and 7 yellow eggs

Assume that the probability that a backup generator will fail a test is $\frac{1}{2}$ and that the college in question has 4 backup generators. After completing Exercise 50, find each probability in Exercises 51–56.

50. Construct a sample space for the test.

51. All generators will fail the test.

52. Exactly 1 generator will fail.

53. Exactly 2 generators will fail.

54. Exactly 3 generators will fail.

55. No generator will fail.

56. Find the sum of the probabilities in Exercises 51–55.

APPLICATIONS *Solve each application.*

57. Quality control In a batch of 10 tires, 2 are known to be defective. If 4 tires are chosen at random, find the probability that all 4 tires are good.

58. Medicine Out of a group of 9 patients treated with a new drug, 4 suffered a relapse. If 3 patients are selected at random from the group of 9, find the probability that none of the 3 patients suffered a relapse.

A survey of 282 people is taken to determine the opinions of doctors, teachers, and lawyers on a proposed piece of legislation, with the results shown in the table. A person is chosen at random from those surveyed. Refer to the table to find each probability.

59. The person favors the legislation.

60. A doctor opposes the legislation.

61. A person who opposes the legislation is a lawyer.

	Number that favor	Number that oppose	Number with no opinion	Total
Doctors	70	32	17	119
Teachers	83	24	10	117
Lawyers	23	15	8	46
Total	176	71	35	282

WRITING ABOUT MATH

62. Explain why all probability values range from 0 to 1.

63. Explain the concept of probability.

SOMETHING TO THINK ABOUT *If $P(A)$ represents the probability of event A, and $P(B|A)$ represents the probability that event B will occur after event A, then*

$$P(A \text{ and } B) = P(A) \cdot P(B|A)$$

64. In a school, 30% of the students are gifted in math and 10% are gifted in art and math. If a student is gifted in math, find the probability that the student is also gifted in art.

65. The probability that a person owns a luxury car is 0.2, and the probability that the owner of such a car also owns a second car is 0.7. Find the probability that a person chosen at random owns both a luxury car and a second car.

Projects

PROJECT 1

Baytown is building an auditorium. The city council has already decided on the layout shown in the illustration. Each of the sections A, B, C, D, E is to be 60 feet in length from front to back. The aisle widths cannot be changed due to fire regulations. The one thing left to decide is how many rows of seats to put in each section. Based on the following information regarding each section of the auditorium, help the council decide on a final plan.

Sections A and C each have four seats in the front row, five seats in the second row, six seats in the third row, and so on, adding one seat per row as we count from front to back.

Section B has eight seats in the front row and adds one seat per row as we count from front to back.

Sections D and E each have 28 seats in the front row and add two seats per row as we count from front to back.

a. One plan calls for a distance of 36 inches, front to back, for each row of seats. Another plan allows for 40 inches (an extra four inches of legroom) for each row. How many seats will the auditorium have under each of these plans?

b. Another plan calls for the higher-priced seats (Sections A, B, and C) to have the extra room afforded by 40-inch rows, but for Sections D and E to have enough rows to make sure that the auditorium holds at least 2,700 seats. Determine how many rows Section D and E would have to contain for this to work. (This answer should be an integer.) How much space (to the nearest tenth of an inch) would be allotted for each row in Sections D and E?

PROJECT 2

Pascal's triangle contains a wealth of interesting patterns. You have seen two in Exercises 88 and 107 of Section 11.1. Here are a few more.

a. Find the hockey-stick pattern in the numbers in the illustration. What would be the missing number in the rightmost hockey stick? Does this pattern work for larger hockey sticks? Experiment.

(Continued)

b. In Illustration 1, find the pattern in the sums of increasingly larger portions of Pascal's triangle. Find the sum of all of the numbers up to and including the row that begins 1 10 45 . . . .

c. In Illustration 2, find the pattern in the sums of the squares of the numbers in each row of the triangle. What is the sum of the squares of the numbers in the row that begins 1 10 45 . . . ? (*Hint*: Calculate $P(2, 1)$, $P(4, 2)$, $P(6, 3)$, ... Do these numbers appear elsewhere in the triangle?)

d. In 1653, Pascal described the triangle in *Treatise on the Arithmetic Triangle*, writing, "I have left out many more properties than I have included. It is remarkable how fertile in properties this triangle is. *Everyone can try his hand.*" Accept Pascal's invitation. Find some of the triangle's patterns for yourself and share your discoveries with your class. Illustration 3 is an idea to get you started.

$$
\begin{aligned}
1^2 &= 1 \\
1^2 + 1^2 &= 2 \\
1^2 + 2^2 + 1^2 &= 6 \\
1^2 + 3^2 + 3^2 + 1^2 &= 20 \\
1^2 + 4^2 + 6^2 + 4^2 + 1^2 &= 70 \\
1^2 + 5^2 + 10^2 + 10^2 + 5^2 + 1^2 &= ? \\
1^2 + 6^2 + 15^2 + 20^2 + 15^2 + 6^2 + 1^2 &= ?
\end{aligned}
$$

ILLUSTRATION 2

ILLUSTRATION 3

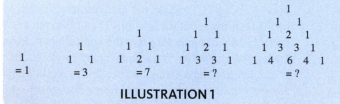

ILLUSTRATION 1

Reach for Success
EXTENSION OF PLANNING FOR YOUR FUTURE

Now that you've considered short-term planning, let's move on and look at your future career goals.

LONG-TERM EDUCATIONAL PLANNING Take advantage of your college resources to help you identify your area of interest. Ultimately, you will need to declare a major.	Your college may have resources to help your choose a major. List a few of these resources. _____ _____
Have you chosen a major? _____ If so, consider careers in that field and list at least three. **1.** _____ **2.** _____ **3.** _____	If not, visit your career services area, see an advisor, or talk with a favorite professor. With whom did you consult? _____ What advice did you receive? _____ _____
What are potential employers looking for in your career interest? _____ _____	Circle True or False for the following statement. Most companies want employees with the problem-solving skills developed in mathematics courses. True False
There are certificate programs and associate's, bachelor's, master's, and doctoral degrees.	What type of certificate or degree is required for your field of interest? _____ What additional mathematics courses (if any) are required for your major? _____ _____

Proper planning can help you avoid taking classes that might not be required for your degree, thus saving you time and money.

11 Review

SECTION 11.1   The Binomial Theorem

DEFINITIONS AND CONCEPTS	EXAMPLES
The symbol $n!$ (**factorial**) is defined as $n! = n(n - 1)(n - 2) \cdot \ldots \cdot (3)(2)(1)$ where n is a natural number. $0! = 1$ $n(n - 1)! = n!$ (n is a natural number)	$7! = 7 \cdot 6 \cdot 5 \cdot 4 \cdot 3 \cdot 2 \cdot 1 = 5{,}040$ $\dfrac{5!}{2!3!} = \dfrac{5 \cdot 4 \cdot \cancel{3!}}{2 \cdot 1 \cdot \cancel{3!}} = \dfrac{20}{2} = 10$ $7 \cdot \mathbf{6!} = 7 \cdot \mathbf{6 \cdot 5 \cdot 4 \cdot 3 \cdot 2 \cdot 1} = 7!$
The binomial theorem: If n is any positive integer, then $(a + b)^n = a^n + \dfrac{n!}{1!(n - 1)!}a^{n-1}b$ $+ \dfrac{n!}{2!(n - 2)!}a^{n-2}b^2 + \cdots$ $+ \dfrac{n!}{r!(n - r)!}a^{n-r}b^r + \cdots + b^n$	$(x - 2y)^4$ $= [x + (-2y)]^4$ $= x^4 + \dfrac{4!}{1!3!}x^3(-2y) + \dfrac{4!}{2!2!}x^2(-2y)^2 + \dfrac{4!}{3!1!}x(-2y)^3 + (-2y)^4$ $= x^4 + 4x^3(-2y) + 6x^2(4y^2) + 4x(-8y^3) + 16y^4$ $= x^4 - 8x^3y + 24x^2y^2 - 32xy^3 + 16y^4$
Specific term of a binomial expansion: The binomial theorem can be used to find a specific term of a binomial expansion. The coefficient of the variables can be found using the formula $\dfrac{n!}{r!(n - r)!}$ where n is the power of the expansion and r is the exponent of the second variable.	To find the third term of $(x + y)^5$, note that the exponent on y will be 2, because the exponent on y is 1 less than the number of the term. The power on x will be 3 because the sum of the powers must be 5. Thus, the variables will be x^3y^2. The coefficient of the variables can be found with the formula. $\dfrac{5!}{2!(5 - 2)!}x^3y^2$ Substitute the values into the formula. $\dfrac{5 \cdot 4 \cdot \cancel{3!}}{2 \cdot 1 \cdot \cancel{3!}}x^3y^2$ Expand the factorials. $10x^3y^2$ Simplify.

REVIEW EXERCISES

Evaluate each expression.

1. $(5!)(2!)$

2. $\dfrac{5!}{3!}$

3. $\dfrac{6!}{2!(6 - 2)!}$

4. $\dfrac{12!}{3!(12 - 3)!}$

5. $(n - n)!$

6. $\dfrac{10!}{6!}$

Use the binomial theorem to find each expansion.

7. $(x + y)^5$

8. $(x - y)^4$

9. $(4x - y)^3$

10. $(x + 4y)^3$

Find the specified term in each expansion.

11. $(x + y)^5$; fifth term

12. $(x - y)^5$; fourth term

13. $(3x - 4y)^3$; second term

14. $(4x + 3y)^4$; third term

SECTION 11.2   Arithmetic Sequences and Series

DEFINITIONS AND CONCEPTS	EXAMPLES
An **arithmetic sequence** is a sequence of the form $$a_1, a_1 + d, a_1 + 2d, \ldots, a_1 + (n-1)d$$ where a_1 is the first term. the nth term is $a_n = a_1 + (n-1)d$. d is the common difference.	Find the first 5 terms of the sequence with $a_1 = 15$ and $d = -2$. $a_1 = 15$ $a_2 = 15 + (-2) = 13$ $a_3 = 15 + 2(-2) = 11$ $a_4 = 15 + 3(-2) = 9$ $a_5 = 15 + 4(-2) = 7$
Arithmetic means: If numbers are inserted between two given numbers a and b to form an arithmetic sequence, the inserted numbers are **arithmetic means** between a and b.	To insert two arithmetic means between 7 and 25, note that $a_1 = 7$ and $a_4 = 25$. Substitute these values into the formula for the nth term and solve for d. $\quad a_n = a_1 + (n-1)d$ $\quad 25 = 7 + (4-1)d$ Substitute values. $\quad 25 = 7 + 3d$ Simplify. $\quad 18 = 3d$ Subtract 7 from both sides. $\quad 6 = d$ Divide both sides by 3. The common difference is 6. The two arithmetic means are $7 + 6 = 13$ and $7 + 2(6) = 19$.
Sum of terms of arithmetic sequences: The sum of the first n terms of an arithmetic sequence is given by $$S_n = \frac{n(a_1 + a_n)}{2} \text{ with } a_n = a_1 + (n-1)d$$ where a_1 is the first term, a_n is the nth term, and n is the number of terms in the sequence.	To find the sum of the first 12 terms of the sequence $4, 12, 20, 28, \ldots$, note that the common difference is 8 and substitute it into the formula for the nth term to find the twelfth term: $$a_{12} = 4 + (12-1)8 = 4 + 88 = 92$$ Then substitute into the formula for the sum of the first n terms: $$S_{12} = \frac{12(4+92)}{2} = \frac{12(96)}{2} = \frac{1,152}{2} = 576$$
Summation notation: $$\sum_{k=1}^{n} f(k) = f(1) + f(2) + \cdots + f(n)$$ Summation notation describes a series.	$$\sum_{k=1}^{3} (3k-1) = [3(1)-1] + [3(2)-1] + [3(3)-1]$$ $$= (3-1) + (6-1) + (9-1)$$ $$= 2 + 5 + 8$$ $$= 15$$

REVIEW EXERCISES

15. Find the eighth term of an arithmetic sequence whose first term is 7 and whose common difference is 5.

16. Write the first five terms of the arithmetic sequence whose ninth term is 242 and whose seventh term is 212.

17. Find two arithmetic means between 8 and 25.

18. Find the sum of the first 20 terms of the sequence 11, 18, 25, ...

19. Find the sum of the first ten terms of the sequence $9, 6\frac{1}{2}, 4, \ldots$.

Find each sum.

20. $\displaystyle\sum_{k=1}^{5} 4k$

21. $\displaystyle\sum_{k=1}^{4} (k^2 + 1)$

22. $\displaystyle\sum_{k=1}^{4} (3k - 4)$

23. $\displaystyle\sum_{k=10}^{10} 36k$

SECTION 11.3 Geometric Sequences and Series

DEFINITIONS AND CONCEPTS	EXAMPLES
Geometric sequences: A geometric sequence is a sequence of the form $$a_1, a_1r, a_1r^2, a_1r^3, \ldots, a_1r^{n-1}$$ where a_1 is the first term. $a_n = a_1r^{n-1}$ is the nth term. r is the common ratio.	The first four terms of the geometric sequence with $a_1 = 8$ and $r = 2$ are as follows: $$a_1 = 8$$ $$a_2 = 8(2) = 16$$ $$a_3 = 8(2)^2 = 32$$ $$a_4 = 8(2)^3 = 64$$
Geometric means: If numbers are inserted between a and b to form a geometric sequence, the inserted numbers are geometric means between a and b.	To insert the positive geometric mean between 4 and 1, note that $a_1 = 4$ and $a_3 = 1$. Then substitute these values into the formula for the nth term and solve for r. $a_n = a_1r^{n-1}$ Formula for the nth term $a_3 = 4r^{3-1}$ Formula for the 3rd term $1 = 4r^2$ Substitute 1 for a_3. $\dfrac{1}{4} = r^2$ Divide both sides by 4. $\dfrac{1}{2} = r$ Take the positive square root of both sides. Since $r = \frac{1}{2}$, $a_2 = a_1r = 4\left(\frac{1}{2}\right) = 2$. The geometric mean between 4 and 1 is 2.
Sum of terms of a geometric sequence: The sum of the first n terms of a geometric sequence is given by $$S_n = \frac{a_1 - a_1r^n}{1 - r} \quad (r \neq 1)$$ where S_n is the sum, a_1 is the first term, r is the common ratio, and n is the number of terms in the sequence.	To find the sum of the first 4 terms of the geometric sequence defined by $a_n = 2(3)^n$, note that the first term and the common ratio are $$a_1 = 2(3)^1 = 2(3) = 6 \quad \text{and} \quad r = 3$$ Then substitute into the formula for the sum of the first n terms and simplify: $$S_n = \frac{a_1 - a_1r^n}{1 - r}$$ $$S_4 = \frac{6 - 6(3)^4}{1 - 3} \quad \text{Substitute values into the formula.}$$ Thus, $$S_4 = \frac{6 - 6(81)}{-2} = \frac{6 - 486}{-2} = 240$$ The sum of the first 4 terms is 240.

REVIEW EXERCISES

24. Write the first five terms of the geometric sequence whose fourth term is 3 and whose fifth term is $\frac{3}{2}$.

25. Find the fifth term of a geometric sequence with a first term of $\frac{1}{16}$ and a common ratio of 2.

26. Find two geometric means between -6 and 384.

27. Find the sum of the first six terms of the sequence 240, 120, 60, ...

28. Find the sum of the first eight terms of the sequence $\frac{1}{8}$, $-\frac{1}{4}$, $\frac{1}{2}$, ...

29. Car depreciation A \$5,000 car depreciates at the rate of 20% of the previous year's value. How much is the car worth after 5 years?

30. Stock appreciation The value of Mia's stock portfolio is expected to appreciate at the rate of 18% per year. How much will the portfolio be worth in 10 years if its current value is \$25,700?

31. Planting corn A farmer planted 300 acres in corn this year. He intends to plant an additional 75 acres in corn in each successive year until he has 1,200 acres in corn. In how many years will that be?

32. Falling objects If an object is in free fall, the sequence 16, 48, 80, . . . represents the distance in feet that the object falls during the first second, during the second second, during the third second, and so on. How far will the object fall during the first 10 seconds?

SECTION 11.4   Infinite Geometric Sequences and Series

DEFINITIONS AND CONCEPTS	EXAMPLES
Sum of an infinite geometric series: If r is the common ratio of an infinite geometric series and if $\lvert r \rvert < 1$, the sum of the series is given by $$S_\infty = \frac{a_1}{1 - r}$$ where a_1 is the first term and r is the common ratio.	To find the sum of the geometric series $8 + 4 + 2 + \cdots$, note that the first term is 8 and the common ratio is $r = \frac{1}{2}$. Since $\left\lvert \frac{1}{2} \right\rvert < 1$, the sum exists. To find it, proceed as follows: $$S_\infty = \frac{a_1}{1 - r}$$ $$S_\infty = \frac{8}{1 - \frac{1}{2}}$$ Substitute values into the formula. Thus, $$S_\infty = \frac{8}{\frac{1}{2}} = 8 \cdot 2 = 16$$ The sum is 16.
To write a repeating decimal as a common fraction, write the decimal as an infinite geometric series and find the sum.	Write $0.\overline{7}$ as a common fraction. Write the infinite geometric series $\frac{7}{10} + \frac{7}{100} + \frac{7}{1,000} + \cdots$ $$S_\infty = \frac{\frac{7}{10}}{1 - \frac{1}{10}} = \frac{7}{10} \div \frac{9}{10} = \frac{7}{10} \cdot \frac{10}{9} = \frac{7}{9}$$

REVIEW EXERCISES

33. Find the sum of the infinite geometric series
$100 + 20 + 4 + \cdots$.

34. Write the decimal $0.\overline{05}$ as a common fraction.

SECTION 11.5   Permutations and Combinations

DEFINITIONS AND CONCEPTS	EXAMPLES
Multiplication principle for events: If E_1 and E_2 are two events and if E_1 can be done in a_1 ways and E_2 can be done in a_2 ways, then the event "E_1 followed by E_2" can be done in $a_1 \cdot a_2$ ways.	If Danielle has 5 shirts, 4 pairs of jeans, and 3 pairs of shoes, she has a choice of $$5 \cdot 4 \cdot 3 = 60$$ different outfits to wear.
Formulas for permutations: $$P(n, r) = \frac{n!}{(n - r)!}$$ $$P(n, n) = n!$$ $$P(n, 0) = 1$$ If order matters, use the permutation formula.	$$P(6, 4) = \frac{6!}{(6 - 4)!} = \frac{6 \cdot 5 \cdot 4 \cdot 3 \cdot 2 \cdot 1}{(2)!} = \frac{720}{2 \cdot 1} = 360$$ $$P(5, 5) = 5! = 5 \cdot 4 \cdot 3 \cdot 2 \cdot 1 = 120$$ $$P(5, 0) = 1$$
Formulas for combinations: $$C(n, r) = \binom{n}{r} = \frac{n!}{r!(n - r)!}$$ $$C(n, n) = \binom{n}{n} = 1 \quad C(n, 0) = \binom{n}{0} = 1$$ If order does not matter, use the combination formula.	$$C(12, 5) = \frac{12!}{5!(12 - 5)!} = \frac{12!}{5!(7)!}$$ $$= \frac{12 \cdot 11 \cdot 10 \cdot 9 \cdot 8 \cdot 7!}{5 \cdot 4 \cdot 3 \cdot 2 \cdot 1 \cdot 7!} = \frac{12 \cdot 11 \cdot 10 \cdot 9 \cdot 8}{5 \cdot 4 \cdot 3 \cdot 2 \cdot 1} = 792$$ $$C(5, 5) = 1 \qquad C(5, 0) = 1$$

The Binomial Theorem (Combinatorial Form):
If n is any positive integer, then

$$(a + b)^n = \binom{n}{0}a^n + \binom{n}{1}a^{n-1}b + \binom{n}{2}a^{n-2}b^2$$
$$+ \cdots + \binom{n}{r}a^{n-r}b^r + \cdots + \binom{n}{n}b^n$$

Use the combinatorial form of the binomial theorem to expand $(x - y)^3$.

$$(x - y)^3 = \binom{3}{0}x^3 - \binom{3}{1}x^2y + \binom{3}{2}xy^2 - \binom{3}{3}y^3$$
$$= 1x^3 - 3x^2y + 3xy^2 - 1y^3$$
$$= x^3 - 3x^2y + 3xy^2 - y^3$$

REVIEW EXERCISES

35. Planning a trip If there are 17 flights from New York to Chicago, and 8 flights from Chicago to San Francisco, in how many different ways could a passenger plan a trip from New York to San Francisco?

Evaluate each expression.

36. $P(7, 7)$

37. $P(5, 0)$

38. $P(8, 6)$

39. $\dfrac{P(9, 6)}{P(10, 7)}$

Evaluate each expression.

40. $C(8, 8)$

41. $C(7, 0)$

42. $\dbinom{8}{6}$

43. $\dbinom{10}{4}$

44. $C(6, 3) \cdot C(7, 3)$

45. Use the combinatorial form of the binomial theorem to expand $(x + 2)^3$.

46. Lining up In how many ways can five people be arranged in a line?

47. Lining up In how many ways can three men and five women be arranged in a line if the women are placed ahead of the men?

48. Choosing people In how many ways can we pick three people from a group of ten?

49. Forming committees In how many ways can we pick a committee of two Democrats and two Republicans from a group containing five Democrats and six Republicans?

SECTION 11.6 Probability

DEFINITIONS AND CONCEPTS	EXAMPLES
An event that cannot occur has a probability of 0. An event that is certain to occur has a probability of 1. All other events have probabilities between 0 and 1.	The probability of rolling a 7 with a single standard die is 0 because there is no 7 on a standard die. The probability of drawing a red card from a stack of red cards is 1 because all the cards are red.
Sample space: The sample space consists of all possibilities of outcomes for an experiment.	The sample space of rolling one standard die is the set of all possible outcomes: $\{1, 2, 3, 4, 5, 6\}$
Probability of an event: If S is the *sample space* of an experiment with n distinct and equally likely outcomes, and E is an event that occurs in s of those ways, then the probability of E is $P(E) = \dfrac{s}{n}$	To find the probability of rolling a prime number with one roll of a standard die, first determine that there are 6 possible outcomes in the sample space: The set of possible outcomes is $\{1, 2, 3, 4, 5, 6\}$. Then determine that there are 3 possible favorable outcomes: The prime numbers on a face of a die are 2, 3, and 5. Thus, $P(\text{prime number}) = \dfrac{3}{6} = \dfrac{1}{2}$ The probability of rolling a prime number is $\dfrac{1}{2}$.

REVIEW EXERCISES

In Exercises 50–52, assume that a dart is randomly thrown at the colored chart.

50. What is the probability that the dart lands in a blue area?

51. What is the probability that the dart lands in an even-numbered area?

52. What is the probability that the dart lands in an area whose number is greater than 2?

53. Find the probability of rolling a sum of 5 on one roll of two dice.

54. Find the probability of drawing a black card from a stack of red cards.

55. Find the probability of drawing a 10 from a standard deck of cards.

56. Find the probability of drawing a 5-card poker hand that has exactly 3 aces.

57. Find the probability of drawing 5 cards, all diamonds, from a standard card deck.

---

11   Test

1. Evaluate: $\dfrac{9!}{7!}$

2. Evaluate: $0!$

3. Find the second term in the expansion of $(x - y)^5$.

4. Find the third term in the expansion of $(x + 2y)^4$.

5. Find the tenth term of an arithmetic sequence with the first three terms of 3, 10, and 17.

6. Find the sum of the first 12 terms of the sequence $-2, 3, 8, \ldots$.

7. Find two arithmetic means between 2 and 98.

8. Evaluate: $\displaystyle\sum_{k=1}^{4}(3k - 4)$

9. Find the seventh term of the geometric sequence whose first three terms are $-\frac{1}{9}$, $-\frac{1}{3}$, and -1.

10. Find the sum of the first six terms of the sequence $\frac{1}{27}, \frac{1}{9}, \frac{1}{3}, \ldots$.

11. Find two geometric means between 3 and 648.

12. Find the sum of all of the terms of the infinite geometric sequence $9, 3, 1, \ldots$.

13. Find the sum of $5, 10, 20, \ldots$

Find the value of each expression.

14. $P(7, 3)$

15. $P(8, 8)$

16. $C(5, 3)$

17. $C(8, 3)$

18. $C(6, 0) \cdot P(6, 5)$

19. $\dfrac{P(6, 4)}{C(6, 4)}$

20. Expand $(x - 3)^4$.

21. **Selecting people**   In how many ways can we select five people from a group of eight?

22. **Selecting committees**   From a group of five men and four women, how many three-person committees can be created that will include two women?

Find each probability.

23. Rolling a 6 on one roll of a die

24. Drawing a jack or a queen from a standard card deck

25. Receiving 5 hearts for a 5-card poker hand

26. Tossing 2 heads in 5 tosses of a fair coin

Cumulative Review

1. Use graphing to solve: $\begin{cases} 2x + y = 5 \\ x - 2y = 0 \end{cases}$

2. Use substitution to solve: $\begin{cases} 3x + y = 4 \\ 2x - 3y = -1 \end{cases}$

3. Use elimination to solve: $\begin{cases} x + 2y = -2 \\ 2x - y = 6 \end{cases}$

4. Use any method to solve: $\begin{cases} \frac{x}{10} + \frac{y}{5} = \frac{1}{2} \\ \frac{x}{2} - \frac{y}{5} = \frac{13}{10} \end{cases}$

5. Evaluate: $\begin{vmatrix} 4 & -3 \\ 2 & -2 \end{vmatrix}$

6. Use Cramer's rule and solve for y only:
$$\begin{cases} 4x - 3y = -1 \\ 3x + 4y = -7 \end{cases}$$

7. Solve:
$$\begin{cases} x + y + z = 1 \\ 2x - y - z = -4 \\ x - 2y + z = 4 \end{cases}$$

8. Solve for z only:
$$\begin{cases} x + 2y + 3z = 6 \\ 3x + 2y + z = 6 \\ 2x + 3y + z = 6 \end{cases}$$

9. Solve by graphing:
$$\begin{cases} 3x - 2y < 6 \\ y < -x + 2 \end{cases}$$

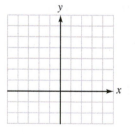

10. Solve by graphing:
$$\begin{cases} y < x + 2 \\ 3x + y \le 6 \end{cases}$$

11. Graph: $f(x) = \left(\frac{1}{2}\right)^x$

12. Write $y = \log_2 x$ as an exponential equation.

Find the value of x.

13. $\log_x 27 = 3$ **14.** $\log_5 125 = x$

15. $\log_2 x = -4$ **16.** $\log_5 x = 0$

17. Find the inverse of $y = \log_2 x$.

18. If $\log_{10} 10^x = y$, then y equals what quantity?

If $\log 7 = 0.8451$ *and* $\log 14 = 1.1461$, *evaluate each expression without using a calculator or tables.*

19. $\log 98$ **20.** $\log 2$

21. $\log 27$

22. $\log \dfrac{7}{5}$ (*Hint:* $\log 10 = 1$.)

23. Solve: $2^{x+5} = 3^x$

24. Solve: $\log 5 + \log x - \log 4 = 1$

Use a calculator for Exercises 25–27.

25. Boat depreciation How much will a \$9,000 boat be worth after 9 years if it depreciates 12% per year?

26. Find $\log_7 9$ to four decimal places.

27. Evaluate: $\dfrac{5!\,7!}{6!}$

28. Use the binomial theorem to expand $(3a - b)^4$.

29. Find the seventh term in the expansion of $(2x - y)^8$.

30. Find the 20th term of an arithmetic sequence with a first term of -11 and a common difference of 6.

31. Find the sum of the first 20 terms of an arithmetic sequence with a first term of 6 and a common difference of 3.

32. Insert two arithmetic means between -3 and 30.

33. Evaluate: $\displaystyle\sum_{k=1}^{4} 2k^2$

34. Evaluate: $\displaystyle\sum_{k=3}^{6} (4k - 5)$

35. Find the seventh term of a geometric sequence with a first term of $\frac{1}{27}$ and a common ratio of 3.

36. Find the sum of the first ten terms of the sequence $\frac{1}{64}, \frac{1}{32}, \frac{1}{16}, \dots$

37. Insert two geometric means between -3 and 192.

38. Find the sum of all the terms of the sequence $9, 3, 1, \dots$

39. Evaluate: $P(9, 3)$

40. Evaluate: $C(10, 7)$

41. Evaluate: $\dfrac{C(8, 4) \cdot C(8, 0)}{P(6, 2)}$

42. If $n > 1$, which is smaller: $P(n, n)$ or $C(n, n)$?

43. Lining up In how many ways can seven people stand in a line?

44. Forming a committee In how many ways can a committee of three people be chosen from a group containing nine people?

45. Cards Find the probability of drawing a red face card from a standard deck of cards.

GLOSSARY

absolute value The distance between a given number and 0 on a number line, usually denoted with $|\ \ |$

absolute value function The function $f(x) = |x|$

acute angle An angle with a measure between 0° and 90°

additive identity The number zero, because for all real numbers a, $a + 0 = 0 + a = a$

additive inverses A pair of numbers, a and b, are additive inverses if $a + b = 0$; also called **negatives** or **opposites**.

algebraic term An expression that is a constant, variable, or a product of constants and variables; for example, 37, xyz, and $32t$ are terms.

algorithm A repeating series of steps or processes

altitude The length of the perpendicular line from a vertex of a triangle to its opposite side

annual growth rate The rate r that a quantity P increases or decreases within one year; the variable r in the exponential growth formula $A = Pe^{rt}$

area The amount of surface enclosed by a two-dimensional geometric figure; expressed as unit²

arithmetic means Numbers that are inserted between two elements in an arithmetic sequence to form a new arithmetic sequence

arithmetic sequence A sequence in which each successive term is equal to the sum of the preceding term and a constant; written $a, a + d, a + 2d, a + 3d, \ldots$

arithmetic series The indicated sum of the terms of an arithmetic sequence

ascending order An ordering of the terms of a polynomial such that a given variable's exponents occur in increasing order

associative properties The properties for addition and multiplication that state for real numbers a, b, and c, $(a + b) + c = a + (b + c)$ and $(ab)c = a(bc)$. *Note:* The associative property does not hold for subtraction or division.

asymptote A straight line that is approached by a graph

augmented matrix A matrix representing a system of linear equations written in standard form, with columns consisting of the coefficients and a column of the constant terms of the equations

axis of symmetry For the parabola $f(x) = ax^2 + bx + c$, a vertical line, $x = -\frac{b}{2a}$

back substitution The process of finding the value of a second (or third, etc.) variable after finding the first by substituting the first into an equation of two variables and, if necessary, substituting both values into an equation of three values to find the third, etc.

base In the expression x^y, x is the base and y is the exponent. The base, x, will be used as a factor y times.

base angles of an isosceles triangle The two congruent angles of an isosceles triangle opposite the two sides that are of the same length

binomial A polynomial with exactly two terms

binomial theorem The theorem used to expand a binomial

boundary line A line that divides the rectangular coordinate plane into two half-planes

center of a circle The point that is equidistant from all points on a circle

center of a hyperbola The midpoint of the segment joining the foci of a hyperbola

change-of-base formula A formula used to convert a logarithm from one base a to some other base b

circle The set of all points in a plane that are the same distance from a fixed point (center)

coefficient matrix A matrix consisting of the coefficients of the variables in a system of linear equations

combinations The number of ways to choose r things taken from a set of n things if order is not important

common difference The constant difference between successive terms in an arithmetic sequence, that is, the number d in the formula $a_n = a_1 + (n - 1)d$

common logarithm A logarithm with a base of 10

common ratio The constant quotient of two successive terms in a geometric sequence

commutative properties The properties for addition and multiplication that state for real numbers a and b, $a + b = b + a$ and $ab = ba$. *Note:* The commutative property does not hold for subtraction and division.

complementary angles Two angles whose measures sum to 90°

completing the square The process of forming a perfect-square trinomial from a binomial of the form $x^2 + bx$

complex fraction A fraction that contains a fraction in its numerator and/or denominator

complex number The sum of a real number and an imaginary number

composite function A new function formed when one function is evaluated in terms of another, denoted by $(f \circ g)(x)$

composite numbers A natural number greater than 1 with factors other than 1 and itself

composition The process of using the output of one function as the input of another function

compound inequality A single statement representing the intersection of two inequalities

compounding periods The number of times per year interest is compounded

compound interest The total interest earned on money deposited in an account where the interest for each compounding period is deposited back into the account, which increases the principle for the next compounding period

conditional equation An equation in one variable that has exactly one solution

cone A three-dimensional surface with a circular base whose cross sections (parallel to the base) are circles that decrease in diameter until it comes to a point

conic section A graph that can be formed by the intersection of a plane and a right-circular cone

conjugate An expression that contains the same two terms as another but with opposite signs between them; for example, $a + \sqrt{b}$ and $a - \sqrt{b}$

consistent system A system of equations with at least one solution

constant A term whose variable factor(s) have an exponent of 0

constant of proportionality If y varies directly with x, that is, $y = kx$, we say k is the constant of proportionality.

constraints Inequalities that limit the possible values of the variables of the objective function

contradiction An equation that is false for all values of its variables; its solution set is $\varnothing$.

coordinate The number that corresponds to a given point on a number line

coordinate plane The plane that contains the x-axis and y-axis

Cramer's rule A method using determinants to solve systems of linear equations

critical points Given a quadratic or rational inequality, the points on the number line corresponding to its critical values

critical values Given a quadratic inequality in standard form, the solutions to the quadratic equation $ax^2 + bx + c = 0$; given a rational inequality in standard form, the values where the denominator is equal to zero and the numerator is equal to zero

cube A rectangular solid whose faces are all congruent squares

cube-root function The function $f(x) = \sqrt[3]{x}$

cubic function A polynomial function whose equation is of third degree; alternately, a function whose equation is of the form $f(x) = ax^3 + bx^2 + cx + d$

cubing function The function $f(x) = x^3$

cylinder A three-dimensional surface whose cross sections (parallel to the base) are the same size and shape as the base of the cylinder

decibel A unit of measure that expresses the voltage gain (or loss) of an electronic device such as an amplifier

decreasing function A function whose value decreases as the independent value increases (graph drops as we move to the right)

degree of a polynomial The largest degree of a polynomial's terms

denominator The expression below the bar of a fraction

dependent equations A system of equations that has infinitely many solutions

dependent variable The variable in an equation of two variables whose value is determined by the independent variable (usually y in an equation involving x and y)

descending order An ordering of the terms of a polynomial, such that a given variable's exponents occur in decreasing order

determinant A calculated value from the elements in a square matrix: For a two-by-two matrix $\begin{bmatrix} a & b \\ c & d \end{bmatrix}$, the determinant is $ad - bc$. For a three-by-three determinant or larger, we use the method of expanding by minors.

difference The result of subtracting two expressions

difference of two cubes An expression of the form $a^3 - b^3$

difference of two squares An expression of the form $a^2 - b^2$

difference quotient The function defined by $\frac{f(x + h) - f(x)}{h}$

direct variation A relationship between two variables x and y of the form $y = kx$, where k is the constant of proportionality

discriminant The part of the quadratic formula, $b^2 - 4ac$, used to determine the number and nature of solutions to the quadratic equation

distributive property The property that states for real numbers a, b, and c, $a(b + c) = ab + ac$ and $(b + c)a = ab + ac$.

domain of a function The set of all permissible input values of a function

domain of a relation The set of all first elements (components) of a relation

eccentricity A measure of the flatness of an ellipse

element of a matrix One of the entries in a matrix

elements of a set The objects in a given set

ellipse The set of all points in a plane the sum of whose distances from two fixed points (foci) is a constant

ellipses A symbol consisting of three dots (. . .) meaning "and so forth" in the same manner

equal ratios Two (or more) ratios that represent equal values

equation A statement indicating that two quantities are equal

equilateral triangle A triangle with all sides of equal length and all angles of equal measure

equivalent equations Two equations that have the same solution set

equivalent fractions Two fractions with different denominators describing the same value; for example, $\frac{3}{11}$ and $\frac{6}{22}$ are equivalent fractions.

equivalent systems Two systems of equations that have the same solution set

even integers The set of integers that can be divided exactly by 2. *Note*: 0 is an even integer.

even root If $b = \sqrt[n]{a}$, b is an even root if n is even.

event A situation that can have several different outcomes or a subset of the sample space of an experiment

exponent In the expression x^y, x is the base and y is the exponent. The exponent y states the number of times that the base x will be used as a factor.

exponential equation An equation that contains a variable in one of its exponents

exponential expression An expression of the form y^x, also called a **power of x**

exponential function A function of the form $f(x) = ab^x$

exponential function (natural) A function of the form $f(t) = Pe^{kt}$

extraneous solution A solution to an equation that does not result in a true statement when substituted for the variable in the original equation

extremes In the proportion $\frac{a}{b} = \frac{c}{d}$, the numbers a and d are called the extremes.

factorial For a natural number n, the product of all the natural numbers less than or equal to n, denoted as $n!$ (*Exception*: 0! is defined as 1.)

factoring The process of finding the individual factors of a product

factoring tree A visual representation of factors of a number, usually used as a tool to express the number in prime-factored form

factors of a number The natural numbers that divide a given number; for example, the factors of 12 are 1, 2, 3, 4, 6, and 12.

factor theorem The theorem that states if $P(x)$ is a polynomial, then $P(r) = 0$ if and only if $(x - r)$ is a factor of $P(r)$

feasibility region The set of points that satisfy all of the constraints of a system of inequalities

Fibonacci sequence The sequence 1, 1, 2, 3, 5, 8, . . . with every successive term the sum of the two previous terms

finite sequence A sequence with a finite number of terms

foci of an ellipse The two fixed points used in the definition of the ellipse

foci of a hyperbola The two fixed points used in the definition of the hyperbola

FOIL method An acronym representing the order for multiplying the terms of two binomials

formula A literal equation in which the variables correspond to defined quantities

fulcrum The point on which a lever pivots

function A relation in which to each first component there corresponds exactly one second component, usually denoted by $f(x)$

fundamental theorem of arithmetic The theorem that states there is exactly one prime factorization for any natural number greater than 1

future value The amount to which an invested amount of money will grow

Gaussian elimination A method of solving a system of linear equations by working with its associated augmented matrix

general form of the equation of a circle The equation of a circle written in the form $x^2 + y^2 + Dx + Ey + F = 0$

general form of a linear equation A linear equation written in the form $Ax + By = C$, A and B not both equal to 0.

general form of the equation of an ellipse The equation of an ellipse written in the form $Ax^2 + Cy^2 + Dx + Ey + F = 0$

general form of the equation of a hyperbola The equation of a hyperbola written in the form $Ax^2 - Cy^2 + Dx + Ey + F = 0$

geometric means Numbers that are inserted between two elements in a geometric sequence to form a new geometric sequence

geometric sequence A sequence in which each successive term is equal to the product of the preceding term and a constant, written $a_1, a_1r, a_1r^2, a_1r^3, \ldots$

geometric series The indicated sum of the terms of a geometric sequence.

graph of a point A point's location in the rectangular or Cartesian plane

greatest integer function The function whose output for a given x is the greatest integer that is less than or equal to x, denoted by $[\ \ ]$

grouping symbol A symbol (such as a radical sign or a fraction bar) or pair of symbols (such as parentheses or braces) that indicate that these operations should be computed before other operations

half-life The time it takes for one-half of a sample of a radioactive element to decompose

half-plane A subset of the coordinate plane consisting of all points on a given side of a boundary line

horizontal line A line parallel to the *x*-axis; a line with the equation $y = b$

horizontal-line test A test used to determine if a given function is one-to-one: If every horizontal line that intersects the graph of the function does so exactly once, then the graph is the graph of a one-to-one function.

horizontal translation A graph that is the same shape as a given graph, except that it is shifted horizontally

hyperbola The set of all points in a plane the difference of whose distances from two fixed points (foci) is a constant

hypotenuse The longest side of a right triangle, the side opposite the 90° angle

identity An equation that is true for all values of its variables; its solution set is $\mathbb{R}$.

identity function The function whose rule is $f(x) = x$

imaginary number The square root of a negative number

improper fraction A fraction whose numerator is greater than or equal to its denominator

inconsistent system A system of equations that has no solution; its solution set is $\varnothing$.

increasing function A function whose value increases as the independent variable increases (graph rises as we move to the right)

independent equations A system of two equations with two variables for which each equation's graph is different

independent variable The variable in an equation of two variables to which we assign input values (usually *x* in an equation involving *x* and *y*)

index If $b = \sqrt[n]{a}$, we say *n* is the index of the radical.

inequality A mathematical statement indicating that two quantities are not necessarily equal

inequality symbols A set of six symbols that are used to describe two expressions that are not equal:

$\approx$ "is approximately equal to"
$\neq$ "is not equal to"
$<$ "is less than"
$>$ "is greater than"
$\leq$ "is less than or equal to"
$\geq$ "is greater than or equal to"

infinite sequence A sequence with infinitely many terms

input value A value substituted for the independent variable

integers The set of numbers given by $\{ \ldots, -4, -3, -2, -1, 0, 1, 2, 3, 4, \ldots \}$

integer square An integer that is the second power of another integer

intercept method The method of graphing a linear equation by first graphing the intercepts, and then connecting them by a straight line

intersection of two intervals A combination of two intervals that contains any point in the first interval that is also in the second interval

interval A set of all real numbers between two given real numbers *a* and *b*; it must be specified whether or not *a* and *b* are included in the interval.

interval notation A way of writing intervals using brackets or parentheses to distinguish between endpoints that are included in a given interval, and endpoints that are not included

inverse function The function, denoted by f^{-1}, obtained by reversing the coordinates of each ordered pair $(x, f(x))$; for example, if $f(2) = 3$, then $f^{-1}(3) = 2$.

irrational numbers The set of numbers that cannot be put in the form of a fraction with integer numerator and nonzero integer denominator; can be expressed only as nonterminating, nonrepeating decimals

isosceles triangle A triangle that has two sides of the same length

least (or lowest) common denominator The smallest expression that is exactly divisible by the denominators of a set of rational expressions

like radicals Radicals that have the same index and radicand

like signs Two numbers that are both positive or both negative

like terms Terms containing the same variables with the same exponents

linear equation in one variable An equation of the form $ax + b = c$, where *a*, *b*, and *c* are real numbers and $a \neq 0$

linear equation in two variables An equation of the form $Ax + By = C$, where *A*, *B*, and *C* are real numbers; *A* and *B* cannot both be 0.

linear inequality in one variable An inequality containing an expression of the form $Ax + B$ $(A \neq 0)$

linear inequality in two variables An inequality containing an expression of the form $Ax + By$ (*A* and *B* not both equal to zero)

linear programming A mathematical technique used to find the optimal allocation of resources

literal equation An equation with more than one variable, usually a formula

logarithm The exponent to which *b* is raised to obtain *x*, denoted by $\log_b x$, $b > 0$, $b \neq 1$

logarithmic equation An equation with logarithmic expressions that contain a variable $y = \log_b x$ if and only if $x = b^y$

logarithmic function The function $f(x) = \log_b x$; $b \neq 1$, $b > 0$

lowest terms A fraction written in such a way so that no integer greater than 1 divides both its numerator and its denominator; also called **simplest form**

major axis (transverse) of an ellipse The line segment passing through the foci joining the vertices of an ellipse

matrix A rectangular array of numbers

mean The sum of a collection of value divided by the number of values

means In the proportion $\frac{a}{b} = \frac{c}{d}$, the numbers b and c are called the means.

median The middle value of a collection of values, obtained by arranging them in increasing order, and either choosing the middle value, or the average of the two middle values

midpoint The point half way between two given points

minor A number associated with an element of a square matrix used to find a determinant of next lower order, by crossing out the elements of the matrix that are in the same row and column as the element

minor axis of an ellipse The line segment passing through the center of an ellipse perpendicular to the major axis

minuend The first expression in the difference of two expressions; for example, in the expression $(3x^2 + 2x - 5) - (4x^2 - x + 5)$, the expression $(3x^2 + 2x - 5)$ is the minuend.

mixed number A rational number expressed as an integer plus a proper fraction

mode The value that occurs most often in a collection of values; there can be more than one mode for a collection of values.

monomial A polynomial with exactly one term

multiplication principle of events The method of determining the number of ways that one event can be followed by another

multiplication property of equality The property that states if a, b, and c are real numbers and $a = b$, then $ac = bc$.

multiplicative identity Is 1, because for all real numbers a, $a \cdot 1 = 1 \cdot a = a$

multiplicative inverse Two numbers whose product is 1; also called **reciprocals**

natural logarithm A logarithm with a base of e

natural numbers The set of numbers given by $\{1, 2, 3, 4, 5, \ldots\}$; also called **positive integers**

negatives A pair of numbers, a and b, represented by points that lie on opposite sides of the origin and at equal distances from the origin. If a and b are negatives, $a + b = 0$. Also called **opposites** or **additive inverses**

number line A line that is used to represent real numbers or sets of real numbers

numerator The expression above the bar of a fraction

numerical coefficient The number factor of a term; for example, the numerical coefficient of $8ab$ is 8.

objective function A function whose value is maximized or minimized

odd integers The set of integers that cannot be divided exactly by 2

odd root If $b = \sqrt[n]{a}$, b is an odd root if n is odd.

one-to-one function A function for which every output corresponds to exactly one input

opposites A pair of numbers, a and b, represented by points that lie on opposite sides of the origin and at equal distances from the origin. If a and b are opposites, $a + b = 0$. Also called **negatives** or **additive inverses**

ordered pair A pair of real numbers, written as (x, y), that describes a unique point in the Cartesian plane

origin The point on a number line that represents the number zero; the point on the rectangular coordinate system that represents the point $(0, 0)$

output value The value of the dependent variable, determined by the choice of input value

parabola The graph of a quadratic function

partial sum The sum of finitely many consecutive terms of a series, starting at the first term

Pascal's triangle A triangular array of numbers in which 1's begin and end each row, and each entry between the 1's is the sum of the closest pair of numbers in the line immediately above it; the first 4 rows of Pascal's triangle are

$$1$$
$$1 \qquad 1$$
$$1 \qquad 2 \qquad 1$$
$$1 \qquad 3 \qquad 3 \qquad 1$$

percent The numerator of a fraction whose denominator is 100

perfect cubes The exact third power of another integer or polynomial; for example, 8 is a perfect cube (2^3) as is $a^3 + 3a^2b + 3ab^2 + b^3$, which is equal to $(a + b)^3$.

perfect squares The exact second power of another integer or polynomial; for example, 9 is a perfect square (3^2) as is $a^2 + 2ab + b^2$, which is equal to $(a + b)^2$.

perfect trinomial square A trinomial of the form $a^2 + 2ab + b^2$ or of the form $a^2 - 2ab + b^2$

perimeter The distance around a non-circular geometric figure

periodic interest rate The interest rate used for each compounding period

permutations The number of ways to choose r things taken from a set of n things if order is important

perpendicular lines Two lines that meet at a 90° angle, two lines whose slopes are negative reciprocals of each other, or a horizontal line and a vertical line

piecewise-defined function A function defined by using different equations for different parts of its domain

plane The graph of an equation of the form $ax + by + cz = d$; a, b, c NOT all equal to 0.

point circle A circle with radius zero; a single point

point-slope form The equation of a line in the form $y - y_1 = m(x - x_1)$, where m is the slope and (x_1, y_1) is a point on the line

polynomial An algebraic expression that is a single term or the sum of several terms containing whole-number exponents on the variables

polynomial function A function whose equation is a polynomial; for example, $f(x) = x^2 - 3x + 6$

positive integers The set of numbers given by {1, 2, 3, 4, 5, . . . }, that is, the set of integers that are greater than zero; also called the **natural numbers**

power of x An expression of the form x^y; also called an **exponential expression**

power rule The rule that states if $a = b$, then $a^n = b^n$ where n is a natural number

prime-factored form A natural number written as the product of factors that are prime numbers

prime number A natural number greater than 1 that can be divided exactly only by 1 and itself

prime polynomial A polynomial that does not factor over the rational numbers

principal square root The positive square root of a number

probability A measure of the likelihood of an event

product The result of multiplying two or more expressions

proper fraction A fraction whose numerator is less than its denominator

proportion A statement that two ratios are equal

pyramid A three-dimensional surface whose cross sections (parallel to the base) are polygons the same shape as the base but decrease in size until it comes to a point

Pythagorean theorem If the length of the hypotenuse of a right triangle is c and the lengths of the two legs are a and b, then $a^2 + b^2 = c^2$.

quadrants The four regions in the rectangular coordinate system formed by the x- and y-axes

quadratic equation An equation that can be written in the form $ax^2 + bx + c = 0$ where $a \neq 0$

quadratic formula A formula that produces the solutions to the general quadratic equation; $x = \dfrac{-b \pm \sqrt{b^2 - 4ac}}{2a}$

quadratic function A polynomial function whose equation in one variable is of second degree; alternately, a function whose equation is of the form $f(x) = ax^2 + bx + c$ $(a \neq 0)$

quadratic inequality An inequality involving a quadratic polynomial in one variable

quotient The result of dividing two expressions

radical sign A symbol used to represent the root of a number

radicand The expression under a radical sign

radius A segment drawn from the center of a circle to a point on the circle; the length of such a line segment

range of a function The set of all output values of a function

range of a relation The set of all second elements of a relation

rate A ratio used to compare two quantities of different units

ratio The comparison of two quantities by their indicated quotient

rational equation An equation that contains one or more rational expressions

rational expression The quotient of two polynomials, where the polynomial in the denominator cannot be equal to 0

rational function A function of the form $f(x) = \dfrac{p(x)}{q(x)}$ where p and q are polynomial functions of x and $q(x) \neq 0$

rationalizing the denominator The process of simplifying a fraction so that there are no radicals in the denominator

rationalizing the numerator The process of simplifying a fraction so that there are no radicals in the numerator

rational numbers The set of fractions that have an integer numerator and a nonzero integer denominator; alternately, any terminating or repeating decimal

real numbers The set that contains the rational numbers and the irrational numbers

reciprocal One number is the reciprocal of another if their product is 1; for example $\frac{3}{11}$ is the reciprocal of $\frac{11}{3}$. Also called **multiplicative inverses**

rectangular coordinate system A grid that allows us to identify each point in a plane with a unique pair of numbers; also called the **Cartesian coordinate system**

rectangular solid A three-dimensional surface whose cross sections (parallel to the base) are the same size and shape as the rectangular base

reflection A graph that is identical in shape to a given graph, except that it is flipped across a line (usually the x- or y-axis)

relation A set of ordered pairs

remainder The computation of $x \div y$ will yield an expression of the form $q + \dfrac{r}{y}$. The quantity r is called the remainder.

remainder theorem The theorem that states if a polynomial function $P(x)$ is divided by $(x - r)$, the remainder is $P(r)$

repeating decimal A decimal expression of a fraction that eventually falls into an infinitely repeating pattern

Richter scale A unit of measure that expresses magnitude (ground motion) of an earthquake

right angle An angle whose measure is 90°

right triangle A triangle that has an angle whose measure is 90°

root A number that makes an equation true when substituted for its variable; if there are several variables, then the set of numbers that make the equation true; also called a **solution**

roster method A method of describing a set by listing its elements within braces

sample space The set of all possible outcomes for an experiment

scientific notation The representation of a number as the product of a number between 1 and 10 (1 included), and an integer power of ten, for example, 6.02×10^{23}

sequence An ordered set of numbers that are defined by the position (1st, 2nd, 3rd, etc.) they hold

set A collection of objects whose members are listed or defined within braces

set-builder notation A method of describing a set that uses a variable (or variables) to represent the elements and a rule to determine the possible values of the variable; for example, we can describe the natural numbers as $\{x \mid x$ is an integer and $x > 0\}$.

similar triangles Two triangles that have the same shape, that is, two triangles whose corresponding angles have the same measure

simplest form The form of a fraction where no expression other than 1 divides both its numerator and its denominator

simultaneous solution Values for the variables in a system of equations that satisfy all of the equations

slope-intercept form The equation of the line in the form $y = mx + b$ where m is the slope and b is the y-coordinate of the y-intercept

solution A number that makes an equation true when substituted for its variable; if there are several variables, then the numbers that make the equation true; also called a **root**

solution set The set of all values that makes an equation true when substituted for the variable in the original equation

sphere A three-dimensional surface in which all points are equidistant from a fixed point (center); a ball shape

square matrix A matrix with the same number of rows as columns

square root If $a = b^2$, then b is a square root of a.

square-root function The function $f(x) = \sqrt{x}$

square-root property The property that states the equation $x^2 = c$ has two solutions: $\sqrt{c}$ and $-\sqrt{c}$

square units The units given to an area measurement

squaring function The function $f(x) = x^2$

standard deviation A measure of how closely a set of data is grouped about its mean

standard form of the equation of a circle The equation of a circle written in the form $(x - h)^2 + (y - k)^2 = r^2$, where (h, k) is the center and r is the radius

standard form of the equation of an ellipse The equation of an ellipse written in the form $\frac{(x - h)^2}{a^2} + \frac{(y - k)^2}{b^2} = 1$, where (h, k) is the center.

standard form of the equation of a hyperbola The equation of a hyperbola written in the form $\frac{(x - h)^2}{a^2} - \frac{(y - k)^2}{b^2} = 1$, where (h, k) is the center.

standard notation The representation of a given number as an integer part, followed by a decimal; for example, 212.3337012

step function A function whose graph is a series of horizontal line segments

straight angle An angle whose measure is 180°

subset A set, all of whose elements are included in a different set; for example, the set $\{1, 3, 5\}$ is a subset of the set $\{1, 2, 3, 4, 5\}$.

subtrahend The second expression in the difference of two expressions; for example, in the expression $(3x^2 + 2x - 5) - (4x^2 - x + 5)$, the expression $(4x^2 - x + 5)$ is the subtrahend.

sum The result of adding two expressions

sum of infinite geometric series The indicated sum of the terms of a geometric sequence for which $|r| < 1$

summation notation A shorthand notation for indicating the sum of a number of consecutive terms in a series, denoted by Σ

sum of two cubes An expression of the form $a^3 + b^3$

sum of two squares An expression of the form $a^2 + b^2$

supplementary angles Two angles whose measures sum to 180°

synthetic division A method used to divide a polynomial by a binomial of the form $x - r$

system of equations Two or more equations that are solved simultaneously

term An expression that is a number, a variable, or a product of numbers and variables; for example, 37, xyz, and $3x$ are terms.

tree diagram A way of visually representing all the possible outcomes of an event

triangular form of a matrix A matrix with all zeros below its main diagonal

trinomial A polynomial with exactly three terms

union of two intervals A combination of two intervals that contains all points in the first interval combined with any additional points in the second interval

unit cost The ratio of an item's cost to its quantity; expressed with a denominator of 1

unlike signs Two numbers have unlike signs if one is positive and one is negative.

unlike terms Terms that are not like terms

variables Letters that are used to represent real numbers

vertex The lowest point of a parabola that opens upward, or the highest point of a parabola that opens downward

vertex angle of an isosceles triangle The angle in an isosceles triangle formed by the two sides that are of the same length

vertical line A line parallel to the y-axis; a line with the equation $x = b$

vertical-line test A method of determining whether the graph of an equation represents a function

vertical translation A graph that is the same shape as a given graph, only shifted vertically

vertices of a hyperbola The endpoints of the axis that passes through the foci

vertices of an ellipse The endpoints of the major axis

volume The amount of space enclosed by a three-dimensional geometric figure

whole numbers The set of numbers given by {0, 1, 2, 3, 4, 5, . . .}; that is, the set of integers that are greater than or equal to zero

x-axis The horizontal number line in the rectangular coordinate system

x-coordinate The first number in an ordered pair

x-intercept The point $(a, 0)$ where a graph intersects the x-axis

y-axis The vertical number line in the rectangular coordinate system

y-coordinate The second number in an ordered pair

y-intercept The point $(0, b)$ where a graph intersects the y-axis

zero-factor property The property of real numbers that states if the product of two quantities is 0, then at least one of those quantities must be equal to 0

There are several ways that a graph can exhibit symmetry about the coordinate axes and the origin. It is often easier to draw graphs of equations if we first find the x- and y-intercepts and find any of the following symmetries of the graph:

1. **y-axis symmetry:** If the point $(-x, y)$ lies on a graph whenever the point (x, y) does, as in Figure I-1(a), we say that the graph is **symmetric about the y-axis**.

2. **Symmetry about the origin:** If the point $(-x, -y)$ lies on the graph whenever the point (x, y) does, as in Figure I-1(b), we say that the graph is **symmetric about the origin**.

3. **x-axis symmetry:** If the point $(x, -y)$ lies on the graph whenever the point (x, y) does, as in Figure I-1(c), we say that the graph is **symmetric about the x-axis**.

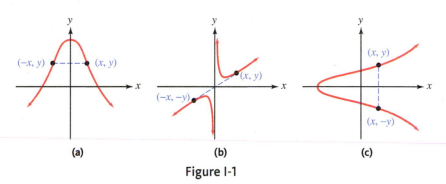

Figure I-1

Tests for Symmetry for Graphs in x and y

- To test a graph for y-axis symmetry, replace x with $-x$. If the new equation is equivalent to the original equation, the graph is symmetric about the y-axis. Symmetry about the y-axis will occur whenever x appears with only even exponents.

- To test a graph for symmetry about the origin, replace x with $-x$ and y with $-y$. If the resulting equation is equivalent to the original equation, the graph is symmetric about the origin.

- To test a graph for x-axis symmetry, replace y with $-y$. If the resulting equation is equivalent to the original equation, the graph is symmetric about the x-axis. The only function that is symmetric about the x-axis is $f(x) = 0$.

EXAMPLE 1 Find the intercepts and the symmetries of the graph of $y = f(x) = x^3 - 9x$. Then graph and state the domain and range.

Solution **x-intercepts:** To find the x-intercepts, we let $y = 0$ and solve for x.

$$y = x^3 - 9x$$
$$0 = x^3 - 9x \qquad \text{Substitute 0 for } y.$$
$$0 = x(x^2 - 9) \qquad \text{Factor out } x, \text{ the GCF.}$$
$$0 = x(x + 3)(x - 3) \qquad \text{Factor } x^2 - 9.$$
$$x = 0 \quad \text{or} \quad x + 3 = 0 \quad \text{or} \quad x - 3 = 0 \qquad \text{Set each factor equal to 0.}$$
$$x = -3 \qquad\qquad x = 3$$

Since the x-coordinates of the x-intercepts are $0, -3$, and 3, the graph intersects the x-axis at $(0, 0), (-3, 0)$, and $(3, 0)$.

y-intercepts: To find the y-intercepts, we let $x = 0$ and solve for y.

$$y = x^3 - 9x$$
$$y = 0^3 - 9(0) \qquad \text{Substitute 0 for } x.$$
$$y = 0$$

Since the y-coordinate of the y-intercept is 0, the graph intersects the y-axis at $(0, 0)$.

Symmetry: We test for symmetry *about the y-axis* by replacing x with $-x$, simplifying, and comparing the result to the original equation.

(1) $\quad y = x^3 - 9x \qquad\qquad$ This is the original equation.
$\qquad y = (-x)^3 - 9(-x) \quad$ Replace x with $-x$.
(2) $\quad y = -x^3 + 9x \qquad\qquad$ Simplify.

Because Equation 2 is not equivalent to Equation 1, the graph is not symmetric about the y-axis.

We test for symmetry *about the origin* by replacing x with $-x$ and y with $-y$, respectively, and comparing the result to the original equation.

(1) $\quad y = x^3 - 9x \qquad\qquad$ This is the original equation.
$\qquad -y = (-x)^3 - 9(-x) \quad$ Replace x with $-x$, and y with $-y$.
$\qquad -y = -x^3 + 9x \qquad\qquad$ Simplify.
(3) $\quad y = x^3 - 9x \qquad\qquad$ Multiply both sides by -1 to solve for y.

Because Equation 3 is equivalent to Equation 1, the graph is symmetric about the origin.

Because the equation is the equation of a nonzero function, there is no symmetry *about the x-axis*.

To graph the equation, we plot the x-intercepts of $(-3, 0), (0, 0)$, and $(3, 0)$ and the y-intercept of $(0, 0)$. We also plot other points for positive values of x and use the symmetry about the origin to draw the rest of the graph, as in Figure I-2(a). (Note that the scale on the x-axis is different from the scale on the y-axis.)

If we graph the equation with a graphing calculator, with window settings of $[-10, 10]$ for x and $[-10, 10]$ for y, we will obtain the graph shown in Figure I-2(b).

From the graph, we can see that the domain is the interval $(-\infty, \infty)$, and the range is the interval $(-\infty, \infty)$.

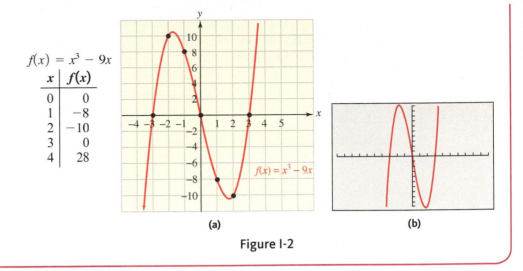

$f(x) = x^3 - 9x$

x	f(x)
0	0
1	−8
2	−10
3	0
4	28

$f(x) = x^3 - 9x$

(a)   (b)

Figure I-2

EXAMPLE 2 Graph the function $f(x) = |x| - 2$ and state the domain and range.

Solution *x*-intercepts: To find the *x*-intercepts, we let $y = 0$ and solve for x.

$$y = |x| - 2$$
$$0 = |x| - 2$$
$$2 = |x|$$
$$x = -2 \quad \text{or} \quad x = 2$$

Since -2 and 2 are solutions, the points $(-2, 0)$ and $(2, 0)$ are the *x*-intercepts, and the graph passes through $(-2, 0)$ and $(2, 0)$.

y-intercepts: To find the *y*-intercepts, we let $x = 0$ and solve for y.

$$y = |x| - 2$$
$$y = |0| - 2$$
$$y = -2$$

Since $y = -2, (0, -2)$ is the *y*-intercept, and the graph passes through the point $(0, -2)$.

Symmetry: To test for *y*-axis symmetry, we replace x with $-x$.

(4) $y = |x| - 2$ This is the original equation.

        $y = |-x| - 2$ Replace x with $-x$.

(5) $y = |x| - 2$ $|-x| = |x|$

Since Equation 5 is equivalent to Equation 4, the graph is symmetric about the *y*-axis. The graph has no other symmetries.

We plot the *x*- and *y*-intercepts and several other points (x, y), and use the *y*-axis symmetry to obtain the graph shown in Figure I-3(a) on the next page.

If we graph the equation with a graphing calculator, with window settings of $[-10, 10]$ for *x* and $[-10, 10]$ for *y*, we will obtain the graph shown in Figure I-3(b).

From the graph, we see that the domain is the interval $(-\infty, \infty)$, and the range is the interval $[-2, \infty)$.

$f(x) = |x| - 2$

x	$f(x)$
0	-2
1	-1
2	0
3	1
4	2

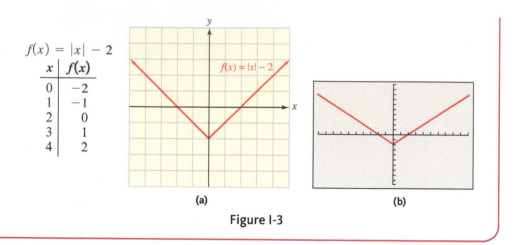

(a)   (b)

Figure I-3

A.1   Exercises

Find the symmetries of the graph of each equation. Do not draw the graph.

1. $y = x^2 - 1$

2. $y = x^3$

3. $y = x^5$

4. $y = x^4$

5. $y = -x^2 + 2$

6. $y = x^3 + 1$

7. $y = x^2 - x$

8. $y^2 = x + 7$

9. $y = -|x + 2|$

10. $y = |x| - 3$

11. $|y| = x$

12. $y = 2\sqrt{x}$

Graph each function and state its domain and range. Check each graph with a graphing calculator.

13. $f(x) = x^4 - 4$

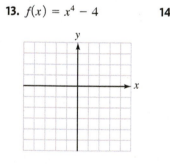

14. $f(x) = \dfrac{1}{2}x^4 - 1$

15. $f(x) = -x^3$

16. $f(x) = x^3 + 2$

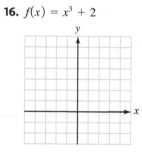

17. $f(x) = x^4 + x^2$

18. $f(x) = 3 - x^4$

19. $f(x) = x^3 - x$

20. $f(x) = x^3 + x$

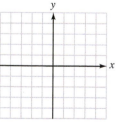

21. $f(x) = \dfrac{1}{2}|x| - 1$

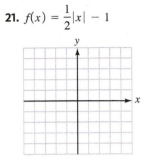

22. $f(x) = -|x| + 1$

23. $f(x) = -|x + 2|$

24. $f(x) = |x - 2|$

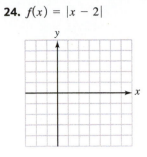

1. How many prime numbers are there in the interval from 40 to 50?
   a. 3   b. 4   c. 5   d. 6   e. none of the above

2. The commutative property of multiplication is written symbolically as
   a. $ab = ba$ b. $(ab)c = a(bc)$
   c. If $a = b$, then $b = a$.   d. $a(b + c) = ab + ac$
   e. none of the above

3. If $a = 3, b = -2$, and $c = 6$, the value of $\dfrac{c - ab}{bc}$ is
   a. 1   b. 0   c. $-\dfrac{11}{12}$ d. -1 e. none of the above

4. Evaluate: $\dfrac{1}{2} + \dfrac{3}{4} \div \dfrac{5}{6}$
   a. $\dfrac{3}{2}$ b. $\dfrac{7}{2}$ c. $\dfrac{9}{8}$ d. $\dfrac{25}{4}$ e. none of the above

5. The expression $\left(\dfrac{a^2}{a^5}\right)^{-5}$ can be written as
   a. 15   b. a^{-15} c. a^{15} d. a^2 e. none of the above

6. Write 0.0000234 in scientific notation.
   a. 2.34×10^{-5} b. 2.34×10^5 c. $234. \times 10^{-7}$
   d. 2.34×10^7 e. none of the above

7. If $P(x) = 2x^2 - x - 1$, find $P(-1)$.
   a. 4   b. 0   c. 2   d. -4 e. none of the above

8. Simplify: $(3x + 2) - (2x - 1) + (x - 3)$
   a. $2x - 2$ b. $2x$ c. 2   d. -2
   e. none of the above

9. Multiply: $(3x - 2)(2x + 3)$
   a. $6x^2 - 6$ b. $6x^2 - 5x - 6$
   c. $6x^2 + 5x + 6$ d. $6x^2 + 5x - 6$
   e. none of the above

10. Divide: $2x + 1 \overline{)2x^2 - 3x - 2}$
    a. $x + 2$ b. $x - 1$ c. $x - 2$ d. $x - 4$
    e. none of the above

11. Solve for x: $5x - 3 = -2x + 10$
    a. $\dfrac{11}{7}$ b. $\dfrac{13}{3}$ c. 1   d. 20   e. none of the above

12. The sum of two consecutive odd integers is 44. The product of the integers is
    a. an even integer   b. 463   c. 46   d. 483
    e. none of the above

13. Solve for x: $2ax - a = b + x$
    a. $\dfrac{a + b}{2a}$ b. $\dfrac{a + b}{2a - 1}$ c. $a + b - 2a$
    d. $-a + b + 1$ e. none of the above

14. The sum of the solutions of $|2x + 5| = 13$ is
    a. 4   b. 8   c. 12   d. 13   e. none of the above

15. Solve for x: $-2x + 5 > 9$
    a. $x > 7$ b. $x < 7$ c. $x < -2$ d. $x > -2$
    e. none of the above

16. Solve for x: $|2x - 5| \leq 9$
    a. $2 \leq x \leq 7$ b. $-2 \geq x$ and $x \leq 7$ c. $x \leq 7$
    d. $-2 \leq x \leq 7$ e. none of the above

17. Factor completely: $3ax^2 + 6a^2x$
    a. $3(ax^2 + 2a^2x)$ b. $3a(x^2 + 2ax)$
    c. $3x(ax + 2a^2)$ d. $3ax(x + 2)$
    e. none of the above

18. The sum of the prime factors of $x^4 - 16$ is
    a. $2x^2$ b. $x^2 + x + 6$ c. $4x + 4$
    d. $x^2 + 2x + 4$ e. none of the above

19. The sum of the factors of $8x^2 - 2x - 3$ is
    a. $8x - 4$ b. $8x - 3$ c. $6x + 2$
    d. $6x - 2$ e. none of the above

20. One of the factors of $27a^3 + 8$ is
    a. $3a - 2$ b. $9a^2 + 12a + 4$ c. $9a^2 + 6a + 4$
    d. $3a^2 - 6a + 4$ e. none of the above

21. The smallest solution of the equation $6x^2 - 5x - 6 = 0$ is
    a. $-\dfrac{3}{2}$ b. $\dfrac{2}{3}$ c. $\dfrac{3}{2}$ d. $-\dfrac{2}{3}$ e. none of the above

22. Simplify: $\dfrac{x^2 + 5x + 6}{x^2 - 9}$
    a. $-\dfrac{2}{3}$ b. $\dfrac{x + 2}{x - 3}$ c. $\dfrac{-5x + 6}{9}$ d. $5x - 3$
    e. none of the above

23. Simplify: $\dfrac{3x + 6}{x + 3} - \dfrac{x^2 - 4}{x^2 + x - 6}$
    a. $\dfrac{3(x + 2)(x + 2)}{(x + 3)(x + 3)}$ b. 3   c. $\dfrac{1}{3}$ d. $\dfrac{x + 2}{x + 3}$
    e. none of the above

24. The numerator of the sum $\dfrac{y}{x + y} + \dfrac{x}{x - y}$ is
    a. $x^2 + 2xy - y^2$ b. $y + x$ c. $y - x$
    d. $x^2 - y^2$ e. none of the above

25. Simplify: $\dfrac{\dfrac{1}{x} + \dfrac{1}{y}}{\dfrac{1}{y}}$
    a. $\dfrac{1}{xy}$ b. 1   c. $\dfrac{y + x}{x}$ d. $y + 1$
    e. none of the above

26. Solve for y: $\dfrac{2}{y + 1} = \dfrac{1}{y + 1} - \dfrac{1}{3}$
    a. 4   b. 3   c. -4 d. -1 e. none of the above

27. The sum of the x- and y-coordinates of the x- and y-intercepts of the graph of $2x + 3y = 6$ is

   a. $-\dfrac{2}{3}$ **b.** 0 **c.** 5 **d.** 6 **e.** none of the above

28. The slope of the line passing through $(3, -2)$ and $(5, -1)$ is

   a. $-\dfrac{1}{2}$ **b.** 2 **c.** -2 **d.** $\dfrac{1}{2}$ **e.** none of the above

29. The graphs of the equations $\begin{cases} 2x - 3y = 4 \\ 3x + 2y = 1 \end{cases}$

   a. are parallel **b.** are perpendicular **c.** do not intersect **d.** are the same line **e.** none of the above

30. An equation of the line passing through $(-2, 5)$ and $(6, 7)$ is

   a. $y = -\dfrac{1}{4}x - \dfrac{11}{2}$ **b.** $y = \dfrac{1}{4}x + \dfrac{11}{2}$

   c. $y = \dfrac{1}{4}x - \dfrac{11}{2}$ **d.** $y = -\dfrac{1}{4}x + \dfrac{11}{2}$

   e. none of the above

31. If $g(x) = x^2 - 3$, then $g(t + 1)$ is

   a. $t^2 - 2$ **b.** -2 **c.** $t - 2$ **d.** $t^2 + 2t - 2$ **e.** none of the above

32. Assume that d varies directly with t. Find the constant of variation if $d = 12$ when $t = 3$.

   a. 36 **b.** 4 **c.** -36 **d.** $\dfrac{1}{4}$ **e.** none of the above

33. The expression $x^{a/2}\, x^{a/5}$ can be expressed as

   a. $x^{a/7}$ **b.** $x^{7a/10}$ **c.** $x^{a/10}$ **d.** $x^{a/5}$ **e.** none of the above

34. The product of $(x^{1/2} + 2)$ and $(x^{-1/2} - 2)$ is

   a. -3 **b.** $-3 + 2x$ **c.** $-3 - 2x^{1/2} + 2x^{-1/2}$ **d.** 0 **e.** none of the above

35. Completely simplify: $\sqrt{112a^3}$ $(a \geq 0)$

   a. $a\sqrt{112a}$ **b.** $4a^2\sqrt{7a}$ **c.** $4\sqrt{7a^2}$ **d.** $16a\sqrt{7a}$ **e.** none of the above

36. Simplify and combine terms: $\sqrt{50} - \sqrt{98} + \sqrt{128}$

   a. $-2\sqrt{2} + \sqrt{128}$ **b.** $6\sqrt{2}$ **c.** $20\sqrt{2}$ **d.** $-4\sqrt{2}$ **e.** none of the above

37. Rationalize the denominator and simplify: $\dfrac{3}{2 - \sqrt{3}}$

   a. $-6 - \sqrt{3}$ **b.** $6 + 3\sqrt{3}$ **c.** $2(2 + \sqrt{3})$ **d.** $6(2 + \sqrt{3})$ **e.** none of the above

38. The distance between the points $(-2, 3)$ and $(6, -8)$ is

   a. $\sqrt{185}$ **b.** $\sqrt{41}$ **c.** $\sqrt{57}$ **d.** $\sqrt{38}$ **e.** none of the above

39. Solve for x: $\sqrt{x + 7} - 2x = -1$

   a. 2 **b.** $-\dfrac{3}{4}$ **c.** $\varnothing$ **d.** $2, -\dfrac{3}{4}$ **e.** none of the above

40. The graph of $y > 3x + 2$ contains no points

   a. in quadrant I **b.** in quadrant II **c.** above the x-axis **d.** in quadrant IV **e.** none of the above

41. The quadratic formula is

   a. $x = \dfrac{b \pm \sqrt{b^2 - 4ac}}{2a}$ **b.** $x = \dfrac{-b \pm \sqrt{b^2 - 4ac}}{a}$

   c. $x = \dfrac{-b \pm \sqrt{b - 4ac}}{a}$ **d.** $x = \dfrac{-b \pm \sqrt{b^2 - 4ac}}{2a}$

   e. none of the above

42. Write the complex number $(2 + 3i)^2$ in $a + bi$ form.

   a. $-5 + 12i$ **b.** -5 **c.** $13 + 12i$ **d.** 13 **e.** none of the above

43. Write the complex number $\dfrac{i}{3 + i}$ in $a + bi$ form.

   a. $\dfrac{1}{3} + 0i$ **b.** $\dfrac{1}{5} + \dfrac{3}{5}i$ **c.** $\dfrac{1}{8} + \dfrac{3}{8}i$

   d. $\dfrac{1}{10} + \dfrac{3}{10}i$ **e.** none of the above

44. The vertex of the parabola determined by $y = 2x^2 + 4x - 3$ is at point

   a. $(0, -3)$ **b.** $(12, 13)$ **c.** $(-1, -5)$ **d.** $(1, 3)$ **e.** none of the above

45. Solve for x: $\dfrac{2}{x} < 3$

   a. $0 < x < \dfrac{2}{3}$ **b.** $x > \dfrac{2}{3}$ **c.** $x < 0$ or $x > \dfrac{2}{3}$

   d. $x < 0$ and $x > \dfrac{2}{3}$ **e.** none of the above

46. If $f(x) = 2x^2 + 1$, then $f(3) =$

   a. 7 **b.** 19 **c.** 17 **d.** 37 **e.** none of the above

47. The inverse function of $y = 3x + 2$ is

   a. $y = \dfrac{x + 2}{3}$ **b.** $x = 3y - 2$ **c.** $y = \dfrac{x - 2}{3}$

   d. $x = \dfrac{y - 2}{3}$ **e.** none of the above

48. The equation of a circle with center at $(-2, 4)$ and radius of 4 units is

   a. $(x + 2)^2 + (y - 4)^2 = 16$
   b. $(x + 2)^2 + (y - 4)^2 = 4$
   c. $(x + 2)^2 + (y - 4)^2 = 2$
   d. $(x - 2)^2 + (y + 4)^2 = 16$
   e. none of the above

49. Find the sum of the solutions of the system

   $\begin{cases} \dfrac{4}{x} + \dfrac{2}{y} = 2 \\[2mm] \dfrac{2}{x} - \dfrac{3}{y} = -1 \end{cases}$

   a. 6 **b.** -6 **c.** 2 **d.** -4 **e.** none of the above

50. Find the y-value of the solution of the system

   $\begin{cases} 4x + 6y = 5 \\ 8x - 9y = 3 \end{cases}$

   a. $\dfrac{3}{4}$ **b.** $\dfrac{1}{3}$ **c.** $\dfrac{1}{2}$ **d.** $\dfrac{2}{3}$ **e.** none of the above

51. The value of the determinant $\begin{vmatrix} 2 & -3 \\ 4 & 4 \end{vmatrix}$ is

   a. 0 **b.** 20 **c.** -4 **d.** -20 **e.** none of the above

52. The value of z in the system $\begin{cases} x + y + z = 4 \\ 2x + y + z = 6 \text{ is} \\ 3x + y + 2z = 8 \end{cases}$
  a. 0   **b.** 1   **c.** 2   **d.** 3   **e.** none of the above

53. $\log_a N = x$ means
  a. $a^x = N$ **b.** $a^N = x$ **c.** $x^N = a$ **d.** $N^a = x$
  e. none of the above

54. Find $\log_2 \dfrac{1}{32}$.
  a. 5   **b.** $\dfrac{1}{5}$ **c.** -5 **d.** $-\dfrac{1}{5}$ **e.** none of the above

55. Find $\log 7 + \log 5$.
  a. $\log 12$ **b.** $\log \dfrac{7}{12}$ **c.** $\log 2$ **d.** $\log 35$
  e. none of the above

56. Find $b^{\log_b x}$.
  a. b **b.** x **c.** 10   **d.** 0   **e.** none of the above

57. Solve for y: $\log y + \log (y + 3) = 1$
  a. $-5, 2$ **b.** 2   **c.** 5   **d.** -5 **e.** none of the above

58. The coefficient of the third term in the expansion of $(a + b)^6$ is
  a. 30   **b.** 2!   **c.** 15   **d.** 120   **e.** none of the above

59. Compute: $P(7, 3)$
  a. 35   **b.** 210   **c.** 21   **d.** 5,040   **e.** none of the above

60. Compute: $C(7, 3)$
  a. 35   **b.** 210   **c.** $7! \cdot 3!$ **d.** 24   **e.** none of the above

61. Find the 100th term of $2, 5, 8, 11, \ldots$.
  a. 301   **b.** 295   **c.** 297   **d.** 299   **e.** none of the above

62. Find the sum of all the terms of $1, \dfrac{1}{3}, \dfrac{1}{9}, \dfrac{1}{27}, \ldots$.
  a. 2   **b.** $\dfrac{2}{3}$ **c.** $\dfrac{3}{2}$ **d.** $\dfrac{5}{3}$ **e.** none of the above

63. Find the probability of rolling a 5 or 6 on one roll of a fair die.
  a. $\dfrac{1}{6}$ **b.** $\dfrac{1}{3}$ **c.** $\dfrac{1}{2}$ **d.** $\dfrac{11}{12}$ **e.** none of the above

APPENDIX 3: Tables

TABLE A Powers and Roots

n	n^2	$\sqrt{n}$	n^3	$\sqrt[3]{n}$	n	n^2	$\sqrt{n}$	n^3	$\sqrt[3]{n}$
1	1	1.000	1	1.000	51	2,601	7.141	132,651	3.708
2	4	1.414	8	1.260	52	2,704	7.211	140,608	3.733
3	9	1.732	27	1.442	53	2,809	7.280	148,877	3.756
4	16	2.000	64	1.587	54	2,916	7.348	157,464	3.780
5	25	2.236	125	1.710	55	3,025	7.416	166,375	3.803
6	36	2.449	216	1.817	56	3,136	7.483	175,616	3.826
7	49	2.646	343	1.913	57	3,249	7.550	185,193	3.849
8	64	2.828	512	2.000	58	3,364	7.616	195,112	3.871
9	81	3.000	729	2.080	59	3,481	7.681	205,379	3.893
10	100	3.162	1,000	2.154	60	3,600	7.746	216,000	3.915
11	121	3.317	1,331	2.224	61	3,721	7.810	226,981	3.936
12	144	3.464	1,728	2.289	62	3,844	7.874	238,328	3.958
13	169	3.606	2,197	2.351	63	3,969	7.937	250,047	3.979
14	196	3.742	2,744	2.410	64	4,096	8.000	262,144	4.000
15	225	3.873	3,375	2.466	65	4,225	8.062	274,625	4.021
16	256	4.000	4,096	2.520	66	4,356	8.124	287,496	4.041
17	289	4.123	4,913	2.571	67	4,489	8.185	300,763	4.062
18	324	4.243	5,832	2.621	68	4,624	8.246	314,432	4.082
19	361	4.359	6,859	2.668	69	4,761	8.307	328,509	4.102
20	400	4.472	8,000	2.714	70	4,900	8.367	343,000	4.121
21	441	4.583	9,261	2.759	71	5,041	8.426	357,911	4.141
22	484	4.690	10,648	2.802	72	5,184	8.485	373,248	4.160
23	529	4.796	12,167	2.844	73	5,329	8.544	389,017	4.179
24	576	4.899	13,824	2.884	74	5,476	8.602	405,224	4.198
25	625	5.000	15,625	2.924	75	5,625	8.660	421,875	4.217
26	676	5.099	17,576	2.962	76	5,776	8.718	438,976	4.236
27	729	5.196	19,683	3.000	77	5,929	8.775	456,533	4.254
28	784	5.292	21,952	3.037	78	6,084	8.832	474,552	4.273
29	841	5.385	24,389	3.072	79	6,241	8.888	493,039	4.291
30	900	5.477	27,000	3.107	80	6,400	8.944	512,000	4.309
31	961	5.568	29,791	3.141	81	6,561	9.000	531,441	4.327
32	1,024	5.657	32,768	3.175	82	6,724	9.055	551,368	4.344
33	1,089	5.745	35,937	3.208	83	6,889	9.110	571,787	4.362
34	1,156	5.831	39,304	3.240	84	7,056	9.165	592,704	4.380
35	1,225	5.916	42,875	3.271	85	7,225	9.220	614,125	4.397
36	1,296	6.000	46,656	3.302	86	7,396	9.274	636,056	4.414
37	1,369	6.083	50,653	3.332	87	7,569	9.327	658,503	4.431
38	1,444	6.164	54,872	3.362	88	7,744	9.381	681,472	4.448
39	1,521	6.245	59,319	3.391	89	7,921	9.434	704,969	4.465
40	1,600	6.325	64,000	3.420	90	8,100	9.487	729,000	4.481
41	1,681	6.403	68,921	3.448	91	8,281	9.539	753,571	4.498
42	1,764	6.481	74,088	3.476	92	8,464	9.592	778,688	4.514
43	1,849	6.557	79,507	3.503	93	8,649	9.644	804,357	4.531
44	1,936	6.633	85,184	3.530	94	8,836	9.695	830,584	4.547
45	2,025	6.708	91,125	3.557	95	9,025	9.747	857,375	4.563
46	2,116	6.782	97,336	3.583	96	9,216	9.798	884,736	4.579
47	2,209	6.856	103,823	3.609	97	9,409	9.849	912,673	4.595
48	2,304	6.928	110,592	3.634	98	9,604	9.899	941,192	4.610
49	2,401	7.000	117,649	3.659	99	9,801	9.950	970,299	4.626
50	2,500	7.071	125,000	3.684	100	10,000	10.000	1,000,000	4.642

TABLE B   Base-10 Logarithms

N	0	1	2	3	4	5	6	7	8	9
1.0	.0000	.0043	.0086	.0128	.0170	.0212	.0253	.0294	.0334	.0374
1.1	.0414	.0453	.0492	.0531	.0569	.0607	.0645	.0682	.0719	.0755
1.2	.0792	.0828	.0864	.0899	.0934	.0969	.1004	.1038	.1072	.1106
1.3	.1139	.1173	.1206	.1239	.1271	.1303	.1335	.1367	.1399	.1430
1.4	.1461	.1492	.1523	.1553	.1584	.1614	.1644	.1673	.1703	.1732
1.5	.1761	.1790	.1818	.1847	.1875	.1903	.1931	.1959	.1987	.2014
1.6	.2041	.2068	.2095	.2122	.2148	.2175	.2201	.2227	.2253	.2279
1.7	.2304	.2330	.2355	.2380	.2405	.2430	.2455	.2480	.2504	.2529
1.8	.2553	.2577	.2601	.2625	.2648	.2672	.2695	.2718	.2742	.2765
1.9	.2788	.2810	.2833	.2856	.2878	.2900	.2923	.2945	.2967	.2989
2.0	.3010	.3032	.3054	.3075	.3096	.3118	.3139	.3160	.3181	.3201
2.1	.3222	.3243	.3263	.3284	.3304	.3324	.3345	.3365	.3385	.3404
2.2	.3424	.3444	.3464	.3483	.3502	.3522	.3541	.3560	.3579	.3598
2.3	.3617	.3636	.3655	.3674	.3692	.3711	.3729	.3747	.3766	.3784
2.4	.3802	.3820	.3838	.3856	.3874	.3892	.3909	.3927	.3945	.3962
2.5	.3979	.3997	.4014	.4031	.4048	.4065	.4082	.4099	.4116	.4133
2.6	.4150	.4166	.4183	.4200	.4216	.4232	.4249	.4265	.4281	.4298
2.7	.4314	.4330	.4346	.4362	.4378	.4393	.4409	.4425	.4440	.4456
2.8	.4472	.4487	.4502	.4518	.4533	.4548	.4564	.4579	.4594	.4609
2.9	.4624	.4639	.4654	.4669	.4683	.4698	.4713	.4728	.4742	.4757
3.0	.4771	.4786	.4800	.4814	.4829	.4843	.4857	.4871	.4886	.4900
3.1	.4914	.4928	.4942	.4955	.4969	.4983	.4997	.5011	.5024	.5038
3.2	.5051	.5065	.5079	.5092	.5105	.5119	.5132	.5145	.5159	.5172
3.3	.5185	.5198	.5211	.5224	.5237	.5250	.5263	.5276	.5289	.5302
3.4	.5315	.5328	.5340	.5353	.5366	.5378	.5391	.5403	.5416	.5428
3.5	.5441	.5453	.5465	.5478	.5490	.5502	.5514	.5527	.5539	.5551
3.6	.5563	.5575	.5587	.5599	.5611	.5623	.5635	.5647	.5658	.5670
3.7	.5682	.5694	.5705	.5717	.5729	.5740	.5752	.5763	.5775	.5786
3.8	.5798	.5809	.5821	.5832	.5843	.5855	.5866	.5877	.5888	.5899
3.9	.5911	.5922	.5933	.5944	.5955	.5966	.5977	.5988	.5999	.6010
4.0	.6021	.6031	.6042	.6053	.6064	.6075	.6085	.6096	.6107	.6117
4.1	.6128	.6138	.6149	.6160	.6170	.6180	.6191	.6201	.6212	.6222
4.2	.6232	.6243	.6253	.6263	.6274	.6284	.6294	.6304	.6314	.6325
4.3	.6335	.6345	.6355	.6365	.6375	.6385	.6395	.6405	.6415	.6425
4.4	.6435	.6444	.6454	.6464	.6474	.6484	.6493	.6503	.6513	.6522
4.5	.6532	.6542	.6551	.6561	.6571	.6580	.6590	.6599	.6609	.6618
4.6	.6628	.6637	.6646	.6656	.6665	.6675	.6684	.6693	.6702	.6712
4.7	.6721	.6730	.6739	.6749	.6758	.6767	.6776	.6785	.6794	.6803
4.8	.6812	.6821	.6830	.6839	.6848	.6857	.6866	.6875	.6884	.6893
4.9	.6902	.6911	.6920	.6928	.6937	.6946	.6955	.6964	.6972	.6981
5.0	.6990	.6998	.7007	.7016	.7024	.7033	.7042	.7050	.7059	.7067
5.1	.7076	.7084	.7093	.7101	.7110	.7118	.7126	.7135	.7143	.7152
5.2	.7160	.7168	.7177	.7185	.7193	.7202	.7210	.7218	.7226	.7235
5.3	.7243	.7251	.7259	.7267	.7275	.7284	.7292	.7300	.7308	.7316
5.4	.7324	.7332	.7340	.7348	.7356	.7364	.7372	.7380	.7388	.7396

TABLE B   (continued)

N	0	1	2	3	4	5	6	7	8	9
5.5	.7404	.7412	.7419	.7427	.7435	.7443	.7451	.7459	.7466	.7474
5.6	.7482	.7490	.7497	.7505	.7513	.7520	.7528	.7536	.7543	.7551
5.7	.7559	.7566	.7574	.7582	.7589	.7597	.7604	.7612	.7619	.7627
5.8	.7634	.7642	.7649	.7657	.7664	.7672	.7679	.7686	.7694	.7701
5.9	.7709	.7716	.7723	.7731	.7738	.7745	.7752	.7760	.7767	.7774
6.0	.7782	.7789	.7796	.7803	.7810	.7818	.7825	.7832	.7839	.7846
6.1	.7853	.7860	.7868	.7875	.7882	.7889	.7896	.7903	.7910	.7917
6.2	.7924	.7931	.7938	.7945	.7952	.7959	.7966	.7973	.7980	.7987
6.3	.7993	.8000	.8007	.8014	.8021	.8028	.8035	.8041	.8048	.8055
6.4	.8062	.8069	.8075	.8082	.8089	.8096	.8102	.8109	.8116	.8122
6.5	.8129	.8136	.8142	.8149	.8156	.8162	.8169	.8176	.8182	.8189
6.6	.8195	.8202	.8209	.8215	.8222	.8228	.8235	.8241	.8248	.8254
6.7	.8261	.8267	.8274	.8280	.8287	.8293	.8299	.8306	.8312	.8319
6.8	.8325	.8331	.8338	.8344	.8351	.8357	.8363	.8370	.8376	.8382
6.9	.8388	.8395	.8401	.8407	.8414	.8420	.8426	.8432	.8439	.8445
7.0	.8451	.8457	.8463	.8470	.8476	.8482	.8488	.8494	.8500	.8506
7.1	.8513	.8519	.8525	.8531	.8537	.8543	.8549	.8555	.8561	.8567
7.2	.8573	.8579	.8585	.8591	.8597	.8603	.8609	.8615	.8621	.8627
7.3	.8633	.8639	.8645	.8651	.8657	.8663	.8669	.8675	.8681	.8686
7.4	.8692	.8698	.8704	.8710	.8716	.8722	.8727	.8733	.8739	.8745
7.5	.8751	.8756	.8762	.8768	.8774	.8779	.8785	.8791	.8797	.8802
7.6	.8808	.8814	.8820	.8825	.8831	.8837	.8842	.8848	.8854	.8859
7.7	.8865	.8871	.8876	.8882	.8887	.8893	.8899	.8904	.8910	.8915
7.8	.8921	.8927	.8932	.8938	.8943	.8949	.8954	.8960	.8965	.8971
7.9	.8976	.8982	.8987	.8993	.8998	.9004	.9009	.9015	.9020	.9025
8.0	.9031	.9036	.9042	.9047	.9053	.9058	.9063	.9069	.9074	.9079
8.1	.9085	.9090	.9096	.9101	.9106	.9112	.9117	.9122	.9128	.9133
8.2	.9138	.9143	.9149	.9154	.9159	.9165	.9170	.9175	.9180	.9186
8.3	.9191	.9196	.9201	.9206	.9212	.9217	.9222	.9227	.9232	.9238
8.4	.9243	.9248	.9253	.9258	.9263	.9269	.9274	.9279	.9284	.9289
8.5	.9294	.9299	.9304	.9309	.9315	.9320	.9325	.9330	.9335	.9340
8.6	.9345	.9350	.9355	.9360	.9365	.9370	.9375	.9380	.9385	.9390
8.7	.9395	.9400	.9405	.9410	.9415	.9420	.9425	.9430	.9435	.9440
8.8	.9445	.9450	.9455	.9460	.9465	.9469	.9474	.9479	.9484	.9489
8.9	.9494	.9499	.9504	.9509	.9513	.9518	.9523	.9528	.9533	.9538
9.0	.9542	.9547	.9552	.9557	.9562	.9566	.9571	.9576	.9581	.9586
9.1	.9590	.9595	.9600	.9605	.9609	.9614	.9619	.9624	.9628	.9633
9.2	.9638	.9643	.9647	.9652	.9657	.9661	.9666	.9671	.9675	.9680
9.3	.9685	.9689	.9694	.9699	.9703	.9708	.9713	.9717	.9722	.9727
9.4	.9731	.9736	.9741	.9745	.9750	.9754	.9759	.9763	.9768	.9773
9.5	.9777	.9782	.9786	.9791	.9795	.9800	.9805	.9809	.9814	.9818
9.6	.9823	.9827	.9832	.9836	.9841	.9845	.9850	.9854	.9859	.9863
9.7	.9868	.9872	.9877	.9881	.9886	.9890	.9894	.9899	.9903	.9908
9.8	.9912	.9917	.9921	.9926	.9930	.9934	.9939	.9943	.9948	.9952
9.9	.9956	.9961	.9965	.9969	.9974	.9978	.9983	.9987	.9991	.9996

TABLE C Base-e Logarithms

N	0	1	2	3	4	5	6	7	8	9
1.0	.0000	.0100	.0198	.0296	.0392	.0488	.0583	.0677	.0770	.0862
1.1	.0953	.1044	.1133	.1222	.1310	.1398	.1484	.1570	.1655	.1740
1.2	.1823	.1906	.1989	.2070	.2151	.2231	.2311	.2390	.2469	.2546
1.3	.2624	.2700	.2776	.2852	.2927	.3001	.3075	.3148	.3221	.3293
1.4	.3365	.3436	.3507	.3577	.3646	.3716	.3784	.3853	.3920	.3988
1.5	.4055	.4121	.4187	.4253	.4318	.4383	.4447	.4511	.4574	.4637
1.6	.4700	.4762	.4824	.4886	.4947	.5008	.5068	.5128	.5188	.5247
1.7	.5306	.5365	.5423	.5481	.5539	.5596	.5653	.5710	.5766	.5822
1.8	.5878	.5933	.5988	.6043	.6098	.6152	.6206	.6259	.6313	.6366
1.9	.6419	.6471	.6523	.6575	.6627	.6678	.6729	.6780	.6831	.6881
2.0	.6931	.6981	.7031	.7080	.7129	.7178	.7227	.7275	.7324	.7372
2.1	.7419	.7467	.7514	.7561	.7608	.7655	.7701	.7747	.7793	.7839
2.2	.7885	.7930	.7975	.8020	.8065	.8109	.8154	.8198	.8242	.8286
2.3	.8329	.8372	.8416	.8459	.8502	.8544	.8587	.8629	.8671	.8713
2.4	.8755	.8796	.8838	.8879	.8920	.8961	.9002	.9042	.9083	.9123
2.5	.9163	.9203	.9243	.9282	.9322	.9361	.9400	.9439	.9478	.9517
2.6	.9555	.9594	.9632	.9670	.9708	.9746	.9783	.9821	.9858	.9895
2.7	.9933	.9969	1.0006	.0043	.0080	.0116	.0152	.0188	.0225	.0260
2.8	1.0296	.0332	.0367	.0403	.0438	.0473	.0508	.0543	.0578	.0613
2.9	.0647	.0682	.0716	.0750	.0784	.0818	.0852	.0886	.0919	.0953
3.0	1.0986	.1019	.1053	.1086	.1119	.1151	.1184	.1217	.1249	.1282
3.1	.1314	.1346	.1378	.1410	.1442	.1474	.1506	.1537	.1569	.1600
3.2	.1632	.1663	.1694	.1725	.1756	.1787	.1817	.1848	.1878	.1909
3.3	.1939	.1969	.2000	.2030	.2060	.2090	.2119	.2149	.2179	.2208
3.4	.2238	.2267	.2296	.2326	.2355	.2384	.2413	.2442	.2470	.2499
3.5	1.2528	.2556	.2585	.2613	.2641	.2669	.2698	.2726	.2754	.2782
3.6	.2809	.2837	.2865	.2892	.2920	.2947	.2975	.3002	.3029	.3056
3.7	.3083	.3110	.3137	.3164	.3191	.3218	.3244	.3271	.3297	.3324
3.8	.3350	.3376	.3403	.3429	.3455	.3481	.3507	.3533	.3558	.3584
3.9	.3610	.3635	.3661	.3686	.3712	.3737	.3762	.3788	.3813	.3838
4.0	1.3863	.3888	.3913	.3938	.3962	.3987	.4012	.4036	.4061	.4085
4.1	.4110	.4134	.4159	.4183	.4207	.4231	.4255	.4279	.4303	.4327
4.2	.4351	.4375	.4398	.4422	.4446	.4469	.4493	.4516	.4540	.4563
4.3	.4586	.4609	.4633	.4656	.4679	.4702	.4725	.4748	.4770	.4793
4.4	.4816	.4839	.4861	.4884	.4907	.4929	.4951	.4974	.4996	.5019
4.5	1.5041	.5063	.5085	.5107	.5129	.5151	.5173	.5195	.5217	.5239
4.6	.5261	.5282	.5304	.5326	.5347	.5369	.5390	.5412	.5433	.5454
4.7	.5476	.5497	.5518	.5539	.5560	.5581	.5602	.5623	.5644	.5665
4.8	.5686	.5707	.5728	.5748	.5769	.5790	.5810	.5831	.5851	.5872
4.9	.5892	.5913	.5933	.5953	.5974	.5994	.6014	.6034	.6054	.6074
5.0	1.6094	.6114	.6134	.6154	.6174	.6194	.6214	.6233	.6253	.6273
5.1	.6292	.6312	.6332	.6351	.6371	.6390	.6409	.6429	.6448	.6467
5.2	.6487	.6506	.6525	.6544	.6563	.6582	.6601	.6620	.6639	.6658
5.3	.6677	.6696	.6715	.6734	.6752	.6771	.6790	.6808	.6827	.6845
5.4	.6864	.6882	.6901	.6919	.6938	.6956	.6974	.6993	.7011	.7029

TABLE C (continued)

N	0	1	2	3	4	5	6	7	8	9
5.5	1.7047	.7066	.7084	.7102	.7120	.7138	.7156	.7174	.7192	.7210
5.6	.7228	.7246	.7263	.7281	.7299	.7317	.7334	.7352	.7370	.7387
5.7	.7405	.7422	.7440	.7457	.7475	.7492	.7509	.7527	.7544	.7561
5.8	.7579	.7596	.7613	.7630	.7647	.7664	.7681	.7699	.7716	.7733
5.9	.7750	.7766	.7783	.7800	.7817	.7834	.7851	.7867	.7884	.7901
6.0	1.7918	.7934	.7951	.7967	.7984	.8001	.8017	.8034	.8050	.8066
6.1	.8083	.8099	.8116	.8132	.8148	.8165	.8181	.8197	.8213	.8229
6.2	.8245	.8262	.8278	.8294	.8310	.8326	.8342	.8358	.8374	.8390
6.3	.8405	.8421	.8437	.8453	.8469	.8485	.8500	.8516	.8532	.8547
6.4	.8563	.8579	.8594	.8610	.8625	.8641	.8656	.8672	.8687	.8703
6.5	1.8718	.8733	.8749	.8764	.8779	.8795	.8810	.8825	.8840	.8856
6.6	.8871	.8886	.8901	.8916	.8931	.8946	.8961	.8976	.8991	.9006
6.7	.9021	.9036	.9051	.9066	.9081	.9095	.9110	.9125	.9140	.9155
6.8	.9169	.9184	.9199	.9213	.9228	.9242	.9257	.9272	.9286	.9301
6.9	.9315	.9330	.9344	.9359	.9373	.9387	.9402	.9416	.9430	.9445
7.0	1.9459	.9473	.9488	.9502	.9516	.9530	.9544	.9559	.9573	.9587
7.1	.9601	.9615	.9629	.9643	.9657	.9671	.9685	.9699	.9713	.9727
7.2	.9741	.9755	.9769	.9782	.9796	.9810	.9824	.9838	.9851	.9865
7.3	.9879	.9892	.9906	.9920	.9933	.9947	.9961	.9974	.9988	2.0001
7.4	2.0015	.0028	.0042	.0055	.0069	.0082	.0096	.0109	.0122	.0136
7.5	2.0149	.0162	.0176	.0189	.0202	.0215	.0229	.0242	.0255	.0268
7.6	.0281	.0295	.0308	.0321	.0334	.0347	.0360	.0373	.0386	.0399
7.7	.0412	.0425	.0438	.0451	.0464	.0477	.0490	.0503	.0516	.0528
7.8	.0541	.0554	.0567	.0580	.0592	.0605	.0618	.0631	.0643	.0656
7.9	.0669	.0681	.0694	.0707	.0719	.0732	.0744	.0757	.0769	.0782
8.0	2.0794	.0807	.0819	.0832	.0844	.0857	.0869	.0882	.0894	.0906
8.1	.0919	.0931	.0943	.0956	.0968	.0980	.0992	.1005	.1017	.1029
8.2	.1041	.1054	.1066	.1078	.1090	.1102	.1114	.1126	.1138	.1150
8.3	.1163	.1175	.1187	.1199	.1211	.1223	.1235	.1247	.1258	.1270
8.4	.1282	.1294	.1306	.1318	.1330	.1342	.1353	.1365	.1377	.1389
8.5	2.1401	.1412	.1424	.1436	.1448	.1459	.1471	.1483	.1494	.1506
8.6	.1518	.1529	.1541	.1552	.1564	.1576	.1587	.1599	.1610	.1622
8.7	.1633	.1645	.1656	.1668	.1679	.1691	.1702	.1713	.1725	.1736
8.8	.1748	.1759	.1770	.1782	.1793	.1804	.1815	.1827	.1838	.1849
8.9	.1861	.1872	.1883	.1894	.1905	.1917	.1928	.1939	.1950	.1961
9.0	2.1972	.1983	.1994	.2006	.2017	.2028	.2039	.2050	.2061	.2072
9.1	.2083	.2094	.2105	.2116	.2127	.2138	.2148	.2159	.2170	.2181
9.2	.2192	.2203	.2214	.2225	.2235	.2246	.2257	.2268	.2279	.2289
9.3	.2300	.2311	.2322	.2332	.2343	.2354	.2364	.2375	.2386	.2396
9.4	.2407	.2418	.2428	.2439	.2450	.2460	.2471	.2481	.2492	.2502
9.5	2.2513	.2523	.2534	.2544	.2555	.2565	.2576	.2586	.2597	.2607
9.6	.2618	.2628	.2638	.2649	.2659	.2670	.2680	.2690	.2701	.2711
9.7	.2721	.2732	.2742	.2752	.2762	.2773	.2783	.2793	.2803	.2814
9.8	.2824	.2834	.2844	.2854	.2865	.2875	.2885	.2895	.2905	.2915
9.9	.2925	.2935	.2946	.2956	.2966	.2976	.2986	.2996	.3006	.3016

Use the properties of logarithms and $\ln 10 \approx 2.3026$ to find logarithms of numbers less than 1 or greater than 10.

Exercises 1.1 (page 13)

1. $5, -5$ **3.** yes **5.** $\frac{3}{4}$ **7.** $\frac{4}{5}$ **9.** $\frac{3}{20}$ **11.** $\frac{8}{49}$
13. 1 **15.** $\frac{31}{20}$ **17.** variable **19.** natural **21.** integers
23. even **25.** natural, 1, itself **27.** 0 **29.** rational
31. $<$ **33.** $\approx$ **35.** inequality, compound inequality
37. union, intersection **39.** 0.875, terminating
41. $-0.4\overline{6}$, repeating **43.** 1, 2, 9 **45.** $-3, 0, 1, 2, 9$
47. $\sqrt{3}$ **49.** 2 **51.** 2 **53.** 9 **55.** $<$ **57.** $>$
59. $<$ **61.** $>$ **63.** $12 < 19$ **65.** $-5 \geq -6$
67. $-3 \leq 5$ **69.** $0 > -3$

71.
73.
75. **77.**
79. **81.**
83. **85.**
87. 25 **89.** -5 **91.** 7 **93.** 20 **95.** -10 or 10
97. $x \geq 0$ **99.**

101. **103.**
105. **107.** **113.** 99

Getting Ready (page 15)

1. 9 **2.** 9 **3.** 12 **4.** 12 **5.** 5 **6.** 5
7. 3 **8.** 6

Exercises 1.2 (page 25)

1. addition **3.** subtraction **5.** division
7. **9.**
11. $77.53 **13.** absolute, common **15.** change, add
17. negative **19.** product, quotient **21.** 0
23. $C = \pi D$; $C = 2\pi r$ **25.** $a + b = b + a$ **27.** reciprocals

29. $5, \frac{1}{7}$ **31.** -9 **33.** -5 **35.** $\frac{1}{6}$ **37.** $\frac{7}{36}$
39. -12 **41.** 0 **43.** $\frac{11}{10}$ **45.** $-\frac{1}{6}$ **47.** -12
49. 27 **51.** $-\frac{4}{7}$ **53.** $-\frac{27}{28}$ **55.** -2 **57.** 7
59. -2 **61.** $\frac{24}{25}$ **63.** 23 **65.** -13 **67.** 1
69. 7 **71.** 1 **73.** 9 **75.** 8 **77.** 9
79. 12 **81.** -8 **83.** $-\frac{5}{4}$ **85.** 19,900 **87.** 100.4 ft
89. comm. prop. of add. **91.** dist. prop.
93. additive identity prop. **95.** mult. inverse prop.
97. assoc. prop. of add. **99.** comm. prop. of mult.
101. assoc. prop. of add. **103.** distrib. prop. **105.** 70
107. 22 **109.** 168 **111.** 4 **113.** -1 **115.** -3
117. -12 **119.** -20 **121.** 11 **123.** 1 **125.** -9
127. -1 **129.** $62 **131.** $+4°$ **133.** $12°$
135. 6,900 gal **137.** $+1,325$ m **139.** $421.88
141. $1,211 **143.** 83 **145.** yes **147.** 30 cm

Getting Ready (page 29)

1. 4 **2.** 27 **3.** -64 **4.** 81 **5.** $\frac{1}{27}$ **6.** $-\frac{16}{625}$

Exercises 1.3 (page 37)

1. 5^3 **3.** $(-y)^2$ **5.** a^2b^3 **7.** $(x^2)^3$ **9.** $\left(\frac{2}{3}\right)^4$
11. $-(ab)^3$ **13.** 7 **15.** 1 **17.** base, exponent, power
19. x^{m+n} **21.** $x^n y^n$ **23.** 1 **25.** x^{m-n} **27.** $A = s^2$
29. $A = \frac{1}{2}bh$ **31.** $A = \pi r^2$ **33.** $V = lwh$
35. $V = Bh$ **37.** $V = \frac{1}{3}Bh$ **39.** base is 5, exponent is 3
41. base is x, exponent is 5 **43.** base is b, exponent is 6
45. base is $-mn^2$, exponent is 3 **47.** 9 **49.** -9 **51.** 9
53. $\frac{16}{625}x^4$ **55.** x^8 **57.** x^9y^8 **59.** $-4x^9y^5$ **61.** p^{10}
63. x^{18} **65.** x^6y^4 **67.** a^{20} **69.** x^{22} **71.** $64x^6$
73. $x^{12}y^8$ **75.** $\frac{a^{28}}{b^{35}}$ **77.** $\frac{1}{16}a^8b^{20}$ **79.** 1 **81.** -1
83. 7 **85.** 50 **87.** $\frac{1}{25}$ **89.** $\frac{1}{81x^4}$ **91.** $\frac{1}{x^4}$ **93.** x^5
95. a^5 **97.** $\frac{1}{x^{11}}$ **99.** $5a^9b^2$ **101.** $\frac{1}{x^{10}}$ **103.** $\frac{a^6}{b^4}$
105. $\frac{16}{y^{16}}$ **107.** a^{n-1} **109.** b^{3n-9} **111.** 108 **113.** $-\frac{1}{216}$
115. $\frac{1}{324}$ **117.** $\frac{27}{8}$ **119.** 15 m^2 **121.** 113 cm^2
123. 343 m^3 **125.** 360 ft^3 **127.** a^4b^5 **129.** $-243p^{10}q^{15}$
131. x^{12} **133.** $\frac{s^3}{r^9}$ **135.** $\frac{1}{b^{72}}$ **137.** $-\frac{1}{d^3}$ **139.** 1
141. $\frac{1}{9x^3}$ **143.** $\frac{1}{a^{n+1}}$ **145.** $\frac{1}{a^{n-2}}$ **147.** $\frac{64b^{12}}{27a^9}$
149. $-\frac{b^3}{8a^{21}}$ **151.** 45 cm^2 **153.** 300 cm^2 **155.** 168 ft^3
157. 2,714 m^3 **159.** 3.462825992 **161.** -244.140625
171. $922,824.13 **177.** $\frac{7}{12}$

Getting Ready (page 40)

1. 10 **2.** 100 **3.** 1,000 **4.** 10,000 **5.** $\frac{1}{100}$
6. $\frac{1}{10,000}$ **7.** 4,000 **8.** $\frac{7}{10,000}$

Exercises 1.4 (page 46)

1. 23,000 **3.** $-5{,}700$ **5.** 0.032 **7.** -0.0051 **9.** 0.75
11. $1.\overline{4}$ **13.** 89 **15.** scientific, 10^n **17.** left
19. 5.2×10^3 **21.** 1.76×10^7 **23.** 5.9×10^{-3}
25. 9.6×10^{-6} **27.** -4.5×10^4 **29.** -3.7×10^{-5}
31. 460 **33.** 796,000 **35.** 6.0×10^{-4} **37.** 5.27×10^3
39. 2×10^{11} **41.** 1.44×10^7 **43.** 0.00526
45. 0.00037 **47.** 23,650,000 **49.** 32,300,000
51. 3.17×10^{-4} **53.** 5.23×10^1 **55.** 4×10^{-2}
57. 6×10^3 **59.** 1.2874×10^{13} **61.** 5.671×10^{10}
63. 0.64 **65.** 3.6×10^{25} **67.** g, x, v, i, r
69. 1.19×10^8 cm/hr **71.** 1.67248×10^{-18} g
73. 1.51×10^{10} in. **75.** 5.9×10^4 mi **77.** 2.54×10^8
79. almost 23 years **81.** 2.5×10^{13} mi **85.** 332

Getting Ready (page 48)

1. 2 **2.** 4 **3.** 3 **4.** 6

Exercises 1.5 (page 57)

1. -5 **3.** 7 **5.** 5 **7.** x **9.** $x + 1$ **11.** $x - 6$
13. 1 **15.** x^8 **17.** $-\frac{1}{64x^3}$ **19.** equation
21. equivalent **23.** c, c **25.** term
27. numerical coefficients, variables **29.** yes **31.** yes
33. 2 **35.** -3 **37.** 9 **39.** 28 **41.** $\frac{2}{3}$ **43.** 4
45. -8 **47.** $-\frac{16}{3}$ **49.** $15x - 27$ **51.** $5x + 10$
53. $-4x + 9$ **55.** $21x - 19$ **57.** -11 **59.** 13
61. -8 **63.** $-\frac{7}{3}$ **65.** 6 **67.** 4 **69.** 3 **71.** 0
73. -2 **75.** 24 **77.** 6 **79.** -6 **81.** 3 **83.** 2,500
85. identity, $\mathbb{R}$ **87.** contradiction, $\varnothing$ **89.** $w = \frac{A}{l}$
91. $B = \frac{3V}{h}$ **93.** $t = \frac{I}{pr}$ **95.** $w = \frac{p - 2l}{2}$
97. $B = \frac{2A}{h} - b$ **99.** $x = \frac{y - b}{m}$ **101.** $l = \frac{a - S + Sr}{r}$
103. $l = \frac{2S - na}{n}$ **105.** $5x + 10$ **107.** $-3x - 5$
109. 3 **111.** $\frac{13}{12}$ **113.** 3 **115.** 9 **117.** -4
119. -6 **121.** 13 **123.** 1.7 **125.** $\varnothing$, contradiction
127. $-\frac{5}{2}$ **129.** $n = \frac{l - a + d}{d}$ **131.** $m = \frac{Fd^2}{(GM)}$
133. $C = \frac{5}{9}(F - 32)$; 0°, 21.1°, 100°
135. $n = \frac{100 \times (C - 6.50)}{7}$; 621, 1,000, 1,692.9 kWh
137. $R = \frac{E}{I}$; $R = 8$ ohms **139.** $n = \frac{360°}{(180° - a)}$; 8

Getting Ready (page 60)

1. 20 **2.** 72 **3.** 41 **4.** 7

Exercises 1.6 (page 66)

1. 195 **3.** $756 **5.** 40 m^2 **7.** $\frac{256x^{20}}{81}$ **9.** a^{n-1}
11. $5x + 4$ **13.** $89x$ **15.** acute **17.** 180°
19. isosceles **21.** equilateral **23.** 48 ft **25.** 7 ft, 15 ft
27. 20% **29.** $333\frac{1}{3}\%$ **31.** 300 shares of BB, 200 shares of SS
33. 35 $15 calculators, 50 $67 calculators **35.** 60°
37. 10 **39.** 72.5° **41.** 56° **43.** 12 m by 24 m
45. 156 ft by 312 ft **47.** 5 ft **49.** 90 lb **51.** 12 roses
53. 60 mi, 120 mi **55.** $578 **57.** 7 in. **59.** 50
61. 10 ft **63.** 4 ft **65.** $-40°$

Getting Ready (page 70)

1. $40 **2.** $0.08x$ **3.** $6,000 **4.** $(15,000 - x)$
5. 200 mi **6.** $20

Exercises 1.7 (page 74)

1. $30 **3.** $(480 - 0.04x)$ **5.** $75 **7.** 1 **9.** 8
11. principal, rate, time **13.** value, price, number
15. $2,000 in a money market, $10,000 in a 5-year CD
17. 10.8% **19.** $8,250 **21.** $3\frac{1}{2}$ hr **23.** $\frac{2}{3}$ hr
25. $\frac{1}{8}$ hr **27.** 20 lb of $2.95 candy; 10 lb of $3.10 candy
29. 15 **31.** 2 gal **33.** 30 **35.** $80 **37.** $1\frac{1}{2}$ hr
39. 125 **41.** 12 mi **43.** 10 oz **45.** 6 in.

Chapter Review (page 79)

1. 0, 1, 2, 4 **2.** 1, 2, 4 **3.** $-4, -\frac{2}{3}, 0, 1, 2, 4$
4. $-4, 0, 1, 2, 4$ **5.** π **6.** $-4, -\frac{2}{3}, 0, 1, 2, \pi, 4$
7. $-4, -\frac{2}{3}$ **8.** $1, 2, \pi, 4$ **9.** 2 **10.** 4 **11.** $-4, 0, 2, 4$
12. 1 **13.**

14.

15.

16.

17.

18.

19.

20.

21.

22. 0 **23.** 1 **24.** -5
25. -6 **26.** 8 **27.** 22 **28.** -28 **29.** -27
30. 3 **31.** -9 **32.** 5 **33.** -4 **34.** -13
35. 5 **36.** -13 **37.** -15 **38.** 39 **39.** 53
40. 21 **41.** 42 **42.** 4 **43.** 5 **44.** -12
45. -24 **46.** -4 **47.** -4 **48.** 12 **49.** 55
50. 0 **51.** -60 **52.** -1 **53.** 4 **54.** 15.4
55. 15 **56.** 15 **57.** yes **58.** $-\frac{19}{6}$ **59.** $\frac{2}{15}$
60. $-\frac{4}{3}$ **61.** $\frac{11}{5}$ **62.** distrib. prop. **63.** comm. prop. of add. **64.** assoc. prop. of add. **65.** add. identity prop. **66.** add. inverse prop. **67.** comm. prop. of mult. **68.** assoc. prop. of mult. **69.** mult. identity prop. **70.** mult. inverse prop. **71.** double negative rule **72.** 625 **73.** -243 **74.** -64
75. $-\frac{1}{625}$ **76.** $-6x^6$ **77.** $-3x^8$ **78.** $\frac{1}{x^3}$
79. $\frac{1}{y^5}$ **80.** $27x^6$ **81.** $256x^{16}$ **82.** $-32x^{10}$
83. $243x^{15}$ **84.** $\frac{1}{x^{10}}$ **85.** x^{20} **86.** $\frac{16}{x^{10}}$ **87.** $\frac{x^{24}}{81}$
88. x^2 **89.** x^5 **90.** $\frac{1}{a^5}$ **91.** $\frac{1}{a^3}$ **92.** $\frac{1}{y^7}$ **93.** y^9
94. $\frac{1}{x}$ **95.** x^3 **96.** $\frac{1}{81a^{12}b^8}$ **97.** $\frac{64y^3}{125}$ **98.** 135
99. -9 **100.** 1.93×10^{10} **101.** 2.73×10^{-8}
102. 8,400,000 **103.** 0.0000000083
104. 5.02×10^{11} gallons **105.** -4 **106.** -5 **107.** 8
108. -8 **109.** 19 **110.** 8 **111.** 12 **112.** $\varnothing$
113. 2 **114.** $\mathbb{R}$ **115.** $r^3 = \frac{3V}{4\pi}$ **116.** $h = \frac{3V}{\pi r^2}$
117. $x = \frac{6v}{ab} - y$ or $x = \frac{6v - aby}{ab}$ **118.** $r = \frac{V}{\pi h^2} + \frac{h}{3}$

119. 5 ft from one end   **120.** 209 m²   **121.** $2\frac{2}{3}$ ft
122. $18,000 at 10%, $7,000 at 9%   **123.** 10 liters   **124.** $\frac{1}{3}$ hr

Chapter 1 Test (page 87)

1. 1, 2, 5   **2.** $\sqrt{7}$
3.
4.
5. **6.**
7. **8.**
9. -8 **10.** 5   **11.** 2   **12.** 20   **13.** -4
14. 1   **15.** 0.3   **16.** 0.5   **17.** -6 **18.** -10
19. 6   **20.** 1   **21.** comm. prop. of add.
22. distrib. prop.   **23.** x^8 **24.** $8x^6y^9$ **25.** $\frac{1}{m^8}$
26. $\frac{m^4}{n^{10}}$ **27.** 4.7×10^6 **28.** 2.3×10^{-7} **29.** 653,000
30. -21 **31.** -12 **32.** -2 **33.** 6   **34.** $\varnothing$
35. $i = \frac{f(P - L)}{s}$ **36.** 36 cm²   **37.** $4,000   **38.** 80 liters

Getting Ready (page 91)

1.
2. **3.**
4.

Exercises 2.1 (page 102)

1. -4 **3.** 2   **5.** 3   **7.** $\frac{3}{2}$ **9.** 9   **11.** $\frac{1}{2}$ **13.** -7
15. ordered pair   **17.** origin   **19.** rectangular or Cartesian
coordinate   **21.** linear   **23.** y-axis
25. vertical   **27.** sub 1   **29–36.**
37. $(2, 4)$ **39.** $(-2, -1)$
41. $(4, 0)$ **43.** $(0, 0)$

45. 5, 4, 2   **47.** $-5, -3, 3$

49.   **51.**

53.   **55.**

57.   **59.**

61. $48   **63.** $3,000   **65.** $(1, 2)$ **67.** $(9, 13)$
69. $\left(\frac{11}{2}, 4\right)$ **71.** $\left(-\frac{7}{2}, 5\right)$

73.   **75.**

77. $(-4, 0)$ **79.** $(9, -2)$ **81.** -1 **83.** -3.933
85. 1.22   **87.** 4.67   **89.** $162,500   **91.** 200
93. 12 mi   **95. a.** $y = 0.25x + 50$ **b.** 60, 70, 80, 90;
$(40, 60), (80, 70), (120, 80), (160, 90)$ **c.** $55

97. a. In 1990, there were 65.5 million swimmers.   **b.** 43 million
101. $a = 0, b > 0$

Getting Ready (page 106)

1. 1   **2.** -1 **3.** $\frac{13}{14}$ **4.** $\frac{3}{14}$

Exercises 2.2 (page 113)

1. $\frac{1}{5}$ 3. 0   5. undefined   7. 5   9. $\frac{x^{12}}{y^8}$

11. 1   13. change   15. rise   17. horizontal

19. positive   21. perpendicular, reciprocals   23. 2

25. $-\frac{8}{9}$ 27. -1 29. $-\frac{1}{3}$ 31. $-\frac{5}{3}$ 33. $\frac{3}{4}$

35. negative   37. positive   39. undefined

41. 0   43. perpendicular   45. neither   47. parallel

49. perpendicular   51. 0   53. undefined   55. $\frac{3}{2}$

57. 0   59. perpendicular   61. neither   63. parallel

65. same line   67. same line   69. not the same line

71. $x = 0$, undefined slope   79. $\frac{1}{4}$ 81. $\frac{1}{25}, \frac{1}{10}, \frac{4}{25}$

83. $\frac{7}{500}$ of a degree increase per year   85. a. $\frac{2}{25}$ b. $\frac{1}{20}, \frac{1}{20}$

87. $525.13 per yr   91. m 93. -4

Getting Ready (page 117)

1. 14   2. $-\frac{5}{3}$ 3. $y = 3x - 4$ 4. $x = \frac{-By - 3}{A}$

Exercises 2.3 (page 126)

1. $y = \frac{2}{3}x - 2$ 3. $\frac{3}{7}$ 5. undefined   7. -3

9. -1 11. 20 oz   13. $y - y_1 = m(x - x_1)$

15. $Ax + By = C$ 17. perpendicular

19. $y - 5 = \frac{1}{2}(x - 1)$ 21. $y = -3(x - 2)$

23. $y = \frac{3}{2}x + 2$ 25. $y = \frac{7}{3}x - 3$ 27. $y = 5x - 12$

29. $y = -7x + 54$ 31. $y = -4$ 33. $y = -\frac{9}{8}x - \frac{29}{8}$

35. $-\frac{5}{2}, (0, 5)$ 37. $-\frac{1}{3}, \left(0, -\frac{5}{6}\right)$

39. $1, (0, -1)$ 41. $\frac{2}{3}, (0, 2)$

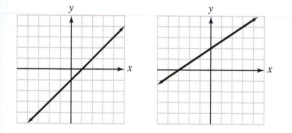

43. parallel   45. perpendicular   47. parallel

49. perpendicular   51. $y = 4x$ 53. $y = 4x - 3$

55. $y = -\frac{1}{7}x - \frac{20}{7}$ 57. $y = -\frac{1}{4}x + \frac{11}{2}$

59. perpendicular   61. parallel   63. perpendicular

65. perpendicular   67. $y = \frac{4}{5}x - \frac{26}{5}$ 69. $y = -\frac{5}{4}x + 3$

71. $\frac{7}{2}, (0, 2)$ 73. $-\frac{2}{3}, (0, 6)$

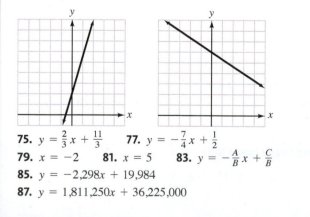

75. $y = \frac{2}{3}x + \frac{11}{3}$ 77. $y = -\frac{7}{4}x + \frac{1}{2}$

79. $x = -2$ 81. $x = 5$ 83. $y = -\frac{A}{B}x + \frac{C}{B}$

85. $y = -2,298x + 19,984$

87. $y = 1,811,250x + 36,225,000$

89. $y = 50,000x + 250,000$ 91. $y = -\frac{950}{3}x + 1,750$

93. $490   95. $154,000   97. $180   107. $a < 0, b > 0$

Getting Ready (page 129)

1. 1   2. 7   3. -20 4. $-\frac{11}{4}$

Exercises 2.4 (page 138)

1. $\{-2, 0, 3\}$ 3. $\{(-2, 0), (0, 3)(3, 9)\}$ 5. $y = \frac{3}{2}x - 4$

7. -2 9. 2   11. relation   13. domain   15. 0

17. cannot   19. slope; y-intercept   21. y 23. y

25. D: $\{-2, -1, 5, 6\}$; R: $\{-4, 1, 3\}$; yes

27. D: $\{-2, 6, 5\}$; R: $\{3, 8, 5, 4\}$; no

29. D: $(-\infty, 1]$; R: $(-\infty, \infty)$; not a function

31. D: $(-\infty, \infty)$; R: $(-\infty, \infty)$; a function   33. $9, -3, 0$

35. $3, -5, \frac{3}{2}$ 37. $4, 9$ 39. $5, 24$ 41. $-1, -1$

43. $2, 2$ 45. $5w, 5w + 5$ 47. $3w - 5, 3w - 2$

49. 12   51. $2b - 2a$ 53. yes   55. yes   57. no

59. no   61. D: $(-\infty, \infty)$ 63. D: $(-\infty, -5) \cup (-5, \infty)$

65. $(-\infty, -3) \cup (-3, \infty)$ 67. $(-\infty, \infty)$

69. D: $(-\infty, \infty)$; R: $(-\infty, \infty)$ 71. D: $(-\infty, \infty)$; R: $(-\infty, \infty)$

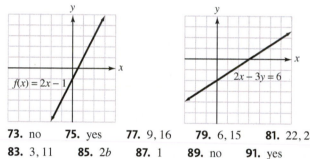

73. no   75. yes   77. $9, 16$ 79. $6, 15$ 81. $22, 2$

83. $3, 11$ 85. $2b$ 87. 1   89. no   91. yes

93. D: $\{2\}$; R: $(-\infty, \infty)$; not a function

95. D: $(-\infty, \infty)$; R: $\{1\}$; function

97. D: $[0, \infty)$; R: $(-\infty, \infty)$; not a function   99. 192

101. a. $I(h) = 0.10h + 50$ b. $61.50   103. 624 ft

105. 12 ft   107. 77°F   111. yes

Getting Ready (page 142)

1. $2, (0, -3)$ 2. $-3, (0, 4)$ 3. $6, -9$ 4. $4, \frac{5}{2}$

Exercises 2.5 (page 149)

1. $5, 0$ 3. $26, -9$ 5. $1, 0$ 7. -13 9. $23, 29$

11. $a \cdot b = b \cdot a$ 13. 1   15. quadratic, parabola

17. absolute value   19. horizontal   21. 2, downward

23. 7, to the left   25. reflected

27.   29.

31.

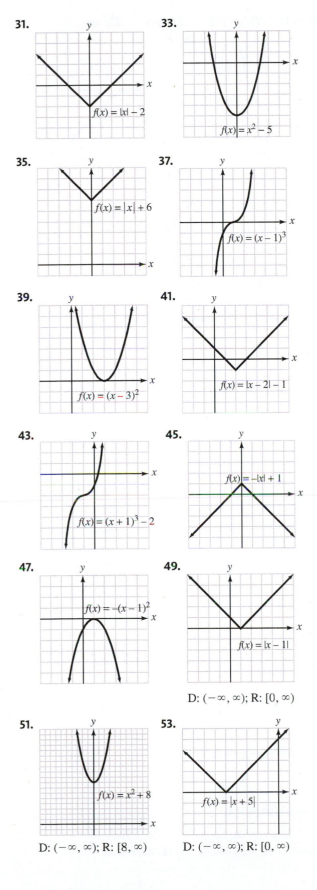

$f(x) = |x| - 2$

33.

$f(x) = x^2 - 5$

35.

$f(x) = |x| + 6$

37.

$f(x) = (x - 1)^3$

39.

$f(x) = (x - 3)^2$

41.

$f(x) = |x - 2| - 1$

43.

$f(x) = (x + 1)^3 - 2$

45.

$f(x) = -|x| + 1$

47.

$f(x) = -(x - 1)^2$

49.

$f(x) = |x - 1|$

D: $(-\infty, \infty)$; R: $[0, \infty)$

51.

$f(x) = x^2 + 8$

D: $(-\infty, \infty)$; R: $[8, \infty)$

53.

$f(x) = |x + 5|$

D: $(-\infty, \infty)$; R: $[0, \infty)$

55.

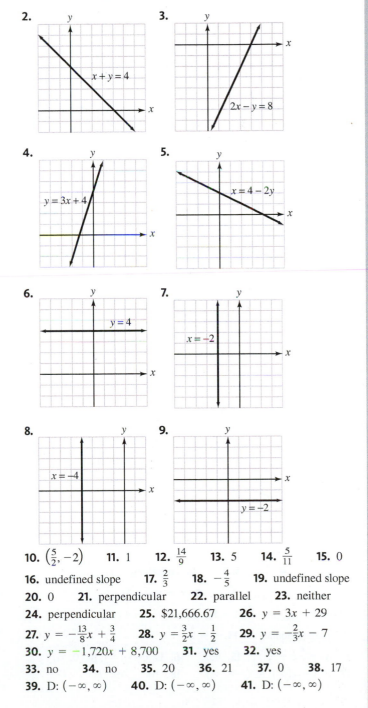

$f(x) = (x - 6)^2$

D: $(-\infty, \infty)$; R: $[0, \infty)$

57.

Chapter Review (page 155)

1. $-10, -6, -6, 0, -2, 6, 9$; $(-9, -10)$, $(-6, -8)$, $(-3, -6)$, $(0, -4)$, $(3, -2)$, $(6, 0)$, $(9, 2)$

2.

$x + y = 4$

3.

$2x - y = 8$

4.

$y = 3x + 4$

5.

$x = 4 - 2y$

6.

$y = 4$

7.

$x = -2$

8.

$x = -4$

9.

$y = -2$

10. $\left(\frac{5}{2}, -2\right)$ **11.** 1   **12.** $\frac{14}{9}$ **13.** 5   **14.** $\frac{5}{11}$ **15.** 0

16. undefined slope   **17.** $\frac{2}{3}$ **18.** $-\frac{4}{5}$ **19.** undefined slope

20. 0   **21.** perpendicular   **22.** parallel   **23.** neither

24. perpendicular   **25.** $21,666.67   **26.** $y = 3x + 29$

27. $y = -\frac{13}{8}x + \frac{3}{4}$ **28.** $y = \frac{3}{2}x - \frac{1}{2}$ **29.** $y = -\frac{2}{3}x - 7$

30. $y = -1,720x + 8,700$ **31.** yes   **32.** yes

33. no   **34.** no   **35.** 20   **36.** 21   **37.** 0   **38.** 17

39. D: $(-\infty, \infty)$ **40.** D: $(-\infty, \infty)$ **41.** D: $(-\infty, \infty)$

42. D: $(-\infty, 2) \cup (2, \infty)$ **43.** D: $(-\infty, 3) \cup (3, \infty)$
44. D: $(-\infty, \infty)$ **45.** a function; D: $(-\infty, \infty)$; R: $(-\infty, \infty)$
46. not a function; D: $[-4, 4]$; R: $[-2, 2]$
47. not a function; D: $(-\infty, \infty)$; R: $[-1, \infty)$
48. a function; D: $(-\infty, \infty)$; R: $[-3, \infty)$ **49.** yes
50. yes **51.** yes **52.** no

53. **54.**

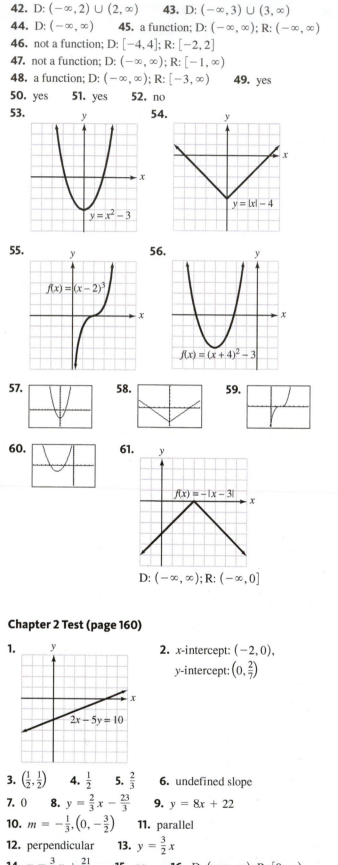

$y = x^2 - 3$

$y = |x| - 4$

55. **56.**

$f(x) = (x - 2)^3$

$f(x) = (x + 4)^2 - 3$

57. **58.** **59.**

60. **61.**

$f(x) = -|x - 3|$

D: $(-\infty, \infty)$; R: $(-\infty, 0]$

Chapter 2 Test (page 160)

1.

$2x - 5y = 10$

2. x-intercept: $(-2, 0)$,
y-intercept: $\left(0, \frac{2}{7}\right)$

3. $\left(\frac{1}{2}, \frac{1}{2}\right)$ **4.** $\frac{1}{2}$ **5.** $\frac{2}{3}$ **6.** undefined slope
7. 0 **8.** $y = \frac{2}{3}x - \frac{23}{3}$ **9.** $y = 8x + 22$
10. $m = -\frac{1}{3}, \left(0, -\frac{3}{2}\right)$ **11.** parallel
12. perpendicular **13.** $y = \frac{3}{2}x$
14. $y = \frac{3}{2}x + \frac{21}{2}$ **15.** no **16.** D: $(-\infty, \infty)$; R: $[0, \infty)$
17. D: $(-\infty, \infty)$; R: $(-\infty, \infty)$ **18.** 10 **19.** 7

20. $3a + 1$ **21.** $x^2 - 2$ **22.** yes **23.** no
24. **25.**

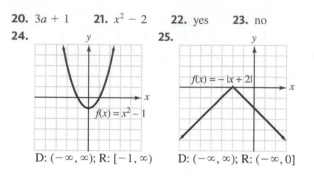

$f(x) = x^2 - 1$

$f(x) = -|x + 2|$

D: $(-\infty, \infty)$; R: $[-1, \infty)$ D: $(-\infty, \infty)$; R: $(-\infty, 0]$

Cumulative Review (page 161)

1. $1, 2, 6, 7$ **2.** $0, 1, 2, 6, 7$ **3.** $-2, 0, 1, 2, \frac{13}{12}, 6, 7$
4. $\sqrt{5}, \pi$ **5.** -2 **6.** $-2, 0, 1, 2, \frac{13}{12}, 6, 7, \sqrt{5}, \pi$
7. $2, 7$ **8.** 6 **9.** $-2, 0, 2, 6$ **10.** $1, 7$
11.

12.

13. 2 **14.** -2 **15.** 23 **16.** -2 **17.** -4
18. -3 **19.** -3 **20.** -5 **21.** assoc. prop. of add.
22. distrib. prop. **23.** comm. prop. of add. **24.** assoc.
prop. of mult. **25.** $-8x^{12}y^3$ **26.** c^2 **27.** $-\dfrac{b^3}{a^2}$
28. $\dfrac{16a^{14}}{b^{14}}$ **29.** 3.82×10^{-5} **30.** 932,000,000
31. 7 **32.** -27 **33.** -1 **34.** 6 **35.** $a = \dfrac{2S}{n} - l$
36. $h = \dfrac{2A}{b_1 + b_2}$ **37.** $28, 30, 32$ **38.** 14 cm by 42 cm
39. yes **40.** $-\frac{7}{5}$ **41.** $y = -\frac{7}{5}x + \frac{11}{5}$
42. $y = -3x - 3$ **43.** 5 **44.** 15 **45.** $2t - 1$
46. $3r^2 + 2$
47. yes; D: $(-\infty, \infty)$; **48.** yes; D: $(-\infty, \infty)$;
R: $(-\infty, 1]$ R: $[0, \infty)$

$y = -x^2 + 1$

$y = \left|\frac{1}{2}x - 3\right|$

Getting Ready (page 165)

1. 2 **2.** -7 **3.** 11 **4.** 3

Exercises 3.1 (page 172)

1. yes **3.** no **5.** 8.5×10^8 **7.** 2.39×10^5
9. system **11.** inconsistent **13.** dependent

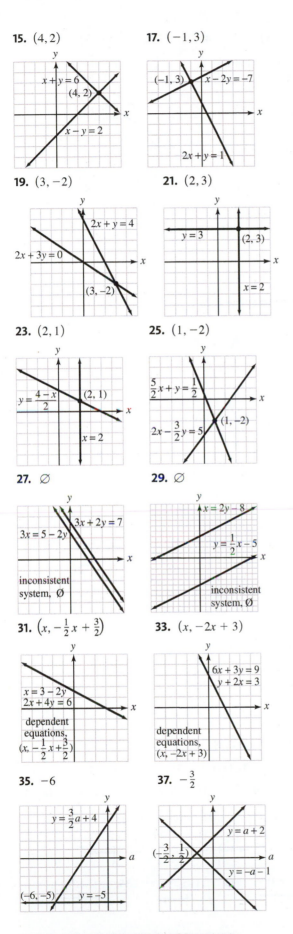

15. $(4, 2)$ **17.** $(-1, 3)$

19. $(3, -2)$ **21.** $(2, 3)$

23. $(2, 1)$ **25.** $(1, -2)$

27. $\varnothing$ **29.** $\varnothing$

31. $\left(x, -\frac{1}{2}x + \frac{3}{2}\right)$ **33.** $(x, -2x + 3)$

35. -6 **37.** $-\frac{3}{2}$

39. no solution   **41.** one solution

43. $\varnothing$ **45.** $\left(x, 2 - \frac{5}{6}x\right)$

47. $\left(3, \frac{5}{2}\right)$ **49.** $\left(-1, \frac{2}{3}\right)$ **51.** $(-0.37, -2.69)$
53. $(-7.64, 7.04)$ **55. a.** \$2 million   **b.** \$3 million
c. 10,000 cameras   **57. a.** yes   **b.** $(3.75, -0.5)$ **c.** no
59. 3 hr   **61.** $(5, 1{,}400)$ **65.** One possible answer is
$$\begin{cases} x + y = -3 \\ x - y = -7 \end{cases}$$

Getting Ready (page 175)

1. $6x - 21$ **2.** $-12x - 20$ **3.** $3x - 9$ **4.** $-2x + 2$

Exercises 3.2 (page 185)

1. 1   **3.** $7x = 14$ **5.** $a^9 b^{14}$ **7.** $\frac{1}{81x^{32}y^4}$ **9.** setup, unit
11. parallelogram   **13.** Opposite   **15.** $(2, 6)$ **17.** $(5, 3)$
19. $(4, 8)$ **21.** $(-2, -3)$ **23.** $(-2, 1)$ **25.** $(4, -2)$
27. $(3, -2)$ **29.** $\left(\frac{3}{4}, \frac{1}{3}\right)$ **31.** $\left(\frac{4}{5}, \frac{3}{4}\right)$ **33.** $(20, -12)$
35. $\varnothing$ **37.** $\varnothing$ **39.** $\left(x, 3x - \frac{3}{2}\right)$ **41.** $(x, 2x + 6)$
43. $\frac{2}{3}$ **45.** $-\frac{691}{1{,}980}$ **47.** $(2, 4)$ **49.** $\left(5, \frac{3}{2}\right)$
51. $(4, -2)$ **53.** $\left(-2, \frac{3}{2}\right)$ **55.** $\left(\frac{1}{2}, -3\right)$ **57.** $(2, -4)$
59. $(2, 3)$ **61.** $\left(-\frac{1}{3}, 1\right)$ **63.** \$65   **65.** 625 Ω, 750 Ω
67. 16 m by 20 m   **69.** \$3,000 at 10%, \$5,000 at 12%
71. 40 oz of 8% solution, 60 oz of 15% solution   **73.** 55 mph
75. 85 racing bikes, 120 mountain bikes   **77.** 200 plates
79. 21   **81.** 750   **83.** 6,500 gal per month
85. A (a smaller loss)   **87.** 590 units per month
89. A (smaller loss)   **91.** A   **93.** $35°, 145°$
95. $x = 22.5, y = 67.5$ **97.** $72°$
99. $f^2 = \frac{1}{4\pi^2 LC}$

Getting Ready (page 190)

1. yes   **2.** yes   **3.** no   **4.** yes

Exercises 3.3 (page 196)

1. yes   **3.** $\frac{8}{3}$ **5.** 1   **7.** $18s^2 + 1$ **9.** plane
11. infinitely   **13.** yes   **15.** $(1, 1, 2)$ **17.** $\left(\frac{3}{4}, \frac{1}{2}, \frac{1}{3}\right)$
19. $\varnothing$ **21.** $(x, 2x - 1, 3x + 1)$ **23.** $(x, 0, 1 - x)$
25. $(3, 2, 1)$ **27.** $(2, 6, 9)$ **29.** $\varnothing$ **31.** $\left(\frac{1}{2}, \frac{5}{3} - \frac{4}{3}z, z\right)$
33. $(0, 2, 2)$ **35.** 30 expensive, 50 middle-priced,
100 inexpensive   **37.** $y = x^2 - 4x$ **39.** 1, 6, 12
41. $A = 50°, B = 60°, C = 70°$ **43.** 1 unit of A, 2 units of
B, and 3 units of C   **45.** 250 \$5 tickets, 375 \$3 tickets, 125 \$2
tickets   **47.** 3 poles, 2 bears, 4 deer   **49.** 78%, 21%, 1%
51. $x^2 + y^2 - 2x - 2y - 2 = 0$ **55.** $(1, 1, 0, 1)$

Getting Ready (page 199)

1. 5   8   13   **2.** 0   3   7   **3.** -1 -1 -2
4. 3   3 -5

Exercises 3.4 (page 205)

1. $\begin{cases} 2x - y = 3 \\ x + 5y = 7 \end{cases}$ **3.** $\begin{cases} 3x - 2y + 4z = 1 \\ 5x + 2y - 3z = -7 \\ -x + 9y + 8z = 0 \end{cases}$

5. 4.7×10^8 **7.** 7.5×10^5 **9.** matrix   **11.** 3, columns
13. augmented, coefficient   **15.** $\begin{bmatrix} 2 & -3 \\ 4 & 2 \end{bmatrix}$ **17.** yes
19. 5   **21.** 2   **23.** $(3,1)$ **25.** $(2,-2)$ **27.** $(1,2,3)$
29. $(2-z, 1-z, z)$ **31.** $(1,2)$ **33.** $\varnothing$
35. $(-z-2, 2-z, z)$ **37.** $(2, z-3, z)$ **39.** $(0,-2)$
41. $(8,8)$ **43.** $(4,5,4)$ **45.** $\varnothing$ **47.** $(2,0)$
49. $(1,2)$ **51.** $(2,3)$ **53.** $\varnothing$ **55.** $(2,1,0)$
57. $(3, 3x-9)$ **59.** $(4-z, 2, z)$ **61.** $(x, 0, 1-x)$
63. $(-1, -1, 2)$ **65.** 20 nickels, 40 dimes, and 4 quarters
67. $y = 2x^2 - x + 1$ **69.** $76°, 104°$ **71.** $59°, 31°$
73. $40°, 65°, 75°$ **77.** $k \neq 0$

Getting Ready (page 207)

1. -22 **2.** 22   **3.** -13 **4.** -13

Exercises 3.5 (page 215)

1. -10 **3.** -16 **5.** $\begin{vmatrix} -21 & -1 \\ 7 & 5 \end{vmatrix}$ **7.** 3   **9.** 0
11. number, square   **13.** $\begin{vmatrix} a_2 & c_2 \\ a_3 & c_3 \end{vmatrix}$ **15.** Cramer's rule
17. consistent, independent   **19.** 14   **21.** 1   **23.** -2
25. -13 **27.** 26   **29.** 0   **31.** $(-1,3)$ **33.** $(4,2)$
35. $\varnothing$ **37.** $\varnothing$ **39.** $(x, -\frac{2}{3}x + 3)$
41. $(x, \frac{4}{3}x - 2)$ **43.** $(1,1,2)$ **45.** $(3,2,1)$ **47.** $\varnothing$
49. $(\frac{1}{2}, \frac{5}{3} - \frac{4}{3}z, z)$ **51.** $4a^2 - b^2$
53. $10a$ **55.** 0   **57.** -23 **59.** 26   **61.** $(-\frac{1}{2}, \frac{1}{3})$
63. $(2,-1)$ **65.** $(5, \frac{14}{5})$ **67.** $(3,-2,1)$ **69.** $(\frac{3}{4}, \frac{1}{2}, \frac{1}{3})$
71. $(-2,3,1)$ **73.** -4 **75.** 2   **77.** $50°, 80°$
79. \$5,000 in HiTech, \$8,000 in SaveTel, \$7,000 in HiGas
85. -4

Chapter Review (page 221)

1. $(3,5)$ **2.** $(-2,3)$

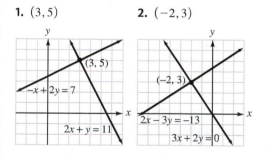

3. $(x, 6 - \frac{3}{2}x)$ **4.** $\varnothing$

5. $(-1,3)$ **6.** $(x, 3x+5)$ **7.** $(3,4)$ **8.** $(-4,2)$
9. $(-2,-3)$ **10.** $(1,-1)$ **11.** $(x, -\frac{2}{3}x - 2)$ **12.** $\varnothing$
13. $(1,2,3)$ **14.** $\varnothing$ **15.** $(x, \frac{1}{5}x - \frac{6}{5}, 0)$ **16.** $(1,3,2)$
17. $(1,2)$ **18.** $(-2-4z, -2-z, z)$ **19.** 14
20. -22 **21.** -3 **22.** 28   **23.** $(2,1)$ **24.** $(-1,3)$
25. $(1,-2,3)$ **26.** $\varnothing$

Chapter 3 Test (page 226)

1. $(2,1)$

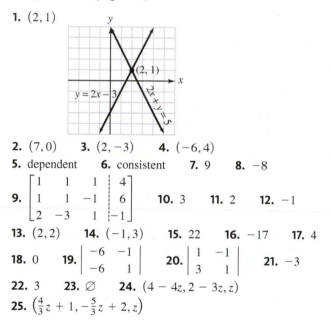

2. $(7,0)$ **3.** $(2,-3)$ **4.** $(-6,4)$
5. dependent   **6.** consistent   **7.** 9   **8.** -8
9. $\begin{bmatrix} 1 & 1 & 1 & | & 4 \\ 1 & 1 & -1 & | & 6 \\ 2 & -3 & 1 & | & -1 \end{bmatrix}$ **10.** 3   **11.** 2   **12.** -1
13. $(2,2)$ **14.** $(-1,3)$ **15.** 22   **16.** -17 **17.** 4
18. 0   **19.** $\begin{vmatrix} -6 & -1 \\ -6 & 1 \end{vmatrix}$ **20.** $\begin{vmatrix} 1 & -1 \\ 3 & 1 \end{vmatrix}$ **21.** -3
22. 3   **23.** $\varnothing$ **24.** $(4-4z, 2-3z, z)$
25. $(\frac{4}{3}z + 1, -\frac{5}{3}z + 2, z)$

Getting Ready (page 231)

4.

Exercises 4.1 (page 238)

1. $\{x | x < 2\}$ **3.** $\{x | x \geq -5\}$ **5.** $(-6, \infty)$
7. $(-\infty, 2]$ **9.** $\dfrac{1}{t^{10}}$ **11.** 320 or more   **13.** $\neq$ **15.** $<$
17. $\geq$ **19.** $a < c$ **21.** reversed   **23.** $c < x, x < d$
25. open   **27.** $\{x | x < 1\}$ **29.** $\{x | x \geq -5\}$
31. $\{x | x < 3\}$ **33.** $\{x | x \geq -2\}$ **35.** $(-\infty, 20]$

37. $(-\infty, 10)$

39. $(-36, \infty)$

41. $(-\infty, 45/7]$

43. $(-2, 4)$

45. $(-2, 5)$

47. $(8, 11)$

49. $[-4, 6)$

51. $(2, 3)$

53. $(-\infty, -15)$

55. $(-\infty, 2) \cup (7, \infty)$

57. $(-\infty, 1)$

59. $\varnothing$

61. $[2, \infty)$

63. $(2, \infty)$

65. $(-3, \infty)$

67. $(-\infty, 3]$

69. $(-\infty, -2)$

71. $(-\infty, 1.5]$

73. $\varnothing$

75. $[-2, 4]$

77. $(-\infty, -2] \cup [-1, \infty)$

79. $(-\infty, \infty)$

81. $x < 1$

83. $x \geq -4$

85. 5 hr   **87.** 4   **89.** more than \$5,000   **91.** 18
93. 88 or higher   **95.** 13   **97.** anything over \$900
99. 139   **103.** no   **105.** a, b, c

Getting Ready (page 241)

1. 5   **2.** 0   **3.** 10   **4.** -5 **5.** $(-5, 4)$
6. $(-\infty, -3) \cup (3, \infty)$

Exercises 4.2 (page 251)

1. $-5, 5$ **3.** $-12, 8$ **5.** $(-8, 8)$
7. $(-\infty, -2] \cup [2, \infty)$ **9.** $(-\infty, \infty)$
11. $(-\infty, 7) \cup (9, \infty)$ **13.** $\frac{3}{4}$ **15.** 6   **17.** $t = \dfrac{A - P}{pr}$
19. x **21.** 0   **23.** reflected   **25.** $a = b$ or $a = -b$
27. $x \leq -k$ or $x \geq k$ **29.**

$f(x) = |x| - 2$

31.

$f(x) = -|x + 4|$

33. $4, -4$ **35.** $9, -3$

37. $4, -1$ **39.** $2, -\frac{1}{2}$ **41.** $7, -8$
43. $0, -6$ **45.** $2, \frac{2}{5}$ **47.** $0, -12$
49. -8 **51.** 5   **53.** $0, -2$ **55.** 0

57. $(-4, 4)$

59. $[-21, 3]$

61. $(-6, 18)$

63. $(-1/5, 3/5)$

65. $(-\infty, -1) \cup (1, \infty)$

67. $(-\infty, -2] \cup [10/3, \infty)$

69. $(-\infty, -10) \cup (14, \infty)$

71. $(-\infty, -1) \cup (15, \infty)$

73. $(-\infty, -4) \cup (14, \infty)$

75. $(-\infty, -24) \cup (-18, \infty)$

77. $(-\infty, \infty)$

79. $(-\infty, \infty)$

81. $\varnothing$ **83.** $\varnothing$ **85.** $8, -4$ **87.** $-4, -28$

89. $4, \frac{1}{2}$ **91.** $\varnothing$ **93.** $(-8/3, 4)$ **95.** $\varnothing$

97. $(-\infty, -16/3) \cup (4, \infty)$

99. $(-\infty, -3/4) \cup (-3/4, \infty)$

101. -5 **103.** $\frac{14}{3}, -6$ **105.** $3, -1$ **107.** $\frac{4}{3}$

109. $[-3/2, 2]$

111. $(-\infty, -4] \cup [-1, \infty)$

113. $(-\infty, -2) \cup (5, \infty)$

115. $(-\infty, \infty)$

117. $(-\infty, -12) \cup (36, \infty)$

119. $\varnothing$ **121.** $\frac{20}{3}$

123.

125. $\varnothing$ **127.** $|x| < 4$

129. $|x + 3| > 6$ **131.** $4 \text{ ft} \leq d \leq 6 \text{ ft}$
133. $70° \leq t \leq 86°$ **135.** $|c - 0.6°| \leq 0.5°$ **141.** $k < 0$
143. x and y must have different signs.

Getting Ready (page 254)

1. yes   **2.** yes   **3.** no   **4.** yes

Exercises 4.3 (page 259)

1. below   **3.** above   **5.** left   **7.** above   **9.** $(4, -1)$
11. $(3, 0)$ **13.** linear   **15.** edge   **17.** dashed

19. $y \geq x$; $y = x$

21. $y \geq -2$; $y = -2$

23. $y > x + 1$; $y = x + 1$

25. $x < 4$; $x = 4$

27. $x = -2$; $-2 \leq x < 0$; $x = 0$

29. $y = 3$; $y < -2$ or $y > 3$; $y = -2$

31. $y \geq 1 - \frac{3}{2}x$; $y = 1 - \frac{3}{2}x$

33. $0.5x + y > 1.5 + x$; $0.5x + y = 1.5 + x$

35. $y = \frac{1}{3}x - 1$; $y < \frac{1}{3}x - 1$

37. $3x \geq -y + 3$; $3x = -y + 3$

39.

41.

43. $x \leq 3$

45. $y > -\frac{3}{2}x + 3$ **47.** $-2 \leq x \leq 3$

49. $y \leq x$ **51.** $y > -1$ or $y \leq -3$

53. $(1,1), (2,1), (2,2)$ **55.** $(2,2), (3,3), (5,1)$

Number of complicated returns / Number of simple returns; $x + 3y \leq 9$

Number of hours second housekeeper works / Number of hours first housekeeper works; $6x + 7y \leq 42$

57. $(40, 20), (60, 40), (80, 20)$

Number of Cleanco shares / Number of Traffico shares; $50x + 60y \leq 6{,}000$

Getting Ready (page 262)

1. yes   **2.** yes   **3.** no   **4.** no

Exercises 4.4 (page 267)

1. $<, >$ **3.** $t = \dfrac{A - p}{pr}$ **5.** $x = z\sigma + \mu$

7. $d = \dfrac{l - a}{n - 1}$ **9.** intersect

11. $y = 3x + 2$; $y = -2x + 3$

13. $x + 3y = 2$; $3x + 2y = 6$

15. $2x + 3y = 6$; $3x + y = 1$; $x = 0$

17. $y = 0$; $x = 0$; $x - y = 4$

19. $y = -x^2 + 2$; $2x - y = 4$

21. $y = -x^2 + 4$; $y = x^2 - 4$

23. $3x + y = 1$; $-x + 2y = 6$

25. $3x + 6y = 18$; $9x + 3y = 18$; $x = 0$; $y = 0$

27.

29.   BRONCOS   PACKERS

← Packers moving this direction

31. 1 $10 CD and 2 $15 CDs, 4 $10 CDs and 1 $15 CD

33. 2 desk chairs and 4 side chairs, 1 desk chair and 5 side chairs

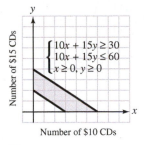

$$\begin{cases} 10x + 15y \geq 30 \\ 10x + 15y \leq 60 \\ x \geq 0, y \geq 0 \end{cases}$$

Number of $15 CDs / Number of $10 CDs

$$\begin{cases} 150x + 100y \\ \quad \leq 900 \\ y > x \\ x \geq 0, y \geq 0 \end{cases}$$

Number of side chairs / Number of desk chairs

37. no

Getting Ready (page 269)

1. 0   **2.** 6   **3.** 10   **4.** 12

Exercises 4.5 (page 277)

1. 20   **3.** $(2, 3)$ **5.** $\frac{9}{4}$ **7.** $y = \frac{9}{4}x - \frac{7}{4}$
9. constraints   **11.** objective   **13.** $P = 12$ at $(0, 4)$
15. $P = \frac{13}{6}$ at $\left(\frac{5}{3}, \frac{4}{3}\right)$ **17.** $P = 0$ at $(0, 0)$ **19.** $P = 0$ at
$(0, 0)$ **21.** $P = \frac{18}{7}$ at $\left(\frac{3}{7}, \frac{12}{7}\right)$ **23.** $P = 3$ at $(1, 0)$
25. $P = -12$ at $(-2, 0)$ **27.** $P = -2$ at $(1, 2)$ and $(-1, 0)$
and the edge joining the vertices   **29.** 3 tables, 12 chairs,
$1,260   **31.** 30 IBMs, 30 Macs, $2,700   **33.** 15 DVD
players, 30 TVs, $1,560   **35.** $150,000 in stocks, $50,000 in
bonds; $17,000

Chapter Review (page 282)

1. $(-\infty, 3]$

2. $(2, \infty)$

3. $(-\infty, -24]$

4. $(-\infty, -51/11)$

5. $(-1/3, 2)$

6. $(2, \infty)$

7. $[-1, \infty)$ **8.** $45,000 or more   **9.** 9

10. 8   **11.** -7 **12.** -6

13.

$f(x) = |x + 1| - 3$

14.

$f(x) = |x - 2| + 1$

15. $1, -\frac{5}{2}$ **16.** $\frac{26}{3}, -\frac{10}{3}$ **17.** $14, -10$ **18.** $-2, \frac{1}{5}$
19. $-1, 1$ **20.** $\frac{13}{12}$ **21.** $(-5, -2)$

22. $[-3, 19/3]$ **23.** $\varnothing$ **24.** $(-\infty, -4) \cup (22/5, \infty)$

25. $(-\infty, 4/3] \cup [4, \infty)$ **26.** $(-\infty, \infty)$

27. $2x + 3y > 6$ $2x + 3y = 6$

28. $y = 4 - x$ $y \leq 4 - x$

29. $x = -2$ $-2 < x < 4$ $x = 4$

30. $y = 1$ $y \leq -2$ or $y > 1$ $y = -2$

31. $3x + 2y = 6$ $y = x + 1$

32. $y = x + 3$ $y = x^2 - 4$

33. max of 12 at $(4, 0)$ **34.** 1,000 bags of X, 1,400 bags of Y

Chapter 4 Test (page 286)

1. $(-\infty, -5]$

2. $(-2, 16)$

3. 4   **4.** $4\pi - 4$ **5.**

$f(x) = |x + 1| - 4$

6.

$f(x) = |x - 2| + 3$

7. $3, -\frac{3}{5}$ **8.** $-5, \frac{23}{3}$

9. $\varnothing$ **10.** $1, 3$ **11.** $[-7, 1]$

12. $(-\infty, -9) \cup (13, \infty)$

13. $(-\infty, 1) \cup (3, \infty)$

14. $\varnothing$

15.

16.

17.

18.

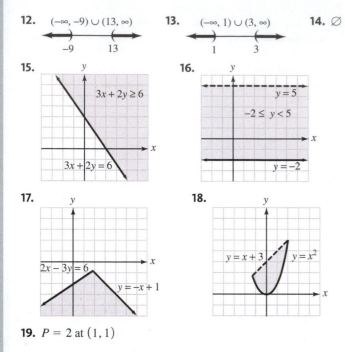

19. $P = 2$ at $(1, 1)$

Cumulative Review (page 287)

1.

2. 7 **3.** 8 **4.** -6

5. x^{24} **6.** x^{14} **7.** x^4 **8.** $a^{2-n}b^{n-2}$ **9.** 4.55×10^8

10. 1.2×10^{-5} **11.** $\frac{11}{2}$ **12.** 3 **13.** 6 **14.** $\varnothing$

15. perpendicular **16.** parallel **17.** $y = \frac{1}{3}x + \frac{11}{3}$

18. $h = \frac{2A}{b_1 + b_2}$ **19.** 4 **20.** 30

21. $(2, 1)$ **22.** $(1, 1)$

23. $(2, -2)$ **24.** $(3, 1)$ **25.** $(-1, -1, 3)$

26. $\left(-z, \frac{1}{2} - z, z\right)$ **27.** -1 **28.** 16 **29.** $(-1, -1)$

30. $(1, 2, -1)$ **31.** $(-\infty, 11]$ **32.** $(-3, 3)$ **33.** $2, \frac{4}{5}$

34. $-5, -\frac{3}{5}$ **35.** $\left[-\frac{2}{3}, 2\right]$ **36.** $(-\infty, -4) \cup (1, \infty)$

37.

38.

39.

40.

41. 24

Getting Ready (page 291)

1. $3a^2b^2$ **2.** $-5x^3y$ **3.** $4p^2 + 7q^2$ **4.** $a^3 - b^2$

Exercises 5.1 (page 298)

1. yes **3.** no **5.** yes **7.** no **9.** y^{14} **11.** $3y^{23}$

13. 1.14×10^8 **15.** sum, whole **17.** binomial

19. descending order **21.** one **23.** monomial

25. trinomial **27.** binomial **29.** monomial

31. 7 **33.** 0 **35.** 2 **37.** 8 **39.** 10

41. $-2x^4 - 5x^2 + 3x + 7$ **43.** $7y + 4y^2 - 5y^3 - 2y^5$

45. 2 **47.** 8 **49.** $50a^2 + 5a + 2$ **51.** $32x^2 - 4x + 2$

53. 13 **55.** 35 **57.** 132 **59.** -48 **61.** 0 ft

63. 64 ft **65.**

67. **69.**

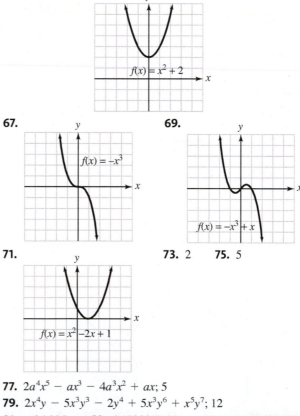

71.

73. 2 **75.** 5

77. $2a^4x^5 - ax^3 - 4a^3x^2 + ax$; 5

79. $2x^4y - 5x^3y^3 - 2y^4 + 5x^3y^6 + x^5y^7$; 12

81. -34.225 **83.** 0.1728317132 **85.** -0.1226844822

87. **89.** **91.** 20 ft

93. 63 ft   **95.** 198 ft   **97.** 10 in.^{2} **99.** 18 in.^{2}
101. 10 m, 42 m, 26 m, 10 m   **105.** no   **107.** 12

Getting Ready (page 301)

1. $2x + 6$ **2.** $-8x + 20$ **3.** $5x + 15$ **4.** $-2x + 6$

Exercises 5.2 (page 303)

1. 26   **3.** 22   **5.** -8 **7.** $(-\infty, -3]$ **9.** $(-1, 9)$
11. exponents   **13.** coefficients   **15.** like terms, $25x$
17. unlike terms   **19.** like terms, $-5r^2t^3$ **21.** unlike terms
23. $16y$ **25.** $2x^3y^2z$ **27.** $6x^3y - xy^3$ **29.** $10x^4y^2$
31. $7a + 11$ **33.** $x^2 - 5x + 6$ **35.** $13ab^2 - 5$
37. $4a^2b^4$ **39.** $9x^2$ **41.** $3x^3y^2$ **43.** $2t + 1$
45. $-5a^2 + 4a + 4$ **47.** $5pq^2 + 5pq$ **49.** $13xy + 2x^2y^2$
51. $3x - 9$ **53.** $-9t + 27$ **55.** $8x^3 - x^2$
57. $14a^2 + 16a - 24$ **59.** $3x^2 + 2$ **61.** $3x^2 - 4x - 4$
63. $-y^3 + 4y^2 + 6$ **65.** $3x^3 - x + 13$
67. $-x^3 - 2x^2 + 2x - 4$ **69.** $4y^2 - 9y + 3$
71. $-5y^2 + 10y + 21$ **73.** $-3p - 14q$
75. $11a^2 - 20b^3 - 9$ **77.** $6x^3 - 6x^2 + 14x - 17$
79. $-9y^4 + 3y^3 - 15y^2 + 20y + 12$ **81.** $x^2 - 8x + 22$
83. $-3y^3 + 18y^2 - 28y + 35$ **85.** $3x + 3$
87. $(x^2 + 5x + 6)$ m   **89.** \$136,000
91. $f(x) = 2{,}500x + 275{,}000$
93. $f(x) = -2{,}100x + 16{,}600$
95. $f(x) = -4{,}800x + 35{,}800$ **99.** $-2x + 9$
101. $-8x^2 - 2x + 2$

Getting Ready (page 306)

1. $12a$ **2.** $4a^4$ **3.** $-7a^4$ **4.** $12a^5$ **5.** $2a + 8$
6. $a^2 - 3a$ **7.** $-4a + 12$ **8.** $-2b^2 - 4b$

Exercises 5.3 (page 313)

1. dist. prop.   **3.** assoc. prop. of mult.
5. comm. prop. of mult.   **7.** 10   **9.** -15 **11.** \$51,025
13. variable   **15.** term   **17.** $x^2 + 2xy + y^2$
19. $x^2 - y^2$; conjugate; difference   **21.** $-24a^3b^3$
23. $-15a^2b^2c^3$ **25.** $-120a^9b^3$ **27.** $25x^{16}y^8$
29. $7x + 28$ **31.** $-a^2 + ab$ **33.** $15x^3 + 20x$
35. $-6x^3 + 6x^2 - 4x$ **37.** $10a^6b^4 - 25a^2b^6$
39. $7r^3st + 7rs^3t - 7rst^3$ **41.** $x^2 + 6x + 8$
43. $z^2 - 11z + 24$ **45.** $6y^3 + 11y^2 + 9y + 2$
47. $6x^3 - x^2 + 3x + 20$ **49.** $2a^2 - 3a - 2$
51. $2x^2 + xy - 6y^2$ **53.** $9x^2 - 6xy - 3y^2$
55. $8a^2 + 14ab - 15b^2$ **57.** $x^2 + 4x + 4$
59. $a^2 - 8a + 16$ **61.** $4a^2 + 4ab + b^2$
63. $4x^2 - 4xy + y^2$ **65.** $x^2 - 4$ **67.** $a^2 - b^2$
69. $2p^3 - 7p^2q + 4pq^2 + 4q^3$ **71.** $a^3 + 3a^2b + 3ab^2 + b^3$
73. $a^3 - 3a^2b - ab^2 + 3b^3$ **75.** $2x^3 + x^2 - 18x - 9$

77. $2x^5 + x$ **79.** $\frac{3x}{y^4z} - \frac{x^6}{y^{10}z^2}$ **81.** $\frac{1}{x^2} - y^2$
83. $2x^3y^3 - \frac{2}{x^3y^3} + 3$ **85.** $x^{3n} - x^{2n}$ **87.** $a^{3n} + a^{4n}$
89. $3x^2 + 12x$ **91.** $13y^2 - 10y + 17$ **93.** $4x^2 - 9y^2$
95. $x^3 - y^3$ **97.** $-405x^7y^4$ **99.** $-12m^4n^2 - 12m^3n^3$
101. $10x^3 + x^2y + 5xy^2 - 4y^3$ **103.** $8a^3 - b^3$
105. $4x^3 - 8x^2 - 9x + 6$ **107.** $x^3 + 8$ **109.** $m^2 - mn + 2n^2$
111. $5x^2 - 36x + 7$ **113.** $24y^2 - 4yz + 21z^2$
115. $x^{2n} - \frac{x^n}{y^n} - x^ny^n + 1$ **117.** $x^{4n} - y^{4n}$
119. $x^{2n} + x^n + x^ny^n + y^n$
121. $9.2127x^2 - 7.7956x - 36.0315$
123. $299.29y^2 - 150.51y + 18.9225$ **125. a.** x^2
b. $x^2 - 2xy + y^2$ **c.** $xy - y^2$ **d.** $xy - y^2$ **e.** y^2
127. 15   **129.** $r = -\frac{1}{5}p^2 + 90p$
131. $(4x^2 - 12x + 9)$ ft^2 **133.** $\frac{1}{2}(b^2 + 3b - 10)$ in.^{2}
137. 2.31×10^7

Getting Ready (page 316)

1. $4a + 8$ **2.** $-5b + 25$ **3.** $a^2 + 5a$ **4.** $-b^2 + 3b$
5. $6a^2 + 12ab$ **6.** $-6p^3 - 10p^2$

Exercises 5.4 (page 322)

1. 2   **3.** 4   **5.** prime   **7.** $a^2 - 81$ **9.** $16r^4 - 9s^2$
11. $n^3 - 125$ **13.** factoring   **15.** greatest common factor
17. $2^3 \cdot 3$ **19.** $3^3 \cdot 5$ **21.** 2^7 **23.** $5^2 \cdot 13$ **25.** 12
27. 2   **29.** $4a^2$ **31.** $6xy^2z^2$ **33.** 4   **35.** 1
37. $2(x + 4)$ **39.** $2x(x - 3)$ **41.** $7x(x + 2)$
43. $9x^2y^2(7x + 9y^2)$ **45.** $3ab(3a^2b^2 - 5ab + 1)$
47. $9x^7y^3(5x^3 - 7y^4 + 9x^3y^7)$ **49.** prime
51. prime   **53.** $-5(a + 2)$ **55.** $-x^3(2x - 1)$
57. $-3x(2x + y)$ **59.** $-12ab^2(3a - 4b)$
61. $x^{3n}(x^{6n} + x^{3n} + 1)$ **63.** $y^n(2y^2 - 3y^3)$
65. $(x + y)(4 + t)$ **67.** $(a - b)(r - s)$
69. $(u + v)(u + v - 1)$ **71.** $-(x + y)(a - b)$
73. $(a + b)(x - y + z)$ **75.** $(a - b + c)(x - y + z)$
77. $(3 - c)(c + d)$ **79.** $(a + b)(a - 4)$
81. $(a + b)(x - 1)$ **83.** $(x + a)(x + y + z)$
85. $x(m + n)(p + q)$ **87.** $y(x + y)(x + y + 2z)$
89. $n(2n - p + 2m)(n^2p - 1)$ **91.** $ab(2a - 5b)(a - 3)$
93. $r_1 = \frac{rr_2}{r_2 - r}$ **95.** $f = \frac{d_1d_2}{d_2 + d_1}$ **97.** $a^2 = \frac{b^2x^2}{b^2 - y^2}$
99. $r = \frac{S - a}{S - l}$ **101.** $3z(9z^2 + 4z + 1)$ **103.** prime
105. $-7u^2v^3z^2(9uv^3z^7 - 4v^4 + 3uz^2)$
107. $4y^{-2n}(2y^{4n} + 3y^{2n} + 4)$ **109.** $(m + n + p)(3 + x)$
111. $(x + y)(x + y + z)$ **113.** $(x + y)(a + b)$
115. $(x + 2)(x + y)$ **117.** $x^{-2}(x^6 - 5x^8)$
119. $a^{-3}b^{-1}(b^2 - 5a^{-3})$ **121.** $a = \frac{Hb}{2b - H}$
123. $y - \frac{x + 3}{3x - 2}$ **125. a.** $12x^3$ in.^{2} **b.** $20x^2$ in.^{2}
c. $4x^2(3x - 5)$ in.^{2} **131.** yes   **133.** no   **135.** yes

Getting Ready (page 324)

1. $a^2 - b^2$ **2.** $25p^2 - q^2$ **3.** $9m^2 - 4n^2$ **4.** $4a^4 - b^4$
5. $a^3 - 27$ **6.** $p^3 + 8$

Exercises 5.5 (page 329)

1. $(6a^2)^2$ **3.** $(y^2)^3$ or $(y^3)^2$ **5.** $(2b^2)^3$
7. $x^2 + 2x + 1$ **9.** $4m^2 + 4mn + n^2$
11. $a^2 + 7a + 12$ **13.** $8r^2 - 10rs + 3s^2$
15. 1, 4, 9, 16, 25, 36, 49, 64, 81, 100   **17.** cannot
19. $(p^2 - pq + q^2)$ **21.** $(x + 2)(x - 2)$
23. $(t + 15)(t - 15)$ **25.** $(b + 9)(b - 9)$
27. $(4x^2 + 9y)(4x^2 - 9y)$ **29.** $(25a + 13b^2)(25a - 13b^2)$
31. $(x^2 + y^2)(x + y)(x - y)$
33. $(4a^2 + 9b^2)(2a + 3b)(2a - 3b)$
35. $(x + y + z)(x + y - z)$ **37.** $(a - b + c)(a - b - c)$
39. $2(x + 12)(x - 12)$ **41.** $2x(x + 4)(x - 4)$
43. $5x(x + 5)(x - 5)$ **45.** $t^2(rs + x^2y)(rs - x^2y)$
47. $(x + 3)(x^2 - 3x + 9)$ **49.** $(p + 4)(p^2 - 4p + 16)$
51. $(t - v)(t^2 + tv + v^2)$ **53.** $(x - 2y)(x^2 + 2xy + 4y^2)$
55. $5(x + 5)(x^2 - 5x + 25)$ **57.** $4x^2(x - 4)(x^2 + 4x + 16)$
59. $2u^2(4v - t)(16v^2 + 4vt + t^2)$
61. $(a + b)(x + 3)(x^2 - 3x + 9)$
63. $(x + y)(x^2 - xy + y^2)(x - y)(x^2 + xy + y^2)$
65. $(1 + a)(1 - a + a^2)(1 - a)(1 + a + a^2)$
67. $(x - y - z)(x^2 + xy + xz + y^2 + 2yz + z^2)$
69. $(a - b - c)(a^2 - 2ab + b^2 + ac - bc + c^2)$
71. $(a + b)(a - b + 1)$ **73.** $(a - b)(a + b + 2)$
75. prime   **77.** $(2y^2 + 1)(2y^2 - 1)$
79. $(m + n)(m^2 - mn + n^2)$ **81.** prime
83. $(x + y - z)(x^2 - xy + xz + y^2 - 2yz + z^2)$
85. $(2ab^2c^3 + 3d^4)(2ab^2c^3 - 3d^4)$
87. $(2x^2 + 5y)(4x^4 - 10x^2y + 25y^2)$
89. $(10a^2 - 7bc^2)(100a^4 + 70a^2bc^2 + 49b^2c^4)$
91. $(2x + y)(1 + 2x - y)$ **93.** $(x^n - 2)(x^{2n} + 2x^n + 4)$
95. $(a^m + b^n)(a^{2m} - a^mb^n + b^{2n})$
97. $2(x^{2m} + 2y^m)(x^{4m} - 2x^{2m}y^m + 4y^{2m})$
99. $0.5g(t_1 + t_2)(t_1 - t_2)$ **101.** $\frac{4}{3}\pi(r_1 - r_2)(r_1^2 + r_1r_2 + r_2^2)$
105. $(x^{16} + y^{16})(x^8 + y^8)(x^4 + y^4)(x^2 + y^2)(x + y)(x - y)$

Getting Ready (page 332)

1. $a^2 - a - 12$ **2.** $6a^2 + 7a - 5$ **3.** $a^2 + 3ab + 2b^2$
4. $2a^2 - ab - b^2$ **5.** $6a^2 - 13ab + 6b^2$
6. $16a^2 - 24ab + 9b^2$

Exercises 5.6 (page 341)

1. 20, 3   **3.** 2, −10   **5.** 15, −2   **7.** 31   **9.** 12
11. −3   **13.** $2xy + y^2$; $2xy + y^2$ **15.** key number
17. $x + 2$ **19.** $x - 3$ **21.** $2a + 1$ **23.** $2m + 3n$
25. $(x + 8)(x + 1)$ **27.** $(x - 2)(x - 5)$ **29.** prime
31. $(x + 5)(x - 6)$ **33.** $(x + 1)^2$ **35.** $(a - 9)^2$

37. $(2y + 1)^2$ **39.** $(3b - 2)^2$ **41.** $(a - 8)(a + 4)$
43. $(x - 3)(x - 2)$ **45.** $b^2(a - 11)(a - 2)$
47. $x^2(b - 7)(b - 5)$ **49.** $(3y + 2)(2y + 1)$
51. $(3x - 4)(2x + 1)$ **53.** $(4a - 3)(2a + 3)$
55. $(2z + 3)(3z + 4)$ **57.** prime   **59.** prime
61. $-2z(2y + 3)(y + 2)$ **63.** $-z(2y^2 - 5)(5y^2 - 1)$
65. $x(x + 12y)(x - 5y)$ **67.** $x(x^2 + 7)(x^2 + 2)$
69. $(x^n + 1)^2$ **71.** $(2a^{3n} + 1)(a^{3n} - 2)$ **73.** $(x + 2)^2$
75. $(a + b + 4)(a + b - 6)$ **77.** $(a - 16)(a - 1)$
79. $(2u + 3)(u + 1)$ **81.** $(5r + 2s)(4r - 3s)$
83. $(5u + v)(4u + 3v)$ **85.** $(x + 2 + y)(x + 2 - y)$
87. $(x + 1 + 3z)(x + 1 - 3z)$
89. $(c + 2a - b)(c - 2a + b)$ **91.** $(a + 4 + b)(a + 4 - b)$
93. $(a + b)(a - 4b)$ **95.** $(3z + 4)^2$
97. $(a + 10)(a - 5)$ **99.** $(y - 7)(y + 3)$
101. prime   **103.** $-3(x - 3)(x - 2)$
105. $3(x + 7)(x - 3)$ **107.** $-4(x - 5)(x + 4)$
109. $-(2x - 3)^2$ **111.** $-(3a + 2b)(a - b)$
113. $(2y - 3t)(y + 2t)$ **115.** $t(3t^2 - 3t + 1)$
117. $-2(9x - 1)(5x + 1)$ **119.** $(x^{2n} + y^{2n})^2$
121. $(3x^n - 1)(2x^n + 3)$ **123.** $x^2(7x - 8)(3x + 2)$
125. $(x^2 + 5)(x^2 + 3)$ **127.** $(y^2 - 10)(y^2 - 3)$
129. $(a + 3)(a - 3)(a + 2)(a - 2)$
131. $(z^2 + 3)(z + 2)(z - 2)$
133. $(x^3 + 3)(x + 1)(x^2 - x + 1)$
135. $(3x + 3y + 4)(2x + 2y - 5)$
137. $(2x + y + z)(2x + y - z)$ **139.** $x + 3$
141. $(2x + 11)$ in.; $(2x - 1)$ in.; 12 in.   **145.** yes

Getting Ready (page 344)

1. $3p^2q - 3pq^2$ **2.** $4p^2 - 9q^2$
3. $p^2 + 5p - 6$ **4.** $6p^2 - 13pq + 6q^2$
5. $2p^2 + 6p - 8$ **6.** $p^3 + 8$

Exercises 5.7 (page 346)

1. 2; GCF 3   **3.** 3; GCF 2   **5.** 2
7. $12a^2 + 4a - 7$ **9.** $4y^2 - 11y + 3$ **11.** $n^2 - 4n - 45$
13. common factors   **15.** trinomial   **17.** $(x - 5)^2$
19. $(4x - 5y)(16x^2 + 20xy + 25y^2)$ **21.** $(x - t)(y + s)$
23. $(6a + 7b)(6a - 7b)$ **25.** $(6x + 5)(2x + 7)$
27. $2(3x - 4)(x - 1)$ **29.** prime
31. $(7x - 1)(x - 8)$ **33.** $(10z + 9t)(10z - 9t)$
35. $2x^2y^2z^2(2 - 13z)$ **37.** prime
39. $(a + b)(f + e)$ **41.** $(5x + 5y - 4)(x + y + 1)$
43. $(25x^2 + 16y^2)(5x + 4y)(5x - 4y)$
45. $2x(4x^3 - 3x^2 + 6x + 12)$ **47.** $(2x - 9)(3x + 7)$
49. $x^4(1 + y^2)(1 + y)(1 - y)$
51. $(x + 1)(x - 1)(x + 4)(x - 4)$
53. $(x + 5 + y^4)(x + 5 - y^4)$
55. $(3x - 1 + 5y)(3x - 1 - 5y)$ **57.** prime
59. $(x + 1)^2(x^2 - x + 1)(x - 1)$

61. $(x + 3)(x - 3)(x + 2)(x^2 - 2x + 4)$

63. $2z(x + y)(x - y)^2(x^2 + xy + y^2)$

65. $(x^m - 3)(x^m + 2)$ **67.** $(a^n - b^n)(a^{2n} + a^n b^n + b^{2n})$

69. $(x - 1)(x - 2)(x - y)$ **73.** $\left(\frac{3}{x} + 2\right)\left(\frac{2}{x} - 3\right)$

75. $(x^2 + x + 1)(x^2 - x + 1)$

Getting Ready (page 347)

1. $2a(a - 2)$ **2.** $(a + 5)(a - 5)$ **3.** $(3a - 2)(2a + 3)$

4. $a(3a - 2)(2a + 1)$

Exercises 5.8 (page 353)

1. $2, 3$ **3.** $2, 3, -1$ **5.** $2, 3, 5, 7$

7. $40{,}081.00 \text{ cm}^3$ **9.** $ax^2 + bx + c = 0$ **11.** $-4, -5$

13. $3, 4$ **15.** $0, -4$ **17.** $0, -1$ **19.** $3, -3$

21. $\frac{5}{2}, -\frac{5}{2}$ **23.** $\frac{1}{3}, -1$ **25.** $2, \frac{1}{2}$ **27.** $0, 7, -3$

29. $0, 0, -1$ **31.** $1, -1, 5, -5$ **33.** $2, -2, 4, -4$

35. $7, -7$ **37.** $\frac{1}{2}, -3$ **39.** $1, 6$ **41.** $-3, -\frac{1}{3}$

43. $2, -\frac{5}{6}$ **45.** $0, -2, -3$ **47.** $1, -1, -3$

49. $3, -3, -\frac{3}{2}$ **51.** $0, 3, -3$ **53.** $\frac{1}{2}, 2$ **55.** $1, -\frac{1}{2}$

57. $2, -\frac{1}{3}$ **59.** $-\frac{7}{5}, \frac{1}{2}$ **61.** $0, 5$ **63.** $3, 3$

65. $-2, -4$ **67.** $\frac{1}{5}, -\frac{5}{3}$ **69.** $0, \frac{5}{6}, -7$ **71.** $2, -2, -\frac{1}{3}$

73. $0, -1, \frac{2}{5}$ **75.** no solution   **77.** 1

79. $14, 16$ or $-16, -14$ **81.** $6, 7$ **83.** 7 cm by 21 cm

85. 12 cm   **87.** 5 ft by 12 ft   **89.** 20 ft by 40 ft

91. 10 sec   **93.** 11 sec and 19 sec   **95.** 50 m   **97.** 3 ft

101. $x^2 - 8x + 15 = 0$ **103.** $x^2 + 5x = 0$

Chapter Review (page 358)

1. 5   **2.** $8; 8x^4y^4 + 13x^3y^2 + 9x^2y$ **3.** 1

4. -6 **5.** $-t^2 - 4t + 6$ **6.** $-z^2 + 4z + 6$

7.   **8.**

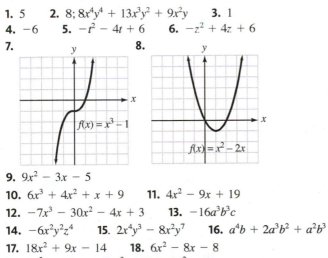

9. $9x^2 - 3x - 5$

10. $6x^3 + 4x^2 + x + 9$ **11.** $4x^2 - 9x + 19$

12. $-7x^3 - 30x^2 - 4x + 3$ **13.** $-16a^3b^3c$

14. $-6x^2y^2z^4$ **15.** $2x^4y^3 - 8x^2y^7$ **16.** $a^4b + 2a^3b^2 + a^2b^3$

17. $18x^2 + 9x - 14$ **18.** $6x^2 - 8x - 8$

19. $9x^2 - 48xy + 64y^2$ **20.** $9x^2 - 4$

21. $3x^4 - 3x^3 + 4x^2 + 2x - 4$

22. $2a^3 - 3a^2b - 3ab^2 + 2b^3$ **23.** $-6(x + 3)$

24. $8x^2(x + 4)$ **25.** $5xy^2(xy - 2)$ **26.** $7a^3b(ab + 7)$

27. $-4x^2y^3z^2(2z^2 + 3x^2)$ **28.** $3a^2b^4c^2(4a^4 + 5c^4)$

29. $9x^2y^3z^2(3xz + 9x^2y^2 - 10z^5)$

30. $-12a^2b^3c^2(3a^3b - 5a^5b^2c + 2c^5)$ **31.** $x^n(x^n + 1)$

32. $y^{2n}(1 - y^{2n})$ **33.** $x^{-2}(x^{-2} - 1)$ **34.** $a^{-3}(a^9 + a^3)$

35. $5x^2(x + y)^3(1 - 3x^2 - 3xy)$

36. $-7a^2b^2(a - b)^3(7a^2 - 7ab - 9b^2)$

37. $(m + 5)(n + 6)$ **38.** $(a + b)(c - 3)$

39. $(x^2 + 4)(x^2 + y)$ **40.** $(a^3 + c)(a^2 + b^2)$

41. $h = \frac{S - 2wl}{2w + 2l}$ **42.** $l = \frac{S - 2wh}{2w + 2h}$ **43.** $(z + 6)(z - 6)$

44. $(y^2 + 4)(y + 2)(y - 2)$ **45.** $(xy^2 + 8z^3)(xy^2 - 8z^3)$

46. prime   **47.** $(x + z + t)(x + z - t)$

48. $(c + a + b)(c - a - b)$ **49.** $2(x^2 + 7)(x^2 - 7)$

50. $3x^2(x^2 + 10)(x^2 - 10)$ **51.** $(x + 7)(x^2 - 7x + 49)$

52. $3(p + 3)(p^2 - 3p + 9)$ **53.** $8(y - 4)(y^2 + 4y + 16)$

54. $4y(x + 3z)(x^2 - 3xz + 9z^2)$ **55.** $(x + 9)(x + 9)$

56. $(a - 7)(a - 7)$ **57.** $(y + 20)(y + 1)$

58. $(z - 5)(z - 6)$ **59.** $-(x + 7)(x - 4)$

60. $(y - 9)(y + 4)$ **61.** $(4a - 1)(a - 1)$

62. prime   **63.** prime   **64.** $-(5x + 2)(3x - 4)$

65. $y(y + 2)(y - 1)$ **66.** $2a^2(a + 3)(a - 1)$

67. $-2(4x - 3)(x - 2)$ **68.** $4(2x + 3)(x - 2)$

69. $3(5x + y)(x - 4y)$ **70.** $5(6x + y)(x + 2y)$

71. $(8x + 3y)(3x - 4y)$ **72.** $(2x + 3y)(7x - 4y)$

73. $5x(2x + 3)(x - 1)$ **74.** $3y(x + 3)(x - 7)$

75. $(z - 2)(z + x + 2)$ **76.** $(x + 2 + y)(x + 2 - y)$

77. $(x + 2 + 2p^2)(x + 2 - 2p^2)$

78. $(2y - 5)(4y^2 + 10y + 25)$ **79.** $(x^m + 3)(x^m - 1)$

80. $(x + 3)(x - 3)(x^2 + 4)$ **81.** $0, \frac{7}{5}$ **82.** $3, -3$

83. $-\frac{5}{6}, \frac{1}{2}$ **84.** $\frac{2}{7}, 5$ **85.** $-\frac{2}{3}, 0, \frac{4}{5}$ **86.** $-\frac{2}{3}, 7, 0$

87. 7 cm   **88.** 17 m by 20 m

Chapter 5 Test (page 363)

1. 6   **2.** 13   **3.** -34 **4.** -6

5.   **6.** $5y^2 + y - 1$

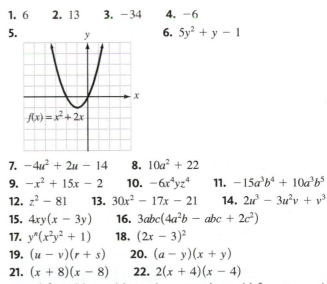

7. $-4u^2 + 2u - 14$ **8.** $10a^2 + 22$

9. $-x^2 + 15x - 2$ **10.** $-6x^4yz^4$ **11.** $-15a^3b^4 + 10a^3b^5$

12. $z^2 - 81$ **13.** $30x^2 - 17x - 21$ **14.** $2u^3 - 3u^2v + v^3$

15. $4xy(x - 3y)$ **16.** $3abc(4a^2b - abc + 2c^2)$

17. $y^n(x^2y^2 + 1)$ **18.** $(2x - 3)^2$

19. $(u - v)(r + s)$ **20.** $(a - y)(x + y)$

21. $(x + 8)(x - 8)$ **22.** $2(x + 4)(x - 4)$

23. $4(y^2 + 4)(y + 2)(y - 2)$ **24.** $(b + 4)(b^2 - 4b + 16)$

25. prime   **26.** $3(u - 2)(u^2 + 2u + 4)$

27. $(a - 10)(a + 1)$ **28.** $(3b + 2)(2b - 1)$

29. $3(u + 2)(2u - 1)$ **30.** $5(4r + 1)(r - 1)$

31. $(x^n + 1)^2$ **32.** $(x + 3 + y)(x + 3 - y)$ **33.** $-3, 2$

34. $-3, \frac{6}{5}$ **35.** $-1, 10$ **36.** $0, -\frac{7}{2}, -\frac{5}{2}$ **37.** $r = \frac{r_1 r_2}{r_2 + r_1}$
38. $6, -1$ **39.** 25 **40.** 11 ft by 22 ft

Getting Ready (page 367)

1. $\frac{3}{4}$ **2.** $\frac{4}{5}$ **3.** $-\frac{5}{13}$ **4.** $\frac{7}{9}$

Exercises 6.1 (page 374)

1. 0 **3.** undefined **5.** -1 **7.** $3x(x-3)$
9. $(3x^2 + 4y)(9x^4 - 12x^2 y + 16y^2)$ **11.** rational
13. asymptote **15.** $a, 1, 0$ **17.** $\frac{a}{b}, 0$
19. -6 **21.** $-\frac{4}{3}$ **23.** $(-\infty, -6) \cup (-6, \infty)$
25. $\left(-\infty, -\frac{4}{3}\right) \cup \left(-\frac{4}{3}, \infty\right)$
27. $(-\infty, 2) \cup (2, \infty)$ **29.** $(-\infty, -2) \cup (-2, 2) \cup (2, \infty)$

31. $-\frac{3}{4}$ **33.** $-\frac{28}{9}$ **35.** $4x^2$ **37.** $-\frac{4y}{3x}$ **39.** $x-9$
41. $\frac{1}{x-y}$ **43.** $\frac{x-3}{x-5}$ **45.** $\frac{a+1}{3}$ **47.** $\frac{x+1}{x+3}$
49. $\frac{x+4}{2(2x-3)}$ **51.** $\frac{(x+1)^2}{(x-1)^3}$ **53.** $\frac{3a+b}{y+b}$ **55.** -1
57. $\frac{b}{3}$ **59.** 20 hr **61.** 12 hr **63.** \$5,555.56
65. \$50,000 **67.** $\frac{12}{13}$ **69.** $-\frac{6a^2}{7}$ **71.** $\frac{5x^3}{3}$
73. $\frac{-2(x+y)}{7x}$ **75.** $\frac{3y}{7(y-z)}$ **77.** $\frac{2x+1}{2-x}$
79. $-\frac{y^2 + 2y + 4}{5(2+y)}$ **81.** $\frac{x+6}{x+4}$ **83.** in lowest terms
85. $\frac{m-2n}{n-2m}$ **87.** $x+2$ **89.** $\frac{1}{x-3}$ **91.** 3
93. $\frac{2a-3b}{a-2b}$ **95.** $\frac{1}{x^2 + xy + y^2 - 1}$ **97.** $\frac{x-y}{x+y}$
99. $-\frac{m+n}{2m+n}$ **101.** $\frac{3(x-y)}{x+2}$ **103.** almost 8 days
105. about 2.55 hr **107. a.** \$33,333.33 **b.** \$116,666.67
109. $c = f(x) = 1.25x + 700$ **111.** \$1,325 **113.** \$1.95
115. $c = f(n) = 0.09n + 7.50$ **117.** \$77.25 **119.** 9.75¢
123. yes **125.** a, d

Getting Ready (page 377)

1. $\frac{3}{2}$ **2.** $\frac{2}{5}$ **3.** $\frac{4}{3}$ **4.** $\frac{14}{3}$

Exercises 6.2 (page 383)

1. $\frac{1}{2}$ **3.** $-\frac{1}{3}$ **5.** 6 **7.** $-28x^7 + 7x^5$ **9.** $m^{2n} - 4$
11. $\frac{ac}{bd}$ **13.** 0 **15.** $\frac{7}{12}$ **17.** $-\frac{4}{3}$ **19.** $\frac{xy^2 d}{c^3}$
21. $-\frac{x^{10}}{y^2}$ **23.** $\frac{x-2}{x}$ **25.** $-\frac{x+1}{x+3}$ **27.** $\frac{y+1}{(y-1)(x-2)}$
29. $-\frac{x+3}{x+2}$ **31.** 1 **33.** $\frac{2p-1}{p+2}$ **35.** $-\frac{x+3}{x+1}$
37. $x-5$ **39.** $\frac{x^2 - 14x + 49}{x^4 + 6x^2 + 9}$
41. $\frac{4m^4 - 4m^3 - 11m^2 + 6m + 9}{x^4 - 2x^2 + 1}$ **43.** $\frac{x-6}{x+9}$ **45.** $\frac{n+2}{n+1}$
47. $\frac{(a+7)^2(a-5)}{12x^2}$ **49.** $\frac{-(x-3)(x-6)}{(x+2)(x+3)}$ **51.** $\frac{x-2}{x}$
53. $\frac{x-5}{x(x+5)}$ **55.** $-\frac{x^7}{18y^4}$ **57.** $\frac{x^3}{x-2}$ **59.** $\frac{3x}{2}$
61. $\frac{t+1}{t}$ **63.** -1 **65.** $\frac{x+y}{x-y}$ **67.** $\frac{x-5}{x+5}$

69. $\frac{x-1}{3x+2}$ **71.** $\frac{x-7}{x+7}$ **73.** 1 **75.** $\frac{(b-2)(b-3)}{2}$ cm²
77. $\frac{(k+2)(k+4)}{k+1}$ **81.** $\div, \cdot$

Getting Ready (page 385)

1. yes **2.** no **3.** yes **4.** yes **5.** no **6.** yes

Exercises 6.3 (page 392)

1. $2x$ **3.** $9x + 3$ **5.** $2y - 2$ **7.** $(-1, 4]$
9. $\frac{V}{\pi r^2} = h$ **11.** $\frac{a+c}{b}$ **13.** subtract, keep **15.** LCD
17. $\frac{2}{3}$ **19.** $-\frac{1}{3}$ **21.** $\frac{4}{3x}$ **23.** $\frac{a-7}{a+b}$ **25.** 2
27. 2 **29.** 3 **31.** $\frac{6x}{(x-3)(x-2)}$ **33.** 72
35. $x(x+2)(x-2)$ **37.** $(x+3)^2(x^2 - 3x + 9)$
39. $(2x+3)^2(x+1)^2$ **41.** 2 **43.** $\frac{3(a-1)}{3a-2}$ **45.** $\frac{9}{20}$
47. $\frac{9a}{10}$ **49.** $\frac{4y + 35x}{14xy}$ **51.** $\frac{9a^2 - 4b^2}{6ab}$ **53.** $\frac{4x-5}{x-3}$
55. $\frac{6y-9}{y-3}$ **57.** $\frac{8x-2}{(x+2)(x-4)}$ **59.** $\frac{7x+29}{(x+5)(x+7)}$
61. $\frac{2x^2 + x}{(x+3)(x+2)(x-2)}$ **63.** $\frac{-x^2 + 11x + 8}{(3x+2)(x+1)(x-3)}$
65. $\frac{y^2 + y - x^2}{(y-2x)^2}$ **67.** $\frac{-3x^2 + x + 1}{(x-1)^2}$ **69.** $\frac{-4x^2 + 14x + 54}{x(x+3)(x-3)}$
71. $\frac{x^3 + x^2 - 1}{x(x+1)(x-1)}$ **73.** $\frac{-y^2 + 48y + 161}{(y+4)(y+3)}$
75. $\frac{4x-1}{(x+2)(x-2)}$ **77.** 0 **79.** 1 **81.** $\frac{2y-5x}{xy}$
83. $\frac{2x^2 + 5x + 4}{x+1}$ **85.** $\frac{-x^2 + 3x + 2}{(x-1)(x+1)^2}$ **87.** $\frac{a^2 + 3a + 1}{a+3}$
89. $-\frac{5x+1}{(2x-1)(x+1)}$ **91.** $\frac{5x-y}{6}$ **93.** $\frac{-x^2 + 11x - 15}{x-9}$
95. $\frac{-5x^3 + 9x^2 - 4x + 23}{x-1}$ **97.** $\frac{3x^2 - 2x - 17}{(x-3)(x-2)}$
99. $\frac{2a}{a-1}$ **101.** $\frac{x^2 - 6x - 1}{2(x+1)(x-1)}$ **103.** $\frac{2b}{a+b}$
105. $\frac{7mn^2 - 7n^3 - 6m^2 + 3mn + n}{(m-n)^2}$ **107.** $\frac{2}{m-1}$
111. $\frac{11x^2 + 4x + 12}{2x}$ ft **113.** $\frac{7x^2 - 11x + 12}{4x}$ cm

Getting Ready (page 395)

1. 3 **2.** -34 **3.** $3 + 2a$ **4.** $5 - 2b$

Exercises 6.4 (page 401)

1. $\frac{2}{3}$ **3.** $\frac{a+b}{a-b}$ **5.** $\frac{3y}{2x}$ **7.** 8 **9.** $-4, 4, -2, 2$
11. complex **13.** $\frac{3}{4}$ **15.** $\frac{2}{3xz}$ **17.** $\frac{2(x+2)}{x-3}$ **19.** $-\frac{1}{7}$
21. $\frac{3+x}{x^2 + 2x}$ **23.** $\frac{y-x}{x^2 y^2}$ **25.** $\frac{5b + 2a}{a^2 b^2}$ **27.** $\frac{5x^2 y^2}{xy + 1}$
29. $p + q$ **31.** $\frac{-1}{a+b}$ **33.** $\frac{y+x}{y-x}$ **35.** $\frac{x^2(xy^2 - 1)}{y^2(x^2 y - 1)}$
37. $\frac{(b+a)(b-a)}{b(b-a-ab)}$ **39.** $\frac{x-1}{x}$ **41.** $x^2 + x - 6$
43. $\frac{y+x}{y-x}$ **45.** $\frac{x+2}{x-3}$ **47.** $\frac{a-1}{a+1}$ **49.** $\frac{y+x}{x^2 y}$
51. $\frac{xy^2}{y-x}$ **53.** $\frac{5b}{5b+4}$ **55.** $\frac{3a^2 + 2a}{2a+1}$ **57.** -1
59. $\frac{10}{3}$ **61.** xy **63.** $\frac{(-x^2 + 2x - 2)(3x+2)}{(2-x)(-3x^2 - 2x + 9)}$
65. $\frac{2(x^2 - 4x - 1)}{-x^2 + 4x + 8}$ **67.** $\frac{-2}{x^2 - 3x - 7}$ **69.** $\frac{k_1 k_2}{k_2 + k_1}$
71. $\frac{4k}{9}$ **75.** $\frac{1}{y+x}$

Getting Ready (page 404)

1. $\frac{9}{2}$ 2. $4, -4$

Exercises 6.5 (page 411)

1. $12(x+3)$ 3. $a(a-5)$ 5. $(x+3)(x-3)$

7. $\frac{n^6}{m^{12}}$ 9. 0   11. rational   13. 10   15. 40

17. $\frac{19}{36}$ 19. $\frac{17}{25}$ 21. $2, -5$ 23. $-4, 3$ 25. $\varnothing$

27. $\varnothing$; 3 is extraneous   29. 2   31. $\frac{1}{2}$ 33. $r_1 = \frac{Rr_2}{r_2 - R}$

35. $f = \frac{pq}{q+p}$ 37. 0   39. $8, \frac{41}{13}$ 41. 5, 3 is extraneous

43. $0, \frac{4}{3}$ 45. $\varnothing$, -2 is extraneous   47. 2

49. $2, -2$ is extraneous   51. $2, -\frac{3}{2}$ 53. $-\frac{11}{6}$

55. $r = \frac{S-a}{S-l}$ 57. $R = \frac{r_1 r_2 r_3}{r_1 r_3 + r_1 r_2 + r_2 r_3}$ 59. $1\frac{7}{8}$ days

61. $5\frac{5}{6}$ min   63. $1\frac{1}{5}$ days   65. $2\frac{4}{13}$ weeks   67. 3 mph

69. 60 mph and 40 mph   71. 3 mph   73. 60 mph

75. 7   77. 12 days   79. $\frac{20}{3}$ cm

Getting Ready (page 414)

1. $\frac{2}{3}$ 2. $-\frac{5}{9}$ 3. 5   4. 16

Exercises 6.6 (page 419)

1. $x^2 + x - 5$ 3. $2x^3 + 5x^2 - x - 6$ 5. $8x^2 + 2x + 4$

7. $-2y^3 - 3y^2 + 6y - 6$ 9. $\frac{1}{b}$ 11. quotient

13. $\frac{x^2}{6y^3}$ 15. $\frac{3b^4}{4a^4}$ 17. $-\frac{5}{7xy^7 f^2}$ 19. $\frac{13a^n b^{2n} c^{3n-1}}{3}$

21. $\frac{2x}{3} - \frac{x^2}{6}$ 23. $\frac{3x^3}{2} + \frac{4x}{y}$ 25. $7xy^2 + 1 - 5x^2 y$

27. $\frac{x^4 y^4}{2} - \frac{x^3 y^9}{4} + \frac{3}{4xy^2}$ 29. $x + 2$ 31. $x - 6$

33. $3x + 4$ 35. $5x^2 + 2x - 3$ 37. $4a^2 - 3a + \frac{7}{a+1}$

39. $3x^2 + x - 2 + \frac{16}{2x+3}$ 41. $x - 7$ 43. $3x^2 - x + 2$

45. $9a^2 + 6ab + 4b^2$ 47. $a^2 + a + 1 + \frac{2}{a-1}$

49. $3x - 5 + \frac{3}{2x+3}$ 51. $x^2 + x + 1$ 53. $2y + 2$

55. $x^2 + x + 2$ 57. $x^4 + x^2 + 4$ 59. $5 - 4x^{4n} y^{4n} - 3x^n y^n$

61. $\frac{b^2}{4a} - \frac{a^3}{2b^4} + \frac{3}{4ab}$ 63. $3x^2 - x + 2$ 65. $5a^2 - 3a - 4$

67. $x - 8y$ 69. $16x^4 - 8x^3 y + 4x^2 y^2 - 2xy^3 + y^4$

71. $(3x+5)$ ft   73. $3x - 2; x + 5$ 77. It is.

Getting Ready (page 421)

1. $x + 1$ with a remainder of $1, 1$

2. $x + 3$ with a remainder of $9, 9$

Exercises 6.7 (page 426)

1. 2   3. 2   5. 0   7. $12a^2 + 4a - 1$

9. $17x^2 - 19x + 14$ 11. $x - r$ 13. $P(r)$

15. $x - 4 + \frac{-6}{x-9}$ 17. $x - 1 + \frac{-5}{x+5}$ 19. $2x^2 + 4x + 3$

21. $4x^2 - 8x + 14 + \frac{8}{x+2}$ 23. $6x^2 - x + 1 + \frac{3}{x+1}$

25. $3x^2 - 2x + 4$ 27. -1 29. -97

31. 23   33. -361 35. -8 37. 44

39. yes; $P(x) = (x-3)(x^2+5)$ 41. no

43. $3x^2 - x + 2$ 45. $x - 7 + \frac{28}{x+2}$ 47. $x + 2$

49. $x + 7$ 51. -3 53. -3

55. yes; $P(x) = x(7x^2 - 5x - 8)$

57. yes; $P(x) = (x-5)(x^2 + 5x + 25)$ 59. $\frac{29}{32}$ 61. 2

63. -1 65. 64   69. 1

Getting Ready (page 428)

1. $\frac{7}{3}$ 2. 1.44×10^9 3. linear function

4. rational function

Exercises 6.8 (page 437)

1. $=$ 3. $\neq$ 5. $k = 200$ 7. $\frac{1}{x^2}$ 9. -2

11. 4.7×10^5 13. 0.0025   15. unit costs, rates

17. extremes, means   19. direct   21. rational

23. joint, constant of proportionality   25. direct

27. neither   29. 7   31. 5   33. 5   35. 5

37. $-\frac{5}{2}, -1$ 39. $\varnothing$ 41. $A = kp^2$ 43. $a = \frac{k}{b^2}$

45. $B = kmn$ 47. $X = \frac{kw}{q}$ 49. $4, -1$ 51. $3, -3$

53. 12   55. $P = \frac{ka^2}{j^3}$ 57. L varies jointly with m and n.

59. E varies jointly with a and the square of b.   61. X varies directly with x^2 and inversely with y^2.   63. R varies directly with L and inversely with d^2.   65. \$62.50   67. $8\frac{1}{2}$ gal

69. 32 ft   71. 80 ft   73. 0.18 g   75. 42 ft   77. $46\frac{7}{8}$ ft

79. 6,750 ft   81. 64π in.^{2} 83. 432 mi   85. 25 days

87. 12 in.^{3} 89. 9   91. \$9,000   93. 0.275 in.

95. 546 Kelvin   97. $85\frac{1}{3}$ 99. 3 ohms

Chapter Review (page 444)

1. 5   2. $-3, \frac{7}{2}$

3. vertical asymptote: $x = 0$

$$f(x) = \frac{3x+2}{x}$$

4. $(-\infty, 1) \cup (1, \infty)$ 5. $\frac{53m}{147n^2}$

6. $\frac{x-9}{x+9}$ 7. $\frac{1}{x-6}$

8. $\frac{1}{2x+4}$ 9. $-\frac{5}{4}$ 10. -2

11. $\frac{m+2n}{2m+n}$ 12. $-\frac{a+b}{c+d}$

13. 1   14. 1   15. $\frac{3x(x-1)}{(x-3)(x+1)}$

16. 1   17. $\frac{3x+7}{x-y}$ 18. $\frac{y+9}{y^2+3}$ 19. $\frac{5x+13}{(x+2)(x+3)}$

20. $\frac{4x^2 + 9x + 12}{(x-4)(x+3)}$ 21. $\frac{5x^2 + 11x}{(x+1)(x+2)}$ 22. $\frac{2(3x+1)}{x-3}$

23. $\frac{5x^2 + 23x + 4}{(x+1)(x-1)^2}$ 24. $\frac{x^2 + 26x + 3}{(x+3)(x-3)^2}$ 25. $\frac{5b - 4a}{a^2 b^2}$

26. $\frac{3q + 4p}{4q - 3p}$ 27. $\frac{2x+1}{x+1}$ 28. $\frac{3x+2}{3x-2}$ 29. $\frac{x-2}{x+3}$

30. $\frac{1}{x}$ 31. $\frac{y-x}{y+x}$ 32. $\frac{x^2 y^2}{(x-y)^2 (y^2 - x^2)}$ 33. 15

34. $-1, -2$ 35. $\varnothing$; 2 and -2 are extraneous

36. $-1, -12$ 37. $a^2 = \frac{x^2 b^2}{b^2 + y^2}$ 38. $b = \frac{Ha}{2a - H}$

39. 50 mph   40. 200 mph   41. $14\frac{2}{5}$ hr   42. $18\frac{2}{3}$ days

43. $-\dfrac{x^2}{3y^3}$ **44.** $-3x^2y + \dfrac{3x}{2} + y$ **45.** $x + 5y$

46. $x^2 + 2x - 1 + \dfrac{6}{2x + 3}$

47. yes; $P(x) = (x - 3)(x^2 - 4x - 3)$ **48.** no   **49.** 5

50. $-4, -12$ **51.** 70.4 ft   **52.** 54   **53.** 6   **54.** 2

55. 16   **56.** $5,460

Chapter 6 Test (page 450)

1. $\dfrac{-2}{3xy}$ **2.** $\dfrac{5}{x - 3}$ **3.** -7

4. a. $x = 3$ **b.** $(-\infty, 3) \cup (3, \infty)$ **5.** $\dfrac{19}{3}$

6. $\dfrac{xz}{y^4}$ **7.** $\dfrac{x + 1}{4x - 1}$ **8.** 1   **9.** $\dfrac{(x + y)^2}{2}$

10. $\dfrac{2}{x + 1}$ **11.** -1 **12.** 6   **13.** 2   **14.** $\dfrac{2s + r^2}{rs}$

15. $\dfrac{2x + 3}{(x + 1)(x + 2)}$ **16.** $\dfrac{u^2}{2vw}$ **17.** $\dfrac{2x + y}{xy - 2}$ **18.** $\dfrac{5}{2}$

19. 5; 3 is extraneous   **20.** $x^2 = \dfrac{a^2b^2 - y^2a^2}{b^2}$ **21.** $r_2 = \dfrac{rr_1}{r_1 - r}$

22. 10 days   **23.** $5,000 at 6% and $3,000 at 10%

24. $\dfrac{-6x}{y} + \dfrac{4x^2}{y^2} - \dfrac{3}{y^3}$ **25.** $3x^2 + 4x + 2$ **26.** -7

27. 47   **28.** 18 ft   **29.** $6, -1$ **30.** $\dfrac{44}{3}$

Cumulative Review 6 (page 451)

1. $\dfrac{x^{11}}{y^2}$ **2.** $\dfrac{b^4}{a^4}$ **3.** $\dfrac{81b^{16}}{16a^8}$ **4.** $x^{21}y^3$ **5.** 7,230,000

6. 0.000712   **7.** 1   **8.** -2 **9.** $\dfrac{5}{6}$ **10.** $-\dfrac{3}{4}$

11. -5 **12.** $-\dfrac{1}{3}$ **13.** 3   **14.** 8   **15.** $-\dfrac{16}{25}$

16. $t^2 - 4t + 3$ **17.** $y = \dfrac{kxz}{r}$ **18.** no

19. $\left[\dfrac{5}{2}, \infty\right)$

20. $(-\infty, 3] \cup \left[\dfrac{11}{3}, \infty\right)$

21. binomial   **22.** 7   **23.** -8

24.

25. $4x^2 - 4x + 14$

26. $-3x^2 - 3$

27. $8x^2 + 22x - 21$

28. $2x^{2n} + 3x^n - 2$

29. $3rs^3(r - 2s)$

30. $(a - b)(x - 4)$

31. $(x + y)(u + v)$

32. $(9x^2 + 4y^2)(3x + 2y)(3x - 2y)$

33. $(3y - 4x)(9y^2 + 12xy + 16x^2)$ **34.** $(2x + 3)(3x - 2)$

35. $(3x - 5)^2$ **36.** $(5x + 3)(3x - 2)$

37. $(3a + 2b)(9a^2 - 6ab + 4b^2)$ **38.** $(2x + 5)(3x - 7)$

39. $(x + 5 + y^2)(x + 5 - y^2)$ **40.** $(y + x - 2)(y - x + 2)$

41. $0, 3, -3$ **42.** $-\dfrac{1}{3}, -\dfrac{7}{2}$ **43.** $\dfrac{2x - 3}{3x - 1}$ **44.** $\dfrac{x - 2}{x + 3}$

45. $\dfrac{4}{x - y}$ **46.** $\dfrac{a^2 + ab^2}{a^2b - b^2}$ **47.** 0   **48.** -17

49. $x + 4$ **50.** $-x^2 + x + 5 + \dfrac{8}{x - 1}$

Getting Ready (page 455)

1. 0   **2.** 16   **3.** 16   **4.** -16 **5.** $\dfrac{8}{125}$ **6.** $\dfrac{81}{256}$

7. $49x^2y^2$ **8.** $343x^3y^3$

Exercises 7.1 (page 464)

1. $5, -5$ **3.** $2x, -2x$ **5.** 2   **7.** $3, -3$ **9.** $\dfrac{x + 5}{x - 2}$

11. 1   **13.** $\dfrac{3(m^2 + 2m - 1)}{(m + 1)(m - 1)}$ **15.** $(5x^2)^2, 6^2 = 36$

17. positive   **19.** 5, left   **21.** radical, index, radicand

23. $|x|$ **25.** x **27.** even   **29.** $7y^2$ **31.** $a^2 + b^3$

33. $10, -10$ **35.** $7, -7$ **37.** 11   **39.** -8 **41.** $-\dfrac{5}{7}$

43. not real   **45.** 0.4   **47.** $5|x|$ **49.** $8y^2$

51. $|x + 3|$ **53.** $|a + 3|$ **55.** 1   **57.** -3 **59.** $-\dfrac{4}{3}$

61. 0.4   **63.** $5y$ **65.** $-10pq$ **67.** -3 **69.** -2

71. 2   **73.** $\dfrac{2}{5}$ **75.** not real   **77.** $\dfrac{1}{2}$ **79.** $3|x|$

81. $2a$ **83.** $\dfrac{1}{2}|x|$ **85.** $|x^3|$ **87.** $-x$ **89.** $-3a^2$

91. D: $[-4, \infty)$, R: $[0, \infty)$ **93.** D: $[0, \infty)$, R: $(-\infty, -3]$

95. D: $(-\infty, \infty)$, R: $(-\infty, \infty)$ **97.** D: $(-\infty, \infty)$, R: $(-\infty, \infty)$

99. 4   **101.** not real   **103.** $5|b|$ **105.** $|t + 12|$

107. $-\dfrac{1}{2}m^2n$ **109.** $0.2z^3$ **111.** $x + 2$ **113.** $0.1x^2|y|$

115. 3.4641   **117.** 26.0624   **119.** about 4 sec   **121.** 1.67

123. 11.8673   **125.** 3 units   **127.** 4 sec

129. about 7.4 amperes

Getting Ready (page 467)

1. 25   **2.** 169   **3.** 18   **4.** 11,236

Exercises 7.2 (page 471)

1. 25   **3.** 25   **5.** 4   **7.** $6x^2 + 7x - 20$

9. $20a^2 - 23ab + 6b^2$ **11.** hypotenuse

13. $a^2 + b^2 = c^2$ **15.** positive   **17.** 10 ft   **19.** 80 m

21. 48 in.   **23.** 4 mi   **25.** 5   **27.** 13   **29.** 10

31. 17   **33.** 10.2   **35.** 13.6   **37.** about 127 ft

39. about 135 ft   **41.** 9.9 cm   **43.** 0.05 ft   **45.** 13 ft

47. yes   **49.** 173 yd   **51.** 24 cm^2 **53.** no

55. no   **63.** about 25   **65.** 16 ft/sec

Getting Ready (page 474)

1. x^7 **2.** a^{12} **3.** a^4 **4.** 1   **5.** $\dfrac{1}{x^4}$ **6.** x^3

7. $\dfrac{b^6}{c^9}$ **8.** a^{10}

Exercises 7.3 (page 480)

1. -49 **3.** 125 **5.** $\frac{1}{16}$ **7.** 8 **9.** 4 **11.** $x \le 1$

13. $r > 28$ **15.** $1\frac{2}{3}$ pints   **17.** $a \cdot a \cdot a \cdot a$ **19.** a^{mn}

21. $\frac{a^n}{b^n}$ **23.** $\frac{1}{a^n}, 0$ **25.** $\left(\frac{b}{a}\right)^n$ **27.** $|x|$ **29.** $\sqrt[3]{5}$

31. $\sqrt[5]{8}$ **33.** $\sqrt[7]{13a}$ **35.** $\sqrt[4]{\frac{1}{2}x^3 y}$ **37.** $\sqrt[4]{6a^3 b}$

39. $\sqrt{x^2 + y^2}$ **41.** 7 **43.** 3 **45.** $\frac{1}{3}$ **47.** $\frac{1}{2}$

49. -3 **51.** not real   **53.** $7^{1/2}$ **55.** $(3a)^{1/4}$

57. $5b^{1/7}$ **59.** $\left(\frac{1}{7}abc\right)^{1/6}$ **61.** $\left(\frac{1}{2}mn\right)^{1/5}$

63. $(x^2 + y^2)^{1/3}$ **65.** $5|y|$ **67.** $3x$ **69.** $|x + 1|$

71. not real   **73.** 125 **75.** 27 **77.** $81x^4$ **79.** $\frac{1}{4}$

81. $\frac{1}{2}$ **83.** $\frac{1}{8}$ **85.** $\frac{1}{64x^3}$ **87.** $\frac{1}{9y^2}$ **89.** 8 **91.** $\frac{16}{81}$

93. $5^{8/9}$ **95.** $4^{3/5}$ **97.** $\frac{1}{36}$ **99.** $9^{1/5}$ **101.** $7^{1/2}$

103. $2^{2/3}$ **105.** x^{14} **107.** $a^{2/9}$ **109.** $y + y^2$

111. $x^2 - x + x^{3/5}$ **113.** $x - y$

115. $x^{4/3} + 2x^{2/3}y^{2/3} + y^{4/3}$ **117.** $\sqrt{p}$ **119.** $\sqrt{5b}$

121. 2 **123.** 2 **125.** 0 **127.** -3 **129.** $125x^6$

131. $\frac{4x^2}{9}$ **133.** $\frac{1}{4p^2}$ **135.** $-\frac{3}{2x}$ **137.** $a^{3/4}b^{1/2}$

139. $\frac{n^{2/5}}{m^{3/5}}$ **141.** $\frac{2x}{3}$ **143.** $\frac{1}{3}x$ **145.** $x^2 + 3x^3 - 4x^{4/3}$

147. $\frac{1}{x} - 2 + x$ **149.** 2.47 **151.** 1.01 **153.** 0.24

155. -1.32 **159.** yes

Getting Ready (page 482)

1. 15 **2.** 24 **3.** 5 **4.** 7 **5.** $4x^2$ **6.** $\frac{8}{11}x^3$

7. $3ab^3$ **8.** $-2a^4$

Exercises 7.4 (page 489)

1. $4 \cdot 7$ **3.** $9 \cdot 2$ **5.** $8 \cdot 2$ **7.** $27 \cdot 2$ **9.** $24x^2$

11. $-\frac{21x^7}{y^4}$ **13.** $9t^2 + 12t + 4$ **15.** $3p + 4 + \frac{-5}{2p - 5}$

17. $\sqrt[n]{a}\sqrt[n]{b}$ **19.** like   **21.** 6 **23.** t **25.** $5x$

27. 10 **29.** $7x$ **31.** $6b$ **33.** 2 **35.** $3a$

37. $2\sqrt{5}$ **39.** $-10\sqrt{2}$ **41.** $2\sqrt[3]{10}$ **43.** $-3\sqrt[3]{3}$

45. $2\sqrt[4]{2}$ **47.** $2\sqrt[5]{3}$ **49.** $\frac{\sqrt{5}}{7a}$ **51.** $\frac{a\sqrt[3]{7}}{4}$

53. $\frac{p\sqrt[3]{3}}{10q}$ **55.** $\frac{m^3\sqrt[5]{3}}{2n^2}$ **57.** $3y\sqrt{7y}$ **59.** $4a^2\sqrt{3a}$

61. $-4a\sqrt{7a}$ **63.** $5ab\sqrt{7b}$ **65.** $-5b^2\sqrt{3}$

67. $-3x^2\sqrt[3]{2}$ **69.** $2x^4 y\sqrt[3]{2}$ **71.** $\frac{z}{4x}$ **73.** $4\sqrt{3}$

75. $-\sqrt{2}$ **77.** $2\sqrt{2}$ **79.** $27\sqrt{5}$ **81.** $4\sqrt{7} - 7\sqrt{6}$

83. $5\sqrt[3]{3}$ **85.** $-\sqrt[3]{4}$ **87.** -10 **89.** $-11\sqrt[3]{2}$

91. $x\sqrt[3]{3x^2}$ **93.** $8z\sqrt{y}$ **95.** $-7y^2\sqrt{y}$ **97.** $-6x\sqrt[3]{2x}$

99. $a = \frac{2}{3}, c = \frac{2}{3}\sqrt{2}$ **101.** $b = 5\sqrt{2}, c = 10$

103. $a = 5, b = 5$ **105.** $a = 7, b = 7$

107. $b = 5\sqrt{3}, c = 10$ **109.** $a = 9, c = 18$

111. $a = 12, b = 12\sqrt{3}$ **113.** $a = \frac{15}{2}, b = \frac{15}{2}\sqrt{3}$

115. $10\sqrt{2x}$ **117.** $2x^3 y\sqrt[4]{2}$ **119.** $\frac{\sqrt[4]{5x}}{2z}$ **121.** $-4\sqrt{2}$

123. 81 **125.** $94\sqrt[4]{3}$ **127.** $10\sqrt[4]{3}$ **129.** $9\sqrt[3]{5y}$

131. $9\sqrt[6]{12xyz}$ **133.** $4x\sqrt[5]{xy^2}$ **135.** $2x + 2$

137. $h = 2.83, x = 2.00$ **139.** $x = 8.66, h = 10.00$

141. $x = 4.69, y = 8.11$ **143.** $x = 12.11, y = 12.11$

145. $10\sqrt{3}$ mm, 17.32 mm   **149.** $12 + 13\sqrt{3}$

Getting Ready (page 492)

1. a^7 **2.** b^3 **3.** $a^2 - 2a$ **4.** $6b^3 + 9b^2$

5. $a^2 - 3a - 10$ **6.** $4a^2 - 9b^2$

Exercises 7.5 (page 499)

1. 5 **3.** xy **5.** 5 **7.** ab **9.** 1 **11.** $\frac{1}{3}$

13. $2, \sqrt{7}, \sqrt{5}$ **15.** $\sqrt{x} - 1$ **17.** conjugate

19. 4 **21.** $5\sqrt{2}$ **23.** $6\sqrt{2}$ **25.** 5 **27.** $r\sqrt[3]{10s}$

29. ab^2 **31.** $30x^2 y\sqrt{2y}$ **33.** $x^2(x + 3)$

35. $12\sqrt{5} - 15$ **37.** $12\sqrt{6} + 6\sqrt{14}$ **39.** $-1 - 2\sqrt{2}$

41. $8 + 5\sqrt{2}$ **43.** $8x - 14\sqrt{x} - 15$

45. $a\sqrt{6} - 3\sqrt{a} + 2\sqrt{3a} - 3\sqrt{2}$ **47.** $\frac{\sqrt{7}}{7}$

49. $\frac{\sqrt{6}}{3}$ **51.** $\frac{2\sqrt{30}}{5}$ **53.** $\frac{\sqrt{21}}{6}$ **55.** $\frac{3\sqrt[3]{4}}{4}$

57. $\frac{\sqrt[3]{21}}{6}$ **59.** $2\sqrt{2x}$ **61.** $\frac{\sqrt{5y}}{y}$ **63.** $\sqrt{2} + 1$

65. $\frac{3\sqrt{2} - \sqrt{10}}{4}$ **67.** $2 + \sqrt{3}$ **69.** $\frac{9 - 2\sqrt{14}}{5}$

71. $\frac{2(\sqrt{x} - 1)}{x - 1}$ **73.** $\frac{x(\sqrt{x} + 4)}{x - 16}$ **75.** $\frac{x - 2\sqrt{xy} + y}{x - y}$

77. $\frac{1}{\sqrt{3} - 1}$ **79.** $\frac{x - 9}{x(\sqrt{x} - 3)}$ **81.** 18 **83.** $5a\sqrt{b}$

85. $2a^2 b^2\sqrt[3]{2}$ **87.** $3x(y + z)\sqrt[3]{4}$

89. $-8x\sqrt{10} + 6\sqrt{15x}$ **91.** $5z + 2\sqrt{15z} + 3$

93. $18r - 12\sqrt{2r} + 4$ **95.** $-6x - 12\sqrt{x} - 6$ **97.** $\sqrt[3]{3}$

99. $\frac{\sqrt[4]{4}}{2}$ **101.** $\frac{\sqrt[5]{2}}{2}$ **103.** $\frac{\sqrt[3]{2ab^2}}{b}$ **105.** $\sqrt{2z} + 1$

107. $\frac{x - y}{x - \sqrt{xy}}$ **109.** $f/4$ **111.** $r = \frac{\sqrt{\pi A}}{\pi}$

113. $4\sqrt{2}$ cm   **115.** $2\sqrt{3}$ ft, $4\sqrt{3}$ ft   **119.** $\frac{x - 9}{4(\sqrt{x} + 3)}$

Getting Ready (page 502)

1. a **2.** $5x$ **3.** $x + 4$ **4.** $y - 3$

Exercises 7.6 (page 507)

1. x **3.** $x + 3$ **5.** $x^2 + 14x + 49$ **7.** 22

9. 17 **11.** $x^n = y^n$, power rule   **13.** square

15. extraneous   **17.** 1 **19.** 11 **21.** 4 **23.** $\frac{5}{2}, \frac{1}{2}$

25. 7; 3 is extraneous   **27.** $4, 3$ **29.** $0, 1$

31. $2, -4$ **33.** 0 **35.** 1 **37.** 1

39. 6; 0 is extraneous   **41.** 5 **43.** 3; -1 is extraneous

45. $g = \frac{v^2}{2h}$ **47.** $l = \frac{8T^2}{\pi^2}$ **49.** $A = P(r + 1)^3$

51. $v^2 = c^2\left(1 - \frac{L_A^2}{L_B^2}\right)$ **53.** 2; 7 is extraneous   **55.** 16

57. 0; 4 is extraneous   **59.** 2; 142 is extraneous   **61.** $1, -4$

63. 2; -8 is extraneous   **65.** -1; 1 is extraneous

67. $1, 9$ **69.** 2; -2 is extraneous   **71.** $\varnothing$; 6 is extraneous

73. $\varnothing$; 8 is extraneous   **75.** 2 **77.** 1

79. $2; -1$ is extraneous   **81.** 2,010 ft   **83.** about 29 mph
85. 19%   **87.** \$5   **89.** $R = \frac{8kl}{\pi r^4}$ **93.** $0, 4$

Getting Ready (page 510)

1. $7x$ **2.** $-x + 10$ **3.** $12x^2 + 5x - 25$ **4.** $9x^2 - 25$

Exercises 7.7 (page 518)

1. not real   **3.** real   **5.** not real   **7.** $3 - \sqrt{5}$
9. $12 + 7x$ **11.** -1 **13.** 20 mph   **15.** i
17. $-i$ **19.** imaginary   **21.** $\frac{\sqrt{a}}{\sqrt{b}}$ **23.** $5, 7$
25. conjugates   **27.** $5i$ **29.** $11i$ **31.** $i\sqrt{7}$
33. $2i\sqrt{2}$ **35.** $9 + 0i$ **37.** $9 + 3i$ **39.** yes
41. no   **43.** $7 + 2i$ **45.** $3 - 5i$ **47.** $15 + 7i$
49. $6 - 8i$ **51.** $24 + 18i$ **53.** $-25 - 25i$
55. $7 + i$ **57.** $32 - 22i$ **59.** $15 - \sqrt{3}i$
61. $7 + 24i$ **63.** 61   **65.** 25   **67.** $2 + i$
69. $\frac{25}{13} + \frac{5}{13}i$ **71.** $\frac{5}{13} - \frac{12}{13}i$ **73.** $\frac{11}{10} + \frac{3}{10}i$
75. $-\frac{42}{25} - \frac{6}{25}i$ **77.** $\frac{35}{34} + \frac{21}{34}i$ **79.** i **81.** -1
83. i **85.** $-i$ **87.** 1   **89.** i **91.** $-2 - 5i$
93. $0 - i$ **95.** $0 + \frac{4}{5}i$ **97.** $\frac{1}{8} - 0i$
99. $0 + \frac{3}{5}i$ **101.** 10   **103.** 13   **105.** $\sqrt{74}$
107. 1   **109.** no   **111.** $-20 - 30i$
113. $-5 + 12i$ **115.** $\frac{1}{4} + \frac{3}{4}i$ **117.** $\frac{1}{4} - \frac{\sqrt{15}}{4}i$
119. $-\frac{5}{169} + \frac{12}{169}i$ **121.** $2 + 9i$ **123.** $-15 + 2\sqrt{3}i$
125. $5 + 5i$ **127.** $16 + 2i$ **129.** $7 - 4i$ volts
131. $1 - 3.4i$ **139.** $\frac{5}{3 + i}$

Chapter Review (page 524)

1. 9   **2.** -13 **3.** -6 **4.** 15   **5.** -3
6. -6 **7.** 5   **8.** -2 **9.** $5|x|$ **10.** $|x + 3|$
11. $3a^2b$ **12.** $4x^2|y|$
13. D: $[-2, \infty)$; R: $[0, \infty)$ **14.** D: $[1, \infty)$; R: $(-\infty, 0]$

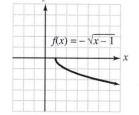

15. D: $[0, \infty)$; R: $(-\infty, 2]$ **16.** D: $(-\infty, \infty)$; R: $(-\infty, \infty)$

17. 12   **18.** about 5.7   **19.** 3 mi   **20.** 8.2 ft
21. 88 yd   **22.** 16,000 yd, or about 9 mi   **23.** 10
24. 2.83 units   **25.** 9   **26.** -7 **27.** 27   **28.** 64
29. -3 **30.** -4 **31.** $\frac{1}{4}$ **32.** $\frac{1}{4}$ **33.** $-16,807$
34. $\frac{1}{1,024}$ **35.** $\frac{25}{9}$ **36.** $\frac{27}{8}$ **37.** $3xy^{1/3}$ **38.** $2x^{1/2}y$
39. $125x^{9/2}y^6$ **40.** $\frac{1}{4u^{4/3}v^2}$ **41.** $5^{3/4}$ **42.** $a^{7/9}$
43. $u - 1$ **44.** $v + v^2$ **45.** $x + 2x^{1/2}y^{1/2} + y$
46. $a^{4/3} - b^{4/3}$ **47.** $\sqrt[3]{5}$ **48.** $\sqrt{x}$ **49.** $\sqrt[3]{3ab^2}$
50. $\sqrt{5ab}$ **51.** $4\sqrt{11}$ **52.** $5\sqrt[3]{2}$ **53.** $2\sqrt[4]{2}$
54. $2\sqrt[5]{3}$ **55.** $2x\sqrt{2x}$ **56.** $2x^3y^2\sqrt{6y}$ **57.** $2xy\sqrt[3]{2x^2y}$
58. $3x^2y\sqrt[3]{2x}$ **59.** $4x$ **60.** $2x$ **61.** $\frac{\sqrt[3]{2a^2b}}{3x}$
62. $\frac{\sqrt{17xy}}{8a^2}$ **63.** $5\sqrt{3}$ **64.** $-\sqrt{2}$ **65.** 0
66. $8\sqrt[4]{2}$ **67.** $29x\sqrt{2}$ **68.** $32a\sqrt{3a}$ **69.** $13\sqrt[3]{2}$
70. $-4x\sqrt[4]{2x}$ **71.** $7\sqrt{2}$ m   **72.** $6\sqrt{3}$ cm, 18 cm
73. 7.07 in.   **74.** 8.66 cm   **75.** $10\sqrt{15}$ **76.** 72
77. $2x\sqrt{5}$ **78.** $3x^2$ **79.** $-2x$ **80.** $-20x^3y^3\sqrt{xy}$
81. $9 - 4\sqrt{3}$ **82.** $5 - 6\sqrt{5}$ **83.** $\sqrt{10} - \sqrt{5}$
84. $3 + \sqrt{6}$ **85.** 4   **86.** $5 + 2\sqrt{6}$ **87.** $x - y$
88. $6u + \sqrt{u} - 12$ **89.** $\frac{\sqrt{3}}{3}$ **90.** $\frac{\sqrt{15}}{5}$ **91.** $\frac{\sqrt{xy}}{y}$
92. $\frac{\sqrt[3]{u^2}}{u^2v^2}$ **93.** $2(\sqrt{2} + 1)$ **94.** $\frac{\sqrt{6} + \sqrt{2}}{2}$
95. $2(\sqrt{x} - 4)$ **96.** $\frac{a + 2\sqrt{a} + 1}{a - 1}$ **97.** $\frac{3}{5\sqrt{3}}$
98. $\frac{1}{\sqrt[3]{3}}$ **99.** $\frac{9 - x}{2(3 + \sqrt{x})}$ **100.** $\frac{a - b}{a + \sqrt{ab}}$ **101.** 9
102. 16, 9   **103.** $5, -1$ is extraneous   **104.** $\frac{9}{16}$ **105.** 2
106. $0, -2$ **107.** $10 - 12i$ **108.** $2 - 50i$
109. $-96 + 3i$ **110.** $-2 - 2\sqrt{2}i$ **111.** $22 + 29i$
112. $-22 + 7i$ **113.** $-12 + 28\sqrt{3}i$ **114.** $118 + 10\sqrt{2}i$
115. $0 - \frac{3}{4}i$ **116.** $0 - \frac{2}{5}i$ **117.** $\frac{12}{5} - \frac{6}{5}i$ **118.** $\frac{21}{10} + \frac{7}{10}i$
119. $\frac{15}{17} + \frac{8}{17}i$ **120.** $\frac{4}{5} - \frac{3}{5}i$ **121.** $\frac{15}{29} - \frac{6}{29}i$
122. $\frac{1}{3} + \frac{1}{3}i$ **123.** 15   **124.** 26   **125.** $-i$ **126.** -1

Chapter 7 Test (page 530)

1. 6   **2.** $5\sqrt{5}$ **3.** $2|x|$ **4.** $2x$
5. D: $[2, \infty)$, R: $[0, \infty)$ **6.** D: $(-\infty, \infty)$, R: $(-\infty, \infty)$

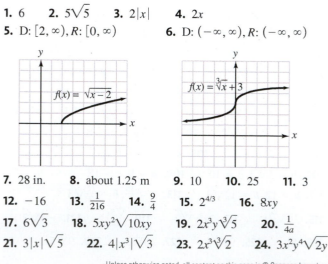

7. 28 in.   **8.** about 1.25 m   **9.** 10   **10.** 25   **11.** 3
12. -16 **13.** $\frac{1}{216}$ **14.** $\frac{9}{4}$ **15.** $2^{4/3}$ **16.** $8xy$
17. $6\sqrt{3}$ **18.** $5xy^2\sqrt{10xy}$ **19.** $2x^3y\sqrt[3]{5}$ **20.** $\frac{1}{4a}$
21. $3|x|\sqrt{5}$ **22.** $4|x^3|\sqrt{3}$ **23.** $2x^3\sqrt[3]{2}$ **24.** $3x^2y^4\sqrt{2y}$

25. $-\sqrt{3}$ **26.** $14\sqrt[3]{5}$ **27.** $2y^2\sqrt{3y}$ **28.** $6z\sqrt[4]{3z}$
29. $-6x\sqrt{y} - 2xy^2$ **30.** $3 - 7\sqrt{6}$ **31.** $\dfrac{\sqrt{5}}{5}$
32. $\sqrt{3t} + 1$ **33.** $\dfrac{3}{\sqrt{21}}$ **34.** $\dfrac{a - b}{a - 2\sqrt{ab} + b}$
35. 10   **36.** $\varnothing$; 4 is extraneous   **37.** $-1 + 11i$
38. $4 - 7i$ **39.** $8 + 6i$ **40.** $-10 - 11i$ **41.** $0 - \dfrac{\sqrt{2}}{2}i$
42. $\dfrac{1}{2} + \dfrac{1}{2}i$

Getting Ready (page 535)

1. $(x + 5)(x - 5)$ **2.** $(b + 9)(b - 9)$
3. $(3x + 2)(2x - 1)$ **4.** $(2x - 3)(2x + 1)$

Exercises 8.1 (page 544)

1. $x^2 - 25$ **3.** $x^2 + 4x + 4$ **5.** $x^2 - 14x + 49$
7. 1   **9.** $t \leq 4$ **11.** $x = \sqrt{c}, x = -\sqrt{c}$
13. positive or negative   **15.** $9, -9$ **17.** $6, -6$
19. $-3, -4$ **21.** $\dfrac{2}{3}, -\dfrac{5}{2}$ **23.** ± 6 **25.** $\pm\sqrt{5}$
27. $0, -6$ **29.** $2 \pm \sqrt{5}$ **31.** $\pm\dfrac{7}{4}i$ **33.** $\pm\dfrac{9}{2}i$
35. 4   **37.** $\dfrac{9}{4}$ **39.** $2, -6$ **41.** $2, 4$ **43.** $-\dfrac{1}{3}, -\dfrac{3}{2}$
45. $\dfrac{5}{3}, -\dfrac{1}{2}$ **47.** $\dfrac{3}{4}, -\dfrac{3}{2}$ **49.** $1, -\dfrac{1}{2}$ **51.** $-\dfrac{7}{10} \pm \dfrac{\sqrt{29}}{10}$
53. $1 \pm \dfrac{\sqrt{6}}{3}$ **55.** $-1 \pm i$ **57.** $-4 \pm i\sqrt{2}$ **59.** $8, 1$
61. $\pm\dfrac{4\sqrt{3}}{3}$ **63.** $4, 10$ **65.** $-5 \pm \sqrt{3}$ **67.** $2, \dfrac{1}{2}$
69. $\dfrac{1}{3} \pm \dfrac{\sqrt{10}}{3}$ **71.** $0, -3$ **73.** $-1, -4$ **75.** $10, -1$
77. $\dfrac{1}{4} \pm \dfrac{3\sqrt{7}}{4}i$ **79.** $d = \dfrac{\sqrt{6h}}{2}$ **81.** $c = \dfrac{\sqrt{Em}}{m}$
83. $-\dfrac{1}{4} \pm \dfrac{\sqrt{41}}{4}$ **85.** $-\dfrac{1}{2} \pm \dfrac{\sqrt{13}}{2}$ **87.** 4 sec   **89.** 72 mph
91. 4%   **93.** width: $7\dfrac{1}{4}$ ft; length: $13\dfrac{3}{4}$ ft   **97.** $\dfrac{3}{4}$

Getting Ready (page 546)

1. $x^2 + 12x + 36, (x + 6)^2$ **2.** $x^2 - 7x + \dfrac{49}{4}, \left(x - \dfrac{7}{2}\right)^2$
3. 7   **4.** 8

Exercises 8.2 (page 551)

1. $5, -3, -4$ **3.** $y = \dfrac{-Ax + C}{B}$ **5.** $3\sqrt{5}$ **7.** $\sqrt{5}$
9. $7, -4, -9$ **11.** $-1, -5$ **13.** $7, -2$ **15.** $-4, -5$
17. $-1, \dfrac{3}{2}$ **19.** $\dfrac{4}{3}, -\dfrac{2}{5}$ **21.** $-\dfrac{3}{2}, -\dfrac{1}{2}$ **23.** $-\dfrac{1}{2} \pm \dfrac{\sqrt{5}}{10}$
25. $-\dfrac{1}{5} \pm \dfrac{\sqrt{6}}{5}$ **27.** $-1 \pm i$ **29.** $-\dfrac{5}{2} \pm \dfrac{\sqrt{3}}{2}i$
31. $\dfrac{2}{3} \pm \dfrac{\sqrt{2}}{3}i$ **33.** $\dfrac{1}{3} \pm \dfrac{2\sqrt{2}}{3}i$ **35.** $N = \dfrac{1}{2} \pm \dfrac{\sqrt{1 + 8C}}{2}$
37. $x = \dfrac{k}{2} \pm \dfrac{\sqrt{k^2 - 4ay}}{2}$ **39.** $-\dfrac{1}{2}, \dfrac{4}{3}$ **41.** $-\dfrac{5}{2} \pm \dfrac{\sqrt{17}}{2}$
43. $\dfrac{1}{3} \pm \dfrac{\sqrt{7}}{3}$ **45.** $\dfrac{1}{4}, -\dfrac{3}{4}$ **47.** $-\dfrac{1}{2} \pm \sqrt{5}$
49. $-\dfrac{1}{3} \pm \dfrac{\sqrt{5}}{3}i$ **51.** $8.98, -3.98$ **53.** $x^2 - 8x + 15 = 0$
55. $x^3 - x^2 - 14x + 24 = 0$ **57.** 8 ft by 12 ft   **59.** 4 units
61. $\dfrac{4}{3}$ cm   **63.** 12, 14 or $-14, -12$ **65.** 6, 7 or $-6, -7$
67. 30 mph   **69.** $4.80 or $5.20   **71.** 4,000

73. 2.26 in.   **75.** 1985   **77.** about 6.13×10^{-3} M
81. $\sqrt{2}, -3\sqrt{2}$ **83.** $i, 2i$

Getting Ready (page 554)

1. 17   **2.** -8

Exercises 8.3 (page 560)

1. -3 **3.** yes   **5.** no   **7.** -2 **9.** $\dfrac{3}{7}$
11. $b^2 - 4ac$ **13.** rational, unequal   **15.** rational, equal
17. complex conjugates   **19.** irrational, unequal
21. rational, unequal   **23.** $8, -8$ **25.** $12, -12$
27. $1, -1, 4, -4$ **29.** $1, -1, \sqrt{2}, -\sqrt{2}$
31. $1, -1, \sqrt{5}, -\sqrt{5}$ **33.** $1, -1, 2, -2$ **35.** $25, 4$
37. $\dfrac{9}{4}$; 1 is extraneous   **39.** $1; \dfrac{9}{4}$ is extraneous
41. $\varnothing$; $\dfrac{4}{9}$ and 1 are both extraneous   **43.** $-8, -27$
45. $-\dfrac{8}{27}, 1$ **47.** $-9, -1$ **49.** $-7, 4$ **51.** $0, 2$
53. $-1, -\dfrac{27}{13}$ **55.** 1 (double root), -1 (double root)
57. $-4, \dfrac{2}{3}$ **59.** $x = \pm\sqrt{r^2 - y^2}$
61. $y = \dfrac{-5x \pm \sqrt{25x^2 - 12x}}{2x}$ **63.** $\dfrac{2}{3}, -\dfrac{1}{4}$ **65.** $-\dfrac{5}{4} \pm \dfrac{\sqrt{17}}{4}$
67. $\dfrac{1}{3} \pm \dfrac{\sqrt{11}}{3}i$ **69.** $-1 \pm 2i$ **71.** yes   **73.** $\dfrac{81}{16}$; 1 is
extraneous   **75.** $-\dfrac{8}{27}, 1$ **77.** $1, -1, 2\sqrt{3}, -2\sqrt{3}$
79. $2, -2, \sqrt{6}i, -\sqrt{6}i$ **81.** $\dfrac{3}{8}, 1$ **83.** $\dfrac{1}{27}, -1$ **85.** $1 \pm i$
87. $-\dfrac{5}{7}, 3$ **89.** $d = \pm\dfrac{\sqrt{kI}}{I}$ **91.** $\mu^2 = \dfrac{\Sigma x^2}{N} - \sigma^2$
93. 5   **95.** $12, -3$ **97.** $k < -\dfrac{4}{3}$ **101.** no

Getting Ready (page 562)

1. -2 **2.** 2   **3.** 0   **4.** 8   **5.** 1   **6.** 4

Exercises 8.4 (page 574)

1. $x = 3$ **3.** $x = -1$ **5.** $3, -3$ **7.** 10
9. $3\dfrac{3}{5}$ hr   **11.** $f(x) = ax^2 + bx + c, a \neq 0$
13. maximum, minimum, vertex   **15.** upward
17. to the right   **19.** upward

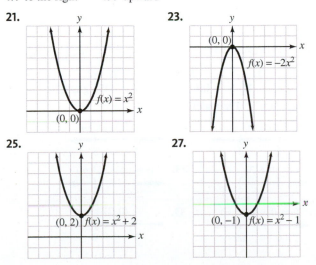

21.   **23.**   **25.**   **27.**

29.

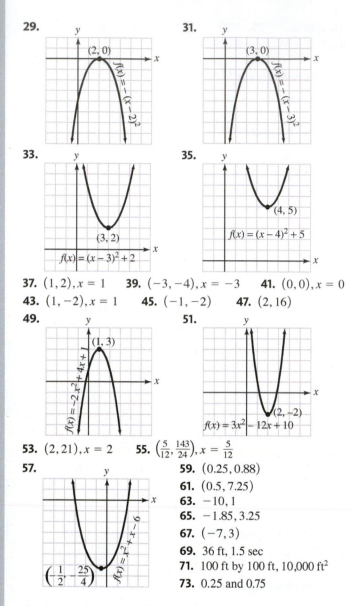

$(2, 0)$, $f(x) = -(x - 2)^2$

31.

$(3, 0)$, $f(x) = -(x - 3)^2$

33.

$(3, 2)$, $f(x) = (x - 3)^2 + 2$

35.

$(4, 5)$, $f(x) = (x - 4)^2 + 5$

37. $(1, 2)$, $x = 1$ **39.** $(-3, -4)$, $x = -3$ **41.** $(0, 0)$, $x = 0$

43. $(1, -2)$, $x = 1$ **45.** $(-1, -2)$ **47.** $(2, 16)$

49.

$(1, 3)$, $f(x) = -2x^2 + 4x + 1$

51.

$(2, -2)$, $f(x) = 3x^2 - 12x + 10$

53. $(2, 21)$, $x = 2$ **55.** $\left(\frac{5}{12}, \frac{143}{24}\right)$, $x = \frac{5}{12}$

57.

$\left(-\frac{1}{2}, -\frac{25}{4}\right)$, $f(x) = x^2 + x - 9$

59. $(0.25, 0.88)$

61. $(0.5, 7.25)$

63. $-10, 1$

65. $-1.85, 3.25$

67. $(-7, 3)$

69. 36 ft, 1.5 sec

71. 100 ft by 100 ft, 10,000 ft^2

73. 0.25 and 0.75

75. 75 ft by 75 ft, 5,625 ft^2 **77.** 14 ft **79.** 5,000

81. 3,276, \$14,742 **83.** \$35

Getting Ready (page 578)

1. $(x + 5)(x - 3)$ **2.** $(x - 2)(x - 1)$

Exercises 8.5 (page 586)

1. $x = 2$ **3.** $x < 2$ **5.** $x > -3$ **7.** -3

9. $y = kx^2$ **11.** $y = kab$ **13.** -7 **15.** greater

17. quadratic **19.** undefined **21.** sign

23. $(1, 4)$

25. $(-\infty, 3) \cup (5, \infty)$

27. $[-4, 3]$

29. $(-\infty, -5] \cup [3, \infty)$

31. $(-\infty, 0) \cup (1/2, \infty)$

33. $(0, 2]$

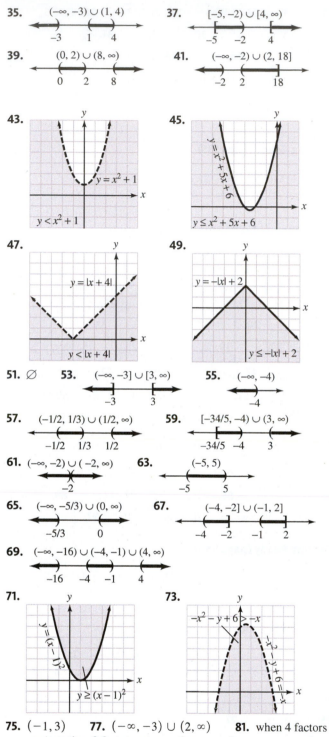

35. $(-\infty, -3) \cup (1, 4)$

37. $[-5, -2) \cup [4, \infty)$

39. $(0, 2) \cup (8, \infty)$

41. $(-\infty, -2) \cup (2, 18]$

43. $y < x^2 + 1$

45. $y = x^2 + 5x + 6$, $y \leq x^2 + 5x + 6$

47. $y = |x + 4|$, $y < |x + 4|$

49. $y = -|x| + 2$, $y \leq -|x| + 2$

51. $\varnothing$ **53.** $(-\infty, -3] \cup [3, \infty)$ **55.** $(-\infty, -4)$

57. $(-1/2, 1/3) \cup (1/2, \infty)$ **59.** $[-34/5, -4) \cup (3, \infty)$

61. $(-\infty, -2) \cup (-2, \infty)$ **63.** $(-5, 5)$

65. $(-\infty, -5/3) \cup (0, \infty)$ **67.** $(-4, -2] \cup (-1, 2]$

69. $(-\infty, -16) \cup (-4, -1) \cup (4, \infty)$

71. $y = (x - 1)^2$, $y \geq (x - 1)^2$

73. $-x^2 - y + 6 > -x$, $-x^2 - y + 6 = -x$

75. $(-1, 3)$ **77.** $(-\infty, -3) \cup (2, \infty)$ **81.** when 4 factors are negative, 2 factors are negative, or no factors are negative

Getting Ready (page 588)

1. $3x - 1$ **2.** $x + 3$ **3.** $2x^2 - 3x - 2$

Exercises 8.6 (page 594)

1. -3 **3.** $4x^2 + 8x + 2$ **5.** $4x^2 + 2x + 1$

7. $-4x^2 + 2x + 5$ **9.** $-\dfrac{5x + 2}{x + 3}$ **11.** $\dfrac{x - 4}{3x^2 - x - 12}$

13. $f(x) + g(x)$ **15.** $f(x)g(x)$ **17.** domain

19. $f(g(x))$ **21.** given point   **23.** $7x, (-\infty, \infty)$

25. $x, (-\infty, \infty)$ **27.** $12x^2, (-\infty, \infty)$

29. $\frac{4}{3}, (-\infty, 0) \cup (0, \infty)$ **31.** $3x - 2, (-\infty, \infty)$

33. $-x - 4, (-\infty, \infty)$ **35.** $2x^2 - 5x - 3, (-\infty, \infty)$

37. $\frac{x-3}{2x+1}, \left(-\infty, -\frac{1}{2}\right) \cup \left(-\frac{1}{2}, \infty\right)$ **39.** 31   **41.** 24

43. -1 **45.** $\frac{-7}{8}$ **47.** $2x^2 - 1$ **49.** $36x^2 + 12x$

51. 4   **53.** $2x + h$ **55.** $4x + 2h$ **57.** $2x + h + 1$

59. $2x + h + 3$ **61.** $4x + 2h + 3$

63. $-2x^2 + 3x - 3, (-\infty, \infty)$

65. $(3x - 2)/(2x^2 + 1), (-\infty, \infty)$ **67.** $3, (-\infty, \infty)$

69. $(x^2 - 4)/(x^2 - 1), (-\infty, -1) \cup (-1, 1) \cup (1, \infty)$

71. 58   **73.** 56   **75.** 2   **77.** $9x^2 - 9x + 2$ **79.** 4

81. $x + a$ **83.** $2x + 2a$ **85.** $x + a + 1$ **87.** $x + a + 3$

89. $2x + 2a + 3$ **95.** $3x^2 + 3xh + h^2$

97. $C(t) = \frac{5}{9}(2,668 - 200t)$

Getting Ready (page 596)

1. $y = \frac{x-2}{3}$ **2.** $y = \frac{2x-10}{3}$

Exercises 8.7 (page 603)

1. $(0, 1)$ **3.** $(3, 7)$ **5.** $(-1, -1)$ **7.** $7 - 7i$

9. $18 - i$ **11.** 10   **13.** one-to-one   **15.** 2

17. x **19.** yes   **21.** no

23.
one-to-one

25.
one-to-one

27.
not one-to-one

29.
one-to-one

31. $\{(3, 4), (2, 3), (1, 2)\}$; yes

33. $\{(2, 1), (3, 2), (3, 1), (5, 1)\}$; no   **35.** $f^{-1}(x) = \frac{1}{4}x - \frac{1}{4}$

37. $f^{-1}(x) = 5x - 4$

39. $f^{-1}(x) = \frac{x-3}{4}$ **41.** $f^{-1}(x) = \frac{x-2}{3}$

43. $y = \pm\sqrt{x - 9}$, no   **45.** $f^{-1}(x) = \sqrt[3]{x}$, yes

47. 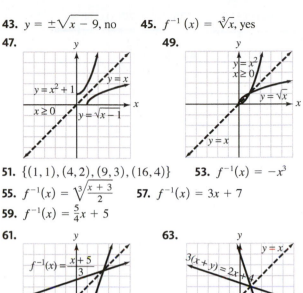   **49.**

51. $\{(1, 1), (4, 2), (9, 3), (16, 4)\}$ **53.** $f^{-1}(x) = -x^3$

55. $f^{-1}(x) = \sqrt[3]{\frac{x+3}{2}}$ **57.** $f^{-1}(x) = 3x + 7$

59. $f^{-1}(x) = \frac{5}{4}x + 5$

61.   **63.**

67. $f^{-1}(x) = \frac{x+1}{x-1}$

Chapter Review (page 608)

1. $\frac{2}{3}, -\frac{3}{4}$ **2.** $-\frac{1}{3}, -\frac{5}{2}$ **3.** $\frac{2}{3}, -\frac{4}{5}$ **4.** $1, -7$

5. $-2, -6$ **6.** $\frac{7}{2}, 1$ **7.** $\frac{1}{4} \pm \frac{\sqrt{41}}{4}$ **8.** $6, -1$

9. $0, 7$ **10.** $\frac{1}{2}, -7$ **11.** $1, -\frac{3}{5}$ **12.** $\frac{1}{4} \pm \frac{\sqrt{17}}{4}$

13. $-\frac{1}{2} \pm \frac{\sqrt{7}}{2}i$ **14.** 4 cm by 6 cm   **15.** 2 ft by 3 ft

16. 7 sec   **17.** 196 ft   **18.** irrational, unequal

19. complex conjugates   **20.** $12, 152$ **21.** $k \geq -\frac{7}{3}$

22. $1, 49$ **23.** $8, -27$ **24.** 1   **25.** $1, -\frac{8}{5}$

26. $\frac{14}{3}$ **27.** 1

28.   **29.**

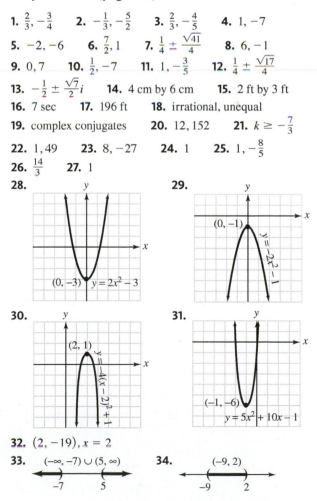

32. $(2, -19), x = 2$

33. $(-\infty, -7) \cup (5, \infty)$ **34.** $(-9, 2)$

35. $(-\infty, 0) \cup [3/5, \infty)$

36. $(-7/2, 1) \cup (4, \infty)$

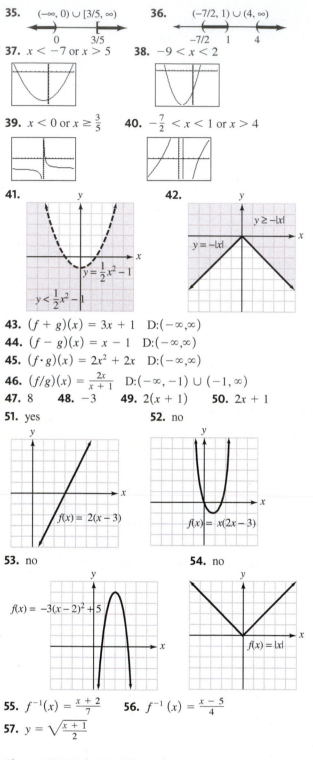

37. $x < -7$ or $x > 5$

38. $-9 < x < 2$

39. $x < 0$ or $x \geq \frac{3}{5}$

40. $-\frac{7}{2} < x < 1$ or $x > 4$

41. **42.**

$y < \frac{1}{2}x^2 - 1$, $y = \frac{1}{2}x^2 - 1$

$y \geq -|x|$, $y = -|x|$

43. $(f+g)(x) = 3x + 1$ D:$(-\infty,\infty)$

44. $(f-g)(x) = x - 1$ D:$(-\infty,\infty)$

45. $(f \cdot g)(x) = 2x^2 + 2x$ D:$(-\infty,\infty)$

46. $(f/g)(x) = \frac{2x}{x+1}$ D:$(-\infty, -1) \cup (-1, \infty)$

47. 8 **48.** -3 **49.** $2(x+1)$ **50.** $2x + 1$

51. yes **52.** no

$f(x) = 2(x-3)$

$f(x) = x(2x-3)$

53. no **54.** no

$f(x) = -3(x-2)^2 + 5$

$f(x) = |x|$

55. $f^{-1}(x) = \frac{x+2}{7}$ **56.** $f^{-1}(x) = \frac{x-5}{4}$

57. $y = \sqrt{\frac{x+1}{2}}$

Chapter 8 Test (page 615)

1. $7, -4$ **2.** $-\frac{3}{2}, -\frac{5}{3}$ **3.** $x = \pm 12$ **4.** $x = 5 \pm \sqrt{6}i$

5. $-3 \pm \sqrt{2}$ **6.** $\frac{5}{2} \pm \frac{\sqrt{37}}{2}$ **7.** $-\frac{5}{4} \pm \frac{\sqrt{17}}{4}$

8. $1 \pm \sqrt{5}i$ **9.** nonreal **10.** 2 **11.** 10 in. **12.** $1, \frac{1}{4}$

13. $f(x) = \frac{1}{2}x^2 - 4$, $(0, -4)$

14. $(2, 7), x = 2$

15. $y = -x^2 + 3$, $y \leq -x^2 + 3$

16. $(-\infty, -2) \cup (4, \infty)$

17. $(-3, 2]$

18. $(g+f)(x) = 5x - 1$ D: $(-\infty, \infty)$

19. $(f-g)(x) = 3x + 1$ D: $(-\infty, \infty)$

20. $(g \cdot f)(x) = 4x^2 - 4x$ D: $(-\infty, \infty)$

21. $(g/f)(x) = \frac{x-1}{4x}$ D: $(-\infty, 0) \cup (0, \infty)$ **22.** 3

23. -4 **24.** -8 **25.** -9 **26.** $4(x-1)$

27. $4x - 1$ **28.** $f^{-1}(x) = \frac{12 - 2x}{3}$

Cumulative Review (page 616)

1. D: $(-\infty, \infty)$; R: $[-2, \infty)$ **2.** D: $(-\infty, \infty)$; R: $(-\infty, 0]$

3. $y = 4x + 7$ **4.** $y = -\frac{2}{3}x - 2$ **5.** $-4a^2 + 12a - 7$

6. $20x^2 - 7x - 6$ **7.** $(x^2 + 4y^2)(x + 2y)(x - 2y)$

8. $(3x + 2)(5x - 4)$ **9.** $9, -1$ **10.** $0, \frac{2}{3}, -\frac{1}{2}$

11. $6ab^2$ **12.** $4t\sqrt{3t}$ **13.** $-4y^2$ **14.** $4x$ **15.** $\frac{1}{2}$

16. 9 **17.** y^2 **18.** $x^{17/12}$

19. $f(x) = \sqrt{x-2}$ domain $= [2, \infty)$ range $= [0, \infty)$

20. $f(x) = -\sqrt{x+2}$ domain $= [-2, \infty)$ range $= (-\infty, 0]$

21. $x^{4/3} - x^{2/3}$ **22.** $\frac{1}{x} + 2 + x$ **23.** $7\sqrt{2}$

24. $-12\sqrt[4]{2} + 10\sqrt[4]{3}$ **25.** $-18\sqrt{6}$ **26.** $\frac{5\sqrt[3]{x^2}}{x}$

27. $\frac{x + 3\sqrt{x} + 2}{x - 1}$ **28.** $\sqrt{xy}$ **29.** $2, 7$ **30.** $\frac{1}{4}$

31. $3\sqrt{2}$ in. **32.** $2\sqrt{3}$ in. **33.** 10 **34.** $\pm 5i$

35. $1, -\frac{3}{2}$ **36.** $-\frac{2}{3} \pm \frac{\sqrt{7}}{3}$

37. $V(0,5); x = 0$

38.

39. $13 - 3i$ **40.** $-5 - 7i$ **41.** 61   **42.** $12 - 6i$

43. $-12 - 10i$ **44.** $\frac{3}{2} + \frac{1}{2}i$ **45.** $\sqrt{13}$ **46.** $\sqrt{61}$

47. -2 **48.** 9, 16

49. $(-\infty, -2) \cup (3, \infty)$ **50.** $(-2, 3]$

51. 14   **52.** 99   **53.** $12x^2 - 12x + 5$ **54.** $6x^2 + 3$

55. $f^{-1}(x) = \frac{x-2}{3}$ **56.** $f^{-1}(x) = \sqrt[3]{x-4}$

Getting Ready (page 621)

1. 8   **2.** 5   **3.** $\frac{1}{25}$ **4.** $\frac{8}{27}$

Exercises 9.1 (page 629)

1. 9   **3.** 108   **5.** $\frac{1}{9}$ **7.** $\frac{1}{12}$ **9.** 40   **11.** 120°

13. exponential   **15.** $(0, \infty)$ **17.** increasing

19. $P\left(1 + \frac{r}{k}\right)^{kt}$ **21.** b   **23.** left

25.   **27.**

29.   **31.**

33. $b = \frac{1}{2}$ **35.** $b = 3$

37.   **39.**

41. $b = 2$

43. increasing function   **45.** decreasing function

47. approximately 1.9 billion   **49.** $\frac{256}{6,561} A_0$

51. $22,080.40   **53.** $7,551.32   **55.** $2,273,996.13

57. 5.0421×10^{-5} coulombs   **59.** $1,115.33

Getting Ready (page 632)

1. 2   **2.** 2.25   **3.** 2.44   **4.** 2.59

Exercises 9.2 (page 637)

1. 1   **3.** 40.17   **5.** 0.22   **7.** $8x^3\sqrt{5x}$ **9.** $23y\sqrt{2y}$

11. 2.72   **13.** increasing   **15.** $A = Pe^{rt}$

17.   **19.**

21. no   **23.** no

25.   **27.**

29. $20,029.87   **31.** $11,809.16   **33.** 10.6 billion

35. 2.6   **37.** 72 yr   **39.** 2.31 grams   **41.** 2,498.27 grams

43. $7,518.28 from annual compounding; $7,647.95 from continuous compounding   **45.** 5,363   **47.** 0.076   **49.** 0 mps

51. this object   **53.** $3,094.15   **57.** 2   **59.** $k = e^{10}$

Getting Ready (page 640)

1. 1   **2.** 25   **3.** $\frac{1}{25}$ **4.** 4

Exercises 9.3 (page 647)

1. 3   **3.** 5   **5.** 2   **7.** 5   **9.** 2

11. $\varnothing$; $\frac{3}{5}$ is extraneous   **13.** 4   **15.** $(0, \infty)$ **17.** x

19. exponent   **21.** $(b, 1), (1, 0)$ **23.** $20 \log \frac{E_O}{E_I}$

25. $4^3 = 64$ **27.** $\left(\frac{1}{2}\right)^3 = \frac{1}{8}$ **29.** $4^{-3} = \frac{1}{64}$ **31.** $\left(\frac{1}{2}\right)^3 = \frac{1}{8}$

33. $\log_7 49 = 2$ **35.** $\log_6 \frac{1}{36} = -2$ **37.** $\log_{1/2} 32 = -5$

39. $\log_x z = y$ **41.** 5   **43.** 3   **45.** 49   **47.** 6

49. 5   **51.** $\frac{1}{25}$ **53.** 5   **55.** $\frac{3}{2}$ **57.** 3   **59.** $\frac{1}{64}$

61. increasing   **63.** decreasing

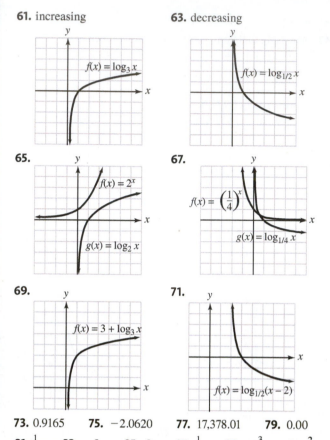

65.   **67.**

69.   **71.**

73. 0.9165   **75.** −2.0620   **77.** 17,378.01   **79.** 0.00

81. $\frac{1}{6}$ **83.** −3   **85.** 8   **87.** $\frac{1}{2}$ **89.** $-\frac{3}{2}$ **91.** $\frac{2}{3}$

93. 8   **95.** 4   **97.** 4   **99.** 4   **101.** 3   **103.** 100

105. 25.25   **107.** 8   **109.** $b = 3$ **111.** no value of b

113. 4.4   **115.** 4   **117.** 4.2 yr old   **119.** 10.8 yr

121. 29.0 dB   **123.** 49.5 dB

Getting Ready (page 650)

1. 2   **2.** −3   **3.** 1   **4.** 0

Exercises 9.4 (page 654)

1. $3^2 = 9$ **3.** $\log_2 8 = 3$ **5.** $y = -7x + 2$

7. $y = -\frac{3}{2}x + \frac{1}{2}$ **9.** $y = 6$ **11.** $\frac{2x^2 + x + 1}{x(x + 1)}$

13. $\frac{x + y}{y - x}$ **15.** $(0, \infty), (-\infty, \infty)$ **17.** 10   **19.** $\frac{\ln 2}{r}$

21. 3.2288   **23.** 2.2915   **25.** 9.9892   **27.** 23.8075

29.   **31.**   **33.** no   **35.** no

37. 0.0408   **39.** no real value   **41.** 0.0089

43. 24.3385   **45.** 5.8 yr   **47.** 13.9 yr   **49.** about 3 hr

55. $t = \frac{\ln 5}{r}$

Getting Ready (page 656)

1. x^{m+n} **2.** 1   **3.** x^{mn} **4.** x^{m-n}

Exercises 9.5 (page 664)

1. 2   **3.** 343   **5.** 3   **7.** $\frac{1}{9}$ **9.** 2   **11.** $\sqrt{85}$

13. $y = -\frac{7}{6}x + \frac{2}{3}$ **15.** 1   **17.** x **19.** −   **21.** x

23. =   **25.** 0   **27.** 2   **29.** 10   **31.** 1

37. $\log_b 7 + \log_b x + \log_b y$ **39.** $\log_b 5 + \log_b x - \log_b y$

41. $3 \log_b x + 2 \log_b y$ **43.** $3 \log_b x + \frac{1}{2} \log_b z - 5 \log_b y$

45. $\log_b \frac{x - 3}{x^5}$ **47.** $\log_b(x^7 \sqrt{z})$ **49.** 1.4472

51. 0.3521   **53.** 0.1339   **55.** 2.4014   **57.** 1.7712

59. −1.0000   **61.** 1.8928   **63.** 2.3219   **65.** 0

67. 7   **69.** 10   **71.** 1   **73.** $\frac{1}{3}(\log_b x + \log_b y - \log_b z)$

75. $\frac{1}{3} \log_b x - \frac{1}{4} \log_b y - \frac{1}{4} \log_b z$ **77.** $\log_b \frac{(x + 2)^3 \sqrt{x + 1}}{y^4}$

79. $\log_b \frac{\frac{x}{z} + x}{\frac{y}{z} + y} = \log_b \frac{x}{y}$ **81.** 2.0493   **83.** 0.4682

87. false, $b^1 = b$ **89.** false, $\log_b xy = \log_b x + \log_b y$

91. true   **93.** false, $\frac{\log_b A}{\log_b B}$ is simplified   **95.** true

97. true   **99.** 4.77   **101.** from 2.5119×10^{-8} to 1.585×10^{-7} **103.** It will increase by $k \ln 2$.

105. The intensity must be cubed.

Getting Ready (page 666)

1. $2 \log x$ **2.** $\frac{1}{2} \log x$ **3.** 0   **4.** $2b \log a$

Exercises 9.6 (page 675)

1. $x = 10^2$ **3.** $(x + 1) = 10^1$ **5.** 1.2619   **7.** 0.1461

9. 0, 16   **11.** $\frac{2}{3}, -4$ **13.** exponential   **15.** $A_0 e^{-kt}$

17. 1.8928   **19.** 3.9120   **21.** 48.9728   **23.** 22.1184

25. 2.7737   **27.** 2.4772   **29.** 1, −3   **31.** −2

33. 1.8   **35.** 3, −1   **37.** 3   **39.** 2   **41.** 10, −10

43. 5; −2 is extraneous   **45.** 20; −5 is extraneous

47. 10; −100 is extraneous   **49.** 4; 0 is extraneous

51. 4; −1 is extraneous   **53.** 20   **55.** 8   **57.** 0

59. ±1.0878   **61.** 0, 1.0566   **63.** −4

65. 10; 1 is extraneous   **67.** 0   **69.** 6; −1 is extraneous

71. 9; −9 is extraneous   **73.** 0.2789   **75.** 100, 1   **77.** 4

79. 1, 7   **81.** 4   **83.** 53 days   **85.** 6.2 yr

87. about 4,200 yr   **89.** 25.3 yr   **91.** 2.828 times larger

93. 13.3   **95.** 42.7 days   **97.** 5.6 yr   **99.** 5.4 yr

101. because $\ln 2 \approx 0.7$ **105.** $x \le 3$

Chapter Review (page 680)

3.   **4.**

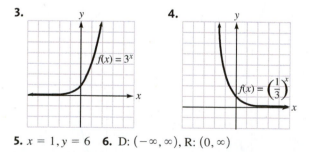

5. $x = 1, y = 6$ **6.** D: $(-\infty, \infty)$, R: $(0, \infty)$

7. $f(x) = \left(\frac{1}{2}\right)^x$, $f(x) = \left(\frac{1}{2}\right)^x - 2$

8. $f(x) = \left(\frac{1}{2}\right)^x$, $f(x) = \left(\frac{1}{2}\right)^{x+2}$

9. \$142,312.97   **10.** \$143,865.07

11. $f(x) = e^x + 1$

12. $f(x) = e^{x-3}$

13. about 582,000,000   **14.** 30.69 grams

15. D: $(0, \infty)$, R: $(-\infty, \infty)$ **17.** 2   **18.** $-\frac{1}{2}$ **19.** 0

20. -2 **21.** $\frac{1}{2}$ **22.** $\frac{1}{3}$ **23.** 81   **24.** 9   **25.** 27

26. -1 **27.** $\frac{1}{8}$ **28.** 2   **29.** 4   **30.** 2   **31.** 10

32. $\frac{1}{25}$ **33.** 5   **34.** 3

35. $f(x) = \log(x - 2)$

36. $f(x) = 3 + \log x$

37. $y = 4^x$, $y = \log_4 x$

38. $y = \left(\frac{1}{3}\right)^x$, $y = \log_{1/3} x$

39. 53 dB   **40.** 4.4   **41.** 5.8916   **42.** -0.1111
43. 10.3398   **44.** 2.5715

45. $f(x) = 1 + \ln x$

46. $f(x) = \ln(x + 1)$

47. about 23 yr   **48.** 0   **49.** 1   **50.** 3   **51.** 5
52. 5   **53.** 0   **54.** 7   **55.** 2   **56.** 6   **57.** 3
58. $5 \log_b x + 4 \log_b y - 2 \log_b z$
59. $\frac{1}{2}(\log_b x - \log_b y - 2 \log_b z)$ **60.** $\log_b \frac{x^3 z^7}{y^5}$
61. $\log_b \frac{y^3 \sqrt{x}}{z^7}$ **62.** 3.36   **63.** 1.56   **64.** 2.64

65. -6.72 **66.** 1.7604
67. about 7.94×10^{-4} gram-ions per liter   **68.** $k \ln 2$ less
69. $\frac{\log 7}{\log 3} \approx 1.7712$ **70.** 2   **71.** 31.0335   **72.** 2
73. $\frac{\log 3}{\log 3 - \log 2} \approx 2.7095$ **74.** $-1, -3$ **75.** 25, 4
76. 4; -2 is extraneous   **77.** 2; -3 is extraneous   **78.** 4, 3
79. 6; -1 is extraneous   **80.** 31   **81.** $\frac{\ln 9}{\ln 2} \approx 3.1699$
82. $\varnothing$ **83.** $\frac{e}{e - 1} \approx 1.5820$ **84.** 1   **85.** about 3,400 yr

Chapter 9 Test (page 686)

1. $f(x) = 2^x + 1$

2. $f(x) = 2^{-x}$

3. $\frac{3}{64}$ gram   **4.** \$1,060.90   **5.** $f(x) = e^x$

6. \$4,451.08   **7.** 2   **8.** 3   **9.** $\frac{1}{27}$ **10.** 7   **11.** 2
12. $\frac{27}{8}$

13. $f(x) = -\log_3 x$

14. $f(x) = \ln x$

15. $\log x + 3 \log y + 4 \log z$ **16.** $\frac{1}{2}(\ln a - 2 \ln b - \ln c)$
17. $\log \frac{b\sqrt{a + 2}}{c^3}$ **18.** $\log \frac{\sqrt[3]{a}}{c\sqrt[3]{b^2}}$ **19.** 1.2552   **20.** 0.4259
21. $\frac{\log 2}{\log 3}$ or $\frac{\ln 2}{\ln 3}$ **22.** $\frac{\log e}{\log \pi}$ or $\frac{\ln e}{\ln \pi}$ **23.** true
24. false, $\frac{\log a}{\log b}$ is simplified   **25.** false, $\log a^{-3} = -3 \log a$
26. false, cannot take logarithm of a negative number
27. 6.4   **28.** 46   **29.** $\frac{\log 4}{\log 7} \approx 0.7124$
30. $\frac{\log 3}{(\log 3) - 2} \approx -0.3133$ **31.** 4   **32.** 10; -1 is extraneous

Getting Ready (page 689)

1. $x^2 - 4x + 4$ **2.** $x^2 + 8x + 16$ **3.** $\frac{81}{4}$ **4.** 36

Exercises 10.1 (page 698)

1. downward   **3.** upward   **5.** $(x + 7)^2$ **7.** $(x - 10)^2$
9. $6, -1$ **11.** $3, -\frac{1}{4}$ **13.** conic   **15.** standard, $(0, 3)$, 4
17. circle, general   **19.** parabola, $(3, 2)$, right

21. $C(0,0); r = 3$

23. $C(2,0); r = 3$

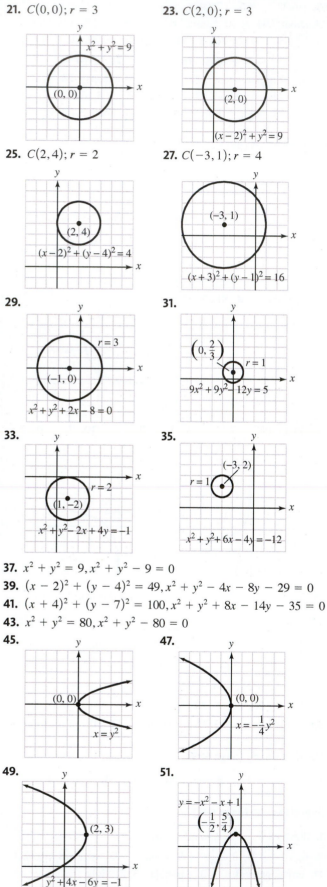

25. $C(2,4); r = 2$

27. $C(-3,1); r = 4$

29.

31.

33.

35.

37. $x^2 + y^2 = 9, x^2 + y^2 - 9 = 0$

39. $(x - 2)^2 + (y - 4)^2 = 49, x^2 + y^2 - 4x - 8y - 29 = 0$

41. $(x + 4)^2 + (y - 7)^2 = 100, x^2 + y^2 + 8x - 14y - 35 = 0$

43. $x^2 + y^2 = 80, x^2 + y^2 - 80 = 0$

45.

47.

49.

51.

53.

55.

57.

59.

61.

63.

65.

67.

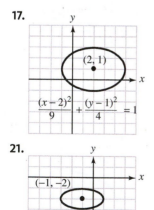

69. $(x - 7)^2 + y^2 = 9$ **71.** no

73. 30 ft away   **75.** 2 AU

Getting Ready (page 702)

1. $y = \pm b$ **2.** $x = \pm a$

Exercises 10.2 (page 709)

1. $(0,3), (0,-3)$ **3.** $(2,0), (-2,0)$ **5.** $35y^4 + \frac{10}{x^3}$

7. $\frac{y^2 + x^2}{y^2 - x^2}$ **9.** ellipse, sum   **11.** center

13. $(0,0)$, major axis, $2b$

15.

17.

19.

21.

23.

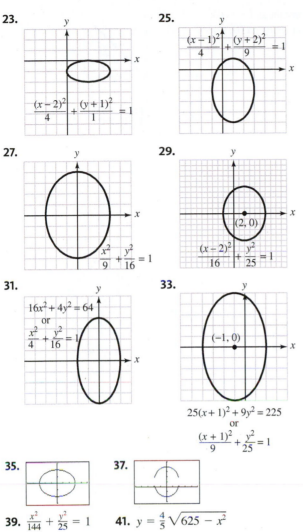

$$\frac{(x-2)^2}{4} + \frac{(y+1)^2}{1} = 1$$

25.

$$\frac{(x-1)^2}{4} + \frac{(y+2)^2}{9} = 1$$

27.

$$\frac{x^2}{9} + \frac{y^2}{16} = 1$$

29.

(2, 0)

$$\frac{(x-2)^2}{16} + \frac{y^2}{25} = 1$$

31.

$$16x^2 + 4y^2 = 64$$
or
$$\frac{x^2}{4} + \frac{y^2}{16} = 1$$

33.

(−1, 0)

$$25(x+1)^2 + 9y^2 = 225$$
or
$$\frac{(x+1)^2}{9} + \frac{y^2}{25} = 1$$

35.

37.

39. $\frac{x^2}{144} + \frac{y^2}{25} = 1$ **41.** $y = \frac{4}{5}\sqrt{625 - x^2}$

43. 10π square units

Getting Ready (page 713)

1. $y = \pm 2.0$ **2.** $y = \pm 2.9$

Exercises 10.3 (page 720)

1. $(2, 0), (-2, 0)$ **3.** $-5x^2(x^2 - 2x + 3)$

5. $(7a + 3b)(2a - 3b)$ **7.** hyperbola, difference

9. center   **11.** $(\pm a, 0)$, y-intercepts

13.

$$\frac{x^2}{9} - \frac{y^2}{4} = 1$$

15.

$$\frac{y^2}{4} - \frac{x^2}{9} = 1$$

17.

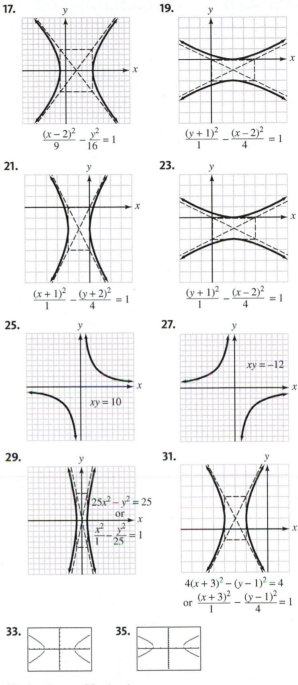

$$\frac{(x-2)^2}{9} - \frac{y^2}{16} = 1$$

19.

$$\frac{(y+1)^2}{1} - \frac{(x-2)^2}{4} = 1$$

21.

$$\frac{(x+1)^2}{1} - \frac{(y+2)^2}{4} = 1$$

23.

$$\frac{(y+1)^2}{1} - \frac{(x-2)^2}{4} = 1$$

25.

$$xy = 10$$

27.

$$xy = -12$$

29.

$$25x^2 - y^2 = 25$$
or
$$\frac{x^2}{1} - \frac{y^2}{25} = 1$$

31.

$$4(x+3)^2 - (y-1)^2 = 4$$
or $\frac{(x+3)^2}{1} - \frac{(y-1)^2}{4} = 1$

33.   **35.**

37. 3 units   **39.** 4 units

Getting Ready (page 723)

1. $-y^2 = 44$ **2.** $5x^2 = 8$

Exercises 10.4 (page 727)

1. 0, 1, 2   **3.** 0, 1, 2, 3, 4   **5.** $-25x\sqrt{5}$ **7.** $\frac{x}{2}$

9. graphing, substitution   **11.** four

13. $(4,2),(-4,-2)$ **15.** $(-1,3),(1,3)$

17. constant on $(-\infty,0)$, increasing on $(0,\infty)$

17. $(3,0),(0,5)$ **19.** $(1,2),(2,1)$ **21.** $(2,4),(-2,4)$
23. $\left(\sqrt{5},5\right),\left(-\sqrt{5},5\right)$
25. $(5,3),(5,-3),(-5,3),(-5,-3)$
27. $(2,4),(2,-4),(-2,4),(-2,-4)$
29. $(-4,3),(-4,-3),(4,3),(4,-3)$

19. decreasing on $(-\infty,0)$, increasing on $(0,2)$, decreasing on $(2,\infty)$

31. $\left(\frac{1}{5},-\frac{18}{5}\right),(3,2)$

21.   **23.**

33. $(0,-5),(3,4),(-3,4)$
35. $\left(-\sqrt{15},5\right),\left(\sqrt{15},5\right),(-2,-6),(2,-6)$
37. $(6,0),(-6,0)$ **39.** $(-2,3),(2,3),(-2,-3),(2,-3)$
41. $(4,4)$ **43.** $\left(-1,\frac{9}{2}\right),\left(3,-\frac{3}{2}\right)$ **45.** $\left(\frac{1}{2},\frac{1}{3}\right),\left(\frac{1}{3},\frac{1}{2}\right)$
47. $(1,0),(5,0),$ **49.** 4 and 8

25.

51. 7 cm by 9 cm   **53.** either $750 at 9% or $900 at 7.5%
55. 68 mph, 4.5 hr   **59.** 0, 1, 2, 3, 4

27. $30; c (in $10)

Getting Ready (page 729)

1. positive   **2.** negative   **3.** 98   **4.** -3

29. After 2 hours, network B is cheaper.

Exercises 10.5 (page 733)

1. 1, 4, 9; increasing   **3.** 9, 4, 1; decreasing   **5.** 20
7. domains   **9.** constant, $f(x)$ **11.** step
13. increasing on $(-\infty,0)$, decreasing on $(0,\infty)$
15. decreasing on $(-\infty,0)$, constant on $(0,2)$, increasing on $(2,\infty)$

Chapter Review (page 738)

1.
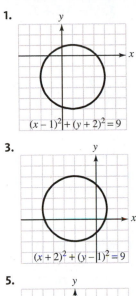
$(x-1)^2 + (y+2)^2 = 9$

2.
(0, 0)
$x^2 + y^2 = 16$

3.
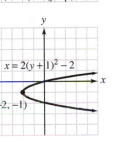
$(x+2)^2 + (y-1)^2 = 9$

4.
(5, 2)
$x = -3(y-2)^2 + 5$

5.
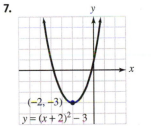
$x = 2(y+1)^2 - 2$
$(-2, -1)$

6.
$y = -2(x-1)^2 + 3$
(1, 3)

7.
$(-2, -3)$
$y = (x+2)^2 - 3$

8.
$9x^2 + 16y^2 = 144$ or
$\dfrac{x^2}{16} + \dfrac{y^2}{9} = 1$
(0, 0)

9.
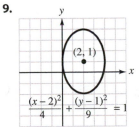
(2, 1)
$\dfrac{(x-2)^2}{4} + \dfrac{(y-1)^2}{9} = 1$

10.
$\dfrac{(x+1)^2}{9} + \dfrac{(y-1)^2}{4} = 1$

11.

$9x^2 - y^2 = -9$
or
$\dfrac{y^2}{9} - \dfrac{x^2}{1} = 1$

12.
$xy = 9$
or
$y = \dfrac{9}{x}$

13. hyperbola

14.
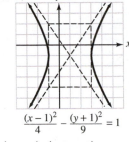
$\dfrac{(x-1)^2}{4} - \dfrac{(y+1)^2}{9} = 1$

15. $(3,4), (3,-4), (-3,4), (-3,-4)$

16. $(\sqrt{6}, 3), (\sqrt{6}, -3), (-\sqrt{6}, 3), (-\sqrt{6}, -3)$

17. increasing on $(-\infty, -2)$, constant on $(-2, 1)$, decreasing on $(1, \infty)$

18.
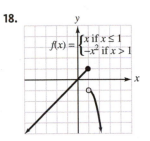
$f(x) = \begin{cases} x \text{ if } x \le 1 \\ -x^2 \text{ if } x > 1 \end{cases}$

19.
$f(x) = 3[\![x]\!]$

Chapter 10 Test (page 745)

1. $(-5, 2); 3$

2. $(-4, 2); 5$

3.
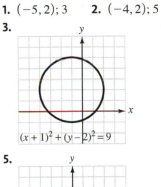
$(x+1)^2 + (y-2)^2 = 9$

4.
$x = (y-2)^2 - 1$

5.
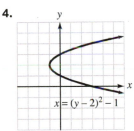
$9x^2 + 4y^2 = 36$
or
$\dfrac{x^2}{4} + \dfrac{y^2}{9} = 1$

6.
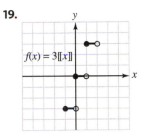
$\dfrac{(x-2)^2}{9} - y^2 = 1$

7.

$\dfrac{(x+3)^2}{1} + \dfrac{(y+1)^2}{4} = 1$

8.
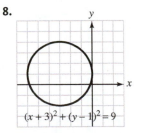
$(x+3)^2 + (y-1)^2 = 9$

9.

$$\frac{(x+1)^2}{9} - \frac{(y-2)^2}{1} = 1$$

10. $(0, 3), (4, 0)$

11. $(3, 4), (-3, 4)$

12. increasing on $(-3, 0)$, decreasing on $(0, 3)$

13.

$$f(x) = \begin{cases} -x^2 & \text{if } x < 0 \\ -x & \text{if } x \ge 0 \end{cases}$$

14.

$f(x) = 2[[x]]$

Cumulative Review (page 746)

1. $8x^2 - 18xy - 5y^2$ **2.** $a^{2n} - 2a^n - 3$ **3.** $\frac{7}{a-5}$

4. $a^2 - 3a + 2$ **5.** 1   **6.** $\frac{4a-1}{(a+2)(a-2)}$ **7.** parallel

8. perpendicular   **9.** $y = -6x + 3$ **10.** $y = -\frac{9}{13}x + \frac{7}{13}$

11.

$2x - 3y < 6$

$2x - 3y = 6$

12.

$y \ge x^2 - 4$

$y = x^2 - 4$

13. $5\sqrt{2}$ **14.** $81x\sqrt[3]{3x}$ **15.** 5; 0 is extraneous

16. 3; 5 is extraneous   **17.** $-\frac{5}{2}, \frac{2}{5}$ **18.** $\frac{1}{2} \pm \frac{3}{2}i$

19. $16x^2 + 8x - 2$ **20.** $f^{-1}(x) = \sqrt[3]{\frac{x+1}{2}} = \frac{\sqrt[3]{4x+4}}{2}$

21.

$y = \left(\frac{1}{2}\right)^x$

22. $5^y = x$ **23.** $\frac{4}{3}, -1$

24. 16

25.

$(0, -1)$

$x^2 + (y+1)^2 = 9$

26.

$x^2 - 9(y+1)^2 = 9$
or
$$\frac{x^2}{9} - \frac{(y+1)^2}{1} = 1$$

27. $(z - 2, -2z + 3, z)$

28. $(2, 1), (2, -1), (-2, 1), (-2, -1)$ **29.** $(3, 4), (-3, 4)$

Getting Ready (page 751)

1. $x^2 + 4x + 4$ **2.** $x^2 - 6x + 9$ **3.** $x^3 + 3x^2 + 3x + 1$

4. $x^3 - 6x^2 + 12x - 8$

Exercises 11.1 (page 757)

1. 720   **3.** 30   **5.** $x^2 + 6xy + 9y^2$ **7.** $x^2 - 6xy + 9y^2$

9. 2   **11.** 5   **13.** one   **15.** Pascal's   **17.** 8!

19. 1   **21.** $a^3 + 3a^2b + 3ab^2 + b^3$

23. $a^4 - 4a^3b + 6a^2b^2 - 4ab^3 + b^4$ **25.** 24

27. $-40,320$ **29.** 30   **31.** 144   **33.** $\frac{1}{110}$

35. 2,352   **37.** 72   **39.** 10   **41.** $x^3 + 3x^2y + 3xy^2 + y^3$

43. $a^6 + 6a^5b + 15a^4b^2 + 20a^3b^3 + 15a^2b^4 + 6ab^5 + b^6$

45. $a^4 - 4a^3b + 6a^2b^2 - 4ab^3 + b^4$

47. $x^3 - 6x^2y + 12xy^2 - 8y^3$ **49.** $8x^3 + 12x^2y + 6xy^2 + y^3$

51. $8x^3 + 36x^2y + 54xy^2 + 27y^3$ **53.** $6a^2b^2$ **55.** $15x^2y^4$

57. $-4xy^3$ **59.** $28x^6y^2$ **61.** $640x^3y^2$ **63.** $-12x^3y$

65. $-70,000x^4$ **67.** $810xy^4$ **69.** 40,320   **71.** 21

73. $\frac{1}{168}$ **75.** 39,916,800   **77.** $2.432902008 \times 10^{18}$

79. $81 + 216y + 216y^2 + 96y^3 + 16y^4$

81. $\frac{x^3}{8} - \frac{x^2y}{4} + \frac{xy^2}{6} - \frac{y^3}{27}$

83. $x^5 - 5x^4y + 10x^3y^2 - 10x^2y^3 + 5xy^4 - y^5$

85. $\frac{x^4}{16} + \frac{x^3y}{6} + \frac{x^2y^2}{6} + \frac{2xy^3}{27} + \frac{y^4}{81}$ **89.** $180x^4y^2$

91. $-\frac{1}{6}x^3y$ **93.** $375x^4$ **95.** $\frac{n!}{3!(n-3)!}a^{n-3}b^3$

97. $\frac{n!}{4!(n-4)!}a^{n-4}b^4$ **99.** $\frac{n!}{(r-1)!(n-r+1)!}a^{n-r+1}b^{r-1}$

107. $1, 1, 2, 3, 5, 8, 13, \dots$; beginning with 2, each number is the sum of the previous two numbers.   **109.** 36

Getting Ready (page 760)

1. 3, 5, 7, 9   **2.** 4, 7, 10, 13

Exercises 11.2 (page 766)

1. 2, 4, 6, 8   **3.** 5, 6, 7, 8   **5.** 5, 7, 9, 11

7. $27x^2 - 12x + 14$ **9.** $\frac{6a^2 + 16}{(a+2)(a-2)}$

11. sequence, finite, infinite   **13.** arithmetic, difference

15. arithmetic mean   **17.** series

19. $1 + 2 + 3 + 4 + 5 + 6 + 7$ **21.** 4   **23.** 88

25. 3, 5, 7, 9, 11   **27.** $-5, -8, -11, -14, -17$

29. 4, 7, 10, 13, 16; 46   **31.** $-7, -9, -11, -13, -15; -65$

33. 247   **35.** -283 **37.** 5, 11, 17, 23, 29

39. $-4, -11, -18, -25, -32$ **41.** $\frac{17}{4}, \frac{13}{2}, \frac{35}{4}$

43. 12, 14, 16, 18   **45.** 1,335   **47.** 459

49. $3 + 6 + 9 + 12$ **51.** $16 + 25 + 36$ **53.** 60

55. 31   **57.** $-118, -111, -104, -97, -90$

59. 34, 31, 28, 25, 22   **61.** 5, 12, 19, 26, 33   **63.** 355

65. -179 **67.** -23 **69.** 12   **71.** $\frac{29}{2}$ **73.** 1.25

75. 354   **77.** 255   **79.** 1,275   **81.** 2,500   **83.** 12

85. $60, $110, $160, $210, $260, $310; $6,010   **87.** 11,325

89. 368 ft   **93.** $\frac{3}{2}, 2, \frac{5}{2}, 3, \frac{7}{2}, 4$

Getting Ready (page 769)

1. 10, 20, 40   **2.** 18, 54, 162

Exercises 11.3 (page 774)

1. 32   **3.** 324   **5.** $\frac{7}{27}$ **7.** $[-1, 5]$

9. $(-\infty, -3) \cup [4, \infty)$ **11.** geometric   **13.** common ratio

15. $S_n = \frac{a_1 - a_1 r^n}{1 - r}$ **17.** 27   **19.** 2.5   **21.** 6

23. 4, 12, 36, 108, 324; 8,748

25. $-5, -1, -\frac{1}{5}, -\frac{1}{25}, -\frac{1}{125}; -\frac{1}{15,625}$

27. 243   **29.** $\frac{1}{32}$ **31.** 2, 8, 32, 128, 512

33. $2, -10, 50, -250, 1,250$ **35.** 6, 18, 54

37. $-20, -100, -500, -2,500$ **39.** -16 **41.** $10\sqrt{2}$

43. 225   **45.** 122   **47.** 2, 10, 50, 250, 1,250

49. $-64, 32, -16, 8, -4$ **51.** 26,244   **53.** $\frac{1}{27}$

55. 3   **57.** No geometric mean exists.   **59.** -255

61. 381   **63.** $\frac{156}{25}$ **65.** $-\frac{21}{4}$ **67.** about 669 people

69. $1,469.74   **71.** $140,853.75

73. $\left(\frac{1}{2}\right)^{11} \approx 0.0005$ sq unit   **75.** $4,309.14

81. arithmetic mean

Getting Ready (page 777)

1. 4   **2.** 4.5   **3.** 3   **4.** 2.5

Exercises 11.4 (page 780)

1. 3   **3.** $\frac{1}{5}$ **5.** $\frac{1}{5}$ **7.** yes   **9.** no   **11.** infinite

13. $S_\infty = \frac{a_1}{1 - r}$ **15.** 64   **17.** 250   **19.** 8   **21.** $\frac{-243}{4}$

23. no sum because $r > 1$ **25.** no sum because $r > 1$

27. $\frac{1}{9}$ **29.** $-\frac{1}{3}$ **31.** $\frac{4}{33}$ **33.** $\frac{25}{33}$ **35.** $-\frac{81}{2}$

37. $-\frac{81}{2}$ **39.** 5,000   **41.** 30 m   **45.** $\frac{4}{5}$

49. no; $0.999999 = \frac{999,999}{1,000,000} < 1$

Getting Ready (page 782)

1. 24   **2.** 120   **3.** 30   **4.** 168

Exercises 11.5 (page 789)

1. 840   **3.** 120   **5.** 56   **7.** 12, -5 **9.** 8

11. $p \cdot q$ **13.** permutation   **15.** $P(n, r) = \frac{n!}{(n - r)!}$

17. combination   **19.** $C(n, r) = \frac{n!}{r!(n - r)!}$ **21.** 120

23. 30   **25.** 240   **27.** $n!$ **29.** 4   **31.** 56

33. 735   **35.** 2   **37.** $x^4 + 4x^3y + 6x^2y^2 + 4xy^3 + y^4$

39. $a^5 + 5a^4b + 10a^3b^2 + 10a^2b^3 + 5ab^4 + b^5$

41. $8x^3 + 12x^2y + 6xy^2 + y^3$

43. $64x^3 - 48x^2y + 12xy^2 - y^3$

45. 1   **47.** 1,260   **49.** $\frac{n!}{2!(n - 2)!}$

51. $81x^4 - 216x^3 + 216x^2 - 96x + 16$ **53.** $-1,250x^2y^3$

55. $-4x^6y^3$ **57.** 35   **59.** 5,040   **61.** 1,000,000

63. 136,080   **65.** 8,000,000   **67.** 2,880   **69.** 13,800

71. 720   **73.** 900   **75.** 364   **77.** 5

79. 1,192,052,400   **81.** 18   **83.** 7,920   **87.** 48

Getting Ready (page 791)

1. 1, 2, 3, 4, 5, 6   **2.** H, T

Exercises 11.6 (page 795)

1. $C(52, 26)$ **3.** -1 **5.** 3, -1 **7.** -2

9. experiment   **11.** $\frac{s}{n}$ **13.** 0   **15. a.** 6   **b.** 52   **c.** $\frac{6}{52}, \frac{3}{26}$

17. {(1, H), (2, H), (3, H), (4, H), (5, H), (6, H), (1, T), (2, T), (3, T), (4, T), (5, T), (6, T)}   **19.** {a, b, c, d, e, f, g, h, i, j, k, l, m, n, o, p, q, r, s, t, u, v, w, x, y, z}   **21.** $\frac{1}{9}$ **23.** $\frac{5}{18}$ **25.** $\frac{1}{4}$

27. $\frac{3}{26}$ **29.** $\frac{33}{391,510}$ **31.** $\frac{9}{460}$ **33.** $\frac{1}{6}$ **35.** $\frac{2}{3}$ **37.** $\frac{19}{42}$

39. $\frac{13}{42}$ **41.** $\frac{3}{8}$ **43.** 0   **45.** $\frac{3}{169}$ **47.** $\frac{1}{6}$ **49.** $\frac{7}{12}$

51. $\frac{1}{16}$ **53.** $\frac{3}{8}$ **55.** $\frac{1}{16}$ **57.** $\frac{1}{3}$ **59.** $\frac{88}{141}$ **61.** $\frac{15}{71}$

65. 0.14

Chapter Review (page 800)

1. 240   **2.** 20   **3.** 15   **4.** 220   **5.** 1   **6.** 5,040

7. $x^5 + 5x^4y + 10x^3y^2 + 10x^2y^3 + 5xy^4 + y^5$

8. $x^4 - 4x^3y + 6x^2y^2 - 4xy^3 + y^4$

9. $64x^3 - 48x^2y + 12xy^2 - y^3$

10. $x^3 + 12x^2y + 48xy^2 + 64y^3$ **11.** $5xy^4$ **12.** $-10x^2y^3$

13. $-108x^2y$ **14.** $864x^2y^2$ **15.** 42

16. 122, 137, 152, 167, 182   **17.** $\frac{41}{3}, \frac{58}{3}$ **18.** 1,550

19. $-\frac{45}{2}$ **20.** 60   **21.** 34   **22.** 14   **23.** 360

24. $24, 12, 6, 3, \frac{3}{2}$ **25.** 1   **26.** 24, -96 **27.** $\frac{945}{2}$

28. $-\frac{85}{8}$ **29.** $1,638.40   **30.** $134,509.57   **31.** 12 yr

32. 1,600 ft   **33.** 125   **34.** $\frac{5}{99}$ **35.** 136   **36.** 5,040

37. 1   **38.** 20,160   **39.** $\frac{1}{10}$ **40.** 1   **41.** 1   **42.** 28

43. 210   **44.** 700   **45.** $x^3 + 6x^2 + 12x + 8$

46. 120   **47.** 720   **48.** 120   **49.** 150   **50.** $\frac{3}{8}$

51. $\frac{1}{2}$ **52.** $\frac{7}{8}$ **53.** $\frac{1}{9}$ **54.** 0   **55.** $\frac{1}{13}$ **56.** $\frac{94}{54,145}$

57. $\frac{33}{66,640}$

Chapter 11 Test (page 805)

1. 72   **2.** 1   **3.** $-5x^4y$ **4.** $24x^2y^2$ **5.** 66

6. 306   **7.** 34, 66   **8.** 14   **9.** -81 **10.** $\frac{364}{27}$

11. 18, 108   **12.** $\frac{27}{2}$ **13.** no sum   **14.** 210

15. 40,320   **16.** 10   **17.** 56   **18.** 720   **19.** 24

20. $x^4 - 12x^3 + 54x^2 - 108x + 81$ **21.** 56   **22.** 30

23. $\frac{1}{6}$ **24.** $\frac{2}{13}$ **25.** $\frac{33}{66,640}$ **26.** $\frac{5}{16}$

Cumulative Review (page 805)

1. $(2, 1)$ **2.** $(1, 1)$ **3.** $(2, -2)$ **4.** $(3, 1)$

5. -2 **6.** -1 **7.** $(-1, -1, 3)$ **8.** 1

9.

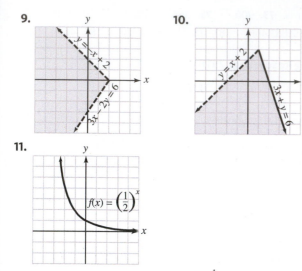

10.

11.

$f(x) = \left(\dfrac{1}{2}\right)^x$

12. $2^y = x$ **13.** 3   **14.** 3   **15.** $\dfrac{1}{16}$ **16.** 1

17. $y = 2^x$ **18.** x **19.** 1.9912   **20.** 0.3010

21. 1.4314   **22.** 0.1461   **23.** $\dfrac{5 \log 2}{\log 3 - \log 2}$ **24.** 8

25. \$2,848.31   **26.** 1.1292   **27.** 840

28. $81a^4 - 108a^3b + 54a^2b^2 - 12ab^3 + b^4$ **29.** $112x^2y^6$

30. 103   **31.** 690   **32.** 8 and 19   **33.** 60   **34.** 52

35. 27   **36.** $\dfrac{1,023}{64}$ **37.** 12, -48 **38.** $\dfrac{27}{2}$ **39.** 504

40. 120   **41.** $\dfrac{7}{3}$ **42.** $C(n, n)$ **43.** 5,040   **44.** 84

45. $\dfrac{3}{26}$

Appendix 1 (page A-4)

1. y-axis   **3.** origin   **5.** y-axis   **7.** none

9. none   **11.** x-axis

13. D: $(-\infty, \infty)$, R: $[-4, \infty)$ **15.** D: $(-\infty, \infty)$, R: $(-\infty, \infty)$,

$f(x) = x^4 - 4$

$f(x) = -x^3$

17. D: $(-\infty, \infty)$, R: $[0, \infty)$ **19.** D: $(-\infty, \infty)$, R: $(-\infty, \infty)$

$f(x) = x^4 + x^2$

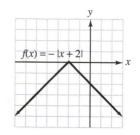

$f(x) = x^3 - x$

21. D: $(-\infty, \infty)$, R: $[-1, \infty)$ **23.** D: $(-\infty, \infty)$, R: $(-\infty, 0]$

$f(x) = \dfrac{1}{2}|x| - 1$

$f(x) = -|x + 2|$

Appendix 2 (page A-7)

1. a   **3.** d   **5.** c   **7.** c   **9.** d   **11.** e   **13.** b

15. c   **17.** e   **19.** d   **21.** d   **23.** e   **25.** c

27. c   **29.** b   **31.** d   **33.** b   **35.** e   **37.** b

39. a   **41.** d   **43.** d   **45.** c   **47.** c   **49.** a

51. b   **53.** a   **55.** d   **57.** b   **59.** b   **61.** d

63. b

INDEX

A

Absolute value
  of complex numbers, 518
  defined, 243
  equations, 243–246
  equations with two, 246
  function on a rectangular coordinate system, 242–243
  inequalities containing, 247
  of real numbers, 12
  solving inequalities, 250
ac method, 339–340
Accent on technology
  approximating square roots, 457
  approximating zeros of polynomials, 425
  checking algebra, 386, 397
  checking an algebraic simplification, 371
  checking solutions of quadratic equations, 541–542
  evaluating determinants, 210–211
  evaluating natural logarithms, 651
  finding base-10 (common) logarithms, 645
  finding powers, 35
  finding the domain and range of a function, 369
  finding the period of a pendulum, 462
  generating tables of solutions, 97
  graphing circles, 693
  graphing ellipses, 706
  graphing exponential function with a base of *e*, 635
  graphing exponential functions, 626
  graphing functions, 144
  graphing hyperbolas, 716
  graphing inequalities, 258
  graphing logarithmic functions, 644
  graphing one-to-one exponential functions, 624
  graphing polynomial functions, 297
  graphing quadratic functions, 570–571
  rational exponents, 477
  solving equations, 351, 406
  solving equations containing radicals, 504
  solving equations graphically, 170–171
  solving exponential equations, 668–669
  solving inequalities, 583–584
  solving investment problems, 628
  solving linear inequalities, 237–238
  solving logarithmic equations, 670–671
  solving quadratic equations, 350
  solving systems by graphing, 169–170
  solving systems of equations, 724
  translations of exponential function, 634–635

using scientific notation, 43
  verifying properties of logarithms, 657
Acute angles, 63
Addend, 16
Addition
  associative properties of, 23
  commutative properties of, 23
  complex numbers, 513
  polynomials, 301–302
  properties of equality, 49–51
  property of inequalities, 232
  radicals, 486–487
  real numbers, 16
Additive identity, 24
Additive inverse property, 24–25
Algebraic expressions, 3
Algebraic terms, 51, 291
Annual growth rate, 633
Areas
  of circle, 37
  of rectangle, 36
  of square, 36
  of trapezoid, 37
  of triangle, 37
Arithmetic means, 763
Arithmetic sequence, 759–768
  defined, 761
  finding specified term of, given the first term and one or more other terms, 762–763
  finding specified term of, given the first term and the common difference, 761
  finding sum of the first *n* terms of, 763–764
  review, 801
  sum of the first *n* terms of, 764
Arithmetic series, 764, 801
Asymptotes, 716
Augmented matrix, 200
Axis of symmetry, 564

B

Babbage, Charles, 387
Back substitution, 200
Base-10 logarithms, A-12
Base angles, 63
Base-*e* exponential functions, 632–639
  graphing, 635
  review, 681–682
Base-*e* logarithms, A-13
Bimodal distribution, 21
Binomial theorem, 751–759
  applying to expand binomial, 754–756
  defined, 788
  factorial notation and, 753–754

finding specified term of a binomial expansion, 756–757
  Pascal's triangle and, 752–753
  raising binomial to a power, 751–752
  review, 800
Binomials
  applying binomial theorem to expand, 754–756
  defined, 292
  finding specified term of expression, 756–757
  multiplication by using distributive property, 308–309
  multiplication by using FOIL method, 308–309
  multiplication of two conjugate, 309–311
  raising to a power, 751–752
  squaring, 309–311
  using synthetic division, 421–424
Boundary line, 255

C

Calculators
  graphing circles, 693
  graphing ellipses, 706–707
  graphing hyperbolas, 716
  graphing lines, 100
  solving logarithmic equations, 651
  solving systems of equations, 724
Careers and Mathematics
  agricultural managers, 1
  atmospheric scientists, 89
  cosmetologists, 163
  craft and fine artists, 289
  diesel service technicians and mechanics, 229
  environmental engineer, 365
  farmers, 1
  financial analysts, 749
  meteorologists, 89
  pest-control workers, 619
  photographers, 453
  police officers and detectives, 533
  ranchers, 1
  water transportation occupations, 687
Carroll, Lewis, 210
Cauchy, Augustin, 511
Cayley, Arthur, 200
Change-of-base formula, 661–662
Circles
  circumference of, 22
  defined, 690
  general form of equation of, 692
  graphing, 693

Circles (*continued*)
radius of, 690
review, 738–740
standard form of equation of a circle with center at (0, 0), 691
standard form of equation of a circle with center at (h, k), 690
Circumference, 22
Closed interval, 234
$C(n, 0)$, 787
$C(n, n)$, 787
$C(n, r)$, 786
Coefficient matrix, 200
Combinations
review, 803–804
using to find coefficients of the terms of a binomial expansion, 788–789
using to find the number of n things taken r at a time, 785–787
Combined variation, 436
Common difference, 761
Common ratio, 769
Complementary angles, 63
Completing the square, 538–542, 608
Complex conjugates, 514
Complex fractions
review, 446–447
simplifying, 395–401
Complex numbers, 510–520
absolute value of, 518
addition, 513
equality of, 513
multiplication, 514
review, 529–530
subtraction, 513
Composite function, 591
Composite numbers, 4
Composition, 590–592
Compound inequalities
defined, 9–11
graphing, 9–11
solving in one variable, 235–238
solving in two variables, 254–259
Compound interest, 627
Compound interest, formula for, 543
Compounding periods (n), 628
Conditional equations, 54
Conic sections, 689
Conjugate binomials, 309–311
Conjugates, 497
Consistent systems, 166
Constant, 51
Constant on the interval, 729
Constraints, 270
Continuous compound interest, 632
Contradiction, 55
Coordinate plane, 92
Coordinates, defined, 4
Cramer, Gabriel, 211
Cramer's rule, 211–215
solving system of linear equations using, 211–215
for three linear equations in three variables, 214
for two linear equations in two variables, 212
Critical points, 579
Critical values, 579

Cube-root function, 460
Cube roots, 458
Cubes
factoring the sum and difference of two, 326–328
perfect, 326

D
Decibel voltage gain, 645
Decimals
repeating, 7, 180
terminating, 7
writing repeating decimal as common fraction, 779–780
Decreasing function, 623
Decreasing on the interval, 729
Denominators
addition of two rational expressions with like, 385–386
addition of two rational expressions with unlike, 388–390
finding the least common, 387
rationalizing, 494–497
subtraction of two rational expressions with like, 385–386
subtraction of two rational expressions with unlike, 388–390
Dependent equations, 168
Dependent variable, 132
Descartes, René, 92, 511, 689
Determinants
array of signs for a three × three, 209
solving systems of linear equations using, 207–218
value of a three × three, 208
value of a two × two, 208
Difference, 16
Difference of two squares, 310
Difference quotient, 592
Direct variation, 432–433
Discount, 62
Discriminant, 554–555, 609–610
Distance formula, 469–471, 524–525
Distributive property, 308–309
Division
division of polynomials by polynomials, 415–417
fractions, 381–382
monomial by a monomial, 414
polynomial by a monomial, 415
polynomial with one or more missing terms by a polynomial, 418
polynomials, 414–419
properties of equality, 49–51
property of radicals, 483
rational expressions, 381–382
real numbers, 18–19
Domain
of cube-root functions, 460–461
of rational functions, 368–369
of relation, 130
of square-root functions, 460–461
Double negative rule, 25
Doubling time formula, 654

E
Edge, 255
Elementary row operations, 200

Elements
of matrix, 199
of sets, 3
Elimination (addition) method
solving system of equations containing second-degree equation by, 725–726
solving systems of linear equations in two variables by, 177–179
Ellipses, 4
defined, 703
general form of the equation of, 707
graphing, 703–707
review, 740–741
standard equation of horizontal ellipse centered at (h, k), 705
standard equation of vertical ellipse centered at (h, k), 705
Empty set, 55
Equality
of complex numbers, 513
logarithmic property of, 659
properties of, 49–51
Equations. *See also* Linear equations
absolute value, 243–246
conditional, 54
containing rational expressions, 404–411
defined, 48
of ellipse centered at the origin, 705
equivalent, 49
exponential, 666–669
finding slope of a line given its, 107–108
as functions, 134
general form of a circle, 692
general form of the equation of ellipses, 707
general form of the equation of hyperbola, 717
general form of the equation of parabola that opens to left/right, 695–696
general form of the equation of parabola that opens upward/downward, 695
of a hyperbola centered at the origin with vertices on the x-axis, 716
of a hyperbola centered at the origin with vertices on the y-axis, 716
of a hyperbola of the form $xy = k$, 719
of hyperbolas centered at (h, k), 716
literal, 55
logarithmic, 666, 669–671
of parabolas, 695
prediction, 124
regression, 124
review, 447–448
solving, 351, 406
solving application involving a logarithmic, 671–674
solving application involving an exponential, 671–674
solving by factoring, 347–353
solving graphically, 170–171
solving higher-degree polynomial, 350–351
solving systems of, 724
standard equation of horizontal ellipse centered at (h, k), 705

standard equation of vertical ellipse centered at (h, k), 705
standard form of a circle with center at $(0, 0)$, 691
standard form of a circle with center at (h, k), 690
systems containing second-degree, 723–729
with two absolute values, 246
writing equations of lines, 117–129
Equilateral triangle, 63
Equivalent equations, 49
Equivalent systems, 166
Euler, Leonhard, 511, 633
Evaluating
algebraic expression when given values for its variables, 22
application containing an exponential function, 626
applying remainder theorem to evaluate polynomial, 424
common logarithm, 644–645
exponential expression using rules for order of operations, 35
logarithm by applying properties of logarithms, 661
natural logarithms, 651
numeric expression following rules for order of operations, 19–20
rational functions, 373–374
Even integers, 5
Events
finding probability of, 792–795
multiplication principle for, 782–783
Everyday connections
astronaut training, 570
focus on conics, 718
fuel oil production, 276
highest and longest jumpers, 430–431
labor force, 113
staffing, 204
stretching rubber bands, 297
United States federal deficit, 45
U.S. population growth, 674
winning the lottery, 794
Experiments, 792
Exponential equations, 666–669, 685
Exponential functions
defined, 622
graphing, 622–624
graphing translation of exponential functions, 625–626
graphing with a base of e, 634
as logarithmic function, 640–642
properties of, 623
review, 680–681
translations of, 634–635
Exponential growth, 633
Exponents
defined, 29
natural-number, 29
negative, 33
power rules of, 32
product rule of, 30
properties of, 35, 478–479
rational, 474–482, 525–526
review, 83–84

rules for order of operations for expressions without, 20
rules of, 474
zero, 32
Expressions, algebraic, 22–23
Extracting the roots, 535
Extraneous solutions, 406
Extremes, 429

F

Factor theorem, 424–425
Factorial notation, 753
Factorials, 753–754
Factoring
defined, 317
difference of two squares, 324–326
greatest common factor from a polynomial, 318–320
polynomial with four or six terms by grouping, 320–321
polynomial with four terms by grouping a trinomial, 341
polynomials by grouping involving a difference of squares, 329
review, 362–363
solving a formula for a specified variable by, 321
solving equations by, 347–353
solving quadratic equations by, 535–536
sum and difference of two cubes, 326–328
techniques, 344–346
testing trinomials for, 336
trinomial by grouping (*ac* method), 339–340
trinomial using substitution, 339
trinomials, 332–341
Factors, 17
Feasibility region, 270
Fibonacci, Leonardo, 549
Fibonacci sequence, 549, 760
Finite sequences, 760
Finite sets, 3
Foci, 703
Focus, 703
FOIL method, 308–309
Formulas
area of circle, 37
area of rectangle, 36
area of square, 36
area of trapezoid, 37
area of triangle, 37
change-of-base, 661–662
circumference of circle, 22
compound interest, 627
doubling time, 654
exponential growth, 633
future value, 628
involving a rational equation for a specified variable, 407
midpoint, 99
perimeter of rectangle, 22
perimeter of square, 22
perimeter of trapezoid, 22
perimeter of triangle, 22
radioactive decay, 636
solving for a specified variable by factoring, 321
solving for indicated variable, 55–56

special product, 310
for a specified variable by factoring, 321
vertex of a parabola, 568
volume of cone, 37
volume of cube, 36
volume of cylinder, 37
volume of pyramid, 37
volume of rectangular solid, 36
volume of sphere, 36, 37
Fractions
division, 381–382
to negative powers, 34
repeating decimal in form of, 180
simplifying complex, 395–401
writing repeating decimal as common, 779–780
Fulcrum, 65
Functions
composite, 591
defined, 130
difference of two, 589–590
difference quotient of, 592
as equations, 134
finding domain of, 135–136
finding inverse of one-to-one, 599
finding the domain and range of, 369
graphing, 144
graphing a reflection of, 147–148
graphing horizontal translation of, 146
graphing quadratic, cubic, and absolute value, 142–144
graphing vertical translation of, 145
inverses of, 595–604
notation, 132–136
objective, 270
one-to-one, 596
operations on, 589
polynomials, 293
product of two, 589–590
quotient of two, 589–590
relation as, 129–130
review, 158–160, 613–615
solving application requiring the composition of two, 593
sum of two, 589–590
Fundamental rectangle, 716
Future value, 628

G

Gauss, Carl Friedrich, 483, 511
Gaussian elimination, 200
General term, 760
Geometric means, 770–771
Geometric sequences, 768–776
defined, 769
finding specified term of, given the first term and common ratio, 769–770
finding sum of the first n terms of, 771–772
review, 802
solving application involving, 772–774
Golden ratio, 549
Graphing
absolute value functions, 142–144
circles, 693
cubic functions, 142–144
ellipses, 706–707

Graphing (*continued*)

ellipses given an equation in general form, 707–708

ellipses given an equation in standard form, 703–707

exponential function with a base of *e*, 634, 635

exponential functions, 622–624, 626

greatest integer function, 731–732

horizontal translations of absolute value function, 146–147

horizontal translations of cubic function, 146–147

horizontal translations of quadratic function, 146–147

hyperbola of the form $xy = k$, 718–719

hyperbolas, 716

inequalities, 258

intercept method, 96

linear equations, 91–105

lines, 100–101

logarithmic functions, 642–643, 644, 652–653

model applications, 98–99

nonlinear inequality in two variables, 584–585

one-to-one exponential functions, 624

parabolas, 694–695

piecewise-defined function, 730–731

polynomial functions, 297

quadratic functions, 142–144, 568–569

reflection of absolute value function about *x*-axis, 147–149

reflection of cubic function about *x*-axis, 147–149

reflection of quadratic function about *x*-axis, 147–149

review, 158–160

solving systems by, 169–170

systems of linear equations, 165–174

translation of exponential functions, 625–626

vertical and horizontal translation of a logarithmic function, 643–644

vertical translation of absolute value function, 145

vertical translation of cubic function, 145

vertical translation of quadratic function, 145

Graphs

absolute value function on a rectangular coordinate system, 242–243

compound inequalities, 9–11

horizontal line and vertical line, 97–98

horizontal translations of, 565–566

intercepts of a line, 95–96

of intervals, 8–9

linear equation in two variables by plotting points, 94–95

linear equation using the slope and *y*-intercept, 120

linear function, 136–137

linear inequalities in two variables, 255–257

polynomial functions, 295–296

of quadratic functions, 562–578

reflection of, 147–149

slope of line, 106–107

in space, 148

symmetries of, A-1–A-5

vertical line test, 131–132

vertical translations of, 564–565

Greatest common factor (GCF)

of two or more monomials, 317–318

of two or more polynomials, 318–320

review, 359–360

Greatest integer function, 731–732, 744–745

Grouping

factoring a polynomial with four or six terms by, 320–321

factoring a trinomial by, 339–340

factoring polynomial with four terms by, 341

greatest common factor and factoring by, 316–321

involving a difference of squares, 329

review, 359–360

symbols, 19

H

Half-open interval, 234

Half-planes, 255

Hopper, Grace Murray, 106

Horizontal line, 97–98

Horizontal lines

slope of, 108–109

test, 597–598

Horizontal translations, 146–147, 565–566

Hypatia, 301

Hyperbolas

defined, 713

equations of, 715, 716

general form of the equation of, 717

graphing, 716

graphing an equation in general form, 717–719

graphing an equation in standard form, 713–717

graphing of the form $xy = k$, 718–719

review, 741–742

solving application involving, 719–720

Hypotenuse, 467

I

Identity, 54

Imaginary numbers, 510

Inconsistent systems, 167–168

Increasing function, 622

Increasing on the interval, 729

Independent equations, 168

Independent variable, 132

Index, 458

Index of the summation, 765

Inequalities

containing one absolute value term, 247

quadratic, 264–266

solving systems of, 263

symbols, 231

Infinite geometric series, 776–781

finding the *n*th partial sum of, 777–778

review, 803

sum of, 778–779

Infinite sequences, 760

Input value, 134

Integers

defined, 4

determining if trinomial is factorable over, 336–339

even, 5

odd, 5

Intercept method, 96

Intersection, 10, 235

Interval notation, 9, 233

Intervals

closed, 234

defined, 8

half-open, 234

interval notation, 9

open, 234

types of, 234

unbounded, 234

union of, 9

ways to describe, 11

Inverse variation, 434–435

Irrational numbers, 6–7

Isosceles triangle, 63

J

Joint variation, 435–436

K

Kovalevskaya, Sonya, 373–374

L

Least common denominator (LCD), 387

Legs, 467

Leibniz, Gottfried Wilhelm, 43

Like denominators

addition of two rational expressions with, 385–386

subtraction of two rational expressions with, 385–386

Like signs

adding real numbers with, 16

dividing real numbers with, 16

multiplying real numbers with, 17

Like terms

combining, 301

defined, 51

Linear equations

defined, 50

determining lines that are parallel, perpendicular, or neither, 120–121

graphing, 91–105

graphing by using the slope and *y*-intercept, 120

in one variable, 48–59, 70–77

review, 84–87, 155–157

solving, 52

using to solve applications, 59–70

Linear functions

graph of, 136–137

that model an application, 137–138

Linear inequalities, 584–586

defined, 233

review, 281–285

solving, 237–238

solving in one variable, 231–238

solving in two variables, 254–259, 262–264

Linear programming

defined, 270

review, 285–286
solving systems using, 269–277
Lines
  graphing, 100–101
  horizontal, 97–98
  point-slope form of, 117–119
  slope-intercept form of, 119
  slope of, 106
  slope of nonvertical, 107
  slopes of horizontal, 108–109
  slopes of parallel lines, 109–110
  slopes of vertical, 108–109
  vertical, 97–98
  writing equations of, 117–129, 157–158
Literal equation, 55
Logarithmic equations, 666, 669–671, 685
Logarithmic functions, 640–649
  defined, 640
  as exponential function, 640–642
  graphing, 642–643
  graphing vertical and horizontal
    translation of a logarithmic function,
    643–644
  review, 682–683
Logarithms, 641
  applying change-of-base formula,
    661–662
  base-10, A-12
  base-e, A-13
  evaluating, 644–645, 651, 661
  expanding logarithmic expression,
    659–660
  finding base-10 (common), 645
  logarithmic property of equality, 659
  natural, 650–656, 683–684
  power rules of, 658–659
  product and quotient properties of, 657
  properties of, 657, 661
  review, 684–685
  solving, 662–663
  solving application involving, 645–646
  writing logarithmic expression
    as single, 660
LORAN (LOng RAnge Navigation), 713

M

Major axis, 705
Malthusian growth model, 672
Markdown, 62
Matrix, 199
Means
  defined, 20–21
  proportions and, 429
Median, defined, 21
Midpoint, 99–100
Midpoint formula, 99
Minor axis, 705
Minors, 208
Mode, defined, 21
Monomials
  defined, 292
  degree of, 292
  division by monomial, 414
  division of polynomials by, 415
  greatest common factor of, 317–318
  multiplication, 306–307
Multiplication
  associative properties of, 23

commutative properties of, 23
complex numbers, 514
distributive property of, 23–24
polynomial by a monomial, 307
polynomial by a polynomial, 307–308
polynomials, 306–313
principle for events, 782–783
properties of equality, 49–51
property of inequalities, 232
property of radicals, 483
radical expressions, 492–494
real numbers, 17–18
two binomials using the distributive
  property, 308–309
two binomials using the FOIL method,
  308–309
Multiplicative inverse property, 25

N

Napier, John, 650
Natural logarithms
  evaluating, 650–651
  graphing, 652–653
  review, 683–684
  solving application involving, 653–654
  solving logarithmic equation with
    calculator, 651
Natural numbers
  defined, 4
  prime-factored form of, 317
Negative exponents, 33
Negative numbers, 4, 24
Negative reciprocals, 110
Newton, Issac, 703
Noether, Amalie, 233
Notation
  function, 132–136
  interval, 9, 234
  mathematical, 311
  scientific, 41–42, 84
  set-builder, 5, 231–235
  standard, 40–42
nth root, 458–460
Numbers. *See also* Real numbers
  complex, 510–518
  composite, 4
  imaginary, 510
  integers, 4
  irrational, 6–7
  natural, 4
  negative, 4, 24
  opposite, 24
  positive, 4
  prime, 4
  rational, 5–6
  whole, 4
Numerators, 497–498
Numerical coefficients, 51

O

Objective functions, 270–272
Odd integers, 5
One-to-one function
  determining, 596
  finding inverse of, 599
Open interval, 234
Opposite numbers, 24
Order, 458

Order of operations, 19–20, 35–36, 51
Ordered pairs
  defined, 93
  dependent linear system as, 179–180
  expressing solutions of dependent
    system, 168–169
  plotting on coordinate plane, 92–94
  relations as, 130
Ordered triple, 190
Origin, 92, A-1
Output value, 134

P

Parabolas, 564, 566
  equations of, 695
  general form of equation of parabola
    that opens to left/right, 695–696
  general form of equation of parabola
    that opens upward/downward, 695
  graphing, 694–695
  review, 738–740
Paraboloids, 694
Parallel lines, 109–110
Partial sum, 777
Pascal, Blaise, 689, 752
Pascal's triangle, 752–753
Perfect cubes, 326
Perfect square trinomials, 310
Perfect squares, 324
Perimeters, 22
Periodic interest rate (i), 628
Permutations, 783, 803–804
Perpendicular lines, 110–111
Perspectives
  Ahmes Papyrus, 42
  complex fractions, 399–400
  complex numbers, 511
  Fibonacci sequence, 549
  graphs in space, 148
  Lewis Carroll, 210
  mathematical notation, 311
  problem-solving approach, 271
pH of a solution, 662–663
Piecewise-defined functions, 729, 744–745
Planes, 190
$P(n, 0)$, 785
$P(n, n)$, 785
$P(n, r)$, 784
Point circle, 690
Point-slope form, 117–119
Polya, George, 271
Polynomial functions
  defined, 293
  evaluating, 293–295
  graphs, 295–296
  review, 358
Polynomials
  addition, 301–302
  applying remainder theorem to
    evaluate, 424
  applying the factor theorem to, 424–425
  approximating zeros of, 425
  defined, 291
  degree of, 293
  division, 414–419
  division by a binomial of the form $(x - r)$
    using synthetic division, 421–424
  division by monomial, 415

Polynomials (*continued*)
  division of polynomials by,
    415–417
  division of polynomials with one or
    more missing terms by, 418
  evaluating, 293–295
  factoring by grouping involving a
    difference of squares, 329
  factoring with four or six terms by
    grouping, 320–321
  factoring with four terms by grouping
    trinomials, 341
  greatest common factor from,
    318–320
  multiplication, 306–313
  multiplication by a monomial, 307
  multiplication by a polynomial,
    307–308
  in one variable, 291
  prime, 319
  review, 358–359, 448
  in several variables, 292
  subtraction, 302–303
Positive numbers, 4
Power rule
  defined, 502
  of logarithms, 658
Powers, A-11
Powers of *i*, 517
Prediction equation, 124
Prime-factored form, 317
Prime numbers, 4
Prime polynomials, 319
Principal square root, 456
Probability, 791–797, 804–805
Problem solving, 61
Product, 17
Properties
  addition of inequalities, 232
  additive inverse, 24–25
  associative of addition, 23
  associative of multiplication, 23
  commutative of addition, 23
  commutative of multiplication, 23
  distributive, 308–309
  distributive of multiplication over
    addition, 23–24
  of equality, 49–51
  of exponents, 35
  of factorials, 754
  of logarithms, 657, 661
  multiplication of inequalities, 232
  multiplicative inverse, 25
  of rational expressions, 370
  of real numbers, 23–25
  transitive, 232
  trichotomy, 15, 232
  of zero and one, 24
  zero-factor, 348
Proportions
  defined, 428
  review, 449–450
  solving, 428–430
  solving an application involving, 430
Pyramid, volume of, 37
Pythagoras of Samos, 469

Pythagorean theorem, 467–473,
  524–525

Q
Quadratic equations
  checking solutions of, 541–542
  defined, 348
  solutions of, 558–559
  solving, 350
  solving an application using, 351–353
  solving by applying the square-root
    property, 537–538
  solving by completing the square,
    538–542
  solving by factoring, 535–536
  solving by using quadratic formula,
    546–553
  using discriminant to determine type of
    solutions to, 554–555
Quadratic formula, 546–553, 608–609
Quadratic functions, 562–578, 610–611
Quadratic inequalities
  review, 284–285, 611–613
  solving in one variable, 578, 579
  solving in two variables, 584–585
Quotient, 18
Quotient rule, 33

R
Radical equations, 501–509
Radical expressions, 455–466
  addition, 486–489
  multiplication, 492–494
  rationalizing the denominator that
    contains, 494–497
  rationalizing the numerator that
    contains, 497–498
  review, 523–524, 526–528
  simplifying and combining, 483–486
  subtraction, 486–489
Radical sign, 456
Radicals
  addition, 486–487
  division property of, 483
  like, 486
  multiplication property of, 483
  properties of, 511
  similar, 486
  subtraction, 486–487
Radicand, 456
Radioactive decay formula, 636
Radius, 690
Ramanujan, Srinivasa, 296
Range
  of rational functions, 369
  of a relation, 130
Rates, 428
Ratio, 428
Rational equations
  defined, 404
  solving, 404–407
  solving an application involving,
    408–410
Rational exponents, 474–482
  applying the properties of exponents to,
    478–479

changing to radicals, 476
  negative, 477–478
  review, 525–526
  using to simplify radicals, 479–480
Rational expressions
  addition, 385–386
  addition with unlike denominators,
    388–390
  applying the properties of radicals,
    483–486
  containing factors that are negatives,
    372–373
  defined, 367
  division, 381–382
  finding all values of a variable for,
    367–368
  finding least common denominator
    of, 387
  multiplication, 378–380
  operations on three or more, 382–383
  properties of, 370
  review, 444–446
  simplified form of, 484
  simplifying, 370–372, 479–480
  squares of, 380
  subtraction, 385–386
  subtraction with unlike denominators,
    388–390
Rational function
  finding the domain of, 368
  review, 444
  that models an application,
    373–374
Rational inequalities, 581–583
Rational numbers, 5
Rationalizing
  denominator, 494–497
  numerator, 497–498
Real numbers
  absolute value of, 12
  addition, 16
  arithmetic and properties of, 15–28
  division, 18–19
  multiplication, 17–18
  properties of, 23–25
  review, 79–83
  set of, 7
  subtraction, 16
  system, 3–15
Reciprocals, 25
Rectangular coordinate system
  absolute value function on, 242–243
  formation of, 92
Reflection, 147
Regression equation, 124
Relations
  domain of, 130
  function as, 130
  ordered pairs as, 130
  range of, 130
Remainder theorem, 424
Repeating decimals, 7
Rhind, Henry, 42
Richter scale, 646
Right angles, 63
Right triangle, 63

Roots, 49, A-11
Roster method, 5
Rules
　　double negative, 25
　　for order of operations, 19–20, 35
　　power rules of exponents, 32
　　product rule of exponents, 30
　　quotient, 33

S
Sample final examination, A-7–A-9
Sample space, 792
Scientific notation
　　converting from standard notation, 41
　　converting to standard notation,
　　　　41–42
　　defined, 41
　　review, 84
　　using, 43
　　using to simplify computation, 42–43
Second-degree equations, 348
　　review, 743–744
　　solving system of equations by
　　　　elimination (addition), 725–726
　　solving system of equations by
　　　　graphing, 723–724
　　solving system of equations by
　　　　substitution, 724–725
　　systems containing, 723–729
Sequences
　　defined, 760
　　examples of, 760
Series, 764
Set-builder notation, 5, 231–235
Sets
　　elements of, 3
　　empty, 55
Significant digits, finding, 44
Similar terms
　　combining, 301
　　defined, 51
Similar triangles, 431–432
Slope-intercept form, 119
Slopes
　　finding through two given points, 107
　　finding with graph, 106–107
　　of horizontal lines, 108–109
　　interpreting in an application,
　　　　111–112
　　nonvertical line, 107
　　nonvertical lines, 107
　　of parallel lines, 109–110
　　of perpendicular lines, 110–111
　　review, 157
　　slopes of perpendicular lines, 110–111
　　of vertical lines, 108–109
Solution set, 49
Solutions, 49
Solving
　　absolute value equations and
　　　　inequalities in one variable,
　　　　241–251
　　absolute value inequalities, 250
　　application by setting up and solving
　　　　a system of three linear equations,
　　　　194–196

application by setting up and solving
　　a system of two linear equations,
　　180–185
application containing a radical
　　expression, 498–499
application involving a logarithm,
　　645–646
application involving a logarithmic
　　equation, 671–674
application involving a natural
　　logarithm, 653–654
application involving a parabola,
　　696–697
application involving a rational
　　equation, 408–410
application involving an ellipse, 708–709
application involving an exponential
　　equation, 671–674
application involving combined
　　variation, 436
application involving direct variation,
　　432–433
application involving exponential
　　growth or decay, 635–637
application involving geometric
　　sequences, 772–774
application involving hyperbolas,
　　719–720
application involving inverse variation,
　　434–435
application involving joint variation,
　　435–436
application involving the greatest
　　integer function, 732
application involving the piecewise-
　　defined function, 732
application requiring answers rounded
　　to the proper significant digit, 44
application requiring use of the square-
　　root property, 542–543
application using a quadratic equation,
　　351–353
application using a quadratic function,
　　571–573
areas and volumes of geometric figures,
　　36–37
equation containing two absolute value
　　terms, 246
equations, 406
equations by factoring, 347–353
equations containing radicals, 504
equations containing rational
　　expressions, 404–411
equations graphically, 170
equations that can be written in
　　quadratic form, 555–558
exponential equations, 667–669
formula containing a radical for a
　　specified variable, 506–507
formula for a specified variable by
　　factoring, 321
formula involving a rational equation
　　for a specified variable, 407
formulas for indicated variable, 55–56
higher-degree polynomial equation by
　　factoring, 350–351

inequalities, 583–584
inequality containing one absolute
　　value term, 247–250
investment problems, 628
linear equation in one variable, 48–59,
　　70–77
linear inequalities in one variable,
　　231–238
linear inequalities in two variables,
　　254–259
logarithmic equation with
　　a calculator, 651
logarithmic equations, 669–671
proportions, 428–430
quadratic equation by factoring, 348–350
quadratic equations by applying the
　　square-root property, 537–538
quadratic equations by completing the
　　square, 538–542
quadratic equations by factoring,
　　535–536
quadratic equations using quadratic
　　formula, 546–553
radical equation containing one radical,
　　502–505
radical equation containing two
　　radicals, 505–506
rational equation, 404–407
solving system of equations containing
　　second-degree equation by
　　elimination (addition), 725–726
system of equations containing
　　a second-degree equation by
　　substitution, 724–725
system of linear equations using
　　Cramer's rule, 211–215
system of linear inequalities in two
　　variables, 262–264
system of quadratic inequalities in two
　　variables, 264–266
system of two linear equations by
　　elimination (addition), 177–179
system of two linear equations by
　　substitution, 175–177
systems of equations, 724
systems of linear equations using
　　matrices, 199–203
systems of three linear equations in
　　three variables, 189–192
systems using linear programming,
　　269–277
using linear equations to solve
　　applications, 59–70
Special product formulas, 310
Square matrix, 200, 208
Square-root function, 460
Square-root property, 535
　　review, 608
　　solving an application requiring use of,
　　　　542–543
　　solving quadratic equations by
　　　　applying, 537–538
Square roots
　　approximating, 457
　　calculating, 456
　　simplifying perfect-square root, 455–457

Square roots (*continued*)
  simplifying perfect-square root
    expression, 457–458
Squares
  difference of two squares, 360
  factoring polynomials by grouping
    involving a difference of, 329
  factoring the difference of two,
    324–326
  perfect, 324
  of rational expressions, 380
  sum of two, 325
Standard deviation, 462
Standard notation
  converting from scientific notation,
    41–42
  converting to scientific notation, 41
  defined, 40
Straight angles, 63
Subsets, 4
Substitution
  factoring a trinomial using, 339
  solving system of equations containing
    second-degree equation by,
    724–725
Substitution method, 175–177
Subtraction
  complex numbers, 513
  polynomials, 302–303
  properties of equality, 49–51
  radicals, 486–487
  real numbers, 16–17
Sum, 16
Summation notation, 765
Supplementary angles, 63
Symmetry
  of graphs, A-1–A-5
  x-axis, A-1
  y-axis, A-1
Synthetic division
  of polynomial by a binomial of the
    form $(x - r)$ using, 421–424
  review, 449
Systems of linear equations
  review, 221–226
  solving by using Cramer's rule,
    211–215
  solving by using determinants,
    207–218
  solving by using matrices,
    199–207
  solving in three variables, 189–198

solving in two variables by elimination
  (addition), 177–179
solving in two variables by graphing,
  165–174
solving in two variables by substitution,
  175–177
solving systems of two, 169

T
Tables
  base-10, A-12
  base-*e*, A-13
  powers, A-11
  roots, A-11
Terminating decimals, 7
Test point, 255
Tests
  for factorability, 336
  vertical line, 131
Theorem
  factor, 424–425
  remainder, 424
Transitive property, 232
Tree diagram, 782
Triangles
  equilateral, 63
  finding length of side, 487–489
  isosceles, 63
  right, 63, 467
  similar, 431–432
Triangular form, 200
Trichotomy property, 15, 232
Trinomials
  defined, 292
  factoring, 332–341
  factoring by using substitution, 339
  factoring polynomial with four terms
    by grouping, 341
  factoring with a leading coefficient of
    one, 332–335
  factoring with a leading coefficient
    other than one using trial and
    error, 335
  perfect square, 310
  review, 360–362
  test for factorability, 336

U
Unbounded interval, 234
Union, 9, 236
Unit costs, 428
Unlike denominators, 388–390

Unlike signs
  adding real numbers with, 16
  dividing real numbers with, 16
  multiplying real numbers with, 17
Unlike terms, 51

V
Variables, 3
Variance, 573
Variation
  combined, 436
  direct, 432–433
  inverse, 434–435
  joint, 435–436
  review, 449–450
Vertex, 564, 566, 567–568
Vertex angles, 63
Vertical line test, 135
Vertical lines, 97–98
  graphing, 97–98
  slope of, 108–109
Vertical translations, 145, 564–565
Vertices, 705
Vieta, François, 7, 311, 511
Volume
  of cone, 37
  of cube, 36
  of cylinder, 37
  of pyramid, 37
  of rectangular solid, 36
  of sphere, 36, 37

W
Weber–Fechner law, 663
Whole numbers, 4

X
x-axis, 92, 147–148
x-axis symmetry, A-1
x-coordinate, 92
x-intercepts, A-2, A-3

Y
y-axis, 92
y-axis symmetry, A-1
y-coordinate, 92
y-intercepts, A-2, A-3
y as a function of *x*, 135

Z
Zero exponents, 32
Zero-factor property, 348